Collins EURO ESSENTIAL ROAD ATLAS

GW00361079

Published by Collins
An imprint of HarperCollins Publishers
Westerhill Road
Bishopbriggs
Glasgow G64 2QT
www.collins.co.uk

HarperCollins Publishers
1st Floor, Watermarque Building,
Ringsend Road, Dublin 4, Ireland

First published 2004

New edition 2022

© HarperCollins Publishers Ltd 2022
Maps © Collins Bartholomew Ltd 2022

A catalogue record for this book is available from the British Library

ISBN 978-0-00-840396-6

10 9 8 7 6 5 4 3 2 1

Printed and bound in India by Replika Press PVT Ltd

All mapping in this atlas is generated from Collins Bartholomew digital databases. Collins Bartholomew, the UK's leading independent geographical information supplier, can provide a digital, custom, and premium mapping service to a variety of markets.
For further information:
e-mail: collinsbartholomew@harpercollins.co.uk
or visit our website at: www.collinsbartholomew.com

If you would like to comment on any aspect of this book, please contact us at the above address or online.
e-mail: collinsmaps@harpercollins.co.uk

f facebook.com/collinsref   🐦 @collins_ref

Contents

	Distance map and chart
ii – iii	Map symbols
iii	Map symbols and country identifiers
iv	International road signs and travel web links
v	Key to map pages
vi – xvii	Route planning maps 1:4 654 000
2 – 181	Road maps 1:1 000 000
182 – 187	Road maps 1:750 000
188 – 191	Index to place names 1:750 000
192 – 207	City through-route maps
192	Amsterdam, Athina, Barcelona, Belfast
193	Beograd, Berlin, Bern, Birmingham
194	Bonn, Bordeaux, Bratislava, Brussel/Bruxelles
195	București, Budapest, Cardiff, Chișinau
196	Dublin, Edinburgh, Firenze, Frankfurt
197	Glasgow, Göteborg, Den Haag, Hamburg
198	Helsinki, İstanbul, København, Köln
199	Leipzig, Lisboa, Ljubljana, London
200	Lyon, Madrid, Manchester, Marseille
201	Milano, München, Napoli, Oslo
202	Palermo, Paris, Podgorica, Praha
203	Rīga, Roma, Rotterdam, Sankt Peterburg
204	Sarajevo, Sevilla, Skopje, Sofia
205	Stockholm, Strasbourg, Tallinn, Torino
206	Toulouse, Valencia, Venezia, Vilnius
207	Warszawa, Wien, Zagreb, Zürich
208 – 211	City street plans
208	Amsterdam, Athina, Berlin, Bern
209	Brussel/Bruxelles, Dublin, Helsinki, København
210	Lisboa, London, Madrid, Oslo
211	Paris, Roma, Stockholm, Wien
212 – 272	Index to place names 1:1 000 000
	Driving information and web links

Map symbols

Road maps	Carte routière	Strassenkarten
E55 Euro route number	Route européenne	Europastrasse
A13 Motorway	Autoroute	Autobahn
Motorway – toll	Autoroute à péage	Gebührenpflichtige Autobahn
Motorway – toll (vignette)	Autoroute à péage (vignette)	Gebührenpflichtige Autobahn (Vignette)
37 Motorway junction – full access	Echangeur d'autoroute avec accès libre	Autobahnauffahrt mit vollem Zugang
12 Motorway junction – restricted access	Echangeur d'autoroute avec accès limité	Autobahnauffahrt mit beschränktem Zugang
Motorway services	Aire de service sur autoroute	Autobahnservicestelle
309 Main road – dual carriageway	Route principale à chaussées séparées	Hauptstrasse – Zweispurig
Main road – single carriageway	Route principale à une seule chaussée	Hauptstrasse – Einspurig
516 Secondary road – dual carriageway	Route secondaire à chaussées séparées	Zweispurige Nebenstrasse
Secondary road – single carriageway	Route secondaire à seule chaussée	Einspurige Nebenstrasse
Other road	Autre route	Andere Strasse
Motorway tunnel	Autoroute tunnel	Autobahntunnel
Main road tunnel	Route principale tunnel	Hauptstrassetunnel
Motorway/road under construction	Autoroute/route en construction	Autobahn/Strasse im Bau
Road toll	Route à péage	Gebührenpflichtige Strasse
Distance marker **16** Distances in kilometres **10** Distances in miles (UK only)	Marquage des distances Distances en kilomètres Distances en miles (GB)	Distanz-Markierung Distanzen in Kilometern Distanzen in Meilen (GB)
Steep hill	Colline abrupte	Steile Strasse
2587 Mountain pass (height in metres)	Col (Altitude en mètres)	Pass (Höhe in Metern)
Scenic route	Parcours pittoresque	Landschaftlich schöne Strecke
International airport	Aéroport international	Internationaler Flughafen
Car transport by rail	Transport des autos par voie ferrée	Autotransport per Bahn
Railway	Chemin de fer	Eisenbahn
Tunnel	Tunnel	Tunnel
Funicular railway	Funiculaire	Seilbahn
Rotterdam Car ferry	Bac pour autos	Autofähre
2587 Summit (height in metres)	Sommet (Altitude en mètres)	Berg (Höhe in Metern)
Volcano	Volcan	Vulkan
Canal	Canal	Kanal
International boundary	Frontière d'Etat	Landesgrenze
Disputed International boundary	Frontière litigieuse	Umstrittene Staatsgrenze
Disputed Territory boundary	Frontière territoriale contestée	Umstrittene Gebietsgrenze
UK Country abbreviation	Abréviation du pays	Regionsgrenze
Urban area	Zone urbaine	Stadtgebiet
28 Adjoining page indicator	Indication de la page contigüe	Randhinweis auf Folgekarte
National Park	Parc national	Nationalpark

1:1 000 000

1 centimetre to 10 kilometres   0   10   20   30   40   50   60   70   80 km   1 inch to 16 miles

0   10   20   30   40   50 miles

City maps and plans	Plans de ville	Stadtpläne
★ Place of interest	Site d'interêt	Sehenswerter Ort
■ Railway station	Gare	Bahnhof
Parkland	Espace vert	Parkland
Woodland	Espace boisé	Waldland
General place of interest	Site d'interêt général	Sehenswerter Ort
Academic/Municipal building	Établissement scolaire/installations municipales	Akademisches/Öffentliches Gebäude
Place of worship	Lieu de culte	Andachtsstätte
Transport location	Infrastructure de transport	Verkehrsanbindung

Places of interest

	English	French	German
🏛	Museum and Art Gallery	Musée / Gallerie d'art	Museum / Kunstgalerie
	Castle	Château	Burg / Schloss
	Historic building	Monument historique	historisches Gebäude
	Historic site	Site historique	historische Stätte
m	Monument	Monument	Denkmal
	Religious site	Site religieux	religiöse Stätte
	Aquarium / Sea life centre	Aquarium / Parc Marin	Aquarium
	Arboretum	Arboretum	Arboretum, Baumschule
	Botanic garden (National)	Jardin botanique national	botanischer Garten
★	Natural place of interest (other site)	Réserve naturelle	landschaftlich interessanter Ort
	Zoo / Safari park / Wildlife park	Parc Safari / Réserve sauvage / Zoo	Safaripark / Wildreservat / Zoo
★	Other site	Autres sites	Touristenattraktion
	Theme park	Parc à thème	Freizeitpark
◆	World Heritage site	Patrimoine Mondial	Weltkulturerbe
	Athletics stadium (International)	Stade international d'athlétisme	internationales Leichtathletik Stadion
	Football stadium (Major)	Stade de football	Fußballstadion
	Golf course (International)	Parcours de golf international	internationaler Golfplatz
	Grand Prix circuit (Formula 1) / Motor racing venue / MotoGP circuit	Circuit auto-moto	Autodrom
	Rugby ground (International - Six Nations)	Stade de rugby	internationales Rugbystadion
	International sports venue	Autre manifestation sportive	internationale Sportanlage
	Tennis venue	Court de tennis	Tennis
Valcotos ⊛	Winter sports resort	Sports d'hiver	Wintersport

Country identifiers

Code	English	French	German	Code	English	French	German
A	Austria	Autriche	Österreich	**IRL**	Ireland	Irlande	Irland
AL	Albania	Albanie	Albanien	**IS**	Iceland	Islande	Island
AND	Andorra	Andorre	Andorra	**L**	Luxembourg	Luxembourg	Luxemburg
B	Belgium	Belgique	Belgien	**LT**	Lithuania	Lituanie	Litauen
BG	Bulgaria	Bulgarie	Bulgarien	**LV**	Latvia	Lettonie	Lettland
BIH	Bosnia and Herzegovina	Bosnie-et-Herzégovine	Bosnien und Herzegowina	**M**	Malta	Malte	Malta
BY	Belarus	Bélarus	Belarus	**MA**	Morocco	Maroc	Marokko
CH	Switzerland	Suisse	Schweiz	**MC**	Monaco	Monaco	Monaco
CY	Cyprus	Chypre	Zypern	**MD**	Moldova	Moldavie	Moldawien
CZ	Czechia (Czech Republic)	République tchèque	Tschechische Republik	**MNE**	Montenegro	Monténégro	Montenegro
D	Germany	Allemagne	Deutschland	**N**	Norway	Norvège	Norwegen
DK	Denmark	Danemark	Dänemark	**NL**	Netherlands	Pays-Bas	Niederlande
DZ	Algeria	Algérie	Algerien	**NMK**	North Macedonia	Macédoine du Nord	Nordmazedonien
E	Spain	Espagne	Spanien	**P**	Portugal	Portugal	Portugal
EST	Estonia	Estonie	Estland	**PL**	Poland	Pologne	Polen
F	France	France	Frankreich	**RKS**	Kosovo	Kosovo	Kosovo
FIN	Finland	Finlande	Finnland	**RO**	Romania	Roumanie	Rumänien
FL	Liechtenstein	Liechtenstein	Liechtenstein	**RSM**	San Marino	Saint-Marin	San Marino
FO	Faroe Islands	Iles Féroé	Färöer-Inseln	**RUS**	Russia	Russie	Russland
GBA	Alderney	Alderney	Alderney	**S**	Sweden	Suède	Schweden
GBG	Guernsey	Guernsey	Guernsey	**SK**	Slovakia	République slovaque	Slowakei
GBJ	Jersey	Jersey	Jersey	**SLO**	Slovenia	Slovénie	Slowenien
GBM	Isle of Man	Île de Man	Insel Man	**SRB**	Serbia	Sérbie	Serbien
GBZ	Gibraltar	Gibraltar	Gibraltar	**TN**	Tunisia	Tunisie	Tunisien
GR	Greece	Grèce	Griechenland	**TR**	Turkey	Turquie	Türkei
H	Hungary	Hongrie	Ungarn	**UA**	Ukraine	Ukraine	Ukraine
HR	Croatia	Croatie	Kroatien	**UK**	United Kingdom GB & NI	Royaume-Uni	Vereinigtes Königreich
I	Italy	Italie	Italien				

International road signs and travel web links

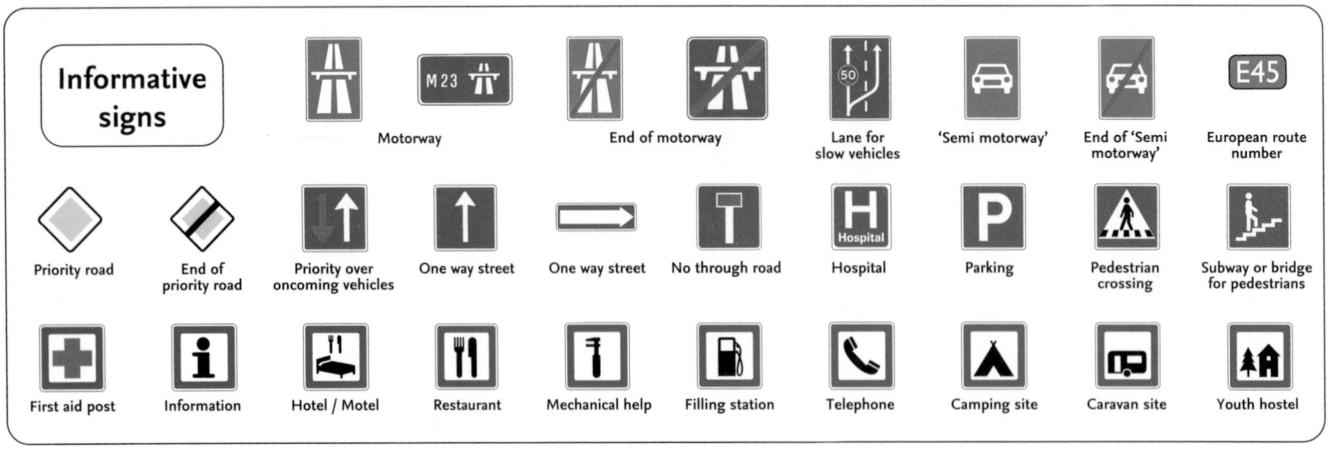
Informative signs
Motorway
End of motorway
Lane for slow vehicles
'Semi motorway'
End of 'Semi motorway'
European route number
Priority road
End of priority road
Priority over oncoming vehicles
One way street
One way street
No through road
Hospital
Parking
Pedestrian crossing
Subway or bridge for pedestrians
First aid post
Information
Hotel / Motel
Restaurant
Mechanical help
Filling station
Telephone
Camping site
Caravan site
Youth hostel

Warning signs
Right bend
Left bend
Double bend
Roundabout
Intersection with non-priority road
Traffic merges from left
Traffic merges from right
Road narrows
Road narrows at left
Road narrows at right
Give way
Slippery road
Uneven road
Steep hill – descent
Tunnel
Opening bridge
Road works
Loose chippings
Level crossing with barrier
Level crossing without barrier
Tram
'Count down' posts
'Danger' level crossing
Low flying aircraft
Falling rocks
Cross wind
Quayside or river bank
Two-way traffic
Traffic signals ahead
Pedestrians
Children
Animals
Wild animals
Other dangers
Width of carriageway
Beginning of regulation
Repetition sign
End of regulation

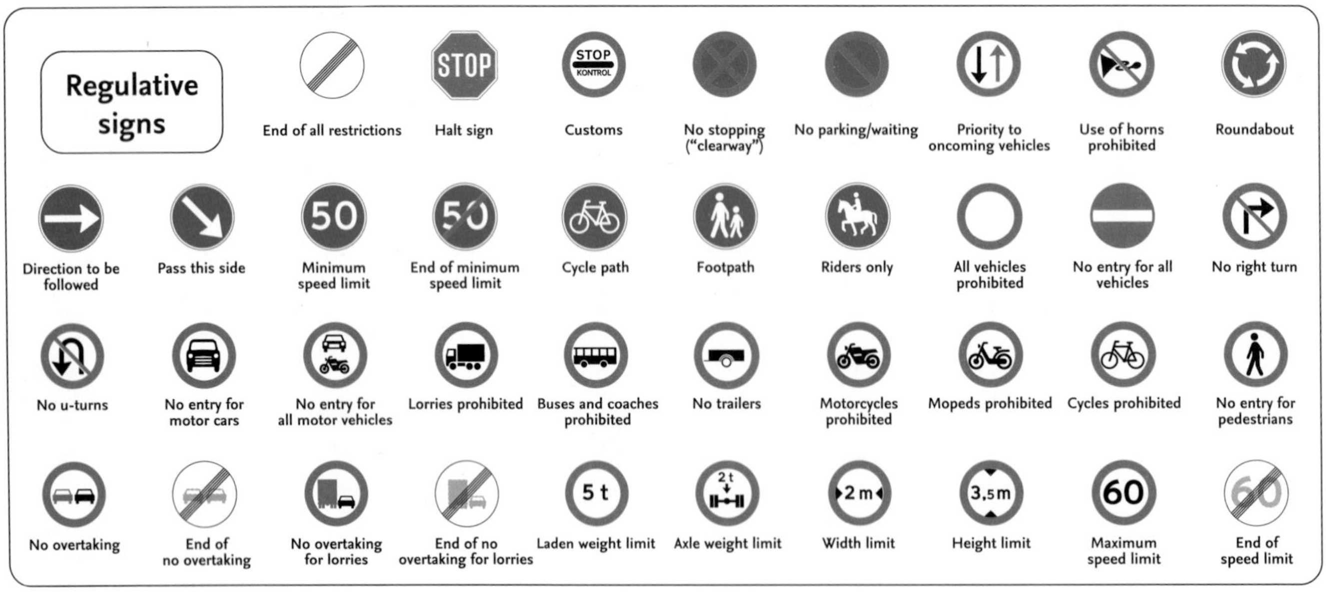
Regulative signs
End of all restrictions
Halt sign
Customs
No stopping ("clearway")
No parking/waiting
Priority to oncoming vehicles
Use of horns prohibited
Roundabout
Direction to be followed
Pass this side
Minimum speed limit
End of minimum speed limit
Cycle path
Footpath
Riders only
All vehicles prohibited
No entry for all vehicles
No right turn
No u-turns
No entry for motor cars
No entry for all motor vehicles
Lorries prohibited
Buses and coaches prohibited
No trailers
Motorcycles prohibited
Mopeds prohibited
Cycles prohibited
No entry for pedestrians
No overtaking
End of no overtaking
No overtaking for lorries
End of no overtaking for lorries
Laden weight limit
Axle weight limit
Width limit
Height limit
Maximum speed limit
End of speed limit

Travel & route planning

Driving information	www.drive-alive.co.uk
The AA	www.theaa.com
The RAC	www.rac.co.uk
ViaMichelin	www.viamichelin.com
Bing Maps	www.bing.com/mapspreview
Motorail information	www.seat61.com/Motorail
Ferry information	www.aferry.com
Eurotunnel information	www.eurotunnel.com/uk/home/

General information

UK Foreign & Commonwealth Office	www.gov.uk/government/organisations/foreign-commonwealth-office
Country profiles	www.cia.gov/the-world-factbook/
World Heritage sites	whc.unesco.org/en/list
World time	www.greenwichmeantime.com
Weather information	www.metoffice.gov.uk

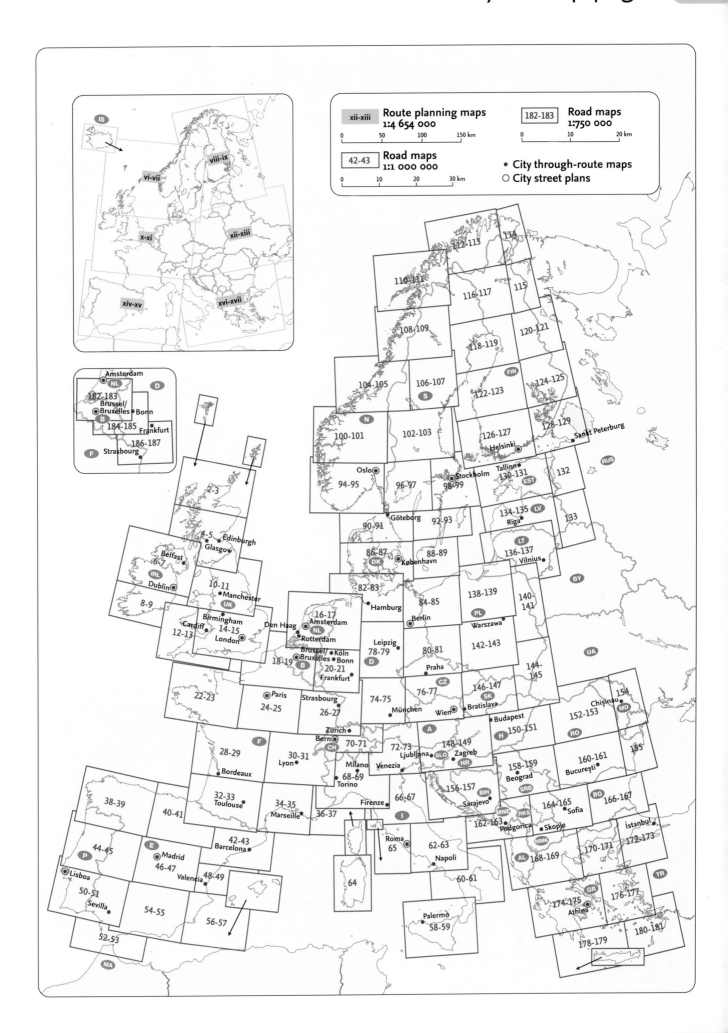

Route planning maps
1:4 654 000
xii-xiii
0 50 100 150 km

Road maps
1:1 000 000
42-43
0 10 20 30 km

Road maps
1:750 000
182-183
0 10 20 km

● City through-route maps
○ City street plans

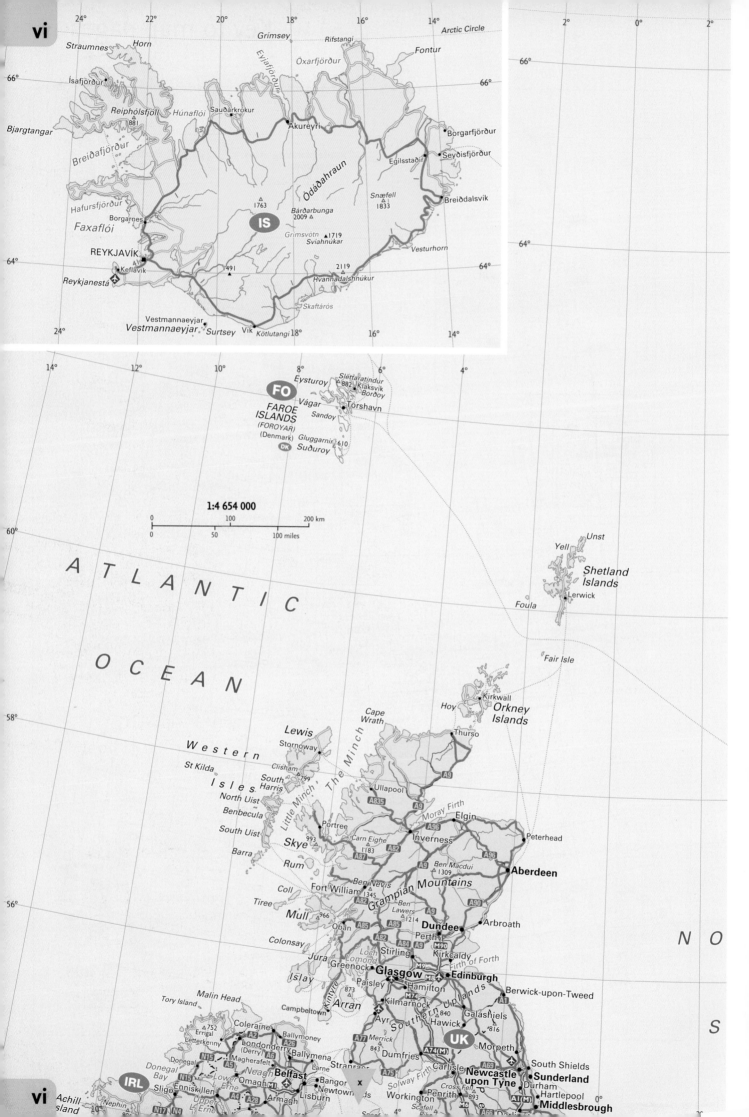

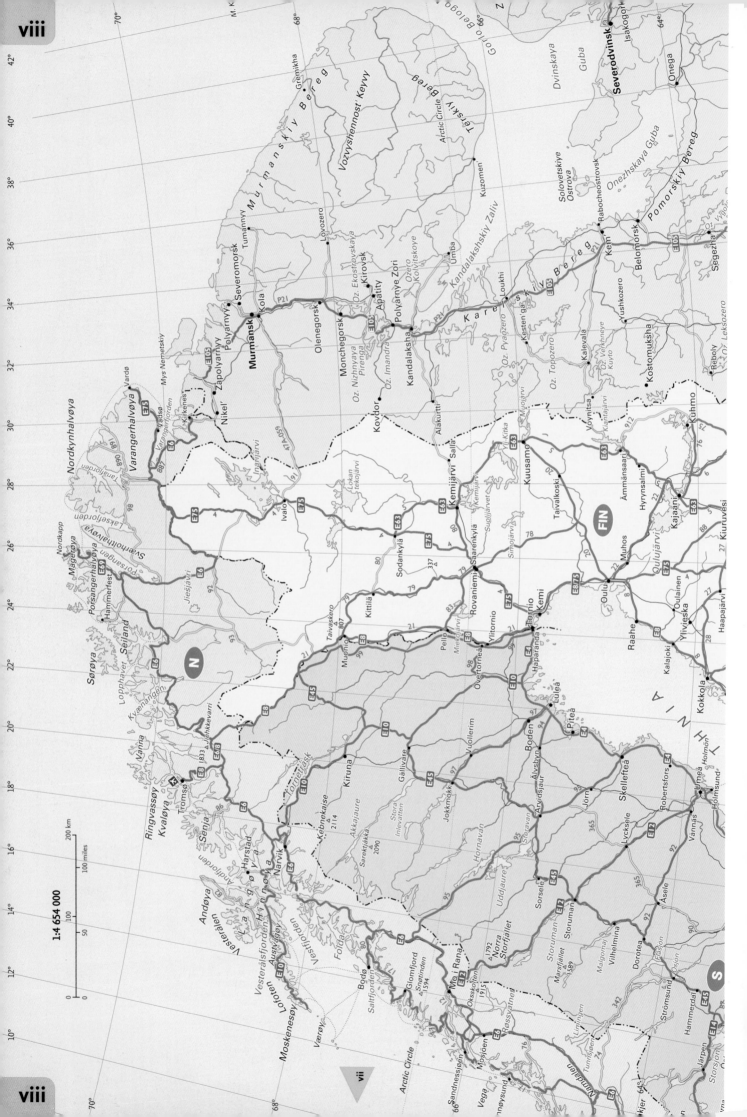

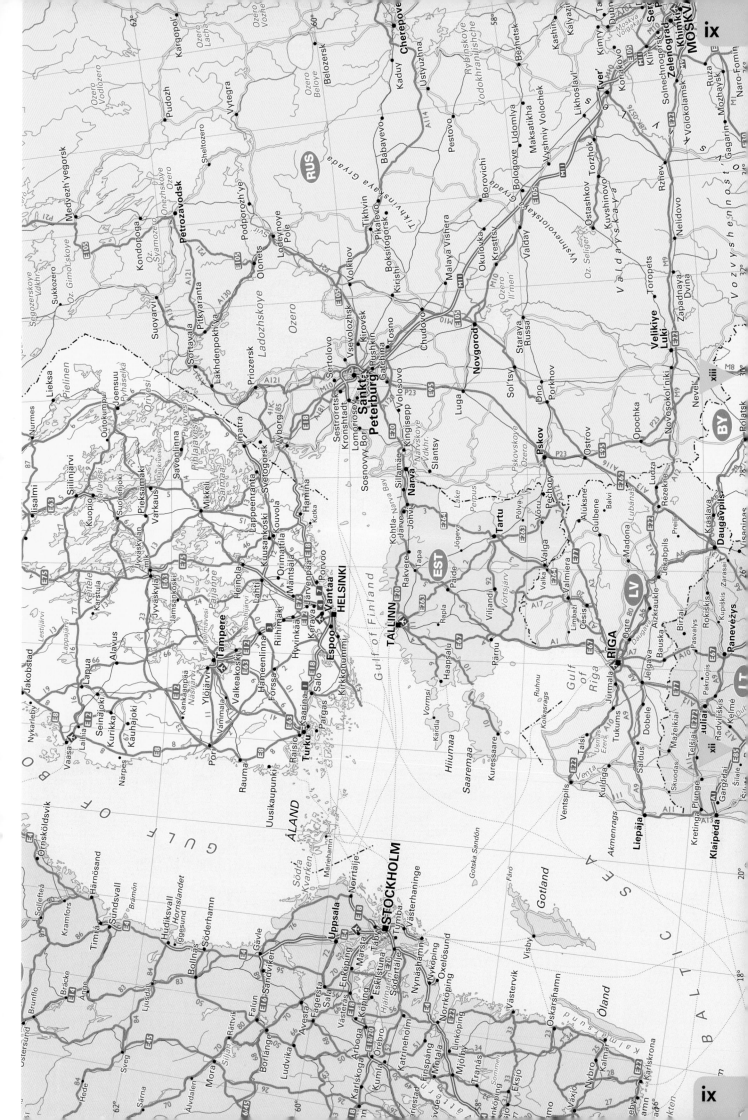

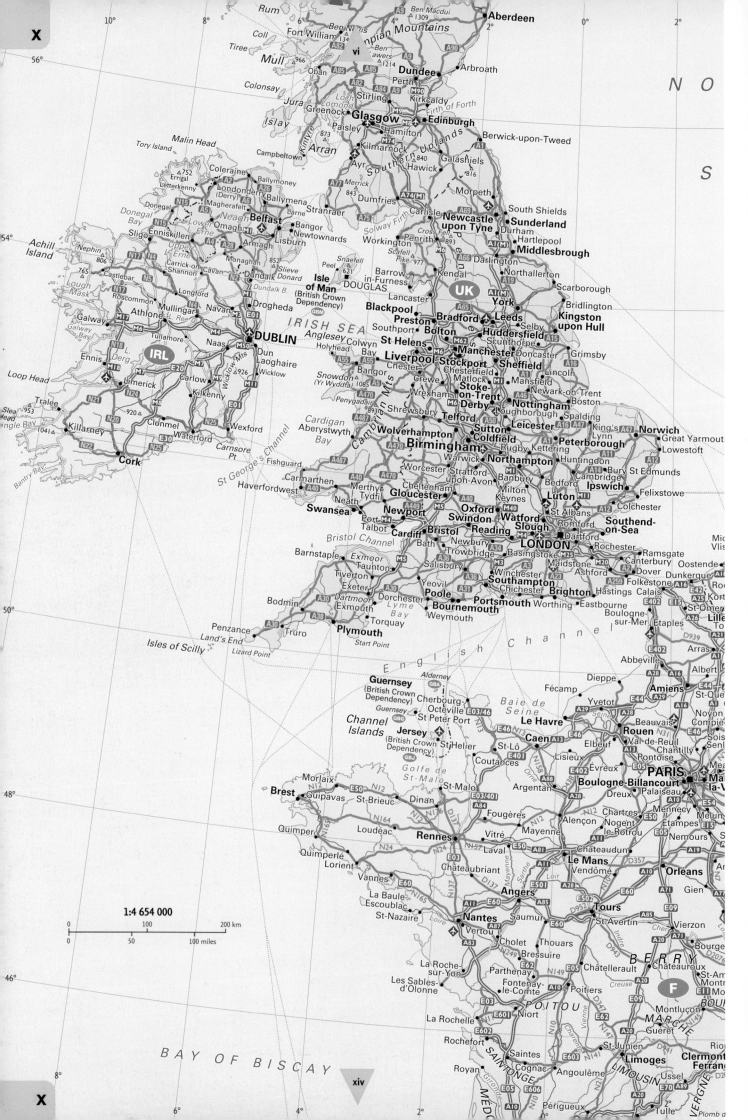

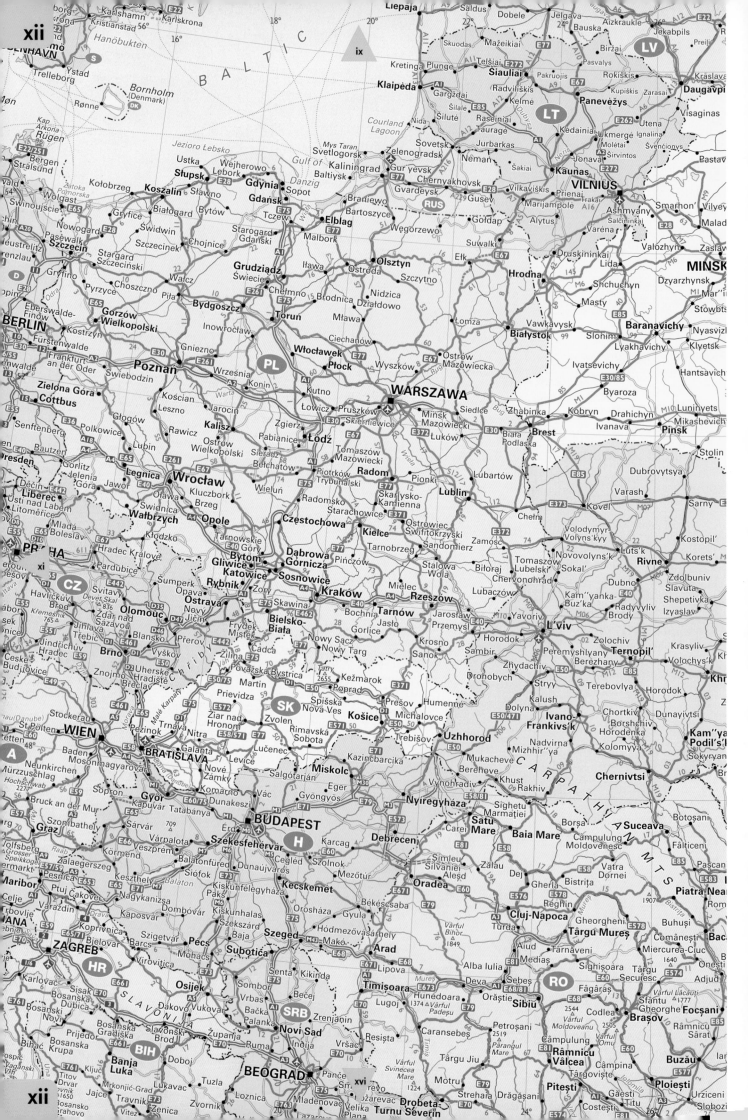

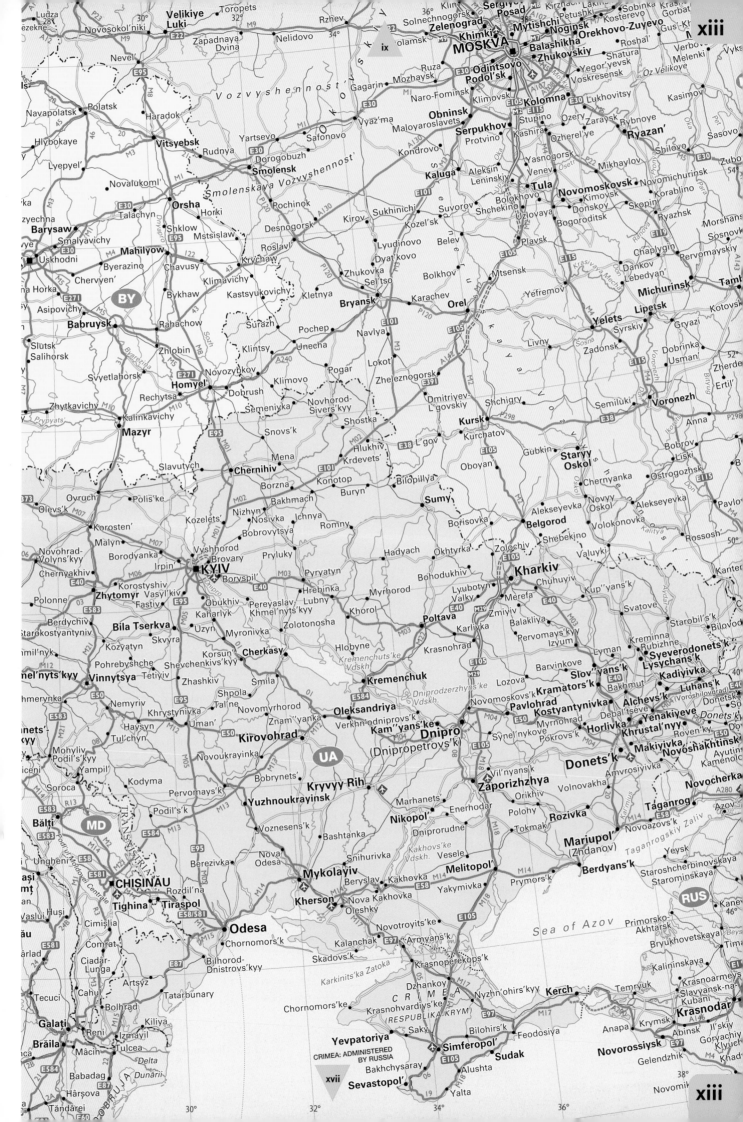

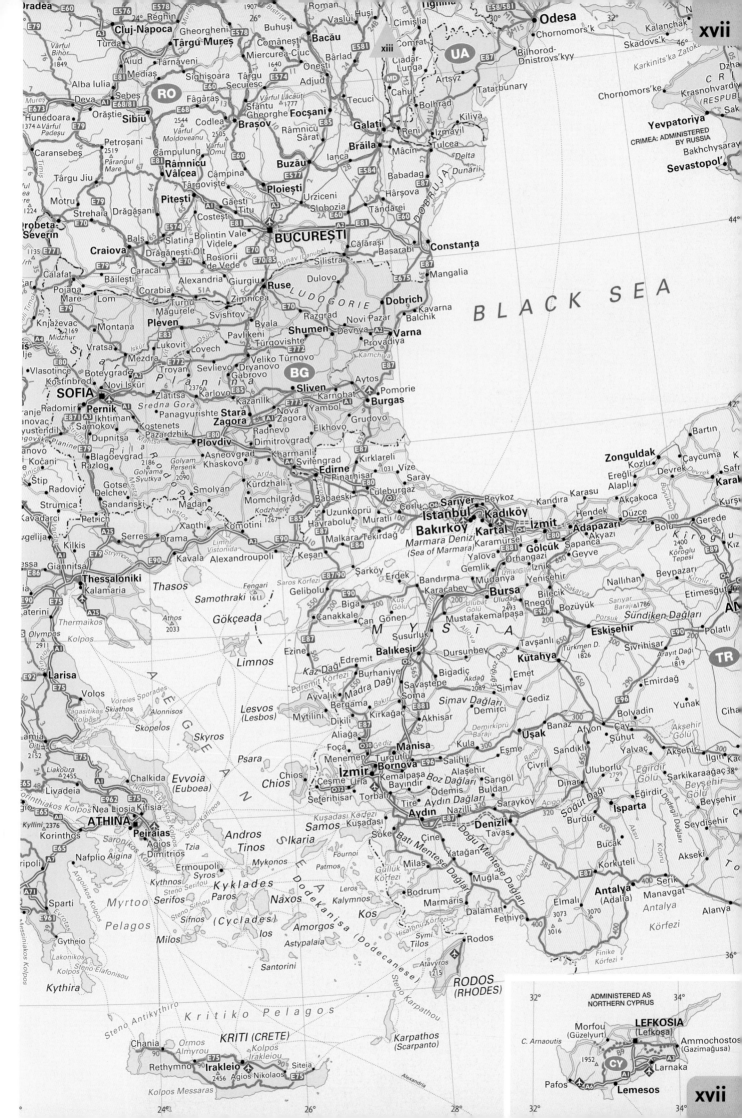

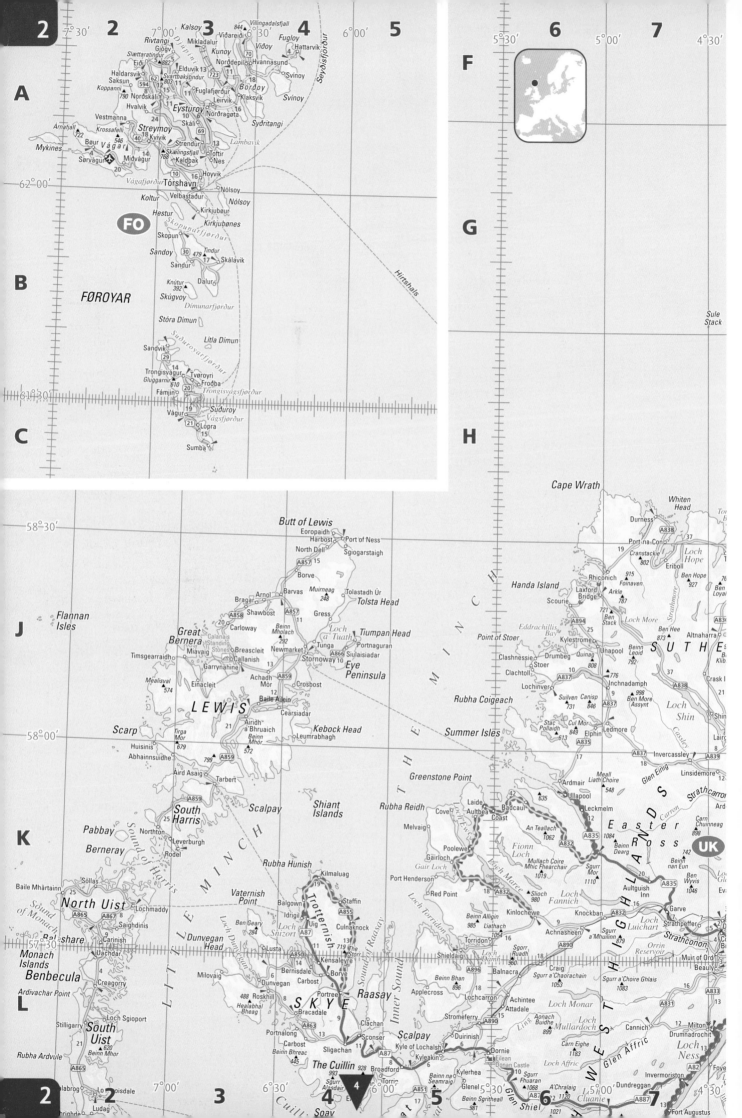

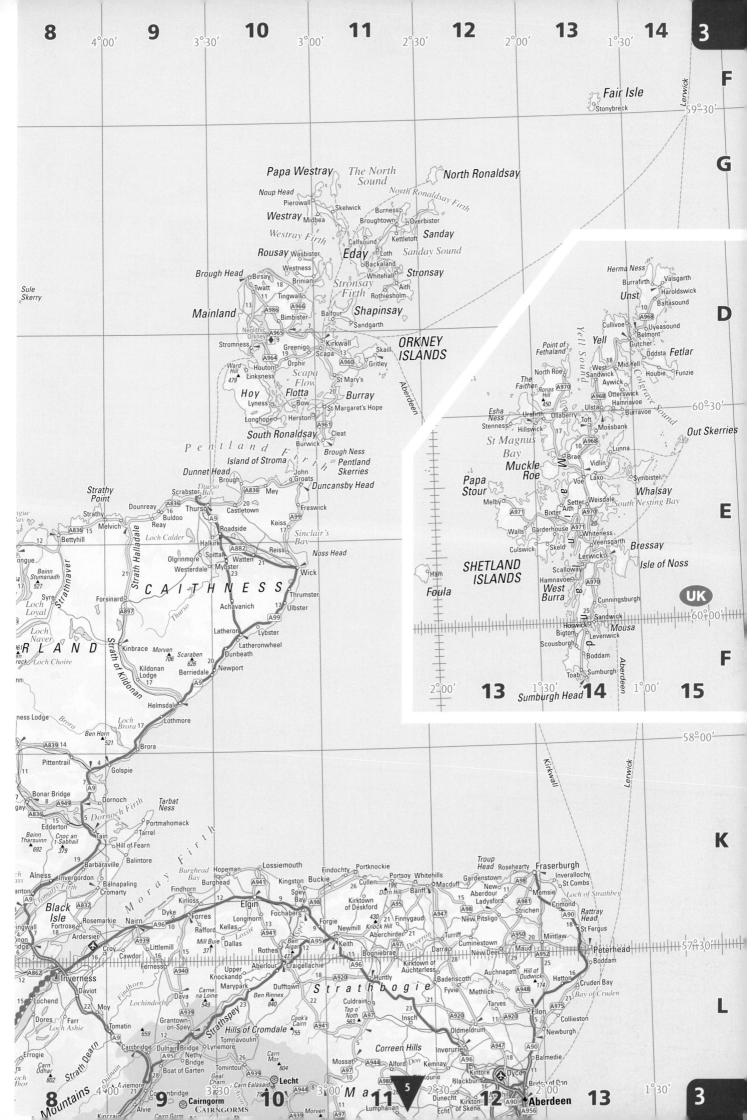

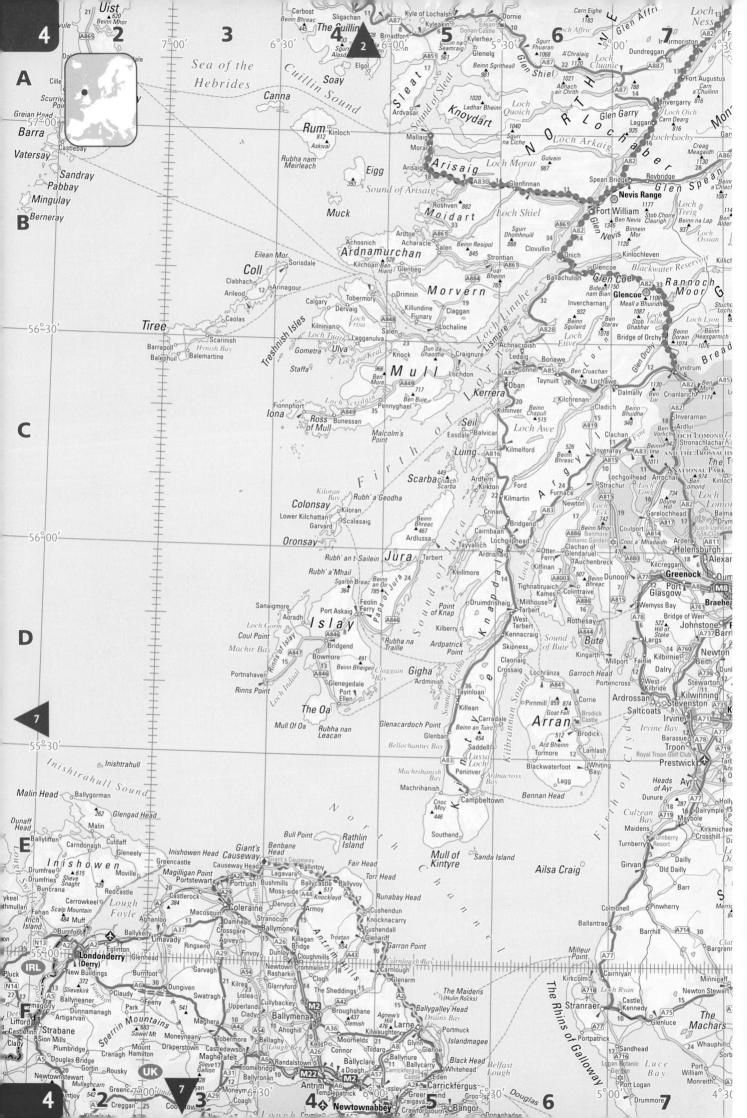

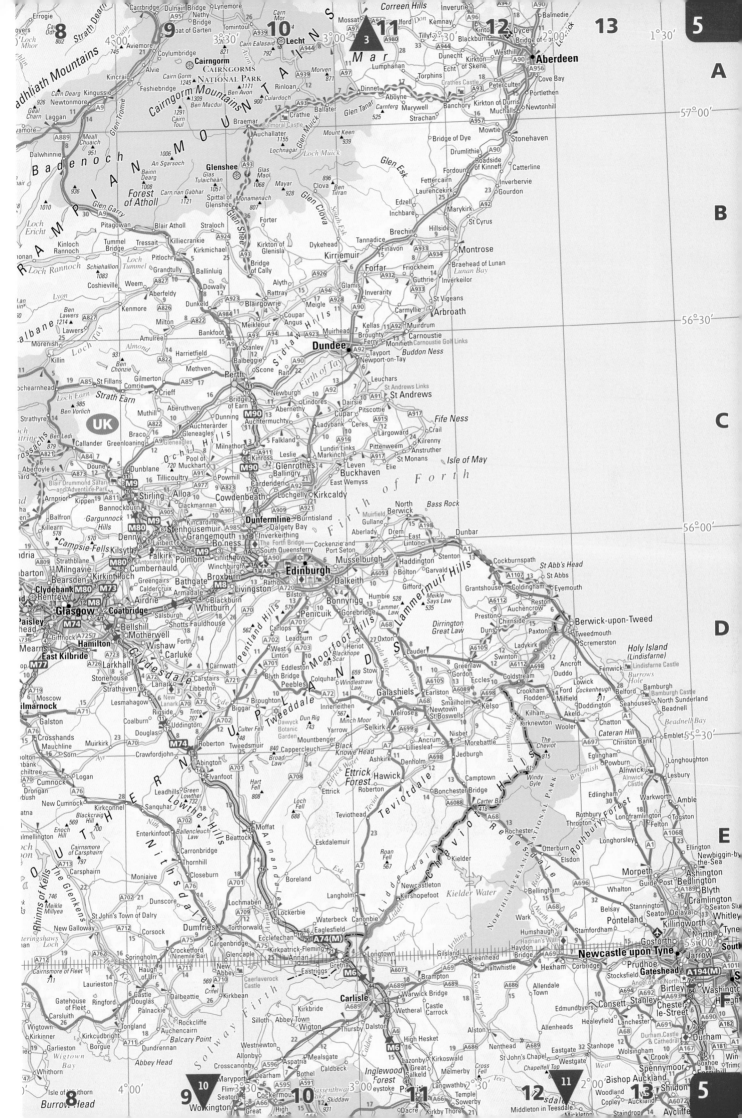

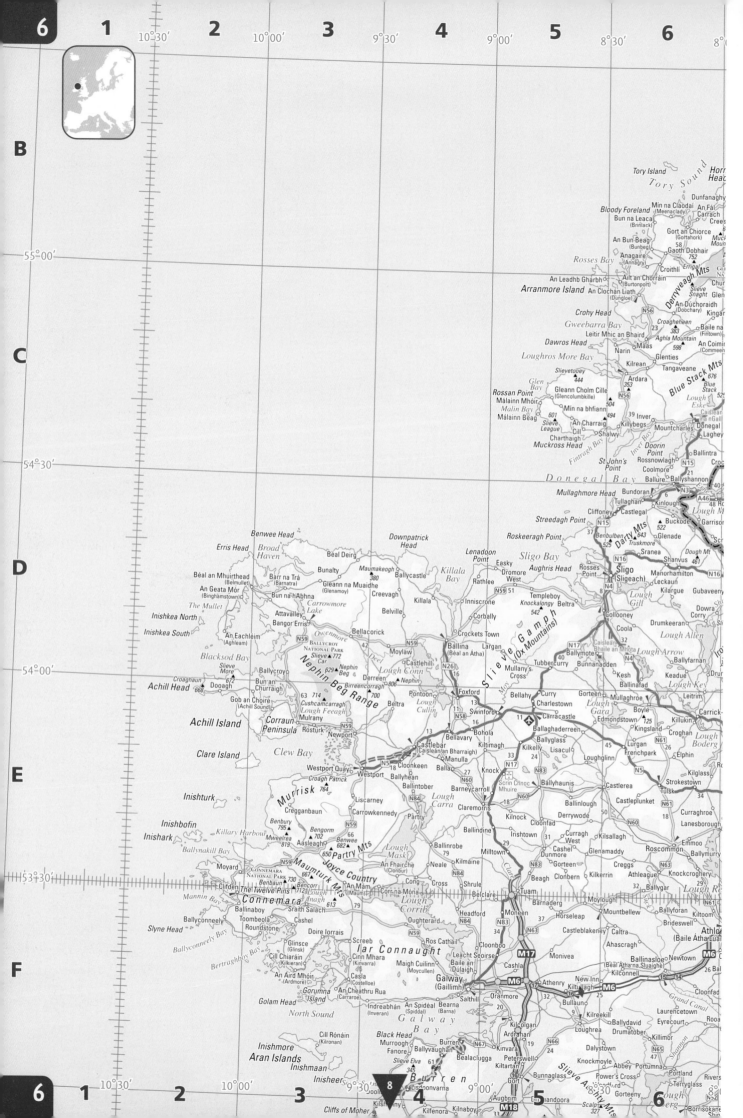

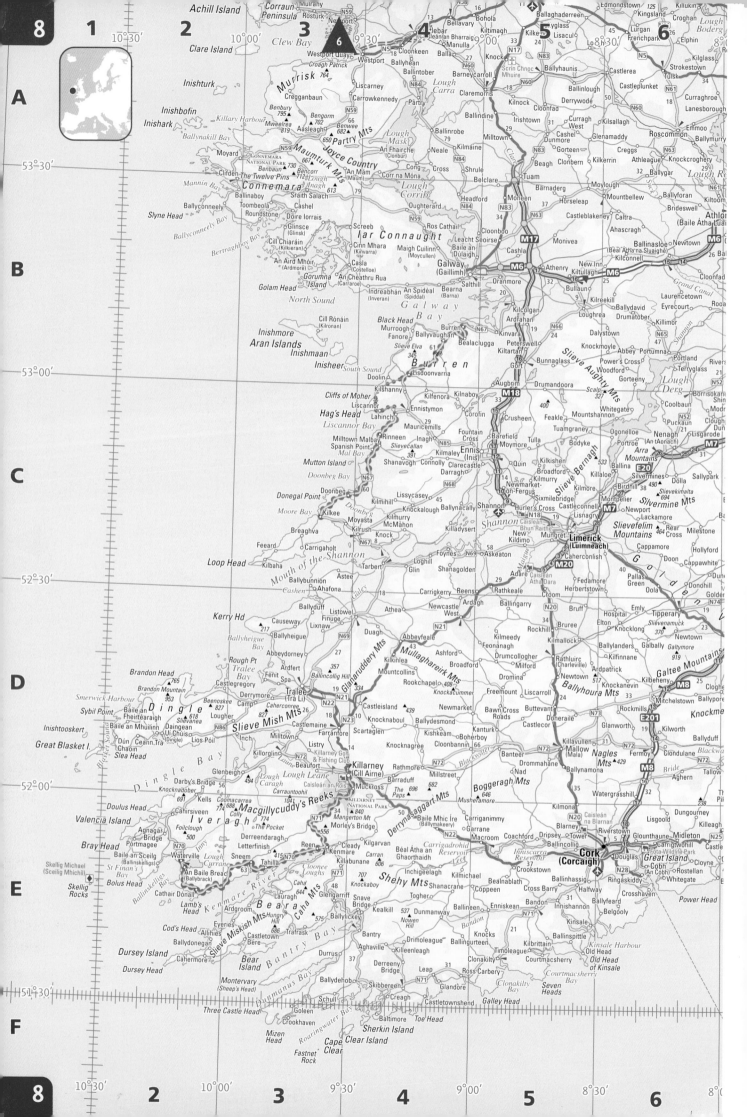

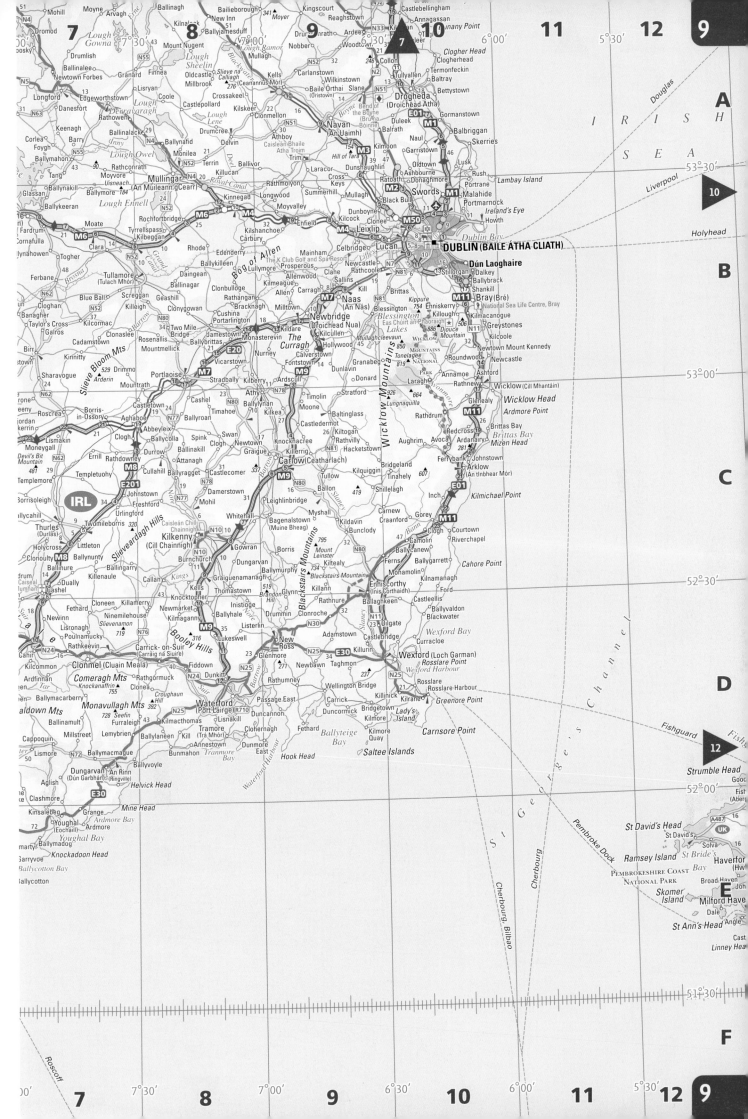

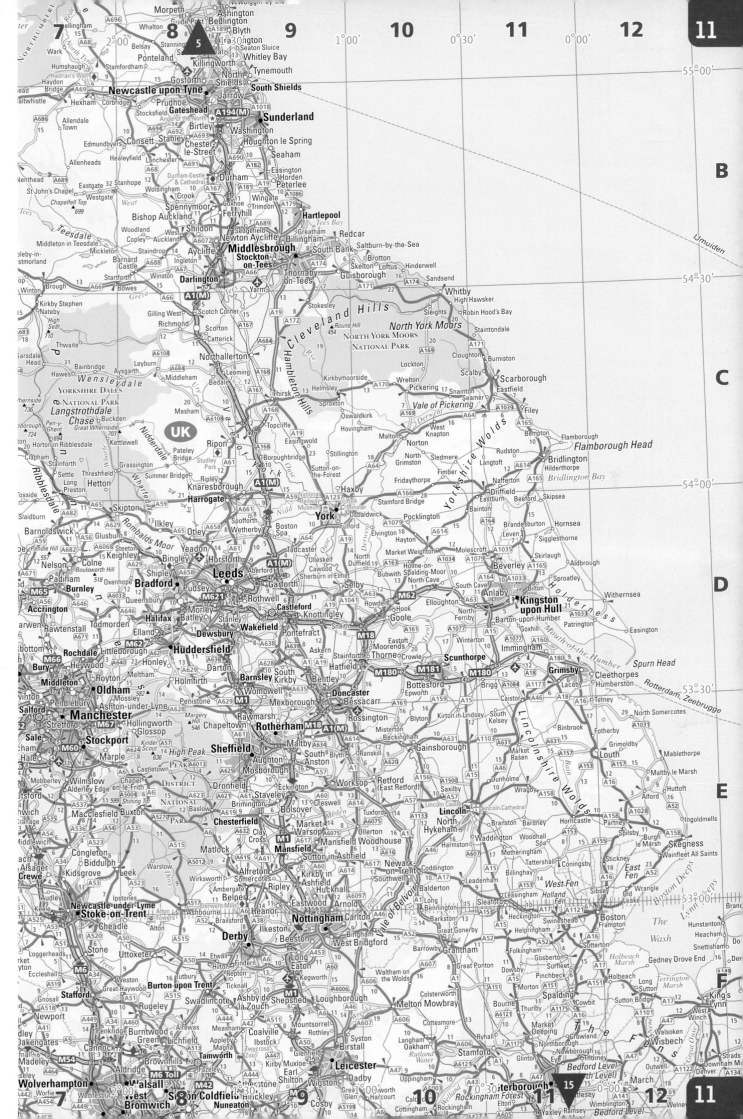

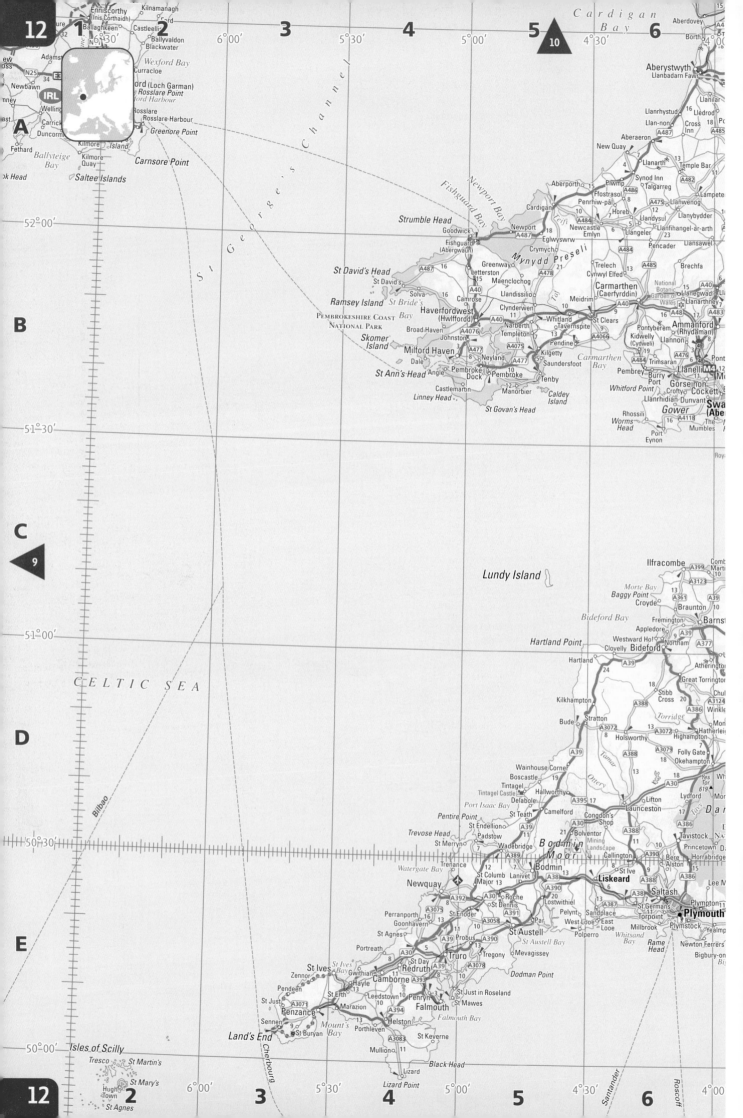

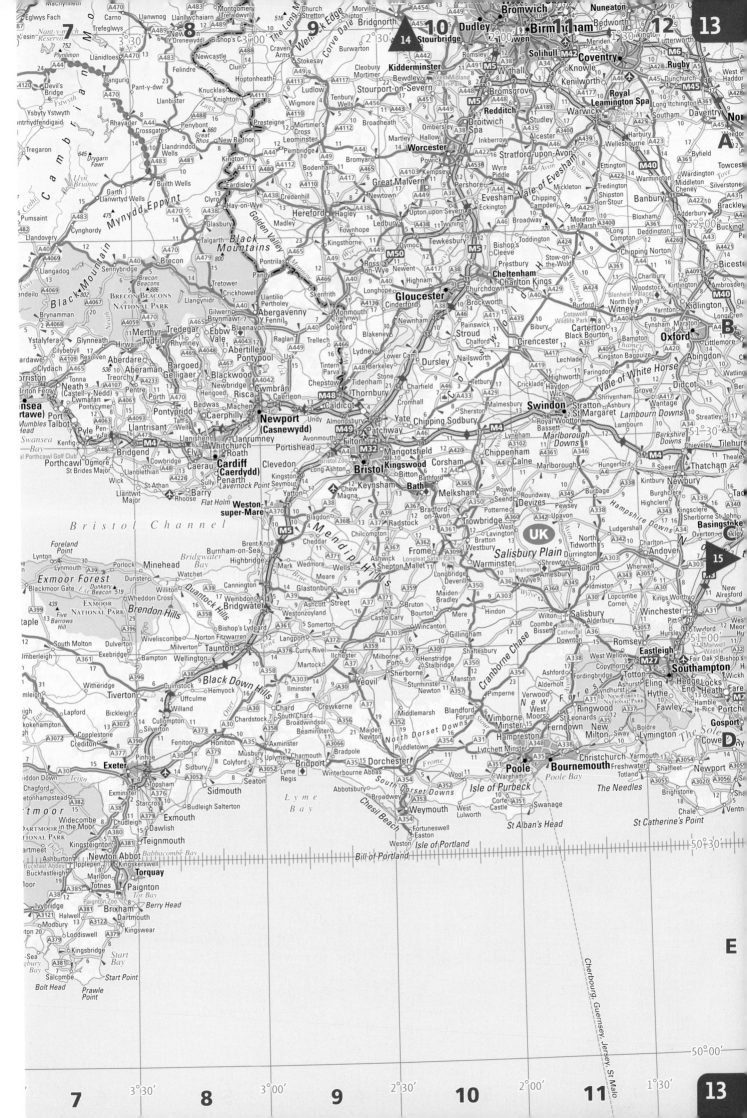

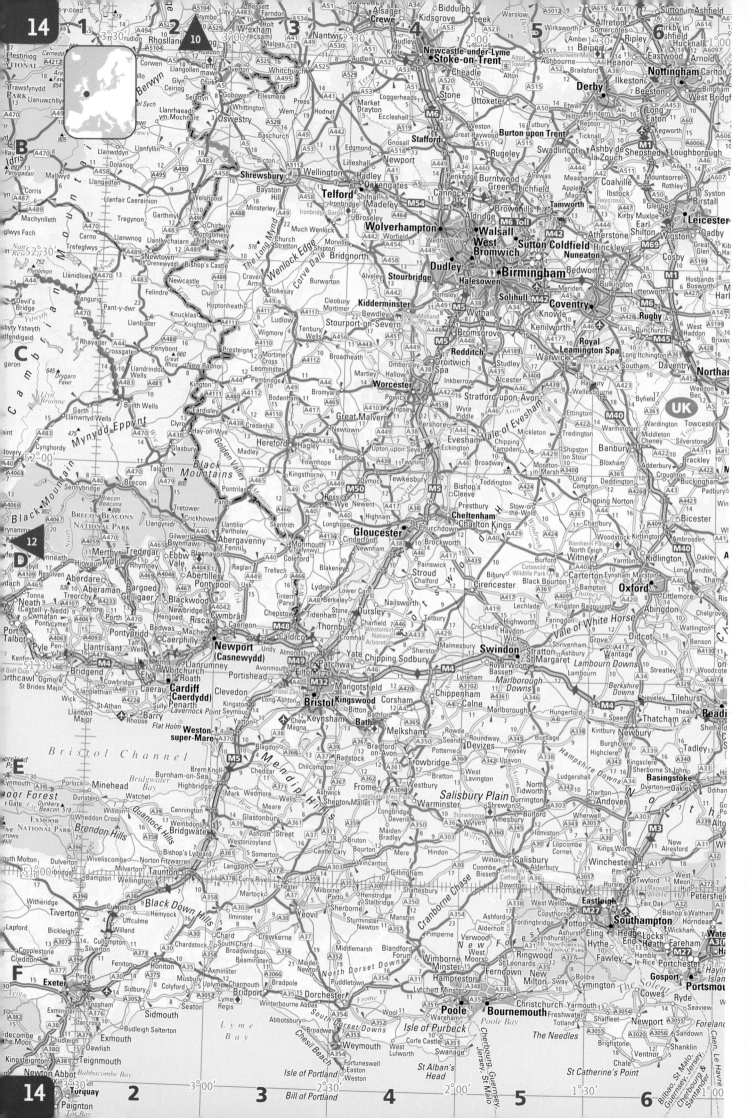

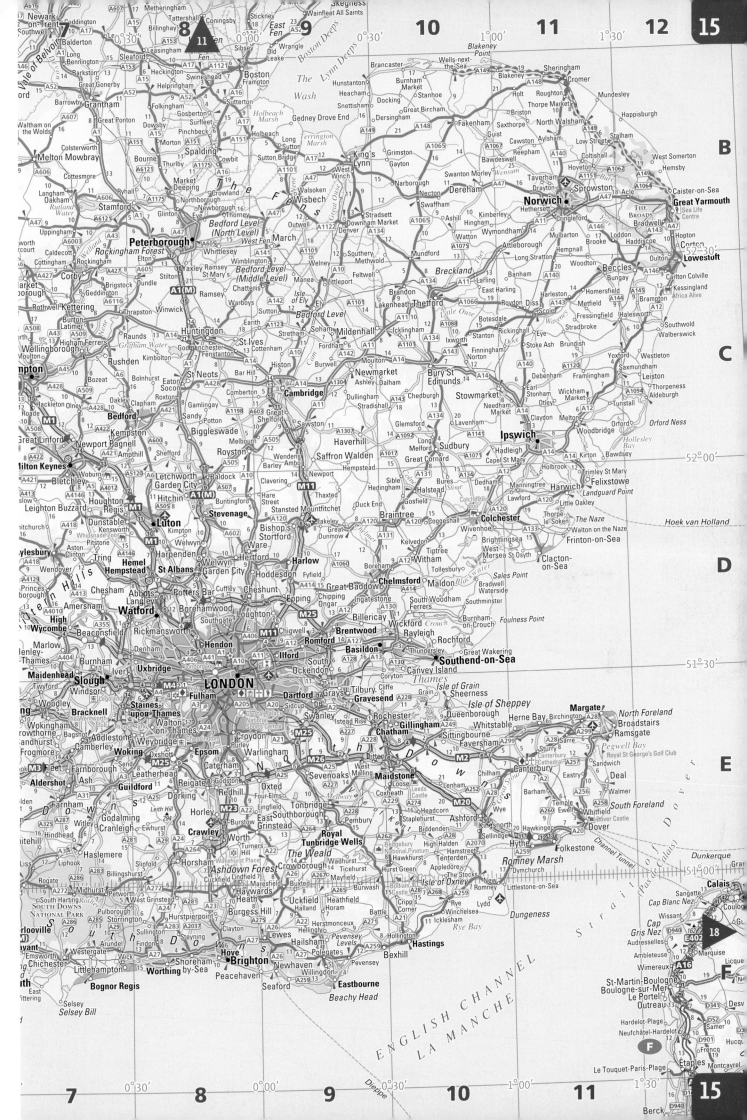

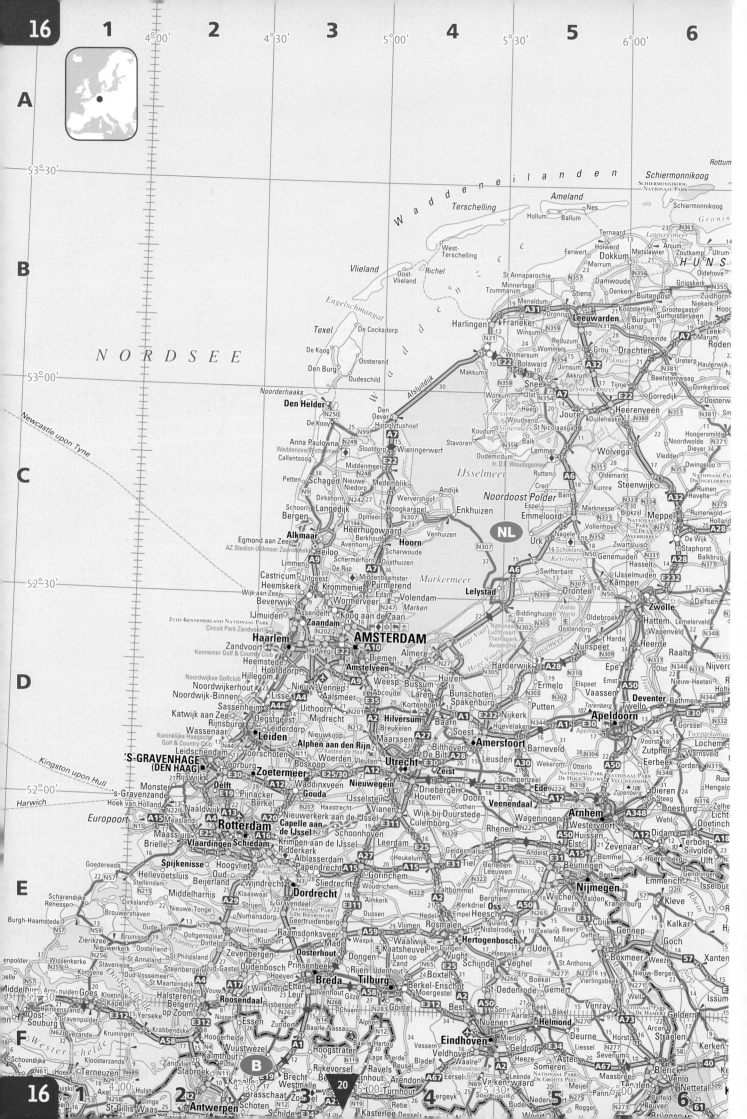

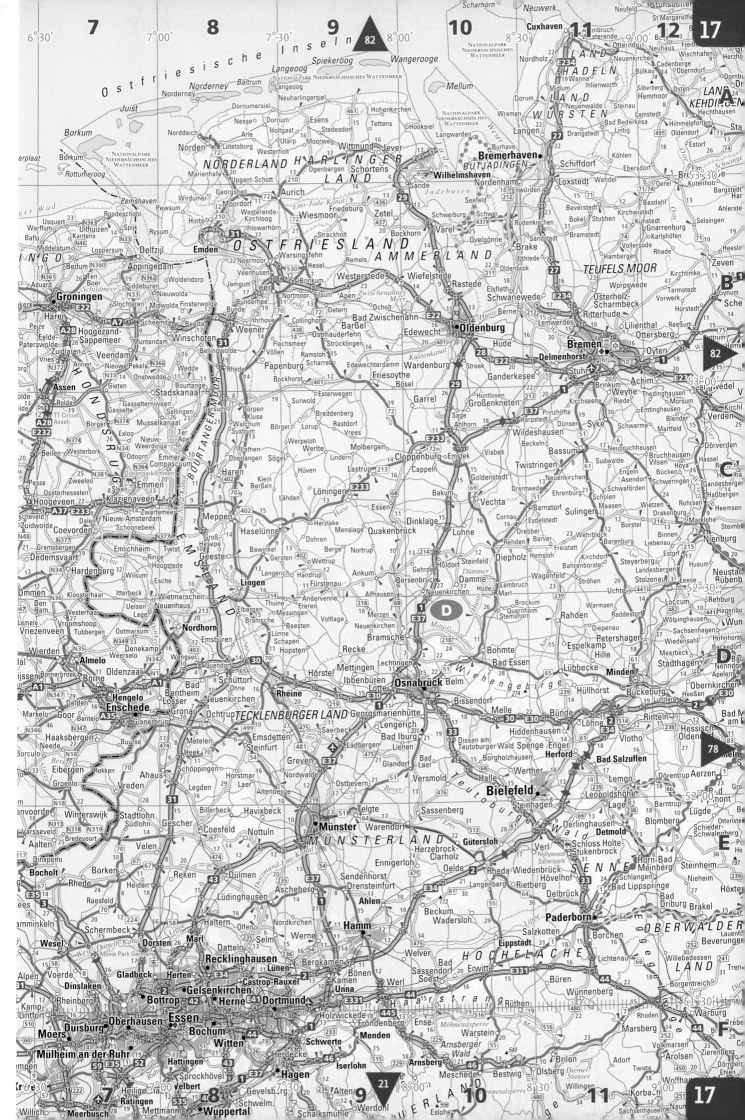

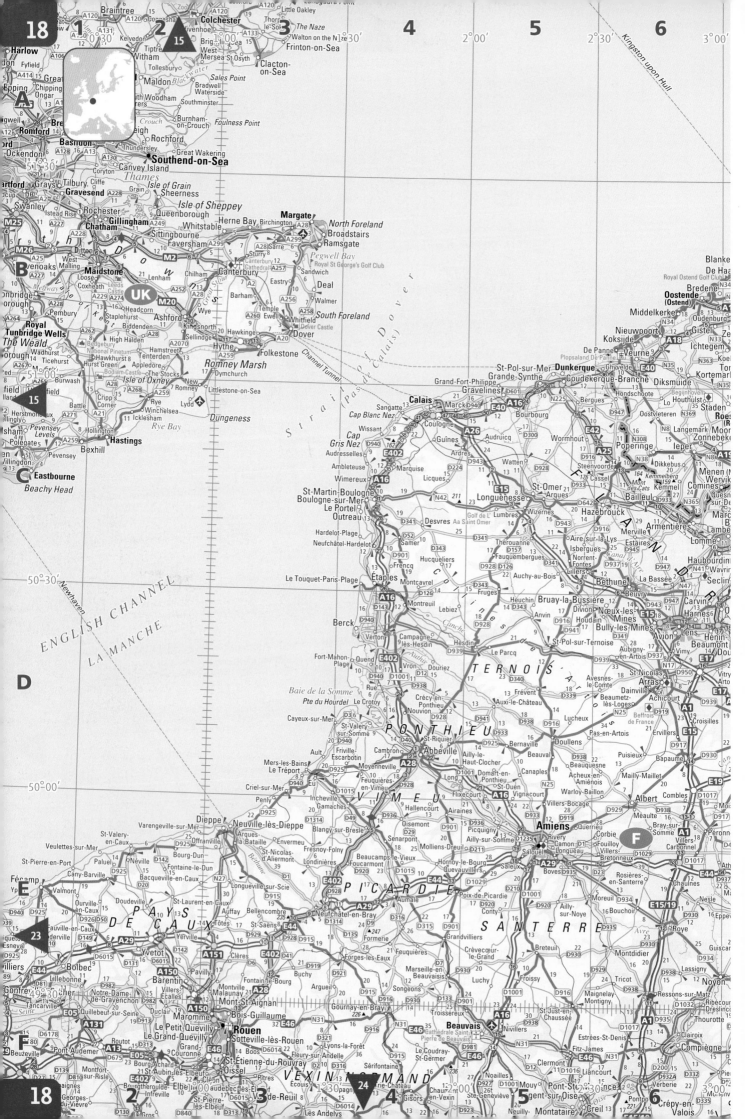

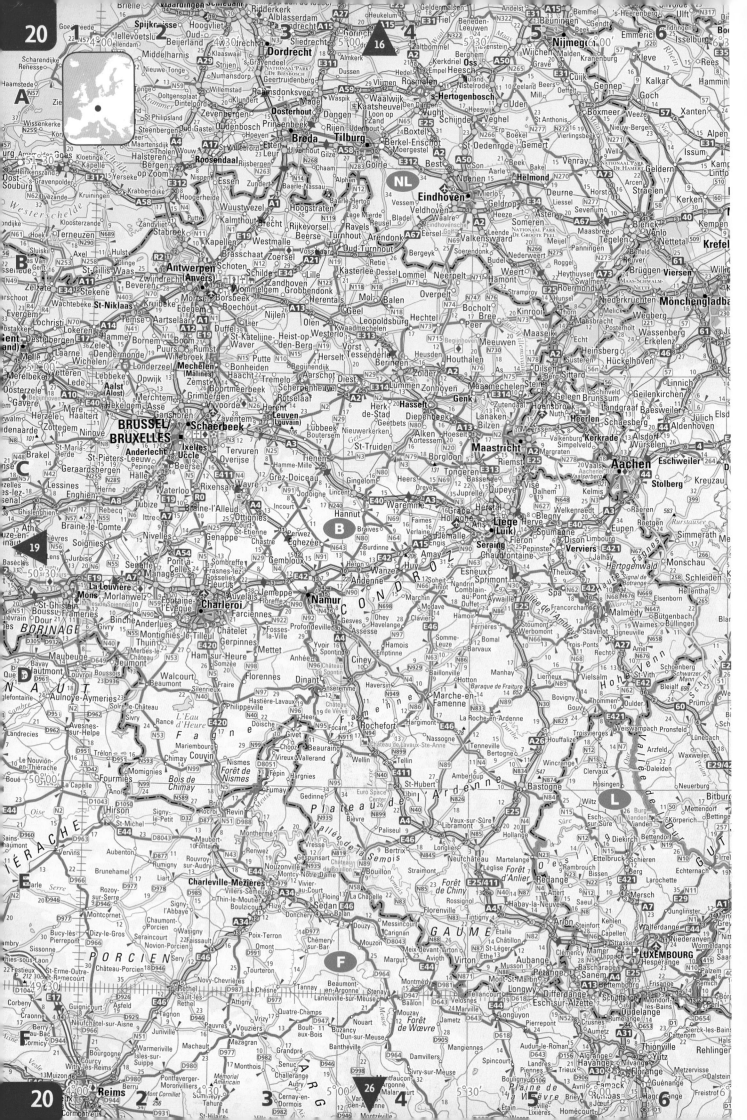

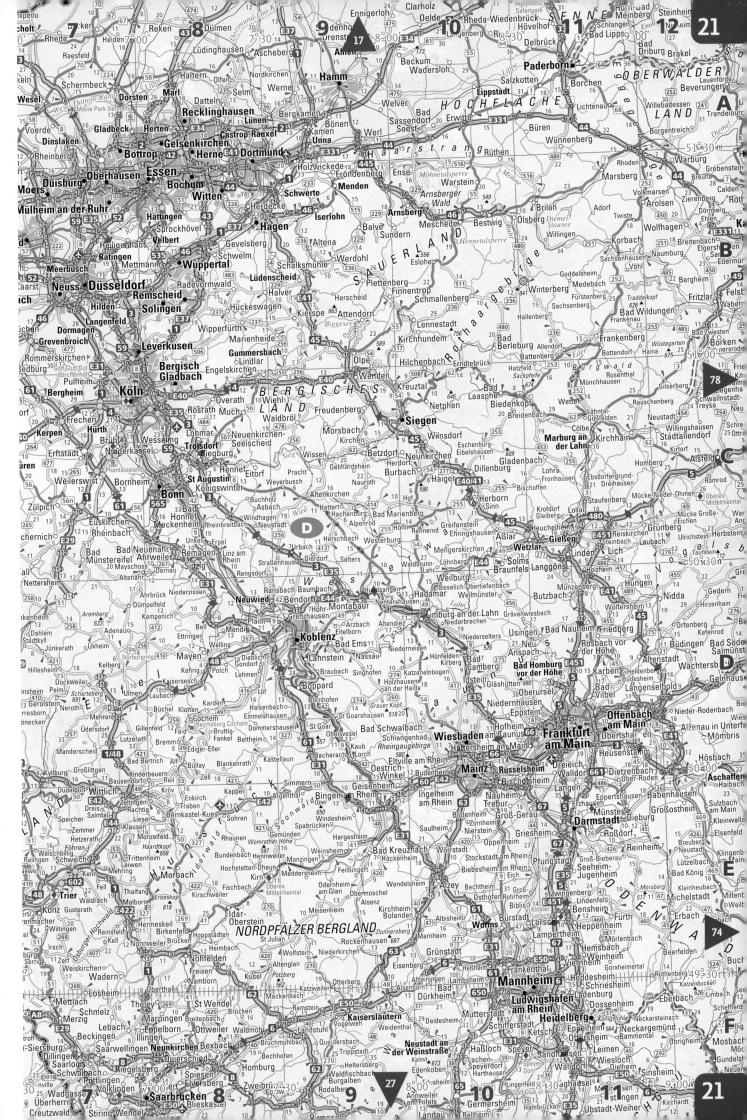

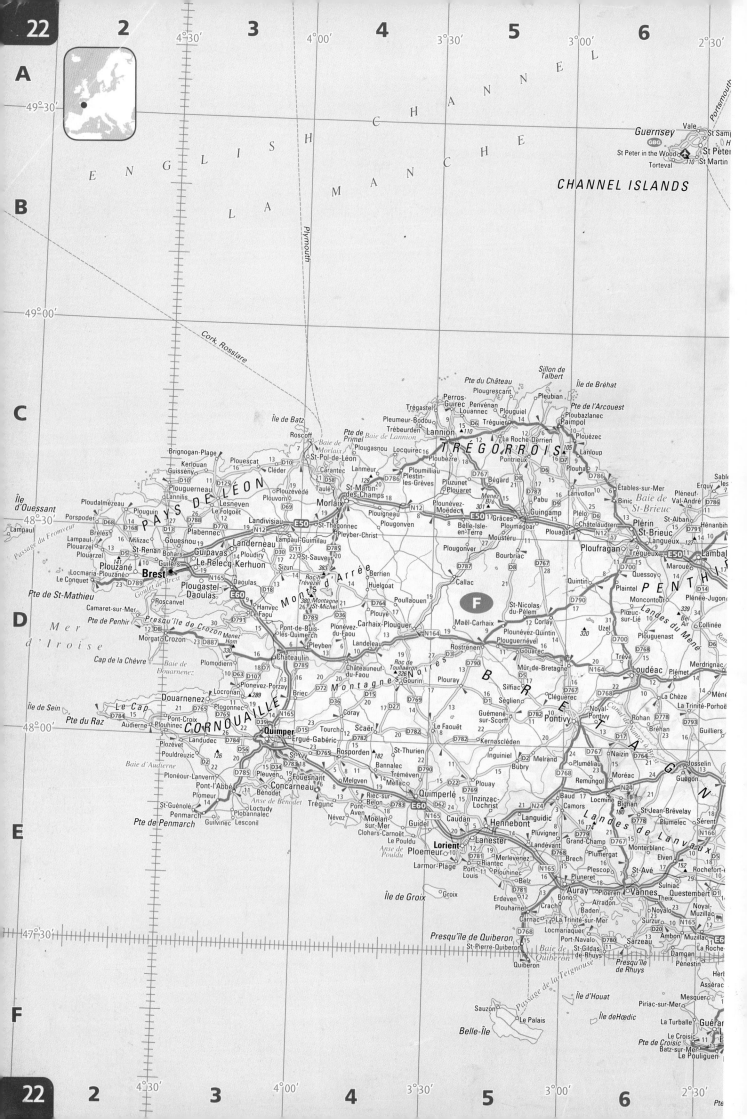

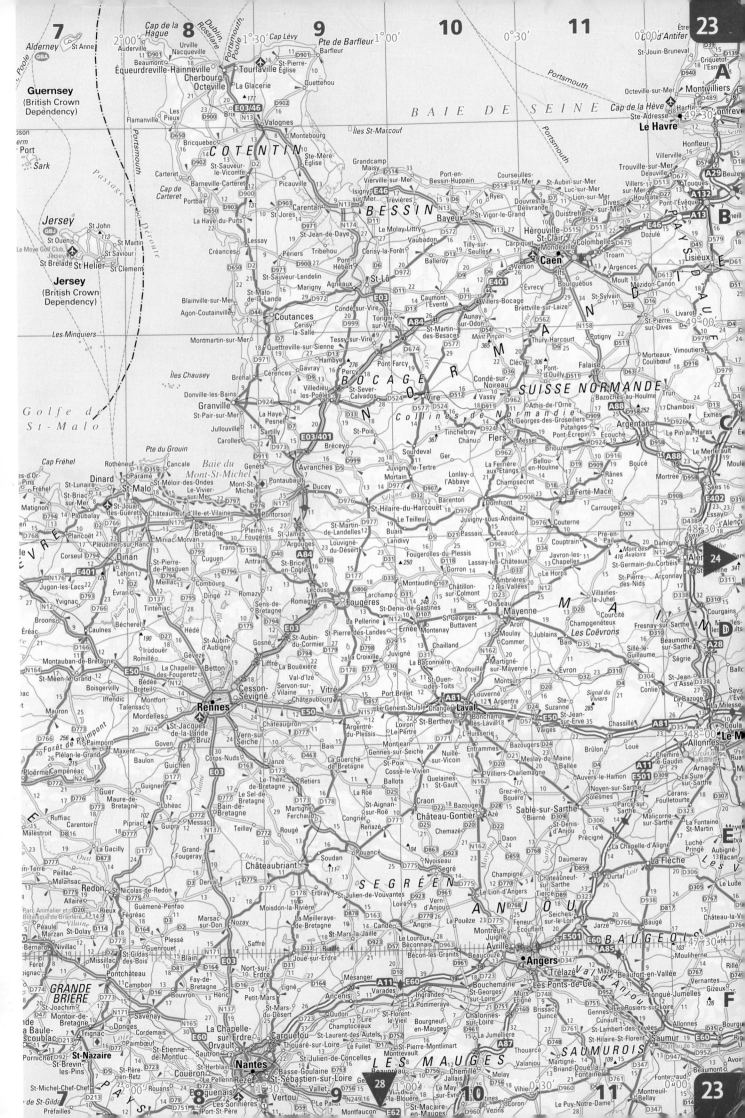

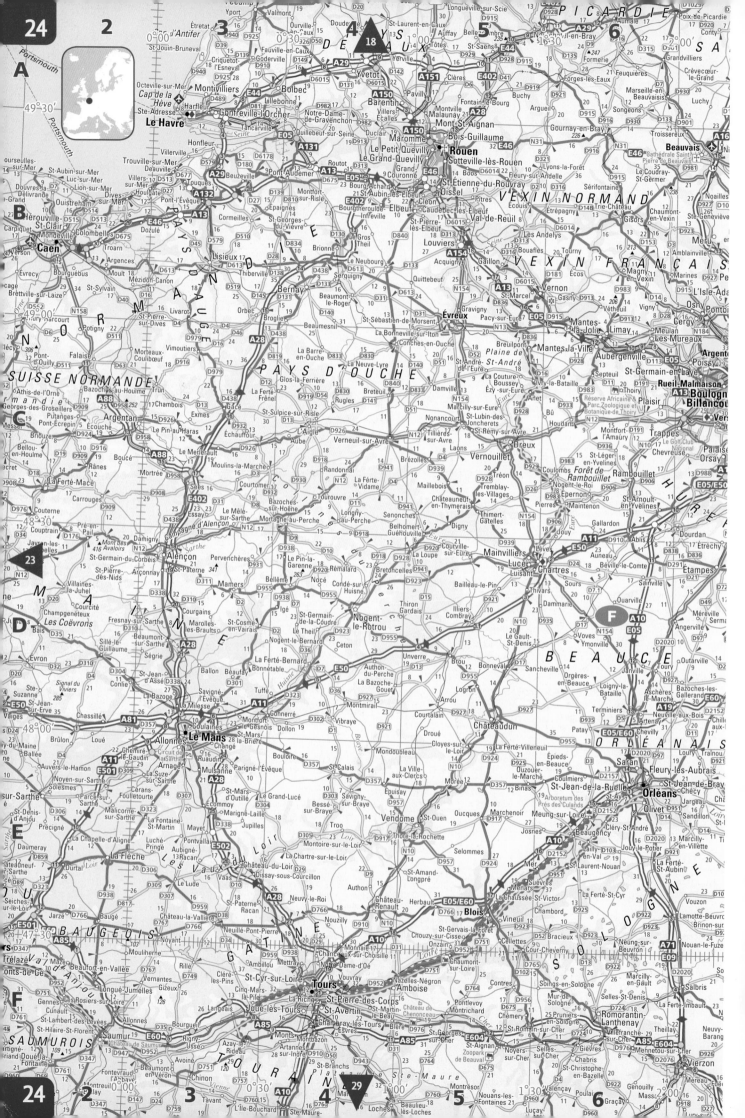

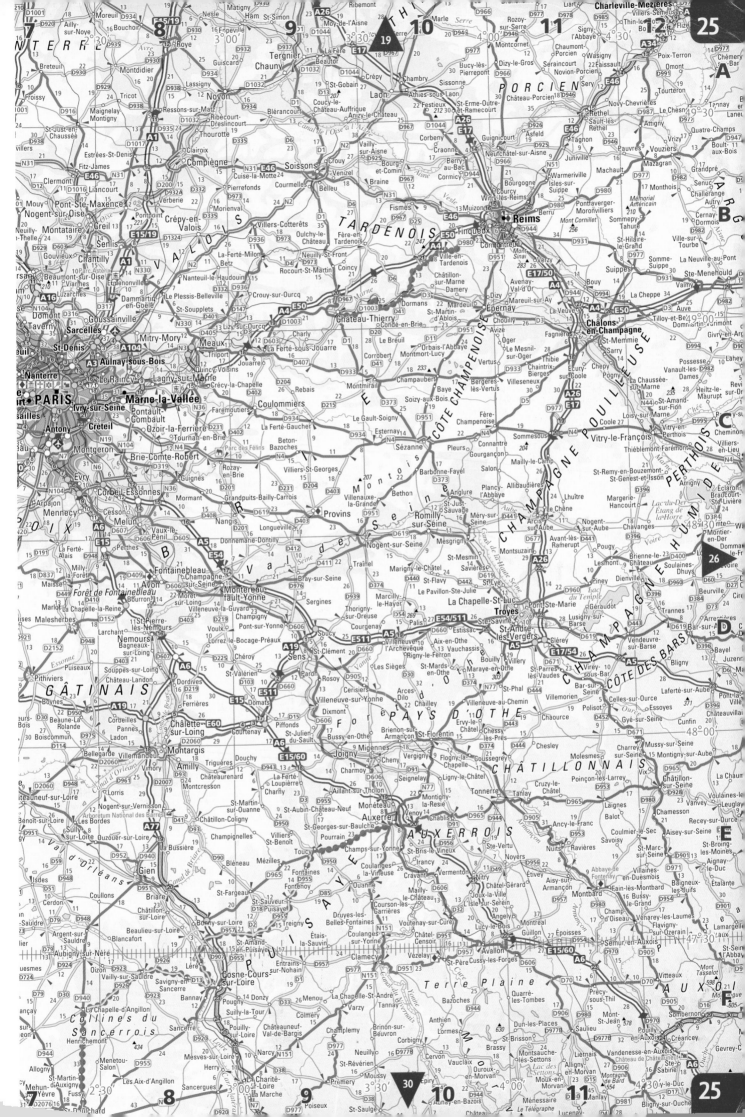

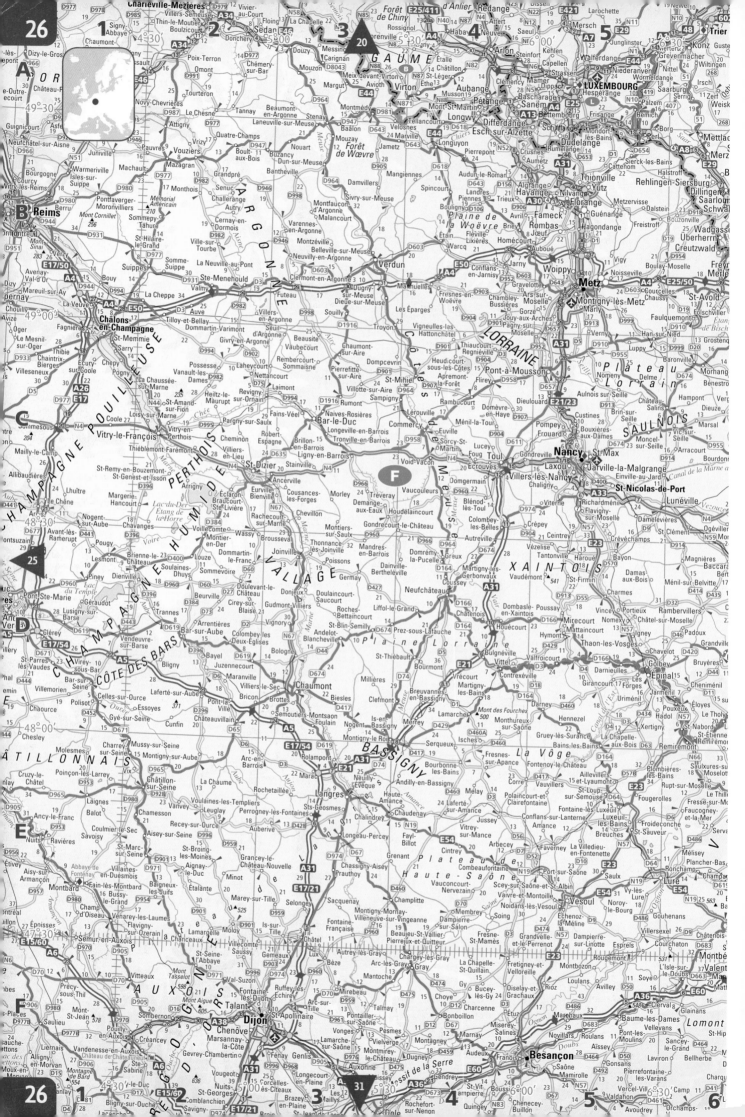

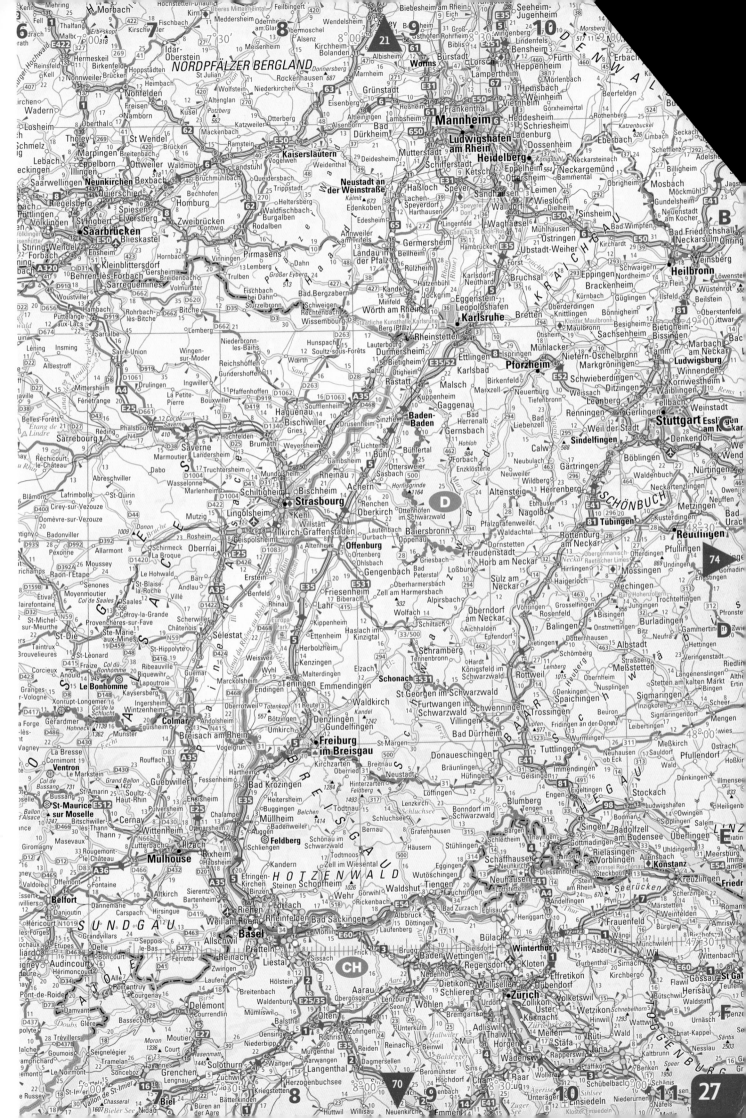

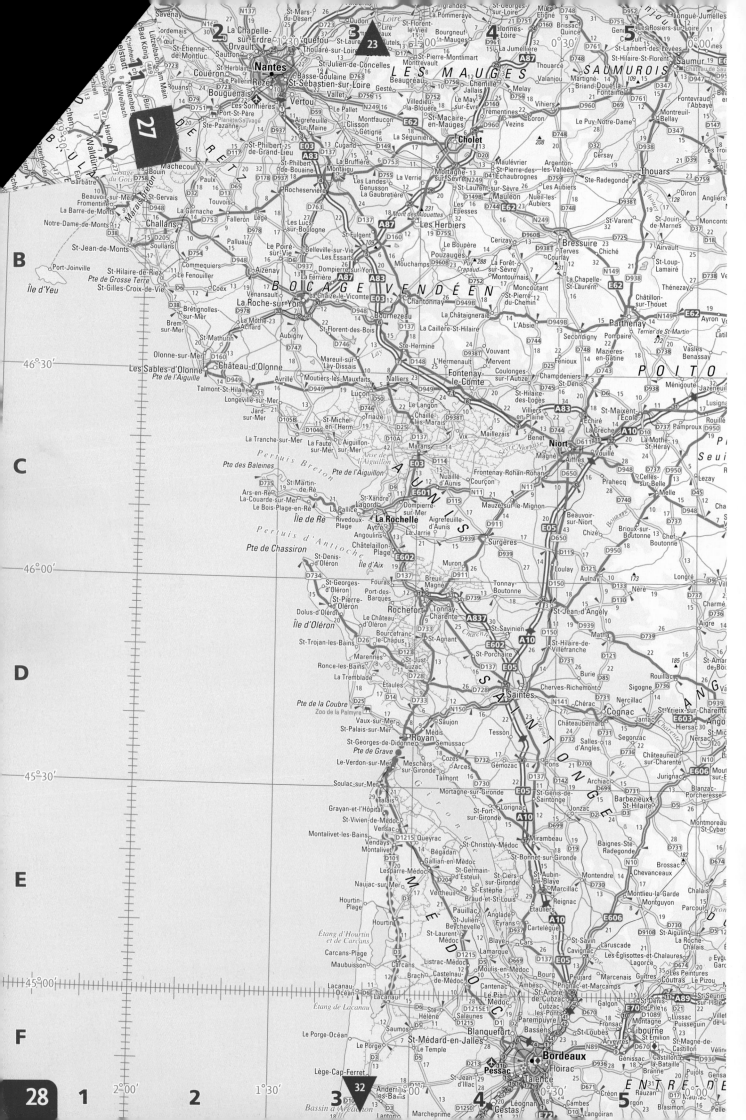

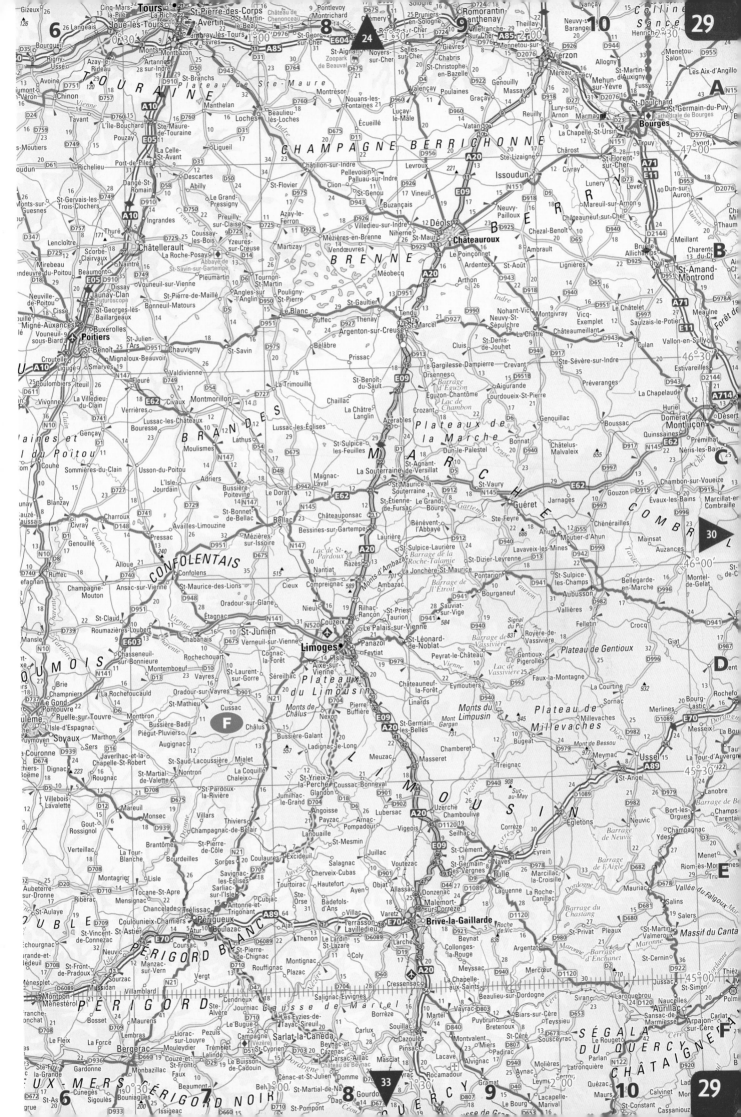

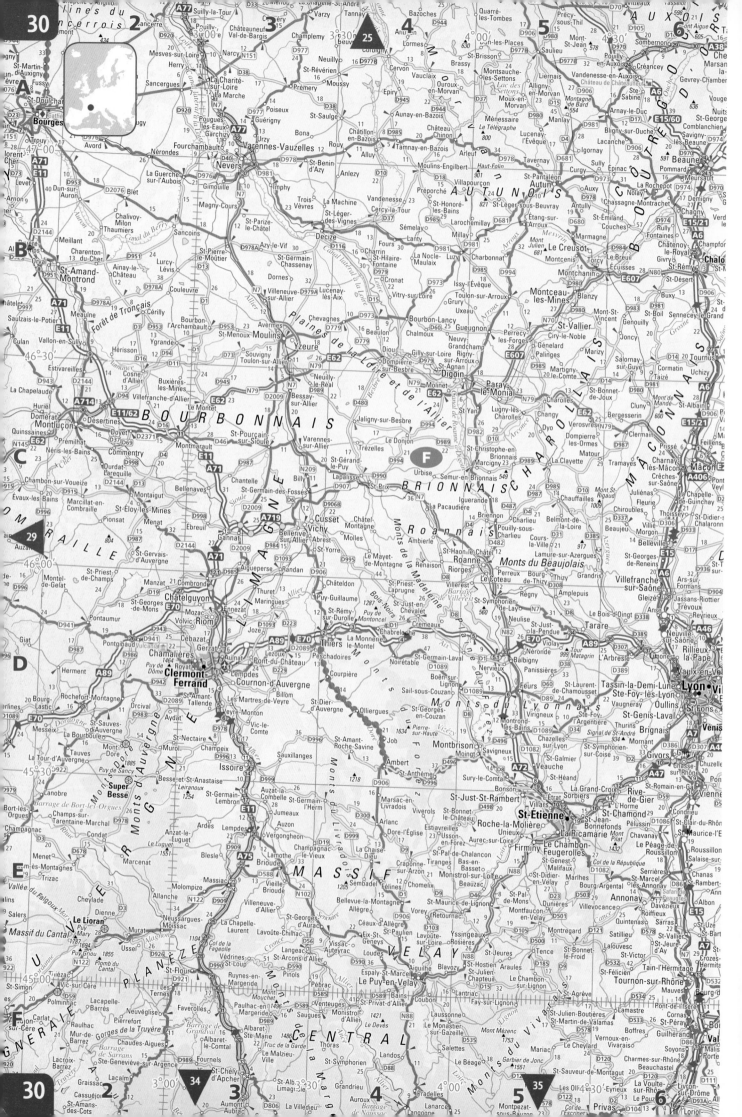

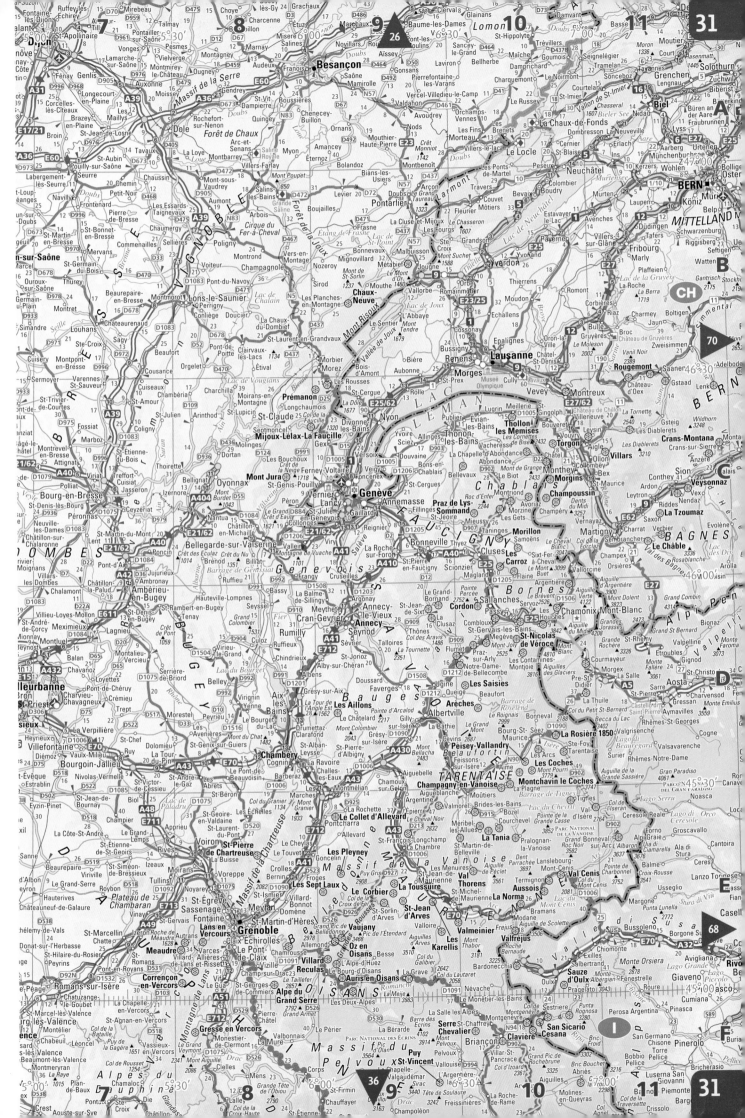

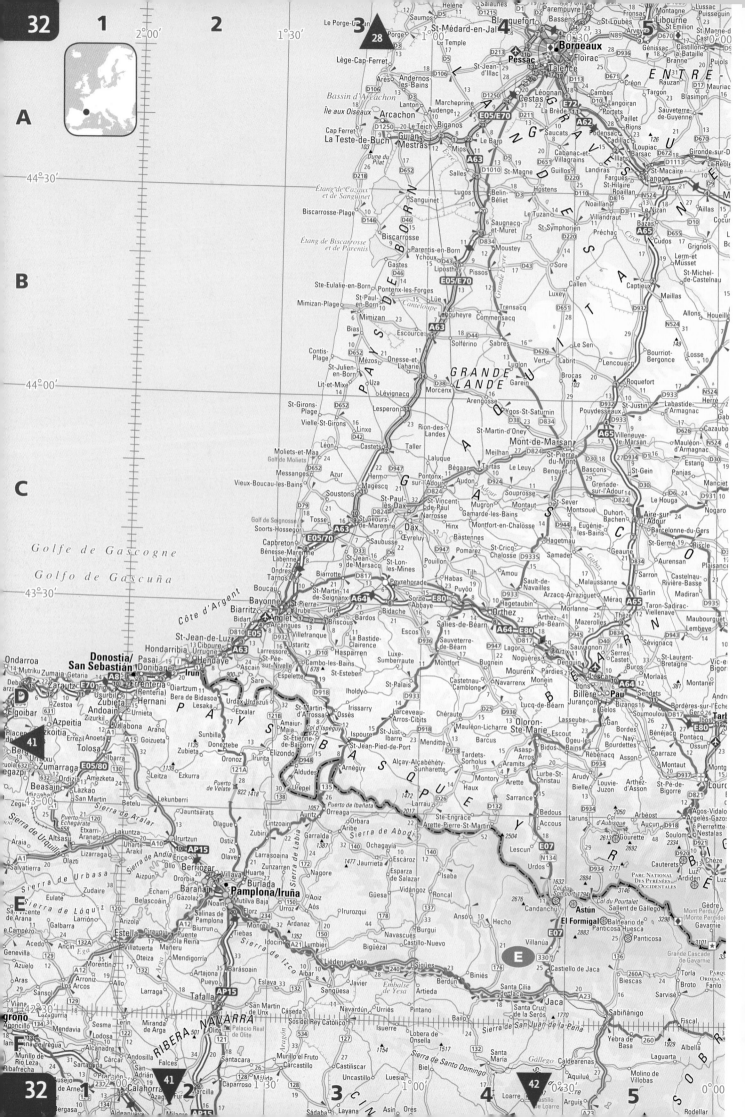

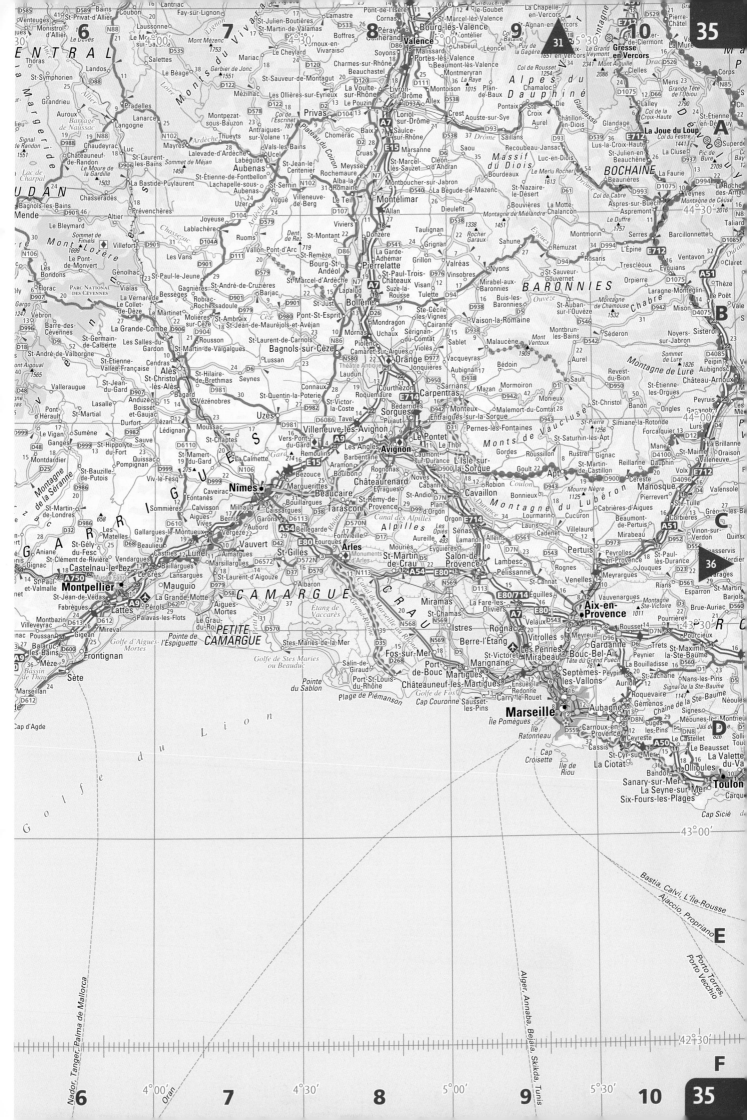

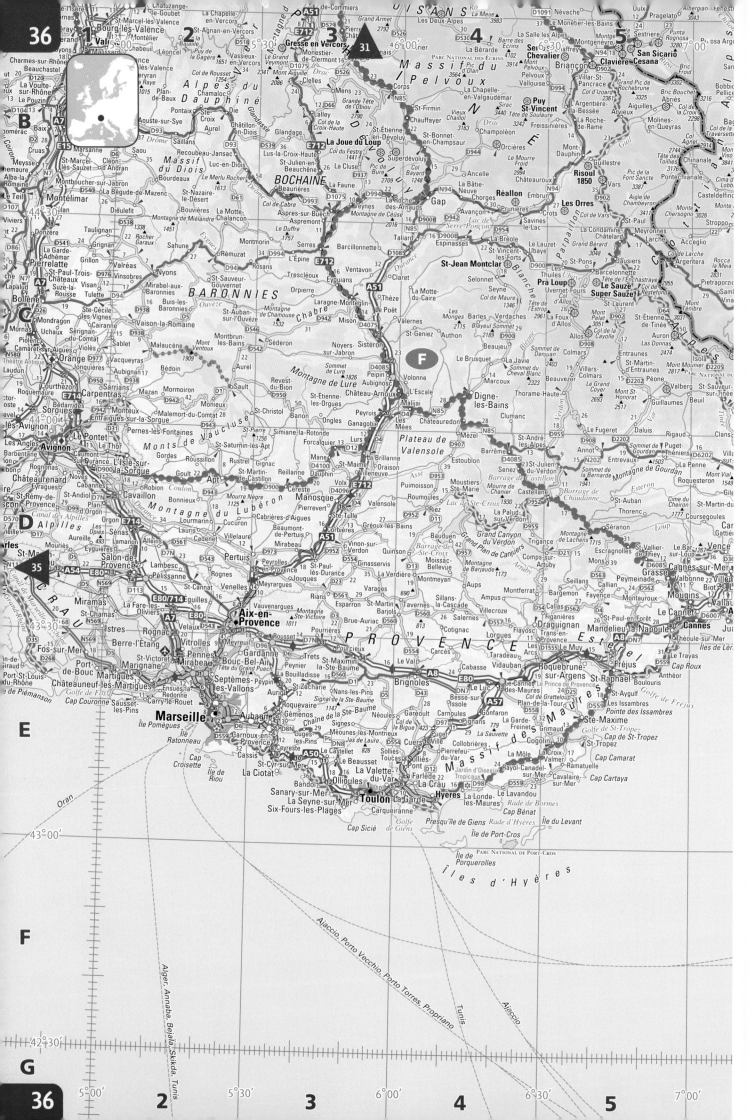

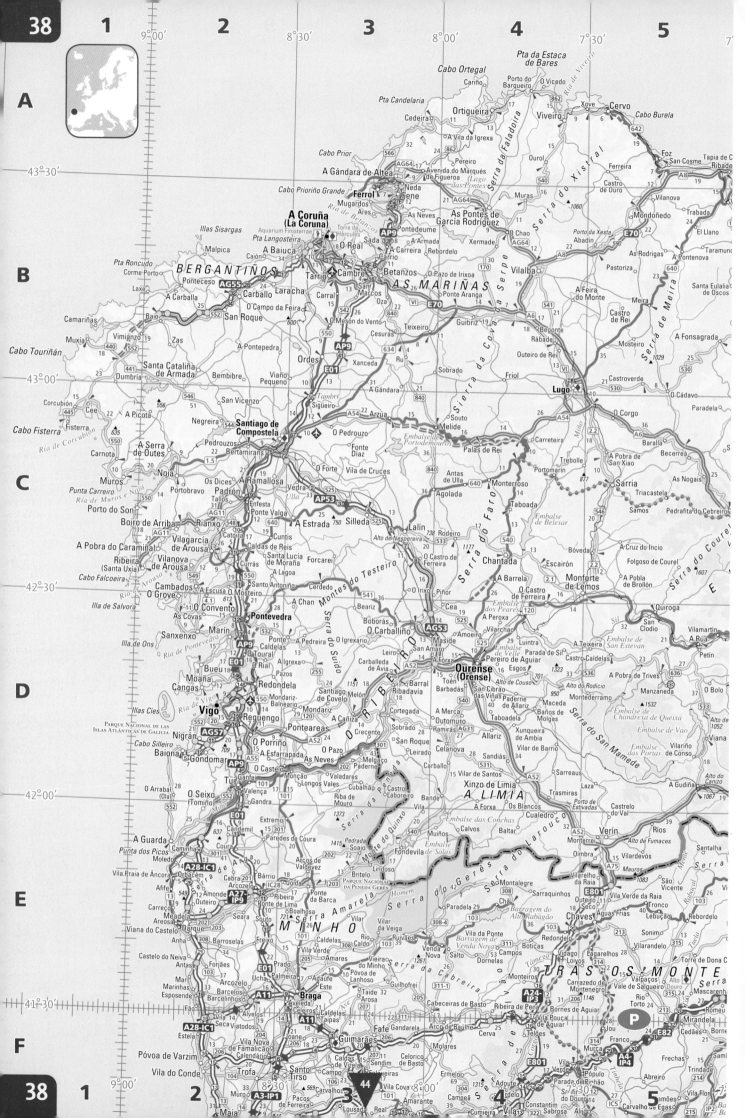

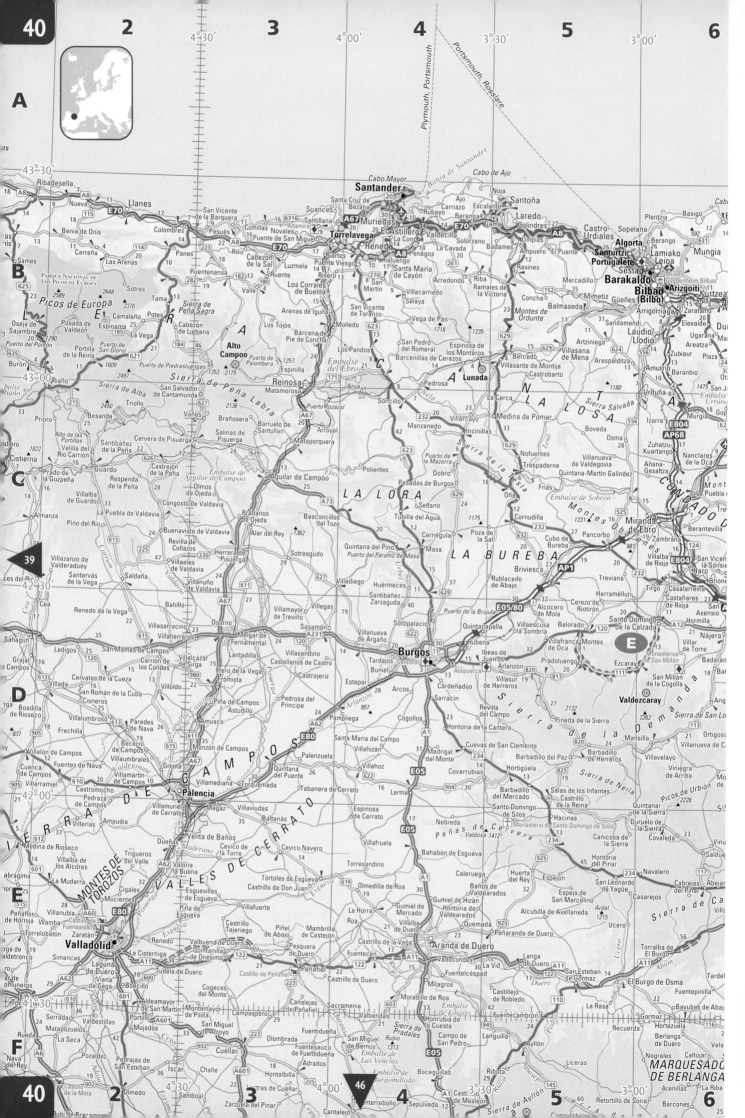

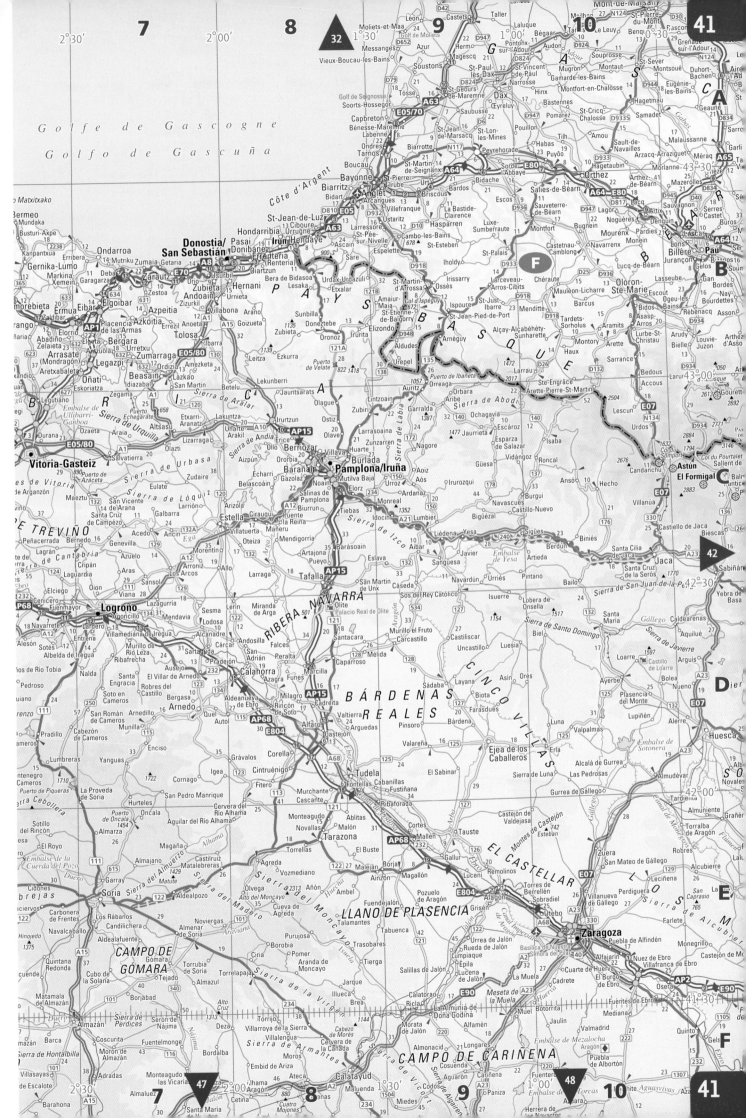

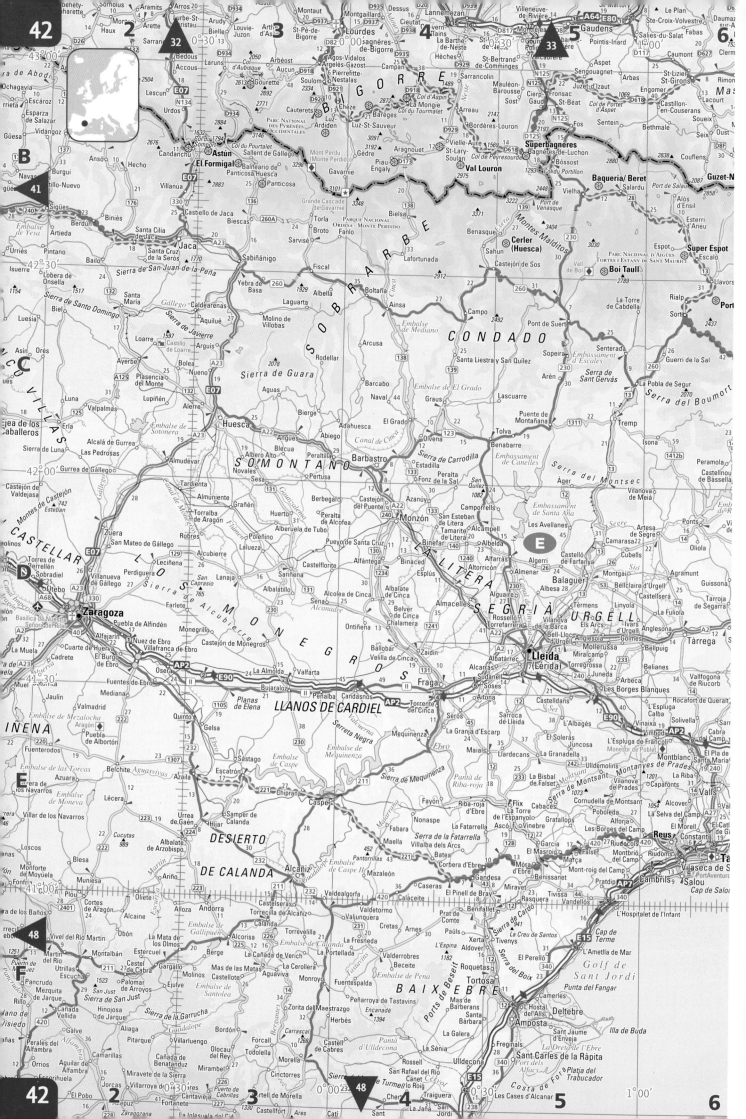

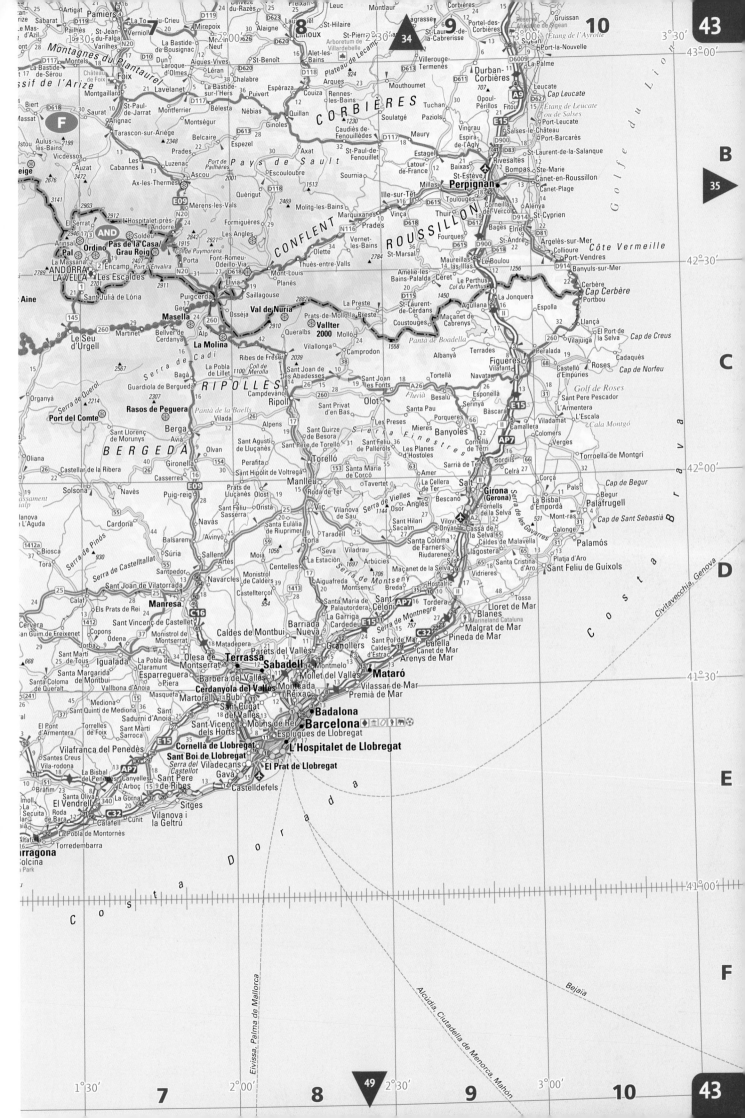

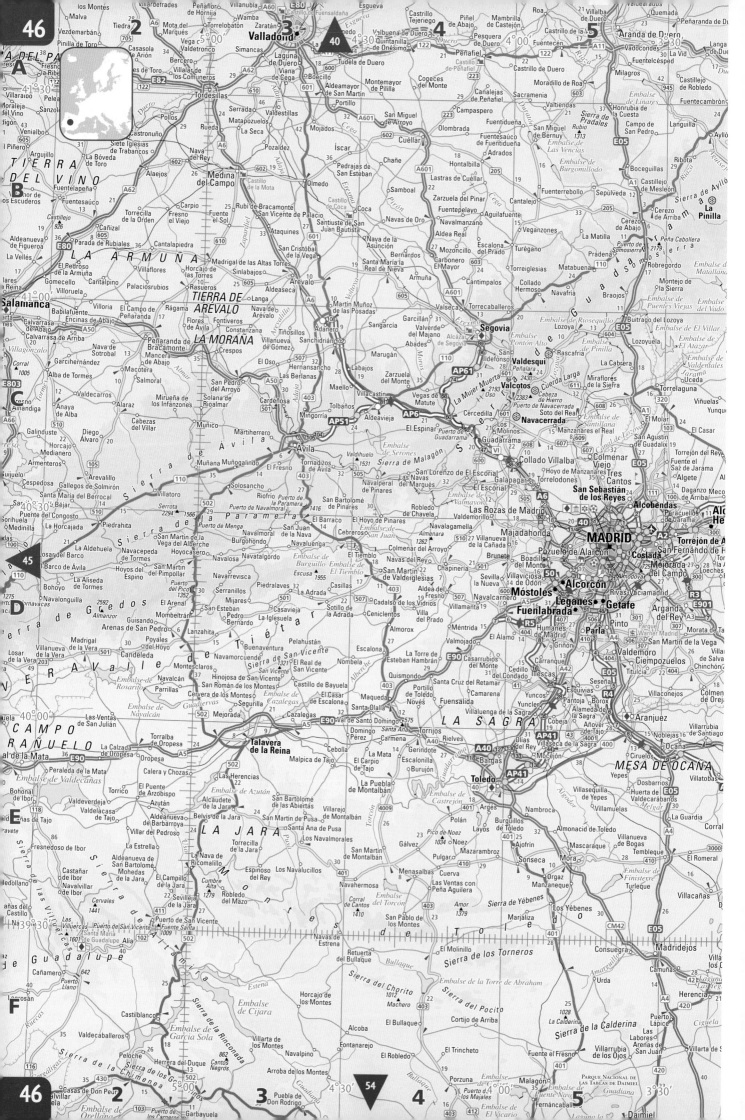

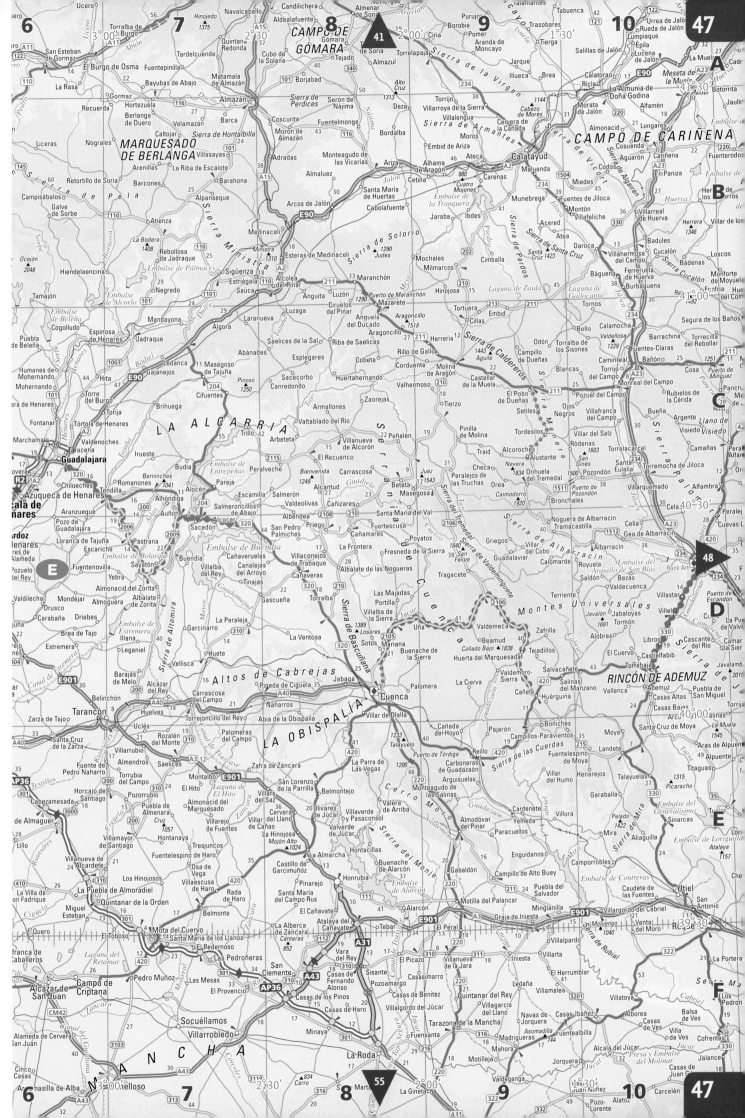

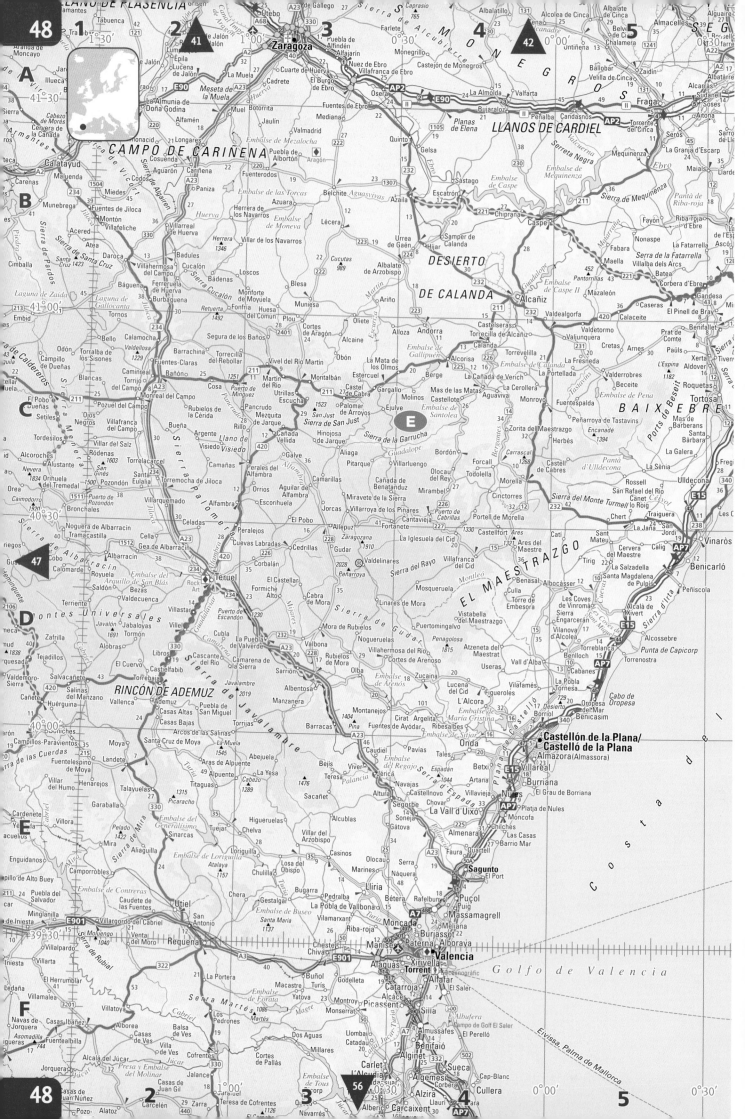

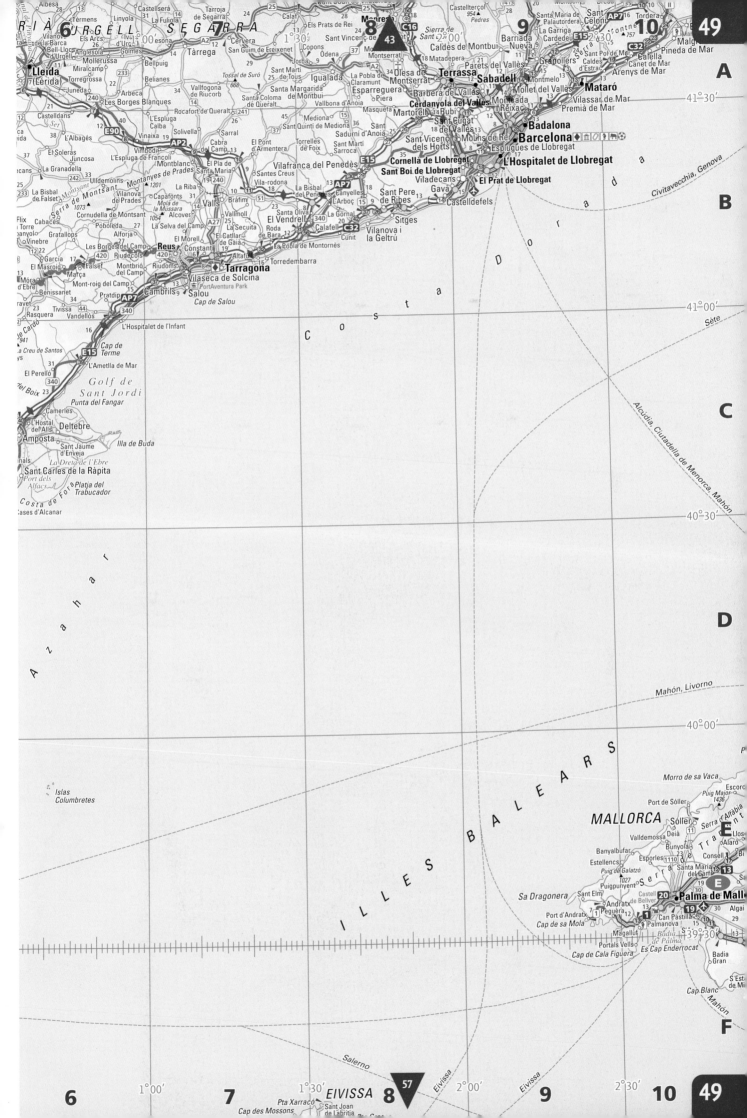

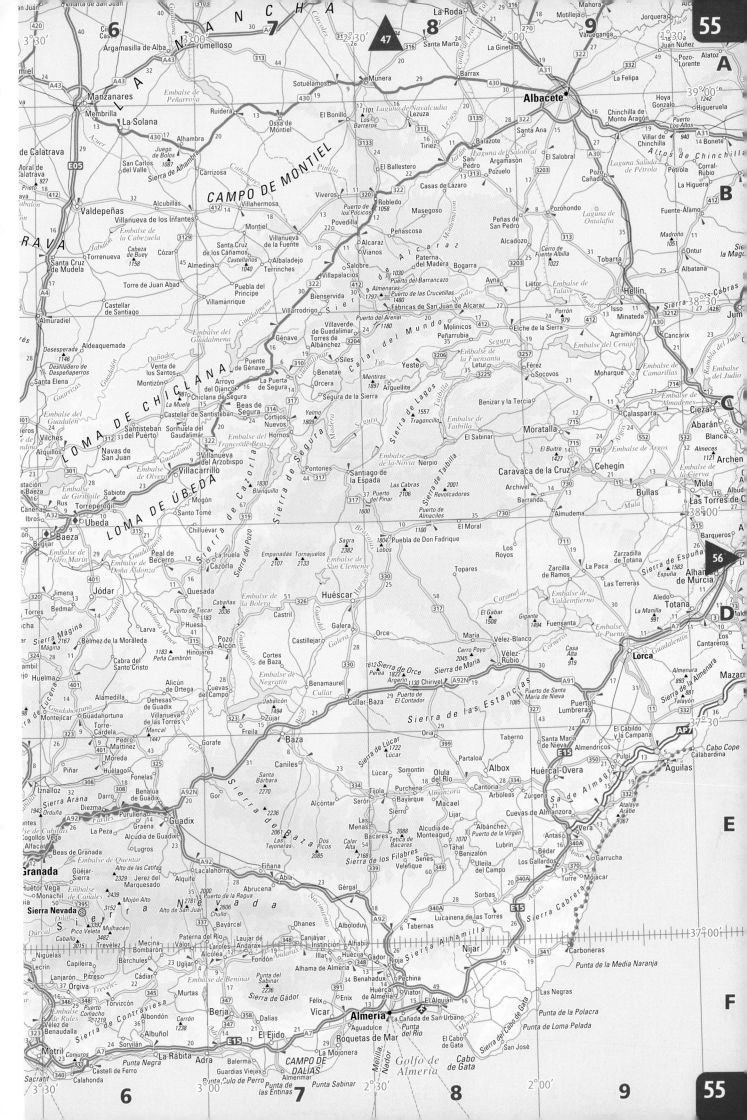

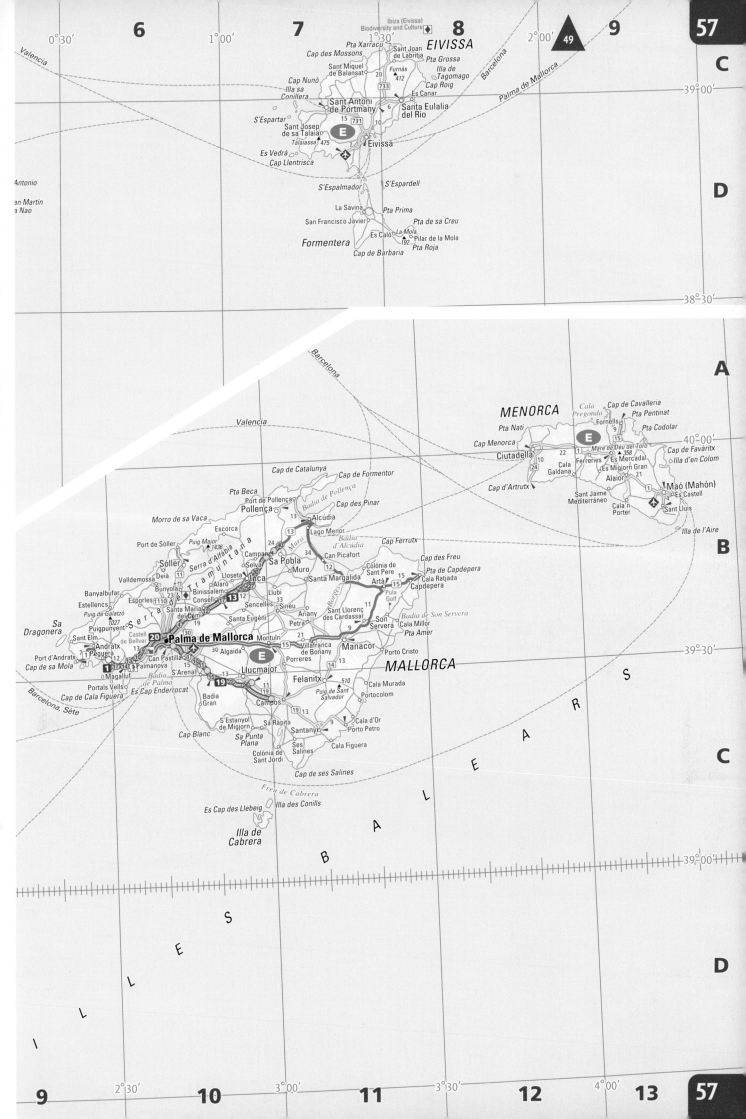

6 0°30' **7** 1°00' **8** 1°30' 2°00' ▲**49** **9**

C

Valencia

Ibiza (Eivissa)
Biodiversity and Culture ◆

EIVISSA

Pta Xarracó
Cap des Mossons
Sant Joan
de Labritja
Pta Grossa
Illa de
Tagomago
Sant Miquel
de Balansat
Furnás
▲412
Cap Nunó
Illa sa
Conillera
733
Cap Roig
Es Canar

39°00'

Santa Eulalia
del Rio

Sant Antoni
de Portmany

S'Espartar
15 731
6
10

Sant Josep
de sa Talaia
(E)
7
Tàlaiassa 475

Eivissa

Es Vedrà
Cap Llentrisca

D

An Antonio
an Martín
e Nao

S'Espardell
S'Espalmador

La Savina
Pta Prima

San Francisco Javier
Pta de sa Creu
La Mola
Formentera
Es Caló 192 Pilar de la Mola
Pta Roja

Cap de Barbaria

38°30'

Barcelona

A

Valencia

MENORCA
Cala
Pregonda
Cap de Cavalleria
Fornells
Pta Pentinat
Pta Nati
9
Pta Codolar
(E)
40°00'
Cap Menorca
15
Mare de Déu del Toro
Cap de Favàritx
Ciutadella
10
Cala
Galdana
24
Ferreries
▲358
Es Mercadal
Illa d'en Colom
Es Migjorn Gran
Alaior
21
1
Maó (Mahón)
Cap d'Artrutx
Sant Jaime
Mediterráneo
Es Castell
Cala n
Porter
Sant Lluís

Illa de l'Aire

Cap de Catalunya
Cap de Formentor

Pta Beca
Badia de Pollença
Port de Pollença
Cap des Pinar
Pollença
13
Alcúdia
Morro de sa Vaca
13
Lago Menor
Escorca
Badia
d'Alcúdia
Cap Ferrutx
Port de Sóller
24
Muro
Can Picafort
Puig Major
▲1436
34
Sóller
Campanet
Sa Pobla
Colònia de
Sant Pere
Cap des Freu
Selva
Muro
Pta de Capdepera
Valldemossa
Deià
11
Lloseta
Inca
12
Santa Margalida
Artà
Cala Ratjada
Bunyola
Alaró
30
Capdepera
Banyalbufar
23
Binissalem
Llubí
Pula
Golf
15
Estellencs
Esporles
1110
Consell
Sencelles
Sineu
11
Puig de Galatzó
Santa Maria
del Camí
33
Son
Servera
Cala Millor
1027
Santa Eugènia
Ariany
Sant Llorenç
des Cardassar
Puigpunyent
19
Petra
Pta Amer
Sa
Dragonera
Sant Elm
Castell
de Bellver
20
Palma de Mallorca
Montuïri
21
Villafranca
de Bonany
Manacor
Badia de Son Servera
Andratx
13
Algaida
Porto Cristo
Port d'Andratx
Peguera
Can Pastilla
30
15
14
13
39°30'
Cap de sa Mola
Magaluf
Palmanova
Llucmajor
Porreres
MALLORCA
S
Portals Vells
S'Arenal
13
Badia
de Palma
Felanitx
510
Cala Murada
Barcelona, Sète
Cap de Cala Figuera
Es Cap Enderrocat
19
22
11
Puig de Sant
Salvador
Portocolom
Badia
Gran
Campos
19
13
Cala d'Or
S'Estanyol
de Migjorn
Sa Ràpita
Santanyí
9
Porto Petro
Cap Blanc
Sa Punta
Plana
Ses
Salines
Cala Figuera
Colònia de
Sant Jordi
Cap de ses Salines
E
A
R
S

Freu de Cabrera

B

Es Cap des Llebeig
Illa des Conills

Illa de
Cabrera

39°00'

C

B

A

L

E

D

A

R

S

I

L

L

E

S

D

9 2°30' **10** 3°00' **11** 3°30' **12** 4°00' **13**

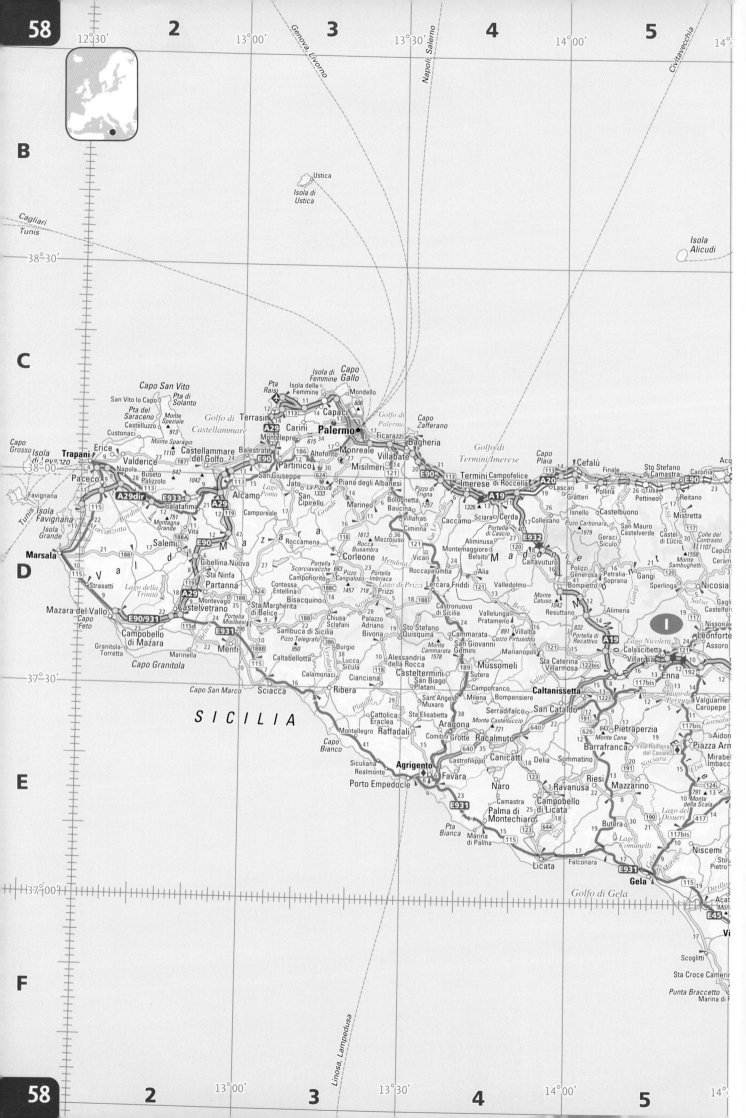

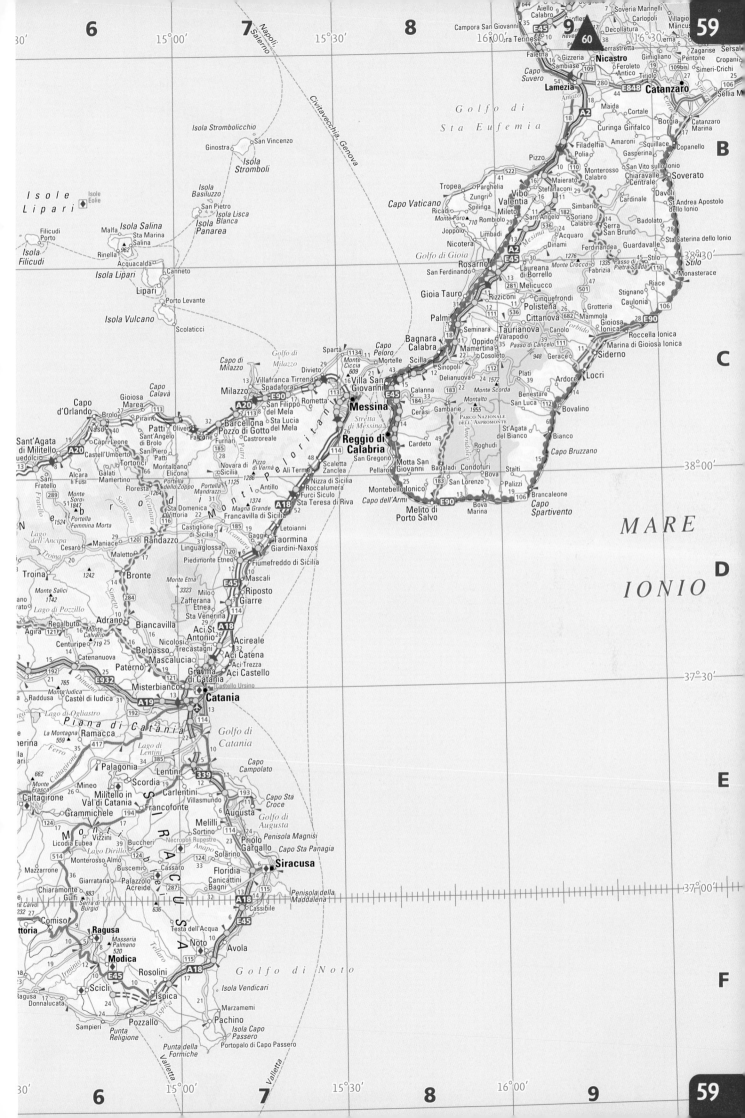

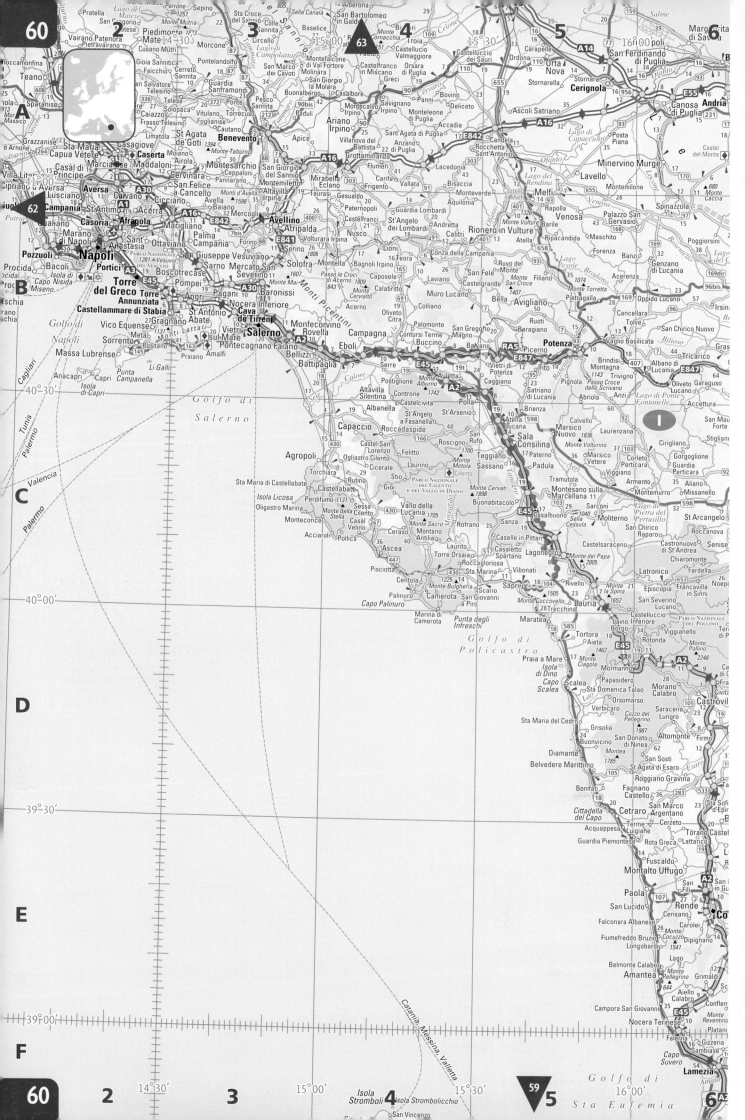

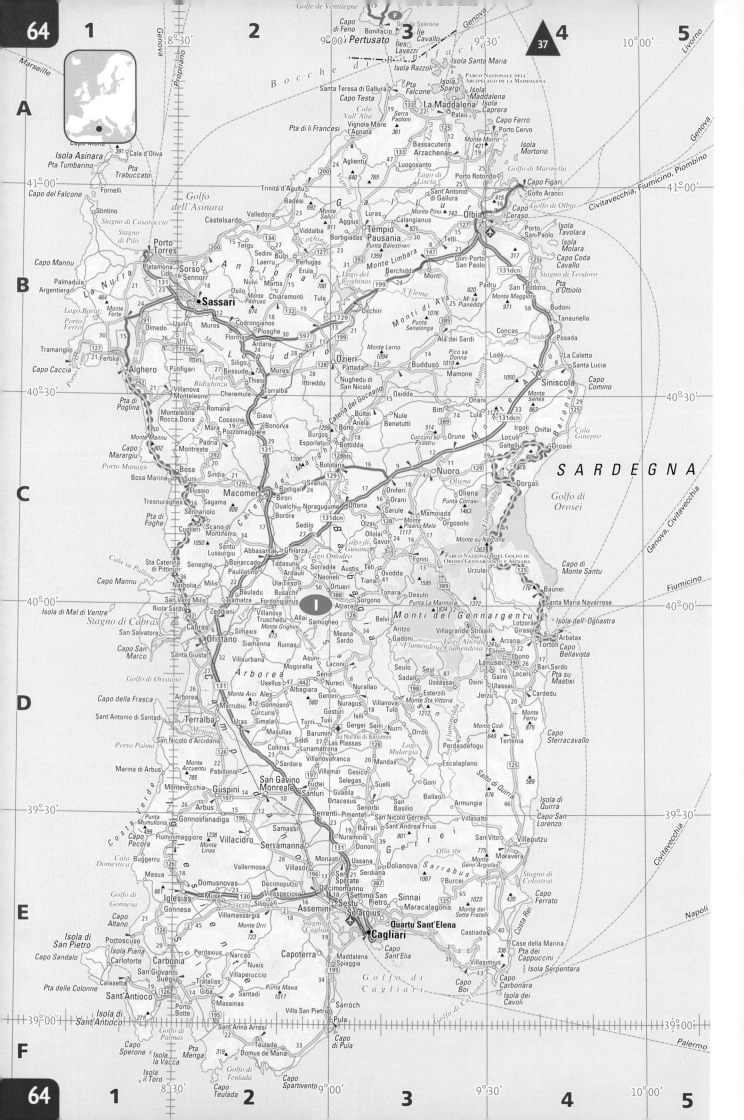

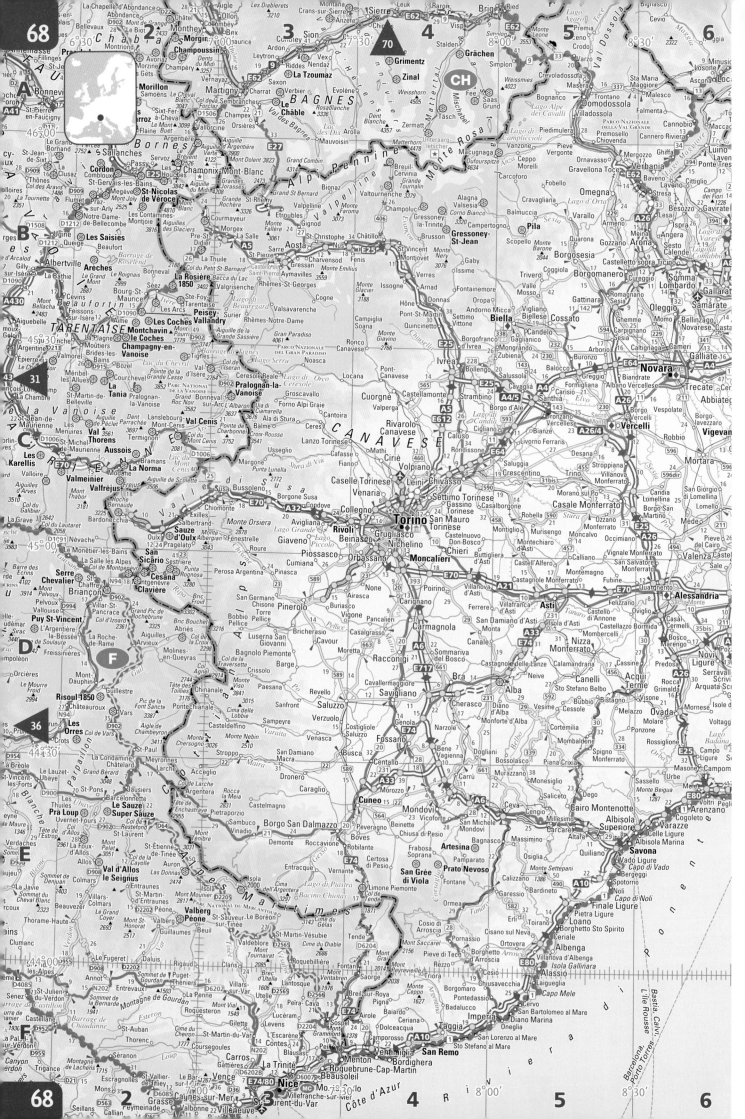

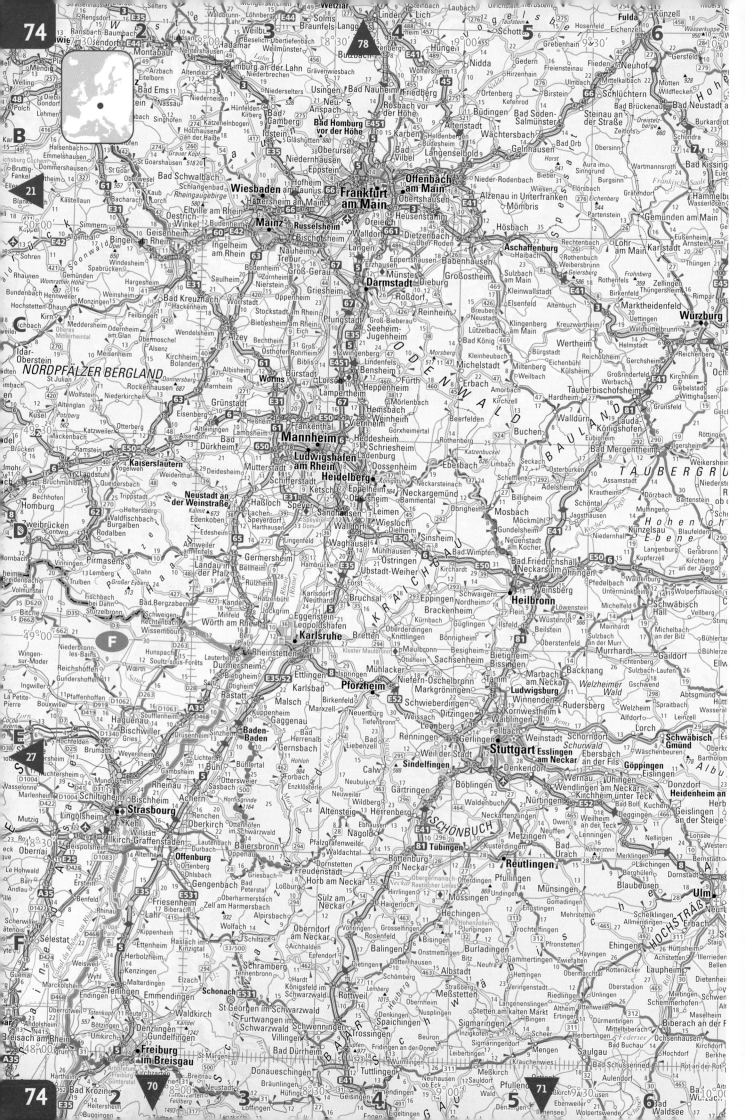

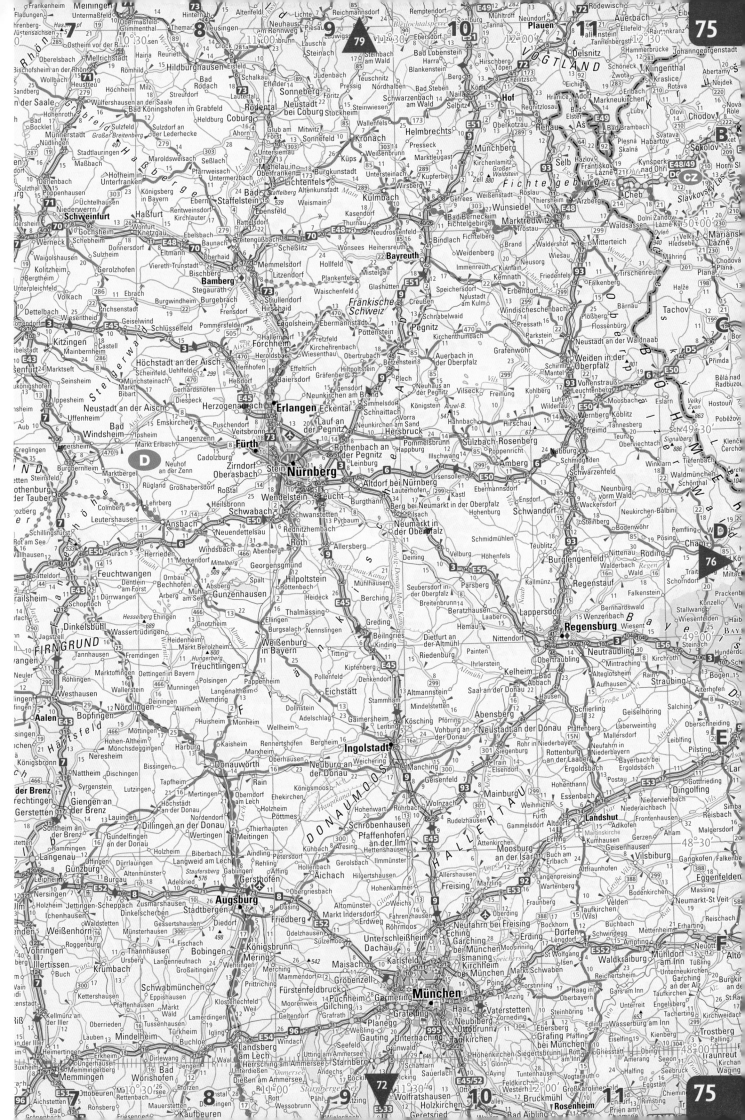

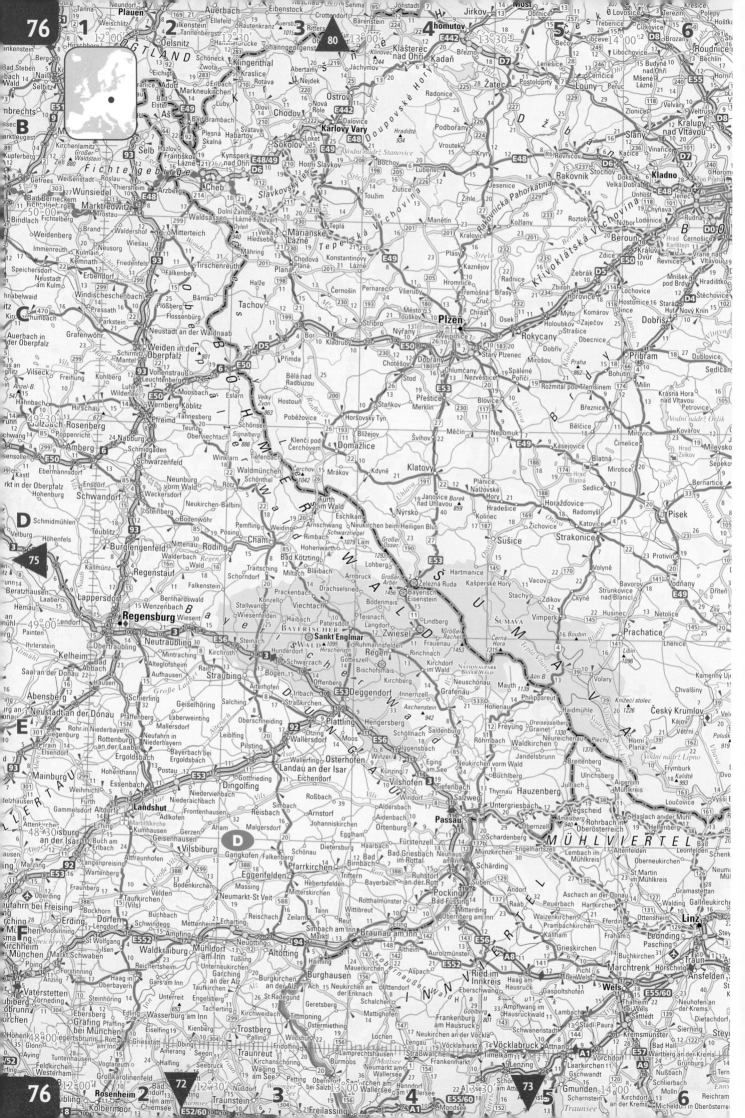

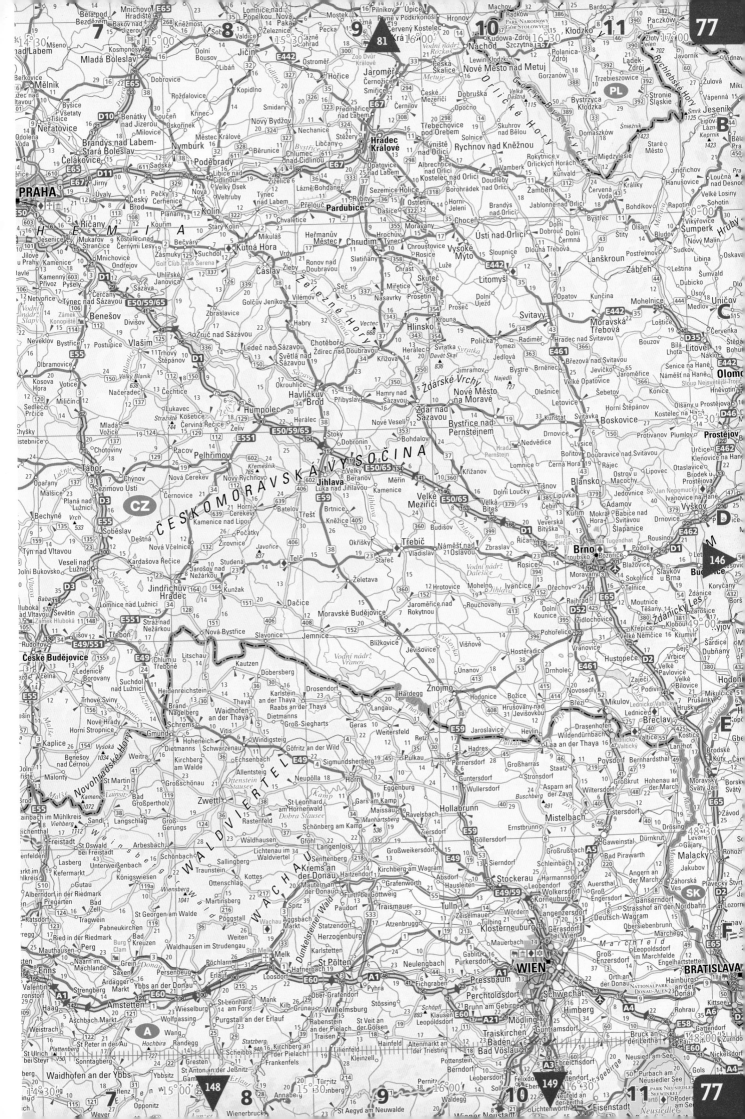

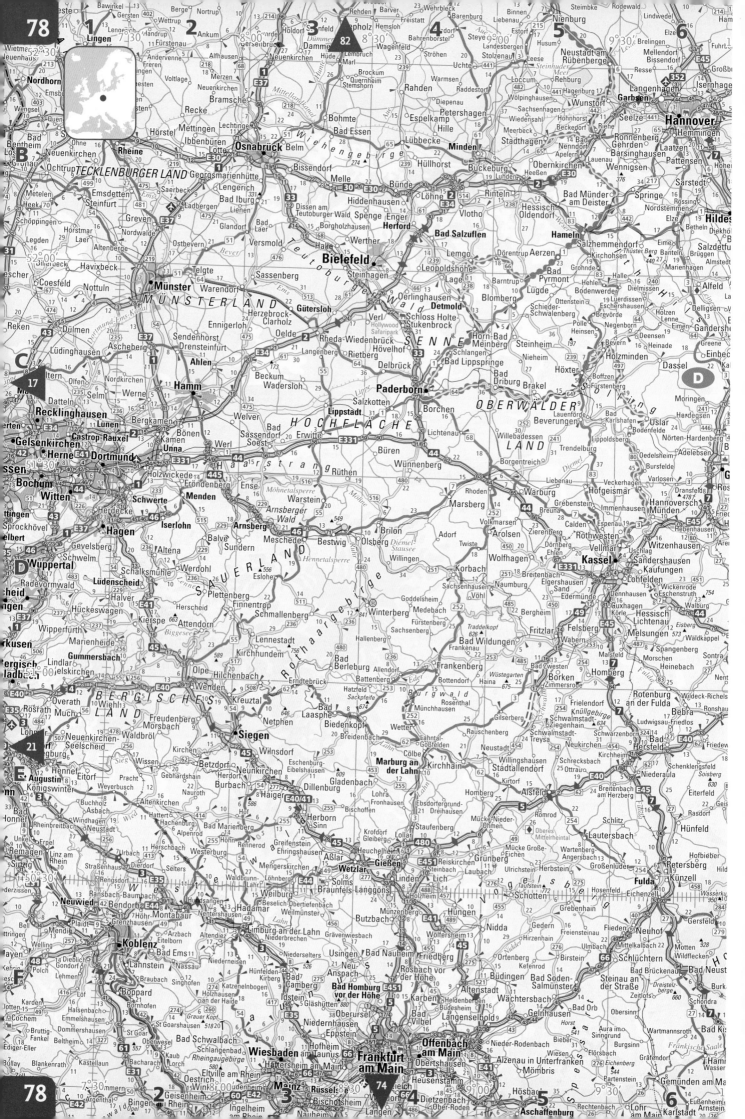

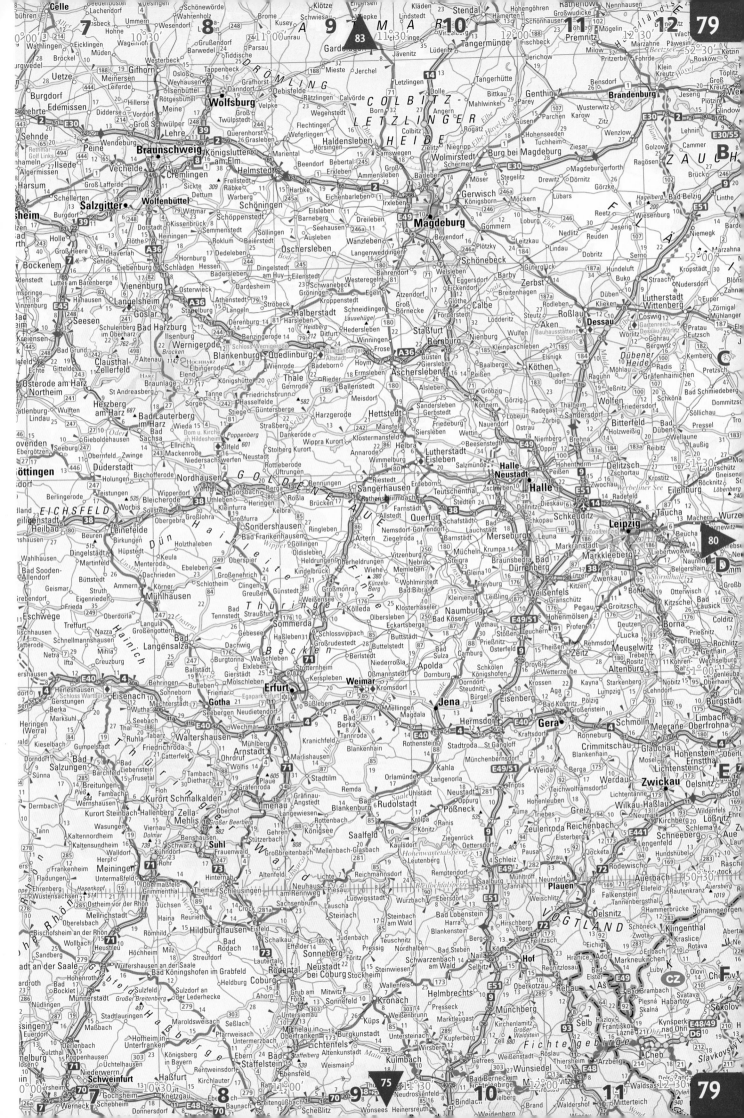

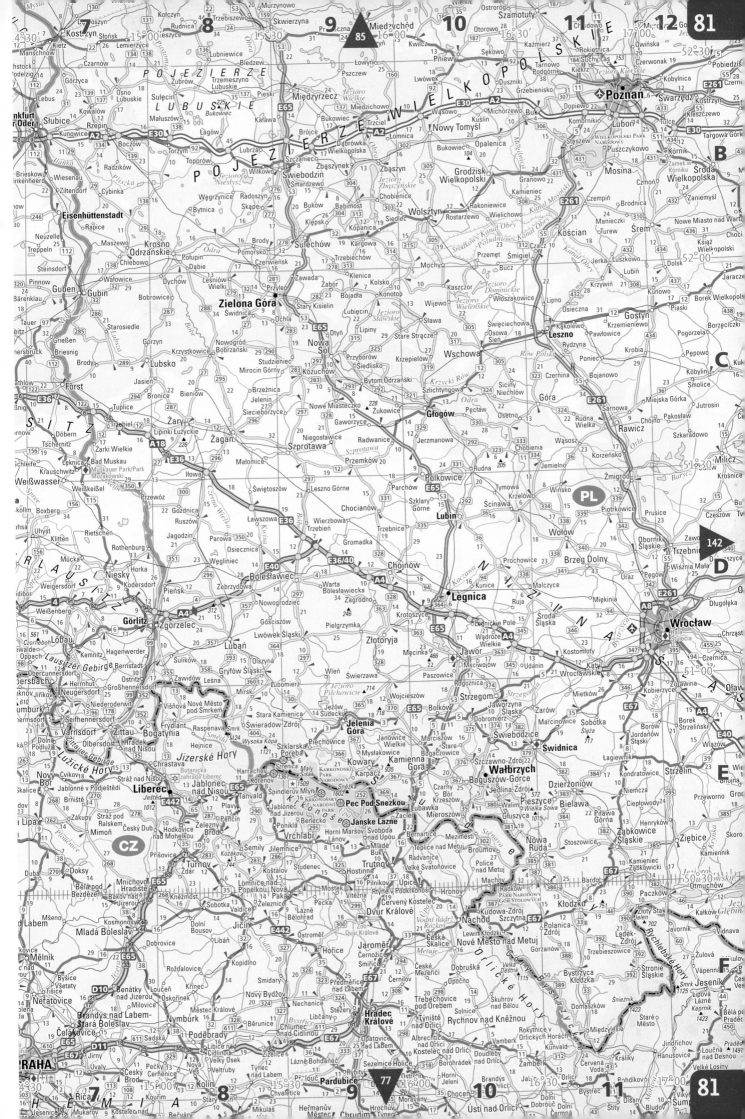

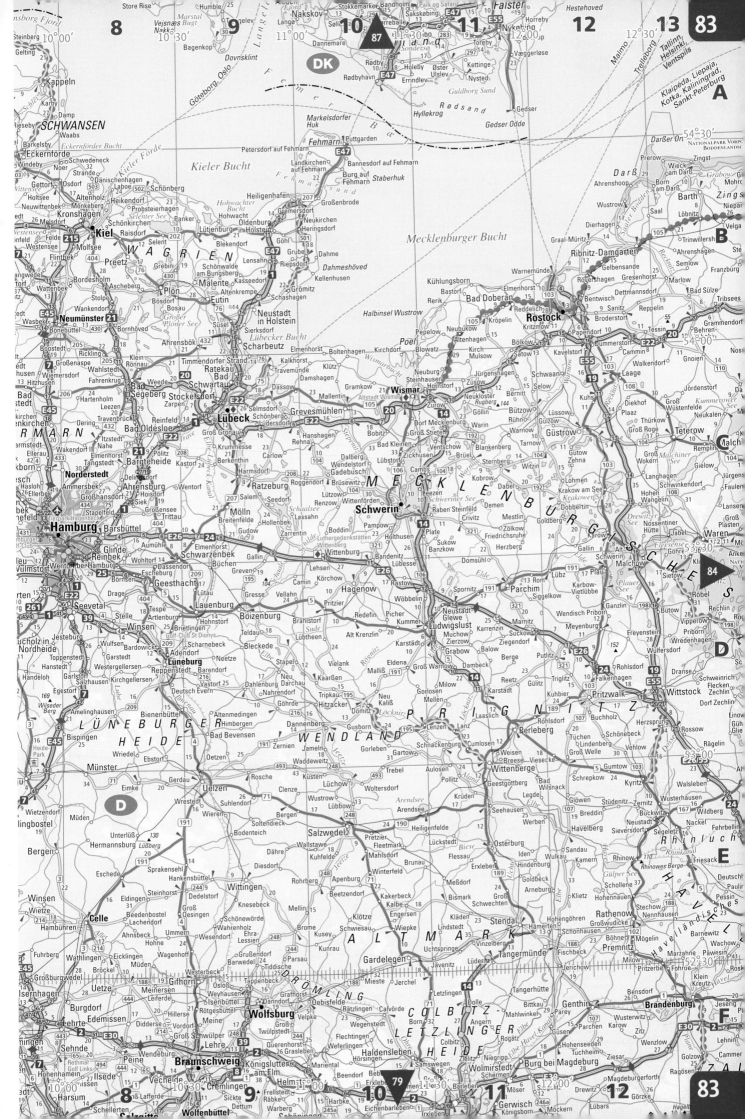

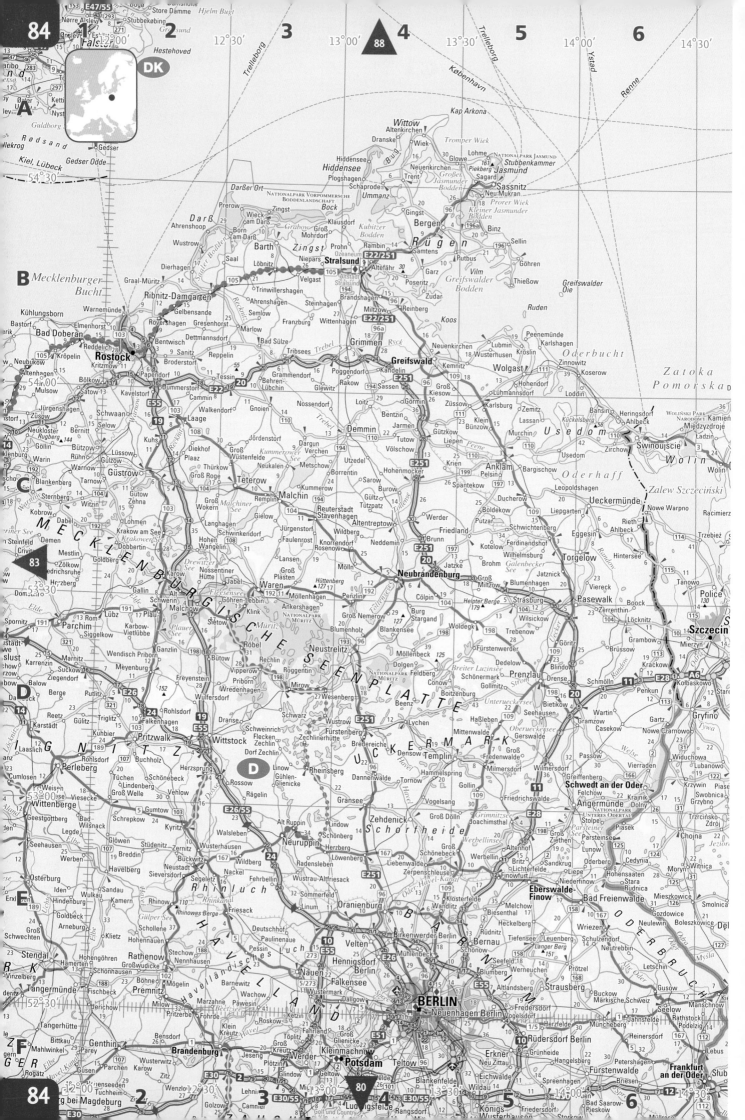

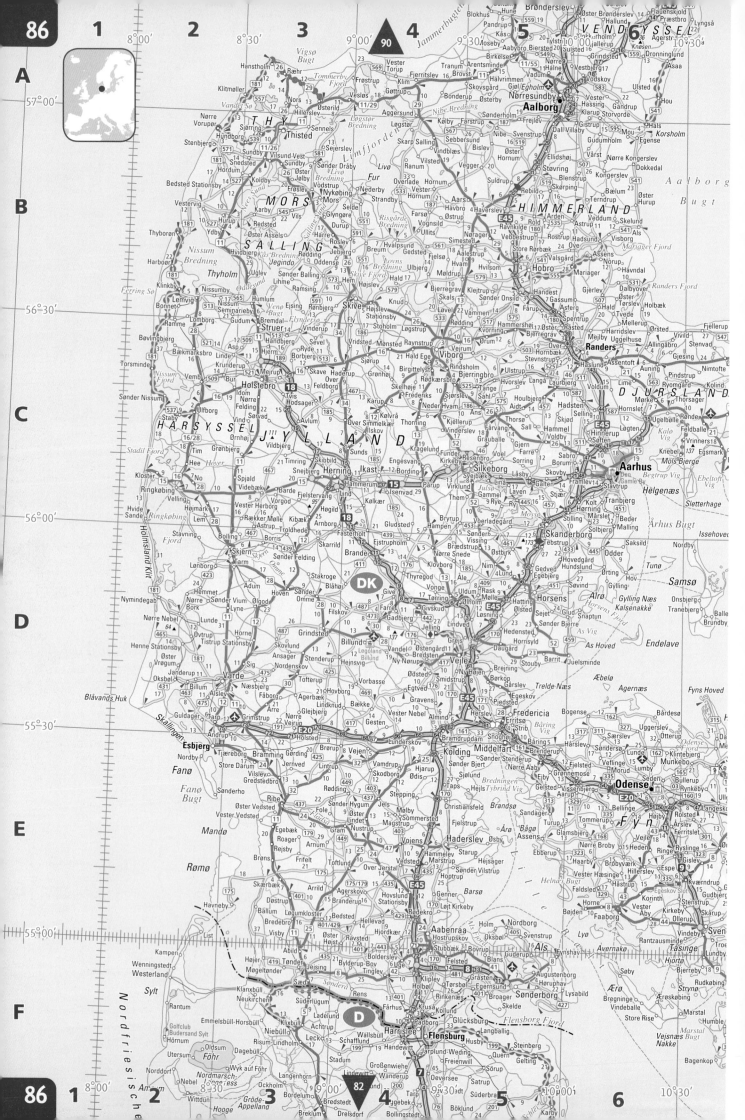

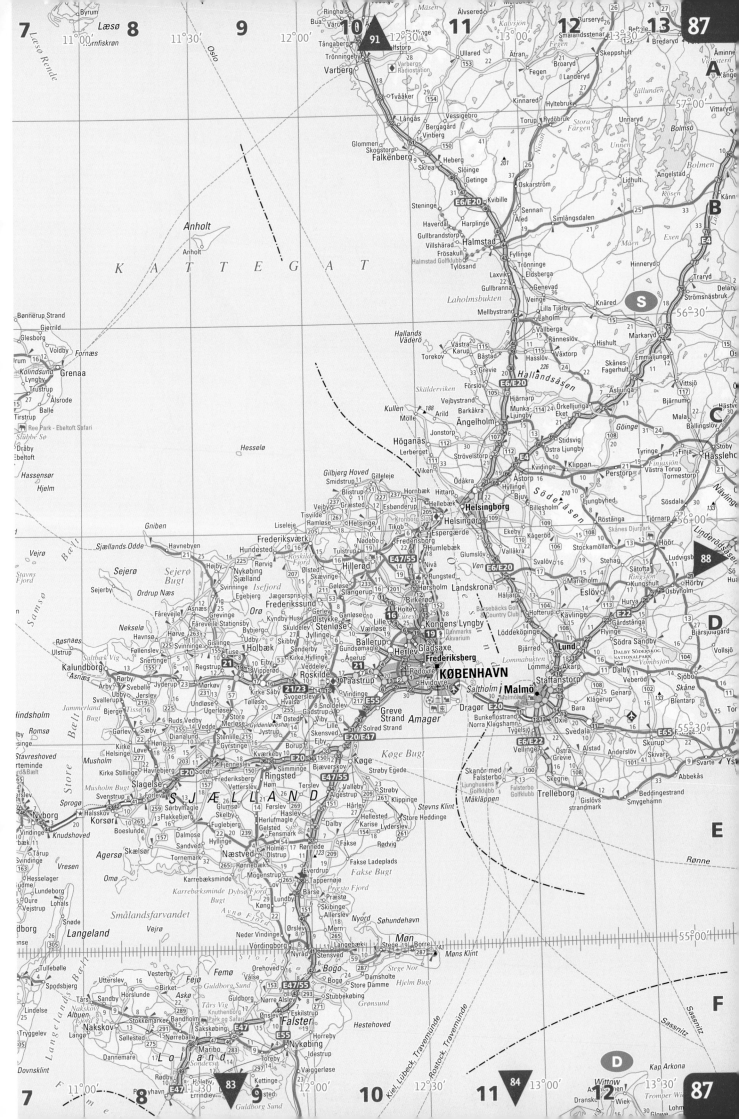

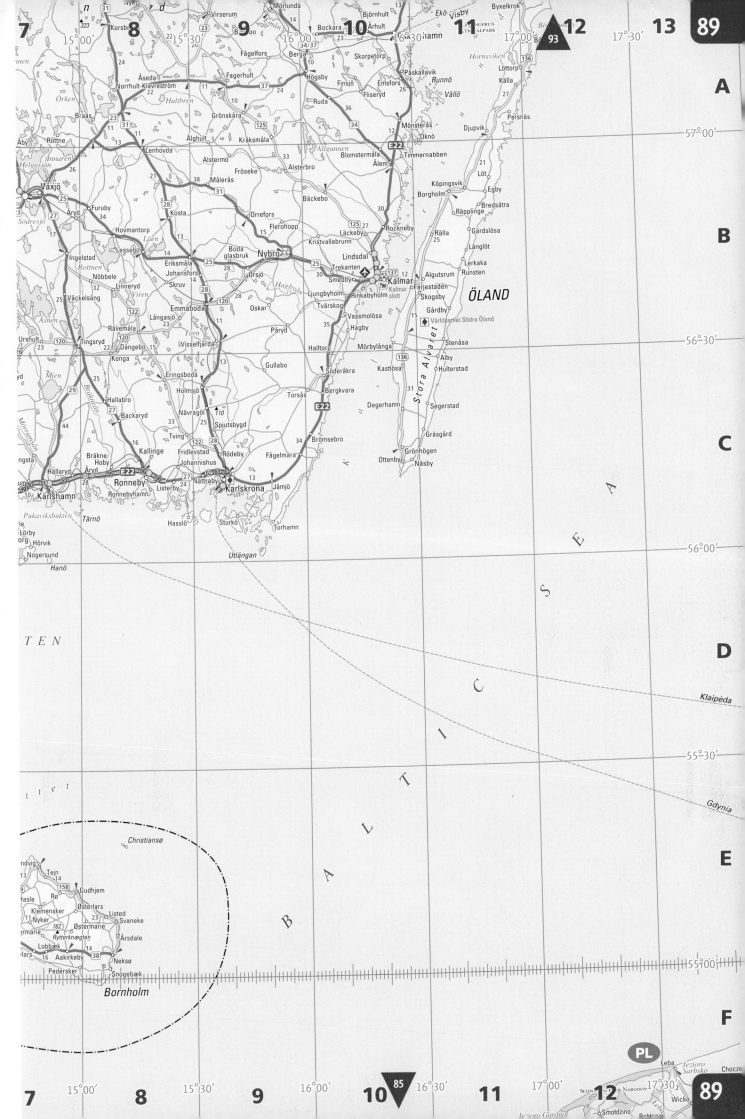

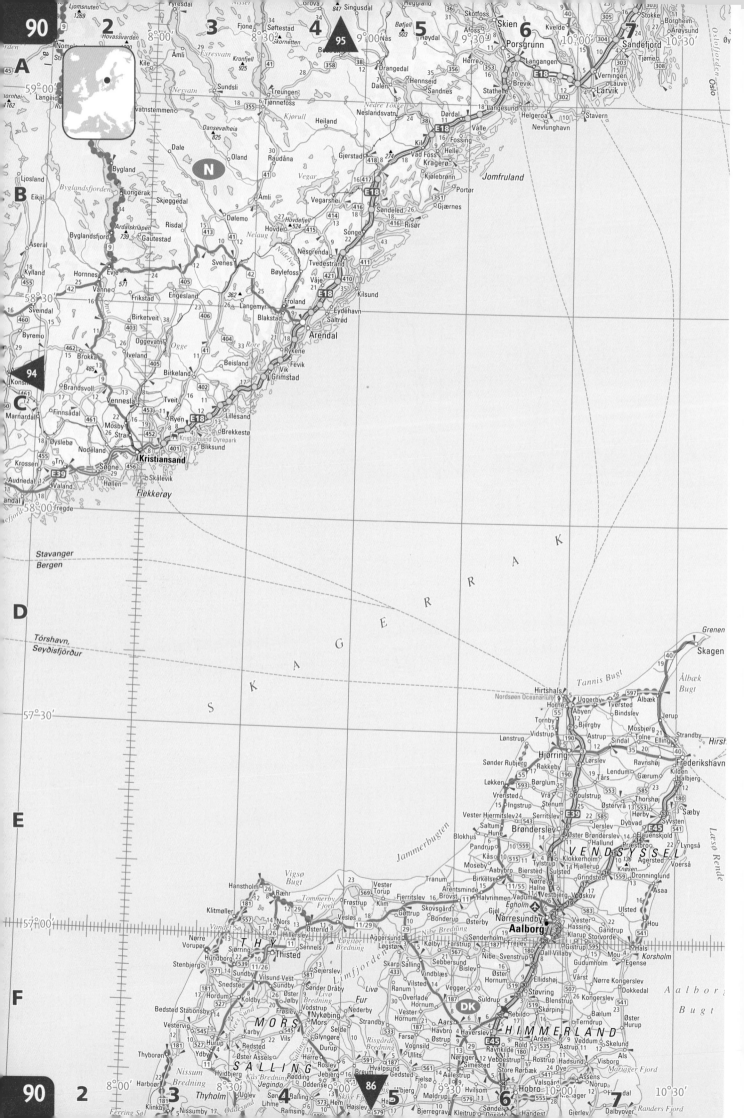

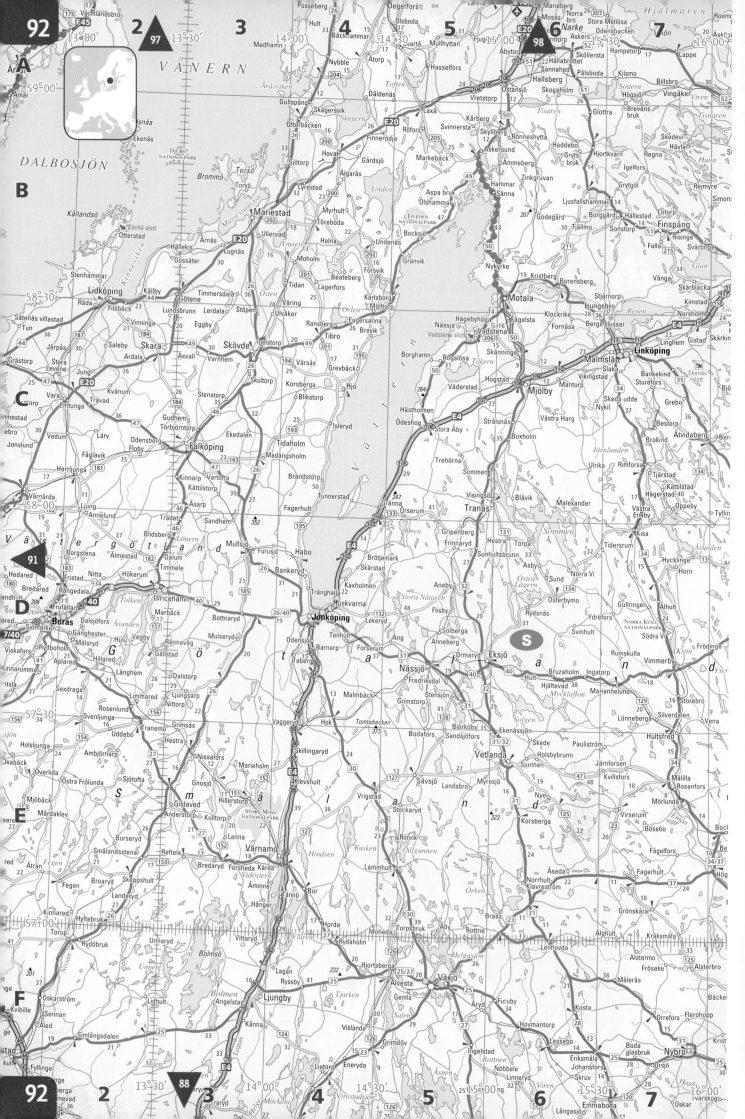

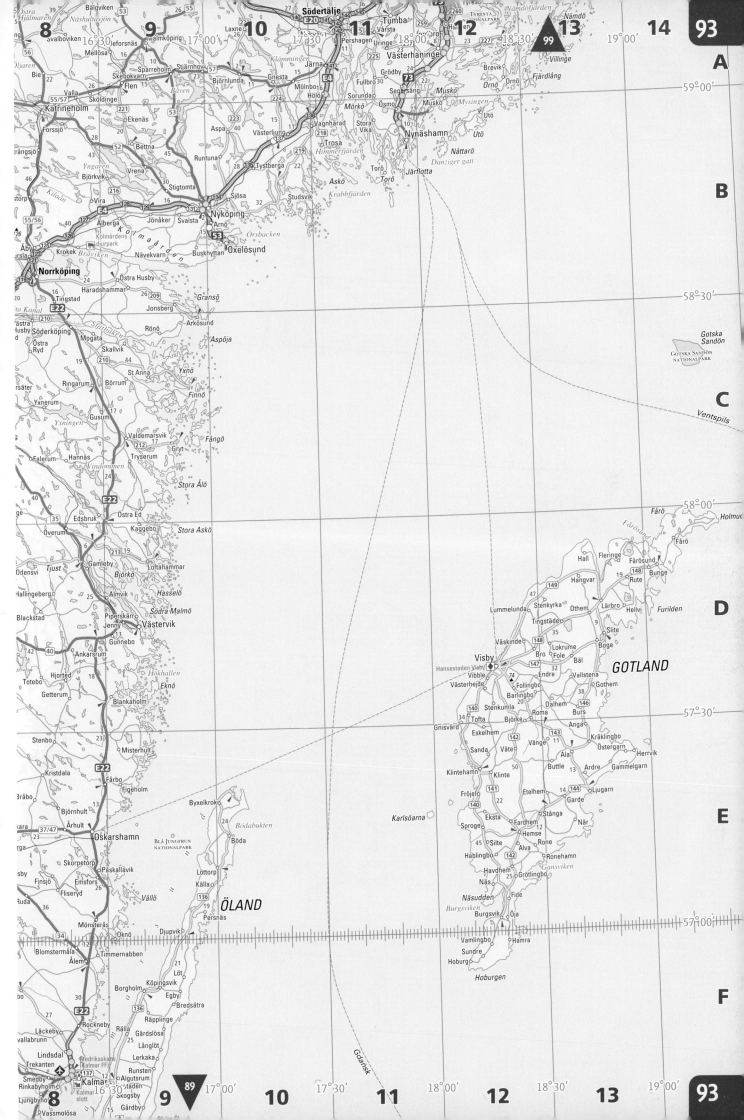

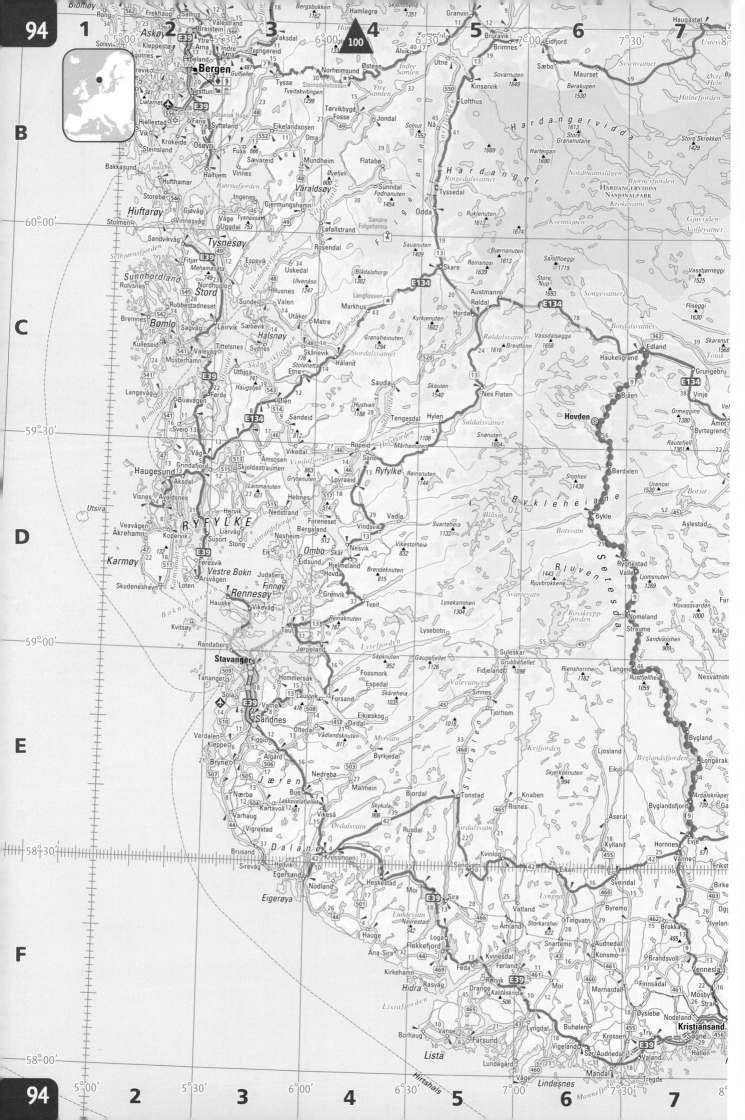

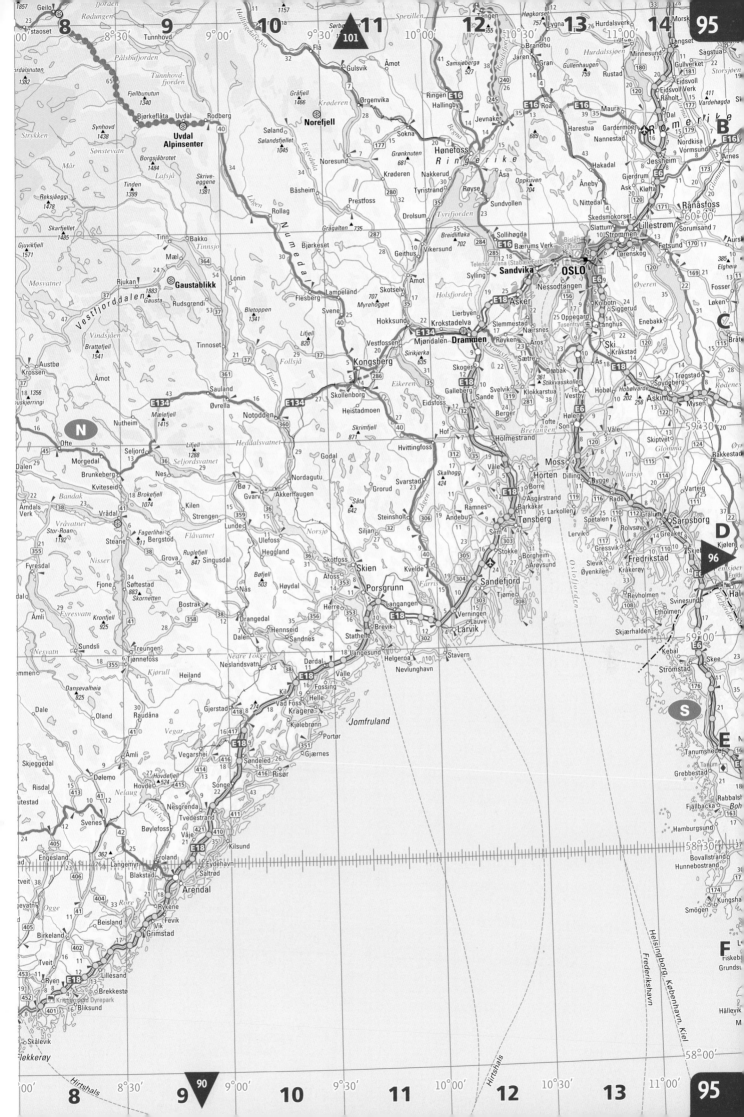

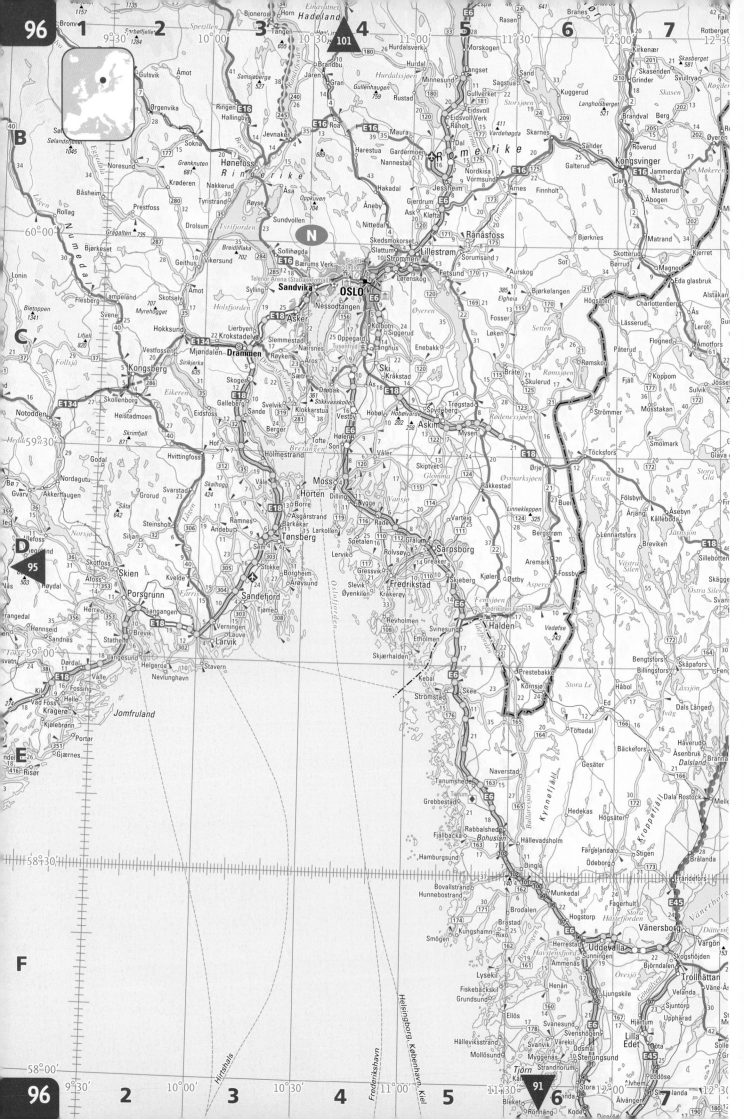

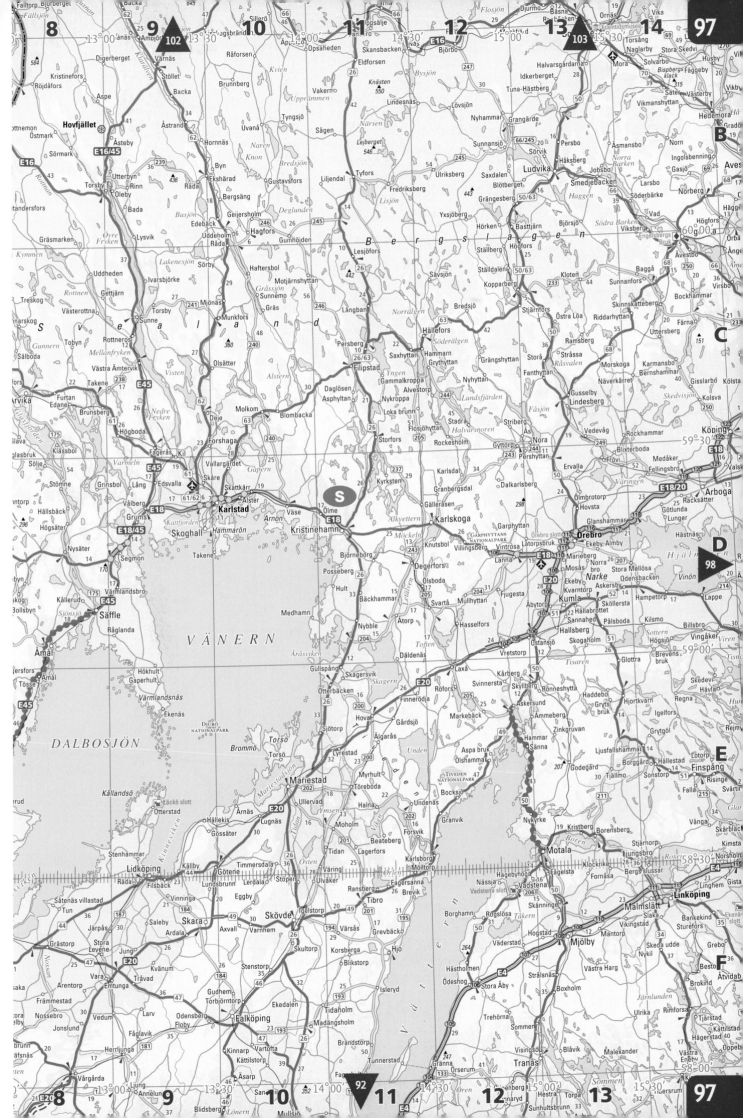

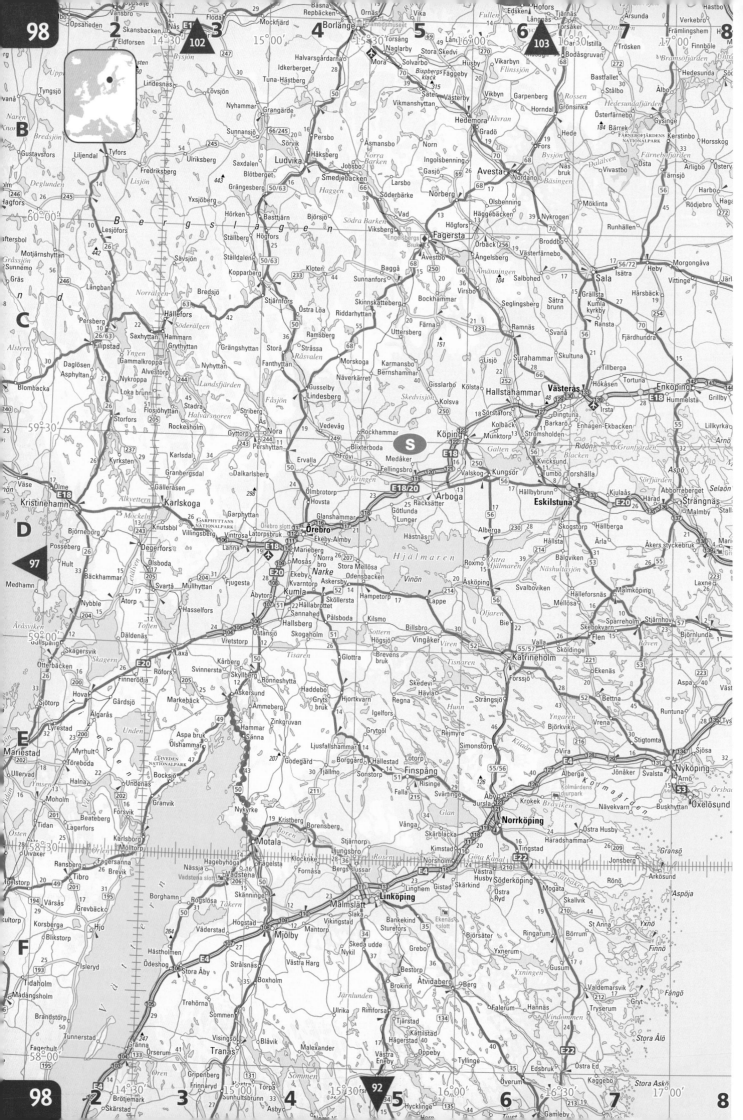

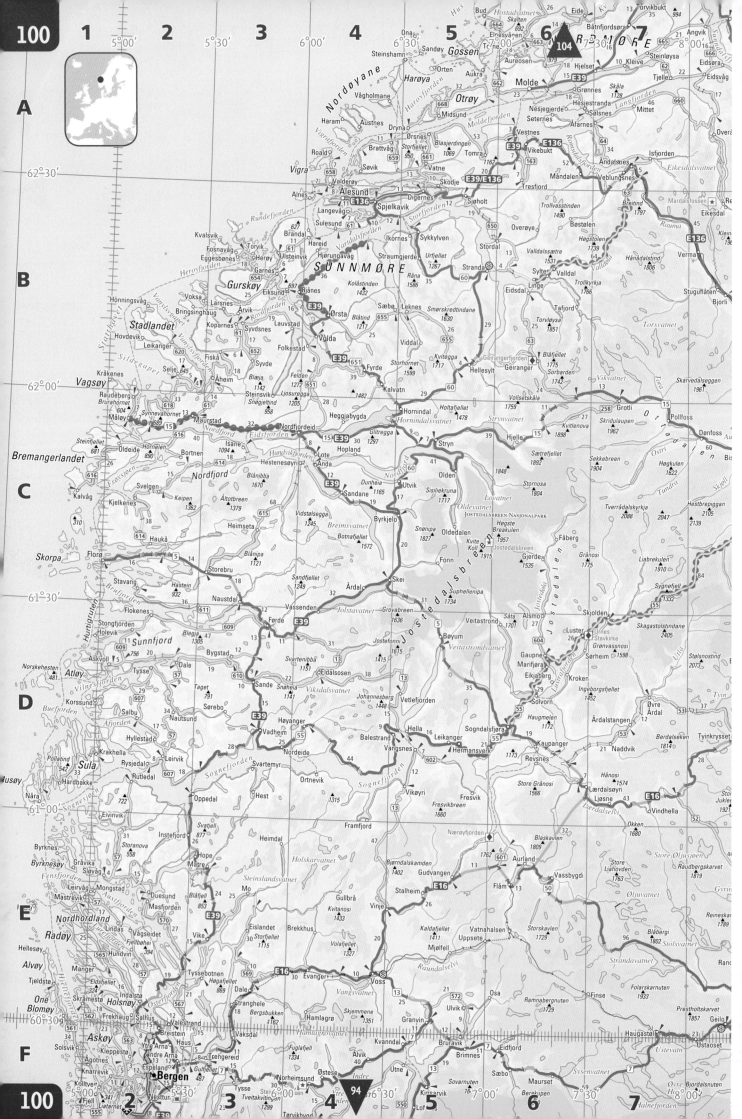

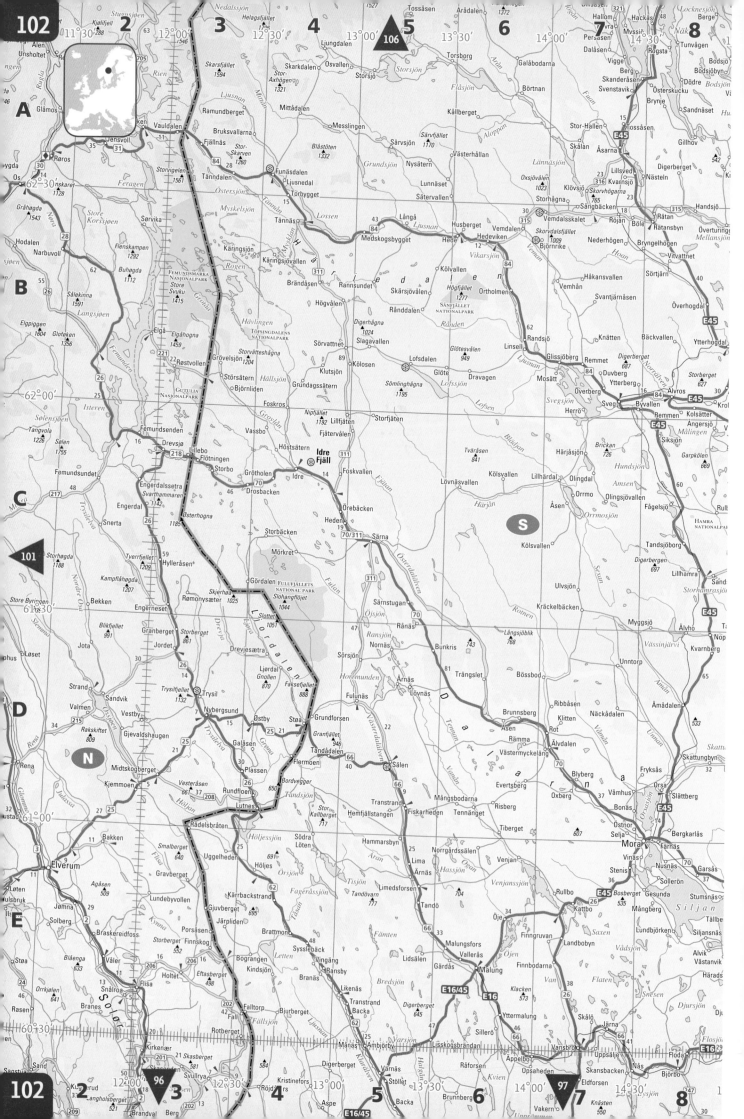

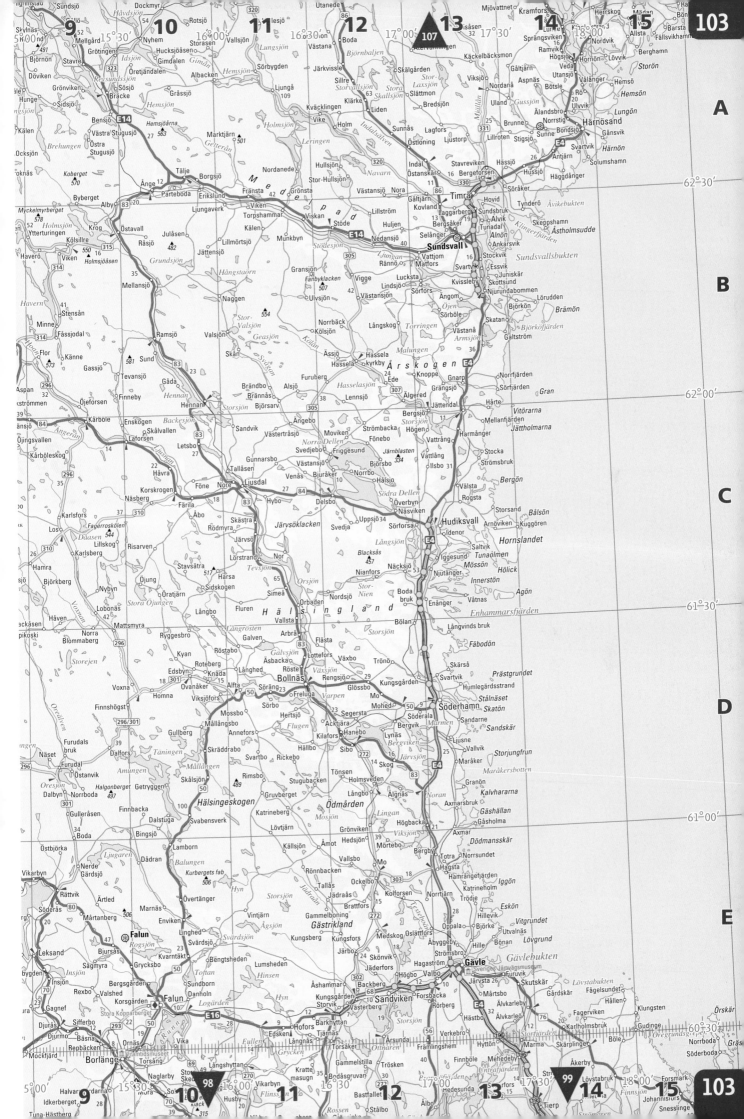

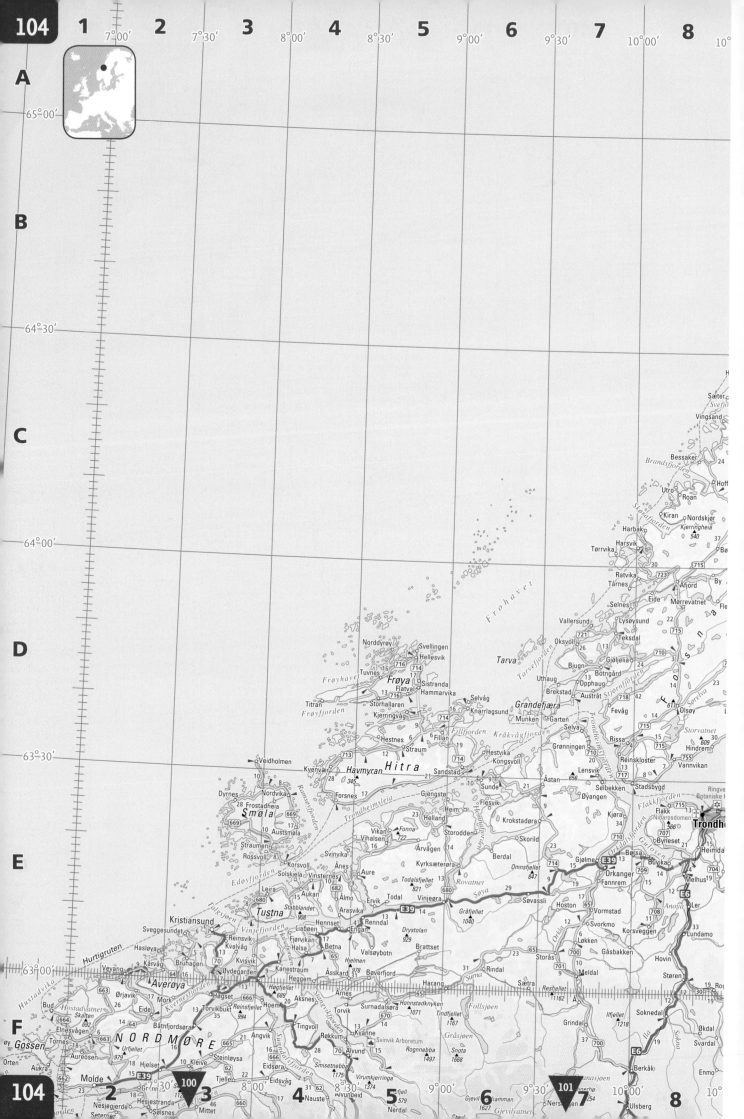

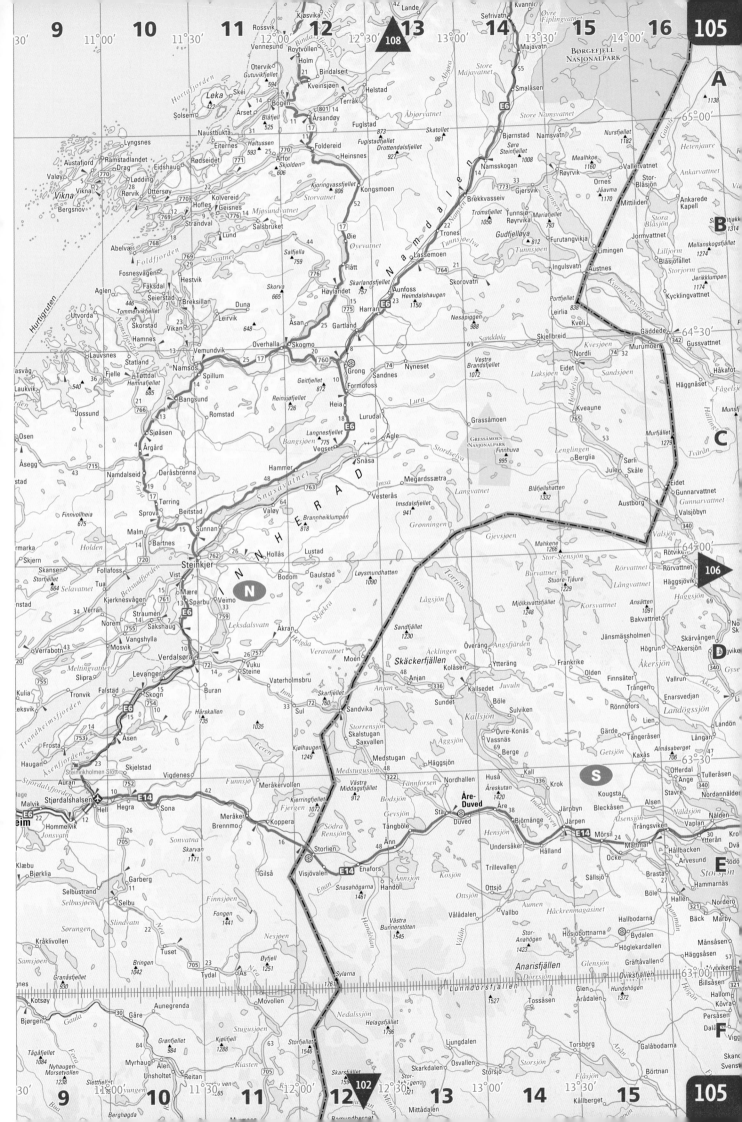

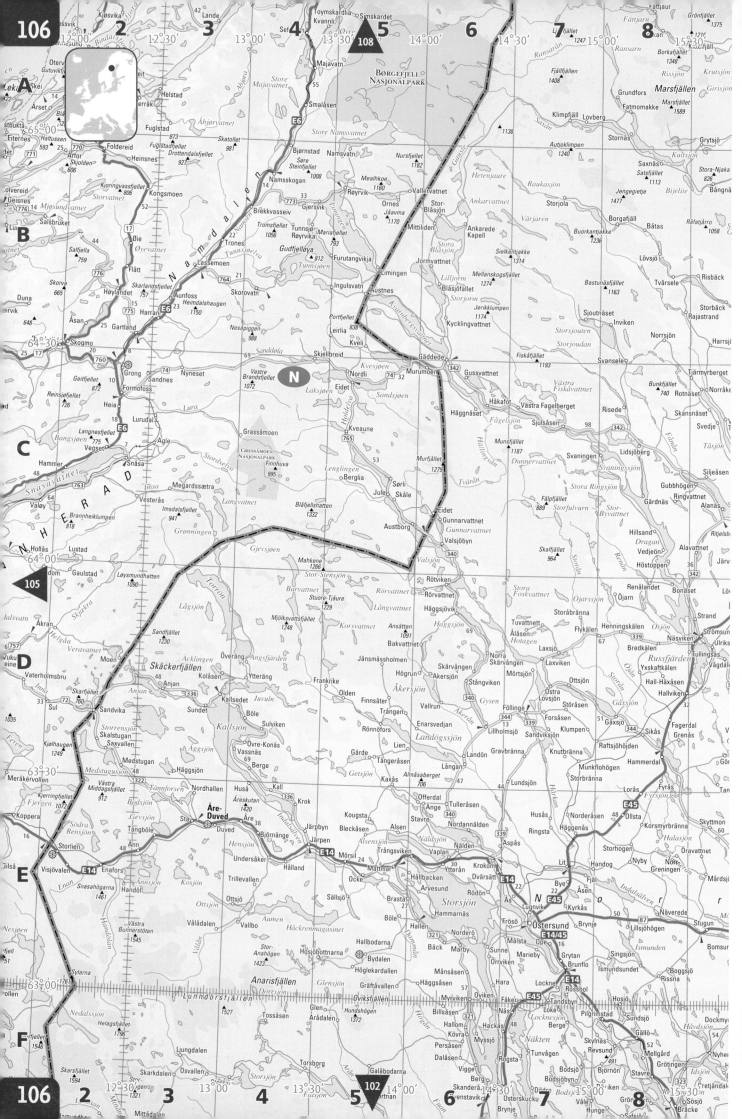

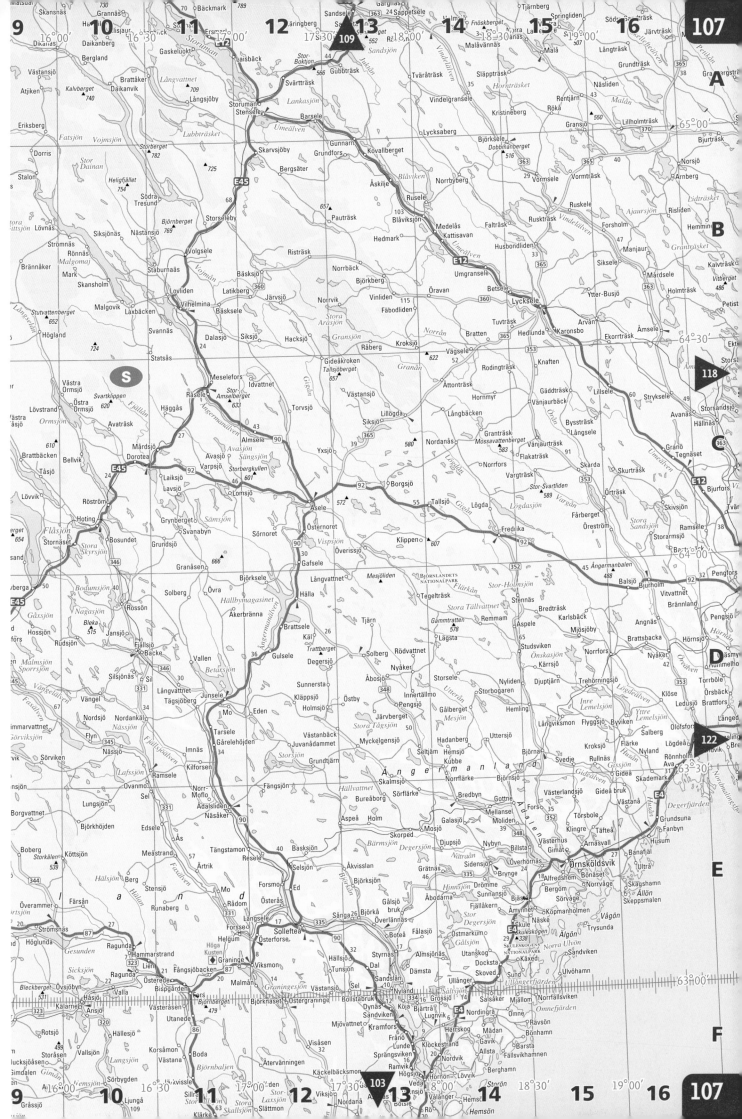

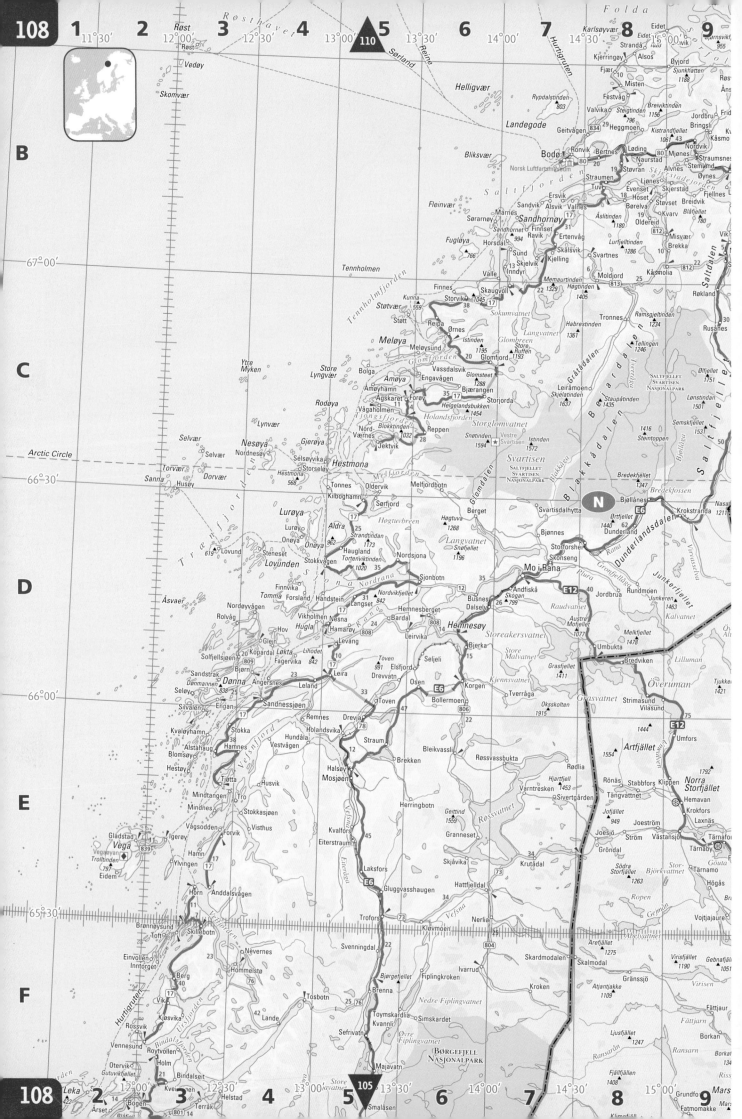

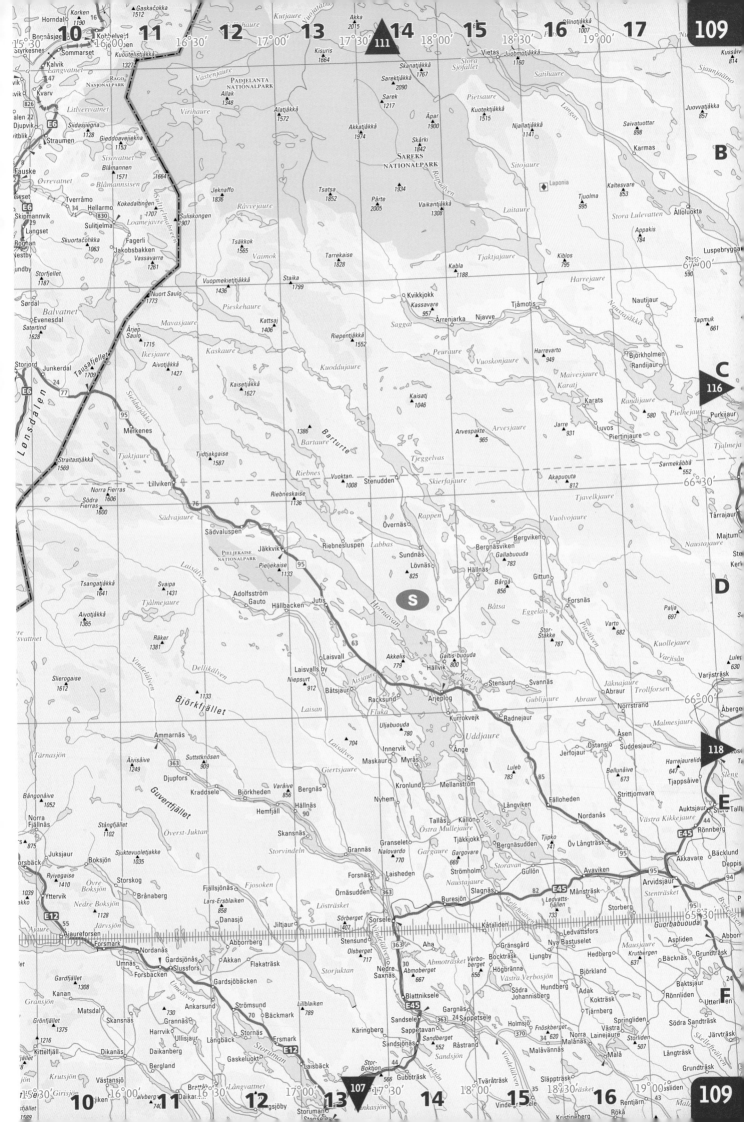

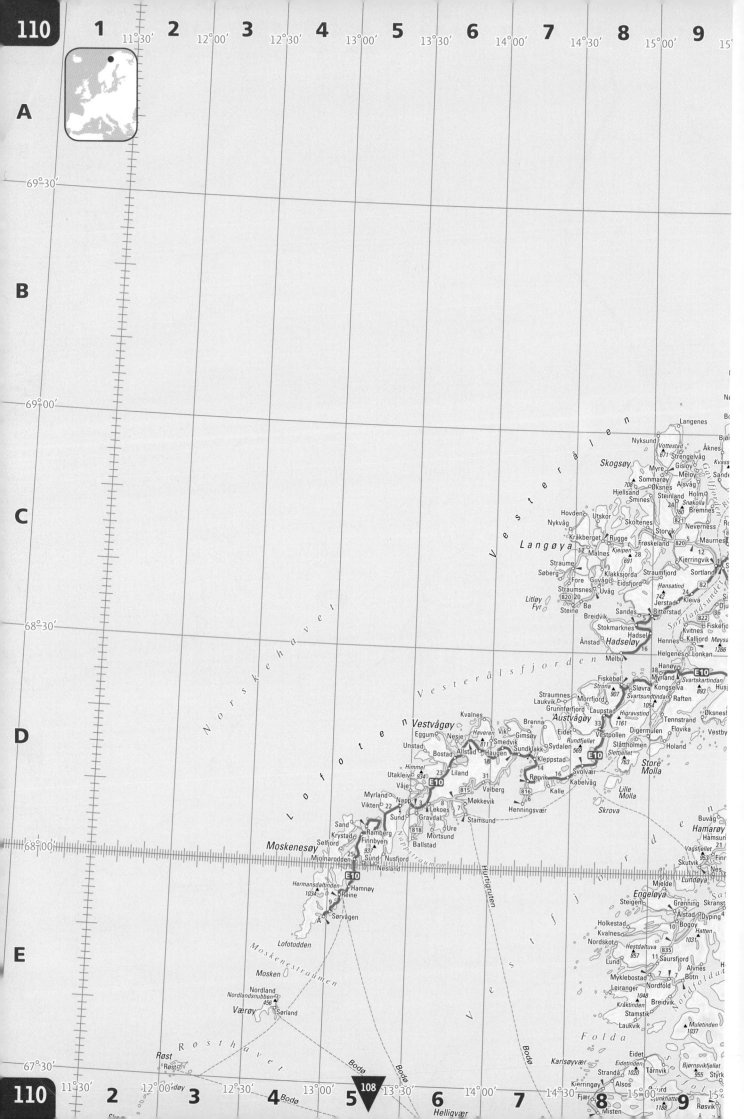

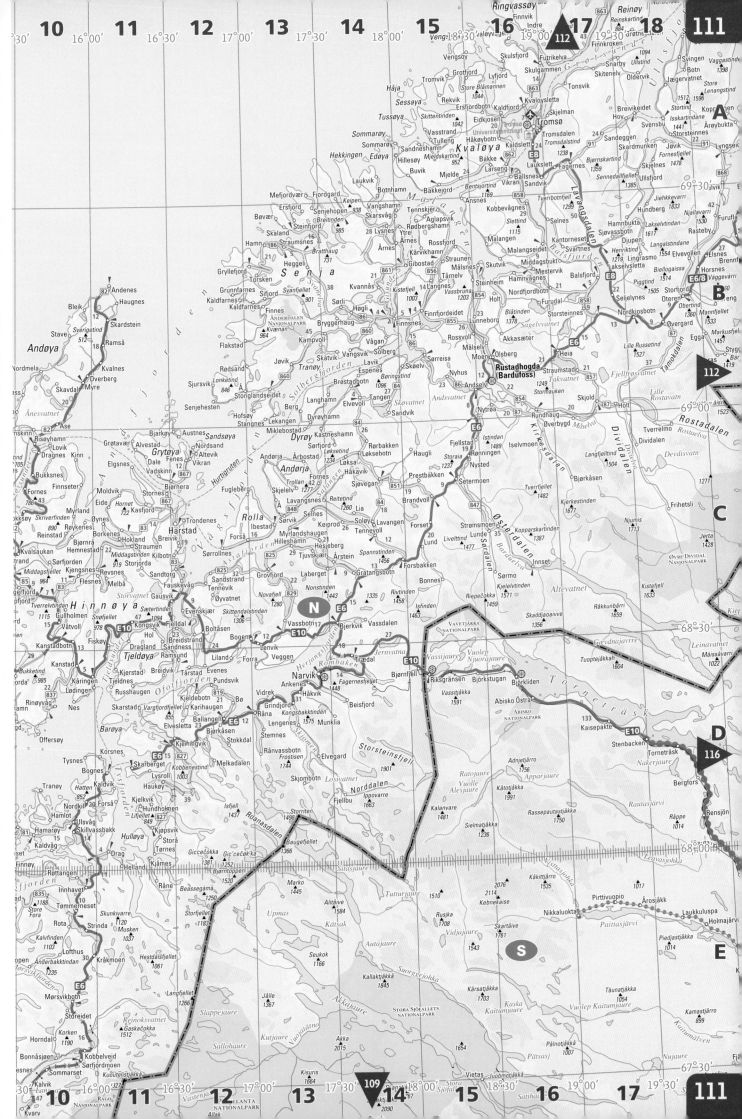

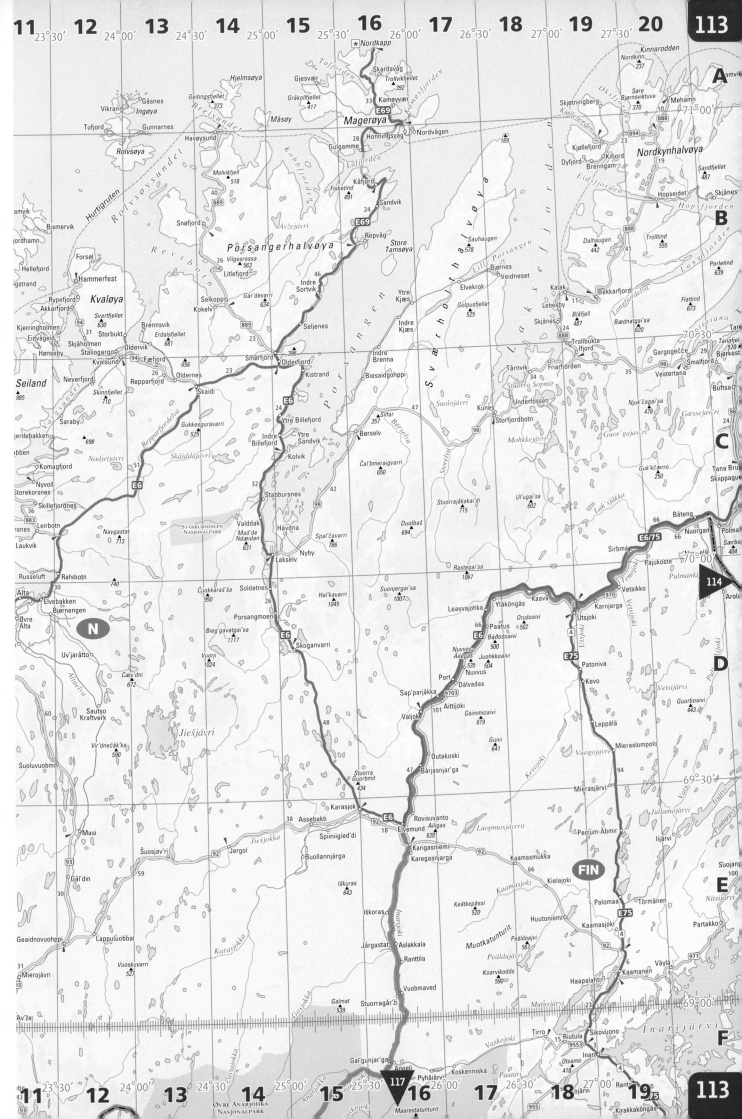

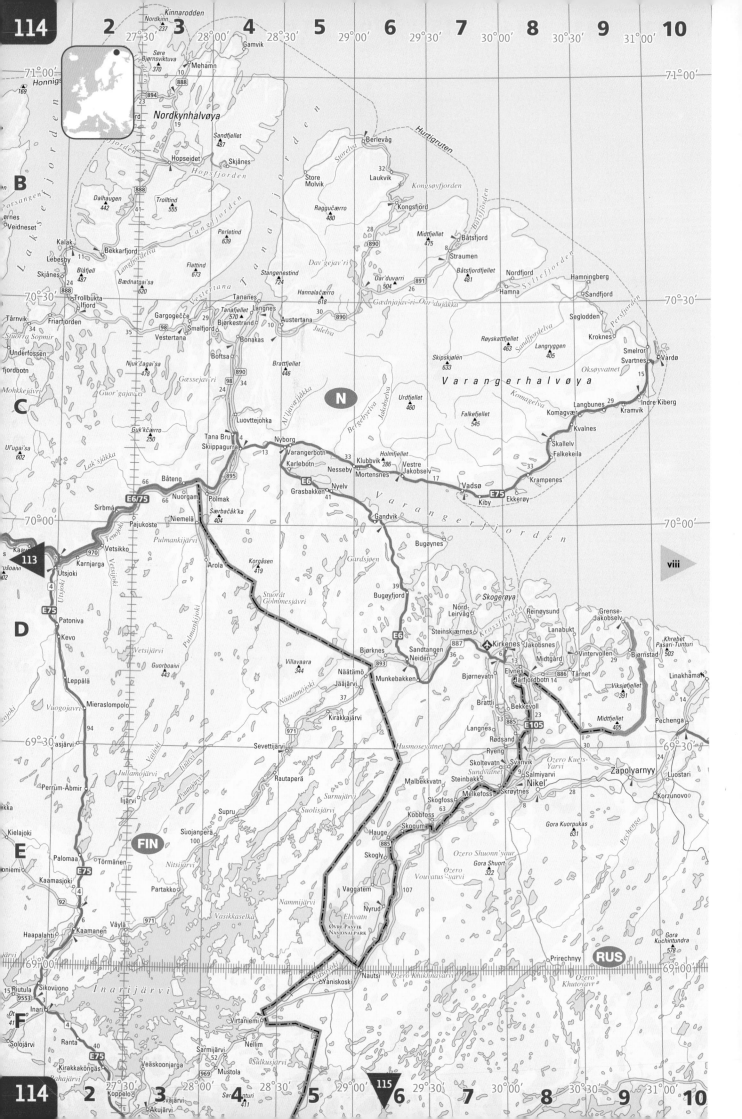

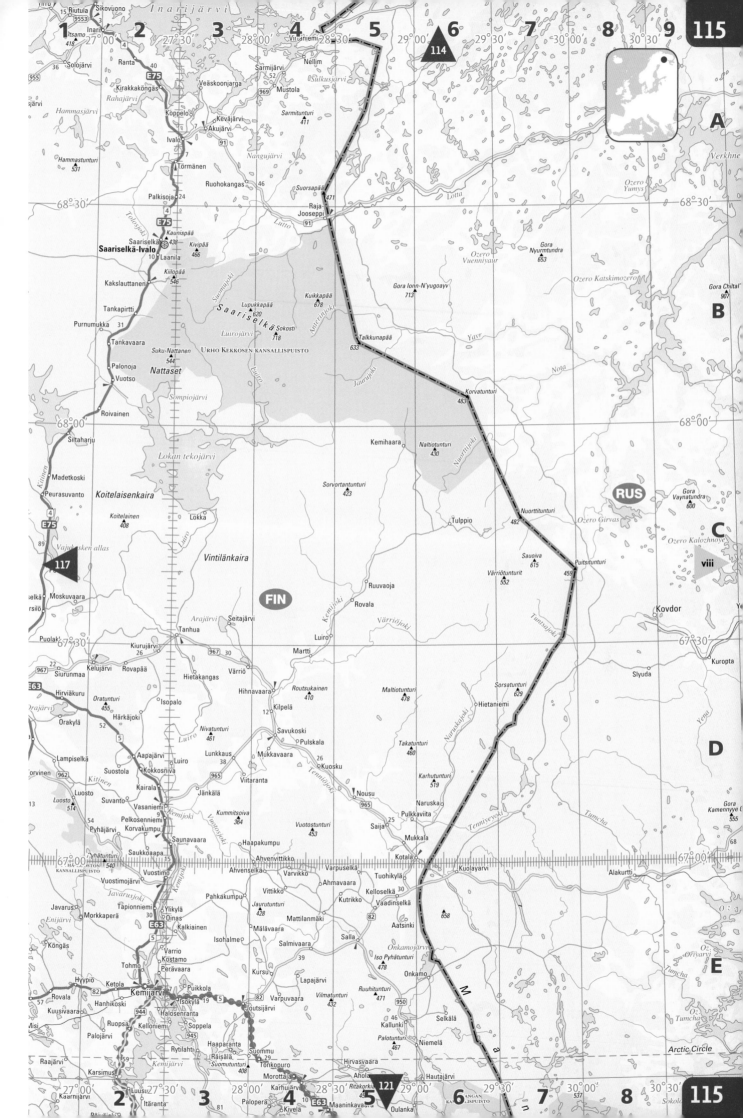

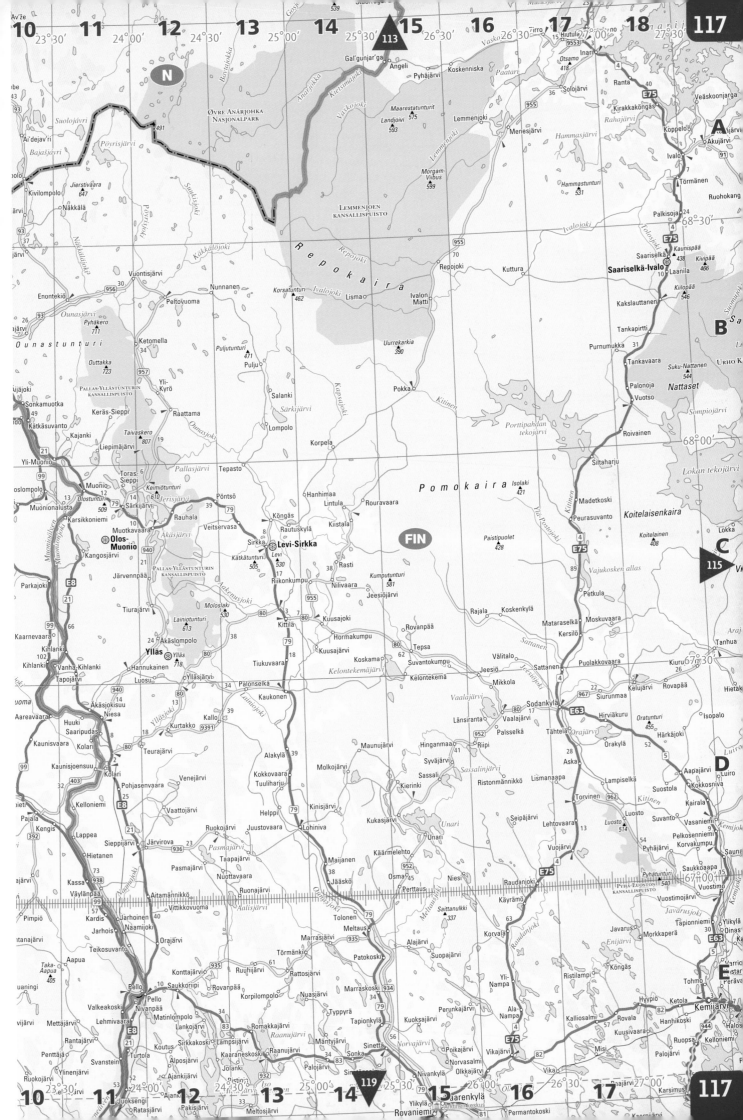

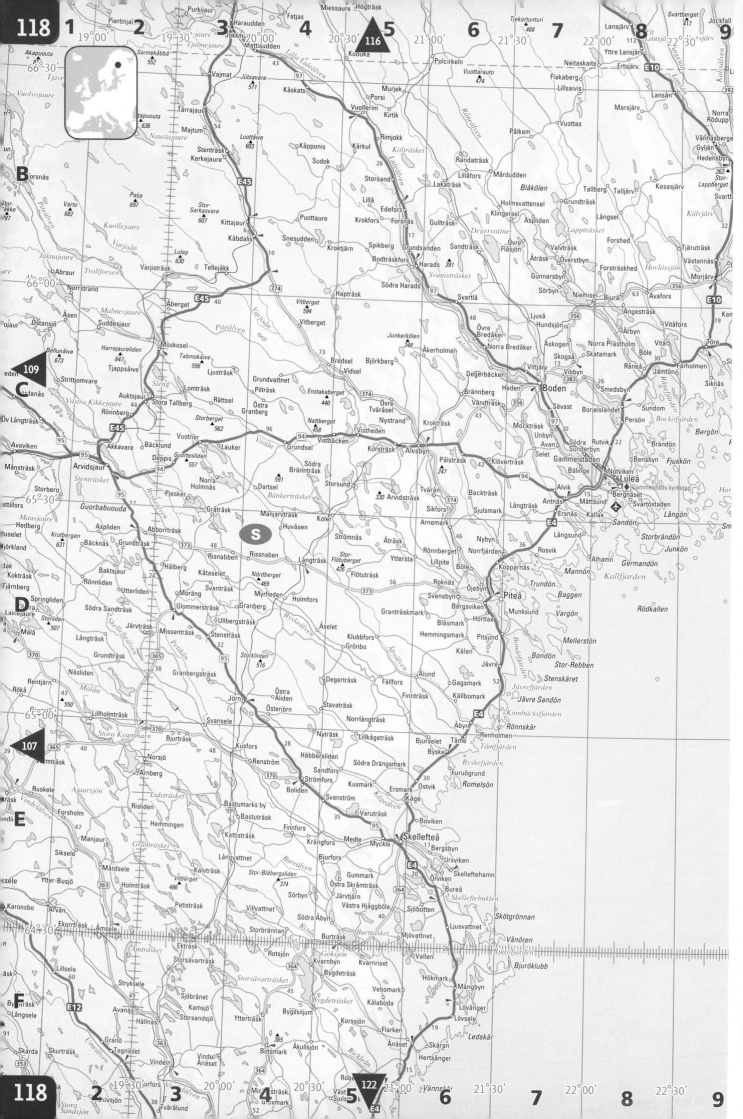

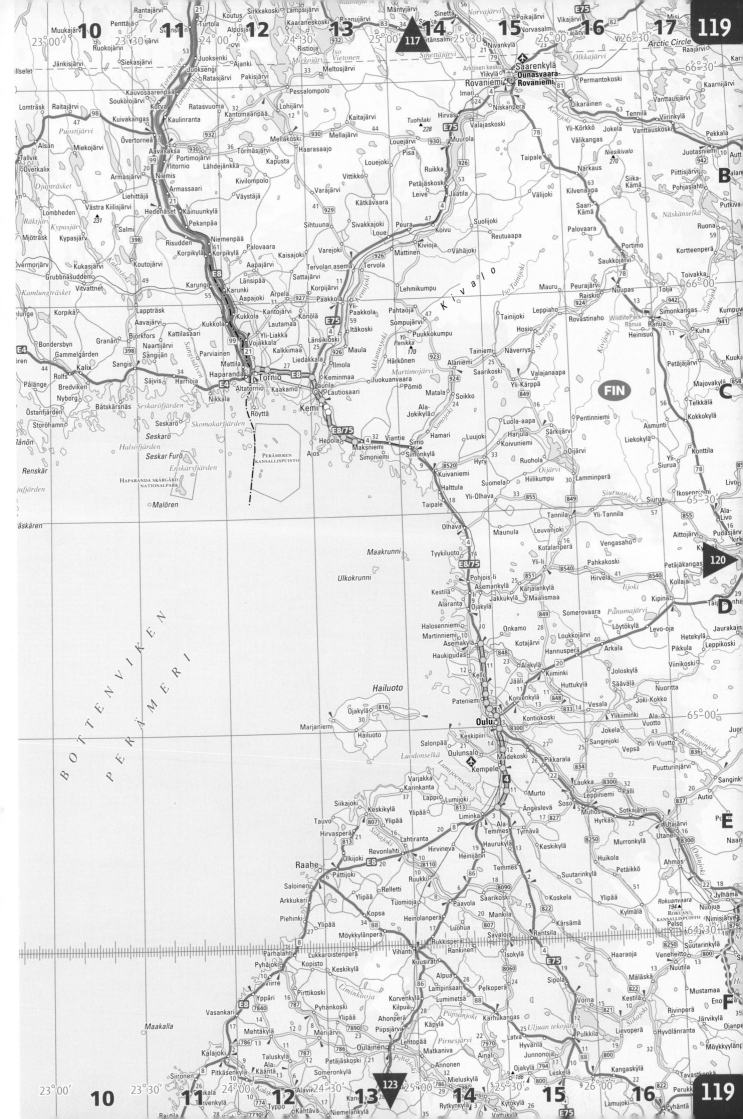

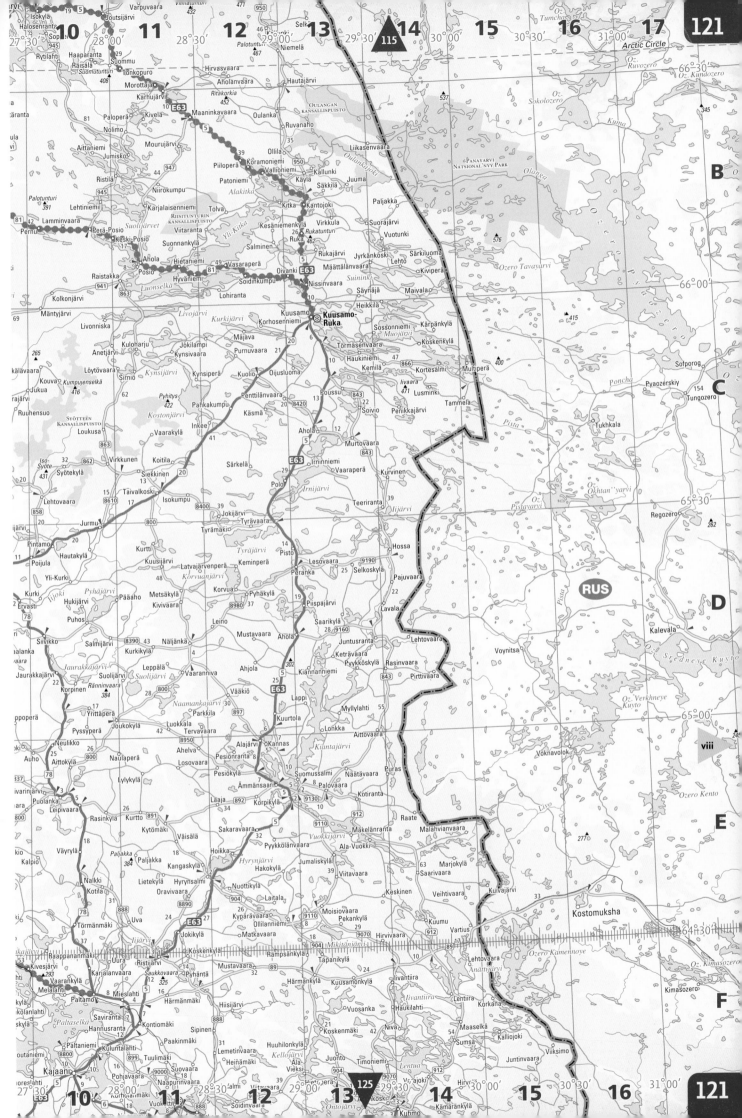

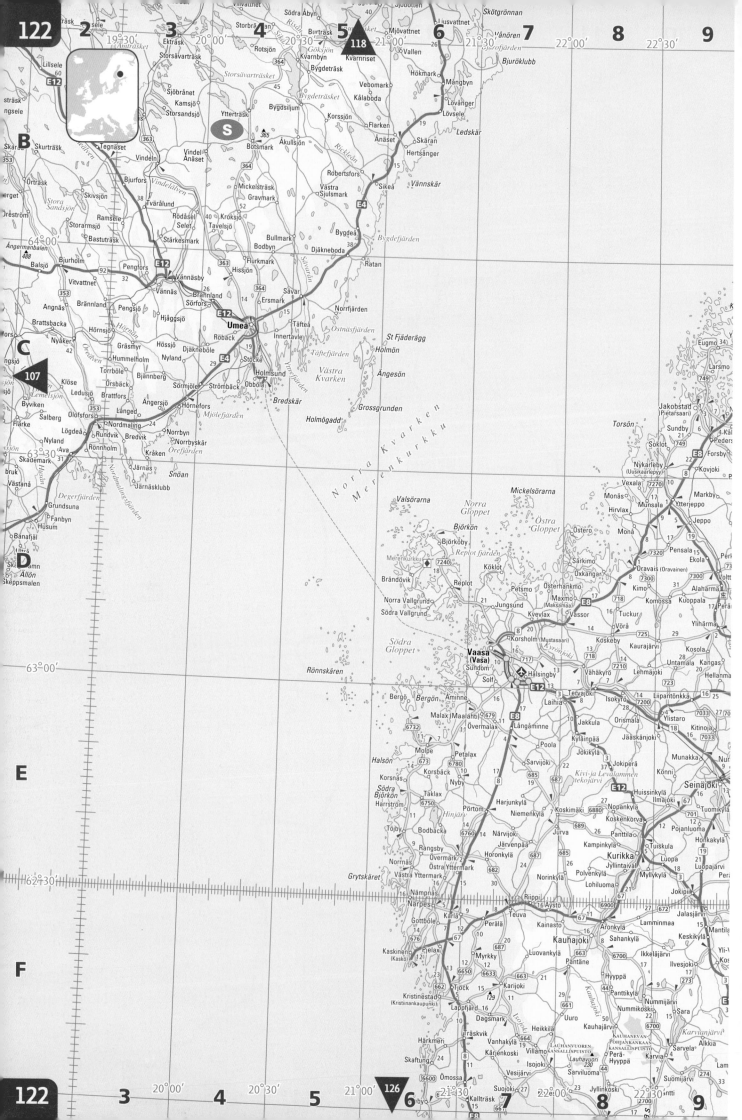

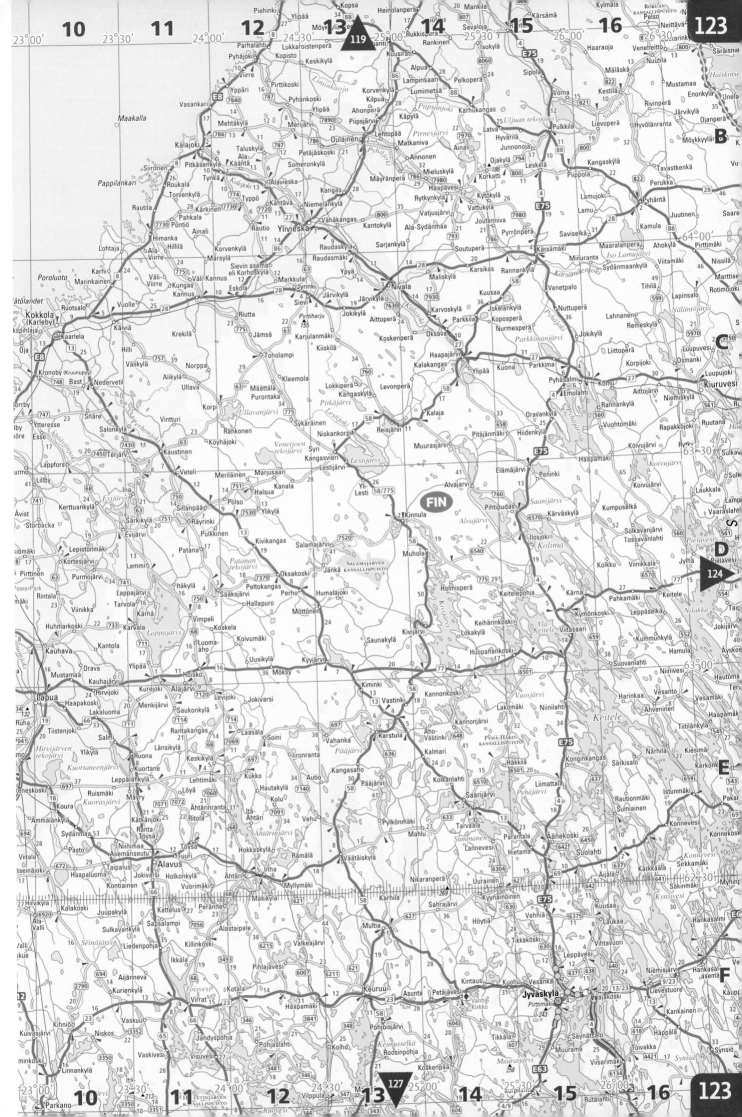

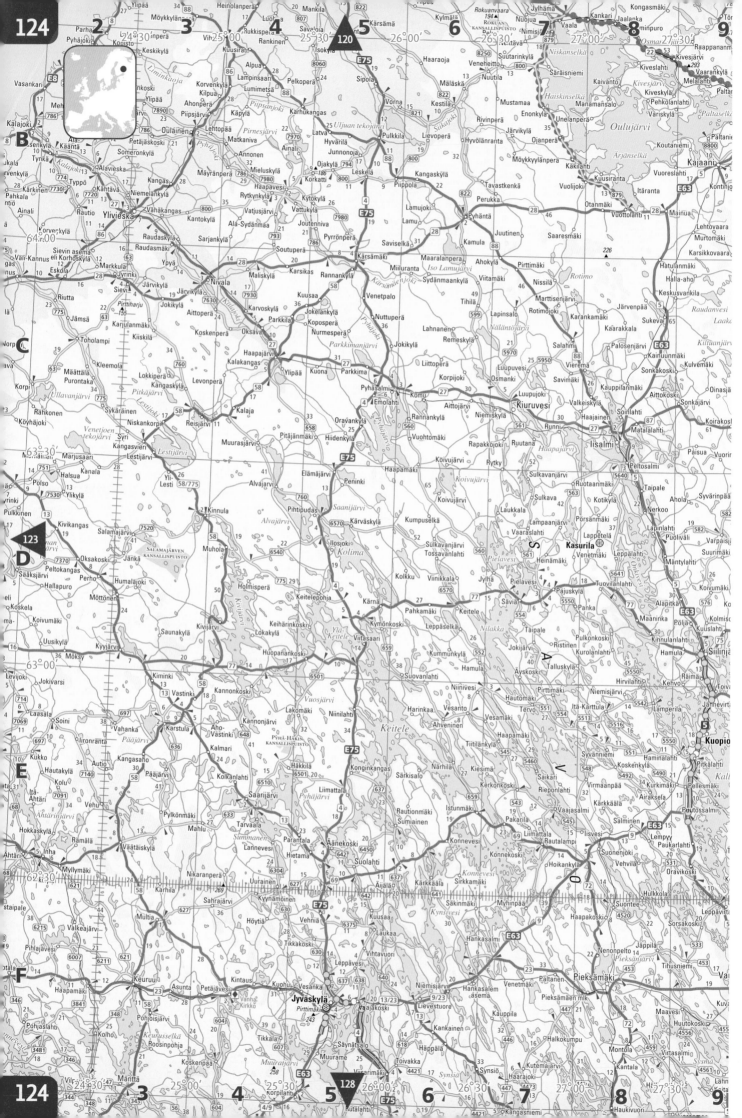

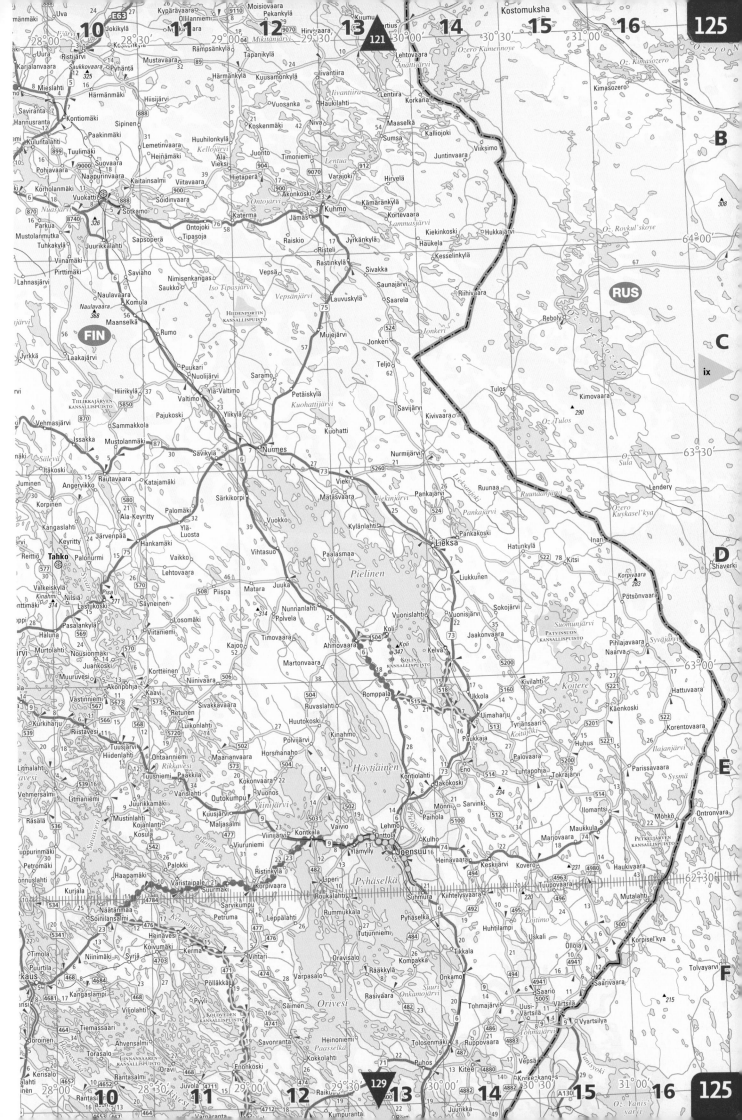

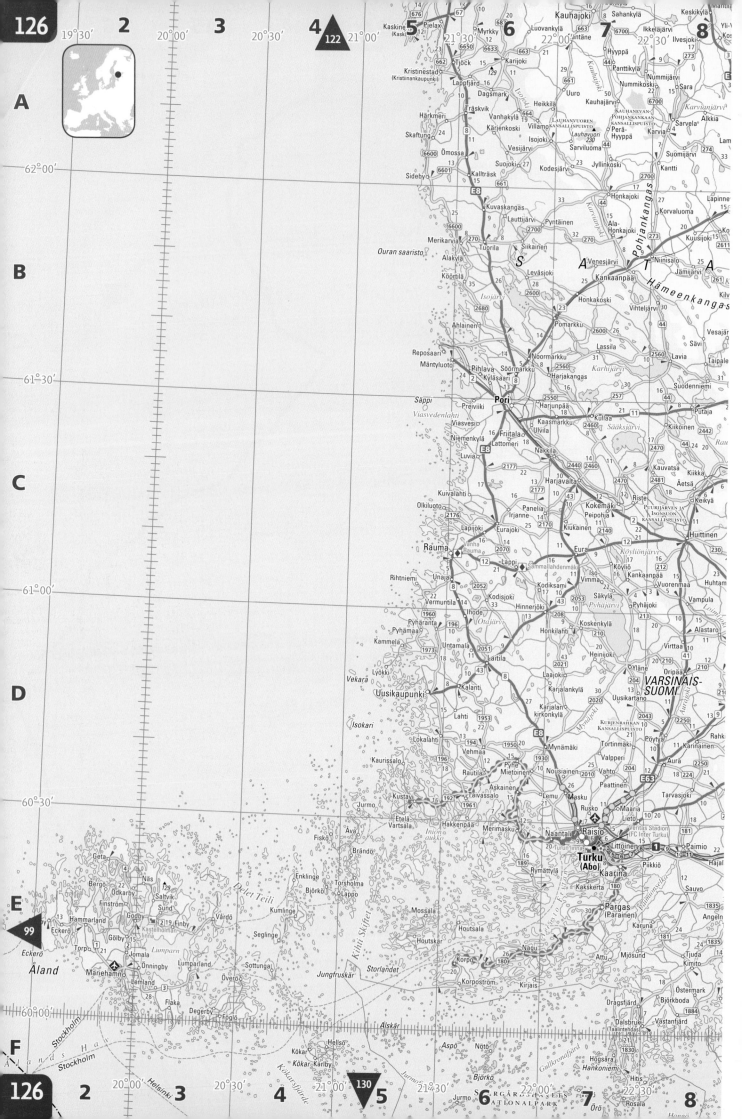

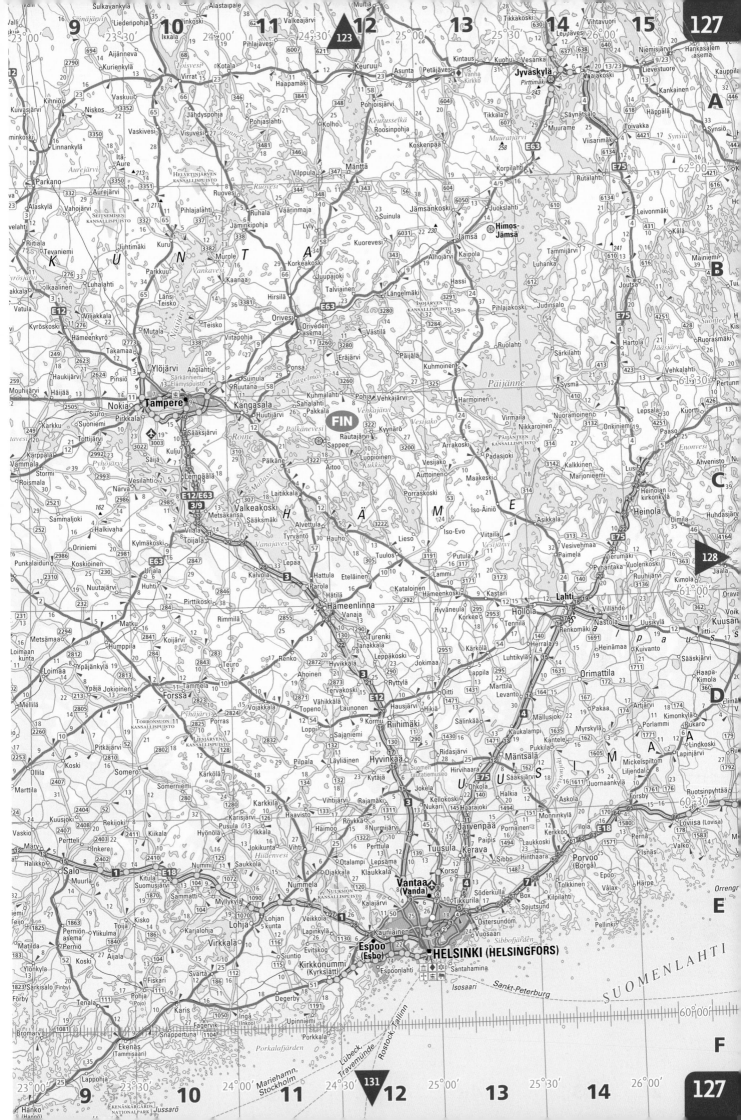

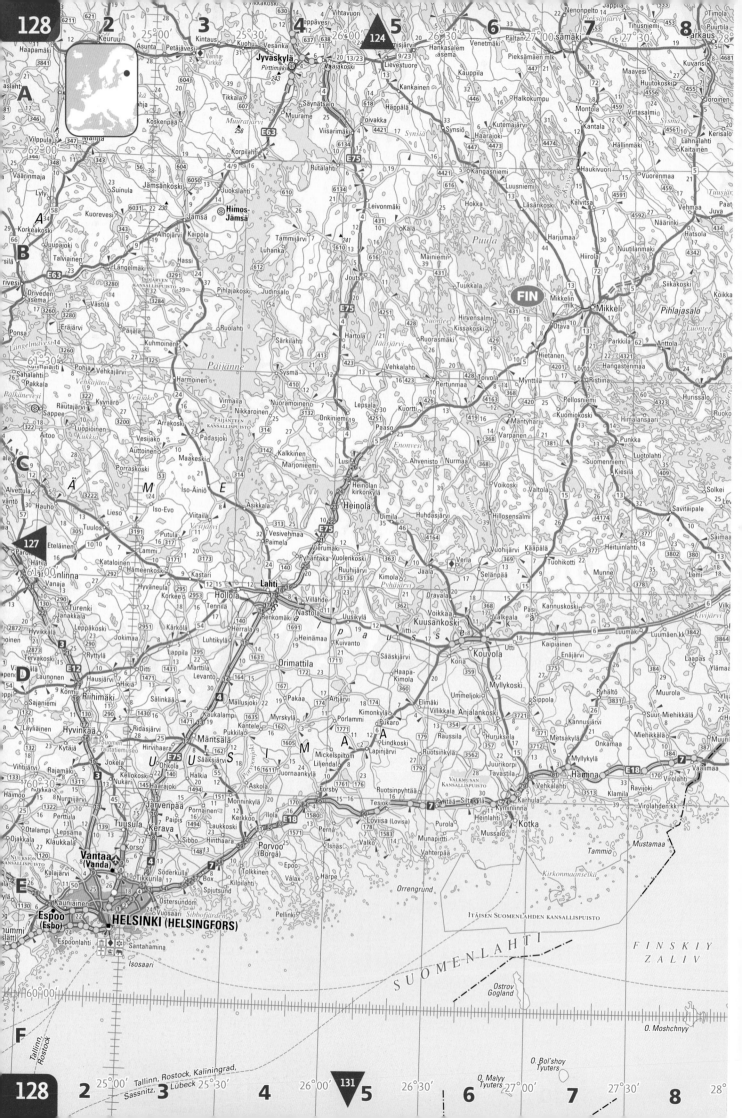

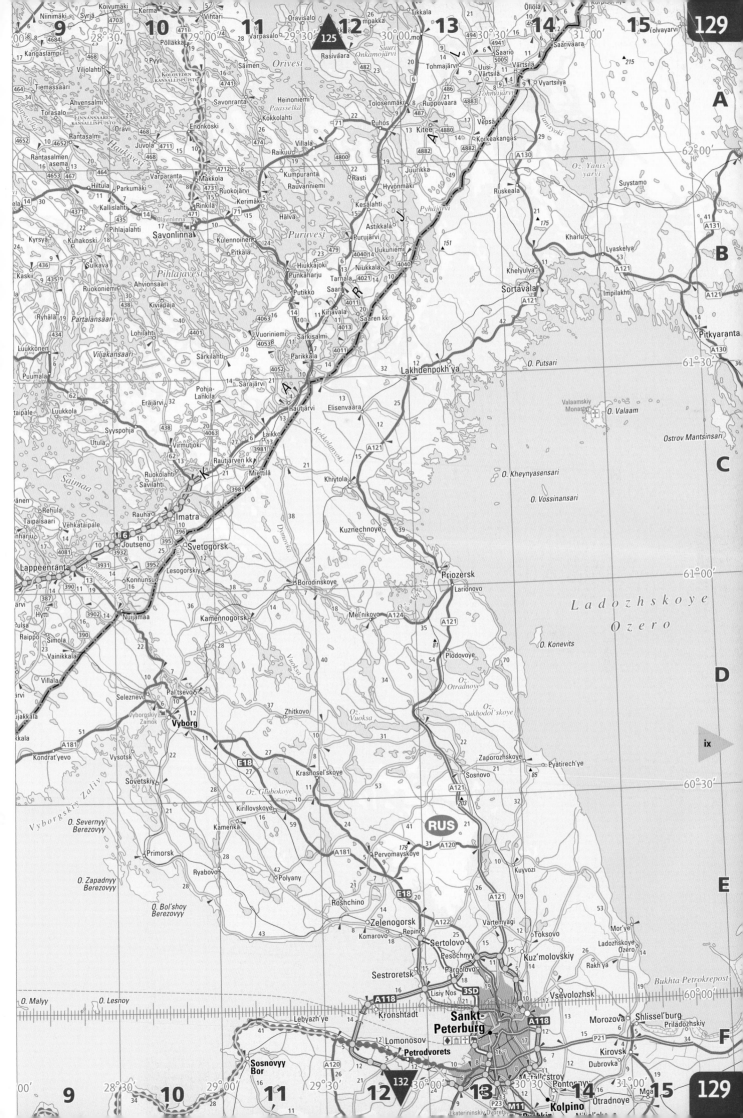

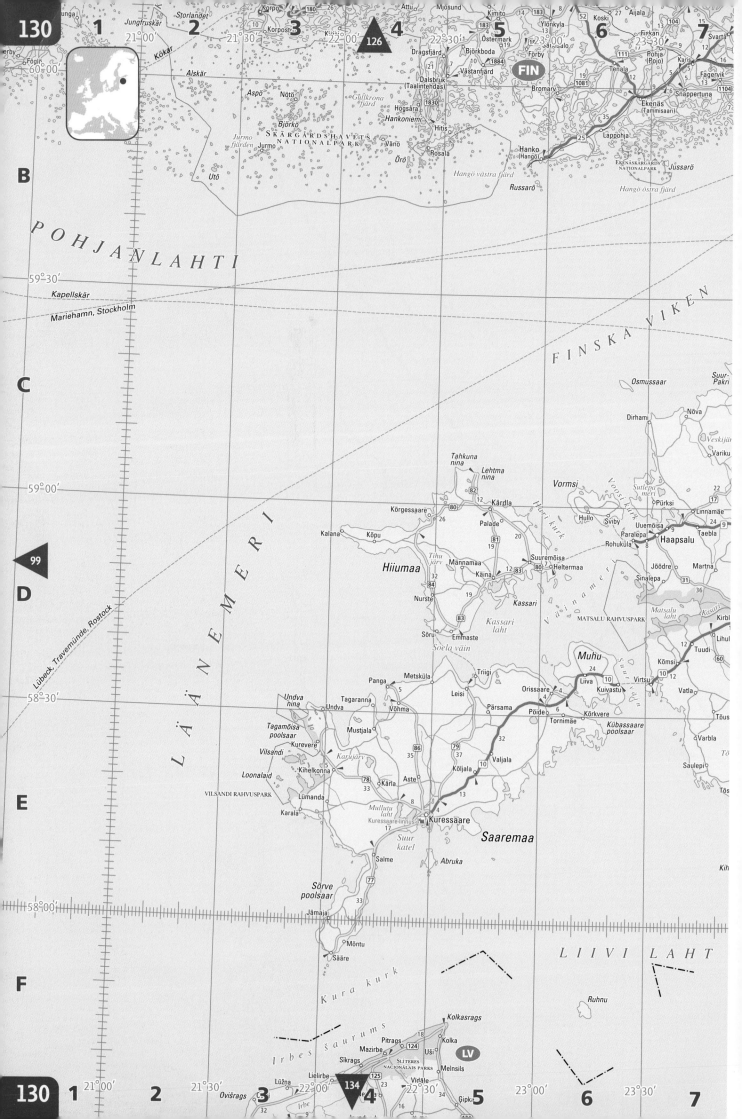

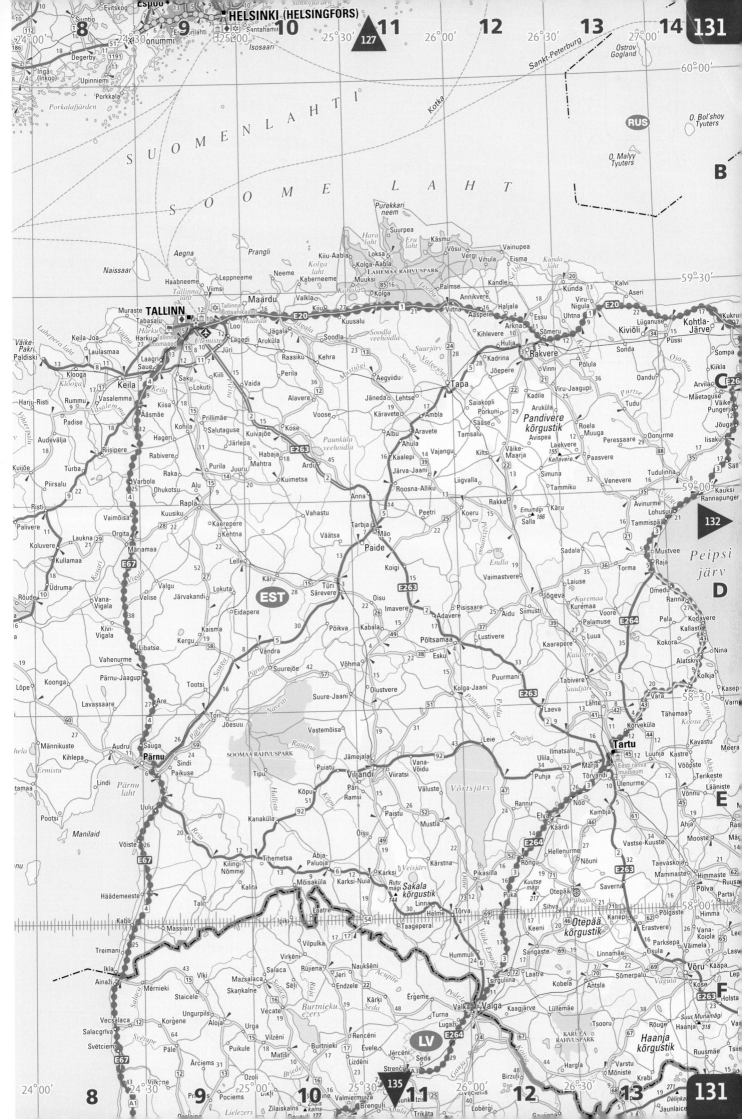

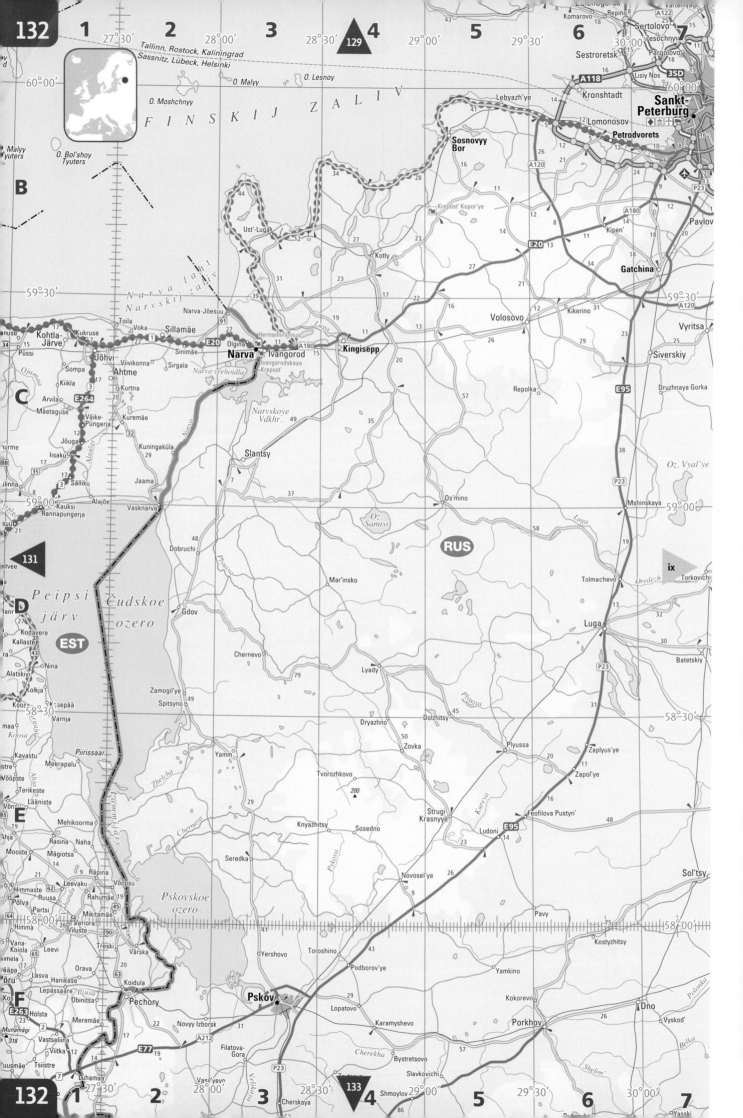

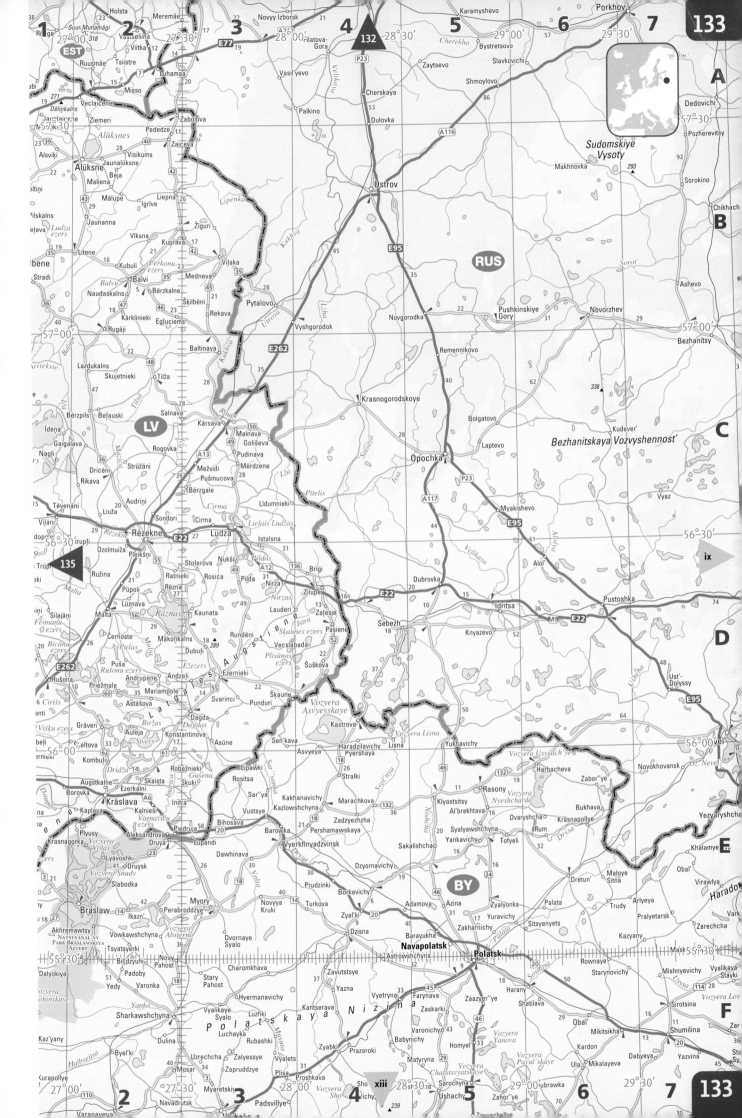

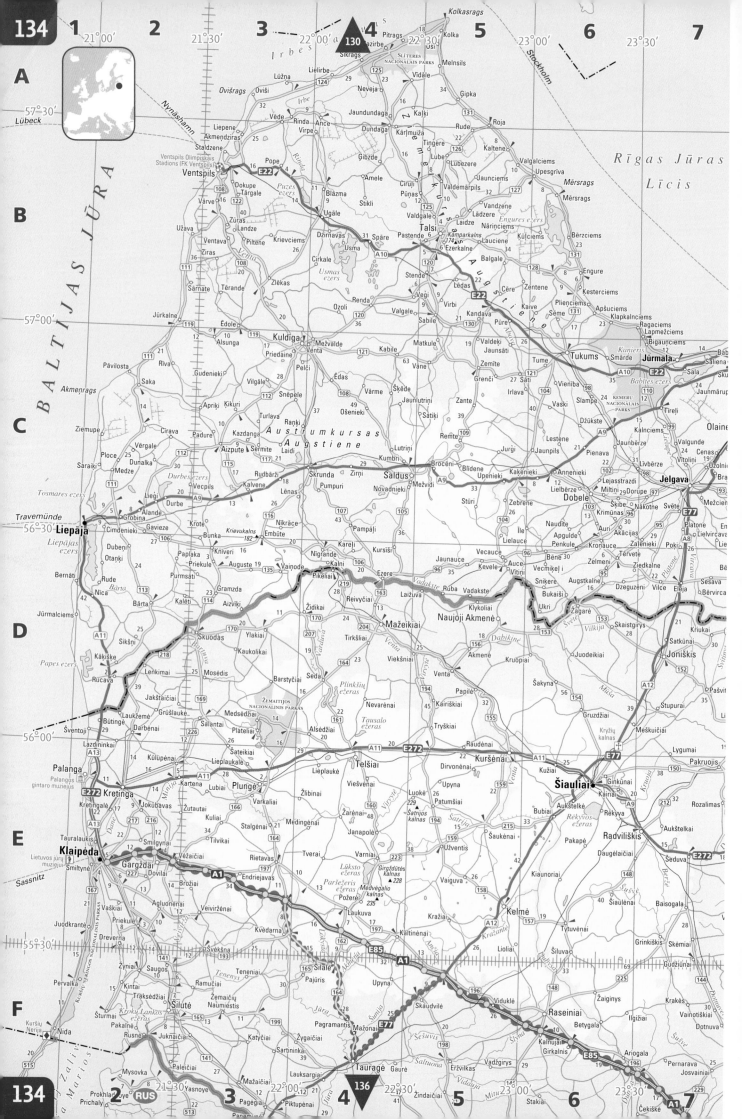

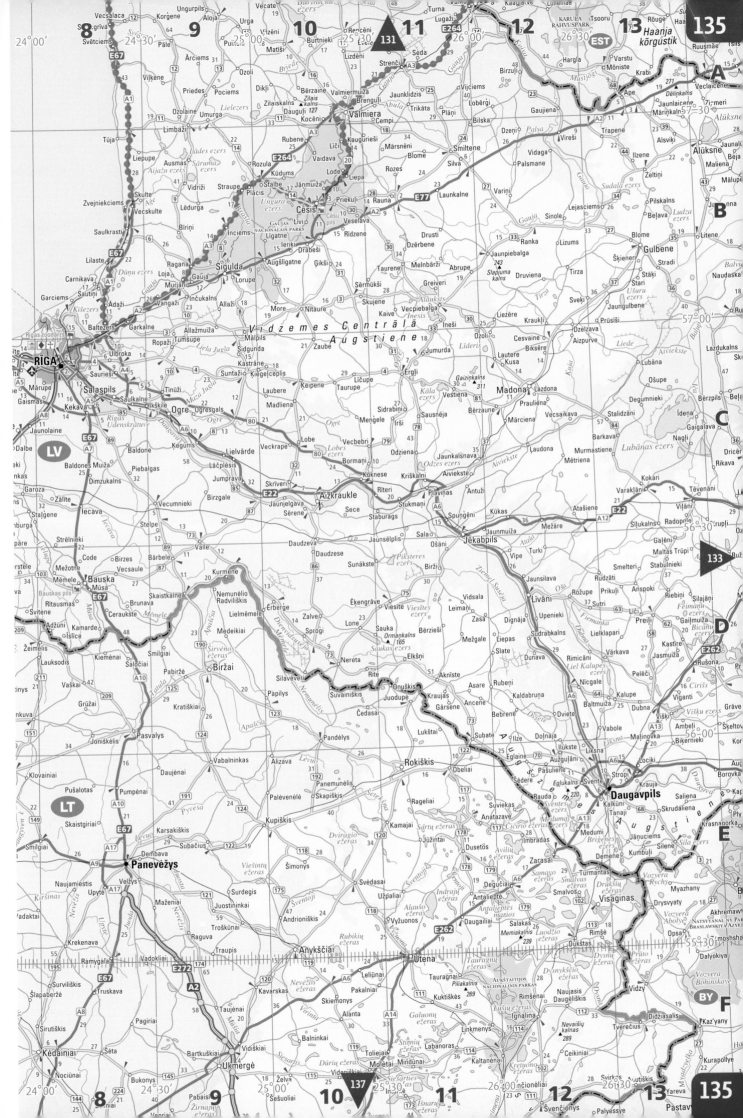

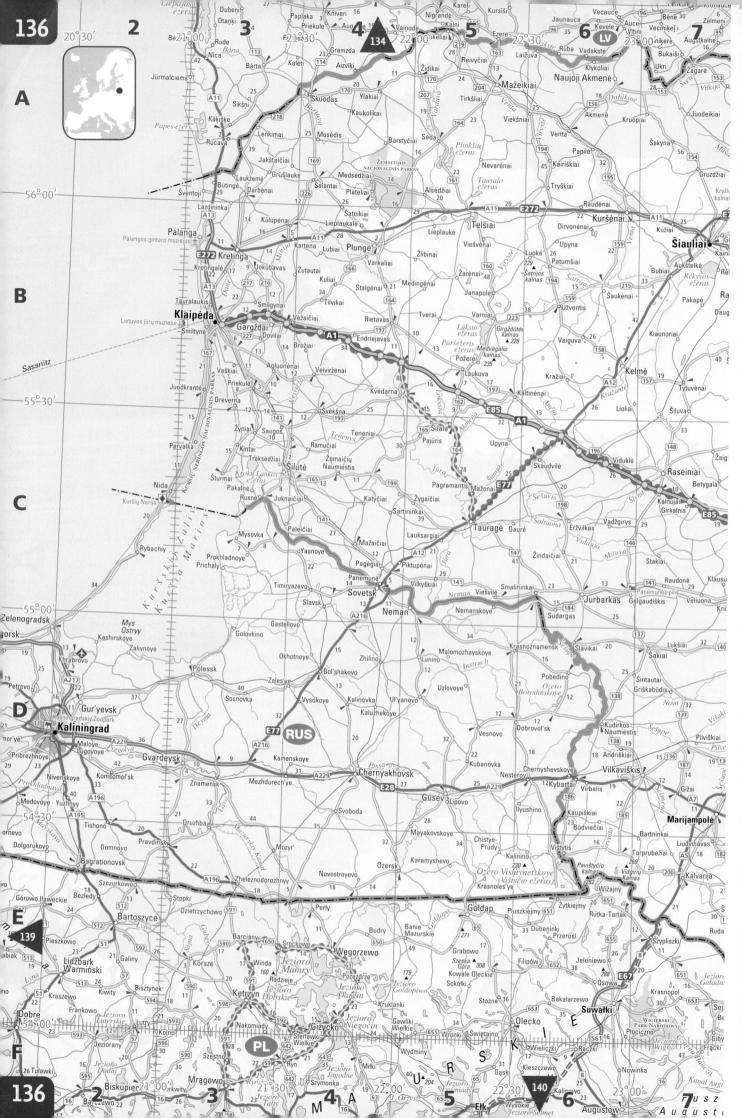

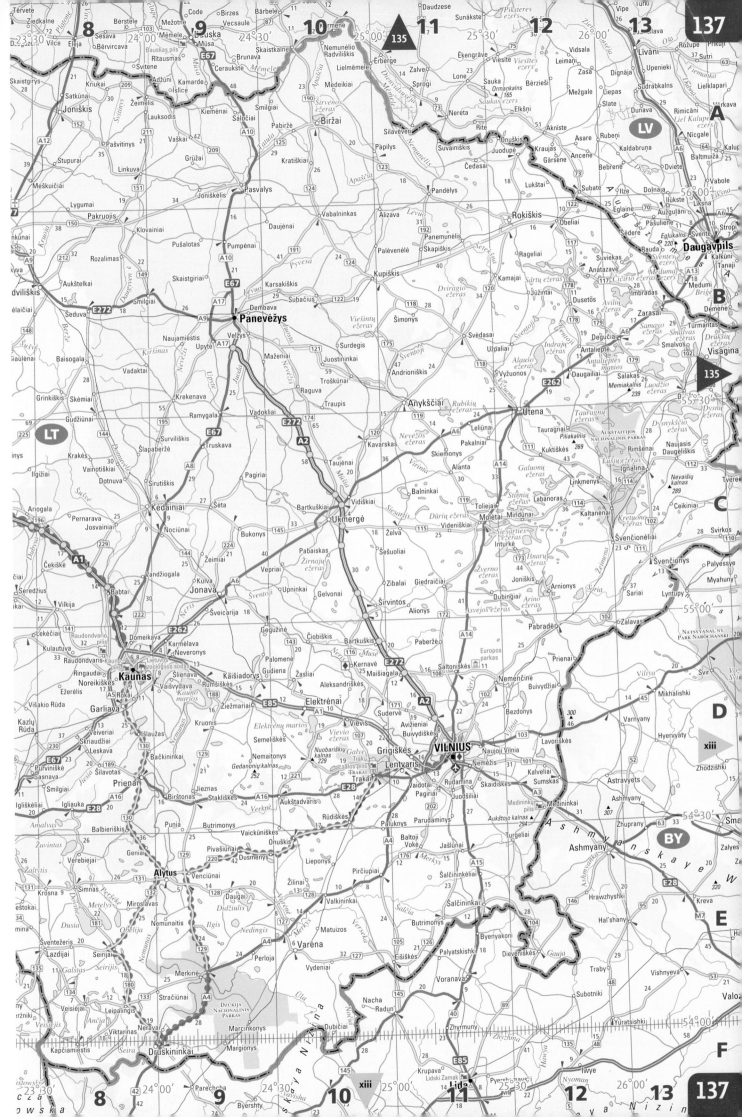

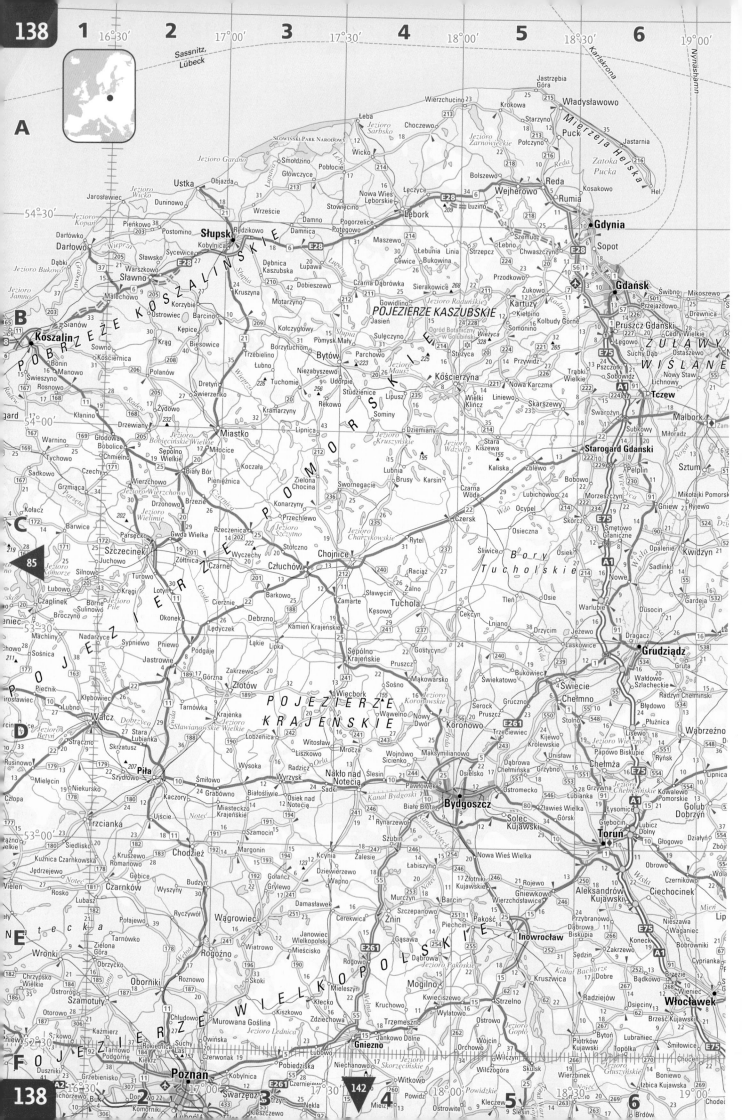

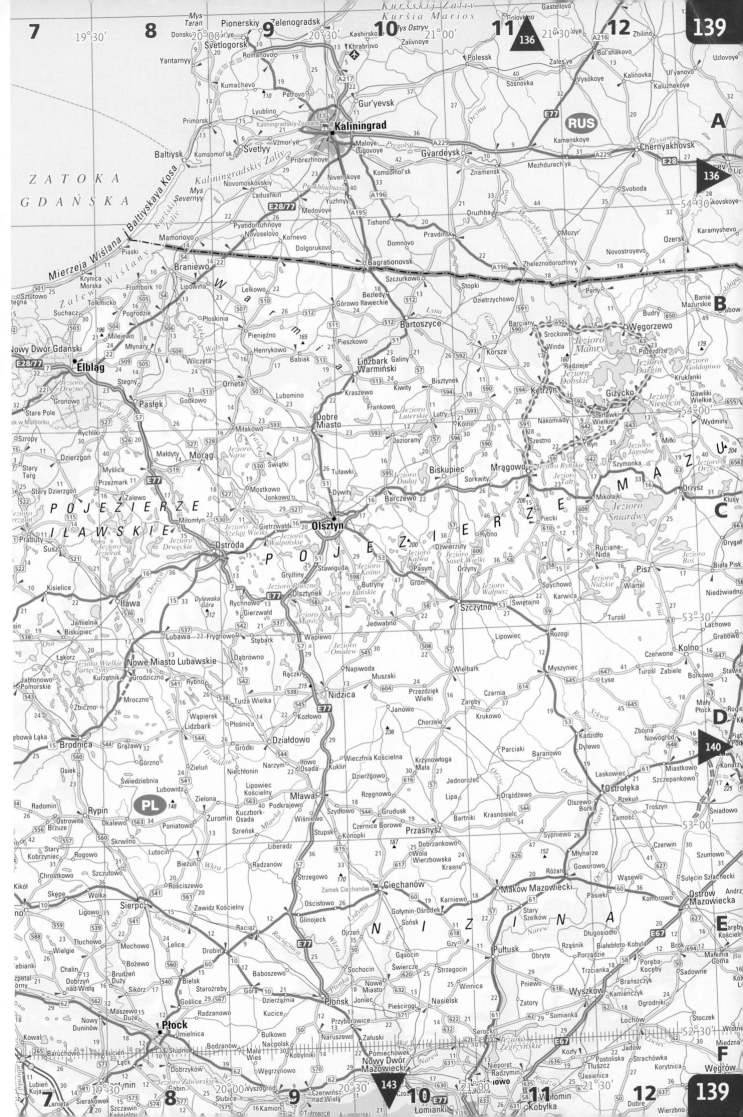

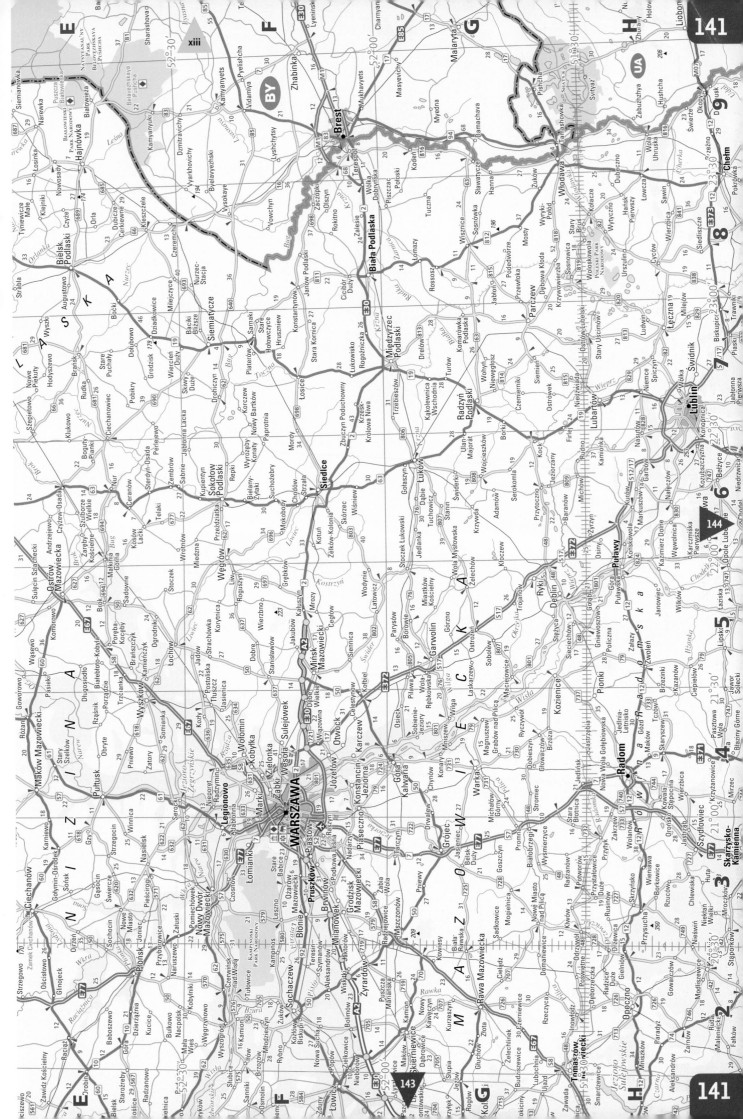

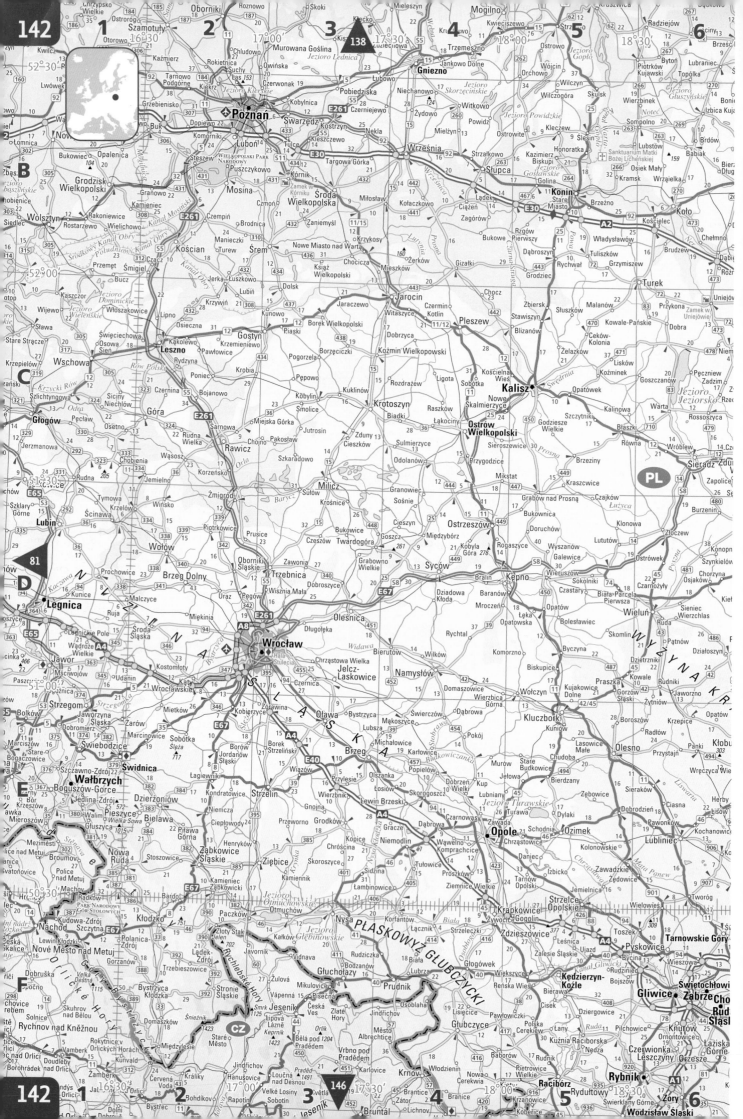

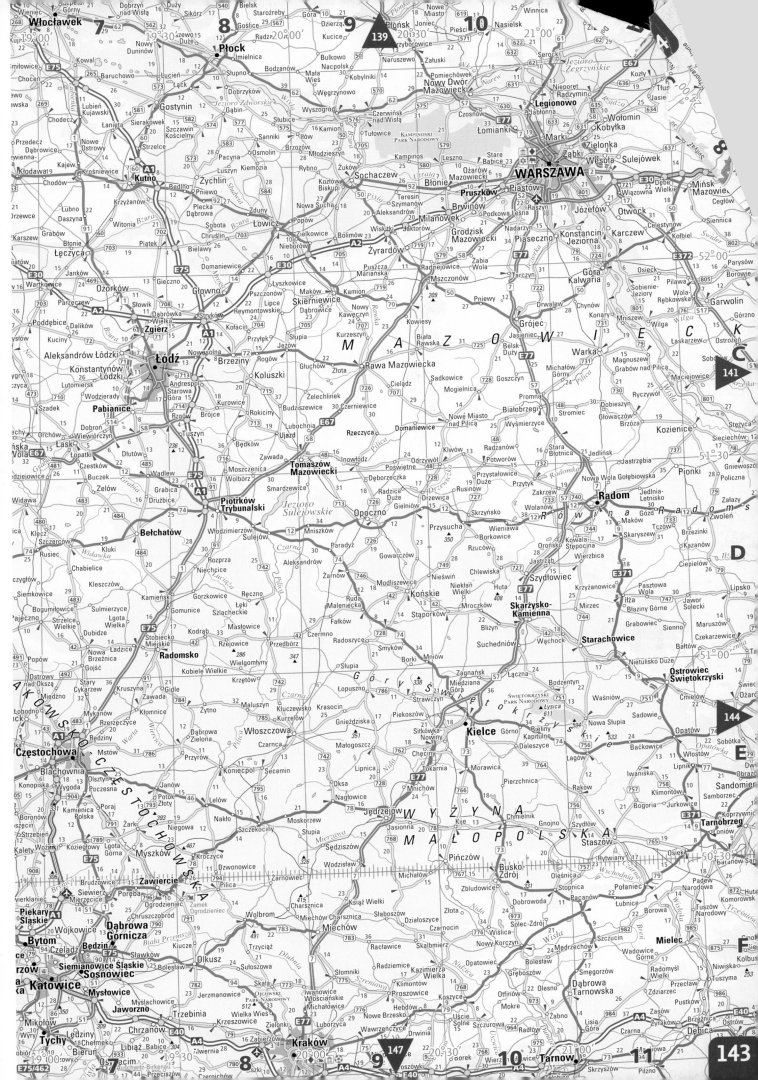

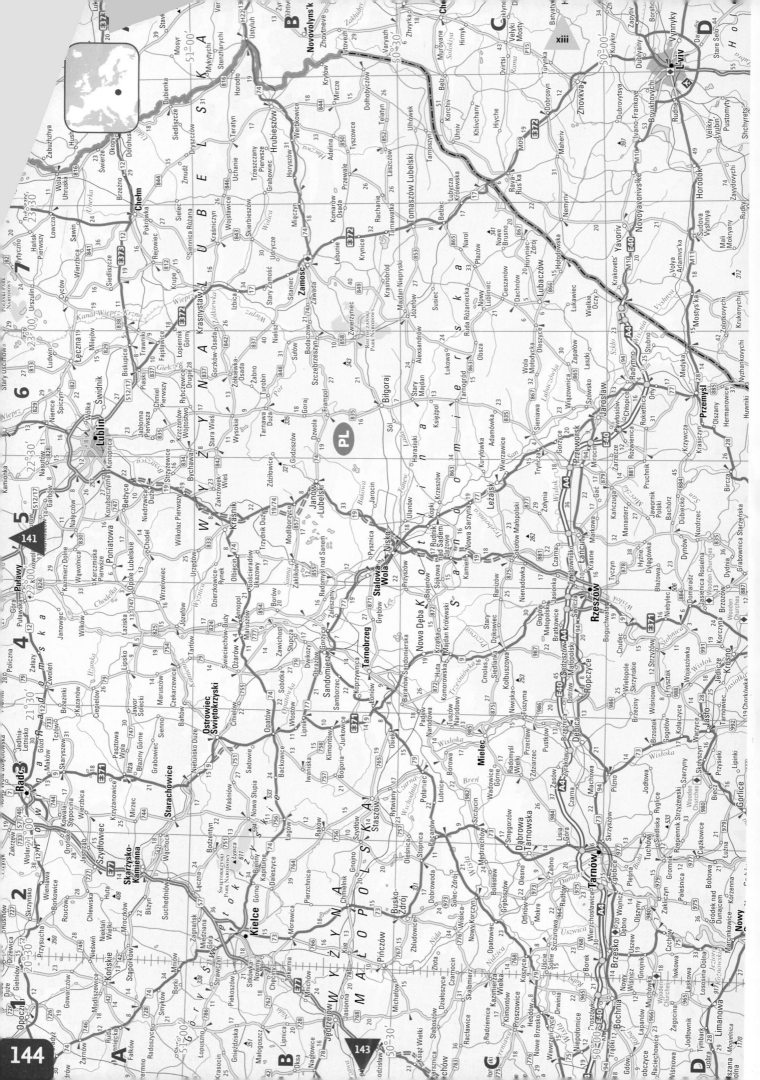

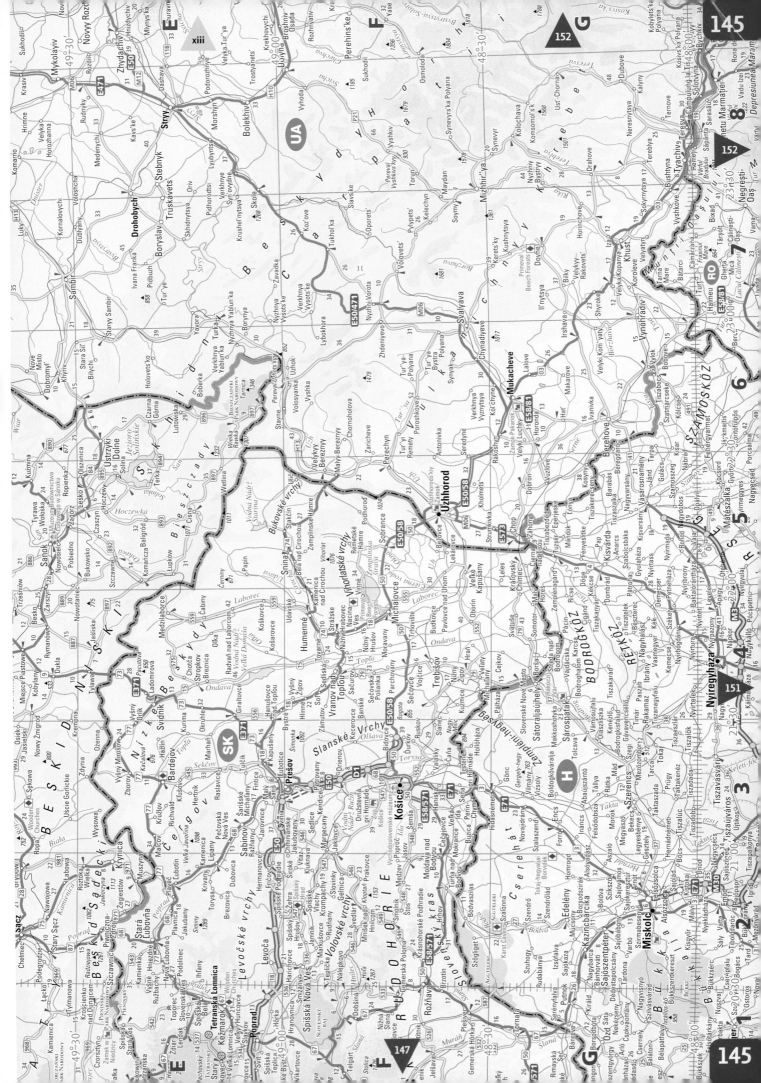

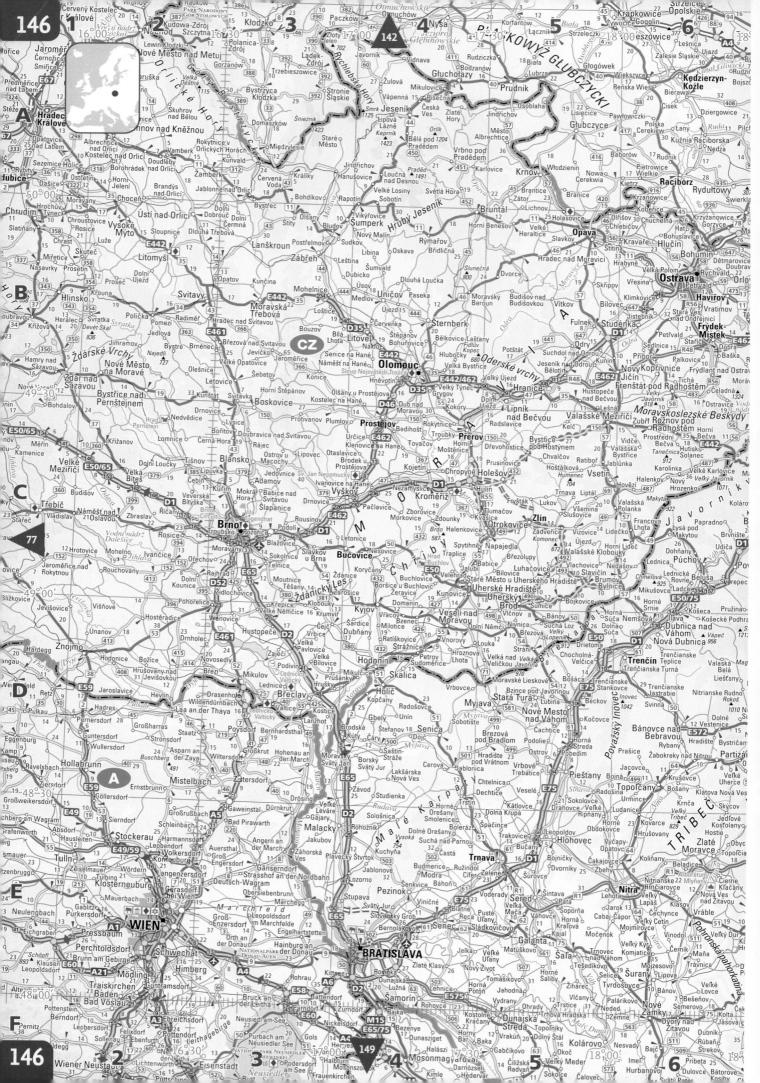

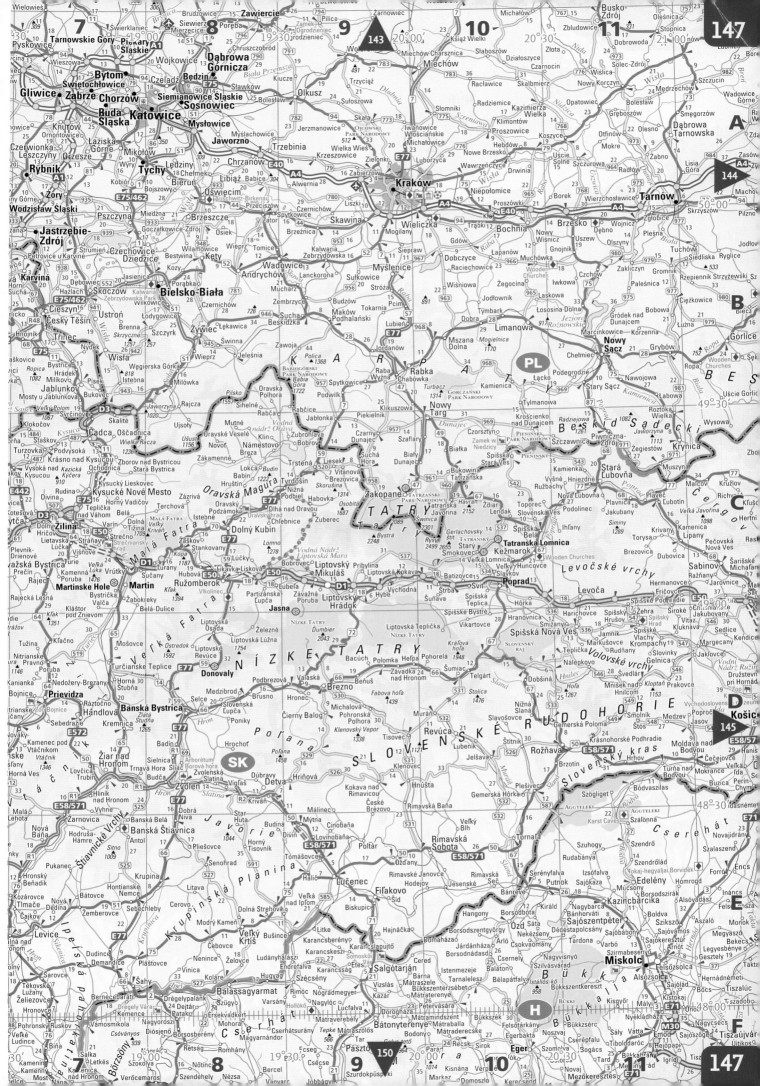

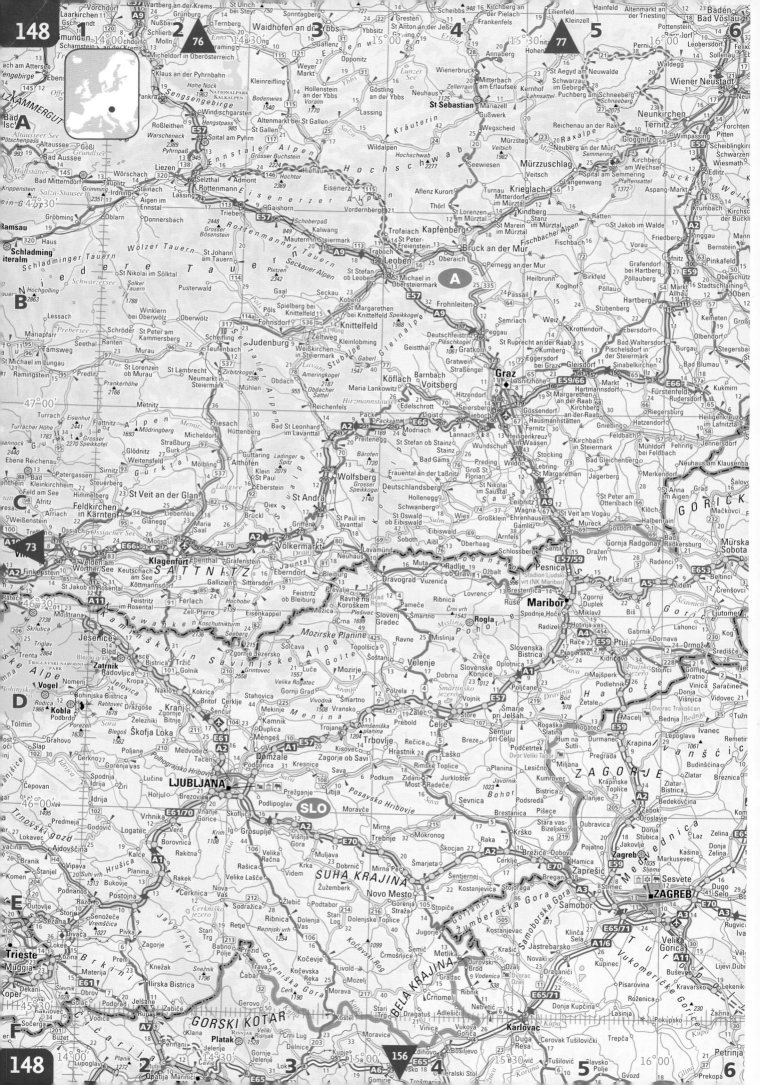

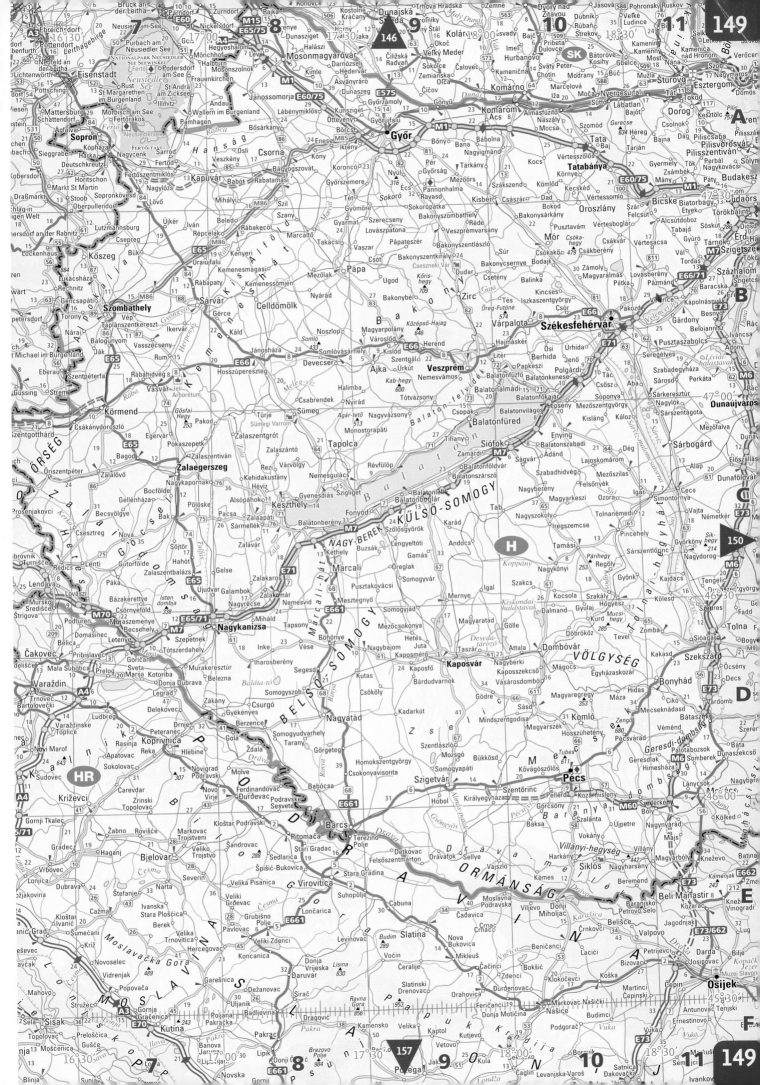

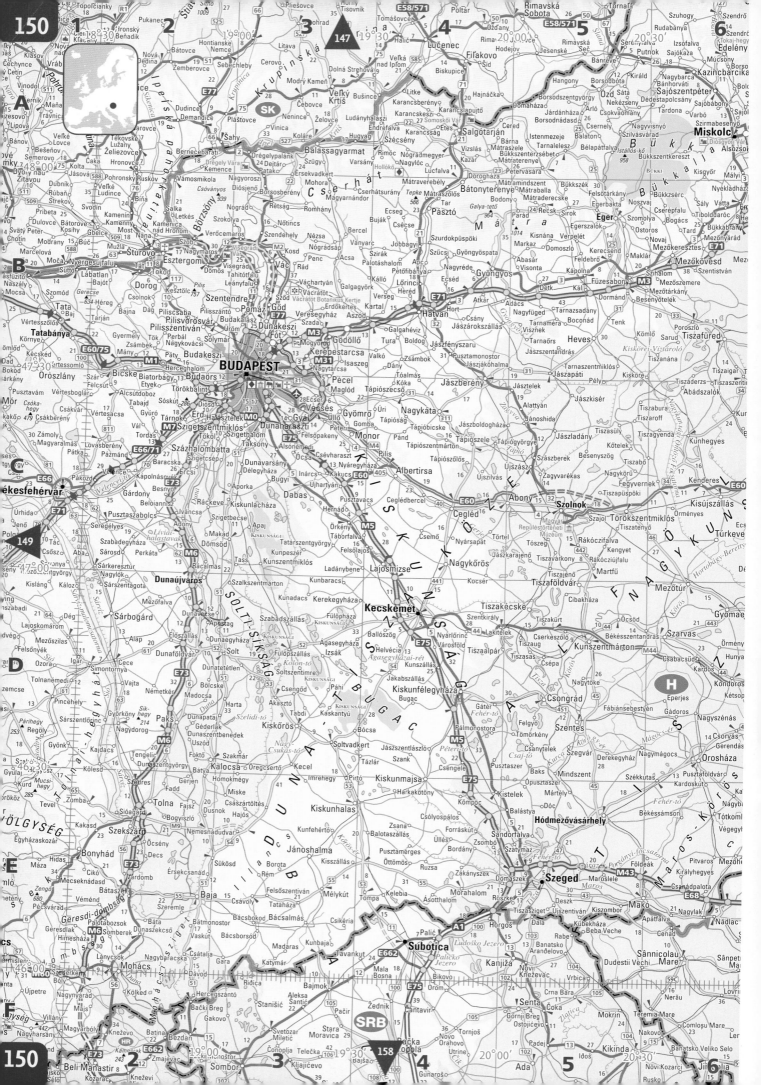

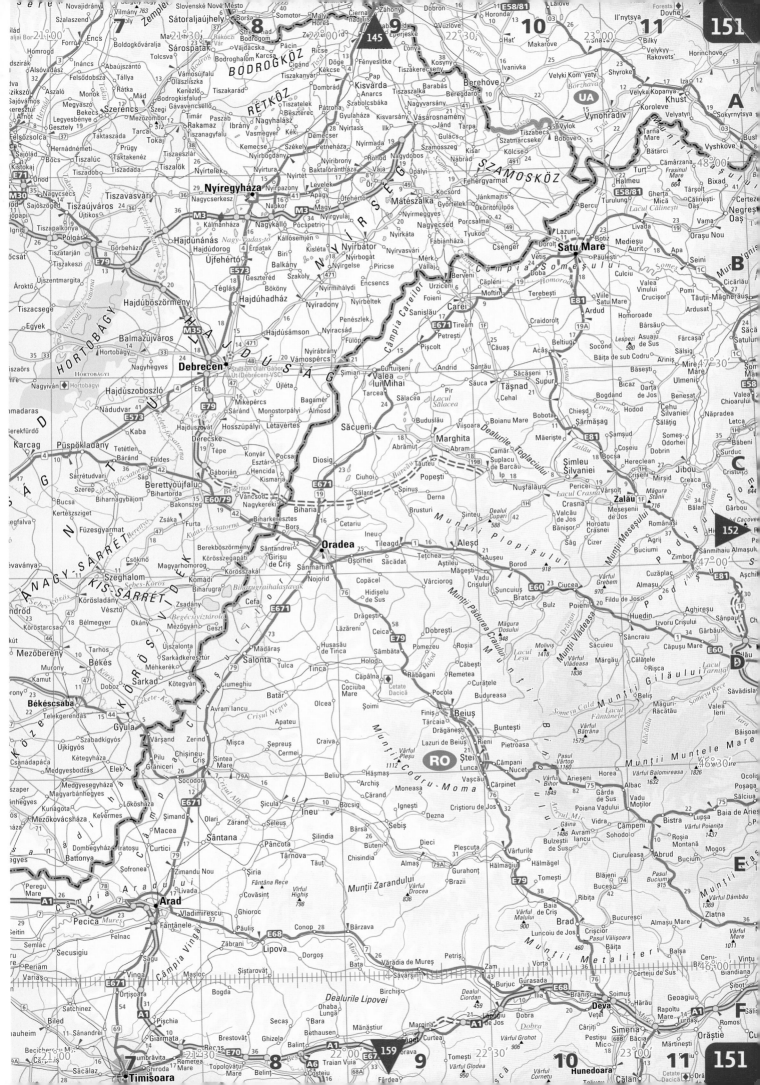

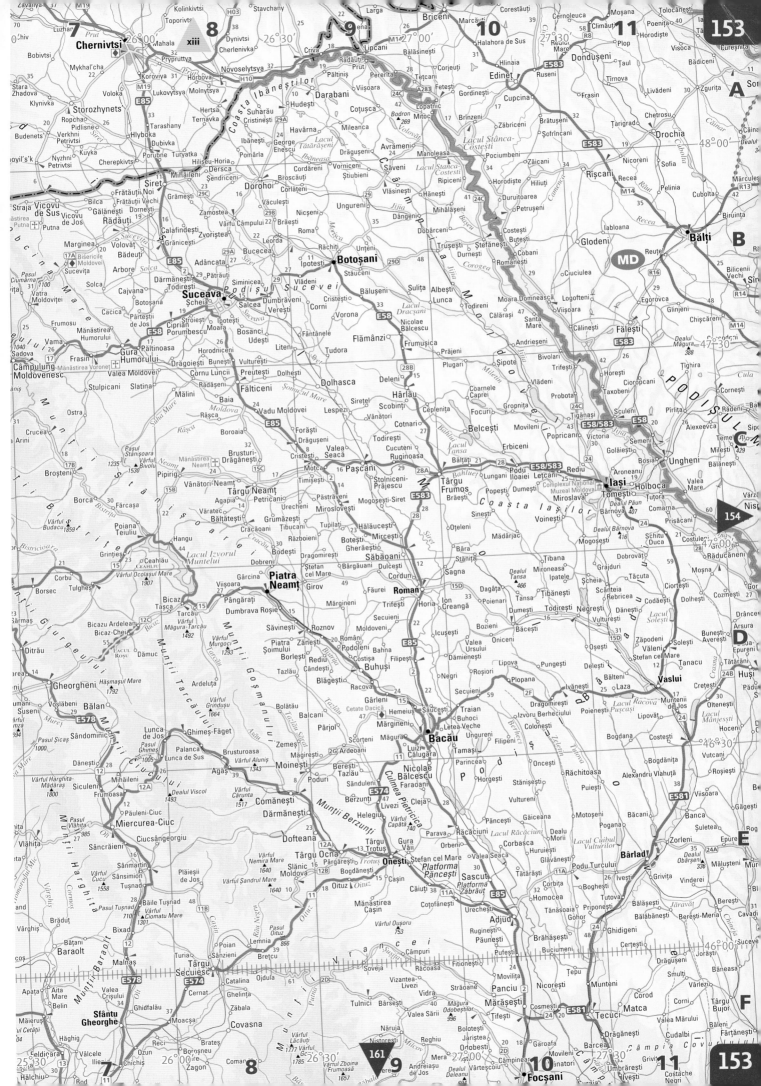

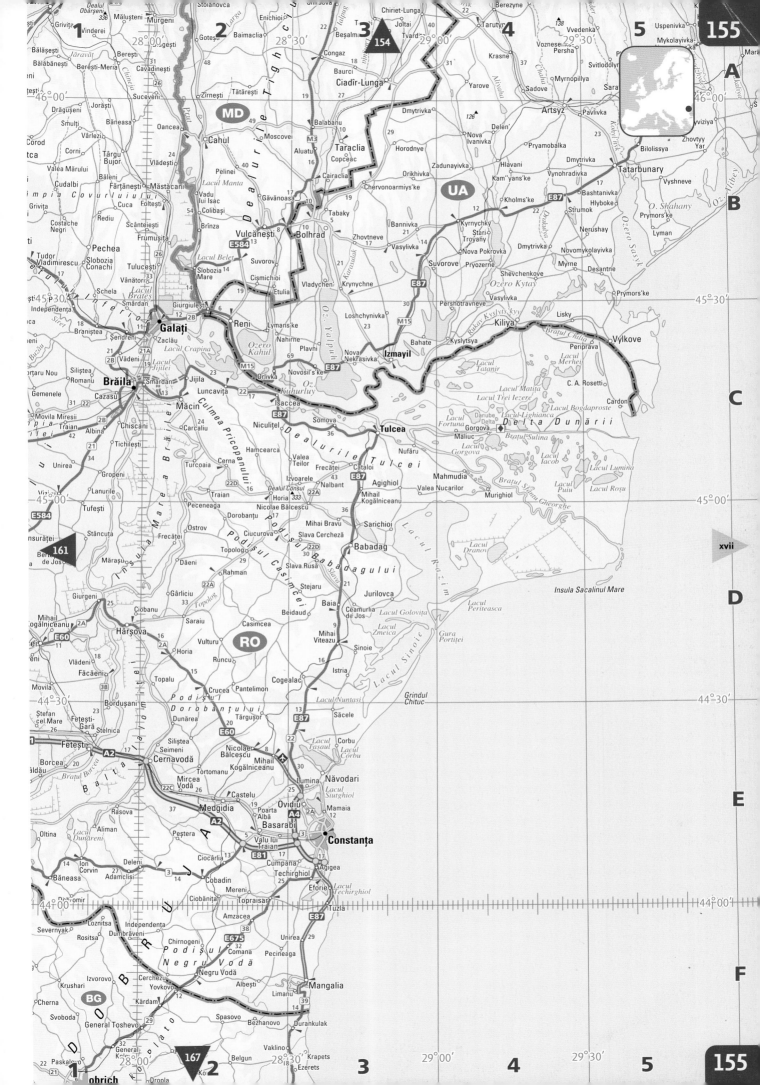

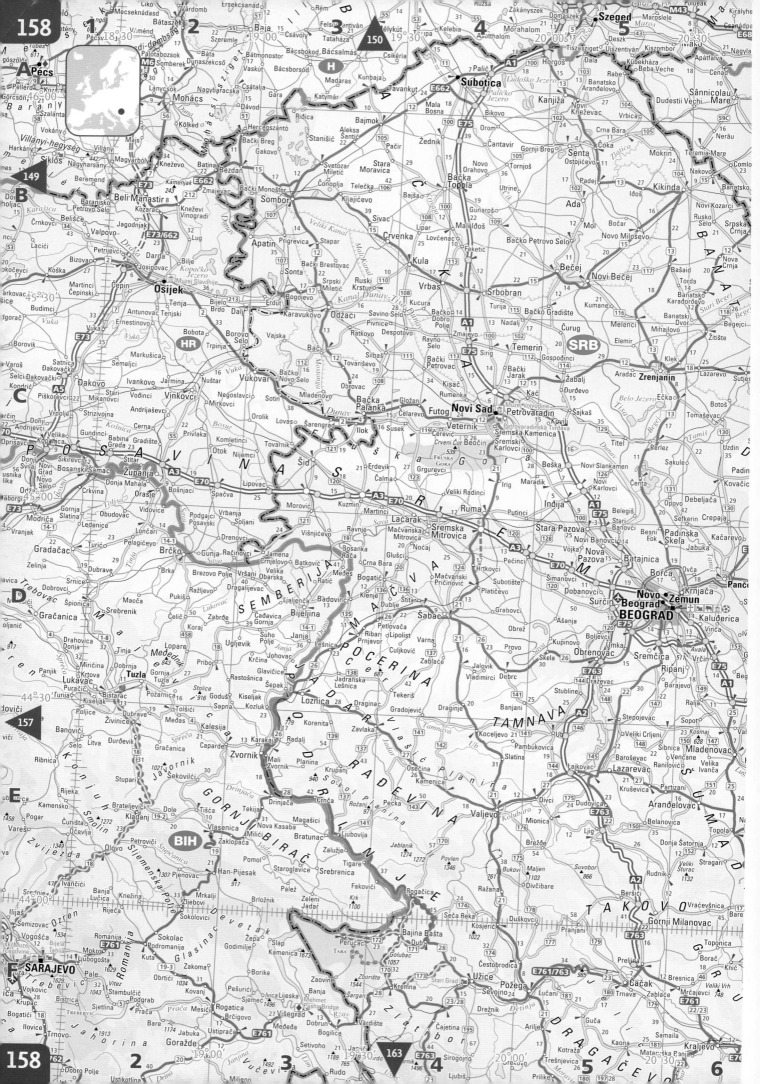

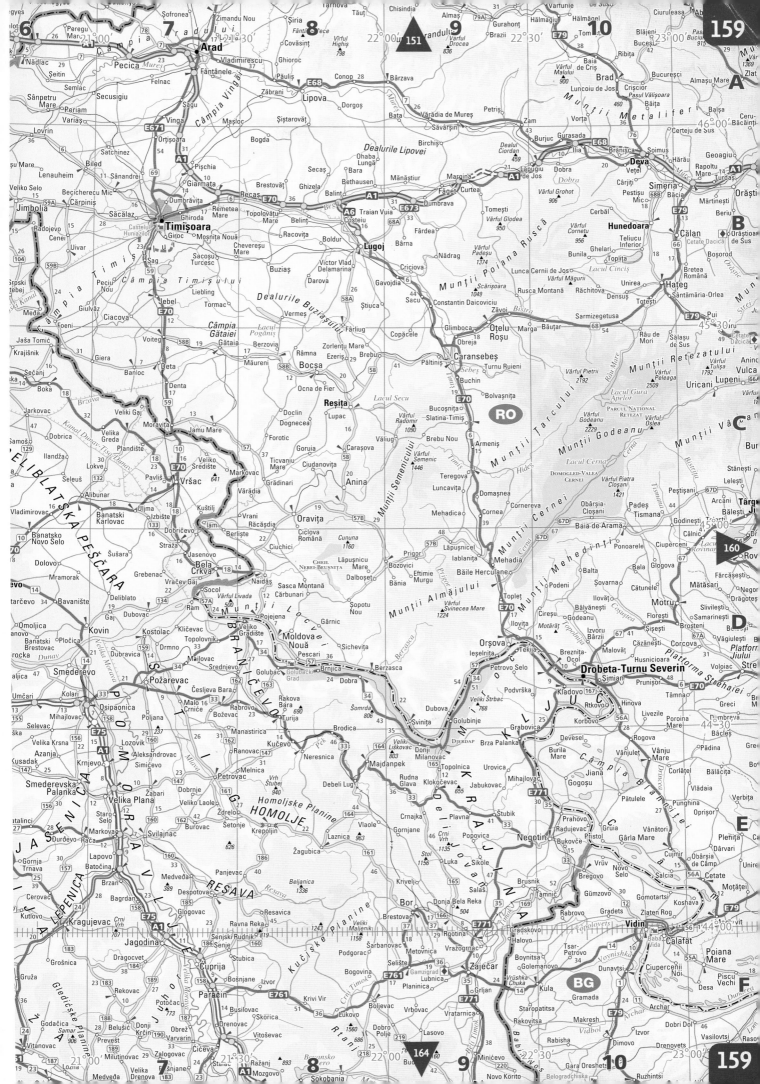

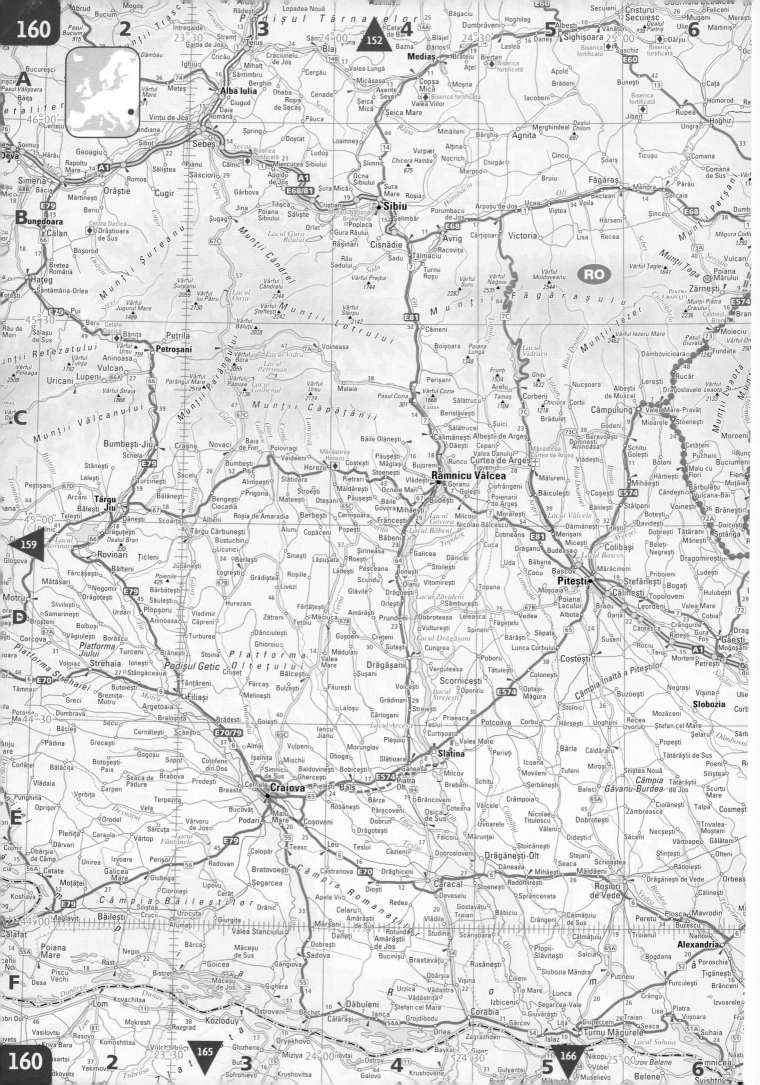

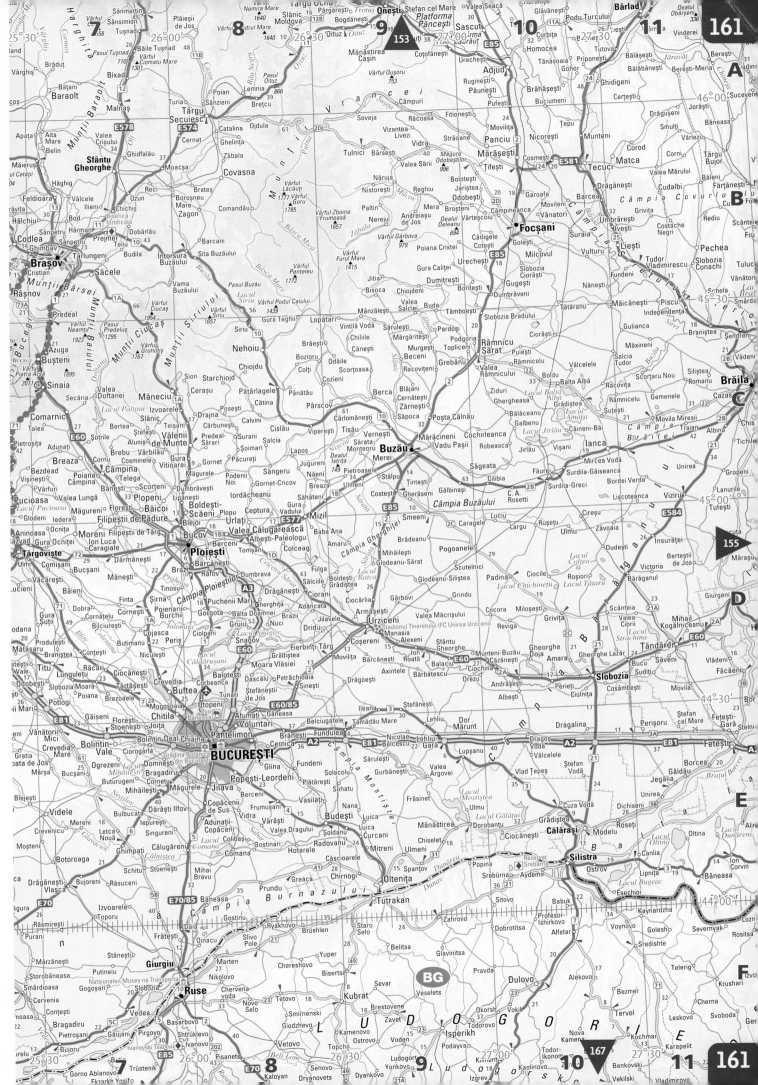

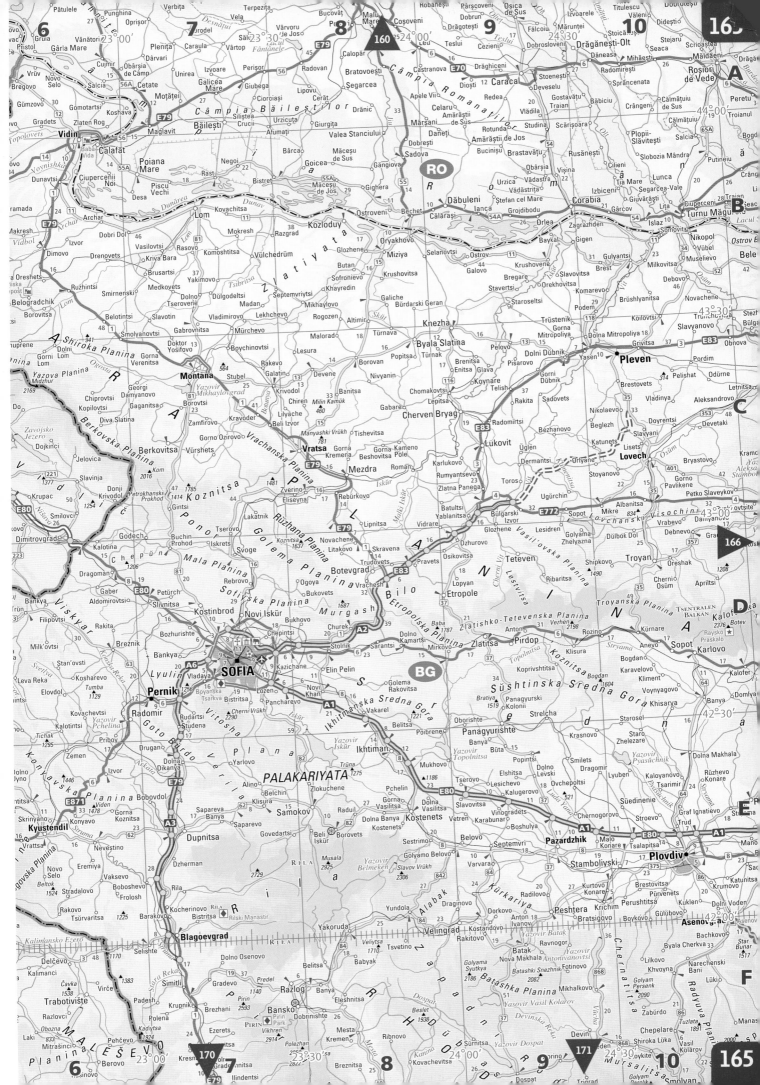

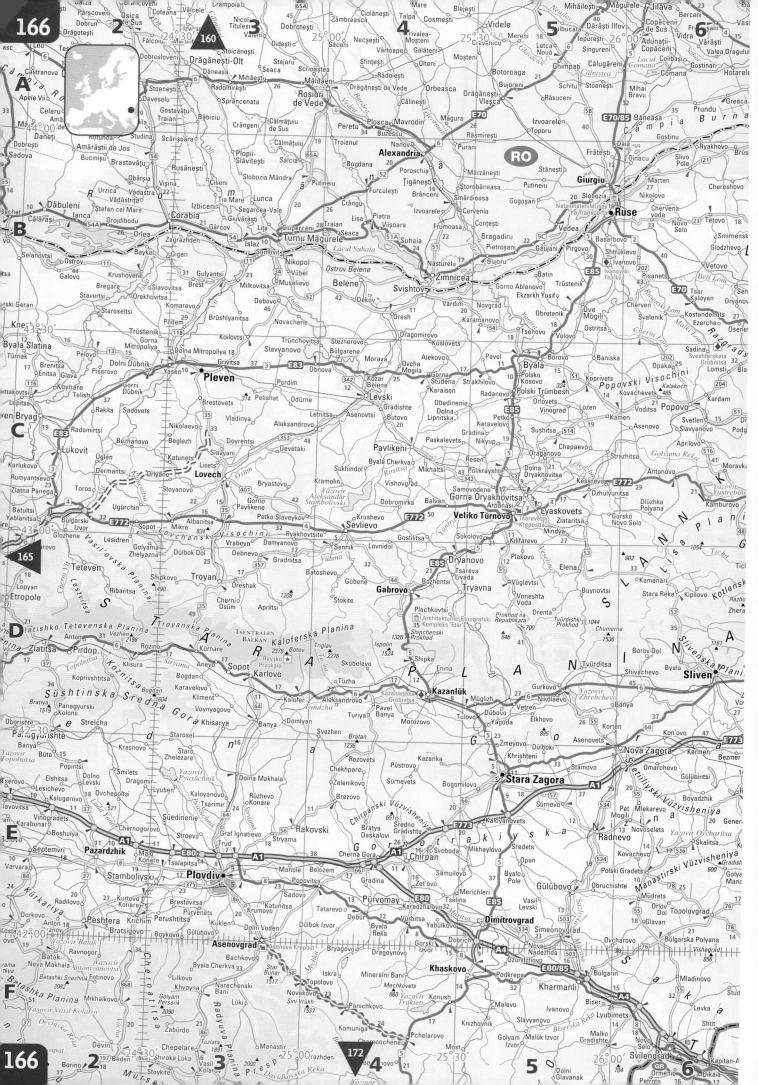

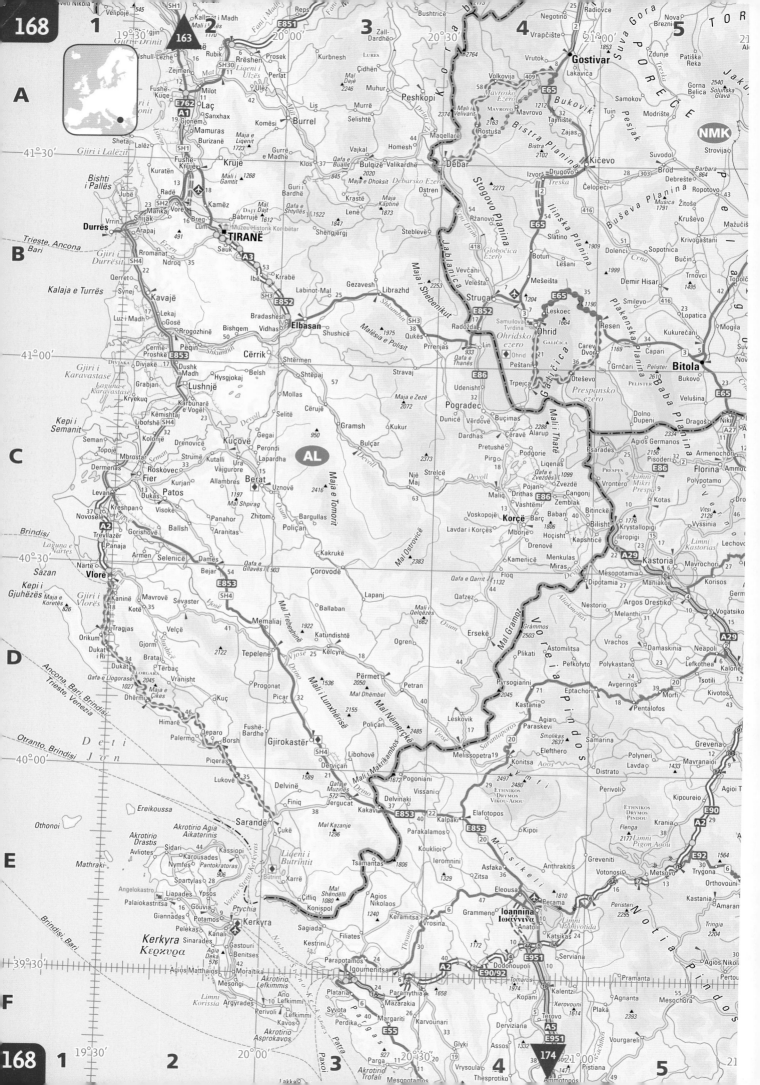

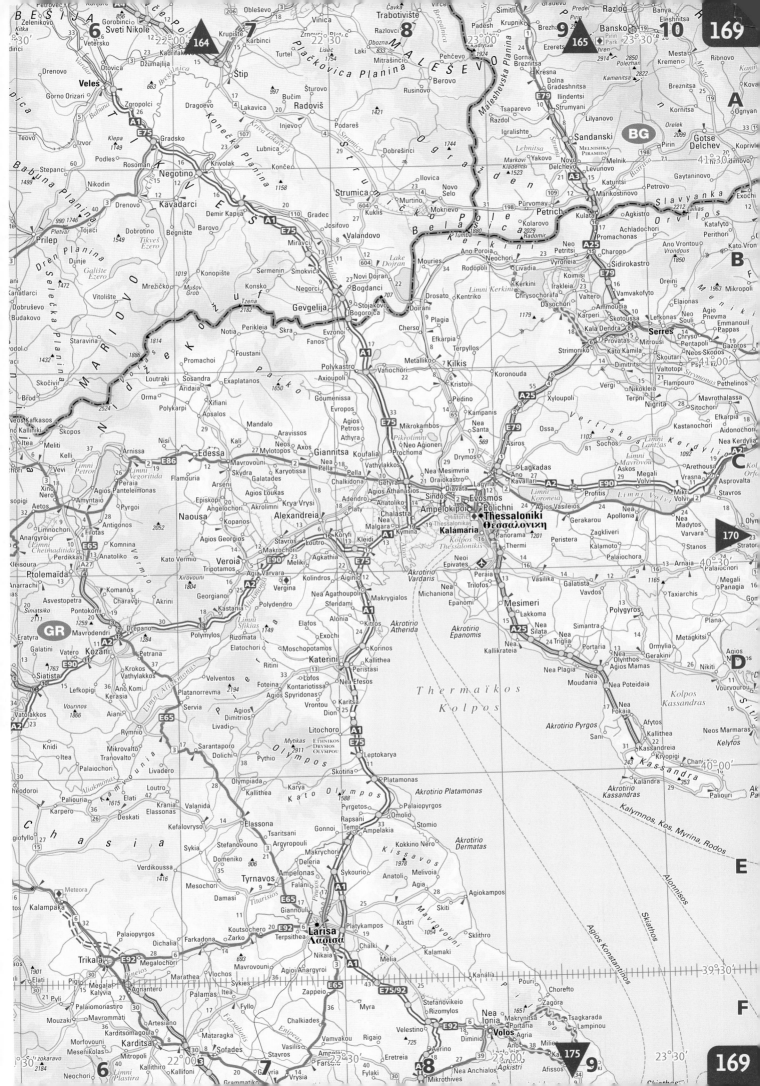

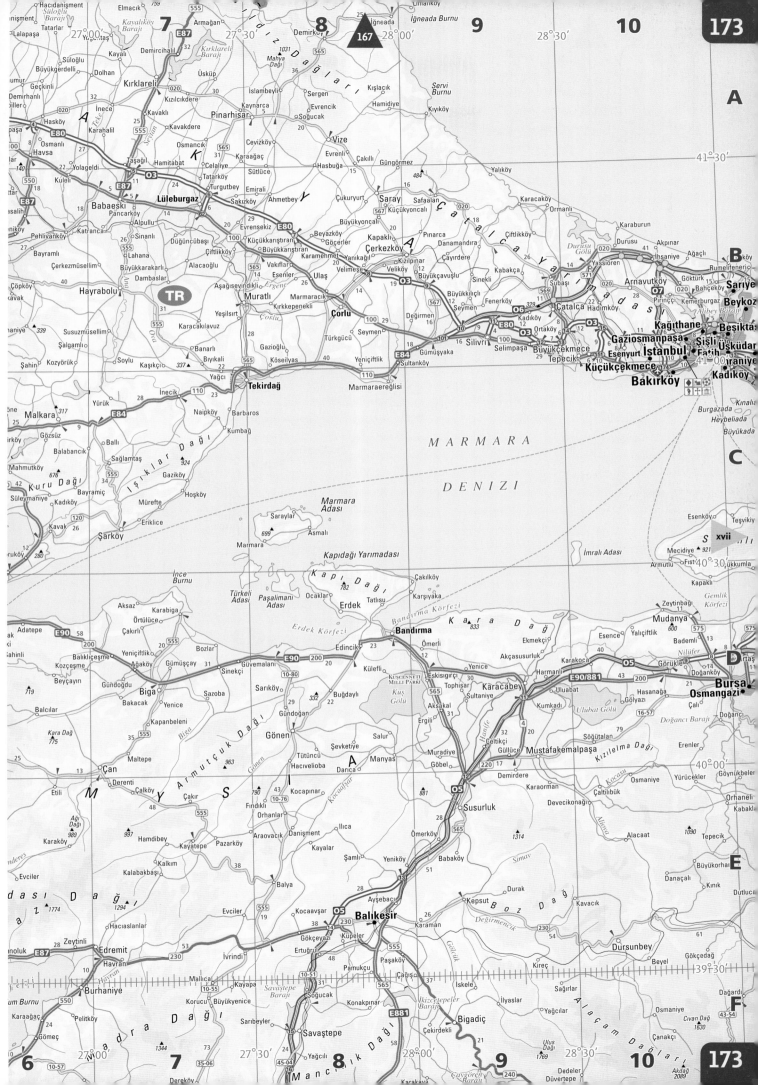

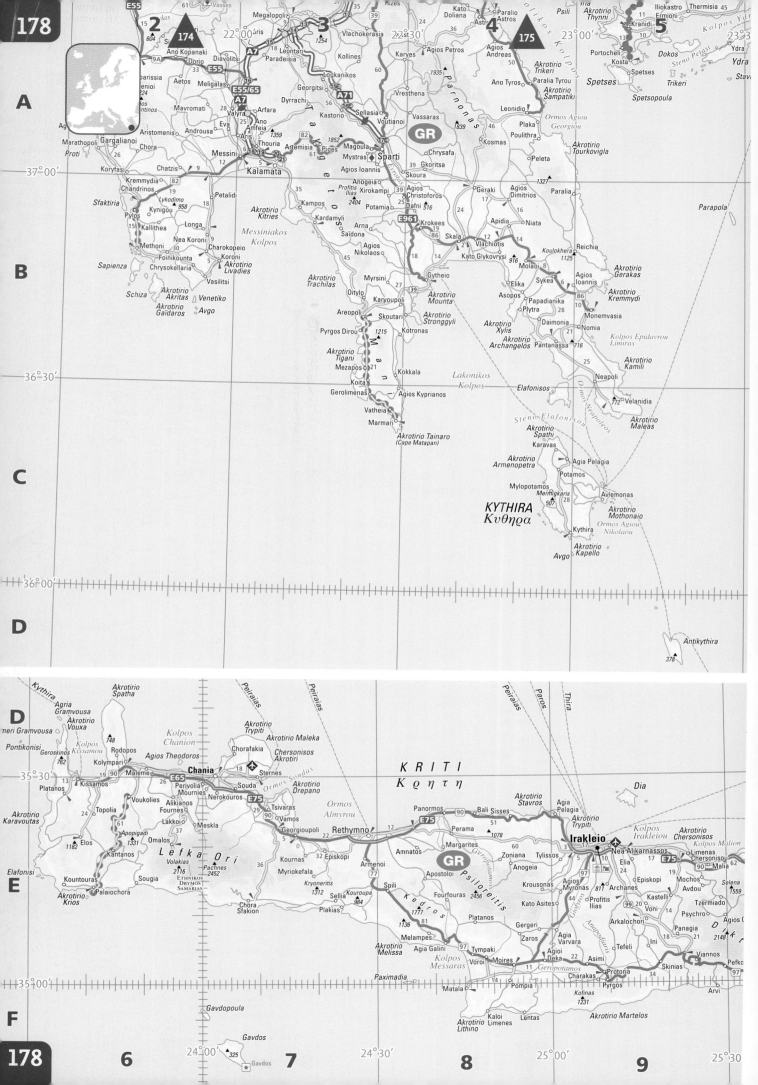

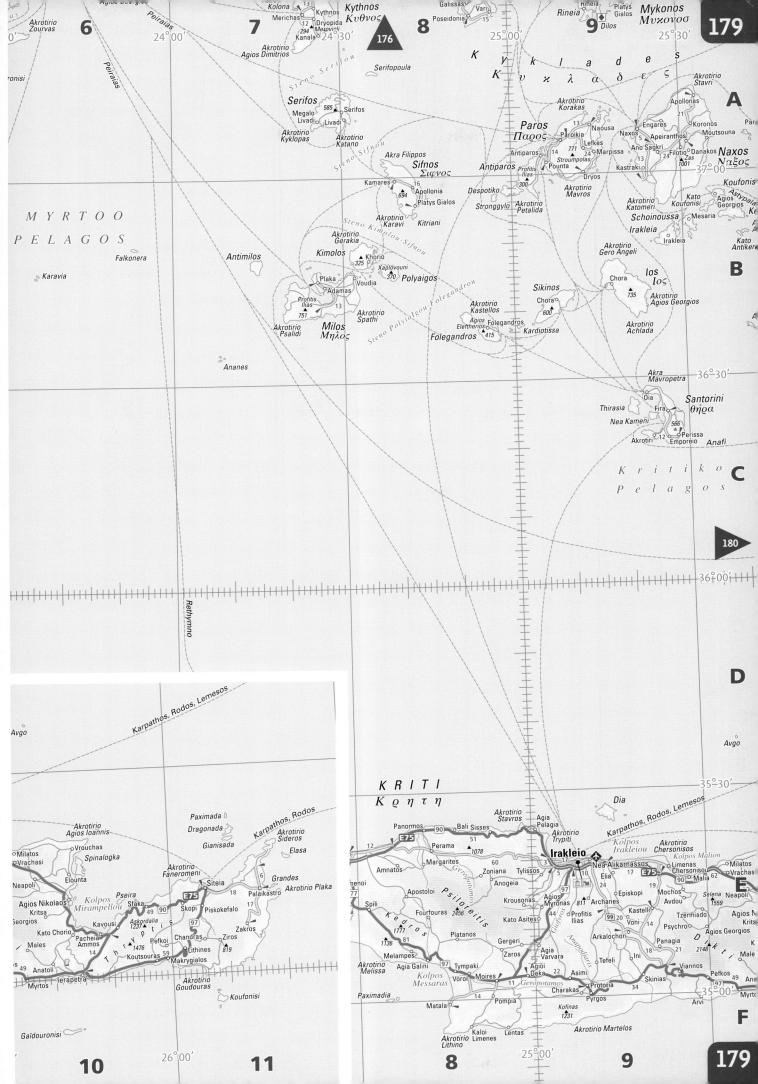

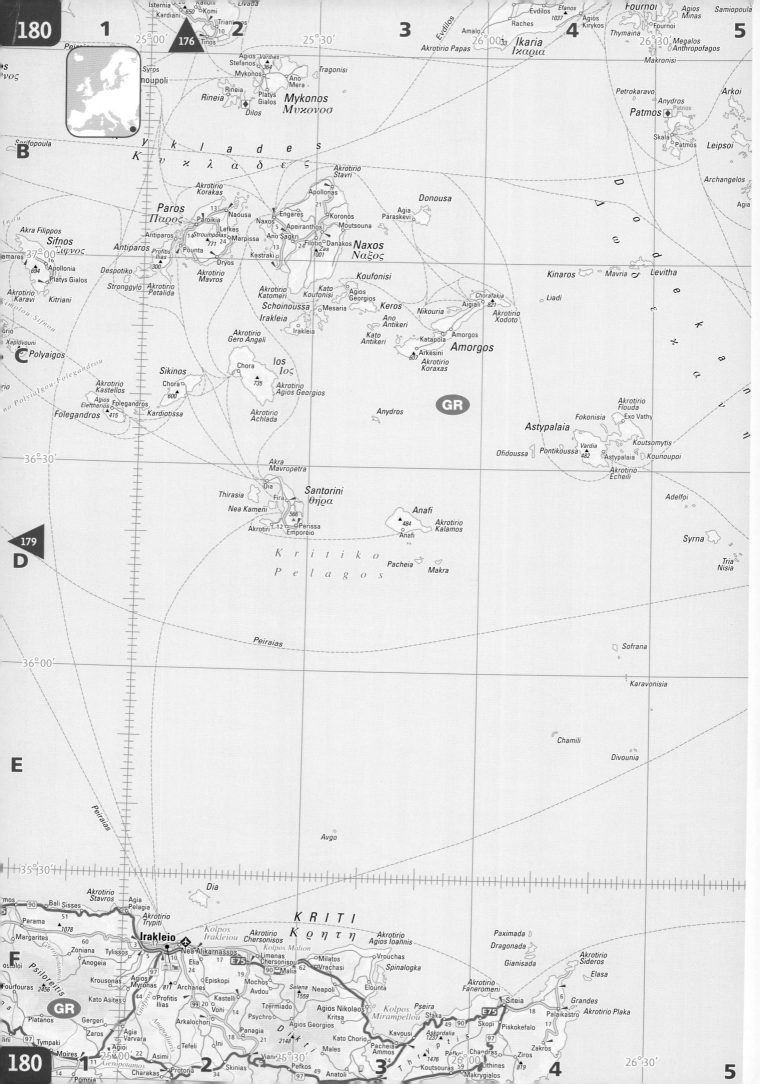

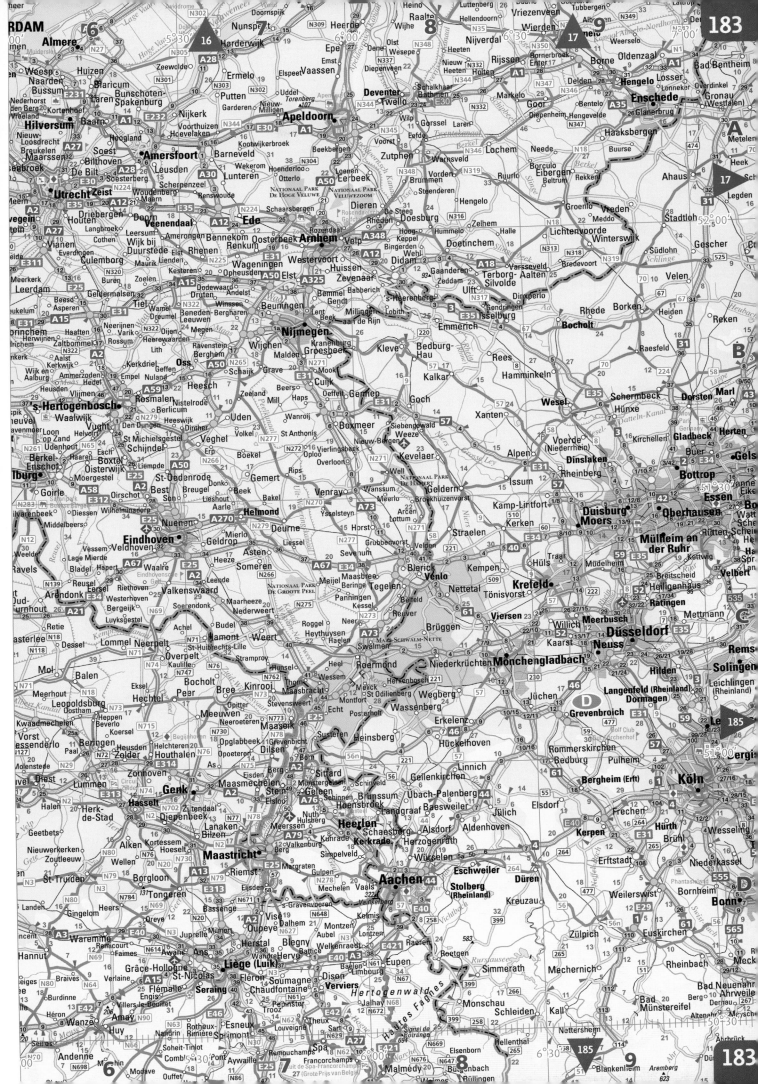

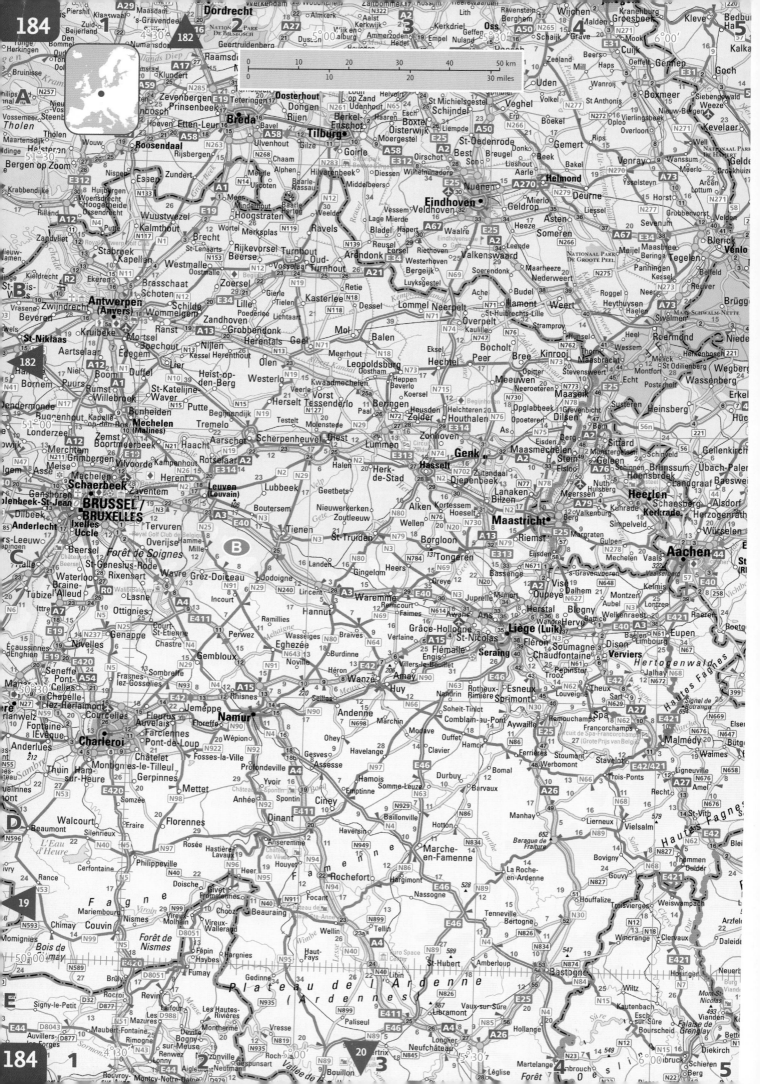

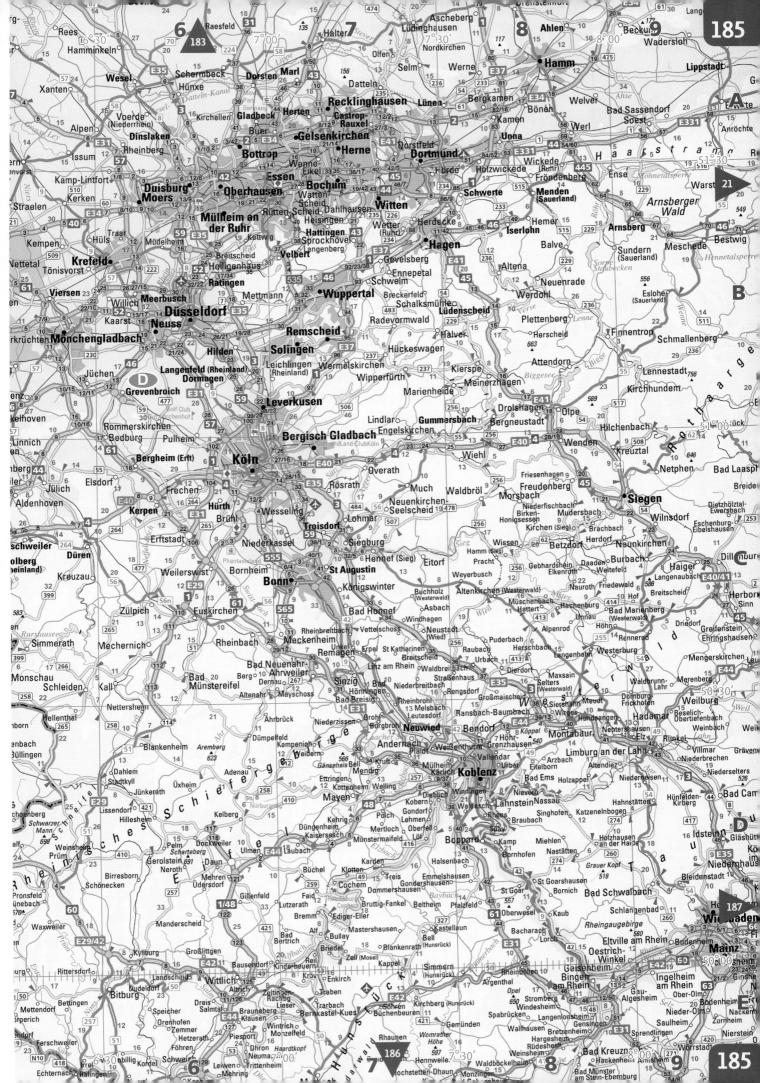

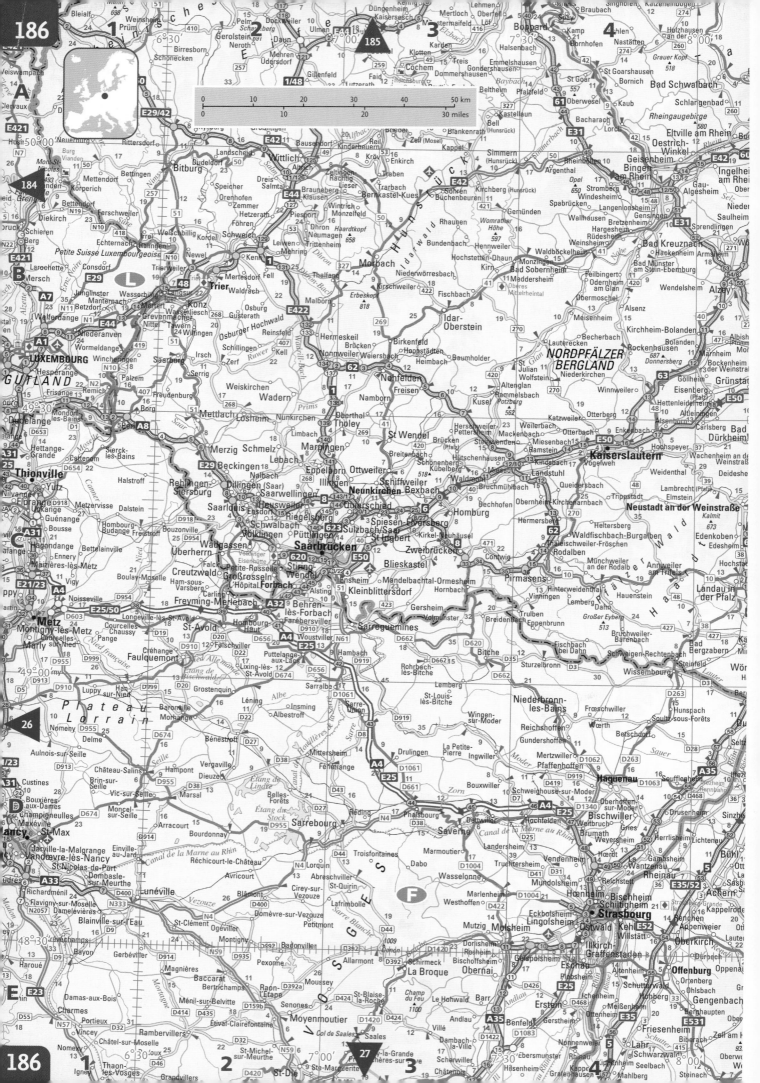

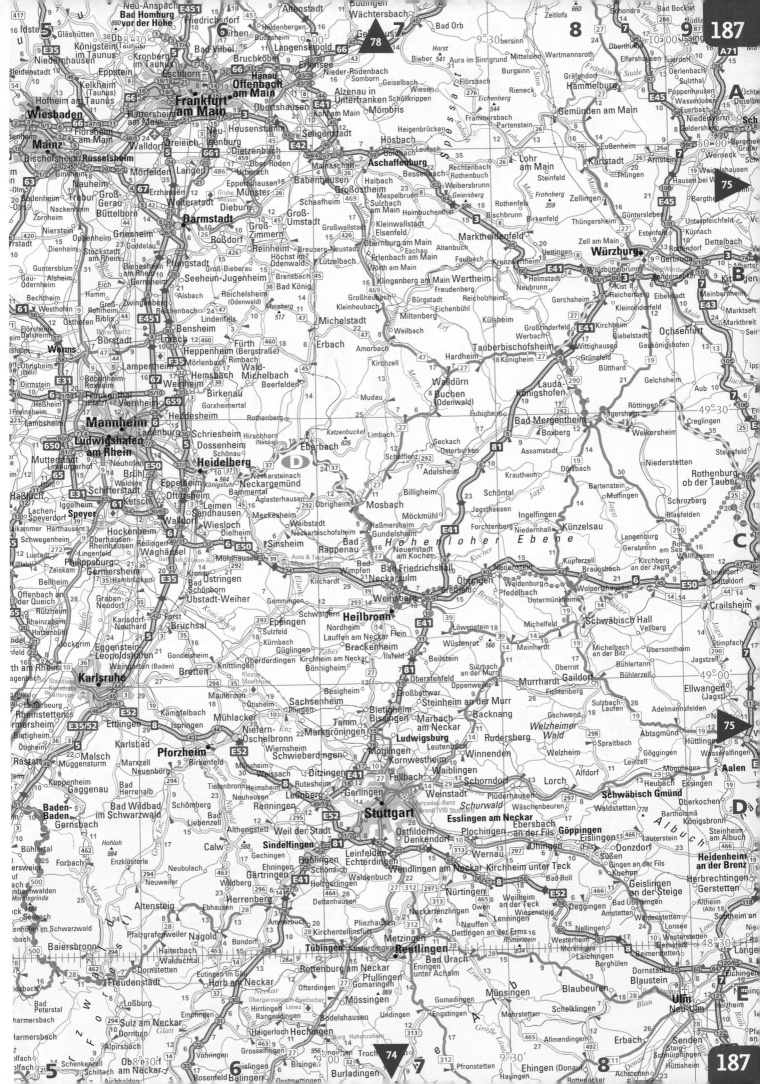

A

Aachen D 183 D8
Aalen D 187 D9
Aalsmeer NL 182 A5
Aalst B 182 D4
Aalst NL 183 B6
Aalten NL 183 B9
Aalter B 182 C2
Aardenburg NL 182 C2
Aarle NL 183 C7
Aarschot B 182 C5
Aartselaar B 182 C4
Abcoude NL 182 A5
Abreschviller F 186 D3
Abtsgmünd D 187 D8
Achel B 183 C6
Achern D 186 D5
Achstetten D 187 E8
Adegem B 182 C2
Adelmannsfelden D 187 D8
Adelsheim D 187 C7
Adenau D 185 D6
Aglasterhausen D 187 C6
Ahaus D 183 A10
Ahlen D 185 A8
Ahrbrück D 183 D6
Aichhalden D 187 E5
Aiglemont F 184 E2
Albergen NL 183 A9
Albertshofen D 187 B9
Albestroff F 186 D2
Albisheim (Pfrimm) D 186 B5
Abblasserdam NL 182 B5
Aldenhoven D 183 D8
Alf D 185 E7
Alfdorf D 187 D8
Alken D 183 D6
Allarmont F 186 E3
Allmendingen D 187 E8
Almelo NL 183 A9
Almere NL 183 A6
Almkerk NL 182 B5
Alost B 182 D4
Alpen D 183 B9
Alpenrod D 185 C8
Alphen NL 182 C5
Alphen aan den Rijn NL 182 A5
Alpirsbach D 187 E5
Alsbach D 187 B6
Alsdorf D 183 D8
Alsenborn D 186 C4
Alsenz D 186 B4
Alsheim D 185 E9
Alsting F 186 C2
Altena D 185 B8
Altenahr D 183 D9
Altenbuch D 187 B7
Altendiez D 185 D8
Altenglan D 186 B3
Altenheim D 186 E4
Altenkirchen (Westerwald) D 185 C8
Altenstadt D 187 A6
Altensteig D 187 D6
Altheim (Alb) D 187 D9
Althengstett D 187 D6
Altleiningen D 186 B5
Altrich D 185 E6
Alzenau in Unterfranken D 187 A7
Alzey D 186 B5
Amay B 183 D6
Amberloup B 184 D4
Ameide NL 182 B5
Amel B 184 D5
Amerongen NL 183 B6
Amersfoort NL 183 A6
Ammerbuch D 187 D6
Ammerzoden NL 183 B6
Amnéville F 186 C1
Amorbach D 187 B7
Amstelveen NL 182 A5
Amsterdam NL 182 A5
Amstetten D 187 D8
Andelst NL 183 B7
Andenne B 183 E6
Anderlecht B 182 D4
Anderlues B 182 E4
Andernach D 185 D7
Andlau F 186 E3
Anhée B 182 E5
Annœullin F 182 D1
Annweiler am Trifels D 186 C4
Anröchte D 185 A9
Ans B 183 D7
Anseremme B 184 D2
Antoing B 182 D2
Antwerpen B 182 C4
Anvers B 182 C4
Anzegem B 182 D2
Apeldoorn NL 183 A7
Appenweier D 186 D4
Arcen NL 183 C8
Ardooie B 182 D2
Arendonk B 183 C6
Argenthal D 185 D7
Armentières F 182 D1
Armsheim D 185 E9
Arnemuiden NL 182 C3
Arnhem NL 183 B7
Arnsberg D 185 B9
Arnstein D 187 B8
Arracourt F 186 D2
Arry F 186 C1
Arzbach D 185 D8
Arzfeld D 185 D5
As B 183 C7
Asbach D 185 C7
Aschaffenburg D 187 B7
Ascheberg D 185 A8
Asperen NL 183 B6
Assamstadt D 187 C8
Asse B 182 D4
Assenede B 182 C3
Assesse B 184 D3
Asten NL 183 C7
Ath B 182 D3
Attendorn D 185 B8
Aub D 187 B9
Aubel B 183 D7
Aubenton F 184 E1
Auby F 182 E2
Aulnois-sur-Seille F 186 D1
Aura im Sinngrund D 187 A8
Auvelais B 182 E5
Auvillers-les-Forges F 184 E1
Avelgem B 182 D2
Avricourt F 186 D2

Awans B 183 D6
Axel NL 182 C3
Aywaille B 183 E7

B

Baarle-Hertog B 182 C5
Baarle-Nassau NL 182 C5
Baarn NL 183 A6
Babberich NL 183 B8
Babenhausen D 187 B6
Baccarat F 186 E2
Bacharach D 185 D7
Backnang D 187 D7
Bad Bentheim D 183 A10
Bad Bergzabern D 186 C4
Bad Berleburg D 185 B9
Bad Bertrich D 185 D7
Bad Boll D 187 D8
Bad Breisig D 185 D7
Bad Camberg D 185 D9
Bad Dürkheim D 186 C5
Bad Ems D 185 D8
Baden-Baden D 187 D5
Bad Friedrichshall D 187 C7
Bad Herrenalb D 187 D5
Badhoevedorp NL 182 A5
Bad Homburg vor der Höhe D 187 A6
Bad Honnef D 185 C7
Bad Hönningen D 185 C7
Bad Kissingen D 187 A9
Bad König D 187 B7
Bad Kreuznach D 185 E8
Bad Laasphe D 185 B9
Bad Liebenzell D 187 D6
Bad Marienberg (Westerwald) D 185 C8
Bad Mergentheim D 187 C8
Bad Münstereifel D 183 D9
Bad Neuenahr-Ahrweiler D 183 D10
Badonviller F 186 D2
Bad Orb D 187 A7
Bad Peterstal D 187 E5
Bad Rappenau D 187 C7
Bad Sassendorf D 185 A9
Bad Schönborn D 187 C6
Bad Schwalbach D 185 D9
Bad Sobernheim D 185 E8
Bad Überkingen D 187 D8
Bad Urach D 187 E7
Bad Vilbel D 187 A6
Bad Wildbad im Schwarzwald D 187 D6
Bad Wimpfen D 187 C7
Baelen B 183 D8
Baesweiler D 183 D8
Baiersbronn D 187 D5
Baillonville B 184 D3
Bakel NL 183 C7
Balen B 183 C6
Balingen D 187 E6
Balve D 185 B8
Bammental D 187 C6
Barneveld NL 183 A7
Baronville F 186 D2
Barr F 186 E3
Bartenstein D 187 C8
Bartholomä D 187 D8
Barvaux B 184 D3
Basècles B 182 D3
Bassenge B 183 D7
Bastogne B 184 D4
Bathmen NL 183 A8
Battice B 183 D7
Baumholder D 186 B3
Bausendorf D 185 D6
Bavel NL 182 B5
Bayon F 186 E2
Beaumont B 184 D1
Beauraing B 184 D2
Becherbach D 186 B4
Bechhofen D 186 C3
Bechtheim D 187 B5
Beckingen D 186 C2
Beckum D 185 A9
Bedburg D 183 D9
Bedburg-Hau D 183 B8
Beek NL 183 B7
Beek NL 183 B7
Beekbergen NL 183 A7
Beerfelden D 187 B6
Beernem B 182 C2
Beers NL 183 B7
Beerse B 182 C5
Beersel B 182 D4
Beerst B 182 C1
Beesd NL 183 B6
Begijnendijk B 182 C5
Behren-lès-Forbach F 186 C2
Beilstein D 187 C7
Beimerstetten D 187 E8
Belfeld NL 183 C8
Bell D 185 D7
Bell (Hunsrück) D 185 D7
Bellenberg D 187 E9
Belles-Forêts F 186 D2
Bellheim D 187 C5
Beloeil B 182 D3
Beltheim D 185 D7
Beltrum NL 183 A9
Bemmel NL 183 B7
Bendorf D 185 D8
Beneden-Leeuwen NL 183 B6
Bénestroff F 186 D2
Benfeld F 186 E4
Bennebroek NL 182 A5
Bennekom NL 183 A7
Bentelo NL 183 A9
Berchem B 182 D3
Berg D 187 D7
Berg L 184 E5
Berg NL 183 D7
Berg (Pfalz) D 187 C5
Bergen op Zoom NL 182 C4
Bergeyk NL 183 C6
Bergharen NL 183 B7
Berghaupten D 186 D4
Bergheim (Erft) D 183 D9
Berghem NL 183 B7
Berghülen D 187 E8
Bergisch Gladbach D 183 D10
Bergkamen D 185 A8
Bergneustadt D 185 B8
Bergschenhoek NL 182 B5
Beringe NL 183 C7
Beringen B 183 C6
Berkel NL 182 B4
Berkel-Enschot NL 183 B6
Berlicum NL 183 B6
Bernissart B 182 E3
Bernkastel-Kues D 185 E7
Bernstadt D 187 D9
Bertogne B 184 D4
Bertrange L 186 B1
Bertrichamps F 186 E2
Bertrix B 184 E3
Beselich-Obertiefenbach D 185 D9
Besigheim D 187 D7
Bessenbach D 187 B7
Best NL 183 C6
Bestwig D 185 B9
Betschdorf F 186 D4
Bettelainville F 186 C1
Bettembourg L 186 B1
Bettendorf L 184 E5
Bettingen D 185 E5
Betzdorf L 186 B1
Betzdorf D 185 C8
Beuningen NL 183 B7
Beuvrages F 182 E2
Beveren B 182 C4
Beverlo B 183 C6
Bexbach D 186 C3
Biberach D 186 D4
Biblis D 187 B5
Bieber D 187 A7
Biebesheim am Rhein D 187 B5
Bietigheim D 187 D5
Bietigheim-Bissingen D 187 D7
Bièvre B 184 E3
Billigheim D 187 C7
Bilthoven NL 183 A6
Bilzen B 183 D7
Binche B 182 E4
Bingen am Rhein D 185 E8
Bingerden NL 183 B8
Birkenau D 187 B6
Birkenfeld D 186 B3
Birkenfeld D 187 B8
Birkenfeld D 187 D6
Birken-Honigsessen D 185 C8
Birresborn D 185 D6
Bischbrunn D 187 B7
Bischheim F 186 D4
Bischoffsheim F 186 E3
Bischofsheim D 185 E9
Bischwiller F 186 D4
Bisingen D 187 E6
Bissen L 184 E5
Bitburg D 185 E6
Bitche F 186 C3
Bladel NL 183 C6
Blainville-sur-l'Eau F 186 D1
Blâmont F 186 D2
Blankenberge B 182 C2
Blankenheim D 183 D9
Blankenrath D 185 D7
Blaricum NL 183 A6
Blaton B 182 D3
Blaubeuren D 187 E8
Blaufelden D 187 C8
Blaustein D 187 E8
Blegny B 183 D7
Bléharies B 182 D2
Bleialf D 185 D5
Bleidenstadt D 185 D9
Bleiswijk NL 182 A5
Blerick NL 183 C8
Bleskensgraaf NL 182 B5
Blieskastel D 186 C3
Bobenheim-Roxheim D 187 B5
Böblingen D 187 D7
Bocholt B 183 C7
Bocholt D 183 B9
Bochum D 185 A7
Bockenheim an der Weinstraße D 187 B5
Bodegraven NL 182 A5
Bodelshausen D 187 E6
Bodenheim D 185 E9
Boechout B 182 C4
Boekel NL 183 B7
Boekhoute B 182 C3
Bogny-sur-Meuse F 184 E2
Böhl D 187 C5
Bolanden D 186 B5
Bolnes NL 182 B5
Bomal B 184 D4
Bondorf D 187 D6
Bondues F 182 D2
Bönen B 185 A8
Bonheiden B 182 C5
Bonn D 183 D10
Bönnigheim D 187 C7
Boom B 182 C4
Boortmeerbeek B 182 D5
Boppard D 185 D7
Borculo NL 183 A9
Borgloon B 183 D6
Borken D 183 B9
Borne NL 183 C7
Borne NL 183 A9
Bornem B 182 C4
Bornerbroek NL 183 A9
Bornheim D 183 D9
Bornhofen D 185 D8
Bornich D 185 D8
Borssele NL 182 C3
Boskoop NL 182 A5
Bottrop D 183 B9
Bouillon B 184 E3
Boulay-Moselle F 186 C1
Bourdonnay F 186 D2
Bourscheid L 184 E5
Bousse F 186 C1
Boussu B 182 E3
Boutersem B 182 D5
Bouxières-aux-Dames F 186 D1
Bouxwiller F 186 D3
Bouzonville F 186 C2
Bovigny B 184 D4
Boxberg D 187 C8
Boxmeer NL 183 B7
Boxtel NL 183 B6
Brachbach D 185 C8
Brackenheim D 187 C7
Braine-l'Alleud B 182 D4
Braine-le-Comte B 182 D4
Braives B 183 D6
Brakel B 182 D3
Brasschaat B 182 C4
Braubach D 185 D8
Brauneberg D 185 E6
Braunfels D 185 C9
Braunsbach D 187 C8
Brecht B 182 C5
Breckerfeld D 185 B7

Breda NL 182 B5
Bredene B 182 C1
Bredevoort NL 183 B8
Bree B 183 C7
Breidenbach F 186 C3
Breitenbach F 186 C3
Breitscheid D 185 C7
Breitscheid D 185 C9
Bremm D 185 D7
Brensbach D 187 B6
Breskens NL 182 C3
Bretten D 187 C6
Bretzenheim D 185 E8
Bretzfeld D 187 C7
Breuberg-Neustadt D 187 B7
Breukelen NL 183 A6
Briedel D 185 D7
Brielle NL 182 B4
Brin-sur-Seille F 186 D1
Broekhuizenvorst NL 183 C8
Brohl D 185 D7
Brouvelieures F 186 E2
Brouwershaven NL 182 B3
Bruchköbel D 187 A6
Bruchmühlbach D 186 C3
Bruchsal D 187 C6
Bruchweiler-Bärenbach D 186 C4
Brücken D 186 B3
Brücken (Pfalz) D 186 C3
Brugelette B 182 D3
Bruges B 182 C2
Brugge B 182 C2
Brüggen D 183 C8
Brühl D 183 D9
Brühl D 187 C6
Bruinisse NL 182 B4
Brûly B 184 E2
Brumath F 186 D4
Brummen NL 183 A8
Brunehamel F 184 E1
Brunssum NL 183 D7
Brussel B 182 D4
Bruxelles B 182 D4
Bruttig-Fankel D 185 D7
Büchel D 185 D7
Buchen (Odenwald) D 187 B7
Büchenbeuren D 185 E7
Buchholz (Westerwald) D 185 C7
Budel NL 183 C7
Büdenheim D 185 E9
Büdesheim D 187 A6
Buer D 185 A7
Buggenhout B 182 C4
Bühl D 186 D5
Bühlertal D 187 D5
Bühlertann D 187 C8
Bühlerzell D 187 C8
Bullay D 185 D7
Büllingen B 184 D5
Bundenbach D 185 E7
Bunschoten-Spakenburg NL 183 A6
Burbach D 185 C9
Burdinne B 183 D6
Buren NL 183 B6
Burgbrohl D 185 D7
Burgh-Haamstede NL 182 B3
Burgsinn D 187 A8
Bürgstadt D 187 B7
Burladingen D 187 E7
Bürstadt D 187 B5
Bussum NL 183 A6
Bütgenbach B 184 D5
Büttelborn D 187 B6
Bütthard D 187 B8
Buurse NL 183 A9

C

Calw D 187 D6
Capelle aan de IJssel NL 182 B5
Carling F 186 C2
Carlsberg D 186 B5
Carvin F 182 E1
Castrop-Rauxel D 185 A7
Cattenom F 186 C1
Ceintrey F 186 D1
Celles B 182 D2
Cerfontaine B 184 D1
Chaam NL 182 C5
Chaligny F 186 D1
Champigneulles F 186 D1
Chapelle-lez-Herlaimont B 182 E4
Charleroi B 182 E4
Charmes F 186 E1
Chastre B 182 D5
Château-Salins F 186 D2
Châtelet B 182 E4
Châtel-sur-Moselle F 186 E1
Châtenois F 186 E3
Chaudfontaine B 183 D7
Chavelot F 186 E1
Chièvres B 182 D3
Chimay B 184 D1
Chooz F 184 D2
Ciney B 184 D3
Cirey-sur-Vezouze F 186 D2
Clavier B 183 E6
Clervaux L 184 D5
Clinge NL 182 C4
Cochem D 185 D7
Coesfeld D 183 B10
Colijnsplaat NL 182 B3
Colroy-la-Grande F 186 E3
Comblain-au-Pont B 183 E7
Comines B 182 D1
Condé-sur-l'Escaut F 182 E3
Consdorf L 186 B1
Contwig D 186 C3
Cothen NL 183 A6
Courcelles B 182 E4
Courcelles-Chaussy F 186 C1
Courcelles-sur-Nied F 186 C1
Courrières F 182 E1
Courtrai B 182 D2
Court-St-Etienne B 182 D5
Couvin B 184 D2
Crailsheim D 187 C9
Creglingen D 187 C9
Créhange F 186 C2
Creutzwald F 186 C2
Crévéchamps F 186 D1
Cuijk NL 183 B7
Culemborg NL 183 B6
Custines F 186 D1
Cysoing F 182 D2

D

Daaden D 185 C8
Dabo F 186 D3
Dahlem D 185 D6
Dahlhausen D 183 C10
Dahn D 186 C4
Daknam B 182 C4
Daleiden D 184 D5
Dalheim D 183 C6
Dalstein F 186 C1
Damas-aux-Bois F 186 E1
Dambach-la-Ville F 186 E3
Damelevières F 186 D1
Damme B 182 C2
Darmstadt D 187 B6
Datteln D 185 A7
Daun D 185 D6
De Bilt NL 183 A6
Deerlijk B 182 D2
Deggingen D 187 D8
De Haan B 182 C1
Deidesheim D 187 C5
Deinze B 182 D3
Delden NL 183 A9
Delft NL 182 A4
Delme F 186 D1
De Meern NL 183 A6
Den Bommel NL 182 B4
Denderleeuw B 182 D4
Dendermonde B 182 C4
Den Dungen NL 183 B6
Denekamp NL 183 A10
Den Haag NL 182 A4
Denkendorf D 187 D7
Dentergem B 182 D2
Dernau D 183 D10
Dessel B 183 C6
Destelbergen B 182 C3
Dettelbach D 187 B9
Dettenhausen D 187 D7
Dettingen an der Erms D 187 D7
Dettwiller F 186 D3
Deurne NL 183 C7
Deventer NL 183 A8
Deville F 184 E2
Dhron D 185 E6
Didam NL 183 B8
Dieblich D 185 D7
Dieburg D 187 B6
Diekirch L 184 E5
Dielheim D 187 C6
Diemen NL 182 A5
Dienheim D 185 E9
Diepenbeek B 183 D6
Diepenheim NL 183 A9
Diepenveen NL 183 A8
Dierdorf D 185 C8
Dieren NL 183 A8
Diessen NL 183 C6
Diest B 183 D6
Dietzenbach D 187 A6
Dietzhölztal-Ewersbach D 185 C9
Dieuze F 186 D2
Dikkebus B 182 D1
Diksmuide B 182 C1
Dilbeek B 182 D4
Dillenburg D 185 C9
Dillingen (Saar) D 186 C2
Dilsen B 183 C7
Dinant B 184 D2
Dinslaken D 183 B9
Dinteloord NL 182 B4
Dinther NL 183 B6
Dinxperlo NL 183 B8
Dirksland NL 182 B4
Dirmstein D 187 B5
Dison B 183 D7
Ditzingen D 187 D7
Dockweiler D 185 D6
Dodewaard NL 183 B7
Doesburg NL 183 A8
Doetinchem NL 183 B8
Doische B 184 D2
Dombasle-sur-Meurthe F 186 D1
Domburg NL 182 B3
Domèvre-sur-Vezouze F 186 D2
Dommershausen D 185 D7
Dongen NL 182 B5
Donk B 183 B7
Donzdorf D 187 D8
Doorn NL 183 A6
Doornspijk NL 183 A7
Dordrecht NL 182 B5
Dorlisheim F 186 D3
Dormagen D 183 C9
Dornburg-Frickhofen D 185 C9
Dornhan D 187 E6
Dornstadt D 187 E8
Dornstetten D 187 E5
Dorsten D 183 B9
Dorstfeld D 185 A7
Dortmund D 185 A7
Dörzbach D 187 C8
Dossenheim D 187 C6
Dreieich D 187 A6
Dreis D 185 E6
Drensteinfurt D 185 A8
Dreumel NL 183 B6
Driebergen NL 183 A6
Driedorf D 185 C9
Drolshagen D 185 B8
Drongen B 182 C3
Drulingen F 186 D3
Drusenheim F 186 D4
Druten NL 183 B7
Dudelange L 186 C1
Dudeldorf D 185 E6
Duffel B 182 C5
Duisburg D 183 C9
Dümpelfeld D 183 D9
Düngenheim D 185 D7
Durbach D 186 D5
Durbuy B 184 D3
Düren D 183 D8
Durmersheim D 187 D5
Düsseldorf D 183 C9
Dussen NL 182 B5

E

Eberbach D 187 C6
Ebersbach an der Fils D 187 D8
Ebersmunster F 186 E4
Ebhausen D 187 D6
Écaussinnes-d'Enghien B 182 D4

Echt NL 183 C7
Echternach L 185 E5
Eckbolsheim F 186 D4
Ede NL 183 A7
Edegem B 182 C4
Edenkoben D 186 C5
Edesheim D 186 C5
Ediger-Eller D 185 D7
Eefde NL 183 A8
Eerbeek NL 183 A8
Eernegem B 182 C1
Eersel NL 183 C6
Eggenstein-Leopoldshafen D 187 C5
Eghezée B 182 D5
Ehingen (Donau) D 187 E8
Ehningen D 187 D6
Ehringshausen D 185 C9
Eibelstadt D 187 B9
Eibergen NL 183 A9
Eich D 187 B5
Eichenbühl D 187 B7
Eijsden NL 183 D7
Eindhoven NL 183 C6
Einville-au-Jard F 186 D1
Eisden B 183 D7
Eisenberg (Pfalz) D 186 B5
Eislingen (Fils) D 187 D8
Eitelborn D 185 D8
Eitorf D 185 C7
Ekeren B 182 C4
Eksaarde B 182 C3
Eksel B 183 C6
Elchingen D 187 E9
Elfershausen D 187 A8
Elkenroth D 185 C8
Ellenberg D 187 D9
Ellezelles B 182 D3
Ellwangen (Jagst) D 187 D9
Elsdorf D 183 D9
Elsenborn B 184 D5
Elsenfeld D 187 B7
Elsloo NL 183 D7
Elspeet NL 183 A7
Elst NL 183 B6
Elst NL 183 B7
Eltville am Rhein D 185 D9
Elz D 185 D9
Emmelshausen D 185 D8
Emmerich D 183 B8
Empel NL 183 B6
Empfingen D 187 E6
Emptinne B 184 D3
Emst NL 183 A7
Engden D 183 A10
Engelskirchen D 185 C7
Enghien B 182 D4
Engis B 183 D6
Engstingen D 187 E7
Eningen unter Achalm D 187 D7
Enkenbach D 186 C4
Enkirch D 185 E7
Ennepetal D 185 B7
Ennery F 186 C1
Enschede NL 183 A9
Ensdorf D 186 C2
Ense D 185 B9
Ensheim D 186 C3
Enter NL 183 A9
Enzklösterle D 187 D5
Epe NL 183 A7
Epfendorf D 187 E6
Eppelborn D 186 C2
Eppelheim D 187 C6
Eppenbrunn D 186 C4
Eppertshausen D 187 B6
Eppingen D 187 C7
Eppstein D 187 A5
Erbach D 187 B7
Erbach D 187 E8
Erftstadt D 183 D9
Erkelenz D 183 C8
Erlenbach am Main D 187 B7
Ermelo NL 183 A7
Erndtebrück D 185 C9
Erp NL 183 B7
Erpel D 185 C7
Erquelinnes B 184 D1
Erstein F 186 E4
Ertvelde B 182 C3
Erwitte D 185 A9
Erzhausen D 187 B6
Esch NL 183 B6
Eschau D 187 B7
Eschau F 186 E4
Eschborn D 187 A6
Eschenburg-Eibelshausen D 185 C9
Esch-sur-Sûre L 184 E4
Eschweiler D 183 D8
Eslohe (Sauerland) D 185 B9
Esneux B 183 E7
Essen B 182 C4
Essen D 183 C10
Essingen D 187 D9
Esslingen am Neckar D 187 D7
Estaimpuis B 182 D2
Estenfeld D 187 B9
Étival-Clairefontaine F 186 E2
Ettelbruck L 184 E5
Ettenheim D 186 E4
Etten-Leur NL 182 B5
Ettlingen D 187 D5
Ettringen D 185 D7
Eubigheim D 187 C8
Euerdorf D 187 A9
Eupen B 183 D8
Euskirchen D 183 D9
Eußenheim D 187 A8
Eutingen im Gäu D 187 E6
Everdingen NL 183 B6
Evergem B 182 C3

F

Faches-Thumesnil F 182 D2
Faid D 185 D7
Faimes B 183 D6
Falck F 186 C1
Fameck F 186 C1
Farciennes B 182 E4
Farébersviller F 186 C2
Faulbach D 187 B7
Faulquemont F 186 C2
Fell D 186 B2
Fellbach D 187 D7
Fénétrange F 186 D3
Fépin F 184 D2
Ferrières B 183 E7

Ferschweiler D 185 E5
Fichtenberg D 187 D8
Fijnaart NL 182 B4
Finnentrop D 185 B8
Fischbach D 186 B3
Fischbach bei Dahn D 186 C4
Flavigny-sur-Moselle F 186 D1
Flein D 187 C7
Flémalle B 183 D6
Fléron B 183 D7
Flers-en-Escrebieux F 182 E2
Fleurus B 182 E5
Flines-lez-Raches F 182 E2
Florange F 186 C1
Floreffe B 182 E5
Florennes B 184 D2
Flörsbach D 187 A7
Flörsheim am Main D 187 A5
Flörsheim-Dalsheim D 187 B5
Focant B 184 D3
Föhren D 185 E6
Folschviller F 186 C2
Fontaine-l'Évêque B 182 E4
Forbach D 186 C2
Forbach F 186 C2
Forchtenberg D 187 C8
Forst D 187 C6
Fosses-la-Ville B 184 D2
Fraire B 182 E5
Frameries B 182 E3
Frammersbach D 187 A7
Francorchamps B 183 E7
Frankenthal (Pfalz) D 187 B5
Frankfurt am Main D 187 A6
Frasnes-lez-Buissenal B 182 D3
Frasnes-lez-Gosselies B 182 D4
Frechen D 183 D9
Freinsheim D 187 B5
Freisen D 186 B3
Freistroff F 186 C1
Fresnes-sur-Escaut F 182 E3
Freudenberg D 185 C8
Freudenberg D 187 B7
Freudenburg D 186 B2
Freudenstadt D 187 E5
Freyming-Merlebach F 186 C2
Friedewald D 185 C8
Friedrichsdorf D 187 A6
Friesenhagen D 185 C8
Friesenheim D 186 E4
Frisange L 186 B1
Frœschwiller F 186 D4
Fromelennes F 184 D2
Fröndenberg D 185 B8
Frouard F 186 D1
Fumay F 184 E2
Fürth D 187 B6

G

Gaanderen NL 183 B8
Gaggenau D 187 D5
Gaildorf D 187 D8
Gambsheim F 186 D4
Gammertingen D 187 E7
Gand B 182 C3
Ganshoren B 182 D4
Garderen NL 183 A7
Gärtringen D 187 D6
Gau-Algesheim D 185 E9
Gaukönigshofen D 187 B9
Gau-Odernheim D 185 E9
Gavere B 182 D3
Gebhardshain D 185 C8
Gechingen D 187 D6
Gedinne B 184 E2
Geel B 183 C6
Geertruidenberg NL 182 B5
Geesteren NL 183 A9
Geetbets B 183 C6
Geffen NL 183 B6
Geilenkirchen D 183 C8
Geiselbach D 187 A7
Geisenheim D 185 E8
Geislingen D 187 E6
Geislingen an der Steige D 187 D8
Geispolsheim F 186 D4
Gelchsheim D 187 B9
Geldermalsen NL 183 B6
Geldern D 183 B8
Geldrop NL 183 C7
Geleen NL 183 D7
Gelnhausen D 187 A7
Gelsenkirchen D 183 B10
Gembloux B 182 D5
Gemert NL 183 B7
Gemmingen D 187 C6
Gemünden B 183 C7
Gemünden am Main D 187 A8
Genappe B 182 D4
Gendringen NL 183 B8
Gendt NL 183 B7
Gengenbach D 186 E5
Genk B 183 D7
Gennep NL 183 B7
Gensingen D 185 E8
Gent B 182 C3
Geraardsbergen B 182 D3
Gerabronn D 187 C8
Gerbéviller F 186 E2
Gerbrunn D 187 B8
Gerchsheim D 187 B8
Gerlingen D 187 D7
Germersheim D 187 C5
Gernsbach D 187 D5
Gernsheim D 187 B5
Gerolstein D 185 D6
Gerpinnes B 184 D2
Gersheim D 186 C3
Gerstetten D 187 D9
Gerstheim F 186 E4
Gescher D 183 B10
Gespunsart F 184 D3
Gesves B 184 D3
Gevelsberg D 185 B7
Ghislenghien B 182 D3
Giebelstadt D 187 B8
Gierle B 182 C5
Gillenfeld D 185 D6
Gilze NL 182 B5
Gingelom B 183 D6
Gingen an der Fils D 187 D8
Ginsheim D 185 E9
Gistel B 182 C1
Givet F 184 D2
Gladbeck D 183 B9
Glanerbrug NL 183 A9
Gläshütten D 187 A5
Goch D 183 B8

Goddelau D 187 B6
Goedereede NL 182 B3
Goes NL 182 B3
Göggingen D 187 D8
Goirle NL 182 C5
Göllheim D 186 B5
Gomadingen D 187 E7
Gomaringen D 187 E7
Gondelsheim D 187 C6
Gondershausen D 185 D7
Gondorf D 185 D7
Goor NL 183 A9
Göppingen D 187 D8
Gorinchem NL 182 B5
Gorssel NL 183 A8
Gorxheimertal D 187 B6
Gouda NL 182 A5
Goudswaard NL 182 B4
Gouvy B 184 D4
Graben-Neudorf D 187 C5
Grâce-Hollogne B 183 D6
Gräfendorf D 187 A8
Grafenrheinfeld D 187 B9
Grandvillers F 186 E2
Grave NL 183 B7
Greifenstein D 185 C9
Grevenbicht NL 183 C7
Grevenbroich D 183 C9
Grevenmacher L 186 B1
Grez-Doiceau B 182 D5
Gries F 186 D4
Griesbach D 187 E5
Griesheim D 187 B6
Grimbergen B 182 D4
Grobbendonk B 182 C5
Groenlo NL 183 A9
Groesbeek NL 183 B7
Gronau (Westfalen) D 183 A10
Groß-Bieberau D 187 B6
Großbottwar D 187 D7
Grosselfingen D 187 E6
Groß-Gerau D 187 B5
Großheubach D 187 B7
Großlangheim D 187 B9
Großlittgen D 185 D6
Großmaischeid D 185 C8
Großostheim D 187 A7
Großrinderfeld D 187 B8
Groß-Rohrheim D 187 B5
Großrosseln D 186 C2
Groß-Umstadt D 187 B6
Großwallstadt D 187 B7
Groß-Zimmern D 187 B6
Grostenquin F 186 D2
Grubbenvorst NL 183 C8
Grünsfeld D 187 B8
Grünstadt D 187 B5
Gschwend D 187 D8
Guénange F 186 C1
Güglingen D 187 C6
Gulpen NL 183 D7
Gummersbach D 185 B8
Gundelsheim D 187 C7
Gundershoffen F 186 D4
Guntersblum D 185 E9
Güntersleben D 187 B8
Gusterath D 186 B2
Gutach (Schwarzwaldbahn)
D 187 E5

H

Haacht B 182 D5
Haaften NL 183 B6
Haaksbergen NL 183 A9
Haaltert B 182 D4
Haaren D 183 B6
Haarlem NL 182 A5
Haastrecht NL 182 A5
Hachenburg D 185 C8
Hackenheim D 185 E8
Hadamar D 185 D9
Haelen NL 183 C7
Hagen D 185 B7
Hagondange F 186 C1
Haguenau F 186 D4
Hahnstätten D 185 D9
Haibach D 187 B7
Haiger D 185 C9
Haigerloch D 187 E6
Haiterbach D 187 D6
Halen B 183 D6
Halfweg NL 182 A5
Halle B 182 D4
Halle NL 183 A8
Halluin B 182 D2
Halsenbach D 185 D7
Halsteren NL 182 B4
Halstroff F 186 C1
Haltern D 183 B10
Halver D 185 B7
Hambach F 186 D3
Hambrücken D 187 C6
Hamm D 185 A8
Hamm (Sieg) D 185 C8
Hamme B 182 C4
Hammelburg D 187 A8
Hamme-Mille B 182 D5
Hamminkeln D 183 B9
Hamoir B 183 E7
Hamois B 184 D3
Hamont B 183 C7
Hampont F 186 D2
Ham-sous-Varsberg F 186 C2
Ham-sur-Heure B 184 D1
Hanau D 187 A6
Handzame B 182 C2
Hannut B 183 D6
Han-sur-Nied F 186 D1
Hapert NL 183 C6
Haps NL 183 B7
Harderwijk NL 183 A7
Hardheim D 187 B7
Hardinxveld-Giessendam
NL 182 B5
Harelbeke B 182 D2
Hargesheim D 185 E8
Hargimont B 184 D3
Hargnies F 184 D2
Harmelen NL 182 A5
Harnes F 182 E1
Haroué F 186 E1
Harthausen D 187 C5
Haslach im Kinzigtal D 186 E5
Hasselt B 183 D6
Haßloch D 187 C5
Haßmersheim D 187 C7
Hastière-Lavaux B 184 D2
Hattersheim am Main D 187 A5
Hattert D 185 C8

Hattingen D 183 C10
Hatzenbühl D 187 C5
Haubourdin F 182 D1
Hauenstein D 186 C4
Hausach D 187 E5
Hausen bei Würzburg D 187 B9
Haut-Fays B 184 D3
Havelange B 184 D3
Haversin B 184 D3
Hayange F 186 C1
Haybes F 184 D2
Hayingen D 187 E7
Hazerswoude-Rijndijk NL
182 A5
Hechingen D 187 E6
Hechtel B 183 C6
Heddesheim D 187 C6
Hedel NL 183 B6
Heek D 183 A10
Heel NL 183 C7
Heemstede NL 182 A5
Heenvliet NL 182 B4
Heer B 184 D2
Heerde NL 183 A8
Heerewaarden NL 183 B6
Heerlen NL 183 D7
Heers B 183 D6
Heesch NL 183 B7
Heeswijk NL 183 B6
Heeten NL 183 A8
Heeze NL 183 C7
Heidelberg D 187 C6
Heiden D 183 B9
Heidenheim an der Brenz
D 187 D9
Heigenbrücken D 187 A7
Heilbronn D 187 C7
Heiligenhaus D 183 C9
Heimbach D 186 B3
Heimbuchenthal D 187 B7
Heimsheim D 187 D6
Heinkenszand NL 182 C4
Heino NL 183 A8
Heinsberg D 183 C8
Heisingen D 183 C10
Heist-op-den-Berg B 182 C5
Hekelgem B 182 D4
Helchteren B 183 C6
Heldenbergen D 187 A6
Hellendoorn NL 183 A8
Hellenthal D 184 C4
Hellevoetsluis NL 182 B4
Helmond NL 183 C7
Helmstadt D 187 B8
Heltersberg D 186 C4
Helvoirt NL 183 B6
Hem F 182 D2
Hemer D 185 B8
Hemsbach D 187 B6
Hengelo NL 183 A8
Hengelo NL 183 A9
Hengevelde NL 183 A9
Hénin-Beaumont F 182 E1
Hennef (Sieg) D 185 C7
Hennweiler D 185 E7
Heppen B 183 C6
Heppenheim (Bergstraße)
D 187 B6
Herbeumont B 184 E3
Herborn D 185 C9
Herbrechtingen D 187 D9
Herdecke D 185 B8
Herdorf D 185 C8
Herent B 182 D5
Herentals B 182 C5
Herenthout B 182 C5
Herk-de-Stad B 183 D6
Herkenbosch NL 183 C8
Herkingen NL 182 B4
Hermersberg D 186 C4
Hermeskeil D 186 B2
Herne B 182 D4
Herne D 185 A7
Héron B 183 D6
Herrenberg D 187 D6
Herrlisheim F 186 D4
Herschbach D 185 C8
Herscheid D 185 B8
Herschweiler-Pettersheim
D 186 C3
Herselt B 182 C5
Herstal B 183 D7
Herten D 183 B10
Herve B 183 D7
Herwijnen NL 183 B6
Herzele B 182 D3
Herzogenrath D 183 D8
Hespérange L 186 B1
Heßheim D 187 B5
Hettange-Grande F 186 C1
Hettenleidelheim D 186 B5
Hetzerath D 185 E6
Heubach D 187 D8
Heukelum NL 183 B6
Heusden B 183 C6
Heusden NL 183 B6
Heusenstamm D 187 A6
Heusweiler D 186 C2
Heythuysen NL 183 C7
Hilchenbach D 185 C9
Hilden D 183 C9
Hillegom NL 182 A5
Hillesheim D 185 D6
Hilsenheim F 186 E4
Hilvarenbeek NL 183 C6
Hilversum NL 183 A6
Hinterweidenthal D 186 C4
Hirrlingen D 187 E6
Hirschhorn (Neckar) D 187 C6
Hochfelden F 186 D4
Hochspeyer D 186 C4
Hochstadt (Pfalz) D 187 C5
Hochstetten-Dhaun D 185 E7
Höchst im Odenwald D 187 B6
Hockenheim D 187 C5
Hoek NL 182 C3
Hoek van Holland NL 182 B4
Hoenderloo NL 183 A7
Hoensbroek NL 183 D7
Hœrdt F 186 D4
Hoeselt B 183 D7
Hoevelaken NL 183 A6
Hoeven NL 182 B5
Hof D 185 C9
Hofheim am Taunus D 187 A5
Hohberg D 186 E4
Höhn D 185 C8
Höhr-Grenzhausen D 185 D8
Hollange B 184 D3
Holten NL 183 A8
Holzappel D 185 D8
Holzgerlingen D 187 D7

Holzhausen an der Haide
D 185 D8
Holzheim D 187 E9
Holzwickede D 185 A8
Hombourg-Budange F 186 C1
Hombourg-Haut F 186 C2
Homburg D 186 C3
Hoofddorp NL 182 A5
Hoogerheide NL 182 B4
Hoog-Keppel NL 183 B8
Hoogland NL 183 A6
Hoogstraten B 182 C5
Hoogvliet NL 182 B4
Hoornaar NL 182 B5
Hoppstädten D 186 B3
Horb am Neckar D 187 E6
Hörde D 185 B7
Hornbach D 186 C3
Horst NL 183 C8
Hösbach D 187 A7
Hosingen L 184 D5
Hotton B 184 D3
Houffalize B 184 D4
Houten NL 183 A6
Houthalen B 183 C6
Houthulst B 182 D1
Houyet B 184 D3
Hückelhoven D 183 C8
Hückeswagen D 185 B7
Huijbergen NL 182 C4
Huissen NL 183 B7
Huizen NL 183 A6
Hüls D 183 B8
Hulsberg NL 183 D7
Hulst NL 182 C4
Hummelo NL 183 A8
Hundsangen D 185 D8
Hünfelden-Kirberg D 185 D9
Hünxe D 183 B9
Hürth D 183 D9
Hütschenhausen D 186 C3
Hüttisheim D 187 E8
Hüttlingen D 187 D9
Huy B 183 D6
Hymont F 186 E1

I

Ichenheim D 186 E4
Ichtegem B 182 C2
Idar-Oberstein D 186 B3
Idstein D 185 D9
Ieper B 182 D1
Iffezheim D 187 D5
Igel D 186 B2
Igersheim D 187 C8
Iggelheim D 187 C5
Igney F 186 E1
IJsselstein NL 183 A6
IJzendijke NL 182 C3
Illingen D 186 C3
Illingen D 187 D6
Illkirch-Graffenstaden F 186 D4
Ilsfeld D 187 C7
Incourt B 182 D5
Ingelfingen D 187 C8
Ingelheim am Rhein D 185 E9
Ingelmunster B 182 D2
Ingwiller F 186 D3
Insming F 186 D2
Iphofen D 187 B9
Ippesheim D 187 B9
Irrel D 185 E5
Irsch D 186 B2
Iserlohn D 185 B8
Ispringen D 187 D6
Isselburg D 183 B8
Issum D 183 B8
Ittre B 182 D4
Ixelles B 182 D4
Izegem B 182 D2

J

Jabbeke B 182 C2
Jagsthausen D 187 C7
Jagstzell D 187 C9
Jalhay B 183 D7
Jarville-la-Malgrange F 186 D1
Jemeppe B 182 E5
Jockgrim D 187 C5
Jodoigne B 182 D5
Jouy-aux-Arches F 186 C1
Jüchen D 183 C8
Jülich D 183 C8
Jungingen D 187 E7
Junglinster L 186 B1
Jünkerath D 185 D6
Juprelle B 183 D7
Jurbise B 182 D3

K

Kaarst D 183 C9
Kaatsheuvel NL 183 B6
Kahl am Main D 187 A7
Kaisersesch D 185 D7
Kaiserslautern D 186 C4
Kalkar D 183 B8
Kall D 183 D9
Kalmthout B 182 C4
Kamen D 185 A8
Kamerik NL 182 A5
Kamp D 185 D8
Kampenhout B 182 D5
Kämpfelbach D 187 D6
Kamp-Lintfort D 183 C9
Kandel D 187 C5
Kapelle NL 182 C3
Kapellen B 182 C4
Kapelle-op-den-Bos B 182 C4
Kappel D 185 D7
Kappel-Grafenhausen D 186 E4
Kappelrodeck D 186 D5
Kaprijke B 182 C3
Karben D 187 A6
Karden D 185 D7
Karlsbad D 187 D6
Karlsdorf-Neuthard D 187 C6
Karlsruhe D 187 C5
Karlstadt D 187 B8
Kastellaun D 185 D7
Kasterlee B 182 C5
Katwijk aan Zee NL 182 A4
Katzenelnbogen D 185 D8

Katzweiler D 186 B4
Kaub D 185 D8
Kaulille B 183 C7
Kautenbach L 184 E5
Kehl D 186 D4
Kehlen L 186 B1
Kehrig D 185 D7
Kelberg D 185 D6
Kelbra (Taunus) D 187 A5
Kell D 186 C2
Kelmis B 183 D7
Kempen D 183 C8
Kempenich D 185 D7
Kenn D 185 E6
Kerkdriel NL 183 B6
Kerken D 183 C8
Kerkrade NL 183 D7
Kerkwijk NL 183 B6
Kerpen D 183 D9
Kessel B 182 C5
Kessel NL 183 C8
Kesteren NL 183 B7
Ketsch D 187 C6
Kettwig D 183 C9
Kevelaer D 183 B8
Kierspe D 185 B8
Kinderbeuern D 185 D7
Kindsbach D 186 C4
Kinrooi B 183 C7
Kippenheim D 186 E4
Kirchardt D 187 C6
Kirchberg (Hunsrück) D 185 E7
Kirchberg an der Jagst D 187 C8
Kirchellen D 183 B9
Kirchen (Sieg) D 185 C8
Kirchheim am Neckar D 187 C7
Kirchheim-Bolanden D 186 B5
Kirchheim unter Teck D 187 D7
Kirchhundem D 185 B9
Kirchzell D 187 B7
Kirkel-Neuhäusel D 186 C3
Kirn D 185 E7
Kirschweiler D 186 B3
Kist D 187 B8
Kitzingen D 187 B9
Klaaswaal NL 182 B5
Klausen D 185 E6
Kleinblittersdorf D 186 C3
Kleinheubach D 187 B7
Kleinrinderfeld D 187 B8
Kleinwallstadt D 187 B7
Kleve D 183 B8
Klingenberg am Main D 187 B7
Kloetinge NL 182 C4
Kloosterzande NL 182 C4
Klotten D 185 D7
Klundert NL 182 B5
Knesselare B 182 C3
Knittlingen D 187 C6
Knokke-Heist B 182 C2
Kobern D 185 D7
Koblenz D 185 D8
Koekelare B 182 C1
Koersel B 183 C6
Koewacht NL 182 C3
Kolitzheim D 187 B9
Köln D 183 D9
Königsbronn D 187 D9
Königstein im Taunus D 187 A5
Königswinter D 185 C7
Konz D 186 B2
Kootwijkerbroek NL 183 A7
Kopstal L 186 B1
Kordel D 185 E6
Kornwestheim D 187 D7
Körperich D 185 E5
Kortemark B 182 C2
Kortenhoef NL 183 A6
Kortessem B 183 D6
Kortgene NL 182 B3
Kortrijk B 182 D2
Kottenheim D 185 D7
Koudekerke NL 182 C3
Krabbendijke NL 182 C4
Kranenburg D 183 B8
Krautheim D 187 C8
Krefeld D 183 C8
Kreuzau D 183 D8
Kreuztal D 185 C8
Kreuzwertheim D 187 B8
Krimpen aan de IJssel NL 182 B5
Kronau D 187 C6
Kronberg im Taunus D 187 A5
Kröv D 185 D7
Kruft D 185 D7
Kruibeke B 182 C4
Kruiningen NL 182 C4
Kruishoutem B 182 D3
Kuchen D 187 D8
Külsheim D 187 B8
Kunrade NL 183 D7
Künzelsau D 187 C8
Kupferzell D 187 C8
Kuppenheim D 187 D5
Kürnach D 187 B9
Kürnbach D 187 C6
Kusel D 186 B3
Kusterdingen D 187 D7
Kuurne B 182 D2
Kwaadmechelen B 183 C6
Kyllburg D 185 D6

Langemark B 182 D1
Langen D 187 B6
Langenau D 187 E9
Langenaubach D 185 C9
Langenberg D 183 C10
Langenberg D 183 A9
Langenburg D 187 C8
Langenfeld (Rheinland) D
183 C9
Langenhahn D 185 C8
Langenlonsheim D 185 E8
Langenselbold D 187 A7
Langsur D 186 B1
Lannoy F 182 D2
La Petite-Pierre F 186 D3
Laren B 182 D4
Laren NL 183 A6
La Roche-en-Ardenne D 184 D4
Larochette L 186 B1
Lasne B 182 D4
Lattrop NL 183 A9
Laubach D 185 D7
Lauda-Königshofen D 187 B8
Lauf D 185 D6
Laufach D 187 A7
Lauffen am Neckar D 187 C7
Lautenbach D 186 D5
Lauterbourg F 187 D5
Lauterecken D 186 B4
Lauterstein D 187 D8
La Wantzenau F 186 D4
Laxou F 186 D1
Lebach D 186 C2
Le Ban-St-Martin F 186 C1
Lebbeke B 182 C4
Lede B 182 D3
Ledegem B 182 D2
Leende NL 183 C7
Leerdam NL 183 B6
Leersum NL 183 A6
Leffinge B 182 C1
Leforest F 182 E1
Legden D 183 A10
Léglise B 184 E4
Lehmen D 185 D7
Le Hohwald F 186 E3
Leichlingen (Rheinland) D
183 C10
Leiden NL 182 A5
Leiderdorp NL 182 A5
Leidschendam NL 182 A4
Leimen D 187 C6
Leimuiden NL 182 A5
Leinfelden-Echterdingen
D 187 D7
Leinzell D 187 D8
Leiwen D 185 E6
Lembeke B 182 C3
Lemberg F 186 C3
Lemberg F 186 D3
Lendelede B 182 D2
Léning F 186 D2
Lennestadt D 185 B9
Lenningen D 187 D7
Lens B 182 D3
Lent NL 183 B7
Leonberg D 187 D7
Leopoldsburg B 183 C6
Les Hautes-Rivières F 184 D2
Les Mazures F 184 D2
Lessines B 182 D3
Leun D 185 C9
Leusden NL 183 A6
Leutesdorf D 185 D7
Leuven B 182 D5
Leuze-en-Hainaut B 182 D3
Leverkusen D 183 C9
L'Hôpital F 186 C2
Libin B 184 E3
Libramont B 184 E3
Lichtaart B 182 C5
Lichtenau D 186 D5
Lichtenvoorde NL 183 B9
Lichtervelde B 182 C2
Liège B 183 D7
Liempde NL 183 B6
Lienden NL 183 B7
Lier B 182 C5
Lierneux B 184 D4
Lieser D 185 D7
Lieshout NL 183 B7
Liessel NL 183 C7
Ligneuville B 184 D5
Lille B 182 C5
Lille F 182 D2
Limbach D 186 C2
Limbach D 187 C7
Limbourg B 183 D7
Limburg an der Lahn D 185 D9
Limburgerhof D 187 C5
Lincent B 183 D6
Lindenfels D 187 B6
Lindlar D 185 B7
Lingenfeld D 187 C5
Lingolsheim F 186 D4
Linnich D 183 C8
Linz am Rhein D 185 C7
Lippstadt D 185 A9
Lisse NL 182 A5
Lissendorf D 185 D6
Lith NL 183 B6
Lixing-lès-St-Avold F 186 C2
Lobith NL 183 B8
Lochem NL 183 A8
Locristi B 182 C3
Loenen NL 183 A8
Löf D 185 D7
Lohmar D 185 C7
Löhnberg D 185 C9
Lohr am Main D 187 B8
Lokeren B 182 C3
Lomme F 182 D1
Lommel B 183 C6
Londerzeel B 182 D4
Lonneker NL 183 A9
Lonny F 184 E2
Lonsee D 187 D8
Lontzen B 183 D8
Loon op Zand NL 183 B6
Loos F 182 D2
Lopik NL 182 B5
Lorch D 185 D8
Lorch D 187 D8
Lorquin F 186 D2
Losheim D 186 B2
Loßburg D 187 E5
Losser NL 183 A9
Lotenhulle B 182 C3
Lottum NL 183 C8
Louvain B 182 D5

Langmark B 182 D1
Louveigné B 183 D7
Lovendegem B 182 C3
Löwenstein D 187 C7
Lubbeek B 182 D5
Lüdenscheid D 185 B8
Lüdinghausen D 185 A7
Ludres F 186 E1
Ludwigsburg D 187 D7
Ludwigshafen am Rhein
D 187 C5
Luik B 183 D7
Lummen B 183 D6
Lünebach D 185 D5
Lünen D 185 A8
Lunéville F 186 D1
Lunteren NL 183 A7
Luppy F 186 D1
Lustadt D 187 C5
Luttenberg NL 183 A8
Lützelbach D 187 B7
Lutzerath D 185 D7
Luxembourg L 186 B1
Luyksgestel NL 183 C6

M

Maarheeze NL 183 C7
Maarn NL 183 A6
Maarssen NL 183 A6
Maarssenbroek NL 183 A6
Maasbracht NL 183 C7
Maasbree NL 183 C8
Maasdam NL 182 B5
Maaseik B 183 C7
Maasland NL 182 B4
Maasmechelen B 183 D7
Maassluis NL 182 B4
Maastricht NL 183 D7
Machelen B 182 D4
Mackenbach D 186 C4
Made NL 182 B5
Magnières F 186 E2
Mahlberg D 186 E4
Maikammer D 187 C5
Mainaschaff D 187 B7
Mainbernheim D 187 B9
Mainhardt D 187 C8
Mainz D 185 D9
Maizières-lès-Metz F 186 C1
Malborn D 186 B2
Maldegem B 182 C2
Malden NL 183 B7
Malmédy B 184 D5
Malsch D 187 D5
Manage B 182 D4
Mandelbachtal-Ormesheim
D 186 C3
Manderscheid D 185 D6
Manhay B 184 D4
Mannheim D 187 C5
Manternach L 186 B1
Marange-Silvange F 186 C1
Marbach am Neckar D 187 D7
Marche-en-Famenne D 184 D4
Marchiennes F 182 E2
Marchin B 183 E6
Marcq-en-Barœul F 182 D2
Margraten D 183 D7
Mariembourg B 184 D2
Marienheide D 185 B8
Markelo NL 183 A9
Markgröningen D 187 D7
Marktbreit D 187 B9
Marktheidenfeld D 187 B8
Marktsteft D 187 B9
Marl D 183 B10
Marlenheim F 186 D4
Marly F 186 C1
Marmoutier F 186 D3
Marnheim D 186 B5
Marpingen D 186 C3
Marsal F 186 D2
Martelange B 184 E4
Marxell D 187 D5
Marxzell D 187 D5
Maßbach D 187 A9
Mastershausen D 185 D7
Mattaincourt F 186 E1
Maubert-Fontaine F 184 E1
Maulbronn D 187 D6
Maurik NL 183 B6
Maxéville F 186 D1
Maxsain D 185 C8
Mayen D 185 D7
Mayschoss D 183 D10
Mechelen B 182 C4
Mechelen NL 183 D7
Mechernich D 183 D9
Meckenheim D 183 D10
Meckesheim D 187 C6
Meddersheim D 186 B4
Meddo NL 183 A9
Meer B 182 C5
Meerbeek B 182 D5
Meerbusch D 183 C9
Meerhout B 183 C6
Meerkerk NL 182 B5
Meerle B 182 C5
Meerlo NL 183 B8
Meerssen NL 183 D7
Meetkerke B 182 C2
Meeuwen B 183 C7
Megen NL 183 B7
Mehren D 185 D6
Mehring D 185 E6
Mehrstetten D 187 E8
Meijel NL 183 C7
Meinerzhagen D 185 B8
Meise B 182 D4
Meisenheim D 186 B4
Meißenheim D 186 E4
Melick NL 183 C8
Meliskerke NL 182 B3
Melle B 182 D3
Menden (Sauerland) D 185 B8
Mendig D 185 D7
Menen B 182 D2
Mengerskirchen D 185 C9
Ménil-sur-Belvitte F 186 E2
Menin B 182 D2
Merbes-le-Château B 184 D1
Merchtem B 182 D4
Mere B 182 D3
Merelbeke B 182 D3
Merenberg D 185 C9
Merklingen D 187 D8
Merksplas B 182 C5
Mersch L 186 B1
Mertert L 186 B1
Mertesdorf D 185 E6
Mertloch D 185 D7
Mertzwiller F 186 D4

Merzig D 186 C2
Meschede D 185 B9
Mespelbrunn D 187 B7
Metelen D 183 A10
Mettendorf D 185 E5
Mettet B 184 D2
Mettlach D 186 C2
Mettmann D 183 C9
Metz F 186 C1
Metzervisse F 186 C1
Metzingen D 187 D7
Meudt D 185 D8
Meulebeke B 182 D2
Michelbach an der Bilz D 187 C8
Michelfeld D 187 C8
Michelstadt D 187 B7
Middelbeers NL 183 C6
Middelburg NL 182 B3
Middelharnis NL 182 B4
Middelkerke B 182 C1
Miehlen D 185 D8
Mierlo NL 183 C7
Miesau D 186 C3
Miesenbach D 186 C4
Mijdrecht NL 182 A5
Mill NL 183 B7
Millingen aan de Rijn NL 183 B8
Milmort B 183 D7
Miltenberg D 187 B7
Minderhout B 182 C5
Minfeld D 186 C5
Mirecourt F 186 E1
Mittelsinn D 187 A8
Mittersheim F 186 D2
Möckmühl D 187 C7
Modave B 183 E6
Moerbeke B 182 C3
Moergestel NL 183 B6
Moerkerke B 182 C2
Moers D 183 C9
Mögglingen D 187 D8
Möglingen D 187 D7
Mol B 183 C6
Molenbeek-St-Jean B 182 D4
Molenstede B 182 C5
Molsheim F 186 D3
Mömbris D 187 A7
Momignies B 184 D1
Moncel-sur-Seille F 186 D1
Mönchengladbach D 183 C8
Mondorf-les-Bains L 186 B1
Mons B 182 E3
Monschau D 183 D8
Monsheim D 187 B5
Mönsheim D 187 D6
Monster NL 182 A4
Montabaur D 185 D8
Montcy-Notre-Dame F 184 E2
Montfoort NL 183 A6
Montfort NL 183 C7
Monthermé F 184 E2
Montignies-le-Tilleul B 184 D1
Montigny F 186 E2
Montigny-lès-Metz F 186 C1
Montzen B 183 D7
Monzelfeld D 185 E7
Monzingen D 185 E8
Mook NL 183 B7
Moorslede B 182 D2
Morbach D 185 E7
Mörfelden D 187 B6
Morhange F 186 D2
Morlanwelz B 182 E4
Mörlenbach D 187 B6
Morsbach D 185 C8
Mortsel B 182 C4
Mosbach D 187 C7
Mössingen D 187 E7
Mouscron B 182 D2
Moussey F 186 E3
Moyenmoutier F 186 E2
Much D 185 C7
Mudau D 187 B7
Müdelheim D 183 C9
Mudersbach D 185 C8
Muggensturm D 187 D5
Mühlacker D 187 D6
Mühlhausen D 187 C6
Mulfingen D 187 C8
Mülheim an der Ruhr D 183 C9
Mülheim-Kärlich D 185 D7
Münchweiler an der Rodalb
D 186 C4
Munderkingen D 187 E8
Mundolsheim F 186 D4
Munkzwalm B 182 D3
Münnerstadt D 187 A9
Münsingen D 187 E7
Münster D 187 B6
Munstergeleen NL 183 D7
Münstermaifeld D 185 D7
Murrhardt D 187 C8
Müschenbach D 185 C8
Mutterstadt D 187 C5
Mutzig F 186 D3

N

Naaldwijk NL 182 B4
Naarden NL 183 A6
Nackenheim D 185 E9
Nagold D 187 D6
Nalbach D 186 C2
Namborn D 186 B3
Namur B 182 E5
Nancy F 186 D1
Nandrin B 183 D6
Nassau D 185 D8
Nassogne B 184 D3
Nastätten D 185 D8
Nauheim D 187 B5
Nauroth D 185 C8
Neckarbischofsheim D 187 C6
Neckargemünd D 187 C6
Neckarsteinach D 187 C6
Neckarsulm D 187 C7
Neckartenzlingen D 187 D7
Nederhorst den Berg NL 183 A6
Nederlangbroek NL 183 A6
Nederweert NL 183 C7
Neede NL 183 A9
Neer NL 183 C8
Neerijnen NL 183 B6
Neeroeteren B 183 C7
Neerpelt B 183 C7
Nellingen D 187 D8
Nentershausen D 185 D8
Neroth D 185 D6
Nersingen D 187 E9
Netphen D 185 C9
Nettersheim D 183 D9

Nettetal *D* 183 C8
Neubrunn *D* 187 B8
Neubulach *D* 187 D6
Neuenbürg *D* 187 D6
Neuenkirchen-Seelscheid
D 185 C7
Neuenrade *D* 185 B8
Neuenstadt am Kocher
D 187 C7
Neuenstein *D* 187 C8
Neuerburg *D* 185 D5
Neufchâteau *B* 184 E3
Neuffen *D* 187 D7
Neufmanil *F* 184 E2
Neufra *D* 187 E7
Neuhausen *D* 187 D6
Neuhofen *D* 187 C6
Neu-Isenburg *D* 187 A6
Neuler *D* 187 D9
Neumagen *D* 185 E6
Neunkirchen *D* 185 C9
Neunkirchen *D* 186 C3
Neuss *D* 183 C9
Neustadt (Wied) *D* 185 C7
Neustadt an der Weinstraße
D 186 C5
Neu-Ulm *D* 187 E9
Neuves-Maisons *F* 186 D1
Neuweiler *D* 187 D6
Neuweid *D* 185 D7
Nevele *B* 182 C3
Newel *D* 185 E6
Niederanven *L* 186 B1
Niederbrechen *D* 185 D9
Niederbreitbach *D* 185 C7
Niederbronn-les-Bains *F* 186 D4
Niederfischbach *D* 185 C7
Niederkassel *D* 183 D10
Niederkirchen *D* 186 B4
Niederkrüchten *D* 183 C8
Niederneisen *D* 185 D9
Niedernhall *D* 187 C8
Niedernhausen *D* 185 D9
Nieder-Olm *D* 185 E9
Nieder-Rodenbach *D* 187 A7
Niederselters *D* 185 D9
Niederstetten *D* 187 C8
Niederwerrn *D* 187 A9
Niederwörresbach *D* 186 B3
Niederzissen *D* 185 D7
Niefern-Öschelbronn *D* 187 D6
Niel *B* 182 C4
Nierstein *D* 185 E9
Nieuw-Bergen *NL* 183 B8
Nieuwegein *NL* 183 A6
Nieuwerkerk *NL* 182 B4
Nieuwerkerk aan de IJssel
NL 182 B5
Nieuwerkerken *B* 183 D6
Nieuwe-Tonge *NL* 182 B4
Nieuw-Heeten *NL* 183 A8
Nieuwkoop *NL* 182 A5
Nieuw-Loosdrecht *NL* 183 A6
Nieuw-Milligen *NL* 183 A7
Nieuw-Namen *NL* 182 C4
Nieuwveen *NL* 182 A5
Nieuw-Vennep *NL* 182 A5
Nieuw-Vossemeer *NL* 182 B4
Nievern *D* 185 D8
Nijkerk *NL* 183 A7
Nijlen *B* 182 C5
Nijmegen *NL* 183 B7
Nijverdal *NL* 183 A8
Nilvange *F* 186 C1
Ninove *B* 182 D4
Nismes *B* 184 D2
Nispen *NL* 182 C4
Nistelrode *NL* 183 B7
Nittel *D* 186 B1
Nivelles *B* 182 D4
Nohfelden *D* 186 B3
Noisseville *F* 186 C1
Nomeny *F* 186 D1
Nomexy *F* 186 E1
Nonnenweier *D* 186 E4
Nonnweiler *D* 186 B2
Noordwijk aan Zee *NL* 182 A4
Noordwijk-Binnen *NL* 182 A4
Noordwijkerhout *NL* 182 A5
Nootdorp *NL* 182 A4
Nordheim *D* 187 C7
Nordkirchen *D* 185 A8
Nouzonville *F* 184 E2
Noville *B* 182 D5
Nüdlingen *D* 187 A9
Nuenen *NL* 183 C7
Nuland *NL* 183 B6
Numansdorp *NL* 182 B4
Nunkirchen *D* 186 C2
Nunspeet *NL* 183 A7
Nürtingen *D* 187 D7
Nuth *NL* 183 D7

O

Oberderdingen *D* 187 C6
Oberfell *D* 185 D7
Oberharmersbach *D* 186 E5
Oberhausen *D* 183 C9
Oberhausen-Rheinhausen
D 187 C5
Oberhoffen-sur-Moder *F* 186 D4
Oberkirch *D* 186 D5
Oberkochen *D* 187 D9
Obermoschel *D* 186 B4
Obernai *F* 186 E3
Obernburg am Main *D* 187 B7
Oberndorf am Neckar *D* 187 E6
Obernheim-Kirchenarnbach
D 186 C4
Ober-Olm *D* 185 E9
Ober-Roden *D* 187 B6
Oberrot *D* 187 C8
Obersinn *D* 187 A8
Obersontheim *D* 187 C8
Oberstenfeld *D* 187 C7
Oberthal *D* 186 B3
Oberthulba *D* 187 A8
Obertshausen *D* 187 C8
Oberursel (Taunus) *D* 187 A6
Oberwesel *D* 185 D8
Oberwolfach *D* 187 E5
Obrigheim *D* 187 C7
Obrigheim (Pfalz) *D* 187 B5
Ochsenfurt *D* 187 B9
Ochtrup *D* 183 A10
Odernheim am Glan *D* 186 B4
Oedelem *B* 182 C2
Oeffelt *NL* 183 B7
Oegstgeest *NL* 182 A5
Oene *NL* 183 A8

Oerlenbach *D* 187 A9
Oestrich-Winkel *D* 185 D8
Offenbach am Main *D* 187 A6
Offenbach an der Queich
D 187 C5
Offenburg *D* 186 E4
Ofterdingen *D* 187 E7
Oftersheim *D* 187 C6
Ogéviller *F* 186 D2
Ohey *B* 183 E6
Ohlsbach *D* 186 E4
Öhringen *D* 187 C7
Oignies *F* 182 E1
Oijen *NL* 183 B6
Oirschot *NL* 183 B6
Oisterwijk *NL* 183 B6
Oldenzaal *NL* 183 A9
Olen *B* 182 C5
Olfen *D* 185 A7
Olpe *D* 185 B8
Olst *NL* 183 A8
Onstmettingen *D* 187 E6
Ooltgensplaat *NL* 182 B4
Oostburg *NL* 182 C2
Oostende *B* 182 C1
Oosterbeek *NL* 183 B7
Oosterhout *NL* 182 B5
Oosterland *NL* 182 B4
Oosterzele *B* 182 D3
Oostham *D* 183 C6
Oostkamp *B* 182 C2
Oostkapelle *NL* 182 B3
Oostmalle *B* 182 C5
Oost-Souburg *NL* 182 B3
Oostvoorne *NL* 182 B4
Ootmarsum *NL* 183 A9
Opglabbeek *B* 183 C7
Opheusden *B* 183 B7
Opitter *B* 183 C7
Oploo *NL* 183 B7
Opoeteren *B* 183 C7
Oppenau *D* 187 E5
Oppenheim *D* 185 E9
Oppenweiler *D* 187 D7
Opwijk *B* 182 D4
Orchies *F* 182 E2
Orenhofen *D* 185 E6
Oreye *B* 183 D6
Ortenberg *D* 186 E4
Osburg *D* 186 B2
Oss *NL* 183 B7
Ossendrecht *NL* 182 C4
Ostend *B* 182 C1
Osterburken *D* 187 C7
Ostfildern *D* 187 D7
Osthofen *D* 187 B5
Ostricourt *F* 182 E2
Östringen *D* 187 C6
Ostwald *F* 186 E4
Ötigheim *D* 187 D5
Ötisheim *D* 187 D6
Ottenheim *D* 186 E4
Ottenhöfen im Schwarzwald
D 187 D5
Otterbach *D* 186 C4
Otterberg *D* 186 B4
Otterlo *NL* 183 A7
Ottersweier *D* 187 D5
Ottignies *B* 182 D4
Ottweiler *D* 186 C3
Oud-Beijerland *NL* 182 B4
Ouddorp *NL* 182 B3
Oudenaarde *B* 182 D3
Oudenbosch *NL* 182 B5
Oudenburg *B* 182 C2
Oude-Tonge *NL* 182 B4
Oudewater *NL* 182 A5
Oud-Gastel *NL* 182 B4
Oud-Turnhout *B* 182 C5
Oud-Vossemeer *NL* 182 B4
Oudzele *B* 182 C2
Ouffet *B* 183 E6
Oulder *B* 184 D5
Oupeye *B* 183 D7
Overath *D* 185 C7
Overdinkel *NL* 183 A10
Overijse *B* 182 D5
Overloon *NL* 183 B7
Overpelt *B* 183 C6
Ovezande *NL* 182 C3
Owen *D* 187 D7

P

Paal *B* 183 C6
Padoux *F* 186 E2
Paliseul *B* 184 E3
Palzem *D* 186 B1
Pange *F* 186 C1
Panningen *NL* 183 C7
Papendrecht *NL* 182 B5
Partenstein *D* 187 A8
Partenstein *D* 187 A8
Pâturages *B* 182 E3
Pecq *B* 182 D2
Peer *B* 183 C6
Pelm *D* 185 D6
Pepingen *B* 182 D4
Pepinster *B* 183 D7
Perl *D* 186 C1
Péruwelz *B* 182 D3
Perwez *B* 182 D5
Petite-Rosselle *F* 186 C3
Petitmont *F* 186 D2
Pexonne *F* 186 E2
Pfaffenhofen an der Roth
D 187 E9
Pfaffenhoffen *F* 186 D4
Pfalzfeld *D* 185 D8
Pfalzgrafenweiler *D* 187 D6
Pfedelbach *D* 187 C7
Pforzheim *D* 187 D6
Pfronstetten *D* 187 E7
Pfullingen *D* 187 E7
Pfungstadt *D* 187 B6
Phalsbourg *F* 186 D3
Philippeville *B* 184 D2
Philippine *NL* 182 C3
Philippsburg *D* 187 C5
Piershil *NL* 182 B4
Piesport *D* 185 E6
Pijnacker *NL* 182 A4
Pirmasens *D* 186 C4
Pittem *B* 182 D2
Plaidt *D* 185 D7
Plettenberg *D* 185 B8
Pliezhausen *D* 187 D7
Plobsheim *F* 186 E4
Plochingen *D* 187 D7
Ploegsteert *B* 182 D1
Plüderhausen *D* 187 D8

Poederlee *B* 182 C5
Polch *D* 185 D7
Polsbroek *NL* 182 B5
Pompey *F* 186 D1
Pont-à-Celles *B* 182 D4
Pont-à-Marcq *F* 182 D2
Pont-de-Loup *B* 182 E5
Poppenhausen *D* 187 A9
Portieux *F* 186 E1
Posterholt *NL* 183 C7
Poussay *F* 186 E1
Pracht *D* 185 C8
Prinsenbeek *NL* 182 B5
Profondeville *B* 184 D2
Pronsfeld *D* 185 D5
Provenchères-sur-Fave *F* 186 E3
Prüm *D* 185 D5
Puderbach *D* 185 C8
Pulheim *D* 183 C9
Putte *B* 182 C5
Putte *NL* 182 C4
Puttelange-aux-Lacs *F* 186 C3
Putten *NL* 183 A7
Püttlingen *D* 186 C2
Puurs *B* 182 C4

Q

Quaregnon *B* 182 E3
Queidersbach *D* 186 C4
Quendorf *D* 183 A10
Quesnoy-sur-Deûle *F* 182 D2
Quierschied *D* 186 C3
Quiévrain *B* 182 E3
Quiévrechain *F* 182 E3

R

Raalte *NL* 183 A8
Raamsdonksveer *NL* 182 B5
Radevormwald *D* 185 B7
Raeren *B* 183 D8
Raesfeld *D* 183 B9
Ralingen *D* 185 E6
Rambervillers *F* 186 E2
Rambrouch *L* 184 E4
Ramillies *B* 182 D5
Rammelsbach *D* 186 B3
Rammingen *D* 187 E9
Ramstein *D* 186 C4
Rance *B* 184 D1
Randersacker *D* 187 B8
Rangendingen *D* 187 E6
Ransbach-Baumbach *D* 185 D8
Ranst *B* 182 C5
Raon-l'Étape *F* 186 E2
Rastatt *D* 187 D5
Ratingen *D* 183 C9
Raubach *D* 185 C8
Ravels *B* 182 C5
Ravenstein *NL* 183 B7
Rebecq *B* 182 D4
Réchicourt-le-Château *F* 186 D2
Recht *B* 184 D5
Rechtenbach *D* 187 B8
Recklinghausen *D* 183 B10
Réding *F* 186 D3
Rees *D* 183 B8
Rehlingen-Siersburg *D* 186 C2
Reichelsheim (Odenwald)
D 187 B6
Reichenbach *D* 187 B6
Reichenberg *D* 187 B8
Reicholzheim *D* 187 B8
Reichshoffen *F* 186 D4
Reichstett *F* 186 D4
Reil *D* 185 D7
Reilingen *D* 187 C6
Reinheim *D* 187 B6
Reinsfeld *D* 186 B2
Reken *D* 183 B10
Rekken *NL* 183 A9
Remagen *D* 185 C7
Remich *L* 186 B1
Remicourt *B* 183 D6
Remouchamps *B* 183 E7
Remscheid *D* 183 C10
Renchen *D* 186 E4
Renesse *NL* 182 B3
Rengsdorf *D* 185 C7
Renkum *NL* 183 B7
Rennerod *D* 185 C9
Renningen *D* 187 D6
Renswoude *NL* 183 A7
Renwez *F* 184 E2
Retie *B* 183 C6
Reusel *NL* 183 C6
Reutlingen *D* 187 E7
Reuver *NL* 183 C8
Revin *F* 184 E2
Rhaunen *D* 185 E7
Rhede *D* 183 B9
Rheden *NL* 183 A8
Rheinau *D* 187 D5
Rheinbach *D* 183 D9
Rheinberg *D* 183 B9
Rheinböllen *D* 185 D8
Rheinbreitbach *D* 185 C7
Rheinbrohl *D* 185 C7
Rheinstetten *D* 187 D5
Rheinzabern *D* 187 C5
Rhenen *NL* 183 B7
Rhens *D* 185 D8
Rhinau *F* 186 E4
Rhisnes *B* 182 D5
Rhoon *NL* 182 B4
Richardménil *F* 186 D1
Ridderkerk *NL* 182 B5
Riegelsberg *D* 186 C2
Riemst *B* 183 D7
Rieneck *D* 187 A8
Riethoven *NL* 183 C6
Rijen *NL* 182 B5
Rijkevorsel *B* 182 C5
Rijnsburg *NL* 182 A4
Rijsbergen *NL* 182 B5
Rijsel *F* 182 D2
Rijssen *NL* 183 A9
Rijswijk *NL* 182 A4
Rilland *NL* 182 C4
Rimbach *D* 187 B6
Rimogne *F* 184 E2
Rips *NL* 183 B7
Rittersdorf *D* 185 D5
Rixensart *B* 182 D5
Rochefort *B* 184 D3
Rochehaut *B* 184 E3
Rochin *F* 182 D2
Rockanje *NL* 182 B4

Rockenhausen *D* 186 B4
Rocroi *F* 184 E2
Rodalben *D* 186 C4
Roermond *NL* 183 C7
Roeselare *B* 182 D2
Roetgen *D* 183 D8
Roggel *NL* 183 C7
Rohrbach-lès-Bitche *F* 186 C3
Rombas *F* 186 C1
Rommerskirchen *D* 183 C9
Ronse *B* 182 D3
Roosendaal *NL* 182 B4
Rosée *B* 184 D2
Rosenfeld *D* 187 E6
Rosheim *F* 186 D3
Rosmalen *NL* 183 B6
Rösrath *D* 185 C7
Roßdorf *D* 187 B6
Rossum *NL* 183 B6
Rot am See *D* 187 C9
Rothenberg *D* 187 B6
Rothenbuch *D* 187 B7
Rothenburg ob der Tauber
D 187 C9
Rothenfels *D* 187 B8
Rotheux-Rimière *B* 183 D6
Rotselaar *B* 182 D5
Rottenacker *D* 187 E8
Rottenburg am Neckar *D* 187 E6
Rottendorf *D* 187 B9
Rotterdam *NL* 182 B5
Röttingen *D* 187 B8
Roubaix *F* 182 D2
Roulers *B* 182 D2
Rouvroy *F* 182 E1
Rouvroy-sur-Audry *F* 184 E2
Rozenburg *NL* 182 B4
Rozendaal *NL* 183 A7
Ruddervoorde *B* 182 C2
Rudersberg *D* 187 D7
Rüdesheim *D* 185 E8
Ruiselede *B* 182 C2
Rülzheim *D* 187 C5
Rumes *B* 182 D2
Rumigny *F* 184 E1
Rumst *B* 182 C4
Runkel *D* 185 D9
Rüsselsheim *D* 187 A5
Rutesheim *D* 187 D6
Rüthen *D* 185 B9
Rütten-Scheid *D* 183 C9
Ruurlo *NL* 183 A8

S

Saales *F* 186 E3
Saarbrücken *D* 186 C2
Saarburg *D* 186 B2
Saarlouis *D* 186 C2
Saarwellingen *D* 186 C2
Sachsenheim *D* 187 D7
St-Amand-les-Eaux *F* 182 E2
St-Avold *F* 186 C3
St-Blaise-la-Roche *F* 186 E3
St-Clément *F* 186 D2
St-Dié *F* 186 E2
Ste-Marguerite *F* 186 E2
Ste-Marie-aux-Mines *F* 186 E3
St-Firmin *F* 186 E1
St-Ghislain *B* 182 E3
St-Hubert *B* 184 D3
St-Louis-lès-Bitche *F* 186 D3
St-Max *F* 186 D1
St-Michel *F* 184 E1
St-Michel-sur-Meurthe *F* 186 E2
St-Nicolas *B* 183 D7
St-Nicolas-de-Port *F* 186 D1
St-Oedenrode *NL* 183 B6
St-Quirin *F* 186 D3
St-Vith *B* 184 D5
Salmtal *D* 185 E6
Sandhausen *D* 187 C6
Sankt Augustin *D* 185 C7
Sankt Goar *D* 185 D8
Sankt Goarshausen *D* 185 D8
Sankt Ingbert *D* 186 C3
Sankt Julian *D* 186 B4
Sankt Katharinen *D* 185 C7
Sankt Wendel *D* 186 C3
Santpoort *NL* 182 A5
Sarralbe *F* 186 D3
Sarrebourg *F* 186 D3
Sarreguemines *F* 186 C3
Sarre-Union *F* 186 D3
Sart *D* 183 D7
Sasbach *D* 186 D5
Sasbachwalden *D* 186 D5
Sassenheim *NL* 182 A5
Sas Van Gent *NL* 182 C3
Satteldorf *D* 187 C9
Saulheim *D* 185 E9
Saverne *F* 186 D3
Schaafheim *D* 187 B7
Schaarsbergen *NL* 183 A7
Schaerbeek *B* 182 D4
Schaesberg *NL* 183 D8
Schaijk *NL* 183 B7
Schalkhaar *NL* 183 A8
Schalksmühle *D* 185 B8
Scharendijke *NL* 182 B3
Schebheim *D* 187 B9
Schefflenz *D* 187 C7
Schelklingen *D* 187 E8
Schenkenzell *D* 187 E5
Schermbeck *D* 183 B9
Scherpenheuvel *B* 182 D5
Scherpenzeel *NL* 183 A6
Scherwiller *F* 186 E3
Schiedam *NL* 182 B4
Schieren *L* 184 D4
Schifferstadt *D* 187 C5
Schiffweiler *D* 186 C3
Schijndel *NL* 183 B6
Schilde *B* 182 C5
Schillingen *D* 186 B2
Schiltach *D* 187 E5
Schiltigheim *F* 186 D4
Schinnen *NL* 183 D7
Schinveld *NL* 183 D7
Schipluiden *NL* 182 B4
Schirmeck *F* 186 E3
Schlangenbad *D* 185 D9
Schleiden *D* 183 D8
Schmallenberg *D* 185 B9
Schmelz *D* 186 C2
Schnelldorf *D* 187 C9
Schnürpflingen *D* 187 E8
Schoenberg *B* 185 D5
Schöllkrippen *D* 187 A7
Schömberg *D* 187 D6

Schönaich *D* 187 D7
Schondra *D* 187 A8
Schönecken *D* 185 D5
Schönenberg-Kübelberg
D 186 C3
Schöntal *D* 187 C8
Schoondijke *NL* 182 C3
Schoonhoven *NL* 182 B5
Schorndorf *D* 187 D8
Schoten *B* 182 C4
Schriesheim *D* 187 C6
Schrozberg *D* 187 C8
Schuttertal *D* 186 E4
Schutterwald *D* 186 E4
Schüttorf *D* 183 A10
Schwäbisch Gmünd *D* 187 D8
Schwäbisch Hall *D* 187 C8
Schwaigern *D* 187 C7
Schwalbach *D* 186 C2
Schwegenheim *D* 187 C5
Schweich *D* 185 E6
Schwieberdingen *D* 187 D7
Seckach *D* 187 C7
Seclin *F* 182 D2
Seebach *D* 187 D5
Seeheim-Jugenheim *D* 187 B6
Seilles *B* 183 E6
Seinsheim *D* 187 B9
Sélestat *F* 186 E3
Seligenstadt *D* 187 A6
Selm *D* 185 A7
Selters (Westerwald) *D* 185 C8
Seltz *F* 186 D5
Senden *D* 187 E9
Seneffe *B* 182 D4
Seraing *B* 183 D7
Serooskerke *NL* 182 B3
Serrig *D* 186 B2
Severnum *D* 183 C8
's-Gravendeel *NL* 182 B5
's-Gravenhage *NL* 182 A4
's-Gravenmoer *NL* 182 B5
's-Gravenpolder *NL* 182 C3
's-Gravenvoeren *B* 183 D7
's-Gravenzande *NL* 182 B4
's-Heerenberg *NL* 183 B8
's-Heerenhoek *NL* 182 C3
's-Hertogenbosch *NL* 183 B6
Siebengewald *NL* 183 B8
Siegburg *D* 185 C7
Siegen *D* 185 C9
Sierck-les-Bains *F* 186 C1
Siershahn *D* 185 D8
Signy-le-Petit *F* 184 E1
Sijsele *B* 182 C2
Silenrieux *B* 184 D2
Simmerath *D* 183 D8
Simmern (Hunsrück) *D* 185 E8
Simpelveld *NL* 183 D8
Sindelfingen *D* 187 D6
Singhofen *D* 185 D8
Sinn *D* 185 C9
Sinsheim *D* 187 C6
Sint Annaland *NL* 182 B4
Sint Anthonis *NL* 183 B7
Sint-Genesius-Rode *B* 182 D4
Sint-Gillis-Waas *B* 182 C4
Sint-Huibrechts-Lille *B* 183 C7
Sint Jansteen *NL* 182 C4
Sint-Katelijne-Waver *B* 182 C5
Sint-Laureins *B* 182 C3
Sint-Lenaarts *B* 182 C5
Sint Maartensdijk *NL* 182 B4
Sint-Margriete *B* 182 C3
Sint-Maria-Lierde *B* 182 D3
Sint-Martens-Latem *B* 182 C3
Sint Michielsgestel *NL* 183 B6
Sint-Niklaas *B* 182 C4
Sint Odilienberg *NL* 183 C8
Sint-Pauwels *B* 182 C4
Sint Philipsland *NL* 182 B4
Sint-Pieters-Leeuw *B* 182 D4
Sint-Truiden *B* 183 D6
Sinzheim *D* 187 D5
Sinzig *D* 185 C7
Sittard *NL* 183 D7
Sivry *B* 184 D1
Sleidinge *B* 182 C3
Sliedrecht *NL* 182 B5
Sluis *NL* 182 C2
Sluiskil *NL* 182 C3
Soerendonk *NL* 183 C7
Soest *D* 185 A9
Soest *NL* 183 A6
Soesterberg *NL* 183 A6
Soheit-Tinlot *B* 183 E6
Sohren *D* 185 E7
Soignies *B* 182 D4
Solingen *D* 183 C10
Sombreffe *B* 182 D5
Someren *NL* 183 C7
Somme-Leuze *B* 184 D3
Somzée *B* 184 D1
Son *NL* 183 B6
Soufflenheim *F* 186 D4
Soultz-sous-Forêts *F* 186 D4
Soumagne *B* 183 D7
Spa *B* 183 E7
Spabrücken *D* 185 E8
Spay *D* 185 D8
Speicher *D* 185 E6
Speyer *D* 187 C5
Spiere *B* 182 D2
Spiesen-Elversberg *D* 186 C3
Spijkenisse *NL* 182 B4
Spontin *B* 184 D3
Spraitbach *D* 187 D8
Sprendlingen *D* 185 E8
Sprimont *B* 183 E7
Sprockhövel *D* 185 B7
Stabroek *B* 182 C4
Staden *B* 182 D2
Stadtkyll *D* 185 D6
Stadtlohn *D* 183 B9
Staig *D* 187 E9
Standdaarbuiten *NL* 182 B5
Stavelot *B* 184 D4
Stavenisse *NL* 182 B4
Steenbergen *NL* 182 B4
Steenderen *NL* 183 A8
Stein *NL* 183 D7
Steinach *D* 186 E5
Steinfeld *D* 186 D3
Steinfeld *D* 187 B8
Steinheim am Albuch *D* 187 D9

Steinheim an der Murr *D*
*187 D7
Steinsfeld *D* 187 C9
Steinwenden *D* 186 C4
Stekene *B* 182 C4
Stellendam *NL* 182 B4
Stevensweert *NL* 183 C7
Stimpfach *D* 187 C9
Stockstadt am Rhein *D* 187 B5
Stolberg (Rheinland) *D* 183 D8
Stoumont *B* 183 D7
Straelen *NL* 183 C8
Straimont *B* 184 E3
Stramproy *NL* 183 C7
Strasbourg *F* 186 D4
Strassen *L* 186 B1
Straßenhaus *D* 185 C8
Strijen *NL* 182 B5
Stromberg *D* 185 E8
Sturzelbronn *F* 186 C4
Stuttgart *D* 187 D7
Suddendorf *D* 183 A10
Südlohn *D* 183 B9
Sulz am Neckar *D* 187 E6
Sulzbach am Main *D* 187 B7
Sulzbach an der Murr *D* 187 C7
Sulzbach-Laufen *D* 187 D8
Sulzbach/Saar *D* 186 C3
Sulzfeld *D* 187 C6
Sulzthal *D* 187 A9
Sundern (Sauerland) *D* 185 B9
Susteren *NL* 183 C7
Swalmen *NL* 183 C8

T

Taintrux *F* 186 E2
Talange *F* 186 C1
Tamm *D* 187 D7
Tantonville *F* 186 E1
Tauberbischofsheim *D* 187 B8
Tawern *D* 186 B2
Tegelen *NL* 183 C8
Tellin *B* 184 D3
Templeuve *F* 182 D2
Temse *B* 182 C4
Tenneville *B* 184 D4
Ter Aar *NL* 182 A5
Terborg-Silvolde *NL* 183 B8
Terheijden *NL* 182 B5
Terneuzen *NL* 182 C3
Tervuren *B* 182 D5
Tessenderlo *B* 183 C6
Testelt *B* 182 D5
Teteringen *NL* 182 B5
Thaleischweiler-Fröschen
D 186 C4
Thalfang *D* 186 B2
Thaon-les-Vosges *F* 186 E1
't Harde *NL* 183 A7
Theux *B* 183 D7
Thionville *F* 186 C1
Tholen *NL* 182 B4
Tholey *D* 186 C3
Thommen *B* 184 D5
Thorn *NL* 183 C7
Thuin *B* 184 D1
Thüngen *D* 187 B8
Thüngersheim *D* 187 B8
Tiefenbronn *D* 187 D6
Tiel *NL* 183 B6
Tielen *B* 182 C5
Tielt *B* 182 D2
Tienen *B* 182 D5
Tilburg *NL* 183 B6
Tongeren *B* 183 D6
Tönisvorst *D* 183 C8
Torhout *B* 182 C2
Tourcoing *F* 182 D2
Tournai *B* 182 D2
Traar *B* 183 C9
Traben-Trarbach *D* 185 E7
Trarbach *D* 185 E7
Trebur *D* 187 B5
Treis *D* 185 D7
Tremelo *B* 182 D5
Trier *D* 186 B2
Trierweiler *D* 186 B2
Trippstadt *D* 186 C4
Trittenheim *D* 185 E6
Trochtelfingen *D* 187 E7
Troisdorf *D* 183 D10
Troisfontaines *F* 186 D3
Trois-Ponts *B* 184 D4
Troisvierges *L* 184 D5
Trooz *D* 183 D7
Truchtersheim *F* 186 D4
Trulben *D* 186 C4
Tubbergen *NL* 183 A9
Tübingen *D* 187 D7
Tubize *B* 182 D4
Turnhout *B* 182 C5
Twello *NL* 183 A8

U

Übach-Palenberg *D* 183 D8
Überherrn *D* 186 C2
Ubstadt-Weiher *D* 187 C6
Uccle *B* 182 D4
Üchtelhausen *D* 187 A9
Uckange *F* 186 C1
Uddel *NL* 183 A7
Uden *NL* 183 B7
Udenhout *NL* 183 B6
Üdersdorf *D* 185 D6
Uettingen *D* 187 B8
Uffenheim *D* 187 B9
Uhingen *D* 187 D8
Uithoorn *NL* 182 A5
Ulft *NL* 183 B8
Ulicoten *NL* 182 C5
Ulm *D* 187 E9
Ulmen *D* 185 D6
Ulvenhout *NL* 182 B5
Undingen *D* 187 E7
Unna *D* 185 A8
Unnau *D* 185 C8
Untermünkheim *D* 187 C8
Unterpleichfeld *D* 187 B9
Urbach *D* 185 C8
Urbar *D* 185 D8
Urberach *D* 187 B6
Utrecht *NL* 183 A6
Üxheim *D* 185 D6

V

Vaals *NL* 183 D8
Vaassen *NL* 183 A7
Valkenburg *NL* 183 D7
Valkenswaard *NL* 183 C6
Vallendar *D* 185 D8
Vandœvre-lès-Nancy *F* 186 D1
Varik *D* 183 B7
Varsseveld *NL* 183 B8
Vaux-sur-Sûre *B* 184 D4
Veenendaal *NL* 183 A7
Veere *NL* 182 B3
Veerle *B* 182 C5
Veghel *NL* 183 B7
Velbert *D* 183 C10
Veldegem *B* 182 C2
Velden *NL* 183 C8
Veldhoven *NL* 183 C6
Velen *D* 183 B9
Vellberg *D* 187 C8
Velp *NL* 183 B7
Vendenheim *F* 186 D4
Venlo *NL* 183 C8
Venray *NL* 183 B7
Vergaville *F* 186 D2
Verlaine *B* 183 D6
Verny *F* 186 C1
Verviers *B* 183 D7
Vessem *NL* 183 C6
Vettelschoss *D* 185 C7
Vianden *L* 184 E5
Vianen *NL* 183 B6
Vic-sur-Seille *F* 186 D2
Vielsalm *B* 184 D4
Vierlingsbeek *NL* 183 B7
Viernheim *D* 187 B6
Viersen *D* 183 C8
Vieux-Condé *F* 182 E3
Vigy *F* 186 C1
Villé *F* 186 E3
Villeneuve-d'Ascq *F* 182 D2
Villers-le-Bouillet *B* 183 D6
Villers-lès-Nancy *F* 186 D1
Villmar *D* 185 D9
Vilvoorde *B* 182 D4
Vincey *F* 186 E1
Vinkt *B* 182 C2
Vinningen *D* 186 C4
Vireux-Molhain *F* 184 E2
Vireux-Wallerand *F* 184 E2
Visé *B* 183 D7
Vlaardingen *NL* 182 B4
Vleuten *NL* 183 A6
Vlijmen *NL* 183 B6
Vlissingen *NL* 182 C3
Voerde (Niederrhein) *D* 183 B9
Vogelenzang *NL* 182 A5
Vogelweh *D* 186 C4
Vöhringen *D* 187 E6
Vöhringen *D* 187 E9
Volkach *D* 187 B9
Volkel *NL* 183 B7
Völklingen *D* 186 C2
Volmunster *F* 186 C3
Voorburg *NL* 182 A4
Voorhout *NL* 182 A4
Voorschoten *NL* 182 A4
Voorst *NL* 183 A8
Voorthuizen *NL* 183 A7
Vorden *NL* 183 A8
Vorst *B* 183 C6
Vosselaar *B* 182 C5
Vrasene *B* 182 C4
Vreden *D* 183 A9
Vreeland *NL* 183 A6
Vresse *B* 184 E2
Vriezenveen *NL* 183 A9
Vrouwenpolder *NL* 182 B3
Vught *NL* 183 B6

W

Waalre *NL* 183 C6
Waalwijk *NL* 183 B6
Waarschoot *B* 182 C3
Wachenheim an der Weinstraße
D 187 C5
Wachtebeke *B* 182 C3
Wächtersbach *D* 187 A7
Waddinxveen *NL* 182 A5
Wadern *D* 186 B2
Wadersloh *D* 185 A9
Wadgassen *D* 186 C2
Wageningen *NL* 183 B7
Waghäusel *D* 187 C5
Waiblingen *D* 187 D7
Waibstadt *D* 187 C6
Waigolshausen *D* 187 B9
Waimes *B* 184 D4
Walcourt *B* 184 D1
Waldachtal *D* 187 E6
Waldböckelheim *D* 185 E8
Waldbreitbach *D* 185 C7
Waldbröl *D* 185 C8
Waldbrunn-Lahr *D* 185 C9
Waldbüttelbrunn *D* 187 B8
Waldenbuch *D* 187 D7
Waldenburg *D* 187 C8
Waldesch *D* 185 D8
Waldfischbach-Burgalben
D 186 C4
Wald-Michelbach *D* 187 B6
Waldmohr *D* 186 C3
Waldrach *D* 186 B2
Waldsee *D* 187 C5
Waldstetten *D* 187 D8
Walferdange *L* 186 B1
Walldorf *D* 187 A6
Walldorf *D* 187 C6
Walldürn *D* 187 B7
Wallhausen *D* 185 E8
Wallhausen *D* 187 C9
Wamel *NL* 183 B6
Wandre *B* 183 D7
Wanne-Eikel *D* 183 B10
Wanroij *NL* 183 B7
Wanssum *NL* 183 B8
Wanze *B* 183 D6
Wapenveld *NL* 183 A8
Waregem *B* 182 D2
Waremme *B* 183 D6
Warmond *NL* 182 A5
Warnsveld *NL* 183 A8
Warstein *D* 185 B9
Wartmannsroth *D* 187 A8
Wäschenbeuren *D* 187 D8
Waspik *NL* 182 B5
Wasseiges *B* 183 D6
Wasselonne *F* 186 D3

Wassenaar *NL* **182** A4
Wassenberg *D* **183** C8
Wasseralfingen *D* **187** D9
Wasserbillig *D* **186** B2
Wasserliesch *D* **186** B2
Wasserlosen *D* **187** A9
Wateringen *NL* **182** A4
Waterland-Oudeman *B* **182** C3
Waterloo *B* **182** D4
Watten-Scheid *D* **183** C10
Wavre *B* **182** D5
Wavrin *F* **182** D1
Waxweiler *D* **185** D5
Waziers *F* **182** E2
Weelde *B* **182** C5
Weerselo *NL* **183** A9
Weert *NL* **183** C7
Weesp *NL* **183** A6
Weeze *D* **183** B8
Wegberg *D* **183** C8
Wehl *NL* **183** B8
Weibern *D* **183** E10
Weibersbrunn *D* **187** B7
Weidenstetten *D* **187** D8
Weidenthal *D* **186** C4
Weiersbach *D* **186** B3
Weikersheim *D* **187** C8
Weilbach *D* **187** B7
Weilburg *D* **185** D9
Weil der Stadt *D* **187** D6
Weilerbach *D* **186** C4
Weilerswist *D* **183** D9
Weilheim an der Teck *D* **187** D8
Weilmünster *D* **185** D9
Weinbach *D* **185** D9

Weingarten (Baden) *D* **187** C6
Weinheim *D* **187** B6
Weinsberg *D* **187** C7
Weinsheim *D* **185** D5
Weinsheim *D* **185** E8
Weinstadt *D* **187** D7
Weiskirchen *D* **186** B2
Weissach *D* **187** D6
Weißenhorn *D* **187** E9
Weißenthurm *D* **185** D7
Weiswampach *L* **184** D5
Weitbruch *D* **186** D4
Weitefeld *D* **185** C8
Weiterstadt *D* **187** B6
Wekerom *NL* **183** A7
Welkenraedt *B* **183** D7
Well *NL* **183** B8
Wellen *B* **183** D6
Wellin *B* **184** D3
Welling *D* **185** D7
Welschbillig *D* **185** E6
Welver *D* **185** A8
Welzheim *D* **187** D8
Wemeldinge *NL* **182** B4
Wendelsheim *D* **186** B5
Wenden *D* **185** C8
Wendlingen am Neckar *D* **187** D7
Wenduine *B* **182** C2
Wengsel (Isterberg) *D* **183** A10
Wépion *B* **182** E5
Werbach *D* **187** B8
Werbomont *B* **184** D4
Werdohl *D* **185** B8
Werkendam *NL* **182** B5

Werl *D* **185** A8
Wermelskirchen *D* **185** B7
Wernau *D* **187** D7
Werne *D* **185** A8
Werneck *D* **187** B9
Wertheim *D* **187** B8
Wervik *B* **182** D2
Wesel *D* **183** B9
Wesepe *NL* **183** A8
Wesseling *D* **183** D9
Wessem *NL* **183** C7
Westerburg *D* **185** C8
Westerheim *D* **187** D8
Westerhoven *NL* **183** C6
Westerlo *B* **182** C5
Westerstetten *D* **187** D8
Westervoort *NL* **183** B7
Westhausen *D* **187** D9
Westhofen *D* **187** B5
Westhoffen *F* **186** D3
Westkapelle *NL* **182** B2
Westmalle *B* **182** C5
Wetter (Ruhr) *D* **185** B7
Wetteren *B* **182** C3
Wevelgem *B* **182** D2
Weyerbusch *D* **185** C8
Weyersheim *F* **186** D4
Wichelen *B* **182** C3
Wickede (Ruhr) *D* **185** B8
Wiehl *D* **185** C8
Wielsbeke *B* **182** D2
Wierden *NL* **183** A9
Wiernsheim *D* **187** D6
Wiesbaden *D* **185** D9
Wiesen *D* **187** A7

Wiesensteig *D* **187** D8
Wiesloch *D* **187** C6
Wijchen *NL* **183** B7
Wijhe *NL* **183** A8
Wijk bij Duurstede *NL* **183** B6
Wijk en Aalburg *NL* **183** B6
Wildberg *D* **187** D6
Wildberg *D* **187** E5
Wilhelminadorp *NL* **183** C6
Willebroek *B* **182** C4
Willemstad *NL* **182** B4
Willich *D* **183** C9
Willstätt *D* **186** D4
Wilnis *NL* **182** A5
Wilnsdorf *D* **185** C9
Wilp *NL* **183** A8
Wiltingen *D* **186** B2
Wiltz *L* **184** E4
Wincheringen *D* **186** B1
Wincrange *L* **184** D4
Windesheim *D* **185** E8
Windhagen *D* **185** C2
Wingene *B* **182** C2
Wingen-sur-Moder *F* **186** D3
Winnenden *D* **187** D7
Winningen *D* **185** D8
Winnweiler *D* **186** B4
Winssen *NL* **183** B7
Winterswijk *NL* **183** B9
Wintrich *D* **185** E6
Wipperfürth *D* **185** B7
Wirges *D* **185** D8
Wissembourg *F* **186** C4
Wissen *D* **185** C8
Wissenkerke *NL* **182** B3
Witten *D* **185** B7

Wittighausen *D* **187** B8
Wittlich *D* **185** E6
Woensdrecht *NL* **182** C4
Woerden *NL* **182** A5
Wœrth *F* **186** D4
Woippy *F* **186** C1
Wolfach *D* **187** E5
Wolfstein *D* **186** B4
Wolpertshausen *D* **187** C8
Wolphaartsdijk *NL* **182** B3
Wommelgem *B* **182** C5
Wormeldange *L* **186** B1
Worms *D* **187** B5
Wörrstadt *D* **185** E9
Wortel *B* **182** C5
Wörth am Main *D* **187** B7
Wörth am Rhein *D* **187** C5
Woudenberg *NL* **183** A6
Woudrichem *NL* **182** B5
Woustviller *F* **186** C3
Wouw *NL* **182** B4
Wuppertal *D* **183** C10
Würselen *D* **183** D8
Würzburg *D* **187** B8
Wüstenrot *D* **187** C7
Wuustwezel *B* **182** C5

Yerseke *NL* **182** C4
Ysselsteyn *NL* **183** C7
Yutz *F* **186** C1
Yvoir *B* **184** D2

Zaamslag *NL* **182** C3
Zaandam *NL* **182** A5
Zaltbommel *NL* **183** B6
Zandhoven *B* **182** C5
Zandvliet *B* **182** C4
Zandvoort *NL* **182** A5
Zarren *B* **182** C1
Zaventem *B* **182** D4
Zeddam *NL* **183** B8
Zedelgem *B* **182** C2
Zeebrugge *B* **182** C2
Zeeland *NL* **183** B7
Zeewolde *NL* **183** A7
Zeiskam *D* **187** C5
Zeist *NL* **183** A6
Zeitlofs *D* **187** A8
Zele *B* **182** C4
Zelhem *NL* **183** A8
Zell (Mosel) *D* **185** D7
Zell am Harmersbach *D* **186** E5
Zell am Main *D* **187** B8
Zellingen *D* **187** B8
Zeltingen-Rachtig *D* **185** E7

Zelzate *B* **182** C3
Zemmer *D* **185** E6
Zemst *B* **182** D4
Zerf *D* **186** B2
Zevenaar *NL* **183** B8
Zevenbergen *NL* **182** B5
Zierikzee *NL* **182** B3
Zoelen *NL* **183** B6
Zoersel *B* **182** C5
Zoetermeer *NL* **182** A4
Zolder *B* **183** C6
Zomergem *B* **182** C3
Zonhoven *B* **183** D6
Zonnebeke *B* **182** D1
Zornheim *D* **185** E9
Zottegem *B* **182** D3
Zoutleeuw *B* **183** D6
Zuid-Beijerland *NL* **182** B4
Zuidland *NL* **182** B4
Zuienkerke *B* **182** C2
Zülpich *D* **183** D9
Zulte *B* **182** D2
Zundert *NL* **182** C5
Zutendaal *B* **183** D7
Zutphen *NL* **183** A8
Zweibrücken *D* **186** C3
Zwevegem *B* **182** D2
Zwevezele *B* **182** C2
Zwiefalten *D* **187** E7
Zwijndrecht *B* **182** C4
Zwijndrecht *NL* **182** B5
Zwingenberg *D* **187** B6

Xanten *D* **183** B8

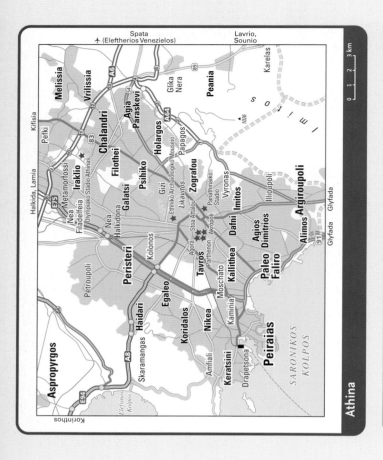

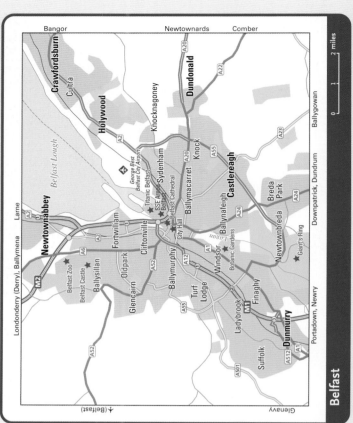

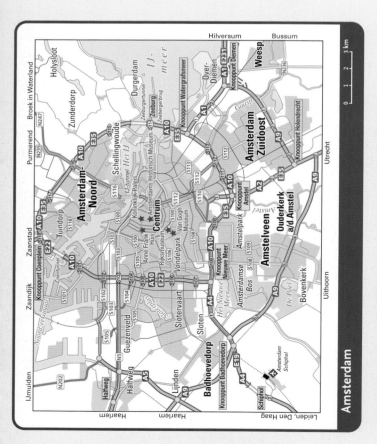

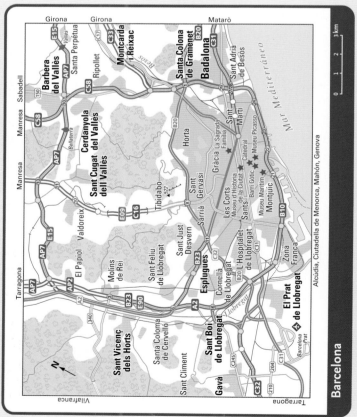

Athina

Belfast

Amsterdam

Barcelona

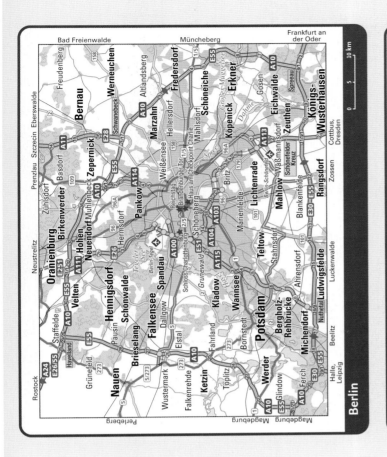

Berlin

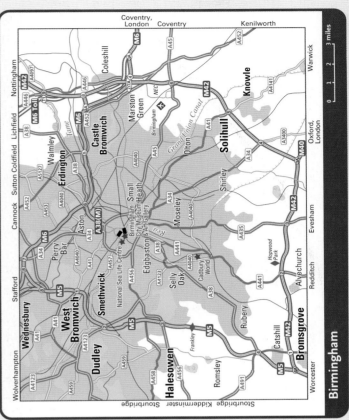

Birmingham

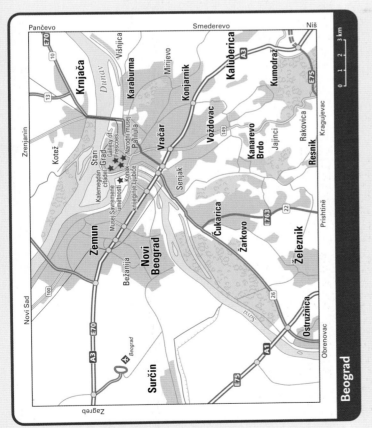

Beograd

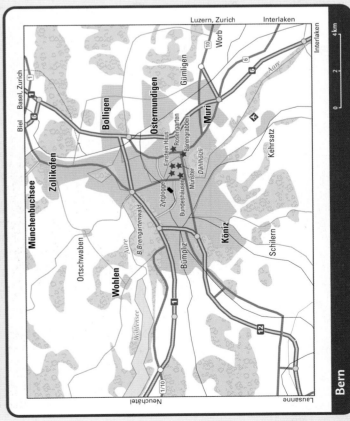

Bern

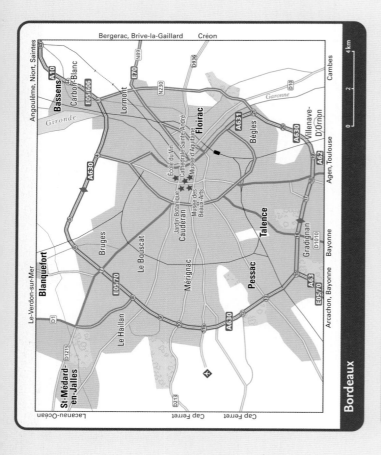

Bordeaux

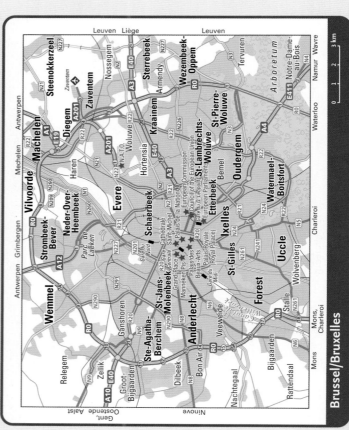

Brussel/Bruxelles

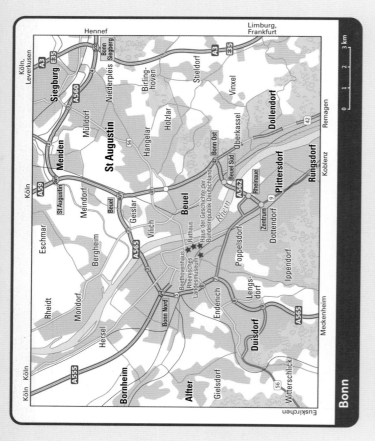

Bonn

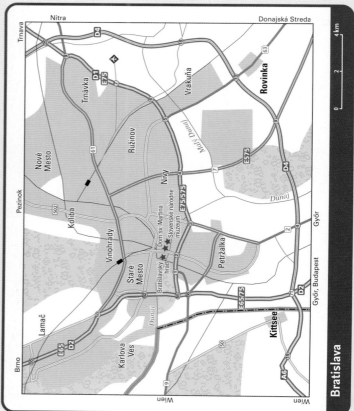

Bratislava

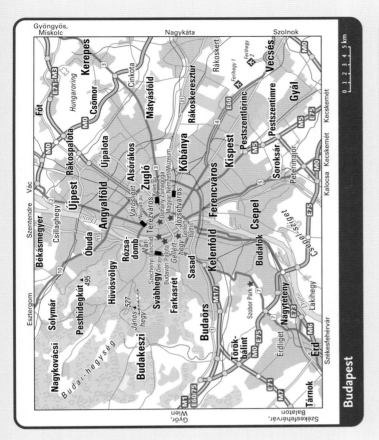

Gyöngyös, Miskolc — Nagykáta — Szolnok

Kerepes · Csömör · Cinkota · Rákoskert · Vecsés
Fót · E71 M3 · Hungaroring · Ferihegy 1 · Ferihegy 2
Szentendre · Mátyásföld · Rákoskeresztúr · Gyál
Vác · Újpalota · Pestszentlőrinc · E60 · Kecskemét
Solymár · Rákospalota · Alsórákos · Zugló · Kőbánya · M5
Békásmegyer · Csillaghegy · Angyalföld · Terézváros · Kispest · E75
Óbuda · Rózsadomb · Józsefváros · Ferencváros · Soroksár · Péterimajor
Hűvösvölgy · Svábhegy · Gellért-hegy · Csepel · Kalocsa
Pesthidegkút · Budai-hegyseg · Farkasrét · Kelenföld · Budafok
Nagykovácsi · Budakeszi · Sasad · Budaörs · Nagytétény · Lakihegy
Törökbálint · Érdliget · Érd · Tárnok
Esztergom · Székesfehérvár

Budapest

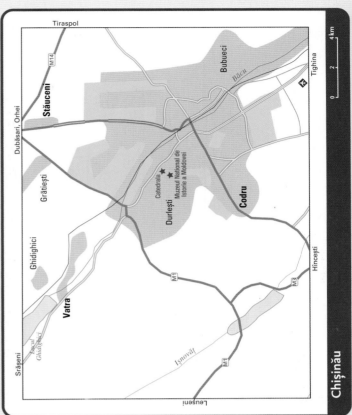

Tiraspol — 4km
Dubăsari, Orhei · M14 · Bubueci · Tighina · Băcu
Stăuceni · Gratieşti · Codru
Ghidighici · Durleşti · Catedrala · Muzeul National de Istorie a Moldovei
Vatra · M1 · M3 · Hinceşti
Străşeni · Iacul Ghidighici · Işnovăţ · M1
Leuşeni

Chişinău

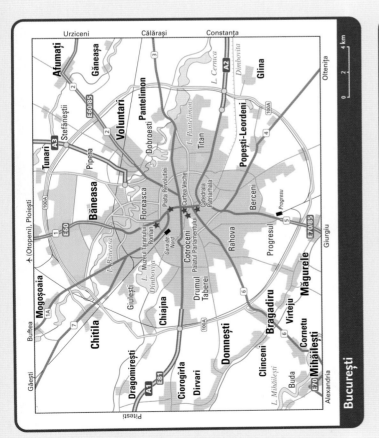

Urziceni — Călăraşi — Constanţa
Afumaţi · Găneasa · Glina
Tunari · Ştefăneştii · Pantelimon · Dîmboviţa · Olteniţa
Voluntari · Dobroeşti · Popeşti-Leordeni
Băneasa · Floreasca · Berceni
Mogoşoaia · Piaţa Revoluţiei · Curtea Veche · Catedrala Patriarhală
Chitila · Giuleşti · Rahova · Progresul
Dragomireşti · Chiajna · Drumul Taberei · Măgurele
Ciorogîrla · Donneşti · Vîrteju
Divari · Clinceni · Bragadiru · Cornetu · Mihăileşti
Găeşti · Alexandria

Bucureşti

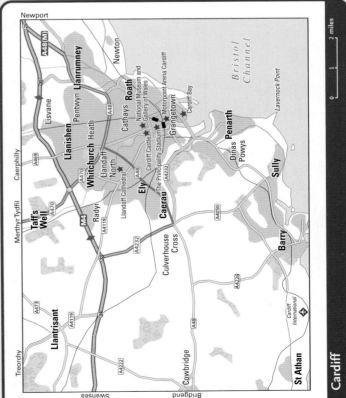

Newport — 2 miles
A48(M) · Llanrumney · Newton · Bristol Channel
Lisvane · Roath · Motorpoint Arena Cardiff · Cardiff Bay
Llanishen · Pentwyn · Cathays · National Museum and Gallery of Wales · Lavernock Point
Caerphilly · Whitchurch · Heath · Grangetown · Penarth
Llandaff North · Cardiff Castle · Dinas Powys
Merthyr Tydfil · Radyr · Llandaff Cathedral · Ely · The Principality Stadium · Sully
Taff's Well · Caerau · Barry
Culverhouse Cross
Llantrisant · Cardiff International
Treorchy · Cowbridge · St Athan
Swansea · Bridgend

Cardiff

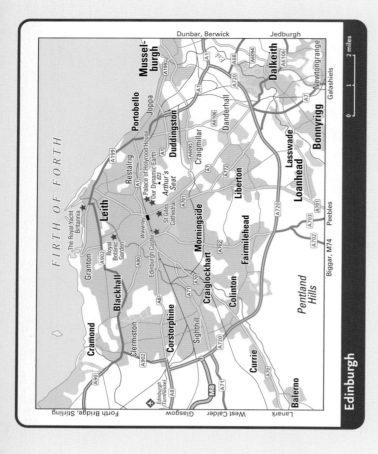

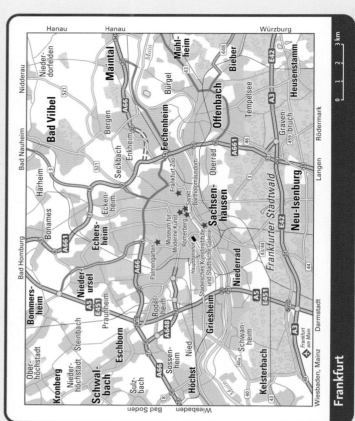

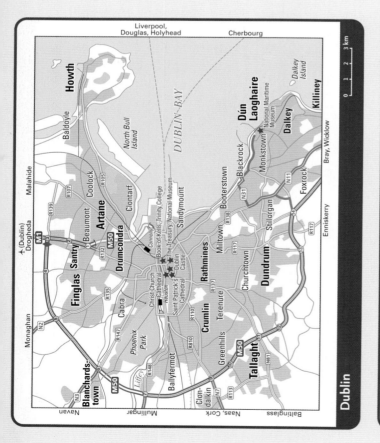

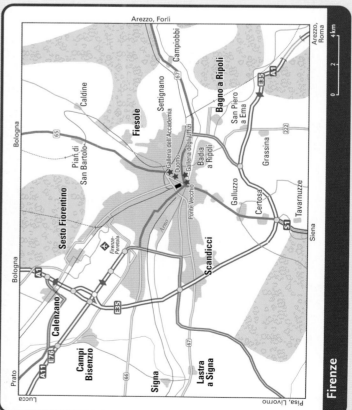

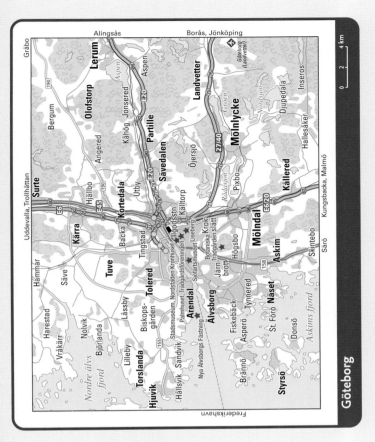

Göteborg

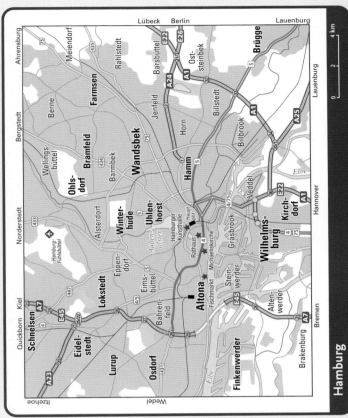

Hamburg

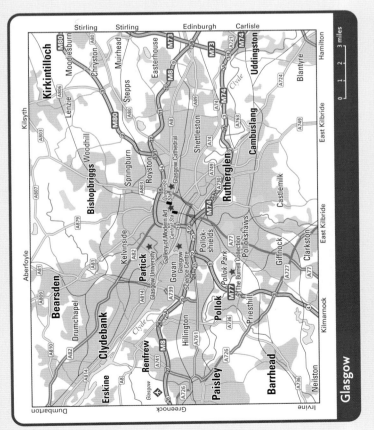

Glasgow

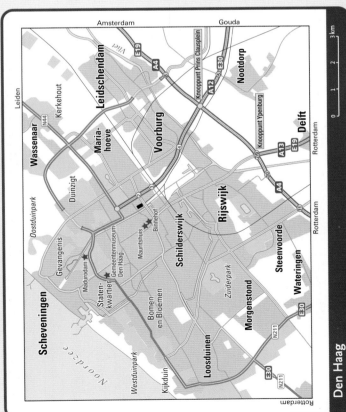

Den Haag

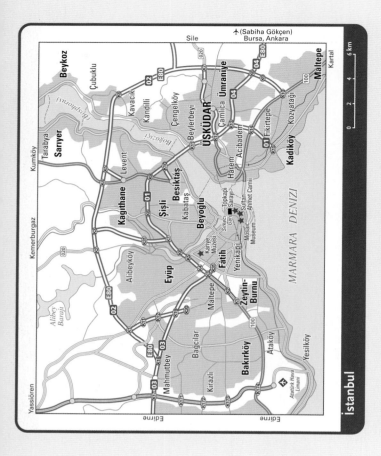

İstanbul

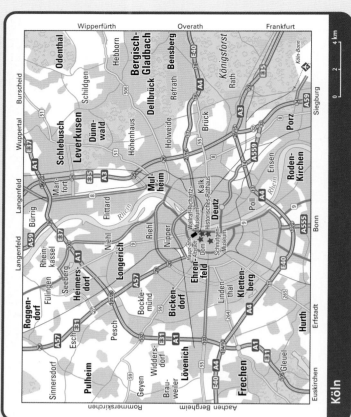

Köln

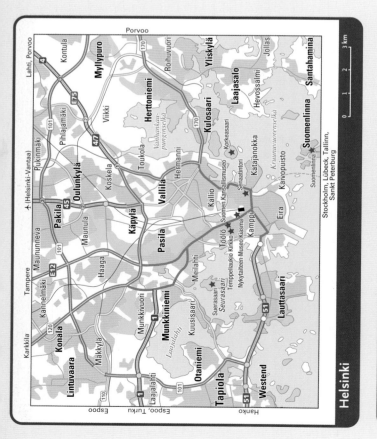

Helsinki

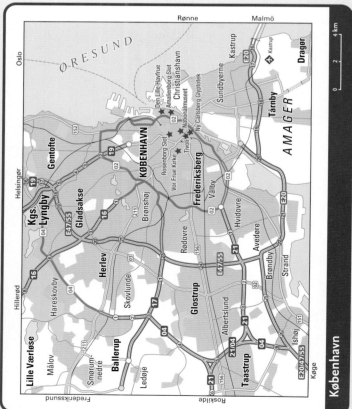

København

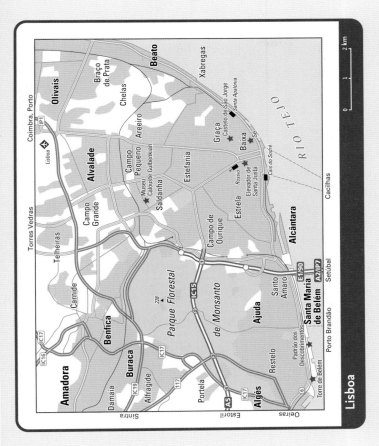

Lisboa

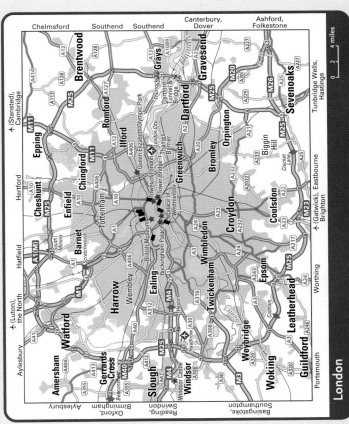

London

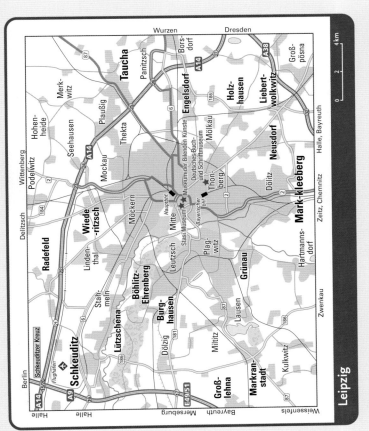

Leipzig

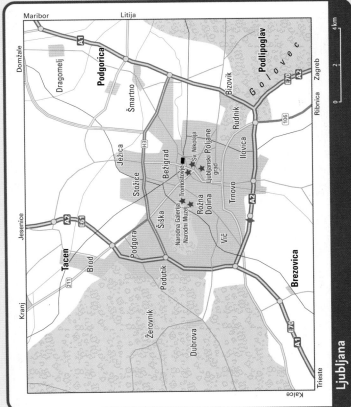

Ljubljana

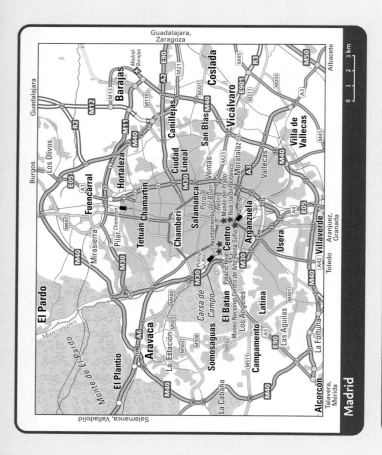

Madrid

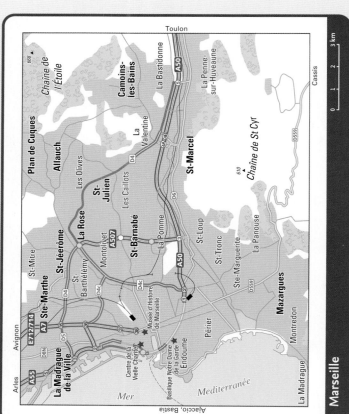

Marseille

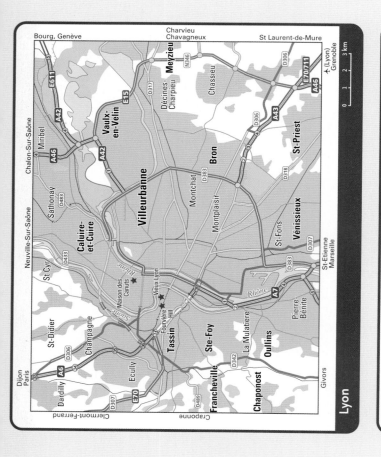

Lyon

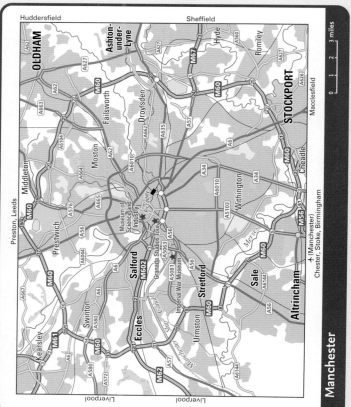

Manchester

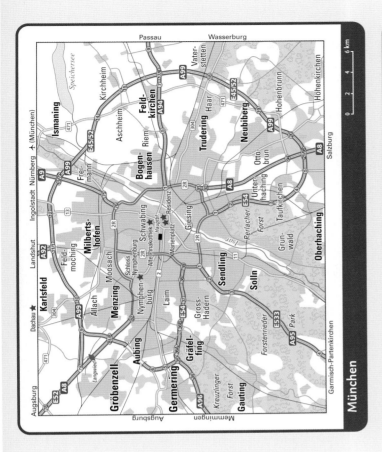

München

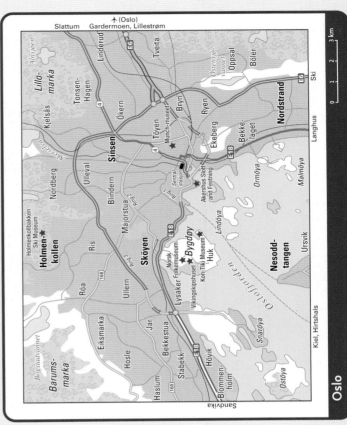

Oslo

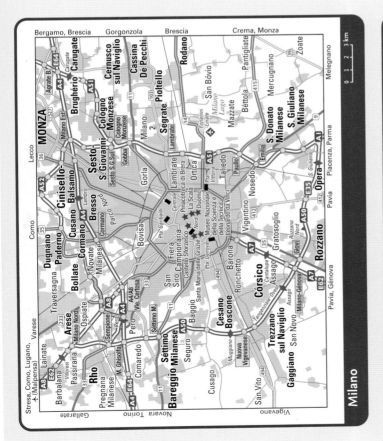

Milano

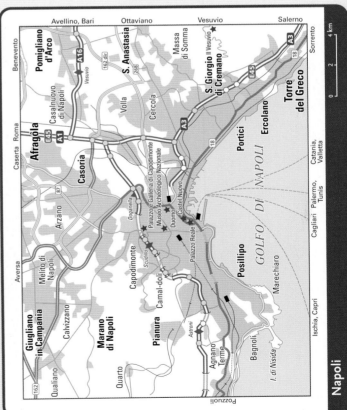

Napoli

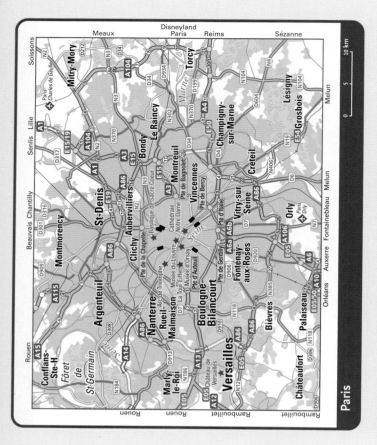

Paris

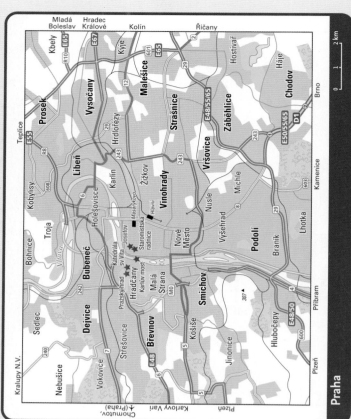

Praha

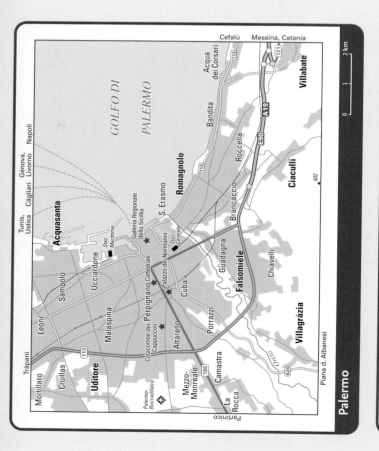

Palermo

Podgorica

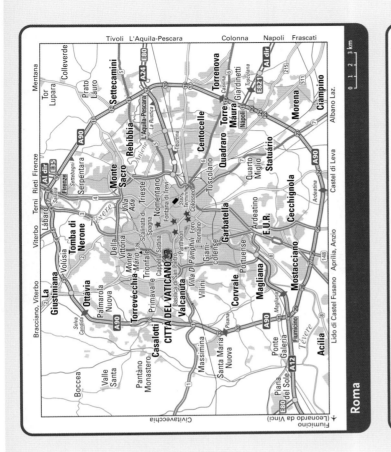

Roma

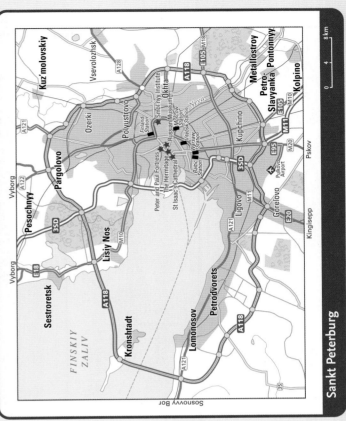

Sankt Peterburg

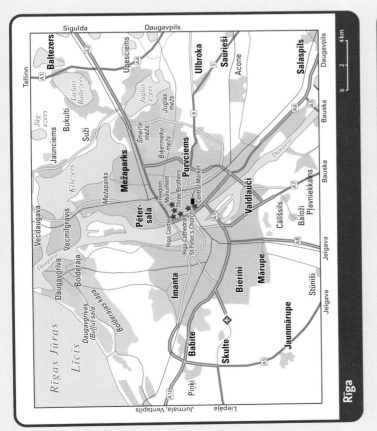

Rīga

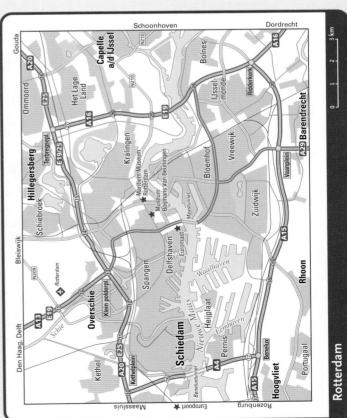

Rotterdam

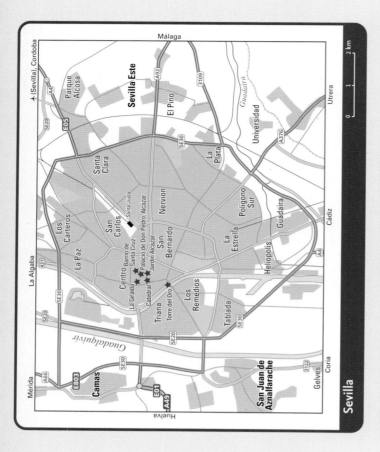

Sevilla

Málaga

Sevilla Este

El Pino

Parque Alcosa

(Sevilla), Cordoba

Universidad

Utrera

Guadaira

Santa Clara

La Plata

Los Carteros

San Carlos

Nervion

La Paz

Centro Barrio de Santa Cruz
La Giralda
Catedral
Palacio de Don Pedro, Alcázar
Jardin Alcázar
San Bernardo

La Estrella

Poligono Sur

Guadaira

Heliopolis

Triana
Torre del Oro

Los Remedios

Tablada

La Algaba

Guadalquivir

Camas

San Juan de Aznalfarache

Mérida

Coria

Gelves

Huelva

Cádiz

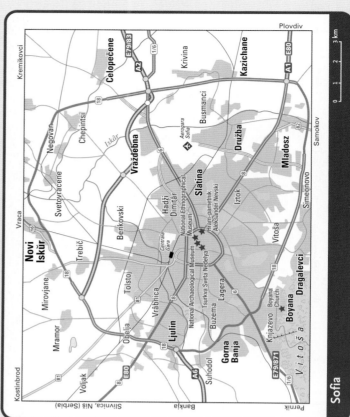

Sofia

Plovdiv

Celopečene

Krivina

Kazichane

Kremikovci

Negován

Chepintsi

Iskâr

Vraždebna

Aerogara Sofia

Družba

Mladosz

Vraca

Svetovracene

Benkovski

Hadži Dimitâr

Slatina

National-Ethnographical Museum

Iztok

Vitoša

Novi Iskâr

Trebič

Hram-pametnik
Aleksânder Nevski

Tsurkva Sveta Nedelya

Simeonovo

Miroyjane

Tolstoi

National Archaeological Museum

Centralna Gara

Lagera

Buzema

Dragalevci

Mramor

Obelja

Vrâbnica

Ljulin

Gorna Banja

Knjazevo

Boyana Church

Boyana

Vitoša

Kostinbrod

Voljak

Suhodol

Banka

Pernik

Silvnica, Niš (Serbia)

E79/871

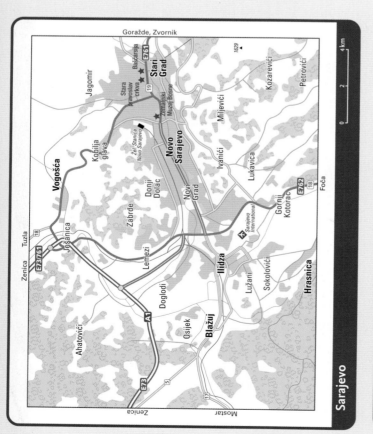

Sarajevo

Goražde, Zvornik

1629

Baščaršija
Stari Grad

E761

Jagomir

Stara Pravoslav crkva

Kozarevići

Petrovići

Zemaljski Muzej Bosne

Miljevići

Kobilja glava

Žel. Stanica Novo Sarajevo

Novo Sarajevo

Ivanići

Lukavica

Vogošća

Donji Dolac

Novi Grad

E762

Foča

Zabrde

Sarajevo International

Gornji Kotorac

Tuzla

Zenica

Lemezi

Ilidža

Jošanica

Lužani

Sokolovići

Hrasnica

Doglodi

A1

Osijek

Blažuj

Ahatovići

E73

Zenica

Mostar

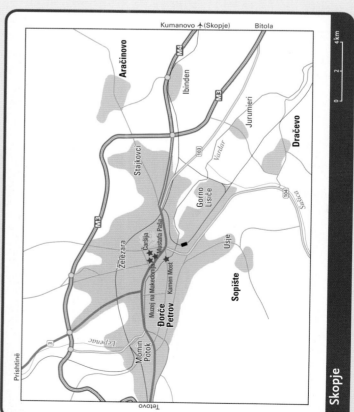

Skopje

Kumanovo (Skopje) Bitola

Aračinovo

M4

Ibinden

M3

Jurumleri

Stajkovci

Vardar

Dračevo

Gorno Lisiče

M3

Železara
Čaršija
Mustafa Paša

Usje

Muzej na Makedonija
Kamen Most

Sopište

Đorče Petrov

Lepenac

Mômin Potok

Prishtinë

Tetovo

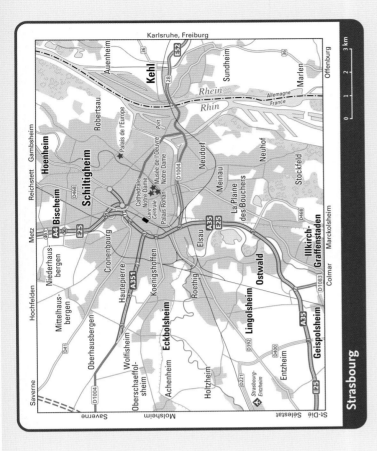

Strasbourg

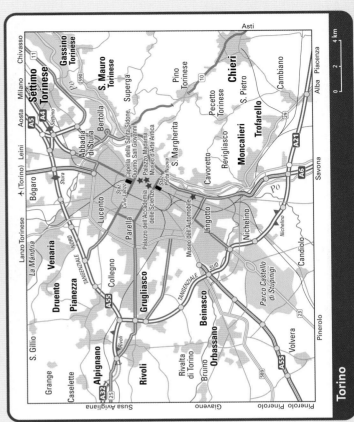

Torino

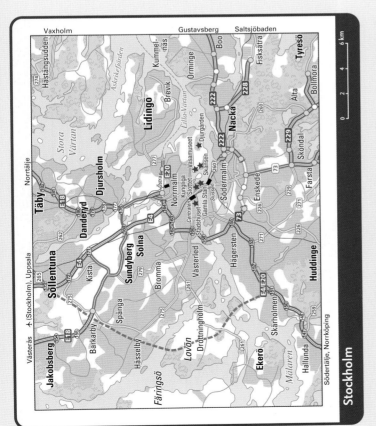

Stockholm

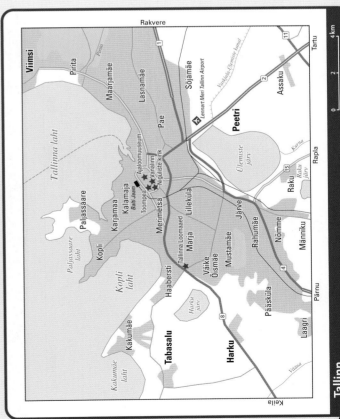

Tallinn

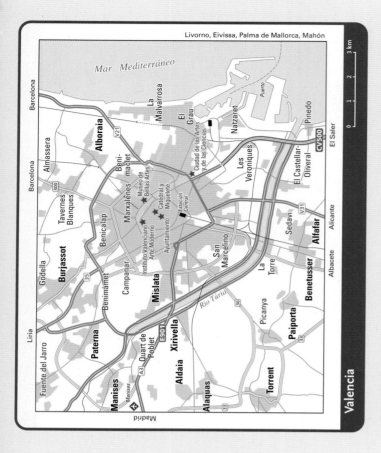

Livorno, Eivissa, Palma de Mallorca, Mahón

Mar Mediterráneo

Barcelona
Barcelona
Barcelona

Almassera
Alboraia
La Malvarrosa
El Grau
Natzaret
Pinedo
El Saler

Tavernes Blanques
Benicalap
Benimàmet
Campanar
Marxalenes
Beni-maclet
Les Veroniques
El Castellar-Oliveral

Museo de Bellas Artes
Ciudad de las Artes y de las Ciencias
Catedral y Miguelete
Estación Central
Museo Valenciano Arte Moderno
Ayuntamiento
Instituto Valenciano

Gódella
Burjassot
Benimàmet

Paterna
Quart de Poblet
Mislata
San Marcelino
La Torre
Sedavi
Alfafar

Fuente del Jarro
Liria
Manises
Aldaia
Xirivella
Rio Túria
Picanya
Benetusser
Paiporta

Madrid
Manises
Alaquas
Torrent

Valencia

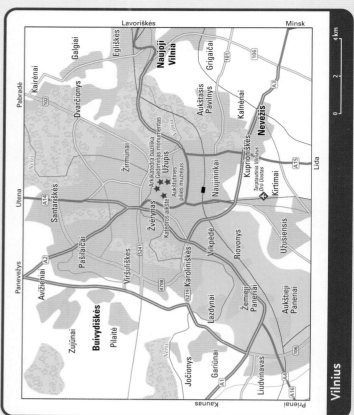

Lavoriškės
Minsk

Galgiai
Egliškės
Naujoji Vilnia
Grigaičiai

Pabradė
Kairėnai
Dvarčionys
Kalnėnai
Nevėžis

Utena
Žirmūnai
Arkikatedra bazilika
Gedimino monumentas
Užupis
Santariškės
Katedros aikštė
Aukštutinės pilies muziejus
Naujininkai
Kuprioniškės
Tartpautinis Vilniaus Oro Uostas
Kirtimai

Panevėžys
Pašilaičiai
Žvėrynas
Vikpėdė
Užusiensis
Avižieniai
Viršuliškės
Karoliniškės
Žemieji Paneriai

Zujūnai
Buivydiškės
Pilaitė
Lazdynai
Aukštieji Paneriai

Jočionys
Gariūnai
Liudvinavas

Kaunas
Prienai
Vilnius

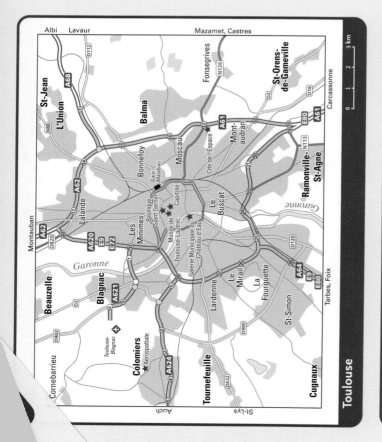

Albi Lavaur
Mazamet, Castres

St-Jean
L'Union
Fonsegrives
St-Orens-de-Gameville

Balma
Carcassonne

Montaudran
Cité de l'Espace
Moscau

Bonnefoy
Ramonville-St-Agne

Basilique Saint Sernin
Capitole
Le Busca
Garonne

Lalande
Musée de Toulouse-Lautrec
Galerie Municipale du Château d'Eau

Beauzelle
Les Minimes
Le Mirail
Lardenne
La Fourguette

Blagnac
Tarbes, Foix

Cornebarrieu
Toulouse-Blagnac Aérospatiale
Colomiers
St-Simon

Tournefeuille
Cugnaux

Auch
St-Lys

Toulouse

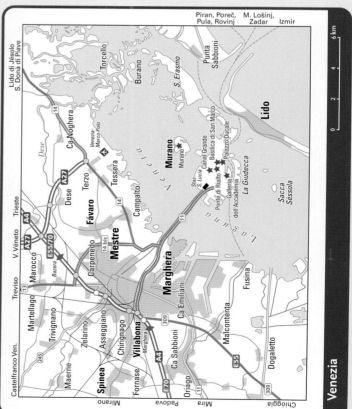

Piran, Poreč, M. Lošinj,
Pula, Rovinj Zadar Izmir

Lido di Jésolo
S. Donà di Piave
Torcello
Punta Sabbioni

Burano
S. Erasmo
Lido

Ca'Noghera
Murano
Murano

Venezia-Marco-Polo
Tessera
Basilica di San Marco
Palazzo Ducale

Dese
Terzo
Staz. S. Lucia
Canal Grande
Ponte di Rialto
Galleria dell'Accademia
Sacca Sessola

Trieste
Fávaro
Campalto
La Giudecca

V. Veneto
Marocco
Carpenedo
Mestre

Treviso
Bazera
Marghera
Fusina

Martellago
Assegiano
Ca'Emiliani

Trivignano
Zelarino
Chirignago
Villabona
Malcontenta

Maerne
Fornase
Ca'Sabbioni
Oriago

Castelfranco Ven.
Spinea
Dogaletto

Mirano
Padova
Mira
Chioggia

Venezia

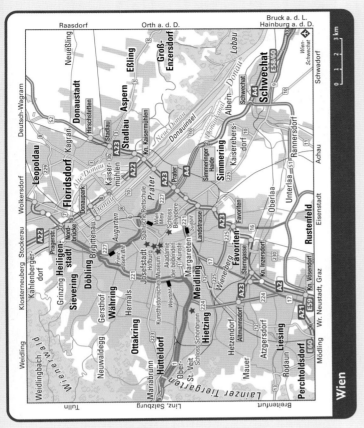

Wien

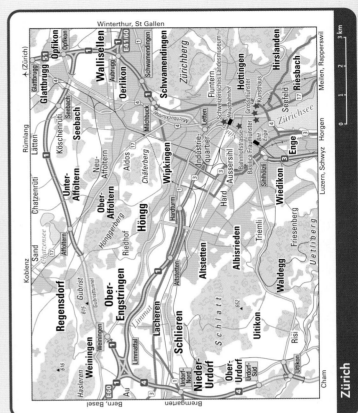

Zürich

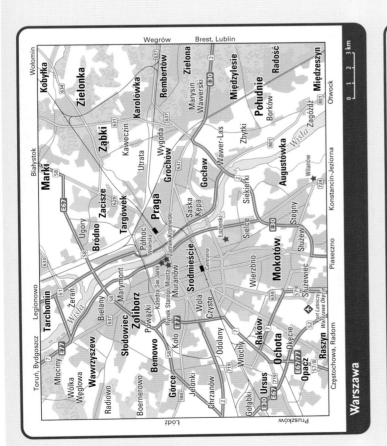

Warszawa

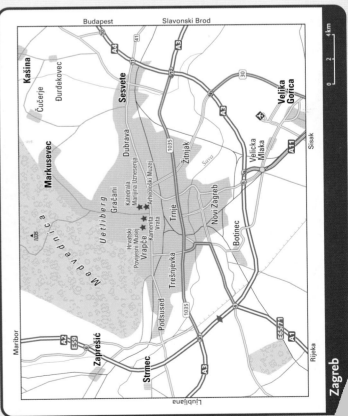

Zagreb

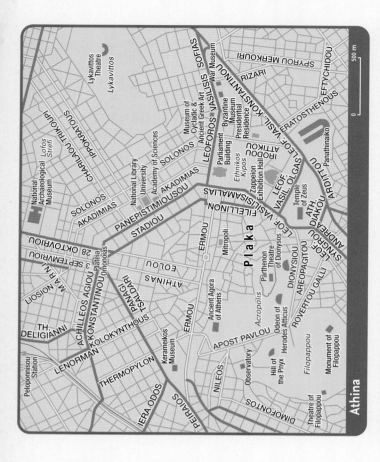

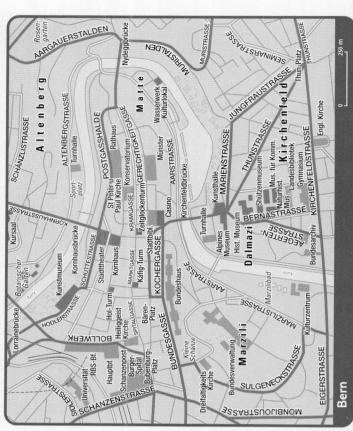

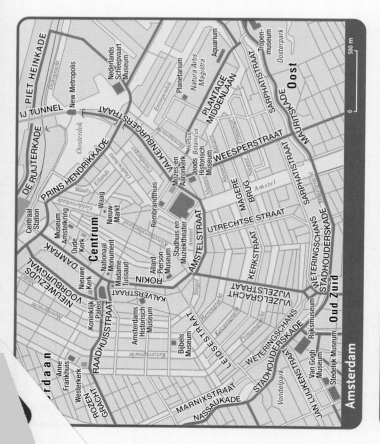

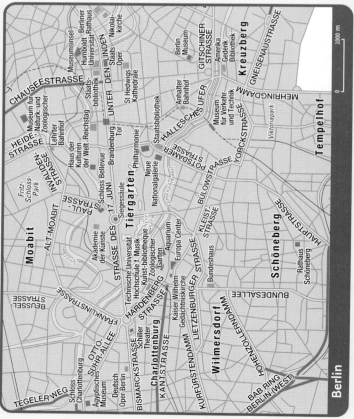

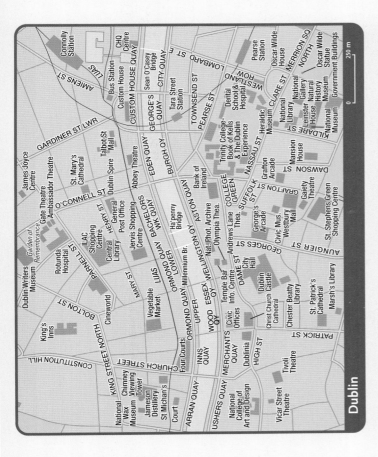

Dublin

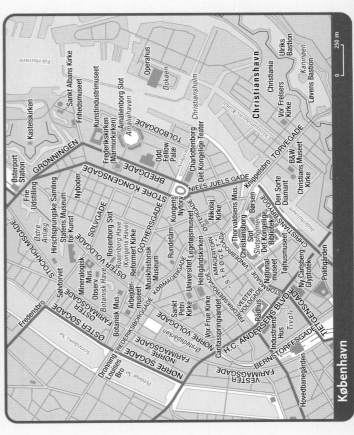

København

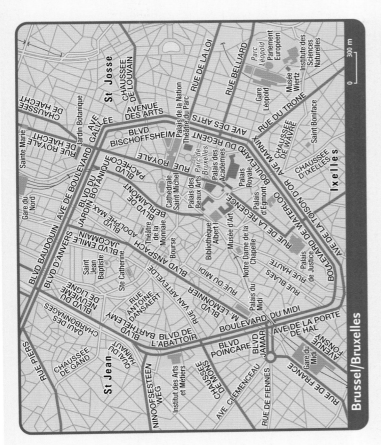

Brussel/Bruxelles

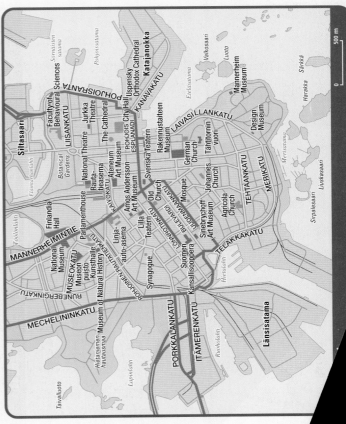

Helsinki

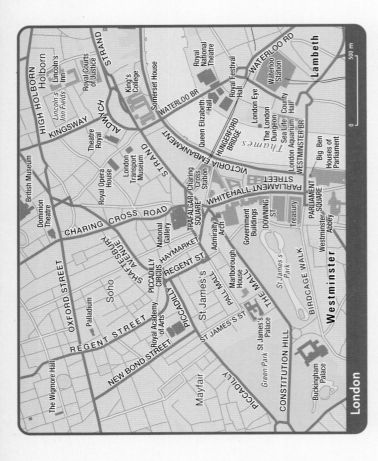

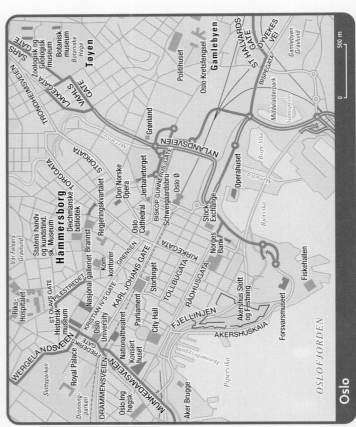

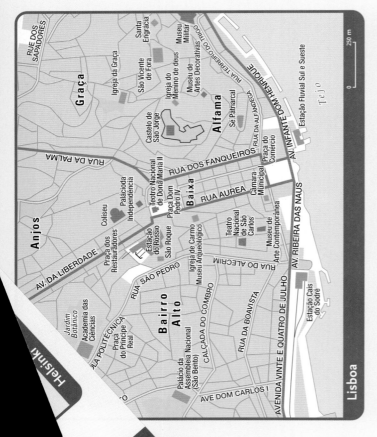

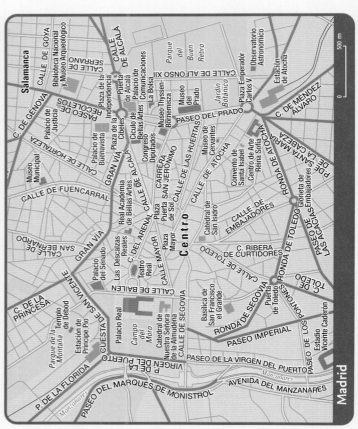

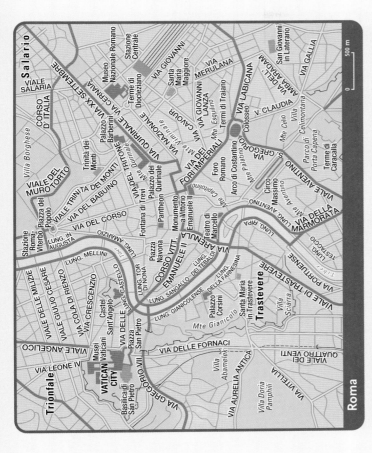

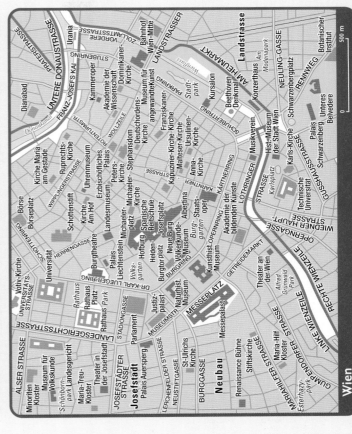

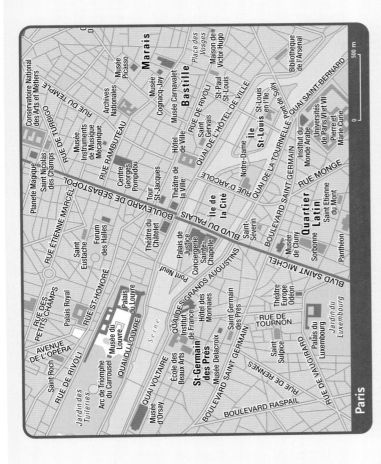

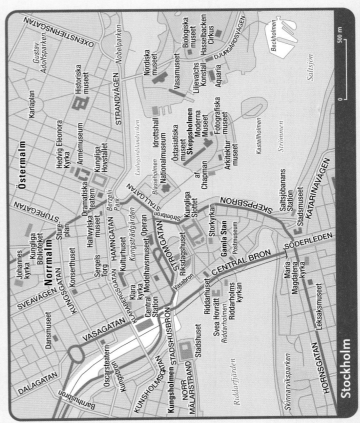

A

Å N 104 F7
Å N 110 E4
Å N 111 B12
Å N 111 C13
Aabenraa DK 86 E4
Aabybro DK 86 A5
Aachen D 20 C6
Aadorf CH 27 F10
Aakirkeby DK 89 E7
Aalborg DK 86 A5
Aalen D 75 E7
Aalestrup DK 86 B4
Aalsmeer NL 16 D3
Aalst B 19 C9
Aalst NL 183 B6
Aalten NL 17 E7
Aalter B 19 B7
Äänekoski FIN 123 E15
Aapajärvi FIN 115 D2
Aapajärvi FIN 119 B12
Aapajoki FIN 119 B12
Aapua S 117 E11
Aarau CH 27 F9
Aarberg CH 31 A11
Aardenburg NL 19 B7
Aareavaara S 117 D10
Aarhus DK 86 C6
Aarle NL 16 F5
A Armada E 38 B3
Aars DK 86 B5
Aarschot B 19 C10
Aartrijke B 182 C2
Aartselaar B 19 B9
Aarup DK 86 E6
Aarwangen CH 27 F8
Aasleagh IRL 6 E3
Ääsmäe EST 131 C9
Aaspere EST 131 C12
Aatsinki FIN 115 E5
Aavajärvi S 119 C11
Aavasaksa FIN 119 B11
Aba H 149 B11
Abaclia MD 154 E3
Abades E 46 C4
Abadín E 38 B5
Abadiño-Zelaieta E 41 B6
Abádszalók H 150 C6
A Baiuca E 38 B3
Abak TR 181 A7
Abalar TR 172 A6
Abánades E 47 C8
Abanilla E 56 E2
Abano Terme I 66 B4
Abarán E 55 C10
A Barrela E 38 C4
Abasár H 150 B5
Abaújszántó H 145 G3
Abbadia San Salvatore I 62 B1
Abbasanta I 64 C2
Abbekås S 87 E13
Abbeville F 18 D4
Abbey IRL 6 F6
Abbeydorney IRL 8 D3
Abbeyfeale IRL 8 D4
Abbeyleix IRL 9 C8
Abbey Town UK 5 F10
Abbiategrasso I 69 C6
Abborrberg S 109 F12
Abborreberget S 98 D8
Abborrträsk S 109 F17
Abbotsbury UK 13 D9
Abbots Langley UK 15 D8
Abcoude NL 16 D3
Abejar E 40 E6
Abejuela E 48 E3
Abela P 50 C2
Abelvær N 105 B10
Abenberg D 75 D8
Abenójar E 54 B4
Abensberg D 75 E10
Aberaeron UK 12 A6
Aberaman UK 13 B8
Aberchirder UK 3 K11
Aberdare UK 13 B8
Aberdaron UK 10 F2
Aberdeen UK 3 L12
Aberdovey UK 10 F3
Aberfeldy UK 5 B9
Aberffraw UK 10 E3
Aberford UK 11 D9
Aberfoyle UK 5 C8
Abergavenny UK 13 B8
Abergele UK 10 E4
Åberget S 109 E18
Abergwaun UK 12 B5
Abergynolwyn UK 10 F4
Aberlady UK 5 C11
Aberlour UK 3 L10
Abernethy UK 5 C10
Aberporth UK 12 A5
Abersoch UK 10 F2
Abertamy CZ 75 B12
Abertawe UK 13 B7
Abertillery UK 13 B8
Abertura E 45 F9
Aberuthven UK 5 C9
Aberystwyth UK 12 A6
Abetone I 66 D2
Abfaltersbach A 72 C6
Abhainnsuidhe UK 2 K2
Abia de la Obispalía E 47 D8
Abiego E 42 C3
Abild DK 86 F3
Abilly F 29 B7
Abingdon UK 13 B12
Abington UK 5 E9
Abisko Östra S 111 D16
Abja-Paluoja EST 131 E10
Abla E 55 E7
Ablis F 24 C6
Ablitas E 41 E8
Abmelaseter N 112 E6
Abo FIN 126 E7
Åbo S 103 C10
Åbodarna S 107 E14
Åbogen N 96 B7
Abondance F 31 C10
Abony H 150 C5
Åbosjö S 107 D13
Aboyne UK 5 A11
Abragão P 44 B4
Abram RO 151 C9
Abrămuț RO 151 C9
Abrantes P 44 F4
Abraur S 109 D16
Abreiro P 38 F5
Abreschviller F 27 C7
Abrest F 30 C3
Abriès F 31 F10
Abrigada P 44 F2
Abriola I 60 B5
Abrucena E 55 E7
Abrud RO 151 E11
Abrupe LV 135 B11
Absam A 72 B4
Absberg D 75 D8
Absdorf A 77 F9
Abtenau A 73 A7
Abtsgmünd D 74 E6
Abukhava BY 140 C10
Åby S 89 A7
Åby S 93 B8
Åbyen DK 90 D7
Åbyggeby S 103 E13
Åbyn S 118 D6
Åbytorp S 92 A6
Acaill IRL 6 E3
A Cañiza E 38 D3
A Carballa E 38 B2
Acarlar TR 177 D10
A Carreira E 38 B3
Acâș RO 151 B10
Acate I 58 E5
Accadia I 60 A4
Acceglio I 36 C5
Accettura I 60 C6
Acciano I 62 C5
Acciaroli I 60 C4
Accous F 32 E4
Accrington UK 11 D7
Accumoli I 62 B4
Acebo E 45 D7
Acedera E 45 F9
Acedo E 32 E1
Acehuche E 45 E7
Aceituna E 45 D8
Acered E 47 B9
Acerenza I 60 B5
Acerno I 60 B4
Acerra I 60 B2
Aceuchal E 51 B7
Ach A 76 F3
Achadh Mòr UK 2 J3
A Chan E 38 D3
Acharacle UK 4 B5
Acharnes GR 175 C8
Achavanich UK 3 J10
Achel B 183 C6
Achenkirch A 72 A4
Achern D 27 C9
Acheux-en-Amiénois F 18 D6
Achicourt F 18 D6
Achill IRL 6 E3
Achilleio GR 175 A6
Achim D 17 B12
Achintee UK 2 L6
Achladochori GR 169 B10
Achladokampos GR 175 D6
Achnacroish UK 4 B6
Achnasheen UK 2 K6
Achosnich UK 4 B4
Achstetten D 71 A9
Achtrup D 82 A6
Aci Castello I 59 D7
Aci Catena I 59 D7
Acireale I 59 D7
Aci Sant'Antonio I 59 D7
Aci Trezza I 59 D7
Acktjära S 103 D11
Acle UK 15 B12
A Coruña E 38 B3
Acquacalda I 59 B6
Acqualagna I 67 E6
Acquanegra sul Chiese I 66 B1
Acquapendente I 62 B1
Acquappesa I 60 E5
Acquaro I 59 B9
Acquarossa CH 71 E7
Acquasanta Terme I 62 B4
Acquasparta I 62 B3
Acquaviva Picena I 62 B5
Acquedolci I 59 C6
Acquigny F 24 B5
Acqui Terme I 37 B8
Acri I 61 E6
A Cruz do Incio E 38 C5
Ács H 149 A10
Acsa H 150 B3
Acuto I 62 D4
Ada SRB 150 F5
Adács H 150 B4
Adahuesca E 42 C3
Adak S 109 F16
Ådalsliden S 107 E11
Adamas GR 179 B7
Adamclisi RO 155 E1
Adamov CZ 77 D11
Adamova BY 133 E4
Adamów PL 141 G6
Adamówka PL 144 C6
Adamstown IRL 9 D9
Adămuș RO 152 E4
Adamuz E 53 A7
Adâncata RO 153 B8
Adâncata RO 161 D8
Ádánd H 149 C10
Adanero E 46 C3
Adão P 45 D6
Adare IRL 8 C5
Adatepe TR 173 D6
Adaúfe P 38 E3
Adavere EST 131 D11
Ādaži LV 135 B8
Adderbury UK 13 A12
Addlestone UK 15 E8
Adegem B 182 C2
Adelboden CH 31 C12
Adelebsen D 78 C6
Adelfia I 61 A7
Adelina PL 144 B8
Adelmannsfelden D 187 D8
Adelschlag D 75 E9
Adelsheim D 27 B11
Adelsried D 75 F8
Ademuz E 47 D10
Adenau D 21 D7
Adendorf D 83 D8
Adendro GR 169 C8
Adenstedt D 78 C6
Adjud RO 153 E10
Adlešiči SLO 148 E4
Adliswil CH 27 F10
Adlkofen D 75 E11
Admont A 73 A9
Adolfsström S 109 D12
Adony H 149 B11
Adorf D 75 B11
Adorf (Diemelsee) D 17 F11
Adoufe P 38 F4
Adra E 55 F6
Adradas E 41 F7
Adrados E 40 F3
Adrano I 59 D6
Adria I 66 B5
Adriani GR 171 B6
Adriers F 29 C7
Aduard NL 17 B6
Adulsbruk N 101 E14
Ådum DK 86 D3
Adunați RO 161 C7
Adunații-Copăceni RO 161 E8
Adutiškis LT 135 F13
Adzaneta de Albaida E 56 D4
Adžūni LV 135 D8
Aegviidu EST 131 C11
Aerino GR 169 F8
Ærøskøbing DK 86 F6
Aerzen D 17 D12
A Escusa E 38 C2
A Esfarrapada E 38 D2
A Estrada E 38 C3
Aetos GR 169 C6
Aetos GR 174 B3
Aetos GR 174 E4
Äetsä FIN 126 C8
Afantou GR 181 D8
Åfarnes N 100 A7
A Feira do Monte E 38 B4
Affing D 75 F8
Afife P 38 E2
Afissos GR 169 F9
Åfjord N 104 D8
Aflenz Kurort A 73 A11
A Fonsagrada E 38 B5
A Forxa E 38 D4
A Forxa E 38 D4
Åfoss N 90 A6
Afragola I 60 B2
Afritz A 73 C8
Afumați RO 160 E2
Afumați RO 161 D8
Afytos GR 169 D9
Aga D 79 E11
Ağaçli TR 173 B10
Ağaköy TR 173 D7
Agalas GR 174 D2
Agallas E 45 D8
A Gándara E 38 B3
A Gándara de Altea E 38 A3
Agapia RO 153 C8
Agăș RO 153 D8
Ágasegyháza H 150 D3
Agde F 34 D5
Agen F 33 B7
Àger E 42 C5
Agerbæk DK 86 D3
Agerskov DK 86 E4
Agersted DK 86 A6
Ågerup DK 87 D10
Ágfalva H 149 A7
Aggersund DK 86 A4
Aggius I 64 B3
Aggsbach Markt A 77 F8
Aghaboe IRL 9 C7
Aghagallon UK 7 C10
Aghalee UK 7 C10
Aghanloo UK 4 E3
Aghaville IRL 8 E4
Aghern IRL 8 D6
Aghione F 37 G10
Aghireșu RO 151 D11
Aghleam IRL 6 D2
Aghnagar Bridge IRL 8 E2
Agia GR 169 E8
Agia Anna GR 175 B7
Agia Anna GR 175 C6
Agia Effimia GR 174 C2
Agia Efthymia GR 174 C5
Agia Galini GR 178 E8
Agia Kyriaki GR 174 E4
Agia Marina GR 175 B6
Agia Marina GR 175 C9
Agia Marina GR 177 E8
Agia Paraskevi GR 168 D4
Agia Paraskevi GR 174 A3
Agia Paraskevi GR 177 E6
Agia Pelagia GR 178 C4
Agia Pelagia GR 178 E9
Agiasma GR 171 C7
Agiasos GR 177 A7
Agia Triada GR 174 D4
Agia Triada GR 175 D6
Agia Varvara GR 169 D7
Agia Varvara GR 178 E9
Agigea RO 155 E3
Agighiol RO 155 C3
Agino Selo BIH 157 C7
Agiofyllo GR 169 E6
Agioi Anargyroi GR 169 E7
Agioi Apostoloi GR 175 C8
Agioi Deka GR 178 E8
Agioi Theodoroi GR 169 E5
Agioi Theodoroi GR 175 A6
Agioi Theodoroi GR 175 D7
Agiokampos GR 169 E8
Agiokampos GR 175 B7
Agionori GR 175 D6
Agios Andreas GR 175 E6
Agios Athanasios GR 169 C8
Agios Athanasios GR 171 B6
Agios Charalampos GR 171 C9
Agios Christoforos GR 175 F5
Agios Dimitrios GR 169 D7
Agios Dimitrios GR 175 D8
Agios Dimitrios GR 175 F6
Agios Efstratios GR 171 E7
Agios Georgios GR 169 C7
Agios Georgios GR 174 B4
Agios Georgios GR 175 C6
Agios Georgios GR 177 F6
Agios Georgios GR 178 E9
Agios Germanos GR 168 C5
Agios Ioannis GR 174 E5
Agios Ioannis GR 175 C7
Agios Ioannis GR 178 B4
Agios Kirykos GR 177 D7
Agios Konstantinos GR 174 B3
Agios Konstantinos GR 174 D5
Agios Konstantinos GR 175 B6
Agios Konstantinos GR 175 D9
Agios Kyprianos GR 178 C3
Agios Leon GR 174 D2
Agios Loukas GR 169 C7
Agios Loukas GR 175 C9
Agios Mamas GR 169 D9
Agios Matthaios GR 168 F2
Agios Myronas GR 178 E9
Agios Nikolaos GR 168 E3
Agios Nikolaos GR 168 E5
Agios Nikolaos GR 169 D10
Agios Nikolaos GR 174 B2
Agios Nikolaos GR 174 B4
Agios Nikolaos GR 174 B4
Agios Nikolaos GR 178 B3
Agios Nikolaos GR 179 E10
Agios Panteleïmonas GR 169 C6
Agios Paraskevi GR 171 F10
Agios Petros GR 169 C8
Agios Petros GR 174 B2
Agios Petros GR 175 E6
Agios Spyridonas GR 169 D7
Agios Spyridonas GR 174 A2
Agios Stefanos GR 175 C8
Agios Stefanos GR 176 E5
Agios Thomas GR 175 C8
Agios Vasileios GR 169 C9
Agios Vasileios GR 175 D6
Agira I 59 D6
Agivey UK 4 E3
Agkathia GR 169 C7
Agkistro GR 169 B9
Aglapsvik N 111 B15
Aglasterhausen D 187 C6
Agle N 105 C13
Aglen N 105 B10
Agliana I 66 E3
Aglientu I 64 A3
Aglish IRL 9 D7
Agluonėnai LT 134 E2
Agnadello I 69 C8
Agnagar Bridge IRL 8 E2
Agnanta GR 168 F5
Agnantero GR 169 F6
Agneaux F 23 B9
Agnita RO 152 F5
Agno CH 69 A6
Agnone I 63 D6
Agolada E 38 C3
Agoncillo E 32 F1
Agon-Coutainville F 23 B8
Agordo I 72 D5
Agost E 56 E3
Agos-Vidalos F 32 D5
Ågotnes N 94 B2
Agra GR 177 A7
Agramón E 55 C9
Agramunt E 42 D6
Agrate Brianza I 69 B7
Agreda E 41 E8
Agrés E 56 D4
Agria GR 169 F9
Agridi GR 174 D4
Agrigento I 58 E4
Agrij RO 151 C11
Agrili GR 174 E4
Agrinio GR 174 B3
Agriovotano GR 175 A7
Agrochão P 39 E5
Agropoli I 60 C3
Ågskaret N 108 C5
Aguadulce E 53 B7
Aguadulce E 55 F7
A Guarda E 38 E2
Aguarón E 41 F9
Aguas E 42 C3
Aguas Belas P 44 E4
Aguas de Busot E 56 D4
Águas de Moura P 50 B2
Águas Frias P 38 E5
Aguaviva E 42 F3
A Gudiña E 38 D5
Agudo E 54 B3
Águeda P 44 C4
Aguessac F 34 B5
Agugliano I 67 E7
Aguiar P 50 C4
Aguiar da Beira P 44 C5
Aguilafuente E 46 B4
Aguilar de Alfambra E 42 F2
Aguilar de Campóo E 40 C3
Aguilar de la Frontera E 53 A7
Aguilar del Río Alhama E 41 E8
Águilas E 55 E9
Agullana E 34 F4
Agullent E 56 D3
Aha S 109 F14
Ahafona IRL 8 C3
Aham D 75 E11
Ahascragh IRL 6 F6
Ahaus D 17 D8
Åheim N 100 B3
Ahelva FIN 121 E11
Ahigal E 45 D8
Ahigal de Villarino E 45 B8
Ahillones E 51 C8
Ahja EST 131 E14
Ahjola FIN 121 D12
Ahlainen FIN 126 B6
Ahlatli TR 167 F8
Ahlbeck D 84 C6
Ahlbeck D 84 C6
Ahlden (Aller) D 82 E7
Ahlen D 17 E9
Ahlerstedt D 17 B12
Ahlhorn D 17 C10
Ahmas FIN 119 E17
Ahmavaara FIN 115 E4
Ahmetbey TR 173 B8
Ahmetbeyli TR 177 C9
Ahmetçeeli TR 172 E6
Ahmetli TR 177 C9
Ahmovaara FIN 125 D13
Ahnsbeck D 79 A7
Ahoghill UK 4 F4
Ahoinen FIN 127 D11
Ahokylä FIN 123 C17
Ahola FIN 121 B11
Ahola FIN 121 C13
Ahola FIN 121 D13
Ahola FIN 124 D9
Aholanvaara FIN 115 E5
Ahonperä FIN 119 F13
Ahorn D 75 B8
Aho-Vastinki FIN 123 E14
Ahrbrück D 21 D7
Ahrensbök D 83 B9
Ahrensburg D 83 C8
Ahrenshagen D 83 B13
Ahrenshoop D 83 B12
Ähtäri FIN 123 E12
Ähtärinranta FIN 123 E12
Ahtme EST 132 C1
Ahula EST 131 C11
Ahun F 29 C10
Åhus S 88 D6
Ahveninen FIN 123 E16
Ahvenisto FIN 127 C16
Ahvensalmi FIN 125 F10
Ahvenselkä FIN 115 E4
Ahvenvittikko FIN 115 D4
Ahvionsaari FIN 129 B10
Aiani GR 169 D6
Aianteio GR 175 D7
Aibar E 32 E3
Aibl A 73 C11
Aichach D 75 F9
Aichhalden D 27 D9
Aichstetten D 71 B10
Ai'dejav'ri N 117 A10
Aidenbach D 76 E4
Aidipsos GR 175 B7
Aidone I 58 E5
Aidonochori GR 169 C10
Aidt DK 86 C5
Aidu EST 131 D12
Aiello Calabro I 59 A9
Aielo de Malferit E 56 D3
Aieta I 60 D5
Aiffres F 28 C5
Aigeira GR 174 C5
Aigen im Ennstal A 73 A9
Aigen im Mühlkreis A 76 E5
Aigiali GR 177 F6
Aigina GR 175 D7
Aiginio GR 169 D8
Aigio GR 174 C5
Aigle CH 31 C10
Aiglemont F 184 E2
Aignan F 33 C6
Aignay-le-Duc F 25 E12
Aigre F 28 D6
Aigrefeuille-d'Aunis F 28 C4
Aigrefeuille-sur-Maine F 28 A3
A Igrexa E 38 D3
Aiguafreda E 43 D8
Aiguebelle F 31 D9
Aigueblanche F 31 D10
Aigueperse F 30 C3
Aigues-Mortes F 35 C7
Aigues-Vives F 33 E9
Aigues-Vives F 34 D4
Aigues-Vives F 35 C7
Aiguilhe F 30 E4
Aiguilles F 31 F10
Aiguillon F 33 B6
Aigurande F 29 C9
Äijäjoki FIN 116 B10
Äijälä FIN 123 E16
Aijala FIN 127 E9
Aijänneva FIN 123 F11
Aillant-sur-Tholon F 25 E9
Aillas F 32 B5
Aillevillers-et-Lyaumont F 26 E5
Ailly-le-Haut-Clocher F 18 D4
Ailly-sur-Noye F 18 E5
Ailly-sur-Somme F 18 E5
Ailt an Chorráin IRL 6 C6
Aimargues F 35 C7
Aime F 31 D10
Ainali FIN 119 F14
Ainali FIN 123 B11
Ainay-le-Château F 29 B11
Ainaži LV 131 F8
Aindling D 75 E8
Ainet A 73 C6
Ainsa E 33 F6
Ainzón E 41 E9
Airaines F 18 E4
Airaksela FIN 124 E8
Airasca I 31 F11
Aird Asaig UK 2 K3
Airdrie UK 5 D9
Aire-sur-l'Adour F 32 C5
Aire-sur-la-Lys F 18 C5
Airidh a'Bhruaich UK 2 J3
Airola I 60 A3
Airole I 37 D7
Airolo CH 71 D7
Airvault F 28 B5
Aisey-sur-Seine F 25 E12
Aïssey F 26 F5
Aisymi GR 171 B9
Aisy-sur-Armançon F 25 E11
Aitamännikkö FIN 117 D12
Aita Mare RO 153 F7
Aiterhofen D 75 E12
Aith UK 3 E14
Aith UK 3 G11
Aitolahti FIN 127 B10
Aitoliko GR 174 C3
Aiton RO 152 D3
Aitona E 42 E4
Aitoo FIN 127 C11
Aitrach D 71 B10
Aitrang D 71 B11
Aittaniemi FIN 121 B10
Aittijoki FIN 113 D17
Aittojärvi FIN 119 D17
Aittojärvi FIN 123 C16
Aittokoski FIN 124 C8
Aittokylä FIN 121 E10
Aittoperä FIN 123 C13
Aittovaara FIN 121 D13
Aiud RO 152 E3
Aiviekste LV 135 C11
Aix-en-Othe F 25 D10
Aix-en-Provence F 35 C9
Aixe-sur-Vienne F 29 D8
Aix-les-Bains F 31 D8
Aizenay F 28 B2
Aizkraukle LV 135 C10
Aizpún E 32 E2
Aizpurve LV 135 C12
Aizpute LV 134 C3
Aizvīķi LV 134 D3
Ajaccio F 37 H9
Ajanki FIN 117 E12
Ajankijärvi FIN 117 E12
Ajat F 29 E8
Ajaureforsen S 109 E10
Ajdovščina SLO 73 E8
Ajka H 149 B9
Ajo E 40 B4
Ajofrín E 46 E5
Ajos FIN 119 C13
Akācijas LV 134 C6
Åkarp S 87 D12
Äkäsjokisuu FIN 117 D11
Äkäslompolo FIN 117 C12
Akasztó H 150 D3
Akçaova TR 181 B7
Akçasusurluk TR 173 D9
Akeld UK 5 D12
Aken D 79 C11
Åkerbränna S 107 D11
Åkerby S 99 B9
Åkerholmen S 118 C6
Åkersberga S 99 D10
Akersjön S 105 D16
Åkers styckebruk S 98 D8
Åkerstrømmen N 101 C14
Akheloy BG 167 D9
Akhremawtsy BY 133 E2
Akhtopol BG 167 E9
Akkan S 109 F12
Akkarfjord N 113 B12
Akkarvik N 112 C5
Akkasæter N 111 B16
Akkavare S 109 E17
Akkerhaugen N 95 D10
Akköy TR 171 E10
Akköy TR 177 E9
Akkrum NL 16 B5
Akmeņdziras LV 134 B3
Akmenė LT 134 D5
Åknes N 110 C9
Aknīste LV 135 D11
Akonkoski FIN 121 F13
Akonpohja FIN 125 D10
Akpinar TR 173 B10
Akraifnio GR 175 C7
Åkran N 105 D12
Akrata GR 174 C5
Åkrehamn N 94 D2
Akrini GR 169 D6
Akrolimni GR 169 C7
Akropotamos GR 170 C6
Akrotiri GR 179 C9
Aksakal TR 173 D9
Aksaz TR 173 D7
Aksdal N 94 D2
Aksnes N 104 F4
Akujärvi FIN 114 F3
Åkullsjön S 118 F5
Åkvisslan S 107 E13
Ål N 101 E9
Ala I 69 B11
Ala S 93 E13
Alacaat TR 173 E10
Alacant E 56 E4
Alacaoğlu TR 173 B7
Alaçati TR 177 C7
Alà dei Sardi I 64 B3
Ala di Stura I 31 E11
Alaejos E 39 F9
Alagna Valsesia I 68 B4
A Lagoa E 38 C2
Alagoa P 44 F5
Alagón E 41 E9
Alahärmä FIN 122 D9
Ala-Honkajoki FIN 126 B7
Alaigne F 33 D10
Alaior E 57 B13
Alájar E 51 D6
Alajärvi FIN 117 E15
Alajärvi FIN 121 E12
Alajärvi FIN 123 D11
Alajõe EST 132 C1
Ala-Jokikylä FIN 119 C14
Ala-Kääntä FIN 119 F12
Ala-Keyritty FIN 125 D10
Alakurtti RUS 115 E8
Alakylä FIN 117 D13
Alakylä FIN 119 D15
Alakylä FIN 126 B6
Ala-Livo FIN 119 D17
Alameda E 53 B7
Alameda de Cervera E 47 F6
Alameda de la Sagra E 46 D5
Alamedilla E 55 D6
Alamillo E 54 B3
Alan HR 67 B10
Ala-Nampa FIN 117 E16
Alanäs S 106 C9
Ālande LV 134 C2
Alandroal P 50 B5
Ålandsbro S 103 A14
Alange E 51 B7
Alaniemi FIN 119 C14
Alanís E 51 C8
Alanta LT 135 F10
Alap H 149 C11
Alapitkä FIN 124 D9
Alaquàs E 48 F4
Alaranta FIN 119 D14
Alaraz E 45 C10
Alarcón E 47 E8
Alar del Rey E 40 C3
Alaró E 49 E10
Alarup AL 168 C4
Ålåsen S 106 D7
Alaskylä FIN 127 B8
Alassio I 37 C8
Alastaipale FIN 123 F11
Alastaro FIN 126 D8
Ala-Sydänmaa FIN 123 B14
Alata F 37 H9
Ala-Temmes FIN 119 E15
Alatoz E 47 F10
Alatri I 62 D4
Alatskivi EST 131 D14
Alattyán H 150 C5
Ala-Valli FIN 123 F9
Alavattnet S 106 C9
Alavere EST 131 C10
Ala-Vieksi FIN 121 F12
Alavieska FIN 119 F12
Ala-Viirre FIN 123 B11
Ala-Vuokki FIN 121 E13
Ala-Vuotto FIN 119 D16
Alavus FIN 123 E11
Alba I 37 B8
Alba Adriatica I 62 B5
Albac RO 151 E10
Albacete E 55 A9
Albacken S 103 A11
Alba de Tormes E 45 C9
Ålbæk DK 90 D7
Albagiara I 64 D2
Albaida E 56 D4
Alba Iulia RO 152 E3
Albaladejo E 55 B7
Alba-la-Romaine F 35 A8
Albalate de Arzobispo E 42 E3
Albalate de Cinca E 42 D4
Albalate de las Nogueras E 47 D8
Albalate de Zorita E 47 D7
Albalatillo E 42 D3
Alban F 33 C10
Albánchez E 55 E8
Albanella I 60 C4
Albanitsa BG 165 C10
Albano di Lucania I 60 B6
Albano Laziale I 62 D3
Albano Vercellese I 68 C5
Albanyà E 43 C9
Albaredo per San Marco I 69 A8
Albaret-le-Comtal F 30 F3
Albareto I 69 E8
Albaret-Ste-Marie F 30 F3
Albaron F 35 C7
Albarracín E 47 D10
Albatana E 55 B9
Albatàrrec E 42 D5
Albatera E 56 E3
Albbruck D 27 E9
Albelda E 42 D4
Albelda de Iregua E 41 D7
Albella E 32 F5
Albendea E 47 C8
Albendín E 53 A8
Albenga I 37 C8
Albeni RO 160 C3
Albens F 31 D8
Albentosa E 48 D3
Ålberga S 93 B9
Alberga S 98 D6
Albergaria-a-Velha P 44 C4
Albergaria dos Doze P 44 E3
Albergen NL 183 A9
Alberic E 48 F4
Alberndorf in der Riedmark A 77 F6
Albernoa P 50 D4
Albero Alto E 41 D11
Alberobello I 61 B8
Alberona I 60 A4
Alberoni I 66 B5
Alberschwende A 71 C9
Albersdorf D 82 B6
Albert F 18 E6
Albertacce F 37 G9
Alberta Ligure I 37 B10
Albertirsa H 150 C4
Albertshofen D 187 B9
Albertville F 31 D9
Alberuela de Tubo E 42 D3
Albesa E 42 D5
Albești RO 152 E5
Albești RO 153 B10
Albești RO 153 D11
Albești RO 155 F2
Albești RO 161 D10
Albeștii de Argeș RO 160 C5
Albeștii de Muscel RO 160 C6
Albești-Paleologu RO 161 D8
Albestroff F 27 C6
Albi F 33 C10
Albias F 33 B8
Albidona I 61 D6
Albignasego I 66 B4
Albina RO 155 C1
Albino I 69 B8
Albires E 39 D9
Albisheim (Pfrimm) D 21 E10
Albisola Marina I 37 C9
Albisola Superiore I 37 C9
Alblasserdam NL 16 E3
Ålbo S 98 B7
Albocàsser E 48 D5
Alboloduy E 55 E7
Albolote E 53 B9
Albon F 30 E6
Albondón E 55 F6
Alboraya E 48 E4
Alborea E 47 F10
Albota RO 160 D5
Albox E 55 E8
Albrechtice nad Orlicí CZ 77 B10
Al'brekhtava BY 133 E5
Albstadt D 27 D11
Albu EST 131 C11
Albudeite E 55 C10
Albufeira P 50 E3
Albujón E 56 F2
Albuñol E 55 F6
Albuñuelas E 53 C9
Alburquerque E 45 F7
Alby S 89 C11
Alby S 103 A9
Alby-sur-Chéran F 31 D9
Alcácer E 48 F4
Alcácer do Sal P 50 C2
Alcáçovas P 50 C3
Alcadozo E 55 B9
Alcafozes P 45 E6
Alcaine E 42 F2
Alcains P 44 E6
Alcalá de Guadaira E 51 E8
Alcalá de Gurrea E 41 D10
Alcalá de Henares E 46 D6
Alcalá del Júcar E 47 F10
Alcalá de los Gazules E 52 D5
Alcalá del Río E 51 D8
Alcalá del Valle E 51 F9
Alcalà de Xivert E 48 D5
Alcalá la Real E 53 B9
Alcalalí E 56 D4
Alcamo I 58 D2
Alcampell E 42 D4
Alcanadre E 32 F1
Alcanede P 44 F3
Alcanena P 44 F3
Alcanhões P 44 F3
Alcañices E 39 E7
Alcañiz E 42 E3
Alcántara E 45 E7
Alcantarilla E 56 F2
Alcantud E 47 C8
Alcaracejos E 54 C3
Alcara li Fusi I 59 C6
Alcaraz E 55 B8
Alcaria Ruiva P 50 D4
Alcarràs E 42 D5
Alcaucín E 53 C8
Alcaudete E 53 A8
Alcaudete de la Jara E 46 E3
Alçay-Alçabéhéty-Sunharette F 32 D4
Alcázar del Rey E 47 D7
Alcázar de San Juan E 47 F6
Alcedar MD 154 B3
Alcester UK 13 A11
Alçitepe TR 171 D10
Alcoba E 46 F4
Alcobaça P 44 E3
Alcobendas E 46 C5
Alcocer E 47 D7
Alcocero de Mola E 40 D5
Alcochete P 50 B2
Alcoentre P 44 F3
Alcoi E 56 D4
Alcolea E 53 A7
Alcolea E 55 F7
Alcolea de Calatrava E 54 B4
Alcolea de Cinca E 42 D4
Alcolea del Pinar E 47 B8
Alcolea del Río E 51 D8
Alcollarín E 45 F9
Alconchel E 51 B5
Alcóntar E 55 E7
Alcorcón E 46 D5
Alcorisa E 42 F3
Alcoroches E 47 C9
Alcossebre E 48 D5
Alcoutim P 50 E5
Alcover E 42 E6
Alcoy E 56 D4
Alcsútdoboz H 149 B11
Alcubierre E 41 E11
Alcubilla de Avellaneda E 40 E5
Alcubillas E 55 B6
Alcublas E 48 E3
Alcúdia E 57 B11
Alcudia de Guadix E 55 E6
Alcudia de Monteagud E 55 E8
Alcuéscar E 45 F8
Aldbrough UK 11 D11
Aldeacentenera E 45 E9

Aldeadávila de la Ribera E 45 B7
Aldea del Cano E 45 F8
Aldea del Fresno E 46 D4
Aldea del Obispo E 45 C7
Aldea del Rey E 51 B8
Aldea de Trujillo E 45 E9
Aldealafuente E 41 E7
Aldealpozo E 41 E7
Aldeamayor de San Martín E 39 E10
Aldeanueva de Barbarroya E 45 E10
Aldeanueva de Ebro E 41 D8
Aldeanueva de Figueroa E 45 B9
Aldeanueva de la Vera E 45 D9
Aldeanueva del Camino E 45 D9
Aldeanueva de San Bartolomé E 45 E10
Aldeaquemada E 55 C6
Aldea Real E 46 B4
Aldearrodrigo E 45 B9
Aldeaseca E 46 B3
Aldeatejada E 45 C9
Aldeavieja E 46 C4
Aldeburgh UK 15 C12
Aldehuela de la Bóveda E 45 C8
Aldehuela de Yeltes E 45 C8
Aldeia da Mata P 44 F5
Aldeia da Ponte P 45 D6
Aldeia de João Pires P 45 D6
Aldeia dos Elvas P 50 D3
Aldeia dos Fernandes P 50 D3
Aldeia dos Palheiros P 50 D3
Aldeia Velha P 44 F4
Aldenhoven D 20 C6
Aldeno I 69 B11
Alderbury UK 13 C11
Alderholt UK 13 D11
Alderley Edge UK 11 E7
Aldersbach D 76 E4
Aldershot UK 15 E7
Aldinac SRB 164 B5
Aldingham UK 10 C5
Aldomirovtsi BG 165 D6
Aldover E 42 F5
Aldridge UK 11 F8
Aldudes F 32 D3
Åle DK 86 D5
Åled S 87 B11
Aledo E 55 D9
Alekovo BG 161 F10
Alekovo BG 166 C4
Aleksandriškės LT 137 D10
Aleksandrova LV 133 E2
Aleksandrovac SRB 159 E7
Aleksandrovac SRB 163 C11
Aleksandrovo BG 165 C10
Aleksandrovo BG 166 D4
Aleksandrovo BG 167 E7
Aleksandrów PL 141 B5
Aleksandrów PL 141 H1
Aleksandrów Kujawski PL 138 E6
Aleksandrów Łódzki PL 143 C7
Aleksa Šantić SRB 150 F3
Aleksinac SRB 164 B4
Ålem S 89 B10
Ålen N 101 A14
Alençon F 23 D12
Alenquer P 44 F3
Alénya F 34 E4
Alerheim D 75 E8
Aléria F 37 G11
Alerre E 41 D11
Alès F 35 B7
Ales I 64 D2
Aleşd RO 151 C9
Alesón E 41 D6
Alessandria I 37 B9
Alessandria del Carretto I 61 D6
Alessandria della Rocca I 58 D3
Alessano I 61 D10
Ålesund N 100 B4
Alet-les-Bains F 33 D10
Alexandria GR 169 C7
Alexandria UK 4 D7
Alexandria RO 160 F6
Alexandroupoli GR 171 C9
Alexandru Vlahuță RO 153 E11
Alexeni RO 161 D9
Alexsandrów PL 144 C6
Alezio I 61 C10
Alf D 21 D8
Alfacar E 53 B9
Alfafar E 48 F4
Alfaiates P 45 D7
Alfajarín E 41 E10
Alfambra E 42 F1
Alfambras P 50 E2
Alfamén E 41 F9
Alfândega da Fé P 39 F6
Alfántega E 42 D4
Alfarim P 50 C1
Alfaro E 41 D8
Alfarràs E 42 D5
Alfatar BG 161 F10
Alfdena I 62 D6
Alfeizerão P 44 E2
Alfeld (Leine) D 78 C6
Alfena P 44 B3
Alferce P 50 E3
Alfonsine I 66 D5
Alford UK 3 L11
Alford UK 11 E12
Alforja E 42 E5
Alfredshem S 107 E15
Alfreton UK 11 E9
Alfta S 103 D11
Alfundão P 50 C3
Algaida S 57 B10
Algajola E 37 F9
Algámitas E 53 B6
Algar E 52 C5
Ålgarås S 92 B4
Ålgård N 94 E3
Algarinejo E 53 B8
Algarrobo E 53 C8
Algatocín E 53 D6
Algeciras E 53 D6
Algemesí E 48 F4
Ålgered S 103 B12
Algermissen D 79 B6
Algerri E 42 D5
Algestrup DK 87 E10
Algete E 46 C6
Alghero I 64 B1
Ålghult S 89 A9
Alginet E 48 F4
Ålgnäs S 103 D12

Algodonales E 51 F9
Algodor P 50 D4
Algora E 47 C7
Algorta E 40 B6
Algoso P 39 F6
Algoz P 50 E3
Algrange F 20 F6
Alguaire E 42 D5
Alguazas E 56 E2
Algueña E 56 E3
Algutsrum S 89 B11
Alhabia E 55 F7
Alhama de Almería E 55 F7
Alhama de Aragón E 41 F8
Alhama de Granada E 53 B9
Alhama de Murcia E 55 D10
Alhambra E 55 B6
Alhamn S 118 C7
Alhaurín de la Torre E 53 C7
Alhaurín el Grande E 53 C7
Alhendín E 53 B9
Alhojärvi FIN 127 B13
Alhóndiga E 47 C7
Ålhult S 92 D7
Alía E 45 F10
Alia I 58 D4
Aliaga E 42 F2
Aliağa TR 177 B8
Aliaguilla E 47 E10
Aliano I 60 C6
Aliartos GR 175 C7
Alibunar SRB 159 C6
Alicante E 56 E4
Alicún de Ortega E 55 D6
Alife I 60 A2
Alija del Infantado E 39 D8
Alijó P 38 F5
Alikianos GR 178 E6
Alikylä FIN 123 C11
Aliman RO 155 E1
Alimena I 58 D4
Aliminusa I 58 D4
Alimpeşti RO 160 C3
Alingsås S 91 D12
Alino BG 165 E6
Alionys 1 LT 137 C11
Aliseda E 45 F7
Alistrati GR 170 B5
Ali Terme I 59 C7
Aliveri GR 175 C9
Alixan E 35 A8
Alizava LT 135 E10
Aljaraque E 51 E5
Aljezur P 50 E2
Ajinovići SRB 163 C8
Aljubarrota P 44 E3
Aljucén E 51 A7
Aljustrel P 50 D3
Alken B 19 C11
Alkkia FIN 122 F9
Alkmaar NL 16 C3
Allai I 64 D2
Allaire F 23 E7
Allambres AL 168 C2
Allan F 35 A8
Allanche F 30 E2
Allariz E 38 D4
Allarmont F 27 D7
Allassac F 29 E8
Allaži LV 135 B9
Allažmuiža LV 135 B9
Alle CH 27 F7
Alleghe I 72 D5
Allègre F 30 E4
Alleins F 35 C9
Allemond F 31 E9
Allen IRL 7 F9
Allendale Town UK 5 F11
Allendorf (Eder) D 21 B11
Allenheads UK 5 F12
Allensbach D 27 E11
Allentsteig A 77 E8
Allenwood IRL 7 F9
Allepuz E 42 F2
Allerona I 62 B1
Allersberg D 75 D9
Allershausen D 75 F10
Allerslev DK 87 E10
Allevard F 31 E9
Allex F 30 F6
Allibaudières F 25 C11
Alligny-en-Morvan F 25 F11
Allihies IRL 8 E2
Allingåbro DK 86 C6
Allinges F 31 C9
Allinge-Sandvig DK 89 E7
Alliste I 61 D10
Allmendingen D 74 F6
Allo E 32 E1
Allogny F 25 F7
Ålloluokta S 109 B17
Allonby UK 5 F10
Allonnes F 23 E12
Allonnes F 23 F12
Allons F 32 B5
Alloue F 29 C7
Alloza E 42 F2
Allschwil CH 27 E8
Allsta S 103 A13
Allstedt D 79 D9
Allumiere I 62 C1
Alluy F 30 A4
Almacelles E 42 D4
Almáchar E 53 C8
Almada P 50 B1
Almadén E 54 B3
Almadén de la Plata E 51 D7
Almadenejos E 54 B3
Almagro E 54 B5
Almaji RO 160 E3
Almajano E 41 E7
Almaluez E 41 F7
Almansa E 56 D2
Almansil P 50 E3
Almanza E 39 C9
Almaraz E 45 E9
Almargen E 53 B6
Almarza E 41 E7
Almaş RO 151 E9
Almásfüzitő H 149 A10
Almaşu RO 151 D11
Almaşu Mare RO 151 E11
Almazán E 41 F6
Almazora E 48 E4
Almazul E 41 E7
Almedina E 55 B7
Almedinilla E 53 B8
Almeida de Sayago E 39 F7

Almelo NL 17 D7
Almenar E 42 D5
Almenara E 48 E4
Almenar de Soria E 41 E7
Almendra E 45 B6
Almendra P 45 B6
Almendral E 51 B6
Almendralejo E 51 B7
Almendricos E 55 E9
Almendros E 47 E7
Almere NL 16 D4
Almería E 55 F8
Almese I 31 E11
Almind DK 86 D4
Almkerk NL 16 E3
Álmo N 104 E4
Almodôvar P 50 D3
Almodóvar del Campo E 54 B4
Almodóvar del Pinar E 47 E9
Almodóvar del Río E 53 A6
Almofala P 44 C5
Almogía E 53 C7
Almograve P 50 D2
Almoguera E 47 D7
Almoharín E 45 F8
Almonacid de la Sierra E 41 F9
Almonacid del Marquesado E 47 E7
Almonacid del Zorita E 47 D7
Almonacid de Toledo E 46 E5
Almonaster la Real E 51 D6
Almondsbury UK 13 B9
Almonte E 51 E6
Almoradí E 56 E3
Almorox E 46 D4
Almsele S 107 C12
Almsjönäs S 107 E14
Älmsta S 99 C11
Almstedt D 79 B6
Almudema E 55 C9
Almudévar E 41 D10
Almuñécar E 53 C9
Almunge S 99 C10
Almuniente E 41 E11
Almünster A 73 A9
Álmuradiel E 55 B6
Almussafes E 48 F4
Almvik S 93 D9
Almyropotamos GR 175 C9
Almyros GR 175 A6
Alnes N 100 B4
Alness UK 3 K8
Alnwick UK 5 E13
Alobras E 47 D10
Alocén E 47 C7
Aloja LV 131 F9
Aloľ' RUS 133 D6
Alonia GR 169 D8
Álora E 53 C7
Alós d'Ensil E 33 E8
Alosno E 51 D5
Alost B 19 C9
Alost B 182 D4
Alovera E 47 C6
Alozaina E 53 C7
Alp E 33 F9
Alpalhão P 44 E5
Alpanseque E 41 F6
Alpbach A 72 A4
Alpedrinha P 44 D6
Alpedriz P 44 E3
Alpen D 17 E7
Alpenrod D 21 C9
Alpens E 43 C8
Alpera E 56 D2
Alphen NL 16 E3
Alphen aan den Rijn NL 16 D3
Alpiarça P 44 F3
Alpirsbach D 27 D9
Alpnach CH 70 D6
Alpua FIN 119 F14
Alpuente E 47 E11
Alpullu TR 173 B7
Alquería de la Condesa E 56 D4
Alquerías E 56 E2
Alqueva P 50 D3
Alquife E 55 E6
Alrance F 34 B4
Alrewas UK 11 F8
Als DK 86 B6
Alsager UK 11 E7
Alsån S 119 B9
Alsbach D 187 B6
Alsdorf D 20 C6
Alsédžiai LT 134 D4
Alsen S 105 E15
Alsenborn D 21 F9
Alseno I 69 D8
Alsenz D 21 E9
Alsfeld D 21 C12
Alsheim D 185 E9
Alsike S 99 C9
Alsjö S 103 D10
Alslev DK 86 D2
Alsleben (Saale) D 79 C10
Alslev DK 86 D2
Alsmo N 100 D6
Alsónémedi H 150 C3
Alsópáhok H 149 C8
Alsos N 108 A8
Alsódász H 145 G2
Alsószolca H 145 G2
Ålsrode DK 87 C7
Alstad N 110 E9
Alstahaug N 108 E3
Alstakan S 97 C8
Alster S 97 D10
Alsterbro S 89 B9
Alstermo S 89 B9
Alsting F 186 C2
Alston UK 5 F11
Alsunga LV 134 C3
Alsvåg N 110 C9
Alsvik N 108 B7
Alsviki LV 133 B1
Älta N 113 D11
Älta S 99 D10
Altafulla E 43 E6
Altamura I 61 B8
Altare I 37 C8
Altatornio FIN 119 C12
Altaussee A 73 A8
Altavilla Irpina I 60 B4
Altavilla Silentina I 60 B4
Altdöbern D 80 C6
Altdorf CH 71 D7
Altdorf D 75 E11

Altdorf bei Nürnberg D 75 D9
Alt Duvenstedt D 82 B7
Alte P 50 E3
Altea E 56 D4
Altefähr D 84 B4
Alteglofsheim D 75 E11
Alteidet N 112 C9
Altena D 17 F9
Altenahr D 21 C7
Altenau D 79 C7
Altenberg D 80 E5
Altenberge D 17 D8
Altenbruch-Westerende D 17 A11
Altenbuch D 74 C5
Altenburg D 79 E11
Altendiez D 21 D9
Altenfeld D 75 A8
Altenfelden A 76 F5
Altenglan D 21 E8
Altenhagen D 83 B11
Altenheim D 27 D8
Altenhof D 84 E5
Altenholz D 83 B8
Altenkirchen D 84 A4
Altenkirchen (Westerwald) D 21 C9
Altenkrempe D 83 B9
Altenkunstadt D 75 B9
Altenmarkt an der Triesting A 77 F9
Altenmarkt bei Sankt Gallen A 73 A10
Altenmarkt im Pongau A 73 B7
Altenmedingen D 83 D9
Altenmünster D 75 F8
Altenstadt D 21 D11
Altenstadt D 71 B11
Altenstadt D 71 B11
Altensteig D 27 D10
Altentreptow D 84 C4
Alter do Chão P 44 F5
Altes Lager D 80 B4
Altevik N 111 C12
Altfraunhofen D 75 F11
Altheim A 76 F4
Altheim D 27 D11
Altheim (Alb) D 187 D9
Althengstett D 187 D6
Althofen A 73 C9
Altier F 35 B6
Altimir BG 165 B8
Altina RO 152 F4
Altindag TR 177 C9
Altinoluk TR 172 E6
Altinova TR 177 A8
Altintaş TR 181 A8
Altkirch F 27 E7
Alt Krenzlin D 83 D10
Altlandsberg D 80 A5
Altleiningen D 21 E10
Altmannstein D 75 E10
Altnaharra UK 2 J8
Altofonte I 58 C3
Altomonte I 60 D6
Altomünster D 75 F9
Alton UK 11 F8
Alton UK 15 E7
Altopascio I 66 E2
Altorricón E 42 D4
Altötting D 75 F12
Altrich D 185 E6
Altrincham UK 11 E7
Alt Ruppin D 84 E3
Altsasu E 32 E1
Alt Schwerin D 83 C12
Altshausen D 71 B9
Altstätten CH 71 C9
Alttäjärvi S 116 C5
Altura E 48 E4
Altusried D 71 B10
Alu EST 131 C9
Aluatu MD 154 F3
Alūksne LV 133 B2
Ålund S 118 D6
Aluniş S 99 B10
Aluniş RO 152 C5
Aluniş RO 152 D5
Aluniş RO 161 C7
Alunu RO 160 C3
Alustante E 47 C9
Alva S 93 E12
Alvaiázere P 44 E4
Alvajärvi FIN 123 D14
Alvalade P 50 D3
Alvaneu CH 71 D9
Alvängen S 91 D11
Alvarenga P 44 C4
Álvares P 44 E4
Álvaro P 44 E5
Alverca da Beira P 45 C6
Alverca P 171 B10
Alvesta S 88 B7
Alvestad N 111 C11
Alvestorp S 97 C12
Alvettula FIN 127 C11
Alvhem S 91 C11
Ålvho S 102 D8
Alviano I 62 B2
Alvie UK 3 L9
Alvignac F 29 F9
Ålvik N 94 B4
Alvik S 103 E9
Alvik S 118 C5
Alvito I 62 D5
Alvito P 50 C4
Älvkarleby S 103 E13
Älvkarleö S 99 A8
Alvnes N 95 B11
Alvnes N 110 E9
Alvocó da Serra P 44 D5
Álvros S 102 B8
Älvsbyn S 118 C6
Älvsered S 87 A11
Ålvund N 101 A9
Ålvundeid N 101 A9
Alwernia PL 143 F8
Alyki GR 174 A3
Alyth UK 5 B10
Alytus LT 137 E9
Alzenau in Unterfranken D 21 D12
Alzey D 21 E10
Alzira E 48 F4
Alzon F 34 C5
Alzonne F 33 D10
Ămădalen S 102 D8

Amadora P 50 B1
Amaiur-Maia E 32 D3
Åmål S 91 A12
Åmål S 91 B12
Amalfi I 60 B3
Amaliada GR 174 D3
Amaliapoli GR 175 A6
Amalo GR 177 D6
Amance F 26 E5
Amancey F 31 A9
Amandola I 62 B4
Amantea I 60 E6
Amara RO 161 D10
Amarante P 38 F3
Amaranto GR 168 E5
Amărăştii RO 160 F4
Amărăştii de Jos RO 160 F4
Amărăştii de Sus RO 160 F4
Amareleja P 51 C5
Amares P 38 E3
Amaroni GR 169 B9
Amaru RO 161 D9
Amarynthos GR 175 C8
Amaseno I 62 E4
Amatrice I 62 B4
Amaxades GR 171 B8
Amay B 19 C11
Ambasaguas E 39 C9
Ambazac F 29 C9
Ambel E 41 E9
Amberg D 75 D10
Ambergate UK 11 E9
Ambérieu-en-Bugey F 31 D7
Amberloup B 19 D12
Ambert F 30 E4
Ambès F 28 E4
Ambierle F 30 D4
Ambillou F 24 F3
Ambjörby S 97 B9
Ambjörnarp S 91 E13
Ambla EST 131 C11
Amblainville F 24 B7
Amble UK 5 E13
Ambleside UK 10 C6
Ambleteuse F 15 F12
Amboise F 24 F4
Ambon F 22 E6
Ambrault F 29 B9
Ambrières-les-Vallées F 23 D10
Ambronay F 31 C7
Ambrosden UK 13 B12
Åmdals Verk N 95 D8
Ameide NL 182 B5
Amele U 16 E5
Amelia I 62 B2
Amélie-les-Bains-Palalda F 34 F4
Amelinghausen D 83 D8
Amêndoa P 44 E4
Amendoeira P 50 D4
Amendolara I 61 D7
Amer E 43 C9
Amerang D 75 G11
A Merca E 38 D4
Amerongen NL 183 B6
Amersfoort NL 16 D4
Amersham UK 15 D7
Amesbury UK 13 C11
Amezketa E 32 D1
Amfikleia GR 175 B6
Amfilochia GR 174 B3
Amfipoli GR 169 C10
Amfissa GR 174 B5
Amieira P 50 C4
Amiens F 18 E5
Amilly F 25 E8
Åminne FIN 122 E7
Åminne S 87 A14
Åmland N 94 F5
Åmli N 90 A3
Åmli N 90 B3
Amlwch UK 10 E3
Ämmälänkylä FIN 123 E9
Ammanford UK 12 B7
Ämmänsaari FIN 121 E12
Ammarnäs S 109 F11
Ämmeberg S 92 B6
Ammenäs S 91 C10
Ammerbuch D 187 D6
Ammern D 79 D7
Ammersbek D 83 C8
Ammerzoden NL 183 B6
Ammochori GR 168 C5
Ammotopos GR 168 F4
Ammoudia GR 168 B9
Amnatos GR 178 E8
Amnéville F 186 C1
Amoeiro E 38 D4
Amonde P 38 E2
Amorbach D 21 E12
Amorebieta E 40 B6
Amorgos GR 177 F6
Amori GR 171 B10
Amorosi I 60 A3
Amos GR 177 F6
Amou F 32 C4
Åmøyhamn N 108 C5
Ampelakia GR 169 D8
Ampeleia GR 169 F7
Ampelokipoi GR 169 C8
Ampelonas GR 169 E7
Ampezzo I 73 D6
Ampfing D 75 F11
Ampflwang im Hausruckwald A 76 F5
Amplepuis F 30 D5
Amposta E 42 F5
Ampthill UK 15 C8
Ampudia E 39 E10
Ampuero E 40 B5
Ampus F 36 D4
Amriswil CH 27 E11
Amrosia GR 171 B8
Amstelveen NL 16 D3
Amsterdam NL 16 D3
Amstetten A 77 F7
Amstetten D 187 D8
Amulree UK 5 B9
Amurrio E 40 B6
Amusco E 40 D3
Amvrosia GR 171 B8
Amygdaleonas GR 171 C6
Amygdalia GR 174 C5
Amýntaio GR 169 C6

Amyntaio GR 169 C6
Amzacea RO 155 F2
Anacapri I 60 B2
Anadia P 44 D4
Anafi GR 180 D3
Anafonitria GR 174 D2
Anagaire IRL 6 B6
Anagni I 62 D4
Ánan'yiv UA 154 B5
Anarcs H 145 G5
Anargyroi GR 169 C6
Anarrachi GR 169 D6
Ånåset S 118 F6
Åna-Sira N 94 F4
Anatazavė LT 135 E11
Anatoli GR 168 E4
Anatoli GR 169 E8
Anatoli GR 179 E10
Anatoliki Fragkista GR 174 B4
Anatoliko GR 169 C6
Anatoliko GR 169 C8
Anavatos GR 177 C7
Anavra GR 174 A5
Anavra GR 175 A6
Anavyssos GR 175 D8
Anaya de Alba E 45 C10
An Baile Breac IRL 8 E2
An Baile Nua IRL 8 D5
An Bhlarna IRL 8 E5
An Bun Beag IRL 6 B6
An Cabhán IRL 7 E8
An Caiseal IRL 6 E3
An Caisleán Riabhach IRL 6 E6
An Caisleán Nua IRL 7 F10
An Caisleán Nua IRL 8 D4
An Cathair IRL 9 D7
Ance LV 134 A4
Ancelle F 36 B4
Ancene LV 135 D11
Ancenis F 23 F9
Ancerville F 26 C3
An Charraig IRL 6 C5
An Cheathru Rua IRL 6 F3
An Chill IRL 9 F9
Ancín E 32 E1
An Cionn Garbh IRL 6 C5
An Clochán IRL 7 C7
An Clochán Liath IRL 6 B6
An Cóbh IRL 8 E6
An Coimín IRL 7 C7
An Coireán IRL 8 E2
Ancona I 67 E8
Ancroft UK 5 D13
Ancrum UK 5 D11
Ancy-le-Franc F 25 E11
Anda N 100 C4
Andalo I 69 A11
Åndalsnes N 100 A7
Andau A 149 A8
Andavías E 39 E8
Anddalsvågen N 108 E3
Andebu N 95 D12
Andechs D 75 G9
Andeer CH 71 D8
Andelfingen CH 27 E9
Andelot-Blancheville F 26 D3
Andelst NL 16 E5
Andenes N 111 B11
Andenne B 19 D11
Anderlecht B 19 C9
Anderlues B 19 D9
Andermatt CH 71 D7
Andernach D 185 D7
Andernos-les-Bains F 28 F3
Andersbo S 99 B9
Anderslöv S 87 E12
Anderstorp S 88 A5
Andervenne D 17 C9
Andfiskå N 108 D7
Andijk NL 16 C4
Andilly-en-Bassigny F 26 E4
Andlau F 27 D7
Andoain E 32 D1
Andocs H 149 C9
Andolsheim F 27 D7
Andorf A 76 F5
Andørja N 111 C13
Andorra E 42 F2
Andorra la Vella AND 33 E9
Andosilla E 32 F2
Andover UK 13 C12
Andrano I 61 D10
Andrăşeşti RO 161 D10
Andratx E 49 E9
Andravida GR 174 D3
Andreas GBM 10 C3
Andreiaşu de Jos RO 153 F9
Andrespol PL 143 C8
Andrest F 33 D6
Andretta I 60 B4
Andria I 60 A6
Andrid RO 151 B10
Andrieşeni RO 153 B10
Andrijaševci HR 157 B10
Andrijevica MNE 163 D8
Andrioniškis LT 135 E10
Andriši LT 136 D6
Andritsaina GR 174 E4
Andronianoi GR 175 B9
Andros GR 176 D4
Androusa GR 174 E4
Andrup DK 86 D3
Andrupene LV 133 D2
Andrychów PL 147 B8
Andrzejewo PL 141 E6
Andselv N 111 B15
Andrup DK 86 D3
An Dúchoraidh IRL 6 C6
Andújar E 53 A8
Anduze F 35 B6
Andželi LV 133 D3
An Eachléim IRL 6 D2
Aneby S 92 D5
Åneby N 95 B13
Anela I 64 C3
Anenii Noi MD 154 D4
Anes F 24 C5
Anet F 24 C5
Anevo BG 165 D10
Aneza GR 174 A2
An Fál Carrach IRL 6 B6
An Fear Bán IRL 7 F7
An Fhairche IRL 6 E4

Ånge S 109 E14
An Geata Mór IRL 6 D2
Ångebo S 103 C11
Angeja P 44 C3
Ångelholm S 87 C11
Angeli FIN 113 F16
Angelniemi FIN 127 E8
Angelochori GR 169 C7
Angelokastro GR 174 B3
Angelokastro GR 175 D7
Ångelsberg S 97 C15
Angelstad S 87 B13
Ångely F 25 E11
Anger D 73 A6
Angera I 68 B6
Angermünde D 84 D6
Angern an der March A 77 F11
Angers F 23 F10
Ångersjö S 102 C8
Ångersjö S 122 C3
Angersnes N 108 E3
Angervikko FIN 125 D10
Angerville F 24 D7
Ångesån S 116 E9
Ångeslevä FIN 119 E15
Ångesträsk S 118 B8
Anghiari I 66 E5
Anglade F 28 E4
Angle UK 9 E12
Anglès F 34 C4
Anglès F 34 C4
Anglesola E 42 D5
Angles-sur-l'Anglin F 29 B7
Anglet F 32 D2
Angliers F 29 B6
Anglure F 25 C10
Angnäs S 107 D16
Angoisse F 29 E8
Ångom S 103 B13
An Gort IRL 6 F5
Angoulême F 29 D6
Angoulins F 28 C3
Angri I 60 B3
Angrie F 23 E10
Angüés E 42 C3
Anguiano E 40 D6
Anguillara Sabazia I 62 C2
Anguillara Veneta I 66 B4
Anguita E 47 B7
Angvik N 100 A8
Anha P 38 E2
Anhée B 19 D10
Anholt DK 87 B9
Aniane F 35 C6
Aniche F 19 D7
Anif A 73 A7
Anina RO 159 C8
Aninoasa RO 160 C2
Aninoasa RO 160 C5
Aninoasa RO 160 D2
Aninoasa RO 161 C6
Anizy-le-Château F 19 E7
Anjalankoski FIN 128 C6
Anjan S 105 D13
Anjum NL 16 B6
Ankaran SLO 73 E8
Ankarede Kapell S 105 B16
Ankarsrum S 93 D8
Ankarsund S 109 F12
Ankarsvik S 103 B13
Ankenes N 111 D13
Ankerlia N 112 E6
Ankershagen D 84 D3
Anklam D 84 C5
Ankum D 17 C9
Anlaby UK 11 D11
An Leadhb Gharbh IRL 6 C5
Anlezy F 30 A4
An Longfort IRL 7 E7
An Mám IRL 6 E3
An Mhala Raithní IRL 6 E3
An Móta F 7 F7
An Muileann gCearr IRL 7 E8
Ånn S 105 E13
Anna E 56 D3
Anna EST 131 C11
Annaberg A 77 F9
Annaberg-Buchholtz D 80 E4
Annaburg D 80 C4
Annagassan IRL 7 E10
Annagry IRL 6 B6
Annahilt UK 7 D10
Annahütte D 80 C5
Annalong UK 7 D11
Annamoe IRL 7 F10
Annan UK 5 F10
Anna Paulowna NL 16 C3
Annarode D 79 C9
An Náis IRL 7 F9
Annayalla IRL 7 D9
Annbank UK 4 E7
Anneberg S 91 D11
Anneberg S 92 D5
Annecy F 31 D9
Annecy-le-Vieux F 31 D9
Annefors S 103 D11
Annelund S 91 D13
Annemasse F 31 C9
Annen NL 17 B7
Annenieki LV 134 C6
Annestown IRL 9 D8
Anneyron F 30 E6
Annfield UK 5 F13
Annikvere EST 131 C12
Annœullin F 182 D1
Annonay F 30 E6
Annonen FIN 119 F14
Annopol PL 144 B5
Annot F 36 D5
Annweiler am Trifels D 21 F9
Ano Agios Vlasios GR 174 B4
Andújar E 53 A8
Anduze F 35 B6
Ano Amfeia GR 174 E5
Ano Chora GR 174 B4
Ano Diakofto GR 174 C5
Ano Drosini GR 171 B9
Anoeta E 32 D1
Anogeia GR 174 C5
Anogeia GR 178 E8
Ano Kalliniki GR 168 C5
Ano Kardamyla GR 177 B7
Ano Kavallari GR 169 C10
Ano Komi GR 169 D6
Ano Kopanaki GR 174 E4
Ano Lechonia GR 169 F9
Ano Lefkimmi GR 168 F3
Ano Mera GR 176 E5
Añón E 41 E8
Ano Poroia GR 169 B9
Anor F 19 E9
Añora E 54 C3
Ano Sagkri GR 176 E5
Ano Steni GR 175 B8

Ano Syros GR 176 E4
Ano Trikala GR 174 D5
Ano Tyros GR 175 E6
Anould F 27 D6
Añover de Tajo E 46 E5
Ano Vrontou GR 169 B10
Anquela del Ducado E 47 C8
Anras A 72 C6
An Rinn IRL 9 D7
Anröchte D 185 A9
An Ros IRL 7 E10
Ans B 19 C12
Ans DK 86 C5
Ansac-sur-Vienne F 29 D7
Ansager DK 86 D3
Ansbach D 75 D8
An Sciobairín IRL 8 E4
Anse F 30 D6
Ansedonia I 65 C4
Anseremme B 19 D10
Ansfelden A 76 F6
Ansião P 44 E4
Ansizca TR 177 C9
Ansjö S 103 A11
Ansnes N 111 B15
Ansó E 32 E4
An Spidéal IRL 6 F4
Anspoki LV 135 D13
Ansprung D 80 E4
Ånstad N 110 C8
Anstruther UK 5 C11
Ånsvik N 109 B9
Antaliepté LT 135 E11
Antanhol P 44 D4
An tAonach IRL 8 C6
Antas E 55 E9
Antas P 38 E2
Antas P 44 C6
Antas de Ulla E 38 C4
An Teampall Mór IRL 9 C7
An Tearmann IRL 7 B7
Antequera E 53 B7
Anterselva di Sopra I 72 C5
Anthéor F 36 E5
Anthien F 25 F10
Anthili GR 175 B5
Anthotopos GR 175 A6
Anthrakitis GR 168 E4
Antibes F 36 D6
Antigonos GR 169 C6
Antikyra GR 175 C6
Antillo I 59 D7
Antimacheia GR 177 F9
An tInbhear Mór IRL 9 C10
Antiparos GR 176 E5
Antirrio GR 174 C4
Antisanti F 37 G10
Antissa GR 171 F9
Antjärn S 103 A14
Antnäs S 118 C7
Antoing B 19 C7
Antol SK 147 E7
Anton BG 165 D9
Anton Ivanov BG 165 E9
Antonivka UA 145 F6
Antonne-et-Trigonant F 29 E7
Antonovo BG 166 C6
Antony F 25 C7
Antraigues-sur-Volane F 30 F5
Antrain F 23 D9
Antrim UK 4 F4
Antrodoco I 62 C4
An tSeanga Mheáin IRL 6 C6
Antsla EST 131 F13
Anttis S 116 D9
Anttola FIN 128 B8
An Tulch IRL 9 C9
Antully F 30 B5
Antunovac Tenjski HR 149 F11
Antuži LV 135 C11
Antwerpen B 19 B9
Antzuola E 32 D1
An Uaimh IRL 7 E9
Anvers B 19 B9
Anvers B 182 C4
Anvin F 18 D5
Anykščiai LT 135 E10
Anzano di Puglia I 60 A4
Anzat-le-Luguet F 30 E3
Anzegem B 19 C7
Anzère CH 31 C11
Anzi I 60 B5
Anzin F 19 D7
Anzing D 75 F10
Anzio I 62 E3
Aoiz E 32 E3
Aoradh UK 4 D4
Aós E 32 E3
Aosta I 31 D11
Aouste-sur-Sye F 31 F7
Apa RO 151 B11
Apagy H 145 H4
Apahida RO 152 D3
Apaj H 150 C2
Apața RO 153 F7
Apateu RO 151 D8
Apátfalva H 150 E6
Apatin SRB 158 F3
Apatovac HR 149 D7
Apc H 150 B4
Ape LV 131 F13
Apecchio I 66 E5
A Pedreira E 38 B3
Apeiranthos GR 177 E6
Apeldoorn NL 16 D5
Apelern D 17 D12
Apele Vii RO 160 E4
Apen D 17 B9
Apenburg D 83 E10
Apensen D 82 D7
Aperi GR 181 E6
A Peroxa E 38 D4
Apgulde LV 134 C6
Apice I 60 A3
A Picota E 38 C2
Apidia GR 175 F6
Apiro I 67 F7
Aplared S 91 D14
A Pobla de Brollón E 38 C5
A Pobla de San Xiao E 38 C5
A Pobra de Trives E 38 D5
A Pobra do Caramiñal E 38 C2
Apoikia GR 176 D4
Apolakkia GR 181 D7
Apold RO 152 E5
Apolda D 79 D10
Apoldu de Jos RO 152 F3
Apollona GR 181 D7
Apollonas GR 177 E6
Apollonia GR 176 F4
A Pontenova E 38 B5
A Pontepedra E 38 B2
Áporka H 150 C3

Apostag H 149 C11
Apostoloi GR 178 E8
Appelbo S 97 B11
Appenweier D 186 D4
Appenzell CH 27 F11
Appiano sulla Strada del Vino I 72 D3
Appietto F 37 G9
Appignano I 67 F7
Appingedam NL 17 B7
Appleby-in-Westmorland UK 10 B7
Appleby Magna UK 11 F8
Applecross UK 2 L5
Appledore UK 12 C6
Appledore UK 15 E10
Apprieu F 31 E7
Apremont-la-Forêt F 26 C4
Aprica I 69 A9
Apricena I 63 D8
Aprigliano I 61 E6
Apriki LV 134 C3
Aprilia I 62 D3
Aprilovo BG 166 C5
Apriltsi BG 165 D10
A Proba E 39 C6
Apsalos GR 169 C7
Apšciems LV 134 B6
Apt F 35 C9
Aquaviva delle Fonti I 61 B7
Aquilonia I 60 B4
Aquilué E 32 F5
Aquino I 62 E5
Aracena E 51 D6
Arachnaio GR 175 D6
Arachova GR 175 C6
Aračinovo NMK 164 E4
Arad RO 151 E7
Aradac SRB 158 C5
Aradeo I 61 C10
Aragnouet F 33 E6
Aragona I 58 E4
Aragoncillo E 47 C8
Araia E 32 E1
A Ramallosa E 38 C2
Aramits F 32 D4
Aramon F 35 C8
Aranda de Duero E 40 E4
Aranda de Moncayo E 41 E8
Aranđelovac SRB 158 E6
Aranhas P 45 D6
Aranitas AL 168 C2
Aranjuez E 46 D5
Arano E 32 D2
Aranzueque E 47 D6
Araovacik TR 173 D6
Arapaj AL 168 B2
Arapiles E 45 C9
Aras E 32 E1
Aras de Alpuente E 47 E10
Araswika N 104 E4
Aratos GR 171 B9
Araules F 30 E5
Aravissos GR 169 C7
Arazede P 44 D3
Arbanasi BG 166 C5
Arbas F 33 E7
Arbatax I 64 D4
Arbeca E 42 D5
Arbecey F 26 E4
Arbedo CH 69 A7
Arbéost F 32 E5
Arberg D 75 D8
Arbesbach A 77 E7
Arbeteta E 47 C8
Arboga S 97 D14
Arbois F 31 B8
Arboleas E 55 E8
Arbon CH 27 E11
Arborea I 64 D2
Arborio I 68 C5
Arbostad N 111 C13
Arbrå S 103 D11
Arbroath UK 5 B11
Arbúcies E 43 D9
Arbus I 64 D2
Árby DK 87 D8
Árbyn S 118 C6
Arcachon F 32 A3
Arcangues F 32 D2
Arcani RO 159 C11
Arce I 62 D5
Arcen NL 16 F6
Arc-en-Barrois F 26 E3
Arces F 28 D4
Arces-Dilo F 25 D10
Arc-et-Senans F 31 A8
Arcevia I 67 E6
Archaia Epidavros GR 175 D7
Archaia Korinthos GR 175 D6
Archaia Olympia GR 174 D4
Archanes GR 178 E9
Archangelos GR 181 D8
Archar BG 159 F10
Archena E 55 C10
Archiac F 28 D5
Archidona E 53 B8
Archipoli GR 181 D8
Archiș RO 151 D9
Archivel E 55 C9
Arcidosso I 65 B5
Ārciems LV 131 F9
Arcis-sur-Aube F 25 C11
Arc-lès-Gray F 26 F4
Arco I 69 B10
Arco de Baúlhe P 38 F4
Arcola I 69 E8
Arçon F 31 B9
Arçonnay F 23 D12
Arcos E 40 C4
Arcos P 50 B4
Arcos de Jalón E 47 B8
Arcos de la Frontera E 52 C5
Arcos de las Salinas E 47 E10
Arcos de Valdevez P 38 E3
Arcozelo E 38 C5
Arcozelo P 38 E2
Arc-sur-Tille F 26 F3
Arcusa E 33 F6
Arda BG 171 B7
Ardagger Markt A 77 F7
Ardagh IRL 8 C4
Ardal N 100 C4
Ardala S 91 C14
Ardales E 53 C7
Ardara I 64 B2

Ardara IRL 6 C6
Ardauli I 64 C2
Ardea I 62 D3
Ardee IRL 7 E9
Ardeluța RO 153 D8
Arden DK 86 B5
Arden UK 4 C7
Ardentes F 29 B9
Ardeoani RO 153 D9
Ardersier UK 3 K8
Ardes F 30 E3
Ardez CH 71 D10
Ardfern UK 4 C5
Ardfert IRL 8 D3
Ardfinnan IRL 9 D7
Ardgay UK 3 K8
Ardglass UK 7 D11
Ardgroom IRL 8 E3
Ardino BG 171 A8
Ardkeen UK 7 D11
Ardlui UK 4 C7
Ardlussa UK 4 C5
Ardmair UK 2 K6
Ardminish UK 4 D5
Ardmore IRL 6 F3
Ardmore IRL 9 E7
Ardon CH 31 C11
Ardooie B 19 C7
Ardore I 59 C9
Arahrain IRL 6 F5
Ardrossan UK 4 D7
Ardscull IRL 9 F8
Ardtoe UK 4 B5
Ardu EST 131 C10
Ardud RO 151 B10
Ardusat RO 151 B11
Ardvasar UK 4 A5
Åre S 105 E14
Áre S 105 E14
Areatza E 40 B6
Arefu RO 160 C5
Areias P 44 E4
Aremark N 96 D6
Arén E 33 F6
Arenas de Iguña E 40 B3
Arenas del Rey E 53 C9
Arenas de San Juan E 46 F5
Arenas de San Pedro E 45 D10
Arendal N 90 C4
Arendonk B 16 F4
Arendsee (Altmark) D 83 E10
Arengosse F 32 B4
Arenillas E 40 F6
Arenshausen D 79 D6
Arentorp S 91 C12
Arentsminde DK 86 A5
Arenys de Mar E 43 D9
Arenzano I 37 C9
Areopoli GR 178 B3
Areosa P 38 E2
Ares E 38 B3
Arès F 28 F3
Ares del Maestre E 48 D4
Aresing D 75 E9
Arespalditza E 40 B5
Arethousa GR 169 C10
Arette F 32 D4
Arette-Pierre-St-Martin F 32 E4
Aretxabaleta E 41 B7
Arevalillo E 45 C10
Arévalo E 46 B3
Arezzo I 66 F4
Arfara GR 174 E5
Arfons F 33 D10
Árfor N 105 B11
Argalasti GR 169 F9
Argallón E 51 C9
Argamasilla de Alba E 47 F6
Argamasilla de Calatrava E 54 B4
Argamasón E 55 B8
Arganda del Rey E 46 D6
Arganil P 44 D4
Arganza E 39 C6
Árgård N 105 D9
Argegno I 69 B7
Argelès-Gazost F 32 D5
Argelès-sur-Mer F 34 E5
Argeliers F 34 D4
Argelita E 48 D4
Argenbühl D 71 B9
Argences F 23 B11
Argenta I 66 C4
Argentan F 23 C11
Argentat F 29 E9
Argente E 47 C10
Argenteuil F 25 C7
Argenthal D 21 E9
Argentiera I 64 B1
Argentière F 31 D10
Argentine I 62 E4
Argenton-du-Plessis F 23 F11
Argenton-les-Vallées F 28 B5
Argenton-sur-Creuse F 29 B9
Argentré F 23 D10
Argentré-du-Plessis F 23 E10
Argent-sur-Sauldre F 25 E7
Argés E 46 E4
Argetoaia RO 160 D2
Arginonta GR 177 E8
Argonay F 31 D9
Argos GR 175 D6
Argos Orestiko GR 168 D5
Argostoli GR 174 C1
Argouges F 23 D9
Arguedas E 41 D8
Argueil F 18 E4
Arguellite E 55 C8
Arguís E 32 F5
Arguisuelas E 47 E9
Argujillo E 39 F8
Argyrades GR 168 F2
Argyropouli GR 169 F6
Arhult S 89 A10
Århus DK 86 C6
Ariano Irpino I 60 A4
Ariano nel Polesine I 66 C5
Ariany E 57 B11
Aribe E 32 E3
Ariccia I 62 D3
Aridaia GR 169 C7
Arieșeni RO 151 E10
Arignac F 33 E9
Arija E 40 C4
Arild S 87 C11
Arileod UK 4 B3
Arilje SRB 158 F5
Arinagour UK 4 B3
Ariniș RO 151 C11
Ariño E 42 E2

Arinsal AND 33 E8
Arinthod F 31 C8
Ariogala LT 134 F6
Aris GR 174 E5
Arisaig UK 4 B5
Aristomenis GR 174 E4
Arisvi GR 171 B9
Aritzo I 64 D3
Ariz P 44 C5
Ariza E 41 F7
Arizgoiti E 40 B6
Arizola E 32 E2
Arjäng S 96 D7
Arjeplog S 109 D14
Arjona E 53 A8
Arjonilla E 53 A8
Arkala FIN 119 D16
Arkalochori GR 178 E9
Arkasa E 181 F6
Arkelstorp S 88 C6
Arkesini GR 180 C3
Arkitsa GR 175 B7
Arkkukari FIN 119 E12
Arklow IRL 9 C10
Arkna EST 131 C12
Arkösund S 93 C9
Árla S 98 D7
Arlanc F 30 E4
Arlanzón E 40 D5
Arle (Großheide) D 17 A8
Arlena di Castro I 62 C1
Arles F 35 C8
Arleuf F 25 F10
Arleux F 19 D7
Arligbo S 98 B8
Arló H 145 G1
Arlon B 19 E12
Arluno I 69 B6
Arlyeva BY 133 E6
Armação de Pêra P 50 E3
Armadale UK 5 D9
Armagh UK 7 D9
Armagnac F 32 D3
Armallones E 47 C8
Armamar P 44 B5
Armășești RO 161 D9
Armasjärvi S 119 B11
Armassaari FIN 119 B11
Armen AL 168 D2
Armenioi GR 175 A7
Armeniș RO 159 C9
Armenistis GR 177 D7
Armeno I 68 B5
Armenochori GR 168 C5
Armenoi GR 178 E7
Armenteros E 45 D9
Armentières F 18 C6
Armento I 60 C5
Armilla E 53 B9
Armissan F 34 D5
Armolia GR 177 C7
Armoy UK 4 E4
Armsheim D 185 E9
Armuña I 64 D3
Armunia E 39 C8
Armutalani TR 181 C8
Armutlu TR 173 C10
Armutlu TR 177 C10
Arna GR 174 F5
Arnac-Pompadour F 29 E8
Arnad I 68 B4
Arnage F 23 E12
Arnaia GR 169 D10
Árnäs S 91 B14
Árnäs S 102 C5
Árnäs S 102 E5
Arnäsvall S 107 E15
Arnavutköy TR 173 B10
Arnay-le-Duc F 25 F11
Arnberg S 107 B16
Arnborg DK 86 C3
Arnbruck D 76 D3
Arneburg D 83 E12
Arnedillo E 41 D7
Arnedo E 41 D7
Arnéguy F 32 D3
Arnemark S 118 D6
Arnemuiden NL 16 F1
Árnes F 24 F4
Árnes N 95 B14
Árnes N 104 F5
Árnes N 111 B15
Arnfels A 148 C4
Arnhem NL 16 E5
Arnionys I LT 137 C12
Arnissa GR 169 C6
Arnö S 93 B10
Arnol UK 2 J3
Arnold UK 11 E9
Arnoldstein A 73 C8
Arnö-viken S 103 C13
Arnoyhamn N 112 C4
Arnprior UK 5 C8
Arnsberg D 17 F10
Arnschwang D 76 D3
Arnstadt D 79 E8
Arnstein D 74 C6
Arnstorf D 76 E3
Arnum DK 86 E3
Aroania GR 174 D5
Aróktő H 151 B6
Arola FIN 113 D20
Arolla CH 31 C11
Arolsen D 17 F12
Aron F 23 D10
Arona I 68 B6
Aroneanu RO 153 C11
Aronkylä FIN 122 F8
Áros N 105 B12
Arosa CH 71 D9
Arosa P 38 E3
Árosjåkk S 111 E17
Árøybukta N 111 A19
Árøysund N 90 A7
Arpajon F 25 C7
Arpajon-sur-Cère F 29 F10
Arpașu de Jos RO 152 F5
Arpela FIN 119 B12
Arpino I 62 D5
Arquata del Tronto I 62 B4
Arques F 18 C5
Arques F 33 E10
Arques-la-Bataille F 18 E3
Arquillos E 55 C6
Arrabalde E 39 D8
Arracourt F 26 C6
Arradon F 22 E6
Arraiolos P 50 B4
Arrankoski FIN 127 C13

Arranhó P 50 B1
Arras F 18 D6
Arrasate E 41 B7
Árre DK 86 D3
Arreau F 33 E6
Arredondo E 40 B4
Arrën AL 163 F9
Árrenjarka S 109 C15
Arrentières F 25 D12
Arriach A 73 C8
Arriana GR 171 B9
Arriate E 53 C6
Arrifana P 45 C6
Arrigny F 25 C12
Arrigorriaga E 40 B6
Arrild DK 86 E3
Arriondas E 39 B9
Arroba de los Montes E 46 F3
Arrochar UK 4 C7
Arronches P 45 F6
Arrone I 62 B3
Arroniz E 32 E1
Arrou F 24 D5
Arroyal E 40 C3
Arroyo de la Luz E 45 E7
Arroyo del Ojanco E 55 C7
Arroyo de San Serván E 51 B7
Arroyomolinos de León E 51 C7
Arruda dos Vinhos P 50 B1
Arry F 186 C1
Ársandøy N 105 A12
Årsdale DK 89 E8
Arseni GR 169 C7
Ars-en-Ré F 28 C2
Årset N 105 A11
Arsiè I 72 E4
Arsiero I 69 B11
Arsière DK 86 E6
Arsoli I 62 C4
Ars-sur-Formans F 30 D6
Ars-sur-Moselle F 26 B5
Árstein N 111 C12
Årsunda S 98 A7
Arsura RO 153 D12
Arsvågen N 94 D2
Artà E 57 B11
Artajona E 32 E2
Artana E 48 E4
Artannes-sur-Indre F 24 F4
Artemisia GR 174 E5
Artemisio GR 174 D5
Artemisio GR 175 A7
Artèn I 72 D4
Artena I 62 D3
Artenay F 24 D6
Artern (Unstrut) D 79 D9
Artés E 43 D7
Artesa de Segre E 42 D6
Artesiano GR 169 F6
Arth CH 71 C7
Arthez-d'Asson F 32 D5
Arthez-de-Béarn F 32 D4
Arthon F 29 B9
Arthon-en-Retz F 23 F8
Artieda E 32 E4
Artigat F 33 D8
Artix F 32 D4
Artjärvi FIN 127 C14
Ärtled S 103 E9
Artlenburg D 83 D8
Artogne I 69 B9
Artotina GR 174 C5
Artziniega E 40 B5
A Rúa E 38 D5
Arudy F 32 D5
Aruküla EST 131 C10
Aruküla EST 131 C12
Arundel UK 15 F7
Árvågen N 104 E5
Arvagh IRL 7 E7
Arvån S 107 B16
Arvesund S 105 E16
Arveyres F 28 F5
Arvi GR 178 F9
Arvidsjaur S 109 E17
Arvidsträsk S 118 C5
Arvieu F 34 B4
Arvieux F 31 F10
Árvik N 100 B3
Arvika S 97 C8
Árviksand N 112 C4
Arvila EST 131 C14
Arzachena I 64 A3
Arzacq-Arraziguet F 32 C5
Arzana I 64 D4
Aržano HR 157 E7
Arzberg D 21 D9
Arzberg D 75 B11
Arzberg D 80 C4
Arzberg D 20 D6
Arzfeld D 20 D6
Arzignano I 72 E3
Arzl im Pitztal A 71 C11
Arzúa E 38 C3
As B 19 B12
Aš CZ 75 B11
Ås N 95 C13
Ås N 105 E11
Ås N 96 C7
Ås S 106 E7
Ås S 107 E11
Åsa N 95 B12
Åsa S 91 E11
Asaa DK 86 A6
Aşağıçobanisa TR 177 B10
Aşağıgüvendikli TR 173 B8
Åsan N 105 B12
Ašanja SRB 158 D5
Åsare LV 135 D11
Åsarna S 102 A7
Åsarp S 91 C14
Asasp-Arros F 32 D4
Asbach D 21 C8
Åsbacka S 103 D12
Åsby S 87 A10
Åsby S 92 D6
Ascain F 32 D2
Asciano I 66 F4
Aschach an der Donau A 76 F6
Aschaffenburg D 21 E12
Aschbach Markt A 77 F7
Ascheberg D 17 E9
Ascheberg (Holstein) D 83 B8
Aschères-le-Marché F 24 D7
Aschersleben D 79 C9
Aşchileu RO 151 D11
Asciano I 66 F4
Ascó E 42 E5

Asco F 37 G10
Ascoli Piceno I 62 B5
Ascoli Satriano I 60 A5
Ascona CH 68 A6
Ascou F 33 E9
As Covas E 38 D2
Åse N 111 B10
Åsebyn S 96 D7
Åseda S 89 A8
Åsegg N 105 C9
Åsele S 107 C12
Åselet S 118 C7
Åsen N 90 B1
Åsemakylä FIN 119 D14
Åsemankylä FIN 119 D14
Åsemanseutu FIN 123 E11
Åsen N 105 D10
Åsen S 102 C7
Åsen S 102 D6
Åsen S 106 C6
Åsen S 109 E16
Åsenbruk S 91 B11
Asendorf D 17 C12
Åsenovets BG 166 D5
Asenovgrad BG 165 E10
Asenovo BG 166 C6
Asenovtsi BG 165 C11
Åseral N 90 B1
Asendelft NL 16 D3
Asfeld F 19 F9
Asfendiou GR 177 F9
Asferg DK 86 B5
Asgårdstrand N 95 D12
Ash UK 15 E7
Åshammar S 103 E12
Ashbourne IRL 7 E10
Ashbourne UK 11 E8
Ashburton UK 13 D7
Ashbury UK 13 B11
Ashby de la Zouch UK 11 F9
Ashcott UK 13 C9
Ashevo RUS 133 B7
Ashford UK 13 D11
Ashford UK 15 E10
Ashford IRL 9 C10
Ashford IRL 8 D4
Ashill UK 15 B10
Ashington UK 5 E13
Ashkirk UK 5 E11
Ashley UK 15 C9
Ashmyany BY 137 D13
Ashmyany BY 137 E12
Ashton-under-Lyne UK 11 E7
Ashurst UK 13 D11
Ashwick UK 13 C9
Asiago I 72 E4
Asikkala FIN 127 C14
Asilah MA 52 F4
Asimi GR 178 E9
Asín E 32 F3
Asiros GR 169 C9
Ask N 95 B14
Aska FIN 117 D17
Askainen FIN 126 D6
Askeaton IRL 8 C5
Asker N 95 C12
Askern UK 11 D9
Askersby S 92 A6
Askersund S 92 B5
Askilje S 107 B13
Askim N 95 C14
Asklipieio GR 177 C14
Asklipieio GR 181 D7
Askogen S 118 C7
Askola FIN 127 D14
Asköping S 92 A8
Askos GR 169 C9
Askvoll N 100 D2
Aslackala FIN 113 E16
Aslestad N 94 D3
Åsli N 101 E11
Åsljunga S 87 C12
Asmali TR 173 C8
Åsmansbo S 97 B13
Åsmarka N 101 D13
Asmunti FIN 119 C17
Asnæs DK 87 D8
As Neves E 38 B3
As Neves E 38 D3
As Nogais E 38 C5
Asola I 66 B1
Asolo I 72 E4
Asopia GR 175 C8
Asopos GR 178 B4
Asos GR 174 C2
Ásotthalom H 150 E4
Asp DK 86 C3
Aspa S 93 B10
Aspa bruk S 92 B5
Aspach A 76 F4
Aspang-Markt A 148 A6
Asparn an der Zaya A 77 E10
Asparukhovo BG 167 D8
Aspås S 106 E6
Aspatria UK 5 F10
Aspe E 56 E3
Aspe S 97 B8
Aspeå S 107 E12
Aspeå S 107 D14
Asperen NL 183 B6
Aspet F 33 D7
Asphyltan S 97 C11
Aspliden S 109 F17
Aspliden S 118 C5
Aspnäs S 99 B8
Aspnäs S 103 A14
As Pontes de García Rodríguez E 38 B4
Aspremont F 35 A10
Aspres-sur-Buëch F 35 A10
Aspropyrgos GR 175 C8
Asprovalta GR 169 C10
Aspudden S 116 E3
As Rodrigas E 38 B5
Assamstadt D 74 D6
Asse B 19 C9
Assebakti N 113 E15
Asseiceira P 44 F4
Assemini I 64 E3
Assen NL 17 B7
Assendelft NL 16 D3
Assenede B 16 F1
Assens DK 86 B6
Assens DK 86 E5
Assesse B 19 D11
Assier F 33 A9
Assisi I 62 A3
Assjö S 103 B12

Åsskard N 104 E5
Aßlar D 21 C10
Assling A 73 C6
Asso I 69 B7
Asson F 32 D5
Assoro I 58 D5
Assos GR 168 D4
Assumar P 44 F6
Ásta N 101 D14
Astaffort F 33 B7
Astakos GR 174 B3
Åstan N 104 E7
Astašova LV 133 D2
Aste EST 130 E4
Åsteby S 97 B9
Astee IRL 8 C3
Asten A 76 F6
Asten NL 16 F5
Asti I 37 B8
Astikkala FIN 129 B12
Aştileu RO 151 C9
Astomilitsa GR 168 D4
Aston Clinton UK 15 D7
Astorga E 39 D7
Åstorp S 87 C11
Åstrand S 97 B9
Astravyets BY 137 D12
Astrazup RKS 163 E10
Astros GR 175 E6
Astrowna BY 133 F7
Astrowshchyna BY 133 F4
Astrup DK 86 B5
Astrup DK 86 C3
Astrup DK 90 E7
Astudillo E 40 D3
Astypalaia GR 180 C4
Asuaju de Sus RO 151 B11
Asūne LV 133 D3
Asuni I 64 D2
Ásványráró H 146 F5
Asvestopetra GR 169 D6
Asvyeya BY 133 G3
Aszaló H 145 G2
Aszód H 150 B3
Atalaia P 44 F4
Atalaia P 44 F5
Atalanti GR 175 B6
Atalaya del Cañavate E 47 E8
Ataquines E 40 F2
Atarfe E 53 B9
Atašiene LV 135 C12
Atea E 47 B9
Åteca E 41 F8
A Teixeira E 38 D5
Åţel RO 152 E5
Ateleta I 63 D6
Atella I 60 B5
Atena Lucana I 60 C5
Återvänningen S 103 A13
Atessa I 63 C7
Ath B 19 C8
Athani GR 174 B2
Áth an Sceire IRL 7 F10
Athboy IRL 7 E8
Athea IRL 8 C4
Athenry IRL 6 F5
Athenstedt D 79 C8
Atheras GR 174 C1
Atherington UK 12 D6
Atherstone UK 11 F8
Athies F 18 E6
Athies-sous-Laon F 19 E8
Athikia GR 175 D6
Athina GR 175 D8
Athis-de-l'Orne F 23 C11
Athleague IRL 6 E6
Athlone IRL 7 F7
Áth Trasna IRL 8 D4
Athy IRL 7 G9
Athyra GR 169 C8
Atienza E 47 B7
Atina I 62 D5
Aţintiş RO 152 E4
Atjiken S 107 A9
Atkár H 150 B4
Atna N 101 C13
Åtorp S 97 D11
Atouguia da Baleia P 44 F2
Åtran S 87 A11
Åträsk S 118 D5
Åträsk S 118 D5
Atri I 62 B5
Atripalda I 60 B3
Atsiki GR 171 E8
Attala H 149 D10
Attali GR 175 B8
Attanagh IRL 9 C8
Attavalley IRL 6 D3
Attendorn D 21 B9
Attenkirchen D 75 E10
Attersee A 73 A8
Attert B 19 E12
Attica UK 7 D10
Attigliano I 62 B2
Attignat F 31 C7
Attigny F 19 F10
Attleborough UK 15 B11
Attnang A 76 F5
Áttonträsk S 107 C14
Attu FIN 126 E7
Atur F 29 E7
Åtvidaberg S 92 C7
Atzara I 64 D3
Atzenbrugg A 77 F9
Atzendorf D 79 C10
Atzeneta del Maestrat E 48 D4
Au A 71 C9
Au CH 71 C9
Aub D 75 C7
Aubagne F 35 D10
Aubange B 19 E12
Aube F 24 C4
Aubel B 183 D7
Aubenas F 35 A7
Aubenton F 19 E9
Aubergenville F 24 C6
Auberive F 26 E3
Aubeterre-sur-Dronne F 29 E6
Aubiet F 33 C7
Aubignan F 35 B9
Aubigné-Racan F 23 E12
Aubignosc F 35 B10
Aubigny F 28 B3
Aubigny-en-Artois F 18 D6
Aubigny-sur-Nère F 25 F7
Aubin F 33 A10
Aubonne CH 31 B9
Aubord F 35 C7
Auboué F 20 F5
Aubusson F 29 D10

Auby F 19 D7
Aucamville F 33 C8
Auce LV 134 D5
Auch F 33 C7
Auchallater UK 5 B10
Auchenbreck UK 4 D6
Auchencairn UK 5 F9
Auchencrow UK 5 D12
Auchnagatt UK 3 L12
Auchterarder UK 5 C9
Auchtermuchty UK 5 C10
Auchy-au-Bois F 18 C5
Aucun F 32 E5
Audenge F 32 A3
Auderville F 23 A8
Audeux F 26 F4
Audevälja EST 131 C9
Audierne F 22 D2
Audincourt F 27 F6
Audlem UK 10 F6
Audley UK 11 E7
Audnedal N 90 C1
Audon F 32 C4
Audresselles F 15 F12
Audrini LV 133 C2
Audru EST 131 E8
Audruicq F 18 C5
Audun-le-Roman F 20 F5
Aue D 79 E12
Auerbach D 75 A11
Auerbach in der Oberpfalz
  D 75 C10
Auersthal A 77 F11
Auffay F 18 E3
Aufhausen D 75 E11
Augbrim IRL 6 F5
Auggen D 27 E8
Augher UK 7 D8
Aughnacloy UK 7 D9
Aughrim IRL 6 F5
Aughrim IRL 9 C10
Aughton UK 11 E9
Auignac F 29 D7
Augsburg D 75 F8
Augšligatne LV 135 B10
Augstkalne LV 133 E2
Augstkalne LV 134 D6
Augusta I 59 E7
Auguste LV 134 D3
Augustenborg DK 86 F5
Augustów PL 136 F6
Augustowo PL 141 E8
Augustusburg D 80 E4
Auho FIN 121 E10
Aukan N 104 E4
Aukra N 100 A5
Aukrug D 82 B7
Aukštadvaris LT 137 D10
Aukštelkai I LT 134 E7
Aukštelk LT 134 E6
Auktsjaur S 109 E17
Auleja LV 133 D2
Aulendorf D 71 B9
Auletta I 60 B4
Aulla I 69 E8
Aullène F 37 H10
Aulnat F 30 D3
Aulnay F 28 C5
Aulnay-sous-Bois F 25 C8
Aulnois sur-Seille F 26 C5
Aulnoye-Aymeries F 19 D8
Aulon F 30 D7
Aulosen D 83 E11
Ault F 18 D3
Aultbea UK 2 K5
Aultguish Inn UK 2 K7
Aulus-les-Bains F 33 E8
Auma D 79 E12
Aumale F 18 E4
Aumetz F 20 F5
Aumont F 31 B8
Aumont-Aubrac F 30 F3
Aumühle D 83 C8
Aunay-en-Bazois F 25 F10
Aunay-sur-Odon F 23 B10
Auneau F 24 D6
Aunegrenda N 105 F10
Aunfoss N 105 B12
Auning DK 86 C6
Auñón E 47 C7
Aups F 36 D4
Aura FIN 126 D8
Aurach D 75 D7
Aurach bei Kitzbühel A 72 B5
Auran N 105 E9
Auray F 22 E6
Aurdal N 101 E10
Aure N 104 E5
Aurec-sur-Loire F 30 E5
Aureilhan F 33 D6
Aureille F 35 C8
Aurejärvi FIN 123 G10
Aurel F 35 A9
Aurel F 35 B9
Aurensan F 32 C5
Aureosen N 100 A6
Auri N 134 C6
Aurich D 17 B8
Aurignac F 33 D7
Aurillac F 29 F10
Auriol F 35 D10
Aurisina I 73 E8
Auritz E 32 E3
Aurland N 100 E6
Aurolzmünster A 76 F4
Auron F 36 C5
Auronzo di Cadore I 72 C5
Auros F 32 A5
Auroux F 30 F4
Aurskog N 95 C14
Ausa-Corno I 73 E7
Ausejo E 32 F1
Auseu RO 151 C10
Ausleben D 79 B9
Ausmas UK 3 B9
Ausonia I 62 E5
Außervillgraten A 72 C5
Aussillon F 33 C10
Aussonne F 33 C8
Austafjord N 105 B9
Austbø N 95 C8
Austbørg N 105 C15
Austertana N 114 C5
Austis I 64 C3
Austmannli N 94 C5
Austnes N 100 A4
Austnes N 105 B15
Austnes N 111 C12
Austräh N 104 D7
Austsmøla N 104 E4
Auterive F 33 D8
Authon F 24 E4

Authon F 36 C4
Authon-du-Perche F 24 D4
Autio FIN 119 E17
Autio FIN 123 E13
Autol E 41 D7
Autrans F 31 E8
Autreville F 26 D4
Autrey-lès-Gray F 26 F3
Autry F 19 F10
Autti FIN 119 B18
Auttoinen FIN 127 C13
Autun F 30 B5
Auve F 25 C11
Auvelais B 19 D10
Auvers-le-Hamon F 23 E11
Auvillar F 33 B7
Auvillers-les-Forges F 184 E1
Auxerre F 25 E10
Auxi-le-Château F 18 D5
Auxonne F 26 F3
Auxy F 30 B5
Auzances F 29 C10
Auzat F 33 E8
Auzat-la-Combelle F 30 E3
Auzon F 30 E3
Ava S 107 D16
Avafors S 118 B8
Availles-Limouzine F 29 C7
Avaldsnes N 94 D2
Avallon F 25 E10
Avan S 118 C7
Avanäs S 107 C17
Avanca F 44 C3
Avançon F 36 B4
Avantas GR 171 C9
Avant-lès-Ramerupt F 25 D11
Avasjö S 107 C11
Avato GR 171 C7
Avaträsk S 107 C10
Avaviken S 109 E16
Avdira GR 171 C7
Avdou GR 178 E9
A Veiga E 39 D6
Aveiras de Cima P 44 F3
Aveiro P 44 C3
Avelar P 44 E4
Avelàs de Caminho P 44 D4
Aveleda P 38 E3
Aveleda P 39 E6
Avelgem B 19 C7
Avella I 60 B3
Avellino I 60 B4
Avenay-Val-d'Or F 25 B11
Avenches CH 31 B11
Avenhorn NL 16 C3
Avenida do Marquês de
  Figueroa E 38 A3
Avermes F 30 B4
Avers CH 71 E9
Aversa I 60 B2
Avesnes-le-Comte F 18 D6
Avesnes-sur-Helpe F 19 D8
Avesta S 98 B6
Åvestbo S 97 C14
Avetrano I 61 C9
Avezzano I 62 D4
Avgerinos GR 168 D5
Avià E 43 C7
Aviano I 73 D6
Aviemore UK 3 L9
Avigliana I 31 E11
Avigliano I 60 B5
Avigliano Umbro I 62 B2
Avignon F 35 C8
Ávila E 46 C3
A Vila da Igrexa E 38 A4
Avilés E 39 A8
Avilley F 26 F5
Avintes P 44 B3
Avinurme EST 131 D13
Avinyó E 43 D7
Avio I 69 B10
Avion F 18 D6
Avioth F 19 F11
Avis P 44 F5
Avispea EST 131 C12
Åvist FIN 122 D9
Avize F 25 C11
Avižieniai LT 137 D11
Avlakia GR 177 D8
Avlemonas GR 178 C5
Avliotes GR 168 E2
Avlonari GR 175 B9
Avlonas GR 175 C8
Avlum DK 86 C3
Avoca IRL 9 C10
Avoine F 23 F12
Avola I 59 F7
Avon F 25 D8
Avonmouth UK 13 B9
Avord F 29 A11
Avoriaz F 31 C10
Avoudrey F 26 F5
Avrämeni RO 153 A9
Avrämeşti RO 152 E6
Avram Iancu RO 151 D8
Avram Iancu RO 151 E10
Avranches F 23 D9
Avren BG 171 B9
Avricourt F 186 D2
Avrig RO 152 F4
Avril F 20 F5
Avrillé F 23 E10
Avrillé F 28 C3
Avtovac BIH 157 F10
Av'ze N 112 E10
Awans B 183 D6
Axams A 72 B3
Axat F 33 E10
Axel NL 16 F1
Axente Sever RO 152 E4
Axintele RO 161 D9
Axioupoli GR 169 C8
Ax-les-Thermes F 33 E9
Axmar S 103 D13
Axmarsbruk S 103 D13
Axminster UK 13 D9
Axos GR 169 C7
Axvall S 91 C14
Ayamonte E 50 E5
Aycliffe UK 5 F13
Aydat F 30 D3
Aydemir BG 161 E10
Ayen F 29 E8
Ayerbe E 32 F4
Aying D 72 A4
Aylesbury UK 15 D7
Ayllón E 40 F5
Aylsham UK 15 B11
Aymavilles I 31 D11
Ayna E 55 B8

Aynac F 29 F9
Ayoo de Vidriales E 39 D7
Ayora E 47 F10
Ayr UK 4 E7
Ayrancilar TR 177 C9
Ayron F 28 B6
Ayşebaci TR 173 E8
Aysgarth UK 11 C8
Ayskoski FIN 123 D17
Äystö FIN 122 F7
Aytos BG 167 D8
Aytré F 28 C3
Ayvacik TR 171 E10
Ayvalik TR 172 F6
Aywaille B 19 D12
Azagra E 32 F2
Azaila E 41 F11
Azambuja P 44 F3
Azanja SRB 159 E6
Azanúy E 42 D4
Azaruja P 50 B4
Azatli TR 172 B6
Azay-le-Ferron F 29 B8
Azay-le-Rideau F 24 F3
Azé F 30 C6
Azerables F 29 C8
Azina BY 133 E5
Azinhaga P 44 F3
Azinheira dos Barros P 50 C3
Azinhoso P 39 F6
Azkoitia E 32 D1
Aznalcázar E 51 E7
Aznalcóllar E 51 D7
Azóia P 50 C1
Azpeitia E 32 D1
Azuaga E 51 C8
Azuara E 41 F10
Azuel E 54 C4
Azuelo E 32 E1
Azuga RO 161 C7
Azuqueca de Henares E 47 C6
Azur F 32 C3
Azután E 45 E10
Azy-le-Vif F 30 B3
Azyory BY 140 C10
Azzano Decimo I 73 E6
Azzone I 69 B9

B

Baalberge D 79 C10
Baâlon F 19 F11
Baar CH 27 F9
Baarle-Hertog B 16 F3
Baarle-Nassau NL 16 F3
Baarn NL 16 D4
Baba Ana RO 161 D8
Babadag RO 155 D3
Babaeski TR 173 B7
Babaköy TR 173 E9
Baban AL 168 C4
Băbana RO 160 D5
Babberich NL 183 B8
Babchyntsi UA 154 A2
Babenhausen D 21 E11
Băbeni RO 151 C11
Băbeni RO 160 D4
Babiak PL 136 E1
Babiak PL 138 F6
Babice CZ 146 C4
Babice CZ 147 C9
Babice nad Svitavou CZ 77 D11
Babići BIH 157 D7
Băbiciu RO 160 E5
Babilafuente E 45 C10
Babimost PL 81 B9
Babin F 34 C8
Babina SRB 164 D5
Babina Greda HR 157 B10
Babino Polje HR 162 D4
Babīte LV 135 C7
Babno Polje SLO 73 E10
Babócsa H 149 D8
Bábolna H 149 A9
Baborów PL 142 F4
Baboszewo PL 139 E9
Babót H 149 A6
Babrujě AL 168 B2
Babtai LT 137 C8
Babuk BG 161 E10
Babušnica SRB 164 C5
Babyak BG 165 F8
Babynichy BY 133 F4
Bać MNE 163 D7
Bač SRB 158 C3
Băcani RO 153 E11
Bacares E 55 D8
Bacău RO 153 D9
Baccarat F 27 D6
Baceno I 68 A5
Băceşti RO 153 D10
Bach A 71 C10
Bacharach D 21 D9
Bachkovo BG 165 F10
Bachórz PL 144 D5
Băcia RO 151 F11
Baciki Bliższe PL 141 F7
Băcioi MD 154 D3
Baciu RO 152 D3
Bäck S 105 E16
Backa S 91 E10
Backa S 97 A9
Backa S 97 B9
Backaland UK 3 G11
Bačka Palanka SRB 158 C3
Backaryd S 89 C7
Bačka Topola SRB 150 F4
Backberg S 103 E12
Backe S 107 D10
Bäckebo S 89 B10
Bäckefors S 91 B11
Bäckhammar S 91 A15
Bački Breg SRB 150 F2
Bački Brestovac SRB 158 B3
Bački Jarak SRB 158 C4
Bački Monoštor SRB 150 F2
Bačkininkai LT 137 D8
Bački Petrovac SRB 158 C4
Bäcklund S 109 E17
Bäckmark S 107 A11
Backnang D 27 C11
Bäcknäs S 109 F17
Bačko Dobro Polje SRB 158 B4
Bačko Gradište SRB 158 C4
Bačko Novo Selo SRB 157 B11
Bačko Petrovo Selo SRB 158 B5
Baćkowice PL 143 E11
Bäckvallen S 102 B8
Baco RO 151 F11
Bâcles PL 159 E11
Bacoli I 60 B2
Bacqueville-en-Caux F 18 E2

Bácsalmás H 150 E3
Bácsbokod H 150 E3
Bácsborsód H 150 E3
Bacúch SK 147 D9
Baczyna PL 85 E8
Bada S 97 B9
Badacsonytomaj H 149 C8
Bad Abbach D 75 E11
Bad Aibling D 72 A5
Badajoz E 51 B6
Badalona E 43 E8
Badalucco I 37 D7
Bádames E 40 B5
Badarán E 40 D6
Bad Aussee A 73 A8
Bad Bayersoien D 71 B12
Bad Bederkesa D 17 A11
Bad Belzig D 79 B12
Bad Bentheim D 17 D8
Bad Bergzabern D 27 B8
Bad Berka D 79 E9
Bad Berleburg D 21 B10
Bad Berneck im Fichtelgebirge
  D 75 B10
Bad Bertrich D 21 D8
Bad Bevensen D 83 D9
Bad Bibra D 79 D10
Bad Birnbach D 76 F4
Bad Blankenburg D 79 E9
Bad Blumau D 148 B6
Bad Bocklet D 75 B7
Bad Boll D 74 E6
Bad Brambach D 75 B11
Bad Bramstedt D 83 C7
Bad Breisig D 185 D7
Bad Brückenau D 74 B6
Bad Buchau D 71 A9
Bad Camberg D 21 D10
Badcaul UK 2 K6
Badderen N 112 D6
Bad Doberan D 83 B11
Bad Driburg D 17 E12
Bad Düben D 79 C12
Bad Dürkheim D 21 F10
Bad Dürrenberg D 79 D11
Bad Dürrheim D 27 D10
Badeborn D 79 C9
Bad Elster D 75 B11
Bademler TR 177 C8
Bademli TR 171 F10
Bademli TR 173 D10
Bademli TR 177 B8
Bad Ems D 21 D9
Baden A 77 F10
Baden CH 27 F9
Baden F 22 E6
Bádenas E 47 B10
Baden-Baden D 27 C9
Bad Endorf D 72 A5
Badenscoth UK 3 L12
Badenweiler D 27 E8
Baderna HR 67 B8
Badersleben D 79 C8
Badesi I 64 B2
Bad Essen D 17 D10
Bädeuți RO 153 B7
Bad Fallingbostel D 82 E7
Bad Feilnbach D 72 A5
Bad Frankenhausen (Kyffhäuser)
  D 79 D9
Bad Freienwalde D 84 E6
Bad Friedrichshall D 21 F12
Bad Füssing D 76 F4
Bad Gams A 73 C11
Bad Gandersheim D 79 C7
Badgastein A 73 B7
Bad Gleichenberg A 148 C5
Bad Goisern A 73 A8
Bad Griesbach im Rottal D 76 F4
Bad Grönenbach D 71 B10
Bad Großpertholz A 77 E7
Bad Grund (Harz) D 79 C7
Bad Hall A 76 F6
Bad Harzburg D 79 C8
Bad Herrenalb D 27 C9
Bad Hersfeld D 78 E6
Bad Hindelang D 71 B10
Badhoevedorp NL 182 A5
Bad Hofgastein A 73 B7
Bad Homburg vor der Höhe
  D 21 D11
Bad Honnef D 21 C8
Bad Hönningen D 185 C7
Badia I 72 C4
Badia Calavena I 69 B11
Badia Gran E 49 F10
Badia Polesine I 66 B3
Badia Tedalda I 66 E5
Bad Iburg D 17 D10
Bădiceni MD 153 A12
Badín SK 147 D8
Bad Ischl A 73 A8
Bad Karlshafen D 78 C5
Bad Kissingen D 75 B7
Bad Kleinen D 83 C10
Bad Kleinkirchheim A 73 C8
Bad König D 21 E12
Bad Königshofen im Grabfeld
  D 75 B7
Bad Kösen D 79 D10
Bad Köstritz D 79 E11
Bad Kötzting D 76 D3
Bądkowo PL 138 E6
Bad Kreuzen A 77 F7
Bad Kreuznach D 21 E9
Bad Krozingen D 27 E8
Bad Laasphe D 21 C10
Bad Laer D 17 D10
Bad Langensalza D 79 D8
Bad Lauchstädt D 79 D10
Bad Lausick D 79 D12
Bad Lauterberg im Harz
  D 79 C7
Bad Leonfelden A 76 E6
Bad Liebenstein D 79 E7
Bad Liebenwerda D 80 C4
Bad Liebenzell D 27 C9
Bad Lippspringe D 17 E11
Badljevina HR 149 E8
Bad Lobenstein D 75 B10
Bad Marienberg (Westerwald)
  D 21 C9
Bad Mergentheim D 74 D6
Bad Mitterndorf A 73 A8
Bad Münder am Deister D 78 B5
Bad Münstereifel D 21 C7
Bad Muskau D 81 C7
Bad Nauheim D 21 D11
Bad Nenndorf D 17 D12
Bad Neuenahr-Ahrweiler
  D 21 C8
Bad Neustadt an der Saale
  D 75 B7

Badolato I 59 B10
Badolatosa E 53 B7
Bad Oldesloe D 83 C8
Badonviller F 27 C6
Bad Orb D 21 D12
Bad Peterstal D 27 C9
Bad Pirawarth A 77 F11
Bad Pyrmont D 17 E12
Bad Radkersburg A 148 C5
Bad Ragaz CH 71 C9
Bad Rappenau D 21 F11
Bad Reichenhall D 73 A6
Bad Rodach D 75 B8
Bad Saarow-Pieskow D 80 B6
Bad Sachsa D 79 C8
Bad Säckingen D 27 E8
Bad Salzdetfurth D 79 B7
Bad Salzuflen D 17 D11
Bad Salzungen D 79 E7
Bad Sankt Leonhard im
  Lavanttal A 73 C10
Bad Sassendorf D 17 E10
Bad Schandau D 80 E6
Bad Schmiedeberg D 79 C12
Bad Schönborn D 187 C6
Bad Schussenried D 71 A9
Bad Schwalbach D 21 D10
Bad Schwartau D 83 C9
Bad Segeberg D 83 C8
Bad Sobernheim D 185 E8
Bad Soden-Salmünster
  D 74 B5
Bad Sooden-Allendorf D 79 D6
Bad Staffelstein D 75 B8
Bad Steben D 75 B10
Bad Sulza D 79 D10
Bad Sülze D 83 B13
Bad Tennstedt D 79 D8
Bad Tölz D 72 A4
Bad Überkingen D 187 D8
Badules E 47 B10
Bad Urach D 27 D11
Bad Vilbel D 21 D11
Bad Vöslau A 77 G10
Bad Waldsee D 71 B9
Bad Waltersdorf A 148 B6
Bad Wildbad im Schwarzwald
  D 187 D6
Bad Wildungen D 21 B12
Bad Wilsnack D 83 E11
Bad Wimpfen D 21 F12
Bad Windsheim D 75 C7
Bad Wörishofen D 71 B11
Bad Wurzach D 71 B9
Bad Zell A 77 E7
Bad Zwesten D 21 B12
Bad Zwischenahn D 17 B10
Bad Zurzach CH 27 E9
Băe RO 151 D11
Baena E 53 A7
Baesweiler D 20 C6
Baeza E 53 C10
Baflo NL 17 B7
Bagà E 43 C7
Băgaciu RO 152 E4
Bagaladi I 59 C8
Bagamér H 151 C8
Bağarasi TR 177 B8
Bağarasi TR 177 D10
Bagard F 35 B7
Bağcilar TR 181 B8
Bâgé-le-Châtel F 30 C6
Bagenalstown IRL 9 C9
Bagenkop DK 83 A9
Bages F 34 E4
Baggå S 97 C14
Bagheria I 58 C4
Bagn N 101 E11
Bagnacavallo I 66 D4
Bagnac-sur-Célé F 33 A10
Bagnara Calabra I 59 C8
Bagnaria I 37 B10
Bagnaria Arsa I 73 E7
Bagnasco I 37 C8
Bagneaux-sur-Loing F 25 D8
Bagnères-de-Bigorre F 33 D6
Bagnères-de-Luchon F 33 E7
Bagni di Lucca I 66 D2
Bagni di Masino I 69 A8
Bagni di Rabbi I 71 D11
Bagno a Ripoli I 66 E3
Bagno di Romagna I 66 E4
Bagnoli del Trigno I 63 D7
Bagnoli di Sopra I 66 B3
Bagnoli Irpino I 60 B4
Bagnolo in Piano I 66 C2
Bagnolo Mella I 66 B1
Bagnolo Piemonte I 31 F11
Bagnolo San Vito I 66 B2
Bagnols-les-Bains F 35 A6
Bagnols-sur-Cèze F 35 B8
Bagnoregio I 62 B2
Bagod H 149 C7
Bagolino I 69 B9
Bagrationovsk RUS 136 F2
Bagrdan SRB 159 E7
Bagshot UK 15 E7
Báguena E 47 B10
Bağyurdu TR 177 C10
Bağyüzü TR 173 F6
Bahabón de Esgueva E 40 E4
Bahate UA 155 C3
Bahçeburun TR 181 B7
Bahçeköy TR 173 B10
Bahçeli TR 171 E10
Bahíllo E 39 D10
Bahmut MD 153 C12
Bahna RO 153 D9
Bahnea RO 152 E4
Báhoň SK 146 E4
Bahrenborstel D 17 C11
Bahrendorf D 79 C10
Baia RO 153 C8
Baia RO 155 D3
Baia de Aramă RO 159 D10
Baia de Arieş RO 151 E11
Baia de Criş RO 151 E10
Baia de Fier RO 160 C3
Baiano I 60 B3
Baiardo I 37 D7
Baia Sprie RO 152 B3
Băicoi RO 161 C7
Băiculeşti RO 160 C5
Baia Mare RO 152 B3
Băiești RO 152 E5
Baiersbronn D 27 C9
Baiersdorf D 75 C9

Baignes-Ste-Radegonde F 28 E5
Baigneux-les-Juifs F 25 E12
Baile an Bhiataigh IRL 7 E10
Baile an Bhuinneánaigh
  IRL 8 C3
Baile an Chinnéidigh IRL 7 F10
Baile an Dúlaigh IRL 6 F4
Baile an Fheirtéaraigh IRL 8 D2
Baile an Mhóta IRL 6 D5
Baile an Mhuilinn IRL 8 E4
Baile an Róba IRL 6 E4
Baile an Sceilg IRL 8 E2
Baile Átha an Rí IRL 6 F5
Baile Átha Buí IRL 7 E8
Baile Átha Cliath IRL 7 F10
Baile Átha Fhirdhia IRL 7 E9
Baile Átha Í IRL 7 G9
Baile Átha Luain IRL 7 F7
Baile Átha Troim IRL 7 E9
Baile Brigín IRL 7 E10
Baile Coimin IRL 7 F9
Băile Govora RO 160 C4
Băile Herculane RO 159 D9
Baile Loch Riach IRL 6 F5
Baile Mhartainn UK 2 K2
Baile Mhic Andáin IRL 9 C8
Baile Mhic Íre IRL 8 E4
Baile Mhistéale IRL 8 D6
Bailén E 54 C5
Baile na Finne IRL 6 C6
Baile na Lorgan IRL 7 D9
Băile Olănești RO 160 C4
Baile Órtha IRL 7 F9
Bálesti RO 160 C2
Băile Tuşnad RO 153 E7
Baile Uílcín IRL 7 E9
Bailieborough IRL 7 E9
Baillargues F 35 C7
Bailleau-le-Pin F 24 D5
Bailleul F 18 C6
Baillonville B 19 D11
Bailo E 32 E4
Baimaclia MD 154 E2
Bainbridge UK 11 C8
Bain-de-Bretagne F 23 E8
Baindt D 71 B9
Bains F 30 E4
Bains-les-Bains F 26 D5
Bainton UK 11 D10
Baio E 38 B2
Baiona E 38 D2
Bais F 23 D9
Bais F 23 D11
Baiso I 66 D2
Băişoara RO 151 D11
Baisogala LT 134 E7
Băiţa RO 151 D10
Băiţa de sub Codru RO 151 B11
Baix E 35 A8
Baixa da Banheira P 50 B1
Baixas F 34 E4
Baj H 149 A10
Baja H 150 E2
Bajč SK 146 F6
Bajgorë RKS 163 D10
Bajina Bašta SRB 158 F4
Bajmok SRB 150 F3
Bajna H 149 A11
Bajót H 149 A11
Bajovo Polje MNE 157 F10
Bajram Curri AL 163 E9
Bajša SRB 150 F4
Bak H 149 C7
Bakacak TR 173 D8
Bakar HR 67 B10
Bakel NL 16 F5
Bakır TR 177 A10
Bakırköy TR 173 C10
Bakkafjörður IS 94 B2
Bakke N 111 A9
Bakkeby N 112 D6
Bakko N 95 C9
Bakonszeg H 151 C7
Bakonybél H 149 B9
Bakonycsernye H 149 B10
Bakonysárkány H 149 B10
Bakonyszentkirály H 149 B9
Bakonyszentlászló H 149 B9
Bakonyszombathely H 149 B9
Bakov nad Jizerou CZ 77 B7
Baks H 150 D5
Baksa H 149 E9
Baktalórántháza H 145 H5
Baktsjaur S 109 F17
Bakum D 17 C10
Bakvattnet S 105 D16
Bâl S 93 D13
Bala RO 159 D10
Balabancik TR 173 C6
Bălăbănești MD 154 C4
Bălăbănești RO 153 E11
Balaci RO 160 E5
Bălăceanu RO 161 C10
Balaciu RO 161 D9
Balaguer E 42 D5
Balallan UK 2 J3
Balan F 31 D7
Bălan RO 151 C11
Bălan RO 153 D7
Bălănești MD 153 C12
Bălănești RO 160 C2
Balanivka UA 154 A5
Balaruc-les-Bains F 35 D6
Bălăşești RO 153 E11
Bălăşești MD 153 A9
Balástya H 150 E5
Balassagyarmat H 147 E8
Balástya H 150 E5
Balatfüred H 149 C9
Balatonberény H 149 C8
Balatonboglár H 149 C9
Balatonföldvár H 149 C9
Balatonfőkajár H 149 B10
Balatonföldvár H 149 C9
Balatonfőzfő H 149 B10
Balatonkenese H 149 B10
Balatonlelle H 149 C9
Balatonszabadi H 149 C10
Balatonszárszó H 149 C9
Balatonvilágos H 149 B10
Bălăuşeri RO 152 E5
Balazote E 55 B8
Balbeggie UK 5 C10

Balbieriškis LT 137 D8
Balbigny F 30 D5
Balbriggan IRL 7 E10
Balcani RO 153 D9
Bălceşti RO 160 D3
Balchik BG 167 C10
Balcilar TR 173 D6
Balçova TR 177 C9
Balderton UK 11 E10
Baldock UK 15 D8
Baldone LV 135 C8
Baldones Muiža LV 135 C8
Baldoveneşti RO 160 E4
Bale HR 67 B8
Baleizão P 50 C4
Balemartine UK 4 C3
Balen B 19 B11
Băleni RO 153 F11
Băleni RO 161 D7
Balephuil UK 4 C3
Balerma E 55 F7
Bălești RO 153 C12
Balestrand N 100 D5
Balestrate I 58 C3
Balfour UK 3 G11
Balfron UK 5 C8
Balgale LV 134 B5
Balgown UK 2 K4
Bălgviken S 98 D6
Bali GR 178 E8
Baligród PL 145 E5
Balikesir TR 173 E8
Balikliçeşme TR 173 D7
Balikliova TR 177 C8
Bălilești RO 160 C5
Bälinge S 99 C7
Balinge S 118 C7
Balingen D 27 D10
Balinka H 149 B10
Balint RO 151 F8
Balintore UK 3 K9
Balje D 17 A12
Baljevac BIH 156 D6
Baljevac SRB 163 C10
Baljvine BIH 157 D7
Balk NL 16 C5
Balkány H 151 B8
Balkbrug NL 16 C6
Balla IRL 6 E4
Ballaban AL 168 D3
Ballabio I 69 B7
Ballachulish UK 4 B6
Ballaghaderreen IRL 6 E5
Ballaghkeen IRL 9 C10
Ballangen N 111 D12
Ballantrae UK 4 E6
Ballao I 64 D3
Ballasalla GBM 10 C2
Ballater UK 5 A10
Ballaugh GBM 10 C2
Balle DK 87 C7
Ballée E 23 E11
Ballee UK 7 C11
Ballen DK 86 D7
Ballenstedt D 79 C9
Balleroy F 23 B10
Ballerup DK 87 D10
Ballesteros de Calatrava E 54 B5
Balli TR 173 C7
Ballina IRL 6 D4
Ballina IRL 8 C6
Ballinaboy IRL 6 F2
Ballinafad IRL 6 D6
Ballinagar IRL 7 F8
Ballinagh IRL 7 E8
Ballinakill IRL 9 C8
Ballinalack IRL 7 E8
Ballinalee IRL 7 E7
Ballinamallard UK 7 D7
Ballinamore IRL 7 D7
Ballinamult IRL 9 C7
Ballinasloe IRL 6 F6
Ballincollig IRL 8 E5
Ballindine IRL 6 E5
Ballindooly IRL 6 F4
Ballineen IRL 8 E5
Ballingarry IRL 8 C5
Ballingarry IRL 9 C7
Ballingeary IRL 8 E4
Ballingry UK 5 C10
Ballinglass Y SRB 87 C13
Ballingurteen IRL 8 E4
Ballinhassig IRL 8 E5
Ballinlough IRL 6 E5
Ballinluig UK 5 B9
Ballinrobe IRL 6 E4
Ballinskelligs IRL 8 E2
Ballintober IRL 6 E4
Ballintoy UK 4 E4
Ballintra IRL 6 C6
Ballintubber IRL 6 E4
Ballivor IRL 7 E9
Ballobar E 42 D4
Ballon F 23 D12
Ballószög H 150 D4
Ballots F 23 E9
Ballsh AL 168 C2
Ballsnes N 111 A16
Ballstad N 110 D6
Ballstädt D 79 D8
Ballum DK 86 E3
Ballum NL 16 B5
Ballure IRL 6 C6
Ballybay IRL 7 D9
Ballybofey IRL 7 F10
Ballybrack IRL 7 F11
Ballybrack IRL 8 F2
Ballybritas IRL 7 F8
Ballybunnion IRL 8 C3
Ballybunion IRL 8 C3
Ballycahill IRL 9 C7
Ballycanew IRL 9 C10
Ballycarry UK 4 F5
Ballycastle IRL 6 D4
Ballycastle UK 4 E4
Ballyclare UK 4 F5
Ballyclogh IRL 8 D5
Ballyconnely IRL 6 F2
Ballyconnell IRL 7 D7
Ballycotton IRL 8 E6
Ballycroy IRL 6 D3
Ballydavid IRL 6 F6
Ballydehob IRL 8 F4
Ballydesmond IRL 8 D4
Ballydonegan IRL 8 E3
Ballyduff IRL 8 C3
Ballyduff IRL 8 D5
Ballyfarnan IRL 6 D6
Ballyfarnon IRL 6 D6
Ballyfeard IRL 8 E6

Ballyferriter IRL 8 D2
Ballyforan IRL 6 F6
Ballygar IRL 6 E6
Ballygarrett IRL 9 C10
Ballygawley UK 7 D8
Ballygeary IRL 9 D10
Ballyglass IRL 6 E5
Ballygorman IRL 4 E2
Ballygowan UK 7 C11
Ballyhaise IRL 7 D8
Ballyhalbert UK 7 D12
Ballyhale IRL 9 D8
Ballyhaunis IRL 6 E5
Ballyhean IRL 6 E4
Ballyheigue IRL 8 D3
Ballyjamesduff IRL 7 E8
Ballykeeran IRL 7 F7
Ballykelly UK 4 E2
Ballykilleen IRL 7 F8
Ballylanders IRL 8 D6
Ballylaneen IRL 9 D8
Ballylickey IRL 8 E4
Ballyliffen IRL 4 E2
Ballyliffin IRL 4 E2
Ballylynan IRL 9 C8
Ballymacarberry IRL 9 D7
Ballymacmague IRL 9 D7
Ballymadog IRL 9 E7
Ballymagorry UK 4 F2
Ballymahon IRL 7 E7
Ballymakeery IRL 8 E4
Ballymartin UK 7 D11
Ballymena UK 4 F4
Ballymoney UK 4 E3
Ballymore IRL 7 F7
Ballymote IRL 6 D5
Ballymurphy IRL 9 C9
Ballymurry IRL 6 E6
Ballynacally IRL 8 C4
Ballynafid IRL 7 E8
Ballynahinch UK 7 D11
Ballynahowen IRL 7 F7
Ballynakill IRL 7 F7
Ballynamona IRL 8 D5
Ballyneaner UK 4 F2
Ballynunty IRL 9 C7
Ballynure UK 4 F5
Ballyporeen IRL 8 D6
Ballyragget IRL 9 C8
Ballyroan IRL 9 C8
Ballyronan UK 4 F3
Ballyshannon IRL 6 C6
Ballyvaldon IRL 9 D10
Ballyvaughan IRL 6 F4
Ballyvoy UK 4 E4
Ballyvoyle IRL 9 D7
Ballywalter UK 7 C12
Ballyward UK 7 D10
Balma F 33 C9
Balmaha UK 4 C7
Balmaseda E 40 B5
Balmazújváros H 151 B7
Balme I 31 E11
Balmedie UK 3 L12
Balmuccia I 68 B5
Balnacra UK 2 L6
Balnapaling UK 3 K8
Balneario de Panticosa Huesca E 32 E5
Balninkai LT 135 F10
Balocco I 68 C5
Balogunyom H 149 B7
Balot F 25 E11
Balotaszállás H 150 E4
Balotești RO 161 D8
Balow D 83 D11
Balrath IRL 7 E10
Balș RO 160 E4
Balșa RO 151 E11
Balsa de Ves E 47 F10
Balsa Pintada E 56 F2
Balsareny E 43 D7
Balsfjord N 111 B17
Balsicas E 56 F3
Balsjö S 107 D16
Balsorano I 62 D5
Bålsta S 99 C9
Balsthal CH 27 F8
Balta RO 159 D10
Balta UA 154 B5
Balta Albă RO 161 C10
Balta Berilovac SRB 164 C5
Balta Doamnei RO 161 D8
Baltanás E 40 E3
Baltar E 38 E4
Baltasound UK 3 D15
Bălțătești RO 153 C8
Bălțați RO 153 C10
Bălteni RO 153 D11
Bâlteni RO 159 D11
Baltezers LV 135 B8
Bălți MD 153 B11
Baltimore IRL 8 F4
Baltinava LV 133 C3
Baltinglass IRL 9 C9
Baltiysk RUS 139 A8
Baltmuiža LV 135 D13
Baltoji Vokė LT 137 E11
Baltora S 99 C11
Bałtów PL 143 D12
Baltray IRL 7 E10
Băluşeni RO 153 B9
Balvan BG 166 C4
Bâlvănești RO 159 D10
Balvano I 60 B5
Balve D 17 F9
Balvi LV 133 B2
Balvicar UK 4 C5
Balya TR 173 E8
Balzers FL 71 C9
Bamberg D 75 C8
Bamburgh UK 5 D13
Bammental D 21 F11
Bampini GR 174 B3
Bampton UK 13 B11
Bampton UK 13 D8
Bana H 149 A9
Banafjäl S 107 E16
Banagher IRL 7 F7
Banarli TR 173 B7
Banassac F 34 B5
Banatski Brestovac SRB 159 D6
Banatski Dvor SRB 158 B6
Banatski Karlovac SRB 159 C7
Banatsko Aranđelovo SRB 150 E5
Banatsko Karađorđevo SRB 158 B6
Banatsko Novo Selo SRB 159 D6
Banatsko Veliko Selo SRB 150 F6
Banbridge UK 7 D10
Banbury UK 13 A12
Banca RO 153 E11
Banchory UK 5 A12
Band RO 152 D4
Bande E 38 D4
Bandenitz D 83 D10
Bandholm DK 83 A10
Bandirma TR 173 D8
Bandol F 35 D10
Bandon IRL 8 E5
Bandurove UA 154 A5
Băneasa RO 153 F11
Băneasa RO 155 E1
Băneasa RO 161 E8
Bănești MD 154 B2
Bănești RO 161 C7
Banevo BG 167 D8
Banff UK 3 K11
Bångnäs S 106 B9
Bangor UK 4 F5
Bangor UK 10 E3
Bangor IRL 6 D3
Bangor Erris IRL 6 D3
Bangsund N 105 C10
Banham UK 15 C11
Bánhorváti H 145 G2
Bănia RO 159 D9
Banie PL 85 D7
Banie Mazurskie PL 136 E5
Baniska BG 166 C5
Bănișor RO 151 C10
Bănița RO 159 C11
Banite BG 171 A8
Banitsa BG 165 C8
Banja BIH 158 F3
Banja SRB 163 B8
Banja Lučica BIH 157 D10
Banja Luka BIH 157 C7
Banjani SRB 158 D4
Banja Vrućia BIH 157 C8
Bankekind S 92 C7
Bankeryd S 92 D4
Bankfoot UK 5 B9
Bankya BG 165 D6
Bankya BG 165 D7
Banloc RO 159 C7
Bannalec F 22 E4
Bannay F 25 F8
Bannesdorf auf Fehmarn D 83 B10
Bannewitz D 80 E5
Bannivka UA 155 B3
Bannockburn UK 5 C9
Bañobárez E 45 C7
Bañón E 47 C10
Banon F 35 B10
Baños de la Encina E 54 C5
Baños de Molgas E 38 D4
Baños de Montemayor E 45 D9
Baños de Río Tobía E 40 D6
Baños de Valdearados E 40 E4
Bánov CZ 146 D5
Bánov SK 146 E6
Banova Jaruga HR 149 F7
Bánovce nad Bebravou SK 146 D6
Banoviči BIH 157 D10
Bánréve H 145 G1
Bansin D 84 C6
Banská Belá SK 147 E7
Banská Bystrica SK 147 D8
Banská Štiavnica SK 147 E7
Banské SK 145 F4
Bansko BG 165 F7
Bant NL 16 C5
Banteer IRL 8 D5
Banteln D 78 B6
Bantheville F 19 F11
Bantry IRL 8 E4
Banya BG 165 D10
Banya BG 165 E9
Banya BG 165 F8
Banya BG 166 D5
Banya BG 167 D9
Banyalbufar E 49 E10
Banyeres de Mariola E 56 D3
Banyliv UA 152 A6
Banyliv-Pidhirnyy UA 152 A7
Banyoles E 43 C9
Banyuls-sur-Mer F 34 F5
Banzi I 60 B6
Banzkow D 83 C11
Bapaume F 18 D6
Bar MNE 163 E7
Bara RO 151 F8
Bâra RO 153 C10
Bara S 87 D12
Barabás H 145 G5
Baracska H 149 B11
Bărăganul RO 161 D11
Baragiano I 60 B5
Barahona E 41 F6
Barajas de Melo E 47 D7
Barajevo SRB 158 D5
Barakaldo E 40 B6
Barakovo BG 165 E7
Baralla E 38 C5
Barañain E 32 E2
Baranbio E 40 B6
Báránd H 151 C7
Baranello I 63 D7
Baranjsko Petrovo Selo HR 149 E10
Barano d'Ischia I 62 F5
Baranów PL 141 G6
Baranów PL 142 D5
Baranowo PL 139 D11
Baranów Sandomierska PL 143 F12
Barão de São João P 50 E2
Baraolt RO 153 E7
Baraqueville F 33 B10
Barásoain E 32 E2
Barassie UK 4 D7
Bărăști RO 160 D5
Baravukha BY 133 E5
Barbacena P 51 B5
Barbadás E 38 D4
Barbadillo de Herreros E 40 D5
Barbadillo del Mercado E 40 D5
Barbadillo del Pez E 40 D5
Barban HR 67 B9
Barbarano Vicentino I 66 B4
Barbaraville UK 3 K8
Barbaros TR 173 C7
Barbaste F 33 B6
Barbastro E 42 C4
Barbat HR 67 C10
Barbate de Franco E 52 D5
Bărbătescu RO 161 D9
Bărbătești RO 160 D3
Barbatovac SRB 164 C3
Barbâtre F 28 B1
Bărbele LV 135 D9
Barbentane F 35 C8
Barberá del Vallès E 43 D8
Barberaz F 31 D8
Barberino di Mugello I 66 D3
Barbezieux-St-Hilaire F 28 E5
Barbonne-Fayel F 25 C10
Bărbulețu RO 160 C6
Barbullush AL 163 F8
Barby (Elbe) D 79 C10
Barç AL 168 C4
Barca E 41 F6
Bârca RO 160 F3
Barcabo E 42 C4
Barcada BIH 157 B8
Barca de Alva P 45 B7
Bărcănești RO 161 D8
Bărcănești RO 161 D9
Barcani RO 161 B8
Barcarrota E 51 B6
Barcea RO 153 F10
Barcelinhos P 38 E2
Barcellona Pozzo di Gotto I 59 C7
Barcelona E 43 E8
Barcelonne-du-Gers F 32 C5
Barcelonnette F 36 C5
Barcelos P 38 E2
Bárcena del Monasterio E 39 B6
Bárcena de Pie de Concha E 40 B3
Barcenillas de Cerezos E 40 B4
Barchfeld D 79 E7
Barciany PL 136 E3
Barcillonnette F 35 B10
Barcin PL 138 E4
Barcino PL 85 B11
Barcis I 72 D6
Barcones E 40 F6
Barcos P 44 B5
Barcs H 149 E8
Barcus F 32 D4
Barczewo PL 136 F2
Bardal N 108 D5
Bardar MD 154 D3
Barde DK 86 C3
Bardejov SK 145 E3
Bárdena E 41 D9
Bardenitz D 80 B3
Bårdesø DK 86 D6
Bardi I 69 D8
Bardineto I 37 C8
Bardney UK 11 E11
Bardo PL 77 B11
Bardolino I 66 A2
Bardonecchia I 31 E10
Bardos F 32 D3
Bardowick D 83 D8
Bardsea UK 10 C5
Bárdudvarnok H 149 D9
Bare BIH 157 E10
Bare SRB 158 E6
Bare SRB 163 C8
Barefield IRL 8 C5
Barèges F 33 E6
Barenburg D 17 C11
Barendorf D 83 D9
Bärenklau D 81 C7
Bärenstein D 76 B4
Bärenstein D 80 E5
Barentin F 18 E2
Barenton F 23 C10
Barevo BIH 157 D7
Barfleur F 23 A9
Barga I 66 D1
Bargagli I 37 C10
Bargas E 46 E4
Bârgăuani RO 153 D9
Barge I 31 F11
Bargemon F 36 D5
Bargen CH 27 E10
Bargenstedt D 82 B6
Barghe I 69 B9
Bârghiș RO 152 F5
Bargischow D 84 C5
Barglów Kościelny PL 140 C7
Bargoed UK 13 B8
Bargrennan UK 4 E7
Bargstedt D 17 B12
Bargteheide D 83 C8
Bargullas AL 168 C3
Barham UK 15 E11
Bar Hill UK 15 C9
Bari I 61 A7
Barić Draga HR 156 D3
Barile I 60 B5
Barilović HR 148 F5
Barinas E 56 E2
Båring DK 86 E5
Barì Sardo I 64 D4
Barisciano I 62 C5
Barjac F 35 B7
Bårjasnjar'ga N 113 D15
Barjols F 35 C11
Barkåkar N 95 D12
Barkåkra S 87 C11
Barkald N 101 C13
Barkarö S 98 C7
Barkava LV 135 C13
Barkelsby D 83 A7
Barkhyttan S 103 E11
Barkowo PL 85 C12
Barkston UK 11 F10
Bârla RO 160 E5
Bârlad RO 153 E11
Barleben D 79 B10
Bar-le-Duc F 26 C3
Barles F 36 C4
Barletta I 60 A6
Barley UK 15 C9
Barlinek PL 85 E8
Barlingbo S 93 D12
Barmouth UK 10 F3
Barmstedt D 82 C7
Bárna H 147 E9
Barna IRL 6 F4
Bârna RO 159 B9
Barnaderg IRL 6 F5
Barnard Castle UK 11 B8
Barnarp S 92 D4
Barnatra IRL 6 D3
Bärnau D 75 C11
Barnbach A 73 B11
Barneberg D 79 B9
Barneveld NL 16 D5
Barneville-Carteret F 23 B8
Barnewitz D 79 A12
Barneycarroll IRL 6 E5
Barnoldswick UK 11 D7
Bârnova RO 153 C11
Barnowko PL 85 E7
Barnsley UK 11 D9
Barnstädt D 79 D10
Barnstaple UK 12 C6
Barnstorf D 17 C11
Barntrup D 17 E12
Baronissi I 60 B3
Baronville F 26 C6
Baroševac SRB 158 E5
Barovo NMK 169 B7
Barowka BY 133 E3
Barqueros E 55 D10
Barr F 27 D7
Barr UK 4 E7
Barracas E 48 D3
Barrachina E 47 C10
Barraduff IRL 8 D4
Barrafranca I 58 E5
Barral E 38 D3
Barrali I 64 E3
Barranco do Velho P 50 E4
Barrancos P 51 C6
Barranda E 55 C9
Barrapoll UK 4 C3
Barrax E 55 A8
Barre-des-Cévennes F 35 B6
Barreiro P 50 B1
Bärrek S 98 B7
Barrême F 36 D4
Barrhead UK 5 D8
Barrhill UK 4 E7
Barriada Nueva E 43 D8
Bário P 38 E2
Barrio del Peral E 56 F3
Barrio Mar E 48 E4
Barrit DK 86 D5
Barr na Trá IRL 6 D3
Barroca P 44 D5
Barroselas P 38 E2
Barrowby UK 11 F10
Barrow-in-Furness UK 10 C5
Barruecopardo E 45 B7
Barruelo de Santullan E 40 C3
Barry UK 13 C8
Barry IRL 7 E7
Bârsa RO 151 E9
Barsac F 32 A5
Bârsana RO 145 H9
Bârsău RO 151 B11
Barsbüttel D 83 C8
Bårse DK 87 E9
Barsele S 107 A12
Bârsești RO 153 F9
Barsinghausen D 78 B5
Barßel D 17 B9
Barsta S 103 A15
Barstyčiai LT 134 D3
Bar-sur-Aube F 25 D12
Bar-sur-Seine F 25 D11
Bārta LV 134 D2
Bartenheim F 27 E7
Bartenstein D 74 D6
Barth D 83 B13
Bartholomä D 74 E6
Bartholomäberg A 71 C9
Bartkuškiai LT 135 F9
Bartkuškis LT 137 D10
Bartnes N 105 C10
Bartniki PL 139 D10
Bartninkai LT 136 E7
Barton UK 10 D6
Barton-upon-Humber UK 11 D11
Bartoszyce PL 136 E2
Baru RO 159 C11
Baruchowo PL 139 F7
Barulho P 45 F6
Barumini I 64 D3
Baruth D 80 B5
Barvas UK 2 J3
Barvaux B 19 D11
Barver D 17 C11
Barwedel D 79 A8
Barwice PL 85 C10
Barxeta E 56 C4
Bârza RO 160 E4
Bârzana E 39 B8
Bârzava RO 151 E9
Barzio I 69 B7
Bašaid SRB 158 B5
Basarabeasca MD 154 E3
Basarabi RO 155 E2
Basarbovo BG 161 F7
Bàscara E 43 C9
Bascharage L 20 E5
Baschi I 62 B2
Baschurch UK 10 F6
Basciano I 62 B5
Basconcillos del Tozo E 40 C4
Bascons F 32 C5
Bascous F 33 C6
Bascov RO 160 D5
Basdahl D 17 B12
Basècles B 19 C8
Basel CH 27 E8
Baselga di Pinè I 69 A11
Baselice I 60 A3
Bas-en-Basset F 30 E5
Băsești RO 151 C11
Båsheim N 95 B10
Bashtanivka UA 154 F4
Basigo E 40 B6
Basildon UK 15 D9
Basiliano I 73 D7
Basingstoke UK 13 C12
Baška CZ 146 B6
Baška HR 67 C10
Baška Voda HR 157 F6
Baskemölla S 88 D6
Bäsksele S 107 B11
Bäsksjö S 107 B12
Basksjön S 107 E12
Baslow UK 11 E8
Bäsna S 97 A13
Bassacutena I 64 A3
Bassano del Grappa I 72 E4
Bassano Romano I 62 C2
Bassecourt CH 27 F7
Basse-Goulaine F 23 F9
Bassenge B 19 C12
Bassens F 28 F4
Bassiano I 62 D4
Bassoues F 33 C6
Bassum D 17 C11
Bassy F 31 D8
Bast FIN 123 C10
Båstad S 87 C11
Bastardo I 62 B3
Bastasi BIH 156 D5
Bastasi BIH 157 F10
Bastelica F 37 G10
Bastelicaccia F 37 H9
Bastennes F 32 C4
Bastfallet S 98 B7
Bastia F 37 F10
Bastia I 62 A3
Bastogne B 19 D12
Bastorf D 83 B11
Basttjärn S 97 B12
Bastumarks by S 118 E3
Bastuträsk S 107 C16
Bastuträsk S 118 E4
Báta H 149 D11
Bata RO 151 E9
Batajnica SRB 158 D5
Batak BG 165 F9
Batalha P 44 E3
Bățani RO 153 E7
Batăr RO 151 D8
Bătarci RO 145 G7
Båtas S 106 B8
Bátaszék H 149 D11
Batea E 42 E4
Batelov CZ 77 D8
Båteng N 113 C20
Baterno E 54 B3
Batetskiy RUS 132 D7
Bath UK 13 C10
Bathford UK 13 C10
Bathgate UK 5 D9
Bathmen NL 16 D6
Batin BG 161 F7
Batina HR 149 E11
Batizovce SK 145 E1
Batković BIH 158 D3
Batley UK 11 D8
Batllavë RKS 164 D3
Bátmonostor H 150 E2
Båtnfjordsøra N 100 A7
Batočina SRB 159 E7
Bátonyterenye H 147 F9
Bátorove Kosihy SK 146 F6
Batoș RO 152 D5
Batoshevo BG 166 D4
Bátovce SK 147 E7
Batovo BG 167 C9
Batrina HR 157 B8
Båtsfjord N 114 B7
Batsi GR 176 D4
Båtsjaur S 109 D13
Båtskärsnäs S 119 C10
Battenberg (Eder) D 21 B11
Bätterkinden CH 27 F8
Battice B 183 D7
Battipaglia I 60 B3
Battle UK 15 F9
Battonya H 151 E7
Batultsi BG 165 C9
Bátya H 150 E2
Batyatychi UA 144 C9
Batz-sur-Mer F 22 F7
Baucina I 58 D4
Baud F 22 E5
Bauduen F 36 D4
Bauen CH 71 D7
Baugé F 23 E11
Baugy F 25 F8
Bauladu I 64 C2
Baulon F 23 E8
Baume-les-Dames F 26 F5
Baumholder D 186 B3
Baunach D 75 C8
Baunei I 64 C4
Baurci MD 154 E3
Bausendorf D 21 D7
Bauska LV 135 D8
Băuțar RO 159 B10
Bautzen D 80 D6
Bavanište SRB 159 D6
Bavay F 19 D8
Bavel NL 182 B5
Baveno I 68 B6
Bavilliers F 27 E6
Bavorov CZ 76 D6
Bawdeswell UK 15 B11
Bawdsey UK 15 C11
Bawinkel D 17 C8
Bawn Cross Roads IRL 8 D5
Bayárcal E 55 E7
Bayarque E 55 E8
Baybuzivka UA 154 A5
Baye F 25 C10
Bayel F 25 D12
Bayerbach D 76 F4
Bayerbach bei Ergoldsbach D 75 E11
Bayerisch Eisenstein D 76 D4
Bayeux F 23 B10
Bayindir TR 177 C10
Bayir TR 181 B8
Bayirköy TR 172 D6
Baykal BG 160 F4
Bayon F 26 D5
Bayonne F 32 D3
Bayramiç TR 172 E6
Bayramiç TR 173 C6
Bayramli TR 173 B6
Bayreuth D 75 C10
Bayrischzell D 72 A5
Bayston Hill UK 10 F6
Baytaly UA 154 B5
Bayubas de Abajo E 40 E6
Baza E 55 D7
Bázakerettye H 149 C7
Bazas F 32 B5
Bazet F 33 D6
Baziège F 33 D9
Bazillac F 33 D6
Bazna RO 152 E4
Bazoches F 25 F10
Bazoches-au-Houlme F 23 C11
Bazoches-les-Gallerandes F 24 D7
Bazoches-sur-Hoëne F 24 C3
Bazougers F 23 D10
Bazouges F 23 E10
Bazzano I 66 D3
Beaconsfield UK 15 D7
Beadnell UK 5 D13
Beagh IRL 6 E5
Bealach A Doirín IRL 6 E5
Bealach Conglais IRL 9 C9
Bealach Feich IRL 7 C7
Bealaclugga IRL 6 F4
Béal an Átha IRL 6 D4
Béal an Mhuirthead IRL 6 D3
Béal Átha an Ghaorthaidh IRL 8 E4
Béal Átha Beithe IRL 7 D9
Béal Átha hAmhnais IRL 6 E5
Béal Átha Liag IRL 7 E7
Béal Átha na Muice IRL 6 E5
Béal Átha na Sluaighe IRL 6 F6
Béal Átha Seanaidh IRL 6 C6
Béal Deirg IRL 6 D3
Béal Easa IRL 6 E4
Bealnablath IRL 8 E5
Beaminster UK 13 D9
Beamud E 47 D9
Beannchar IRL 7 F7
Beantraí IRL 8 E4
Beariz E 38 D3
Bearna IRL 6 F4
Bearsden UK 5 D8
Beas E 51 E6
Beasain E 32 D1
Beas de Granada E 53 B10
Beas de Segura E 55 C7
Beateberg S 92 B4
Beattock UK 5 E10
Beaucaire F 35 C8
Beaucamps-le-Vieux F 18 E4
Beauchastel F 30 F6
Beaucouzé F 23 F10
Beaufay F 24 D3
Beaufort F 31 B7
Beaufort F 31 D10
Beaufort IRL 8 D3
Beaufort-en-Vallée F 23 F11
Beaugency F 24 E6
Beaujeu F 30 C6
Beaujeu F 36 C4
Beaujeu-St-Vallier-Pierrejux-et-Quitteur F 26 E4
Beaulieu F 35 C7
Beaulieu-lès-Loches F 24 F5
Beaulieu-sur-Dordogne F 29 F9
Beaulieu-sur-Loire F 25 E8
Beaulon F 30 B4
Beauly UK 2 L8
Beaumarchés F 33 C6
Beaumaris UK 10 E3
Beaumesnil F 24 B4
Beaumetz-lès-Loges F 18 D6
Beaumont B 19 D9
Beaumont F 23 A8
Beaumont F 29 B6
Beaumont F 29 F7
Beaumont-de-Lomagne F 33 C7
Beaumont-de-Pertuis F 35 C10
Beaumont-en-Argonne F 19 E11
Beaumont-en-Véron F 23 F12
Beaumont-le-Roger F 24 B4
Beaumont-lès-Valence F 30 F6
Beaumont-sur-Oise F 25 B7
Beaumont-sur-Sarthe F 23 D12
Beaune F 30 A6
Beaune-La Rolande F 25 D7
Beaupréau F 23 F10
Beauquesne F 18 D5
Beauraing B 19 D10
Beaurepaire F 31 E7
Beaurepaire-en-Bresse F 31 B7
Beaurières F 35 A10
Beausite F 26 C3
Beausoleil F 37 D6
Beautor F 19 E7
Beauvais F 18 F5
Beauval F 18 D5
Beauvezer F 36 C5
Beauville F 33 B7
Beauvoir-sur-Mer F 28 B1
Beauvoir-sur-Niort F 28 C5
Beauzac F 30 E5
Beauzelle F 33 C8
Beba Veche RO 150 E5
Bebertal D 79 B9
Bebington UK 10 E5
Bebra D 78 E6
Bebrene LV 135 D12
Bebrina HR 157 B8
Beccles UK 15 C12
Becedas E 45 D9
Beceite E 42 F4
Bečej SRB 158 B5
Beceni RO 161 C9
Becerreá E 38 C5
Becerril de Campos E 39 D10
Becherbach D 186 B4
Bécherel F 23 D8
Bechet RO 160 F3
Bechhofen D 21 F8
Bechhofen D 75 D8
Bechlín CZ 76 B6
Bechtheim D 21 E10
Bechyně CZ 77 D6
Becicherecu Mic RO 151 F7
Bečići MNE 163 E6
Becilla de Valderaduey E 39 D9
Beçin TR 181 B7
Beckdorf D 82 D7
Beckedorf D 17 D12
Beckeln D 17 C11
Beckingen D 21 F7
Beckingham UK 11 E10
Beckov SK 146 D5
Beckum D 17 E10
Beclean RO 152 C4
Beclean RO 152 F5
Bécon-les-Granits F 23 F10
Bečov CZ 76 B5
Becsehely H 149 D7
Becsvölgye H 149 C7
Bečváry CZ 77 C8
Bedale UK 11 C8
Bédar E 55 E9
Bédarieux F 34 C5
Bédarrides F 35 B8
Bedburg D 21 C7
Bedburg-Hau D 183 B8
Beddgelert UK 10 E3
Beddingestrand S 87 E12
Bédée F 23 D8
Bedekovčina HR 148 D5
Beden BG 165 F9
Beder DK 86 C6
Bedford UK 15 C8
Bedihošť CZ 77 D12
Bjdków PL 143 C8
Bedlington UK 5 E13
Bedlno PL 143 B8
Bedmar E 53 A10
Bednja HR 148 D5
Bédoin F 35 B9
Bedollo I 69 A11
Bedonia I 37 B11
Bedous F 32 D4
Bedsted DK 86 E4
Bedsted Stationsby DK 86 B2
Béduer F 33 A9
Bedum NL 17 B7
Bedwas UK 13 B8
Bedworth UK 13 A12
Bjdzin PL 143 F7
Bjdzino PL 85 B9
Beedenbostel D 79 A7
Beeford UK 11 D11
Beek NL 16 E5
Beek NL 16 E5
Beek NL 183 D7
Beekbergen NL 183 A7
Beelitz D 80 B3
Beendorf D 79 B9
Beenz D 84 D4
Beerfelden D 21 E11
Beernem B 19 B7
Beers NL 16 E5
Beerse B 16 F3
Beersel B 19 C9
Beerst B 182 C1
Beerta NL 17 B8
Beesd NL 183 B6
Beesenstedt D 79 C10
Beeskow D 80 B6
Beesten D 17 D9
Beeston UK 11 F9
Beetsterzwaag NL 16 B6
Beetzendorf D 83 E10
Bégaar F 32 C4
Bégadan F 28 E4
Begaljica SRB 158 D6
Bégard F 22 C5
Begejci SRB 158 B6
Beğendik TR 172 C6
Begíjar E 53 A9
Begijnendijk B 19 B10
Beglezh BG 165 C10
Begnište NMK 169 B6
Begonte E 38 B4
Begur E 43 D10
Behramkale TR 171 F10
Behren-lès-Forbach F 27 B6
Behren-Lübchin D 83 B13
Behringen D 79 D8
Beica de Jos RO 152 D5
Beidaud RO 155 D3
Beierfeld D 79 E12
Beierstedt D 79 B8
Beilen NL 17 C7
Beilngries D 75 D9
Beilstein D 27 B11
Beimerstetten D 187 E8
Beinasco I 37 A7
Beinette I 37 C7
Beinwil CH 27 F9
Beirã P 44 F6
Beisfjord N 111 D14
Beisland N 90 C3
Beith UK 4 D7
Beitostølen N 101 D9
Beitstad N 105 C10
Beiuș RO 151 D9
Beja LV 133 B2
Beja P 50 C4
Bejar AL 168 D2
Béjar E 45 D9
Bejís E 48 E3
Bekecs H 145 G3
Békés H 151 D7
Békéscsaba H 151 D7
Békéssámson H 150 E6
Békésszentandrás H 150 D5
Bekkarfjord N 113 B19
Bekken N 101 C15
Bekkevoll N 114 D8
Belá SK 147 C7
Bélâbre F 29 B8
Bela Crkva SRB 159 D7
Beladice SK 146 E6
Belá-Dulice SK 147 C7
Belalcázar E 51 B9
Belá nad Cirochou SK 145 F5
Bělá nad Radbuzou CZ 75 C12
Belanovce NMK 164 E4
Belanovica SRB 158 E5
Bela Palanka SRB 164 C5
Bélapátfalva H 145 G1
Bělá pod Bezdězem CZ 77 A7
Bělá pod Pradědem CZ 77 B12
Belascoáin E 32 E2
Belauski LV 133 C2
Belava LV 135 B13
Belazaima do Chão P 44 C4
Belcaire F 33 E9
Belcastel F 33 B10
Belcești RO 153 C10
Bełchatów PL 143 D7
Belchin BG 165 E7
Belchite E 41 F10
Bělčice CZ 76 C5
Belciugatele RO 161 E8
Belclare IRL 6 F5
Belcoo UK 7 D7
Belderg IRL 6 D3
Beldibi TR 181 C8
Beled H 149 B8
Belegiš SRB 158 C5
Belej HR 67 C9
Belene BG 160 F6
Bélesta F 33 E9
Beleți-Negrești RO 160 D6
Belevi TR 177 C9
Belezna H 149 D7
Belfast UK 7 C11
Belfeld NL 16 F6
Belford UK 5 D13
Belfort F 27 E6
Belfort-du-Quercy F 33 B9
Belforte del Chienti I 67 F7
Belgern D 80 D4
Belgershain D 79 D12
Belgioioso I 69 C7
Belgodère F 37 F10
Belgooly IRL 8 E6
Belgun BG 155 F2
Belhomert-Guéhouville F 24 C5
Beli HR 67 B9
Belianes E 42 D6
Belica HR 149 D7
Beli Iskŭr BG 165 E8
Beli Izvor BG 165 C7
Beli Manastir HR 149 E11
Belin RO 153 F7
Belin-Béliet F 32 B4
Belinchón E 47 D6
Belinț RO 151 F8
Beli Potok SRB 164 B5
Beliș RO 151 D11
Belišće HR 149 E10
Belitsa BG 161 F9
Belitsa BG 165 D8
Belitsa BG 165 F8
Beliu RO 151 E9
Bělkovice-Lašťany CZ 146 B4
Bell D 21 D8
Bell (Hunsrück) D 185 D7
Bella I 60 B5
Bellac F 29 C8
Bellacorick IRL 6 D3
Bellaghy UK 4 F3
Bellaghy IRL 6 E5
Bellagio I 69 B7
Bellahy IRL 6 E5
Bellanicë RKS 163 E10
Bellano I 69 A7

Bellante I 62 B5
Bellaria I 66 D5
Bellavary IRL 6 E4
Bellavista E 51 E8
Bellclaire d'Urgell E 42 D5
Belleek UK 7 D10
Bellegarde F 25 E7
Bellegarde F 35 C8
Bellegarde-en-Marche F 29 D10
Bellegarde-sur-Valserine F 31 C8
Belle-Isle-en-Terre F 22 C5
Bellême F 24 D4
Bellenaves F 30 C3
Bellenberg D 71 A10
Bellencombre F 18 E3
Bellerive-sur-Allier F 30 C3
Belles-Forêts F 27 C6
Belleu F 19 F7
Bellevaux F 31 C10
Belleville F 30 C6
Belleville-sur-Meuse F 26 B3
Belleville-sur-Vie F 28 B3
Bellevue-la-Montagne F 30 E4
Belley F 31 D8
Bellheim D 21 F10
Bellherbe F 26 F6
Bellignat F 31 C8
Bellinge DK 86 E6
Bellingham UK 5 E12
Bellingwolde NL 17 B8
Bellinzago Novarese I 68 B6
Bellinzona CH 69 A7
Bellizzi I 60 B3
Bell-Lloc d'Urgell E 42 D5
Bello E 47 C10
Bellobradë RKS 163 C10
Bellopojë RKS 163 D10
Bellou-en-Houlme F 23 C11
Bellpuig E 42 D6
Bellreguard E 56 D4
Bellshill UK 5 D8
Belluno I 72 D5
Bellver de Cerdanya E 33 F9
Bellvik S 107 C10
Belm D 17 D10
Bélmegyer H 151 D7
Bélmez E 51 C9
Bélmez de la Moraleda E 53 A10
Belmont UK 3 D15
Belmont-de-la-Loire F 30 C5
Belmonte E 39 B7
Belmonte E 47 E7
Belmonte P 45 D6
Belmonte Calabro I 60 E6
Belmonte del Sannio I 63 D6
Belmonte in Sabina I 62 C3
Belmontejo E 47 E8
Belmont-sur-Rance F 34 C4
Belmullet IRL 6 D3
Belo Brdo RKS 163 C10
Beloci MD 154 B4
Belœil B 19 C8
Belogradchik BG 159 F10
Beloiannisz H 149 B11
Belojin SRB 164 C3
Belorado E 40 D5
Beloslav BG 167 C9
Bělotín CZ 146 B5
Belotintsi BG 165 B6
Belovo BG 165 E9
Belozem BG 166 E4
Belp CH 31 B11
Belpasso I 59 D6
Belpech F 33 D9
Belper UK 11 E9
Belsay UK 5 E13
Belsh AL 168 C2
Belsk Duży PL 141 G3
Beltheim D 21 D8
Beltinci SLO 148 C6
Beltiug RO 151 B10
Beltra IRL 6 D5
Beltra IRL 6 E4
Beltrum NL 183 A9
Belturbet IRL 7 D8
Beluša SK 146 C6
Belušić SRB 159 F7
Belvédère-Campomoro F 37 H9
Belvedere Marittimo I 60 D5
Belvedere Ostrense I 67 E7
Belver de Cinca E 42 D4
Belver de los Montes E 39 E9
Belvès F 29 F8
Belvèze-du-Razès F 33 D10
Belvì I 64 D3
Belville IRL 6 D4
Belvís de la Jara E 46 E3
Belz F 22 E5
Belz UA 144 C9
Bełżec PL 144 C7
Bełżyce PL 141 H6
Bembibre E 38 B2
Bembibre E 39 C7
Bemmel NL 16 E5
Bemposta P 39 F7
Bemposta P 44 F4
Bempton UK 11 C11
Benabarre E 42 C4
Benacazón E 51 E7
Benahadux E 55 F8
Benahavís E 53 C6
Benalmádena E 53 C7
Benalúa de Guadix E 55 E6
Benalúa de las Villas E 53 B9
Benalup de Sidonia E 52 D5
Benamargosa E 53 C8
Benamaurel E 55 D7
Benamejí E 53 B7
Benamocarra E 53 C8
Benaocaz E 53 C6
Benaoján E 53 C6
Benasal E 48 D4
Benasau E 56 D4
Benasque E 33 E7
Benassay F 28 B6
Benatae E 55 C7
Benátky nad Jizerou CZ 77 B7
Benavente E 39 D8
Benavente P 50 B2
Benavides de Orbigo E 39 D8
Benavila P 44 F5
Benburb UK 7 D9
Bencatel P 50 B5
Bendorf D 21 D9
Bēne LV 134 D6
Benecko CZ 81 E9
Beneden-Leeuwen NL 16 E4
Benediktbeuren D 72 A3
Benedita P 44 F3
Benegiles E 39 E8
Bénéjacq F 32 D5
Benejama E 56 D3
Benejúzar E 56 E3
Benesat RO 151 C11
Benešov CZ 77 C7
Benešov nad Černou CZ 77 E7
Benešov nad Ploučnicí CZ 80 E6
Bénesse-Maremne F 32 C3
Benestare I 59 C9
Bénestroff F 27 C6
Benet F 28 C4
Benetutti I 64 C3
Bene Vagienna I 37 B7
Bénévent-l'Abbaye F 29 C9
Benevento I 60 A3
Benfeld F 27 D8
Benfica do Ribatejo P 44 F3
Bengeşti-Ciocadia RO 160 C3
Bengtsfors S 91 A11
Bengtsheden S 103 E10
Benia de Onís E 39 B10
Beniarbeig E 56 D5
Beniarrés E 56 D4
Beničanci HR 149 E10
Benicarló E 48 D5
Benicasim E 48 D5
Benidorm E 56 D4
Beniel E 56 E3
Benifaió E 48 F4
Benifallet E 42 F5
Benifallim E 56 D4
Benigánim E 56 D4
Benilloba E 56 D4
Benissa E 56 D5
Benissanet E 42 E5
Benitachell E 56 D5
Benitses GR 168 E2
Benizalón E 55 E8
Benizar y la Tercia E 55 C9
Benken CH 27 F11
Benkovac HR 156 D4
Benkovski BG 161 F10
Benkovski BG 171 B8
Benllech UK 10 E3
Benlloch E 48 D5
Bennebroek NL 182 A5
Bennekom NL 183 A7
Bennewitz D 79 D12
Bennungen D 79 D9
Bénodet F 22 E3
Benquerença P 45 D6
Benquerenças P 44 E5
Benquerencia de la Serena E 51 B9
Benquet F 32 C4
Bensafrim P 50 E2
Bensbyn S 118 C8
Bensdorf D 79 B11
Benshausen D 79 E8
Bensheim D 21 E11
Bensjö S 103 A9
Benson UK 13 B12
Bentelo NL 17 D7
Bentivoglio I 66 C3
Bentley UK 11 D9
Bentwisch D 83 B12
Bentzin D 84 C4
Beňuš SK 147 D9
Benzingerode D 79 C8
Beočin SRB 158 C4
Beograd SRB 158 D5
Bera de Bidasoa E 32 D2
Beragh UK 7 C8
Berane MNE 163 D8
Beranga E 40 B4
Berango E 40 B6
Berantevilla E 40 C6
Berat AL 168 C2
Bérat F 33 D8
Beratzhausen D 75 D10
Berbegal E 42 D3
Berbeşti RO 160 C3
Berca RO 161 C9
Bercedo E 40 B5
Bercel H 147 F8
Berceni RO 161 D8
Berceni RO 161 E8
Bercero E 39 E9
Berceto I 69 D8
Berchem B 19 C8
Berchidda I 64 B3
Berching D 75 D9
Berchtesgaden D 73 A7
Bérchules E 55 F6
Bercianos del Páramo E 39 D8
Berck F 15 G12
Bercu RO 145 H6
Berdal N 104 E6
Berdalen N 94 D6
Berducedo E 39 B6
Berdún E 32 E4
Bere Alston UK 12 E6
Beregdaróc H 145 G6
Bereguardo I 69 C7
Berehomet UA 152 A6
Berehove UA 145 G6
Berek HR 149 E7
Berekböszörmény H 151 C8
Berekfürdő H 150 C6
Beremend H 149 E10
Bereşti RO 153 E11
Bereşti-Meria RO 153 E11
Bereşti-Tazlău RO 153 E9
Berettyóújfalu H 151 C8
Berevoeşti RO 160 C5
Berezeni RO 154 E2
Berezyne UA 154 E4
Berg D 71 B9
Berg D 75 B10
Berg D 183 D9
Berg L 20 E6
Berg N 96 B7
Berg N 108 F3
Berg N 111 B13
Berg NL 19 C12
Berg NL 183 D7
Berg S 92 C8
Berg S 102 A7
Berg S 107 E10
Berg (Pfalz) D 27 C9
Berga D 79 D9
Berga D 79 E11
Berga E 43 C7
Berga S 89 A10
Bergagård S 87 B11
Bergaland N 94 D4
Bergama TR 177 A9
Bergamo I 69 B8
Bergantino I 66 B3
Bergara E 32 D1
Bergasa E 32 F1
Bergatreute D 71 B9
Berg bei Neumarkt in der Oberpfalz D 75 D9
Bergby S 103 E13
Berge D 17 C9
Berge D 83 D11
Berge E 42 F3
Berge S 102 A8
Berge S 105 D14
Bergeforsen S 103 A13
Bergeggi I 37 C8
Bergen D 73 A6
Bergen D 83 E7
Bergen D 84 B4
Bergen N 94 B2
Bergen NL 16 C3
Bergen (Dumme) D 83 E9
Bergen op Zoom NL 16 F2
Berger N 95 C12
Bergerac F 29 F6
Bergères-lès-Vertus F 25 C11
Bergesserin F 30 C6
Berget N 108 D6
Bergeyk NL 16 F4
Bergfors S 111 D18
Berghamn S 103 A15
Bergharen NL 183 B7
Berghaupten D 186 E4
Bergheim A 73 A7
Bergheim D 75 E9
Bergheim (Edertal) D 21 B12
Bergheim (Erft) D 21 C7
Berghem NL 16 E5
Berghem S 91 E12
Berghin RO 152 E3
Berghülen D 74 F6
Berg im Drautal A 73 C7
Bergisch Gladbach D 21 C8
Bergkamen D 17 E9
Bergkarlås S 102 D8
Bergkvara S 89 C10
Bergland S 107 A10
Berglia N 105 C15
Bergmo N 112 D7
Bergnäs S 109 E13
Bergnäset S 118 C8
Bergnäsudden S 109 E15
Bergnäsviken S 109 D15
Bergneustadt D 185 B8
Bergö FIN 99 B13
Bergö FIN 122 E6
Bergom S 107 E15
Bergsåker S 103 B13
Bergsäng S 97 B10
Bergsäter S 107 B12
Bergsbyn S 118 E6
Bergschenhoek NL 182 B5
Bergsgården S 103 E10
Bergshamra S 99 C11
Bergsjö S 103 C13
Bergsnov N 105 B9
Bergs slussar S 92 C7
Bergstod N 95 D9
Bergstrøm N 96 D6
Bergsviken S 118 D6
Bergtheim D 75 C7
Bergues F 18 C5
Bergün CH 71 D9
Bergvik S 103 D12
Bergviken S 109 D16
Bergwitz D 79 C12
Berhida H 149 B10
Beringe NL 183 C7
Beringel P 50 C4
Beringen B 19 B11
Berislăveşti RO 160 C4
Beriu RO 151 F11
Berja E 55 F7
Berka D 79 E7
Berkåk N 101 A12
Berkel NL 16 E2
Berkel-Enschot NL 16 E4
Berkeley UK 13 B10
Berkenthin D 83 C9
Berkheim D 71 A10
Berkhout NL 16 C4
Berkoviči BIH 157 F9
Berkovitsa BG 165 C7
Berlanga E 51 C8
Berlanga de Duero E 40 F6
Berleşti RO 160 D3
Berlevåg N 114 B6
Berlicum NL 183 B6
Berlin D 80 A4
Berlingerode D 79 D7
Berlişte RO 159 D7
Berlstedt D 79 D9
Bermatingen D 27 E11
Bermeo E 40 B6
Bermillo de Sayago E 39 F7
Bern CH 31 B11
Bernac-Dessus F 33 D6
Bernalda I 61 C7
Bernardos E 46 B4
Bernartice CZ 76 D6
Bernartice CZ 81 E9
Bernāti LV 134 D1
Bernau D 27 E9
Bernau D 84 E5
Bernau am Chiemsee D 72 A5
Bernaville F 18 D5
Bernay F 24 B4
Bernbeuren D 71 B11
Bernburg (Saale) D 79 C10
Berndorf A 77 G10
Berne D 17 B10
Bernecebaráti H 147 E7
Bernedo E 41 C7
Bernhardsthal A 77 E11
Bernhardswald D 75 D11
Bernin F 31 E8
Bernis F 35 C7
Bernisdale UK 2 L4
Bernissart B 19 D8
Bernitt D 83 C11
Bernkastel-Kues D 21 E8
Bernolákovo SK 146 E4
Bernsdorf D 80 D6
Bernshammar S 97 C14
Bernstadt D 74 E7
Bernstadt D 81 D7
Bernstein A 148 B6
Beromünster CH 27 F9
Beronovo BG 167 D7
Beroun CZ 76 C6
Berovo NMK 165 F6
Berra I 66 C4
Berre-l'Étang F 35 D9
Berriedale UK 3 J9
Berrien F 22 D4
Berriozar E 32 E2
Berru F 19 F9
Berry-au-Bac F 19 F8
Bersenbrück D 17 C9
Bershad' UA 154 A5
Beršići SRB 158 E5
Bersone I 69 B10
Bērstele LV 135 D7
Bertamirans E 38 C2
Bertea RO 161 C7
Berteştii de Jos RO 155 D1
Bertinoro I 66 D5
Bertnes N 108 B8
Bertogne B 19 D12
Bertrange L 186 B1
Bertrichamps F 27 D6
Bertrix B 19 E11
Běrunice CZ 77 B8
Berveni RO 151 B9
Bērvircava LV 134 D7
Berwick-upon-Tweed UK 5 D12
Bērzaine LV 131 F10
Berzasca RO 159 D8
Bērzaune LV 135 C12
Bērzciems LV 134 B6
Berzence H 149 D8
Bērzgale LV 133 C3
Bērzieši LV 135 D11
Bērzkalne LV 133 B2
Berźniki PL 137 E7
Berzovia RO 159 C8
Bērzpils LV 133 C2
Berzunți RO 153 E9
Beşalma MD 154 E3
Besalú E 43 C9
Besançon F 26 F5
Besande E 39 C10
Bescanó E 43 D9
Beselich-Obertiefenbach D 21 D10
Bešeňov SK 146 E6
Besenyőtelek H 150 B5
Besenyszög H 150 C5
Besigheim D 27 C11
Beşiktaş TR 173 B11
Beška SRB 158 C5
Besko PL 145 D4
Besni Fok SRB 158 D5
Besnyő H 149 B11
Besozzo I 68 B6
Bessacarr UK 11 D9
Bessaker N 104 C8
Bessan F 34 D5
Bessay-sur-Allier F 30 C3
Bessbrook UK 7 D10
Besse F 31 E9
Besse-et-St-Anastaise F 30 D2
Bessèges F 35 B7
Bessenbach D 187 B7
Bessé-sur-Braye F 24 E4
Besse-sur-Issole F 36 E4
Bessières F 33 C9
Bessines-sur-Gartempe F 29 C8
Bessude I 64 B2
Best NL 16 E4
Bestensee D 80 B5
Bestorp S 92 C7
Bestwig D 17 F10
Bestwina PL 147 B8
Beszterec H 145 G4
Betanzos E 38 B3
Betelu E 32 D2
Bétera E 48 E4
Beteta E 47 C8
Bethausen RO 151 F8
Betheln D 78 B6
Bethesda UK 10 E3
Bethmale F 33 E8
Bethon F 25 C10
Béthune F 18 C6
Betna N 104 E4
Beton-Bazoches F 25 C9
Betschdorf F 186 D4
Betsele S 107 B15
Bettelainville F 186 C1
Bettembourg L 20 E6
Bettendorf L 20 E6
Bettingen D 20 E6
Bettna S 93 B9
Bettola I 37 B11
Betton F 23 D8
Bettyhill UK 3 H8
Bettystown IRL 7 E10
Betws-y-coed UK 10 E4
Betxi E 48 E4
Betygala LT 134 F6
Betz F 25 B8
Betzdorf D 21 C9
Betzdorf L 186 B1
Betzenstein D 75 C9
Betzigau D 71 B10
Beucha D 79 D12
Beuil F 36 C5
Beuningen NL 16 E5
Beuren D 79 D7
Beuron D 27 D10
Beurville F 25 D12
Beuvrages F 182 E2
Beuvry F 18 C6
Beuzeville F 18 F1
Bevagna I 62 B3
Bevaix CH 31 B10
Beveren B 19 B9
Beverley UK 11 D11
Beverlo B 183 C6
Bevern D 78 C5
Beverstedt D 17 B11
Beverungen D 21 A12
Beverwijk NL 16 D3
Béville-le-Comte F 24 D6
Bewdley UK 13 A10
Bex CH 31 C11
Bexbach D 21 F8
Bexhill UK 15 F9
Beyağaç TR 181 B9
Beyazköy TR 173 B8
Beyçayiri TR 173 D6
Beyel TR 173 E10
Beyendorf D 79 B10
Beykoz TR 173 B11
Beynac-et-Cazenac F 29 F8
Beynat F 29 E9
Beynost F 31 D7
Beyoba TR 177 B10
Beyobasi TR 181 C9
Bezas E 47 D10
Bezau A 71 C9
Bezdan SRB 150 F2
Bezdead RO 161 C6
Bezdonys LT 137 D12
Bèze F 26 F3
Bezenye H 146 F4
Bezhanitsy RUS 133 C7
Bezhanovo BG 155 F2
Bezhanovo BG 165 C9
Béziers F 34 D5
Bezledy PL 136 E2
Bezmer BG 161 F10
Bezmer BG 166 E6
Bezouce F 35 C8
Bežovce SK 145 F5
Biadki PL 142 C4
Biała PL 142 F4
Biała-Parcela Pierwsza PL 142 D5
Biała Piska PL 139 C13
Biała Podlaska PL 141 F8
Biała Rawska PL 141 G2
Białe Błota PL 138 D4
Białebłoto-Kobyla PL 139 E12
Białka PL 145 E1
Białobrzegi PL 141 G3
Białogard PL 85 C9
Białośliwie PL 85 D12
Białowieża PL 141 E9
Biały Bór PL 85 C11
Biały Dunajec PL 147 C10
Białystok PL 140 D8
Biancavilla I 59 D6
Bianco I 59 C9
Biandrate I 68 C5
Bians-les-Usiers F 31 B9
Bianzè I 68 C5
Biar E 56 D3
Biarritz F 32 D2
Biarrotte F 32 C3
Biars-sur-Cère F 29 F9
Bias F 32 B3
Bias F 33 B7
Biasca CH 71 E7
Biatorbágy H 149 B11
Bibbiena I 66 E4
Bibbona I 66 F2
Biberach D 27 D9
Biberach an der Riß D 71 A9
Biberbach D 75 E8
Biberist CH 27 F8
Bibinje HR 156 D3
Bibione I 73 E7
Biblis D 21 E10
Bibury UK 13 B11
Bicaz RO 151 C11
Bicaz RO 153 D8
Bicaz-Chei RO 153 D7
Bicazu Ardelean RO 153 D7
Biccari I 60 A4
Bicester UK 13 B12
Bichiş RO 152 E4
Bichl D 72 A3
Bichlbach A 71 C11
Bickleigh UK 13 D7
Bicorp E 48 F3
Bicos P 50 D3
Bicske H 149 A11
Bicton UK 10 F6
Bidache F 32 D3
Bidart F 32 D2
Biddenden UK 15 E10
Biddinghuizen NL 16 D5
Biddulph UK 11 E7
Bideford UK 12 C6
Bidjovagge N 112 E10
Bidos F 32 D4
Bidovce SK 145 F3
Bie S 93 A8
Bieber D 21 D12
Biebesheim am Rhein D 21 E10
Biecz PL 144 D3
Biedenkopf D 21 C11
Biel CH 27 F7
Biel E 32 F4
Bielany-Żyłaki PL 141 F6
Bielawa PL 81 E11
Bielawy PL 143 B8
Bielefeld D 17 D11
Bielice PL 85 D7
Bieliny Kapitulne PL 143 E10
Biella I 68 B5
Bielle F 32 D5
Bielsa E 33 E6
Bielsk PL 139 E8
Bielsko-Biała PL 147 B8
Bielsk Podlaski PL 141 E8
Bienenbüttel D 83 D8
Bieniów PL 81 C8
Bienno I 69 B9
Bienservida E 55 B7
Bienvenida E 51 C7
Bierawa PL 142 F5
Bierdzany PL 142 E5
Bière CH 31 B9
Bierge E 42 C3
Bierné F 23 E10
Biersted DK 86 A5
Biert F 33 E8
Biertan RO 152 E5
Bieruń PL 143 F7
Bierutów PL 142 D4
Bierzwienna-Długa PL 142 B6
Bierzwnik PL 85 D9
Biesaidgohppi N 113 C16
Biescas E 32 E5
Biesenthal D 84 E5
Biesiekierz PL 85 B10
Biesles F 26 D3
Biesowice PL 85 B11
Biessenhofen D 71 B11
Bietigheim D 27 C9
Bietigheim-Bissingen D 27 C11
Bietikow D 84 D5
Bièvre B 19 E11
Bieżuń PL 139 E8
Biga TR 173 D7
Bigadiç TR 173 F9
Biganos F 32 A4
Bigastro E 56 E3
Bigauņciems LV 134 C7
Bigbury-on-Sea UK 13 E7
Biggar UK 5 D9
Biggleswade UK 15 C8
Bignan F 22 E6
Bignasco CH 71 E7
Bigor MNE 163 E7
Bigüézal E 32 E3
Biguglia F 37 F10
Bihać BIH 156 C4
Biharia RO 151 C8
Biharkeresztes H 151 C8
Biharnagybajom H 151 C7
Bihartorda H 151 C7
Biharugra H 151 D8
Bihosava BY 133 E3
Bijela MNE 163 E6
Bijele Poljane MNE 163 D6
Bijeljani BIH 157 F9
Bijeljina BIH 158 D3
Bijelo Brdo HR 149 E11
Bijelo Bučje BIH 157 D8
Bijelo Polje MNE 163 C8
Bikernieki LV 135 E13
Bikovo SRB 150 F4
Biksēre LV 135 C12
Bílá Lhota CZ 77 C11
Bilalovac BIH 157 D9
Bilbao E 40 B6
Bilbo E 40 B6
Bilbor RO 153 C7
Bilca RO 153 B7
Bilciureşti RO 161 D7
Bil'dzyuhi BY 133 F2
Bileća BIH 162 D5
Biled RO 151 F6
Biłgoraj PL 144 B6
Bilicenii Vechi MD 153 B12
Bílina CZ 80 E5
Bilisht AL 168 C4
Biljača SRB 164 E4
Bilje HR 149 E11
Bilky UA 145 G7
Billdal S 91 D10
Billerbeck D 17 E8
Billère F 32 D5
Billericay UK 15 D9
Billesholm S 87 C11
Billiat F 31 C8
Billigheim D 21 F12
Billingham UK 11 B9
Billinghay UK 11 E11
Billingsfors S 91 B11
Billingshurst UK 15 E8
Billom F 30 D3
Billsåsen S 105 F16
Billsbro S 92 A7
Billsta S 107 E15
Billum DK 86 D2
Billund DK 86 D4
Billy F 30 C3
Bilolissya UA 154 F5
Bílovec CZ 146 B6
Bílovice CZ 146 C5
Bilska LV 135 B11
Bilston UK 5 D10
Bilthoven NL 16 D4
Bilto N 112 E7
Bilton UK 11 D11
Bilyayivka UA 154 E6
Bilychi UA 145 E6
Bilyn UA 152 A4
Bilyne UA 154 B5
Bilzen B 19 C12
Bimbister UK 3 G10
Bíňa SK 147 F7
Binaced E 42 D4
Binas F 24 E5
Binasco I 69 C7
Binbrook UK 11 E11
Binche B 19 D9
Bindalseit N 105 A12
Bindlach D 75 C10
Bindslev DK 90 D7
Binefar E 42 D4
Bingen D 27 D11
Bingen am Rhein D 21 E9
Bingerden NL 183 B8
Bingham UK 11 F10
Binghamstown IRL 6 D2
Bingley UK 11 D8
Bingsjö S 103 D10
Binic F 22 C6
Biniés E 32 E4
Binissalem E 49 E10
Binnen D 17 C12
Binz D 84 B5
Binzen D 27 E8
Biograd na Moru HR 156 E3
Biokovina BIH 157 D7
Biol F 31 E7
Bionaz I 31 D11
Biorra IRL 7 F7
Biosca E 43 D6
Biot F 36 D6
Biota E 41 D9
Birchington UK 15 E11
Birchiş RO 151 F9
Bircza PL 144 D5
Birdhill IRL 8 C6
Birgittelyst DK 86 C4
Biri H 151 B8
Biri N 101 E13
Bīriņi LV 135 B9
Biristrand N 101 D12
Birkeland N 90 C3
Birkelse DK 86 A5
Birkenau D 187 B6
Birkenfeld D 21 E8
Birkenfeld D 27 C10
Birkenfeld D 74 C6
Birkenhead UK 10 E5
Birken-Honigsessen D 185 C8
Birkenwerder Berlin D 84 E4
Birkerød DK 87 D10
Birket DK 87 F8
Birketveit N 90 C2
Birkfeld A 148 B5
Birkungen D 79 D7
Birmingham UK 13 A11
Birnbaum A 73 C6
Biron F 33 A7
Birori I 64 C2
Birr IRL 7 F7
Birresborn D 21 D7
Birsay UK 3 G10
Birstall UK 11 F9
Birstein D 21 D12
Birštonas LT 137 D9
Birtavarre N 112 E6
Birtley UK 5 F13
Biruința MD 153 B12
Biržai LT 135 D9
Birzes LV 135 D8
Birzgale LV 135 C9
Birži LV 135 D11
Birzuli LV 131 F12
Biš SLO 148 C5
Bisaccia I 60 A4
Bisacquino I 58 D3
Biscarrosse F 32 B3
Biscarrosse-Plage F 32 B3
Bisceglie I 61 A7
Bischberg D 75 C8
Bischbrunn D 187 B7
Bischhausen (Waldkappel) D 78 D6
Bischheim F 27 C8
Bischoffen D 21 C10
Bischofferode D 79 D7
Bischoffsheim F 186 E3
Bischofsheim D 21 E10
Bischofsheim an der Rhön D 74 B7
Bischofshofen A 73 B7
Bischofsmais D 76 E4
Bischofswerda D 80 D6
Bischofswiesen D 73 A6
Bischofszell CH 27 E11
Bischwiller F 27 C8
Bisenti I 62 B5
Biser BG 166 F5
Bisertsi BG 161 F9
Bishop Auckland UK 5 F13
Bishop's Castle UK 10 G6
Bishop's Cleeve UK 13 B10
Bishop's Lydeard UK 13 C8
Bishop's Stortford UK 15 D9
Bishop's Waltham UK 13 D12
Bishqem AL 168 B2
Bishtazhin RKS 163 E10
Bisignano I 60 D6
Bisingen D 27 D10
Biskupice PL 141 H7
Biskupice PL 142 D5
Biskupice SK 147 E9
Biskupiec PL 136 F2
Biskupiec PL 139 C7
Bislev DK 86 B5
Bismark (Altmark) D 83 E11
Bismervik N 113 B11
Bismo N 101 C8
Bisoca RO 161 B9
Bispgården S 107 E11
Bispingen D 83 D8
Bissen L 20 E6
Bissendorf D 17 D10
Bissendorf (Wedemark) D 78 A6
Bissingen D 75 E8
Bistagno I 37 B8
Bistarac BIH 157 C10
Bistra RO 151 E11
Bistra RO 152 B4
Bistreț RO 160 F3
Bistrets BG 167 E8
Bistrica BIH 157 B7
Bistrica BIH 157 C7
Bistrica BIH 157 E8
Bistrica BIH 157 E10
Bistrica SLO 73 D9
Bistrica SLO 148 D5
Bistričak BIH 157 D8
Bistrița RO 152 C5
Bistrița Bârgăului RO 152 C5
Bistritsa BG 165 D7
Bistritsa BG 165 E7
Bisztynek PL 136 E2
Bitburg D 21 E7
Bitche F 27 B7
Bitetto I 61 A7
Bitinckë AL 168 C4
Bitola NMK 168 B5
Bitonto I 61 A7
Bitritto I 61 A7
Bitschwiller-lès-Thann F 27 E7
Bitterfeld D 79 C11
Bitterstad N 110 C9
Bitti I 64 C3
Bittkau D 79 B10
Bitton UK 13 C10
Bitz D 27 D11
Biurrun E 32 E2
Bivio CH 71 E9
Bivolari RO 153 B10
Bivona I 58 D3
Bixad RO 145 H7
Bixad RO 153 E7
Bixter UK 3 E14
Biyikali TR 173 B7
Biyikli TR 177 D10
Bizanet F 34 D4
Bizanos F 32 D5
Bizovac HR 149 E10
Bjåen N 94 C6
Bjærangen N 108 C6
Bjæverskov DK 87 E10
Bjännberg S 122 C3
Bjarkøy N 111 C12
Bjärnum S 87 C13
Bjärred S 87 D12
Bjärsjölagård S 87 D13
Bjärtrå S 107 F13
Bjästa S 107 E15
Bjela BIH 157 F10
Bjelajci BIH 157 B6
Bjelopolje HR 156 C4
Bjelovar HR 149 E7
Bjergby DK 90 D7
Bjerge DK 87 D8
Bjerka N 108 D6
Bjerkvik N 111 C14
Bjerreby DK 86 F7
Bjerregrav DK 86 B5
Bjerringbro DK 86 C5
Bjøllånes N 108 C8
Bjoneroa N 95 A12
Bjønnes N 108 D7
Björbo S 97 B12
Björboholm S 91 D11
Bjordal N 94 E4
Bjørgen N 105 F9
Bjørgo N 101 E10
Björkå S 107 E13
Bjørkåsen N 111 C17
Bjørkåsen N 111 D12
Björkberg S 103 C9
Björkberg S 107 B13
Björkberg S 118 C5
Björkboda FIN 126 E8
Björke S 93 D12
Björke S 103 E13
Bjørkeflåta N 95 B9
Bjørkelangen N 95 C15
Bjørkeset N 95 C10
Bjørkestrand N 114 C4
Björketorp S 91 E12
Björkfors S 119 C10
Björkheden S 109 E12
Björkhöjden S 107 E10
Björkholmen S 109 C17
Björkland S 109 F16
Bjørklia N 105 E9
Björkliden S 111 D16
Björklinge S 99 B9
Björknäset S 107 F12
Bjørknes N 96 B6
Bjørknes N 114 D6
Björkö FIN 126 E4
Björköby FIN 122 D6
Björköby S 92 D5
Björkön S 103 B14
Björksele S 107 B15
Björksele S 107 D12
Björksjön S 107 E13
Björkstugan S 111 D15
Björkvik S 93 B9
Bjorli N 100 B8
Bjørn N 108 D4
Björna S 107 D15
Björnänge S 105 E14
Björndalen S 91 C11
Björneborg S 97 D11
Bjørnengen N 113 D11

Bjørnera N 111 C12
Bjørnes N 113 B18
Bjørnevatn N 114 D7
Bjørnfjell N 111 D15
Björnhult S 89 A10
Björnliden S 102 B3
Björnlunda S 93 A10
Björnön S 103 A9
Bjørnrå N 111 C10
Björnrike S 102 B6
Björnsjö S 107 D14
Björnskinn N 111 B10
Bjørnstad N 105 B14
Bjørnstad N 114 D9
Björsarv S 103 B11
Björsbo S 103 C12
Björsjö S 97 B13
Bjugn N 104 D7
Bjurå S 118 B8
Bjuråker S 103 C12
Bjurberget S 97 A8
Bjurfors S 107 C17
Bjurfors S 118 E5
Bjurholm S 107 D16
Bjursås S 103 E9
Bjurselet S 118 E6
Bjurträsk S 107 B17
Bjuv S 87 C11
Blace NMK 164 E3
Blace SRB 164 C3
Blachownia PL 143 E6
Black Bourton UK 13 B11
Black Bull IRL 7 F4
Blackburn UK 3 L12
Blackburn UK 5 D9
Blackburn UK 10 D7
Blackmoor Gate UK 13 C7
Blackpool UK 10 D5
Blackrock IRL 7 E10
Blackstad S 93 D8
Blacktown UK 7 C7
Blackwater IRL 9 D10
Blackwaterfoot UK 4 D6
Blackwood UK 13 B8
Bladel NL 16 F4
Blaenau Ffestiniog UK 10 F4
Blaenavon UK 13 B8
Blagaj BIH 157 F8
Blagaj Japra BIH 156 B5
Blagdon UK 13 C9
Blăgești RO 153 D9
Blăgești RO 153 E12
Blagnac F 33 C8
Blagoevgrad BG 165 E7
Blagoevo BG 166 C6
Blåhøj DK 86 D4
Blaibach D 75 D12
Blain F 23 F8
Blainville-sur-l'Eau F 186 D1
Blainville-sur-Mer F 23 B8
Blair Atholl UK 5 B9
Blairgowrie UK 5 B10
Blaj RO 152 E3
Blajan F 33 C10
Blăjani RO 161 C9
Blăjel RO 152 E4
Blăjeni RO 151 D10
Blakeney UK 13 B10
Blakeney UK 15 B11
Blakstad N 90 C4
Blåmont F 27 C6
Blan F 33 C10
Blanca S 55 C10
Blancafort F 25 E8
Blancas E 47 C10
Blandford Forum UK 13 D10
Blandiana RO 151 F11
Blanes E 43 D9
Blaney UK 7 D7
Blangy-sur-Bresle F 18 E4
Blankaholm S 93 D9
Blankenberg D 83 C11
Blankenberge D 19 B7
Blankenburg (Harz) D 79 C8
Blankenfelde D 80 B4
Blankenhain D 79 E9
Blankenhain D 79 E11
Blankenheim D 21 C10
Blankensee D 84 D4
Blankensee D 75 B10
Blanquefort F 28 F4
Blans DK 86 F5
Blansko CZ 77 D11
Blanzac-Porcheresse F 28 E6
Blanzay F 29 C6
Blanzy F 30 B5
Blaricum NL 183 A6
Blarney IRL 8 E5
Blasimon F 28 F5
Blåsjöfallet S 105 B16
Blåsmark S 118 D6
Błaszki PL 142 C5
Blatec NMK 164 F6
Blatets BG 167 D7
Blatna BIH 156 C5
Blatná CZ 76 D5
Blatné SK 146 E4
Blatnica BIH 157 D8
Blato HR 162 D2
Blaton B 182 D3
Blattniksele S 109 F14
Blaubeuren D 74 F6
Blaufelden D 74 D6
Blausasc F 37 D6
Blaustein D 187 E8
Blåvik S 92 C6
Blåviksjön S 107 B14
Blavozy F 30 E4
Blaye F 28 E4
Blaževo SRB 163 C10
Błazny Górne PL 141 H4
Błazma LV 134 B4
Blažovce CZ 77 D11
Błazowa PL 144 D5
Blázquez E 51 C9
Blažuj BIH 157 E9
Blecķāsen S 105 E15
Bleckede D 83 D9
Blecua E 42 C3
Bled SLO 73 D9
Błędowo PL 138 D5
Bledzew PL 81 A8
Blegny B 19 C12
Bléharies B 19 C7
Bleialf D 20 D6
Bleicherode D 79 D8
Bleidenstadt D 185 D9
Bleik N 111 B10
Bleikvassli N 108 E6
Bleiswijk NL 182 A5

Blejești RO 161 E6
Blejoi RO 161 D8
Blekendorf D 83 B9
Bleket S 91 D10
Blender D 17 C12
Bléneau F 25 E8
Blénod-lès-Toul F 26 C4
Blentarp S 87 D13
Blera I 62 C2
Blérancourt F 19 E7
Blerick NL 16 F6
Blesa E 42 F3
Bleskensgraaf NL 182 B5
Blesle F 30 E3
Blessington IRL 7 F9
Blet F 29 B11
Bletchley UK 15 D7
Blidene LV 134 C5
Blidsberg S 91 D14
Blieskastel D 21 F8
Bligny F 25 D12
Bligny-sur-Ouche F 25 F12
Blikstorp S 91 C15
Bliksund N 90 C3
Blimea E 39 B8
Blindow D 84 D5
Blinja HR 149 F6
Blistrup DK 87 C10
Blixterboda S 97 D13
Blizanów PL 142 C6
Blíževov CZ 76 D3
Blížkovice CZ 77 E9
Bliznatsi BG 155 C9
Blizyn PL 141 H3
Bllacë RKS 163 E10
Blois F 24 E5
Blokhus DK 86 A5
Blokzijl NL 16 C5
Blombacka S 97 C10
Blomberg D 17 E12
Blomberg D 80 A5
Blumenhagen D 84 C5
Blumenholz D 84 D4
Blūskovo BG 167 C8
Blyberg S 102 D7
Blyth UK 5 E13
Blyth UK 11 E9
Blyth Bridge UK 5 D10
Blyton UK 11 E10
Bø N 95 D10
Bø N 110 C8
Bo N 111 B10
Bø N 111 D12
Boada E 45 C8
Boadilla del Monte E 46 D5
Boadilla de Rioseco E 39 D10
Boal E 39 B6
Boalhosa P 38 E3
Boan MNE 163 D7
Boara Pisani I 66 B3
Boat of Garten UK 3 L9
Boa Vista P 44 E3
Boavista P 44 F2
Bobadilla E 53 B7
Bobâlna RO 152 C3
Bobbio I 37 B10
Bobbio Pellice I 31 F11
Bobenheim-Roxheim D 187 B5
Boberg S 107 E19
Boberka UA 145 E6
Bobicești RO 160 E4
Böbing D 71 A11
Bobingen D 71 A11
Bobitz D 83 C10
Bobivtsi UA 153 A7
Böblingen D 27 C11
Bobolice PL 85 B10
Boborás E 38 D3
Boboshevo BG 165 E6
Bobota HR 149 F11
Bobota RO 151 C10
Bobovdol BG 165 E7
Bobove UA 145 G6
Bobowa PL 144 D2
Bobowo PL 138 C6
Bobrov SK 147 C9
Bobrovec SK 147 C9
Bobrówko PL 85 E8
Bobrówko PL 81 C8
Bobrowniki PL 138 E6
Bobrowniki PL 140 D9
Bobryk-Druhyy UA 154 B6
Bocacara E 45 C8
Bocairent E 56 D3
Bočar SRB 150 F5
Bocchigliero I 61 E7
Boceguillas E 40 F4
Bocfölde H 149 C7
Bochnia PL 144 D1
Bocholt B 19 B12
Bocholt D 17 E7
Bochov CZ 76 B4
Bochum D 17 E8
Bockara S 89 A10
Bockenem D 79 B7
Bockenheim an der Weinstraße
D 187 B5
Bockhammar S 97 C14
Bockhorn D 17 B10
Bockhorn D 75 F10
Bockhorst D 17 C9
Bočki PL 141 E8
Bocksjö S 92 B5
Bockträsk S 109 F15
Bocognano F 37 G10
Boconád H 150 B5
Bőcs H 145 G3
Bócsa H 150 D3
Bocşa RO 151 C10
Bocşa RO 159 C8
Bocsig RO 151 E8
Boczów PL 81 B7
Bod RO 153 F7
Boda S 103 A12

Boda S 103 D9
Boda bruk S 103 C12
Bodaczów PL 144 B7
Bodafors S 92 D5
Boda glasbruk S 89 B9
Bodajk H 149 B10
Bodåsgruvan S 98 B6
Bodbacka FIN 122 E6
Bodbyn S 122 C3
Boddam UK 3 F14
Boddam UK 3 L13
Boddin D 83 C10
Boddin D 83 D10
Bodelshausen D 187 E6
Boden S 118 C7
Bodenfelde D 78 C6
Bodenheim D 21 E10
Bodenkirchen D 75 F11
Bodenmais D 76 D4
Bodenteich D 83 A9
Bodenwerder D 78 C6
Bodenwöhr D 75 D11
Bodești RO 153 C8
Bodman D 27 E11
Bodmin UK 12 E5
Bodø N 108 B7
Bodom N 105 D11
Bodonal de la Sierra E 51 C6
Bodoney UK 7 C7
Bodony H 147 F10
Bodrež SRB 158 D5
Bodroghalom H 145 G4
Bodrogkisfalud H 145 G3
Bodrum TR 177 E9
Bodsjö S 102 A8
Bodsjöbyn S 102 A8
Bodträskfors S 118 B5
Bódvaszilas H 145 F2
Bodyke IRL 8 C5
Bodzanów PL 139 E9
Bodzanów PL 142 E5
Bodzentyn PL 143 E10
Bölge S 99 B9
Böle S 102 B7
Böle S 105 D14
Böle S 105 E15
Böle S 118 C7
Böle S 118 D6
Bolea E 41 D10
Bolekhiv UA 145 E8
Boleráz SK 146 E5
Bolesław PL 143 F7
Bolesław PL 143 F8
Bolesławiec PL 81 D9
Bolesławiec PL 142 D5
Bolga N 108 C5
Bolgatovo RUS 133 C5
Bolhás H 149 D8
Bolhrad UA 155 B3
Bolholt UK 13 B11
Boliden S 118 E4
Bolimów PL 141 F2
Bolintin-Deal RO 161 E7
Bolintin-Vale RO 161 E7
Boliqueime P 50 E3
Boljanić BIH 157 C8
Boljanići MNE 163 C7
Boljevac SRB 159 F8
Boljevci SRB 158 D5
Bölkow D 83 B11
Bolków PL 81 E9
Bollebygd S 91 D12
Bollène F 35 B8
Bollengo I 68 C4
Bollermoen N 108 D6
Bolligen CH 31 B11
Bolling DK 86 D3
Bollingstedt D 82 A6
Bollnäs S 103 D11
Bollstabruk S 107 E13
Bollullos Par del Condado
E 51 E6
Bolnes NL 182 B5
Bolnhurst UK 15 C8
Bologna I 66 D3
Bologne F 26 D3
Bolognola I 62 B4
Bolotana I 64 C2
Boloteşti RO 153 F10
Bolsena I 62 B1
Bol'shakovo RUS 136 D4
Bolsfjäll S 106 B8
Bolsover UK 11 E9
Bolsward NL 16 B5
Bolszewo PL 138 A5
Boltaña E 33 F6
Boltåsen S 111 C12
Boltenhagen D 83 B10
Boltigen CH 31 B11
Bolton UK 11 D7
Bölüntü TR 181 A7
Bolvaşniţa RO 159 C9
Bolventor UK 12 D5
Bóly H 149 E11
Bolyarovo BG 167 E7
Bolzano I 72 D3
Bomal B 19 D12
Bomba I 63 C6
Bombarral P 44 F2
Bomlitz D 82 E7
Bompas F 34 E4
Bompensiere I 58 E4
Bompietro I 58 D5
Bomporto I 66 C3
Bomsund S 107 E9
Bona F 30 A3
Bonaduz CH 71 D8
Bonakas N 113 C11
Bönan S 103 E13
Bonanza E 52 C4
Boñar E 39 C9
Bonar Bridge UK 3 K8
Bonarcado I 64 C2
Bonares E 51 E6
Bonäs S 102 D7
Bönäset S 107 E15
Bonassola I 37 C11
Bonawe UK 4 C6
Bonboillon F 26 F4
Bonchamp-lès-Laval
F 23 D10
Bonchester Bridge UK 5 E11
Boncourt CH 27 F7
Bondeno I 66 C3
Bondersbyn S 119 C11
Bonderup DK 86 A4
Bondorf D 187 D6
Bondsjö S 103 A14
Bondues F 182 D2
Bönebüttel D 83 B8
Bonefro I 63 D7
Bönen D 17 E9
Bo'ness UK 5 C9
Bonete E 55 B10
Bönhamn S 103 A15
Boniches E 47 E9
Boniewo PL 138 E6
Bonifacio F 37 J10

Bojnice SK 147 D7
Bojničky SK 146 E5
Bojnik SRB 164 C4
Bojszów PL 142 F5
Bojszowy PL 143 F7
Boka SRB 159 C6
Bokel D 17 B11
Böklund D 82 A7
Bokod H 149 A10
Bököny H 151 B8
Bokšić HR 149 E10
Boksjön S 109 E10
Bol HR 156 F6
Bolanden D 186 B5
Bolandoz F 31 A9
Bolaños de Calatrava E 54 B5
Bolătău RO 153 D8
Bolayir TR 173 C6
Bolbaite E 56 C3
Bolbec F 18 E1
Bolboşi RO 159 D11
Bölcske H 150 D2
Boldekow D 84 C5
Bolderslev DK 86 F4
Boldeşti-Grădiştea RO 161 D9
Boldeşti-Scăeni RO 161 D8
Boldog H 150 B4
Boldogkőváralja H 145 G3
Boldre UK 13 D11
Boldu RO 161 C10
Boldur RO 159 B8
Boldva H 145 G2
Bole I 41 D10
Bolekhiv UA 145 E8
Boleráz SK 146 E5
Bolesław PL 143 F7

Bonifati I 60 D5
Bönigen CH 70 D5
Bonin PL 85 B10
Bonnåsjøen N 109 A10
Bonn D 21 C8
Bonnåsjøen N 109 A10
Bonnat F 29 C9
Bonndorf im Schwarzwald
D 27 E9
Bønnerup Strand DK 87 B7
Bonnes F 111 C15
Bonnet DK 86 B2
Bonnétable F 24 D3
Bonneuil-Matours F 29 B7
Bonneval F 24 D5
Bonneval-sur-Arc F 31 E11
Bonneville F 31 C9
Bonnieux F 35 C9
Bönnigheim D 27 B11
Bonnyrigg UK 5 D10
Bonny-sur-Loire F 25 E8
Bono I 22 E6
Bono I 64 C3
Bonorva I 64 C2
Bons-en-Chablais F 31 C9
Bonson F 30 D5
Bonţida RO 152 D3
Bőny H 149 A9
Bonyhád H 149 D11
Boo S 99 D10
Boock D 84 D6
Boom B 19 B9
Boortmeerbeek B 19 C10
Boos F 18 F3
Boostedt D 83 B8
Bootle UK 10 E6
Bopfingen D 75 E7
Boppard D 21 D9
Boqueixón E 38 C2
Bor CZ 75 C12
Bor S 88 A3
Bor SRB 159 E9
Borač SRB 158 F5
Borås S 91 D12
Borăscu RO 160 D2
Borba P 50 B4
Borbona I 62 B3
Borča SRB 158 D5
Borcea RO 155 E1
Borchen D 17 E11
Borci BIH 157 C7
Borci BIH 157 F8
Börcs H 149 A9
Borculo NL 17 D7
Bordalba E 41 F8
Bordány H 150 E4
Bordeaux F 28 F4
Bordeira P 50 E2
Bordei Verde RO 161 C11
Bordelum D 82 A5
Bordères-Louron F 33 E6
Bordères-sur-l'Échez F 33 D6
Bordes F 32 D5
Bordes F 33 D6
Bordesholm D 83 B8
Bordeşti RO 161 B10
Bordighera I 37 D7
Bordils E 43 C9
Bording DK 86 C4
Bordón E 42 F3
Borduşani RO 155 E1
Bore I 69 D8
Boreham UK 15 D9
Borehamwood UK 15 D8
Borek PL 143 F10
Borek Strzeliński PL 81 E12
Borek Wielkopolski PL 81 C12
Boreland UK 5 E10
Børelva N 108 B8
Borensberg S 92 B6
Borg D 20 E6
Børga FIN 127 E14
Borgafjäll S 106 B8
Borgentreich D 17 E12
Börger D 17 C9
Borger NL 17 C7
Borggård S 92 B7
Borghamn S 92 C5
Borgheim N 90 A7
Borghetto d'Arroscia I 37 C7
Borghetto di Borbera I 37 B9
Borghetto di Vara I 69 E8
Borghetto Santo Spirito I 37 C8
Borgholm S 89 B11
Borgholzhausen D 17 D10
Borgia I 59 B10
Borgloon B 19 C11
Børglum DK 90 E6
Borgo F 37 F10
Borgo a Mozzano I 66 E2
Borgo d'Ale I 68 C5
Borgofonte I 66 B2
Borgofranco d'Ivrea I 68 B4
Borgohmanero I 68 B5
Borgomaro I 37 D7
Borgone Susa I 31 E11
Borgonovo Val Tidone I 69 C7
Borgo Pace I 66 E5
Borgorose I 62 C4
Borgo San Dalmazzo I 37 C6
Borgo San Lorenzo I 66 E3
Borgo San Martino I 68 C5
Borgosesia I 68 B5
Borgo Tossignano I 66 D4
Borgo Val di Taro I 69 E8
Borgo Valsugana I 72 D3
Borgo Velino I 62 C4
Borgo Vercelli I 68 C5
Borgsjö S 103 A10
Borgsjö S 107 C13
Borgstedt D 83 B7
Borgstena S 91 D13
Borgue UK 5 F8
Borgvattnet S 107 D10
Borhaug N 94 F5
Boriaeva BG 181 B7
Borike BIH 157 E11
Borino BG 165 F9
Bofitov CZ 77 D11
Borja E 41 E9
Borja E 41 E9
Borjana SLO 73 D7
Borjelslandet S 118 C8
Borkan S 106 A8
Borkavichy BY 133 E4
Borken D 17 E7
Borken (Hessen) D 21 B12
Borkenes N 111 C11
Borki PL 141 G7
Borki PL 143 D9

Børkop DK 86 D5
Borkowice PL 141 H3
Borkowo PL 139 D12
Borkum D 17 A7
Borlänge S 97 B13
Borlești RO 153 D9
Børmarka N 104 C8
Bormio I 71 E10
Bórmarka N 104 C8
Bormujos E 51 E7
Born D 79 B9
Born NL 183 C7
Borna D 79 D12
Borna D 80 D4
Born am Darß D 83 B13
Borne NL 17 D7
Bornem B 19 B9
Bornerbroek NL 17 D7
Bornes P 39 F6
Bornes de Aguiar P 38 E4
Borne Sulinowo PL 85 C11
Bornheim D 21 C7
Bornhofen D 21 D9
Bornhöved D 83 B8
Bornich D 185 D8
Bornos E 51 F8
Bornova TR 177 C9
Boronów PL 143 E6
Borore I 64 C2
Boroşneu Mare RO 153 F8
Boroszów PL 142 E5
Borota H 150 E3
Boroughbridge UK 11 C9
Borovan BG 165 C8
Borovany CZ 77 E7
Borovo Dol BG 166 D6
Borovnica BIH 157 D9
Borovnica SLO 73 E9
Borovo BG 166 C5
Borovo Selo HR 157 B10
Borovtsi BG 165 C7
Borów PL 81 E11
Borów PL 144 B4
Borowa PL 143 F11
Borowie PL 141 G5
Borox E 46 D5
Borrby S 88 E6
Borre DK 87 F10
Borre N 95 D12
Borrentin D 84 C3
Borris IRL 9 C9
Borris IRL 9 C9
Borrisokane IRL 6 G6
Borrisoleigh IRL 9 C7
Borrowdale UK 10 B5
Børrud N 96 C7
Börrum S 93 C9
Borş RO 151 C8
Borşa RO 152 B4
Borşa RO 152 D3
Borša SK 145 G4
Borsbeek B 19 B9
Borsec RO 153 D7
Børselv N 113 C16
Borsh AL 168 D2
Borshchiv UA 154 A5
Borshchovychi UA 144 D9
Borsio GR 174 E5
Borský Svätý Jur SK 77 E12
Borsodbóta H 145 G1
Borsodnádasd H 145 G1
Borsodszentgyörgy H 145 G1
Borsodszirák H 145 G2
Borsosberény H 147 F8
Borssele NL 182 C3
Borstel D 17 C11
Bortigali I 64 C2
Bortigiadas I 64 B3
Borth UK 10 G3
Bort-les-Orgues F 29 E10
Börtnan S 102 A6
Bortnen N 100 C2
Borum DK 86 C6
Borup DK 86 C6
Borve UK 2 L4
Borynya UA 145 E6
Boryslav UA 145 E7
Borzęcin PL 143 G12
Borzonasca I 37 C10
Borzytuchom PL 85 B12
Bosa I 64 C2
Bošáca SK 146 D5
Bosa Marina I 64 C1
Bosanci RO 153 B8
Bosanska Dubica BIH 157 B6
Bosanska Gradiška BIH 157 B7
Bosanska Kostajnica BIH 156 B6
Bosanska Krupa BIH 156 C5
Bosanski Brod BIH 157 B8
Bosanski Kobas BIH 157 B8
Bosanski Novi BIH 156 B5
Bosanski Petrovac BIH 156 C5
Bosanski Šamac BIH 157 B8
Bosanско Grahovo BIH 156 D5
Bošany SK 146 D6
Bősárkány H 149 A8
Bosau D 83 B8
Boscastle UK 12 D5
Bosco I 64 C3
Bosco Chiesanuova I 69 B11
Bosco Marengo I 37 B9
Boscotrecase I 60 B2
Bösdorf D 83 B8
Bösebo S 89 A9
Bösel D 17 B9
Boshulya BG 165 E9
Bosia RO 153 C11
Bosilegrad SRB 164 E5
Bosiljevo HR 148 F4
Boskoop NL 16 D3
Boskovice CZ 77 D11
Bošnjace SRB 164 D4
Bošnjaci HR 157 B10
Bošnjani SRB 159 F7
Bošnjane SRB 159 F7
Boşorod RO 159 B11

Bössbod S 102 D7
Bosset F 29 F6
Bossolasco I 37 B8
Bôssost E 33 E7
Bostad N 110 D6
Bostan BIH 157 F7
Bostance NL 184 E5
Bøstølen N 100 C8
Boston UK 11 F11
Boston Spa UK 11 D9
Bostrak N 90 A4
Bosundet S 107 C10
Botão P 44 D4
Boteå S 107 E13
Boteni RO 160 C6
Botesdale UK 15 C11
Boteşti RO 153 C9
Boteşti RO 160 C6
Botevgrad BG 165 D8
Botevo BG 167 C9
Bothel UK 5 F10
Boticas P 38 E4
Botiz RO 151 B10
Botiza RO 152 B4
Botn N 110 C9
Botn N 111 A18
Botn N 112 D4
Botngård N 104 D7
Botnhamn N 111 A14
Botoroaga RO 161 E7
Botorrita E 41 E9
Botoš SRB 158 C5
Botoşana RO 153 B8
Botoşani RO 153 B9
Botoşeşti-Paia RO 159 E11
Botricello I 61 F7
Bötsle S 103 A14
Botsmark S 118 E4
Bottendorf (Burgwald) D 21 B11
Bottesford UK 11 F10
Bottidda I 64 C3
Bottnaryd S 91 D14
Bottrop D 17 E7
Botun NMK 168 B4
Bötzingen D 27 D8
Bouafles F 24 B5
Boucau F 32 C3
Bouc-Bel-Air F 35 D9
Boucé F 23 D11
Bouchain F 19 D7
Bouchemaine F 23 F10
Bouchoir F 18 E6
Boudry CH 31 B10
Boué F 19 E8
Bouglon F 33 B6
Bouguenais F 23 F8
Bouillargues F 35 C7
Bouillon B 19 E11
Bouilly F 25 D10
Bouin F 28 B2
Boujailles F 31 B9
Boujan-sur-Libron F 34 D5
Boulay-Moselle F 26 B5
Boulazac F 29 E7
Boulbon F 35 C8
Bouligny F 19 F12
Bouloc F 33 C10
Boulogne-Billancourt F 25 C7
Boulogne-sur-Gesse F 33 D7
Boulogne-sur-Mer F 15 F12
Bouloire F 24 E4
Boulouris F 36 E5
Boult-aux-Bois F 19 F10
Boulzicourt F 19 E10
Bouniagues F 29 F7
Bøur FO 2 A2
Bourbon-Lancy F 30 B4
Bourbon-l'Archambault F 30 B3
Bourbonne-les-Bains F 26 E4
Bourbourg F 15 F12
Bourbriac F 22 D5
Bourcefranc-le-Chapus F 28 D3
Bourdeaux F 35 A9
Bourdeilles F 29 E7
Bourdonnay F 27 C6
Bouresse F 29 C7
Bourg F 28 E4
Bourg-Achard F 18 F2
Bourganeuf F 29 D9
Bourg-Argental F 30 E6
Bourg-de-Péage F 31 E7
Bourg-de-Thizy F 30 C5
Bourg-de-Visa F 33 B7
Bourg-Dun F 18 E1
Bourg-en-Bresse F 31 C7
Bourges F 29 A10
Bourg-et-Comin F 19 F8
Bourg-Lastic F 29 D11
Bourg-lès-Valence F 30 F6
Bourgneuf-en-Mauges F 23 F10
Bourgneuf-en-Retz F 28 A2
Bourgogne F 19 F9
Bourgoin-Jallieu F 31 D7
Bourg-St-Andéol F 35 B8
Bourg-St-Bernard F 33 C9
Bourg-St-Maurice F 31 D10
Bourgtheroulde-Infreville
F 18 F2
Bourguébus F 23 B11
Bourgueil F 23 F12
Bourmont F 26 D4
Bourne UK 11 F11
Bournemouth UK 13 D11
Bournezeau F 28 B3
Bourran F 33 B6
Bourriot-Bergonce F 32 B5
Bourron-Marlotte F 25 D8
Bourscheid L 184 E5
Bourtange NL 17 C8
Bourton UK 13 C10
Boussac F 29 C10
Bousse F 186 C1
Boussens F 33 D7
Boussois F 19 D9
Boussu B 19 D8
Boutersem B 19 C10
Bouveret CH 31 C10
Bouverie UK 35 A9
Bouvron F 23 F8
Bouxières-aux-Dames F
26 C5
Bouxwiller F 27 C7
Bouy F 25 B11
Bouzonville F 21 F7
Bouzov CZ 77 C11
Bova I 59 D8
Bøvær N 111 B13
Bovalino I 59 C9
Bovallstrand S 91 C9
Bova Marina I 59 D8
Bovan SRB 159 F8
Bovec SLO 73 D8
Bóveda E 38 C5

Boveda E 40 C5
Bovegno I 69 B9
Bovenden D 78 C6
Bøverfjord N 104 E5
Boves F 18 E5
Boves I 37 C7
Bovigny B 20 D5
Boviken S 118 E6
Boville Ernica I 62 D4
Bovino I 60 A4
Bø;lingbjerg DK 86 C2
Bovolone I 66 B3
Bovrup DK 86 F5
Bow UK 3 H10
Bowes UK 11 B7
Bowmore UK 4 D4
Box FIN 127 E13
Boxberg D 81 D7
Boxberg D 187 C8
Boxdorf D 80 D5
Boxholm S 92 C6
Boxtel NL 16 E4
Boyadzhik BG 166 E6
Boyanovo BG 167 E7
Boychinovtsi BG 165 C7
Boykovo BG 165 E10
Boyle IRL 6 E6
Bø;lefoss N 90 B4
Boynes F 25 D7
Boynitsa BG 159 F9
Bø;um N 100 D5
Božava HR 67 D9
Bozburun TR 181 C8
Bozeat UK 15 C7
Bozel F 31 E10
Boževac SRB 159 D7
Boževo PL 139 E8
Bozhentsi BG 166 D4
Bozhurishte BG 165 D7
Božice CZ 77 E10
Bozieni MD 154 D3
Bozieni RO 153 D10
Bozioru RO 161 C9
Božjakovina HR 148 E6
Bozlar TR 173 D7
Bozouls F 34 B4
Bozovici RO 159 D9
Bozveliysko BG 167 C8
Bozzolo I 66 B1
Bra I 37 B7
Braås S 89 A8
Bråbo S 89 A10
Brabova RO 160 E2
Bracadale UK 2 L4
Bracciano I 62 C2
Brach F 28 E4
Brachbach D 21 C9
Bracieux F 24 E6
Bräcke S 103 A9
Brackenheim D 27 B11
Brackley UK 13 A12
Bracknagh IRL 7 F8
Bracknell UK 15 E7
Braco UK 5 C9
Brad RO 151 E10
Bradashesh AL 168 B3
Brădeanu RO 161 D9
Brădeni RO 152 E5
Brădeşti RO 152 E6
Brădeşti RO 160 D3
Bradford UK 11 D8
Bradford-on-Avon UK 13 C10
Bradpole UK 13 D9
Bradu RO 160 D5
Brădulet RO 160 C5
Brădut RO 153 E7
Bradwell UK 15 B12
Bradwell Waterside UK 15 D10
Brae UK 3 E14
Brædstrup DK 86 D5
Braemar UK 5 A10
Brăeşti RO 153 B8
Brăeşti RO 153 C10
Brăeşti RO 161 C9
Bråfirm E 43 E6
Braga P 38 E3
Bragadiru RO 161 E7
Bragadiru RO 161 F7
Bragança P 39 E6
Bragar UK 2 J3
Brăhăşeşti RO 153 E10
Brahlstorf D 83 D9
Brăila RO 155 C1
Brailsford UK 11 F8
Braine F 19 F8
Braine-l'Alleud B 19 C9
Braine-le-Comte B 19 C9
Braintree UK 15 D10
Braives B 19 C11
Brajković BIH 157 D8
Brake (Unterweser) D 17 B10
Brakel B 19 C8
Brakel D 17 E12
Bräkne-Hoby S 89 C8
Brålanda S 91 B11
Bralin PL 142 D4
Brallo di Pregola I 37 B10
Bralos GR 174 B5
Braloştiţa RO 160 D3
Bram F 33 D10
Bramans F 31 E10
Bramberg am Wildkogel
    A 72 B5
Bramdrupdam DK 86 D4
Bramming DK 86 E3
Brampton UK 5 F11
Brampton UK 15 C12
Bramsche D 17 D9
Bramsche D 17 D10
Bramstedt D 17 B11
Bran RO 160 B6
Brånaberg S 109 E11
Branäs S 102 E4
Brancaleone I 59 D9
Brancaster UK 15 B10
Brânceni RO 160 F6
Brâncoveneşti RO 152 D5
Brâncoveni RO 160 E4
Brand D 75 C10
Brandal N 100 B4
Brändåsen S 102 B4
Brändbo S 103 B11
Brandbu N 95 B13
Brande DK 86 D4
Brande-Hörnerkirchen D 82 C7
Brandenberg A 72 B4
Brandenburg D 79 B11
Brand-Erbisdorf D 80 E4
Branderup DK 86 E4

Brandesburton UK 11 D11
Brandis D 80 C4
Brand-Nagelberg A 77 E8
Brando F 37 F10
Brändö FIN 126 E5
Brandon UK 15 C10
Brändön S 118 C8
Brändövik FIN 122 D6
Brandshagen D 84 B4
Brandstorp S 92 C4
Brandsvoll N 90 C2
Brandval N 96 B7
Brandvoll N 111 C15
Brandýs nad Labem-Stará
    Boleslav CZ 77 B7
Brandýs nad Orlicí CZ 77 B10
Branes N 96 A6
Brănești MD 154 C3
Brănești RO 160 C6
Brănești RO 160 D2
Brănești RO 161 E8
Branice PL 142 F4
Braniewo PL 139 B8
Branik SLO 73 E8
Brănișca RO 151 F10
Brănișora RO 152 C4
Brănistea RO 155 C1
Brănistea RO 152 C4
Brankas LV 134 C7
Brănna S 91 B11
Brännåker S 107 B9
Brännäs S 103 B11
Brännberg S 118 C6
Branne F 28 F5
Brännland S 107 D16
Brännland S 122 C4
Brañosera E 40 C3
Brańsk PL 141 E7
Branston UK 11 E11
Brańszczyk PL 139 E12
Brantevik S 88 D6
Brantice CZ 142 F4
Brantôme F 29 E7
Branzi I 69 A8
Braojos E 46 B5
Braone I 69 B9
Braskereidfoss N 101 E15
Braslaw BY 133 E2
Braşov RO 161 B7
Brassac F 33 C10
Brasschaat B 16 F2
Brassy F 25 F10
Brasta S 105 E16
Brastad S 91 C10
Brastavătu RO 160 F4
Brasy CZ 76 C5
Brataj AL 168 D2
Bratca RO 151 D10
Bråte N 95 C14
Bratelu RO 152 E4
Brateljevici BIH 157 D10
Brateş RO 153 F8
Bratislava SK 77 F12
Bratkowice PL 144 C4
Bratovoeşti RO 160 E3
Bratsigovo BG 165 E9
Brattäker S 107 A11
Brattbäcken S 107 C9
Bratten S 107 B14
Brattfors S 103 A12
Brattfors S 107 D17
Brattli N 114 D7
Brattmon S 102 E4
Bratton UK 13 C10
Brattsbacka S 107 D16
Brattsele S 107 C13
Brattset N 104 E5
Brattvåg N 100 A4
Bratunac BIH 158 E3
Brăţuşeni MD 153 A10
Bratya Daskalovi BG 166 E4
Braubach D 21 D9
Braud-et-St-Louis F 28 E4
Braunau am Inn A 76 F4
Brauneberg D 185 E6
Braunfels D 21 C10
Braunlage D 79 C8
Bräunlingen D 27 E9
Braunsbach D 187 C8
Braunsbedra D 79 D10
Braunschweig D 79 B8
Braunton UK 12 C6
Bravicea MD 154 C2
Bravnica BIH 157 D7
Bray IRL 7 F10
Bray-sur-Seine F 25 D9
Bray-sur-Somme F 18 E6
Brazatortas E 54 B4
Brazey-en-Plaine F 26 F3
Brazi RO 161 D8
Brazii RO 151 E9
Brazii RO 161 D8
Brbinj HR 67 D11
Brčigovo BIH 157 E11
Brčko BIH 157 C10
Brdów PL 138 F6
Bré IRL 7 F10
Brea E 41 E8
Brea de Tajo E 47 D6
Breaghva IRL 8 C3
Breascleit UK 2 J3
Breasta RO 160 E3
Bresternica SLO 148 C5
Brestova HR 67 B9
Brestovac SRB 159 E9
Brestovac SRB 164 C4
Brestovac Požeški HR 149 F9
Brestovăţ RO 151 F8
Brestovene BG 161 F9
Brestovets BG 165 C10
Brestovitsa BG 165 E10
Bretagne-d'Armagnac F 33 C6
Breţcu RO 153 E8
Bretea Română RO 159 B11
Breteil F 23 D8
Bretenoux F 29 F9
Breteuil F 18 E5
Breteuil F 24 C4
Brétignolles-sur-Mer F 28 B2
Bretnig D 80 D6
Bretocino S 89 D8
Bretoncelles F 24 D4
Bretten D 27 B10
Bretten UK 10 E6
Brettville-sur-Laize F 23 B11
Bretzenheim D 185 E8
Bretzfeld D 187 C7
Breuberg-Neustadt D 21 E12
Breuches F 26 E5
Breugel NL 183 B6
Breuil-Cervinia I 68 B4
Breuil-Magné F 28 D4

Bredsjön S 103 A13
Bredstedt D 82 A5
Bredsten DK 86 D4
Bredträsk S 107 D15
Bredvik S 107 D17
Bredviken S 108 D8
Bredviken S 119 C10
Bree B 19 B12
Breese D 83 D11
Bregana HR 148 E5
Breganze I 72 E4
Bregare BG 165 B9
Bregeni MD 153 B12
Bregenz A 71 B9
Breg-Lum AL 163 E9
Breg-Lum AL 168 B2
Bregninge DK 86 F6
Bregovo BG 159 E10
Breguzzo I 69 B10
Bréhal F 23 C8
Bréhan F 22 D6
Brehna D 79 C11
Breidenbach D 21 C10
Breidenbach F 27 B7
Breidstrand N 111 C12
Breidvik N 108 B9
Breidvik N 110 C8
Breidvik N 110 E9
Breidvik N 111 D11
Breidvik N 111 D11
Breihölz D 82 B7
Breil CH 71 D8
Breil-sur-Roya F 37 D7
Breisach am Rhein D 27 D8
Breistein N 94 B2
Breitenbach CH 27 F8
Breitenbach D 21 F8
Breitenbach (Schauenburg)
    D 17 F12
Breitenbach am Herzberg
    D 78 E6
Breitenbach am Inn A 72 B4
Breitenberg D 76 E5
Breitenbrunn D 75 D10
Breitenburg D 82 C7
Breitenfelde D 83 C8
Breitengüßbach D 75 C8
Breitenhagen D 79 C10
Breitnau D 27 E9
Breitscheid D 183 C9
Breitscheid D 185 C7
Breitscheid D 185 C9
Breitungen D 79 E7
Breivik N 111 C12
Breivik N 112 B9
Breivikbotn N 112 B9
Breivikeidet N 111 A18
Brejning DK 86 D5
Brekka N 108 D8
Brekken N 101 A15
Brekken N 108 D5
Brekkestø N 90 C3
Brekkhus N 100 E4
Brekkvasselv N 105 B14
Breklum D 82 A5
Brekova SRB 158 F5
Breksillan N 105 B10
Brekstad N 104 D7
Brelès F 22 D2
Brelingen (Wedemark) D 78 A6
Bremdal DK 86 B3
Bremen D 17 B11
Bremerhaven D 17 A11
Bremervörde D 17 B12
Bremgarten CH 27 F9
Bremm D 21 D8
Bremnes N 94 C2
Bremnes N 110 C8
Brem-sur-Mer F 28 B2
Brenderup DK 86 E5
Brenes E 51 D8
Brenguļi LV 131 F11
Brenitsa BG 165 C9
Brenna N 108 F5
Brenna N 110 D8
Brenna PL 147 B7
Brennero I 72 C4
Brennes N 112 D5
Brennfjell N 112 E5
Brenngam N 113 B19
Brennhaug N 101 C10
Brennmo N 105 E11
Brennsvik N 113 B13
Breno I 69 B9
Brénod F 31 C8
Brens F 33 C9
Brensbach D 187 B6
Brent Knoll UK 13 C8
Brentwood UK 15 D9
Brenzone I 69 B10
Brimnes N 94 B5
Brinches P 50 C4
Brincones E 45 B8
Brindisi I 61 B9
Brindisi Montagna I 60 B5
Bringsinghaug N 100 B2
Bringsli N 108 B8
Brinian UK 3 G11
Brinje HR 67 B11
Brinkum D 17 B9
Brinkum D 17 B11
Brinlack IRL 6 B6
Brinon-sur-Beuvron F 25 F9
Brinon-sur-Sauldre F 25 E7
Brin-sur-Seille F 26 C5
Brínza MD 155 B2
Brînzeni MD 153 A10
Brion F 30 F3
Briones E 40 C6
Brionne F 24 B4
Brioude F 30 E3
Brioux-sur-Boutonne F 28 C5
Briouze F 23 C11
Briscous F 32 D3
Brisighella I 66 D4
Brissac-Quincé F 23 F11
Bristol UK 13 C9
Bristol UK 15 B11
Britelo P 38 E3
Briton Ferry UK 13 B7
Brittas IRL 7 F10
Brittas Bay IRL 9 C11
Britvica BIH 157 F8
Britz D 84 E5
Brive-la-Gaillarde F 29 E9
Briviesca E 40 C5
Brix F 23 A8
Brixen im Thale A 72 B5
Brixham UK 13 E7
Brixworth UK 15 C7
Brka BIH 157 C10
Brložnik BIH 157 D10
Brna HR 162 D2

Brnaze HR 157 E6
Brněnec CZ 77 C11
Brništĕ CZ 81 E7
Brnjica SRB 159 D8
Brnjica SRB 163 C9
Brno CZ 77 D11
Bro S 93 D12
Bro S 99 C9
Broadford IRL 8 C5
Broadford IRL 8 D5
Broad Haven UK 9 E12
Broadheath UK 13 A10
Broadstairs UK 15 E11
Broadway UK 13 A11
Broadwey UK 13 D10
Broadwindsor UK 13 D9
Broager DK 86 F5
Broaryd S 87 A12
Broby S 88 C6
Broby S 99 C11
Brobyværk DK 86 E6
Broc CH 31 B11
Broćanac BIH 157 F7
Brocas F 32 B4
Broceni LV 134 C5
Bröckel D 79 A7
Brockum D 17 D10
Brockworth UK 13 B10
Broczyno PL 85 C10
Brod NMK 168 A5
Brod NMK 169 C6
Brod RKS 163 F10
Brod RKS 164 E3
Brodalen S 91 C10
Brodarevo SRB 163 C8
Broddbo S 98 C6
Brodek u Prostějova CZ 77 D12
Broderstorf D 83 B12
Brodica SRB 159 E8
Brodick UK 4 D6
Brodilovo BG 167 E9
Brodina RO 152 B6
Brodnica PL 81 B11
Brodnica PL 139 D7
Brodosané RKS 163 E10
Brodské SK 77 E12
Brody PL 81 C7
Brody PL 81 C7
Brok PL 139 E12
Brokdorf D 17 A12
Brokind S 92 C7
Brokka N 90 C2
Brokstedt D 83 C7
Brolo I 59 C6
Bromarv FIN 127 F9
Brome D 79 A8
Bromma N 95 B10
Bromnes N 112 C2
Bromölla S 88 C6
Bromsgrove UK 13 A10
Bromyard UK 13 A9
Bron F 30 D6
Bronchales E 47 D9
Brønderslev DK 86 A5
Broni I 69 C7
Bronice PL 81 C7
Brønnøysund N 108 F3
Bronnitsya UA 154 A1
Brøns DK 86 E3
Bronte I 59 D6
Bronzani Majdan BIH 157 C6
Brooke UK 15 B11
Brookeborough UK 7 D8
Broons F 23 D7
Broquiès F 34 B4
Brora UK 3 J9
Brørup DK 86 E4
Brösarp S 88 D6
Broscăuţi RO 153 B8
Broseley UK 10 F7
Broshniv Osada UA 145 F9
Brossac F 28 E5
Brøstadbotn N 111 B14
Broşteni RO 153 C7
Broşteni RO 153 F10
Broşteni RO 159 D10
Brotas P 50 B3
Brötjemark S 92 D4
Broto E 32 F5
Brottes F 26 D3
Brotton UK 11 B10
Brøttum N 101 D13
Brou F 24 D5
Brough UK 3 H10
Brough UK 11 B7
Broughshane UK 4 F4
Broughton UK 5 D10
Broughton UK 15 D11
Broughton UK 3 G11
Broughton in Furness UK 10 C5
Broughton UK 3 G11
Broughty Ferry UK 5 C11
Broumov CZ 81 E10
Brousseval F 26 D2
Broutzaiika GR 175 C6
Brouvelieures F 26 D5
Brouwershaven NL 16 E1
Brovst DK 86 A5
Brownhills UK 11 F8
Broxburn UK 5 D10
Brozany CZ 76 B6
Brozas E 45 E7
Brožići LT 134 E3
Brozzo I 69 B9
Brštanovo HR 156 E5
Brtnice CZ 77 D9
Bruay-la-Bussière F 18 D6
Bruchhausen-Vilsen D 17 C12
Bruchköbel D 187 A6
Bruchsal D 27 B10
Bruchweiler-Bärenbach
    D 186 C4
Brück D 79 B12
Bruck an der
    Großglocknerstraße A 73 B6
Bruck an der Leitha A 77 F11
Bruck an der Mur A 73 B11
Brücken D 21 E8
Brücken (Helme) D 79 D9
Brücken (Pfalz) D 21 F8

Brückl A 73 C10
Bruckmühl D 72 A4
Brudzeń Duży PL 139 E8
Brudzew PL 142 B6
Brudzowice PL 143 E7
Brue-Auriac F 35 C10
Brüel D 71 A10
Brüere-Allichamps F 29 B10
Bruff IRL 8 D5
Brugelette B 182 D3
Bruges B 19 B7
Bruges B 182 C5
Brugg CH 27 F9
Brugge B 19 B7
Brüggen D 16 F6
Brüggen D 78 B6
Brugnato I 69 E8
Brugnera I 72 E6
Bruguières I 33 C8
Bruhagen N 104 E3
Brühl D 21 C7
Brühl D 187 C6
Bruinisse NL 16 E2
Bruiu RO 152 F5
Brûlon F 23 E11
Brûly B 19 E10
Brumath F 27 C8
Brummen NL 183 A8
Brumovice CZ 77 E11
Brumov-Bylnice CZ 146 C6
Brumunddal N 101 E13
Brumundsag N 101 E13
Brunau D 83 E10
Brunava LV 135 D8
Brundby DK 86 D7
Brundish UK 15 C11
Brunehamel F 19 E9
Brunete E 46 D5
Brunflo S 106 E7
Brunico I 72 C4
Bruniquel F 33 B9
Brunkeberg N 95 D8
Brunn D 84 C4
Brunna S 99 C8
Brunna S 99 C9
Brunn am Gebirge A 77 F10
Brunnberg S 97 B10
Brunne S 103 A14
Brunnen CH 71 D8
Brunnsberg S 102 D6
Brunsbüttel D 17 A12
Brunssum NL 20 C5
Bruravik N 94 B5
Bruree IRL 8 D5
Brus SRB 163 C11
Brusago I 69 A11
Brusand N 94 E3
Brušane HR 156 C3
Brusarci BG 159 F11
Brüsewitz D 83 C10
Brushlen BG 161 F8
Brushlyanitsa BG 165 B10
Brusio CH 69 A9
Brusnik SK 147 E8
Brusnika Velika BIH 157 B9
Brusno SK 147 D9
Brusque F 34 C4
Brusson I 68 B4
Brüssow D 84 D6
Brusturi RO 151 C9
Brusturi-Drăgăneşti RO 153 C8
Brusturoasa RO 153 D8
Brusy PL 138 C5
Bruton UK 13 C10
Bruttig-Fankel D 21 D8
Bruvno HR 156 C3
Bruxelles B 19 C9
Bruyères F 26 D5
Bruz F 23 D8
Bruzaholm S 92 D6
Brvnište SK 146 C6
Brwinów PL 141 F3
Bryagovo BG 166 F4
Bryastovo BG 165 C10
Bryggerhaug N 111 B14
Brymbo UK 10 E5
Brynamman UK 13 B7
Bryne N 94 E3
Brynford UK 10 E5
Bryngė S 107 E14
Bryngelhögen S 102 B7
Brynje S 102 A3
Brynje S 106 E7
Brynmawr UK 13 B8
Bryrup DK 86 D5
Bryukhovychi UA 144 D8
Brzan SRB 159 E7
Brza Palanka SRB 159 E9
Brzece SRB 163 C10
Brzeg PL 142 E4
Brzeg Dolny PL 81 D11
Brześć Kujawski PL 138 E6
Brzesko PL 143 G10
Brzeszcze PL 143 G7
Brzezie PL 85 C11
Brzezie PL 138 E6
Brzezinki PL 141 H5
Brzeziny PL 142 C5
Brzeziny PL 143 C8
Brzeziny PL 144 C3
Brzeźnica PL 81 C8
Brzeźnica PL 147 B9
Brzeźno PL 85 C9
Brzeźno PL 142 B5
Brzostek PL 144 D4
Brzotin SK 145 F1
Brzóza PL 141 G4
Brzozów PL 141 F2
Brzozów PL 144 D5
Brzozowiec PL 81 A8
Bruzze PL 139 D7
Bů F 24 C5
Bua S 87 A10
Buais F 23 C10
Buarcos P 44 D3
Buavågen N 94 C2
Bubbio I 37 B8
Bubiai LT 134 E6
Bubry F 22 E5
Bubwith UK 11 D10
Buca TR 177 C9
Buçany SK 146 E5
Buccheri I 59 E6
Bucchianico I 63 C6
Buccinasco I 69 C7
Buccino I 60 B4

Bucecea RO 153 B8
Bucelas P 50 B1
Buceş RO 151 E10
Bucy-lès-Gy F 26 F4
Buch D 71 A10
Buch am Erlbach D 75 F11
Buchbach D 75 F11
Buchboden A 71 C9
Büchel D 21 D7
Büchen D 83 D9
Buchen (Odenwald) D 27 A11
Büchenbeuren D 185 E7
Buchholz D 83 D12
Buchholz (Aller) D 82 E7
Buchholz (Westerwald) D 21 C9
Buchin RO 159 C9
Buchin Prohod BG 165 D7
Buch in Tirol A 72 B4
Buchkirchen A 76 F6
Büchlberg D 76 E5
Buchloe D 71 A11
Buchlovice CZ 146 C4
Buchoiz in der Nordheide
    D 83 D8
Buchs CH 71 C8
Buchy F 18 E3
Bučim NMK 164 F5
Buçimas AL 168 C4
Bucine I 66 F4
Bucinişu RO 160 F4
Bučişte NMK 164 F5
Buciumi RO 151 C11
Buciumeni RO 153 F10
Buciumeni RO 161 C6
Buciumi RO 151 C11
Bučje SRB 159 F9
Bučje SRB 163 C7
Buckden UK 11 C7
Bückeburg D 17 D12
Bücken D 17 C12
Buckfastleigh UK 13 D7
Buckhaven UK 5 C10
Buckie UK 3 K11
Buckingham UK 14 D7
Buckley UK 10 E5
Buckode IRL 6 D6
Buckow Märkische Schweiz
    D 80 A6
Bückwitz D 83 E12
Bucoşniţa RO 159 C9
Bucov RO 161 D8
Bucovăţ MD 154 C2
Bucovăţ RO 160 E3
Bucovica BIH 157 E7
Bučovice CZ 77 D12
Bucsa H 151 C7
Bucşani RO 161 D7
Bucşani RO 161 E7
Bucu RO 161 D9
Bucureşti RO 151 E10
Bucureşti RO 161 E8
Bucy-lès-Pierrepont F 19 E8
Bucz PL 81 B9
Buczek PL 143 D7
Bud N 100 A5
Budacu de Jos RO 152 C5
Budakalász H 150 B3
Budakeszi H 149 A11
Budakovo NMK 168 B5
Budaörs H 149 B11
Budapest H 150 C3
Budča SK 147 D8
Buddusò I 64 B3
Bude UK 12 D5
Budeasa RO 160 D5
Budel NL 16 F5
Büdelsdorf D 82 B7
Budenets' UA 153 A7
Budenheim D 21 D11
Budeşti RO 152 B3
Budeşti RO 152 D4
Budeşti RO 160 C4
Budeşti RO 161 E8
Budeyi UA 154 A4
Budia E 47 C7
Budila RO 161 B7
Budimci HR 149 F10
Budimlić Japra BIH 156 C5
Budiná SK 147 E8
Büdingen D 21 D12
Budinščina HR 148 D6
Budišov CZ 77 D9
Budišov nad Budišovkou
    CZ 146 B5
Budkovce SK 145 F5
Budmerice SK 146 E4
Budoia I 72 D6
Budoni I 64 B4
Budrio I 66 C4
Budry PL 136 E4
Budureasa RO 151 D10
Buduslău RO 151 C9
Budva MNE 163 E6
Büdvcčiai LT 136 D6
Budyně nad Ohří CZ 76 B6
Budziszewice PL 141 G1
Budzów PL 147 B9
Budzyń PL 85 E11
Bue N 94 E3
Bueña E 47 C10
Buenache de Alarcón E 47 E8
Buenache de la Sierra E 47 D8
Buenaventura E 46 D3
Buenavista de Valdavia E 39 C10
Buendía E 47 D7
Buer D 183 B10
Buer N 96 D6
Bueu E 38 D2
Bufleben D 79 D8
Buftea RO 161 D7
Bugac H 150 D4
Bugarra E 48 E3
Bugeat F 29 D9
Buğdayli TR 173 D8
Bugeat F 29 D9
Buggenhout B 182 C4
Buggerru I 64 E1
Buggingen D 27 E8
Bugiac MD 154 E3
Bugnara I 62 C5
Bugnein F 32 D4
Bugøyfjord N 114 D7
Bugøynes N 114 D7
Bugyi H 150 C3
Bühl D 27 C9
Bühlertal D 27 C9
Bühlertann D 74 D6
Bühlerzell D 74 D6

Buhoci RO 153 D10
Buhølen N 90 C1
Buhuşi RO 153 D9
Buia I 73 D7
Builth Wells UK 13 A8
Buis-les-Baronnies F 35 B9
Buitenpost NL 16 B6
Buitrago del Lozoya E 46 C5
Buivydiškės LT 137 D11
Buivydžiai LT 137 D12
Buják H 147 F9
Bujalance E 53 A8
Bujan AL 163 E9
Bujanovac SRB 164 E4
Bujaraloz E 42 E3
Buje HR 67 B8
Bujor MD 154 D2
Bujoreni RO 160 C4
Bujoreni RO 161 E7
Bujoru RO 161 F7
Bük H 149 B7
Buk PL 81 B11
Bukaiši LV 134 D6
Bukhava BY 133 E6
Bukhovo BG 165 D8
Bukhovtsi BG 167 C7
Bükkábrány H 145 H2
Bükkösd H 149 D9
Bukksnes N 111 C10
Bükkszék H 145 H1
Bükkszentkereszt H 145 G2
Bükkzsérc H 145 H2
Buko D 79 C11
Bukonys LT 135 F8
Bukova Gora BIH 157 E7
Bukovče SRB 159 E10
Bukovec CZ 147 B7
Bukovets BG 165 D8
Bukovje SLO 73 E9
Bukovo NMK 168 C5
Buków PL 81 B9
Bukowe PL 142 B4
Bukowice PL 81 D12
Bukowiec PL 81 B9
Bukowiec PL 81 B10
Bukowiec PL 138 D5
Bukowina PL 138 D12
Bukowina Tatrzańska PL 145 E1
Bukownica PL 142 D5
Bukowsko PL 145 E5
Bülach CH 27 E10
Bulboaca MD 154 D4
Bulbucata RO 161 E7
Bulçar AL 168 C3
Buldoo UK 3 H9
Bulford UK 13 C11
Bŭlgarene BG 165 C11
Bŭlgarevo BG 167 C10
Bŭlgari BG 167 E9
Bŭlgarin BG 166 F5
Bŭlgarovo BG 167 D8
Bŭlgarska Polyana BG 166 E6
Bulgnéville F 26 D4
Bŭlhaç D 17 A11
Bulkington UK 13 A12
Bulkowo PL 139 E9
Bullas E 55 C9
Bullaun IRL 6 F5
Bullay D 21 D8
Bulle CH 31 B11
Bullerup DK 86 E6
Büllingen B 20 D6
Bullmark S 122 B4
Bully-les-Mines F 18 D6
Bulqizë AL 168 A3
Bultei I 64 C3
Bulz RO 151 D10
Bulzeşti RO 160 D3
Bulzeştii de Sus RO 151 E10
Bulzi I 64 B2
Bumbeşti-Jiu RO 160 C2
Bumbeşti-Piţic RO 160 C3
Buna BIH 157 F8
Bunacurry IRL 6 E4
Bunalty IRL 6 D3
Bun an Churraigh IRL 6 E3
Bun an Phobail IRL 4 E2
Bun an Tábhairne IRL 8 E6
Bunbeg IRL 6 B6
Bunclody IRL 9 C9
Buncrana IRL 4 E2
Bun Cranncha IRL 4 E2
Bünde D 17 B8
Bünde D 17 D11
Bundenbach D 21 E8
Bunderhee D 17 B8
Bun Dobhrain IRL 6 D6
Bundoran IRL 6 D6
Bunessan UK 4 C4
Buneşti RO 152 E6
Buneşti RO 153 B8
Buneşti-Avereşti RO 153 D11
Bungay UK 15 C11
Bunge S 93 D14
Bunić HR 156 C4
Buniel E 40 D4
Bunila RO 159 B10
Bunka LV 134 D3
Bunkeflostrand S 87 D11
Bunkris S 102 C4
Bunmahon IRL 9 D8
Bunnaglass IRL 6 F5
Bun na hAbhna IRL 6 D3
Bunnahowen IRL 6 D3
Bun na Leaca IRL 6 B6
Bunnaddan IRL 6 D5
Bunnanaddan IRL 6 D5
Buñol E 48 F3
Bunschoten-Spakenburg
    NL 16 D4
Bunteşti RO 151 D10
Buntingford UK 15 D8
Bunyola E 49 E10
Buoač BIH 157 C7
Buochs CH 71 D6
Buollannjárga N 113 E15
Buonabitacolo I 60 C5
Buonalbergo I 60 A3
Buonconvento I 66 F3
Buonvicino I 60 D5
Bur DK 86 C2
Buran N 105 D11
Burbach D 21 C10
Burbage UK 13 C11
Burbáguena E 47 B10
Burcei I 64 E3
Bŭrdarski Geran BG 165 B8
Burdinne B 19 C11
Bureå S 118 E6
Bureåborg S 107 E13
Büren D 17 E11

Buren NL 183 B6
Büren an der Aare CH 27 F7
Bures UK 15 D10
Buresjön S 109 E14
Burfjord N 112 D7
Burford UK 13 B11
Burg D 80 C6
Burg (Dithmarschen) D 82 C6
Burganes de Valverde E 39 E8
Burgas BG 167 E8
Burgau A 148 B6
Burgau D 75 F7
Burgau P 50 E2
Burg auf Fehmarn D 83 B10
Burg bei Magdeburg D 79 B10
Burgberg im Allgäu D 71 B10
Burgbernheim D 75 D7
Burgbrohl D 185 D7
Burgdorf CH 70 C5
Burgdorf D 78 C6
Burgdorf D 79 B7
Burgebrach D 75 C8
Bürgel D 79 E10
Burgess Hill UK 15 F8
Burghausen D 76 F3
Burghclere UK 13 C12
Burghead UK 3 K10
Burgh-Haamstede NL 16 E1
Burgh le Marsh UK 11 E12
Burgio I 58 D3
Burgkirchen an der Alz D 75 F12
Burgkunstadt D 75 B9
Bürglen CH 27 E11
Bürglen CH 71 D7
Burglengenfeld D 75 D11
Burgohondo E 46 D3
Burgos E 40 D4
Burgos I 64 C2
Burgsalach D 75 D9
Burgsinn D 74 B6
Burgstädt D 79 E12
Burgstädt D 79 E12
Burg Stargard D 84 D4
Burgsvik S 93 E12
Burgthann D 75 D9
Burgtonna D 79 D8
Burgui E 32 E3
Burguillos E 51 E8
Burguillos del Cerro E 51 C6
Burguillos de Toledo E 46 E5
Burgum NL 16 B6
Burgwindheim D 75 C8
Burhaniye TR 173 F6
Burhave (Butjadingen) D 17 A10
Buriasco I 31 F11
Burie F 28 E5
Burila Mare RO 159 E10
Burizane AL 168 A2
Burjassot E 48 E4
Burjuc RO 151 E9
Burkardroth D 74 B6
Burkhardtsdorf D 80 E3
Burlada E 32 E2
Burladingen D 27 D11
Burlats F 33 C10
Bürmoos A 76 G3
Burnchurch IRL 9 C8
Burness UK 3 G11
Burnfoot UK 4 F3
Burnfoot IRL 4 E2
Burnham UK 15 D7
Burnham Market UK 15 B10
Burnham-on-Crouch UK 15 D10
Burnham-on-Sea UK 13 C9
Burniston UK 11 C11
Burnley UK 11 D7
Burntisland UK 5 C10
Burntwood Green UK 11 F8
Burón E 39 B9
Buronzo I 68 C5
Buros F 32 D5
Burow D 84 C4
Burøysund N 112 C4
Burrafirth UK 3 E14
Burrel AL 168 A3
Burren UK 7 D10
Burren IRL 6 F4
Burriana E 48 E4
Burry Port UK 12 B6
Bürs A 71 C9
Burs S 93 D13
Bursa TR 173 E11
Burscough Bridge UK 10 D6
Burseryd S 87 A12
Bursfelde D 78 C6
Bürstadt D 21 E11
Burstow UK 15 E8
Bursuc MD 154 C2
Burtnieki LV 131 F10
Burton-in-Kendal UK 10 C6
Burton Latimer UK 15 C7
Burtonport IRL 6 C6
Burton upon Trent UK 11 F8
Burträsk S 118 E5
Buru RO 152 D3
Burujón E 46 E4
Burwarton UK 13 A9
Burwash UK 15 F9
Burwell UK 15 C9
Burwick UK 3 H11
Bury UK 11 D7
Bury St Edmunds UK 15 C10
Burzenin PL 142 D6
Busachi I 64 C2
Busalla I 37 B9
Busana I 66 D1
Busanski Dubočac BIH 157 B8
Busca I 37 B6
Buscemi I 59 E6
Busdorf D 82 B7
Buseto Palizzolo I 58 C2
Buševec HR 148 E6
Busha UA 153 A12
Bushmills UK 4 E3
Bushtricë AL 163 F9
Bushtyna UA 145 G7
Busigny F 19 D7
Busilovac SRB 159 F7
Bušince SK 147 E9
Buskhyttan S 93 B9
Byssträsk S 107 C15
Buślary PL 85 C10
Bušletić BIH 157 C9
Bušnevo HR 148 E6
Busot E 56 E4
Busovača BIH 157 D8
Bussang F 27 E6
Busseto I 69 D10
Bussière-Badil F 29 D7
Bussière-Galant F 29 D8
Bussière-Poitevine F 29 C7
Bussigny CH 31 B10

Bussi sul Tirino I 62 C5
Büßleben D 79 E9
Bussolengo I 66 B2
Bussoleno I 31 E11
Bussum NL 16 D4
Bussy-le-Grand F 25 E12
Bussy-en-Othe F 25 D10
Buşteni RO 161 C7
Bustillo del Páramo E 39 D8
Busto Arsizio I 69 B6
Bustuchin RO 160 D3
Busturi-Axpe E 41 B6
Büsum D 82 B5
Bŭta BG 165 E9
Butan BG 160 F3
Buteni RO 151 E9
Butera I 58 E5
Buteşti MD 153 B10
Bütgenbach B 20 D6
Butimanu RO 161 D7
Bütingė LT 134 D2
Butler's Bridge IRL 7 D8
Butoieşti RO 160 D2
Butor MD 154 C4
Bütow D 83 D12
Butrimonys LT 137 E9
Butrimonys LT 137 E11
Butryny PL 139 C10
Bütschwil CH 27 F11
Büttelborn D 187 B6
Buttelstedt D 79 D9
Butterme UK 10 B5
Bütthard D 74 C6
Buttigliera d'Asti I 37 A8
Buttle S 93 E11
Büttstädt D 79 D9
Büttstedt D 79 D7
Butuceni MD 154 B3
Butuceni MD 154 C3
Buturugeni RO 161 E7
Butzbach D 21 D11
Bützow D 83 C11
Buurse NL 17 D8
Buvåg N 110 D9
Buvik N 111 A15
Buvika N 104 E8
Buxerolles F 29 B6
Buxerolles-les-Mines F 30 C2
Buxted UK 15 F9
Buxtehude D 82 D7
Buxton UK 11 E8
Buxy F 30 B6
Buynovtsi BG 166 D5
Büyükada TR 173 C11
Büyükaltiağaç TR 171 B10
Büyükanafarta TR 171 D10
Büyükbelen TR 177 B10
Büyükçavuşlu TR 173 B10
Büyükçekmece TR 173 B10
Büyükçiğli TR 177 C9
Büyükdöllük TR 167 F7
Büyük Evren TR 171 C10
Büyükgerdelli TR 167 F7
Büyükkarakarli TR 173 B8
Büyükkariştiran TR 173 B8
Büyükkiliçli TR 173 B9
Büyükorhan TR 173 E10
Büyükyenice TR 173 F8
Büyükyoncali TR 173 B8
Buza RO 152 D4
Buzançais F 29 B8
Buzancy F 19 F10
Buzău RO 161 C9
Buzescu RO 160 E6
Buzet HR 67 B8
Buzet-sur-Baïse F 33 B6
Buzet-sur-Tarn F 33 C9
Buziaş RO 159 B8
Büzim BIH 156 B5
Buzoeşti RO 160 D5
Buzsák H 149 C9
By N 104 D8
Byala BG 166 C5
Byala BG 166 D6
Byala BG 167 D9
Byala Cherkva BG 165 F10
Byala Cherkva BG 166 C4
Byala Reka BG 166 E4
Byala Reka BG 167 C9
Byala Slatina BG 165 C8
Byal Izvor BG 171 A8
Byalo Pole BG 166 E5
Byberget S 103 A9
Bybjerg DK 87 D9
Bychawa PL 144 A6
Bycina PL 142 F6
Byczyna PL 142 D5
Bydalen S 105 E15
Bydgoszcz PL 138 D5
Bye S 106 E7
Byelavyezhski BY 141 F8
Byel'ki BY 133 F2
Byenyakoni BY 137 E11
Byershty BY 137 F9
Byfield UK 13 A12
Bygdeå S 122 B5
Bygdeträsk S 118 E5
Bygdsiljum S 118 E5
Bygland N 90 B2
Byglandsfjord N 90 B2
Bygstad N 100 D3
Byhleguhre D 80 B6
Bykle N 94 D6
Bylchau UK 10 E4
Bylderup-Bov DK 86 F4
Byn S 97 B10
Byneset N 104 E8
Byremo N 90 C1
Byrkjedal N 94 E4
Byrkjelo N 100 D5
Byrknes N 100 E1
Byrtegrend N 94 C7
Byrum DK 87 A7
Byske S 118 E6
Bysshytten S 93 B9
Bystré CZ 77 C10

Bystrička SK 147 C7
Bystrzyca PL 142 E3
Bystrzyca Kłodzka PL 77 B11
Bytča SK 147 C7
Bytnica PL 81 B8
Bytom PL 143 F6
Bytom Odrzański PL 81 C9
Bytoń PL 138 E6
Bytów PL 85 B12
Byvallen S 102 B7
Byviken S 107 D16
Byxelkrok S 89 A12
Bzenec CZ 146 D4
Bzince pod Javorinou SK 146 D5

C

Cabacés E 42 E5
Cabaj-Čápor SK 146 E6
Cabanac-et-Villagrains F 32 A4
Cabañaquinta E 39 B8
Cabanas del Castillo E 45 E9
Cabanas de Viriato P 44 D5
Cabañas Raras E 39 C6
Cabanes E 48 D5
Cabanillas E 41 D8
Cabannes F 35 C8
Čabar HR 73 E10
Cabasse F 36 E4
Cabeça Gorda P 50 D4
Cabeção P 50 B3
Cabeceiras de Basto P 38 E4
Cabeço de Vide P 44 F5
Cabella Ligure I 37 B10
Căbeşti RO 151 D9
Cabeza de Framontanos E 45 B8
Cabeza del Buey E 51 B9
Cabeza del Caballo E 45 B7
Cabeza la Vaca E 51 C7
Cabezamesada E 47 E6
Cabezarados E 54 B4
Cabezarrubias del Puerto
    E 54 B4
Cabezas del Villar E 45 C10
Cabezas Rubias E 51 D5
Cabezón de Cameros E 41 D6
Cabezón de la Sal E 40 B3
Cabezón de Liébana E 39 B10
Cabezuela del Valle E 45 D9
Čabiny SK 145 E4
Cabo de Palos E 56 F3
Cabolafuente E 47 B8
Cabra E 53 B8
Cabração P 38 E2
Cabra del Camp E 42 E6
Cabra del Santo Cristo E 55 D6
Cabra de Mora E 48 D3
Cabras I 64 D2
Cabrejas del Pinar E 40 E6
Cabrela P 50 B3
Cabrerets F 33 A9
Cabrières-d'Aigues F 35 C10
Cabril P 44 D5
Cabrillanes E 39 C7
Cabrillas E 45 C8
Cabuna RO 149 E10
Cacabelos E 39 C6
Cacaricata I 60 A2
Čačak SRB 158 F5
Caçarelhos P 39 E7
Caccamo I 58 D4
Caccuri I 61 E7
Cacém P 50 B1
Cáceres E 45 F8
Cachopo P 50 E4
Cacica RO 153 B7
Čachtice SK 146 D5
Cacín E 53 B9
Čačinci HR 149 E9
Cadalso de los Vidrios E 46 D4
Cadamstown IRL 7 F7
Cadaqués E 43 C10
Cadaval P 44 F2
Čadavica BIH 157 D6
Čadavica HR 149 E9
Čadavica Gornja BIH 157 C11
Čadca SK 147 C7
Cadelbosco di Sopra I 66 C2
Cadenazzo CH 69 A6
Cadenberge D 17 A12
Cadenet F 35 C9
Cadeo I 69 D8
Cádiar E 55 F6
Cadillac F 32 A5
Cádiz E 52 C4
Cadolzburg D 75 D8
Cadoneghe I 66 B4
Cadours F 33 C8
Cadrete E 41 E10
Caen F 23 B11
Caerau UK 13 C8
Caerdydd UK 13 C8
Caerfyrddin UK 12 B6
Caergybi UK 10 E2
Caerhun UK 10 E4
Caerleon UK 13 B9
Caernarfon UK 10 E3
Caerphilly UK 13 B8
Cafasse I 31 E12
Caggiano I 60 B5
Çağış TR 173 E8
Cagli I 66 E6
Cagliari I 64 E3
Čaglin HR 149 F9
Cagnac-les-Mines F 33 C10
Cagnano Varano I 63 D9
Cagnes-sur-Mer F 36 D6
Caherconlish IRL 8 C6
Cahermore IRL 8 E2
Cahir IRL 9 C7
Cahirsiveen IRL 8 E2
Cahors F 33 B8
Cahuzac-sur-Vère F 33 C9
Căianu RO 152 D3
Căianu Mic RO 152 C4
Caiazzo I 60 A2
Căinari MD 154 D4
Căinarii Vechi MD 153 A12
Căineni RO 160 C4
Căineni-Băi RO 161 C10
Caión E 38 B2
Cairaclia MD 154 F3
Cairanne F 35 B8
Cairnbaan UK 4 C6
Cairnryan UK 4 F6
Cairo Montenotte I 37 C8
Caiseal IRL 9 C7
Caisleán an Bharraigh IRL 6 E4
Caisleán an Chomair IRL 9 C8
Caisleán Uí Chonaill IRL 8 C5
Caissargues F 35 C7
Caister-on-Sea UK 15 B12

Caistor UK 11 E11
Căiuţi RO 153 E9
Caivano I 60 B2
Cajarc F 33 B9
Čajetina SRB 158 F4
Čajić BIH 157 E7
Čajkov SK 147 E7
Čajniče BIH 157 E11
Cajvana RO 153 B7
Çaka SK 146 E6
Čakajovce SK 146 E6
Çakılköy TR 173 D9
Çakıllı TR 173 A8
Čakirli TR 173 E7
Çakırlı TR 173 F10
Čakirli TR 173 F10
Čakovec HR 149 D6
Cala E 51 D7
Calabardina E 55 E9
Calabritto I 60 B4
Calaceite E 42 F4
Calacuccia I 37 G10
Cala d'Oliva I 64 A1
Cala d'Or E 57 C11
Calaf E 43 D7
Calafat RO 159 F10
Calafell E 43 E7
Cala Figuera E 57 C11
Calafíndeşti RO 153 B8
Cala Galdana E 57 B12
Calahonda E 53 C6
Calahorra E 32 F2
Calais F 15 F12
Calalzo di Cadore I 72 D5
Calamandrana I 37 B8
Cala Millor E 57 B11
Calamocha E 47 C10
Calamonaci I 58 D3
Calamonte E 51 B7
Cala Murada E 57 C11
Çalan RO 151 F11
Calañas E 51 D6
Calanda E 42 F3
Calangianus I 64 B3
Calanna I 59 C8
Cala'n Porter E 57 B13
Calaraşauca MD 154 A3
Călăraşi RO 152 D4
Călăraşi RO 153 B10
Călăraşi RO 160 F4
Călăraşi RO 161 E10
Cala Ratjada E 57 B11
Calascibetta I 58 D5
Calasetta I 64 E1
Calasparra E 55 C9
Calatafimi I 58 D2
Calatayud E 41 F8
Călăţele RO 151 D11
Calatorao E 41 F9
Calau D 80 C5
Căldăraru RO 160 E6
Caldarola I 67 F7
Caldaro sulla Strada del Vino
    I 72 D3
Caldas da Rainha P 44 F2
Caldas de Reis E 38 C2
Caldas de Vizela P 38 F3
Caldbeck UK 5 F10
Caldearenas E 32 F4
Caldecott UK 11 F10
Caldelas E 38 E3
Caldelas Taipas P 38 F3
Calden D 21 B12
Caldercruik UK 5 D9
Caldes de Malavella E 43 D9
Caldes de Montbui E 43 D8
Caldes d'Estrac E 43 D9
Caldicot UK 13 B9
Caldogno I 72 E4
Caldonazzo I 69 A11
Calella E 43 D9
Calendário P 38 F2
Calenzana F 37 F9
Calenzano I 66 E3
Calera de León E 51 C7
Calera y Chozas E 45 E10
Caleruega E 40 E5
Caleruela E 45 E10
Calfsound UK 3 G11
Calgary UK 4 B4
Çali TR 173 D10
Calig E 42 G4
Călimăneşti RO 160 C4
Calimera I 61 C10
Callas F 36 D5
Callen F 32 B5
Callian F 36 D5
Callington UK 12 E6
Callosa d'En Sarrià E 56 D4
Callosa de Segura E 56 E3
Calma SRB 158 C4
Călmăţiu RO 160 F5
Călmăţuiu de Sus RO 160 E5
Calmont F 33 B11
Calmont F 33 D9
Calne UK 13 C10
Câlnic RO 152 F3
Câlnic RO 159 D11
Caloloziocorte I 69 B7
Calomarde E 47 D9
Calonge E 43 D10
Calopăr RO 160 E3
Calovec SK 146 F5
Calpe E 56 D5
Caltabellotta I 58 D3
Caltagirone I 59 E6
Caltanissetta I 58 E5
Caltavuturo I 58 D4
Caltignaga I 68 B6
Căltilibük TR 173 E10
Caltojar E 40 F6
Caltra IRL 6 F6
Călugăreni RO 161 E7
Caluso I 68 C4

Calvão P 44 D3
Calvarrasa de Abajo E 45 C9
Calvarrasa de Arriba E 45 C9
Calvello I 60 C5
Calverstown IRL 7 F9
Calvi F 37 F9
Calvi dell'Umbria I 62 C3
Calvignac F 33 B9
Calvinet F 29 F10
Calvini RO 161 C8
Calvi Risorta I 60 A2
Calverd F 35 C7
Calvörde D 79 B8
Calvos E 38 E4
Calw D 27 C10
Calzada de Calatrava E 54 B5
Calzada de Valdunciel E 45 B9
Calzadilla E 45 D7
Calzadilla de los Barros E 51 C7
Camaldoli I 66 E4
Camaleño E 39 B10
Camallera E 43 C9
Camañas E 47 C10
Camar RO 151 C10
Camarasa E 42 D5
Cămăraşu RO 152 D4
Camarena E 46 D4
Camarena de la Sierra E 47 D10
Camarès F 34 C4
Camaret-sur-Aigues F 35 B8
Camaret-sur-Mer F 22 D2
Camarillas E 42 F2
Camariñas E 38 B1
Cămârzana RO 145 H7
Camarzana de Tera E 39 D7
Camas E 51 E7
Camastra I 58 E4
Cambados E 38 C2
Camberley UK 15 E7
Cambes F 28 F5
Cambil E 53 A9
Cambo-les-Bains F 32 D3
Camborne UK 12 E4
Cambra de Baixo P 44 C4
Cambre E 38 B3
Cambrai F 19 D7
Cambridge UK 15 C9
Cambrils E 42 E6
Cambron F 18 D4
Cambs D 83 C11
Camburg D 79 D10
Camelford UK 12 D5
Camenca MD 154 A3
Camerano I 67 E8
Camerata Cornello I 69 B8
Cameri I 68 B6
Camerino I 67 F7
Camerles F 42 F5
Camerota I 60 C4
Camici TR 177 E10
Camigliatello Silano I 61 E6
Camin D 83 D9
Caminha P 38 E2
Caminomorisco E 45 D8
Caminreal E 47 C10
Çamlica TR 173 D10
Çamlik TR 177 D9
Cammarata I 58 D4
Cammer D 79 B12
Cammin D 83 C12
Camolin IRL 9 C10
Camon F 18 D5
Camors F 22 E6
Camp IRL 8 D3
Campagna I 60 B4
Campagnano di Roma I 62 C2
Campagnatico I 65 B4
Campagne F 29 F7
Campagne-lès-Hesdin F 15 G12
Campan F 33 D6
Companario E 51 B8
Campanet E 57 B10
Campana I 61 E7
Campano E 52 D4
Campaspero E 40 F3
Campbeltown UK 4 E5
Campbon F 23 F7
Campdevànol E 43 C8
Campeã P 38 F4
Campello E 56 E4
Campello sul Clitunne I 62 B3
Campelo P 44 F2
Campelos P 44 F2
Campénéac F 23 E7
Câmpeni RO 151 E11
Campertogno I 68 B5
Câmpia Turzii RO 152 D3
Campi Bisenzio I 66 E3
Campiglia Marittima I 65 A3
Campiglia Soana I 31 D12
Campillo de Alto Buey E 47 E9
Campillo de Arenas E 53 A9
Campillo de Dueñas E 47 C9
Campillo de Llerena E 51 B8
Campillos E 53 B7
Campillos-Paravientos E 47 E9
Câmpina RO 161 C7
Câmpineanca RO 153 F10
Campinho P 50 B5
Campisábalos E 40 F5
Campi Salentina I 61 C10
Campli I 62 B5
Campo E 33 F6
Campo P 38 F3
Campobasso I 63 D7
Campobello di Licata I 58 E4
Campobello di Mazara I 58 D2
Campodarsego I 66 A4
Campo de Caso E 39 B9
Campo de Criptana E 47 F6
Campo de San Pedro E 40 F4
Campodimele I 62 E5
Campodolcino I 71 D8
Campofelice di Roccella I 58 D5
Campofiorito I 58 D3
Campoformido I 73 D7
Campofranco I 58 E4
Campofrío E 51 D6
Campoli Appennino I 62 D5
Campo Ligure I 37 B9
Campo Lugar E 45 E9
Campolongo Maggiore I 66 B5
Campo Maior P 51 A5
Campomanes E 39 B7
Campomarino I 63 D8
Campomorone I 37 C9
Componogara I 66 B5
Campora San Giovanni I 59 A9
Camporeale I 58 D3
Campo Redondo P 50 D2

Camporgiano I 66 D1
Camporosso I 37 D7
Camporrells E 42 D4
Camporrobles E 47 E10
Campos E 57 C11
Campos P 38 E4
Camposampiero I 66 A4
Camposanto I 66 C3
Campotéjar E 53 B9
Campotosto I 62 B4
Campo Tures I 72 C4
Camprodon E 33 F10
Camptown UK 5 E11
Câmpulung RO 160 C6
Câmpulung la Tisa RO 145 H8
Câmpulung Moldovenesc
    RO 153 B7
Câmpuri RO 153 E9
Camrose UK 12 B4
Camuñas E 46 F6
Çamyayla TR 181 B8
Çan TR 173 D7
Caña SK 145 F3
Cañada E 56 D3
Cañada de Benatanduz E 42 F2
Cañada del Hoyo E 47 E9
Cañada Vellida E 42 F2
Canak HR 156 C3
Çanakçi TR 173 F10
Çanakkale TR 171 D10
Canale I 37 B7
Canale-di-Verde F 37 G10
Canalejas del Arroyo E 47 D8
Canalejas de Peñafiel E 40 E3
Canals E 56 D3
Canals I 56 D3
Canàl San Bovo I 72 D4
Cañamares E 47 D8
Cañamero E 45 F10
Canaples F 18 D5
Canari F 37 F10
Canaro I 66 C3
Cañaveral E 45 E8
Cañaveral de León E 51 C6
Cañaveras E 47 D8
Cañaveruelas E 47 D7
Canazei I 72 D4
Cancale F 23 C8
Cancellara I 60 B5
Cancello ed Arnone I 60 A2
Cancon F 33 A7
Candanchú E 32 E4
Çandarli TR 177 B8
Candás E 39 A8
Candasnos E 42 D4
Candé F 23 E9
Candela I 60 A5
Candelario E 45 D9
Candeleda E 45 D10
Candelo I 68 B5
Candemil P 38 E2
Cândeşti RO 153 D9
Cândeşti RO 160 C6
Candia Lomellina I 68 C4
Candilichera E 41 E7
Candín E 39 C6
Canedo P 44 B4
Canelli I 37 B8
Canena E 55 C6
Canepina I 62 C2
Canero E 39 A7
Căneşti RO 161 C9
Canet F 34 C5
Canet I 34 D4
Canet de Mar E 43 D9
Canet d'en Berenguer E 48 E4
Cañete E 47 D9
Cañete de las Torres E 53 A8
Cañete la Real E 53 C7
Canet-en-Roussillon F 34 E5
Canet lo Roig E 42 F4
Canet-Plage F 34 E5
Cangas E 38 D2
Cangas del Narcea E 39 B6
Cangas de Onís E 39 B9
Cangonj AL 168 C4
Canha P 50 B2
Canhestros P 50 C3
Cania MD 154 E2
Canicattì I 58 E4
Canicattini Bagni I 59 E7
Canicosa de la Sierra E 40 E5
Caniles E 55 E7
Canillas de Aceituno E 53 C8
Canino I 62 C1
Cañizal E 45 B10
Cañizares E 47 C8
Cañizo E 39 E9
Canjáyar E 55 E7
Canlia RO 161 E11
Canna I 61 C7
Cannara I 62 B3
Cannero Riviera I 68 A6
Cannes F 36 D6
Canneto I 59 C6
Canneto I 66 F2
Canneto sull'Oglio I 66 B1
Cannich UK 2 L7
Canningstown IRL 7 E8
Cannington UK 13 C8
Cannobio I 68 A6
Cannock UK 11 F7
Cano P 50 B4
Canonbie UK 5 E10
Canosa di Puglia I 60 A6
Cánovas E 56 F2
Can Pastilla E 49 E10
Can Picafort E 57 B11
Canredondo E 47 C8
Cansano I 62 D6
Cantagallo I 66 D3
Cantalapiedra E 45 B10
Cantalejo E 40 F4
Cantalice I 62 C4
Cantalpino E 45 B10
Cantanhede P 44 D3
Cantavieja E 42 F3
Čantavir SRB 150 F4
Cantemir MD 154 E2
Cantenac F 28 E4
Canteras E 56 F2
Canterbury UK 15 E11
Cantiano I 66 E6
Cantillana E 51 D8
Cantoira I 31 E11
Cantoria E 55 E8
Cantù I 69 B7
Canvey Island UK 15 D10
Cany-Barville F 18 E2
Canyelles E 43 E7
Caolas UK 4 B3

Caorle I 73 E6
Capaccio I 60 C4
Capaci I 58 C3
Capafonts E 42 E6
Capalbio I 65 C4
Căpâlna RO 151 D9
Căpâlniţa RO 153 E6
Capannoli I 66 E2
Capannori I 66 E2
Caparde BIH 157 D10
Capari NMK 168 B5
Caparica P 50 B1
Caparrosa P 44 C4
Caparroso E 32 F2
Cap-Blanc E 48 F4
Capbreton F 32 C3
Cap d'Agde F 34 D6
Capdenac F 33 A10
Capdenac-Gare F 33 A10
Capdepera E 57 B11
Capel Curig UK 10 E4
Capelins P 51 B5
Capelle aan de IJssel NL 16 E3
Capellen L 20 E6
Capel St Mary UK 15 C11
Capendu F 34 D4
Capestang F 34 D5
Capestrano I 62 C5
Cap Ferret F 32 A3
Capileira E 55 F6
Capilla E 54 B2
Capinha P 44 D6
Capistrello I 62 D4
Capizzi I 58 D5
Căpleni RO 151 B10
Čaplje BIH 157 C6
Čapljina BIH 157 F8
Capodimonte I 62 B1
Capo di Ponte I 69 A9
Capo d'Orlando I 59 C6
Capoliveri I 65 B2
Capolona I 66 E4
Caposele I 60 B4
Capoterra I 64 E2
Cappadocia I 62 C4
Cappagh White IRL 8 C6
Cappamore IRL 8 C6
Cappawhite IRL 8 C6
Cappeen IRL 8 E5
Cappelle sul Tavo I 62 C6
Capeln (Oldenburg) D 17 C10
Cappercleuch UK 5 E10
Cappoquin IRL 9 D7
Capracotta I 63 D6
Capraia Isola I 65 A1
Capranica I 62 C2
Caprarola I 62 C2
Căpreni RO 160 D3
Capri I 60 B2
Căpriana MD 154 C2
Capriati a Volturno I 63 E6
Capri Leone I 59 C6
Caprino Bergamasco I 69 B7
Caprino Veronese I 69 B10
Captieux F 32 B5
Capua I 60 A2
Capurso I 61 A7
Căpuşu Mare RO 151 D11
Capvern-les-Bains F 33 D6
Carabaña E 47 D6
Caracal RO 160 E4
Caracuel de Calatrava E 54 B4
Caragaş MD 154 D5
Caragele RO 161 D10
Caraglio I 37 C6
Caraman F 33 D9
Caramanico Terme I 62 C6
Caramulo P 44 C4
Cărand RO 151 E9
Caranga E 39 B7
Caranguejeira P 44 E3
Caransebeş RO 159 C9
Carantec F 22 C4
Carapelle I 60 A6
Carapinheira P 44 D3
Carasco I 37 C10
Caraşova RO 159 C8
Caraula RO 159 E11
Caravaca de la Cruz E 55 C9
Caravaggio I 69 C8
Carbajales de Alba E 39 E8
Carballeda de Avia E 38 D3
Carballo E 38 B2
Carballo E 38 B2
Carbellino E 45 B8
Carbonera de Frentes E 41 E6
Carboneras E 55 F9
Carboneras de Guadazaón E 47 E9
Carbonero El Mayor E 46 B4
Carboneros E 54 C5
Carbonia I 64 E2
Carbonìn I 72 C5
Carbonne F 33 D8
Carbost UK 2 L4
Carbost UK 2 L4
Cărbunari RO 159 D8
Cărbuneşti RO 161 C8
Carbury IRL 7 F9
Carcaboso E 45 D8
Carcabuey E 53 B8
Carcaixent E 48 F4
Carcaliu RO 155 C2
Carcans F 28 E3
Carcans-Plage F 28 E3
Carção P 39 E6
Cárcar E 32 F2
Carcare I 37 C8
Carcassonne F 33 D10
Carcastillo E 32 F3
Carcelén E 47 F10
Carcès F 36 D4
Carchelejo E 53 A9
Carcoforo I 68 B5
Cardaillac F 29 F9
Çardak TR 172 D6
Cardedeu E 43 D8
Cardedu E 64 D4
Cardeña E 54 C4
Cardeñadijo E 40 D4
Cardenden UK 5 C10
Cardenete E 47 E9
Cardeñosa E 46 C3
Cardeto I 59 C8
Cardiff UK 13 C8
Cardigan UK 12 A5
Cardigos P 44 E4
Cardinale I 59 B9
Cardito I 62 D5
Cardon RO 155 C6
Cardona E 43 D7
Cardosas P 50 E2
Carei RO 151 B9

Carenas E 41 F8
Carentan F 23 B9
Carentoir F 23 E7
Carevdar HR 149 D7
Carev Dvor NMK 168 B5
Cargenbridge UK 5 E9
Cargèse F 37 G9
Carhaix-Plouguer F 22 D4
Caria P 44 D6
Cariati I 61 E7
Caridade P 50 B4
Carife I 60 A4
Carignan F 19 E11
Carignano I 37 B7
Cariñena E 41 F9
Carini I 58 C3
Carinish UK 2 K2
Cariño E 38 A4
Carinola I 60 A1
Carisio I 68 C5
Carisolo I 69 A10
Cârjiţi RO 151 F10
Carland UK 7 C9
Carlanstown IRL 7 E9
Carlantino I 63 D7
Carlat F 29 F11
Carlentini I 59 E7
Carlet E 48 F3
Cârlibaba RO 152 B6
Cârligele RO 153 F10
Carling F 186 C2
Carlingford IRL 7 D10
Carlisle UK 5 F11
Carloforte I 64 E1
Cârlogani RO 160 D4
Carlopoli I 59 A9
Carlops UK 5 D10
Carlow D 83 C9
Carlow IRL 9 C9
Carloway UK 2 J3
Carlsberg D 186 C5
Carlton UK 11 F9
Carlton Colville UK 15 C12
Carluke UK 5 D9
Carlux F 29 F8
Carmagnola I 37 B7
Carmanova MD 154 C5
Carmarthen UK 12 B6
Carmaux F 33 B10
Carmena E 46 E4
Cármenes E 39 C8
Carmiano I 61 C10
Carmona E 51 E8
Carmonita E 45 F8
Carmyllie UK 5 B11
Carnac F 22 E5
Carnagh UK 7 D9
Carndonagh IRL 4 E2
Carnew IRL 9 C10
Carnforth UK 10 C6
Carnières F 19 D7
Carnikava LV 135 B8
Carnlough UK 4 F5
Carno UK 10 F4
Carnota E 38 C1
Carnoules F 36 E4
Carnoustie UK 5 B11
Carnoux-en-Provence F 35 D10
Carnteel UK 7 D9
Carnwath UK 5 D9
Carolei I 60 E6
Carolles F 23 C8
Carona I 69 A8
Caronia I 58 C5
C. A. Rosetti RO 155 C5
C. A. Rosetti RO 161 C10
Carosino I 61 C8
Carovigno I 61 B9
Carovilli I 63 D6
Carpaneto Piacentino I 69 D8
Carpegna I 66 E5
Carpen RO 159 E11
Carpenedolo I 66 B1
Carpentras F 35 B9
Carpi I 66 C2
Carpignano Salentino I 61 C10
Carpignano Sesia I 68 B5
Cărpineni MD 154 D2
Cârpinet RO 151 E10
Carpineti I 66 D1
Carpineto Romano I 62 D4
Cârpiniş RO 151 F6
Carpino I 63 D9
Carpinone I 63 D6
Carpio E 39 F9
Carpiquet F 23 B11
Carquefou F 23 F9
Carqueiranne F 36 E4
Carracastle IRL 6 E5
Carracedelo E 39 C6
Carradale East UK 4 D6
Carragh IRL 7 F9
Carraig Airt IRL 7 B7
Carraig na Siuire IRL 9 D8
Carraig Thuathail IRL 8 E6
Carral E 38 B3
Carralevë RKS 163 E10
Carranque E 46 D5
Carrapateira P 50 E2
Carrapichana P 44 C6
Carrara I 69 E9
Carraroe IRL 6 F3
Carreña E 39 B10
Carreteira E 38 C4
Carriazo E 40 B4
Carrick IRL 6 C5
Carrick IRL 9 D9
Carrickart IRL 7 B7
Carrickfergus UK 4 F5
Carrickmacross IRL 7 E9
Carrickmore UK 4 F2
Carrick-on-Shannon IRL 6 E6
Carrick-on-Suir IRL 9 D8
Carrigaholt IRL 8 C3
Carrigallen IRL 7 E7
Carriganimy IRL 8 E4
Carriganimmy IRL 8 E4
Carrigart IRL 7 B7

Carrigkerry IRL 8 D4
Carrig Mhachaire IRL 7 E9
Carrigtohill IRL 8 E6
Carrigtwohill IRL 8 E6
Carrio E 38 B3
Carrión de Calatrava E 54 A5
Carrión de los Céspedes E 51 E7
Carrión de los Condes E 39 D10
Carrizo de la Ribera E 39 C8
Carrizosa E 55 B7
Carronbridge UK 5 E9
Carros F 37 D6
Carrouges F 23 C11
Carrowkeel IRL 4 E2
Carrowkeel IRL 7 B7
Carrowkennedy IRL 6 E3
Carrù I 37 C7
Carryduff UK 7 C11
Carry-le-Rouet F 35 D9
Cars F 28 E4
Carsac-Aillac F 29 F8
Carsluith UK 5 F8
Carsoli I 62 C4
Carspach F 27 E7
Carsphairn UK 5 E8
Carstairs UK 5 D9
Cartagena E 56 F3
Cártama E 53 C7
Cartaxo P 44 F3
Cartaya E 51 E5
Cartelègue F 28 E4
Carteret F 23 B8
Carterton UK 13 B11
Cartes E 40 B3
Cartoceto I 67 E6
Carucedo E 39 D6
Carunchio I 63 D7
Carvalhal P 44 E4
Carvalhal P 50 C2
Carvalho de Egas P 38 F5
Carvalhosa P 38 F3
Carviçais P 45 B7
Carvin F 18 D6
Carvoeiro P 44 E2
Carvoeiro P 50 E3
Čáry SK 77 E12
Casabermeja E 53 C8
Casabona I 61 E7
Casa Branca P 50 B3
Casa Branca P 50 B4
Casacalenda I 63 D7
Casagiove I 60 A2
Casaglione F 37 G9
Casa l'Abate I 61 C10
Casalanguida I 63 C6
Casalarreina E 40 C6
Casalbordino I 63 C7
Casalbore I 60 A4
Casalborgone I 68 C4
Casalbuono I 60 C5
Casalbuttano ed Uniti I 69 C8
Casàl Cermelli I 37 B9
Casàl di Principe I 60 A2
Casalecchio di Reno I 66 D3
Casaletto Spartano I 60 C5
Casalfiumanese I 66 D4
Casalgrande I 66 D2
Casalgrasso I 37 B7
Casalmaggiore I 66 C1
Casalnuovo Monterotaro I 63 D8
Casalpusterlengo I 69 C8
Casalvecchio di Puglia I 63 D8
Casàl Velino I 60 C4
Casamassima I 61 B7
Casamozza I 37 F10
Casarabonela E 53 C7
Casarano I 61 C10
Casar de Cáceres E 45 E8
Casar de Palomero E 45 D8
Casarejos E 40 E5
Casares E 39 E7
Casares E 53 D6
Casares de las Hurdes E 45 D8
Casariche E 53 B7
Casarrubios del Monte E 46 D4
Casarsa della Delizia I 73 E6
Casarza Ligure I 37 C10
Casas Altas E 47 D10
Casas Bajas E 47 D10
Casas de Benítez E 47 F8
Casas de Don Pedro E 45 F10
Casas de Fernando Alonso E 47 F8
Casas de Haro E 47 F8
Casas de Juan Gil E 47 F10
Casas de Juan Núñez E 47 F10
Casas de Lázaro E 55 B8
Casas del Monte E 45 D9
Casas de los Pinos E 47 F8
Casas del Puerto E 56 E2
Casas de Millán E 45 E8
Casas de Reina E 51 C8
Casas de Ves E 47 F10
Casas-Ibáñez E 47 F10
Casasimarro E 47 F8
Casas Novas de Mares P 50 B5
Casasola de Arión E 39 E9
Casatejada E 45 E10
Casatenovo I 69 B7
Casavieja E 46 D3
Cascais P 50 B1
Cascante E 41 D8
Cascante del Río E 47 D10
Cascia I 62 B4
Casciana Terme I 66 E2
Cascina I 66 E2
Căscioarele RO 161 E8
Căsebres P 50 B2
Cáseda E 32 E3
Casei Gerola I 37 A9
Casekow D 84 D6
Casella I 37 B9
Caselle in Pittari I 60 C5
Caselle Torinese I 68 C4
Case Perrone I 61 B7
Caseras E 42 E4
Caserta I 60 A2
Casével P 50 D3
Cashel IRL 6 F4
Cashel IRL 6 F5
Cashel IRL 7 F7
Cashel IRL 7 G8
Cashel IRL 9 C7
Cashla IRL 6 F5
Casillas E 46 D3
Casillas de Flores E 45 D7
Casimcea RO 155 D2

Caşin RO 153 E9
Casina I 66 C1
Casinos E 48 E3
Casla IRL 6 F3
Casnewydd UK 13 B9
Casola in Lunigiana I 66 D1
Casola Valsenio I 66 D4
Casole d'Elsa I 66 F3
Casoli I 63 C6
Casoria I 60 B2
Caspe E 42 E3
Casperia I 62 C3
Cassà de la Selva I 43 D9
Cassagnes-Bégonhès F 33 B11
Cassaniouze F 29 F10
Cassano allo Ionio I 61 D6
Cassano delle Murge I 61 B7
Cassano Magnano I 68 B6
Cassano Spinola I 37 B9
Cassaro I 59 E6
Cassel F 18 C5
Casseneuil F 33 B7
Casserres E 43 C7
Cassibile I 59 F7
Cassine I 37 B9
Cassino I 62 E5
Cassis F 35 D10
Cassola I 72 E4
Cassuéjouls F 30 F2
Castà SK 146 E4
Castagliona Olona I 69 B6
Castagnaro I 66 B3
Castagneto Carducci I 65 B3
Castagnole delle Lanze I 37 B8
Castagnole Monferrato I 37 B8
Castalla I 56 D3
Castañar de Ibor I 45 E10
Castañares de Rioja E 40 C5
Castanet-Tolosan F 33 C9
Castanheira P 44 C6
Castanheira de Pêra P 44 D4
Castano Primo I 68 B6
Castasegna CH 69 A7
Casteggio I 37 A10
Castejón E 41 D8
Castejón del Puente E 42 D4
Castejón de Monegros E 42 D3
Castejón de Sos E 33 E6
Castejón de Valdejasa E 41 E10
Castelbellino I 67 F7
Castèl Bolognese I 66 D4
Castelbuono I 58 D5
Castelcivita I 60 C4
Castèl d'Ario I 66 B2
Castel del Monte I 62 B5
Castèl del Piano I 65 B5
Castèl del Rio I 66 D4
Castèl di Iudica I 59 E6
Castèl di Lama I 62 B5
Castèl di Lucio I 58 D5
Castèl di Sangro I 62 D6
Casteleiro P 45 D6
Castelfidardo I 67 F8
Castelfiorentino I 66 E2
Castelflorite E 42 D3
Castèl Focognano I 66 E4
Castelforte I 62 E5
Castelfranci I 60 B4
Castelfranco di Sopra I 66 E4
Castelfranco di Sotto I 66 E2
Castelfranco Emilia I 66 C3
Castelfranco in Miscano I 60 A4
Castelfranco Veneto I 72 E4
Castèl Frentano I 63 C6
Castèl Gandolfo I 62 D3
Castelginest F 33 C8
Castèl Giorgio I 62 B1
Castèl Goffredo I 66 B1
Castelgrande I 60 B4
Casteljaloux F 33 B6
Castèll D 75 C7
Castèll'Alfero I 37 B8
Castellalto I 62 B5
Castellammare del Golfo I 58 C2
Castellammare di Stabia I 60 B2
Castellamonte I 68 C4
Castellana Grotte I 61 B8
Castellane F 36 D5
Castellaneta I 61 B7
Castellanos de Castro E 40 D3
Castellarano I 66 C2
Castellar de la Frontera E 53 D6
Castellar de la Muela E 47 C9
Castellar de la Ribera E 43 C6
Castellar de Santiago E 55 B6
Castellar de Santisteban E 55 C6
Castell'Arquato I 69 D8
Castell'Azzara I 62 B1
Castellazzo Bormida I 37 B8
Castelldans E 42 E5
Castell de Cabres E 42 F4
Castell de Castells E 56 D4
Castelldefels E 43 E7
Castell de Ferro E 55 F6
Castelleone I 69 C8
Castelletto sopra Ticino I 68 B6
Castellfort E 42 F3
Castellina in Chianti I 66 F3
Castellina Marittima I 66 F2
Castelliri I 62 D5
Castellnou de Bassella E 43 C6
Castelló E 48 E4
Castello d'Argile I 66 C3
Castelló de Farfanya E 42 D5
Castelló de la Plana E 48 E4
Castelló d'Empúries E 43 C10
Castello di Annone I 37 B8
Castellón de la Plana E 48 E4
Castellote E 42 F3
Castell Tesino I 72 D4
Castellserà E 42 D5
Castellterçol E 43 D8
Castelluccio I 66 B2
Castelluccio dei Sauri I 60 A4
Castelluccio Inferiore I 60 C5
Castelluccio Valmaggiore I 60 A4
Castell'Umberto I 59 C6
Castelluzzo I 58 C2
Castèl Madama I 62 D3
Castèl Maggiore I 66 C3
Castelmagno I 37 C6
Castelmassa I 66 B3
Castelmauro I 63 D7
Castelmoron-sur-Lot F 33 B6
Castelnau-Barbarens F 33 C7
Castelnaudary F 33 D9
Castelnau-d'Auzan F 33 C6

Castelnau-de-Médoc F 28 E4
Castelnau-de-Montmiral F 33 C9
Castelnau d'Estréfonds F 33 C8
Castelnau-le-Lez F 35 C6
Castelnau-Magnoac F 33 D7
Castelnau-Montratier F 33 B8
Castelnau-Rivière-Basse F 32 C5
Castelnovo di Sotto I 66 C2
Castelnovo ne'Monti I 66 D1
Castelnuovo Berardenga I 66 F4
Castelnuovo della Daunia I 63 D8
Castelnuovo di Garfagnana I 66 D1
Castelnuovo di Porto I 62 C3
Castelnuovo di Val di Cecina I 66 F2
Castelnuovo Don Bosco I 37 B7
Castelnuovo Rangone I 66 C2
Castelnuovo Scrivia I 37 B9
Castelo Bom P 45 C7
Castelo Branco P 39 F6
Castelo Branco P 44 E6
Castelo de Paiva P 44 B4
Castelo de Vide P 44 F6
Castelo do Neiva P 38 E2
Castelões P 44 C4
Castelplanio I 67 F7
Castelraimondo I 67 F7
Castèl Ritaldi I 62 B3
Castelrotto I 72 C4
Castelsagrat F 33 B7
Castèl San Giovanni I 69 C7
Castèl San Lorenzo I 60 C4
Castèl San Niccolò I 66 E4
Castèl San Pietro Terme I 66 D4
Castelsantangelo sul Nera I 62 B4
Castelsaraceno I 60 C5
Castelsardo I 64 B2
Castelsarrasin F 33 B8
Castelseras E 42 F3
Casteltermini I 58 D4
Castelu RO 155 E2
Castelverde I 69 C8
Castelvetere in Val Fortore I 60 A3
Castelvetrano I 58 D2
Castelvetro Piacentino I 69 C8
Castèl Viscardo I 62 B2
Castèl Volturno I 60 A1
Castenaso I 66 C3
Castéra-Verduzan F 33 C7
Castetnau-Camblong F 32 D4
Castets F 32 C3
Castiadas I 64 E3
Castielfabib E 47 D10
Castiello de Jaca E 32 E4
Castiglioncello I 66 F1
Castiglione dei Pepoli I 66 D3
Castiglione del Lago I 66 F5
Castiglione della Pescaia I 65 B3
Castiglione della Stiviere I 66 B1
Castiglione di Sicilia I 59 D7
Castiglione d'Orcia I 65 A5
Castiglione Messer Marino I 63 D6
Castiglion Fiorentino I 66 F4
Castignano I 62 B5
Castilblanco E 45 F10
Castilblanco de los Arroyos E 51 D8
Castiliscar E 32 F3
Castillejar E 55 D7
Castillejo de Martin Viejo E 45 C7
Castillejo de Mesleón E 40 F4
Castillejo de Robledo E 40 E5
Castillo de Bayuela E 46 D3
Castillo de Garcimuñoz E 47 E8
Castillo de Locubín E 53 A9
Castillon-en-Couserans F 33 E8
Castillon-la-Bataille F 28 F5
Castillonnès F 33 A7
Castillo-Nuevo E 32 E3
Castilruiz E 41 E7
Castione della Presolana I 69 B9
Castions di Strada I 73 E7
Castlebar IRL 6 E4
Castlebay UK 4 B2
Castlebellingham IRL 7 E10
Castleblakeney IRL 6 F6
Castleblayney IRL 7 D9
Castlebridge IRL 9 D10
Castle Carrock UK 5 F11
Castle Cary UK 13 C9
Castlecomer IRL 9 C8
Castleconnell IRL 8 C5
Castlecor IRL 8 D5
Castledawson UK 4 F3
Castlederg UK 4 F1
Castledermot IRL 9 C9
Castle Douglas UK 5 F9
Castleellis IRL 9 D10
Castlefinn IRL 4 F1
Castleford UK 11 D9
Castlegal IRL 6 D6
Castlegregory IRL 8 D2
Castlehill IRL 6 D4
Castleisland IRL 8 D3
Castle Kennedy UK 4 F7
Castlellis IRL 9 D10
Castlemaine IRL 8 D3
Castlemartyr IRL 8 E6
Castleplunket IRL 6 E6
Castlepollard IRL 7 E8
Castlerea IRL 6 E6
Castlereagh UK 7 C11
Castlerock UK 4 E3
Castletown UK 3 H10
Castletown UK 4 E3
Castletown GBM 10 C2
Castletown IRL 9 D6
Castletown Bere IRL 8 E3
Castletownshend IRL 8 F4
Castlewellan UK 7 D11
Castranova RO 160 E4
Castrejón de la Peña E 39 C10
Castrelo do Val E 38 E5
Castres F 33 C10
Castricum NL 16 D3
Castries F 35 C6
Castrignano del Capo I 61 D10
Castril E 55 D7
Castrillo de Don Juan E 40 E3
Castrillo de Duero E 40 E3
Castrillo de la Reina E 40 E5
Castrillo de la Vega E 40 E4
Castrillo Tejeriego E 40 E3

Castro I 61 C10
Castrobarto E 40 B5
Castrocalbón E 39 D8
Castro Caldelas E 38 D5
Castrocaro Terme I 66 D4
Castro Daire P 44 C5
Castro dei Volsci I 62 D5
Castro del Río E 53 A8
Castro de Ouro E 38 A5
Castro de Rei E 38 B5
Castrofilippo I 58 E4
Castrogonzalo E 39 E8
Castrojeriz E 40 D3
Castro Laboreiro P 38 D3
Castro Marim P 50 E5
Castromocho E 39 E10
Castromonte E 39 E9
Castronuevo E 39 E9
Castronuño E 39 F9
Castronuovo di Sant'Andrea I 60 C5
Castronuovo di Sicilia I 58 D4
Castropignano I 63 D7
Castropodame E 39 C7
Castropol E 39 A5
Castrop-Rauxel D 17 E8
Castroreale I 59 C7
Castro-Urdiales E 40 B5
Castroverde E 38 B5
Castroverde de Campos E 39 E8
Castrovillari I 60 D6
Castuera E 51 B9
Caţa RO 152 E6
Catadau E 48 F3
Çatalca TR 173 B9
Catalina RO 153 F8
Cataloi RO 155 C3
Catania I 59 E7
Catanzaro I 59 B10
Catanzaro Marina I 59 B10
Catarroja E 48 F4
Câtcău RO 152 C3
Căteasca RO 160 D6
Catenanuova I 59 D6
Caterham UK 15 E8
Cateri F 37 F9
Cathair Dónall IRL 8 E2
Cathair na Mart IRL 6 E3
Cathair Saidhbhín IRL 8 E2
Catherdaniel IRL 8 E2
Catì E 42 G4
Čatići BIH 157 D9
Catignano I 62 C5
Câtina RO 152 D4
Câtina RO 161 C8
Cativelos P 44 C5
Catoira E 38 C2
Caton UK 10 C6
Catral E 56 E3
Cattenom F 20 F6
Catterfeld D 79 E8
Catterick UK 5 B12
Catterline UK 5 B12
Cattolica I 67 E6
Cattolica Eraclea I 58 E3
Cătunele RO 159 D10
Catus F 33 A8
Căuaş RO 151 B10
Caudan F 22 E5
Caudebec-lès-Elbeuf F 18 F3
Caudecoste F 33 B7
Caudete E 56 D3
Caudete de las Fuentes E 47 E10
Caudiel E 48 E3
Caudiès-de-Fenouillèdes F 33 E10
Caudry F 19 D7
Caujac F 33 D8
Caulnes F 23 D7
Caulonia I 59 C9
Caumont I 33 D8
Caumont-l'Éventé F 23 B10
Caumont-sur-Durance F 35 C8
Caunes-Minervois F 34 D4
Cauro F 37 H9
Căuşeni MD 154 D3
Caussade F 33 B9
Cautano I 60 A3
Cauterets F 32 E5
Cava de'Tirreni I 60 B3
Cavadineşti RO 153 E12
Cavaglià I 68 C5
Cavaillon F 35 C9
Cavalaire-sur-Mer F 36 E5
Cavaleiro P 50 D2
Cavalese I 72 D4
Cavallermaggiore I 37 B7
Cavallino I 66 B6
Cava Manara I 69 C7
Cavan IRL 7 E8
Cavanagarven IRL 7 D9
Cavarzere I 66 B5
Cavargna I 69 A7
Cavazzo Carnico I 73 D7
Cave I 62 D3
Cave del Predil I 73 D8
Caveirac F 35 C7
Cavezzo I 66 C3
Cavignac F 28 E5
Cavnic RO 152 B3
Cavour I 68 D3
Cavriago I 66 C2
Cavriglia I 66 F3
Cavtat HR 162 D5
Çavuşköy TR 171 C10
Cawdor UK 3 K9
Cawood UK 11 D9
Cawston UK 15 B11
Caxarias P 44 E4
Çaybaşı TR 177 C9
Çayboyu TR 181 B8
Cayeux-sur-Mer F 18 D4
Çayırdere TR 173 B9
Caylus F 33 B9
Cazalegas E 46 D3
Cazalilla E 53 A9
Cazalla de la Sierra E 51 D8
Cazals F 33 A8
Cazals F 33 B9
Căzăneşti RO 159 D10
Căzăneşti RO 161 D10
Cazasu RO 155 C1
Cazaubon F 32 C5
Cazères F 33 D8
Cazes-Mondenard F 33 B8
Cazilhac F 34 C5
Cazin BIH 156 C4

Cazis CH 71 D8
Cažma HR 149 E7
Cazorla E 55 D7
Cazoulès F 29 F8
Cea E 38 B4
Cea E 39 D10
Ceahlău RO 153 C7
Ceamurlia de Jos RO 155 D3
Ceanannus Mór IRL 7 E9
Ceann Toirc IRL 8 D5
Ceann Trá IRL 8 D2
Ceanu Mare RO 152 D3
Cearsiadar UK 2 J3
Ceatharlach IRL 9 C9
Ceaucé F 23 D10
Ceauşu de Câmpie RO 152 D5
Ceaux-d'Allègre F 30 E4
Cébazat F 30 D3
Čebín CZ 77 D10
Cebolla E 46 E3
Cebovce SK 147 E8
Cebreros E 46 D4
Ceccano I 62 D4
Cece H 149 C11
Čečejovce SK 145 F3
Čechtice CZ 77 D7
Čechynce SK 146 E6
Cecina I 66 F2
Ceclavín E 45 E7
Cecuni MNE 163 D8
Cedães P 38 F5
Čedasai LT 135 D10
Cedegolo I 69 A9
Cedeira E 38 A3
Cedillo E 44 E6
Cedillo del Condado E 46 D5
Cedrillas E 48 D3
Cedry Wielkie PL 138 B6
Cedynia PL 84 E6
Cee E 38 C1
Cefa RO 151 D8
Cefalù I 58 C5
Cefn-mawr UK 10 F5
Ceggia I 73 E6
Cegléd H 150 C4
Ceglédbercel H 150 C4
Ceglie Messapica I 61 B9
Cegłów PL 141 F5
Čegrane NMK 163 F10
Cehal RO 151 C10
Cehegín E 55 C9
Cehu Silvaniei RO 151 C11
Ceica RO 151 D9
Ceikiniai LT 135 F12
Ceilhes-et-Rocozels F 34 C5
Ceinos de Campos E 39 D9
Ceintrey F 26 C5
Ceira P 44 D4
Čejč CZ 77 E11
Čejkov SK 145 G4
Cekcyn PL 138 C5
Çekirdekli TR 177 B9
Čekiškė LT 134 F7
Ceków-Kolonia PL 142 C5
Celadas E 47 D10
Čeladná CZ 146 B6
Čelákovice CZ 77 B7
Celaliye TR 173 A7
Celano I 62 D4
Celanova E 38 D4
Čelarevo SRB 158 C4
Celaru RO 160 E4
Celbridge IRL 7 F9
Čelebić BIH 157 E6
Čelebići BIH 157 F10
Celeiros P 38 E3
Celenza Valfortore I 63 D7
Celestynów PL 141 F4
Čelić BIH 157 C10
Celico I 61 E6
Čelinac Donji BIH 157 C7
Celje SLO 148 D4
Cella E 47 D10
Celldömölk H 149 B8
Celle D 79 A7
Celle Ligure I 37 C9
Cellere I 62 C1
Celles B 19 C7
Celles-sur-Belle F 28 C5
Celles-sur-Ource F 25 D11
Cellettes F 24 E5
Cellino Attanasio I 62 B5
Cellole I 62 E5
Čelopeci NMK 168 B5
Čelopek NMK 164 F3
Colorico da Beira P 44 C6
Colorico de Basto P 38 F4
Celrà E 43 C9
Çeltikçi TR 173 D9
Cembra I 69 A11
Čemerno BIH 157 F10
Cempi LV 135 A11
Cénac-et-St-Julien F 29 F8
Cenad RO 150 E6
Cenade RO 152 E4
Cenas LV 134 C7
Cencenighe Agordino I 72 D4
Cendras F 35 B7
Cendrieux F 29 F7
Cenei RO 159 B6
Ceneselli I 66 B3
Cengio I 37 C8
Cenicero E 41 D6
Cenicientos E 46 D4
Čenta SRB 158 C5
Centallo I 37 B7
Centelles E 43 D8
Cento I 66 C3
Centola I 60 C4
Centuri F 37 F10
Centuripe I 59 D6
Cepagatti I 62 C6
Cepari RO 160 C5
Cepin HR 149 E11
Ceplenița RO 153 C9
Čepovan SLO 73 D8
Ceppaloni I 60 A3
Ceppo Morelli I 68 B5
Ceprano I 62 D5
Ceptura RO 161 C9
Čeralije HR 149 E9
Cerami I 59 D6
Cerani BIH 157 F8
Cerano I 68 C6
Ceranów PL 141 E6
Cérans-Foulletourte F 23 E12
Cerasi I 59 C8
Ceraso I 60 C4
Cerașu RO 161 C8
Cerât RO 160 E3
Ceraukste LV 135 D8
Çëravië AL 168 C4
Cerbăl RO 151 F10

Cerbère F 34 F5
Cercal P 44 F3
Cercal P 50 D2
Čerčany CZ 77 C7
Cercedilla E 46 C4
Cercemaggiore I 63 E7
Cerchezu RO 155 F2
Cerchiara di Calabria I 61 D6
Cerchio I 62 C5
Cercy-la-Tour F 30 B4
Cerda I 58 D4
Cerdanyola del Vallès E 43 E8
Cerdedo E 38 C3
Cerdeira P 44 D5
Cerdon F 25 E7
Cēre LV 134 B5
Cerea I 66 B3
Cered H 147 E9
Ceregnano I 66 B4
Cerekwica PL 85 E13
Cérences F 23 C9
Ceres UK 5 C11
Ceres I 31 E11
Ceresole Reale I 31 E11
Céreste F 35 C10
Céret F 34 F4
Čerević SRB 158 C4
Cerezo de Abajo E 46 B5
Cerezo de Arriba E 40 F4
Cerezo de Ríotirón E 40 C5
Cerfontaine B 184 D1
Cergău RO 152 E3
Cergy F 24 B7
Cerhenice CZ 77 B8
Ceriale I 37 C8
Ceriana I 37 D7
Cerignola I 60 A5
Cérilly F 30 B2
Čerin BIH 157 F8
Cerisano I 60 E6
Cerisiers F 25 D9
Cerisy-la-Forêt F 23 B10
Cerisy-la-Salle F 23 B9
Cerizay F 28 B4
Çerkezköy TR 173 B9
Çerkezmüsellim TR 173 B7
Cerklje SLO 73 D9
Cerklje SLO 148 E5
Cerknica SLO 73 E9
Cerkno SLO 73 D8
Cerkwica PL 85 B8
Cermei RO 151 D8
Çermë-Proshkë AL 168 B2
Cermignano I 62 B5
Cërmjan RKS 163 E9
Cerna HR 157 B10
Cerna RO 155 C2
Cernache do Bonjardim P 44 E4
Černá Hora CZ 77 D11
Cernat RO 153 F8
Cernătești RO 160 E2
Cernătești RO 161 C9
Cernavodă RO 155 E2
Cernay F 27 E7
Cernay-en-Dormois F 19 F10
Černčice CZ 76 B5
Cernégula E 40 C4
Cernele RO 160 E3
Cernești RO 152 C3
Cernica RO 161 E8
Cernier CH 31 A10
Černik HR 157 B7
Černík SK 146 E6
Černilov CZ 77 B9
Cernișoara RO 160 C4
Cernobbio I 69 B7
Cernoleuca MD 153 A11
Černošice CZ 76 C6
Černošín CZ 76 C3
Černilov LV 133 D2
Černovice CZ 77 D7
Černožice CZ 77 B9
Cerová SK 146 D4
Cerovac SRB 159 E6
Cerovak Tušilovički HR 148 F5
Ceroviljani BIH 157 B7
Cerovlje HR 67 B9
Cerovo SK 147 E8
Cerralbo E 45 C7
Cerreto d'Esi I 67 F6
Cerreto di Spoleto I 62 B3
Cerreto Sannita I 60 A3
Cerrigydrudion UK 10 E4
Cërrik AL 168 B2
Cerro al Volturno I 62 D6
Cersay F 28 A5
Cersosimo I 61 C6
Certaldo I 66 E3
Certeju de Sus RO 151 F10
Certești RO 153 E11
Certeze RO 145 H7
Certosa di Pavia I 69 C7
Certosa di Pesio I 37 C7
Ceru-Băcăinți RO 151 F11
Cërujë AL 168 C3
Cerva P 38 F4
Cervaro I 62 E5
Cervatos de la Cueza E 39 D10
Červená Voda CZ 77 B11
Cervenia RO 161 F6
Červená Voda CZ 77 C12
Červený Kostelec CZ 77 B10
Cervera E 42 D6
Cervera de la Cañada E 41 F8
Cervera del Llano E 47 E8
Cervera del Maestre E 48 D5
Cervera de los Montes E 46 D3
Cervera del Río Alhama E 41 D8
Cervera de Pisuerga E 40 C3
Cerveteri I 62 D2
Cervia I 66 D5
Cervignano del Friuli I 73 E7
Cervinara I 60 A3
Červiná Řečice CZ 77 C8
Cervione F 37 G10
Cervo E 38 A5
Cervo I 37 D8
Cervon F 25 F10
Cerzeto I 60 E6
Cesano Boscone I 69 C7
Cesano Maderno I 69 B7
Cesarò I 59 D6
Cesena I 66 D5
Cesenatico I 66 D5
Cesio I 37 C7
Cēsis LV 135 B10
Česká Kamenice CZ 80 E6
Česká Lípa CZ 81 E7
Česká Skalice CZ 77 B10
Česká Ves CZ 77 B12
České Brezovo SK 147 E9
České Budějovice CZ 77 E6
České Mezeříčí CZ 77 B10
Český Brod CZ 77 B7
Český Dub CZ 81 E7
Český Krumlov CZ 76 E6
Český Těšín CZ 147 B7
Česljeva Bara SRB 159 D7
Çeşme TR 177 C7
Cespedosa E 45 C9
Cessalto I 73 E6
Cessenon-sur-Orb F 34 D5
Cessole I 37 B8
Cesson F 25 C8
Cesson-Sévigné F 23 D8
Cestas F 28 F4
Čestobrodica SRB 158 F5
Cesuras E 38 B3
Cesvaine LV 135 C12
Cetariu RO 151 C9
Cetate RO 152 C5
Cetate RO 159 E11
Cetatea de Baltă RO 152 E4
Cetățeni RO 160 C6
Cetina E 41 F8
Cetingrad HR 156 B4
Cetinje MNE 163 E6
Ceto I 69 A9
Ceton F 24 D4
Cetona I 62 B1
Cetraro I 60 D5
Ceuta WHAT 53 E6
Ceutí E 56 E2
Ceva I 37 C8
Cevico de la Torre E 40 E3
Cevico Navero E 40 E3
Cevins F 31 D9
Cevio CH 68 A6
Çevizköy TR 173 A8
Čevo MNE 163 D6
Cewice PL 85 B13
Ceyreste F 35 D10
Ceyzériat F 31 C7
Cezieni RO 160 E4
Chã P 38 E4
Chaam NL 16 F3
Chabanais F 29 D7
Chabeuil F 31 F7
Chabielice PL 143 D7
Chablis F 25 E10
Chabówka PL 147 B9
Chabreloche F 30 D4
Chabris F 24 F6
Chacim P 39 F6
Chagford UK 13 D7
Chagny F 30 B6
Chaïkali GR 174 C4
Chaillac F 29 C8
Chailland F 23 D10
Chaillé-les-Marais F 28 C3
Chailley F 25 D10
Chaintrix-Bierges F 25 C11
Chaironeia GR 175 C6
Chalabre F 33 E10
Chalais CH 31 C12
Chalais F 28 E6
Chalamera E 42 D4
Chalamont F 31 C7
Chalampé F 27 E8
Chalandri GR 175 C8
Chalandritsa GR 174 C4
Chalastra GR 169 C8
Chale UK 13 D12
Chaleix F 29 D7
Châlette-sur-Loing F 25 D8
Chalford UK 13 B10
Chalgrove UK 13 B12
Chaligny F 26 C5
Chalin PL 139 E7
Chalindrey F 26 E3
Chalivoy-Milon F 29 B11
Chalki GR 169 E8
Chalki GR 181 D7
Chalkiades GR 169 F7
Chalkiades GR 174 A2
Chalkida GR 175 C8
Chalkidona GR 169 C8
Chalkio GR 175 D6
Challans F 28 B2
Challerange F 19 F10
Challes-les-Eaux F 31 D8
Chalmoux F 30 B4
Chalonnes-sur-Loire F 23 F10
Châlons-en-Champagne
F 25 C11
Chalon-sur-Saône F 30 B6
Châlus F 29 D7
Cham CH 27 F9
Cham D 75 D12
Chamalières F 30 D3
Chamaloc F 31 F7
Chamberet F 29 D9
Chambéria F 31 C8
Chambéry F 31 D8
Chambley-Bussières F 26 B4
Chambois F 23 C12
Chambon-sur-Voueize F 29 C10
Chambord F 24 E6
Chamboulive F 29 E9
Chambray-lès-Tours F 24 F4
Chambry F 19 E8
Chamesson F 25 E12
Chamonix-Mont-Blanc F 31 D10
Chamoux-sur-Gelon F 31 D9
Champagnac F 29 E10
Champagnac-de-Belair F 29 E7
Champagnac-le-Vieux F 30 E4
Champagne-Mouton F 29 D6
Champagne-sur-Seine F 25 D8
Champagney F 26 E6
Champagnole F 31 B8
Champaubert F 25 C10
Champdeniers-St-Denis F 28 C5
Champ-d'Oiseau F 25 E11
Champeix F 30 D3
Champéry CH 31 C10
Champex CH 31 C11
Champforgeuil F 30 B6
Champgenéteux F 23 D11
Champier F 31 E7
Champigné F 23 E10
Champignelles F 25 E9
Champigneulles F 186 D1
Champigny F 25 D9
Champlemy F 25 F9
Champlitte F 26 E4
Champniers F 29 D6
Champoléon F 31 F9
Champoluc I 68 B4
Champsecret F 23 C10
Champs-sur-Tarentaine-
Marchal F 29 E11
Champs-sur-Yonne F 25 E10
Champ-sur-Drac F 31 E8
Champtoceaux F 23 F9
Chamrousse F 31 E8
Chamusca P 44 F4
Chanac F 34 B5
Chanas F 30 E6
Chança P 44 F5
Chanceaux F 25 E12
Chanceaux-sur-Choisille F 24 F4
Chancelade F 29 E7
Chancelaria P 44 E3
Chandras GR 179 E11
Chandrinos GR 174 F4
Chañe E 40 F3
Changé F 23 D10
Changé F 23 E12
Changy F 30 C5
Chania GR 178 D7
Chaniotis GR 169 D10
Chantada E 38 C4
Chantelle F 30 C3
Chantilly F 25 B7
Chantonnay F 28 B3
Chanu F 23 C10
Chaource F 25 D11
Chapaevo BG 166 C5
Chapayevka UA 154 A5
Chapel-en-le-Frith UK 11 E8
Chapelle-lez-Herlaimont
B 182 E4
Chapeltown UK 11 E9
Chapinería E 46 D4
Charakas GR 178 E9
Charavgi GR 169 D6
Charbonnat F 30 B5
Charcenne F 26 F4
Charchilla F 31 C8
Chard UK 13 D9
Chardstock UK 13 D9
Charenton-du-Cher F 29 B11
Charfield UK 13 B10
Chargey-lès-Gray F 26 F4
Charlbury UK 13 B12
Charlemont UK 7 D9
Charleroi B 19 D9
Charlestown IRL 6 E5
Charleville IRL 8 D5
Charleville-Mézières F 19 E10
Charlieu F 30 C5
Charlottenberg S 96 C7
Charlton UK 13 C12
Charlton Kings UK 13 B10
Charly F 25 C9
Charmé F 29 D6
Charmes F 26 D5
Charmes-sur-Rhône F 30 F6
Charmey CH 31 B11
Charmouth UK 13 D9
Charmoy F 25 E9
Charnay-lès-Mâcon F 30 C6
Charny F 25 E9
Charokopeio GR 178 B2
Charolles F 30 C5
Charomkhava BY 133 F3
Charopo GR 169 B9
Chârost F 29 B10
Charquemont F 27 F6
Charrat CH 31 C11
Charrey-sur-Seine F 25 E12
Charrin F 30 B4
Charroux F 29 C6
Charsznica PL 143 F8
Chartres F 24 D5
Charvensod I 31 D11
Charvieu-Chavagneux F 31 D7
Charzyno PL 85 B9
Chassagne-Montrachet F 30 B6
Chasseneuil-sur-Bonnieure
F 29 D6
Chasseradès F 35 A6
Chasse-sur-Rhône F 30 D6
Chassigny-Aisey F 26 E3
Chassillé F 23 D11
Chastre B 19 C10
Château-Arnoux F 35 B10
Châteaubernard F 28 D5
Châteaubourg F 23 D9
Châteaubriant F 23 E9
Château-Chinon F 30 A4
Château-d'Oex CH 31 C11
Château-d'Olonne F 28 B2
Château-du-Loir F 24 E3
Châteaudun F 24 D5
Châteaugiron F 23 D8
Château-Gontier F 23 E10
Château-Landon F 25 D8
Château-la-Vallière F 24 E3
Châteaulin F 22 D3
Châteaumeillant F 29 B10
Châteauneuf-de-Galaure
F 31 E6
Châteauneuf-de-Randon
F 35 A6
Châteauneuf-d'Ille-et-Vilaine
F 23 C8
Châteauneuf-du-Faou F 22 D4
Châteauneuf-en-Thymerais
F 24 C5
Châteauneuf-la-Forêt F 29 D9
Châteauneuf-les-Martigues
F 35 D9
Châteauneuf-sur-Charente
F 28 D5
Châteauneuf-sur-Cher F 29 B10
Châteauneuf-sur-Loire F 25 E7
Châteauneuf-sur-Sarthe
F 23 E11
Châteauneuf-Val-de-Bargis
F 25 F9
Châteauponsac F 29 C8
Château-Porcien F 19 E9
Châteauredon F 36 C4
Châteaurenard F 25 E8
Châteaurenard F 35 C8
Châteaurenaud F 31 B7
Château-Renault F 24 E4
Châteauroux F 29 B9
Châteauroux F 36 B5
Château-Salins F 26 C6
Château-Thierry F 25 B9
Châteauvillain F 25 D12
Châtel F 31 C10
Châtelaillon-Plage F 28 C3
Châtelaudren F 22 C6
Châtel-Censoir F 25 E10
Châteldon F 30 D4
Châtelet B 19 D10
Châtel-Gérard F 25 E11
Châtelguyon F 30 D3
Châtellerault F 29 B7
Châtel-Montagne F 30 C4
Châtel-St-Denis CH 31 B10
Châtel-sur-Moselle F 26 D5
Châtelus-Malvaleix F 29 C10
Châtenois F 26 D4
Châtenois F 27 D7
Châtenois-les-Forges F 27 E6
Châtenoy-le-Royal F 30 B6
Chatham UK 15 E10
Châtillon B 19 E12
Châtillon I 68 B4
Châtillon-Coligny F 25 E8
Châtillon-en-Bazois F 30 A4
Châtillon-en-Diois F 35 A9
Châtillon-en-Michaille F 31 C8
Châtillon-la-Palud F 31 D7
Châtillon-sur-Chalaronne
F 31 C6
Châtillon-sur-Colmont F 23 D10
Châtillon-sur-Indre F 29 B8
Châtillon-sur-Loire F 25 E8
Châtillon-sur-Marne F 25 B10
Châtillon-sur-Seine F 25 E12
Châtillon-sur-Thouet F 28 B5
Chatte F 31 E7
Chatteris UK 15 C9
Chatton UK 5 D13
Chatuzange-le-Goubet F 31 F7
Chatzis GR 174 E4
Chauchina E 53 B9
Chaudenay F 26 E4
Chaudes-Aigues F 30 F3
Chaudeyrac F 35 A6
Chaudfontaine B 19 C12
Chauffailles F 30 C5
Chauffayer F 31 F9
Chaulnes F 18 E6
Chaumergy F 31 B7
Chaumont F 26 D3
Chaumont-en-Vexin F 18 F4
Chaumont-Porcien F 19 E9
Chaumont-sur-Aire F 26 C3
Chaumont-sur-Loire F 24 F5
Chaunay F 29 C6
Chauny F 19 E7
Chaussin F 31 B7
Chauvigny F 29 B7
Chavanay F 30 E6
Chavanges F 25 C12
Chavanoz F 31 D7
Chavari GR 174 D3
Chavelot F 26 D5
Chaves P 38 E5
Chazelles-sur-Lyon F 30 D5
Cheadle UK 11 F8
Cheb CZ 75 B11
Checa E 47 C9
Chechel'nyk UA 154 A4
Chęciny PL 143 E9
Chécy F 24 E7
Chedburgh UK 15 C10
Cheddar UK 13 C9
Chef-Boutonne F 28 C5
Chekhpare BG 166 E4
Cheles E 51 B5
Chella E 56 C3
Chełm PL 141 H8
Chełmek PL 143 F7
Chełmiec PL 145 D2
Chełmno PL 138 D5
Chełmno PL 142 B6
Chelmsford UK 15 D9
Chełmża PL 138 D6
Cheltenham UK 13 B10
Chelva E 47 E11
Chemazé F 23 E10
Chémery F 24 F5
Chémery-sur-Bar F 19 E10
Chemillé F 23 F10
Chemin F 31 B7
Cheminon F 25 C12
Chemiré-le-Gaudin F 23 E11
Chemnitz D 80 E3
Chenecey-Buillon F 26 F4
Chénérailles F 29 C10
Cheniménil F 26 D6
Chenôve F 26 F3
Cheny F 25 E10
Chepelare BG 165 F10
Chepintsi BG 165 D7
Chepstow UK 13 B9
Chepy F 25 C11
Chera E 48 E3
Chérac F 28 D5
Cherasco I 37 B7
Chéraute F 32 D4
Cherbourg-Octeville F 23 A8
Cheremule I 64 B2
Cherepkivtsi UA 153 A7
Chereshovo BG 161 F8
Cherlenivka UA 153 A8
Cherna BG 155 F1
Cherna Gora BG 166 E4
Chernevo BG 167 C9
Chernevo RUS 132 D3
Cherni Osŭm BG 165 D10
Cherni Vrŭkh BG 167 E8
Chernivtsi UA 153 A7
Chernogorovo BG 165 E9
Cherno More BG 167 D8
Chernoochene BG 166 F4
Chernyakhovsk RUS 136 D4
Chernyshevskoye RUS 136 D6
Chéroy F 25 D9
Cherskaya RUS 132 F3
Cherso GR 169 B8
Chert E 42 F4
Cherveix-Cubas F 29 E8
Cherven BG 161 F8
Chervena voda BG 161 F8
Cherven Bryag BG 165 C9
Cherventsi BG 167 B8
Cherves-Richemont F 28 D5
Chervona Hreblya UA 154 A4
Chervonoarmiys'ke UA 154 F3
Chervonohrad UA 144 C9
Chesham UK 15 D7
Cheshunt UK 15 D8
Chesley F 25 E11
Chessy-les-Prés F 25 D10
Cheste E 48 F3
Chester UK 10 E6
Chesterfield UK 11 E9
Chester-le-Street UK 5 F13
Chețani RO 152 E4
Chetrosu MD 153 A11
Chevagnes F 30 B4
Chevanceaux F 28 E5
Chevereșu Mare RO 159 B7
Chevillon F 26 C3
Chevilly F 24 D6
Chevreuse F 24 C7
Chew Magna UK 13 C9
Cheylade F 30 E2
Chezal-Benoît F 29 B10
Chèze F 32 E5
Chiajna RO 161 E8
Chiampo I 66 A3
Chianciano Terme I 62 A1
Chianni I 66 E2
Chiaramonte Gulfi I 59 E6
Chiaramonti I 64 B2
Chiaravalle I 67 E7
Chiaravalle Centrale I 59 B9
Chiari I 69 B8
Chiaromonte I 60 C6
Chiasso CH 69 B7
Chiavari I 37 C10
Chiavenno I 69 A7
Chiché F 28 B5
Chichester UK 15 F7
Chichiș RO 153 F7
Chiclana de la Frontera E 52 D4
Chiclana de Segura E 55 C6
Chieming D 72 A6
Chienes I 72 C4
Chieri I 37 A7
Chiesa in Valmalenco I 69 A8
Chieșd RO 151 C10
Chiesina Uzzanese I 66 E2
Chieti I 63 C6
Chieuti I 63 D8
Chieveley UK 13 C12
Chièvres B 19 C8
Chigwell UK 15 D9
Chiheru de Jos RO 152 D5
Chikhachevo RUS 133 B7
Chilches E 48 E4
Chilcompton UK 13 C9
Chilham UK 15 E10
Chiliile RO 161 C9
Chiliomodi GR 175 D6
Chilleurs-aux-Bois F 24 D7
Chillón E 54 B3
Chilluévar E 55 C6
Chiloeches E 47 C6
Chimay B 19 D9
Chimeneas E 53 B9
Chinanale I 36 B6
Chinchilla de Monte Aragón
E 55 B9
Chinchón E 46 D6
Chindrieux F 31 D8
Chinon F 23 F12
Chinteni RO 152 D3
Chiochiș RO 152 D4
Chioggia I 66 B5
Chiojdeni RO 161 B9
Chiojdu RO 161 C8
Chiomonte I 31 E10
Chiona GR 174 C4
Chionata GR 174 C2
Chios GR 177 C7
Chiperceni MD 154 B3
Chipiona E 52 C4
Chippenham UK 13 C10
Chipping Campden UK 13 A11
Chipping Norton UK 13 B11
Chipping Ongar UK 15 D9
Chipping Sodbury UK 13 B10
Chippis CH 31 C12
Chiprana E 42 E3
Chiprovtsi BG 165 C6
Chirbury UK 10 F5
Chiren BG 165 C8
Chiriet-Lunga MD 154 E3
Chirivel E 55 D8
Chirk UK 10 F5
Chirnogeni RO 155 F2
Chirnogi RO 161 E9
Chirnside UK 5 D12
Chiroubles F 30 C6
Chirpan BG 166 E4
Chirpăr RO 152 F5
Chirsova MD 154 E3
Chiscani RO 155 C1
Chișcăreni MD 153 B12
Chișcăreni MD 154 C3
Chiselet RO 161 E9
Chișinău MD 154 C3
Chisindia RO 151 E9
Chișineu-Criș RO 151 D8
Chistye-Prudy RUS 136 E5
Chițcani MD 154 D5
Chitignano I 66 E4
Chitila RO 161 D7
Chiuiești RO 152 C3
Chiuro I 69 A8
Chiusa I 72 C4
Chiusa di Pesio I 37 C7
Chiusaforte I 73 D7
Chiusa Sclafani I 58 D3
Chiusavecchia I 37 D7
Chiusdino I 66 F3
Chiusi I 62 A1
Chiusi della Verna I 66 E4
Chiuza RO 152 C4
Chiva E 48 F3
Chivasso I 68 C4
Chizé F 28 C5
Chlebičov CZ 146 B5
Chlebnice SK 147 C8
Chlebowo PL 81 B7
Chlewiska PL 141 H3
Chłopice PL 144 D6
Chludowo PL 81 A11
Chlumčany CZ 76 C4
Chlumec CZ 80 E5
Chlumec nad Cidlinou CZ 77 B8
Chlum u Třeboně CZ 77 E7
Chmielnik PL 143 E10
Chmielno PL 85 C10
Chmiel Pierwszy PL 144 A6
Chminianske Jakubovany
SK 145 F3
Chobienia PL 81 C10
Chobienice PL 81 B9
Choceň CZ 77 B10
Choceń PL 138 F7
Chocholná-Velčice SK 146 D5
Chocianów PL 81 D9
Chocicza PL 81 B12
Chociwel PL 85 D8
Chocz PL 142 C4
Choczewo PL 138 A4
Chodecz PL 138 F7
Chodel PL 141 H6
Chodov CZ 75 B12
Chodová Planá CZ 75 C12
Chodów PL 141 F6
Chodów PL 143 B7
Chodzież PL 85 E11
Chojna PL 84 E6
Chojnice PL 85 C13
Chojno PL 81 C12
Chojnów PL 81 D9
Cholet F 28 A4
Chomakovtsi BG 165 C9
Chomelix F 30 E4
Chomérac F 35 A8
Chomutov CZ 76 B4
Chonikas GR 175 D6
Chooz F 19 D10
Chop UA 145 G5
Chora GR 174 E4
Chora GR 175 A8
Chora GR 177 D8
Chora GR 179 B9
Chora GR 179 B9
Chorafakia GR 178 D7
Chora Sfakion GR 178 E7
Chorefto GR 169 F9
Chorges F 36 B4
Chorio GR 177 F8
Choristi GR 171 B6
Chorkówka PL 145 D4
Chorley UK 10 D6
Chorna UA 154 B4
Chorna Tysa UA 152 A4
Chornoholova UA 145 F6
Chornomyn UA 154 A4
Chorzele PL 139 D10
Chorzów PL 143 F6
Chorzyna PL 142 D6
Choszczno PL 85 D8
Chotča SK 145 E4
Chotěboř CZ 77 C9
Chotěšov CZ 76 C4
Chotín SK 149 A10
Chouilly F 25 B11
Chouni GR 174 B4
Chouto P 44 F4
Chouzy-sur-Cisse F 24 E5
Chovar E 48 E4
Choye F 26 F4
Chozas de Abajo E 39 C8
Chrást CZ 76 C4
Chrast CZ 77 C9
Chrastava CZ 81 E7
Chřibská CZ 81 E6
Christchurch UK 13 D11
Christiansfeld DK 86 E4
Christon Bank UK 5 E13
Chropyně CZ 146 C4
Chróścina PL 142 E3
Chrostkowo PL 139 E7
Chroustovice CZ 77 C9
Chrudim CZ 77 C9
Chruślin PL 143 B8
Chruszczobród PL 143 F7
Chrysa GR 171 B7
Chrysafa GR 175 E6
Chryso GR 169 B10
Chrysochorafa GR 169 B9
Chrysochori GR 171 C7
Chrysokellaria GR 178 B2
Chrysoupoli GR 171 C7
Chrzanów PL 143 F7
Chrząstowa Wielka PL 81 D12
Chrząstowice PL 142 E5
Chrzypsko Wielkie PL 81 A10
Chtelnica SK 146 D5
Chuchelná CZ 142 G5
Chudleigh UK 13 D7
Chudoba PL 142 E5
Chulilla E 48 E3
Chulmleigh UK 13 D7
Chuprene BG 165 B6
Chur CH 71 D9
Churchdown UK 13 B10
Church Hill IRL 7 C7
Church Stretton UK 10 F6
Churek BG 165 D8
Churwalden CH 71 D9
Chuzelles F 30 D6
Chvalčov CZ 146 C5
Chvaletice CZ 77 B8
Chvalšiny CZ 76 E6
Chwaszczyno PL 138 B5
Chynadiyeve UA 145 F6
Chyňava CZ 76 B6
Chýnov CZ 77 D7
Chynów PL 141 G4
Chyšky CZ 77 C6
Ciacova RO 159 B7
Ciadîr MD 154 D2
Ciadîr-Lunga MD 154 E3
Ciadoux F 33 D7
Ciampino I 62 D3
Cianciana I 58 D3
Ciasna PL 142 E6
Ciążeń PL 142 B4
Cibakháza H 150 D5
Ciborro P 50 B3
Ciboure F 32 D2
Cicagna I 37 C10
Cicărlău RO 151 B11
Cicciano I 60 B3
Cicerale I 60 C4
Ćićevac SRB 159 F7
Cicibór Duży PL 141 F8
Ciclova Română RO 159 C8
Čičov SK 149 A9
Çidhën AL 163 F9
Cidones E 41 E6
Ciechanów PL 139 E10
Ciechanowiec PL 141 E7
Ciechocinek PL 138 E6
Cielądz PL 141 G2
Ciempozuelos E 46 D5
Ciepielów PL 141 H5
Ciepłowody PL 81 E11
Čierna nad Tisou SK 145 G5
Čierne Kľačany SK 146 E6
Čierny Balog SK 147 D9
Cierp-Gaud F 33 E7
Cierznie PL 85 C12
Cieszanów PL 144 C7
Cieszków PL 81 C12
Cieszyn PL 142 D4
Cieszyn PL 147 B7
Cieutat F 33 D6
Cieux F 29 D8
Cieza E 55 C10
Ciężkowice PL 144 D2
Cífer SK 146 E5
Çifliq AL 168 E3
Çiftlikköy TR 173 B7
Çiftlikköy TR 173 B9
Cifuentes E 47 C7
Cigales E 39 E10
Cigánd H 145 G4
Cigliano I 68 C5
Cikó H 149 D11
Cilibia RO 161 C10
Cilieni RO 160 F5
Čilipi HR 162 D5
Čilizská Radvaň SK 146 F5
Cill Airne IRL 8 D4
Cillas E 47 C9
Cill Chainnigh IRL 9 C8
Cill Chaoi IRL 8 C3
Cill Charthaigh IRL 6 C5
Cill Chiaráin IRL 6 F3
Cill Choca IRL 7 F9
Cill Chomhghaill IRL 7 F10
Cill Chormaic IRL 7 F7
Cill Chuillin IRL 7 F9
Cill Dalua IRL 8 C6
Cill Dara IRL 7 F9
Cill Dhéagláin IRL 7 E10
Cill Droichid IRL 7 F9
Cille Bhrighde UK 2 L2
Cilleros E 45 D7
Cill Mhantáin IRL 7 G10
Cill Mocheallóg IRL 8 D5
Cill na Mallach IRL 8 D5
Cill Orglan IRL 8 D3
Cill Rois IRL 8 C4
Cill Rónáin IRL 6 F3
Cilybebyll UK 13 B7
Cimanes del Tejar E 39 C8
Cimballa E 47 B9
Cimdenieki LV 134 C2
Čimelice CZ 76 D6
Ciminna I 58 D4
Cimișlia MD 154 D3
Cimolais I 72 D5
Çınarcık TR 173 C11
Cinco Casas E 47 F6
Cinctorres E 42 F3
Cincu RO 152 F5
Cinderford UK 13 B10
Çine TR 181 A8
Ciñera E 39 C8
Ciney B 19 D11
Cinfães P 44 B4
Cinge TR 177 A9
Cingoli I 67 F7
Cinigiano I 65 B4
Cinisello Balsamo I 69 B7
Cinn Mhara IRL 6 F3
Cinobaňa SK 147 E9
Cinq-Mars-la-Pile F 24 F3
Cinquefrondi I 59 C9
Cintegabelle F 33 D9
Cintrey F 26 E4
Cintruénigo E 41 D8
Ciobănița RO 155 E2
Ciobanu RO 155 D1
Čiobiškis LT 137 D10
Ciocănești RO 161 D7
Ciocănești RO 161 E10
Ciocârlia RO 155 E2
Ciocârlia RO 161 D9
Ciochina RO 161 D10
Ciocile RO 161 D10
Cioc-Maidan MD 154 E3
Ciofrângeni RO 160 C5
Cióirtheach IRL 7 F7
Ciolănești RO 160 E6
Ciolpani RO 161 D8
Cionn tSáile IRL 8 E5
Ciorani RO 161 D8
Ciorăști RO 161 C10
Ciorogârla RO 161 E7
Cioroiași RO 160 E2
Cioropcani MD 153 C11
Ciortești RO 153 D11
Ciprian Porumbescu RO 153 B8
Cirat E 48 D4
Cirauqui E 32 E2
Cīrava LV 134 C2
Circello I 60 A3
Cirencester UK 13 B11
Cireșu RO 159 D10
Cireșu RO 161 D10
Cirey-sur-Blaise F 26 D2
Cirey-sur-Vezouze F 27 C6
Ciria E 41 E8
Ciriè I 68 C4
Cirigliano I 60 C6
Ciripcău MD 154 A2
Cirkale LV 134 B4
Cirma LV 133 C3
Cirò I 61 E8
Cirò Marina I 61 E8
Çirpi TR 177 C10
Ciruelos E 46 E5
Ciruelos del Pinar E 47 B8
Cīruļi LV 134 B4
Ciry-le-Noble F 30 B5
Cisano sul Neva I 37 C8
Cisek PL 142 F5
Cislău RO 161 C8
Cișmichioi MD 155 B2
Cisna PL 145 E5
Cisnădie RO 160 B4
Cisneros E 39 D10
Cisòn di Valmarino I 72 E5
Cissé F 29 B6
Cista Provo HR 157 E6
Cisterna di Latina I 62 D3
Cisternino I 61 B8
Cistierna E 39 C9
Citerna I 66 F5
Čitluk BIH 157 B6
Čitluk BIH 157 F8
Cittadella I 72 E4
Città della Pieve I 62 B2
Città di Castello I 66 F5
Cittanova I 59 C9
Cittareale I 62 B4
Città Sant'Angelo I 62 B6
Cittiglio I 68 B6
Ciucea RO 151 D10
Ciuchici RO 159 D8
Ciuciulea MD 153 B10
Ciucsângeorgiu RO 153 E7
Ciucur-Mingir MD 154 E3
Ciucurova RO 155 D2
Ciudad Real E 54 B5
Ciudad Rodrigo E 45 C7
Ciudanovița RO 159 C8
Ciugud RO 152 E3
Ciuhoi RO 151 C9
Ciulnița RO 161 D10
Ciumani RO 153 D7
Ciumeghiu RO 151 D8
Ciuperceni RO 159 D11
Ciuperceni RO 160 F5
Ciupercenii Noi RO 159 F10
Ciurila RO 152 D3
Ciuruleasa RO 151 E11
Ciutadella E 57 A12
Civaux F 29 C7
Cividale del Friuli I 73 D7
Civita I 61 D6
Civita Castellana I 62 C2
Civita d'Antino I 62 D4
Civitanova del Sannio I 63 D6
Civitanova Marche I 67 F8
Civitaquana I 62 C5
Civitavecchia I 62 C1
Civitella Casanova I 62 C5
Civitella del Tronto I 62 B5
Civitella di Romagna I 66 D4
Civitella in Val di Chiana I 66 F4

Civitella Roveto I 62 D4
Civray F 29 B10
Civray F 29 C6
Cizer RO 151 C10
Čížkovice CZ 76 B6
Čkyně CZ 76 D5
Clabhach UK 4 B3
Clachan UK 2 L4
Clachan UK 4 C7
Clachan of Glendaruel UK 4 C6
Clachtoll UK 2 J6
Clackmannan UK 5 C9
Clacton-on-Sea UK 15 D11
Cladich UK 4 C6
Clady UK 4 F1
Clady UK 4 F3
Claggan UK 4 B5
Clairac F 33 B6
Clairoix F 18 F6
Clairvaux-les-Lacs F 31 B8
Claix F 31 E8
Clamecy F 25 F10
Clane IRL 7 F9
Clans F 36 C4
Claonadh IRL 7 F9
Claonaig UK 4 D6
Clapham UK 11 C7
Clapham UK 15 C8
Clara IRL 7 F7
Clár Chlainne Mhuiris IRL 6 E5
Clarecastle IRL 8 C5
Claremorris IRL 6 E5
Claret F 35 B10
Claro CH 31 D9
Clary F 19 D7
Clashmore IRL 9 D7
Clashnessie UK 2 J6
Claudy UK 4 F2
Clausnitz D 80 E4
Claußnitz D 80 E3
Clausthal-Zellerfeld D 79 C7
Claut I 72 D6
Clauzetto I 73 D6
Clavering UK 15 D9
Clavier B 19 D11
Clay Cross UK 11 E9
Claydon UK 15 C11
Clayton UK 5 F8
Cleady IRL 8 E3
Cleat UK 3 H11
Cleator Moor UK 10 B4
Clécy F 23 C11
Cléder F 22 C3
Cleethorpes UK 11 D11
Clefmont F 26 D4
Cléguérec F 22 D5
Cleja RO 153 E9
Clelles F 31 E8
Clémency L 20 E5
Clenze D 83 E10
Cleobury Mortimer UK 13 A10
Cléon F 18 F3
Cléon-d'Andran F 35 A8
Cléré-les-Pins F 24 F3
Clères F 18 E3
Clérey F 25 D11
Clermain F 30 C6
Clermont F 18 F5
Clermont F 33 D8
Clermont-en-Argonne F 26 B3
Clermont-Ferrand F 30 D3
Clermont-l'Hérault F 34 C5
Clerval F 26 F6
Clervaux L 20 D6
Cléry-St-André F 24 E6
Cles I 72 D3
Clevedon UK 13 C9
Cleveleys UK 10 D5
Clifden IRL 6 F2
Cliffe UK 15 E9
Cliffoney IRL 6 D6
Climăuți MD 153 A11
Clinge NL 16 F2
Clingen D 79 D8
Clion F 29 B8
Clisson F 28 A3
Clitheroe UK 11 D7
Cloch na Rón IRL 6 F3
Clogh IRL 7 F8
Clogh IRL 9 C7
Clogh IRL 9 F8
Clogh IRL 9 C10
Cloghan IRL 7 F7
Cloghan IRL 7 F7
Clogheen IRL 8 D7
Clogher UK 7 D8
Clogherhead IRL 7 E10
Cloghy UK 7 D12
Clohars-Carnoët F 22 E4
Clohernagh IRL 8 D6
Cloich na Coillte IRL 8 E5
Clonakilty IRL 8 E5
Clonaslee IRL 7 F7
Clonbern IRL 6 E5
Clonbulloge IRL 7 F8
Clonbur IRL 6 E4
Clondrohid IRL 8 E4
Clondulane IRL 8 D6
Clonea IRL 9 D8
Clonee IRL 7 F10
Cloneen IRL 9 D7
Clones IRL 7 E8
Clonmel IRL 9 D7
Clonmellon IRL 7 E8
Clonoulty IRL 9 C7
Clonroche IRL 9 D9
Clontibret IRL 7 E8
Clonygowan IRL 7 F8
Cloonbannin IRL 8 D4
Cloonboo IRL 6 F4
Cloonfad IRL 6 E5
Cloonfad IRL 6 F6
Cloonkeen IRL 6 E4
Cloppenburg D 17 C10
Closeburn UK 5 E9
Clough UK 7 D11
Cloughjordan IRL 8 C6
Cloughmills UK 4 E4
Cloughton UK 11 C11
Clova UK 5 B10
Clovelly UK 12 D6
Clovullin UK 4 B6
Cluain Bú IRL 8 C5
Cluainín IRL 6 D6
Cluain Eois IRL 7 D8
Cluainín IRL 6 D6
Cluain Meala IRL 9 D7
Cluis F 29 B9
Cluj-Napoca RO 152 D3
Clumanc F 36 C4
Clun UK 13 A8
Cluny F 30 C6

Cluses F 31 C10
Clusone I 69 B8
Clydach UK 13 B7
Clydebank UK 5 D8
Clynderwen UK 12 B5
Clyro UK 13 A8
Čmielów PL 143 E12
Čmolas PL 143 F12
Coachford IRL 8 E5
Coagh UK 4 F3
Coalburn UK 5 D9
Coalisland UK 7 C9
Coalville UK 11 F9
Coaña E 39 A6
Coarnele Caprei RO 153 C10
Coarraze F 32 D5
Coast UK 2 K5
Coatbridge UK 5 D8
Cobadin RO 155 E2
Cobani MD 153 B10
Cobeja E 46 D5
Cobeta E 47 C8
Cobh IRL 8 E6
Cobia RO 160 D6
Coburg D 75 B8
Coca E 46 B3
Cocentaina E 56 D4
Cochem D 21 D8
Cochirleanca RO 161 C10
Cociuba Mare RO 151 D9
Cockburnspath UK 5 D12
Cockenzie and Port Seton UK 5 D11
Cockerham UK 10 D6
Cockermouth UK 5 F10
Cockett UK 12 B7
Cocora RO 161 D10
Cocu RO 160 D5
Codaeşti RO 153 D11
Coddington UK 11 E10
Code LV 135 D12
Codevigo I 66 B5
Codigoro I 66 C5
Codlea RO 161 B6
Codogno I 69 C7
Codos E 41 F9
Codreanca MD 154 C3
Codroipo I 73 E6
Codrongianos I 64 B2
Codru MD 154 D3
Coesfeld D 17 E8
Coevorden NL 17 C7
Coëx F 28 B2
Cofrentes E 47 F10
Cogealac RO 155 D3
Cogeces del Monte E 40 E3
Coggeshall UK 15 D10
Coggia F 37 G9
Cogiglia I 68 B5
Cogîlniceni MD 154 B3
Çöğmen TR 181 C10
Cognac F 28 D5
Cognac-la-Forêt F 29 D8
Cogne I 31 D11
Cognin F 31 D8
Cogolin F 36 E5
Cogollos E 40 D4
Cogollos Vega E 53 B9
Cogolludo E 47 C6
Cogula F 45 C6
Coillan Chollaigh IRL 7 E9
Coimbra P 44 D4
Coimbrão P 44 E3
Coín E 53 C7
Coincy F 25 B9
Coja P 44 D5
Cojasca RO 161 D7
Cojocna RO 152 D3
Čoka SRB 150 F5
Colares P 50 B1
Colayrac-St-Cirq F 33 B7
Cölbasna MD 154 B4
Cölbe D 21 C11
Colbitz D 79 B10
Colbordolo I 67 E6
Colceag RO 161 D8
Colchester UK 15 D10
Coldingham UK 5 D12
Colditz D 79 D11
Coldstream UK 5 D12
Coleford UK 13 B9
Coleraine UK 4 E3
Colibaşi MD 155 B11
Colibaşi RO 160 D5
Colibaşi RO 161 C8
Colibiţa RO 152 C5
Colico I 69 A7
Coligny F 31 C7
Colindres E 40 B5
Colintraive UK 4 D6
Collado Hermoso E 46 B5
Collado Villalba E 46 C5
Collagna I 66 D1
Collanzo E 39 B8
Collarmele I 62 C5
Collazzone I 62 B2
Collecchio I 66 C1
Collecorvino I 62 C5
Colledara I 62 B5
Collegno I 31 E11
Collepardo I 62 D4
Collepasso I 61 C10
Collesalvetti I 66 E1
Colle Sannita I 60 A3
Collesano I 58 D4
Colletorto I 63 D7
Colliano I 60 B4
Colli a Volturno I 62 D6
Collinas I 64 E2
Collinée F 22 D6
Collingham UK 11 E10
Collinghorst (Rhauderfehn) D 17 B9
Collio I 69 B9
Collioure F 34 E5
Collobrières F 36 E4
Collon IRL 7 E10
Collonges F 31 C8
Collonges-la-Rouge F 29 E9
Collooney IRL 6 D6
Colmar F 27 D8
Colmars F 36 C5
Colmberg D 75 D7
Colmeal P 44 D5
Colmenar E 53 C8

Colmenar del Arroyo E 46 D4
Colmenar de Montemayor E 45 D9
Colmenar de Oreja E 46 D6
Colmenar Viejo E 46 C5
Colméry F 25 F9
Colmonell UK 4 E7
Colne UK 11 D7
Colobraro I 61 C6
Cologna Veneta I 66 B3
Cologne F 33 C7
Cologno al Serio I 69 B8
Colombelles F 23 B11
Colombey-les-Belles F 26 C4
Colombey-les-Deux-Églises F 26 D2
Colombier CH 31 B10
Colombiès F 33 B10
Colombres E 39 B10
Colomers E 43 C9
Colomiers F 33 C8
Coloneşti RO 153 D10
Coloneşti RO 160 D5
Colònia de Sant Jordi E 57 C11
Colònia de Sant Pere E 57 B10
Colonnella I 62 B5
Colorno I 66 C1
Colos P 50 D3
Cölpin D 84 C4
Colquhar UK 5 D10
Colroy-la-Grande F 27 D7
Colsterworth UK 11 F10
Colți RO 161 C8
Coltishall UK 15 B11
Colunga E 39 B9
Colwyn Bay UK 10 E4
Coly F 29 E8
Colyford UK 13 D8
Colyton UK 13 D8
Comacchio I 66 C5
Comana RO 161 F7
Comana RO 155 F2
Comana RO 161 E8
Comana de Sus RO 152 F6
Comandău RO 153 E8
Comănești RO 153 E8
Comares E 53 C8
Comarna RO 153 C11
Comarnic RO 161 C7
Combeaufontaine F 26 E4
Combe Martin UK 12 C6
Comber UK 7 C11
Comberton UK 15 C9
Comblain-au-Pont B 19 D12
Comblanchien F 26 F2
Combles F 18 D6
Combloux F 31 D10
Combourg F 23 D8
Combronde F 30 D3
Comeglians I 73 C6
Comèlico Superiore I 72 C6
Comillas E 40 B3
Comines B 18 C6
Comişani RO 161 D7
Comiso I 59 F6
Comitini I 58 E4
Comloşu Mare RO 150 F6
Commeen IRL 7 C7
Commenailles F 31 B7
Commentry F 30 C2
Commequiers F 28 B2
Commer F 23 D10
Commercy F 26 C4
Como I 69 B7
Cómpeta E 53 C9
Compiano I 37 B11
Compiègne F 18 F6
Compolibat F 33 B10
Comporta P 50 C2
Compreignac F 29 C8
Comps-sur-Artuby F 36 D5
Comrat MD 154 E3
Comrie UK 5 C9
Comunanza I 62 B4
Cona I 66 B5
Conca F 37 H10
Concarneau F 22 E4
Concas I 64 B4
Conceição P 50 E4
Concha E 40 B5
Conches-en-Ouche F 24 C4
Concordia Sagittaria I 73 E6
Concorès F 33 A8
Concots F 33 B9
Condat F 30 E2
Condé-en-Brie F 25 B10
Condé-sur-Huisne F 24 D4
Condé-sur-l'Escaut F 182 E3
Condé-sur-Noireau F 23 C10
Condino I 69 B10
Condofuri I 59 C9
Condom F 33 C6
Condove I 31 E11
Condrieu F 30 E6
Condrița MD 154 C3
Conegliano I 72 E5
Conflans-en-Jarnisy F 26 B4
Conflans-sur-Lanterne F 26 E5
Confolens F 29 C7
Cong IRL 6 E4
Congaz MD 154 E3
Congdon's Shop UK 12 D6
Congleton UK 11 E7
Congosto E 39 C7
Congosto de Valdavia E 39 C10
Congrier F 23 E9
Conil de la Frontera E 52 D4
Coningsby UK 11 E11
Coniston UK 10 C5
Conlie F 23 D11
Conliège F 31 B8
Conlig UK 4 F5
Connah's Quay UK 10 E5
Connantre F 25 C10
Connaux F 35 B8
Connel UK 4 C6
Connerré F 24 D3
Connolly IRL 8 C4
Connor UK 4 F4
Conon Bridge UK 2 K8
Conop RO 151 E9
Conow D 84 D4
Conques F 33 A10
Conques-sur-Orbiel F 33 D10
Conquista E 54 C3
Consandolo I 66 C4
Consdorf L 186 B1
Conselice I 66 C4

Consell E 49 E10
Conselve I 66 B4
Consett UK 5 F13
Constância P 44 F4
Constanța RO 155 E3
Constantí E 42 E6
Constantim P 38 F4
Constantim P 39 E7
Constantina E 51 D8
Constantin Daicoviciu RO 159 B9
Constanzana E 46 C3
Consuegra E 46 F5
Contarina I 66 B5
Contes F 37 D6
Contessa Entellina I 58 D3
Conthey CH 31 C11
Contigliano I 62 C3
Contis-Plage F 32 B3
Contres F 24 F5
Contrexéville F 26 D4
Controne I 60 B4
Contwig D 21 F8
Conty F 18 E5
Conversano I 61 B8
Convoy IRL 7 C7
Conwy UK 10 E4
Conza della Campania I 60 B4
Cookstown UK 4 F3
Coola IRL 6 D6
Coolbaun IRL 6 G6
Coole I 25 C11
Coole IRL 7 E8
Coolmore IRL 6 C6
Coombe Bissett UK 13 C11
Cootehill IRL 7 D8
Copăcani RO 161 E7
Copăcel RO 159 C9
Copăceni MD 153 C12
Copăceni RO 160 D3
Copăcenii de Sus RO 161 E8
Copanca MD 154 D5
Copanello I 59 B9
Copceac MD 155 B11
Copertino I 61 C10
Çöpköy TR 173 B6
Copley UK 5 F13
Copons E 43 D7
Copparo I 66 C4
Coppeen IRL 8 E5
Copplestone UK 13 D7
Coppull UK 10 D6
Copşa Mică RO 152 E4
Copythorne UK 13 D12
Corabia RO 160 F4
Cora Chaitlín IRL 8 C5
Cora Droma Rúisc IRL 6 E6
Čoralići BIH 156 C5
Corato I 61 A6
Coray F 22 D4
Corbalán E 47 D11
Corbally IRL 6 D4
Corbasca RO 153 E10
Corbeanca RO 161 D8
Corbeil-Essonnes F 25 C7
Corbeilles F 25 D8
Corbeni RO 160 C5
Corbeny F 19 F8
Corbera E 48 F4
Corbera d'Ebre E 42 E4
Corbi RO 160 C5
Corbie F 18 E6
Corbières CH 31 B11
Corbières F 35 C10
Corbigny F 25 F10
Corbii Mari RO 161 D6
Corbița RO 153 E10
Corbridge UK 5 F12
Corbu RO 155 E3
Corbu RO 153 E3
Corbu RO 160 D5
Corby UK 11 G10
Corcaigh IRL 8 E6
Corchiano I 62 C2
Corciano I 66 F5
Corcieux F 27 D6
Corcova RO 159 D11
Corcubión E 38 C1
Córdoba E 53 A7
Cordemais F 23 F8
Cordenòns I 73 E6
Cordes-sur-Ciel F 33 B9
Córdoba E 53 A7
Cordobilla de Lácara E 45 F8
Corduente E 47 C9
Cordun RO 153 D9
Coreglia Antelminelli I 66 D2
Corella E 41 D8
Coreses E 39 E8
Corestăuți MD 153 A10
Corfe Castle UK 13 D10
Corfinio E 62 C5
Cori I 62 D3
Coria E 45 E7
Coria del Río E 51 E7
Coriano I 66 E6
Corigliano Calabro I 61 D7
Corinaldo I 67 E7
Coripe E 51 F9
Corjeuți MD 153 A10
Corlăţel RO 159 E10
Corlățeni RO 161 B10
Corlay F 22 D5
Corlea IRL 7 F7
Corleone I 58 D3
Corleto Perticara I 60 C6
Cormatin F 30 B6
Cormeilles F 18 F1
Corme Porto E 38 B1
Cormicy F 19 F8
Cormons I 73 E7
Cormontreuil F 20 F2
Cornafulla IRL 6 F7
Cornago E 41 D7
Cornamona IRL 6 E4
Cornaredo I 69 B6
Cornas F 30 F6
Cornățelu RO 161 D7
Cornau D 17 C10
Cornea RO 159 C9
Cornedo all'Isarco I 72 C3
Cornedo Vicentino I 69 B11
Cornella-del-Vercol F 34 E4
Cornellà de Llobregat E 43 E8
Cornellà de Terri E 43 C9

Cornellana E 39 B7
Cornereva RO 159 C9
Corneşti MD 153 C12
Corneşti RO 152 C3
Corneşti RO 161 D6
Cornetu RO 161 E7
Corni RO 153 C9
Corni RO 153 F11
Corniglio I 69 E9
Cornimont F 27 E6
Cornu RO 161 C7
Cornuda I 72 E5
Cornudella de Montsant E 42 E5
Cornudilla E 40 C5
Cornu Luncii RO 153 C8
Cornus F 34 C5
Corod RO 153 F11
Corofin IRL 8 C4
Coroieni RO 152 C3
Coroisânmărtin RO 152 E4
Coron F 23 F10
Çorovodë AL 168 C3
Corps F 31 E8
Corps-Nuds F 23 E8
Corral de Almaguer E 47 E6
Corral de Calatrava E 54 B4
Corrales E 39 E8
Corral-Rubio E 55 B10
Corre F 26 E5
Correggio I 66 C2
Corrèze F 29 E9
Corridonia I 67 F8
Corrie UK 4 D6
Corris UK 10 F4
Corr na Móna IRL 6 E4
Corrobert F 25 C10
Corsano I 61 D10
Corseul F 23 D7
Corsham UK 13 C10
Corsock UK 5 E9
Cortale I 59 B9
Corte F 37 G10
Corte de Peleas E 51 B6
Cortegada E 38 D3
Cortegana E 51 D6
Cortemilia I 37 B8
Corteno Golgi I 69 A9
Corteolona I 69 C7
Cortes E 41 E9
Cortes de Aragón E 42 F2
Cortes de Arenoso E 48 D3
Cortes de Baza E 55 C7
Cortes de la Frontera E 53 C6
Cortes de Pallás E 48 F3
Cortiçadas do Lavre P 50 B3
Cortijo de Arriba E 46 F4
Cortijos Nuevos E 55 C7
Cortina d'Ampezzo I 72 C5
Corton UK 15 B12
Cortona I 66 F4
Coruche P 50 B2
Corullón E 39 C6
Corund RO 152 E6
Corvara in Badia I 72 C4
Corvera E 56 F2
Corwen UK 10 F5
Coryton UK 15 D10
Cosa E 47 D10
Cosâmbeşti RO 161 D10
Cosăuţi MD 154 A2
Coşbuc RO 152 C4
Cosby UK 11 F9
Coşcodeni MD 153 B12
Coscurita E 41 F7
Coşeiu RO 151 C11
Cosenza I 60 E6
Coşereni RO 161 D9
Coşeşti RO 160 C5
Coshieville UK 5 B9
Çösici BIH 157 D8
Cosio d'Arroscia I 37 C7
Coslada E 46 D5
Cosmeşti RO 153 F10
Cosmeşti RO 160 E6
Cosminele RO 161 C7
Cosne-Cours-sur-Loire F 25 F8
Cosne-d'Allier F 30 C2
Coşniţa MD 154 C4
Cosoleto I 59 C8
Coşoveni RO 160 E3
Cossato I 68 B5
Cossé-le-Vivien F 23 E10
Cossoine I 64 C2
Cossonay CH 31 B10
Costacciaro I 67 F6
Costa da Caparica P 50 B1
Costa di Rovigo I 66 B4
Costa Volpino I 69 B8
Coşteiu RO 151 F9
Costelloe IRL 6 F3
Costeşti MD 153 C11
Costeşti MD 154 D3
Costeşti RO 160 C4
Costeşti RO 160 D4
Costeşti RO 161 D5
Costeştii din Vale RO 161 D6
Costigliole Saluzzo I 37 B6
Costişa RO 153 D9
Costuleni RO 153 C11
Cosuenda E 41 F9
Coswig D 79 C11
Coswig D 80 D5
Coteana RO 160 E4
Coteşti RO 161 D10
Cothen NL 16 D4
Coti-Chiavari F 37 H9
Cotignac F 36 D4
Cotignola I 66 D4
Cotofăneşti RO 153 E9
Coţofenii din Dos RO 160 E3
Cotronei I 61 E7
Cottanello I 62 C3
Cottbus D 80 C6
Cottenham UK 15 C9
Cottingham UK 11 D10
Cottmore UK 11 F10
Cottbus D 80 C6
Cotu-Chiavari... [Cotignac]
Coubon F 30 F5
Couches I 30 B6
Couço P 50 B3
Coucy-le-Château-Auffrique F 19 F7
Coudekerque-Branche F 18 B5

Couëron F 23 F8
Couflens F 33 E8
Coufouleux F 33 C9
Couiza F 33 E10
Coulaines F 23 D12
Coulanges-la-Vineuse F 25 E10
Coulanges-sur-Yonne F 25 E10
Coulaures F 29 E7
Couleuvre F 30 B2
Coullons F 25 E7
Coulmier-le-Sec F 25 E11
Coulmiers F 24 E6
Coulogne F 15 F12
Coulombiers F 29 C6
Coulombs F 24 C5
Coulommiers F 25 C9
Coulonges-sur-l'Autize F 28 C4
Coulounieix-Chamiers F 29 E7
Coulport F 4 D6
Coupar Angus UK 5 B10
Couptrain F 23 D11
Cour-Cheverny F 24 E5
Courcelles B 19 D9
Courcelles-Chaussy F 26 B5
Courcelles-sur-Nied F 186 C1
Courchaton F 26 E6
Courchevel F 31 D10
Cour-Cheverny F 24 E5
Courcité F 23 D11
Courçon F 28 C4
Courcy F 19 F9
Courgains F 23 D12
Courgenay CH 27 F7
Courgenay F 25 D9
Courlay F 28 B4
Courmayeur I 31 D10
Courmelles F 19 F7
Cournon-d'Auvergne F 30 D3
Courpière F 30 D4
Courrendlin CH 27 F7
Courrensan F 33 C6
Courrières F 182 E1
Coursac F 29 E7
Coursan F 34 D5
Coursegoules F 36 D6
Courseulles-sur-Mer F 23 B11
Cours-la-Ville F 30 C5
Courson-les-Carrières F 25 E9
Court CH 27 F7
Courtalain F 24 D5
Courtelary CH 27 F7
Courtenay F 25 D9
Courthézon F 35 B8
Courtmacsherry IRL 8 E5
Courtomer F 23 C12
Courtown IRL 9 C10
Courtrai B 182 D2
Cousance F 31 B7
Cousances-les-Forges F 26 C3
Coussac-Bonneval F 29 D8
Coussay-les-Bois F 29 B7
Coussegrey F 25 E11
Coussey F 26 D4
Coustouges F 34 F4
Coutances F 23 B9
Couterne F 23 C11
Couto de Cima P 44 C5
Couvet CH 31 B10
Couvin B 19 D9
Couze-et-St-Front F 29 F7
Couzeix F 29 D8
Covaleda E 40 E6
Covarrubias E 40 D5
Covăsinţ RO 151 E8
Covasna RO 153 E8
Cove UK 2 K5
Cove Bay UK 3 L12
Coventry UK 13 A11
Covilhã P 44 D6
Cowbit UK 11 F11
Cowbridge UK 13 C8
Cowdenbeath UK 5 C10
Cowes UK 13 D12
Cox E 56 E3
Cox F 33 C8
Coxheath UK 15 E9
Coxhoe UK 5 F13
Coylumbridge UK 3 L9
Cózar E 55 B6
Cozes F 28 E4
Cozieni RO 161 C9
Cozmeşti RO 153 D11
Cozzano F 37 H10
Craanford IRL 9 C10
Crăciunelu de Jos RO 152 E3
Crăciuneşti RO 152 E3
Craco I 61 C6
Craidorolţ RO 151 B10
Craig UK 2 L6
Craigavad UK 4 F5
Craigavon UK 7 D10
Craigellachie UK 3 L10
Craignure UK 4 C5
Crail UK 5 C11
Crailsheim D 75 D7
Craiova RO 160 E3
Cramlington UK 5 E13
Cranage UK 11 E7
Cranagh UK 4 F2
Cranford IRL 7 B7
Crângeni RO 160 E5
Cran-Gevrier F 31 D9
Crângu RO 160 F6
Crângurile RO 160 D6
Cranleigh UK 15 E8
Crans-sur-Sierre CH 31 C11
Craon F 23 E10
Craonne F 19 F8
Craponne-sur-Arzon F 30 E4
Crask Inn UK 2 J7
Crasna RO 151 C11
Crasna RO 160 C3
Crasnoe MD 154 D5
Crathie UK 5 A10
Crato P 44 F5
Cravagliana I 68 B5
Cravant F 25 E10
Craven Arms UK 13 A9
Crawfordjohn UK 5 E9
Crawley UK 15 E8
Creaca RO 151 C11
Creagh IRL 8 E4

Creagorry UK 2 L2
Creamhghort IRL 7 B7
Créances F 23 B9
Créancey F 25 F12
Crecente E 38 D3
Crèches-sur-Saône F 30 C6
Crécy-en-Ponthieu F 18 D4
Crécy-la-Chapelle F 25 C8
Crécy-sur-Serre F 19 E8
Credenhill UK 13 A9
Crediton UK 13 D7
Creeslough IRL 7 B7
Creevagh IRL 6 D4
Creggan UK 4 F2
Creggan UK 7 D9
Cregganbaun IRL 6 E3
Creggs IRL 6 E6
Creglingen D 75 D7
Cregneash GBM 10 C2
Créhange F 186 C2
Creil F 18 F5
Creissels F 34 B5
Crema I 69 C8
Cremeaux F 30 D4
Crémenes E 39 C9
Crémieu F 31 D7
Cremlingen D 79 B8
Cremona I 69 C9
Crenševci SLO 148 C6
Créon F 28 F5
Crepaja SRB 158 C6
Crépey F 26 C4
Crépy F 19 E8
Crépy-en-Valois F 18 F6
Cres HR 67 C9
Crescentino I 68 C5
Crespina I 66 E2
Crespino I 66 C4
Crespos E 46 C3
Cressensac F 29 E9
Crest F 31 F7
Creswell UK 11 E9
Cretas E 42 F4
Créteil F 25 C7
Creţeni RO 160 D4
Creţeşti RO 153 D11
Creußen D 75 C10
Creutzwald F 21 F7
Creuzburg D 79 D7
Crevalcore I 66 C3
Crevant F 29 C9
Crévechamps F 26 C5
Crèvecœur-le-Grand F 18 F4
Crevedia RO 161 D7
Crevedia Mare RO 161 E7
Crevenicu RO 161 E7
Crevillente E 56 E3
Crevoladossola I 68 A5
Crewe UK 10 E7
Crewkerne UK 13 D9
Crianlarich UK 4 C7
Cricău RO 152 E3
Criccieth UK 10 F3
Criciova RO 159 B9
Crickhowell UK 13 B8
Cricklade UK 13 B11
Cricova MD 154 C3
Crieff UK 5 C9
Criel-sur-Mer F 18 D3
Crikvenica HR 67 B10
Crimmitschau D 79 E11
Crimond UK 3 K13
Crinan UK 4 C5
Cringleford UK 15 B11
Crinitz D 80 C5
Crişan E 41 F7
Cripp's Corner UK 15 F10
Criquetot-l'Esneval F 18 E1
Crişcior RO 151 E10
Crişeni RO 151 C11
Crispiano I 61 B8
Crissolo I 31 F11
Cristeşti RO 152 D4
Cristeşti RO 153 B9
Cristeşti RO 153 C9
Cristian RO 152 F4
Cristian RO 161 B6
Cristineşti RO 153 A8
Criştioru de Jos RO 151 E10
Cristóbal E 45 D9
Cristolţ RO 151 C11
Cristuru Secuiesc RO 152 E6
Criuleni MD 154 C4
Criva MD 153 A9
Crivitz D 83 C11
Crkvice BIH 157 D8
Crkvice MNE 163 D6
Crkvine SRB 163 C9
Crljivica BIH 157 C11
Črmošnjice SLO 73 E11
Črna SLO 73 D10
Crna Bara SRB 150 F5
Crna Bara SRB 158 D3
Crnac HR 149 E9
Crnajka SRB 159 E9
Crna Trava SRB 164 D5
Crnča SRB 158 E3
Crni Lug BIH 156 D6
Crni Lug HR 67 B10
Crnjelovo BIH 157 C11
Črnkovci HR 149 E10
Crnoklište SRB 164 C5
Črnomelj SLO 148 E5
Crock D 75 B8
Crocketford UK 5 E9
Crockets Town IRL 6 D4
Crockmore UK 5 D12
Crocmaz MD 154 E5
Crocq F 29 D10
Crodo I 68 A5
Crofty UK 12 B6
Croghan IRL 6 E6
Crognaleto I 62 B4
Croisilles F 18 D6
Croithlí IRL 6 B6
Crolles F 31 E8
Crolly IRL 6 B6
Cromadh IRL 8 C5
Cromarty UK 3 K8
Cromer UK 15 B11
Cromhall UK 13 B10
Cronat F 30 B4
Crook UK 5 F13
Crookham UK 5 D12
Crookhaven IRL 8 F3
Crookstown IRL 8 E5
Croom IRL 8 C5
Cropalati I 61 D7
Cropani I 59 B10
Crosbost UK 2 J4

Crosby UK 10 E5
Crosia I 61 D7
Cross IRL 6 E4
Crossaig UK 4 D6
Crossakeel IRL 7 E8
Cross Barry IRL 8 E5
Crosscanonby UK 5 F10
Crossdoney IRL 7 D8
Crossgar UK 7 D11
Crossgare UK 4 E3
Crossgates UK 13 A8
Crosshands UK 5 D8
Crosshaven IRL 8 E6
Crosshill UK 4 E7
Cross Inn UK 12 A6
Cross Keys UK 7 E9
Crossmaglen UK 7 D9
Crotone I 61 E8
Crots F 36 B4
Crottendorf D 76 B3
Croughton UK 13 B12
Croutelle F 29 B6
Crouy F 19 F7
Crouy-sur-Ourcq F 25 B9
Crowborough UK 15 E9
Crowland UK 11 F11
Crowle UK 11 D10
Crowthorne UK 15 E7
Croy UK 3 K8
Croyde UK 12 C6
Crozant F 29 C9
Crozes-Hermitage F 30 E6
Crozon F 22 D3
Cruas F 35 A8
Crucea RO 153 C7
Crucea RO 155 D2
Crucișor RO 151 B11
Crucoli I 61 E7
Cruden Bay UK 3 L13
Cruglic MD 154 C4
Crumlin UK 4 F4
Cruseilles F 31 C9
Crușet RO 160 D3
Crusheen IRL 8 C5
Crusnes F 20 F5
Cruzy F 34 D4
Cruzy-le-Châtel F 25 E11
Crvenka SRB 158 B3
Crymych UK 12 B5
Csabacsűd H 150 D6
Csabrendek H 149 B8
Csákánydoroszló H 149 C7
Csákberény H 149 B10
Csákvár H 149 B10
Csanádapáca H 151 D6
Csanádpalota H 150 E6
Csány H 150 B4
Csanytelek H 150 D5
Császár H 149 A10
Császártöltés H 150 E3
Csátalja H 150 E2
Csávoly H 150 E3
Csécse H 147 F9
Csemő H 150 C4
Csengele H 150 D4
Csenger H 145 G7
Csengőd H 150 D3
Csépa H 150 D5
Csepreg H 149 B7
Cserépfalu H 145 H2
Cserhátsurány H 147 F8
Cserkeszőlő H 150 D5
Csernely H 145 G1
Csesztreg H 149 C7
Csetény H 149 B9
Csévharaszt H 150 C3
Csíkéria H 150 E3
Csókakő H 149 B10
Csököly H 149 D9
Csokonyavisonta H 149 D8
Csokvaomány H 145 G1
Csolnok H 149 A11
Csólyospálos H 150 E4
Csongrád H 150 D5
Csopak H 149 C9
Csór H 149 B10
Csorna H 149 A8
Csörnyeföld H 149 D7
Csorvás H 150 D6
Csősz H 149 B10
Csót H 149 B9
Csurgó H 149 D8
Cuacos de Yuste E 45 D9
Cuadros E 39 C8
Cualedro E 38 E4
Cuarte de Huerva E 41 E10
Cuba P 50 C4
Cubalhão P 38 D3
Cubells E 42 D5
Cubjac F 29 E7
Cubla E 47 D10
Cubo de Bureba E 40 C5
Cubo de la Solana E 41 E7
Cubolta MD 153 B12
Cubzac-les-Ponts F 28 F5
Cuca RO 153 F11
Cuca RO 160 D5
Cucalón E 47 E8
Cucerdea RO 152 E4
Cuci RO 152 E4
Cuckfield UK 15 E8
Cucuron F 35 C9
Cucuteni RO 153 C9
Cudalbi RO 153 F11
Cudillero E 39 A7
Čudnići BIH 157 D8
Cudos F 32 B5
Cuéllar E 40 F3
Cuenca E 47 D8
Cuenca de Campos E 39 D9
Cuers F 36 E4
Cuerva E 46 E4
Cueva de Agreda E 41 E8
Cuevas Bajas E 53 B8
Cuevas de Almanzora E 55 E9
Cuevas del Becerro E 53 C6
Cuevas del Campo E 55 D7
Cuevas de San Clemente E 40 D4
Cuevas de San Marcos E 53 B8
Cuevas Labradas E 42 G1
Cuffley UK 15 D8
Cugand F 28 A3
Cuges-les-Pins F 35 D10
Cugir RO 151 F11
Cuglieri I 64 C2
Cugnaux F 33 C8
Cuguen F 23 D8
Cuhureștii de Sus MD 154 B3
Cuijk NL 16 E5
Cúil an tSúdaire IRL 7 F8
Cuiseaux F 31 C7

Cuise-la-Motte F 18 F7
Cuisery F 31 B6
Cujmir RO 159 E10
Çukë AL 168 E3
Çukuryurt TR 173 B8
Culan F 29 B10
Culciu RO 151 B11
Culdaff IRL 4 E2
Culdrain UK 3 L11
Culemborg NL 16 E4
Culjković SRB 158 D4
Culla E 48 D4
Cullahill IRL 9 C8
Cúllar-Baza E 55 D7
Cullen UK 3 K11
Cullera E 48 F4
Cullivoe UK 3 D14
Cullompton UK 13 D8
Cully CH 31 C10
Cullybackey UK 4 F4
Cullyhanna UK 7 D9
Culmstock UK 13 D8
Culnaknock UK 2 K4
Culoz F 31 D8
Culswick UK 3 E13
Cumbernauld UK 5 D8
Cumbres de San Bartolomé E 51 C6
Cumbres Mayores E 51 C6
Cumeada P 50 E3
Cumiana I 31 F11
Cumieira P 38 F4
Cuminestown UK 3 K12
Cumlosen D 83 D11
Cumnock UK 5 E8
Cumpăna RO 155 E3
Cunault F 23 F11
Cunegès F 29 F6
Cuneo I 37 C7
Cunewalde D 81 D7
Cunfin F 25 D12
Cungrea RO 160 D4
Cunha P 44 C6
Cunica MD 154 B3
Čuništa BIH 157 D10
Cunit E 43 E7
Cunningburgh UK 3 E14
Cuntis E 38 C2
Cuorgnè I 68 C4
Cupar UK 5 C10
Cupcina MD 153 A10
Cupello I 63 D7
Cupra Marittima I 62 A5
Cupramontana I 67 F7
Ćuprija SRB 159 F7
Cupșeni RO 152 B3
Cuq-Toulza F 33 C9
Curăţele RO 151 D10
Curcani RO 161 E9
Curcuris I 64 D2
Cureggio I 68 B5
Cureșnița MD 153 A12
Curgy F 30 B5
Curinga I 59 B9
Curon Venosta I 71 D11
Curracloe IRL 9 D10
Curraghroe IRL 6 E6
Curragh West IRL 6 E5
Currás E 38 C2
Curry IRL 6 E5
Curry Rivel UK 13 C9
Cursi I 61 C10
Curtea RO 151 F9
Curtea de Argeș RO 160 C5
Curtici RO 151 E7
Curtișoara RO 160 E4
Curtuișeni RO 151 B9
Ćurug SRB 158 C5
Cusano Mutri I 60 A3
Cushendall UK 4 E4
Cushendun UK 4 E4
Cushina IRL 7 F8
Cussac F 29 D7
Cusset F 30 C3
Cussy-les-Forges F 25 F11
Custines F 26 C5
Custonaci I 58 C2
Cutro I 61 E7
Cutrofiano I 61 C10
Cuxac-Cabardès F 33 D10
Cuxac-d'Aude F 34 D4
Cuxhaven D 17 A11
Cuzăplac RO 151 D10
Cuza Vodă RO 161 E10
Cuzmin MD 154 A3
Cuzorn F 33 A7
Cvikov CZ 81 E7
Cwmafan UK 13 B8
Cwmbrân UK 13 B8
Cybinka PL 81 B7
Cybowo PL 85 D9
Cychry PL 85 E7
Cyców PL 141 H8
Cydweli UK 12 B6
Cynghordy UK 13 A7
Cynwyl Elfed UK 12 B6
Cyprianka PL 138 E7
Cysoing F 182 D2
Czacz PL 81 B11
Czajków PL 142 D5
Czaplinek PL 85 C10
Czarna PL 143 F11
Czarna PL 144 C5
Czarna Białostocka PL 140 D8
Czarna Dąbrówka PL 85 B13
Czarna Górna PL 145 E6
Czarna Woda PL 138 C5
Czarncza PL 143 E8
Czarne PL 85 C11
Czarnia PL 139 D11
Czarnków PL 85 E11
Czarnocin PL 143 F10
Czarnów PL 81 A7
Czarnowąsy PL 142 E4
Czarnożyły PL 142 D6
Czarny Bór PL 81 E10
Czarny Dunajec PL 147 C9
Czastary PL 142 D5
Czaszyn PL 145 E5
Czchów PL 144 D2
Czechowice-Dziedzice PL 147 B8
Czechy PL 85 C11
Czechy PL 143 C6
Czekarzewice PL 143 D12
Czeladź PL 143 F7
Czemierniki PL 141 G7
Czempiń PL 81 B11
Czermin PL 142 C4
Czermin PL 143 F12
Czermno PL 141 H2
Czernica PL 81 D12
Czernice Borowe PL 139 D10

Czernichów PL 143 G8
Czernichów PL 147 B8
Czerniejewo PL 85 F12
Czerniewice PL 141 G2
Czernikowo PL 138 E6
Czernina PL 81 C11
Czersk PL 138 C4
Czerwień PL 81 B8
Czerwin PL 139 E12
Czerwińsk nad Wisłą PL 139 F9
Czerwionka-Leszczyny PL 142 F6
Czerwonak PL 81 B11
Czerwone PL 139 D12
Czestków PL 143 C7
Częstochowa PL 143 E7
Czeszów PL 81 D12
Człopa PL 85 D10
Człuchów PL 85 C12
Czmoń PL 81 B12
Czorsztyn PL 145 E1
Czosnów PL 139 F10
Czudec PL 144 D4
Czyże PL 141 E8
Czyżew-Osada PL 141 E6

D

Daaden D 185 C8
Dăbâca RO 152 D3
Dabar HR 156 C3
Dabar HR 156 E6
Dabas H 150 C3
Dabel D 83 C11
Dąbie PL 81 B8
Dąbie PL 141 G6
Dąbie PL 142 B6
Dąbki PL 85 B10
Dabo F 27 C7
Daborgrad BIH 157 C9
Dabrac BIH 157 D6
Dabravolya BY 140 E10
Dabrica BIH 157 F9
Dąbroszyn PL 81 A7
Dąbroszyn PL 142 B6
Dąbrowa PL 138 E4
Dąbrowa PL 142 E4
Dąbrowa PL 142 E4
Dąbrowa Białostocka PL 140 C8
Dąbrowa Biskupia PL 138 E6
Dąbrowa Chełmińska PL 138 D5
Dąbrowa Górnicza PL 143 F7
Dąbrowa Tarnowska PL 143 F10
Dąbrowa Zielona PL 143 E8
Dąbrowice PL 141 G2
Dąbrowice PL 143 B7
Dąbrówka Wielka PL 143 C7
Dąbrówka Wielkopolska PL 81 B9
Dąbrówno PL 139 D9
Dăbuleni RO 160 F4
Dabyeya BY 133 F7
Dachau D 75 F9
Dachnów PL 144 C7
Dachrieden D 79 D7
Dachwig D 79 D8
Dačice CZ 77 D8
Dacre UK 5 F11
Dad H 149 A10
Dadia GR 171 B10
Dădran S 103 E10
Dăeni RO 155 D2
Dăești RO 160 C4
Dáfjord N 112 D3
Dafnes GR 174 C4
Dafni GR 171 D6
Dafni GR 174 D5
Dafni GR 174 C5
Dafni GR 175 C8
Dág H 149 A11
Daganzo de Arriba E 46 C6
Dagârdi TR 173 F11
Dagda LV 133 D3
Dagebüll D 82 A5
Dagenham UK 15 D9
Dagestan Şahinler TR 173 B9
Daglan F 29 F8
Daglösen S 97 C11
Dagmersellen CH 27 F8
Dagpınar TR 181 A9
Dagsmark FIN 122 F7
Dahlem D 21 D7
Dahlen D 80 D4
Dahlenburg D 83 D9
Dahlhausen D 183 D10
Dahme D 80 C4
Dahme D 83 B10
Dahn D 27 B8
Dähre D 83 E9
Daia RO 161 F7
Daia Română RO 152 E3
Daikanberg S 107 A10
Daikanvik S 107 A10
Dailly UK 4 E7
Daimiel E 46 F5
Daimonia GR 178 B4
Daingean IRL 7 F8
Daingean Uí Chúis IRL 8 D2
Dainville F 18 D6
Dainville-Bertheléville F 26 D4
Dairsie UK 5 C11
Daknam B 182 C3
Đakovo HR 149 F10
Dal N 95 B14
Dal S 107 E13
Đala SRB 150 E5
Dalaas A 71 C10
Dalabrog UK 2 L2
Dalaman TR 181 C9
Dalarö S 93 A12
Dalarö S 93 A12
Dalasjö S 107 B11
Dalbe LV 134 C7
Dalbeattie UK 5 F9
Dalberg-Wendelstorf D 83 C10
Dalboșeț RO 159 D8
Dalby DK 86 B5
Dalby DK 87 E10
Dalby GBM 10 C2
Dalby S 87 D12
Dalbyn S 103 D9
Dalbyover DK 86 B6
Dăldenăs S 92 A4
Dale UK 9 E12
Dale N 90 B3
Dale N 100 D2
Dale N 100 B8
Dale N 111 C11
Daleiden D 20 D6
Dalen N 90 A5
Dalen N 94 D7
Dalen NL 17 C7
Daleszyce PL 143 E10

Dalfors S 103 D9
Dalfsen NL 16 C6
Dalgety Bay UK 5 C10
Dalham UK 15 C10
Dalhem B 19 C12
Dalhem S 93 D13
Dalholen N 101 B11
Dalías E 55 F7
Dalików PL 143 C7
Dalj HR 157 B10
Dalkarlsberg S 97 C12
Dalkeith UK 5 D10
Dalkey IRL 7 F11
Dallas UK 3 K10
Dallgow D 80 A4
Dall Villaby DK 86 B5
Dalmally UK 4 C7
Dalmand H 149 D10
Dalmellington UK 5 E8
Dalmine I 69 B8
Dalmose DK 87 E8
Dalovice CZ 76 B3
Dalry UK 4 D7
Dalrymple UK 4 E7
Dalsbruk FIN 126 E8
Dalselv N 108 D6
Dalsjöfors S 91 D13
Dals Långed S 91 B11
Dalstein F 20 F6
Dalston UK 5 F11
Dalstorp S 91 D14
Dalstuga S 103 D10
Dalton-in-Furness UK 10 C5
Daluis F 36 C5
Dalum S 91 D13
Dalur FO 2 B3
Dàlvadas FIN 113 D17
Dalwhinnie UK 5 B8
Dalyan TR 181 C9
Dalyokya BY 135 F13
Dalystown IRL 6 F6
Damachava BY 141 G9
Damaras BG 165 D10
Damasi GR 169 E7
Damaskinia GR 168 D5
Damasławek PL 85 E13
Damazan F 33 B6
Dambach-la-Ville F 186 E3
Dambaslar TR 173 B8
Dambeck D 83 D11
Dâmbovicioara RO 160 C6
Damelevières F 26 C5
Damerstown IRL 9 C8
Damery F 25 B10
Damès AL 168 C3
Damgan F 22 E6
Damhead UK 4 E3
Damhliag IRL 7 E10
Dămienești RO 153 D9
Damigny F 23 C12
Damjan RKS 163 E10
Dammarie F 24 D6
Dammartin-en-Goële F 25 B8
Damme B 19 B7
Damme D 17 C10
Damnica PL 85 B12
Damno PL 85 A12
Damp D 83 A7
Dampierre F 26 F4
Dampierre-sur-Linotte F 26 E5
Dampierre-sur-Salon F 26 E4
Damprichard F 27 F6
Damshagen D 83 C10
Damsholte DK 87 F10
Dämsta S 107 E13
Damüls A 71 C9
Damville F 24 C5
Damvillers F 19 F11
Damwoude NL 16 B6
Damyanovo BG 165 D9
Danaçali TR 173 E8
Danakos GR 177 E6
Danamandira TR 173 B9
Danasjö S 109 E12
Dănciulești RO 160 D3
Dăneasa RO 160 E5
Daneș RO 152 E5
Danesfort IRL 7 E7
Dănești RO 153 D11
Dănești RO 153 D11
Dănești RO 160 D2
Daneți RO 160 F4
Dângebo S 89 B8
Dângeni RO 153 B9
Dangé-St-Romain F 29 B7
Danholn S 103 E10
Dănicei RO 160 D4
Daniec PL 142 E5
Danilovgrad MNE 163 D7
Dänischenhagen D 83 B8
Danişment TR 173 E8
Danjoutin F 27 E6
Dankerode D 79 C9
Danndorf D 79 B8
Dannemare DK 83 A10
Dannemora S 99 B9
Dannenberg (Elbe) D 83 D10
Dannenwalde D 84 D4
Dány H 150 B4
Daon F 23 E10
Daoulas F 22 D3
Darabani RO 153 A9
Dărăști Ilfov RO 161 E7
Darbénai LT 134 D2
Dar Ben Karricha el Behri MA 53 E6
Darby's Bridge IRL 8 D2
Dar Chaoui MA 52 E5
Darda HR 149 E11
Dardesheim D 79 C8
Dardhas AL 168 C4
Darè I 69 A10
Darfo Boario Terme I 69 B9
Dargun D 83 C13
Darica TR 173 B10
Darıcı TR 173 F10
Darkley UK 7 D9
Darlington UK 11 B8
Dârlos RO 152 E4
Darłówko PL 85 B10
Darłowo PL 85 B10
Dărmănești RO 153 D8
Dărmănești RO 153 E8
Dărmănești RO 160 D5
Dărmănești RO 161 D7
Darmstadt D 21 E11
Darney F 26 D5
Darnieulles F 26 D5

Darnózseli H 146 F4
Daroca E 47 B10
Darova RO 159 B8
Darque P 38 E2
Darra UK 3 K12
Darragh IRL 8 C5
Dartford UK 15 E9
Dartmeet UK 13 D7
Dartmouth UK 13 E7
Darton UK 11 D8
Dartsel S 118 C4
Daruvar HR 149 E8
Darwen UK 10 D7
Dascălu RO 161 D8
Dashava UA 145 E9
Dašice CZ 77 B9
Dasing D 75 F9
Dasochori GR 169 B9
Dasochori GR 171 C7
Dassel D 78 C6
Dassendorf D 83 D8
Dassow D 83 C9
Daszyna PL 143 B7
Datça TR 181 C7
Datteln D 17 E8
Datterode (Ringgau) D 79 D7
Daudzeva LV 135 D10
Daudzeva LV 135 C10
Daugai LT 137 E9
Daugailiai LT 135 E11
Daugård DK 86 D5
Daugavpils LV 135 E13
Daugėlaičiai LT 134 E7
Dauguļi LV 135 A10
Daujėnai LT 135 E9
Daumazan-sur-Arize F 33 D8
Daumeray F 23 E11
Daun D 21 D7
Dauphin F 35 C10
Dava UK 3 L9
Daventry UK 13 A12
Davézieux F 30 E6
Davidești RO 160 C6
Davidkovo BG 165 F10
Daviot UK 3 L8
Davle CZ 76 C6
Davleia GR 175 B6
Dávod H 150 E2
Davor HR 157 B7
Davos CH 71 D9
Davutlar TR 177 D9
Davydiv UA 144 D9
Dawlish UK 13 D8
Dax F 32 C3
Deal UK 15 E11
Dealu RO 152 E6
Dealu Morii RO 153 E10
Dearham UK 5 F10
Deauville F 23 B12
Deba E 32 C1
Debar NMK 168 A4
Debeli Lug SRB 159 E8
Debeljača SRB 158 C6
Debenham UK 15 C11
Dębe Wielkie PL 141 F4
Dębica PL 143 F11
De Bilt NL 16 D4
Dęblin PL 141 G5
Debnevo BG 165 D10
Dębnica Kaszubska PL 85 B12
Dębno S 88 E7
Dębno PL 143 G10
Dęborzeczka PL 141 H2
Debovo BG 165 B10
Dębowa Kłoda PL 141 G8
Dębowa Łąka PL 138 D7
Dębowiec PL 147 B7
Debrc SRB 158 D4
Debrecen H 151 B8
Debrešte NMK 168 B5
Debrzno PL 85 C11
Dębsko PL 85 D9
Debür BG 166 E4
Deçan RKS 163 D9
Decazeville F 33 A10
Dechtice SK 146 D5
Decimomannu I 64 E2
Decimoputzu I 64 E2
Děčín CZ 80 E6
Decize F 30 B3
De Cocksdorp NL 16 B3
Decollatura I 59 A9
Decs H 149 D11
Deda RO 152 D5
Dedeler TR 173 F9
Dedelow D 84 D5
Dedelstorf D 83 E9
Dedemsvaart NL 17 C6
Dedham UK 15 D10
Dedovici RUS 132 F6
Deensen D 78 C6
Deerlijk B 19 C7
Dég H 149 C10
Degaña E 39 C6
Degeberga S 88 D6
Degerbäcken S 118 C6
Degerby FIN 99 B9
Degerby FIN 127 F11
Degerfors S 92 A4
Degerhamn S 89 C10
Degersjö S 107 D12
Degerträsk S 118 C4
Deggendorf D 76 E3
Deggingen D 74 F6
Değirmen TR 173 B9
Değirmenci TR 177 C9
Değirmendüzü TR 171 D10
Değirmenyeni TR 167 F7
Dego I 37 C8
Degučiai LT 135 E12
Degunieki LV 135 C13
De Haan B 19 B7
Dehesa de Campoamor E 56 F3
Dehesas de Guadix E 55 D6
Deià E 49 E10
Deidesheim D 21 F10
Deilão P 39 E6
Deining D 75 D10
Deiningen D 75 E8
Deinze B 19 C8
Deiva Marina I 37 C10
Dej RO 152 C3
Deje S 97 C9
Dejtár H 147 E8

Dekani SLO 67 A8
Dekeleia GR 175 C8
De Koog NL 16 B3
De Kooy NL 16 C3
Dekov BG 165 B11
Delabole UK 12 D5
Delary S 87 B13
Delbrück D 17 E11
Delčevo NMK 165 F6
Delden NL 17 D7
Delden D 17 D7
Délegyháza H 150 C3
Deleitosa E 45 E9
Đelekovec HR 149 D7
Delémont CH 27 F7
Delen' UA 154 F4
Deleni RO 153 C9
Deleni RO 155 E2
Deleria GR 169 E7
Delești RO 153 D11
Delfoi GR 175 C6
Delft NL 16 D2
Delfzijl NL 17 B7
Delia I 58 E4
Delianuova I 59 C8
Deliblato SRB 159 D7
Deliceto I 60 A4
Deligrad SRB 159 F7
Deliktaş TR 177 B8
Delingsdorf D 83 C8
Delitzsch D 79 C11
Delle F 27 E7
Delligsen D 78 C6
Dello I 66 B1
Delme F 26 C5
Delmenhorst D 17 B11
Delnice HR 67 B10
Delsbo S 103 C12
Deltebre E 42 F5
Delvin IRL 7 E8
Delvinaki GR 168 E3
Delvinë AL 168 E3
Delvine AL 168 E3
Demandice SK 147 E7
Demange-aux-Eaux F 26 C3
Demecser H 145 G4
De Meern NL 188 A6
Demen D 83 C11
Demene LV 135 E13
Demigny F 30 B6
Demiraj AL 163 E7
Demirci TR 177 A10
Demirdere TR 173 D9
Demirhanli TR 167 F7
Demir Hisar NMK 168 B5
Demir Kapija NMK 169 B7
Demirköy TR 167 F9
Demirli TR 181 C9
Demirtaş TR 173 D11
Demmin D 84 C4
Demonte I 37 C6
Dému F 33 C6
Denain F 19 D7
Denbigh UK 10 E5
Den Bommel NL 182 B4
Den Burg NL 16 B3
Denderleeuw B 182 D4
Dendermonde B 19 B9
Den Dungen NL 16 E4
Denekamp NL 17 D8
Denguin F 32 D5
Den Haag NL 16 D2
Den Haag NL 182 A4
Denham UK 15 D8
Den Ham NL 17 D7
Den Helder NL 16 C3
Denholm UK 5 E11
Denia E 56 D5
Denkendorf D 27 C11
Denkendorf D 75 E9
Denkingen D 27 D10
Denkingen D 27 E11
Denklingen D 71 B11
Dennewitz D 80 C4
Dennhausen (Fuldabrück) D 78 D5
Denny UK 5 C9
Den Oever NL 16 C4
Densow D 84 D4
Densuș RO 159 B10
Denta RO 159 C7
Dentergem B 19 C7
Dentlein am Forst D 75 D7
Denver UK 11 F12
Denzlingen D 27 D8
Déols F 29 B9
De Panne B 18 B6
Deppis S 109 E18
Derăsbrenna N 105 C10
Derby UK 11 F9
Derecske H 151 C8
Dereköy TR 167 F8
Dereköy TR 173 C12
Derenburg D 79 C8
Derenti TR 173 E7
De Rijp NL 16 C3
Dermantsi BG 165 C9
Dermbach D 79 E7
Dermenas AL 168 C2
Derna RO 151 C9
Dernau D 21 C8
Derreen IRL 6 D6
Derreendaragh IRL 8 E3
Derreeny Bridge IRL 8 E4
Derry UK 4 E2
Derryconnelly UK 7 D7
Derrygonnelly UK 7 D7
Derrylin UK 7 D7
Derrymore IRL 8 D3
Derrynawilt UK 7 D8
Derryrush IRL 6 F3
Derrywode IRL 6 E5
Dersca RO 153 B8
Dersingham UK 11 F13
Dersum D 17 C8
Deruta I 62 A2
Dervaig UK 4 B4
Dervenakia GR 175 D6
Derventa BIH 157 C8
Derviçan AL 168 E3
Dervio I 69 A7
Derviziana GR 168 F4
Dervock UK 4 E4
Desa RO 159 F11
Desana I 68 C5
Descantar UA 155 B5
Descargamaría E 45 D8
Descartes F 29 B7
Desenzano del Garda I 66 B2
Désertines F 29 C11
Deseşti RO 152 B3

Desfina GR 175 C6
Deskati GR 169 E8
Deskle SLO 73 D8
Despotovac SRB 159 E7
Despotovo SRB 158 C4
Dessau D 79 C11
Dessel B 16 F4
De Steeg NL 16 D6
Destelbergen B 19 B8
Deštná CZ 77 D7
Destriana E 39 D7
Desulo I 64 C3
Desvres F 15 F12
Deszczno PL 85 E8
Deszk H 150 E5
Deta RO 159 C7
Detern D 17 B8
Detk H 150 B5
Detkovac HR 149 E9
Dětmarovice CZ 146 B6
Detmold D 17 E11
Dettelbach D 75 C7
Dettenhausen D 187 D7
Dettingen an der Erms D 187 D7
Dettmannsdorf D 83 B13
Dettum D 79 B8
Dettwiller F 186 D3
Detva SK 147 D8
Deurne NL 16 F5
Deutsch Evern D 83 D8
Deutschfeistritz A 148 B4
Deutschhof D 84 E3
Deutschkreutz A 149 A7
Deutschlandsberg A 73 C11
Deutschneudorf D 80 E4
Deutsch-Wagram A 77 F11
Deutzen D 79 D11
Deva RO 151 F11
Dévaványa H 151 C6
Devecikonaği TR 173 E10
Devecser H 149 B8
Devene BG 165 C8
Deventer NL 16 D6
Devesel RO 159 E10
Deveselu RO 160 E4
Devetaki BG 165 C10
Deville F 184 E2
Devil's Bridge UK 13 A7
Devin BG 165 F9
Devizes UK 13 C11
Devletliağaç TR 167 F8
Devnya BG 167 C9
Dévrske HR 156 E4
De Wijk NL 11 D8
Dewsbury UK 11 D8
Deza E 41 F7
Dežanovac HR 149 E8
Dezghingea MD 154 E3
Dezna RO 151 E9
Dhërmi AL 168 D2
Dhron D 185 E6
Diafani GR 181 E6
Diakofto GR 174 C5
Dialampi GR 171 B9
Diamante I 60 D5
Dianalund DK 87 D9
Diano d'Alba I 37 B8
Diano Marina I 37 D8
Diavata GR 169 C8
Diavolitsi GR 174 E4
Dichiseni RO 161 E11
Dicomano I 66 E4
Didam NL 16 E6
Didcot UK 13 B12
Didderse D 79 B7
Dideşti RO 160 E5
Didyma GR 175 E7
Didymoteicho GR 171 B10
Didžiasalis LT 135 F13
Die F 31 F7
Dieblich D 21 D8
Dieburg D 21 E11
Dieci RO 151 E9
Diedorf D 75 F8
Diego Álvaro E 45 C10
Diekhof D 83 C12
Diekholzen D 78 B6
Diekirch L 20 E6
Dielheim D 21 F11
Dielmissen D 78 C6
Dielsdorf CH 27 F9
Diemen NL 16 D3
Diémoz F 31 D7
Dienheim D 185 E9
Dienne F 30 E2
Dienville F 25 D12
Diepenau D 17 D11
Diepenbeek B 19 C11
Diepenheim NL 183 A9
Diepenveen NL 183 A8
Diepholz D 17 C10
Dieppe F 18 E3
Dierdorf D 21 C9
Dieren NL 16 D6
Dierhagen D 83 B12
Diesdorf D 83 E9
Dieskau D 79 D11
Dießen am Ammersee D 72 A3
Diessenhofen CH 27 E10
Diest B 19 C11
Dietachdorf A 77 F6
Dietenheim D 71 A10
Dietersburg D 76 F3
Dietfurt an der Altmühl D 75 D10
Dietikon CH 27 F9
Dietingen D 27 D10
Dietmanns A 77 E7
Dietmanns A 77 E8
Dietmannsried D 71 B10
Dietzenbach D 21 D11
Dietzhölztal-Ewersbach D 185 C9
Dieue-sur-Meuse F 26 B3
Dieulefit F 35 A9
Dieulouard F 26 C5
Dieuze F 26 C6
Dieveniškes LT 137 E12
Diever NL 16 C6
Diex A 73 C10
Diezma S 163 D6
Differdange L 20 E5
Digerberget S 103 A8
Digerberget S 102 A8
Digermulen N 110 D8
Digernes N 100 B3
Dignac F 29 D6
Dignano I 73 D6
Dignja LV 135 D12
Dignano I 73 D6
Digne-les-Bains F 36 C4

Digny F 24 C5
Digoin F 30 C4
Dihtiv UA 144 B9
Dijon F 26 F3
Dikaia GR 166 F6
Dikanäs S 107 A10
Dikancë RKS 163 E10
Dikili TR 177 A8
Dikkebus B 18 C6
Dikļi LV 131 F10
Diksmuide B 18 B6
Dilar E 53 B9
Dilbeek B 182 D4
Dilesi GR 175 C8
Dilinata GR 174 C2
Dillenburg D 21 C10
Dilling N 95 D13
Dillingen (Saar) D 21 F7
Dillingen an der Donau D 75 E7
Dilove UA 145 H9
Dilsen B 19 B12
Dimaro I 71 E11
Diminio GR 169 F8
Dimitrie Cantemir RO 153 D12
Dimitritsi GR 169 C9
Dimitrovgrad BG 166 E5
Dimitrovgrad SRB 165 C6
Dimitsana GR 174 D5
Dimovo BG 159 F10
Dimzukalns LV 135 C8
Dinami I 59 B9
Dinan F 23 D7
Dinant B 19 D10
Dinard F 23 C7
Dingé F 23 D8
Dingelstädt D 79 D7
Dingelstedt am Huy D 79 C8
Dingle IRL 8 D2
Dingle S 91 B10
Dingolfing D 75 E12
Dingtuna S 98 C6
Dingwall UK 2 K8
Dinjiška HR 67 D11
Dinkelsbühl D 75 D7
Dinkelscherben D 75 F8
Dinklage D 17 C10
Dinnet UK 5 A11
Dinslaken D 17 E7
Dinteloord NL 16 E2
Dinther NL 183 B6
Dinxperlo NL 17 E6
Diö S 88 B6
Dion GR 169 D7
Diósd H 149 B11
Diosig RO 151 C9
Diósjenő H 147 F8
Dioști RO 160 E4
Diou F 30 B4
Dipignano I 60 E6
Dipotama GR 171 B7
Dipotamia GR 168 D4
Dippach L 20 E6
Dippoldiswalde D 80 E5
Dirdal N 94 E4
Dirhami EST 130 C7
Dirivaara S 116 E8
Dirkshorn NL 16 C3
Dirksland NL 16 E2
Dirlewang D 71 A11
Dirmstein D 187 B5
Dirvonėnai LT 134 E5
Dischingen D 75 E7
Disentis Muster CH 71 D7
Diseröd S 91 D11
Dison B 19 C12
Diss UK 15 C11
Dissay F 29 B6
Dissay-sous-Courcillon F 24 E3
Dissen am Teutoburger Wald D 17 D10
Distington UK 10 B4
Distomo GR 175 C6
Distrato GR 168 D5
Ditfurt D 79 C9
Ditrău RO 153 D7
Ditton UK 15 E9
Ditzingen D 27 C11
Divača SLO 73 E8
Divarata GR 174 C2
Diva Slatina BG 165 C6
Divci SRB 158 E5
Divčibare SRB 158 E5
Dives-sur-Mer F 23 B11
Dividalen N 111 C18
Divieto I 59 C7
Divín SK 147 E9
Divina SK 147 C7
Divion F 18 D6
Divišov CZ 77 C7
Divjakë AL 168 C2
Divonne-les-Bains F 31 C9
Divuša HR 156 B5
Dixmont F 25 D9
Dizy F 25 B10
Dizy-le-Gros F 19 E9
Djäkneboda S 122 B5
Djäkneböle S 122 C4
Djupen N 111 B18
Djupfjord N 110 C9
Djupfors S 109 E11
Djupsjö S 107 E14
Djuptjärn S 107 D15
Djupvik N 109 B10
Djupvik N 112 D5
Djupvik S 89 A11
Djura S 103 E8
Djurås S 97 A13
Djurmo S 97 A13
Djurö S 99 D11
Dlhá nad Oravou SK 147 C8
Dlouhá Loučka CZ 77 C12
Dlouhá Třebová CZ 77 C10
Długołęka PL 81 D12
Długołęka PL 140 D7
Długosiodło PL 139 E12
Dłutów PL 143 C7
Dlŭzhka Polyana BG 166 C6
Dmytrivka UA 154 F3
Dmytrivka UA 154 F4
Dmytrivka UA 155 B4
Dnestrovsc MD 154 D5
Dno RUS 132 F6
Doagh UK 4 F4
Doba RO 151 B10
Dobanovci SRB 158 D5
Dobârceni RO 153 B10
Dobârlău RO 153 F7
Dobbertin D 83 C12
Dobbiaco I 72 C5
Dobczyce PL 144 D1
Dobele LV 134 C6
Döbeln D 80 D4
Doberçan RKS 164 E4
Doberlug-Kirchhain D 80 C5
Döbern D 81 C7
Dobersberg A 77 E8
Doberschütz D 79 D12
Dobiegniew PL 85 E9
Dobieszewo PL 85 B12
Dobieszyn PL 141 G4
Doboj BIH 157 C9
Dobova SLO 148 E5
Doboz H 151 D7
Dobrá CZ 146 B6
Dobra PL 85 C8
Dobra PL 142 C6
Dobra PL 144 D1
Dobra RO 151 F10
Dobra RO 161 D7
Dobra SRB 159 D8
Dobrá Niva SK 147 E8
Dobřany CZ 76 C4
Dobre PL 138 E6
Dobre PL 139 F12
Dobre Miasto PL 136 F1
Dobreni RO 153 D8
Dobrešinci NMK 169 A8
Dobrești RO 151 D9
Dobrești RO 160 D6
Dobrești RO 160 F3
Dobrevo NMK 164 E5
Dobrica SRB 159 C6
Dobričevo SRB 159 D7
Dobrich BG 155 F1
Dobrich BG 166 E5
Dobri Do SRB 164 D3
Dobri Dol BG 159 F11
Dobrin RO 151 C11
Dobrinishte BG 165 F8
Dobříš CZ 76 C6
Dobritz D 79 B11
Dobřív CZ 76 C5
Dobrljin BIH 156 B5
Dobrna SLO 73 D11
Dobrnič SLO 73 E10
Dobrnja BIH 157 C7
Dobrnja BIH 157 C10
Dobrnje SRB 159 E7
Dobro E 40 C4
Dobrodzień PL 142 E5
Döbrököz H 149 D10
Dobromierz PL 81 E10
Dobromir RO 155 E1
Dobromirka BG 166 C4
Dobromirtsi BG 171 B8
Dobromyl' UA 145 D6
Dobroń PL 143 C7
Dobron' UA 145 G5
Dobronín CZ 77 D9
Dobro Polje BIH 157 E10
Dobro Polje SRB 159 F9
Dobrošane NMK 164 E4
Dobrosloveni RO 160 E4
Dobrosyn UA 144 C8
Dobroszyce PL 81 D12
Dobroteasa RO 160 D4
Dobrotești RO 160 E5
Dobrotić SRB 164 C4
Dobrotich BG 167 C8
Dobrotino NMK 169 B6
Dobrotitsa BG 161 F9
Dobrovăț RO 153 D11
Dobrovci BIH 157 C9
Dobrovice CZ 77 B7
Dobrovnik SLO 149 C6
Dobrovol'sk RUS 136 D5
Dobrowoda PL 143 F10
Dobruchi RUS 132 D2
Dobrun BIH 158 F3
Dobrun RO 160 E4
Dobrușa MD 154 B3
Dobruševo NMK 169 B5
Dobruška CZ 77 B10
Dobrzankowo PL 139 E10
Dobrzany PL 85 D8
Dobrzeń Wielki PL 142 E4
Dobrzyca PL 142 C4
Dobrzyków PL 139 F8
Dobrzyń nad Wisłą PL 139 E7
Dobšiná SK 145 F1
Dóc H 150 E5
Docking UK 15 B10
Dockmyr S 103 A10
Docksta S 107 E14
Dockweiler D 21 D7
Doclin RO 159 C8
Doddington UK 5 D12
Dodewaard NL 183 B7
Dodonoupoli GR 168 E4
Dödre S 102 A8
Doesburg NL 16 D6
Doetinchem NL 16 E6
Dofteana RO 153 E9
Doğanbey TR 177 C8
Doğanbey TR 177 D9
Doğanci TR 173 D10
Doğanköy TR 173 D10
Döge H 145 G5
Dogliani I 37 B7
Dognecea RO 159 C8
Doğüşbelen TR 181 C9
Dohna D 80 E5
Dohňany SK 146 C6
Dohren D 17 C9
Doicești RO 160 D6
Doïrani GR 169 B8
Doire Iorrais IRL 6 F3
Doische B 19 D10
Dojč SK 146 D4
Dojkinci SRB 165 C6
Dokka N 101 E12
Dokkas S 116 D6
Dokkedal DK 86 B6
Dokkum NL 16 B5
Doksy CZ 76 B6
Doksy CZ 81 E7
Doktor Yosifovo BG 165 C7
Dokupe LV 134 B3
Dolanog UK 10 F5
Dolbenmaen UK 10 F3
Dolceacqua I 37 D7
Dol-de-Bretagne F 23 C8
Dole BIH 157 D10
Dole F 26 F3
Dølemo N 90 B3
Dolenci NMK 168 B5
Dolenja Vas SLO 73 E10
Dolenjske Toplice SLO 73 E11
Dolgarrog UK 10 E4
Dolgellau UK 10 F4
Dolgen D 84 D4
Dolgorukovo RUS 136 E2
Dolhan TR 167 F8
Dolhasca RO 153 C9
Dolhești RO 153 C9
Dolhești RO 153 D11
Dołhobyczów PL 144 B9
Dolianova I 64 E3
Dolice PL 85 D8
Dolichi GR 169 D7
Doljani BIH 157 E8
Doljani HR 156 D5
Doljevac SRB 164 C4
Dolla IRL 8 C6
Dolle D 79 B10
Dollern D 82 C7
Döllnitz D 79 D11
Dollnstein D 75 E9
Dollon F 24 D4
Dolna MD 154 C2
Dolna Banya BG 165 E8
Dolna Dikanya BG 165 E7
Dolna Gradeshnitsa BG 165 F7
Dolnaja LV 135 D12
Dolná Krupá SK 146 E5
Dolna Lipnitska BG 166 C4
Dolna Makhala BG 165 E10
Dolna Melna BG 164 D6
Dolna Mitropoliya BG 165 C10
Dolna Oryakhovitsa BG 166 C5
Dolná Strehová SK 147 E8
Dolná Súča SK 146 D6
Dolná Tižina SK 147 C7
Dolna Vasilitsa BG 165 E8
Dolné Orešany SK 146 E4
Dolné Vestenice SK 146 D6
Dolni Bousov CZ 77 B8
Dolní Bukovsko CZ 77 D7
Dolní Čermná CZ 77 C11
Dolni Chiflik BG 167 D9
Dolní Dobrouč CZ 77 C11
Dolni Dŭbnik BG 165 C9
Dolní Dvořiště CZ 77 E6
Dolni Glavanak BG 166 F5
Dolní Kounice CZ 77 D10
Dolni Lom BG 165 C6
Dolní Loučky CZ 77 D10
Dolní Němčí CZ 146 D5
Dolní Podluží CZ 81 E7
Dolní Újezd CZ 77 C10
Dolní Újezd CZ 146 B5
Dolni Voden BG 165 E10
Dolní Šandov CZ 75 B12
Dolno Dupeni NMK 168 C5
Dolno Ezerovo BG 167 D8
Dolno Kamartsi BG 165 D8
Dolno Konjare NMK 164 F4
Dolno Levski BG 165 E9
Dolno Osenovo BG 165 F7
Dolno Selo BG 164 E5
Dolno Tserovene BG 159 F11
Dolno Uyno BG 164 E6
Dolný Hričov SK 147 C7
Dolný Kubín SK 147 C8
Dolný Pial SK 146 E6
Dolný Štál SK 146 F5
Dolo I 66 B5
Dolomieu F 31 D8
Dolores E 56 E3
Dolovo SRB 159 D6
Dölsach A 73 C6
Dolsk PL 81 C12
Dołubowo PL 141 E7
Dolus-d'Oléron F 28 D3
Dolyna UA 145 F8
Dolynivka UA 154 A5
Dolyns'ke UA 154 B5
Dolzhitsy RUS 132 D5
Domaháza H 147 E10
Domaniewice PL 141 G2
Domaniewice PL 143 B8
Domanín CZ 146 C4
Domaradz PL 144 D4
Domart-en-Ponthieu F 18 D5
Domaševo BIH 162 D5
Domašinec HR 149 D7
Domașnea RO 159 C9
Domaszek H 150 E5
Domaszków PL 77 B11
Domaszowice PL 142 D4
Domat Ems CH 71 D8
Domats F 25 D9
Domažlice CZ 76 D3
Dombás N 101 B10
Dombasle-en-Xaintois F 26 D4
Dombasle-sur-Meurthe F 186 D1
Dombegyház H 151 E7
Dombóvár H 149 D10
Dombrád H 145 G4
Dombresson CH 31 A10
Domburg NL 16 E1
Domegge di Cadore I 72 D5
Domeikava LT 137 D8
Domène F 31 E8
Domeniko GR 169 E7
Domérat F 29 C11
Domèvre-en-Haye F 26 C4
Domèvre-sur-Vezouze F 27 C6
Domfront F 23 C10
Domgermain F 26 C4
Domiće HR 162 D3
Domingo Pérez E 46 E4
Dömitz D 83 D10
Domlyan BG 165 D10
Dommartin-le-Franc F 26 D2
Dommartin-Varimont F 25 C12
Domme F 29 F8
Dommershausen D 21 D8
Dommitzsch D 80 C3
Domnești RO 160 C5
Domnești RO 161 E7
Domnitsa GR 174 B4
Domnovo RUS 136 E2
Domodossola I 68 A5
Domokos GR 174 A5
Domont F 25 B7
Domoroc RKS 164 D4
Dömös H 149 A11
Domoszló H 147 F10
Dompcevrin F 26 C3
Dompierre-les-Ormes F 30 C5
Dompierre-sur-Besbre F 30 B4
Dompierre-sur-Mer F 28 C3
Dompierre-sur-Yon F 28 B3
Domrémy-la-Pucelle F 26 D4
Dömsöd H 150 C3
Domsühl D 83 D11
Domus de Maria I 64 F2
Domusnovas I 64 E2
Domvaina GR 175 C6
Domžale SLO 73 D10
Donagh UK 7 D8
Donaghadee UK 4 F5
Donaghmore UK 7 C9
Donaghmore IRL 7 F10
Don Álvaro E 51 B7
Doña Mencía E 53 A8
Donard IRL 7 F9
Donaueschingen D 27 E9
Donauwörth D 75 E8
Don Benito E 51 B8
Doncaster UK 11 D9
Donchery F 19 E10
Donduşeni MD 153 A11
Donegal IRL 6 C6
Doneraile IRL 8 D5
Doneztebe E 32 D2
Dønfoss N 100 C8
Dongen NL 16 E3
Donges F 23 F7
Dongo I 69 A7
Donici MD 154 C3
Doñinos de Salamanca E 45 C9
Donja Bela Reka SRB 159 E9
Donja Brela HR 157 F6
Donja Bukovica MNE 163 C7
Donja Dubrava HR 149 D7
Donja Kupčina HR 148 E5
Donja Lepenica BIH 157 B8
Donja Mahala BIH 157 B10
Donja Motičina HR 149 F10
Donja Šatornja SRB 158 E6
Donja Stubica HR 148 E5
Donja Višnjica HR 148 D6
Donja Vrijeska HR 149 E8
Donja Zelina HR 148 E6
Donje Pazarište HR 67 C11
Donjeux F 26 D3
Donji Andrijevci HR 157 B9
Donji Čaglić HR 149 F8
Donji Dubovnik BIH 156 C5
Donji Dušnik SRB 164 C5
Donji Kosinj HR 156 C3
Donji Krčin SRB 159 F7
Donji Krivodol SRB 165 C6
Donji Lapac HR 156 C4
Donji Miholjac HR 149 E10
Donji Milanovac SRB 159 E9
Donji Proložac HR 157 F7
Donji Rujani BIH 157 E6
Donji Seget HR 156 E5
Donji Srb HR 156 D5
Donji Striževac SRB 164 C5
Donji Svilaj BIH 157 B9
Donji Vakuf BIH 157 D7
Donji Vijačani BIH 157 C8
Donji Zemunik HR 156 D3
Donji Širovac HR 156 B5
Donk NL 183 B7
Donkerbroek NL 16 B6
Donnalucata I 59 F6
Donnas I 68 B4
Donnemarie-Dontilly F 25 D9
Donnersbach A 73 B9
Donnersdorf D 75 C7
Donohill IRL 8 C6
Donori I 64 E3
Donostia E 32 D2
Donskoye RUS 139 A8
Donville-les-Bains F 23 C8
Donzdorf D 74 E6
Donzenac F 29 E9
Donzère F 35 B8
Donzy F 25 F9
Dooagh IRL 6 E2
Doochary IRL 6 C6
Dooish UK 4 F2
Doolin IRL 6 F4
Doon IRL 8 C6
Doonbeg IRL 8 C3
Doorn NL 16 D4
Doornspijk NL 183 A7
Dopiewo PL 81 B11
Đorče Petrov NMK 164 E3
Dorchester UK 13 D10
Dørdal N 90 B5
Dordives F 25 D8
Dordrecht NL 16 E3
Dore-l'Église F 30 E4
Dörentrup D 17 D12
Dores UK 3 L8
Dorfen D 75 F11
Dorfgastein A 73 B7
Dorfmark D 82 E7
Dorf Mecklenburg D 83 C10
Dorf Zechlin D 83 D13
Dorgali I 64 C4
Dorgoș RO 151 E8
Dorio GR 174 E4
Dorking UK 15 E8
Dorkovo BG 165 E9
Dorlisheim F 186 D3
Dormagen D 21 B7
Dormánd H 150 B5
Dormans F 25 B10
Dor Mărunt RO 161 E9
Dorna-Arini RO 152 C6
Dorna Candrenilor RO 152 C6
Dornava SLO 148 D5
Dörnberg (Habichtswald) D 17 F12
Dornbirn A 71 C9
Dornburg (Saale) D 79 D10
Dornburg-Frickhofen D 185 C9
Dornbusch D 17 A12
Dorndorf D 79 E7
Dorndorf-Steudnitz D 79 D10
Dornelas P 38 E4
Dornes F 30 B3
Dornești RO 153 B8
Dornhan D 187 E6
Dornie UK 2 L5
Dornișoara RO 152 C6
Dörnitz D 79 B11
Dorno I 69 C6
Dornoch UK 3 K8
Dornstadt D 74 F6
Dornstetten D 27 D9
Dornum D 17 A8
Dornumersiel D 17 A8
Dorobanțu RO 155 D2
Dorobanțu RO 161 E9
Dorog H 149 A11
Dorogháza H 147 F9
Dorohoi RO 153 B8
Dorohusk PL 141 H9
Dorolț RO 151 B10
Doroțcaia MD 154 C4
Dorotea S 107 C10
Dörpen D 17 C8
Dorras N 112 D8
Dorris S 107 B9
Dorstadt D 79 B8
Dorsten D 17 E7
Dortsfeld D 185 A7
Dortan F 31 C8
Dortmund D 17 E8
Dörttepe TR 177 E10
Doruchów PL 142 D5
Dorum D 17 A11
Dorupe LV 134 C7
Dörverden D 17 C12
Dörzbach D 74 D6
Dos Aguas E 48 F3
Dosbarrios E 46 E6
Dos Hermanas E 51 E8
Dospat BG 165 F9
Dossenheim D 21 F11
Doștat RO 152 F3
Dos Torres E 54 C3
Døstrup DK 86 E3
Dotnuva LT 134 F7
Dotternhausen D 27 D10
Döttingen CH 27 E9
Douai F 19 D7
Douarnenez F 22 D3
Doubrava CZ 147 B7
Doubravice nad Svitavou CZ 77 D11
Doubs F 31 B9
Douchy F 25 E9
Douchy-les-Mines F 19 D7
Doucier F 31 B8
Doudeville F 18 E2
Doudleby nad Orlicí CZ 77 B10
Doué-la-Fontaine F 23 F11
Douglas UK 5 D9
Douglas GBM 10 C3
Douglas IRL 8 E6
Douglas Bridge UK 4 F2
Doulaincourt-Saucourt F 26 D3
Doulevant-le-Château F 25 D12
Doullens F 18 D5
Dounaiika GR 174 D3
Doune UK 5 C8
Dounreay UK 3 H9
Dour B 19 D8
Dourdan F 24 C7
Dourgne F 33 D10
Douriez F 18 D4
Doussard F 31 D9
Douvaine F 31 C9
Douvres-la-Délivrande F 23 B11
Douzy F 19 E11
Dovadola I 66 D4
Dover UK 15 E11
Dovhe UA 145 G7
Döviken S 103 A9
Dovilai LT 134 E2
Dovre N 101 C10
Dowally UK 5 B9
Downham Market UK 11 F12
Downpatrick UK 7 D11
Downton UK 13 D11
Dowra IRL 6 D6
Dowsby UK 11 F11
Doxato GR 171 B6
Doyet F 30 C2
Doyrentsi BG 165 C10
Dozulé F 23 B11
Drabeši LV 135 B10
Drâby DK 86 C7
Dračevo BIH 162 D5
Dračevo NMK 164 F4
Drachhausen D 80 C6
Drachselsried D 76 D4
Drachten NL 16 B6
Drag N 105 B10
Drag N 111 D11
Dragacz PL 138 C6
Dragalijevac BIH 157 C11
Dragalina RO 161 E10
Dragalovci BIH 157 C8
Drăgănești RO 151 D9
Drăgănești RO 153 F10
Drăgănești RO 161 D8
Drăgănești de Vede RO 160 E6
Drăgănești-Olt RO 160 E5
Drăgănești-Vlașca RO 161 E7
Draganići HR 148 E5
Draganovo BG 166 C5
Drăganu RO 160 D5
Drăgășani RO 160 D4
Dragash RKS 163 E10
Dragatuš SLO 67 A11
Drage D 83 D8
Drage HR 156 E4
Drăgești RO 151 D9
Draghiceni RO 160 E4
Draginac SRB 158 D3
Draginje SRB 158 D4
Draginovo BG 165 E8
Dragland N 111 C11
Dragnes N 111 C10
Dragnić BIH 157 D7
Dragobi AL 163 E8
Dragočaj BIH 157 C7
Dragocvet SRB 159 F7
Dragodana RO 160 D6
Drăgoești RO 160 D4
Drăgoești RO 161 D9
Dragoevo BG 167 C7
Dragoevo NMK 164 F5
Drăgoiești RO 153 B8
Dragoman BG 165 D6
Dragomance NMK 164 E4
Dragomir BG 165 E9
Dragomirești RO 152 B4
Dragomirești RO 153 C9
Dragomirești RO 153 D10
Dragomirești RO 160 D6
Dragomirovo BG 166 B4
Dragoni I 60 A2
Dragør DK 87 D11
Dragoš NMK 168 C5
Dragoslavele RO 160 C6
Dragoș Vodă RO 161 E10
Drăgotești RO 159 D11
Drăgotești RO 160 E4
Dragotina HR 156 B5
Dragović HR 149 F8
Dragovishtitsa BG 165 E6
Dragoychintsi BG 164 D5
Dragoynovo BG 166 F4
Dragsfjärd FIN 126 E7
Draguignan F 36 D4
Drăgușeni RO 153 A9
Drăgușeni RO 153 C8
Drăgușeni RO 153 F11
Drăgutești RO 159 D11
Drahnsdorf D 80 C5
Drahovce SK 146 D5
Drahove UA 145 G8
Drahovica Donja BIH 157 C9
Drajna RO 161 C8
Draka BG 167 E8
Drakenburg D 17 C12
Draksenič BIH 157 B6
Dralfa BG 167 C6
Drama GR 170 B6
Drammen N 95 C12
Drânceni RO 153 D12
Drange N 94 F5
Drangedal N 90 A5
Drangstedt D 17 A11
Drănic RO 160 E3
Dranse D 83 D13
Dransfeld D 78 D6
Dranske D 84 A4
Draperstown UK 4 F3
Drasenhofen A 77 E11
Draßmarkt A 149 A6
Drávafok H 149 E9
Dravagen S 102 B6
Draviskos GR 170 C5
Dravograd SLO 73 C11
Drawno PL 85 D9
Drawsko PL 85 E10
Drawsko Pomorskie PL 85 C9
Drayton UK 15 B11
Drążdżewo PL 139 D11
Dražen Vrh SLO 148 C5
Draževac SRB 158 D5
Dražgoše SLO 73 D9
Drebber D 17 C10
Drebkau D 80 C6
Drégelypalánk H 147 E8
Dreieich D 21 D11
Dreileben D 79 B9
Dreis D 21 E7
Drelów PL 141 G7
Drelsdorf D 82 A6
Drem UK 5 C11
Drenchia I 73 D8
Drenovac SRB 159 F7
Drenovci HR 157 C10
Drenovë AL 168 C4
Drenovets BG 159 F10
Drenovicë AL 168 C2
Drenovo NMK 164 F4
Drenovo NMK 169 B6
Drense D 84 D5
Drensteinfurt D 17 E9
Drenta BG 166 D5
Drentwede D 17 C11
Drepano GR 169 D6
Drepano GR 175 D6
Dresden D 80 D5
Dretun' BY 133 E6
Dretyń PL 85 B11
Dreumel NL 183 B6
Dreux F 24 C5
Dreverna LT 134 E2
Drevja N 108 E5
Drevjesætra N 102 D4
Dřevohostice CZ 146 C5
Drevsjø N 102 C3
Drevvatn N 108 D5
Drewitz D 79 B11
Drewnica PL 138 B6
Drezdenko PL 85 E9
Drežnica BIH 157 E8
Dreżnik SRB 158 F4
Driceni LV 133 C2
Dridu RO 161 D8
Driebergen NL 16 D4
Driebes E 47 D6
Driedorf D 185 C9
Drienov SK 145 F3
Drietoma SK 146 D5
Driffield UK 11 C11
Drimmo IRL 7 F7
Drimnin UK 4 B5
Drimoleague IRL 8 E4
Drinić BIH 156 C5
Drinjača BIH 157 D11
Drinovci BIH 157 F7
Dripsey IRL 8 E5
Drisht AL 163 E8
Dříteň CZ 76 D6
Drithas AL 168 C4
Driva N 101 A11
Drivstua N 101 B11
Drmno SRB 159 D7
Drnholec CZ 77 E11
Drniš HR 156 E5
Drnje HR 149 D7
Drnovice CZ 77 D11
Drnovice CZ 77 D11
Dro I 69 B10
Drøbak N 95 C13
Drobeta-Turnu Severin RO 159 D10
Drobin PL 139 E8
Drochia MD 153 A11
Drochtersen D 17 A12
Drogheda IRL 7 E10
Drohiczyn PL 141 F7
Drohobych UA 145 E8
Droichead Abhann IRL 8 C5
Droichead Átha IRL 7 E10
Droichead na Bandan IRL 8 E5
Droichead Nua IRL 7 F9
Droitwich Spa UK 13 A10
Drolshagen D 185 B8
Drolsum N 95 B12
Dromara UK 7 D10
Dromina IRL 8 D5
Drommahane IRL 8 D5
Drömme S 107 E14
Dromod IRL 7 E7
Dromore UK 7 C8
Dromore UK 7 D10
Dromore West IRL 6 D5
Dronero I 37 C6
Dronfield UK 11 E9
Drongan UK 5 E8
Drongen B 182 C3
Dronninglund DK 86 A6
Dronrijp NL 16 B5
Dronten NL 16 C5
Dropla BG 155 F2
Drosato GR 169 B8
Drosbacken S 102 C3
Drosendorf A 77 E9
Drosia GR 175 C8
Drösing A 77 E11
Drosopigi GR 168 C5
Droué F 24 D5
Droyßig D 79 D11
Drugan BG 165 E7
Drugovo NMK 168 B4
Druid UK 10 F5
Druimdrishaig UK 4 D5
Drulingen F 27 C7
Drumandoora IRL 6 G5
Drumanespick IRL 7 E8
Drumatober IRL 6 F6
Drumbeg UK 2 J6
Drumbilla IRL 7 D10
Drumcard UK 7 D7
Drumcollogher IRL 8 D5
Drumcondra IRL 7 E9
Drumconrath IRL 7 E9
Drumcree IRL 7 E8
Drumettaz-Clarafond F 31 D8
Drumevo BG 167 C8
Drumfree IRL 4 E2
Drumkeeran IRL 6 D6
Drumlea IRL 7 D7
Drumlish IRL 7 E7
Drumlithie UK 5 B12
Drummin IRL 9 D9
Drummore UK 4 F7
Drumnadrochit UK 2 L8
Drumquin UK 4 F2
Drumshanbo IRL 6 D6
Drung IRL 7 D8
Drusenheim F 27 C8
Druskininkai LT 137 F9
Drusti LV 135 B11
Druten NL 183 B7
Druviena LV 135 B12
Druya BY 133 E2
Druyes-les-Belles-Fontaines F 25 E9
Druysk BY 133 E2
Družbice PL 143 D7
Druzhba RUS 136 E3
Druzhnaya Gorka RUS 132 C7
Družstevná pri Hornáde SK 145 F3
Drvenik HR 157 F7
Drwalew PL 141 G4
Drwinia PL 143 F9
Dryanovets BG 161 F8
Dryanovo BG 166 D4
Dryazhno RUS 132 E4
Drygały PL 139 C13
Drymaia GR 175 B6
Drymen UK 5 C8
Drymos GR 169 C8
Dryna N 100 A5
Dryopida GR 175 E9
Dryos GR 176 E5
Drysvyaty BY 135 E13
Dryszczów PL 144 A8
Drzewce PL 142 B6
Drzewiany PL 85 C11
Drzewica PL 141 H2
Drzonowo PL 85 C11
Drzycim PL 138 C5
Duagh IRL 8 D4
Dualchi I 64 C2
Dually IRL 9 C7
Duas Igrejas P 39 F7
Dub SRB 158 F4
Dubá CZ 77 A7
Dubăsari MD 154 C4
Dubăsarii Vechi MD 154 C4
Duba Stonska HR 162 D4
Dubău MD 154 C4
Dubeczno PL 141 H8
Düben D 79 C11
Duben D 80 C5
Dübendorf CH 27 F10
Dubeņi LV 134 D2
Dubeninki PL 136 E6
Dubí CZ 80 E5
Dubičiai LT 137 E10
Dubicko CZ 77 C11
Dubicze Cerkiewne PL 141 E8
Dubidze PL 143 D7
Dubiecko PL 144 D5
Dubienka PL 144 A8
Dubingiai LT 137 C11
Dubino I 69 A7
Dubivka UA 153 A8
Dublin IRL 7 F10
Dublje SRB 158 D4
Dublovice CZ 76 C6
Dublyany UA 144 D9
Dublyany UA 145 E8
Dubna LV 135 D13
Dub nad Moravou CZ 146 C4
Dubňany CZ 77 E12
Dubnica nad Váhom SK 146 D6
Dubník SK 146 F6
Dubočnica BIH 157 D9
Dubova RO 159 D9
Dubovac SRB 159 D7
Dubove UA 145 G8
Dubovets BG 166 F5
Dubovica SK 145 E2
Dŭbovo BG 166 D5
Dubovo SRB 164 C4
Dubovsko BIH 156 C5
Dubrava BIH 157 E8
Dubrava HR 149 E7
Dubrave BIH 157 C10
Dubrave BIH 157 D6
Dubrave BIH 157 D10
Dubravica BIH 157 D8
Dubravica HR 148 E5
Dubravica SRB 159 D7
Dúbravy SK 147 D8
Dubrawka BY 133 F6
Dubrovka RUS 129 F14
Dubrovka RUS 132 F6
Dubrovka RUS 133 D5
Dubrovnik HR 162 D5
Dubrovytsya UA 144 D8
Dubuļi LV 133 D3
Dubynove UA 154 A6
Ducey F 23 C9
Ducherow D 84 C5
Duchov CZ 80 E5
Duck End UK 15 D9
Duclair F 18 F2
Duda-Epureni RO 153 D12
Dudar H 149 B9
Duddo UK 5 D12
Dudelange L 20 F6
Dudeldorf D 21 E7
Duderstadt D 79 C7
Dudești RO 161 D10
Dudestii Vechi RO 150 E5
Dudince SK 147 E7
Düdingen CH 31 B11
Dudley UK 11 F7
Dudovica SRB 158 E5
Dueñas E 40 E2
Duesund N 100 E2
Dueville I 72 E4
Duffel B 19 B10
Dufftown UK 3 L10
Duga Poljana SRB 163 C9
Duga Resa HR 148 F5
Dugi Rat HR 156 F6
Dugny-sur-Meuse F 26 B3
Dugopolje HR 156 E6
Dugo Selo HR 148 E6
Düğüncübaşi TR 173 B7
Duhort-Bachen F 32 C5
Duino I 73 E8
Duirinish UK 2 L5
Duisburg D 17 F7
Dukas AL 168 C2
Dukat AL 168 D2
Dukat i Ri AL 168 D1

Dukla *PL* 145 D4
Dūkštas *LT* 135 D10
Dŭlbok Dol *BG* 165 D10
Dŭlboki *BG* 166 E5
Dŭlbok Izvor *BG* 166 E4
Dulcești *RO* 153 D9
Duleek *IRL* 7 E10
Dŭlgodeltsi *BG* 160 F2
Dŭlgopol *BG* 167 C8
Dulina *BY* 133 F12
Duljci *BIH* 157 D7
Dūlmen *D* 17 E8
Dulovce *SK* 146 F6
Dulovka *RUS* 133 A4
Dulovo *BG* 161 F10
Dulverton *UK* 13 C7
Dumbarton *UK* 4 C7
Dumbrava *RO* 151 F9
Dumbrava *RO* 161 D8
Dumbrava Roșie *RO* 153 D8
Dumbrăveni *RO* 152 E5
Dumbrăveni *RO* 153 B8
Dumbrăveni *RO* 155 F1
Dumbrăveni *RO* 161 B10
Dumbrăvești *RO* 161 C12
Dumbrăvița *RO* 151 F7
Dumbrăvița *RO* 152 B3
Dumbrăvița *RO* 153 F6
Dumbría *E* 38 B1
Dumești *RO* 153 C10
Dumești *RO* 153 D10
Dumfries *UK* 5 E9
Dumitra *RO* 152 C4
Dumitrești *RO* 161 B9
Dummerstorf *D* 83 C12
Dumnicë e Poshtme *RKS* 164 D3
Dümpelfeld *D* 21 D7
Dun *F* 33 D9
Duna *N* 105 B11
Dunaegyháza *H* 149 C11
Dunaföldvár *H* 149 C11
Dunaharaszti *H* 150 C3
Dunajská Lužná *SK* 146 E4
Dunajská Streda *SK* 146 F5
Dunakeszi *H* 150 B3
Dunalka *LV* 134 C2
Dún an Rí *IRL* 7 E9
Dunapataj *H* 150 D3
Dunărea *RO* 155 E2
Dunaszeg *H* 149 A9
Dunaszekcső *H* 149 D11
Dunaszentbenedek *H* 149 C11
Dunaszentgyörgy *H* 149 C11
Dunasziget *H* 146 F4
Dunatetétlen *H* 150 D3
Dunaújváros *H* 149 C11
Dunava *LV* 135 D12
Dunavarsány *H* 150 C3
Dunavecse *H* 150 D2
Dunavtsi *BG* 159 F10
Dunbar *UK* 5 C11
Dunbeath *UK* 3 J10
Dunblane *UK* 5 B9
Dunboyne *IRL* 7 F10
Dún Búinne *IRL* 7 F10
Duncannon *IRL* 9 D9
Dún Chaoin *IRL* 8 D2
Dunchurch *UK* 13 A12
Duncormick *IRL* 9 D9
Dundaga *LV* 134 A4
Dundalk *IRL* 7 D10
Dún Dealgan *IRL* 7 D10
Dun Dealgan *IRL* 7 D10
Dundee *UK* 5 C11
Dunderland *N* 108 D8
Dundonald *UK* 7 C11
Dundreggan *UK* 2 L7
Dundrennan *UK* 5 F9
Dundrum *IRL* 9 C8
Dundrum *UK* 7 D11
Dunecht *UK* 3 L12
Dunes *F* 33 B7
Dunfanaghy *IRL* 7 B7
Dunfermline *UK* 5 C10
Dungannon *UK* 7 C9
Dún Garbhán *IRL* 9 D7
Dungarvan *IRL* 9 D7
Dungarvan *IRL* 9 D7
Düngenheim *D* 185 D7
Dungiven *UK* 4 F3
Dungloe *IRL* 6 C6
Dungourney *IRL* 8 E6
Dunholme *UK* 11 E11
Dunicë *AL* 168 C4
Dunières *F* 30 E5
Duninowo *PL* 85 A11
Dunje *NMK* 169 B6
Dunkeld *UK* 5 B9
Dunkerque *F* 18 B5
Dunkerrin *IRL* 9 C7
Dunkitt *IRL* 9 D8
Dún Laoghaire *IRL* 7 F10
Dunlavin *IRL* 7 F9
Dunleer *IRL* 7 E10
Dún Léire *IRL* 7 E10
Dun-le-Palestel *F* 29 C10
Dun-les-Places *F* 25 F11
Dunlop *UK* 4 D7
Dunloy *UK* 4 E4
Dún Mánmhaí *IRL* 8 E4
Dunmanway *IRL* 8 E4
Dún Mór *IRL* 9 B9
Dunmore *IRL* 6 E5
Dunmore East *IRL* 9 D9
Dunmurry *UK* 7 C11
Dunnamanagh *UK* 4 F2
Dún na nGall *IRL* 6 C6
Dunning *UK* 5 C9
Dunoon *UK* 4 D7
Dunquin *IRL* 8 D2
Duns *UK* 5 D12
Dunscore *UK* 5 E9
Dünsen *D* 17 C11
Dunshaughlin *IRL* 7 E9
Dunstable *UK* 15 D7
Dunster *UK* 13 C7
Dun-sur-Auron *F* 29 B11
Dun-sur-Meuse *F* 19 F11
Dunure *UK* 4 E7
Dunvant *UK* 12 B6
Dunvegan *UK* 2 L3
Duplica *SLO* 73 D10
Dupnitsa *BG* 165 E7
Durach *D* 71 B10
Durak *TR* 173 E9
Duras *F* 33 A6

Durbach *D* 27 C9
Durban-Corbières *F* 34 E4
Durbe *LV* 134 C2
Durbuy *B* 184 D3
Dúrcal *E* 53 C9
Durdat-Larequille *F* 29 C11
Đurđenovac *HR* 149 E8
Đurđevac *HR* 149 D8
Đurđevića Tara *MNE* 163 C7
Đurđevik *BIH* 157 D10
Đurđevo *SRB* 158 C5
Đurđevo *SRB* 159 E6
Düren *D* 20 C6
Durfort *F* 35 C6
Durham *UK* 5 F13
Đurkov *SK* 145 F3
Durlas *IRL* 9 C7
Durleşti *MD* 154 C3
Durmanec *HR* 148 D5
Dürmentingen *D* 71 A9
Durmersheim *D* 27 C9
Durnești *RO* 153 B10
Dürnkrut *A* 77 F11
Durrës *AL* 168 B1
Durrington *UK* 13 C11
Dürrlauingen *D* 75 F7
Durrow *IRL* 9 C8
Durrus *IRL* 8 E3
Dürrwangen *D* 75 D7
Dursley *UK* 13 B10
Dursunbey *TR* 173 E10
Durtal *F* 23 E11
Duruelo de la Sierra *E* 40 E6
Duruitoarea *MD* 153 B10
Durup *DK* 86 B3
Durusu *TR* 173 B10
Dusetos *LT* 135 E11
Dushk i Madh *AL* 168 C2
Dusina *BIH* 157 E8
Dūškotna *BG* 167 D8
Duškovci *SRB* 158 F3
Dusnok *H* 150 E2
Dusocin *PL* 138 C6
Düsseldorf *D* 21 B7
Dussen *NL* 16 E3
Dußlingen *D* 187 E7
Düßnitz *D* 79 C12
Dusznik *PL* 81 B10
Dutluca *TR* 173 E11
Dutovlje *SLO* 73 E8
Duved *S* 105 E13
Duvberg *S* 102 B7
Düvertepe *TR* 173 F9
Duži *BIH* 162 D5
Dvärsätt *S* 106 E6
Dvaryshcha *BY* 133 E6
Dve Mogili *BG* 161 F7
Dverberg *N* 111 B10
Dviete *LV* 135 D12
Dvirtsi *UA* 144 C9
Dvor *HR* 156 B5
Dvorce *CZ* 146 B5
Dvornaye Syalo *BY* 133 E3
Dvorníky *SK* 146 E5
Dvory nad Žitavou *SK* 146 F6
Dvůr Králové *CZ* 77 B9
Dwikozy *PL* 144 B4
Dwingeloo *NL* 16 C6
Dyan *UK* 7 D9
Dyankovo *BG* 161 F9
Dybvad *DK* 90 E7
Dyce *UK* 3 L12
Dychów *PL* 81 C8
Dydnia *PL* 144 D5
Dyfjord *N* 113 B19
Dygowo *PL* 85 B9
Dyke *UK* 3 K9
Dykehead *UK* 5 B11
Dykhtynets' *UA* 152 A6
Dylągówka *PL* 144 D5
Dylaki *PL* 142 E5
Dylewo *PL* 139 D11
Dymchurch *UK* 15 E10
Dymock *UK* 13 B10
Dynäs *S* 107 F13
Dynivtsi *UA* 153 A8
Dynów *PL* 144 D5
Dyo *F* 30 C5
Dyping *N* 110 E9
Dyrnes *N* 104 E3
Dyrøyhamn *N* 111 B13
Dyrrachi *GR* 174 E5
Dyulevo *BG* 167 E8
Dyulino *BG* 167 D9
Dyviziya *UA* 154 F5
Dywity *PL* 136 F1
Dzęgużany *LV* 134 D6
Dzelzava *LV* 135 C12
Dzeni *LV* 135 B10
Dzērbene *LV* 135 B11
Dzhebel *BG* 171 A8
Dzherman *BG* 165 E7
Dzhulynka *UA* 154 A6
Dzhulyunitsa *BG* 166 C5
Dzhurivr *UA* 152 A6
Dzhurovo *BG* 165 D9
Dziadkowice *PL* 141 E7
Dziadowa Kłoda *PL* 142 D4
Działdowo *PL* 139 D9
Działoszyce *PL* 143 F9
Działoszyn *PL* 142 D6
Działyń *PL* 138 D7
Dziemiany *PL* 138 B4
Dziergowice *PL* 142 F5
Dzierzążna *LV* 135 D6
Dzierzążno Wielkie *PL* 85 E10
Dzierzgoń *PL* 139 C7
Dzierzgowo *PL* 139 D10
Dzierzkowice-Rynek *PL* 144 B5
Dzierżoniów *PL* 81 E11
Dzietrzniki *PL* 142 D6
Dzietrzychowo *PL* 136 E3
Dziewierzewo *PL* 85 E13
Dżigolj *SRB* 164 C4
Dzirnavas *LV* 134 B4
Dzisna *BY* 133 E4
Dziwnów *PL* 85 B7
Dzūkste *LV* 134 C6
Džumajlija *NMK* 164 F5
Dźwierzuty *PL* 139 C10
Dźwirzyno *PL* 85 B8
Dzwola *PL* 144 B6
Dzwonowice *PL* 143 E8
Dźwierzuty *PL* 133 F4

Earith *UK* 15 C9
Earl Shilton *UK* 11 F9
Earlston *UK* 5 D11
Earl Stonham *UK* 15 C11
Earsairidh *UK* 4 B2
Easdale *UK* 4 C5
Easington *UK* 5 F14
Easington *UK* 11 D12
Easingwold *UK* 11 C9
Easky *IRL* 6 D5
Eastbourne *UK* 15 F9
Eastburn *UK* 11 D11
Eastfield *UK* 11 C11
Eastgate *UK* 5 F12
East Grinstead *UK* 15 E8
East Harling *UK* 15 C10
East Kilbride *UK* 5 D8
Eastleigh *UK* 13 D12
East Linton *UK* 5 D11
East Looe *UK* 12 E6
Eastoft *UK* 11 D10
Easton *UK* 3 L10
East Retford *UK* 11 E10
Eastriggs *UK* 5 F10
East Wemyss *UK* 5 C10
East Wittering *UK* 15 F7
Eastwood *UK* 11 E9
Eaton Socon *UK* 15 C8
Eaunes *F* 33 D8
Eauze *F* 33 C6
Ebberup *DK* 86 E5
Ebbs *A* 72 A5
Ebbw Vale *UK* 13 B8
Ebecik *TR* 181 A9
Ebeleben *D* 79 D8
Ebelsbach *D* 75 C8
Ebeltoft *DK* 87 C7
Ebene Reichenau *A* 73 C8
Ebenfurth *A* 77 G10
Eben im Pongau *A* 73 B7
Ebensee *A* 73 A9
Ebensfeld *D* 75 B8
Ebenthal *A* 73 C9
Eberau *A* 148 B6
Eberbach *D* 21 F11
Ebergötzen *D* 79 C7
Ebermannsdorf *D* 75 D10
Ebermannstadt *D* 75 C9
Ebern *D* 75 B8
Eberndorf *A* 73 C10
Ebersbach *D* 80 D5
Ebersbach *D* 81 D7
Ebersbach an der Fils *D* 74 E6
Ebersberg *D* 75 F10
Eberschwang *A* 76 F5
Ebersdorf *A* 148 B5
Ebersdorf *D* 17 A12
Ebersdorf *D* 75 B10
Ebersmunster *F* 186 E4
Eberstein *A* 73 C10
Eberswalde-Finow *D* 84 E5
Ebes *H* 151 C8
Ebhausen *D* 27 C10
Ebikon *CH* 27 F9
Ebnat-Kappel *CH* 27 F11
Eboli *I* 60 B4
Ebrach *D* 75 C7
Ebreichsdorf *A* 77 G10
Ébreuil *F* 30 C3
Ebstorf *D* 83 D8
Écaussinnes-d'Enghien *B* 182 D4
Ecclefechan *UK* 5 E10
Eccles *UK* 5 D12
Eccleshall *UK* 11 F7
Eceabat *TR* 171 D9
Echallens *CH* 31 B10
Echarri *E* 32 E2
Échauffour *F* 24 C3
Échenoz-la-Méline *F* 26 E5
Eching *D* 75 E11
Eching *D* 75 F10
Echinos *GR* 171 B7
Echiré *F* 28 C5
Échirolles *F* 31 E8
Échourgnac *F* 29 E6
Echsenbach *A* 77 E8
Echt *UK* 3 L12
Echt *NL* 19 B12
Echte *D* 79 C7
Echternach *L* 20 E6
Ečka *SRB* 158 C5
Eckartsberga *D* 79 D10
Eckbolsheim *F* 186 D4
Eckental *D* 75 C9
Eckernförde *D* 83 B7
Eckerö *FIN* 99 B13
Eckington *UK* 11 E9
Eckington *UK* 13 A10
Éclaron-Braucourt-Ste-Livière *F* 25 C12
Écommoy *F* 23 E12
Écos *F* 24 B6
Écouché *F* 23 C11
Écouflant *F* 23 E10
Écouis *F* 18 F3
Écrouves *F* 26 C4
Écs *H* 149 A9
Ecséd *H* 150 B4
Ecseg *H* 147 F9
Ecsegfalva *H* 151 C6
Ecser *H* 150 C3
Écueillé *F* 29 A8
Écuisses *F* 30 B6
Écury-sur-Coole *F* 25 C11
Ed *S* 91 B10
Ed *S* 107 E12
Eda glasbruk *S* 96 C7
Edam *NL* 16 C4
Edane *S* 97 C9
Ėdas *LV* 134 C4
Eddelak *D* 82 C6
Edderton *UK* 3 K8
Eddleston *UK* 5 D10
Ede *NL* 16 D5
Ede *S* 103 B13
Edebäck *S* 97 B10
Edefors *S* 118 B5
Edegem *B* 19 B9
Edelény *H* 145 G2
Edelschrott *A* 73 B11
Edemissen *D* 79 B7
Eden *UK* 4 F5
Eden *S* 107 D12
Edenderry *IRL* 7 F8
Edenkoben *D* 21 F10
Edermünde *D* 78 D5
Ederny *UK* 7 C7

Edesheim *D* 21 F10
Edessa *GR* 169 C7
Edewecht *D* 17 B9
Edewechterdamm *D* 17 B9
Edgeworthstown *IRL* 7 E7
Edgmond *UK* 11 F7
Ediger-Eller *D* 21 D8
Edinburgh *UK* 5 D10
Edincik *TR* 173 D9
Edineț *MD* 153 A10
Edirne *TR* 171 A11
Edland *N* 94 C7
Edling *D* 75 F11
Edlingham *UK* 5 E13
Edlitz *A* 148 A6
Edmondstown *IRL* 6 E5
Edmundbyers *UK* 5 F13
Édole *LV* 134 B3
Edolo *I* 69 A9
Edremit *TR* 173 E7
Edrosa *P* 39 E6
Edroso *P* 39 E6
Edsbro *S* 99 C11
Edsbruk *S* 93 C8
Edsbyn *S* 103 D10
Edsele *S* 107 E11
Edsken *S* 98 A6
Edsvalla *S* 97 D9
Edzell *UK* 5 B11
Eefde *NL* 16 D6
Eeklo *B* 19 B8
Eelde-Paterswolde *NL* 17 B7
Eerbeek *NL* 16 D6
Eernegem *B* 182 C2
Eersel *NL* 16 F4
Eferding *A* 76 F6
Effelder *D* 75 B9
Effeltrich *D* 75 C9
Efforie *RO* 155 E3
Efpalio *GR* 174 C4
Efringen-Kirchen *D* 27 E8
Efxeinoupoli *GR* 175 A6
Egby *S* 89 B11
Egebæk *DK* 86 E3
Egebjerg *DK* 86 D5
Egebjerg *DK* 87 D9
Egeln *D* 79 C9
Egense *DK* 86 B6
Eger *H* 145 H1
Egerbakta *H* 145 H1
Egersund *N* 94 F3
Egersund *N* 94 F3
Egerszalók *H* 145 H1
Egervár *H* 149 C7
Egeskov *DK* 86 D5
Egestorf *D* 83 D8
Egga *N* 111 B19
Eggby *S* 91 C14
Eggebek *D* 82 A6
Eggenburg *A* 77 E9
Eggenfelden *D* 75 F12
Eggesin *D* 84 C6
Eggingen *D* 27 E9
Eggiwil *CH* 70 D5
Egglham *D* 76 E4
Eggolsheim *D* 75 C9
Eggstätt *D* 72 A5
Eghezée *B* 19 C10
Eging am See *D* 76 E4
Eglaine *LV* 135 E12
Égletons *F* 29 E10
Egling *D* 72 A4
Eglingham *UK* 5 E13
Eglinton *UK* 4 E2
Eglisau *CH* 27 E9
Eglish *UK* 7 D9
Egluceims *LV* 134 B5
Eglwys Fach *UK* 10 F4
Eglwyswrw *UK* 12 A5
Egmond aan Zee *NL* 16 C3
Egorovca *MD* 153 B11
Egremont *UK* 10 C4
Egsmark *DK* 86 C7
Egton *UK* 11 C10
Éguilles *F* 35 C9
Eguisheim *F* 27 D7
Éguzon-Chantôme *F* 29 C9
Ehekirchen *D* 75 E8
Ehingen *D* 75 D8
Ehingen (Donau) *D* 71 A9
Ehlen (Habichtswald) *D* 17 F12
Ehningen *D* 187 D6
Ehra-Lessien *D* 79 A8
Ehrenberg-Wüstensachsen *D* 74 B7
Ehrenburg *D* 17 C11
Ehrenfriedersdorf *D* 80 E3
Ehrenhausen *A* 148 C5
Ehringshausen *D* 21 C10
Eibar *E* 32 D1
Eibelstadt *D* 74 C7
Eibenstock *D* 75 B12
Eibergen *NL* 17 D7
Eibiswald *A* 148 D4
Eich *D* 21 E10
Eichenbarleben *D* 79 B9
Eichenbühl *D* 27 A11
Eichendorf *D* 76 E3
Eichenzell *D* 74 B6
Eichgraben *A* 77 F9
Eichigt *D* 75 B11
Eichstätt *D* 75 E9
Eichwalde *D* 80 B5
Eicklingen *D* 79 A7
Eide *N* 100 A6
Eide *N* 111 C11
Eidet *N* 105 C15
Eidet *N* 105 C16
Eidet *N* 108 A9
Eidet *N* 111 B10
Eidfjord *N* 94 B6
Eiði *FO* 2 A2
Eidkjosen *N* 111 A16
Eidsbugarden *N* 101 D8
Eidsdal *N* 100 B6
Eidsfjord *N* 110 C8

Eidsfoss *N* 95 C12
Eidshaug *N* 105 B10
Eidsnes *N* 112 C10
Eidstranda *N* 111 A19
Eidsund *N* 94 D3
Eidsvåg *N* 100 A8
Eidsvåg *N* 100 B8
Eidsvoll *N* 95 B14
Eidsvoll Verk *N* 95 B14
Eidvågeid *N* 113 B12
Eigeltingen *D* 27 E10
Eigenrieden *D* 79 D7
Eigersund *N* 94 F3
Eiholt *N* 100 B8
Eik *N* 94 D3
Eikelandsosen *N* 94 B3
Eiken *N* 94 F6
Eikesdal *N* 100 B8
Eikjaberg *N* 100 D6
Eikjeskog *N* 94 F4
Eikjil *N* 90 B1
Eiksund *N* 100 B3
Eilenburg *D* 79 D12
Eilenstedt *D* 79 C9
Eilsleben *D* 79 B9
Eime *D* 78 B6
Eimen *D* 78 C6
Eimke *D* 83 E8
Eina *N* 101 E13
Einacleit *UK* 2 J3
Einbeck *D* 78 C6
Eindhoven *NL* 16 F4
Einsiedeln *CH* 27 F10
Einville-au-Jard *F* 26 C5
Einvollen *N* 108 F3
Eisden *D* 183 D7
Eiselfing *D* 75 F11
Eisenach *D* 79 E7
Eisenberg *D* 79 E10
Eisenberg (Pfalz) *D* 21 E10
Eisenerz *A* 73 A10
Eisenhüttenstadt *D* 81 B7
Eisenkappel *A* 73 D10
Eisenstadt *A* 77 G11
Eisfeld *D* 75 B8
Eišiškės *LT* 137 E10
Eislandet *N* 100 A8
Eisleben, Lutherstadt *D* 79 C10
Eislingen (Fils) *D* 74 E6
Eitelborn *D* 21 D8
Eiterfeld *D* 78 E6
Eiternes *N* 105 B11
Eiterstraum *N* 108 D7
Eitorf *D* 21 C8
Eivinvik *N* 100 E2
Eivissa *E* 57 D7
Eja *P* 44 B4
Ejby *DK* 86 E5
Ejby *DK* 87 D9
Ejsing *DK* 86 B4
Ejstrupholm *DK* 86 D4
Ejulve *E* 42 F2
Ekeby *S* 87 C11
Ekeby *S* 92 A6
Ekeby *S* 99 C9
Ekeby-Almby *S* 97 D13
Ekedalen *S* 91 C14
Ekenäs *FIN* 127 F9
Ekenäs *S* 91 B11
Ekenässjön *S* 92 E6
Ékengrave *LV* 135 D11
Ekeren *B* 182 C4
Ekerö *S* 99 D9
Eket *S* 87 C12
Ekkerøy *N* 114 C8
Ekmekçi *TR* 173 D9
Ekola *FIN* 122 D9
Ekorrträsk *S* 107 B16
Eksaarde *B* 182 C3
Eksel *B* 183 C6
Eksharad *S* 97 B9
Eksjö *S* 92 D5
Eksta *S* 93 E12
Ekträsk *S* 118 F3
Ekzarh-Antimovo *BG* 167 D7
Ekzarkh Yosif *BG* 161 F7
Elafos *GR* 169 D7
Elafotopos *GR* 168 E4
Elaiochori *GR* 171 C6
Elaionas *GR* 174 B5
Elaionas *GR* 174 B5
El Álamo *E* 46 D5
El Algar *E* 56 F3
El Almendro *E* 51 D5
El Alquián *E* 55 F8
Elämäjärvi *FIN* 123 D15
El Arahal *E* 51 E8
El Arenal *E* 45 D10
Elassona *GR* 174 F5
El Astillero *E* 40 B4
Elateia *GR* 175 B6
Elati *GR* 169 E6
Elati *GR* 169 E6
Elatochori *GR* 169 D7
Elatou *GR* 174 B4
El Ballestero *E* 55 B8
El Barco de Ávila *E* 45 D9
El Barraco *E* 46 D3
Elbasan *AL* 168 B3
Elbergen *D* 17 D8
Elbeuf *F* 18 F3
Elbigenalp *A* 71 C10
Elbingerode (Harz) *D* 79 C8
Elblag *PL* 139 B7
El Bodón *E* 45 D7
El Bonillo *E* 55 B7
El Bosque *E* 53 C6
El Bullaque *E* 46 F4
El Burgo *E* 53 C7
El Burgo de Ebro *E* 41 E10
El Burgo de Osma *E* 40 E5
El Burgo Ranero *E* 39 D9
El Buste *E* 41 E8
El Cabaco *E* 45 C8
El Cabildo y la Campana *E* 55 E9
El Cabo de Gata *E* 55 F8
El Campillo *E* 51 D6
El Campillo de la Jara *E* 45 E10
El Campo de Peñaranda *E* 45 C10
El Cañavate *E* 47 E8
El Carpio *E* 53 A8
El Carpio de Tajo *E* 46 E4
El Casar *E* 46 C6
El Casar de Escalona *E* 46 D3
El Castellar *E* 48 D3
El Castillo de las Guardas *E* 51 D7
El Catllar de Gaià *E* 43 E6

El Centenillo *E* 54 C5
El Cerro de Andévalo *E* 51 D6
Elche *E* 56 E3
Elche de la Sierra *E* 55 C8
Elchingen *D* 187 E9
Elciego *E* 41 C6
Elçili *TR* 172 B6
El Coronil *E* 51 E8
El Cubo de Don Sancho *E* 45 C8
El Cubo de Tierra del Vino *E* 39 F8
El Cuervo *E* 47 D10
El Cuervo *E* 51 F7
Elda *E* 56 E3
Eldalsosen *N* 100 D4
Eldena *D* 83 D9
Eldforsen *S* 97 B11
Eldingen *D* 83 E8
Eldsberga *S* 87 B12
Elduvik *FO* 2 A3
Elefsina *GR* 175 C8
Eleftheres *GR* 171 C6
Eleftherio *GR* 168 D4
Eleftheroupoli *GR* 171 C6
Eleja *LV* 134 D7
Elek *H* 151 D7
Elektrėnai *LT* 137 D10
Elemir *SRB* 158 C5
Elena *BG* 166 D5
Elend *D* 79 C8
Eleousa *GR* 168 E4
El Escorial *E* 46 C4
Eleshnitsa *BG* 165 F8
El Espinar *E* 46 C4
El Estrecho *E* 56 F3
Elexalde *E* 40 B6
El Fendek *MA* 53 E5
Elfershausen *D* 187 A8
El Fresno *E* 46 C3
Elgå *N* 101 B15
Elgin *UK* 3 K10
Elgoibar *E* 32 D1
Elgol *UK* 2 L4
El Grado *E* 42 C4
El Granado *E* 50 D5
El Grau de Borriana *E* 48 E4
Elgsnes *N* 111 C11
El Herrumblar *E* 47 F9
El Hito *E* 47 E7
El Hoyo de Pinares *E* 46 C4
Elia *GR* 178 B3
Elice *I* 62 B5
Elie *UK* 5 C11
Elika *GR* 178 B4
Elimäki *FIN* 127 D15
Elini *I* 64 D4
Elin Pelin *BG* 165 D8
Elisenvaara *RUS* 129 C12
Eliseyna *BG* 165 C8
Elizondo *E* 32 D2
Eljas *E* 45 D7
Ełk *PL* 136 F5
Elkenroth *D* 185 C8
Elkhovo *BG* 166 D5
Elkhovo *BG* 167 E7
Elkšņi *LV* 135 D11
Ellan *S* 99 B11
Elland *UK* 11 D8
Ellefeld *D* 75 B11
Ellenberg *D* 187 C9
Ellerau *D* 83 C7
Ellerbek *D* 83 C7
Ellesmere *UK* 10 F6
Ellesmere Port *UK* 10 E6
Ellezelles *B* 19 C8
Ellidshøj *DK* 86 B5
Elling *DK* 90 E7
Ellingen *D* 75 D8
Ellington *UK* 5 E13
El Llano *E* 38 B6
Ellmau *A* 72 A5
Ellös *S* 91 C9
Ellrich *D* 79 C8
Elloughton *UK* 11 D10
Ellwangen (Jagst) *D* 75 E7
Ellwürden *D* 17 B10
Elm *CH* 71 D8
Elm *D* 17 A12
Elmacik *TR* 167 F8
El Masroig *E* 42 E5
Elmen *A* 71 C11
Elmenhorst *D* 83 B12
Elmenhorst *D* 83 C8
Elmenhorst *D* 83 C10
El Molar *E* 46 C5
El Molinillo *E* 46 F4
El Moral *E* 55 D8
Elmshorn *D* 82 C7
Elne *F* 34 E4
Elnesvågen *N* 100 A6
Elorrio *E* 41 B6
Elorz *E* 32 E2
Elos *GR* 178 E6
El Oso *E* 46 C3
Els Prats de Rei *E* 43 D7
Elsa *EST* 131 E12
Elsdorf *D* 21 C7
Elsenborn *B* 20 D6
Elsendorf *D* 75 E11
Elsenfeld *D* 21 E12
El Serrat *AND* 33 E9
Elsfjord *N* 108 D6
Elsfleth *D* 17 B10
Elshitsa *BG* 165 D9
Elsloo *NL* 183 D7
Elsnes *N* 111 B19
Elsnig *D* 80 C3
Elsnigk *D* 79 C11
El Solerás *E* 42 E5
Elspeet *NL* 16 D5
Elst *NL* 16 E5
Elst *NL* 183 B6
Elster *D* 79 C12
Elsterberg *D* 79 E11
Elsterwerda *D* 80 D5
Elstra *D* 80 D6
El Tiemblo *E* 46 D4
Eltmann *D* 75 C8
El Toboso *E* 47 E7
Elton *UK* 11 F11
Elton *IRL* 8 D6
El Torno *E* 45 D9
El Trinchete *E* 46 F4
El Tumbalejo *E* 51 E6
Eltville am Rhein *D* 21 D10
Elva *EST* 131 E12
Elvål *N* 101 C14
Elvanfoot *UK* 5 E9
Elvas *P* 51 B5
Elvebakken *N* 113 C11
Elvegard *N* 111 D13
Elvekrok *N* 113 B17
Elvelund *N* 112 C16
Elvemund *N* 113 E16
Elven *F* 22 E6
El Vendrell *E* 43 E7
El Verger *E* 56 D5
Elverum *N* 101 E15
Elvesletta *N* 111 D12
Elvevoll *N* 111 B14
Elvevollen *N* 111 B18
El Villar de Arnedo *E* 32 F1
Elvington *UK* 11 D10
El Viso *E* 54 C3
El Viso del Alcor *E* 51 E8
Elvnes *N* 114 D8
Elx *E* 56 E3
Elxleben *D* 79 D8
Ely *UK* 13 D8
Ely *UK* 15 C9
Elz *D* 185 D9
Elzach *D* 27 D9
Elze *D* 78 B6
Elze (Wedemark) *D* 78 A6
Embid *E* 47 C9
Embid de Ariza *E* 41 F8
Embleton *UK* 5 E13
Embrun *F* 36 B5
Emburga *LV* 135 C7
Embüte *LV* 134 C3
Emden *D* 17 B8
Emecik *TR* 181 C7
Emiralem *TR* 177 B9
Emirali *TR* 173 B6
Emlichheim *D* 17 C7
Emly *IRL* 8 D6
Emmaboda *S* 89 B9
Emmaljunga *S* 87 C13
Emmanouil Pappas *GR* 169 C10
Emmaste *EST* 130 D7
Emmeloord *NL* 16 C5
Emmelshausen *D* 21 D9
Emmen *CH* 27 F9
Emmen *NL* 17 C7
Emmendingen *D* 27 D8
Emmer-Compascuum *NL* 17 C8
Emmerich *D* 16 E6
Emmoo *IRL* 6 E6
Emőd *H* 145 H2
Emolahti *FIN* 123 C15
Empel *NL* 16 E4
Empesos *GR* 174 A3
Empfingen *D* 187 E6
Empoli *I* 66 E2
Emponas *GR* 181 D7
Emporeio *GR* 179 C9
Emporeios *GR* 177 C9
Emptinne *B* 19 D11
Emsbüren *D* 17 D8
Emsdetten *D* 17 D9
Emsfors *S* 89 A10
Emskirchen *D* 75 C8
Emst *NL* 16 D5
Emstek *D* 17 C10
Emsworth *UK* 15 F7
Emtinghausen *D* 17 C11
Emtunga *S* 91 C12
Emyvale *IRL* 7 D9
Enafors *S* 105 E13
Enäjärvi *FIN* 128 D7
Enånger *S* 103 C13
Enarsvedjan *S* 105 D16
Encamp *AND* 33 E9
Encinas de Abajo *E* 45 C10
Encinasola *E* 51 C6
Encinas Reales *E* 53 B8
Encinedo *E* 39 D6
Enciso *E* 41 D7
Encs *H* 145 G3
Encsencs *H* 151 B9
Endingen *D* 27 D8
Endre *S* 93 D12
Endrefalva *H* 147 E9
Endriejavas *LT* 134 E3
Endrinal *E* 45 C9

Endzele LV 131 F10
Enebakk N 95 C14
Enego I 72 E4
Eneryda S 88 B6
Enese H 149 A8
Enez TR 171 C10
Enfesta E 38 C2
Enfield IRL 7 F9
Engan N 101 B11
Engan N 104 E4
Engan N 108 D3
Engares GR 176 E5
Engavågen N 108 C6
Engden D 183 A10
Engelberg CH 71 D6
Engelhartstetten A 77 F11
Engelhartszell A 76 E5
Engeln D 17 C11
Engelsberg D 75 F12
Engelskirchen D 21 C8
Engen D 27 E10
Enger D 17 D11
Engerdal N 101 C15
Engerdalssetra N 101 C15
Engerneset N 102 C3
Engersen D 79 A9
Engesland N 90 B3
Engesvang DK 86 C4
Enghien B 19 C9
Engis B 19 C11
Englefontaine F 19 D8
Engomer F 33 E8
Engstingen D 27 D11
Enguera E 56 D3
Enguidanos E 47 E9
Engure LV 134 B6
Engvik N 112 C2
Enhagen-Ekbacken S 98 C7
Enichioi MD 154 E2
Enina BG 166 D4
Eningen unter Achalm D 187 E7
Enitsa BG 165 C9
Enix E 55 F7
Enkenbach D 186 C4
Enkhuizen NL 16 C4
Enkirch D 21 E8
Enklinge FIN 99 B15
Enköping S 98 C8
Enmo N 101 A12
Enna I 58 D5
Ennepetal D 185 B7
Ennery F 186 C1
Ennezat F 30 D3
Ennigerloh D 17 E10
Ennis IRL 8 C5
Enniscorthy IRL 9 C9
Enniscrone IRL 6 D4
Enniskean IRL 8 E5
Enniskerry IRL 7 F10
Enniskillen UK 7 D7
Ennistymon IRL 8 C4
Enns A 77 F6
Eno FIN 125 E14
Enonkoski FIN 125 F11
Enonkylä FIN 119 F17
Enontekiö FIN 117 B11
Ens NL 16 C5
Enschede NL 17 D7
Ensdorf D 75 D10
Ensdorf D 186 C2
Ense D 17 F10
Ensheim D 21 F8
Ensisheim F 27 E7
Enskogen S 103 C9
Ensuès-la-Redonne F 35 D9
Enter NL 183 A9
Enterkinfoot UK 5 E9
Entlebuch CH 70 D6
Entracque I 37 C6
Entradas P 50 D3
Entraigues-sur-la-Sorgue F 35 B8
Entrains-sur-Nohain F 25 F9
Entrammes F 23 E10
Entraunes F 36 C5
Entraygues-sur-Truyère F 33 A11
Entrena E 41 D6
Entrevaux F 36 D5
Entroncamento P 44 F4
Envendos P 44 E5
Envermeu F 18 E3
Enviken S 103 E10
Enying H 149 C10
Enzklösterle D 27 C9
Eochaill IRL 9 E7
Eoropaidh UK 2 H4
Épaignes F 18 F1
Epalinges CH 31 B10
Epano Fellos GR 176 D4
Epanomi GR 169 D8
Epe NL 16 D5
Eperjes H 150 D6
Eperjeske H 145 G5
Épernay F 25 B10
Épernon F 24 C6
Epfendorf D 27 D10
Épieds-en-Beauce F 24 E6
Épierre F 31 E9
Épila E 41 E9
Épinac F 30 B6
Épinal F 26 D5
Epiry F 25 F10
Episcopia I 60 C6
Episkopi GR 169 C7
Episkopi GR 178 E7
Episkopi GR 178 E9
Epitalio GR 174 D3
Époisses F 25 E11
Epoo FIN 127 E14
Eppelborn D 21 F7
Eppelheim D 21 F11
Eppenbrunn D 186 C4
Eppertshausen D 21 E11
Eppeville F 19 E7
Epping UK 15 D9
Eppingen D 27 B10
Eppishausen D 71 A11
Eppstein D 21 D10
Epsom UK 15 E8
Eptachori GR 168 D5
Épuisay F 24 E4
Epureni RO 153 E11
Epworth UK 11 D10
Équeurdreville-Hainneville F 23 A8
Eraclea I 73 E6
Eraclea Mare I 67 A6
Eräjärvi FIN 127 B12
Eräjärvi FIN 129 C12
Erastvere EST 131 F13
Erateini GR 174 C5
Eratyra GR 169 D6

Erba I 69 B7
Erbach D 21 E12
Erbach D 74 F6
Erbendorf D 75 C11
Érberge LV 135 D10
Erbiceni RO 153 C10
Erbray F 23 E9
Erchie I 61 C9
Ercsi H 149 B11
Érd H 149 B11
Erdeborn D 79 D10
Erdek TR 173 D8
Erdelli TR 177 B10
Erdeven F 22 E5
Erdevik SRB 158 C3
Erding D 75 F10
Erdőkertes H 150 B3
Erdut HR 157 A11
Erdweg D 75 F9
Erdzelija NMK 164 F5
Éréac F 23 D7
Eremitu RO 152 D5
Eremiya BG 165 E6
Erenler TR 173 D10
Eresos GR 177 A6
Eretreia GR 169 F8
Eretria GR 175 C8
Erfde D 82 B6
Erftstadt D 21 C7
Erfurt D 79 E9
Érgeme LV 131 F11
Ergersheim D 75 C7
Ergili TR 173 D9
Ērgļi LV 135 C11
Ergoldsbach D 75 E11
Ergué-Gabéric F 22 D3
Erharting D 75 F12
Eriboll UK 2 J7
Erice E 32 E2
Erice I 58 C2
Ericeira P 50 B1
Eriklice TR 173 C7
Eriksberg S 107 D9
Erikslund S 103 A10
Eriksmåla S 89 B8
Erikssund S 99 C9
Ering D 76 F4
Eringsboda S 89 C8
Eriswil CH 70 C5
Erjova MD 154 B4
Erkelenz D 20 B6
Erkheim D 71 A10
Erkner D 80 B5
Erla E 41 D10
Erlach CH 31 A11
Erlangen D 75 C9
Erlauf A 77 F8
Erlbach D 75 B11
Erlenbach am Main D 187 B7
Erli I 37 C8
Erlsbach A 72 C5
Ermelo NL 16 D5
Ermelo P 38 F4
Ermenonville F 25 B8
Ermesinde P 44 B3
Ermidas do Sado P 50 C3
Ermioni GR 175 E7
Ermoclia MD 154 D5
Ermoupoli GR 176 E4
Ermsleben D 79 C9
Ermua E 41 B6
Erndtebrück D 21 C10
Ernée F 23 D10
Ernei RO 152 D5
Ernestinovo HR 149 F11
Ernstbrunn A 77 E10
Erp NL 16 E5
Érpatak H 151 B8
Erpel D 21 C8
Erquelinnes B 184 D1
Erquy F 22 C7
Erra P 50 A3
Errenteria E 32 D1
Errezil E 32 D1
Erribera E 41 B7
Errill IRL 9 C7
Errindlev DK 83 A11
Erritsø DK 86 D5
Errogie UK 3 L8
Érsekcsanád H 150 E2
Érsekė AL 168 D4
Érsekvadkert H 147 E8
Ersfjord N 111 B13
Ersmark S 118 E5
Ersmark S 122 C4
Ersnäs S 118 C7
Erstein F 27 D8
Erstfeld CH 71 D7
Ersvik N 108 B7
Ertenvåg N 108 B7
Ertingen D 71 A8
Erto I 72 D5
Ertsjärv S 116 E8
Ertuğrul TR 173 E8
Ertvelde B 182 C3
Erula I 64 B2
Ervalla S 97 D13
Ervasti FIN 121 D9
Ervedosa do Douro P 44 B6
Ervenik HR 156 D5
Ervidel P 50 D3
Ervik N 104 E5
Ervillers F 18 D6
Ervy-le-Châtel F 25 D10
Erwitte D 17 E11
Erxleben D 79 B9
Erxleben D 83 E11
Erythres GR 175 C7
Erzhausen D 21 E11
Erzvilkas LT 134 F5
Esanatoglia I 67 F6
Esbjerg DK 86 E2
Esbo FIN 127 E12
Esbønderup DK 87 C10
Escairón E 38 C4
Escalante E 40 B5
Escalaplano I 64 D3
Escalhão P 45 C7
Escaló E 33 E8
Escalona E 46 D4
Escalona del Prado E 46 B4
Escalonilla E 46 E4
Escalos de Baixo P 44 E6
Escalos de Cima P 44 E6
Escamilla E 47 C8
Es Canar E 57 C8
Escañuela E 53 A8
Escariche E 47 D6
Escárigo P 45 C6
Escároz E 32 E3
Es Castell E 57 B13

Escatrón E 42 E3
Esch NL 183 B6
Eschau D 187 B7
Eschau F 186 E4
Eschborn D 187 A6
Esche D 17 C7
Escheburg D 83 D8
Eschede D 83 E8
Eschenbach CH 27 F9
Eschenburg-Eibelshausen D 21 C10
Eschenstruth (Helsa) D 78 D6
Eschershausen D 78 C6
Eschlkam D 76 D3
Escholzmatt CH 70 D5
Esch-sur-Alzette L 20 F5
Esch-sur-Sûre L 20 E5
Eschwege D 79 D7
Eschweiler D 20 C6
Escombreras E 56 F3
Escorca E 49 E10
Escorihuela E 42 F2
Escos F 32 D3
Escoublac F 22 F6
Escource F 32 B3
Escout F 32 D4
Escragnolles F 36 D5
Escucha E 42 F2
Escurial E 45 F9
Escúzar E 53 B9
Esechioi RO 161 E10
Esence TR 173 D10
Esenköy TR 173 B8
Esenler TR 173 B10
Esens D 17 A9
Esenyurt TR 173 B10
Esgos E 38 D4
Esgueira P 44 C3
Esguevillas de Esgueva E 40 E3
Eskdalemuir UK 5 E10
Eskelhem S 93 E12
Eskiçine TR 181 A8
Eskilstorp S 87 E13
Eskilstrup DK 84 A11
Eskilstuna S 98 D7
Eskioba TR 177 C10
Eskişehir TR 173 D9
Eskisigirçi TR 173 D9
Eskogen S 99 C10
Eskola FIN 123 C12
Eskoriatza E 41 B6
Esku EST 131 D11
Eslarn D 75 C12
Eslava E 32 E2
Eslohe (Sauerland) D 17 F10
Eslöv S 87 D12
Es Mercadal E 57 B13
Es Migjorn Gran E 57 B13
Esnes F 19 D7
Esnouveaux B 19 C12
Espa N 101 E14
Espadañedo E 39 D7
Espaly-St-Marcel F 30 E4
Esparragalejo E 51 B7
Esparragosa E 51 B9
Esparragosa de la Serena E 51 B8
Esparreguera E 43 D7
Esparron F 35 C10
Esparza de Salazar E 32 E3
Espe DK 86 E6
Espedal N 94 E4
Espeja E 45 C7
Espeja de San Marcelino E 40 E5
Espejo E 53 A7
Espejón E 40 E5
Espel NL 16 C5
Espeland N 94 B2
Espelette F 32 D3
Espelkamp D 17 D11
Espenau D 78 D5
Espenes N 111 B14
Espera E 51 E8
Esperança P 45 F6
Espéraza F 33 E10
Espergærde DK 87 D11
Esperia I 62 E5
Espevik N 94 C2
Espezel F 33 E10
Espiel E 54 C3
Espinama E 39 B10
Espinasses F 36 B4
Espinhal P 44 D4
Espinho P 44 B3
Espinilla E 40 B3
Espinosa de Cerrato E 40 E4
Espinosa de Henares E 47 C6
Espinosa de los Monteros E 40 B4
Espinoso del Rey E 46 E3
Espira-de-l'Agly F 34 E4
Espírito Santo P 50 D4
Espite P 44 E3
Esplantas F 30 F4
Esplegares F 47 C8
Esplús E 42 D4
Espolla E 34 F5
Espondeilham F 34 D5
Esponellà E 43 C9
Espoo FIN 127 E12
Espoonlahti FIN 127 E12
Esporlatu I 64 C2
Esporles E 49 E10
Esposende P 38 E2
Espot E 33 E8
Esprels F 26 E5
Esquivias E 46 D5
Esrange S 116 C6
Esrum DK 87 C10
Essay F 23 C12
Essel D 82 E7
Essen B 16 F2
Essen D 17 F8
Essen (Oldenburg) D 17 C9
Essenbach D 75 E11
Essert F 27 E6
Essingen D 74 E7
Essingen S 97 B12
Essoyes F 25 D12
Essu EST 131 C12
Essvik S 103 B13
Estación de Baeza E 54 C5
Estadilla E 42 C4
Estagel F 34 E4
Estaimpuis B 19 C7
Estaing F 33 A11
Estaires F 18 C5
Estang F 32 C5
Estarreja P 44 C3
Estavayer-le-Lac CH 31 B10
Este I 66 B4
Este P 38 E3
Estela P 38 F2

Estella E 32 E1
Estellencs E 49 E10
Estenfeld D 187 B9
Estepa E 53 B7
Estepar E 40 D4
Estepona E 53 D6
Esteras de Medinaceli E 47 B8
Estercuel E 42 F2
Esternay F 25 C10
Esterri d'Aneu E 33 E8
Esterwegen D 17 C9
Esterzili I 64 D3
Esztergom H 149 A11
Étables-sur-Mer F 22 C6
Étagnac F 29 D7
Étain F 19 F12
Étais-la-Sauvin F 25 E9
Étalle B 19 E12
Étampes F 24 D7
Étang-sur-Arroux F 30 B5
Étaples F 15 F12
Étaules F 28 D3
Étauliers F 28 E4
Eteläinen FIN 127 C12
Etelä-Vartsala FIN 126 D5
Etelhem S 93 E13
Éternoz F 31 A9
Etes H 147 E9
Ethe B 19 E12
Etholmen N 91 A9
Etili TR 173 E6
Étival F 31 B8
Étival-Clairefontaine F 27 D6
Étivey F 25 E11
Etna N 101 D11
Etne N 94 C3
Étréchy F 24 D7
Étrépagny F 18 F4
Étretat F 23 A12
Étreux F 19 E8
Étrín Bajo E 51 B6
Etropole BG 165 D9
Etroubles I 31 D11
Ettelbruck L 20 E6
Ettenheim D 27 D8
Etten-Leur NL 16 E3
Ettington UK 13 A11
Ettrick UK 5 E10
Ettringen D 21 D8
Etulia MD 155 B2
Étupes F 27 E6
Étuz F 26 F4
Etwall UK 11 F8
Etxalar E 32 D2
Etxarri-Aranatz E 32 E1
Etyek H 149 B11
Eu F 18 D3
Eubigheim D 74 C6
Euerdorf D 74 B7
Eugénie-les-Bains F 32 C5
Eugmo FIN 122 C9
Eulate E 32 E1
Eupen B 20 C6
Euper D 79 C12
Eura FIN 126 C7
Eurajoki FIN 126 C6
Eurasburg D 72 A3
Eurville-Bienville F 26 C3
Euskirchen D 21 C7
Eußenheim D 74 B6
Eutin D 83 B9
Eutingen im Gäu D 187 E6
Eutzsch D 79 C12
Eva FIN 126 C4
Eva DK 174 E4
Evanger N 100 E4
Evanton UK 3 K8
Évaux-les-Bains F 29 C10
Evciler TR 173 E6
Evciler TR 173 E7
Evdilos GR 177 D7
Évele LV 131 F11
Evenes N 111 D12
Evenesdal N 109 C10
Evenset N 108 B8
Evenskjær N 111 C12
Evenstad N 101 C14
Everdingen NL 183 B6
Everdrup DK 87 E9
Evergem B 19 B8
Everöd S 88 D6
Eversberg S 102 D6
Evertsberg S 102 D6
Evesham UK 13 A11
Évian-les-Bains F 31 C10
Evijärvi FIN 123 D10
Evitskog FIN 127 E11
Evje N 90 B2
Évlalo GR 171 C7
Evolène CH 31 C12
Evora P 50 B4
Évora-Monte P 50 B4
Evosmos GR 169 C8
Évran F 23 D8
Évrecy F 23 B10
Evrencik TR 173 A8
Evrenli TR 173 A8
Evrensekiz TR 173 B7
Évron F 23 D11
Evropos GR 169 C7
Evrostina GR 174 C5
Évry F 25 C7
Evzonoi GR 169 B8
Ewhurst UK 15 E8
Examilia GR 175 D6
Exaplatanos GR 169 C7
Exarchos GR 175 B6
Excideuil F 29 E8
Exebridge UK 13 C7
Exeter UK 13 D7
Exilles I 31 E10
Exloo NL 17 C7

Exmes F 23 C12
Exminster UK 13 D8
Exmouth UK 13 D8
Exochi GR 169 B10
Exochi GR 169 D7
Exo Chora GR 174 D2
Exo Vathy GR 180 C4
Extremo P 38 E3
Eydehavn N 90 B4
Eydelstedt D 17 C11
Eye UK 11 F11
Eye UK 15 C11
Eyemouth UK 5 D12
Eyeries IRL 8 E3
Eyguians F 35 B10
Eyguières F 35 C8
Eygurande-et-Gardedeuil F 29 E6
Eymet F 33 A6
Eymoutiers F 29 D9
Eynsham UK 13 B12
Eyragues F 35 C8
Eyrans F 28 E4
Eyrecourt IRL 6 F6
Eyrein F 29 E9
Eyzahut F 35 A8
Eyzin-Pinet F 31 E6
Ezcaray E 40 D6
Ezerche BG 166 B6
Ezere LV 134 D4
Ezerėlis LT 137 D8
Ezerets BG 155 F3
Ezerets BG 165 F7
Ezeriş RO 159 C8
Ezerkalne LV 134 B5
Ezerkalni LV 133 D1
Ezernieki LV 133 D3
Ezine TR 171 E10
Ezkurra E 32 D2
Ézy-sur-Eure F 24 C5

F

Faaborg DK 86 E6
Fabara E 42 E4
Fabas F 33 D8
Fåberg N 100 C6
Fåberg N 101 D12
Fabero E 39 C6
Fábiánháza H 145 H5
Fabianki PL 138 E7
Fábiánsebestyén H 150 D5
Fåbodliden S 107 B13
Fåborg DK 86 D3
Fabrègues F 35 C6
Fabrezan F 34 D4
Fabriano I 67 F6
Fabrica di Roma I 62 C2
Fábricas de San Juan de Alcaraz E 55 B8
Fabrizia I 59 C9
Fabro I 62 B2
Făcăeni RO 155 D1
Faches-Thumesnil F 182 D2
Facinas E 52 D5
Fadd H 149 D11
Faedis I 73 D7
Faenza I 66 D4
Faeto I 60 A4
Fafe P 38 F3
Fagagna I 73 D7
Făgăraș RO 152 F5
Fågelfors S 89 A9
Fågelmara S 89 C9
Fågelsjö S 102 C8
Fågelsta S 92 C6
Fågelsundet S 103 E14
Fagerås S 97 D9
Fagerdal S 106 D8
Fagerhaug N 101 A11
Fagerhult S 89 A9
Fagerhult S 92 C11
Fagerhult S 92 C4
Fagerli N 109 B11
Fagernes N 101 E10
Fagernes N 111 A17
Fagersanna S 92 C4
Fagersta S 97 C14
Fagervik FIN 127 E11
Fagervika N 108 D4
Fagerviken S 99 A9
Făget RO 151 F9
Făgetelu RO 160 D5
Fäggeby S 97 B14
Fåglavik S 91 C13
Fagnano Castello I 60 D6
Fagnières F 25 C11
Fahan IRL 4 E2
Fahrdorf D 82 A7
Fahrenkrug D 83 C8
Fahrenzhausen D 75 F10
Fahrland D 80 B4
Faicchio I 60 A2
Faid D 21 D8
Faido CH 71 E7
Failde P 39 E6
Faimes B 19 C11
Fain-lès-Montbard F 25 E11
Fains-Véel F 26 C3
Fairlie UK 4 D7
Fair Oak UK 13 D12
Faissault F 19 E10
Fajão P 44 D5
Fajarda P 50 B2
Fajsławice PL 144 A6
Fajsz H 149 D11
Fakenham UK 15 B10
Fåker S 106 F7
Fakija BG 167 E8
Fakovići BIH 158 E4
Faksdal N 105 B10
Fakse DK 87 E10
Fakse Ladeplads DK 87 E10
Falaise F 23 C11
Falani GR 169 E7
Fålasjö S 107 E13
Falces E 32 F2
Fălciu RO 154 E2
Falck F 186 C2
Fălcoiu RO 160 E5
Falconara I 59 C7
Falconara Albanese I 60 E6
Falconara Marittima I 67 E7
Falcone I 59 C7
Faldsled DK 86 E6
Fale N 101 A9
Faleria I 62 C2
Falerna I 59 A9
Falerone I 62 A4
Falerum S 93 C8
Fălești MD 153 B11
Faux F 29 F7

Falkekeila N 114 C8
Falkenberg D 75 C11
Falkenberg D 75 F12
Falkenberg D 80 C4
Falkenberg S 87 B11
Falkenhagen D 83 D12
Falkenhain D 79 D12
Falkensee D 80 A4
Falkenstein D 75 D11
Falkenstein D 75 D11
Falkirk UK 5 D9
Falkland UK 5 C10
Falköping S 91 C14
Fall N 97 A8
Falla S 92 B7
Falleron F 28 B2
Fällfors S 118 D5
Fålloheden S 109 E15
Falltorp S 97 A8
Falmenta I 68 A6
Falmouth UK 12 E4
Falset E 42 E5
Falstad N 105 D10
Fälticeni RO 153 C8
Faltrisk S 107 B15
Falun S 103 E10
Falur S 103 E10
Famalicão P 44 D6
Fambach D 79 E7
Fameck F 20 F6
Fámjin FO 2 B3
Fana N 94 B2
Fanano I 66 D2
Fanari GR 171 C8
Fanbyn S 107 E11
Fångsjöbacken S 107 E11
Fångsjön S 107 E12
Fanjeaux F 33 D10
Fanlo E 32 E5
Fannrem N 104 E7
Fano I 67 E7
Fanore IRL 6 F4
Fânos GR 168 C5
Fântânele RO 151 E7
Fântânele RO 152 E5
Fântânele RO 153 B9
Fanthyttan S 97 C13
Fåo P 38 E2
Felde D 83 B7
Feldebrő H 147 F10
Feldioara RO 153 F7
Feldkirch A 71 C9
Feldkirchen in Kärnten A 73 C9
Feldkirchen-Westerham D 72 A4
Feldru RO 152 C5
Feleacu RO 152 D3
Felgar P 45 B7
Felgueiras P 38 F3
Felgyő H 150 D5
Feliceni RO 152 E6
Felindre UK 13 A8
Felino I 66 C1
Felitto I 60 C4
Félix E 55 F7
Felixdorf A 77 G10
Felixstowe UK 15 D11
Felizzano I 37 B8
Fell D 21 E7
Fellbach D 27 C11
Felletin F 29 D10
Fellingsbro S 97 D14
Felnac RO 151 E7
Felsberg D 21 B12
Felsődobsza H 145 G3
Felsőlajos H 150 C4
Felsőnyék H 149 C10
Felsőpakény H 150 C3
Felsőszentiván H 150 E3
Felsőszentmárton H 149 E9
Felsőtárkány H 145 H1
Felsőzsolca H 145 G2
Felsted DK 86 F4
Felton UK 5 E13
Feltre I 72 D4
Feltwell UK 11 F12
Femundsenden N 101 C15
Femundsundet N 101 C15
Fénay F 26 F3
Fendeille F 33 D9
Fene E 38 B3
Fenerköy TR 173 B9
Fenes N 111 C12
Fenestrelle I 31 E11
Fénétrange F 27 C7
Feneu F 23 E10
Fengersfors S 91 B11
Fenioux F 28 B5
Fenis I 31 D11
Fenit IRL 8 D3
Feniton UK 13 D8
Fensmark DK 87 E9
Fenstanton UK 15 C8
Fenwick UK 5 D13
Fényeslitke H 145 G5
Feofilova Pustyn' RUS 132 E5
Feolin Ferry UK 4 D4
Feonanagh IRL 8 D5
Fépin F 19 D10
Ferbane IRL 7 F7
Ferdinandea I 59 B9
Ferdinandovac HR 149 D8
Ferdinandshof D 84 C5
Fère-Champenoise F 25 C10
Fère-en-Tardenois F 25 B10
Férel P 23 E7
Ferentillo I 62 B3
Ferentino I 62 D4
Feres GR 171 C10
Férez E 55 C9
Feria E 51 B6
Feričanci HR 149 E9
Ferizaj RKS 164 E3
Ferlach A 73 C9
Fermignano I 66 E6
Fermo I 67 F8
Fermoselle E 39 F7
Fermoy IRL 8 D6
Fernáncaballero E 46 F5
Fernán Núñez E 53 A7
Ferndown UK 13 D11
Ferness UK 3 K9
Ferney-Voltaire F 31 C9
Fernitz A 148 C5
Ferns IRL 9 C9
Feroleto Antico I 59 B9
Ferrals-les-Corbières F 34 D4
Ferrandina I 61 C6
Ferrara I 66 C4
Ferrazzano I 63 D7
Ferreira E 38 A5
Ferreira do Alentejo P 50 C3

Ferreira do Zêzere P 44 E4
Ferrel P 44 F2
Ferreras de Abajo E 39 E7
Ferreras de Arriba E 39 E7
Ferrere I 37 B7
Ferreries E 57 B13
Ferreruela de Huerva E 47 B10
Ferreruela de Tábara E 39 E7
Ferrette F 27 E7
Ferrière I 37 B11
Ferrières B 19 D12
Ferrières F 25 D8
Ferritslet DK 86 E7
Ferrol E 38 B3
Ferrybank IRL 9 C10
Ferryhill UK 5 F13
Ferschweiler D 185 E5
Fertilia I 64 B1
Fertőd H 149 A7
Fertőrákos H 149 A7
Fertőszentmiklós H 149 A7
Ferwert NL 16 B5
Feshiebridge UK 3 L9
Fessenheim F 27 E8
Festieux F 19 E8
Festvåg N 108 B8
Fetești MD 153 A10
Fetești RO 155 E1
Fetești-Gară RO 155 E1
Fethard IRL 9 D7
Fethard IRL 9 D9
Fetsund N 95 C14
Fettercairn UK 5 B11
Feucht D 75 D9
Feuchtwangen D 75 D7
Feugarolles F 33 B6
Feuquières F 18 E4
Feuquières-en-Vimeu F 18 D4
Feurs F 30 D5
Fevåg N 104 D7
Fevik N 90 C4
Feytiat F 29 D8
Fiac F 33 C9
Fiamignano I 62 C4
Fiano I 31 E12
Fiano Romano I 62 C3
Fiavè I 69 B10
Ficarazzi I 58 C3
Ficarolo I 66 C3
Fichtelberg D 75 C10
Fichtenberg D 74 E6
Fichti GR 175 D6
Ficulle I 62 B2
Fiddown IRL 9 D8
Fide S 93 E12
Fidenza I 69 D9
Fidjeland N 94 E5
Fiè allo Sciliar I 72 C4
Fieberbrunn A 72 B6
Fieni RO 160 C6
Fier AL 168 C2
Fiera di Primiero I 72 D4
Fierbinți-Târg RO 161 D8
Fierzë AL 163 E9
Fiesch CH 70 E6
Fiesole I 66 E3
Fiesso Umbertiano I 66 C4
Figaniéres F 36 D4
Figari F 37 J10
Figeac F 33 A10
Figeholm S 93 E9
Figgjo N 94 E3
Figline Valdarno I 66 E3
Figueira da Foz P 44 D3
Figueira de Castelo Rodrigo
P 45 C7
Figueira e Barros P 44 F5
Figueiredo de Alva P 44 C5
Figueiró da Granja P 44 C6
Figueiró dos Vinhos P 44 E4
Figueres E 43 C9
Figueroles E 48 D4
Figueruela de Arriba E 39 E7
Fijnaart NL 182 B5
Filadelfia I 59 B9
Filakio GR 171 A10
Fiľakovo SK 147 E9
Filattiera I 69 E8
Fildu de Jos RO 151 D11
Filettino I 62 D4
Filey UK 11 C11
Filiano I 60 B5
Filiași RO 160 D3
Filiates GR 168 E3
Filicudi Porto I 59 B6
Filipeni RO 153 D10
Filipești RO 153 D9
Filipeștii de Pădure RO 161 D7
Filipeștii de Târg RO 161 D7
Filipjakov HR 156 E3
Filipovtsi BG 165 B6
Filipów PL 136 E6
Filippiada GR 174 A2
Filipstad S 97 C11
Filisur CH 71 D9
Fillan N 104 D5
Fillinges F 31 C9
Fillyra GR 171 B9
Filotas GR 169 C6
Filotio GR 177 E6
Filottrano I 67 F7
Filsbäck S 91 C13
Filskov DK 86 D4
Filsum D 17 B9
Filton UK 13 B9
Filzmoos A 73 B8
Fimber UK 11 C10
Finale I 58 C5
Finale Emilia I 66 C3
Finale Ligure I 37 C8
Fiñana E 55 E7
Finavon UK 5 B11
Finby FIN 99 B14
Finby FIN 127 E8
Findern UK 11 F8
Findhorn UK 3 K9
Findikli TR 172 D6
Findikli TR 173 E8
Findochty UK 3 K11
Findon UK 15 F8
Finestrat E 56 D4
Finhan F 33 C8
Finiq AL 168 E3
Finiș RO 151 D9
Finja S 87 C13
Finkenstein A 73 C8
Finnbacka S 103 D10
Finnbodarna S 102 E7
Finnböle S 98 B8
Finnbyen N 110 D5
Finnea IRL 7 E8
Finneby S 103 B9
Finnentrop D 21 B9
Finnerödja S 92 B4
Finnes N 108 C6
Finnes N 111 B13
Finnfjordeidet N 111 B15
Finnfors S 118 E4
Finngruvan S 102 E7
Finnholt N 96 B6
Finningham UK 15 C11
Finnkroken N 111 A17
Finnøy N 111 D9
Finnsådal N 90 C2
Finnsäter S 105 D15
Finnset N 101 B8
Finnset N 108 B7
Finnset N 111 C10
Finnseter N 111 C10
Finnshögst S 103 D9
Finnskog N 102 E3
Finnsnes N 111 B14
Finntorp S 96 D7
Finnträsk S 118 D6
Finnvik N 111 A16
Finnvika N 108 D4
Finnygaud UK 3 K11
Fino Mornasco I 69 B7
Finowfurt D 84 E5
Finse N 100 E7
Finsjö S 89 A10
Finspång S 92 B7
Finsterwalde D 80 C5
Finsterwolde NL 17 B8
Finström FIN 99 B13
Finta RO 161 D7
Fintel D 82 D7
Fintice SK 145 E3
Fintona UK 7 D8
Fintown IRL 6 C6
Finuge IRL 8 D3
Finvoy UK 4 E3
Fiodh Ard IRL 9 D7
Fionnphort UK 4 C4
Fiorano Modense I 66 C2
Fiorenzuola d'Arda I 69 D8
Fira GR 179 C9
Firenze I 66 E3
Firenzuola I 66 D3
Firlej PL 141 G7
Firmi F 33 A10
Firminy F 30 E5
Firmo I 60 D6
Fiscal E 32 E5
Fischach D 75 F8
Fischbach A 148 B5
Fischbach D 21 E8
Fischbachau D 72 A4
Fischbach bei Dahn D 27 B8
Fischbeck D 79 A11
Fischen im Allgäu D 71 C10
Fiskå N 100 B3
Fiskardo GR 174 C2
Fiskari FIN 127 E10
Fiskarheden S 102 D5
Fiskebäckskil S 91 C9
Fiskebøl N 110 D8
Fiskefjord N 110 C9
Fiskö FIN 126 E4
Fiskøy N 111 C11
Fisksätra S 99 D10
Fismes F 19 F8
Fisterra E 38 C1
Fitero E 41 D8
Fitionești RO 153 F10
Fitjar N 94 C2
Fitou F 34 E4
Fitz-James F 18 F5
Fiuggi I 62 D4
Fiumalbo I 66 D2
Fiumefreddo Bruzio I 60 E6
Fiumefreddo di Sicilia I 59 D7
Fiumicino I 62 D2
Fiunary UK 4 B5
Fivemiletown UK 7 D8
Fivizzano I 66 D1
Fix-St-Geneys F 30 E4
Fizeșu Gherlii RO 152 C3
Fjær N 108 B8
Fjäl S 106 E7
Fjälkinge S 88 C6
Fjäll S 96 C7
Fjällåkern S 107 E14
Fjällåsen S 116 C4
Fjällbacka S 91 B9
Fjällnäs S 102 A3
Fjällsjö S 107 E10
Fjärås Kyrkby S 91 E11
Fjätervålen S 102 C5
Fjell N 94 B2
Fjellbu N 111 D14
Fjelldal N 111 C12
Fjelle N 111 C15
Fjellerup DK 86 B7
Fjellnes N 108 B9
Fjellstad N 111 C15
Fjelsø DK 86 B4
Fjelsted DK 86 E6
Fjelstervang DK 86 C3
Fjelstrup DK 86 E5
Fjennesev DK 86 E4
Fjerritslev DK 86 A4
Fjone N 90 A3
Fjordgard N 111 A14
Fjørvikan N 104 E4
Fjugesta S 92 A5
Flå N 95 B10
Flachau A 73 B7
Flachsmeer D 17 B8
Fladungen D 75 A7
Flagnac F 33 A10
Flaine F 31 C10
Flaka FIN 99 B14
Flakaberg S 116 E7
Flakaträsk S 107 D16
Flakaträsk S 109 F12
Flakk N 104 E8
Flakkebjerg DK 87 E8
Flakstad N 111 D9
Flåm N 100 E6
Flamanville F 23 A8
Flămânzi RO 153 B9
Flamborough UK 11 C11
Flamouria GR 169 C10
Flampouro GR 169 C10
Flärke S 107 D16
Flärken S 118 F5
Flästa S 103 D11
Flatabø N 94 B4
Flåten N 112 D6
Flatøydegard N 101 E11
Flått N 105 B12
Flattach A 73 C7
Flattnitz A 73 C9
Flatval N 104 D5
Flauenskjold DK 86 A6
Flavigny-sur-Moselle F 26 C5
Flavigny-sur-Ozerain F 25 E12
Flavin F 34 B4
Flawil CH 27 F11
Flayosc F 36 D4
Flechtingen D 79 B9
Flecken Zechlin D 83 D13
Fleet UK 15 E7
Fleetmark D 83 E10
Fleetwood UK 10 D5
Flein D 27 B11
Flekkefjord N 94 F5
Flémalle B 19 C11
Flen S 93 A9
Flensburg D 82 A6
Flenstad N 104 D8
Fleringe S 93 D13
Flermoen N 102 D4
Flerohopp S 89 B9
Fléron B 19 C12
Flers F 23 C10
Flers-en-Escrebieux F 182 E2
Flesberg N 95 C10
Flesnes N 111 C10
Flessau D 83 E11
Flesvik N 104 E6
Fleurance F 33 C7
Fleuré F 29 C7
Fleurie F 30 C6
Fleurier CH 31 B10
Fleurus B 19 D10
Fleury F 34 D5
Fleury-les-Aubrais F 24 E6
Fleury-sur-Andelle F 18 F3
Fléville-Lixières F 19 F12
Flieden D 74 B6
Fließ A 71 C11
Flimby UK 5 F9
Flims CH 71 D8
Flines-lez-Raches F 182 E2
Flint UK 10 E5
Flintbek D 83 B8
Flirey F 26 C4
Flisa N 102 E3
Flisby S 92 D5
Fliseryd S 89 A10
Flix E 42 E5
Flixecourt F 18 D5
Flize F 19 E10
Floby S 91 C13
Floda S 91 D11
Floda S 97 A12
Flodden UK 5 D12
Flogned S 96 C7
Floh D 79 E7
Flöha D 80 E4
Floing F 19 E10
Floirac F 28 F4
Flokenes N 100 D2
Floq AL 168 C4
Flor S 103 B9
Florac F 35 B6
Florange F 20 F6
Floreffe B 19 D10
Florennes B 19 D10
Florenville B 19 E11
Flores de Ávila E 45 C10
Floresta I 59 D6
Florești MD 154 B2
Florești RO 152 D3
Florești RO 159 D10
Florești RO 161 C7
Florești-Stoenești RO 161 E7
Floriada GR 174 A3
Florida de Liébana E 45 B9
Floridia I 59 E7
Florina GR 168 C5
Florinas I 64 B2
Florø N 100 C2
Flörsbach D 74 B5
Flörsheim am Main D 187 A5
Flörsheim-Dalsheim D 187 B5
Flosjöhyttan S 97 C12
Flossenbürg D 75 C11
Flöthe D 79 B7
Flötningen S 102 C3
Flötuträsk S 118 D5
Floure F 33 D10
Flovika N 110 D9
Fluberg N 101 E12
Flüelen CH 71 D7
Flühli CH 70 D6
Flumeri I 60 A4
Flumet F 31 D10
Fluminimaggiore I 64 E2
Flums CH 27 F11
Fluren S 103 C11
Flurkmark S 122 C4
Flussio I 64 C2
Fly DK 86 C4
Flyggsjö S 107 D15
Flyinge S 87 D12
Flykälen S 106 D8
Flyn S 107 E10
Fobello I 68 B5
Foča BIH 157 F10
Foça TR 177 B8
Focant B 19 D11
Fochabers UK 3 K10
Focșani RO 153 F10
Focuri RO 153 C10
Foeçy F 24 F7
Foeni RO 159 B6
Foggia I 63 E9
Foglianise I 60 A3
Föglö FIN 99 B14
Fohnsdorf A 73 C10
Fohrde D 79 B11
Föhren D 21 E7
Foiano della Chiana I 66 F4
Foieni RO 151 B9
Foinikounta GR 178 B2
Foix F 33 E8
Fojnica BIH 157 E8
Foktő H 149 E11
Földeák H 150 E6
Folderid N 105 B12
Földes H 151 C7
Fole DK 86 E3
Fole S 93 D13
Folegandros GR 179 B8
Folgaria I 69 B11
Folgoso de Courel E 38 C5
Folgoso de la Ribera E 39 C7
Folia GR 170 C6
Folignano I 62 B5
Foligno I 62 B3
Folkestad N 100 B4
Folkestone UK 15 E11
Folkingham UK 11 F11
Follafoss N 105 D10
Folldal N 101 B12
Follebu N 101 D12
Follina I 72 E5
Follingbo S 93 D12
Föllinge S 106 D7
Follonica I 65 B3
Folly Gate UK 12 D6
Fölsbyn S 96 D7
Foltești RO 155 B2
Fonãîtika GR 174 D3
Fondevila E 38 E3
Fondi I 62 E4
Fondón E 55 F7
Fône S 103 C10
Fonebo S 103 C12
Fonelas E 55 E6
Fonfría E 39 E7
Fonfría E 42 E3
Fønhus N 101 E11
Foni I 64 C3
Fonni I 64 C3
Fonroque F 29 F6
Fonsorbes F 33 C8
Fontaine F 27 E6
Fontaine F 31 E8
Fontainebleau F 25 D8
Fontaine-Française F 26 E3
Fontaine-le-Bourg F 18 E3
Fontaine-le-Dun F 18 E2
Fontaine-lès-Dijon F 26 E3
Fontaine-l'Évêque B 19 D9
Fontaine-lès-Luxeuil F 26 E5
Fontainemore I 68 B4
Fontaines F 25 D8
Fontaines F 30 B6
Fontaine F 37 C7
Fontanar E 47 C6
Fontanarejo E 46 F3
Fontane I 37 C7
Fontanelice I 66 D4
Fontanellato I 66 C1
Fontanès F 35 C7
Fontanigorda I 37 B10
Fontanosas E 54 B3
Fontcecchio I 62 C5
Fonte Díaz E 38 C3
Fontellas E 41 D8
Fontenay-le-Comte F 28 C4
Fontenilles F 33 C8
Fontenoy F 25 D8
Fontenoy-le-Château F 26 E5
Fontes P 44 B4
Fontevraud-l'Abbaye F 23 F12
Fontiveros E 46 C3
Font-Romeu-Odeillo-Via
F 33 F10
Fontstown IRL 7 F9
Fontvieille F 35 C8
Fonyód H 149 C8
Fonz E 42 C4
Fonzaso I 72 D4
Foppolo I 69 A8
Forano I 62 C3
Forăști RO 153 B8
Forbach D 27 C9
Forbach F 27 B6
Förby FIN 126 E8
Forcall E 42 F3
Forcalquier F 35 C10
Forcarei E 38 C3
Force I 62 B4
Forchheim D 75 C9
Forchtenberg D 187 C8
Forchtenstein A 149 A6
Ford UK 4 C6
Ford UK 5 D12
Ford IRL 9 C10
Førde N 94 C2
Førde N 100 D3
Fordham UK 15 C9
Fordingbridge UK 13 D11
Fordongianus I 64 D2
Fordoun UK 5 B12
Fore N 110 C7
Foreneset N 94 D4
Forenza I 60 B5
Føresvik N 94 D2
Forfar UK 5 B11
Forges-les-Eaux F 18 E4
Forgie UK 3 K10
Forino I 60 B3
Forio I 62 F5
Föritz D 75 B9
Forjães P 38 E2
Forkhill UK 7 D10
Førland N 94 F6
Forlev DK 87 E8
Forlì I 66 D5
Forlimpopoli I 66 D5
Formazza I 71 E6
Formby UK 10 D5
Formello I 62 C2
Formerie F 18 E4
Formia I 62 E5
Formiche Alto I 48 D3
Formicola I 60 A2
Formigine I 66 C2
Formigliana I 68 C5
Formignana I 66 C4
Formigueres F 33 E10
Formofoss N 105 C12
Fornäs S 92 C6
Fornelli I 64 A1
Fornells UK 4 D7
Fornells de la Selva E 43 D9
Forni di Sopra I 72 D6
Forni di Sotto I 73 D6
Forno Alpi Graie I 31 E11
Forno di Zoldo I 72 D5
Fornos P 45 B7
Fornos de Algodres P 44 C5
Fornovo di Taro I 66 C1
Foros do Arrão P 44 F4
Foros de Vale Figueira P 50 B3
Forotic RO 159 C8
Forra N 111 D13
Forráskút H 150 E4
Forres UK 3 K9
Forró H 145 G3
Fors S 98 B6
Fors S 107 E12
Fors S 107 E15
Forså N 111 C12
Forså S 111 D11
Forsand N 94 E4
Forsåsen S 106 D7
Forsbacka S 103 E12
Forsbacken S 109 F11
Forsbakken N 111 C14
Forsby FIN 122 C9
Forsby FIN 127 E14
Forserum S 92 D4
Forshaga S 97 C9
Forshed S 118 B4
Forsheda S 87 A13
Forsholm S 107 B16
Forsinard UK 3 J9
Forsland N 108 D4
Førslev DK 87 E9
Förslöv S 87 C11
Forsmark S 99 B10
Forsmark S 109 F10
Forsmo S 107 E13
Forsnäs S 109 D16
Forsnäs S 109 E13
Forsnäs S 118 B5
Forsnes N 104 D4
Forsøl N 113 B12
Forssa FIN 127 D10
Forsse S 107 E11
Forssjö S 93 B8
Forst D 81 C7
Forstinning D 75 F10
Forstråskhed S 118 B8
Forsvik S 92 B4
Fortanete I 42 F2
Fort Augustus UK 2 L7
Forte dei Marmi I 66 E1
Fortezza I 72 C4
Forth UK 5 D9
Fortios P 44 F6
Fortrose UK 3 K8
Fortuna E 56 E2
Fortuneswell UK 13 D10
Fort William UK 4 B6
Forvik N 108 E3
Fos F 33 E7
Fosdinovo I 69 E9
Foskros I 102 B4
Foskvallen S 102 C4
Fosnavåg N 100 B3
Fosnavågen N 105 B10
Fossacesia I 63 C6
Fossalta di Portogruaro I 73 E6
Fossano I 37 B7
Fossato di Vico I 67 F6
Fossby N 91 A10
Fosse N 94 B6
Fosser N 95 C14
Fosses-la-Ville B 19 D10
Fossheim N 112 D5
Fossing F 90 B5
Fossmork N 94 E4
Fossombrone I 67 E6
Fos-sur-Mer F 35 D8
Fót H 150 B3
Foteina GR 169 D6
Fotherby UK 11 E11
Fotinovo BG 165 F9
Fotolivos GR 170 B6
Foucarmont F 18 E4
Fouesnant F 22 E3
Foug F 26 C4
Fougères F 23 D9
Fougerolles F 26 E5
Fougerolles-du-Plessis F 23 D10
Fouilloy F 18 E6
Foulain F 26 D3
Foulayronnes F 33 B7
Fountain Cross IRL 8 C5
Fouras F 28 D3
Fourcès F 33 C6
Fourchambault F 30 A3
Four Elms UK 15 E9
Fourfouras GR 178 E8
Fourmies F 19 D9
Fourna GR 174 A4
Fournels F 30 F3
Fournes GR 178 E6
Fournoi GR 177 D7
Fourques F 34 E4
Fourques F 35 C8
Fours F 30 B4
Foustani GR 169 C7
Fownhope UK 13 A9
Foxdale GBM 10 C2
Foxford IRL 6 E4
Foyers UK 2 L8
Foygh IRL 7 E7
Foynes IRL 8 C4
Foz E 38 A5
Foz P 44 F2
Foz do Arelho P 44 F2
Foz Giraldo P 44 E5
Frabosa Soprana I 37 C7
Fraga E 42 E4
Fragagnano I 61 C8
Fraham A 76 F6
Frailes E 53 A9
Fraire B 19 D10
Fraize F 27 D7
Frameries B 19 D8
Framley DK 86 B5
Framlingham UK 15 C11
Främlingshem S 98 A7
Frammersbach D 187 A7
Främmestad S 91 C12
Frampol PL 144 B6
Frampton UK 13 B9
Francavilla al Mare I 63 C6
Francavilla di Sicilia I 59 D7
Francavilla Fontana I 61 B9
Francavilla in Sinni I 60 C6
Francescas F 33 B6
Frâncești RO 160 C4
Franco P 38 F5
Francofonte I 59 E6
Francorchamps B 20 D5
Francoules F 33 A8
Francova Lhota CZ 146 C6
Frändefors S 91 C11
Frânești D 79 C10
Frangy F 31 C8
Frankenau D 21 B11
Frankenberg D 80 E4
Frankenberg (Eder) D 21 B11
Frankenburg am Hausruck
A 76 F4
Frankenfels A 77 G8
Frankenheim D 75 A7
Frankenmarkt A 76 G4
Frankenthal (Pfalz) D 21 E10
Frankfurt am Main D 21 D11
Frankfurt an der Oder D 81 B7
Frankowo PL 136 E6
Frankrike S 105 D15
Frånö S 103 A14
Franois F 26 F4
Fränsta S 103 A11
Františkovy Lázně CZ 75 B11
Franzburg D 83 B13
Frascati I 62 D3
Frascineto I 60 D6
Fraserburgh UK 3 K12
Frasin MD 153 A11
Frasin RO 153 B7
Frasne F 31 B9
Frasnes-lez-Buissenal B 19 C8
Frasnes-lez-Gosselies B 19 C9
Frassinoro I 66 D2
Frasso Telesino I 60 A3
Frastanz A 71 C9
Frata RO 152 D4
Frătăutii Noi RO 153 B7
Frătăutii Vechi RO 153 B7
Fratel P 44 E5
Frătești RO 161 F7
Fratta Polesine I 66 B4
Fratta Todina I 62 B2
Fraubrunnen CH 31 A12
Frauenau D 76 D4
Frauenfeld CH 27 E10
Frauenkirchen A 77 G11
Frauental an der Laßnitz
A 148 C4
Frauenwald D 79 E8
Fraunberg D 75 F10
Frayssinet-le-Gélat F 33 A8
Frecăței RO 155 C3
Frecăței RO 155 D2
Frechas P 38 F5
Frechen D 21 C7
Frechilla E 39 D10
Freckenhorst D 17 E9
Freckleton UK 10 D6
Fredderica DK 86 D5
Frederiksberg DK 87 D11
Frederiksberg DK 87 E8
Frederikshavn DK 90 E8
Frederikssund DK 87 D10
Frederiksværk DK 87 D10
Fredrika S 107 C14
Fredriksberg S 97 B11
Fredriksdal S 92 D5
Fredrikstad S 91 A8
Fredropol PL 145 D5
Free N 110 C7
Freeland UK 13 B12
Freemount IRL 8 D5
Freenagh de la Sierra E 51 C8
Fregene I 62 D2
Freginals E 42 F5
Fréhel F 23 C7
Freiberg D 80 E4
Freiburg (Elbe) D 17 A12
Freiburg im Breisgau D 27 E8
Freienbach CH 27 F10
Freienstein D 17 A12
Freienwill D 82 A6
Freihung D 75 C10
Freila E 55 D7
Freilassing D 73 A6
Freimersheim D 187 B5
Freisen D 21 E8
Freising D 75 F10
Freissinières F 31 F10
Freistadt A 77 E7
Freistatt D 17 C11
Freiston D 12 F6
Freital D 80 E5
Freixedas P 45 C6
Freixial E 44 E4
Freixiosa P 44 C5
Freixo P 38 E2
Freixo de Espada à Cinta
P 45 B7
Fréjairolles F 33 C10
Frejlev DK 86 A5
Fréjus F 36 E5
Frekhaug N 94 A2
Frellstedt D 79 B8
Freluga S 103 D11
Fremdingen D 75 E7
Fremington UK 12 C6
Frenchpark IRL 6 E6
Frencq F 15 F12
Frensdorf D 75 C8
Frenštát pod Radhoštěm
CZ 146 B6
Frepa D 17 D9
Fresach A 73 C8
Fresagrandinaria I 63 D7
Freshford IRL 9 C8
Freshwater UK 13 D11
Fresnay-sur-Sarthe F 23 D12
Fresneda de la Sierra E 47 D8
Fresnedoso de Ibor E 45 E9
Fresnes-en-Woëvre F 26 B4
Fresnes-sur-Apance F 26 E4
Fresnes-sur-Escaut F 182 E3
Fresne-St-Mamès F 26 E4
Fresno de la Ribera E 39 E9
Fresno de Sayago E 39 F8
Fresno el Viejo E 45 B10
Fresnoy-Folny F 18 E3
Fresnoy-le-Grand F 19 E7
Fresse-sur-Moselle F 27 E6
Fressingfield UK 15 C11
Fresvik N 100 D5
Fretigney-et-Velloreille F 26 F4
Freudenberg D 21 C9
Freudenberg D 187 B7
Freudenburg D 186 B2
Freudenstadt D 27 D9
Frévent F 18 D5
Freyburg (Unstrut) D 79 D10
Freyenstein D 83 D12
Freyming-Merlebach F 27 B6
Freyung D 76 E5
Friarfjorden N 113 C18
Frías E 40 C5
Fribourg CH 31 B11
Frick D 21 C7
Frick-Fjord N 111 C11
Fričovce SK 145 E2
Fridafors S 88 C7
Fridalen N 109 B9
Fridaythorpe UK 11 C10
Fridingen an der Donau
D 27 D10
Fridlevstad S 88 C9
Fridolfing D 76 G3
Frieda (Meinhard) D 79 D7
Friedberg A 148 B5
Friedberg D 75 F8
Friedberg (Hessen) D 21 D11
Friedeburg D 17 B9
Friedeburg (Saale) D 79 C10
Friedenfels D 75 C11
Friedersdorf D 79 C11
Friedersdorf D 80 B6
Friedewald D 185 C8
Friedland D 78 E6
Friedland D 80 B6
Friedland D 84 C5
Friedrichroda D 79 E8
Friedrichsbrunn D 79 C9
Friedrichsdorf D 187 A6
Friedrichshafen D 71 B9
Friedrichskoog D 82 B5
Friedrichsruhe D 83 C11
Friedrichswalde D 84 D5
Frielendorf D 21 C12
Friemar D 79 E8
Friesach A 73 C9
Friesack D 83 E13
Friesenhagen D 185 C8
Friesenheim D 27 D8
Friesenried D 71 B11
Friesoythe D 17 B9
Frifelt DK 86 E3
Frigento I 60 A4
Friggesund S 103 C12
Frigiliana E 53 C9
Frihetsli N 111 C18
Friitala FIN 126 C6
Frikes GR 174 C2
Frikstad N 90 B2
Frillesås S 91 E11
Frinnaryd S 92 D5
Frinton-on-Sea UK 15 D11
Friockheim UK 5 B11
Friol E 38 B3
Frisange I 20 E6
Frístad S 91 D13
Frist.la S 91 D12
Fritzens A 72 B4
Fritzlar D 21 B12
Friville-Escarbotin F 18 D4
Frizington UK 10 B5
FroŠha FO 2 B3
Fröðinge S 92 D6
Frodsham UK 10 E6
Frœschwiller F 186 D4
Froges N 31 E8
Frogmore UK 15 E7
Frohburg D 79 D12
Frohnleiten A 148 B4
Froideconche F 26 E5
Froissy F 18 F5
Fröjel S 93 E12
Froland N 90 B4
Frolosh BG 165 E6
Frombork PL 139 B8
Frome UK 13 C10
Fromelennes F 184 D2
Fromentine F 28 B1
Frómista E 40 D3
Fröndenberg D 17 F9
Fronhausen D 21 C11
Fronsac F 28 F5
Fronsac F 33 E7
Fronteira P 44 F5
Frontenard F 31 B7
Frontenay-Rohan-Rohan
F 28 C4
Frontenhausen D 75 E12
Frontignan F 35 D6
Fronton F 33 C8
Frösakull S 87 B11
Frose D 79 C9
Fröseke S 89 B9
Frosinone I 62 D4
Frøskeland N 110 C9
Frøslev DK 86 B3
Frøslev DK 86 A3
Frøstrup DK 86 A3
Frouard F 26 C5
Frövi S 97 D13
Frufällan S 91 D12
Fruges F 18 C5
Frumoasa RO 153 E7
Frumoasa RO 161 F6
Frumosu RO 153 B7
Frumușani RO 161 E8
Frumușeni RO 151 E6
Frumușița RO 155 B2
Frunzivka UA 154 C5
Frutigen CH 70 D5
Fry GR 181 F6
Frýdek-Místek CZ 146 B6
Frýdlant CZ 81 E8
Frýdlant nad Ostravicí CZ
146 B6
Frygnowo PL 139 C9
Fryksås S 102 D8
Frymburk CZ 76 E6
Fryšták CZ 146 C5
Frysztak PL 144 D4
Fubine I 37 B8
Fucecchio I 66 E2
Fuchstal D 71 B11
Fuencaliente E 54 C4
Fuendejalón E 41 E9
Fuengirola E 53 C7
Fuenlabrada E 46 D5
Fuenmayor E 41 D6
Fuensalida E 46 D4
Fuensanta E 55 E9
Fuensanta de Martos E 53 A9
Fuente-Álamo E 53 B10
Fuente-Álamo E 55 E10
Fuente-Álamo E 56 F2
Fuentealbilla E 47 F9
Fuentecambrón E 40 E5
Fuentecén E 40 E4
Fuente de Cantos E 51 C7
Fuente del Arco E 51 C8
Fuente del Maestre E 51 B7
Fuente de Pedro Naharro
E 47 E6
Fuente de Piedra E 53 B7
Fuente el Fresno E 46 F5
Fuente el Saz de Jarama E 46 C6

Fuente el Sol E 46 B3
Fuenteguinaldo E 45 D7
Fuentelapeña E 39 F9
Fuentelcésped E 40 E4
Fuentelespino de Haro E 47 E7
Fuentelespino de Moya E 47 E10
Fuentelmonge E 41 F7
Fuentenovilla E 47 D6
Fuente Obejuna E 51 C9
Fuente Palmera E 53 A6
Fuentepelayo E 40 F4
Fuentepinilla E 40 E6
Fuenterodos E 41 F10
Fuenterrebollo E 40 F4
Fuenterroble de Salvatierra
E 45 C9
Fuentesaúco E 45 B10
Fuentesaúco de Fuentidueña
E 40 F3
Fuentes-Claras E 47 C10
Fuentes de Andalucía E 51 E9
Fuentes de Ebro E 41 E10
Fuentes de Jiloca E 47 B9
Fuentes de León E 51 C6
Fuentes de Nava E 39 D10
Fuentes de Oñoro E 45 C7
Fuentes de Ropel E 39 E8
Fuentespalda E 42 F4
Fuente-Tójar E 53 A8
Fuente Vaqueros E 53 B9
Fuentidueña E 40 F4
Fuerte del Rey E 53 A9
Fuertescusa E 47 D8
Fügen A 72 B4
FuglafjørSur FO 2 A3
Fugleberg N 111 C12
Fuglebjerg DK 87 E9
Fuglstad N 105 A13
Fuhrberg (Burgwedel) D 78 A6
Fulda D 74 A6
Fulford UK 5 E9
Fulga RO 161 D8
Fulham UK 15 E8
Fullbro S 93 A11
Fulnek CZ 146 B5
Fülöp H 151 B9
Fülöpháza H 150 D3
Fülöpszállás H 150 D3
Fulunäs S 102 D5
Fumay F 19 E10
Fumel F 33 B7
Fumone I 62 D4
Funäsdalen S 102 A4
Fundada P 44 E4
Fundão P 44 D6
Fundata RO 160 C6
Fundeni RO 161 B11
Fundeni RO 161 E9
Funder Kirkeby DK 86 C4
Fundulea RO 161 E9
Fundu Moldovei RO 152 B6
Funes E 32 F2
Funes I 72 C4
Funzie UK 3 D15
Furadouro P 44 C3
Furceni MD 154 C3
Furci I 63 C7
Furci Siculo I 59 D7
Furculeşti RO 160 F6
Furiani F 37 F10
Furnace UK 4 C6
Furnari I 59 C7
Furralleigh IRL 9 D8
Fürstenau D 17 C9
Fürstenberg D 21 A12
Fürstenberg D 84 D4
Fürstenberg (Lichtenfels)
D 21 B11
Fürstenfeld A 148 B6
Fürstenfeldbruck D 75 F9
Fürstenwalde D 80 B6
Fürstenwerder D 84 D5
Fürstenzell D 76 E4
Furta H 151 C7
Furtan S 97 C8
Furtei I 64 D2
Fürth D 21 E11
Fürth D 75 D8
Furth D 75 E11
Furth bei Göttweig A 77 F9
Furth im Wald D 76 D3
Furtwangen im Schwarzwald
D 27 D9
Furuberg S 103 B12
Furuby S 89 B8
Furudal N 111 B19
Furudal S 103 D9
Furudals bruk S 103 D9
Furuflaten N 111 B19
Furuögrund S 118 E6
Furusjö S 91 D14
Furusund S 99 C11
Furutangvika N 105 B14
Furuvik S 103 E13
Fusa N 94 B3
Fuscaldo I 60 E6
Fushë-Arrëz AL 163 E9
Fushë-Bardhe AL 168 C4
Fushë Kosovë RKS 164 D3
Fushë-Krujë AL 168 B2
Fushë-Kuqe AL 168 A2
Fusignano I 66 D4
Fußach A 71 C9
Füssen D 71 B11
Fussy F 25 F7
Fustiñana E 41 D9
Futeau F 26 B3
Futog SRB 158 C4
Futrikelva N 111 A17
Füzesabony H 150 B5
Füzesgyarmat H 151 C7
Fužine HR 67 B10
Fyfield UK 15 D9
Fylaki GR 169 F8
Fyli GR 175 C8
Fyllinge S 87 B11
Fyllo GR 169 F7
Fynshav DK 86 F5
Fyrås S 106 D8
Fyrde N 100 B4
Fyresdal N 90 A3
Fyteies GR 174 B3
Fyvie UK 3 L12

G

Gaal A 73 B10
Gaanderen NL 183 B8
Gabaldón E 47 E9
Gabare BG 165 C8
Gabarret F 32 C6
Gabčíkovo SK 146 F5
Gabela BIH 157 F8
Gaber BG 165 D6
Gabia la Grande E 53 B9
Gabicce Mare I 67 E6
Gąbin PL 139 F8
Gablingen D 75 F8
Gablitz A 77 F10
Gáborján H 151 C8
Gabriac F 34 B4
Gabrnik SLO 148 D5
Gabrovnitsa BG 165 B7
Gabrovo BG 166 D4
Gaby I 68 B4
Gać PL 144 C5
Gacé F 23 C12
Gacko SLO 148 D5
Gåda S 103 B10
Gadbjerg DK 86 D4
Gäddede S 105 B16
Gäddträsk S 107 C15
Gadmen CH 70 D6
Gadoni I 64 D3
Gádor E 55 F8
Gádoros H 150 D6
Gadstrup DK 87 D10
Gadžin Han SRB 164 C5
Gaeiras P 44 F2
Gærum DK 90 E7
Gãeşti RO 160 D6
Gaeta I 62 E5
Gafanha da Boa Hora P 44 C3
Gafanha da Nazaré P 44 C3
Gafanha do Carmo P 44 C3
Gafanhoeira P 50 B3
Gáfete P 44 F5
Gaflenz A 73 A10
Gafsele S 107 C12
Gaganitsa BG 165 C7
Gaggenau D 27 C9
Gaggi I 59 D7
Gaggio Montano I 66 D2
Gaglianico I 68 B5
Gagliano Castelferrato I 59 D6
Gagliano del Capo I 61 D10
Gagnef S 103 E9
Gagnières F 35 B7
Gagsmark S 118 D6
Găiceana RO 153 E10
Gaigalava LV 133 C2
Gaildorf D 74 E6
Gaillac F 33 C9
Gaillac-d'Aveyron F 34 B4
Gaillard F 31 C9
Gaillimh IRL 6 F4
Gaillon F 24 B5
Gailmui LV 135 D13
Gaimersheim D 75 E9
Gainsborough UK 11 E10
Gaiole in Chianti I 66 F3
Gaios GR 174 A1
Gairloch UK 2 K5
Gairo I 64 D4
Gais CH 71 C8
Gais I 72 C4
Găiseni RO 161 D7
Gaishorn A 73 A10
Gaismas LV 135 C8
Gaj SRB 159 D7
Gakovo SRB 150 F3
Galåbodarna S 102 A6
Galambok H 149 C8
Galan F 33 D6
Gălăneşti RO 153 B7
Gálániuta N 112 F10
Galanta SK 146 E5
Galapagar E 46 C5
Galaroza E 51 D6
Galåsen N 102 D4
Galashiels UK 5 D11
Galasjö S 107 E14
Galata BG 167 C9
Galatades GR 169 C7
Galatas GR 175 E7
Galatas GR 175 E7
Gălăteni RO 160 E6
Galati RO 155 C2
Galatii Bistriţei RO 152 D4
Galati Mamertino I 59 C6
Galatin BG 165 C8
Galatina I 61 C10
Galatini GR 169 D6
Galatista GR 169 D9
Galatone I 61 C10
Galaxidi GR 174 C5
Galbally IRL 8 D6
Galbarra E 32 E1
Galbenu RO 161 C10
Gålberget S 107 D14
Galda de Jos RO 152 E3
Gâldãu RO 161 E11
Gâl'din N 113 E12
Galeata I 66 E4
Galende E 39 D6
Galēni LV 135 D13
Galera E 55 D7
Galéria F 37 G9
Gălești RO 152 E5
Galewice PL 142 D5
Galgagyörk H 150 B3
Galgahéviz H 150 B4
Gálgău RO 152 C3
Galgon F 28 F5
Gal'gunjar'ga N 113 F16
Galicea RO 160 D4
Galicea Mare RO 160 E2
Galiche BG 165 B8
Galinduste E 45 C9
Galiny PL 136 E3
Galissas GR 176 E4
Galisteo E 45 E8
Gallardon F 24 C5
Gallargues-le-Montueux
F 35 C7
Galleberg N 95 C12
Gallegos de Argañan E 45 C7
Gallegos de Solmirón E 45 C10
Galleno I 66 E2
Gälleråsen S 97 D11
Gallese I 62 C2
Gallian-en-Médoc F 28 E4
Galliate I 68 B6
Gallicano I 66 D1
Gallicano nel Lazio I 62 D3
Gallin D 83 C10
Gallin D 83 D12
Gallio I 72 E4
Gallipoli I 61 C10
Gällivare S 116 D5
Gallizien A 73 C10
Gallneukirchen A 76 F6
Gällö S 103 A9
Gällstad S 91 D13
Galluccio I 60 A1
Gallur E 41 E9
Galovo BG 160 F4
Galros IRL 7 F8
Gålsjö bruk S 107 E13
Galston UK 5 D8
Galtelli I 64 C4
Galten DK 86 C5
Galterud N 96 B6
Gåltjärn S 103 A13
Gåltjärn S 103 A14
Galtström S 103 B13
Galtür A 71 D10
Galve E 42 F2
Galve de Sorbe E 47 B6
Galveias P 44 F5
Galven S 103 D11
Gálvez E 46 E4
Galway IRL 6 F4
Gama E 40 B5
Gamaches F 18 E4
Gamache-les-Bains F 32 C4
Gämás H 149 C9
Gambara I 66 B1
Gambarie I 59 C8
Gambassi Terme I 66 E2
Gambatesa I 63 D7
Gambettola I 66 D5
Gambolò I 68 C6
Gambsheim F 27 C8
Gaming A 77 G8
Gamleby S 93 D8
Gamlingay UK 15 C8
Gamlitz A 148 C5
Gammalkroppa S 97 C11
Gammelboning S 103 E11
Gammelgården S 119 C10
Gammelgarn S 93 E13
Gammel Rye DK 86 D5
Gammelsdorf D 75 E10
Gammelstaden S 118 C8
Gammelstilla S 98 B7
Gammertingen D 27 D11
Gams CH 71 C8
Gamvik N 104 D7
Gamvik N 113 A17
Gamvik N 113 A21
Gamvik N 113 A13
Gan F 32 D5
Ganagobie F 35 B10
Gand B 19 B8
Gand B 19 B8
Gandarela P 38 F3
Gandellino I 69 A8
Ganderkesee D 17 B11
Gandesa E 42 E4
Gandesbergen D 17 C12
Gandía E 56 D4
Gandino I 69 B8
Gandra P 38 D2
Gandrup DK 86 A6
Gandvik N 114 C6
Găneasa RO 160 E4
Găneasa RO 161 D8
Gănești RO 152 E4
Ganfei P 38 D2
Ganges F 35 C6
Gånghester S 91 D13
Gangi I 58 D5
Gângiova RO 160 F3
Gangkofen D 75 F12
Ganløse DK 87 D10
Gannat F 30 C3
Gänserndorf A 77 F11
Ganshoren B 19 C9
Gânsvik S 103 A15
Ganzlin D 83 D12
Gaoth Dobhair IRL 6 B6
Gap F 36 B4
Gaperhult S 91 B13
Gara H 150 E3
Garaballa E 47 E10
Garagarza E 32 D1
Garaguso I 60 B6
Gara Khitrino BG 167 C7
Garbayuela E 46 F3
Garberg N 105 E9
Gârbou RO 151 C11
Gârbova RO 152 F3
Gârbovi RO 161 D9
Garbów PL 141 H6
Garbsen D 78 B6
Garching an der Alz D 75 F12
Garching bei München D 75 F10
Garchizy F 30 A3
Garcia E 42 E5
Garciaz E 45 E9
Garciems LV 135 B8
Garcihernández E 45 C10
Garcillán E 46 C4
Garčin HR 157 B9
Gârcina RO 153 D8
Garcinarro E 47 D7
Gârcov RO 160 F5
Garda I 69 B10
Gârda de Sus RO 151 E10
Gardanne F 35 D9
Gârdãs S 102 E6
Gårdby S 89 B11
Garde I 72 E4
Garde S 105 D15
Gardeja PL 138 C6
Gardelegen D 79 B9
Gardenstown UK 3 K12
Garderen NL 183 A7
Garderhouse UK 3 E14
Gardermoen N 95 B14
Gardiki GR 174 A5
Garding D 82 B5
Gärdnäs S 106 C8
Gardone Riviera I 69 B10
Gardone Val Trompia I 69 B9
Gardonne F 29 F6
Gárdony H 149 B11
Gardouch F 33 D9
Gärdsjö S 92 A3
Gardsjöbäcken S 109 F12
Gårdskär S 103 E14
Gärdslösa S 89 B11
Gårdstånga S 87 D12
Gärdnäs S 106 C8
Gare N 102 A1
Gâre F 27 E7
Garein F 32 B4
Gårelehöjden S 107 D11
Garelochhead UK 4 C7
Gareóult F 36 E4
Gareśnica HR 149 E7
Garforth UK 11 D9
Gargaliani GR 174 E4
Gargáligas E 45 F9
Gargallo E 42 F2
Garganta la Olla E 45 D9
Gargazzone I 72 C3
Gargilesse-Dampierre F 29 B9
Gargnano I 69 B10
Gargnäs S 107 A13
Gargogečče N 113 C20
Gargždai LT 134 F2
Garijp NL 16 B5
Garkalne LV 135 B8
Gârla Mare RO 159 E10
Garlasco I 69 C6
Gârleni RO 153 D9
Garliava LT 137 D8
Gârliciu RO 155 D2
Garlieston UK 5 F8
Garlin F 32 C5
Garlitos E 54 B2
Garlstorf D 83 D8
Garmisch-Partenkirchen
D 72 B3
Garnes N 100 B3
Gârnic RO 159 D8
Garoaffa RO 153 F10
Garons F 35 C7
Garoza LV 135 C7
Garpenberg S 98 B6
Garphyttan S 97 D12
Garrafe de Torío E 39 C8
Garralda E 32 E3
Garrane IRL 8 E4
Garray E 41 E7
Garrel D 17 C10
Garrison UK 6 D6
Garristown IRL 7 E10
Garrovillas E 45 E7
Garrucha E 55 E9
Garryhine UK 2 J3
Garryvoe IRL 8 E6
Garsås S 102 E8
Garsdale Head UK 11 C7
Gârsene LV 135 D11
Gârslev DK 86 D5
Gârsnas S 88 D6
Garstang UK 10 D6
Garsten A 77 A9
Gärstedt D 83 D8
Garten N 104 D7
Garth UK 13 A7
Garthmyl UK 10 F5
Gartow D 83 D10
Gärtringen D 27 C10
Gartz D 84 D6
Garvagh UK 4 F3
Garvagh IRL 7 E7
Garvaghy UK 7 D8
Garvald UK 5 D11
Garvamore UK 5 A8
Garvão P 50 D3
Garvard UK 4 C4
Garvary UK 7 D7
Garve UK 2 K7
Garwolin PL 141 G5
Garz D 84 B4
Gąsawa PL 138 E4
Gåsbakken N 104 E7
Gaschurn A 71 D10
Gascueña E 47 D7
Gåsholma S 103 D13
Gasjö S 97 B14
Gaskeluokt S 107 A11
Gąski PL 136 F5
Gâsnes N 112 D18
Gâsnes N 113 A13
Gasny F 24 B6
Gąsocin PL 139 E10
Gaspoltshofen A 76 F5
Gasperina I 59 B10
Gasselte NL 17 C7
Gasselternijveen NL 17 C7
Gassino Torinese I 68 C4
Gassjö S 103 B9
Gassum DK 86 B6
Gastellovo RUS 136 C4
Gastes F 32 B3
Gastoúni GR 174 D3
Gastouri GR 168 E2
Gata E 45 D7
Gata de Gorgos E 56 D5
Gâtaia RO 159 C7
Gatchina RUS 132 B7
Gatehouse of Fleet UK 5 F8
Gáter H 150 D4
Gateshead UK 5 F13
Gátova E 48 E3
Gattendorf A 77 F11
Gatteo a Mare I 66 D5
Gattières F 37 D6
Gattinara I 68 B5
Gau-Algesheim D 185 E8
Gauchy F 19 E7
Gaucín E 53 D6
Gauja LV 135 B9
Găujani RO 161 F7
Gaujiena LV 131 F12
Gaukönigshofen D 74 C6
Gaulstad N 105 D12
Gau-Odernheim D 185 E9
Gaupne N 100 D6
Gauré N 90 A7
Gausvik N 111 C11
Gautestad N 90 B2
Gauting D 75 F9
Gauto N 109 D12
Gavà E 43 E8
Gavalou GR 174 B4
Găvănoasa MD 155 B3
Gavardo I 69 B9
Gavarnie F 32 E5
Gávavencsellő H 145 G4
Gavere B 19 C8
Gavi I 37 B9
Gavião P 44 F5
Gaviere LV 134 C2
Gavik S 103 A15
Gavirate I 68 B6
Gavoi I 64 C3
Gavojdia RO 159 B9
Gavorrano I 65 B3
Gâre N 102 A1
Gavray F 23 C9
Gavrio GR 176 D4
Gavrolimni GR 174 C4
Gävsta S 99 B10
Gaweinstal A 77 F11
Gawliki Wielkie PL 136 E5
Gaworzyce PL 81 C9
Gaytaninovo BG 169 B10
Gayton UK 11 F13
Gazalos D 185 C10
Gaziantep TR 177 C9
Gazikoy TR 173 C7
Gazioğlu TR 173 B8
Gaziosmanpaşa TR 173 B10
Gazlaz E 32 E2
Gazoldo degli Ippoliti I 66 B2
Gazoros SRB 169 B10
Gazzo Veronese I 66 B3
Gazzuolo I 66 B2
Gbelce SK 147 F7
Gbely SK 77 E12
Gdansk PL 138 B6
Gdinj HR 157 F6
Gdov RUS 132 D2
Gdynia PL 138 A6
Geaca RO 152 D4
Gea de Albarracín E 47 D10
Geaidnovuohppi N 113 E12
Geamãna MD 154 D4
Geashill IRL 7 F8
Geaune F 32 C5
Gebesee D 79 D8
Gebhardshain D 21 C9
Gechingen D 187 D6
Geçkinli TR 167 F7
Geddington UK 15 C7
Géderlak H 149 C11
Gedern D 21 D12
Gédinne B 19 E10
Gedser DK 83 A11
Gedsted DK 86 B4
Gedved DK 86 D5
Geel B 19 B11
Geertruidenberg NL 16 E3
Geeste D 17 C8
Geesteren NL 183 A9
Geestgottberg D 83 E11
Geesthacht D 83 D8
Geetbets B 19 C11
Geffen NL 183 B6
Gefrees D 75 B10
Gefyra GR 169 C8
Gefyri GR 169 F7
Gegai GR 174 B2
Gegužinė LT 137 D9
Gehrde D 17 C10
Gehrden D 78 B6
Gehren D 79 E9
Geijersholm S 97 B10
Geilenkirchen D 20 C6
Geilo N 95 A8
Geirangr N 100 B6
Geisa D 78 E6
Geiselbach D 187 A7
Geiselhöring D 75 E11
Geiselwind D 75 C7
Geisenhausen D 75 F11
Geisenheim D 21 E9
Geising D 80 E5
Geisingen D 27 E10
Geislingen D 187 E6
Geislingen an der Steige
D 74 E6
Geismar D 79 D7
Geisnes N 105 B11
Geispolsheim F 27 C8
Geisthal A 73 B11
Geithain D 79 D12
Geithus N 95 C11
Gela I 58 E5
Gelbensande D 83 B12
Gelchsheim D 74 C7
Geldermalsen NL 16 E4
Geldern D 16 E6
Geldersheim D 75 B7
Geldrop NL 16 F5
Geleen NL 19 C12
Gelelagar UK 13 A8
Gelenau D 79 E12
Gelgaudiškis LT 136 D5
Gelibolu TR 172 D6
Gellénháza H 149 C7
Gelligaer UK 13 B8
Gelnhausen D 21 D12
Gelnica SK 145 F2
Gelos F 32 D5
Gelsa E 41 E11
Gelse H 149 C7
Gelsenkirchen D 17 E8
Gelsted DK 86 E5
Gelsted DK 87 E9
Geltendorf D 71 A12
Gelting D 83 A7
Geltow D 80 B3
Gelu RO 151 E6
Gelvonai LT 137 D9
Gembloux B 19 C10
Gemeaux F 26 F3
Gemenele RO 155 C1
Gemert NL 16 E5
Gemerská Hôrka SK 145 F1
Gemerská Poloma SK 145 F1
Gemert NL 16 E5
Gemla S 88 B7
Gemmingen D 187 C6
Gemona del Friuli I 73 D7
Gemünd D 21 C7
Gémozac F 28 D4
Gemünden D 21 E8
Gemünden am Main D 74 B6
Genappe B 19 C9
Genarp S 87 D12
Génave E 55 C7
Genazzano I 62 D3
Gençay F 29 C7
Gencsapáti H 149 B7
Gendrey F 26 F4
Gendringen NL 16 E6
Géneland P 30 B5
Genemuiden NL 16 C6
General-Inzovo BG 167 E6
General-Kolevo BG 155 F1
Generalski Stol HR 148 F4
General Toshevo BG 155 F2
Genestoso E 39 C7
Genève CH 31 C9
Genévila E 32 E1
Genga I 67 F6
Gengenbach D 27 D9
Geniai TR 171 E10
Geniej TR 177 D10
Génisea GR 171 B7
Genk B 19 C12
Genlis F 26 F3
Gennadi GR 181 D7
Gennep NL 16 E5
Genner DK 86 E4
Gennes F 23 F11
Gennes-sur-Seiche F 23 D9
Genola I 37 B7
Génolhac F 35 B6
Genoni I 64 D3
Genouillac F 29 C9
Genouilly F 24 F6
Genouilly F 30 B6
Genova I 37 C9
Genovés E 56 D3
Gensac F 28 F6
Gensingen D 185 E8
Gent B 19 B8
Genthin D 79 B11
Gentioux-Pigerolles F 29 D9
Genzano di Lucania I 60 B6
Genzano di Roma I 62 D3
Geoagiu RO 151 F10
George Enescu RO 153 A9
Georgensgmünd D 75 D9
Georgianoi GR 169 D7
Georgianoi GR 170 C6
Georgi Damyanovo BG 165 C7
Georgioupoli GR 178 E7
Georgitsi GR 174 E5
Georgsheil D 17 B8
Georgsmarienhütte D 17 D9
Ger F 23 C10
Ger F 32 D5
Gera D 79 E11
Geraardsbergen B 19 C8
Gerabronn D 74 D6
Gerace I 59 C9
Geraci Siculo I 58 D5
Gerakarou GR 169 C9
Geraki GR 175 F6
Gerakini GR 169 D9
Geras A 77 E9
Gerasdorf bei Wien A 77 F10
Géraudot F 25 D11
Gerbéviller F 186 E2
Gerbrunn D 187 B8
Gerbstedt D 79 C10
Gérce H 149 B8
Gerchsheim D 74 C6
Gerdau D 83 E8
Gerena E 51 D7
Gerendals H 151 D6
Geredlak H 149 D11
Geretsberg A 76 F3
Geretsried D 72 A3
Gérgal E 55 E7
Gergei I 64 D3
Gergeri GR 178 E8
Gerhardshofen D 75 C8
Gerindote E 46 E4
Geringswalde D 80 D3
Gerjen H 149 D11
Gerlev DK 87 D10
Gerlingen D 27 C11
Gerlos A 72 B5
Germas GR 168 D5
Germay F 26 D3
Germencik TR 177 D10
Gernika-Lumo E 41 B6
Gernrode D 79 C9
Gernsbach D 27 C9
Gernsheim D 187 B5
Gerola Alta I 69 A8
Gerolimenas GR 178 C3
Gerolsbach D 75 F9
Geroldstein D 21 D7
Gerolzhofen D 75 C7
Gerovo HR 67 A10
Gerpinnes B 19 D10
Gersau CH 71 C7
Gersfeld (Rhön) D 74 B6
Gersheim D 27 B7
Gersten D 17 C9
Gerstetten D 74 E7
Gerstheim F 186 E4
Gersthofen D 75 F8
Gerstungen D 79 D7
Gerswalde D 84 D5
Gerwisch D 79 B10
Gerzat F 30 D3
Gerzen D 75 F11
Gesäter S 91 B10
Gescher D 17 E8
Gesico I 64 D3
Gespunsart F 19 E10
Gessertshausen D 75 F8
Gestalgar E 48 E3
Gesté F 23 F9
Gesten DK 86 D4
Gesturi I 64 D3
Gesualdo I 60 A4
Gesunda S 102 E8
Gesves B 19 D11
Geszt H 151 D8
Geta FIN 99 B13
Getafe E 46 D5
Getaria E 32 D1
Getinge S 87 B11
Getryggen S 103 D10
Getterum S 93 D8
Gettjärn S 97 C9
Gettorf D 83 B7
Gevelsberg D 17 F8
Gévezé F 23 D8
Gevgelija NMK 169 B7
Gevrey-Chambertin F 26 F2
Gex F 31 C9
Geyer D 80 E3
Geyikli TR 171 E10
Gezaveh AL 168 B3
Gföhl A 77 E8
Ghedi I 66 A1
Ghelari RO 159 B10
Ghelinţa RO 153 F8
Ghemme I 68 B5
Gheorghe Doja RO 161 D10
Gheorghe Lazăr RO 161 D10
Gheorgheni RO 152 D6
Gherăeşti RO 153 D9
Gherăseni RO 161 C9
Ghergheasa RO 161 C10
Gherghiţa RO 161 D8
Gherla RO 152 C3
Gherta Mică RO 145 H7
Ghidfalãu RO 153 F7
Ghidigeni RO 153 E10
Ghidighiei MD 154 C3
Ghiffa I 68 B6
Ghilarza I 64 C2
Ghimbav RO 161 B7
Ghimeş-Făget RO 153 D8
Ghimpaţi RO 161 E7
Ghindari RO 152 E5
Ghiorac RO 151 E8
Ghioroiu RO 160 D3
Ghiroda RO 151 F7
Ghislenghien B 19 C8
Ghisonaccia I 37 G10
Ghisoni I 37 G10
Ghizela RO 151 F8
Ghyvelde F 18 B5
Gialtra GR 175 B6
Giannades GR 168 E2
Giannitsa GR 169 C7
Giano dell'Umbria I 62 B3
Giardinello I 58 C3
Giardini-Naxos I 59 D7
Giarmata RO 151 F7
Giarratana I 59 E6
Giarre I 59 D7
Giat F 29 D10
Giave I 64 C2
Giaveno I 31 E11
Giba I 64 E2
Gibellina Nuova I 58 D2
Gibostad N 111 B15
Gibraleón E 51 E6
Gibraltar GBZ 53 D6
Giby PL 136 E7
Gibzde LV 134 B4
Gideå S 107 E15
Gideå bruk S 107 E16
Gideåkroken S 107 C12
Gidle PL 143 E7
Giebelstadt D 74 C6
Gieboldehausen D 79 C7
Gieczno PL 143 C7
Giedraičiai LT 137 C11
Gielniów PL 141 H12
Gielow D 83 C13
Gien F 25 E8
Giengen an der Brenz D 75 E7
Giera RO 159 C6
Gierle B 182 C5
Giersleben D 79 C10
Gierstädt D 79 D8
Gierzwałd PL 139 C9
Gießen D 21 C11
Gieten NL 17 B7
Gietrzwałd PL 139 C9
Gièvres F 24 F5
Giffnock UK 5 D8
Gifford UK 5 D11
Gifhorn D 79 B8
Gigean F 35 C6
Gigen BG 160 F4
Gighera RO 160 F3
Giglio Castello I 65 C3
Gignac F 35 C5
Gignac F 35 C10
Gignod I 31 D11
Gijón E 39 A8
Gikši LV 135 B10
Gilău RO 151 D11
Gilching D 75 F9
Gilena E 53 B7
Gilette F 37 D6
Gilford UK 7 D10
Gilleleje DK 87 C10
Gillenfeld D 21 D7
Gillhov S 102 A8
Gillingham UK 13 C10
Gilling West UK 11 C8
Gilly-sur-Isère F 31 D9
Gilly-sur-Loire F 30 B4
Gilmorton UK 13 C12
Gilserberg D 21 C12
Gilsland UK 5 F11
Gilten D 82 E7
Gilwern UK 13 B8
Gilze NL 16 E3
Gimåt S 107 E16
Gimdalen S 103 A10
Gimigliano I 59 B10
Gimo S 99 B10
Gimont F 33 C7
Gimouille F 30 B3
Gimsøy N 110 D7
Ginasservis F 35 C10
Ginestas F 34 D4
Gingelom B 19 C11
Gingen an der Fils D 187 D6
Gingst D 84 B4
Ginkūnai LT 134 E6
Ginoles F 33 E10
Ginosa I 61 B7
Ginostra I 59 B7
Ginsheim D 185 E9
Gintsi BG 165 D7
Gioi I 60 C4
Gioia dei Marsi I 62 D5
Gioia del Colle I 61 B7
Gioia Sannitica I 60 A2
Gioia Tauro I 59 C8
Gioiosa Ionica I 59 C9
Gioiosa Marea I 59 C6
Giornico CH 71 E7
Giovinazzo I 61 A7
Gipka LV 135 A9
Giraltovce SK 145 E4
Girancourt F 26 D5
Girasole I 64 D4
Girifalco I 59 B9
Girişu de Criş RO 151 C8
Girkalnis LT 134 F6
Giroc RO 159 B7
Giromagny F 27 E6
Girona E 43 C9
Gironde-sur-Dropt F 32 A5
Gironella E 43 C7
Giroussens F 33 C9
Girov RO 153 D8
Girvan UK 4 E7
Gislaved S 88 A5
Gislev DK 86 E7
Gislinge DK 87 D9
Gislövs strandmark S 87 E12
Gisløy N 110 C9
Gisors F 18 F4
Gissi I 63 C7
Gisslarbo S 97 C14
Gistad S 92 C7
Gistel B 18 B6
Gistrup DK 86 B5
Giswil CH 70 D6

Gittelde D 79 C7
Gittun S 109 D16
Giubega RO 160 E2
Giubiasco CH 69 A7
Giugliano in Campania I 60 B2
Giuleşti RO 152 B3
Giuliano di Roma I 62 D4
Giulianova I 62 B5
Giulvăz RO 159 B6
Giuncugnano I 66 D1
Giurgeni RO 155 D1
Giurgita RO 160 E3
Giurgiu RO 161 F7
Giurgiuleşti MD 155 C2
Giuvărăşti RO 160 F5
Give DK 86 D4
Givet F 19 D10
Givors F 30 D6
Givry B 19 D9
Givry F 30 B6
Givry-en-Argonne F 25 C12
Givskud DK 86 D4
Gižai LT 136 D7
Gizałki PL 142 B4
Gizeux F 23 F12
Giżycko PL 136 E4
Gizzeria I 59 B9
Gjakovë RKS 163 E9
Gjærnes N 90 B5
Gjegjan AL 163 E9
Gjegjani AL 163 F9
Gjengstø N 104 E6
Gjerde N 100 C6
Gjerdebakken N 113 C11
Gjerdrum N 95 B14
Gjerlev DK 86 B6
Gjermungshamn N 94 B3
Gjern DK 86 C5
Gjerrild DK 87 C7
Gjerstad N 90 B5
Gjersvik N 105 B14
Gjesing DK 86 D4
Gjesvær N 113 A15
Gjevaldshaugen N 101 D15
Gjilan RKS 164 E3
Gjirokastër AL 168 D3
Gjógy FO 2 A3
Gjøl DK 86 A5
Gjøljelia N 104 D7
Gjølme N 104 E7
Gjønem AL 168 A2
Gjøra N 101 A10
Gjorm AL 168 D2
Gjøvåg N 94 B2
Gjøvik N 101 E13
Gjurakoc RKS 163 D9
Gjuvberget N 102 E3
Gkoritsa GR 175 E6
Gladbeck D 17 E7
Gladenbach D 21 C11
Gladsaxe DK 87 D10
Gladstad N 108 E2
Glainans F 26 F6
Glamis UK 5 B11
Glamoč BIH 157 D6
Glämos N 101 A14
Glamsbjerg DK 86 E6
Glandage F 35 A10
Glandon F 29 E8
Glandore IRL 8 E4
Glandorf D 17 D10
Glanegg A 73 C9
Glanerbrug NL 17 D7
Glanshammar S 97 D13
Glanworth IRL 8 D6
Glarryford UK 4 F4
Glarus CH 71 C8
Glasbury UK 13 A8
Glasgow UK 5 D8
Gläshütten D 21 D10
Glashütten D 75 C9
Glaslough IRL 7 D9
Glassan IRL 7 F7
Glastonbury UK 13 C9
Glaubitz D 80 D4
Glauchau D 79 E12
Glava BG 165 C9
Glava S 97 C8
Glava glasbruk S 96 C7
Glavan BG 166 E6
Glăvăneşti RO 153 E10
Glavanovtsi BG 164 D6
Glavice HR 157 E6
Glavičice BIH 157 C11
Glâvile RO 160 D4
Glavinitsa BG 161 F9
Glavki SRB 171 B7
Gleann Cholm Cille IRL 6 C5
Gleann Doimhin IRL 7 C7
Gleann na Muaidhe IRL 6 D3
Gledica SRB 163 C9
Glein N 108 D4
Gleina D 79 D10
Gleisdorf A 148 B5
Gleizé F 30 C6
Glejbjerg DK 86 D3
Glemsford UK 15 C10
Glen S 102 A6
Glenade IRL 6 D6
Glenamoy IRL 6 D3
Glenariff UK 4 E4
Glenarm UK 4 F5
Glenavy UK 4 F5
Glenbarr UK 4 D5
Glenbeigh IRL 8 D3
Glencaple UK 5 E9
Glencar IRL 6 D6
Glencoe UK 4 B6
Glencolumbkille IRL 6 C5
Glendowan IRL 7 C7
Gleneagles UK 5 C9
Glenealy IRL 7 G10
Gleneely IRL 4 E2
Glenegedale UK 4 D4
Glenelg UK 2 L5
Glenfield UK 11 F9
Glenfinnan UK 4 B6
Glengarriff IRL 8 E3
Glengavlen IRL 7 D7
Glenhead UK 4 F5
Glenluce UK 4 F7
Glenmore IRL 9 D8
Glenoe UK 4 F5
Glenrothes UK 5 C10
Glenties IRL 6 C6
Glère F 27 F6
Glesborg DK 87 C7
Gletsch CH 71 D6
Glewitz D 84 B3

Glibaći MNE 163 C7
Glimåkra S 88 C6
Glimboca RO 159 C9
Glin IRL 8 C4
Glina HR 148 F6
Glina RO 161 E8
Glinde D 83 C8
Glindow D 79 B12
Glinjeni MD 153 B11
Glinojeck PL 139 E9
Glinsce IRL 6 F3
Glinsk IRL 6 F3
Göhren-Lebbin D 83 D13
Goian MD 154 C4
Goicea RO 160 F3
Goieşti RO 160 E3
Goirle NL 16 E4
Góis P 44 D4
Goito I 66 B2
Goizueta E 32 D2
Gojsć PL 143 D7
Gökçealan TR 177 D9
Gökçebayir TR 171 E10
Gökçedağ TR 173 C8
Gökçeören TR 181 B9
Gökçeyazi TR 173 E8
Gokels D 82 B6
Gökova TR 181 B8
Göktepe TR 181 B9
Göktürk TR 173 B10
Gol N 101 E9
Gola HR 149 D8
Gołąb PL 141 H5
Golac SLO 67 A9
Golãieşti RO 153 C11
Golaj AL 163 E9
Gołańcz PL 85 E12
Gołaszyn PL 141 G6
Golbey F 26 D5
Gølby FIN 99 B13
Gölcük TR 173 F8
Golczewo PL 85 C7
Goldach CH 71 C8
Goldbeck D 83 E11
Goldberg D 83 C12
Goldegg A 73 B7
Goldelund D 82 A6
Golden IRL 9 D7
Golden Pot UK 14 E7
Goldenstedt D 17 C10
Goleen IRL 8 F3
Golegã P 44 F4
Golemanovo BG 159 F9
Golema Rakovitsa BG 165 E8
Golema Rečica NMK 163 F10
Golenice PL 85 D7
Golesh BG 161 F11
Goleşti RO 153 D10
Goleşti RO 160 C4
Golfo Aranci I 64 B4
Golina PL 142 B5
Golinhac F 33 A11
Goliševa LV 133 C3
Gölköy TR 177 E9
Gölle H 149 D9
Göllersdorf A 77 F10
Göllheim D 186 B5
Göllin D 83 C11
Gollin D 84 D5
Golling an der Salzach A 73 A7
Göllingen D 79 D9
Gollmitz D 84 D5
Golmes F 42 D5
Golnik SLO 73 D9
Gølstrup DK 86 A4
Golubac SRB 159 D8
Golub-Dobrzyń PL 138 D7
Golubić HR 156 D5
Golubinje SRB 159 E9
Gołymin Izvor BG 166 F5
Golyam Manastir BG 166 E6
Golyamo Belovo BG 165 E8
Gölyazi TR 173 D10
Gołymin-Ośrodek PL 139 E10
Golzow D 79 B11
Gomadingen D 27 D11
Gómara E 41 E7
Gomaringen D 187 E7
Gomba H 150 C4
Gömeç TR 173 F6
Gomecello E 45 B9
Gomes Aires P 50 D3
Gomirje HR 67 B10
Gommern D 79 B10
Gomotartsi BG 159 E11
Gomunice PL 143 D8
Gönc H 145 G3
Gonçalo P 44 D4
Goncelin F 31 B10
Gondershausen D 185 D7
Gondomar E 38 D2
Gondomar P 44 B3
Gondrecourt-le-Château F 26 C4
Gondreville F 26 C4
Gondrin F 33 C6
Gönen TR 173 D8
Gonfaron F 36 E4
Goni I 64 D3
Goniądz PL 140 D7
Gonnesa I 64 E1
Gonnoi SRB 169 E7
Gonnosfanadiga I 64 E2
Gonnosnò I 64 D2
Gonsans F 26 F5
Gônyû H 149 A9
Gonzaga I 66 C2
Goodwick UK 12 A5
Goole UK 11 D10
Goonhaven UK 12 E4
Goor NL 17 D7
Gopegi E 40 C6
Görzke D 79 B11
Gorzkowice PL 143 D8
Gorzków-Osada PL 144 B7
Górzna PL 85 D11
Górzno RO 139 D8
Górzno PL 141 G5
Goraj PL 144 B6
Góra Kalwaria PL 141 G4
Goranu RO 160 C4
Góra Puławska PL 141 H5
Gorawino PL 85 C9

Gogolin PL 142 F5
Gogołów PL 144 D4
Gogoşari RO 161 F7
Gogoşu RO 159 E10
Gogoşu RO 160 E2
Göhl D 83 B9
Gohor RO 153 E10
Gohrau D 79 C11
Göhrde D 83 D9
Göhren D 84 B5
Göhren-Lebbin D 83 D13
Gorażde BIH 157 E10
Gorban RO 153 D12
Görbeháza H 151 B7
Görcsöny H 149 E10
Gördalen S 102 C3
Gordaliza del Pino E 39 C9
Gordes F 35 C9
Gordineşti MD 153 A10
Gørding DK 86 E3
Gordola CH 69 A6
Gordon UK 5 D11
Gordona I 69 A7
Gorebridge UK 5 D10
Gorenja Straža SLO 73 E11
Gorenja vas SLO 73 D9
Gorey IRL 9 C10
Gőrgeteg H 149 D8
Gorgonzola I 69 B7
Gorgota RO 161 D8
Gorgova RO 155 C4
Gorinchem NL 16 E3
Gorino I 66 C5
Gorishovë AL 168 C2
Görisried D 71 B11
Goritsa BG 167 D8
Goritsa BG 167 D9
Göritz D 84 C5
Gorizia I 73 E8
Gorleben D 83 D10
Gørlev DK 87 D8
Gorlice PL 144 D3
Görlitz D 81 D7
Gorlosen D 83 D10
Gormanston IRL 7 E10
Gormanstown IRL 7 E10
Gormaz E 40 E6
Görmin D 84 C4
Gorna Belica NMK 164 F3
Gorna Beshovitsa BG 165 C9
Gorna Breznitsa BG 165 F7
Gorna Koznitsa BG 165 E7
Gorna Kremena BG 165 C8
Gorna Oryakhovitsa BG 166 C5
Gorna Studena BG 166 C4
Gornau D 80 E4
Gørnæs RO 152 D5
Gornet RO 161 C8
Gornet-Cricov RO 161 C8
Gorni Dŭbnik BG 165 C9
Gorni Lom BG 165 C8
Gornja Golubinja BIH 157 C6
Gornja Grabovica BIH 157 E8
Gornja Gračenica HR 149 E7
Gornja Ljubata SRB 164 E5
Gornja Ljuta BIH 157 F8
Gornjane SRB 159 E8
Gornja Ploča HR 156 D4
Gornja Radgona SLO 148 C6
Gornja Sanica BIH 157 C6
Gornja Slatina BIH 157 C9
Gornja Toponica SRB 164 C4
Gornja Trnava SRB 159 E6
Gornja Tuzla BIH 157 C10
Gornje Jelenje HR 67 B10
Gornje Peulje BIH 156 D6
Gornje Ratkovo BIH 157 C7
Gornje Vratno HR 148 D6
Gornji Babin Potok HR 156 C4
Gornji Breg SRB 150 F5
Gornji Grad SLO 73 D10
Gornji Humac HR 157 F6
Gornji Kosinj HR 156 C3
Gornji Lapac HR 156 C5
Gornji Lukavac BIH 157 F9
Gornji Majkovi HR 162 D5
Gornji Malovan BIH 157 E7
Gornji Matejevac SRB 164 C4
Gornji Milanovac SRB 158 E5
Gornji Oččauš BIH 157 D8
Gornji Podgradci BIH 157 B7
Gornji Rajič HR 157 B7
Gornji Ribnik BIH 157 D6
Gornji Tkalec HR 148 E6
Gornji Vakuf BIH 157 E8
Górno PL 143 E10
Gorno Ablanovo BG 161 F7
Gorno Aleksandrovo BG 167 D7
Gorno Orizari NMK 164 F5
Gorno Ozirovo BG 165 C8
Gorno Pavlikene BG 165 C10
Goro I 66 C5
Gorobinci NMK 164 F4
Górowo Iławeckie PL 136 E2
Gorredijk NL 16 B6
Gorreto I 37 B10
Gorron F 23 D10
Görsbach D 79 D9
Gorseinon UK 12 B6
Gorska Polyana BG 167 E7
Gorski Izvor BG 166 E4
Gorsko Novo Selo BG 166 C5
Gorssel NL 16 D6
Gort IRL 6 F5
Gortahork IRL 6 B6
Gort an Choirce IRL 6 B6
Gortavoy Bridge UK 7 C9
Gorteen IRL 6 E5
Gorteen IRL 6 E6
Gorteeny IRL 6 F6
Gortin UK 4 F2
Goruia RO 159 C8
Görükle TR 173 D10
Gorun BG 167 B10
Görvik S 106 D8
Görwihl D 27 E9
Gorxheimertal D 21 E11
Gorzanów PL 77 B11
Gorze F 26 B5
Görzig D 79 C10

Gorzyce PL 144 C6
Gorzyce PL 146 B6
Gorzyń PL 81 A9
Górzyn PL 81 C7
Gosaldo I 72 D4
Gosau A 73 A8
Gosberton UK 11 F11
Gościeradów Ukazowy PL 144 B4
Gościm PL 85 E9
Gościno PL 85 B9
Gósciszów PL 81 D8
Gosë AL 168 B2
Gosfield UK 15 D10
Gosforth UK 5 E13
Gosforth UK 10 C5
Goslar D 79 C7
Gosławice PL 139 E9
Goślice PL 139 E8
Gosné F 23 D9
Gospić HR 156 C3
Gospođinci SRB 158 C4
Gosport UK 13 D12
Gössäter S 91 B13
Gossau CH 27 F11
Gossendorf A 148 C5
Gößl A 73 A8
Gostavățu RO 160 E5
Gostil AL 163 E9
Gostilitsa BG 166 C4
Gostinari RO 161 E8
Gostinu RO 161 E8
Gostivar NMK 163 F10
Gostling an der Ybbs A 73 A10
Gostovići BIH 157 D9
Gostycyn PL 138 D4
Gostyń PL 81 C12
Gostyń PL 85 B7
Gostynin PL 139 F7
Goszcz PL 142 D3
Goszczanów PL 142 C6
Goszczyn PL 141 G3
Göta S 91 C11
Göteborg S 91 D10
Götene S 91 B13
Goteşti MD 154 E2
Gotha D 79 E8
Gothem S 93 D13
Götlunda S 97 C14
Gotse Delchev BG 169 A10
Gottböle FIN 122 F6
Gotteszell D 76 E3
Gottfrieding D 75 E11
Göttingen D 78 C6
Gottmadingen D 27 E10
Gottne S 107 E14
Gøttrup DK 86 A4
Götzis A 71 C9
Gouarec F 22 D5
Gouda NL 16 D3
Goudswaard NL 182 B4
Gouesnou F 22 D3
Gouhenans F 26 E5
Goult F 35 C9
Goumenissa GR 169 C7
Goumois F 27 F6
Gourdan-Polignan F 33 D7
Gourdon F 29 F8
Gourdon UK 5 B12
Gourette F 32 E5
Gourgançon F 25 C11
Gouria GR 174 C3
Gourin F 22 D4
Gournay-en-Bray F 18 F4
Goussainville F 25 B7
Gout-Rossignol F 29 E6
Gouveia P 44 D5
Gouvia GR 168 E2
Gouvieux F 25 B7
Gouvy B 20 D5
Gouzon F 29 C10
Gove P 44 B4
Govedari HR 162 D3
Govedartsi BG 165 E7
Goven F 23 D8
Gowarczów PL 141 H2
Gowidlino PL 138 B5
Goworowo PL 139 E12
Gowran IRL 9 C8
Goxhill UK 11 D11
Göynükbelen TR 173 E11
Gózd PL 141 H4
Gozdnica PL 81 D8
Gozdowice PL 84 E6
Gözsüz TR 173 C10
Gozzano I 68 B5
Graal-Müritz D 83 B12
Grab BIH 157 F10
Grab BIH 162 D5
Graben-Neudorf D 187 C5
Grabica PL 143 D8
Grabjan AL 168 C2
Gråbo S 91 D11
Grabovac HR 157 F7
Grabovci SRB 158 D4
Grabovica SRB 159 E10
Grabow D 83 D11
Grabów PL 143 B7
Grabowiec PL 143 D11
Grabowiec PL 144 B8
Grabówka PL 140 D8
Grabów nad Pilicą PL 141 G4
Grabów nad Prosną PL 142 C5
Grabownica Starzeńska PL 144 D5
Grabówno PL 85 D12
Grabowno Wielkie PL 142 D3
Grabowo PL 136 E5
Grabowo PL 139 D13
Grabs CH 71 C8
Gračac HR 156 D4
Gračanica BIH 157 C9
Gračanica BIH 157 D10
Gračanica BIH 157 E8
Gaçay F 24 F6
Grâce-Hollogne B 19 C11
Grâces F 22 C5
Gračianica BIH 157 E8
Gracze PL 142 E4
Grad SLO 148 C6
Gradac BIH 162 D5
Gradac SLO 148 E4
Gradačac BIH 157 C9
Gradara I 67 E6
Gräddö S 99 C12
Gradec HR 149 E6
Gradec NMK 169 B7
Gradefes E 39 C9
Gradets BG 159 E10
Gradets BG 167 D7
Gradevo BG 165 F7
Gradina BG 166 F5
Gradina BG 166 E4

Gorzyce PL 144 C6
Gradina BIH 157 F9
Gradina SRB 163 C9
Grădinari RO 159 C9
Grădinari RO 160 D4
Grădiniţa MD 154 D5
Gradishte BG 166 C4
Gradište SRB 164 B5
Gradište SRB 164 C5
Grădiştea RO 160 D3
Grădiştea RO 161 D8
Grădiştea RO 161 D8
Gradište Bekteško HR 149 F8
Gradnitsa BG 165 D10
Grado E 39 B7
Grado I 73 E7
Gradø S 97 B15
Gradojević SRB 158 D4
Gradoli I 62 B1
Gradsko NMK 169 A6
Gradskovo SRB 159 E9
Graena E 55 D6
Græsted DK 87 C10
Gräfelfing D 75 F9
Grafenau D 76 E4
Gräfenberg D 75 C9
Gräfendorf D 74 B6
Grafenhausen D 27 E9
Gräfenhainichen D 79 C11
Gräfenhausen D 27 E9
Grafenreinfeld D 187 B9
Gräfenroda D 79 E8
Grafenstein A 73 C9
Grafenwöhr A 77 F9
Gräfenwohr D 75 C10
Grafenwörth A 77 F9
Gräffignano I 62 B2
Grafhorst D 79 B8
Graf Ignatievo BG 165 E10
Gräfinau-Angstedt D 79 E9
Grafing bei München D 75 F10
Grafrath D 75 F9
Gräfsnäs S 91 C12
Gräftåvallen S 105 E15
Gragnano I 60 B3
Grahovac HR 157 E6
Grahovo MNE 163 D6
Grahovo SLO 73 D8
Grái g na Manach IRL 9 C9
Graigue IRL 9 C9
Graiguenamanagh IRL 9 C9
Grain UK 15 E10
Grainau D 71 C10
Grainet D 76 E5
Grajal de Campos E 39 D9
Grajduri RO 153 D11
Grajewo PL 140 C6
Gråmanstorp A 76 F6
Gramastetten A 76 F6
Gramat F 29 F9
Gramatikovo BG 167 E9
Grambow D 84 D6
Grămeşti RO 153 B8
Gramkow D 83 C10
Grammatiko GR 169 C7
Grammendorf D 83 B13
Grammeniza GR 174 A2
Grammeno GR 168 E4
Grammichele I 59 E6
Gramsbergen NL 17 C7
Gramsh AL 168 C3
Gramzda LV 134 D2
Gramzow D 84 D6
Gran N 95 B13
Granabeg IRL 7 F10
Granada E 53 B9
Granån S 119 C10
Granard IRL 7 E8
Granarolo dell'Emilia I 66 C3
Gränsälen S 102 B3
Granátula de Calatrava E 54 B5
Granberg S 118 D3
Granberget N 102 D3
Granbergsdal S 97 D12
Granbergsträsk S 118 D3
Grančarevo BIH 162 D5
Grancey-le-Château-Nouvelle F 26 E3
Grandas E 39 B6
Grandcamp-Maisy F 23 B9
Grand-Champ F 22 E6
Grand-Couronne F 18 F3
Grande-Synthe F 18 B5
Grande-Vabre F 33 A10
Grand-Fort-Philippe F 18 B5
Grand-Fougeray F 23 E8
Grândola P 50 C2
Grandpré F 19 F10
Grandpuits-Bailly-Carrois F 25 C8
Grandrieu F 30 F4
Grandris F 30 C5
Grandson CH 31 B10
Grandtully UK 5 B9
Grandvelle-et-le-Perrenot F 26 E5
Grandvillars F 27 E6
Grandvilliers F 18 E6
Grañén E 41 E11
Grängärde S 97 B12
Grange IRL 9 E7
Grangemouth UK 5 C9
Grange-over-Sands UK 10 C6
Grängesberg S 97 B13
Granges-sur-Vologne F 27 D6
Grängshyttan S 97 C12
Grängsjö S 103 B13
Granhult S 116 E7
Granica BIH 162 D5
Grănicerii RO 151 D8
Grănicești RO 153 B8
Graninge S 107 E12
Granitola-Torretta I 58 D2
Granitsa GR 174 A4
Granja P 51 C3
Granja de Moreruela E 39 E8
Granja de Torrehermosa E 51 C8
Granja do Ulmeiro P 44 D3
Gränna S 92 C4
Grannäs S 107 A10
Grannäs S 109 E13
Granneset N 108 D7
Granö S 107 C16

Granollers E 43 D8
Granön S 103 D13
Granowiec PL 142 D4
Granowo PL 81 B11
Granschütz D 79 D11
Gransee D 84 D4
Granselet S 109 E14
Gränsgård S 109 F15
Gransjö S 107 A15
Gransjön S 103 B11
Grantham UK 11 F10
Grantown-on-Spey UK 3 L9
Granträsk S 107 C14
Granträskmark S 118 D6
Grantshouse UK 5 D12
Gränum S 88 C7
Granvik S 92 B5
Granville F 23 C8
Granvin N 94 A5
Grapska BIH 157 C9
Gräs S 97 C10
Grasbakken N 114 C5
Gräsgård S 89 C11
Grasleben D 79 B9
Gräsmarken S 97 B8
Grasmere UK 10 C5
Gräsmyr S 107 D17
Grassano I 60 B6
Grassau D 72 A5
Grasse F 36 D5
Grassington UK 11 C8
Gråssjö S 103 A13
Gråsten DK 86 F5
Gråstorp S 91 C12
Gratallops E 42 E5
Gratangsbotn N 111 C14
Gratens F 33 D8
Gratia RO 161 E6
Gratini GR 171 B9
Gratkorn A 148 B4
Grätnäs S 107 E14
Gräträsk S 118 C3
Gratteri I 58 D4
Gratwein A 148 B4
Grauballe DK 86 C5
Graulhet F 33 C9
Graupa D 80 D5
Graus E 42 F4
Grávalos E 41 D8
Gravberget N 102 E3
Gravbränna S 106 D7
Gravdal N 110 D6
Grave NL 16 E5
Gravedona I 69 A7
Gravelines F 18 C5
Gravellona Toce I 68 B5
Gravelotte F 26 B5
Gravens DK 86 D4
Grävenwiesbach D 21 D10
Gräveri LV 133 D2
Gravesend UK 15 E9
Gravia GR 174 B5
Gravigny F 24 B5
Gråvika N 100 E1
Gravina di Catania I 59 D7
Gravina in Puglia I 61 B6
Gravmark S 118 C4
Gray F 26 F4
Grayan-et-l'Hôpital F 28 E3
Grays UK 15 E9
Graz A 148 B4
Grazalema E 53 C6
Grążawy PL 139 D8
Grazzanise I 60 A2
Grčak SRB 163 C10
Grdelica SRB 164 D5
Greaca RO 161 E8
Greåker N 95 D14
Gréalou F 33 A9
Great Baddow UK 15 D10
Great Bircham UK 15 B10
Great Clifton UK 5 B9
Great Cornard UK 15 C10
Great Dunmow UK 15 D9
Great Glen UK 11 F9
Great Gonerby UK 11 F10
Greatham UK 11 B9
Great Harwood UK 11 D7
Great Haywood UK 11 F8
Great Linford UK 15 C7
Great Malvern UK 13 A10
Great Ponton UK 11 F10
Great Salkeld UK 5 F11
Great Shelford UK 15 C9
Great Torrington UK 12 D6
Great Wakering UK 15 D10
Great Yarmouth UK 15 B12
Grebănu RO 161 C9
Grebaštica HR 156 E4
Grebbestad S 91 B9
Grebci BIH 162 D5
Grebenac SRB 159 D7
Grebendorf (Meinhard) D 79 D7
Grebenhain D 21 D12
Grebenişu de Câmpie RO 152 D4
Grebenstein D 21 B12
Grebin D 83 B8
Grębków PL 141 F5
Grebo S 92 C7
Grębocin PL 138 D6
Grójbości n PL 143 F10
Grębów PL 144 B4
Greccio I 62 C3
Grecești RO 160 E2
Greci I 60 A4
Greci RO 155 D2
Gredelj BIH 157 F10
Greding D 75 D9
Gredstedbro DK 86 E3
Greencastle UK 4 F2
Greencastle UK 4 E2
Greencastle UK 4 E3
Greene D 78 C6
Greengairs UK 5 D9
Greenhead UK 5 F11
Greenigo UK 3 H10
Greenlaw UK 5 D11
Greenloaning UK 5 C9
Greenock UK 4 D7
Greenodd UK 10 C5
Greenore IRL 7 D10
Greenway UK 12 B5
Greggio I 68 C5
Greifenburg A 73 C7
Greifendorf D 80 D4
Greifenstein D 21 C10
Greiffenberg D 84 D5
Greifswald D 84 B4
Grein A 77 F7

Greiveri LV 135 B11
Greiz D 79 E11
Grejs DK 86 D5
Gremersdorf D 83 B9
Grenaa DK 87 C7
Grenade F 33 C8
Grenade-sur-l'Adour F 32 C5
Grenant F 26 E4
Grenås S 106 D8
Grenchen CH 27 F7
Grenči LV 134 C5
Grenoble F 31 E8
Grense-Jakobselv N 114 D9
Gréoux-les-Bains F 35 C10
Gresenhorst D 83 B12
Gress UK 2 J4
Gressan I 31 D11
Gresse D 83 D9
Gressoney-la-Trinite I 68 B4
Gressvik N 91 A8
Gresten A 77 G8
Grésy-sur-Aix F 31 D8
Grésy-sur-Isère F 31 D9
Gretna UK 5 F10
Greußen D 79 D8
Greux F 26 D4
Grevbäck S 92 C4
Greve in Chianti I 66 E3
Greven D 17 D9
Greven D 83 D9
Grevena GR 168 D5
Grevenbicht NL 183 C7
Grevenbroich D 21 B7
Greveniti GR 168 E5
Grevenmacher L 20 E6
Grevesmühlen D 83 C10
Greve Strand DK 87 D10
Grevie S 87 C11
Grevinge DK 87 D9
Greyabbey UK 7 C11
Greystoke UK 5 F11
Greystones IRL 7 F10
Grez-Doiceau B 19 C10
Grez-en-Bouère F 23 E10
Grgar SLO 73 D8
Grgurevci SRB 158 C4
Griegos E 47 D9
Gries F 27 C8
Gries am Brenner A 72 B3
Griesbach D 27 D9
Griesheim D 21 E11
Grieskirchen A 76 F5
Grießen D 81 C7
Griesstätt D 75 G11
Griffen A 73 C10
Grigiškes LT 137 D11
Grignan F 35 B8
Grigno I 72 D4
Grignols F 32 B5
Grigny F 30 D6
Grigoriopol MD 154 C4
Grijota E 40 D2
Grijpskerk NL 16 B6
Grillby S 98 C8
Grillon F 35 B8
Grimaldi I 60 E6
Grimaud F 36 E5
Grimbergen B 19 C9
Grimma D 79 D12
Grimmen D 84 B4
Grimoldby UK 11 E12
Grimsås S 91 E14
Grimsby UK 11 D11
Grimslöv S 88 B7
Grimstad N 90 C4
Grimston UK 11 F13
Grimstorp S 92 D5
Grimstrup DK 86 D3
Grindafjord N 94 D2
Grindaheim N 101 A11
Grindal N 101 A11
Grinder N 96 B7
Grindjord N 111 D13
Grindsted DK 86 A6
Grindsted DK 86 D3
Grindu RO 161 D9
Grinkiškis LT 134 E7
Griñón E 46 D5
Grins A 71 C11
Grinsbol S 97 C8
Grințies RO 153 C7
Gripenberg S 92 D5
Grisén E 41 E9
Griškabūdis LT 136 D7
Grisolia I 60 D5
Grisolles F 33 C8
Grisslehamn S 99 B11
Gritley UK 3 H11
Grivița RO 153 E11
Grivița RO 153 F11
Grivița RO 161 D10
Grivitsa BG 165 C10
Grkinja SRB 164 C4
Grljan SRB 159 F9
Grnčari NMK 168 B5
Grobbendonk B 19 B10
Gröbenzell D 75 F9
Grobiņa LV 134 C2
Gröbming A 73 B8
Gröbzig D 79 C10
Grocka SRB 158 D6
Gródby S 93 A11
Gródek PL 140 D9
Gródek nad Dunajcem PL 144 D2
Gröden D 80 D5
Grödig A 73 A7
Gröditz D 80 D4
Gródki PL 139 D9
Grodków PL 142 E3
Grodziczno PL 139 D8
Grodziec PL 142 B5
Grodzisk PL 141 E7
Grodzisk Mazowiecki PL 141 F3
Grodzisk Wielkopolski PL 81 B10
Groenlo NL 17 D7
Groesbeek NL 183 B7
Grohnde (Emmerthal) D 78 B5
Groitzsch D 79 D11
Groix F 22 E5
Grojdibodu RO 160 F4
Grójec PL 141 G3
Grom PL 139 C10
Gromadka PL 81 D9
Grömitz D 83 B9
Gromnik PL 144 D2
Gromo I 69 B8
Gronau (Westfalen) D 17 D8
Grønbjerg DK 86 C3
Grønbo S 118 D5
Gröndal S 108 D8

Grong N 105 C12
Grönhögen S 89 C10
Grønhøj DK 86 C4
Gröningen D 79 C9
Groningen NL 17 B7
Grønnemose DK 86 E6
Grønnes N 100 A6
Grønning N 110 E9
Grønningen N 104 D7
Gronowo PL 139 B9
Grönsinka S 98 B7
Grönskåra S 89 A9
Grønsta S 103 A11
Grønviken N 94 D4
Grønviken S 103 A9
Grönviken S 103 D12
Groomsport UK 4 F5
Grootegast NL 16 B6
Gropello Cairoli I 37 B7
Gropeni RO 155 C1
Gropnița RO 153 C10
Grorud N 95 D11
Groscavallo I 31 E11
Grosio I 69 A9
Grošnica SRB 159 F6
Grosotto I 69 A9
Großaitingen D 71 A11
Groß Ammensleben D 79 B10
Großarl A 73 B7
Groß-Bieberau D 21 E11
Groß Börnecke D 79 C9
Großbothen D 79 D12
Großbottwar D 27 C11
Großbreitenbach D 79 E9
Großburgwedel (Burgwedel) D 78 B6
Groß Dölln D 84 E5
Grosselfingen D 27 D10
Großenaspe D 83 C7
Großenbrode D 83 B10
Großenehrich D 79 D8
Groß Engersdorf A 77 F11
Großengottern D 79 D8
Großenhain D 80 D5
Großenkneten D 17 C10
Großenlüder D 78 E6
Großensee D 83 C8
Großenstein D 79 E11
Großenwiehe D 82 A6
Groß-Enzersdorf A 77 F11
Grosseto I 65 B4
Grosseto-Prugna F 37 H9
Groß Fredenwalde D 84 D5
Großfurra D 79 D8
Groß-Gerau D 21 E10
Groß-Gerungs A 77 E7
Groß Glienicke D 80 B4
Großmain D 73 A6
Groß Grönau D 83 C9
Großhabersdorf D 75 C8
Großhansdorf D 83 C8
Großharras A 77 E10
Großhartmannsdorf D 80 E4
Groß Heere (Heere) D 79 B7
Großhennersdorf D 81 E7
Groß-Hesepe D 17 C8
Großheubach D 187 B7
Grössjö S 107 F14
Großkarolinenfeld D 72 A5
Groß Kiesow D 84 B4
Großklein A 148 C4
Groß Köris D 80 B5
Groß Kreutz D 79 B12
Großkrut A 77 E11
Groß Lafferde (Lahstedt) D 79 B7
Großlangheim D 187 B9
Groß Leine D 80 B6
Groß Leuthen D 80 B6
Großlittgen D 21 D7
Großlohra D 79 D8
Großmaischeid D 185 C8
Großmehlen D 80 D5
Groß Miltzow D 84 C5
Groß Mohrdorf D 84 B3
Großmonra D 79 D9
Groß Naundorf D 80 C3
Groß Nemerow D 84 D4
Groß Oesingen D 79 A7
Großolbersdorf D 80 E4
Groß Oßnig D 80 C6
Großostheim D 21 E12
Großpetersdorf A 149 B6
Groß Plasten D 83 C13
Großraming A 73 A10
Großräschen D 80 C6
Großrinderfeld D 74 C6
Groß Roge D 83 C13
Groß-Rohrheim D 21 E10
Großröhrsdorf D 80 D6
Großrosseln D 186 C2
Großrudestedt D 79 D9
Großrußbach A 77 F11
Groß Sankt Florian A 148 C4
Groß Särchen D 80 D6
Großschirma D 80 E4
Großschönau A 77 E7
Groß Schönebeck D 84 E5
Groß Schwechten D 83 E11
Groß Schwülper (Schwülper) D 79 B7
Groß-Siegharts A 77 E8
Groß Stieten D 83 C10
Großthiemig D 80 D5
Großträben D 80 D3
Groß Twülpstedt D 79 B8
Groß-Umstadt D 187 B6
Großwallstadt D 187 B7
Groß Warnow D 83 D11
Großweikersdorf A 77 F9
Groß Welle D 83 D12
Groß Wittensee D 82 B7
Groß Wokern D 83 C12
Großwudicke D 79 A11
Groß Wüstenfelde D 83 C13
Groß-Zimmern D 187 B6
Groß Ziethen D 84 D5
Grostenquin F 27 C6
Grosuplje SLO 73 E10
Grøtavær N 111 C11
Grotfjord N 111 A16
Grötholen S 102 C4
Grötingen S 103 A9
Grötlingbo S 93 E13
Grøtnesdalen N 111 A18
Grottaferrata I 62 D3
Grottaglie I 61 C9
Grottaminarda I 60 A4
Grottammare I 62 B5
Grottazzolina I 62 A5

Grotte I 58 E4
Grotte di Castro I 62 B1
Grotteria I 59 C9
Grottole I 61 B6
Grou NL 16 B5
Grove UK 13 B12
Grövelsjön S 102 B3
Grovfjord N 111 C13
Grozd'ovo BG 167 C9
Grozești RO 153 D12
Grub am Forst D 75 B9
Grubbenvorst NL 183 C8
Grubbnäsudden S 119 B10
Grube D 83 B10
Grubišno Polje HR 149 E8
Gruczno PL 138 D5
Gruda HR 162 D5
Grude BIH 157 F7
Grudusk PL 139 D10
Grudziądz PL 138 D6
Gruey-lès-Surance F 26 D5
Gruglasco I 68 C4
Gruia RO 159 E10
Gruissan F 34 D5
Gruiu RO 161 D8
Grullos E 39 B7
Grumăzești RO 153 C8
Grumbach D 80 D5
Grumo Appula I 61 A7
Grums S 97 D9
Grünau I 43 H8
Grünau A 73 H9
Grünberg D 21 C11
Grünburg A 76 G6
Grundagssätern S 102 B4
Grundfors S 106 A9
Grundfors S 107 B13
Grundforsen S 102 D4
Grundsanden S 118 C4
Grundsel S 118 C4
Grundsjö S 107 C11
Grundsuna S 107 E16
Grundsund S 91 C9
Grundtjärn S 107 D12
Grundträsk S 107 A16
Grundträsk S 109 F17
Grundträsk S 118 B7
Grundvattnet S 118 C4
Grünendeich D 82 C7
Grünewalde D 80 C5
Grungebru N 94 C7
Grünheide D 80 B5
Grünkraut D 71 B9
Grunnfarnes N 111 B12
Grunnfjord N 112 C4
Grunnførfjord N 110 D8
Grunow D 80 B6
Grünsfeld D 74 C6
Grünstadt D 21 E10
Grünwald D 75 F10
Grupčin NMK 164 F3
Grury F 30 B4
Grüšlauke LT 134 D2
Grußendorf (Sassenburg) D 79 A8
Gruta PL 138 D6
Gruvberget S 103 D11
Gruyères CH 31 B11
Gruža SRB 159 F6
Grūžai LT 134 D6
Grybów PL 145 D2
Grycksbo S 103 E9
Gryfice PL 85 C8
Gryfino PL 84 D6
Gryfów Śląski PL 81 D8
Grygov CZ 146 B4
Grylewo PL 85 E12
Gryllefjord N 111 B13
Grynberget S 107 C11
Gryt S 93 C9
Grytan S 106 E7
Grytgöl S 92 B7
Gryts bruk S 92 B6
Grytsjö S 106 A9
Gryttjom S 99 B8
Gryźliny PL 139 C9
Grzebienisko PL 81 B11
Grzmiąca PL 85 C10
Grzybno PL 84 D7
Grzybno PL 138 D5
Grzymiszew PL 142 B5
Grzywna PL 138 D6
Gschnitz A 72 B3
Gschwandt A 73 A8
Gschwend D 74 E6
Gstaad CH 31 C11
Gsteig CH 31 C11
Guadahortuna E 53 A10
Guadajoz E 51 D8
Guadalajara E 47 C6
Guadalaviar E 47 D9
Guadalcanal E 51 C8
Guadalcázar E 53 A7
Guadalest E 56 D4
Guadalmez E 54 B3
Guadarrama E 46 C4
Guadassuar E 48 F4
Guadix E 55 E6
Guaire IRL 9 C10
Guájar-Faraguit E 53 C10
Gualav S 88 C6
Gualdo Cattaneo I 62 B3
Gualdo Tadino I 67 F6
Gualtieri I 66 C2
Guarcino I 62 D4
Guarda P 45 C6
Guardamar del Segura E 56 E3
Guardavalle I 59 B9
Guardea I 62 B2
Guardiagrele I 63 C6
Guardia Lombardi I 60 B4
Guardia Perticara I 60 C6
Guardia Piemontese I 60 E6
Guardia Sanframondi I 60 A3
Guardias Viejas E 55 F7
Guardiola de Berguedà E 43 C7
Guardo E 39 C10
Guardramiro E 45 B8
Guareña E 51 B7
Guaro E 53 C7
Guarromán E 54 C5
Guasila I 64 D3
Guastalla I 66 C2
Gubaveeny IRL 7 C7
Gubbhögen S 106 C9
Gubbio I 66 F6
Gubbträsk S 107 A13
Guben D 81 C7
Gübene BG 166 D4

Gubin PL 81 C7
Guča SRB 158 F5
Guča Gora BIH 157 D8
Gudar E 48 D3
Gudavac BIH 156 C5
Güdesberge D 78 D6
Gudenieki LV 134 C3
Guderup DK 86 F5
Gudhem S 91 C14
Gudhjem DK 89 E7
Gudiena LT 137 D9
Gudinge S 99 B8
Gudme DK 87 E7
Gudow D 83 D9
Gudum DK 86 B2
Gudumholm DK 86 B6
Gudvangen N 100 E5
Gudžiūnai LT 134 E7
Guebwiller F 27 E7
Guégon F 22 E6
Guéjar-Sierra E 53 B10
Guémar F 27 D7
Guémené-Penfao F 23 E8
Guémené-sur-Scorff F 22 D5
Guénange F 20 F6
Güenes E 40 B5
Guenrouet F 23 E7
Guer F 23 E7
Guérande F 22 F7
Guéret F 29 C9
Guérigny F 25 F9
Guerri de la Sal E 33 F8
Güesa E 32 E3
Gueugnon F 30 B5
Gugești RO 161 B10
Güglingen D 27 B10
Guglionesi I 63 D7
Gugutka BG 171 B9
Gühlen-Glienicke D 83 D13
Guia P 44 E3
Guia P 50 E3
Guichen F 23 E8
Guidel F 22 E5
Guide Post UK 5 E13
Guidizzolo I 66 B2
Guidonia-Montecelio I 62 D3
Guiglia I 66 D2
Guignen F 23 E8
Guignes F 25 C8
Guignicourt F 19 F8
Guijo de Coria E 45 D8
Guijo de Galisteo E 45 D8
Guijo de Granadilla E 45 D8
Guijuelo E 45 C9
Guildford UK 15 E8
Guilers F 22 D2
Guilherand F 30 F6
Guilhofrei P 38 E3
Guilhovai P 44 C3
Guillena E 51 D7
Guillestre F 36 B5
Guilliers F 22 D6
Guillon F 25 E11
Guillos F 32 A4
Guilvinec F 22 E3
Guimarães P 38 F3
Guînes F 15 F12
Guingamp F 22 C5
Guipavas F 22 D3
Guipry F 23 E8
Guisando E 45 D10
Guisborough UK 11 B9
Guiscard F 19 E7
Guise F 19 E8
Guiseley F 22 D3
Guissona E 42 D6
Guist UK 15 B10
Guitiriz E 38 B4
Guîtres F 28 E5
Gujan-Mestras F 32 A3
Gulács F 145 G5
Gülbahçe TR 177 C8
Gulbene LV 135 B12
Guldager DK 86 D2
Guldborg DK 84 A1
Gulgamme N 113 B16
Gulianca RO 161 C11
Gullabo S 89 C9
Gullane UK 5 C11
Gullberg S 103 D11
Gullbrå N 100 E4
Gullbrandstorp S 87 B11
Gullbranna S 87 B11
Gulleråsen S 103 D9
Gullholmen N 111 C10
Gullön S 109 F15
Gullringen S 92 D7
Gullspång S 91 B15
Gullträsk S 118 B6
Güllübahçe TR 177 C9
Güllüce TR 173 D9
Güllük TR 177 E10
Gulpen NL 183 D7
Gülpınar TR 171 E10
Gulsele S 107 D11
Gulsvik N 95 B11
Gültz D 84 C4
Gülübintsi BG 166 E6
Gülübovo BG 165 E10
Gülübovo BG 166 E5
Gulyantsi BG 160 F5
Gülzow D 83 C12
Gumhöjden S 97 B10
Gumiel de Hizán E 40 E4
Gumiel de Mercado E 40 E4
Gummark S 118 E5
Gummersbach D 21 B9
Gumpelstadt D 79 E7
Gumtow D 83 E12
Gümüşçay TR 173 D7
Gümüşsuyu TR 177 C9
Gümüşyaka TR 173 B9
Gümzovo BG 159 E10
Gunaroš SRB 150 F4
Gundelfingen D 27 D8
Gundelfingen an der Donau D 75 E7
Gundelsheim D 21 F12
Gundershoffen F 27 C8
Gundinci HR 157 B9
Gündoğan TR 177 E9
Gündoğdu TR 173 D7
Gündsømagle DK 87 D10
Güneyli TR 172 D6
Güngörmez TR 173 C10
Gunja HR 157 C10
Gunnarn S 107 B13
Gunnarnes N 113 A13
Gunnarsbo S 103 C11

Gunnarsbyn S 118 B7
Gunnarskog S 97 C8
Gunnarvattnet S 105 C16
Gunnebo S 93 D9
Günsburg D 75 D9
Güntersberge D 79 C8
Guntersblum D 185 E9
Guntersdorf A 77 E10
Güntersleben D 187 B8
Gunzenhausen D 75 D8
Gura-Bicului MD 154 D4
Gura Caliței RO 161 B9
Gura Camencii MD 154 B2
Gura Foii RO 160 D6
Gura Galbenei MD 154 D3
Gurahonț RO 151 E9
Gura Humorului RO 153 B7
Gura Ocniței RO 161 D7
Gurasada RO 151 F10
Gura Râului RO 152 F3
Gura Teghii RO 161 C8
Gura Văii RO 159 E10
Gura Vadului RO 161 D8
Gura Vitioarei RO 161 D8
Gurbănești RO 161 E9
Gurghiu RO 152 D5
Guri i Bardhë AL 168 B3
Guri i Zi AL 163 E8
Gurk A 73 C9
Gurkovo BG 166 D5
Gurkovo BG 167 C10
Gürlyano BG 164 D5
Gurrea de Gállego E 41 D10
Gurrë e Madhe AL 168 A3
Gurteen IRL 6 E6
Gur'yevsk RUS 136 D2
Gusborn D 83 D10
Gušče HR 149 F7
Güsen D 79 B10
Gusev RUS 136 D5
Gusinje MNE 163 D8
Gușoeni RO 160 D4
Gusow D 80 A6
Guspini I 64 D2
Gussago I 69 B9
Gusselby S 97 C13
Güssing A 149 B6
Gussola I 66 B1
Gussvattnet S 105 C16
Gußwerk A 148 A4
Gustavsberg S 99 D10
Gustavsberg S 107 B10
Gustavsfors S 97 B10
Güsten D 79 C10
Gusterath D 21 E7
Güstrow D 83 C12
Gusum S 93 C8
Gutach (Schwarzwaldbahn) D 187 E5
Gutau A 77 F7
Gutcher UK 3 D14
Güterfelde D 80 B4
Güterglück D 79 C10
Gütersloh D 17 E10
Guthrie UK 5 B11
Gutorfölde H 149 C7
Gutow D 83 C12
Guttannen CH 70 D6
Guttaring A 73 C10
Gützkow D 84 C4
Guvåg N 110 C8
Güvemalani TR 173 D7
Güvercinlik TR 177 E10
Guxhagen D 78 D5
Güzelbahçe TR 177 C8
Güzelçamli TR 177 D9
Gvardeysk RUS 136 D3
Gvarv N 95 D10
Gvozd HR 148 F5
Gvozd MNE 163 D7
Gvozdansko HR 156 B5
Gwda Wielka PL 85 C11
Gweedore IRL 6 B6
Gwithian UK 12 E4
Gy F 26 F4
Gyál H 150 C3
Gyarmat H 149 B8
Gyékényes H 149 D8
Gyenesdiás H 149 C8
Gyermely H 149 A11
Gyé-sur-Seine F 25 D11
Gyhum D 17 B12
Gyljen S 118 B9
Gylling DK 86 D6
Gymno GR 175 C8
Gyomaendrőd H 150 D6
Gyömöre H 149 A9
Gyömrő H 150 C3
Gyón H 150 C3
Györköny H 149 C11
Györság H 149 A9
Györszemere H 149 A9
Györtelek H 145 H5
Györújbarát H 149 A9
Györújfalu H 149 A9
Györzámoly H 149 A9
Gyrstinge DK 87 E9
Gysinge S 98 B7
Gytheio GR 178 B4
Gyttorp S 97 C12
Gyueshevo BG 164 E5
Gyula H 151 D7
Gyulaj H 149 C11
Gyúró H 149 B11
Gzy PL 139 E10

H

Haabneeme EST 131 B9
Haacht B 19 C10
Häädemeeste EST 131 E8
Haaften NL 183 B6
Haag A 77 F7
Haag am Hausruck A 76 F5
Haag in Oberbayern D 75 F11
Haajainen FIN 124 C7
Haaksbergen NL 17 D7
Haaltert B 19 C8
Haanja EST 131 F14
Haapajärvi FIN 123 E15
Haapakoski FIN 123 E16
Haapakoski FIN 124 F8
Haapakumpu FIN 115 D3
Haapalahti FIN 113 C10
Haapaluoma FIN 123 E10
Haapamäki FIN 123 C16
Haapamäki FIN 123 E17
Haapamäki FIN 125 E10
Haaparanta FIN 115 E3
Haapavesi FIN 119 F14
Haapsalu EST 130 D7
Haar D 75 F10
Haarajoki FIN 123 F17
Haarajoki FIN 127 E13
Haarajoa FIN 119 F16
Haarasajo FIN 119 B12
Haarbach D 76 E4
Haarby DK 86 E6
Haaren NL 183 B6
Haarlem NL 16 D3
Haastrecht NL 16 E3
Haavisto FIN 127 D11
Habaja EST 131 C10
Habartov CZ 75 B12
Habas F 32 C4
Habay-la-Neuve B 19 E12
Häbbersliden S 118 E4
Habiller D 187 B8
Hablingbo S 93 E12
Habo S 91 D15
Håbol S 91 B11
Habovka SK 147 C9
Habry CZ 77 C8
Habsheim F 27 E7
Hachenburg D 21 C9
Haciahmetli TR 177 B10
Haciköy TR 172 C6
Hacılar TR 173 D8
Hackås S 102 A7
Hackenheim D 21 E9
Hacketstown IRL 9 C9
Hackleton UK 15 C7
Hacksjö S 107 B12
Hadamar D 21 D10
Hadanberg S 107 D14
Haddebo S 92 B6
Haddington UK 5 D11
Haddiscoe UK 15 B12
Haderslev DK 86 E4
Haderup DK 86 C3
Hadimköy TR 173 B10
Hadleigh UK 15 D10
Hadley UK 10 F7
Hadol F 26 D5
Hadres A 77 E10
Hadsel N 110 C8
Hadsten DK 86 C6
Hadsund DK 86 B6
Hadžići BIH 157 E9
Haelen NL 183 C7
Hafnerbach A 77 F8
Haftersbol S 97 C10
Haga S 98 B8
Haganj HR 149 E7
Hagaström S 103 E13
Hagby S 89 B9
Hagebyhöga S 92 C5
Hagen D 17 F9
Hagenbach D 187 C5
Hagenburg D 17 D12
Hagenow D 83 D10
Hagenwerder D 81 D7
Hageri EST 131 C9
Hägerstad S 92 C7
Hagetaubin F 32 C4
Hagetmau F 32 C4
Hagfors S 97 B10
Häggås S 107 C11
Häggdånger S 103 A14
Häggbäcken S 98 A14
Häggenås S 106 E7
Häggnäset S 105 C16
Häggsåsen S 105 D13
Häggsjön S 105 D13
Häggsjövik S 105 D16
Häghig RO 153 F7
Hagley UK 13 A8
Hagondange F 20 F6
Hagsta S 103 E13
Haguenau F 27 C8
Hahausen D 79 C7
Hahnbach D 75 C10
Hahnstätten D 185 D9
Hahót H 149 C7
Haibach D 187 B7
Haibach D 75 D12
Haidmühle D 76 E5
Haiger D 21 C10
Haigerloch D 27 D10
Häijää FIN 127 B9
Hailsham UK 15 F9
Hailuoto FIN 119 D13
Haiming D 76 F3
Haimoo FIN 127 E11
Haina (Kloster) D 21 B11
Hainburg an der Donau A 77 F11
Hainfeld A 77 F9
Hainichen D 80 E4
Haiterbach D 187 D6
Haitzendorf A 77 F9
Hajala FIN 126 E8
Hajdúböszörmény H 151 B8
Hajdúdorog H 151 B8
Hajdúhadház H 151 B8
Hajdúnánás H 151 B8
Hajdúsámson H 151 C7
Hajdúszoboszló H 151 C7
Hajdúszovát H 151 C7
Hajmáskér H 149 B10
Hajmel AL 163 F8
Hajnáčka SK 147 E9
Hajnówka PL 141 E9
Hajós H 150 E3
Hakadal N 95 B13
Håkafot S 105 C15
Håkansvallen S 102 B3
Hakenstedt D 79 B9
Häkkilä FIN 123 E14
Hakokylä FIN 121 E12
Håkøybotn N 111 A16
Hakstabben N 112 C11
Håkvik N 111 D13
Halahora de Sus MD 153 A10
Håland N 94 C4
Halapić BIH 157 D6
Hålászi H 146 F4
Halásztelek H 150 C2

Hălăucești RO 153 C9
Halbe D 80 B5
Halbenrain A 148 C5
Hålberg S 109 F18
Halberstadt D 79 C9
Halbjerg DK 90 E7
Halblech D 71 B11
Halbturn A 77 G11
Hălchiu RO 153 F7
Hald DK 86 B4
Hald DK 86 B6
Haldarsvik FO 2 A2
Halden N 91 A10
Haldensleben D 79 B9
Hale UK 11 E7
Halen B 183 D6
Halenkov CZ 146 C6
Halenkovice CZ 146 C4
Halesowen UK 13 A10
Halesworth UK 15 C12
Halfing D 72 A5
Halfway IRL 8 E5
Halfweg NL 16 D3
Halhjem N 94 B2
Halič SK 147 E9
Halifax UK 11 D8
Halikko FIN 127 E9
Halimba H 149 B9
Halitpaşa TR 177 B10
Håljarp S 87 D11
Halkia FIN 127 D13
Halkirk UK 3 H10
Halkivaha FIN 127 C9
Halkokumpu FIN 124 F7
Hall S 93 D13
Hälla S 107 C12
Halla-aho FIN 124 C9
Hallabro S 89 C8
Hällabrottet S 92 A6
Halla Heberg S 91 E10
Halland UK 15 F9
Hålland S 105 E14
Hallapuro FIN 123 D12
Hällaryd S 89 C7
Hällbacken S 105 E16
Hällbacken S 109 F12
Hällberga S 98 D7
Hällbo S 103 D11
Hallbodarna S 105 E15
Hällbybrunn S 98 D6
Halle B 19 C9
Halle D 78 C6
Halle NL 183 B8
Halle S 87 A11
Hadsten DK 86 C6
Halle (Saale) D 79 D10
Halle (Westfalen) D 17 D10
Hällefors S 103 A11
Hälleforsnäs S 93 A9
Hallein A 73 A7
Hällekis S 91 C13
Hällen S 103 E14
Hallen S 105 D16
Hallenberg D 21 B11
Hallencourt F 18 E4
Halle-Neustadt D 79 D10
Hallerndorf D 75 C9
Hällesjö S 103 A11
Hällestad S 92 B7
Hällevadsholm S 91 B10
Hällevik S 88 C7
Hälleviksstrand S 91 C9
Hall-Håxåsen S 106 D8
Hallingby N 95 B13
Hallingeberg S 93 D9
Hällinmäki FIN 124 F9
Hall in Tirol A 72 B4
Hällnäs S 107 C17
Hällnäs S 109 D11
Hällnäs S 109 E13
Hallom S 102 A7
Hallow UK 13 A10
Hällsbäck S 97 D8
Hallsberg S 92 A6
Hällsjö S 107 E13
Hällsta S 98 D6
Hallstahammar S 98 C6
Hallstatt A 73 A8
Hallstavik S 99 B11
Halltorp S 89 C10
Halluin B 19 C7
Hallund DK 86 A6
Hällvik S 109 D14
Hallviken S 106 D8
Hallworthy UK 12 D5
Hälmägel RO 151 E10
Hälmagiu RO 151 E10
Halmeu RO 145 H7
Halmstad S 87 B11
Halna S 92 B4
Halosenniemi FIN 119 D14
Halosenranta FIN 115 E2
Halovo SRB 159 F9
Hals DK 86 B6
Halsa N 104 E4
Halsenbach D 21 D9
Hal'shany BY 137 E13
Hälsingby FIN 122 D7
Hälsjö S 103 C12
Halsøy N 108 D5
Halskov DK 87 E8
Halstead UK 15 D10
Halstenbek D 83 C7
Halsteren NL 16 E2
Halstroff F 20 F6
Halsua FIN 123 D12
Haltern D 17 E8
Halttula FIN 119 C14
Haluna FIN 125 D10
Hälvä FIN 129 B10
Halvarsgårdarna S 97 B13
Halver D 21 B9
Halvrimmen DK 86 A5
Halwell UK 13 E7
Halže CZ 75 C12
Ham F 19 E7
Ham UK 3 E12
Hamar N 101 E14
Hamari FIN 119 C14
Hamarøy N 108 D5
Hamarøy N 111 D10
Hambach D 27 B7
Hambergen D 17 B11
Hamble-le-Rice UK 13 D12
Hambrücken D 21 F11
Hambühren D 79 A6
Hamburg F 23 C9
Hamburgsund S 91 B9
Hambye F 23 C9
Hamcearca RO 155 C2
Hamdibey TR 173 E7
Hämeenkoski FIN 127 C13

Hämeenkyrö FIN 127 B9
Hämeenlinna FIN 127 D11
Hämelhausen D 17 C12
Hameln D 17 D12
Hämerten D 79 A10
Hamica HR 148 E5
Hamidiye TR 167 F9
Hamidiye TR 172 B6
Hamilton UK 5 D8
Hamilton's Bawn UK 7 D9
Hamina FIN 128 D7
Haminalahti FIN 124 E9
Hamit TR 181 C9
Hamitabat TR 173 A7
Hamlagrø N 94 A4
Hamlot N 111 D10
Hamm D 17 E9
Hamm (Sieg) D 185 C8
Hammar S 92 B5
Hammarland FIN 99 B13
Hammarn S 97 C12
Hammarnäs S 105 E16
Hammarsbyn S 102 E5
Hammarstrand S 107 E10
Hammarvika N 104 D5
Hamme B 19 B9
Hammel DK 86 C5
Hammelburg D 74 B6
Hammelev DK 86 E4
Hammelspring D 84 D4
Hamme-Mille B 19 C10
Hammenhög S 88 D6
Hammer N 105 C12
Hammerbrücke D 75 B11
Hammerdal S 106 D8
Hammerfest N 113 B12
Hammershøj DK 86 C5
Hammerum DK 86 C4
Hamminkeln D 17 E7
Hamn N 108 E3
Hamn N 111 B13
Hamna N 114 B8
Hamnavoe UK 3 D14
Hamnavoe UK 3 E14
Hamnbukt N 112 C8
Hamnbukta N 111 B18
Hamneidet N 112 D6
Hamnes N 105 B10
Hamnes N 108 E4
Hamnes N 112 D6
Hamningberg N 114 B9
Hamnøy N 110 E5
Hamnvågnes N 111 B16
Hamoir B 19 D12
Hamois B 19 D11
Hamont B 16 F5
Hampen DK 86 C4
Hampetorp S 92 A7
Håmpjåkk S 116 D4
Hampont F 26 C6
Hampreston UK 13 D11
Hamra S 93 F12
Hamra S 103 C9
Hamrångefjärden S 103 E13
Hamre N 112 C4
Hamry nad Sázavou CZ 77 C9
Ham-sous-Varsberg F 186 C2
Hamstreet UK 15 E10
Hamsund N 111 D10
Ham-sur-Heure B 19 D9
Hamula FIN 123 D16
Hamula FIN 124 D9
Hamzabeyli TR 167 F7
Hanaskog S 88 C6
Hanau D 187 A6
Handbjerg DK 86 C3
Handeloh D 83 D7
Handen S 93 A12
Handest DK 86 B5
Handewitt D 82 A6
Handlová SK 147 D7
Handog S 106 E7
Handöl S 105 E12
Handrabury UA 154 B5
Handrup D 17 C9
Handsjö S 102 B8
Handstein N 108 D4
Handzame B 182 C2
Hanebo S 103 D12
Hanerau-Hademarschen D 82 B6
Hanestad N 101 C13
Hănești RO 153 B9
Hangastenmaa FIN 128 B7
Hangelsberg D 80 B5
Hånger S 87 A13
Hangö FIN 127 F8
Hangony H 145 G1
Hangu RO 153 C8
Hangvar S 93 D13
Hanhikoski FIN 115 E2
Hanhimaa FIN 117 C14
Han i Elezit RKS 164 E3
Hanikase EST 132 F1
Haniska SK 145 F3
Hankamäki FIN 125 D10
Hankasalem asema FIN 123 F16
Hankasalmi FIN 123 F16
Hankensbüttel D 83 E9
Han Knežica BIH 157 B6
Hanko FIN 127 F8
Hanna PL 141 G9
Hannäs S 93 C8
Hannover D 78 B6
Hannoversch Münden D 78 D6
Hannukainen FIN 117 C11
Hannusperä FIN 119 D15
Hannusranta FIN 121 F10
Hannut B 19 C11
Hanøy N 110 D9
Han-Pijesak BIH 157 D10
Hanshagen D 83 C10
Hańsk Pierwszy PL 141 H8
Hansnes N 112 D4
Hanstedt D 83 D8
Hanstholm DK 86 A3
Han-sur-Nied F 26 C5
Hanušovce nad Topľou SK 145 E4
Hanušovice CZ 77 B11
Hanvec F 22 D3
Haparanda S 119 C12
Hapert NL 183 C6
Häppälä FIN 123 F16
Happburg D 75 D9
Happisburgh UK 15 B12
Haps NL 183 B7
Hapträsk S 118 B4
Hara S 106 E6
Härad S 98 D7
Haradok BY 133 F7
Harads S 118 B5
Häradsbäck S 88 B6
Häradsbygden S 103 E9
Haradshammar S 93 B9
Haradzilavichy Pyershaya BY 133 D4
Haraldseng N 112 B9
Haram N 100 A4
Harang N 104 E6
Harany BY 133 F6
Harasiuki PL 144 C5
Hărău RO 151 F10
Haraudden S 116 E3
Harbacheva BY 133 E6
Harbak N 104 C8
Harbke D 79 B9
Harbo S 98 B8
Harboør DK 86 B2
Harbost UK 2 J4
Harburg (Schwaben) D 75 E8
Harbury UK 13 A12
Hard A 71 C9
Hardbakke N 100 D1
Hardegg A 77 E9
Hardegsen D 78 C6
Hardelot-Plage F 15 F12
Hardenberg NL 17 C7
Harderwijk NL 16 D5
Hardheim D 27 A11
Hardinxveld-Giessendam NL 182 B5
Hardt D 27 D9
Hareid N 100 B4
Harelbeke B 19 C7
Haren NL 17 B7
Haren (Ems) D 17 C8
Hare Street UK 15 D9
Harestua N 95 B13
Harfleur F 23 A12
Harg S 99 B10
Hargesheim D 21 E9
Hargimont B 19 D11
Hargla EST 131 F12
Hargnies F 19 D10
Hargshamn S 99 B10
Harichovce SK 145 F2
Harinkaa FIN 123 E16
Harjakangas FIN 126 B6
Härjäro S 99 D8
Härjåsjön S 102 C7
Harjavalta FIN 126 C7
Harjula FIN 119 C15
Harjumaa FIN 128 B7
Harjunkylä FIN 122 E7
Harjunpää FIN 126 C6
Harju-Risti EST 131 C7
Harka H 149 A7
Härkäjoki FIN 115 D2
Harkakötöny H 150 E4
Harkány H 149 E10
Härkmeri FIN 122 F6
Härkönen FIN 119 C13
Harku EST 131 C9
Hârlău RO 153 C9
Harlech UK 10 F3
Harleston UK 15 C11
Hårlev DK 87 E10
Harlingen NL 16 B4
Harlow UK 15 D9
Harly F 19 E7
Hărman RO 153 F7
Harmånger S 103 C13
Härmänkylä FIN 121 F13
Harmanli TR 173 D9
Härmänmäki FIN 121 F11
Harmannsdorf A 77 F10
Harmelen NL 182 A5
Harmoinen FIN 127 C13
Harmsdorf D 83 C9
Harmston UK 11 E10
Harnes F 18 D6
Härnösand S 103 A14
Haro E 40 C6
Harodz'ki BY 137 E13
Haroldswick UK 3 D15
Haroué F 26 D5
Härpe FIN 127 E14
Harpefoss N 101 C11
Harpenden UK 15 D8
Harplinge S 87 B11
Harpstedt D 17 C11
Harra D 75 B10
Harrachov CZ 81 E8
Harran N 105 B13
Harre DK 86 B3
Harridslev DK 86 C6
Harrietfield UK 5 C9
Harrioja S 119 C11
Harrislee D 82 A6
Harrogate UK 11 D8
Harrsjö S 106 B9
Harrström FIN 122 E6
Harrvik S 107 A10
Harsa S 103 C10
Hårsbäck S 98 C7
Harsefeld D 82 D7
Hârseni RO 152 F6
Hârsești RO 160 D5
Harsleben D 79 C9
Hårslev DK 86 D6
Hârșova RO 155 D1
Harsprånget S 116 E3
Harstad N 111 C12
Harsum D 78 B6
Harsvik N 104 C8
Harta H 150 D3
Hartberg A 148 B5
Hårte S 103 C13
Hartenholm D 83 C8
Hartha D 80 D3
Harthausen D 21 F10
Hartheim D 27 E8
Hârtiești RO 160 C6
Hart im Zillertal A 72 B4
Hartkirchen A 76 F5
Hartland UK 12 D6
Hartlepool UK 11 B9
Hartmanice CZ 76 D4
Hartola FIN 127 B15
Harwich UK 15 D11
Harzgerode D 79 C9
Hasanağa TR 173 D10
Hasbuğa TR 173 A8
Haselünne D 17 C8
Hasirciarnavutköy TR 171 B10
Håsjö S 107 F10
Hasköy TR 173 A6
Hasköy TR 177 C9
Haslach an der Mühl A 76 E6
Haslach im Kinzigtal D 27 D9
Hasle CH 70 C5
Hasle DK 88 E7
Haslemere UK 15 E7
Haslev DK 87 E9
Hasloh D 83 C7
Hasløya N 104 E2
Haslund DK 86 C6
Hășmaș RO 151 D9
Hasparren F 32 D3
Haßbergen D 17 C12
Hassel (Weser) D 17 C12
Hassela S 103 B12
Hassela kyrkby S 103 B12
Hasselfelde D 79 C8
Hasselfors S 92 A5
Hasselt B 19 C11
Hasselt NL 16 C6
Haßfurt D 75 B8
Hassi FIN 127 B13
Hässjö S 103 A14
Haßleben D 79 D9
Haßleben D 84 D5
Hässleholm S 87 C13
Hasslö S 89 C8
Haßloch D 21 F10
Hasslöv S 87 C12
Haßmersheim D 187 C7
Hästbo S 103 E13
Hästholmen S 92 C5
Hastière-Lavaux B 19 D10
Hastings UK 15 F10
Hästnäs S 97 D14
Håstrup DK 86 E6
Hästveda S 87 C13
Hasvåg N 105 C9
Hasvik N 112 C9
Hať CZ 146 B6
Hat' UA 145 G6
Hațeg RO 159 B10
Hatfield UK 11 D10
Hatherleigh UK 12 D6
Hätilä FIN 127 C11
Hatipkişlasi TR 181 B7
Hatsola FIN 128 B8
Hattarvík FO 2 A4
Hattem NL 16 D6
Hattersheim am Main D 21 D10
Hattert D 21 C9
Hattfjelldal N 108 E6
Hatting DK 86 D5
Hattingen D 17 F8
Hatton UK 3 L13
Hattstedt D 82 A6
Hattula FIN 127 C11
Hattuvaara FIN 125 E16
Hatulanmäki FIN 124 C8
Hatunkylä FIN 125 D15
Hatvan H 150 B4
Hatzenbühl D 27 B9
Hatzendorf A 148 C6
Hatzfeld (Eder) D 21 C11
Haubourdin F 18 C6
Hauenstein D 186 C4
Haugan N 105 D9
Haugastøl N 94 A7
Hauge N 94 F4
Hauge N 114 E6
Haugen N 110 D7
Haugesund N 94 D2
Haugh of Urr UK 5 F9
Haugland N 108 D5
Haugli N 111 C15
Haugnes N 111 B11
Haugnes N 112 C6
Haugset N 112 D7
Hauho FIN 127 C12
Haukå N 100 C2
Haukela FIN 125 B14
Haukeligrend N 94 C7
Haukijärvi FIN 127 B9
Haukilahti FIN 121 F13
Haukiniemi FIN 121 C13
Haukipudas FIN 119 D14
Haukivaara FIN 125 E15
Haukivuori FIN 124 F8
Haukøy N 111 D11
Haulerwijk NL 16 B6
Haurukylä FIN 119 E15
Haus A 73 B8
Haus N 94 B2
Hausach D 187 E5
Hausen D 75 E11
Hausen bei Würzburg D 187 B9
Häusern D 27 E9
Hausham D 72 A4
Hausjärvi FIN 127 D12
Hauske N 94 D3
Hausleiten A 77 F10
Hausmannstätten A 148 C5
Hautajärvi FIN 115 E6
Hautakylä FIN 121 D10
Hautakylä FIN 123 E12
Haute-Amance F 26 E4
Hautefort F 29 E8
Hauterives F 31 E7
Hauteville-Lompnes F 31 D8
Haut-Fays B 184 D3
Hautmont F 19 D8
Hautomäki FIN 123 E17
Haux F 32 D4
Hauzenberg D 76 E5
Havant UK 14 F7
Havârna RO 153 A9
Havbro DK 86 B4
Havdhem S 93 E12
Hávdna N 113 C15
Havelange B 19 D11
Havelberg D 83 E12
Havelte NL 16 C6
Håven S 103 C9
Haverdal S 87 B11
Haverfordwest UK 12 B5
Haverhill UK 15 C9
Haverlah D 79 B7
Haverö S 103 B9
Haversin B 19 D11
Haverslev DK 86 B5
Håverud S 91 B11
Havířov CZ 146 B6
Havixbeck D 17 E8
Hävla S 92 B7
Havlíčkův Brod CZ 77 C9
Havndal DK 86 B6
Havneby DK 86 E3
Havnebyen DK 87 D8
Havnsø DK 87 D8
Havøysund N 113 A14
Håvra S 103 C10
Havran TR 173 E7
Havrebjerg DK 87 E8
Havsa TR 173 A6
Havsskogen S 99 B11
Havtun N 94 B2
Hawarden UK 10 E5
Hawes UK 11 C7
Hawick UK 5 E11
Hawkhurst UK 15 E10
Hawkinge UK 15 E11
Haxby UK 11 C9
Hayange F 20 F6
Haybes F 184 D2
Haydarli TR 177 D10
Haydere TR 181 A8
Haydon Bridge UK 5 F12
Haydon Wick UK 13 B11
Hayingen D 71 A8
Hayle UK 12 E4
Hay-on-Wye UK 13 A8
Hayrabolu TR 173 B7
Hayton UK 11 D10
Hayvoron UA 154 A5
Haywards Heath UK 15 F8
Hazebrouck F 18 C6
Hazerswoude-Rijndijk NL 182 A5
Haźlach PL 147 B7
Hažlín SK 145 E3
Hazlov CZ 75 B11
Heacham UK 11 F12
Headcorn UK 15 E10
Headford IRL 6 F4
Healeyfield UK 5 F13
Heanor UK 11 E9
Heathfield UK 15 F9
Hebdów PL 143 F9
Hebenhausen (Neu-Eichenberg) D 78 D6
Heberg S 87 B11
Hebertsfelden D 76 F3
Hebnes N 94 D3
Heby S 98 C7
Hèches F 33 D6
Hechingen D 27 D10
Hecho E 32 E4
Hechtel B 19 B11
Hechthausen D 17 A12
Heckelberg D 84 E5
Heckington UK 11 F11
Hedared S 91 D12
Hedberg S 109 F16
Heddesheim D 21 F11
Hédé F 23 D8
Hede S 98 B6
Hede S 102 B6
Hedekas S 91 B10
Hedel NL 16 E4
Hedemora S 97 B14
Heden DK 86 E6
Heden S 102 C4
Heden S 118 C7
Hedenäset S 119 B11
Hedensbyn S 118 B9
Hedensted DK 86 D5
Hedersleben D 79 C9
Hédervár H 146 F4
Hedesunda S 98 B8
Hedeviken S 102 B6
Hedge End UK 13 D12
Hedlunda S 107 B15
Hedmark S 107 B13
Hedsjön S 103 E12
Hee DK 86 C2
Heeg NL 16 C5
Heek D 17 D8
Heel NL 183 C7
Heemsen D 17 C12
Heemskerk NL 16 C3
Heemstede NL 16 D3
Heenvliet NL 182 B4
Heer B 19 D10
Heerde NL 16 D6
Heerenveen NL 16 C5
Heerewaarden NL 183 B6
Heerhugowaard NL 16 C3
Heerlen NL 20 C5
Heers B 19 C11
Heesch NL 16 E5
Heeslingen D 17 B12
Heeßen D 17 D12
Heeswijk NL 16 E4
Heeten NL 183 A8
Heeze NL 16 F5
Heggeli N 111 B13
Heggem N 104 E4
Heggenes N 101 D10
Heggjabygda N 100 C4
Heggland N 90 A5
Heggmoen N 108 B8
Hegra N 105 E10
Hegyeshalom H 146 F4
Hehlen D 78 C5
Heia N 105 C12
Heia N 111 B17
Heide D 82 B6
Heideck D 75 D9
Heidelberg D 21 F11
Heiden D 17 E7
Heidenau D 80 E5
Heidenheim D 75 D8
Heidenheim an der Brenz D 75 E7
Heidenreichstein A 77 E8
Heigenbrücken D 187 A7
Heikendorf D 83 B8
Heikkilä FIN 121 C14
Heikkilä FIN 122 F7
Heiland N 90 B4
Heilbronn D 27 B11
Heilbrunn A 148 B5
Heiligenberg D 27 E11
Heiligenfelde D 83 E10
Heiligenhafen D 83 B9
Heiligenhaus D 17 F7
Heiligenkreuz am Waasen A 148 C5
Heiligenkreuz im Lafnitztal A 148 C6
Heiligenstadt Heilbad D 79 D7
Heiligenstedten D 82 C6
Heiloo NL 16 C3
Heilsbronn D 75 D8
Heiltz-le-Maurupt F 25 C12
Heim N 104 E6
Heimbach D 21 E8
Heimbuchenthal D 187 B7
Heimdal N 100 E3
Heimdal N 104 E8
Heimertingen D 71 A10
Heimseta N 100 C3
Heimsheim D 187 D6
Heinade D 78 C6
Heinämaa FIN 127 D14
Heinämäki FIN 121 F12
Heinämäki FIN 123 D17
Heinävaara FIN 125 E14
Heinävesi FIN 125 F11
Heinebach (Alheim) D 78 D6
Heinersbrück D 81 C7
Heinersdorf D 80 B6
Heinersreuth D 75 C10
Heinijärvi FIN 119 E14
Heinijoki FIN 126 D7
Heiningen D 79 B8
Heinisuo FIN 119 C16
Heinkenszand NL 16 F1
Heinlahti FIN 128 E6
Heino NL 183 A8
Heinola FIN 127 C15
Heinolan kirkonkylä FIN 127 C15
Heinolanperä FIN 119 E14
Heinoniemi FIN 125 F13
Heinsberg D 20 B6
Heinsen D 78 C5
Heinsnes N 105 B12
Heisingen D 183 C10
Heistadmoen N 95 C11
Heist-op-den-Berg B 19 B10
Heitersheim D 27 E8
Heituinlahti FIN 128 C8
Hejls DK 86 E5
Hejnice CZ 81 E8
Hejnsvig DK 86 D4
Hejőpapi H 145 H2
Hejsager DK 86 E5
Hekelgem B 19 C9
Hel PL 138 A6
Helbra D 79 C9
Helchteren B 183 C6
Heldburg D 75 B8
Heldenbergen D 21 D11
Heldrungen D 79 D9
Helechal E 51 B9
Helegiu RO 153 E9
Helensburgh UK 4 C7
Helfenberg A 76 E6
Helgenes N 110 C9
Helgeroa N 90 B6
Helgum S 107 E11
Hell N 105 E9
Hella N 100 D5
Helland N 104 E5
Helland N 111 D11
Hellanmaa FIN 122 D9
Hellarmo N 109 B10
Helle N 90 B5
Hellebæk DK 87 C11
Hellefjord N 113 B11
Hellendoorn NL 183 A8
Hellenthal D 20 D6
Hellenurme EST 131 E12
Hellesøy N 100 E1
Hellested DK 87 E10
Hellesvik N 104 D5
Hellesylt N 100 B5
Hellevad DK 86 E4
Hellevoetsluis NL 16 E2
Helligskogen N 112 E6
Hellín E 55 B9
Hellingly UK 15 F9
Hellnes N 112 C6
Hellsö FIN 126 F4
Hellvi S 93 D13
Hellvik N 94 F3
Helmbrechts D 75 B10
Helme EST 131 E11
Helmond NL 16 F5
Helmsdale UK 3 J9
Helmsley UK 11 C9
Helmstadt D 74 C6
Helmstedt D 79 B9
Heľpa SK 147 D9
Helppi FIN 117 D13
Helpringham UK 11 F11
Helse D 82 C6
Helsingborg S 87 C11
Helsinge DK 87 C10
Helsingfors FIN 127 E12
Helsingør DK 87 C11
Helsinki FIN 127 E12
Helstad N 105 A12
Helston UK 12 E4
Heltermaa EST 130 D6
Heltersberg D 21 F9
Helvaci TR 177 B9
Helvécia H 150 D4
Helvoirt NL 183 B6
Hem DK 86 B3
Hem F 182 D2
Hemau D 75 D10
Hemavan S 108 E9
Hemeiuș RO 153 D9
Hemel Hempstead UK 15 D8
Hemer D 185 B8
Hemfjäll S 109 E12
Hemfjällstangen S 102 D5
Hemhofen D 75 C8
Hemling S 107 D15
Hemme D 82 B6
Hemmet DK 86 D2
Hemmingen D 78 B6
Hemmingen S 107 B17
Hemmingsmark S 118 D6
Hemmoor D 17 A12
Hemnesberget N 108 D6
Hemnestad N 111 C11
Hempnall UK 15 C11
Hempstead UK 15 C9
Hemsbach D 21 E11
Hemsbünde D 82 D6
Hemsby UK 15 B12
Hemse S 93 E12
Hemsedal N 101 E9
Hemsjö S 107 D14
Hemslingen D 82 D7
Hemsloh D 17 C11
Hemsö S 103 A15
Hemyock UK 13 D8
Henån S 91 C10
Hénanbihen F 23 C7
Henarejos E 47 E10
Hencida H 151 C8
Hendaye F 32 D2
Hendon UK 15 D8
Hengelo NL 16 D6
Hengelo NL 17 D7
Hengersberg D 76 E4
Hengevelde NL 183 A9
Henggart CH 27 E10
Hengoed UK 13 B8
Hénin-Beaumont F 18 D6
Henley-on-Thames UK 15 D7
Hennan S 103 B10
Henndorf am Wallersee A 73 A7
Hennebont F 22 E5
Hennef (Sieg) D 21 C8
Hennes N 110 C9
Henne Stationsby DK 86 D2
Hennezel F 26 D5
Hennickendorf D 80 B4
Hennigsdorf Berlin D 80 A4
Henningskälen S 106 D8
Henningsvær N 110 D7
Hennseid N 90 A5
Hennset N 104 E4
Hennstedt D 82 B6
Hennweiler D 21 E8
Henrichemont F 25 F8
Henryków PL 81 E12
Henrykowo PL 139 B9
Henstridge UK 13 D10
Hepola FIN 119 C13
Heppen B 183 C6
Heppenheim (Bergstraße) D 21 E11
Herálec CZ 77 C8
Herálec CZ 77 C9
Herbault F 24 E5
Herbertingen D 27 D11
Herbertstown IRL 8 C6
Herbés E 42 F3
Herbeumont B 184 E3
Herbignac F 23 F7
Herbolzheim D 27 D8
Herborn D 21 C10
Herbrechtingen D 75 E7
Herbstein D 21 C12
Herby PL 142 E6
Herceghalom H 149 A11
Herceg-Novi MNE 162 E6
Hercegovac HR 149 E8
Hercegszántó H 150 F2
Herdecke D 17 F8
Herdorf D 21 C9
Hereclean RO 151 C11
Heréd H 150 B4
Hereford UK 13 A9
Héreg H 149 A11
Herencia E 46 F6
Herend H 149 B9
Herent B 19 C10
Herentals B 19 B10
Herenthout B 182 C5
Hérépian F 34 C5
Herford D 17 D11
Hergatz D 71 B9
Hergiswil CH 70 D6
Herguijuela E 45 F9
Héric F 23 F8
Héricourt F 27 E6
Hérimoncourt F 27 F6
Heringen (Helme) D 79 D8
Heringen (Werra) D 79 E7
Heringsdorf D 83 B10
Heringsdorf D 84 C6
Heriot UK 5 D11
Herisau CH 27 F11
Hérisson F 29 B11
Herk-de-Stad B 19 C11
Herkenbosch NL 183 C8
Herkingen NL 182 B4
Herleshausen D 79 D7
Herlev DK 87 D10
Herlufmagle DK 87 E9
Herm F 32 C3
Hermagor A 73 C7
Hermannsburg D 83 E8
Hermanovce SK 145 E3
Hermanowice PL 144 D6
Hermansverk N 100 D5
Heřmanův Městec CZ 77 C9
Herment F 29 D11
Hermersberg D 186 C4
Hermeskeil D 21 E7
Hermisende E 39 E6
Hermsdorf D 79 E10
Hernád H 150 C3
Hernádnémeti H 145 G2
Hernani E 32 D2
Hernansancho E 46 C3
Herne B 19 C9
Herne D 17 E8
Herne Bay UK 15 E11
Herning DK 86 C3
Heroldsbach D 75 C8
Héron B 19 C11
Hérouville-St-Clair F 23 B11
Herøy N 100 B3
Herpf D 79 E7
Herrala FIN 127 D13
Herramélluri E 40 C5
Herräng S 99 B11
Herré F 32 C5
Herre N 90 A6
Herrenberg D 27 C10
Herrera E 53 B7
Herrera del Duque E 45 F10
Herrera de los Navarros E 42 E1
Herrera de Pisuerga E 40 C3
Herrería E 47 C9
Herreruela E 45 F7
Herrestad S 91 C10
Herrieden D 75 D8
Herringbotn N 108 E6
Herrlisheim F 186 D4
Herrljunga S 91 C13
Herrnhut D 81 D7
Herrö S 102 B7
Herrsching am Ammersee D 75 G9
Herrskog S 103 A15
Herrvik S 93 E13
Herry F 25 F8
Hersbruck D 75 C9
Herschbach D 21 C9
Herscheid D 21 B9
Herschweiler-Pettersheim D 186 C3
Herselt B 19 B10
Herslev DK 86 D5
Herstal B 19 C12
Herstmonceux UK 15 F9
Herston UK 3 H11
Herten D 17 E8
Hertford UK 15 D8
Hertník SK 145 E3
Hertsa UA 153 A8
Hertsånger S 118 F6
Hertsjö S 103 D11
Herve B 19 C12
Hervik N 94 D3
Herwijnen NL 183 B6
Herzberg D 80 C4
Herzberg D 83 C11
Herzberg D 84 E3
Herzberg am Harz D 79 C7
Herzebrock-Clarholz D 17 E10
Herzele B 19 C8
Herzfelde D 80 B5
Herzhorn D 17 A12
Herzlake D 17 C9
Herzogenaurach D 75 C8
Herzogenbuchsee CH 27 F8
Herzogenburg A 77 F9
Herzogenrath D 183 D8
Herzsprung D 83 D12
Hesdin F 18 D5
Hesel D 17 B9
Hesjeberg N 111 C13
Hesjestranda N 100 A6
Heskestad N 94 F4
Hespérange L 20 E6
Heßdorf D 75 C8
Hesselager DK 87 E7
Hessen D 79 B8
Hessfjorden N 112 D3
Heßheim D 21 E10
Hessisch Lichtenau D 78 D6
Hessisch Oldendorf D 17 D12
Hest N 100 D3
Hestenesøyri N 100 C4
Hestnes N 104 D5
Hestøy N 108 E3
Hestra S 91 E14
Hestra S 92 D6
Hestvik N 105 B10
Hestvika N 104 D6
Heswall UK 10 E5
Hetekylä FIN 119 D17
Hetés H 149 D9
Hethersett UK 15 B11
Hetlingen D 82 C7
Hettange-Grande F 186 C1
Hettenleidelheim D 186 B5
Hettenshausen D 75 E10
Hettingen D 27 D11
Hetton UK 11 C7
Hettstedt D 79 C10
Hetzerath D 21 E7
Heubach D 187 D8
Heuchelheim D 21 C11
Heuchin F 18 D5
Heudicourt-sous-les-Côtes F 26 C4
Heukelum NL 16 E4
Heusden B 19 B11
Heusden NL 183 B6
Heusenstamm D 21 D11
Heustreu D 75 B7
Heusweiler D 186 C2
Heves H 150 B5
Hévíz H 149 C8
Hevlín CZ 77 E10
Hexham UK 5 F12
Heyrieux F 31 D7
Heysham UK 10 C6
Heythuysen NL 19 B12
Heywood UK 11 D7
Hida RO 151 C11
Hidas H 149 D10
Hidasnémeti H 145 G3
Hiddenhausen D 17 D11
Hiddensee D 84 A4
Hidirköylü TR 177 D10
Hidișelu de Sus RO 151 D9
Hieflau A 73 A10
Hiendelaencina E 47 B7
Hiersac F 28 D5
Hietakangas FIN 115 D3
Hietama FIN 123 E15
Hietanen FIN 117 D11
Hietanen FIN 128 B7
Hietaniemi FIN 115 D6
Hietaniemi FIN 121 B11
Hietaperä FIN 121 F13
Higham Ferrers UK 15 C7
Highampton UK 12 D6
High Bentham UK 10 C6
Highbridge UK 13 C9
Highclere UK 13 C12
High Halden UK 15 E10
High Hawsker UK 11 C10
High Hesket UK 5 F11
High Lorton UK 5 F10
Highnam UK 13 B10
Highworth UK 13 B11
High Wycombe UK 15 D7
Higuera de Arjona E 53 A9
Higuera de la Serena E 51 B8
Higuera de la Sierra E 51 D7
Higuera de Llerena E 51 C8
Higuera de Vargas E 51 C6
Higuera la Real E 51 C6
Higueruela E 55 B10
Higueruelas E 48 E3
Hihnavaara FIN 115 D4
Hiidenkylä FIN 123 C15
Hiidenlahti FIN 125 E10
Hiilikumpu FIN 119 C15
Hiirikylä FIN 125 C10
Hiirola FIN 128 B7
Hiisijärvi FIN 121 F12
Híjar E 42 E3
Hikiä FIN 127 D12
Hilbersdorf D 80 E4
Hilchenbach D 21 C10
Hildburghausen D 75 B8
Hilden D 21 B7
Hilders D 79 E7
Hilderthorpe UK 11 C11
Hildesheim D 78 B6
Hilgertshausen D 75 F9
Hilișeu-Horia RO 153 A8
Hiliuți MD 153 B10
Hillared S 91 D13
Hille D 17 D11
Hille S 103 E13
Hillebola S 99 B9
Hillegom NL 16 D3
Hillerød DK 87 D10
Hillerse D 79 B7
Hillerslev DK 86 A3
Hillerslev DK 86 E6
Hillerstorp S 88 A5
Hilleshamn N 111 C13
Hillesheim D 21 D7
Hillesøy N 111 A15
Hillevik S 103 E13
Hilli FIN 123 C11
Hillilä FIN 123 B11
Hill of Fearn UK 3 K9
Hillosensalmi FIN 128 C6
Hillsand S 106 C8
Hillsborough UK 7 D10
Hillside UK 5 B12
Hillswick UK 3 E14
Hilltown UK 7 D10
Hilpoltstein D 75 D9
Hilsenheim F 186 E4
Hilton UK 11 F8
Hiltpoltstein D 75 C9
Hiltula FIN 129 B9
Hilvarenbeek NL 183 C6
Hilversum NL 16 D4
Himalansaari FIN 128 C8
Himanka FIN 123 B11
Himarë AL 168 D2
Himberg A 77 F10
Himbergen D 83 D9
Himesháza H 149 D11
Himma EST 131 F14

Himmaste *EST* 131 E14
Himmelberg *A* 73 C9
Himmelpforten *D* 17 A12
Hînceşti *MD* 154 D3
Hinckley *UK* 11 F9
Hindås *S* 91 D11
Hindenburg *D* 83 E11
Hinderwell *UK* 11 B10
Hindhead *UK* 15 E7
Hindley *UK* 10 D6
Hindon *UK* 13 C10
Hindrem *N* 104 D8
Hinganmaa *FIN* 117 D15
Hingham *UK* 15 B10
Hinnerjöki *FIN* 126 C6
Hinnerup *DK* 86 C6
Hinneryd *S* 87 B13
Hinojal *E* 45 E8
Hinojales *E* 51 C6
Hinojares *E* 55 C7
Hinojosas de Calatrava *E* 54 B4
Hinojos *E* 51 E7
Hinojosa *E* 47 B8
Hinojosa de Duero *E* 45 C7
Hinojosa de Jarque *E* 42 F2
Hinojosa del Duque *E* 51 B9
Hinojosa del Valle *E* 51 C7
Hinojosa de San Vicente
E 46 D3
Hinova *RO* 159 D10
Hinte *D* 17 B8
Hinterhermsdorf *D* 80 E6
Hinternah *D* 75 A8
Hinterrhein *CH* 71 D8
Hintersee *A* 73 A7
Hintersee *D* 84 C6
Hinterweidenthal *D* 186 C4
Hinterzarten *D* 27 E9
Hinthaara *FIN* 127 E13
Hinwil *CH* 27 F10
Hinx *F* 32 C4
Hippolytushoef *NL* 16 C3
Hîrbovăţ *MD* 154 C3
Hîrjauca *MD* 154 C2
Hirka *TR* 181 B9
Hirnyk *UA* 144 C9
Hirrlingen *D* 27 D10
Hirschaid *D* 75 C9
Hirschau *D* 75 C10
Hirschberg *D* 75 B10
Hirschfeld *D* 80 D5
Hirschhorn (Neckar) *D* 187 C6
Hirsila *FIN* 127 B11
Hirsingue *F* 27 E7
Hirson *F* 19 E9
Hîrtop *MD* 154 D3
Hîrtopul Mare *MD* 154 C3
Hirtshals *DK* 90 D6
Hirvas *FIN* 119 B14
Hirvaskoski *FIN* 120 D9
Hirvasperä *FIN* 119 E13
Hirvasvaara *FIN* 115 E5
Hirvelä *FIN* 119 D16
Hirvelä *FIN* 121 F14
Hirvensalmi *FIN* 128 B6
Hirviäkuru *FIN* 115 C1
Hirvihaara *FIN* 127 D13
Hirvijoki *FIN* 123 F9
Hirvikylä *FIN* 123 F9
Hirvilahti *FIN* 124 D8
Hirvineva *FIN* 119 E14
Hirvivaara *FIN* 121 E13
Hirvlax *FIN* 122 D8
Hirwaun *UK* 13 B7
Hirzenhain *D* 21 D12
Hisarönü *TR* 181 C8
Hishult *S* 87 C12
Hissjön *S* 122 C4
Histon *UK* 15 C9
Hita *E* 47 C6
Hitchin *UK* 15 D8
Hitis *FIN* 126 F8
Hittarp *S* 87 C11
Hittisau *A* 71 C9
Hitzacker *D* 83 D10
Hitzendorf *A* 148 B4
Hitzhusen *D* 83 C7
Hiukkajoki *FIN* 129 B12
Hiyche *UA* 144 C8
Hjägsjö *S* 122 C3
Hjällerup *DK* 86 A6
Hjältevad *S* 92 C6
Hjärnarp *S* 87 C11
Hjärsås *S* 88 C6
Hjärtum *S* 91 C11
Hjarup *DK* 86 E4
Hjelle *N* 100 C6
Hjellestad *N* 94 B2
Hjellsand *N* 110 C8
Hjelmeland *N* 94 D4
Hjelset *N* 100 A7
Hjerkinn *N* 101 B11
Hjerm *DK* 86 C3
Hjo *S* 92 C4
Hjordkær *DK* 86 E4
Hjørring *DK* 90 E7
Hjorted *S* 93 D8
Hjortkvarn *S* 92 B6
Hjortsberga *S* 88 B6
Hjørungavåg *N* 100 B4
Hjuvik *S* 91 D10
Hlavani *UA* 154 F4
Hlebine *HR* 149 D7
Hligeni *MD* 154 B3
Hlinaia *MD* 153 A10
Hlinaia *MD* 154 C1
Hlinaia *MD* 154 D5
Hlinik nad Hronom *SK* 147 D7
Hlinné *SK* 145 F4
Hlinsko *CZ* 77 C9
Hlohovec *SK* 146 E5
Hlubočky *CZ* 146 B4
Hluboká nad Vltavou *CZ* 77 D6
Hlučín *CZ* 146 B6
Hlyboka *UA* 153 A7
Hlybokaye *BY* 133 F3
Hlyboke *UA* 155 B5
Hnatkiv *UA* 154 A3
Hněvotín *CZ* 77 C12
Hniezdne *SK* 145 E2
Hnizdychiv *UA* 145 E9
Hnojník *CZ* 147 B7
Hnúšťa *SK* 147 D9
Hobol *H* 149 D9
Hobøl *N* 95 C13
Hobro *DK* 86 B5
Hoburg *S* 93 F12
Hoçe e Qytetit *RKS* 163 E10
Hoceni *RO* 153 D12
Hochdonn *D* 82 B6
Hochdorf *CH* 27 F9
Hochdorf *D* 71 A9
Hochfelden *F* 27 C8

Höchheim *D* 75 B7
Hochspeyer *D* 186 C4
Hochstadt (Pfalz) *D* 187 C5
Höchstadt an der Aisch *D* 75 C8
Höchstädt an der Donau
D 75 E8
Hochstetten-Dhaun *D* 21 E8
Höchst im Odenwald *D* 187 B6
Hoçişti *AL* 168 C4
Hockenheim *D* 187 C6
Hoczew *PL* 145 E5
Hodac *RO* 152 D5
Hodalen *N* 101 B14
Hoddesdon *UK* 15 D8
Hodejov *SK* 147 E10
Hodenhagen *D* 82 E7
Hodkovice nad Mohelkou
CZ 81 E8
Hódmezővásárhely *H* 150 E5
Hodnet *UK* 10 F6
Hodod *RO* 151 C11
Hodonice *CZ* 77 E11
Hodonín *CZ* 77 E12
Hodoşa *RO* 152 D5
Hodruša-Hámre *SK* 147 E7
Hodsager *DK* 86 C3
Hodyszewo *PL* 141 E7
Hoek *NL* 16 F1
Hoek van Holland *NL* 16 E2
Hoem *N* 104 F4
Hoenderloo *NL* 183 A7
Hœnheim *F* 186 D4
Hoensbroek *NL* 19 C12
Hœrdt *F* 186 D4
Hoeselt *B* 19 C11
Hoevelaken *NL* 16 D4
Hoeven *NL* 16 E3
Hof *D* 21 C7
Hof *D* 75 B10
Hof *N* 95 C12
Hofbieber *D* 78 E6
Hoffstad *N* 104 C8
Hofgeismar *D* 21 B12
Hofheim am Taunus *D* 21 D10
Hofheim in Unterfranken
D 75 B8
Hofles *N* 105 B11
Hofors *S* 98 A16
Hofsøy *N* 111 B13
Hofterup *D* 87 D11
Höganäs *S* 87 C11
Högås *S* 108 E9
Högbo *S* 103 D12
Högbränna *S* 109 F15
Högen *S* 103 C13
Högfors *S* 97 B15
Högfors *S* 97 C13
Hoghilag *RO* 152 E5
Hoghiz *RO* 152 F6
Høgli *N* 111 B14
Höglunda *S* 103 A10
Hogoşt *RKS* 164 D4
Högrun *S* 105 D16
Högsåra *FIN* 126 F7
Högsäter *S* 91 B11
Högsäter *S* 96 C6
Högsäter *S* 97 D8
Högsby *S* 89 A10
Høgset *N* 104 F3
Högsjö *S* 92 A7
Högsjö *S* 103 A14
Hogstad *S* 92 C6
Hogstorp *S* 91 C11
Högträsk *S* 116 E5
Högvålen *S* 102 B4
Høgyész *H* 149 C10
Hohberg *D* 186 E4
Hohen-Altheim *D* 75 E8
Hohenaspe *D* 82 C7
Hohenau *D* 76 E5
Hohenau an der March *A* 77 E11
Hohenberg *A* 77 G9
Hohenbocka *D* 80 D6
Hohenbucko *D* 80 C4
Hohenburg *D* 75 D10
Hohendorf *D* 84 B5
Hoheneich *A* 77 E8
Hohenems *A* 71 C9
Hohenfels *D* 75 D10
Hohenfurch *D* 71 B11
Hohengöhren *D* 79 A11
Hohenhameln *D* 79 B7
Hohenkammer *D* 75 F10
Hohenkirchen *D* 17 A9
Höhenkirchen-Siegertsbrunn
D 75 F10
Hohenleuben *D* 79 E11
Hohenlockstedt *D* 82 C7
Hohenmocker *D* 84 C4
Hohenmölsen *D* 79 D11
Hohennauen *D* 83 E12
Hohenpeißenberg *D* 71 B12
Hohenroth *D* 75 B7
Hohensaaten *D* 84 E6
Hohenseeden *D* 79 B11
Hohenstein-Ernstthal *D* 79 E12
Hohenthann *D* 75 E11
Hohenthurm *D* 79 C11
Hohen Wangelin *D* 83 C12
Hohenwart *D* 75 F9
Hohenwarth *D* 76 D3
Hohenwestedt *D* 82 B7
Höhn *D* 21 C9
Hohn *D* 82 B7
Hohne *D* 79 A7
Hohnhorst *D* 17 D12
Höhr-Grenzhausen *D* 21 D9
Hohwacht (Ostsee) *D* 83 B9
Hoikankylä *FIN* 124 E7
Hoikka *FIN* 121 E12
Hoisdorf *D* 83 C8
Hoisko *FIN* 123 D11
Højby *DK* 86 E5
Højby *DK* 87 D9
Højer *DK* 86 F3
Højmark *DK* 86 C2
Højslev *DK* 86 B4
Højslev Stationsby *DK* 86 B4
Hok *S* 92 D5
Hökåsen *S* 98 C7
Hökerum *S* 91 D13
Hökhult *S* 91 A13
Hökhuvud *S* 99 B9
Hokka *FIN* 128 B6
Hokkaskylä *FIN* 123 E12
Hokksund *N* 95 C11
Hokland *N* 111 C11

Hökön *S* 88 C6
Hol *N* 101 E8
Hol *N* 111 C11
Holand *N* 110 D9
Holandsvika *N* 108 E5
Holasovice *CZ* 142 G4
Holbæk *DK* 86 B6
Holbæk *DK* 87 D9
Holbeach *UK* 11 F12
Holboca *RO* 153 C11
Holbøl *DK* 86 F4
Holbrook *UK* 15 D11
Holdorf *D* 17 C10
Holeby *DK* 83 A10
Hølen *N* 95 C13
Holešov *CZ* 146 C5
Holevik *N* 100 D1
Holguera *E* 45 E8
Holíč *SK* 77 E12
Holice *CZ* 77 B9
Höljes *S* 102 E4
Holkestad *N* 110 E8
Holkonkylä *FIN* 123 E11
Hollabrunn *A* 77 E10
Hollandscheveld *NL* 17 C7
Hollange *B* 19 E12
Hollås *N* 105 C11
Holle *D* 79 B7
Hollen *N* 90 C2
Hollenbach *D* 75 F9
Hollenbek *D* 83 C9
Hollenegg *A* 73 C11
Hollenstedt *D* 82 D7
Hollenstein an der Ybbs
A 73 A10
Hollern-Twielenfleth *D* 82 C7
Hollfeld *D* 75 C9
Hollingstedt *D* 82 B6
Hollington *UK* 15 F10
Hollingworth *UK* 11 E8
Holláháza *H* 145 F3
Hollola *FIN* 127 D14
Hollum *NL* 16 B5
Hollybush *UK* 4 E7
Hollyford *IRL* 8 C6
Hollywood *IRL* 7 F9
Holm *D* 82 C7
Holm *DK* 86 E5
Holm *N* 105 A12
Holm *N* 110 C9
Holm *S* 103 A12
Holm *S* 107 E13
Hol'ma *UA* 146 B5
Holmajärvi *S* 111 E18
Holme-Olstrup *DK* 87 E10
Holmenkollen *N* 95 C12
Holmfirth *UK* 11 D8
Holmfors *S* 118 C4
Holmisperä *FIN* 123 D14
Holmsjö *S* 89 C7
Holmsjö *S* 103 A14
Holmsjö *S* 107 D12
Holmsund *S* 122 C4
Holmsvattensel *S* 118 B6
Holmsveden *S* 103 D12
Holmträsk *S* 107 B16
Hölö *S* 93 A11
Holod *RO* 151 D9
Hołodowska *PL* 144 C7
Holoşniţa *MD* 153 A12
Holoubkov *CZ* 76 C5
Holovets'ko *UA* 145 E6
Holovne *UA* 141 H10
Holsbybrunn *S* 92 E6
Holsljunga *S* 91 E12
Holsta *EST* 131 F14
Holstebro *DK* 86 C3
Holsted *DK* 86 D3
Hölstein *CH* 27 F8
Holsworthy *UK* 12 D6
Holt *UK* 10 E6
Holt *UK* 15 B11
Holt *N* 111 B16
Holt *N* 111 B17
Holte *DK* 87 D10
Holten *NL* 17 D6
Holtet *N* 102 E3
Holtgast *D* 17 A9
Holthusen *D* 83 C10
Holtsee *D* 83 B7
Holungen *D* 79 D7
Holwerd *NL* 16 B5
Holycross *IRL* 9 C7
Holyhead *UK* 10 E2
Holywell *UK* 7 D7
Holywell *UK* 10 E5
Holywood *UK* 4 F5
Holzappel *D* 185 D8
Holzen *D* 78 C6
Holzgerlingen *D* 187 D7
Holzhausen an der Haide
D 21 D9
Holzheim *D* 21 D11
Holzheim *D* 75 E8
Holzheim *D* 75 E8
Holzheim *D* 75 F7
Holzkirchen *D* 72 A4
Holzminden *D* 21 A12
Holzthaleben *D* 79 D8
Holzweißig *D* 79 C11
Holzwickede *D* 17 E9
Høm *DK* 87 E9
Homberg (Efze) *D* 21 B12
Homberg (Ohm) *D* 21 C12
Hombourg-Budange *F* 186 C1
Hombourg-Haut *F* 186 C2
Homburg *D* 21 F8
Homécourt *F* 20 F5
Homersfield *UK* 15 C11
Homesh *AL* 168 A3
Hommelstø *N* 108 E4
Hommelvik *N* 105 E9
Hommersåk *N* 94 E3
Homna *S* 103 D10
Homocea *RO* 153 E10
Homokmégy *H* 150 E3
Homokszentgyörgy *H* 149 D9
Homorode *RO* 151 B11
Homorod *RO* 152 E6
Homorod *H* 145 G2
Homyel' *BY* 133 F5
Hondarribia *E* 32 D2
Hondón de las Nieves *E* 56 E3
Hondón de los Frailes *E* 56 E3
Hondschoote *F* 18 B5
Hône *I* 68 B4
Hønefoss *N* 95 B12
Honfleur *F* 23 B12
Høng *DK* 87 D8
Honiton *UK* 13 D8

Honkajoki *FIN* 122 G8
Honkakoski *FIN* 126 B7
Honkakylä *FIN* 122 D9
Honkilahti *FIN* 126 D7
Honley *UK* 11 D8
Honningsvåg *N* 100 B2
Honningsvåg *N* 113 B16
Hönö *S* 91 D10
Honoratka *PL* 142 B5
Honrubia *E* 47 E8
Honrubia de la Cuesta *E* 40 F4
Hønseby *N* 113 B11
Hontalbilla *E* 40 F3
Hontanaya *E* 47 E7
Hontianske Nemce *SK* 147 E7
Hontoria de la Cantera *E* 40 D4
Hontoria del Pinar *E* 40 E5
Hontoria de Valderados
E 40 E4
Hoofddorp *NL* 16 D3
Hoogerheide *NL* 16 F2
Hoogersmilde *NL* 16 C6
Hoogezand-Sappemeer
NL 17 B7
Hoogkarspel *NL* 16 C4
Hoog-Keppel *NL* 183 A8
Hoogkerk *NL* 17 B6
Hoogland *NL* 183 A6
Hoogstede *D* 17 C7
Hoogstraten *B* 16 F3
Hoogvliet *NL* 16 E2
Hook *UK* 11 D10
Hook *UK* 14 E7
Hooksiel *D* 17 A10
Höör *S* 87 D13
Hoorn *NL* 16 C4
Hoornaar *NL* 182 B5
Hopârta *RO* 152 E3
Hope *UK* 10 E5
Hope *N* 100 D3
Hopeman *UK* 3 K10
Hopen *N* 111 E10
Hopfgarten im Brixental
A 72 B5
Hopfgarten in Defereggen
A 72 C6
Hopland *N* 100 C4
Hoppstädten *D* 21 E8
Hoppula *FIN* 115 F2
Hopseidet *N* 113 B20
Hopsten *D* 17 D9
Hopton *UK* 15 C12
Hoptonheath *UK* 13 A9
Hoptrup *DK* 86 E4
Horam *UK* 15 F9
Horažd'ovice *CZ* 76 D5
Horb am Neckar *D* 27 D10
Horbova *UA* 153 A8
Hörbranz *A* 71 B9
Hørby *DK* 90 E7
Hörby *S* 87 D13
Horcajo de las Torres *E* 45 B10
Horcajo de los Montes *E* 46 F3
Horcajo de Santiago *E* 47 E7
Horcajo Medianero *E* 45 C9
Horda *N* 94 C5
Horda *S* 88 A6
Hörde *D* 185 B7
Horden *UK* 11 B9
Hordum *DK* 86 B3
Horea *RO* 151 E10
Horeb *UK* 12 A6
Horeşti *MD* 153 C11
Horezu *RO* 160 C3
Horgen *CH* 27 F10
Horgenzell *D* 71 B9
Horgeşti *RO* 153 E10
Horgoš *SRB* 150 E4
Horia *RO* 153 D9
Horia *RO* 155 C2
Horia *RO* 155 D1
Hořice *CZ* 77 B9
Horincea *UA* 153 E10
Horincove *UA* 145 G7
Horitschon *A* 149 A7
Horjul *SLO* 73 D9
Horka *D* 81 D7
Hörken *S* 97 B12
Hørken *S* 97 B12
Horka *D* 81 D7
Horley *UK* 15 E8
Hörlitz *D* 80 C5
Hormakumpu *FIN* 117 C14
Hormilla *E* 40 D6
Horn *A* 77 E9
Horn *N* 95 A12
Horn *S* 92 D7
Hornachos *E* 51 B7
Hornachuelos *E* 51 D9
Horná Kraľová *SK* 146 E5
Horná Potôň *SK* 146 E4
Horná Streda *SK* 146 D5
Horná Štubňa *SK* 147 D7
Horná Súča *SK* 146 D5
Horná Ves *SK* 146 D6
Hornbach *D* 21 F8
Horn-Bad Meinberg *D* 17 E11
Hornbæk *DK* 87 C10
Hornburg *D* 79 B8
Hornby *UK* 10 C6
Horncastle *UK* 11 E11
Horndal *N* 109 A10
Horndal *S* 98 B6
Horndean *UK* 14 F6
Horne *DK* 86 D3
Horne *DK* 86 E6
Horne *DK* 90 D6
Horneburg *D* 82 C7
Hörnefors *S* 122 C3
Hornindal *N* 100 C5
Hørning *DK* 86 C6
Horní Planá *CZ* 76 E6
Horní Slavkov *CZ* 75 B12
Horní Štěpánov *CZ* 77 C11
Horní Stropnice *CZ* 77 E7
Horní Suchá *CZ* 146 B6
Horní Jelení *CZ* 77 B9
Horní Jiřetín *CZ* 80 E5
Horní Lideč *CZ* 146 C6
Horní Moštěnice *CZ* 146 C5
Horní Planá *CZ* 76 E6
Horní Slavkov *CZ* 75 B12
Horní Štěpánov *CZ* 77 C11
Horní Stropnice *CZ* 77 E7
Horní Suchá *CZ* 146 B6
Hornnäs *S* 122 A4
Hornnes *N* 90 B2

Hornön *S* 103 A14
Hornos *E* 55 C7
Hornow *D* 81 C7
Hornoy-le-Bourg *F* 18 E4
Hornsea *UK* 11 D11
Hörnsjö *S* 107 D17
Hornslet *DK* 86 C6
Hornstorf *D* 83 C11
Hornsyld *DK* 86 D5
Hörnum *D* 82 A4
Hornum *DK* 86 B4
Horný Bar *SK* 146 F4
Horný Tisovník *SK* 147 E8
Horný Vadičov *SK* 147 C7
Horoatu Crasnei *RO* 151 C10
Horodişte *MD* 153 A11
Horodişte *MD* 153 B11
Horodiştea *MD* 154 A3
Horodka *UA* 154 A3
Horodkiv *PL* 144 B9
Horodniceni *RO* 153 B8
Horodnye *UA* 154 D5
Horodok *UA* 144 D4
Horoměřice *CZ* 76 B6
Horonda *UA* 145 G6
Horonkylä *FIN* 122 G9
Hořovice *CZ* 76 C5
Horrabridge *UK* 12 D6
Horreby *DK* 83 A11
Horred *S* 91 E11
Hörsching *A* 76 F6
Horsdal *N* 108 B7
Horseleap *IRL* 6 F5
Horsens *DK* 86 D5
Horsforth *UK* 11 D8
Horsham *UK* 15 E8
Hørsholm *DK* 87 D10
Hörsingen *D* 79 B9
Horslunde *DK* 87 F9
Horsmanaho *FIN* 125 E12
Horsnes *N* 111 B18
Horšovský Týn *CZ* 76 C3
Horssko *S* 98 B8
Horst *NL* 16 F6
Horst (Holstein) *D* 82 C7
Horstedt *D* 82 B12
Hörstel *D* 17 D9
Horstmar *D* 17 D8
Hort *H* 150 B4
Horten *N* 95 C12
Hortezuela *E* 40 F6
Hortigüela *E* 40 D5
Hortlax *S* 118 D6
Hortobágy *H* 151 B8
Horton in Ribblesdale *UK* 11 C7
Høruphav *DK* 86 F5
Hørve *DK* 87 D8
Hörvik *S* 89 C7
Horw *CH* 70 C6
Horwich *UK* 10 D6
Horyniec-Zdrój *PL* 144 C7
Horyszów *PL* 144 B8
Hösbach *D* 21 D12
Hosena *D* 80 D6
Hosenfeld *D* 74 A5
Hoset *N* 108 B8
Hosingen *L* 20 D6
Hosio *FIN* 119 C15
Hosjö *S* 98 B6
Hosjöbottnarna *S* 105 E15
Hoşköy *TR* 173 C7
Hospital *IRL* 8 C6
Hospital *IRL* 9 C7
Hossa *FIN* 121 D14
Hóssjö *S* 122 C3
Hossjön *S* 107 D9
Hoşfkirch *D* 27 E11
Hosszúhetény *H* 149 D10
Hosszúpályi *H* 151 C8
Hosszúpereszteg *H* 149 B8
Hošťálková *CZ* 146 C5
Hostalric *E* 43 D9
Hostens *F* 32 B4
Hošt̕eradice *CZ* 77 E10
Hostie *SK* 146 E6
Hostinné *CZ* 81 E9
Hostka *CZ* 76 B6
Hostomice *CZ* 76 C5
Hostomice *CZ* 80 E5
Hostoň *CZ* 75 C12
Hostrupskov *DK* 86 E4
Höstsätern *S* 102 C4
Hot *AL* 163 E7
Hotarele *RO* 161 E8
Hoting *S* 107 C10
Hotinja vas *SLO* 148 D5
Hotonj *BIH* 162 D4
Hotton *B* 19 D11
Hou *DK* 86 A6
Hou *DK* 86 D6
Houbie *UK* 3 D15
Houdain *F* 18 D6
Houdan *F* 24 C6
Houdelaincourt *F* 26 C3
Houécourt *F* 26 D4
Houeillès *F* 32 B6
Houeydets *F* 33 D6
Houffalize *B* 19 D12
Houghton le Spring *UK* 5 F11
Houghton Regis *UK* 15 D7
Houlberg *DK* 86 C5
Houlgate *F* 23 B11
Hourtin *F* 28 E3
Hourtin-Plage *F* 28 E3
Hougulia *N* 101 D11
Houten *NL* 16 D4
Houthalen *B* 19 C11
Houthulst *B* 18 C6
Houton *UK* 3 H10
Houtsala *FIN* 126 E6
Houtskär *FIN* 126 E5
Houyet *B* 184 D3
Hov *DK* 86 D6
Hov *N* 101 E12
Hov *N* 108 B7
Hov *N* 111 A18
Hova *S* 92 B4
Hovborg *DK* 86 D3
Hovda *N* 94 D2
Hovde *N* 90 B4
Hovden *N* 110 D8
Hovdevik *N* 100 B2
Hove *UK* 15 F8
Hovedgård *DK* 86 D5
Hovelhof *D* 17 E11
Hoven *DK* 86 D3
Hoveton *UK* 15 B11
Hovézí *CZ* 146 C6
Hovid *S* 103 B12
Hovin *N* 104 E8
Hovmantorp *S* 89 B8
Hovringen *N* 101 C10
Høvringen *N* 101 C10
Hovsherad *N* 94 F4
Hovslund Stationsby *DK* 86 E4
Hovsta *S* 97 D13

Howden *UK* 11 D10
Howth *IRL* 7 F10
Hoya *D* 17 C12
Hoya Gonzalo *E* 55 B9
Høyanger *N* 100 D4
Høydal *N* 90 A5
Hoyerswerda *D* 80 D6
Høylake *UK* 10 E5
Høylandet *N* 105 B12
Hoym *D* 79 C9
Hoyocasero *E* 46 D3
Hoyo de Manzanares *E* 46 C5
Hoyos *E* 45 D7
Hoyos del Espino *E* 45 D10
Hoyti̊ş *FIN* 123 F14
Høyvik *FO* 2 A3
Hozha *BY* 137 F8
Hrabove *UA* 154 A4
Hrabyně *CZ* 146 B6
Hradec Králové *CZ* 77 B9
Hradec nad Moravicí *CZ* 146 B5
Hradec nad Svitavou *CZ* 77 C10
Hrádek *CZ* 77 E12
Hrádek nad Nisou *CZ* 81 E7
Hradenytsi *UA* 154 D5
Hradešice *CZ* 76 D5
Hradiště *SK* 146 D6
Hradiště pod Vrátnom *SK*
146 D5
Hradištko *CZ* 76 C6
Hraň *SK* 145 F4
Hranice *CZ* 75 B11
Hranice *CZ* 146 B5
Hranovnica *SK* 145 F1
Hrasnica *BIH* 157 E9
Hrastnik *SLO* 73 D11
Hrawzhyshki *BY* 137 E12
Hrebenky *UA* 154 D5
Hreljin *HR* 67 B10
Hrhov *SK* 145 F2
Hrimne *UA* 145 D8
Hriňová *SK* 147 D9
Hristovaia *MD* 154 A3
Hrnjadi *BIH* 156 D5
Hrob *CZ* 80 E5
Hrochot *SK* 147 D8
Hrochův Týnec *CZ* 77 C9
Hrodna *BY* 140 C9
Hromnice *CZ* 76 C4
Hronec *SK* 147 D8
Hronov *CZ* 77 B10
Hronovce *SK* 147 E7
Hronský Beň Beňx *SK* 147 E7
Hrotovice *CZ* 77 D10
Hroznová Lhota *CZ* 146 D4
Hrtkovci *SRB* 158 D4
Hrubieszów *PL* 144 B8
Hruşca *MD* 154 A3
Hrušky *CZ* 77 E11
Hrušova *MD* 154 C6
Hrušovany nad Jevišovkou
CZ 77 E10
Hruštín *SK* 147 C8
Hrusznew *PL* 141 F7
Hrvaćani *BIH* 157 C7
Hrvace *HR* 156 E6
Hrvatska Dubica *HR* 157 B6
Hrvatska Kostajnica *HR* 156 B6
Hrynyava *UA* 152 B5
Huaröd *S* 88 D5
Huarte *E* 32 E2
Hubová *SK* 147 C8
Hückelhoven *D* 20 B6
Hückeswagen *D* 21 B8
Hucknall *UK* 11 E9
Hucksjöåsen *S* 103 A10
Hucqueliers *F* 15 F12
Huddersfield *UK* 11 D8
Hüde *D* 17 C11
Hude (Oldenburg) *D* 17 B10
Hudeşti *RO* 153 A9
Hudiksvall *S* 103 C13
Huécija *E* 55 F7
Huedin *RO* 151 D11
Huélago *E* 55 E6
Huelgoat *F* 22 D3
Huelma *E* 53 A10
Huelva *E* 51 E6
Huelves *E* 47 D7
Huércal de Almería *E* 55 F8
Huércal-Overa *E* 55 E9
Huérguina *E* 47 D9
Huérmeces *E* 40 C4
Huerta del Marquesada *E* 47 D9
Huerta del Rey *E* 40 E5
Huerta de Valdecarábanos
E 46 E5
Huertahernando *E* 47 C8
Huerto *E* 42 D3
Huesa *E* 55 D7
Huesa del Común *E* 42 F2
Huéscar *E* 55 D7
Huete *E* 47 D7
Huétor-Tájar *E* 53 B8
Huétor-Vega *E* 53 B9
Huévar *E* 51 E7
Hüfingen *D* 27 E9
Hufthamar *N* 94 B2
Hugh Town *UK* 12 F2
Huglfing *D* 72 A3
Huguag *H* 147 E8
Huhdasjärvi *FIN* 127 C16
Huhmarkoski *FIN* 123 D10
Huhtamo *FIN* 126 C7
Huhti *FIN* 127 C10
Huhtilampi *FIN* 125 F14
Huhus *FIN* 125 E15
Huijbergen *NL* 182 C4
Huikola *FIN* 119 E16
Huisheim *D* 75 E8
Huisinis *UK* 2 K2
Huissen *NL* 16 E5
Huissinkylä *FIN* 122 E8
Huittinen *FIN* 126 C7
Huizen *NL* 16 D4
Hujakkala *FIN* 128 D8
Hukjärvi *FIN* 121 D10
Hukkajärvi *FIN* 121 E14
Hulín *CZ* 146 C4
Hulja *EST* 131 C12
Huljen *S* 103 B12
Hulkkola *FIN* 124 E8
Hull *UK* 11 D11
Hüllhorst *D* 17 D11
Hullo *EST* 130 D6
Hullsjön *S* 103 A13
Hüls *D* 183 C8
Hulsberg *NL* 183 D7
Hulst *NL* 16 F2
Hult *S* 91 A15

Hult *S* 92 D6
Hulterstad *S* 89 C11
Hultsfred *S* 92 D7
Hulu *S* 91 D13
Hulubeşti *RO* 160 D6
Hulyanka *UA* 154 C4
Hum *BIH* 157 F10
Hum *BIH* 162 D5
Humaljoki *FIN* 123 D13
Humanes de Madrid *E* 46 D5
Humanes de Mohernando
E 47 C6
Humberston *UK* 11 D11
Humbie *UK* 5 D11
Humble *DK* 83 A9
Humenné *SK* 145 F4
Humilladero *E* 53 B7
Humlebæk *DK* 87 D11
Humlegårdsstrand *S* 103 D13
Humlum *DK* 86 B3
Hummelholm *S* 107 D17
Hummelo *NL* 183 A8
Hummelsta *S* 98 C7
Hummuli *EST* 131 F12
Hum na Sutli *HR* 148 D5
Humpolec *CZ* 77 C8
Humppila *FIN* 127 D9
Humshaugh *UK* 5 E12
Huncovce *SK* 145 E1
Hundåla *N* 108 E4
Hundberg *N* 111 B18
Hundberg *S* 109 F16
Hundborg *DK* 86 B3
Hundeluft *D* 79 C11
Hunderdorf *D* 75 E12
Hundested *DK* 87 D9
Hundholmen *N* 111 D11
Hundorp *N* 101 C11
Hundsangen *D* 21 D9
Hundshübel *D* 79 E12
Hundsjö *S* 118 C7
Hundslund *DK* 86 D6
Hundvin *N* 100 E2
Hune *DK* 86 A5
Hunedoara *RO* 151 F10
Hünfeld *D* 78 E6
Hünfelden-Kirberg *D* 21 D10
Hunge *S* 103 A9
Hungen *D* 21 D11
Hungerford *UK* 13 C11
Hunnebostrand *S* 91 C9
Hunsel *NL* 19 B12
Hunspach *F* 27 C8
Hunstanton *UK* 11 F12
Huntingdon *UK* 15 C8
Huntlosen *D* 17 C10
Huntly *UK* 3 L11
Hünxe *D* 183 B9
Hunya *H* 151 D6
Huopanankoski *FIN* 123 D15
Hüpstedt *D* 79 D7
Hurbanovo *SK* 146 F6
Hurdal *N* 95 B14
Hurdalsverk *N* 95 B14
Hurezani *RO* 160 D3
Huriel *F* 29 C10
Hurissalo *FIN* 128 B7
Hurler's Cross *IRL* 8 C5
Hursley *UK* 13 C12
Hurst Green *UK* 15 E9
Hurstpierpoint *UK* 15 F8
Hurteles *E* 41 D7
Hürth *D* 21 C7
Huruieşti *RO* 153 E10
Huruksela *FIN* 128 D6
Hurup *DK* 86 B2
Hurva *S* 87 D12
Huså *S* 105 E14
Husabø *S* 91 D10
Husaru *S* 102 B6
Husbondliden *S* 107 B15
Husby *D* 82 A7
Husby *S* 97 B15
Husby *S* 99 B9
Huschka *UA* 141 H9
Huşi *RO* 153 D12
Husinec *CZ* 76 D5
Husjorda *N* 111 D9
Huskvarna *S* 92 D4
Husnes *N* 94 C3
Husnicioara *RO* 159 D10
Husøy *N* 108 D3
Hussjö *S* 103 A13
Hustopeče *CZ* 77 E11
Hustopeče nad Bečvou *CZ*
146 B5
Husum *D* 17 C12
Husum *D* 82 B6
Husum *S* 107 E16
Husvik *N* 108 H3
Huta *DK* 86 E4
Huta Komorowska *PL* 143 F12
Hutisko-Solanec *CZ* 146 C6
Hutovo *BIH* 162 D4
Hütschenhausen *D* 186 C3
Hüttau *A* 73 B7
Hüttenberg *A* 73 C10
Hüttisheim *D* 74 F6
Hüttlingen *D* 75 E7
Huttoft *UK* 11 E12
Hüttschlag *A* 73 B7
Huttukylä *FIN* 119 D15
Huttwil *CH* 27 F8
Huuhilonkylä *FIN* 121 F13
Huuki *S* 117 C11
Huutijärvi *FIN* 127 C11
Huutokoski *FIN* 124 E8
Huutokoski *FIN* 125 E12
Huutoniemi *FIN* 113 E18
Huvåsen *S* 118 D4
Hüven *D* 17 C9
Huwniki *PL* 145 D6
Huy *B* 19 D11
Hvalpsund *DK* 86 B4
Hvalsø *DK* 87 D9
Hvalvík *FO* 2 A2
Hvam *DK* 86 B5
Hvannasund *FO* 2 A3
Hvar *HR* 156 F5
Hvidbjerg *DK* 86 B3
Hvide Sande *DK* 86 C2
Hvidovre *DK* 87 D10
Hvilsom *DK* 86 B5
Hvittingfoss *N* 95 D11
Hvorslev *DK* 86 C5
Hwlffordd *UK* 12 B5
Hybe *SK* 147 C9
Hybo *S* 103 C11
Hycklinge *S* 92 D7
Hyères *F* 36 E4
Hyermanavichy *BY* 133 F3

Hyervvaty *BY* 137 D13
Hylen *N* 94 C5
Hylleråsen *N* 102 C3
Hyllestad *N* 100 D2
Hyllinge *DK* 87 E9
Hyllinge *S* 87 C11
Hyltebruk *S* 87 A12
Hymont *F* 26 D5
Hyönölä *FIN* 127 E10
Hyrkäs *FIN* 119 E16
Hyry *FIN* 119 C14
Hyrynsalmi *FIN* 121 E11
Hysgjokaj *AL* 168 C2
Hyssna *S* 91 D12
Hythe *UK* 13 D12
Hythe *UK* 15 E11
Hytti *FIN* 129 D9
Hyttön *S* 99 B8
Hyväneula *FIN* 127 D13
Hyväniemi *FIN* 121 B11
Hyvärilä *FIN* 119 F15
Hyvikkälä *FIN* 127 E10
Hyvinkää *FIN* 127 E10
Hyvölänranta *FIN* 119 F16
Hyvönmäki *FIN* 129 B12
Hyypiö *FIN* 115 E1
Hyyppä *FIN* 122 F8
Hyżne *PL* 144 D5

Iablaniţa *RO* 159 D9
Iabloana *MD* 153 B11
Iacobeni *RO* 159 C7
Iacobeni *RO* 152 C6
Ialoveni *MD* 154 D3
Iam *RO* 159 C7
Ianca *RO* 160 F4
Ianca *RO* 161 C10
Iancu Jianu *RO* 160 E4
Iara *RO* 152 D3
Iargara *MD* 154 E2
Iarova *MD* 153 A12
Iaşi *RO* 153 C11
Iasmos *GR* 171 B8
Ibahernando *E* 45 F9
Iballë *AL* 163 E9
Ibăneşti *RO* 152 D5
Ibăneşti *RO* 153 A8
Ibarra *E* 32 D1
Ibbenbüren *D* 17 D9
Ibdes *E* 47 B9
Ibë *AL* 168 A2
Ibeas de Juarros *E* 40 D4
Ibestad *N* 111 C13
Ibi *E* 56 D3
Ibos *F* 32 D5
Ibrány *H* 145 G4
Ibriktepe *TR* 171 B10
Ibros *E* 53 A10
Ibstock *UK* 11 F9
Ichenhausen *D* 75 F7
Ichenheim *D* 186 E4
Ichtegem *B* 18 B7
Icking *D* 72 A3
Icklesham *UK* 15 F10
Icklingham *UK* 15 C10
Iclänzel *RO* 152 D4
Iclod *RO* 152 D3
Icoana *RO* 160 E5
Icuşeşti *RO* 153 D9
Idanha-a-Nova *P* 45 E6
Idanha-a-Velha *P* 45 E6
Idar-Oberstein *D* 21 E8
Ideciu de Jos *RO* 152 D5
Iden *D* 83 E11
Ideţa *LV* 133 C1
Idenor *S* 103 C13
Idestrup *DK* 83 A11
Idiazabal *E* 32 D1
Idivuoma *S* 116 A6
Idkerberget *S* 97 B13
Idmiston *UK* 13 C11
Idocin *E* 32 E3
Idom *DK* 86 C2
Idoš *SRB* 150 F5
Idre *S* 102 B4
Idrigill *UK* 2 K4
Idrija *SLO* 73 D9
Idritsa *RUS* 133 D5
Idro *I* 69 B9
Idstedt *D* 82 A7
Idstein *D* 21 D10
Idvattnet *S* 107 C12
Iecava *LV* 135 C8
Iedera *RO* 161 C7
Ieper *B* 18 C6
Iepureşti *RO* 161 E7
Ierapetra *GR* 179 E10
Ieriķi *LV* 135 B10
Ierissos *GR* 170 D5
Iernut *RO* 152 E4
Ieromnini *GR* 168 E4
Ieropigi *GR* 168 C5
Ieşelniţa *RO* 159 D9
Ifaistos *GR* 171 B8
Iffendic *F* 23 D7
Iffezheim *D* 187 D5
Ifjord *N* 113 C19
Ifs *F* 23 B11
Ifta *D* 79 D7
Ig *SLO* 73 E10
Igal *H* 149 C9
Igalo *MNE* 162 E6
Igar *H* 149 C11
Igé *F* 24 D4
Igea *E* 41 D7
Igel *D* 186 B2
Igelfors *S* 92 B7
Igelstorp *S* 91 C14
Igelsdorf *D* 75 C9
Igerøy *N* 108 E3
Igersheim *D* 74 D6
Iggelheim *D* 187 C5
Iggensbach *D* 76 E4
Iggesund *S* 103 C13
Ighiu *RO* 152 E3
Igis *D* 71 D9
Iglesias *I* 64 E2
Igliauka *LT* 137 D8
Igling *D* 71 A11
Igliškeliai *LT* 137 D8
Ignalina *LT* 135 F12
Ignatievo *BG* 167 C9
Iğneada *TR* 167 F9
Igneşti *RO* 151 E9
Igney *F* 26 D5
Igornay *F* 30 A5
Igoumenitsa *GR* 168 E3
Igralishte *BG* 169 A9
Igrejinha *P* 50 B4
Igrici *H* 145 H2
Igrīve *LV* 133 B2

Igualada *E* 43 D7
Igualeja *E* 53 C6
Igueña *E* 39 C7
Iguerande *F* 30 C5
Iharosberény *H* 149 D8
Ihľany *SK* 145 E2
Ihlienworth *D* 17 A11
Ihlow/erhörn (Ihlow) *D* 17 B9
Ihode *FIN* 126 D6
Iholdy *F* 32 D3
Ihrhove *D* 17 B8
Ihrlerstein *D* 75 E10
Ihsaniye *TR* 173 B10
Ii *FIN* 119 D14
Iijärvi *FIN* 113 E20
Iinattijärvi *FIN* 121 D9
Iironranta *FIN* 123 E12
Iisalmi *FIN* 124 C7
Iisvesi *FIN* 124 E8
Iitti *FIN* 127 D15
Iitto *FIN* 116 A6
Iivantiira *FIN* 121 F13
IJlst *NL* 16 B5
IJmuiden *NL* 16 D3
IJsselmuiden *NL* 16 C5
IJsselstein *NL* 16 D4
IJzendijke *NL* 16 F1
Ikaalinen *FIN* 127 B9
Ikast *DK* 86 C4
Ikazn' *BY* 133 E2
Ikervár *H* 149 B8
Ikhtiman *BG* 165 E8
Ikizdere *TR* 177 D10
Ikkala *FIN* 127 E11
Ikkala *FIN* 127 E11
Ikkeläjärvi *FIN* 122 F9
Ikla *EST* 131 F8
Ikornes *N* 100 B5
Ikosenniemi *FIN* 119 C17
Ikrény *H* 149 A9
Ikškile *LV* 135 C8
Ilandža *SRB* 159 C6
Ilanz *CH* 71 D8
Ilava *SK* 146 D6
Iława *PL* 139 C8
Ilbono *I* 64 D4
Ilchester *UK* 13 C9
Ildir *TR* 177 C7
Île *LV* 134 C6
Ileana *RO* 161 D9
Ileanda *RO* 152 C3
Ilfeld *D* 79 C8
Ilford *UK* 15 D9
Ilfracombe *UK* 12 C6
Ilgiżiai *LT* 134 F6
Ílhavo *P* 44 C3
Ilia *RO* 151 F10
Ilica *TR* 173 E8
Ilidza *BIH* 157 E9
Ilieni *RO* 153 F7
Ilijaš *BIH* 157 E9
Ilindentsi *BG* 165 F7
Iliokastro *GR* 175 E7
Ilirska Bistrica *SLO* 73 E8
Ilk *H* 145 G5
Ilkeston *UK* 11 F9
Ilkley *UK* 11 D8
Illana *E* 47 D7
Illar *E* 55 F7
Illats *F* 32 A5
Illerrieden *D* 71 A10
Illertissen *D* 71 A10
Illescas *E* 46 D5
Ille-sur-Têt *F* 34 E4
Illiers-Combray *F* 24 D5
Illingen *D* 21 F8
Illingen *D* 187 D6
Illkirch-Graffenstaden *F* 27 C8
Illmensee *D* 27 E11
Illmitz *A* 149 A7
Íllora *E* 53 B9
Illschwang *D* 75 D10
Illueca *E* 41 E8
Illzach *F* 27 E7
Ilmajoki *FIN* 122 E9
Ilmatsalu *EST* 131 E13
Ilmenau *D* 79 E8
Ilminster *UK* 13 D9
Ilmmünster *D* 75 F10
Ilmola *FIN* 119 C13
Il'nytsya *UA* 145 G7
Ilok *HR* 158 C3
Ilola *FIN* 127 E14
Ilomantsi *FIN* 125 E15
Ilosjoki *FIN* 123 D15
Ilovăţ *RO* 159 D10
Ilovica *NMK* 169 B8
Ilovice *BIH* 157 E9
Ilovița *RO* 159 D10
Iłów *PL* 139 F9
Iłowa *PL* 81 C8
Iłowo Osada *PL* 139 D9
Ilsbo *S* 103 C13
Ilsede *D* 79 B7
Ilsenburg (Harz) *D* 79 C8
Ilseng *N* 101 E14
Ilsfeld *D* 27 B11
Ilskov *DK* 86 C4
Ilükste *LV* 135 E11
Ilva Mare *RO* 152 C5
Ilva Mică *RO* 152 C5
Ilvesjoki *FIN* 122 F9
İlyaslar *TR* 173 F9
Ilyushino *RUS* 136 D6
Ilz *A* 148 B5
Iłża *PL* 141 H4
Ilze *LV* 135 D12
Ilzene *LV* 135 B13
Imari *FIN* 119 B15
Imatra *FIN* 129 C10
Imavere *EST* 131 D11
Imbradas *LT* 135 E12
Imeľ *SK* 146 F6
Imèr *I* 72 D4
Imeros *GR* 171 C8
Imielnica *PL* 139 E8
Immeln *S* 88 C6
Immendingen *D* 27 D11
Immenhausen *D* 78 D5
Immenreuth *D* 75 C11
Immenstaad am Bodensee *D* 27 E11
Immenstadt im Allgäu *D* 71 B10
Immingham *UK* 11 D11
Imnäs *S* 107 D11
Imola *I* 66 D4
Imotski *HR* 157 F7
Imperia *I* 37 D8
Imphy *F* 30 B3
Impilakhti *RUS* 129 B15
Impiö *FIN* 120 C9
Impruneta *I* 66 E3
Imrehegy *H* 150 E3

İmroz *TR* 171 D9
Imst *A* 71 C11
Ina *FIN* 123 D14
Inagh *IRL* 8 C4
Ináncs *H* 145 G3
Inarcs *H* 150 C3
Inari *FIN* 113 F19
Inari *FIN* 125 D15
Inca *E* 57 B10
Inch *IRL* 8 D3
Inch *IRL* 9 C10
Inchbare *UK* 5 B11
Incheville *F* 18 D3
Inchigeelagh *IRL* 8 E4
Inchnadamph *UK* 2 J7
Inciems *LV* 135 B9
Incinillas *E* 40 C4
Încircliova *RO* 177 D10
Incisa in Val d'Arno *I* 66 E3
Incourt *B* 19 C10
Inčukalns *LV* 135 B9
Indal *S* 103 A13
Indalstø *N* 100 E2
Independenţa *RO* 155 C1
Independenţa *RO* 155 C1
Indija *SRB* 158 C5
Indra *LV* 133 E3
Indreabhán *IRL* 6 F4
Indre Arna *N* 94 B2
Indre Billefjord *N* 113 C15
Indre Brenna *N* 113 B16
Indre Kårvik *N* 111 A16
Indre Kiberg *N* 114 C9
Indre Kjæs *N* 113 B16
Indre Sortvik *N* 113 B15
Indura *BY* 140 D9
Indzhe Voyvoda *BG* 167 E8
Inece *TR* 167 F8
Inecik *TR* 173 C7
Ineši *LV* 135 B11
Ineu *RO* 151 C9
Ineu *RO* 151 E8
Infiesto *E* 39 B9
Ingå *FIN* 127 E11
Ingared *S* 91 D11
Ingatestone *UK* 15 D9
Ingatorp *S* 92 D6
Ingelfingen *D* 187 C8
Ingelheim am Rhein *D* 21 E10
Ingelmunster *B* 19 C7
Ingelstad *S* 89 B7
Ingenes *N* 94 B3
Ingersheim *F* 27 D7
Ingleton *UK* 5 E11
Ingleton *UK* 10 C7
Ingoldmells *UK* 11 E12
Ingolstadt *D* 75 E9
Ingrandes *F* 23 F10
Ingrandes *F* 29 B2
Ingstrup *DK* 90 E6
Inguiniel *F* 22 E5
Ingulsvatn *N* 105 B14
Ingwiller *F* 27 C7
Inha *FIN* 123 E12
Ini *GR* 178 E9
Iniesta *E* 47 F9
Inis *IRL* 8 C5
Inis Córthaidh *IRL* 9 C9
Inis Diomáin *IRL* 8 C4
Inistioge *IRL* 9 D8
Injevo *NMK* 169 A7
Inkberrow *UK* 13 A11
Inke *H* 149 D8
Inkee *FIN* 121 C12
Inkere *FIN* 127 E9
Inkoo *FIN* 127 E11
Inndyr *N* 108 B7
Innerbraz *A* 71 C9
Innerleithen *UK* 5 D10
Innernzell *D* 76 E4
Innertällmo *S* 107 D13
Innertavle *S* 122 C4
Innerkirchen *CH* 70 D6
Innervik *S* 109 F14
Innervillgraten *A* 72 C5
Innhavet *N* 111 D10
Inniscrone *IRL* 6 D4
Innishannon *IRL* 8 E5
Innsbruck *A* 72 B3
Innset *N* 111 C16
Inntorget *N* 108 B7
Inowłódz *PL* 141 G2
Inowrocław *PL* 138 E5
Ins *CH* 31 A11
Insch *UK* 3 L11
Insjön *S* 103 E10
Ińsko *PL* 85 D9
Insming *F* 27 C6
Instefjord *N* 100 D2
Instinción *E* 55 F7
Însurăţei *RO* 161 D11
Intepe *TR* 171 D10
Întorsura Buzăului *RO* 161 B8
Întregalde *RO* 151 E11
Introbio *I* 69 B7
Inturkė *LT* 135 F11
Inver *IRL* 6 C6
Inverallochy *UK* 3 K13
Inveran *IRL* 6 F4
Inveraray *UK* 4 C6
Inverarity *UK* 5 B11
Inverbervie *UK* 5 B12
Invercassley *UK* 2 K7
Invercharnan *UK* 4 B6
Invergarry *UK* 4 A7
Invergordon *UK* 3 K8
Inverkeilor *UK* 5 C10
Inverkeithing *UK* 5 C10
Invermoriston *UK* 2 L7
Inverness *UK* 3 K8
Inverurie *UK* 3 L12
Inviken *S* 106 B8
Inzell *D* 73 A6
Inzigkofen *D* 27 D11
Inzing *A* 72 B3
Inzinzac-Lochrist *F* 22 E5
Ioannina *GR* 168 E4
Ion Corvin *RO* 155 E1
Ion Creangă *RO* 153 D9
Ioneşti *RO* 160 D4
Ioneşti *RO* 160 D4
Ion Luca Caragiale *RO* 161 D7
Ion Roată *RO* 161 D9
Iordăcheanu *RO* 161 C8
Ioulis *GR* 175 D9
Ip *RO* 151 C10
Ipatele *RO* 153 D10
Iphofen *D* 187 B9
Ipiķi *LV* 131 E10
Ipoteşti *RO* 153 B8

Ipoteşti *RO* 153 B9
Ippesheim *D* 75 C7
Ipplepen *UK* 13 E7
Ipsala *TR* 171 C10
Ipsheim *D* 75 C7
Ipstones *UK* 11 E8
Ipswich *UK* 15 C11
Irakleia *GR* 169 B9
Irakleia *GR* 174 B5
Irakleia *GR* 176 F5
Irakleio *GR* 178 E9
Irancy *F* 25 E10
Iratoşu *RO* 151 E7
Irdning *A* 73 A9
Irechekovo *BG* 167 E7
Iregszemcse *H* 149 C10
Irgoli *I* 64 C4
Iği *GR* 175 E7
Irig *SRB* 158 C4
Irishtown *IRL* 6 E5
Irissarry *F* 32 D3
Irjanne *FIN* 126 C6
Irlava *LV* 134 C5
Irlbach *D* 75 E12
Irninniemi *FIN* 121 C13
Irodouër *F* 23 D8
Ironbridge *UK* 10 F7
Irrel *D* 20 E6
Irsch *D* 21 E7
Irsee *D* 71 B11
Irshava *UA* 145 G7
Irši *LV* 135 C11
Irsina *I* 60 B6
Irsta *S* 98 C7
Irueste *E* 47 C7
Irun *E* 32 D2
Iruña *E* 32 E2
Irurita *E* 32 D2
Irurozqui *E* 32 E3
Irurtzun *E* 32 E2
Irvine *UK* 4 D7
Irvinestown *UK* 7 D7
Irxleben *D* 79 B9
Isaba *E* 32 E4
Isaccea *RO* 155 C2
Isačić *BIH* 156 D5
Işalniţa *RO* 160 E3
Isane *N* 100 C3
Isaris *GR* 174 E5
Isaszeg *H* 150 B3
Isàtra *S* 98 B3
Isbergues *F* 18 C5
Isbister *UK* 3 D14
Íscar *E* 40 F2
Isches *F* 26 D4
Ischgl *A* 71 C10
Ischia *I* 60 B1
Ischia di Castro *I* 62 B1
Ischitella *I* 63 D9
Isdes *F* 25 E7
Iselvmoen *N* 111 C16
Isen *D* 75 F11
Is-en-Bassigny *F* 26 D3
Isenbüttel *D* 79 B8
Isenvad *DK* 86 C4
Iseo *I* 69 B9
Iserlohn *D* 17 F9
Isernhagen *D* 78 B6
Isernia *I* 63 D6
Isfjorden *N* 100 A7
Ishakçelebi *TR* 177 B10
Ishull-Lezhë *AL* 163 F8
İshakli *I* 64 D3
İskele *TR* 173 F9
İskender *TR* 172 A6
İşkoras *N* 113 E16
Iskra *BG* 166 F4
Iskrets *BG* 165 D7
Isla Cristina *E* 51 E5
İslambeyli *TR* 167 F9
Isla Plana *E* 56 F2
Išlaužas *LT* 137 D8
Islaz *RO* 160 F5
Isle *F* 29 D8
Isle of Whithorn *UK* 5 F8
Isleryd *S* 92 C4
Isles-sur-Suippe *F* 19 F9
İslice *LV* 135 D8
İsmaili *TR* 177 B9
Ismaning *D* 75 F10
Ismundsundet *S* 106 E8
Isna *P* 44 E5
Isnäs *FIN* 127 E15
Isnello *I* 58 D5
İşnovăţ *MD* 154 C3
Isny im Allgäu *D* 71 B10
Iso-Evo *FIN* 127 D12
Isohalme *FIN* 115 E3
Isojoki *FIN* 122 F7
Isokumpu *FIN* 121 C11
Isokylä *FIN* 115 E3
Isokylä *FIN* 119 F14
Isokyrö *FIN* 122 D8
Isola *I* 36 C6
Isola 2000 *F* 37 C6
Isola *I* 59 C7
Isola del Gran Sasso d'Italia *I* 62 B5
Isola della Scala *I* 66 B3
Isola delle Femmine *I* 58 C3
Isola del Liri *I* 62 D5
Isola di Capo Rizzuto *I* 61 F8
Isole del Cantone *I* 37 B9
Isona *E* 42 C6
Isopalo *FIN* 115 D2
Isorella *I* 66 B1
Iso-Vimma *FIN* 126 C7
Ispagnac *F* 35 B6
Isperikh *BG* 161 F9
Ispica *I* 59 F6
Ispoure *F* 32 D3
Ispra *I* 68 B5
Ispringen *D* 27 C10
Issakka *FIN* 125 C10
Isselburg *D* 17 E6
Issigeac *F* 29 F7
Issime *I* 68 B4
Isso *E* 55 C9
Issogne *I* 68 B4
Issoire *F* 30 D3
Issoudun *F* 29 B9
Íssum *D* 17 E6
Is-sur-Tille *F* 26 E3
Issy-l'Évêque *F* 30 B4
Istalsnä *LV* 133 D3
Istán *E* 53 C7
İstanbul *TR* 173 B10
Istead Rise *UK* 15 E9
Istebna *PL* 147 B7
Istebné *SK* 147 C8
Istenmezeje *H* 147 E10

Isternia *GR* 176 D5
Isthmia *GR* 175 D7
Istiaia *GR* 175 B7
Istibanja *NMK* 164 F5
Istog *RKS* 163 D9
Istres *F* 35 C8
Istria *RO* 155 D3
Istrio *GR* 181 D7
Istunmäki *FIN* 123 E16
Isuerre *E* 32 F3
Iszkaszentgyörgy *H* 149 B10
Itä-Ähtäri *FIN* 123 E12
Itä-Aure *FIN* 123 E10
Itä-Karttula *FIN* 124 E8
Itäkoski *FIN* 119 C13
Itäkoski *FIN* 125 C9
Itäkylä *FIN* 123 D11
Itäranta *FIN* 115 F2
Itäranta *FIN* 120 F9
Itea *GR* 169 C6
Itea *GR* 169 D6
Itea *GR* 169 F7
Itea *GR* 174 C5
Itero de la Vega *E* 40 D3
Iteuil *F* 29 C6
Ithaki *GR* 174 C2
Itrabo *E* 53 C9
Itri *I* 62 E5
Itterbeck *D* 17 C7
Ittireddu *I* 64 B2
Ittiri *I* 64 B2
Ittre *B* 19 C8
Ituero de Azaba *E* 45 D7
Itzehoe *D* 82 C7
Itzstedt *D* 83 C8
Iurceni *MD* 154 C2
Ivalo *FIN* 115 A3
Ivalon Matti *FIN* 117 B15
Iván *H* 149 B7
Ivana Franka *UA* 145 E7
Ivancea *MD* 154 C3
Ivančice *CZ* 77 D10
Ivančići *BIH* 157 D9
Iváncsa *H* 149 B11
Ivanec *HR* 148 D6
Ivăneşti *RO* 153 D10
Ivangorod *RUS* 132 C3
Ivangrad → Berane *HR* 156 E5
Ivanić-Grad *HR* 148 E6
Ivanivka *UA* 145 G6
Ivanjica *SRB* 163 B9
Ivanjska *BIH* 157 C7
Ivankovo *HR* 149 F11
Ivano-Frankove *UA* 144 D8
Ivanovice na Hané *CZ* 77 D12
Ivanovo *BG* 161 F7
Ivanovo *BG* 166 F5
Ivanovo *BG* 167 C7
Ivanovo *SRB* 158 D6
Ivanska *HR* 149 E7
Ivanski *BG* 167 C8
Ivarrud *N* 108 F6
Ivarsbjörke *S* 97 C9
Ivars d'Urgell *E* 42 D5
Ivaylovgrad *BG* 171 A10
Iveland *N* 90 C2
Iver *UK* 15 D9
Iveşti *RO* 153 E11
Iveşti *RO* 153 F11
Ivrea *I* 68 C4
Ivrindi *TR* 173 E7
Ivry-la-Bataille *F* 24 C5
Ivry-sur-Seine *F* 25 C7
Ivybridge *UK* 13 E7
Iwaniska *PL* 143 E11
Iwanowice Włościańskie *PL* 143 F8
Iwkowa *PL* 144 D2
Iwye *BY* 137 F12
Ixelles *B* 19 C9
Ixworth *UK* 15 C10
İyaslar *TR* 177 A10
İža *SK* 149 A10
Iza *UA* 145 G7
Izarra *E* 40 C6
Izbica *PL* 144 B7
Izbica Kujawska *PL* 138 F5
Izbiceni *RO* 160 F5
Izbicko *PL* 142 E5
Izbişte *SRB* 159 C7
Izbižno *BIH* 157 E10
Izeaux *F* 31 E7
Izeda *P* 39 E7
Izegem *B* 19 C7
Izernore *F* 31 C8
Izeron *F* 31 E7
Izgrev *BG* 167 E9
Izgrev *BG* 167 E9
Iž Mali *HR* 67 D11
İzmayil *UA* 155 C3
İzmir *TR* 177 C7
İznájar *E* 53 B8
Iznalloz *E* 53 B9
Izola *SLO* 67 A8
Izsák *H* 150 D3
Izsófalva *H* 145 G2
Izvoare *RO* 160 E2
Izvoarele *RO* 155 C2
Izvoarele *RO* 160 F6
Izvoarele *RO* 161 D6
Izvoarele *RO* 161 E7
Izvoarele Sucevei *RO* 152 B6
Izvor *BG* 165 F10
Izvor *BG* 165 E6
Izvor *NMK* 168 B4
Izvor *NMK* 169 A6
Izvor *SRB* 159 F8
Izvorovo *BG* 155 F1
Izvoru *RO* 160 E6
Izvoru Bârzii *RO* 159 D10
Izvoru Berheciului *RO* 153 D10
Izvoru Crişului *RO* 151 D11

Jablanica *BIH* 157 E8
Jablanica *NMK* 164 F5
Jablanica *SK* 146 D4
Jablonec nad Jizerou *CZ* 81 E8
Jablonec nad Nisou *CZ* 81 E8
Jablonica *SK* 146 D4
Jablonka *PL* 147 C9
Jabłonka *PL* 139 F10
Jabłonka Kościelna *PL* 140 E6
Jabłonna Lacka *PL* 141 F6
Jabłonna *PL* 139 F10
Jabłonna *PL* 141 F6
Jabłonna Pierwsza *PL* 141 H7
Jablonné nad Orlicí *CZ* 77 B11
Jablonné v Podještědí *CZ* 81 E7
Jablonové *SK* 77 F12
Jabłonowo Pomorskie *PL* 139 D7
Jabłuńka *CZ* 146 C6
Jablunkov *CZ* 147 B7
Jabugo *E* 51 D6
Jabuka *BIH* 157 E10
Jabuka *SRB* 158 D6
Jabuka *SRB* 163 C7
Jabukovac *HR* 149 F6
Jabukovac *SRB* 159 E9
Jabukovik *SRB* 164 D5
Jaca *E* 32 E4
Jachenau *D* 72 A3
Jáchymov *CZ* 76 B3
Jacovce *SK* 146 D6
Jaderberg *S* 103 E12
Jaderfors *S* 103 E12
Jadów *PL* 139 F12
Jadraås *S* 103 E11
Jadraque *E* 47 C7
Jadranska Lešnica *SRB* 158 D3
Jægerspris *DK* 87 D9
Jægervatnet *N* 111 A18
Jaén *E* 53 A9
Jägala *EST* 131 C10
Jagare *BIH* 157 C7
Jagerberg *A* 148 C5
Jagodina *SRB* 159 F7
Jagodnjak *HR* 149 E11
Jagodzin *PL* 81 D8
Jagsthausen *D* 27 B11
Jagstzell *D* 75 D7
Jähdyspohja *FIN* 123 F11
Jahnsfelde *D* 80 A6
Jahodná *SK* 146 E5
Jah-Salih *AL* 168 A1
Jajce *BIH* 157 D7
Ják *H* 149 B7
Jakabszállás *H* 150 D4
Jäkälävaara *FIN* 121 C10
Jakkukylä *FIN* 119 D15
Jakkula *FIN* 122 E8
Jäkkvik *S* 109 D12
Jaklovce *SK* 145 F3
Jakobsbakken *N* 109 B10
Jakobsnes *N* 114 D9
Jakobstad *FIN* 122 C9
Jakokoski *FIN* 125 E13
Jakovlje *HR* 148 E5
Jakšić *HR* 149 F9
Jakštaičiai *LT* 134 D2
Jaktorów *PL* 141 F3
Jakubany *SK* 145 E2
Jakubov *SK* 77 F11
Jakubów *PL* 141 F5
Jalance *E* 47 F10
Jalasjärvi *FIN* 122 F8
Jalhay *B* 20 C5
Jaligny-sur-Besbre *F* 30 C4
Jallais *F* 23 F10
Jalovik *SRB* 158 D4
Jałówka *PL* 140 D9
Jalubí *CZ* 146 D6
Jämaja *EST* 130 E4
Jämäs *FIN* 125 B12
Jämejala *EST* 131 D11
Jameln *D* 83 D10
Jamena *SRB* 157 C11
Jametz *F* 19 F11
Jamielnik *PL* 139 C7
Jämijärvi *FIN* 126 B8
Jamilena *E* 53 A9
Jäminkipohja *FIN* 127 B11
Jämjö *S* 89 C9
Jammerdal *N* 96 B7
Jamnik *SK* 145 F2
Jämsä *FIN* 123 C12
Jämsä *FIN* 123 B13
Jämsänkoski *FIN* 127 B13
Jämshög *S* 88 C7
Jämtön *S* 118 C8
Jamu Mare *RO* 159 C8
Janakkala *FIN* 127 D12
Janapole *LT* 134 E4
Jänd *H* 145 G5
Jandelsbrunn *D* 76 E5
Janderup *DK* 86 D2
Jäneda *EST* 131 C11
Jänickendorf *D* 80 B4
Janja *BIH* 158 D3
Janjević *BIH* 157 D9
Janjina *HR* 162 D3
Jänkä *FIN* 123 D13
Jankä *FIN* 115 D3
Jänkisjärvi *S* 116 E10
Jánkmajtis *H* 145 H6
Janków *PL* 143 E7
Jankowo Dolne *PL* 138 E4
Jänmuiža *LV* 135 B10
Jänneberg *FIN* 125 E9
Jánosháza *H* 149 B8
Jánoshida *H* 150 C5
Jánossomorja *H* 149 A8
Janów *PL* 140 D8
Janów *PL* 143 E7
Janowice Wielkie *PL* 81 E9
Janowiec *PL* 141 H5
Janowiec Wielkopolski *PL* 85 E12
Janowo *PL* 139 D10
Janów Lubelski *PL* 144 B5
Janów Podlaski *PL* 141 F8

Jaren *N* 95 B13
Jarfjordbotn *N* 114 D8
Jårgastat *N* 113 E16
Jargeau *F* 24 E7
Jarhoinen *FIN* 117 E11
Jarhois *S* 117 E11
Jariştea *RO* 153 F10
Järkvissle *S* 103 A12
Järlåsa *S* 98 C8
Jarmen *D* 84 C4
Jarménil *F* 26 D5
Jarmina *H* 149 F11
Järna *S* 93 A11
Järna *S* 97 A11
Jarnac *F* 28 D5
Jarnages *F* 29 C10
Järnäs *S* 107 E17
Järnäsklubb *S* 107 E17
Järnforsen *S* 92 E7
Jarny *F* 26 B4
Jarocin *PL* 142 C4
Jarocin *PL* 144 B5
Jarok *SK* 146 E5
Jaroměř *CZ* 77 B9
Jaroměřice *CZ* 77 C11
Jaroměřice nad Rokytnou *CZ* 77 D9
Jaroslavice *CZ* 77 E10
Jarosław *PL* 144 C6
Jarosławiec *PL* 85 A11
Jaroslov nad Nežárkou *CZ* 77 D8
Jarovnice *SK* 145 E3
Järpås *S* 91 C12
Järpbyn *S* 105 E14
Järpen *S* 105 E14
Järpliden *S* 102 E3
Jarplund-Weding *D* 82 A6
Jarque *E* 41 E8
Jarrow *UK* 5 F14
Järva-Jaani *EST* 131 C11
Järvberget *S* 107 D13
Järvenpää *FIN* 117 C19
Järvenpää *FIN* 122 E7
Järvenpää *FIN* 124 C8
Järvenpää *FIN* 125 D10
Järvenpää *FIN* 127 E13
Järvikylä *FIN* 119 F17
Järvikylä *FIN* 123 C13
Järvikylä *FIN* 123 C13
Järville-la-Malgrange *F* 26 C5
Järvirova *FIN* 117 D12
Järvsand *S* 107 C9
Järvsjö *S* 103 C11
Järvsö *S* 103 D11
Järvsta *S* 103 E13
Järvtjärn *S* 118 E5
Järvträsk *S* 107 A16
Jarzé *F* 23 E11
Jaša Tomić *SRB* 159 C6
Jasen *BIH* 162 D3
Jasenak *HR* 67 B11
Jasenica *BIH* 156 C5
Jasenovac *HR* 157 B6
Jasenovo *SRB* 159 D7
Jasenovo *SRB* 163 B8
Jasień *PL* 81 C8
Jasień *PL* 85 B13
Jasienica *PL* 139 F11
Jasienica *PL* 147 B7
Jasienica Rosielna *PL* 144 D4
Jasieniec *PL* 141 G3
Jasika *SRB* 159 F7
Jasionka *PL* 144 C5
Jasionna *PL* 143 G9
Jasionówka *PL* 140 D8
Jaśliska *PL* 145 E4
Jasło *PL* 144 D3
Jašľúnai *LT* 137 D11
Jasmuiža *LV* 135 D13
Jasov *SK* 145 F2
Jásová *SK* 146 F6
Jassans-Riottier *F* 30 D6
Jasseron *F* 31 C7
Jastarnia *PL* 138 A6
Jastrabá *HR* 148 S4
Jastrowie *PL* 85 D11
Jastrząb *PL* 141 H3
Jastrzębia *PL* 141 H4
Jastrzębia Góra *PL* 138 A5
Jastrzębie-Zdrój *PL* 147 B7
Jászapáti *H* 150 B5
Jászárokszállás *H* 150 B4
Jászberény *H* 150 C4
Jászboldogháza *H* 150 C4
Jászfényszaru *H* 150 B4
Jászkarajenő *H* 150 C5
Jászkisér *H* 150 C5
Jászladány *H* 150 C5
Jászszentandrás *H* 150 B5
Jászszentlászló *H* 150 D4
Jásztelek *H* 150 C5
Jatar *E* 53 C9
Jättendal *S* 103 C13
Jättensjö *S* 103 B10
Jatuni *FIN* 116 B9
Jatzke *D* 84 C5
Jatznick *D* 84 C5
Jaulín *E* 41 E9
Jaunalūksne *LV* 133 B2
Jaunanna *LV* 133 B2
Jaunberze *LV* 134 C6
Jaunciems *LV* 134 C5
Jaundundaga *LV* 130 F4
Jaungulbene *LV* 135 B12
Jaunjelgava *LV* 135 C10
Jaunkalsnava *LV* 135 C11
Jaunklidzis *LV* 131 F13
Jaunlaicene *LV* 131 F13
Jaunlutriņi *LV* 134 C4
Jaunmārupe *LV* 135 C8
Jaunmuiža *LV* 135 C8
Jaunolaine *LV* 135 C7
Jaunpiebalga *LV* 135 B12
Jaunpils *LV* 134 C6
Jaunsāti *LV* 134 C6
Jaunsēlpils *LV* 135 C11
Jaunsmilte *LV* 135 D12
Jauntsarats *E* 32 E2
Jaurakainen *FIN* 119 D17
Jaurakkajärvi *FIN* 121 C10
Jaurrieta *E* 32 E3
Jausiers *F* 36 C5
Javalí Viejo *E* 56 F2
Javarus *FIN* 115 E1
Jávea-Xàbia *E* 56 D5

Jävenitz *D* **79** A10
Javerlhac-et-la-Chapelle-St-Robert *F* **29** D7
Javgur *MD* **154** D3
Javier *E* **32** E3
Javorani *BIH* **157** C7
Javorník *CZ* **77** B12
Jävre *S* **118** D6
Javron-les-Chapelles *F* **23** D11
Jawor *PL* **81** D10
Jawornik Polski *PL* **144** D5
Jawor Solecki *PL* **141** H4
Jaworzno *PL* **142** D6
Jaworzno *PL* **143** F7
Jaworzyna Śląska *PL* **81** E10
Jayena *E* **53** C9
Jazeneuil *F* **28** C6
Jebel *RO* **159** B7
Jebjerg *DK* **86** B4
Jedburgh *UK* **5** E11
Jedlanka *PL* **141** G6
Jedlicze *PL* **144** D4
Jedlina-Zdrój *PL* **81** E10
Jedlińsk *PL* **141** G4
Jedlnia-Letnisko *PL* **141** H4
Jedlová *CZ* **77** C10
Jedľové Kostolany *SK* **146** E6
Jednorożec *PL* **139** D11
Jedovnice *CZ* **77** D11
Jjdrzejewo *PL* **85** E10
Jjdrzejów *PL* **143** E9
Jédula *E* **52** C5
Jedwabne *PL* **140** D6
Jedwabno *PL* **139** C10
Jeesiö *FIN* **117** D16
Jeesiöjärvi *FIN* **117** C14
Jegălia *RO* **161** E11
Jegun *F* **33** C6
Jegunovce *NMK* **164** E3
Jejsing *DK* **86** F3
Jēkabpils *LV* **135** C11
Jektvik *N* **108** C5
Jelah *BIH* **157** C8
Jelašca *BIH* **157** F9
Jelcz-Laskowice *PL* **81** D12
Jelenia Góra *PL* **81** E9
Jeleniewo *PL* **136** E6
Jelenin *PL* **81** C8
Jelenje *HR* **67** B9
Jeleśnia *PL* **147** B8
Jelgava *LV* **134** C7
Jelka *SK* **146** E5
Jelling *DK* **86** D4
Jeloboc *MD* **154** C3
Jelovica *SRB* **165** C6
Jełowa *PL* **142** E5
Jels *DK* **86** E4
Jelsa *HR* **157** F6
Jelšane *SLO* **67** A9
Jelšava *SK* **145** F1
Jelsi *I* **63** D7
Jemeppe *B* **19** D10
Jemgum *D* **17** B8
Jemielnica *PL* **142** E5
Jemielno *PL* **81** C11
Jemnice *CZ* **77** D9
Jena *D* **79** E10
Jenbach *A* **72** B4
Jeneč *CZ* **76** B6
Jengen *D* **71** B11
Jenikowo *PL* **85** C8
Jennersdorf *A* **148** C6
Jenny *S* **93** D9
Jenő *H* **149** B10
Jensvoll *N* **101** A15
Jeppo *FIN* **122** D9
Jērcēni *LV* **131** F11
Jerchel *D* **79** B9
Jerez de la Frontera *E* **52** C4
Jerez del Marquesado *E* **55** E6
Jerez de los Caballeros *E* **51** C6
Jerfojaur *S* **109** E16
Jergol *N* **113** E14
Jergucat *AL* **168** E3
Jeri *LV* **131** F10
Jérica *E* **48** E3
Jerichow *D* **79** A11
Jerka *PL* **81** C11
Jernved *DK* **86** E3
Jerslev *DK* **87** D8
Jerslev *DK* **90** E7
Jerstad *N* **110** C9
Jerte *E* **45** D9
Jerup *DK* **90** D7
Jerzens *A* **71** C11
Jerzmanowa *PL* **81** C10
Jerzmanowice *PL* **143** F8
Jerzu *I* **64** D4
Jesenice *CZ* **76** B4
Jesenice *CZ* **77** C7
Jesenice *HR* **156** D4
Jesenice *HR* **156** F6
Jesenice *SLO* **73** D9
Jeseník *CZ* **77** B12
Jeseník nad Odrou *CZ* **146** B5
Jesenské *SK* **147** E10
Jeserig *D* **79** B11
Jeserig *D* **79** B12
Jesi *I* **67** E7
Jesionowo *PL* **85** D8
Jesolo *I* **66** A6
Jessen *D* **80** C3
Jessheim *N* **95** B14
Jeßnitz *D* **79** C11
Jesteburg *D* **83** D7
Jettingen-Scheppach *D* **75** F7
Jeumont *F* **19** D9
Jevenstedt *D* **82** B7
Jever *D* **17** A9
Jevíčko *CZ* **77** C11
Jevišovice *CZ* **77** E10
Jevnaker *N* **95** B12
Jezera *BIH* **157** D8
Jezerane *HR* **67** B11
Jezerc *RKS* **164** E3
Jezero *BIH* **157** D7
Jezero *HR* **156** B3
Jeżewo *PL* **138** C5
Jeżewo *PL* **140** D7
Jeziorany *PL* **136** F2
Jeziorzany *PL* **141** G6
Jeżów *PL* **141** G1
Jeżowe *PL* **144** C5
Jeżów Sudecki *PL* **81** E9
Jiana *RO* **159** E10
Jibert *RO* **152** E6
Jibou *RO* **151** C11
Jichișu de Jos *RO* **152** C3
Jičín *CZ* **77** B8
Jidvei *RO* **152** E4
Jieznas *LT* **137** D9
Jihlava *CZ* **77** D9
Jijila *RO* **155** C2
Jijona-Xixona *E* **56** D4
Jilava *RO* **161** E8
Jilavele *RO* **161** D9
Jilemnice *CZ* **81** E9
Jílové *CZ* **80** E6
Jílové u Prahy *CZ* **77** C7
Jiltjaur *S* **109** E12
Jimbolia *RO* **150** F6
Jimena *E* **53** A10
Jimena de la Frontera *E* **53** D6
Jimramov *CZ* **77** C10
Jina *RO* **152** F3
Jince *CZ* **76** C5
Jindřichov *CZ* **77** B12
Jindřichov *CZ* **142** F4
Jindřichův Hradec *CZ* **77** D8
Jiříkov *CZ* **81** E7
Jirkov *CZ* **76** A4
Jirlău *RO* **161** C10
Jirnsum *NL* **16** B5
Jirny *CZ* **77** B7
Jistebnice *CZ* **77** D7
Jistebník *CZ* **146** B6
Jitia *RO* **161** B9
Jlajkovci *SRB* **163** C10
Joachimsthal *D* **84** E5
Joane *P* **38** F3
Job *F* **30** D4
Jobbágyi *H* **147** F9
Jobsbo *S* **97** B13
Jochberg *A* **72** B5
Jocketa *D* **79** E11
Jockfall *S* **116** E9
Jockgrim *D* **27** B9
Jódar *E* **55** D6
Jodłowa *PL* **144** D3
Jodłownik *PL* **144** D1
Jodoigne *B* **19** C10
Joensuu *FIN* **125** E13
Jõepere *EST* **131** C12
Joesjö *S* **108** E8
Joeström *S* **108** E8
Jõesuu *EST* **131** E9
Jœuf *F* **20** F6
Jõgeva *EST* **131** D12
Jõgua *EST* **131** C14
Johanngeorgenstadt *D* **75** B12
Johannisfors *S* **99** B10
Johannishus *S* **89** C8
Johanniskirchen *D* **76** E3
Johansfors *S* **89** B9
John o'Groats *UK* **3** H10
Johnston *UK* **12** B5
Johnstone *UK* **4** D7
Johnstown *IRL* **9** C7
Johnstown *IRL* **9** C10
Johovac *BIH* **158** D3
Jöhstadt *D* **76** A4
Jõhvi *EST* **131** C14
Joigny *F* **25** E9
Joinville *F* **26** D3
Joița *RO* **161** E7
Jokela *FIN* **119** B16
Jokela *FIN* **119** E16
Jokela *FIN* **127** D12
Jokelankylä *FIN* **123** C14
Jøkelfjordeidet *N* **112** C9
Jokijärvi *FIN* **121** C12
Jokijärvi *FIN* **123** D17
Joki-Kokko *FIN* **119** D16
Jokikunta *FIN* **127** E11
Jokikylä *FIN* **121** E11
Jokikylä *FIN* **122** E8
Jokikylä *FIN* **123** C13
Jokikylä *FIN* **123** C15
Jokilampi *FIN* **121** C12
Jokimaa *FIN* **127** D12
Jokioinen *FIN* **127** D9
Jokiperä *FIN* **122** E8
Jokipii *FIN* **122** E9
Jokivarsi *FIN* **123** E11
Jokivarsi *FIN* **123** E12
Jokkmokk *S* **116** E3
Jokūbavas *LT* **134** E2
Jolanda di Savoia *I* **66** C4
Jolanki *FIN* **117** E13
Jolda *P* **38** E3
Joloskylä *FIN* **119** D16
Joltai *MD* **154** E3
Jomala *FIN* **99** B13
Jømna *N* **101** E15
Jona *CH* **27** F10
Jönåker *S* **93** B9
Jonava *LT* **137** C9
Joncy *F* **30** B6
Jondal *N* **94** B4
Jonesborough *UK* **7** D10
Joniec *PL* **139** E10
Joniškėlis *LT* **135** D8
Joniškis *LT* **134** D7
Joniškis *LT* **137** C12
Jonkeri *FIN* **125** C13
Jönköping *S* **92** D4
Jonkowo *PL* **139** C9
Jonku *FIN* **120** D9
Jonquières *F* **35** B8
Jonsberg *S* **93** B9
Jonsered *S* **91** D11
Jonslund *S* **91** C12
Jonstorp *S* **87** C11
Jonzac *F* **28** E5
Jõõdre *EST* **130** D7
Joppolo *I* **59** B8
Jorăști *RO* **153** F11
Jorba *E* **43** D7
Jorcas *E* **42** F2
Jordanów *PL* **147** B9
Jordanów Śląski *PL* **81** E11
Jordbro *S* **99** D10
Jordbru *N* **108** B9
Jordbrua *N* **108** D8
Jördenstorf *D* **83** C13
Jordet *N* **102** D3
Jork *D* **82** C7
Jörlanda *S* **91** D10
Jormvattnet *S* **105** B16
Jörn *S* **118** D4
Joroinen *FIN* **125** F9
Jørpeland *N* **94** D4
Jorquera *E* **47** F9
Jørstadmoen *N* **101** D12
Jošanica *BIH* **157** E9
Jošanička Banja *SRB* **163** C10
Jošavka *BIH* **157** C7
Joseni *RO* **153** D6
Josenii Bârgăului *RO* **152** C5
Josifovo *NMK* **169** B7
Josipdol *HR* **156** B3
Josipovac *HR* **149** E11
Josnes *F* **24** E6
Jössefors *S* **96** C8
Josselin *F* **22** E6
Jossund *N* **105** C9
Josvainiai *LT* **134** F7
Jota *N* **101** D15
Jou *P* **38** F5
Jouarre *F* **25** C9
Joué-lès-Tours *F* **24** F4
Joué-sur-Erdre *F* **23** E9
Jougne *F* **31** B9
Joukokylä *FIN* **121** D10
Jouques *F* **35** C10
Joure *NL* **16** C5
Journiac *F* **29** F7
Joutenniva *FIN* **123** B15
Joutsa *FIN* **127** B15
Joutseno *FIN* **129** C10
Joutsijärvi *FIN* **115** E3
Joux-la-Ville *F* **25** E10
Jouy *F* **24** C6
Jouy-aux-Arches *F* **26** B5
Jouy-le-Potier *F* **24** E6
Jøvik *N* **111** A18
Jøvik *N* **111** B13
Joyeuse *F* **35** B7
Joze *F* **30** D3
Józefów *PL* **141** F4
Józefów *PL* **144** A4
Józefów *PL* **144** C7
Juankoski *FIN* **125** D10
Juan-les-Pins *F* **36** D6
Juban *AL* **163** E8
Jubë *AL* **168** B1
Jübek *D* **82** A6
Jublains *F* **23** D11
Jubrique *E* **53** C6
Jüchen *D* **20** B6
Juchowo *PL* **85** C10
Jüchsen *D* **75** B8
Jucu *RO* **152** D3
Judaberg *N* **94** D3
Judenbach *D* **75** B9
Judenburg *A* **73** B10
Judinsalo *FIN* **127** B14
Juelsminde *DK* **86** D6
Jugon-les-Lacs *F* **23** D7
Jugorje *SLO* **148** E4
Jugureni *RO* **161** C8
Juhonpieti *S* **116** D10
Juhtimäki *FIN* **127** B9
Juillac *F* **29** E8
Juillan *F* **32** D6
Jujurieux *F* **31** C7
Jukkasjärvi *S* **116** C5
Juknaičiai *LT* **134** F3
Juksjaur *S* **109** E10
Jukua *FIN* **121** C9
Julåsen *S* **103** B10
Julbach *A* **76** E5
Jule *N* **105** C15
Jülich *D* **20** C6
Juliénas *F* **30** C6
Jullouville *F* **23** C8
Jumaliskylä *FIN* **121** E13
Jumeaux *F* **30** E3
Jumilhac-le-Grand *F* **29** E8
Jumilla *E* **55** C10
Juminen *FIN* **125** D9
Jumisko *FIN* **121** B11
Jumprava *LV* **135** C9
Jumurda *LV* **135** C11
Juncal *P* **44** E3
Juncosa *E* **42** E5
Juneda *E* **42** D5
Jung *S* **91** C13
Jungingen *D* **27** D11
Junglinster *L* **20** E6
Jungsund *FIN* **122** D7
Junik *RKS* **163** E9
Juniskär *S* **103** B13
Juniville *F* **19** F9
Jünkerath *D* **21** D7
Junkerdal *N* **109** C10
Junnonoja *FIN* **119** F15
Junosuando *S* **116** D9
Junqueira *P* **50** E5
Junsele *S* **107** D11
Juntinvaara *FIN* **121** F15
Juntusranta *FIN* **121** D13
Juodeikiai *LT* **134** D6
Juodkrantė *LT* **134** E2
Juodšiliai *LT* **137** D11
Juodupė *LT* **135** D11
Juoksengi *S* **117** E11
Juoksenki *FIN* **117** E11
Juokslahti *FIN* **127** B13
Juokuanvaara *FIN* **119** C13
Juonto *FIN* **121** F13
Juorkuna *FIN* **120** E8
Juornaankylä *FIN* **127** D14
Juostininkai *LT* **135** E9
Juotasniemi *FIN* **119** B17
Jupânești *RO* **160** D3
Jupilles *F* **24** E3
Juprelle *B* **19** C12
Jurançon *F* **32** D5
Jurbarkas *LT* **136** C6
Jurbise *B* **19** C8
Jürgenshagen *D* **83** C11
Jürgenstorf *D* **84** C3
Jurģi *LV* **134** C5
Jüri *EST* **131** C9
Jurignac *F* **28** D5
Jurilovca *RO* **155** D3
Jurjevo *HR* **67** C10
Jūrkalne *LV* **134** B2
Jurklošter *SLO* **148** D4
Jurkowice *PL* **143** E11
Jūrmala *LV* **134** C7
Jūrmalciems *LV* **134** D2
Jurmo *FIN* **126** D5
Jurmo *FIN* **126** F6
Jurmu *FIN* **121** D10
Jurovski Brod *HR* **148** E4
Jursla *S* **93** B8
Jurva *FIN* **122** E7
Jussac *F* **29** F10
Jussey *F* **26** E4
Juta *H* **149** D9
Jüterbog *D* **80** C4
Jutis *S* **109** D13
Jutrosin *PL* **81** C12
Jutsajaure *S* **116** D3
Juujärvi *FIN* **120** B9
Juuka *FIN* **125** D12
Juuma *FIN* **121** B13
Juupajoki *FIN* **127** B11
Juupakylä *FIN* **123** F10
Juurikka *FIN* **125** G14
Juurikkalahti *FIN* **125** B10
Juurikkamäki *FIN* **125** E10
Juurikorpi *FIN* **128** D6
Juuru *EST* **131** C9
Juustovaara *FIN* **117** D13
Juutinen *FIN* **123** B17
Juva *FIN* **128** B8
Juvanådammet *S* **107** D12
Juvigné *F* **23** D9
Juvigny-le-Tertre *F* **23** C9
Juvigny-sous-Andaine *F* **23** C10
Juvola *FIN* **125** F11
Juzennecourt *F* **26** D2
Juzet-d'Izaut *F* **33** E7
Jūžintai *LT* **135** E11
Jyderup *DK* **87** D8
Jylhä *FIN* **123** D17
Jylhämä *FIN* **119** E17
Jyllinge *DK* **87** D10
Jyllinkoski *FIN* **122** F8
Jyllintaival *FIN* **122** E8
Jyrinki *FIN* **123** C12
Jyrkänkoski *FIN* **121** B14
Jyrkänkylä *FIN* **125** B13
Jyrkkä *FIN* **125** C9
Jystrup *DK* **87** D9
Jyväskylä *FIN* **123** F15

K

Kaagjärve *EST* **131** F12
Kaakamo *FIN* **119** C12
Kaalepi *EST* **131** C11
Kaamanen *FIN* **113** E19
Kaamasjoki *FIN* **113** E19
Kaamasmukka *FIN* **113** E18
Kaanaa *FIN* **127** B11
Kääntöjärvi *S* **116** D7
Kääpa *EST* **131** F14
Kääpälä *FIN* **128** C6
Kaarakkala *FIN* **124** C8
Kaaraneskoski *FIN* **117** E12
Käärdi *EST* **131** E12
Kaarepere *EST* **131** D13
Kaaresuvanto *FIN* **116** B9
Kaarina *FIN* **126** E7
Kaarlela *FIN* **123** C10
Käärmelehto *FIN* **117** D15
Kaarnevaara *S* **117** C10
Kaarnijärvi *FIN* **115** F1
Kaarßen *D* **83** D10
Kaarst *D* **21** B7
Kaasmarkku *FIN* **126** C7
Kaatsheuvel *NL* **16** E4
Kaava *FIN* **113** D18
Kaavi *FIN* **125** E10
Kaba *H* **151** C7
Kabakça *TR* **173** B9
Kabaklar *TR* **173** E11
Kabakum *TR* **177** A8
Kabala *EST* **131** D11
Kåbdalis *S* **118** B4
Kabelvåg *N* **110** D7
Kaberneeme *EST* **131** B10
Kabile *LV* **134** C4
Kableshkovo *BG* **167** D9
Kabli *EST* **131** E8
Kać *SRB* **158** C4
Kaçanik *RKS* **164** E3
Kačarevo *SRB* **158** D6
Kachkivka *UA* **154** A2
Kachurivka *UA* **154** B5
Kačice *CZ* **76** B5
Käckelbäcksmon *S* **103** A13
Kaczory *PL* **85** D11
Kadaň *CZ* **76** B4
Kadarkút *H* **149** D9
Kadıköy *TR* **173** B9
Kadıköy *TR* **173** C6
Kadıköy *TR* **173** C11
Kadıköy *TR* **177** A9
Kadila *EST* **131** C12
Kadrifakovo *NMK* **164** F5
Kadrina *EST* **131** C12
Kadzidło *PL* **139** D11
Käenkoski *FIN* **125** E15
Kaerepere *EST* **131** D9
Kåfjord *N* **112** D11
Kåfjord *N* **113** B16
Kåfjorddalen *N* **112** E6
Kåge *S* **118** E5
Kågeröd *S* **87** D12
Kaggebo *S* **93** D9
Kağithane *TR* **173** B10
Kagkadi *GR* **174** C3
Kahla *D* **79** E10
Kahl am Main *D* **187** A7
Kåhög *S* **91** D11
Kahraman *TR* **181** A8
Kähtävä *FIN* **119** F12
Kaïafa *GR* **174** D4
Käina *EST* **130** D5
Kainasto *FIN* **122** F8
Kainourgio *GR* **174** B3
Kainulasjärvi *S* **116** D9
Kainuunkylä *FIN* **119** B11
Kainuunmäki *FIN* **124** C8
Kaipiainen *FIN* **128** D7
Kaipola *FIN* **127** B13
Kairala *FIN* **115** D2
Kairiai *LT* **134** E6
Kairiškiai *LT* **134** D5
Kaisajoki *FIN* **119** B12
Kaisepakte *S* **111** D17
Kaisersesch *D* **21** D8
Kaiserslautern *D* **21** F9
Kaisheim *D* **75** E8
Kaišiadorys *LT* **137** D9
Kaisma *EST* **131** D9
Kaitainen *FIN* **124** F9
Kaitainsalmi *FIN* **121** F11
Kaitajärvi *FIN* **119** B13
Kaitum *S* **116** C4
Kaivanto *FIN* **120** F9
Kaive *LV* **134** B6
Kaive *LV* **135** B11
Kajaani *FIN* **121** F10
Kajal *SK* **146** E5
Kajanki *FIN* **117** B11
Kajdacs *H* **149** C11
Kajew *PL* **143** B7
Kajoo *FIN* **125** D12
Kájov *CZ* **76** E6
Kakanj *BIH* **157** D9
Kakasd *H* **149** D11
Kakavi *AL* **168** E3
Kaķenieki *LV* **134** C6
Kakenstorf *D* **82** D7
Kakerbeck *D* **83** E10
Kakhanavichy *BY* **133** E4
Käkilahti *FIN* **120** F8
Kāķišķe *LV* **134** D2
Kakliç *TR* **177** C8
Kąkolewnica Wschodnía *PL* **141** G7
Kąkolewo *PL* **81** C11
Kakovatos *GR* **174** E4
Kakrukë *AL* **168** C3
Kakskerta *FIN* **126** E7
Kakslauttanen *FIN* **115** B2
Kakucs *H* **150** C3
Kál *H* **150** B5
Käl *S* **107** D12
Kälä *FIN* **127** B15
Kalabakbaşi *TR* **173** E7
Kålaboda *S* **118** F5
Kalače *MNE* **163** D9
Kala Dendra *GR* **169** B9
Kalaja *FIN* **123** C14
Kalajärvi *FIN* **127** E12
Kalajoki *FIN* **119** F11
Kalak *N* **113** B19
Kalakangas *FIN* **123** C14
Kalakoski *FIN* **123** F10
Kalamaki *GR* **169** E8
Kalamaki *GR* **174** D2
Kalamaki *GR* **175** D8
Kalamaria *GR* **169** C8
Kalamata *GR* **174** E5
Kalamos *GR* **175** C8
Kalamoti *GR* **177** C7
Kalamoto *GR* **169** C9
Kalampaka *GR* **169** E6
Kalampaki *GR* **171** B6
Kalana *EST* **130** D4
Kalandra *GR* **169** E9
Kala Nera *GR* **169** F9
Kalanistra *GR* **174** C4
Kalanti *FIN* **126** D6
Kälarne *S* **107** F10
Kalathos *GR* **181** D8
Kalavarda *GR* **181** D7
Kalavryta *GR* **174** C5
Kaława *PL* **81** B9
Kalbe (Milde) *D* **83** E10
Kalce *SLO* **73** E9
Kalchevo *BG* **167** E7
Káld *H* **149** B8
Kaldabruņa *LV* **135** D12
Kaldbak *FO* **2** A3
Kaldfarnes *N* **111** B12
Kaldfarnes *N* **111** B13
Kaldfjord *N* **111** A16
Kaldslett *N* **111** A16
Kaldvåg *N* **111** D10
Kaldvik *N* **111** D11
Kale *TR* **181** B9
Kalefeld *D* **79** C7
Kaleköy *TR* **171** D9
Kälen *S* **103** A8
Kälen *S* **103** B11
Kälen *S* **118** D6
Kalentzi *GR* **168** F4
Kalesija *BIH* **157** D10
Kalesmeno *GR* **174** B4
Kalēti *LV* **134** D2
Kalety *PL* **143** E6
Kalevala *RUS* **121** D17
Kali *GR* **169** C7
Kali *HR* **156** D3
Kalianoi *GR* **175** D5
Kalimanci *NMK* **165** F6
Kalimash *AL* **163** E9
Kaliningrad *RUS* **136** D2
Kalinino *RUS* **136** E6
Kalinovik *BIH* **157** F9
Kalinovka *RUS* **136** D4
Kalinowa *PL* **142** C5
Kalinowo *PL* **136** F6
Kaliska *PL* **138** C5
Kalisz *PL* **142** C5
Kalisz Pomorski *PL* **85** D9
Kalita *EST* **131** E9
Kali Vrysi *GR* **170** B5
Kalix *S* **119** C10
Kalixforsen *S* **116** C4
Kalkar *D* **16** E6
Kalkhorst *D* **83** C10
Kalķi *LV* **130** F4
Kalkiainen *FIN* **115** E3
Kalkim *TR* **173** E7
Kalkkimaa *FIN* **119** C12
Kalkkinen *FIN* **127** C14
Kałków *PL* **77** B12
Kalkūni *LV* **135** E12
Kall *D* **21** C7
Kall *S* **105** E14
Källa *S* **89** A11
Kallaste *EST* **131** D14
Kallax *S* **118** C8
Kållberget *S* **102** A6
Källbomark *S* **118** D6
Kållby *FIN* **122** C9
Källby *S* **91** B13
Kalle *N* **110** D7
Kålleboda *S* **96** D7
Kållekärr *S* **91** C10
Kållered *S* **91** D11
Kållerud *S* **91** A12
Kallham *A* **76** F5
Kallifoni *GR* **169** F6
Kallifytos *GR* **171** B6
Kallimasia *GR* **177** C7
Kallinge *S* **89** C8
Kalliojoki *FIN* **121** F15
Kalliopi *GR* **171** E8
Kalliosalmi *FIN* **117** E17
Kallirachi *GR* **171** C7
Kallislahti *FIN* **129** B10
Kallithea *GR* **169** D8
Kallithea *GR* **169** D9
Kallithea *GR* **169** E7
Kallithea *GR* **174** D4
Kallithea *GR* **174** F4
Kallithea *GR* **175** D8
Kallithea *GR* **177** D8
Kallithiro *GR* **169** F6
Kalljord *N* **110** C9
Kallmet i Madh *AL* **163** F8
Kallmünz *D* **75** D10
Kallo *FIN* **117** D12
Kálló *H* **150** B3
Kallön *S* **109** E15
Kalloni *GR* **171** F10
Kalloni *GR* **175** D7
Kalloni *GR* **176** D5
Kállósemjén *H* **145** H4
Kallsedet *S* **105** D13
Källsjön *S* **103** E11
Kallträsk *FIN* **122** F7
Kallunki *FIN* **115** E5
Kallunki *FIN* **121** B13
Kálmánháza *H* **145** H4
Kalmar *S* **89** B10
Kalmari *FIN* **123** E14
Kalmthout *B* **16** F2
Kalna *SRB* **164** C5
Kalna *SRB* **164** D5
Kalná nad Hronom *SK* **147** E7
Kalnciems *LV* **134** C7
Kalni *LV* **134** D4
Kalnieši *LV* **133** E2
Kalnujai *LT* **134** F6
Kalocsa *H* **150** D2
Kalofer *BG* **165** D10
Kaloi Limenes *GR* **178** F8
Kaloneri *GR* **169** D5
Kalo Nero *GR* **174** E4
Kalos Agros *GR* **170** B6
Kalotina *BG* **165** C6
Kalotintsi *BG* **165** D6
Kaloyanovets *BG* **166** E5
Kaloyanovo *BG* **165** E10
Káloz *H* **149** C10
Kalpaki *GR* **168** E4
Kalpio *FIN* **121** E10
Kals am Großglockner *A* **73** B6
Kaltanėnai *LT* **135** F11
Kaltbrunn *CH* **27** F11
Kaltene *LV* **134** B5
Kaltenkirchen *D* **83** C7
Kaltennordheim *D* **79** E7
Kaltensundheim *D* **79** E7
Kaltinėnai *LT* **134** E4
Kaluđerica *SRB* **158** D6
Kalugerovo *BG* **165** E9
Kalundborg *DK* **87** D8
Kalupe *LV* **135** D13
Kałuszyn *PL* **141** F5
Kaluzhekoye *RUS* **136** D4
Kalvåg *N* **100** C1
Kalvarija *LT* **136** E7
Kalvatn *N* **100** B4
Kalveliai *LT* **137** D12
Kalvene *LV* **134** C3
Kalvi *EST* **131** C13
Kälviä *FIN* **123** C10
Kalvik *N* **109** A10
Kalvitsa *FIN* **128** B7
Kalvola *FIN* **127** C11
Kalvträsk *S* **107** B17
Kalwang *A* **73** B10
Kalwaria Zebrzydowska *PL* **147** B9
Kalymnos *GR* **177** F8
Kalyny *UA* **145** G8
Kalythies *GR* **181** D8
Kalyves *GR* **171** C7
Kalyvia *GR* **174** B3
Kalyvia *GR* **174** B3
Kalyvia *GR* **174** D5
Kalyvia Thorikou *GR* **175** D8
Kamajai *LT* **135** E11
Kämäränkylä *FIN* **121** F14
Kamarde *LV* **135** D8
Kamares *GR* **174** C4
Kamares *GR* **175** F10
Kamariotissa *GR* **171** D8
Kambja *EST* **131** E13
Kamburovo *BG* **166** C6
Kamen *BG* **166** C5
Kamen *D* **17** E9
Kamenari *BG* **166** D6
Kamena Vourla *GR* **175** B6
Kamen Bryag *BG* **167** C11
Kamencia *SRB* **158** F6
Kamenec pod Vtáčnikom *SK* **147** D7
Kamengrad *BIH* **156** C6
Kamenica *BIH* **156** D5
Kamenica *BIH* **157** D9
Kamenica *BIH* **158** F3
Kamenica *NMK* **164** E6
Kamenica *SK* **145** E2
Kamenica *SRB* **158** E4
Kamenica *SRB* **164** C5
Kamenica nad Cirochou *SK* **145** F4
Kamenica nad Hronom *SK* **147** F7
Kamenicë *AL* **168** C4
Kamenice *CZ* **77** C7
Kamenice *CZ* **77** D9
Kamenicë *RKS* **164** D4
Kamenice nad Lipou *CZ* **77** D8
Kameničná *SK* **146** F6
Kamenín *SK* **147** F7
Kamenka *RUS* **129** E11
Kamenná Poruba *SK* **147** C7
Kamennogorsk *RUS* **129** D11
Kamenný Most *SK* **147** F7
Kamenný Přívoz *CZ* **77** C7
Kamenný Újezd *CZ* **77** E6
Kameno *BG* **167** D8
Kameno Pole *BG* **165** C8
Kamenovo *BG* **161** F8
Kamensko *BIH* **157** D9
Kamensko *HR* **149** F8
Kamensko *HR* **157** E6
Kamenskoye *RUS* **136** D4
Kamenz *D* **80** D6
Kamerik *NL* **182** A5
Kamern *D* **83** E12
Kames *UK* **4** D6
Kamëz *AL* **168** B2
Kamičak *BIH* **157** C6
Kamicë-Flakë *AL* **163** E7
Kamień *PL* **144** C5
Kamieńczyk *PL* **139** E12
Kamienica *PL* **145** D1
Kamienica Polska *PL* **143** E7
Kamieniec *PL* **81** B10
Kamieniec Ząbkowicki *PL* **77** A11
Kamienka *SK* **145** E2
Kamień Krajeńskie *PL* **85** C13
Kamienna Gora *PL* **81** E10
Kamiennik *PL* **81** E12
Kamień Pomorski *PL* **85** C7
Kamieńsk *PL* **143** D8
Kamień Wielkie *PL* **81** A7
Kamilski Dol *BG* **171** A10
Kamion *PL* **139** F9
Kamion *PL* **141** G2
Kamionka *PL* **141** H6
Kamiros *GR* **181** D7
Kamlunge *S* **119** C9
Kammela *FIN* **126** D5
Kammen *N* **112** C4
Kamnik *SLO* **73** D10
Kamno *SLO* **73** D8
Kamøyvær *N* **113** A16
Kamp *D* **185** D8
Kampanis *GR* **169** C8
Kampen *D* **86** F2
Kampen *NL* **16** C5
Kampenhout *B* **182** D5
Kämpfelbach *D* **187** D6
Kampi *GR* **174** A2
Kampia *GR* **177** B6
Kampinkylä *FIN* **122** E8
Kampinos *PL* **141** F2
Kamp-Lintfort *D* **17** F7
Kampos *GR* **174** C5
Kampos *GR* **174** F5
Kampvoll *N* **111** B14
Kamsjö *S* **118** F3
Kamula *FIN* **123** B16
Kamut *H* **151** D6
Kam"yane *UA* **154** A5
Kam"yans'ke *UA* **154** F4
Kamyanyets *BY* **141** F9
Kamyanyuki *BY* **141** E9
Kanaküla *EST* **131** E10
Kanal *SLO* **73** D8
Kanala *FIN* **123** D12
Kanala *GR* **175** E9
Kanali *GR* **168** E2
Kanali *GR* **174** A2
Kanalia *GR* **169** F8
Kanallaki *GR* **168** F4
Kanan *S* **109** F10
Kanatlarci *NMK* **169** B6
Kańczuga *PL* **144** D5
Kandava *LV* **134** B5
Kandel *D* **27** B9
Kandelin *D* **84** B4
Kandern *D* **27** E8
Kandersteg *CH* **70** E5
Kandila *GR* **174** B2
Kandila *GR* **174** D5
Kandle *EST* **131** B12
Kanepi *EST* **131** F13
Kanestraum *N* **104** E4
Kanfanar *HR* **67** B8
Kangas *FIN* **119** F13
Kangas *FIN* **122** D9
Kangasaho *FIN* **123** E13
Kangasala *FIN* **127** C11
Kangaskylä *FIN* **119** F16
Kangaskylä *FIN* **121** E11
Kangaskylä *FIN* **123** C13
Kangaslahti *FIN* **125** D10
Kangaslampi *FIN* **125** F10
Kangasniemi *FIN* **123** G17
Kangasvieri *FIN* **123** C13
Kangos *S* **116** D9
Kangosjärvi *FIN* **117** C11
Kanianka *SK* **147** D7
Kaninë *AL* **168** D2
Kanjiža *SRB* **150** E5
Kankaanpää *FIN* **126** B7
Kankaanpää *FIN* **126** C7
Kankainen *FIN* **123** F16
Kankari *FIN* **120** E8
Kånna *S* **87** B13
Kannas *FIN* **121** E12
Känne *S* **103** B9
Kannonjärvi *FIN* **123** E14
Kannonkoski *FIN* **123** E14
Kannus *FIN* **123** C11
Kannusjärvi *FIN* **128** D7
Kannuskoski *FIN* **128** D7
Kanpantxua *E* **41** B6
Kanstad *N* **111** D10
Kanstadbotn *N* **111** C10
Kantala *FIN* **124** F8
Kantanos *GR* **178** E6
Kantele *FIN* **127** D14
Kantens *NL* **17** B7
Kantia *GR* **175** D6
Kantojärvi *FIN* **119** C12
Kantojoki *FIN* **121** B13
Kantokylä *FIN* **123** B13
Kantola *FIN* **123** D10
Kantomaanpää *FIN* **119** B12
Kantorneset *N* **111** B17
Kantserava *BY* **133** F4
Kantti *FIN* **122** F9
Kanturk *IRL* **8** D5
Kaolinovo *BG* **161** F10
Kaona *SRB* **158** F5
Kaonik *BIH* **157** D8
Kaonik *SRB* **164** B3
Kapakli *TR* **173** B8
Kapakli *TR* **173** D10
Kapanbeleni *TR* **173** D7
Kapandriti *GR* **175** C8
Kaparelli *GR* **175** C7
Kapčiamiestis *LT* **137** F8
Kapelle *NL* **16** F1
Kapellen *B* **16** F2
Kapelle-op-den-Bos *B* **182** C4
Kapellskär *S* **99** C12
Kapfenberg *A* **148** B4
Kapikargin *TR* **181** C9
Kapitan-Andreevo *BG* **166** F6
Kapiz *TR* **181** B9
Kaplava *LV* **133** E2
Kaplice *CZ* **77** E7
Kapljuh *BIH* **156** C5
Kápolna *H* **150** B5
Kápolnásnyék *H* **149** B11
Kaposfő *H* **149** D9
Kaposmérő *H* **149** D9
Kaposszekcső *H* **149** D10
Kaposvár *H* **149** D9
Kapp *N* **101** E13
Kappel *D* **21** E8
Kappel-Grafenhausen *D* **186** E4
Kappeln *D* **83** A7
Kappelrodeck *D* **186** D5
Kappl *A* **71** C10
Kåpponis *S* **118** B4
Kaprijke *B* **16** F1
Kaprun *A* **73** B6
Kapshticë *AL* **168** C5
Kapsia *GR* **174** D5
Kaptol *HR* **149** F9
Kaptsyowka *BY* **140** C9
Kapušany *SK* **145** E3
Kapusta *FIN* **119** B12
Kapuvár *H* **149** A8
Käpylä *FIN* **119** F14
Karaağaç *TR* **173** A7
Karaağaç *TR* **173** F6
Karaağaçli *TR* **177** B9
Karabanove *UA* **154** C5
Karabiga *TR* **173** D7
Karaböğürtlen *TR* **181** B9
Karabunar *BG* **165** E9
Karaburun *TR* **173** B10
Karaburun *TR* **177** B7
Karaca *TR* **181** C8
Karacabey *TR* **173** D9
Karacadağ *TR* **167** F9
Karacakilavuz *TR* **173** B7
Karacaköy *TR* **173** B9
Karád *H* **149** C9
Karadzhalovo *BG* **166** E4
Karahalil *TR* **173** A7
Karaincirli *TR* **171** C10
Karainebeyli *TR* **171** D10
Karaisen *BG* **166** C4
Karakaja *BIH* **157** D11
Karakasim *TR* **172** A6
Karakaya *TR* **173** F9
Karakoca *TR* **173** D10
Karaköy *TR* **173** E6
Karaköy *TR* **177** B9

Karaköy TR 181 B9
Karakurt TR 177 A10
Karala EST 130 E3
Karaman TR 173 E9
Karamanovo BG 166 B5
Karamehmet TR 173 B8
Karamyshevo RUS 132 F4
Karamyshevo RUS 136 E5
Karancsberény H 147 E9
Karancskeszi H 147 E9
Karancslapujtő H 147 E9
Karancsság H 147 E9
Karankamäki FIN 124 C8
Karaoğlanli TR 177 B10
Karaorman TR 173 E9
Karaova TR 177 E10
Karapchiv UA 152 A6
Karapelit BG 161 F11
Karasjok N 113 E15
Karatoulas GR 174 D4
Karats S 109 C16
Karavas GR 178 C4
Karavelovo BG 165 D10
Karavelovo BG 167 E7
Käravete EST 131 C11
Karavomylos GR 175 B6
Karavukovo SRB 157 B11
Karben D 21 D11
Kårberg S 92 B5
Karbinci NMK 164 F5
Kårböle S 103 C9
Kårböleskog S 103 C9
Karbow-Vietlübbe D 83 D12
Karbunarë e Vogël AL 168 C2
Karby D 83 A7
Karby DK 86 B3
Karby S 99 C10
Karcag H 151 C6
Karcsa H 145 G4
Karczew PL 141 F4
Karczmiska Pierwsze PL 141 H6
Kärda S 87 A13
Kardakata GR 174 C1
Kardam BG 155 F2
Kardam BG 166 C6
Kardamaina GR 177 F9
Kardamyli GR 174 F5
Kardašova Řečice CZ 77 D7
Karden D 21 D8
Kardiani GR 176 D5
Kardis S 117 E11
Karditsa GR 169 F6
Karditsomagoula GR 169 F6
Kärdla EST 130 D5
Kardon BY 133 F6
Kardos H 150 D6
Kardoskút H 150 E6
Karegasnjarga FIN 113 E16
Kareļi LV 134 D4
Karesuando S 116 B8
Kargowa PL 81 B9
Karhi FIN 123 C10
Karhila FIN 123 F13
Karhujärvi FIN 115 F4
Karhukangas FIN 119 F14
Karhula FIN 128 D6
Kariani GR 170 C5
Karigasniemi FIN 113 E16
Karihaugen N 111 D12
Karijoki FIN 122 F7
Karinainen FIN 126 D8
Käringberg S 107 A12
Kåringen N 111 D10
Käringsjön S 102 B3
Käringsjövallen S 102 B4
Karinkanta FIN 119 E13
Karis FIN 127 E10
Karise DK 87 E10
Karisjärvi FIN 127 D11
Karitaina GR 174 E5
Karitsa GR 169 D7
Karjalaisenniemi FIN 121 B11
Karjalan kirkonkylä FIN 126 D7
Karjalankylä FIN 119 D15
Karjalankylä FIN 126 D7
Karjalanvaara FIN 120 B9
Karjalanvaara FIN 121 F10
Karjalohja FIN 127 E10
Kärjenkoski FIN 122 F7
Karjulanmäki FIN 123 C13
Karkalou GR 174 D5
Kārķi LV 131 F11
Kärkinen FIN 119 F12
Kärkkääla FIN 123 E16
Kärkkäälä FIN 124 E8
Karkkila FIN 127 D11
Karkku FIN 127 C9
Kārklinieki LV 133 B2
Kärkölä FIN 127 D10
Kärkölä FIN 127 D13
Karksi EST 131 E11
Karksi-Nuia EST 131 E11
Kårkul S 118 B5
Kärla EST 130 E4
Karlå FIN 122 F7
Karlby FIN 126 F4
Karlebotn N 114 C5
Karleby FIN 123 C10
Karlholmsbruk S 99 A9
Karlino PL 85 B9
Kārļmuiža LV 134 B4
Karlobag HR 67 C11
Karlovac HR 148 F5
Karlovasi GR 177 D8
Karlovice CZ 142 F3
Karlovo BG 165 D10
Karlovy Vary CZ 76 B3
Karłowice PL 142 E4
Karlsbäck S 107 D15
Karlsbad D 27 C10
Karlsberg S 103 C9
Karlsborg S 92 B5
Karlsburg D 84 C5
Karlsdal S 97 D12
Karlsdorf-Neuthard D 27 B10
Karlsfeld D 75 F9
Karlsfors S 103 C9
Karlshagen D 84 B5
Karlshamn S 89 C7
Karlshöfen D 17 B12
Karlskoga S 97 D12
Karlskrona S 89 C9
Karlsøy N 112 C4
Karlsruhe D 27 B9
Karlstad S 97 D10
Karlstadt D 74 C6
Karlstein an der Thaya A 77 E8
Karlstetten A 77 F9
Karlukovo BG 165 C9
Karmansbo S 97 C14
Karmas S 109 B17
Karmėlava LT 137 D9
Kärnä FIN 123 D11
Kärnä FIN 123 D15
Kärna S 91 D10
Karnaliyivka UA 154 E6
Karnice PL 85 B8
Karniewo PL 139 E10
Karnjarga FIN 113 D19
Karnobat BG 167 D7
Karojba HR 67 B8
Karolinka CZ 146 C6
Karonsbo S 107 B15
Karoti GR 171 B10
Karousades GR 168 E2
Karow D 79 B11
Karow D 83 C12
Karpacz PL 81 E9
Kärpänkylä FIN 121 C14
Karpathos GR 181 E6
Karpenisi GR 174 B4
Karperi GR 169 B9
Karpero GR 169 E6
Kärppälä FIN 127 C9
Karpuzlu TR 171 C10
Karpuzlu TR 181 A7
Kärrbackstrand S 102 E4
Karrebæksminde DK 87 E9
Karrenzin D 83 D11
Kärrsjö S 107 D15
Karsakiškis LT 135 E9
Kärsämä FIN 119 E15
Kärsämäki FIN 123 C15
Kārsava LV 133 C3
Karsikas FIN 123 C14
Karsikkoniemi FIN 117 C11
Karsikkovaara FIN 124 B9
Karsimus FIN 115 E2
Karsin PL 138 C4
Karşiyaka TR 173 D9
Karşiyaka TR 177 C9
Karsko PL 85 E8
Kårsta S 99 C10
Karstädt D 83 D10
Karstädt D 83 D11
Kärstna EST 131 E11
Karstula FIN 123 E13
Karszew PL 142 B6
Kartal H 150 B4
Kartavoll N 94 E3
Kartena LT 134 E2
Kartitsch A 72 C6
Kartuzy PL 138 B5
Käru EST 131 D10
Käru EST 131 D12
Karuna FIN 126 E8
Karungi S 119 B11
Karunki FIN 119 B12
Karup DK 86 C4
Kårvåg N 104 E3
Karvala FIN 123 D11
Kärväskylä FIN 123 D15
Karvia FIN 122 F9
Kårvikhamn N 111 B15
Karviná CZ 147 B7
Karvoskylä FIN 123 C14
Karvounari GR 168 F3
Karwica PL 139 C11
Karya GR 169 E7
Karya GR 174 B2
Karya GR 175 D6
Karyes GR 171 D6
Karyes GR 175 E5
Karyotissa GR 169 C7
Karyoupoli GR 178 B3
Karystos GR 175 C9
Kås DK 86 A5
Kašalj SRB 163 C10
Kasejovice CZ 76 D5
Kasendorf D 75 B9
Kasepää EST 131 D14
Kasfjord N 111 C11
Kashirskoye RUS 136 D2
Kaşikçi TR 173 B7
Kašina HR 148 E6
Kaskantyú H 150 D3
Kåskats S 116 F4
Kaskii FIN 129 B9
Kaskinen FIN 122 F6
Kaskö FIN 122 F6
Käsmä FIN 121 C12
Kåsmo N 108 B9
Kåsmolia N 108 B9
Käsmu EST 131 B11
Kaspakas GR 171 E8
Kašperské Hory CZ 76 D5
Kaspichan BG 167 C8
Kassa S 117 D11
Kassandreia GR 169 D9
Kasseedorf D 83 B9
Kassel D 78 D6
Kassiopi GR 168 E2
Kastania GR 168 D4
Kastania GR 168 E5
Kastania GR 169 D7
Kastania GR 169 F6
Kastanies GR 171 A10
Kastanochori GR 169 C10
Kastari FIN 127 C13
Kastellaun D 21 D8
Kastelli GR 178 E9
Kastellia GR 174 B5
Kaštel Stari HR 156 E5
Kaštel Sućurac HR 156 E5
Kaštel Šegarski HR 156 D4
Kasterlee B 16 F3
Kastīre LV 135 D13
Kastl D 75 D10
Kastlösa S 89 C10
Kastneshamn N 111 C13
Kastorf D 83 C9
Kastoria GR 168 C5
Kastorio GR 174 E5
Kastraki GR 174 B3
Kastraki GR 176 E5
Kastrāne LV 135 C10
Kastre EST 131 E14
Kastri GR 169 E8
Kastri GR 175 E6
Kastro GR 175 C7
Kastrosykia GR 174 A2
Kastrova BY 133 D4
Kaszaper H 150 E6
Kaszczor PL 81 C10
Katafyto GR 169 B10
Katajamäki FIN 125 D10
Katakolo GR 174 D3
Kåtaliden S 109 F15
Kataloinen FIN 127 C12
Katapola GR 177 F6
Katarraktis GR 177 C7
Kåtaselet S 118 D3
Katastari GR 174 D2
Katerini GR 169 D8
Katerma FIN 125 B12
Kathenoi GR 175 B8
Kathlow D 81 C6
Kätkänjoki FIN 123 E11
Kätkäsuvanto FIN 117 B10
Kätkävaara FIN 119 B13
Kätkesuando S 117 B10
Katlenburg-Lindau D 79 C7
Kátlovce SK 146 D5
Kato Achaïa GR 174 C4
Kato Alepochori GR 175 C7
Kato Asites GR 178 E9
Katochi GR 174 C3
Kato Chorio GR 179 E10
Kato Doliana GR 175 E6
Kato Glykovrysi GR 178 B4
Kato Kamila GR 169 B9
Kato Makrinou GR 174 C4
Kato Nevrokopi GR 169 B10
Kato Sounio GR 175 D9
Kato Tithorea GR 175 B6
Katouna GR 174 B3
Kato Vermio GR 169 C7
Katovice CZ 76 D5
Kato Vlasia GR 174 C4
Kato Vrontou GR 169 B10
Katowice PL 143 F7
Katranca TR 173 B7
Katrineberg S 103 D11
Katrineholm S 93 B8
Katrineholm S 103 E13
Katsdorf A 77 F6
Katsikas GR 168 E4
Kattavia GR 181 E7
Kattbo S 102 E7
Kattelus FIN 123 F11
Kattilasaari S 119 C11
Kättilstad S 92 C7
Kättilstorp S 91 C14
Kattisavan S 107 B14
Kattisträsk S 118 E3
Katundishtë AL 168 D3
Katunets BG 165 C10
Katuņitsa BG 165 E10
Katuntsi BG 169 B9
Katwijk aan Zee NL 16 D2
Katyčiai LT 134 F3
Katymár H 150 E3
Kąty Wrocławskie PL 81 D11
Katzenelnbogen D 21 D9
Katzweiler D 21 E9
Kaub D 21 D9
Kaufbeuren D 71 B11
Kaufungen D 78 D6
Kauguriešі LV 135 B11
Kauhajärvi FIN 122 F8
Kauhajärvi FIN 123 D10
Kauhajoki FIN 122 F8
Kauhava FIN 123 D10
Kaukalampi FIN 127 D13
Kaukolikai LT 134 D3
Kaukonen FIN 117 D13
Kauksi EST 131 C14
Kaulille B 183 C7
Kaulinranta FIN 119 B11
Kaulsdorf D 79 E9
Kaunas LT 137 D8
Kaunata LV 133 D3
Kauniainen FIN 127 E12
Kaunisjoensuu S 117 D11
Kaunisvaara S 117 D10
Kaupanger N 100 D6
Kaupiškiai LT 136 D6
Kauppila FIN 123 F17
Kauppilanmäki FIN 124 C8
Kaurajärvi FIN 122 D9
Kaurissalo FIN 126 D5
Kaustinen FIN 123 C11
Kautenbach L 184 E5
Kautokeino N 112 E11
Kautzen A 77 E8
Kauvatsa FIN 126 C8
Kauvosaarenpää FIN 119 B11
Kavacik TR 172 B6
Kavacik TR 173 E9
Kavadarci NMK 169 B7
Kavajë AL 168 B2
Kavak TR 173 C6
Kavakdere TR 173 A7
Kavakli TR 173 A7
Kavaklidere TR 181 B8
Kavala GR 171 C6
Kavarna BG 167 C10
Kavarskas LT 135 F9
Kavasilas GR 174 D3
Kavastu EST 131 E14
Kavelstorf D 83 C12
Kävlinge S 87 D12
Kavos GR 168 F3
Kavousi GR 179 E10
Kavs'ke UA 145 E8
Kavyli GR 171 A11
Kaxås S 105 D15
Käxed S 107 E14
Kaxholmen S 92 D4
Kayabaşi TR 181 B9
Kayalar TR 173 E8
Kayali TR 167 F8
Kayalioğlu TR 177 B10
Kayapa TR 173 F7
Kayatepe TR 173 E7
Käylä FIN 121 B13
Käymäjärvi S 116 D9
Kayna D 79 E11
Kaynaklar TR 177 C9
Kaynarca TR 173 A7
Kaynardzha BG 161 F10
Käyrämö FIN 117 E16
Kaysersberg F 27 D7
Kazanka BG 166 E4
Kazanlŭk BG 166 D4
Kazanów PL 141 H4
Kazár H 147 E9
Kazdanga LV 134 C3
Kazichane BG 165 D7
Kazikli TR 177 E9
Kazimierza Wielka PL 143 F9
Kazimierz Biskupi PL 142 B5
Kazimierz Dolne PL 141 H5
Kazincbarcika H 145 G2
Kazlowshchyna BY 133 E4
Kazlų Rūda LT 137 D7
Kaźmierz PL 81 A11
Kaznějov CZ 76 C4
Kazyany BY 133 E7
Kaz'yany BY 135 F13
Kçirë AL 163 E8
Kcynia PL 85 E12
Kdyně CZ 76 D4
Keadue IRL 6 D6
Keady UK 7 D9
Kealkill IRL 8 E4
Kebal S 91 B9
Kecel H 150 D3
Kecerovce SK 145 F3
Kechrokampos GR 171 B7
Kechros GR 171 B9
Kecskéd H 149 A10
Kecskemét H 150 D4
Kėdainiai LT 135 F7
Kedros GR 174 A5
Kędzierzyn-Koźle PL 142 F5
Keele UK 11 E7
Keenagh IRL 7 E7
Keeni EST 131 F12
Kefalos GR 181 C5
Kefalovryso GR 169 E7
Kefenrod D 21 D12
Kefermarkt A 77 F7
Ķegums LV 135 C9
Kegworth UK 11 F9
Kehidakustány H 149 C8
Kehl D 27 C8
Kehlen L 20 E6
Kehra EST 131 C10
Kehrig D 21 D8
Kehtna EST 131 D9
Kehvo FIN 124 D9
Keighley UK 11 D8
Keihärinkoski FIN 123 D15
Keikyä FIN 126 C8
Keila EST 131 C8
Keila-Joa EST 131 C8
Keillmore UK 4 D5
Keinäsperä FIN 119 D17
Ķeipene LV 135 C10
Keiprod N 111 C14
Keiss UK 3 H10
Keitele FIN 123 D16
Keitelepohja FIN 123 D15
Keith UK 3 K11
Kék H 145 G4
Ķekava LV 135 C8
Kékcse H 145 G5
Kelankylä FIN 119 C18
Kelberg D 21 D7
Kelbra (Kyffhäuser) D 79 D9
Kelč CZ 146 C5
Këlcyrë AL 168 D3
Kelebia H 150 E4
Kelechyn UA 145 F7
Kelheim D 75 E10
Kelkheim (Taunus) D 187 A5
Kell D 21 E7
Kellas UK 3 K10
Kellas UK 5 B11
Kellenhusen D 83 B10
Kelli GR 169 C6
Kellinghusen D 82 C7
Kellmünz an der Iller D 71 A10
Kello FIN 119 D14
Kellokoski FIN 127 D13
Kelloniemi FIN 115 E3
Kelloniemi FIN 117 D11
Kelloselkä FIN 115 E5
Kells UK 4 F4
Kells IRL 7 E9
Kells IRL 8 E2
Kells IRL 9 C8
Kelmė LT 134 E5
Kelmis B 20 C6
Kelontekemä FIN 117 C15
Kelottijärvi FIN 116 A8
Kelso UK 5 D12
Kelujärvi FIN 115 D2
Kelvä FIN 125 D14
Kelvedon UK 15 D10
Kemalpaşa TR 177 C9
Kemberg D 79 C12
Kemecse H 145 G4
Kemence H 147 E7
Kemenesmagasi H 149 B8
Kemenessömjén H 149 B8
Kemerburgaz TR 173 B10
Kémes H 149 E10
Kemeten A 148 B6
Kemi FIN 119 C13
Kemihaara FIN 115 C5
Kemijärvi FIN 115 E2
Kemilä FIN 121 C14
Keminmaa FIN 119 C13
Keminperä FIN 121 D12
Këmishtaj AL 168 C2
Kemmel B 18 C6
Kemnath D 75 C10
Kemnay UK 3 L12
Kemnitz D 81 D7
Kemnitz D 84 B5
Kempele FIN 119 E15
Kempen D 16 F6
Kempenich D 21 D8
Kempsey UK 13 A10
Kempston UK 15 C8
Kempten (Allgäu) D 71 B10
Kendal UK 10 C6
Kenderes H 150 C6
Kendice SK 145 F3
Kenézlő H 145 G4
Kenfig UK 13 B7
Kengis S 117 D10
Kengyet H 150 C5
Kenilworth UK 13 A11
Kenmare IRL 8 E3
Kenmore UK 5 B9
Kenn D 21 E7
Kennacraig UK 4 D6
Kensaleyre UK 2 L4
Kensworth UK 15 D7
Kentavros GR 171 B7
Kentriko GR 169 B8
Kentro GR 174 D3
Kenyeri H 149 B8
Kenzingen D 27 D8
Kepez TR 171 D10
Kępice PL 85 B11
Kępno PL 142 D4
Kepsut TR 173 E9
Keqekollë RKS 164 D3
Keramitsa GR 168 E3
Keramoti GR 171 C7
Keräntöjärvi S 116 C9
Kerasia GR 169 D6
Kerasona GR 168 F4
Keräs-Sieppi FIN 117 B11
Keratea GR 175 D8
Kerava FIN 127 E13
Kerecsend H 147 F10
Kerekegyháza H 150 D3
Kerepestarcsa H 150 B3
Kerets'ky UA 145 G7
Kergu EST 131 D9
Keri GR 174 D2
Kerimäki FIN 129 B11
Kerisalo FIN 125 F9
Kerkdriel NL 16 E4
Kerkejaure S 118 B3
Kerken D 16 F6
Kerkini GR 169 B9
Kerkkoo FIN 127 E14
Kerkonkoski FIN 123 E17
Kerkrade NL 20 C6
Kerkwijk NL 183 B6
Kerkyra GR 168 E2
Kerlouan F 22 C3
Kerma FIN 125 F11
Kermen BG 166 D6
Kernascléden F 22 D5
Kernavė LT 137 D10
Kernhof A 148 A5
Kerns CH 70 D6
Kerpen D 21 C7
Kerrykeel IRL 7 B7
Kershopefoot UK 5 E11
Kersilö FIN 117 C17
Kerspleben D 79 D9
Kerstinbo S 98 B7
Kerteminde DK 86 E7
Kertészsziget H 151 C7
Kertezi GR 174 D4
Kerttuankylä FIN 123 D10
Kerzers CH 31 B11
Kesälahti FIN 129 B12
Keşan TR 172 C6
Kesäniemenkylä FIN 121 B12
Kesarevo BG 166 C5
Kesasjärv S 118 B8
Kesh UK 7 C7
Kesh IRL 6 D6
Kesik TR 177 B8
Keskijärvi FIN 125 E14
Keskikylä FIN 119 E13
Keskikylä FIN 119 E15
Keskikylä FIN 119 F13
Keskikylä FIN 122 F9
Keskikylä FIN 123 E11
Keskinen FIN 121 E13
Keskipiiri FIN 119 E14
Keski-Posio FIN 121 B11
Keskusvankila FIN 124 C8
Kęsowo PL 138 C4
Kessel B 182 C5
Kessel NL 183 C8
Kesselinkylä FIN 125 C14
Kessingland UK 15 C12
Ķesterciems LV 134 B6
Kesteren NL 183 B7
Kestilä FIN 119 D14
Kestilä FIN 119 F16
Kestrini GR 168 E3
Keswick UK 10 B5
Keszthely H 149 C8
Kesztölc H 149 A11
Kétegyháza H 151 D7
Kéthely H 149 C8
Ketola FIN 115 E2
Ketomella FIN 117 B12
Keträvaara FIN 121 D13
Kętrzyn PL 136 E3
Ketsch D 21 F11
Kétsoprony H 151 D6
Kettering UK 15 C7
Kettershausen D 71 A10
Kettinge DK 83 A11
Kettletoft UK 3 G11
Kettlewell UK 11 C7
Kettwig D 183 C9
Kęty PL 147 B8
Ketzin D 79 B12
Keula D 79 D8
Keuruu FIN 123 F13
Keutschach am See A 73 C9
Keväjärvi FIN 114 F3
Kevelaer D 16 E6
Ķevele LV 134 D5
Kevermes H 151 E7
Kevo FIN 113 D19
Keynsham UK 13 C9
Keyritty FIN 125 D10
Kežmarok SK 145 E1
Khadzhidimovo BG 169 A10
Khalamyer"ye BY 133 E7
Kharlu RUS 129 B14
Kharmanli BG 166 F5
Khaskovo BG 166 F5
Khayredin BG 160 F3
Khelyulya RUS 129 B14
Khisarya BG 165 D10
Khiytola RUS 129 C12
Khlivchany UA 144 C8
Kholmets' UA 145 F5
Kholms'ke UA 155 B4
Khorio GR 179 B8
Khrabrovo RUS 136 D2
Khrishteni BG 166 E5
Khust UA 145 G7
Khvoyna BG 165 F10
Khyriv UA 145 D6
Kiannanniemi FIN 121 D13
Kiato GR 175 C6
Kiaunoriai LT 134 E6
Kibæk DK 86 C3
Kibworth Harcourt UK 11 F10
Kiby N 114 C7
Kičevo NMK 168 A4
Kichenitsa BG 161 F8
Kichevo BG 167 C9
Kidderminster UK 13 A10
Kidlington UK 13 B12
Kidričevo SLO 148 D5
Kidsgrove UK 11 E7
Kidwelly UK 12 B6
Kiefersfelden D 72 A5
Ķieģeļceplis LV 135 C10
Kiekinkoski FIN 125 B14
Kiekrz PL 81 B11
Kiel D 83 B8
Kielajoki FIN 113 E18
Kielce PL 143 E10
Kiełczygłów PL 143 D6
Kielder UK 5 E11
Kieldrecht B 182 C4
Kiełpino PL 138 B5
Kiemėnai LT 135 D8
Kiemozia PL 141 F1
Kienberg D 75 F11
Kierinki FIN 117 D15
Kierspe D 21 B9
Kieselbach D 79 E7
Kiesilä FIN 128 C7
Kiesimä FIN 123 E17
Kietrz PL 142 F5
Kietz D 81 A7
Kifisia GR 175 C8
Kifjord N 113 B19
Kihelkonna EST 130 E4
Kihlanki FIN 117 C11
Kihlanki S 117 C10
Kihlepa EST 131 E8
Kihlevere EST 131 C12
Kihniö FIN 123 F10
Kiihtelysvaara FIN 125 F14
Kiikala FIN 127 E10
Kiikka FIN 126 C8
Kiikla EST 131 C14
Kiikoinen FIN 126 C8
Kiili EST 131 C9
Kiiminki FIN 119 D15
Kiisa EST 131 C9
Kiiskilä FIN 123 C13
Kiistala FIN 117 C14
Kiiu EST 131 C10
Kiiu-Aabla EST 131 B11
Kije PL 143 E10
Kijevë RKS 163 D10
Kijevo HR 156 E5
Kijewo Królewskie PL 138 D5
Kikerino RUS 132 C6
Kikinda SRB 150 F5
Kikół PL 138 E7
Kikorze PL 85 C8
Ķikuri LV 134 C3
Kil N 90 B5
Kil S 97 C9
Kilafors S 103 D12
Kilargue IRL 6 D6
Kilb A 77 F8
Kilbaha IRL 8 C3
Kilbeggan IRL 7 F8
Kilbeheny IRL 8 D6
Kilberry UK 4 D5
Kilberry IRL 7 F8
Kilbirnie UK 4 D7
Kilboghamn N 108 D5
Kilbotn N 111 C12
Kilbrittain IRL 8 E5
Kilby S 99 B10
Kilcar IRL 6 C5
Kilchoan UK 4 B4
Kilchrenan UK 4 C6
Kilcock IRL 7 F9
Kilcolgan IRL 6 F5
Kilcommon IRL 9 D7
Kilconnell IRL 6 F6
Kilconney IRL 7 D8
Kilcoole IRL 7 F10
Kilcormac IRL 7 F7
Kilcreggan UK 4 D7
Kilcullen IRL 7 F9
Kildare IRL 7 F9
Kildavin IRL 9 C9
Kilden DK 90 E7
Kildimo New IRL 8 C5
Kildonan Lodge UK 3 J9
Kile N 90 A2
Kilen N 95 D9
Kilfenora IRL 6 G4
Kilfinan UK 4 D6
Kilfinnane IRL 8 D6
Kilforsen S 107 D11
Kilgarvan IRL 8 E4
Kilgetty UK 12 B5
Kilglass IRL 6 E6
Kilglass IRL 6 F6
Kilham UK 5 D12
Kilidülbahir TR 171 D10
Kilifarevo BG 166 C5
Kilingi-Nõmme EST 131 E9
Kiliya UA 155 C4
Kilkea IRL 9 C9
Kilkee IRL 8 C3
Kilkeel UK 7 D11
Kilkelly IRL 6 E5
Kilkenny IRL 9 C8
Kilkerrin IRL 6 E5
Kilkhampton UK 12 D6
Kilkieran IRL 6 F3
Kilkinlea IRL 8 D4
Kilkis GR 169 C8
Kilkishen IRL 8 C5
Kill IRL 7 F9
Kill IRL 9 D8
Killabunane IRL 8 E3
Killadysert IRL 8 C4
Killagan Bridge UK 4 E4
Killala IRL 6 D4
Killaloe IRL 8 C6
Killamerry IRL 9 D8
Killann IRL 9 C9
Killarga IRL 6 D6
Killarney IRL 8 D4
Killavullen IRL 8 D5
Killeagh IRL 8 E7
Killean UK 4 D5
Killearn UK 5 C8
Killeberg S 88 C6
Killeenleagh IRL 8 E4
Killeigh IRL 7 F8
Killen UK 7 C7
Killenaule IRL 9 C7
Killerrig IRL 9 C9
Killeshandra IRL 7 D7
Killichonan UK 5 B8
Killiecrankie UK 5 B9
Killimor IRL 6 F6
Killin UK 5 C8
Killinaboy IRL 6 G4
Killinchy UK 7 D11
Killinge S 116 C4
Killingworth UK 5 E13
Killinick IRL 9 D10
Killinkoski FIN 123 F11
Killorglin IRL 8 D3
Killough UK 7 D11
Killough IRL 7 F10
Killucan IRL 7 E8
Killukin IRL 6 E6
Killundine UK 4 B5
Killurin IRL 9 D9
Killybegs IRL 6 C6
Killyclogher UK 4 F2
Killylea UK 7 D9
Kilmacanogue IRL 7 F10
Kilmacrenan IRL 7 B7
Kilmacthomas IRL 9 D8
Kilmaganny IRL 9 D8
Kilmaine IRL 6 E4
Kilmaley IRL 8 C4
Kilmallock IRL 8 D5
Kilmaluag UK 2 K4
Kilmarnock UK 4 D8
Kilmartin UK 4 C6
Kilmeague IRL 7 F9
Kilmeedy IRL 8 D5
Kilmelford UK 4 C6
Kilmichael IRL 8 E4
Kilmihill IRL 8 C4
Kilmona IRL 8 E5
Kilmoon IRL 7 E10
Kilmore IRL 8 C5
Kilmore IRL 9 D9
Kilmore Quay IRL 9 D9
Kilmuckridge IRL 9 C10
Kilmurry IRL 8 C5
Kilmurry McMahon IRL 8 C4
Kilnaboy IRL 6 G4
Kilnaleck IRL 7 E8
Kilnamanagh IRL 9 C10
Kilninian UK 4 B4
Kilninver UK 4 C5
Kilnock IRL 6 E5
Kiloran UK 4 C4
Kilpelä FIN 115 D4
Kilpilahti FIN 127 E14
Kilpisjärvi FIN 112 E6
Kilpua FIN 119 F13
Kilquiggin IRL 9 C9
Kilrane IRL 9 D10
Kilrea UK 4 F3
Kilrean IRL 6 C6
Kilreekill IRL 6 F6
Kilrenny UK 5 C11
Kilronan IRL 6 F3
Kilrush IRL 8 C4
Kilsallagh IRL 6 E5
Kilsaran IRL 7 E10
Kilshanchoe IRL 7 F9
Kilshanny IRL 6 G4
Kilskeer IRL 7 E9
Kilsmo S 92 A7
Kilsund N 90 B4
Kilsyth UK 5 D8
Kiltartan IRL 6 F5
Kiltealy IRL 9 C9
Kiltegan IRL 9 C9
Kiltimagh IRL 6 E5
Kiltogan IRL 9 C9
Kiltoom IRL 6 F6
Kiltsi EST 131 C12
Kiltullagh IRL 6 F5
Kilvakkala FIN 127 B9
Kilvenaapa FIN 119 B16
Kilvo S 116 E6
Kilwaughter UK 4 F5
Kilwinning UK 4 D7
Kilworth IRL 8 D6
Kimasozero RUS 121 F17
Kimberley UK 15 B11
Kimbolton UK 15 C8
Kiminki FIN 123 E13
Kimito FIN 126 E8
Kimle H 146 F4
Kimmeria GR 171 B7
Kimo FIN 122 D8
Kimola FIN 127 C15
Kimonkylä FIN 127 D15
Kimovaara RUS 125 C16
Kimpton UK 15 D8
Kimstad S 92 B7
Kinahmo FIN 125 E13
Kinbrace UK 3 J9
Kincardine UK 5 C9
Kincraig UK 3 L9
Kincses H 149 B10
Kindberg A 148 A4
Kindelbrück D 79 D9
Kinderbeuern D 21 D8
Kinding D 75 E9
Kindsbach D 186 C4
Kindsjön S 102 E4
Kineta GR 175 D7
Kingarrow IRL 6 C6
Kingarth UK 4 D6
Kingisepp RUS 132 C4
Kingsbridge UK 13 E7
Kingsclere UK 13 C12
Kingscourt IRL 7 E9
Kingskerswell UK 13 E7
Kingsland IRL 6 E6
King's Lynn UK 11 F12
Kingsnorth UK 15 E10
Kingsteignton UK 13 D7
Kingsthorne UK 13 B9
Kingston UK 3 K10
Kingston Bagpuize UK 13 B12
Kingston Seymour UK 13 C9
Kingston upon Hull UK 11 D11
Kingswear UK 13 E7
Kingswood UK 13 C9
Kings Worthy UK 13 C12
Kington UK 13 A8
Kingussie UK 5 A8
Kingwilliamstown IRL 8 D4
Kinik TR 173 E10
Kinik TR 177 A9
Kinisjärvi FIN 117 D14
Kinloch UK 4 A4
Kinlochard UK 4 C8
Kinlochewe UK 2 K6
Kinlochleven UK 4 B7
Kinloch Rannoch UK 5 B8
Kinloss UK 3 K9
Kinlough IRL 6 D6
Kinn N 111 C10
Kinna S 91 D12
Kinnared S 87 A12
Kinnarp S 91 C14
Kinnarumma S 91 D12
Kinnegad IRL 7 F8
Kinnitty IRL 7 F7
Kinnula FIN 123 D13
Kinnulanlahti FIN 124 D8
Kinrooi B 19 B12
Kinross UK 5 C10
Kinsale IRL 8 E5
Kinsalebeg IRL 9 E7
Kinsarvik N 94 B5
Kintai LT 134 F2
Kintaus FIN 123 F14
Kintbury UK 13 C12
Kintore UK 3 L12
Kinvara IRL 6 F5
Kinvarra IRL 6 F3
Kioni GR 174 C2
Kipen' RUS 132 B6
Kipfenberg D 75 E9
Kipilovo BG 166 D6
Kipinä FIN 119 D17
Kipoi GR 168 E4
Kipoureio GR 168 E5
Kippel CH 70 E5
Kippen UK 5 C8
Kippenheim D 27 D8
Kir AL 163 E8
Kirakkajärvi FIN 114 D5
Kirakkaköngäs FIN 113 F19
Királd H 145 G1
Királyegyháza H 149 D9
Királyhegyes H 150 E6
Kiran N 104 C8
Kiran TR 181 B8
Kirazli TR 172 D6
Kirbla EST 131 D7
Kirby Muxloe UK 11 F9
Kircasalih TR 173 B6
Kirchanschöring D 73 A6
Kirchardt D 21 F11

Kirchbach A 73 C7
Kirchbach in Steiermark A 148 C5
Kirchberg CH 27 F11
Kirchberg CH 31 A12
Kirchberg D 76 E4
Kirchberg D 79 E12
Kirchberg (Hunsrück) D 185 E7
Kirchberg am Wagram A 77 F9
Kirchberg am Walde A 77 E8
Kirchberg am Wechsel A 148 A6
Kirchberg an der Pielach A 77 F8
Kirchberg an der Raab A 148 C5
Kirchbichl A 72 A5
Kirchdorf D 17 C11
Kirchdorf D 83 C10
Kirchdorf an der Iller D 71 A10
Kirchdorf an der Krems A 73 A9
Kirchdorf im Wald D 76 E4
Kirchdorf in Tirol A 72 A5
Kirchehrenbach D 75 C9
Kirchellen D 183 B9
Kirchen (Sieg) D 21 C9
Kirchenlamitz D 75 B11
Kirchenthumbach D 75 C10
Kirchgellersen D 83 D8
Kirchhain D 21 C11
Kirchheim D 74 C6
Kirchheim D 78 E6
Kirchheim am Neckar D 187 C7
Kirchheim bei München D 75 F10
Kirchheim-Bolanden D 21 E10
Kirchheim unter Teck D 27 C11
Kirchhundem D 21 B10
Kirchlauter D 75 B8
Kirchlinteln D 17 C12
Kirch Mulsow D 83 C11
Kirchohsen (Emmerthal) D 78 B5
Kirchroth D 75 E12
Kirchschlag in der Buckligen Welt A 148 A6
Kirchseelte D 17 C11
Kirchtimke D 17 B12
Kirchwalsede D 17 B12
Kirchweidach D 75 F12
Kirchwistedt D 17 B11
Kirchzarten D 27 E8
Kirchzell D 21 E12
Kircubbin UK 7 D11
Kireç TR 173 E9
Kiremitçisalih TR 172 B6
Kirillovskoye RUS 129 E11
Kirjais FIN 126 E6
Kirjavala FIN 129 B12
Kırkağaç TR 177 A10
Kirkbean UK 5 F9
Kirkbride UK 5 F11
Kirkby UK 10 E6
Kirkby in Ashfield UK 11 E9
Kirkby Lonsdale UK 10 C6
Kirkbymoorside UK 11 C10
Kirkby Stephen UK 11 C7
Kirkby Thore UK 5 F11
Kirkcaldy UK 5 C10
Kirkcolm UK 4 F6
Kirkconnel UK 5 E8
Kirkcudbright UK 5 F8
Kirkeby DK 86 E7
Kirkehamn N 94 F5
Kirke Helsinge DK 87 D8
Kirke Hyllinge DK 87 D9
Kirkel-Neuhäusel D 186 C3
Kirkenær N 96 B7
Kirkenes N 114 D8
Kirke Såby DK 87 D9
Kirke Stillinge DK 87 E8
Kirkham UK 10 D6
Kirkinner UK 5 F8
Kirkintilloch UK 5 D8
Kirkjubøur FO 2 B3
Kirkkavak TR 173 B6
Kirkkepenekli TR 173 B8
Kirkkonummi FIN 127 E11
Kirklareli TR 167 F8
Kirkmichael UK 4 E7
Kirkmichael UK 5 B9
Kirk Michael GBM 10 C2
Kirknewton UK 5 D12
Kirkoswald UK 5 F11
Kirkpatrick-Fleming UK 5 E10
Kirkton UK 4 C5
Kirkton of Durris UK 5 A12
Kirkton of Glenisla UK 5 B10
Kirkton of Skene UK 3 L12
Kirktown of Auchterless UK 3 L11
Kirktown of Deskford UK 3 K11
Kirkwall UK 3 H11
Kirn D 21 E8
Kirnujärvi S 116 D10
Kirovsk RUS 129 F14
Kirriemuir UK 5 B10
Kirschweiler D 21 E8
Kirtik S 118 B5
Kirtlington UK 13 B12
Kirton UK 15 C11
Kirton in Lindsey UK 11 E10
Kirtorf D 21 C12
Kiruna S 116 C4
Kisa S 92 D7
Kisač SRB 158 C4
Kisar H 145 G6
Kisbér H 149 A10
Kirovsk UK 5 D12
Kişlacik TR 167 F9
Kisláng H 149 C10
Kislöd H 149 B9
Kismarja H 151 C8
Kisnána H 147 F10
Kisovec SLO 73 D10
Kissakoski FIN 128 B6
Kissamos GR 178 E6
Kissenbrück D 79 B8
Kißlegg D 71 B9
Kisszállás H 150 E3
Kist D 187 B8

Kistanje HR 156 E4
Kistelek H 150 E4
Kistokaj H 145 G2
Kistrand N 113 C15
Kisújszállás H 150 C6
Kisvárda H 145 G5
Kisvarsány H 145 G5
Kiszkowo PL 81 A12
Kiszombor H 150 E5
Kitee FIN 125 F14
Kiten BG 167 E9
Kitinoja FIN 122 E9
Kitka FIN 121 B12
Kitkiöjärvi S 116 C10
Kitros GR 169 D8
Kitsi FIN 125 D15
Kittajaur S 118 B3
Kittelfjäll S 106 A8
Kittilä FIN 117 C13
Kittsee A 77 F12
Kitula FIN 127 E10
Kitzbühel A 72 B5
Kitzen D 79 D11
Kitzingen D 75 C7
Kitzscher D 79 D12
Kiucze PL 143 F8
Kiukainen FIN 126 C7
Kiurujärvi FIN 115 D2
Kiuruvesi FIN 123 C17
Kivarinjärvi FIN 121 E10
Kivelä FIN 115 F4
Kiveri GR 175 D6
Kivesjärvi FIN 121 F9
Kiveskylä FIN 121 F9
Kiveslahti FIN 120 F9
Kiviapaja FIN 129 B10
Kivijärvi FIN 123 D14
Kivijärvi S 116 E10
Kivioja FIN 119 B14
Kiviöli EST 131 C13
Kiviperä FIN 121 B14
Kivivaara FIN 121 D11
Kivivaara FIN 125 C14
Kivi-Vigala EST 131 D8
Kivotos GR 168 D5
Kiwity PL 136 E2
Kiyiköy TR 173 A9
Kızılcaova TR 177 C10
Kızılcikdere TR 167 F8
Kızılpınar TR 173 B9
Kızılyaka TR 181 B8
Kjækan N 112 D9
Kjåmes N 111 D11
Kjeldebotn N 111 D12
Kjelkvik N 111 C11
Kjellerup DK 86 C4
Kjelling N 108 B7
Kjemmoen N 102 D2
Kjengsnes N 111 C11
Kjerknesvågen N 105 D14
Kjerret N 96 B7
Kjerringdal N 112 C9
Kjerringholmen N 113 B12
Kjerringøy N 108 A8
Kjerringvåg N 104 D5
Kjerringvik N 110 C9
Kjerringvik N 111 D11
Kjerstad N 111 D11
Kjølebrønn N 90 B5
Kjølen N 91 A9
Kjøllefjord N 113 B19
Kjølstad N 101 C10
Kjøpsvik N 111 D11
Kjøra N 104 E7
Kjøsvika N 105 A12
Kjulaås S 98 D7
Klaaswaal NL 16 E2
Kľačno SK 147 D7
Kladanj BIH 157 D10
Kläden D 79 A10
Kladnica SRB 163 C9
Kladnice HR 156 E5
Kladno CZ 76 B6
Kladruby CZ 76 C3
Klæbu N 104 E7
Klagenfurt A 73 C9
Klågerup S 87 D12
Klaipėda LT 134 E2
Kłaj PL 143 G9
Klakksjorda N 110 C8
Klaksvík FO 2 A3
Klamila FIN 128 D7
Klana HR 67 B9
Klanac HR 156 C3
Kłanino PL 85 B10
Klanjec HR 148 D5
Klanxbüll D 86 F3
Klapkalnciems LV 134 B6
Kläppsjö S 107 D12
Klärke S 103 A12
Klarup DK 86 A6
Klašnice BIH 157 C7
Klasov SK 146 E6
Klässbol S 97 C8
Kláštěrec nad Ohří CZ 76 B4
Kláštor pod Znievom SK 147 D7
Klátova Nová Ves SK 146 D6
Klatovy CZ 76 D4
Klaukkala FIN 127 E12
Klaus an der Pyhrnbahn A 73 A9
Klausdorf D 80 B4
Klausdorf D 84 B4
Klausen A 21 E7
Klausen Leopoldsdorf A 77 F10
Klausučiai LT 136 C7
Klauvnes N 112 D6
Klazienaveen NL 17 C7
Kłecko PL 81 A12
Klecza Dolna PL 143 D7
Kleczew PL 138 F5
Kleidi GR 169 C8
Kleinarl A 73 B7
Klein Berßen D 17 C8
Kleinblittersdorf D 27 B7
Klein Bünzow D 84 C5
Kleinfurra D 79 D8
Kleinheubach D 21 E12
Kleinjena D 79 D11
Klein Kreutz D 79 B12
Kleinlobming A 73 B10
Kleinmachnow D 80 B4
Kleinpaschleben D 79 C10
Kleinreifling A 73 A10

Kleinrinderfeld D 187 B8
Klein Rönnau D 83 C8
Klein Sankt Paul A 73 C10
Kleinwallstadt D 21 E12
Kleinwelka D 80 D6
Kleinzell A 77 G9
Kleio GR 171 F10
Kleisoura GR 168 C5
Kleitoria GR 174 D5
Kleiva N 110 C9
Kleive N 100 A7
Klejniki PL 141 E8
Klejtrup DK 86 B5
Klek SRB 158 C5
Klembivka UA 154 A2
Klenica PL 81 C9
Klenike SRB 164 E4
Klenje SRB 158 D3
Klenovec SK 147 D9
Klenovice na Hané CZ 77 D12
Kleosin PL 140 D8
Klepacze PL 140 D8
Kleppe N 94 E3
Kleppestø N 94 B2
Kleppstad N 110 D7
Kljpsk PL 81 B9
Kleszczele PL 141 E8
Kleszczewo PL 81 B12
Kleszczewo PL 136 F6
Kleszczów PL 143 C7
Klettwitz D 80 C5
Kleve D 16 E6
Klevshult S 92 E4
Kličevac SRB 159 D7
Kličevo MNE 163 D6
Klieken D 79 C11
Klietz D 83 E12
Klikuszowa PL 147 B9
Klim DK 86 A4
Kliment BG 161 F10
Kliment BG 165 C10
Klimentovo BG 167 C8
Klimkovice CZ 146 B6
Klimontów PL 143 E11
Klimontów PL 143 F9
Klimpfjäll S 106 A7
Klin SK 147 C9
Klinča Sela HR 148 E5
Klinë RKS 163 D10
Klinë e Epërme RKS 163 D10
Klingenberg D 80 E5
Klingenberg am Main D 21 E12
Klingenthal D 75 B11
Klingersel S 118 B7
Klingre S 107 E15
Klink D 81 D7
Klitten D 81 D7
Klixbüll D 82 A5
Kljajićevo SRB 150 F3
Ključ BIH 157 C6
Kllokot RKS 164 E3
Klobouky CZ 77 E11
Kłobuck PL 143 E6
Klobuk BIH 157 F7
Klöch A 148 C5
Klockestrand S 103 A14
Klockrike S 92 C6
Kłoczew PL 141 G5
Kłodawa PL 85 E8
Kłodawa PL 143 B6
Kłodzko PL 77 B11
Kloetinge NL 16 F1
Kløfta N 95 B14
Klokkarstua N 95 C12
Klokkerholm DK 86 A6
Klokočevac SRB 159 E9
Klokočevci HR 149 E10
Klokočov SK 147 C7
Kłomnice PL 143 E7
Klonowa PL 142 D5
Klooga EST 131 C8
Kloosterhaar NL 17 C7
Kloosterzande NL 16 F2
Klos AL 163 F9
Klos AL 168 B3
Klöse S 107 D16
Klošter Ivanić HR 149 E6
Klošter Podravski HR 149 E8
Kloster DK 86 C2
Klosterfelde D 84 E4
Klosterhäseler D 79 D10
Kloster le A 71 C10
Klosterlechfeld D 71 A11
Klostermansfeld D 79 C9
Klosterneuburg A 77 F10
Klosters CH 71 D9
Kloster Zinna D 80 B4
Kloten CH 27 F10
Kloten S 97 C13
Klotten D 21 D8
Klötze (Altmark) D 79 A9
Klovainiai LT 135 E7
Klovborg DK 86 D4
Klöverträsk S 118 C6
Klövsjö S 102 A7
Klubbfors S 118 E4
Klubbvik N 114 C6
Kluczbork PL 142 D5
Kluczewsko PL 143 E8
Kluki PL 143 D7
Kluknava SK 145 F2
Klukowo PL 141 E7
Klumpen S 106 D7
Kog SLO 148 D5
Klöt DK 86 D6
Kluse D 17 C8
Kľúšov SK 145 E3
Klusy PL 139 C13
Klutsjön S 102 B4
Klütz D 83 C10
Klwów PL 141 G3
Klyastsitsy BY 133 E3
Klykoliai LT 134 D5

Klynivka UA 153 A7
Knaben N 94 E6
Kňáda S 103 D10
Knaften S 107 C15
Knäred S 87 B12
Knaresborough UK 11 C9
Knarrevik N 94 B2
Knarrlagsund N 104 D6
Knätten S 102 B7
Knebel DK 86 C6
Knesebeck D 83 E9
Knesselare B 19 B7
Knetzgau D 75 C8
Knežak SLO 73 E9
Kneževi Vinogradi HR 149 E11
Kneževo HR 149 E11
Knezha BG 165 C9
Kněžice CZ 77 D9
Knežina BIH 157 D10
Kněžmost CZ 77 B8
Knić SRB 158 F5
Knidi GR 169 D6
Knighton UK 13 A10
Knin HR 156 D5
Knislinge S 88 C7
Knittelfeld A 73 B10
Knittlingen D 27 B10
Kniveri LV 135 C13
Knivsta S 99 C9
Knizhovnik BG 166 F5
Knjaževac SRB 164 B5
Knock UK 4 C5
Knock UK 4 C5
Knock IRL 6 E5
Knock IRL 8 C4
Knockalough IRL 8 C4
Knockanevin IRL 8 D6
Knockban UK 2 K7
Knockbrack IRL 7 C7
Knockbridge IRL 7 E10
Knockcroghery IRL 6 E6
Knocklong IRL 8 D6
Knockmoyle IRL 6 F6
Knocknaboul IRL 8 D4
Knocknacarry UK 4 E4
Knocknacree IRL 9 C9
Knocknagree IRL 8 D4
Knocks IRL 8 E5
Knocktopher IRL 9 D8
Knokke-Heist B 19 B7
Knoppe S 103 B12
Knorrendorf D 84 C4
Knottingley UK 11 D9
Knowle UK 13 A11
Knucklas UK 13 A8
Knurów PL 142 F6
Knutbränna S 106 D7
Knutby S 99 C10
Knutsbol S 97 D12
Knutsford UK 11 E7
Knyazevo RUS 133 D5
Knyazhe UA 152 A6
Knyazhevo BG 167 E6
Knyazhitsy RUS 133 E4
Knyszyn PL 140 D7
Kobaky UA 152 A5
Kobarid SLO 73 D8
Kobbelveid N 109 A10
Kobbevågnes N 111 B16
Kobbfoss N 114 E7
Kobela BIH 157 D8
København DK 87 D11
Kobenz A 73 B10
Kobeřice CZ 142 G5
Kobern D 185 D7
Kobersdorf A 149 A6
Kobiele Wielkie PL 143 D8
Kobierzyce PL 81 E11
Kobilyane PL 171 A8
Kobiór PL 143 F6
Koblenz D 21 D9
Kobrow D 83 C11
Kobryn BY 141 F9
Kobyla Góra PL 142 D4
Kobylanka PL 85 D7
Kobylany PL 141 F8
Kobylec PL 141 G5
Kobylin PL 81 C12
Kobyłka PL 139 F11
Kobylnica PL 81 B12
Kobylnica PL 85 B12
Kobylniki PL 139 F9
Kocaavşar TR 173 E8
Kocahidir TR 171 C10
Kočani NMK 164 F5
Kocapinar TR 173 E8
Koçarli TR 177 D10
Koçayazı TR 167 F8
Koçayazı TR 167 F8
Koceljevo SRB 158 E4
Kočevje SLO 73 E10
Kočevska Reka SLO 73 E10
Kochanowice PL 142 E6
Kochel am See D 72 A3
Kocherinovo BG 165 E7
Kochmar BG 161 F10
Kochovo BG 167 C7
Kock PL 141 G6
Kočovce SK 146 D5
Kocs H 149 A10
Kocsér H 150 C4
Kocsola H 149 C10
Kocsord H 145 H5
Koczała PL 85 C12
Kodavere EST 131 D14
Kodeń PL 141 G9
Kodersdorf D 81 D7
Kodesjärvi FIN 122 F8
Kodiksami FIN 126 C6
Kodisjoki FIN 126 C6
Koditz D 75 B10
Kodyma UA 154 A4
Koekelare B 18 B6
Koersel B 183 C6
Koeru EST 131 D11
Koewacht NL 182 C3

Koikkala FIN 128 B8
Koilovtsi BG 165 C10
Koimisi GR 169 B9
Koirakoski FIN 124 C9
Koita GR 178 B3
Koitila FIN 121 C11
Koivu FIN 119 B14
Koivujärvi FIN 123 C16
Koivujärvi FIN 123 D16
Koivukylä FIN 124 D9
Koivumäki FIN 124 D9
Koivumäki FIN 125 F11
Koivuniemi FIN 119 C15
Köja S 107 F13
Kojanlanti FIN 125 E11
Kojetín CZ 146 C4
Kōka H 150 C4
Kökar FIN 126 F5
Kokari LV 133 C2
Kokava nad Rimavicou SK 147 D9
Kokelv N 113 B14
Kokemäki FIN 126 C7
Kokkala GR 178 B3
Kokkari GR 177 D8
Kokkinochoma GR 171 C6
Kokkino Nero GR 169 E8
Kokkokylä FIN 119 C17
Kokkola FIN 123 C10
Kokkolahti FIN 125 F12
Kokkosniva FIN 115 D2
Kokkotoi GR 175 D6
Kokkovaara FIN 117 D13
Kokora FIN 131 D14
Kokořin CZ 77 B11
Kokrica SLO 73 D9
Koksijde B 18 B6
Kokträsk S 109 F16
Kola BIH 157 C7
Kolari FIN 117 D11
Kolari S 117 D11
Kolari S 117 D11
Kolárovice SK 147 C7
Kolarovo BG 169 B9
Kolárovo SK 146 F5
Kolašin MNE 163 D8
Kolbäck S 98 C6
Kolbacz PL 85 D7
Kolbaskowo PL 84 D6
Kolbermoor D 72 A5
Kolbiel PL 141 F4
Kolbnitz D 73 C7
Kolbotn N 95 C13
Kolbu N 101 E13
Kolbudy Górne PL 138 B5
Kolbuszowa PL 144 C4
Kølby DK 86 B4
Kol'chyne UA 145 G6
Kölcse H 145 G6
Kołczygłowy PL 85 B12
Kołczyn PL 81 A8
Koldby DK 86 B3
Koldere TR 177 B10
Kolding DK 86 E4
Koler S 118 D5
Kolešd H 149 C11
Kolesjan AL 163 F9
Kolforsen S 103 E12
Kolga EST 131 B11
Kolga-Aabla EST 131 B11
Kolga-Jaani EST 131 D11
Kolho FIN 123 F13
Koli FIN 125 D13
Kolín CZ 77 B8
Kolind DK 86 C7
Kolindros GR 169 D7
Kolinec CZ 76 D4
Kolinkivtsi UA 153 A8
Koliri GR 174 D3
Kolisne UA 154 C5
Kolitzheim D 75 C7
Kõljala EST 130 E5
Koljane HR 156 E5
Kolka LV 130 F5
Kølkær DK 86 C4
Kolkanlahti FIN 123 E14
Kölked H 149 E11
Kolkja EST 131 D14
Kolku FIN 115 D3
Kolkonjärvi FIN 121 D9
Kolkwitz D 80 C6
Kollaja FIN 119 D17
Kolleda D 79 D9
Kollines GR 174 D5
Kölln-Reisiek D 82 C7
Kõljala EST 130 E5
Kolmisoppi FIN 124 D9
Köln D 21 C7
Kolnica PL 140 C8
Kolno PL 139 D12
Kolno PL 139 D12
Kolo PL 142 B6
Kołobrzeg PL 85 B9
Kolochau D 80 C4
Kolochava UA 145 G8
Kolonjë AL 168 C2
Kolonowskie PL 142 E5
Kolosjoki FIN 115 E4
Kölosen S 102 B4
Kolpino RUS 129 F14
Kolsätter S 102 B8
Kolsh AL 163 E9
Kolsillre S 103 B9
Kölsjön S 103 B11
Kolsko PL 81 C9
Kolsva S 97 C14
Kolsvollen S 102 C6
Kölsvallen S 102 C7
Kolt DK 86 C6
Kolta SK 146 E6
Köthen D 17 A11
Kohlberg D 75 C11
Koluič BIH 156 C5
Koluszki PL 141 H2
Koluvere EST 131 D8
Kolvereid N 105 B11
Kolvik N 113 C15

Kølvrå DK 86 C4
Kolymbari GR 178 D6
Kolympia GR 181 D8
Komádi H 151 C7
Komagfjord N 113 C11
Komagvær N 114 C9
Komańcza PL 145 E5
Komanos GR 169 C7
Komar BIH 157 D7
Komara GR 171 A10
Komarani SRB 163 C8
Komarevo BG 165 B10
Komarica BIH 157 C8
Komárno SK 146 F6
Komarno UA 145 D8
Komárom H 149 A10
Komárov CZ 76 C5
Komarovo RUS 129 E12
Komarów-Osada PL 144 B7
Komba H 150 C4
Kömbe FIN 120 F4
Komen SLO 73 E8
Komi GR 176 D5
Komi GR 177 C7
Komin HR 148 D5
Komin HR 157 F8
Komiža HR 63 A10
Komjatice SK 146 E6
Komletinci BIH 157 B10
Komló H 149 D10
Kömlő H 150 B5
Kömlőd H 149 A10
Kommeno GR 174 A3
Komnina GR 169 C6
Komoran RKS 163 D10
Komorniki PL 81 B11
Komorowo PL 139 E11
Komorzno PL 142 D5
Komoshtitsa BG 165 B7
Komossa FIN 122 D9
Komotini GR 171 B8
Kompakka FIN 125 F13
Kompelusvaara S 116 D8
Kömpöc H 150 E4
Kompoti GR 174 A3
Komprachcice PL 142 E4
Kömsi EST 130 D7
Komsomol'sk RUS 136 D2
Komsomol'sk RUS 139 A9
Komsomol'sk UA 145 G8
Komu FIN 123 C16
Komula FIN 125 C10
Komunari BG 167 C8
Komuniga BG 166 F4
Komylio GR 174 B2
Konak TR 177 C9
Konakpınar TR 173 F8
Konare BG 155 F2
Konary PL 141 G4
Konarzyce PL 139 D13
Konarzyny PL 85 C12
Koncanica HR 149 E8
Konče NMK 164 F5
Kondolovo BG 167 E9
Kondoros H 150 D6
Kondratowice PL 81 E11
Kondrat'yevo RUS 129 D10
Kondrić HR 157 B9
Koneck PL 138 E5
Kong DK 87 E9
Konga S 89 B8
Köngäs FIN 115 E1
Köngäs FIN 117 C13
Kongasmäki FIN 121 E9
Kongens Lyngby DK 87 D11
Konginkangas FIN 123 E15
Kongsberg N 95 C11
Kongselva N 110 D9
Kongsfjord N 114 B6
Kongslia N 101 C9
Kongsmoen N 105 B12
Kongsvik N 111 C11
Kongsvinger N 96 B7
Kongsvoll N 104 D6
Konice CZ 77 C11
Koniecpol PL 143 E8
Königheim D 187 B8
Königsberg in Bayern D 75 B8
Königsborn D 79 B10
Königsbronn D 75 E7
Königsbrück D 80 D5
Königsbrunn D 71 A11
Königsdorf D 72 A3
Königsee D 79 E8
Königsfeld im Schwarzwald D 27 D9
Königshofen D 79 D10
Königshütte D 79 C8
Königslutter am Elm D 79 B8
Königsmoos D 75 E9
Königssee D 73 A6
Königstein D 75 C10
Königstein D 80 E6
Königstein im Taunus D 187 A5
Königswartha D 80 D6
Königswiesen A 77 F7
Königswinter D 21 C7
Königs Wusterhausen D 80 B5
Konin PL 142 B6
Konispol AL 168 E3
Konitsa GR 168 D4
Könitz D 79 E10
Köniz CH 70 D5
Konjevrate HR 156 E5
Konjic BIH 157 E8
Konjsko BIH 157 E7
Konjsko BIH 162 D5
Könni FIN 122 E9
Konnersreuth D 75 C11
Konnevesi FIN 123 E16
Konnunsuo FIN 129 C10
Konnuslahti FIN 125 E10
Könölä FIN 119 C12
Konopiska PL 143 E7
Konopište NMK 169 B7
Konopnica PL 142 D6
Konopnica PL 142 H6
Konoša RUS 133 A9
Konotop BG 165 B8
Konotop PL 85 D9
Końskie PL 141 H2
Konsko NMK 169 B7
Kōnski PL 141 H2
Konsmo N 90 C2
Konstancin-Jeziorna PL 141 F4
Konstantinova UA 139 D8
Konstantinovy Lázně CZ 76 C3
Konstantynów PL 141 F8

Konstantynów Łódzki PL 143 C7
Konstanz D 27 E11
Kontariotissa GR 169 D7
Kontiainen FIN 123 E10
Kontias GR 171 E8
Kontinjoki FIN 121 F10
Kontiokoski FIN 119 D15
Kontiolahti FIN 125 E13
Kontiomäki FIN 121 F11
Kontkala FIN 125 E12
Kontopouli GR 171 E8
Kontovazaina GR 174 D4
Konttajärvi FIN 117 E12
Konttila FIN 119 C17
Konttimäki FIN 123 E16
Konush BG 166 F4
Kóny H 149 A8
Konyár FIN 151 C8
Konyavo BG 165 E6
Konz D 21 E7
Konzell D 75 D12
Koog aan de Zaan NL 16 D3
Koonga EST 131 D8
Köörtilä FIN 126 B6
Koosa EST 131 D14
Kootstertille NL 16 B6
Kootwijkerbroek NL 183 A7
Kopani GR 168 F4
Kopanica PL 81 B9
Kopanos GR 169 C7
Kopardal N 108 D4
Koparnes N 100 B3
Kopčany SK 77 E12
Koper SLO 67 A8
Kopervik N 94 D2
Kóphàza H 149 A7
Kopice PL 142 E3
Kopidlno CZ 77 B8
Kopilovtsi BG 165 B10
Köping S 97 C15
Köpingebro S 88 E5
Köpingsvik S 89 B11
Kopisto FIN 119 F12
Kopki PL 144 C5
Koplik i Sipërm AL 163 E7
Köpmanholm S 99 C11
Köpmanholmen S 107 E15
Koporiće RKS 163 C10
Koporiçë RKS 163 C10
Koposperä FIN 123 C14
Koppang N 101 C14
Koppangen N 111 A19
Kopparberg S 97 C13
Kopparmora S 99 D11
Koppraeås S 118 D6
Koppelo FIN 115 E4
Koppera N 105 E11
Koppl A 73 A7
Koppom S 96 C7
Koprivets BG 166 C5
Koprivlen BG 169 A10
Koprivnica HR 149 D7
Koprivnice CZ 146 B6
Koprivshtitsa BG 165 D9
Koprzywnica PL 143 E12
Kopsa FIN 119 C13
Kopstal L 186 B1
Kõpu EST 130 D4
Kõpu EST 131 C10
Koraj BIH 157 C10
Koramonieni FIN 121 B12
Korbach D 17 F11
Korbevac SRB 164 D5
Korbovo SRB 159 D10
Korčanica BIH 156 C5
Korçë AL 168 C4
Korchiv UA 144 C8
Korchow D 83 D10
Korčula HR 162 D3
Korczew PL 141 F7
Korczyna PL 144 D4
Kordel D 185 E6
Korenita SRB 158 E3
Kořenov CZ 81 E8
Korentokylä FIN 120 D9
Korentovaara FIN 125 E16
Koretin RKS 164 D4
Korfantów PL 142 F4
Korfos GR 175 D7
Korgen N 108 D6
Korgene LV 131 F9
Körgessaare EST 130 D4
Korholanmäki FIN 121 F10
Korhosenniemi FIN 121 C12
Koria FIN 128 D6
Korinos GR 169 D7
Korinth DK 86 E6
Korinthos GR 175 D6
Korisia GR 175 D9
Korisos GR 168 D5
Korita BIH 156 D5
Korita BIH 157 F7
Korita HR 162 D4
Korita MNE 163 E8
Korithi GR 174 D2
Korkana FIN 121 F15
Korkatti FIN 119 F15
Korkeakangas FIN 125 F14
Korkeakoski FIN 127 B11
Korkee FIN 123 E17
Körkvere EST 130 E6
Körle D 78 D6
Körmend H 149 B7
Kormista GR 170 C6
Kormu FIN 127 D12
Korňa SK 147 C7
Kornalovychi UA 145 D9
Körner D 79 D8
Kornelimünster D 183 D8
Kornevo RUS 136 E1
Kórnik PL 81 B12
Kornitsa BG 165 F8
Kornofolia GR 171 B10
Kornos CY 181 C2
Kornsjø N 91 B10
Kornwestheim D 27 C11
Környe H 149 A10
Koroleve UA 145 G7
Koromačno HR 67 C9
Koroncó H 149 A9
Koroneia GR 175 C6
Koroni GR 178 C3
Koronisia GR 174 A2
Koronos GR 177 E7
Koronouda GR 169 C9
Koropi GR 175 D8
Korosladány H 151 D7
Korostarcsa H 151 D7
Körösszegapáti H 151 C8
Koroviya UA 153 A7

Korpela FIN 117 B14
Körperich D 20 E6
Korpi FIN 123 C12
Korpijärvi FIN 119 B13
Korpijoki FIN 123 C16
Korpikå S 119 C10
Korpikylä FIN 119 B13
Korpikylä FIN 121 E12
Korpikylä S 119 B11
Korpilahti FIN 123 F15
Korpilombolo S 116 E10
Korpilompolo FIN 117 E13
Korpinen FIN 121 D10
Korpinen FIN 125 D10
Korpisel'kya RUS 121 E15
Korpivaara FIN 125 E12
Korpo FIN 126 E6
Korpoström FIN 126 E6
Korsåmon S 103 A12
Korsbäck FIN 122 E6
Korsberga S 89 A8
Korsberga S 91 C15
Korsgården S 103 E10
Korsholm FIN 122 D7
Korskrogen S 103 C10
Korsmyrbränna S 106 E8
Korsnäs FIN 122 E6
Korsnes N 111 D11
Korso FIN 127 E13
Korsør DK 87 E8
Korssjön S 118 F5
Korssund N 100 D1
Korsträsk S 118 C5
Korsveggen N 104 E8
Korsvoll N 104 E4
Korsze PL 136 E3
Kortemark B 19 B7
Korten BG 166 D5
Kortenhoef NL 16 D4
Kortesalmi FIN 121 C14
Kortesjärvi FIN 123 D10
Kortessem B 19 C11
Kortevaara FIN 125 B13
Kortgene NL 16 E1
Korthi GR 176 D4
Kortrijk B 19 C7
Kortteenperä FIN 119 B17
Kortteinen FIN 125 D11
Korubaşı TR 171 E10
Korucu TR 173 F7
Koruköy TR 177 E10
Korva S 119 B11
Korvakumpu FIN 115 D2
Korvala FIN 117 E16
Korvaluoma FIN 126 B8
Kõrveküla EST 131 E13
Korvenkylä FIN 119 D15
Korvenkylä FIN 119 F14
Korvenkylä FIN 123 B12
Korvua FIN 121 D12
Koryčany CZ 77 D12
Korycin PL 140 D8
Koryfasi GR 174 E4
Koryfi GR 169 C8
Korytnica PL 139 F12
Korzenna PL 144 D2
Korzeńsko PL 81 C11
Korzybie PL 85 B11
Kos GR 177 F9
Kosakowo PL 138 A6
Kosanica MNE 163 C7
Košarovce SK 145 E4
Kösching D 75 E10
Koscian PL 81 B11
Kościelec PL 142 B6
Kościelna Wieś PL 142 C5
Kościernica PL 85 B10
Kościerzyna PL 138 B4
Kosd H 150 B3
Kose EST 131 C10
Kose EST 131 F14
Košeca SK 146 C6
Košecké Podhradie SK 146 D6
Kösedere TR 171 E10
Kösedere TR 177 B8
Köseilyas TR 173 B8
Kosel D 82 A7
Koselji BIH 157 E9
Koserow D 84 B6
Košetice CZ 77 C8
Kosharevo BG 165 D6
Koshava BG 159 E11
Košice SK 145 F3
Kosiv UA 152 A6
Kosiv's'ka Polyana UA 145 G9
Kosjerić SRB 158 F4
Koška HR 149 E10
Koskama FIN 117 C14
Koskeby FIN 122 D8
Koskela FIN 119 E15
Koskela FIN 123 D11
Koskenkorva FIN 122 E8
Koskenkylä FIN 117 C16
Koskenkylä FIN 121 C14
Koskenkylä FIN 121 F11
Koskenkylä FIN 124 E8
Koskenkylä FIN 125 D10
Koskenmäki FIN 121 F13
Koskenniska FIN 113 F17
Koskenpää FIN 123 F14
Koskenperä FIN 123 C14
Koski FIN 127 D9
Koski FIN 127 D10
Koskimäki FIN 122 E7
Koskinou GR 181 D8
Koskioinen FIN 127 C9
Koškovce SK 145 E4
Koskue FIN 122 F9
Koskullskulle S 116 D5
Koslovets BG 166 C4
Kosmach UA 152 A5
Kosman BIH 157 F10
Kosmas GR 175 E6
Kosmio GR 171 B8
Kosmonosy CZ 77 B7
Kosola FIN 122 D9
Kosova Hora CZ 77 C6
Kosovo NMK 156 E5
Kosovska Kamenicë RKS 164 D4
Kosów Lacki PL 141 E6
Koßdorf D 80 D4
Kössen A 72 A5
Kosta S 89 B8
Kosta S 89 B8
Kostakioi GR 174 A2
Kostálov CZ 81 E8
Kostamo FIN 115 E2
Kostandenets BG 166 B6
Kostandovo BG 165 E9
Kostanjevac HR 148 E4
Kostanjevica SLO 148 E4

Kostel SLO 67 A10
Kostelec nad Černými Lesy
  CZ 77 C7
Kostelec nad Orlicí CZ 77 B10
Kostelec na Hané CZ 77 C12
Kostenets BG 165 E8
Kostenets BG 165 D7
Kostice CZ 77 E11
Kostinbrod BG 165 D7
Kostolac SRB 159 D7
Kostolné Kračany SK 146 F5
Kostomfoty PL 81 D11
Kostomuksha RUS 121 E16
Kostrzyn PL 81 A7
Kostrzyn PL 81 B12
Kostyzhitsy RUS 132 F6
Kosula FIN 125 E11
Košute HR 157 E6
Kosy UA 145 B5
Kosyny UA 145 G5
Koszalin PL 85 B10
Koszęcin PL 142 E6
Kőszeg H 149 B7
Koszyce PL 143 F10
Kotajärvi FIN 119 D15
Kotala FIN 115 C5
Kotala FIN 123 F12
Kotaljeva BIH 157 D9
Kotlenka HR 67 B10
Kotljevo SRB 158 F6
Kralovice CZ 76 B4
Kotë AL 168 D2
Kötegyán H 151 D7
Kotel BG 167 D6
Kotělek H 150 C5
Kotelow D 84 C5
Kotešová SK 147 C7
Kotikylä FIN 124 D8
Kotila FIN 121 D10
Kotiranta FIN 121 E13
Kotka FIN 128 E6
Kotlin PL 142 C4
Kotly RUS 132 B4
Kotor MNE 163 E6
Kotoriba HR 149 D7
Kotorsko BIH 157 C9
Kotor Varoš BIH 157 D7
Kotraža SRB 158 F5
Kotronas GR 178 B3
Kötschach A 73 C7
Kotsøy N 105 F9
Kottenheim D 185 D7
Kottes A 77 E8
Köttmannsdorf A 73 C9
Köttsjön S 107 D12
Kotuń PL 141 F6
Koudekerke NL 16 F1
Koudum NL 16 C4
Koufalia GR 169 C8
Koufovouno GR 171 B10
Kougsta S 105 E15
Kouklioi GR 168 E4
Koumanis GR 174 D4
Koura FIN 123 E10
Kouřim CZ 77 C7
Kourkouloi GR 175 B7
Kournas GR 178 E7
Koutaniemi FIN 121 F10
Koutojärvi S 119 B10
Koutsochero GR 169 E7
Koutsopodi GR 175 D6
Koutsouras GR 179 E10
Koutus FIN 117 E12
Kouva FIN 121 C10
Kouvola FIN 128 D6
Kovachevets BG 166 C6
Kovachevitsa BG 165 F8
Kovachevo BG 166 E6
Kovachevtsi BG 165 D6
Kovachitsa BG 160 F2
Kovači BIH 157 E7
Kovačica SRB 158 C6
Kővágószőlős H 149 D10
Kovallberget S 107 B13
Kovanj BIH 157 E9
Kovarce SK 146 E6
Kovářov CZ 76 C6
Kovdor RUS 115 C8
Kovelahti FIN 126 B8
Kovero FIN 125 E15
Kovilj SRB 158 C5
Kovin SRB 159 D6
Kovjoki FIN 122 C9
Kovland S 103 A13
Kövra S 102 A7
Kovren MNE 163 C8
Kowal PL 139 E7
Kowala-Stjpocina PL 141 H4
Kowale PL 142 D5
Kowale Oleckie PL 136 E5
Kowale-Pańskie PL 142 C6
Kowalewo Pomorskie PL 138 D6
Kowalów PL 81 B7
Kowary PL 81 E9
Kowiesy PL 141 G2
Köyceğiz TR 181 C9
Köyhäjoki FIN 123 C11
Köyliö FIN 126 C7
Koynare BG 165 C9
Koyundere TR 177 B9
Koyunyeri TR 171 C10
Kozani GR 169 D6
Kozarac BIH 157 C6
Kozarac HR 149 E11
Kozar Belene BG 166 C4
Kozármisleny H 149 D10
Kozárovce SK 147 E7
Kozçeşme TR 173 D6
Kozica HR 157 F7
Koziegłowy PL 143 E7
Kozielice PL 85 D7
Kozienice PL 141 G5
Kozina SLO 73 E8
Kožlany CZ 76 C5
Kozloduy BG 160 F3
Kozłów Biskupi PL 141 F2
Kozłowo PL 139 D9
Kozluk BIH 157 C11
Kozły PL 139 F11
Koźminek PL 142 C5
Koźmin Wielkopolski PL 142 C3
Koz'ove UA 145 F7
Kozubszczyzna PL 141 H6
Kożuchów PL 81 C9
Kožuhe BIH 157 C9
Kozy PL 147 B8
Kozyörük TR 173 B6
Krabbendijke NL 16 F2
Krabi EST 131 F13
Kräckelbäcken S 102 C7
Krackow D 84 D6
Kraddsele S 109 E12
Kraftsdorf D 79 E10
Krąg PL 85 B11

Kragelund DK 86 C4
Kragerø N 90 B5
Krägi PL 85 C11
Kragujevac SRB 159 E6
Kraj HR 156 E6
Kraj S 105 C9
Krajenka PL 85 D12
Krajišnik SRB 159 C6
Krajnik Dolny PL 84 D6
Krakača BIH 156 B4
Kråkberget N 100 C5
Kråken S 122 C3
Kråkenes N 100 B2
Kråkerøy N 91 A8
Krakes LT 134 F7
Krakhella N 100 D2
Kråklingbo S 93 E13
Kråklivollen N 105 E9
Kråkmoen N 111 E10
Kraków PL 143 F8
Krakow am See D 83 C12
Kråksmåla S 89 A9
Kråkstad N 95 C13
Kralíky CZ 77 B11
Kraljeva BIH 157 D9
Kraljevica HR 67 B10
Kraljevo SRB 158 F6
Kralovice CZ 76 B4
Kráľov Dvůr CZ 76 C6
Kramarzyny PL 85 B12
Kramfors S 103 A14
Kramolin BG 166 D4
Krampenes N 114 C8
Kramsach A 72 B4
Kramsk PL 142 B5
Kramvik N 114 C9
Kranenburg D 16 E6
Kranevo BG 167 C10
Krångfors S 118 E4
Krania GR 168 E5
Krania Elassonas GR 169 E6
Kranichfeld D 79 E8
Kranidi GR 175 E7
Kranj SLO 73 D9
Kranjska Gora SLO 73 D8
Krapanj HR 156 E4
Krapets BG 155 F3
Krapiel PL 85 D8
Krapina HR 148 D5
Krapinske Toplice HR 148 D5
Krapkowice PL 142 F4
Krašič HR 148 E5
Krasiczyn PL 144 D6
Krasiv UA 145 D8
Kráslava LV 133 E2
Kraslice CZ 75 B12
Krasnagorka BY 133 E2
Krásná Hora nad Vltavou
  CZ 76 C6
Krasnapollye BY 133 E6
Krasne PL 139 E10
Krasne PL 144 C5
Krasne UA 154 E4
Kraśniczyn PL 144 B7
Kraśnik PL 144 B5
Krasni Okny UA 154 B4
Krasnobród PL 144 B7
Krasnogorodskoye RUS 133 C2
Krásnohorské Podhradie
  SK 145 F2
Krasnoles'ye RUS 136 E5
Krásno nad Kysucou SK 147 C7
Krasnopol PL 136 E7
Krasno Polje HR 67 C11
Krasnosel'skoye RUS 129 D11
Krasnosielc PL 139 E10
Krasnovo BG 165 E9
Krasnoyil's'k UA 153 A7
Krasnoznamensk RUS 136 D5
Krasnystaw PL 144 B7
Krasocin PL 143 E9
Krastë AL 168 B3
Kraszewice PL 142 D5
Kraszewo PL 139 E9
Kratigos GR 177 B7
Kratiškiai LT 135 D9
Kratovo NMK 164 E5
Kratte masugn S 98 B6
Krauchenwies D 27 D11
Krauja LV 135 E13
Kraujas LV 135 D11
Kraukli LV 135 B12
Kraulshavīm N 95 C12
Krautheim D 74 D6
Kravaře CZ 146 B6
Kravarsko HR 148 E6
Kravoder BG 165 C7
Kraymorie BG 167 E8
Kražiai LT 134 E5
Krčina BIH 157 C11
Krekhovychi UA 145 E9
Krečišta HR 162 D3
Krefeld D 17 F7
Kreiensen D 78 C6
Krekenava LT 135 E8
Krekhovychi UA 145 E9
Krekilä FIN 123 C11
Kremasti GR 181 D8
Kremen BG 165 F8
Kremena BG 155 F2
Kremintsi UA 152 A5
Kremmen D 84 E5
Kremmydia GR 174 F4
Kremna SRB 158 F4
Kremnica SK 147 D7
Krempe D 82 C7
Kremperheide D 17 A12
Krempna PL 145 D4
Krems an der Donau A 77 F9
Kremsbrücke A 73 C8
Kremsmünster A 76 F6
Křepice CZ 77 E11
Krepoljin SRB 159 E8
Kreševo BIH 157 E9
Kreshpan AL 168 C2
Krešice CZ 76 A6
Kresna BG 165 F7
Kresnice SLO 73 D10
Kressbronn am Bodensee
  D 71 B9
Krestena GR 174 D4
Kretinga LT 134 E2
Kretingalė LT 134 E2
Kreuth D 72 A4
Kreuzau D 20 C6
Kreuzlingen CH 27 E11
Kreuztal D 21 C9
Kreuzwertheim D 74 C6
Kreva BY 137 E13
Krezluk BIH 157 D7
Krichim BG 165 E9
Krieglach A 148 A5

Kriegstetten CH 27 F8
Krien D 84 C4
Kriens CH 70 C6
Krievciems LV 134 B3
Krieza GR 175 C9
Krikello GR 174 B4
Krikellos GR 174 B3
Krimpen aan de IJssel NL 16 E3
Křimūnas LV 134 C6
Krinida D 170 C5
Krinides GR 171 B6
Krini D 72 A3
Krinitzmow D 83 C9
Kriukai LT 134 D7
Kriūkai LT 137 C7
Kriva Bara BG 159 F11
Krivača BIH 157 G9
Kriva Feja BG 165 C7
Krivaň SK 147 D8
Krivi Vir SRB 159 F8
Krivogaštani NMK 168 A5
Krivolak NMK 169 A7
Križ HR 149 E7
Križanov CZ 77 D10
Križevci HR 149 D7
Křížová CZ 77 D9
Križpolje HR 67 B11
Krk HR 67 B10
Krka SLO 73 E10
Krmčina HR 156 D3
Krnča SK 146 D6
Krnica HR 67 C9
Krnjača SRB 158 D6
Krnja Jela NMK 163 D7
Krnjak HR 148 F5
Krnjeuša BIH 156 C5
Krnjevo SRB 159 E7
Krnov CZ 142 F4
Krobia PL 81 C11
Kroczyce PL 143 E8
Krøderen N 95 B11
Krog S 103 B9
Krögis D 80 D4
Krok S 105 E14
Krokees GR 175 F6
Krokeide N 94 B2
Krokek S 93 B8
Kroken N 100 D6
Kroken N 108 F7
Krokio GR 175 A6
Kroknäs S 102 A8
Kroknes N 114 C9
Krokom S 106 E6
Krokos GR 169 D6
Krokowa PL 138 A5
Kroksjö S 107 B13
Kroksjö S 107 D14
Kroksjö S 118 F4
Krokstadelva N 95 C12
Krokstadøra N 104 E7
Krokstranda N 108 D9
Krokströmmen S 102 B8
Kroktjärn S 118 B4
Krokträsk S 118 C6
Królowy Most D 140 D8
Krolpa D 79 E10
Kroměříž CZ 146 C4
Krommenie NL 16 C3
Krompachy SK 145 F2
Kromsdorf D 79 E9
Kronach D 75 B9
Kronau D 187 C6
Kronauce LV 134 C6
Kronberg im Taunus D 187 A5
Kronoby FIN 123 C10
Kronprinzenkoog D 82 C5
Kronshagen D 83 B8
Kronshtadt RUS 129 F12
Kronstorf A 77 F6
Kropa SLO 73 D9
Kröpelin D 83 B11
Kropp D 82 B7
Kroppenstedt D 79 C9
Kropstädt D 79 C10
Krościenko nad Dunajcem
  PL 145 D1
Krośba A 73 A7
Krośnice PL 81 D12
Krośniewice PL 143 B7
Krosno PL 144 D4
Krosno Odrzańskie PL 81 B8
Krossen D 79 E10
Krossen N 90 C2
Krossen N 95 C8
Krossmoen N 94 E4
Krostitz D 79 D11
Krote LV 134 C3
Krotoszyce PL 81 D10
Krotoszyn PL 142 C3
Krottendorf A 148 B5
Krouna CZ 77 C10
Krousonas GR 178 E8
Kröv D 21 E8
Krovik S 111 E19
Krovyli GR 171 C9
Krrabë AL 168 B2
Krš HR 156 C3
Kršan HR 67 B9
Krško SLO 148 E4
Krstac MNE 156 B4
Krstinja HR 156 B4
Krtova BIH 157 C9
Kruchowo PL 138 E5
Krüden D 83 E11
Kruft D 185 D7
Kruibeke B 19 B9

Kruiningen NL 16 F2
Kruishoutem B 19 C8
Krujë AL 168 A2
Krukenychi UA 144 D7
Kruklanki PL 136 E5
Krukowo PL 139 D11
Krumbach A 71 C9
Krumbach A 148 A6
Krumbach (Schwaben) D 71 A10
Krumë AL 163 E9
Krummesse D 83 C9
Krumovgrad BG 171 B9
Krumovo BG 165 E10
Krumovo BG 166 E6
Krumpa (Geiseltal) D 79 D10
Krumvíř CZ 77 E11
Krün D 72 A3
Krunderup DK 86 C3
Kruonis LT 137 D9
Kruopiai LT 134 D6
Krupac BIH 157 E9
Krupac SRB 165 C6
Krupanj SRB 158 E3
Krupava BY 137 F11
Krupe PL 144 A7
Krupina SK 147 D8
Krupište NMK 164 F5
Krupka CZ 80 E5
Krupnik BG 165 F7
Kruså DK 82 A6
Kruševac SRB 159 F7
Kruševica SRB 158 E5
Kruševo NMK 168 A5
Kruševo Grdo BIH 157 D8
Krushari BG 155 F1
Krushe e Madhe RKS 164 E2
Krushel'nytsya UA 145 E8
Krushevets BG 167 E8
Krusheno BG 166 C4
Krushovene BG 160 F4
Krushovitsa BG 160 F3
Kruščica SK 146 D6
Kruszewo PL 85 E11
Kruszwica PL 138 E5
Kruszyna PL 85 B12
Kruszyna PL 143 E7
Krutådal N 108 E7
Krute MNE 163 E7
Kruunupyy FIN 123 C10
Kružlov SK 145 E3
Krya Vrysi GR 169 C7
Kryekuq AL 168 C2
Kryezi AL 163 E9
Kryłów PL 144 B9
Krynica PL 145 E2
Krynica Morska PL 139 B7
Krynice PL 144 B7
Krynki PL 140 D9
Krynychne UA 155 B3
Kryoneri GR 175 D6
Kryoneritis GR 175 B7
Kryopigi GR 169 D9
Krypno Kościelne PL 140 D7
Kryry CZ 76 B4
Krystad N 110 D5
Krystallopigi GR 168 C5
Kryva Balka UA 154 C5
Kryve Ozero UA 154 A6
Kryvorivnya UA 152 A5
Kryzhopil' UA 154 A3
Krzanowice PL 142 F5
Krząstka PL 144 C4
Krzcin PL 85 D9
Krzelów PL 81 D11
Krzemienieo PL 81 C11
Krzepice PL 142 E6
Krzepielów PL 81 C10
Krzesk-Królowa Niwa PL 141 F7
Krzeszów PL 81 E9
Krzeszów PL 144 C5
Krzeszowice PL 143 F8
Krzeszyce PL 81 A8
Krzętów PL 143 E8
Krzosy PL 81 B12
Krzynowłoga Mała PL 139 D10
Krzystkowice PL 81 C8
Krzywcza PL 144 D6
Krzywda PL 141 G6
Krzywiń PL 81 C11
Krzywin PL 84 D6
Krzywowierzba PL 141 G8
Krzyżanów CZ 143 B7
Krzyżanowice PL 141 H4
Krzyżanowice PL 146 B6
KrzyżKrzyż PL 85 E10
Ksar Sghir MA 53 E5
Książ Wielki PL 143 F9
Książ Wielkopolski PL 81 B12
Ksijżpol PL 144 C6
Kubanovka RUS 136 D5
Kubbe S 107 D14
Kübekháza H 150 E5
Kübelmäki FIN 124 C9
Küblis CH 71 D9
Kubrat BG 161 F8
Kubuli LV 133 B2
Kuç AL 168 D2
Kučevo SRB 159 E8
Kuchen D 74 E6
Kuchl A 73 A7
Kuchurhan UA 154 D5
Kuchyňa SK 77 F12
Kucice PL 139 E9
Kuciny PL 143 C7
Kučište HR 162 D3
Kuçovë AL 168 C2
Kučurhan UA 154 D5
Kuczbork-Osada PL 139 D9
Kudever' RUS 133 C4
Kudirkos Naumiestis LT 136 D5
Kudowa-Zdrój PL 77 B10
Kūdums LV 135 B10
Kufstein A 72 A5
Kuggerud N 96 B6
Küggören S 103 C12
Kuha FIN 119 C17
Kuhakoski FIN 129 B9
Kühbach D 75 F9
Kuhbier D 83 E11
Kuhfelde D 83 E10
Kühlungsborn D 83 B11

Kuhmalahti FIN 127 B12
Kuhmo FIN 121 F14
Kuhmoinen FIN 127 B13
Kühndorf D 79 E7
Kühren D 79 D12
Kuhs D 83 C12
Kuhstedt D 17 B11
Kuijõe EST 131 C7
Kuimetsa EST 131 C10
Kuinre NL 16 C5
Kuivajärvi FIN 121 E15
Kuivajõe EST 131 C10
Kuivakangas S 119 B11
Kuivalahti FIN 126 C6
Kuivaniemi FIN 119 C14
Kuivanto FIN 127 D15
Kuivasjärvi FIN 123 F10
Kuivastu EST 130 D6
Kujakowice Dolne PL 142 D5
Kūkas LV 135 C12
Kukkaro FIN 123 D15
Kukkia FIN 127 B12
Kukko FIN 123 E12
Kukkola FIN 119 C12
Kukkola S 119 C12
Kuklen BG 165 E10
Kuklin PL 139 D9
Kuklinów PL 81 C11
Kukljica HR 156 D3
Kukljin SRB 164 B3
Kuko FIN 119 F9
Kukurečani NMK 168 B5
Kula BG 159 F9
Kula HR 149 F9
Kula S 145 E3
Kula SRB 158 B4
Kulaši BIH 157 C8
Kulata BG 169 B9
Kulautuva LT 137 D9
Külçiems LV 134 B6
Kuldīga LV 134 C3
Külefli TR 173 D6
Kuleli TR 173 B6
Kulennoinen FIN 129 B11
Kulen Vakuf BIH 156 C5
Kulesze PL 140 D7
Kulho FIN 125 E13
Kulia N 105 D9
Kuliai LT 134 E3
Kulju FIN 127 C10
Kullaa FIN 126 C7
Kullamaa EST 131 D8
Kulleseid N 94 C2
Kulltorp S 88 A5
Kulmain D 75 C10
Kulmbach D 75 B9
Kuloharju FIN 121 C11
Külsheim D 27 A12
Kultima FIN 116 B9
Kuluntalahti FIN 121 F10
Kulva LT 137 C9
Kulvemäki FIN 124 C9
Kulykiv UA 154 D5
Kumachevo RUS 139 A9
Kumane SRB 158 B5
Kumanica SRB 163 C9
Kumanovo NMK 164 E4
Kumbağ TR 173 C7
Kumberg A 148 B5
Kumbri LV 134 C4
Kumbuli LV 135 E13
Kumburun TR 171 E10
Kumhausen D 75 E11
Kumkadı TR 173 D11
Kumkale TR 171 E10
Kumköy TR 173 B11
Kumla S 92 A6
Kumla kyrkby S 98 C7
Kumle FIN 126 E3
Kumlinge FIN 99 B15
Kummer D 83 D10
Kummerow D 83 C13
Kummersdorf-Alexanderdorf
  D 80 B4
Kummersdorf Gut D 80 B4
Kummunkylä FIN 123 D16
Kumpuranta FIN 125 G12
Kumpuselkä FIN 123 D16
Kumpuvaara FIN 119 C18
Kumrovec HR 148 D5
Kunadacs H 150 D3
Kunágota H 151 E7
Kunbaja H 150 E3
Kunbaracs H 150 D3
Kunčina CZ 77 C11
Kunda EST 131 B13
Kunda S 103 E12
Kunes N 113 C18
Kunfehértó H 150 E3
Kungälv S 91 D10
Kungas FIN 123 C11
Kungsängen S 99 D9
Kungsäter S 88 A3
Kungsbacka S 91 D11
Kungsberg S 103 E11
Kungsfors S 103 E12
Kungsgården S 103 D12
Kungsgården S 103 E12
Kungshamn S 91 C9
Kungshult S 87 D12
Kungsör S 98 D6
Kunhegyes H 150 C6
Kunice PL 81 D10
Kunín CZ 146 B6
Kuningaküla EST 132 C2
Kuniów PL 142 E5
Kunmadaras H 150 C6
Kunovice CZ 146 C4
Kunowice PL 81 B7
Kunowo PL 81 C12
Kunpeszér H 150 C3
Kunrade NL 183 D7
Kunrau D 79 A9
Kunštát CZ 77 C10
Kunszállás H 150 D4
Kunszentmárton H 150 D6
Kunszentmiklós H 150 C3
Kunsziget H 149 A8
Kunvald CZ 77 B11
Kunžak CZ 77 D8
Künzell D 74 B6
Künzelsau D 74 D6
Künzing D 76 E4
Kuohatti FIN 125 C12
Kuohu FIN 123 F14
Kuoksajärvi FIN 117 E15
Kuoksu S 116 C7
Kuomo FIN 121 C11
Kuopio FIN 124 E9
Kuopio FIN 129 E13
Kuosku FIN 115 D4
Kuossakåbba S 116 D4
Kuouka S 116 E5
Kup PL 142 E4
Kupari FIN 162 D3
Küpeler TR 173 E8
Kupferberg D 75 B9
Kupferzell D 74 D6
Kupientyn PL 141 F6
Kupinec HR 148 E5
Kupinovo SRB 158 D5
Kupiškis LT 135 E9
Kupjak HR 67 B10
Küplü TR 171 B10
Kuppenheim D 27 C9
Kuprava LV 133 B3
Kupres BIH 157 D7
Kuprešani BIH 157 D7
Küps D 75 B9
Kurapollye BY 135 F13
Kuratën AL 168 B2
Kurbnesh AL 163 F9
Kurd H 149 D10
Kürdzhali BG 171 A8
Kurejoki FIN 123 D11
Kuremaa EST 131 D13
Kuremäe EST 132 C2
Kuressaare EST 130 E4
Kurevere EST 130 D4
Kurienkylä FIN 123 F10
Kurikka FIN 122 E8
Kuřim CZ 77 D11
Kurima SK 145 E3
Kurjala FIN 125 E9
Kurkela AL 168 C2
Kurki FIN 121 D10
Kurkimäki FIN 124 E9
Kurmene LV 135 D9
Kürnach D 187 B9
Kürnare BG 165 D10
Kürnbach D 27 B10
Kurolanlahti FIN 124 D8
Kuropta RUS 115 D9
Kurort Bad Gottleuba D 80 E5
Kurort Kipsdorf D 80 E5
Kurort Oberwiesenthal D 76 B3
Kurort Schmalkalden D 79 E7
Kurort Steinbach-Hallenberg
  D 79 E8
Kurów PL 141 H6
Kurowice PL 143 C8
Kurravaara S 116 C4
Kurrokvejk S 109 E14
Kuršėnai LT 134 D6
Kuršiši LV 134 C4
Kursu FIN 115 E4
Kurśumlija SRB 164 C3
Kuršumlijska Banja SRB 164 C3
Kurtakko FIN 117 D12
Kurtbey TR 172 B6
Kurtna EST 132 C2
Kurtovo Konare BG 165 E10
Kurtti FIN 121 D11
Kurtto FIN 121 E11
Kurtzea E 40 B6
Kuru FIN 127 B10
Kurvinen FIN 121 B10
Kuryłówka PL 144 C5
Kurzelów PL 143 E8
Kurzeszyn PL 141 G2
Kurzętnik PL 139 D8
Kusa LV 135 C12
Kusadak SRB 159 E6
Kuşadası TR 177 D9
Kuşçayır TR 172 E6
Kusel D 21 E8
Kusey D 79 A9
Kushnytsya UA 145 G7
Kušići SRB 163 C9
Kuside MNE 163 D6
Kušlin PL 81 B10
Kusmark S 118 E5
Küsnacht CH 27 F9
Küssnacht CH 27 F9
Kustavi FIN 126 D5
Küsten D 83 E10
Kusterdingen D 27 C11
Kuštilj SRB 159 C7
Kuta BIH 157 E9
Kutalli AL 168 C2
Kutas H 149 D8
Kutemajärvi FIN 124 F7
Kutenholz D 17 B12
Kuti MNE 163 D8
Kutina HR 149 F7
Kutjevo HR 149 F9
Kutlovo SRB 159 E6
Kutná Hora CZ 77 C8
Kutno PL 143 B7
Kutrikko FIN 115 E5
Kuttainen S 116 B9
Kuttanen FIN 116 B9
Kuttura FIN 117 B16
Kúty SK 77 E12
Kuty UA 152 A6
Kuukasjärvi FIN 119 C17
Kuumu FIN 121 E14
Kuurne B 19 C7
Kuurtola FIN 121 D12
Kuusaa FIN 123 C15
Kuusaa FIN 123 F15
Kuusajärvi FIN 117 C14
Kuusajoki FIN 117 C12
Kuusalu EST 131 C10
Kuusamo FIN 121 C11
Kuusankoski FIN 128 D6
Kuusijärvi FIN 121 D11
Kuusijoki FIN 126 B6
Kuusiranta FIN 119 F14
Kuusivaara FIN 115 E2
Kuusjoki FIN 127 D9
Kuuslahti FIN 124 D9
Kuvansi FIN 125 F9
Kuvaskangas FIN 126 B6
Kuyka UA 153 A7
Kuyvozi RUS 129 E13
Kužiai LT 134 E6
Kuzmice SK 145 F4

Kuzmin SRB 158 C3
Kuźmina PL 145 D5
Kuz'molovskiy RUS 129 E13
Kuznechnoye RUS 129 C12
Kuźnia Raciborska PL 142 F5
Kuźnia PL 140 C9
Kuźnica Czarnkowska PL 85 E11
Kväcklingen S 103 A12
Kværkeby DK 87 E9
Kværndrup DK 86 E7
Kvalfors N 108 E5
Kvalnes N 110 D6
Kvalnes N 110 E8
Kvalnes N 111 B11
Kvalnes N 114 C8
Kvaløyhamn N 108 E3
Kvaløysletta N 111 A16
Kvaløyvågen N 111 A16
Kvalsaukan N 111 C10
Kvalsvik N 100 B3
Kvalvåg N 104 E3
Kvalvik N 111 A19
Kvam N 101 C11
Kvannås N 111 B14
Kvanndal N 94 B5
Kvanne N 101 A9
Kvannli N 105 A14
Kvänum S 91 C13
Kvarnberg S 102 D8
Kvarnbyn S 118 F4
Kvarnriset S 118 F5
Kvarnsand N 99 B11
Kvarnsjö S 102 A7
Kvarntäkt S 103 E10
Kvarntorp S 92 A6
Kvarv N 108 B8
Kvarv N 109 B10
Kvasice CZ 146 C4
Kvasy UA 152 A4
Kveaune N 105 C15
Kvédarna LT 134 E3
Kveinsjøen N 105 A12
Kvelde N 90 A6
Kveli N 105 B15
Kvenvær N 104 D4
Kvevlax FIN 122 D7
Kvibille S 87 B11
Kvicksund S 98 D6
Kvidinge S 87 C12
Kvikkjokk S 109 C14
Kvillsfors S 92 E7
Kvinesdal N 94 F5
Kvinlog N 94 F5
Kvissleby S 103 B13
Kvisvik N 104 E3
Kvitblik N 109 B10
Kviteberg N 112 D8
Kviteseid N 95 D8
Kvitnes N 110 C9
Kvitnes N 112 C5
Kvitsøy N 94 D2
Kvívík FO 2 A2
Kvorning DK 86 C5
Kwaadmechelen B 19 B11
Kwidzyn PL 138 C6
Kwieciszewo PL 138 E5
Kwilcz PL 81 A10
Kyan S 103 D10
Kybartai LT 136 D6
Kycklingvattnet S 105 B16
Kydrasivka UA 154 A5
Kyjov CZ 77 D12
Kylänpää FIN 122 E8
Kylänlahti FIN 125 D13
Kyläsaari FIN 126 B6
Kyleakin UK 2 L5
Kyle of Lochalsh UK 2 L5
Kylerhea UK 2 L5
Kylestrome UK 2 J6
Kylland N 90 B1
Kyllburg D 21 D7
Kyllini GR 174 D3
Kylmäkoski FIN 127 C10
Kylmälä FIN 119 E16
Kymi GR 175 B9
Kymina GR 169 C8
Kyminlinna FIN 128 D6
Kymönkoski FIN 123 D15
Kyndby Huse DK 87 D9
Kyngäs FIN 119 D17
Kynsiperä FIN 121 C11
Kynsivaara FIN 121 C11
Kynšperk nad Ohří CZ 75 B12
Kypärävaara FIN 121 E12
Kyparissia GR 174 E4
Kypasjärv S 119 B10
Kyprinos GR 171 A10
Kypseli GR 175 D7
Kyrgia GR 171 B6
Kyriaki GR 175 C6
Kyritz D 83 E12
Kyrkås S 106 E7
Kyrkhult S 88 C7
Kyrksæterøra N 104 E6
Kyrkslätt FIN 127 E11
Kyrksten S 97 D11
Kyrnychky UA 155 B4
Kyröskoski FIN 127 B9
Kyrsyä FIN 129 B9
Kyslytsya UA 155 C4
Kysucké Nové Mesto SK 147 C7
Kysucký Lieskovec SK 147 C7
Kytäjä FIN 127 D12
Kythira GR 178 C4
Kythnos GR 175 E7
Kytökylä FIN 119 F15
Kytömäki FIN 121 E11
Kyustendil BG 165 E6
Kyyjärvi FIN 123 D13
Kyynämöinen FIN 123 F14
Kyynärö FIN 127 C12

La

Laa an der Thaya A 77 E10
Laaber D 75 D10
La Adrada E 46 D3
Laage D 83 C12
Laagri EST 131 C9
Laaja FIN 121 E12
Laajoki FIN 126 D7
Laakajärvi FIN 125 C10
La Alberca E 45 C8
La Alberca E 47 F10
La Alberca de Záncara E 47 E8
La Alberguería de Argañán
  E 45 D7
La Albuera E 51 B6
La Aldehuela E 45 D10
La Algaba E 51 E7
La Aljorra E 56 F2

La Almarcha E 47 E8
La Almolda E 42 D3
La Almunia de Doña Godina
  E 41 F9
Laanila FIN 115 B2
Lääniste EST 131 E14
La Antilla E 51 E5
Laapas FIN 128 D8
Laarkirchen A 76 G5
Laarne B 19 B9
Laasala FIN 123 E12
Laaslich D 83 D11
Laatre EST 131 E10
Laatre EST 131 F12
Laatzen D 78 B6
La Azohía E 56 F2
La Baconnière F 23 D10
Labajos E 46 C3
La Bañeza E 39 D8
Labanoras LT 135 F11
La Barre-de-Monts F 28 B1
La Barre-en-Ouche F 24 C4
La Barthe-de-Neste F 33 E6
Labarthe-Rivière F 33 D7
La Bassée F 18 C6
La Bastide-Clairence F 32 D3
Labastide-d'Anjou F 33 D9
La Bastide de Bousignac
  F 33 D9
La Bastide-de-Sérou F 33 D8
La Bastide-l'Évêque F 33 B10
Labastide-Murat F 33 A9
La Bastide-Puylaurent F 35 A6
Labastide-Rouairoux F 34 C4
La Bastide-St-Pierre F 33 C8
La Bastide-sur-l'Hers F 33 E9
La Bâthie F 31 D9
La Bâtie-Neuve F 36 B4
Lábatlan H 149 A10
Labatut-Rivière F 32 C6
La Baule-Escoublac F 23 F7
La Bazoche-Gouet F 24 D4
La Bazoge F 23 D12
L'Abbaye CH 31 B9
Labégude F 35 A7
La Bégude-de-Mazenc F 35 A8
Labenne F 32 C2
La Bérarde F 31 F9
Labergement-lès-Seurre F 31 A7
Laberweinting D 75 E11
Labin HR 67 B9
La Bisbal de Falset E 42 E5
La Bisbal del Penedès E 43 E6
La Bisbal d'Empordà E 43 D10
Łabiszyn PL 138 E4
Lablachère F 35 A7
Laboe D 83 B8
La Bouëxière F 23 D9
Laboureyre F 32 B4
La Bouilladisse F 35 D10
La Bourboule F 30 D2
Laboutarie F 33 C10
La Bóveda de Toro E 39 F9
Ladi GR 171 B10
Ladignac-le-Long F 29 D8
Ladinhac F 29 F11
Ladispoli F 62 D2
Ladoeiro P 45 E6
Ladomirová SK 145 E4
Ladon F 25 E8
Ladozhskoye Ozero RUS 129 E15
Ladushkin RUS 139 A9
Ladybank UK 5 C10
Ladykirk UK 5 D12
Ladysford UK 3 K12
Lädzere LV 134 B6
Ładzice PL 143 D7
Ładzin PL 84 C7
Laekvere EST 131 C13
Laer D 17 D8
La Ercina E 39 C9
La Hinojosa E 47 E8
Laerma GR 181 D7
Laerru I 64 B2
La Espina E 39 B7
La Estación E 43 D8
La Estrella E 45 E10
Laeva EST 131 D12
La Fare-les-Oliviers F 35 C9
La Farlède F 36 E4
La Fatarella E 42 E4
La Faurie F 35 A10
La Faute-sur-Mer F 28 C3
La Felipa E 47 F9
La Fère F 19 E7
La Ferrière F 28 B3
La Ferrière-aux-Étangs F 23 C10
La Ferté-Alais F 25 D7
La Ferté-Bernard F 24 D4
La Ferté-Frênel F 24 C4
La Ferté-Gaucher F 25 C9
La Ferté-Imbault F 24 F6
La Ferté-Loupière F 25 E9
La Ferté-Macé F 23 C11
La Ferté-Milon F 25 B9
La Ferté-sous-Jouarre F 25 C9
La Ferté-St-Aubin F 24 E6
La Ferté-St-Cyr F 24 E6
Laferté-sur-Amance F 26 E4
Laferté-sur-Aube F 25 D12
La Férte-Vidame F 24 C4
La Ferté-Villeneuil F 24 E5
Lafitole F 33 D6
La Chaise-Dieu F 30 E4
La Chaize-le-Vicomte F 28 B3
La Chambre F 31 E9
Lachania GR 181 E7
La Chapelaude F 29 C11
La Chapelle F 19 E11
La Chapelle-aux-Bois F 26 D5
La Chapelle-aux-Saints F 29 F9
La Chapelle-d'Abondance
  F 31 C10
La Chapelle-d'Aligné F 23 E11
La Chapelle-d'Angillon F 25 F7
La Chapelle-de-Guinchay
  F 30 C6
La Chapelle-des-Fougeretz
  F 23 D8
La Chapelle-en-Valgaudémar
  F 31 F9
La Chapelle-en-Vercors F 31 F8
La Chapelle-la-Reine F 25 D8
La Chapelle-Laurent F 30 E3
Lachapelle-sous-Aubenas
  F 35 A7

La Chapelle-St-André F 25 F9
La Chapelle-St-Laurent F 28 B5
La Chapelle-St-Luc F 25 D11
La Chapelle-St-Quillain F 26 F4
La Chapelle-St-Ursin F 29 A10
La Chapelle-sur-Erdre F 23 F8
Lachar E 53 B9
La Charité-sur-Loire F 25 F9
La Chartre-sur-le-Loir F 24 E4
La Châtaigneraie F 28 B5
La Châtre F 29 B9
La Châtre-Langlin F 29 C8
La Chaume F 25 E12
La Chaussée-St-Victor F 24 E5
La Chaussée-sur-Marne
  F 25 C12
La Chaux-de-Fonds CH 27 F6
La Chaux-du-Dombief F 31 B8
Lachendorf D 79 B7
La Cheppe F 25 B12
La Chèze F 22 D6
Lachowo PL 139 D13
Lacíci HR 146 F6
La Cierva E 47 E8
La Ciotat F 35 D10
La Cisterniga E 39 E10
Lack UK 7 C7
Łąck PL 139 F8
Lackamore IRL 8 C6
Lackeby S 89 B10
Lackendorf A 148 B6
La Clayette F 30 C5
La Clusaz F 31 D9
La Cluse F 35 A10
La Cluse-et-Mijoux F 31 B9
La Codosera E 45 F6
La Concha E 40 B4
La Condamine-Châtelard
  F 36 C5
La Coquille F 29 D7
La Coronada E 51 B8
La Coronada E 51 C8
La Côte-St-André F 31 E7
La-Couarde-sur-Mer F 28 C3
La Couronne F 28 D6
Lacourt F 33 E8
La Courtine F 29 D10
La Couture-Boussey F 24 C5
Lāčplēsis LV 135 C9
Lacq F 32 D5
La Crau F 36 E4
La Crèche F 28 C5
Lacroix-Barrez F 29 F11
La Croixille F 23 D9
La Croix-Valmer F 36 E5
Lacrouzette F 33 C10
La Cumbre E 45 F9
Łączna PL 143 E10
Łącznik PL 142 F4
Ladánybene H 150 C3
Ladbergen D 17 D9
Ladce SK 146 C6
Lądek PL 142 B4
Lądek-Zdrój PL 77 B11
Ladelund D 82 A6
Ladenburg D 21 F11
Lądeşti RO 160 D4
La Guardia de Jaén E 53 A9
La Guerche-de-Bretagne
  F 23 E9
La Guerche-sur-l'Aubois F 30 B2
Laguiole F 30 F2
Laguna Dalga E 39 D8
Laguna de Duero E 39 E10
Laguna de Negrillos E 39 D8
Lagundo I 72 C3
Lagunilla E 45 D9
Lagyna GR 169 C9
La Haba E 51 B8
Lahana TR 173 B7
La Haye-du-Puits F 23 B8
La Haye-Pesnel F 23 C9
Lähdejänkkä FIN 119 B12
Lähden D 17 C9
Lahenpää S 116 E10
Laheycourt F 26 C3
La Higuera E 55 B10
Lahinch IRL 8 C4
La Hiniesta E 39 E8
La Hinojosa E 47 E8
Lahnajärvi S 116 E9
Lahnajärvi FIN 124 D7
Lahnanen FIN 123 D16
Lahnasjärvi FIN 125 C10
Lahnstein D 185 D8
Laholm S 87 C12
La Horcajada E 45 D10
La Horra E 40 E4
Lahr (Schwarzwald) D 27 D8
Lähte EST 131 E13
Lahti FIN 126 D6
Lahti FIN 127 D14
Lahtiranta FIN 119 E13
Laichingen D 74 E6
Laide UK 2 K5
Laidi LV 134 C3
Laidze LV 134 B5
Laifour F 184 E2
L'Aigle F 24 C4
La Iglesuela E 46 D3
La Iglesuela del Cid E 42 G3
Laignes F 25 E11
Laiguéglia I 37 D8
Laiko S 107 D14
Laimont F 26 C3
Lainate I 69 B7
Lainio S 116 C9
Laino Borgo I 60 D5
Lairg UK 3 J8
La Iruela E 55 D7
Laisbäck S 107 A12
Laisheden S 109 E14
Laissac F 34 B4
Laisvall S 109 D13
Laisvalls by S 109 D13
Laitala FIN 121 E12
Laitikkala FIN 127 C11
Laitila FIN 126 D6
Laives I 72 C3
Laize EST 131 D13
Laizuva LT 134 D5
La Jana E 42 F3
La Jarrie F 28 C3
La Javie F 36 C4
Lajkovac SRB 158 E5
Lajosmizse H 150 C3
La Jumellière F 23 F10
Lakaluoma FIN 123 E10
Lakatnik BG 165 C7
Lakaträsk S 118 B6
Lakavica NMK 163 F10
Lakavica NMK 169 A7
Lakenheath UK 15 C10
Lakhdenpokh'ya RUS 129 B13
Laki NMK 165 F6
Łąkie PL 85 C11
Lakitelek H 150 D4
Lakka GR 168 F3
Lakki GR 177 E8
Lakkoi GR 178 E6
Lakkoma GR 171 D9
Lakkopetra GR 174 C3
Łąkociny PL 142 C4
Lakomäki FIN 123 E15
Lakorz PL 139 D7
Lakšárska Nová Ves SK 146 D4
Lakselv N 113 C14
Lakselvsletta N 111 B18
Laksfors N 108 E5
Laktaši BiH 157 C7
Lakuntza E 32 E1
La Lantejuela E 51 E9
La Lapa E 51 B7
Lalapaşa TR 167 F7
Lalas GR 174 D4
L'Albagés E 42 E5
Lalbenque F 33 B9
L'Alcora E 48 D4
L'Alcudia E 48 F3
L'Alcudia de Crespins E 56 D3
Lalevade-d'Ardèche F 35 A7
Lalín E 38 C3
L'Alfàs del Pi E 56 D4
Lalín E 38 C3
Lalinac SRB 164 C5
Lalinde F 29 F7
La Línea de la Concepción
  E 53 D6
Lalioti GR 175 C6
Lallaing F 19 D7
Lalley F 31 F8
Lalm N 101 C10
La Londe-les-Maures F 36 E4
Laloşu RO 160 E4
Laloubère F 33 D6
La Loupe F 24 D5
La Louvière B 19 D9
Lalove UA 145 G6
La Loye F 31 A8
L'Alpe-d'Huez F 31 E9
Lalueza E 42 D3
La Luisiana E 51 D9
Laluque F 32 C4
Lam D 76 D4
La Machine F 30 B3
La Maddalena I 64 A3
Lama dei Peligni I 63 C6
La Magdalena E 39 C8
Lamagistère F 33 B7
La Malène F 34 B5
Lamalou-les-Bains F 34 C5
Lama Mocogno I 66 D2
La Manga del Mar Menor
  E 56 F3
Lamanon F 35 C9
La Marche F 26 F4
Lamarche F 26 D4
Lamarche-sur-Saône F 26 F3
Lamargelle F 25 E12
Lamarque F 28 E4
Lamas P 44 C3
La Massana AND 33 E9
Lamastre F 30 F6
La Mata E 46 E5
La Mata de los Olmos E 42 F3
La Matilla E 46 B5
Lambach A 76 F5
Lamballe F 22 D6
Lambersart F 18 C7
Lambesc F 35 C9
Lambia GR 174 D4
Lambinowice PL 142 F4
Lamborn S 103 D10
Lamborn UK 13 B11
Lambourn UK 13 B11
Lambrecht (Pfalz) D 186 C5
Lambsheim D 21 E10
Lamego P 44 B5
La Meilleraye-de-Bretagne
  F 23 E9
Lameiras P 45 C6
Lamerdingen D 71 A11
L'Ametlla de Mar E 42 F5
Lamesberg D 183 C10
Lamia GR 174 B5
Lamego P 44 B5
Lamiako E 40 B6
Lamim FIN 123 D11
Lammi FIN 127 C13
Lamminkoski FIN 123 F10
Lamminperä FIN 119 C16
Lamminvaara FIN 121 D13
La Mojonera E 55 F7
La Môle F 36 E5
Lamon I 72 D4
La Monnerie-le-Montel F 30 D4
La Morera E 51 B6
La Mothe-Achard F 28 B2
La Mothe-Cassel F 33 A8
La Mothe-St-Héray F 28 C5
Lamotte-Beuvron F 24 E7
La Motte-Chalancon F 35 B9
La Motte-du-Caire F 35 B11
Lamovita BiH 157 C7
Lampaanjärvi FIN 124 D7
Lampaul F 22 C2
Lampaul-Guimiliau F 22 D3
Lampaul-Plouarzel F 22 C1
Lampeland N 95 C11
Lampertheim D 21 E10
Lampertswalde D 80 D5
Lampeter UK 12 A6
Lampinsaari FIN 119 F14
Lampiselkä FIN 115 D1
Lamprechtshausen A 76 G3
Lamprero I 66 C1
Lampsijärvi FIN 117 E13

Lajeosa do Mondego P 45 C6

La Garde F 36 E4
La Garde-Adhémar F 35 B8
La Garde-Freinet F 36 E4
Lagardelle-sur-Lèze F 33 D8
Lagares da Beira P 44 D5
La Garnache F 28 B2
La Garriga E 43 D8
Lagarrigue F 33 C10
La Garrovilla E 51 B7
La Gaubretière F 28 B3
Lagavara UK 4 E4
Lagedi EST 131 C9
Lage D 17 D7
Lage D 17 E11
Lage Mierde NL 16 F4
Lagerfors S 91 B15
Lagfors S 103 A13
Lagg UK 4 A7
Laggan UK 5 A9
Lagganulva UK 4 C4
Laggarberg S 103 B13
Laghey IRL 6 C6
Laghtgeorge IRL 6 F5
Łagiewniki PL 81 E11
La Gineta E 47 F9
Lagkada GR 177 C7
Lagkadas GR 169 C9
Lagkadia GR 174 D5
La Glacerie F 23 A8
Lagnieu F 31 D7
Lagny-sur-Marne F 25 C8
Lago I 60 E6
Lago P 39 F6
Lagoa P 50 E3
Lagoaça P 45 B7
Lago Menor E 57 B11
La Gomera E 46 D3
Lagor F 32 D4
Lagorce F 28 B5
Lagord F 28 C3
La Gornal E 43 E7
Lagos GR 171 B8
Lagos P 50 E2
Lagosanto I 66 C5
Łagów PL 81 B8
Łagów PL 143 E11
Lagrán E 41 C6
La Granada de Riotinto E 51 D7
La Granadella E 42 E5
La Grand-Croix F 30 D6
La Grande-Combe F 35 B7
La Grande-Motte F 35 D7
La Granja d'Escarp E 42 E4
La Granjuela E 51 C9
Lagrasse F 34 D4
La Grave F 31 E9
Lägsta S 107 D14
La Guardia E 46 E5
La Guardia de Jaén E 53 A9
Laguarta E 32 F5
La Guerche F 29 E9
Laguenne F 29 E9
Laguépie F 33 B9

Lamsfeld-Groß Liebitz D 80 C6
Lamspringe D 79 C7
Lamstedt D 17 A12
Lamu FIN 119 F16
La Mudarra E 39 E10
La Muela E 41 E9
La Mure F 31 F8
Lamure-sur-Azergues F 30 C5
Lana I 72 C3
Lanabukt N 114 D8
Lanaja E 41 E11
Lanaken B 19 C12
Lanarce F 30 F5
Lanark UK 5 D9
La Nava E 51 D6
La Nava de Ricomalillo E 46 E3
La Nava de Santiago E 45 F8
Lancaster UK 10 C6
Lanchester UK 5 F13
Lanciano I 63 C6
Lanckorona PL 147 B9
Lancy CH 31 C9
Landau an der Isar D 75 E12
Landau in der Pfalz D 21 F10
Landbobyn S 102 E7
Lande N 105 A13
Landeck A 71 C11
Landeleau F 22 D4
Landen B 183 D6
Landerneau F 22 D3
Landersheim F 27 C7
Landeryd S 87 A12
Landete E 47 E10
Landévant F 22 E5
Landgraaf NL 20 C6
Landiras F 32 A5
Landivisiau F 22 D3
Landivy F 23 D9
Landkirchen auf Fehmarn
  D 83 B10
Landön S 105 D16
Landos F 30 F4
Landquart CH 71 D9
Landrecies F 19 D8
Landres F 19 F12
Landsberg am Lech D 71 A11
Landsbro S 92 E5
Landscheid D 185 E6
Landshut D 75 E11
Landskrona S 87 D11
Landstuhl D 21 F8
Landvetter S 91 D11
Landzhene BG 169 A9
Lane N 110 B9
Lanesborough IRL 7 E7
Lanesøyra N 112 C4
Lanester F 22 E5
La Neuve-Lyre F 24 C4
La Neuveville CH 31 A11
La Neuville-au-Pont F 25 B12
Laneuville-sur-Meuse F 19 F11
Lâng S 97 D9
Langå DK 86 C5
Långa E 46 B3
Långå S 102 B5
Langa de Duero E 40 E5
Långalma S 99 B10
Långamínne FIN 122 E7
Långan S 105 D16
Langangen N 90 A6
Långås S 87 B10
Långasjö S 89 B8
Langau A 77 E9
Långbäck S 107 A11
Långbäcken S 107 C14
Langballig D 82 A7
Långban S 97 C11
Långbo S 103 C10
Långbo S 103 D10
Langbunes N 114 C9
Langburkersdorf D 80 D6
Langdorf D 76 D4
Langeac F 30 E3
Langeais F 24 F3
Langebæk DK 87 F10
Långed S 107 D18
Langedijk NL 16 C3
Langeid N 90 B2
Längelmäki FIN 127 B12
Langeln D 79 C7
Langelsheim D 79 C7
Langemark B 18 C6
Langemyr N 90 B3
Langen D 17 A11
Langen D 21 C11
Langenaltheim D 75 E8
Langenau D 75 F7
Langenaubach D 185 C9
Langenberg D 17 E10
Langenburg D 183 C10
Langenburg D 74 D6
Langenenslingen D 27 D11
Langenes N 110 B9
Längenfeld A 71 C11
Langenfeld (Rheinland) D 21 B7
Langenhagen D 78 B6
Langenhahn D 185 C8
Langenhorn D 82 A6
Langenleuba-Oberhain D 79 D12
Langenlois A 77 E9
Langenlonsheim D 185 E8
Langennaundorf D 80 C4
Langenneufnach D 71 A11
Langenorla D 79 E10
Langenpreising D 75 F10
Langenselbold D 21 D12
Langenthal CH 27 E8
Langenwang A 148 A5
Langenweddingen D 79 B10
Langenzenn D 75 C8
Langenzersdorf A 77 F10
Langeoog D 17 A8
Langesund N 90 A6
Langevåg N 90 C2
Langevåg N 100 B4
Langewiesen D 79 E8
Langfjordbotn N 112 C9
Langgöns D 21 C11
Langhagen D 83 C12
Langham UK 11 F10
Langhamn N 111 B14
Langhirano I 66 C1
Langholm UK 5 E10
Langhus N 95 C13
Långlöt S 89 B11
Långnäs S 98 A6

Langnau CH 70 D5
Langnes N 111 B15
Langnes N 114 C4
Langnes N 114 D7
Langø DK 83 A10
Langogne F 30 F4
Langoiran F 28 F5
Langon F 32 A5
Langport UK 13 C9
Langreo E 39 B8
Langres F 26 E3
Langschlag A 77 E7
Längsel S 118 B8
Långsele S 107 E12
Långsele S 107 C15
Langset N 95 B14
Langset N 108 D5
Langset N 109 B9
Långshyttan S 97 B15
Långsjöby S 107 A11
Långskog S 103 B12
Langslett N 112 D6
Langstrand N 112 B11
Långsund S 118 D7
Langsur D 186 B2
Langtoft UK 11 C11
Långträsk S 107 A16
Långträsk S 118 C3
Långträsk S 118 D4
Langueux F 22 D6
Languidic F 22 E5
Languilla E 40 F5
Langula D 79 D7
Långvattnet S 107 D11
Långvattnet S 107 D13
Långvattnet S 118 E4
Långviken S 109 E15
Långvikshamn S 107 D15
Långvinds bruk S 103 D13
Langwarden (Butjadingen)
  D 17 A10
Langwathby UK 5 F11
Langwedel D 17 C12
Langwedel D 83 B7
Langweid am Lech D 75 F8
Łania PL 139 F7
Lanivet UK 12 E5
Lanjarón E 53 C9
Lankojärvi FIN 117 E12
Lanloup F 22 C6
Lanmeur F 22 C4
Lanna S 87 A13
Lanna S 92 A5
Länna S 99 C9
Lannach A 148 C4
Lannavaara S 116 B7
Lannemezan F 33 D6
Lannepax F 33 C6
Lannevesi FIN 123 E14
Lannilis F 22 C2
Lannion F 22 C5
Lannoy F 19 C7
Lanobre F 29 E11
La Nocle-Maulaix F 30 B4
Lanouaille F 29 E8
Lánov CZ 81 E9
Lansân S 116 F3
Lansargues F 35 C7
Lansen D 83 C13
Länsikoski FIN 119 C13
Länsikylä FIN 123 E11
Länsiranta FIN 115 B3
Länsiranta FIN 117 D16
Länsi-Teisko FIN 127 B11
Lansjärv S 118 B6
Lanškroun CZ 77 C11
Lanslebourg-Mont-Cenis
  F 31 E10
Lanta F 33 D9
Lantadilla E 40 D3
Lanton F 28 F3
Lantosque F 37 D6
Lanty F 30 B4
La Nucía E 56 D4
Lanulejouls F 33 B10
Lanurile RO 155 C1
Lanusei I 64 D4
Lanuvio I 62 D3
Lanvollon F 22 C6
Łany PL 142 F5
Lányvacska UA 155 B4
Lánycsók H 149 D11
Lanz D 83 D11
Lanzahita E 46 D3
Lanžhot CZ 77 E11
Lanzo Torinese I 31 E11
Laon F 19 E8
Laons F 24 C5
La Paca E 55 D9
La Pacaudière F 30 C4
Lapajärvi FIN 115 D4
Lapalisse F 30 C4
La Pallice F 28 C3
La Palma E 56 F3
La Palma del Condado E 51 E6
La Palme F 34 E4
Lapalud F 35 B8
La Palud-sur-Verdon F 36 D4
Lapanj AL 168 C3
Łapanów PL 144 D1
Lapardha AL 168 C2
La Parra E 51 B7
La Parra de las Vegas E 47 E8
Lapas GR 174 C3
La Pellerine F 23 D9
La Penne F 36 D5
La Peraleja F 47 D7
La Pesga E 45 D8
La Petite-Pierre F 27 C7
La Peza E 55 E6
Lapford UK 13 A8
Lapiejoki FIN 126 C6
La Pinilla E 56 F2
Lapinjärvi FIN 127 D15
Lapinkylä FIN 127 E11
Lapinlahti FIN 124 D8
Lapinneva FIN 123 C9
Lapinsalo FIN 123 C17
La Plagne F 31 E10
La Plaza E 39 B7
Laplume F 33 B6
Lapmežciems LV 134 B7
La Pobla de Claramunt E 43 D7
La Pobla del Duc E 56 D4
La Pobla de Lillet E 43 C7
La Pobla de Montornès E 43 E6
La Pobla de Segur E 42 C5
La Pobla de Vallbona E 48 E3
La Pola de Gordón E 39 C8
La Pommeraye F 23 F10
La Porta F 37 G10
La Portellada E 42 F4
La Portera E 47 F10

Lapoş RO 161 C8
La Pouëze F 23 E10
Lapoutroie F 27 D7
Lapovo SRB 159 E7
Lappajärvi FIN 123 D11
Lappe S 92 A7
Lappea FIN 117 D11
Lappersdorf D 75 D11
Lappetelä FIN 124 D8
Lappfjärd FIN 122 F7
Lappfors FIN 123 C10
Lappi FIN 119 E14
Lappi FIN 121 D13
Lappi FIN 126 C6
Lappila FIN 127 D13
Lappo FIN 126 E5
Lappohja FIN 127 F9
Lappträsk S 119 B11
Lappuluobbal N 113 E13
La Preste F 33 F10
La Proveda de Soria E 41 D7
Läpseki TR 172 D6
Laptevo RUS 138 C7
Lapua FIN 123 E10
La Puebla de Almoradiel E 47 E6
La Puebla de Arganzón E 40 C6
La Puebla de Cazalla E 51 E9
La Puebla de los Infantes E 51 D9
La Puebla del Río E 51 E7
La Puebla de Montalbán E 46 E4
La Puebla de Valdavia E 39 C10
La Puebla de Valverde E 48 D3
La Puerta de Segura E 55 C7
Lăpugiu de Jos RO 151 F10
La Punt CH 71 D9
Lăpuş RO 152 C4
Lăpuşata RO 160 D4
Lăpuşna MD 154 D2
Lăpuşnicu Mare RO 159 D8
Lăpuşnicel RO 159 D9
Łapy PL 140 E7
L'Aquila I 62 C4
Laracha E 38 B2
Laracor IRL 7 E9
Laragh IRL 7 F10
Laragne-Montéglin F 35 B10
La Rambla E 53 A7
Laranueva E 47 C7
La Rasa E 40 E5
La Ravoire F 31 D8
Larbert UK 5 C9
L'Arboç E 43 E7
L'Arbresle F 30 D6
Lårbro S 93 D13
Larceveau-Arros-Cibits F 32 D3
Larchamp F 23 D9
Larchant F 25 D8
Larche F 29 E8
Larche F 36 C5
Lardaro I 69 B10
Lardero E 41 D7
Lardos GR 181 D8
Laredo E 40 B5
La Redorte F 34 D4
Laren NL 16 D4
Laren NL 183 A8
La Réole F 32 A5
Larga MD 153 A9
Largan IRL 6 D5
L'Argentière-la-Bessée F 31 F10
Largoward UK 5 C11
Largs UK 4 D7
Largu RO 161 D10
Lari I 66 E2
Lariano I 62 D3
La Riba E 42 E6
La Riba de Escalote E 40 F6
La Ricamarie F 30 E5
La Riche F 24 F4
La Rinconada E 51 E8
Larino I 63 D7
Larionovo RUS 129 D13
Larisa GR 169 E7
Larkhall UK 5 D9
Larkollen N 95 D13
Larling UK 15 C10
L'Armentera E 43 C10
Larmor-Plage F 22 E5
Larne UK 4 F5
La Robla E 39 C8
La Roca de la Sierra E 45 F7
La Roche CH 31 B11
La Roche-Bernard F 23 F7
La Roche-Canillac F 29 E9
La Roche-Chalais F 28 E6
La Roche-de-Rame F 31 F10
La Roche-Derrien F 22 C5
La Roche-des-Arnauds F 35 A10
La Roche-en-Ardenne B 19 D12
La Rochefoucauld F 29 D6
La Rochelle F 28 C3
Larochemillay F 30 B5
La Roche-Posay F 29 B7
La Rochepot F 30 B6
La Roche-sur-Foron F 31 C9
La Roche-sur-Yon F 28 B3
La Rochette F 31 E9
Larochette L 20 E6
La Roda E 47 F8
La Roda de Andalucía E 53 B7
La Roë F 23 E9
Laroles E 55 E7
La Romana E 56 E3
La Romieu F 33 C6
Laroquebrou F 29 F10
Laroque-d'Olmes F 33 E9
La Roque-Ste-Marguerite F 34 B5
Laroque-Timbaut F 33 B7
La Rouquette F 33 B9
Larraga E 32 E2
Larrasoaina E 32 E2
Larrau F 32 D4
Larrazet F 33 C7
Larressore F 32 D3
Larrión E 32 E1
Larsbo S 97 B14
Larseng N 111 A16
Larsmo FIN 122 C9
Larsnes N 100 B3
Laruns F 32 E5
Laruscade F 28 E4
Larva E 55 D6
Larvik N 90 A7
Larymna GR 175 B7
Lasà I 71 D11
Lasalle F 35 B6
La Salle I 31 D11
La Salle les Alpes F 31 F10

La Salvetat-Peyralès F 33 B10
La Salvetat-sur-Agout F 34 C4
La Salzadella E 48 D5
Läsänkoski FIN 128 B6
Las Arenas E 39 B10
La Savina E 57 D7
Las Berlanas E 46 C3
Låsby DK 86 C5
Las Cabezas de San Juan E 51 F8
Lascari I 58 D4
Las Casas E 48 E4
Lascuarre E 42 C5
La Seca E 39 F10
La Secuita E 43 E6
La Séguinière F 28 A4
La Selva del Camp E 42 E6
La Selve F 33 B11
La Sènia E 42 F4
La Seyne-sur-Mer F 35 D10
Las Herencias E 46 E3
Łasin PL 138 C6
Lasinja HR 148 E5
Łask PL 143 C7
Łaskarzew PL 141 G5
Laško SLO 148 D4
Laskowa PL 144 D1
Laskowice PL 138 D5
Laskowiec PL 139 D12
Las Labores E 46 F5
Laslea RO 152 E5
Las Majadas E 47 D8
Las Menas E 55 E8
Las Mesas E 47 F7
Las Navas de la Concepción E 51 D9
Las Navas del Marqués E 46 C4
Lasne B 182 D4
Las Negras E 55 F9
La Solana E 55 B6
La Souterraine F 29 C8
Lasovo SRB 159 F9
Las Palas E 56 F2
Las Pedroñeras E 47 F7
Las Pedrosas E 41 D10
La Spezia I 69 E8
Las Plassas I 64 D2
Las Rotas E 56 D5
Las Rozas de Madrid E 46 C5
Lassahn D 83 C9
Lassan D 84 C5
Lassay-les-Châteaux F 23 D11
Lassemoen N 105 B13
Lässerud S 96 B12
Lasseube F 32 D5
Lassigny F 18 E6
Lassila FIN 126 B7
Lassing A 73 A9
Lassing A 73 A10
Laßnitzhöhe A 148 B5
Lastebasse I 69 B11
Las Terreras E 55 D8
Las Torres de Cotillas E 56 E2
Lastours F 33 D10
Lastovo HR 162 D2
Lastra a Signa I 66 E3
Lastras de Cuéllar E 40 F3
Lastrup D 17 C9
Lastukoski FIN 125 D10
Lastva BIH 162 D5
Lašva BIH 157 D8
Lasva EST 131 F14
La Suze-sur-Sarthe F 23 E12
La Teste-de-Buch F 32 A3
Lathen D 17 C8
Latheron UK 3 J10
Latheronwheel UK 3 J10
La Thuile I 31 D10
Lathus F 29 C7
Latiano I 61 B9
Latikberg S 107 B12
Latillé F 28 B6
Latina I 62 E3
Latisana I 73 E7
Latorpsbruk S 97 D12
La Torre de Cabdella E 33 F7
La Torre de Esteban Hambrán E 46 D4
La Torre de l'Espanyol E 42 E5
La Tour-Blanche F 29 E6
La Tour-d'Auvergne F 29 D11
Latour-de-France F 34 E4
La Tour-du-Crieu F 33 D9
La Tour-du-Pin F 31 D7
La Tour-sur-Orb F 34 C5
Latowicz PL 141 F5
La Tranche-sur-Mer F 28 C3
La Tremblade F 28 D3
La Trimouille F 29 C7
La Trinité F 37 D6
La Trinité-Porhoët F 22 D6
La Trinité-sur-Mer F 22 E5
Latronico I 60 C6
Latronquière F 29 F10
Lattes F 35 C6
Lattomeri FIN 126 C6
Lattrop NL 183 A9
La Turballe F 22 F5
Latva FIN 119 F15
Latva FIN 120 D7
Latvajärvenperä FIN 121 D11
Laubach D 21 C11
Laubach D 21 D8
Lauben D 71 A10
Laubere LV 135 C10
Laubusch D 80 D5
Lauchhammer D 80 D5
Lauchringen D 27 E9
Lauciene LV 134 B5
Lauda-Königshofen D 74 C6
Lauder UK 5 D11
Lauderi LV 133 D4
Laudio E 40 B6
Laudona LV 135 C12
Laudun F 35 B8
Lauenau D 17 D12
Lauenbrück D 82 D7
Lauenburg (Elbe) D 83 D9
Lauenförde D 21 A12
Lauf D 186 D5

Laufach D 187 A7
Lauf an der Pegnitz D 75 C9
Laufen CH 27 E8
Laufenberg CH 27 E9
Lauffen am Neckar D 187 C7
Laugnac F 33 B7
Lauingen (Donau) D 75 E7
Laujar de Andarax E 55 F7
Laukaa FIN 123 F15
Lauker S 118 C3
Laukka FIN 119 E15
Laukkala FIN 123 D17
Laukkoski FIN 127 E13
Laukkuluspa S 111 E18
Laukna EST 131 D8
Lauksargiai LT 134 F4
Laukslett N 111 A17
Lauksodis LT 135 D8
Laukuva LT 134 F4
Laukvik N 105 C9
Laukvik N 110 D8
Laukvik N 110 E8
Laukvik N 111 A14
Laukvik N 113 C11
Laukvik N 114 B6
Laukžemė LT 134 D2
Laulasmaa EST 131 C8
Launac F 33 C8
La Unión E 56 F3
Launkalne LV 135 B11
Launonen FIN 127 D12
Laupen CH 31 B11
Laupheim D 71 A9
Laupstad N 110 D8
Lauragh IRL 8 E3
Lauraguel F 33 D10
Laureana di Borrello I 59 C9
Laurencekirk UK 5 B12
Laurencetown IRL 6 F6
Laurens F 34 C5
Laurenzana I 60 C5
Lauria I 60 C5
Laurière F 29 C8
Laurieston UK 5 F8
Laurila FIN 119 C13
Laurino I 60 C4
Lauris F 35 C9
Lausanne CH 31 B10
Lauscha D 75 B9
Laußig D 79 C12
Laußnitz D 80 D5
Laussonne F 30 F5
Laussou F 33 A7
Lauta D 80 D6
Lautamaa FIN 119 C22
Lautenbach D 27 C9
Lauter D 79 E12
Lauterbourg F 27 C9
Lauterbrunnen CH 70 D5
Lautere LV 135 C12
Lauterecken D 186 B4
Lauterhofen D 75 D10
Lautersbach (Hessen) D 78 E5
Lauterstein D 187 D8
Lautertal D 75 B8
Lautiosaari FIN 119 C13
Lautrec F 33 C10
Lauttavaara FIN 121 E10
Lauttijärvi FIN 126 B6
Lauve N 90 A7
Lauvsnes N 105 C9
Lauvstad N 100 B3
Lauvuskylä FIN 125 C13
Lauvvik N 94 E4
Lauzerte F 33 B8
Lauzès F 33 A9
Lauzun F 33 A6
Lavagna I 37 C10
Laval F 23 D10
Lavala FIN 121 D14
Lavalette F 33 D10
La Valette-du-Var F 35 D10
La Vall d'Uixó E 48 E4
La Valle I 72 C4
Lavamünd A 73 C10
Lavangen N 111 C14
Lavangsnes N 111 C14
Lavara GR 171 B10
Lavardac F 33 B6
Lavassaare EST 131 D8
Lavaur F 33 C9
Lavaveix-les-Mines F 29 C10
Lavda GR 168 D5
Lavdar i Korçës AL 168 C4
La Vecilla E 39 C9
La Vega E 39 B8
La Vega E 39 B8
La Vega E 39 B9
Lavelanet F 33 E9
La Vellés E 45 B9
Lavello I 60 A5
Laven DK 86 C5
Lavena Ponte Tresa I 69 B6
Lavenham UK 15 C10
Laveno I 68 B6
La Ventosa E 47 D8
La Verdière F 35 C10
La Vernarède F 35 B6
La Verpillière F 31 D7
La Verrie F 28 B4
La Veuve F 25 B11
Lavia FIN 126 B8
Laviano I 60 B4
La Victoria E 53 A7
La Vid E 40 E5
Lavik N 111 B14
Lavik N 100 D2
Lavinio-Lido di Enea I 62 E3
Laviron F 26 F6
Lavis I 69 A11
Lavit F 33 C7
Lavoriškės LT 137 D12
La Voulte-sur-Rhône F 30 F6
Lavoûte-Chilhac F 30 E4
Lavoûte-sur-Loire F 30 E4
Lavradio P 50 B3
Lavre P 50 B3
Lavrio GR 175 D9
Lavsjö S 107 C13
La Wantzenau F 186 D4
Lawers UK 5 B8
Lawford UK 15 D11

Lawrencetown UK 7 D10
Ławsk PL 140 D6
Ławszowa PL 81 D8
Ławy PL 85 E7
Laxå S 92 B5
Laxbäcken S 107 B10
Laxe E 38 B2
Laxey GBM 10 C3
Laxford Bridge UK 2 J6
Laxnäs S 108 E9
Laxne N 112 D9
Laxsjö S 106 D7
Laxvik S 87 B11
Laxviken S 106 D7
Layana E 32 F3
La Yesa E 48 E3
Läyliäinen FIN 127 D11
Layos E 46 E4
Layrac F 33 B7
Laz HR 148 E6
Laza E 38 D5
Laza RO 153 D11
Lazagurría E 32 F1
Lazarata GR 174 B2
Lăzarea RO 153 D7
Lăzăreni RO 151 D9
Lazarevac SRB 158 E5
Lazarev Krst MNE 163 D7
Lazarevo SRB 158 C6
Lazdijai LT 137 E8
Lazdininkai LT 134 E2
Lazdona LV 135 C12
Lazdukalns LV 133 C2
Lazeshchyna UA 152 A4
Lazise I 66 B2
Łaziska PL 141 H5
Łaziska Górne PL 142 F6
Lazkao E 32 D1
Lázně Bělohrad CZ 77 B9
Lázně Bohdaneč CZ 77 B9
Lázně Kynžvart CZ 75 B12
Lazonby UK 5 F11
Lazuri RO 151 B10
Lazuri de Beiuş RO 151 D10
Leabgarrow IRL 6 C5
Leachd Seoirse IRL 6 F5
Leadburn UK 5 D10
Leadenham UK 11 E10
Leadhills UK 5 E9
Leamhcán IRL 7 F10
Leamington Spa, Royal UK 13 A11
Leányfalu H 150 B3
Leap IRL 8 E4
Leasingham UK 11 E11
Leatherhead UK 15 E8
Leavvajohka N 113 D17
Łeba PL 85 A13
Lebach D 21 F7
Lebane SRB 164 D4
Le Ban-St-Martin F 186 C1
Le Barp F 32 A4
Le Bar-sur-Loup F 36 D5
Lebbeke B 19 B9
Le Béage F 30 F5
Le Beausset F 35 D10
Le Bez F 33 C10
Lebedivka UA 154 F6
Lébénymiklós H 149 A8
Lebesby N 113 B18
Le Bez F 33 C10
Lebiaż F 18 C5
Le Biot F 31 C10
Le Blanc F 29 B8
Le Bleymard F 35 B6
Łebno PL 138 B5
Le Bois-d'Oingt F 30 D6
Le Bois-Plage-en-Ré F 28 C3
Le Boréon F 37 C6
Le Boulou F 34 E4
Le Boupère F 28 B4
Le Bourg F 29 F9
Le Bourg-d'Oisans F 31 E9
Le Bourget-du-Lac F 31 D8
Le Bousquet-d'Orb F 34 C5
Le Breuil F 30 C5
Le Breuil F 30 B5
Lebrija E 51 F7
Le Brusquet F 36 C4
Lebucquière F 18 D6
Le Bugue F 29 F7
Le Buisson F 34 A5
Le Buisson-de-Cadouin F 29 F7
Łebunia PL 138 B4
Lebus D 81 B7
Lebusa D 80 C4
Lebyazh'ye RUS 129 F11
Leça da Palmeira P 44 B3
Leça do Bailio P 44 B3
Le Cannet F 36 D6
Le Cannet-des-Maures F 36 E4
Le Castellet F 35 D10
Le Cateau-Cambrésis F 19 D8
Le Catelet F 19 D7
Le Caylar F 34 C5
Le Cayrol F 34 A4
Lecce I 61 C10
Lecco I 69 B7
Lece SRB 164 D4
Lécera E 42 E2
Lech A 71 C10
Lechaina GR 174 D3
Lechaio GR 175 D6
Le Chambon-Feugerolles F 30 E5
Le Chambon-sur-Lignon F 30 E5
Lechaschau A 71 C11
Le Château-d'Oléron F 28 D3
Le Châtelard F 31 D9
Le Châtelet F 29 B10
Le Châtelet-en-Brie F 25 C8
Lechbruck D 71 B11
Le Chêne F 25 D11
Le Chesne F 19 E10
Le Cheylard F 30 F5
Le Cheylas F 31 E8
Lechinta RO 152 C5
Lechlade UK 13 B11
Lechovo GR 169 C5
Lechtingen (Wallenhorst) D 17 D10
Leciñena E 41 E10
Leck D 82 A5
Leckaun IRL 6 D6
Leckmelm UK 2 K6
Le Clapier F 34 C5
Le Collet-de-Dèze F 35 B6

Le Conquet F 22 D2
Le Coteau F 30 C5
Le Coudray-St-Germer F 18 F4
Lécousse F 23 D9
Le Crès F 35 C6
Le Creusot F 30 B5
Lecrín E 53 C9
Le Croisic F 22 F6
Le Crotoy F 18 D4
Lectoure F 33 C7
Łęczna PL 141 H7
Łęczyca PL 143 B7
Łęczyce PL 138 A4
Ledaña E 47 E9
Ledanca E 47 C7
Lēdas LV 134 B5
Ledbury UK 13 A10
Lede B 19 C8
Ledeč nad Sázavou CZ 77 C8
Ledegem B 182 D2
Ledenice BIH 157 C9
Ledenice CZ 77 E7
Lédergues F 33 B10
Ledesma E 45 B8
Lédignan F 35 C7
Ledigos E 39 D10
Ledmore UK 2 J7
Lednica SK 146 C6
Lednice CZ 77 E11
Lednické Rovne SK 146 C6
Le Donjon F 30 C4
Le Dorat F 29 C8
Ledrada E 45 D9
Lędurga LV 135 B9
Ledusjö S 107 D16
Ledyczek PL 85 C11
Lidzin PL 85 B8
Ledžiny PL 143 F7
Leeds UK 11 D8
Leedstown UK 12 E4
Leek UK 11 E7
Leek NL 16 B3
Leeming UK 11 C8
Lee Moor UK 12 E6
Leende NL 16 F5
Leer (Ostfriesland) D 17 B8
Leerdam NL 16 E4
Leersum NL 183 A6
Leese D 17 C12
Leeuwarden NL 16 B5
Leevaku EST 131 E14
Leevi EST 131 F14
Leezen D 83 C8
Leezen D 83 C11
Le Faou F 22 D3
Le Faouët F 22 D5
Le Fenouiller F 28 B2
Leffinge B 182 C1
Lefka GR 176 E5
Lefkes GR 176 E5
Lefkimmi GR 168 F3
Lefkonas GR 169 B10
Lefkopigi GR 169 D6
Lefkothea GR 168 D5
Lefktra GR 175 C7
Le Fleix F 29 F6
Le Folgoët F 22 C3
Leforest F 182 E2
Le Fossat F 33 D8
Le Fousseret F 33 D8
Le Fugeret F 36 C5
Le Fuilet F 23 F10
Léganés E 46 D5
Leganiel E 47 D7
Legau D 71 B10
Le Gault-Soigny F 25 C10
Le Gault-St-Denis F 24 D5
Legazpi E 32 D1
Legde D 83 E11
Legden D 17 D8
Legé F 28 B3
Lège-Cap-Ferret F 28 F3
Le Genest-St-Isle F 23 D10
Leggs UK 7 C7
Legionowo PL 139 F10
Léglise B 19 E12
Legnago I 66 B3
Legnano I 69 B6
Legnica PL 81 D10
Legnickie Pole PL 81 D10
Łęgowo PL 138 B6
Legrad HR 149 D7
Le Grand-Bornand F 31 D9
Le Grand-Bourg F 29 C9
Le Grand-Lemps F 31 E7
Le Grand-Lucé F 24 E3
Le Grand-Pressigny F 29 B7
Le Grand-Quevilly F 18 F3
Le Grand-Serre F 31 E7
Le Grau-du-Roi F 35 C7
Le Gros Theil F 18 F2
Le Gua F 31 E8
Legutiano E 41 C6
Legyesbénye H 145 G3
Le Havre F 23 A12
Lehe D 82 B6
Lehliu RO 161 E9
Lehliu-Gară RO 161 E9
Lehmäjoki FIN 122 D8
Lehmen D 21 D8
Lehmikumpu FIN 119 B14
Lehmo FIN 125 E13
Lehndorf D 79 E11
Lehnice SK 146 E4
Lehnin D 79 B12
Le Hohwald F 27 D7
Léhon F 23 D7
Le Horps F 23 D11
Le Houga F 32 C5
Lehota SK 146 E5
Lehre D 79 B8
Lehrberg D 75 D8
Lehrte D 78 B6
Lehsen D 83 D10
Lehtimäki FIN 123 E11
Lehtiniemi FIN 121 B10
Lehto FIN 121 B14
Lehtopää FIN 119 F13
Lehtovaara FIN 121 D14
Lehtovaara FIN 121 D15
Lehtovaara FIN 124 B9
Lehtovaara FIN 125 D11
Lehtse EST 131 C11
Leianokladi GR 174 B5
Leibertingen D 27 D11
Leiblfing D 75 E11

Leibnitz A 148 C5
Leicester UK 11 F9
Leichlingen (Rheinland) D 183 C10
Leiden NL 16 D3
Leiderdorp NL 16 D3
Leidschendam NL 16 D2
Leie EST 131 E12
Leifear IRL 4 F2
Leiferde D 79 B7
Leigh UK 10 E6
Leighlinbridge IRL 9 C9
Leighton Buzzard UK 15 D7
Leikanger N 100 B3
Leikanger N 100 C5
Léim an Bhradáin IRL 7 F10
Leimani LV 135 D12
Leimen D 21 F11
Leimuiden NL 182 A5
Leinburg D 75 D9
Leine N 101 D13
Leinefelde D 79 D7
Leinelelden-Echterdingen D 187 D7
Leini I 68 C4
Leino FIN 121 D12
Leinzell D 74 E6
Leioa E 40 B6
Leipalingis LT 137 E8
Leipheim D 75 F7
Leipivaara FIN 121 E10
Leipojärvi S 116 D6
Leipzig D 79 D11
Leira N 101 E10
Leira N 104 E4
Leira N 108 D5
Leirado E 38 D3
Leirâmoen N 108 C6
Leiranger N 110 E8
Leirbotn N 113 C11
Leiria P 44 E3
Leirlia N 105 B15
Leiro E 38 D3
Leirvåg N 100 E2
Leirvik FO 2 A3
Leirvik N 94 C2
Leirvik N 100 D2
Leirvik N 105 B8
Leirvika N 108 D6
Leisach A 73 C6
Leisi EST 130 D5
Leisnig D 80 D3
Leißling D 79 D10
Leiston UK 15 C12
Leitir Ceanainn IRL 7 C7
Leitir Mhic an Bhaird IRL 6 C6
Leitrim IRL 6 E6
Leitza E 32 D2
Leitzkau D 79 B10
Leive FIN 119 B14
Leivonmäki FIN 127 B15
Leivset N 108 B9
Leiwen D 21 E7
Leixlip IRL 7 F10
Lejasciems LV 135 B13
Lejazstrazdi LV 134 C6
Leka GR 177 D8
Lekaj AL 168 B2
Lekangen N 111 B13
Lekani GR 171 B7
Lekărovce SK 145 F5
Leknes N 100 B5
Lekowo PL 85 C9
Leksand S 103 E9
Leksdal N 105 D10
Leksvik N 105 D9
Lekunberri E 32 D2
Leland N 108 D4
Le Langon F 28 C4
Le Lardin-St-Lazare F 29 E8
Le Lauzet-Ubaye F 36 C4
Le Lavandou F 36 E4
Leleasca RO 160 D4
Leles SK 145 G5
Lelese RO 151 F10
Leleşti RO 159 C11
Le Leuy F 32 C4
Lelice PL 139 E8
Le Lion-d'Angers F 23 E10
Leliūnai LT 135 F10
Lelkowo PL 139 B9
Lelle EST 131 D9
Le Locle CH 31 A10
Le Louroux-Béconnais F 23 E10
Lelów PL 143 E8
Le Luc F 36 E4
Le Lude F 23 E12
Lelystad NL 16 C4
Lem DK 86 C2
Le Malzieu-Ville F 30 F3
Le Mans F 23 E12
Le Markstein F 27 E7
Le Martinet F 35 B7
Le Mas-d'Agenais F 33 B6
Le Mas-d'Azil F 33 D8
Le Masnau-Massuguiès F 34 C4
Le Massegros F 34 B5
Le Mayet-de-Montagne F 30 C4
Le May-sur-Èvre F 23 F10
Lembach im Mühlkreis A 76 E5
Lembeke B 182 C3
Lemberg D 27 B8
Lemberg F 27 B7
Lembeye F 32 D5
Lembras F 29 F7
Lembruch D 17 C10
Lemele NL 16 D6
Lemelerveld NL 16 D6
Le Mêle-sur-Sarthe F 23 D12
Le Merlerault F 23 C12
Le Mesnil-sur-Oger F 25 C11
Lemetinvaara FIN 121 F12
Lemgo D 17 D11
Lemi FIN 128 C8
Lemland FIN 99 B14
Lemmenjoki FIN 117 A16
Lemmer NL 16 C5
Lemnia RO 153 E8
Le Molay-Littry F 23 B10
Le Monastier F 34 A5
Le Monastier-sur-Gazeille F 30 F4
Le Monêtier-les-Bains F 31 F10
Le Montet F 30 C3

Lempäälä FIN 127 C10
Lempdes F 30 D3
Lempdes F 30 E3
Lempuy FIN 124 C9
Lemu FIN 126 D6
Le Muy F 36 E5
Lemvig DK 86 B2
Lemwerder D 17 B11
Lembybrien IRL 9 D7
Lena N 101 E13
Lenart SLO 148 C5
Lēnas LV 134 C3
Lenauheim RO 150 F6
Lencloître F 29 B6
Lencouacq F 32 B5
Lend A 73 B7
Lendak SK 145 E1
Lendava SLO 149 C6
Lendelede B 182 D2
Lendery RUS 125 D16
Lendinara I 66 B4
Lendum DK 90 E7
Lenė AL 168 B3
Leneshice CZ 76 B5
Le Neubourg F 24 B4
Lengau A 72 B6
Lengau A 76 F4
Lengdorf D 75 F11
Lengefeld D 80 E4
Lengenes N 111 D13
Lengenfeld D 79 E11
Lengenwang D 71 B11
Lengerich D 17 D9
Lengerich D 17 D9
Lenggries D 72 A4
Lengnau CH 27 F7
Lengyeltóti H 149 C9
Lenham UK 15 E10
Lenhovda S 89 B8
Lenine UA 154 C5
Léning F 27 C6
Le Nizan F 32 B5
Lenk CH 31 C11
Lenkimai LT 134 D2
Lenna I 69 B8
Lennartsfors S 96 D9
Lennartsnäs S 99 D9
Lenne D 78 C6
Lennestadt D 21 B10
Lenningen D 27 C11
Lennsjö S 103 B12
Leno I 66 B1
Le Noirmont CH 27 F6
Lenola I 62 E5
Le Nouvion-en-Thiérache F 19 D8
Lens B 19 C8
Lens F 18 D6
Lensahn D 83 B9
Lensvik N 104 D7
Lent F 31 C7
Lent NL 183 B7
Lentas GR 178 F8
Lenti H 149 C7
Lentia I 72 D5
Lentiira FIN 121 F14
Lenting D 75 E9
Lentini I 59 E7
Lentvaris LT 137 D11
Lenvik N 111 C13
Lenzburg CH 27 F9
Lenzen D 83 D10
Lenzing A 76 G5
Lenzkirch D 27 E9
Leoben A 73 B11
Leoben A 73 C8
Leobendorf A 77 F10
Leobersdorf A 77 G10
Leogang A 73 B6
Léognan F 28 F4
Leominster UK 13 A9
León E 39 C8
Léon F 32 C3
Leonberg D 27 C11
Léoncel F 31 F7
Leonding A 76 F6
Leonessa I 62 B3
Leonforte I 58 D5
Leonidio GR 175 E6
Leontari GR 174 A5
Leontari GR 174 E5
Leopoldov SK 146 E5
Leopoldsburg B 19 B11
Leopoldsdorf im Marchfelde A 77 F11
Leopoldshagen D 84 C5
Leopoldshöhe D 17 D11
Leorda RO 153 B8
Leordeni RO 160 D6
Leordina RO 152 B4
Leova MD 154 E2
Lepa FIN 127 C15
Le Palais F 22 F5
Le Palais-sur-Vienne F 29 D8
Le Pallet F 23 F9
Le Parcq F 18 D5
Lepassaare EST 132 F1
Le Passage F 33 B7
Le Pavillon-Ste-Julie F 25 D10
Lepe E 51 E5
Le Péage-de-Roussillon F 30 E6
Le Pellerin F 23 F8
Le Périer F 31 F8
Le Perthus F 34 F4
Le Pertre F 23 D9
Le Petit-Quevilly F 18 F3
Le Pian-Médoc F 28 F4
Le Pin-au-Haras F 23 C12
L'Épine F 35 B10
Le Pin-la-Garenne F 24 D3
Lepistönmäki FIN 123 C10
Lepitsa BG 165 C9
Le Pizou F 28 E6
Le Plan F 33 D8
Le Plessis-Belleville F 25 B8
Le Poët F 35 B10
Lepoglava HR 148 D6
Le Poinçonnet F 29 B9
Le Poiré-sur-Vie F 28 B2
Le Pont-de-Beauvoisin F 31 D8
Le Pont-de-Claix F 31 E8
Le Pont-de-Monvert F 35 B6
Leporano I 61 C8
Le Porge F 28 F3
Le Porge-Océan F 28 F3
Le Portel F 15 F12
Leposavić RKS 163 C10
Leposaviq RKS 163 C10
Le Pouldu F 22 E4
Le Pouliguen F 22 F7
Lepoura GR 175 C9

Le Pouzin *F* 30 F6
Leppäjärvi *FIN* 117 B10
Leppäkoski *FIN* 127 D12
Leppälä *FIN* 113 D19
Leppälä *FIN* 121 D11
Leppälahti *FIN* 124 D8
Leppälähkylä *FIN* 123 E11
Leppälahti *FIN* 125 F12
Leppäselkä *FIN* 123 D16
Leppävesi *FIN* 123 F15
Leppävirta *FIN* 119 C14
Leppiaho *FIN* 119 C15
Leppikoski *FIN* 119 D17
Leppiniemi *FIN* 119 E16
Leppneeme *EST* 131 B9
Lepsala *FIN* 127 C15
Lepsäma *FIN* 127 E12
Lepsény *H* 149 B10
Leptokarya *GR* 169 D8
Le Puy-en-Velay *F* 30 E4
Le Puy-Notre-Dame *F* 23 F11
Le Quesnoy *F* 19 D8
Lequile *I* 61 C10
Ler *N* 104 E8
Le Raincy *F* 25 C8
Léran *F* 33 E9
Lerberget *S* 87 C11
Lercara Friddi *I* 58 D4
Lerdala *S* 91 C14
Léré *F* 25 F8
Le Relecq-Kerhuon *F* 22 D3
Lerești *RO* 160 C6
Lerici *I* 69 E8
Lerín *E* 32 F2
Lerkaka *S* 89 B11
Lerma *E* 40 D4
Lerm-et Musset *F* 32 B5
Lermoos *A* 71 C11
Lerot *S* 96 C7
Le Rouget *F* 29 F10
Lérouville *F* 26 C4
Le Rozier *F* 34 B5
Lerum *S* 91 D11
Le Russey *F* 26 F6
Lervik *N* 95 D13
Lerwick *UK* 3 E14
Lesa *I* 68 B6
Les Abrets *F* 31 D8
Les Aix-d'Angillon *F* 25 F8
Lešak *RKS* 163 C10
Lesaka *E* 32 D2
Les Andelys *F* 18 F3
Les Angles *F* 33 E10
Les Angles *F* 35 C8
Lešani *NMK* 168 B4
Les Arcs *F* 31 D10
Les Arcs *F* 36 E4
Les Aubiers *F* 28 B4
Les Avellanes *E* 42 D5
Les Avenières *F* 31 D8
Les Bondons *F* 35 B6
Les Bordes *F* 25 E7
Les Borges Blanques *E* 42 D5
Les Borges del Camp *E* 42 E6
Les Bouchoux *F* 31 C8
Les Brenets *CH* 31 A10
Lesbury *UK* 5 E13
Les Cabannes *F* 33 E9
L'Escala *E* 43 C10
L'Escale *F* 35 B11
Les Cammazes *F* 33 D10
Lescar *F* 32 D5
L'Escarène *F* 37 D6
Les Cases d'Alcanar *E* 42 F5
Lesce *SLO* 73 D9
Lesconil *F* 22 E3
Les Contamines-Montjoie *F* 31 D10
Les Coves de Vinromà *E* 48 D5
Lescun *F* 32 E4
Lescure-d'Albigeois *F* 33 C10
Les Deux-Alpes *F* 31 E9
Les Diablerets *CH* 31 C11
Les Échelles *F* 31 E8
Les Églisottes-et-Chalaures *F* 28 E5
Le Sel-de-Bretagne *F* 23 E8
Le Sen *F* 32 B4
Le Sentier *CH* 31 B9
Les Éparges *F* 26 B4
Les Epesses *F* 28 B4
Les Escaldes *AND* 33 E9
Les Essards-Taignevaux *F* 31 B7
Les Essarts *F* 28 B3
Le Seu d'Urgell *E* 33 F8
Les Eyzies-de-Tayac-Sireuil *F* 29 F8
Les Fins *F* 31 A10
Les Forges *F* 26 D5
Les Fourgs *F* 31 B9
Les Gets *F* 31 C10
Leshak *RKS* 163 C10
Les Hautes-Rivières *F* 184 E2
Les Herbiers *F* 28 B3
Les Houches *F* 31 D10
Lesichevo *BG* 165 E9
Lesično *SLO* 148 D5
Lesidren *BG* 165 D9
Lesina *I* 63 D8
Les Issambres *F* 36 E5
Lesja *N* 101 B9
Lesjöfors *S* 97 C11
Leskava *LT* 137 D8
Leskelä *FIN* 119 F15
Lesko *PL* 145 E5
Leskoec *NMK* 168 B4
Leskova *SRB* 163 C9
Leskovac *SRB* 164 D4
Leskovik *AL* 168 D4
Leskovo *BG* 161 F11
Les Landes-Genusson *F* 28 B3
Leslie *UK* 5 C10
Les Lucs-sur-Boulogne *F* 28 B3
Lesmahagow *UK* 5 D9
Les Maillys *F* 25 C11
Les Marches *F* 31 E8
Les Martres-de-Veyre *F* 30 D3
Les Matelles *F* 35 C6
Les Mazures *F* 184 E2
Les Mées *F* 35 B10
Les Menuires *F* 31 E10
Lesmont *F* 25 D11
Les Mureaux *F* 24 C6
Leśna *PL* 81 B10
Lesneven *F* 22 C3
Leśnica *PL* 142 F5
Leśnica *SRB* 158 D3
Leśniów Wielki *PL* 81 C8
Lesogorskiy *RUS* 129 C10
Les Ollières-sur-Eyrieux *F* 30 F6

Lesovaara *FIN* 121 D13
Lesovo *BG* 167 F7
Lesparre-Médoc *F* 28 E4
Les Peintures *F* 28 E5
Les Pennes-Mirabeau *F* 35 D9
Lesperon *F* 32 C3
Lespezi *RO* 153 C9
Les Pieux *F* 23 A8
Lespignan *F* 34 D4
Lespinassière *F* 34 D4
Les Planches-en-Montagne *F* 31 B9
Les Planes d'Hostoles *E* 43 C9
L'Espluga Calba *E* 42 E6
L'Espluga de Francolí *E* 42 E6
Les Ponts-de-Cé *F* 23 F10
Les Ponts-de-Martel *CH* 31 A10
Les Preses *E* 43 C8
Lespugue *F* 33 D7
Les Rosiers-sur-Loire *F* 23 F11
Les Rousses *F* 31 C9
Les Sables-d'Olonne *F* 28 C2
Lessach *A* 73 B8
Lessay *F* 23 B8
Lessebo *S* 89 B8
Les Sièges *F* 25 D10
Lessines *B* 19 C8
Les Sorinières *F* 23 F8
Lestene *LV* 134 C6
Les Ternes *F* 30 E3
Les Thuiles *F* 36 C5
Lestijärvi *FIN* 123 C13
Leština *CZ* 77 C11
Lestrade *F* 34 B4
Les Trois-Moutiers *F* 24 F3
Leşu *RO* 152 C5
Lesura *BG* 165 C8
Les Vans *F* 35 B7
Les Vignes *F* 34 B5
Leszczyn *PL* 85 C9
Leszno *PL* 81 C11
Leszno *PL* 141 F3
Leszno Górne *PL* 81 D9
Létavértes *H* 151 C8
Letca *RO* 151 C11
Letçani *RO* 153 C10
Letca Nouă *RO* 161 E12
Letchworth Garden City *UK* 15 D8
Letea Veche *RO* 153 D9
Le Teich *F* 32 A3
Le Teil *F* 35 A8
Le Teilleul *F* 23 C10
Le Temple *F* 28 F4
Letenye *H* 149 D7
Le Theil *F* 24 D4
Le-Theil-de-Bretagne *F* 23 E9
Le Thillot *F* 27 E6
Le Tholy *F* 27 D6
Le Thor *F* 35 C8
Letino *I* 63 E7
Letkés *H* 147 F7
Letnitsa *BG* 165 C10
Letoianni *I* 59 D7
Letonice *CZ* 77 D11
Le Touquet-Paris-Plage *F* 15 F12
Le Touvet *F* 31 E8
Letovice *CZ* 77 C11
L'Étrat *F* 30 E5
Le Trayas *F* 36 E5
Le Tréport *F* 18 D3
Letsbo *S* 103 C10
Letschin *D* 80 A6
Letterfinish *IRL* 8 E3
Letterkenny *IRL* 7 C7
Lettermacaward *IRL* 6 C6
Letterston *UK* 12 B5
Letur *E* 55 C7
Le Tuzan *F* 32 B4
Letzlingen *D* 79 B9
Leu *RO* 160 D6
Leuc *F* 33 D10
Leucate *F* 34 E5
Leuchars *UK* 5 C11
Leuchtenberg *D* 75 C11
Leud *RO* 152 B4
Leuenberg *D* 84 E5
Leuglay *F* 25 E12
Leuk *CH* 68 A4
Leukerbad *CH* 70 E5
Leumrabhagh *UK* 2 J4
Leun *D* 185 C9
Leuna *D* 79 D11
Leusden *NL* 16 D4
Leuşeni *MD* 154 D2
Leutasch *A* 72 B3
Leutenberg *D* 79 E9
Leutershausen *D* 75 D7
Leutesdorf *D* 185 D7
Leutkirch im Allgäu *D* 71 B10
Leuvanjoki *FIN* 119 D15
Leuven *B* 19 C10
Leuze-en-Hainaut *B* 19 C8
Le Val *F* 36 E4
Levan *AL* 168 C2
Levänen *FIN* 128 C8
Levang *N* 105 D10
Levanger *N* 105 D10
Levanjska Varoš *HR* 149 F10
Levanto *FIN* 127 D13
Levanto *I* 37 C11
Leva Reka *BG* 165 C9
Levašjoki *FIN* 126 B6
Levél *H* 146 F4
Levelek *H* 151 H4
Leven *UK* 5 C11
Leven *UK* 11 D11
Levens *F* 37 D6
Levens *UK* 10 C6
Levenwick *UK* 3 F14
Leverano *I* 61 C10
Leverburgh *UK* 2 K2
Le-Verdon-sur-Mer *F* 28 D3
Leverkusen *D* 21 B7
Le Vernet *F* 33 D9
Lèves *F* 24 D5
Levet *F* 29 F10
Levice *SK* 147 E7
Levico Terme *I* 69 A11
Lévignac *F* 33 C8
Lévignacq *F* 32 B3
Levijoki *FIN* 123 D11
Le Ville *I* 66 F5
Levinovac *HR* 149 E8
Le-Vivier-sur-Mer *F* 23 C8
Levka *BG* 166 F6
Levkogeia *GR* 170 B5
Levoča *SK* 145 E2

Levonperä *FIN* 123 C14
Levo-oja *FIN* 119 D16
Levroux *F* 29 B9
Levski *BG* 166 C4
Levunovo *BG* 169 B9
Lewes *UK* 15 F9
Lewin Brzeski *PL* 142 E4
Lewin Kłodzki *PL* 77 B10
Leyburn *UK* 11 C8
Leyland *UK* 10 D6
Leyme *F* 29 F9
Leysdown-on-Sea *UK* 15 E10
Leysin *CH* 31 C11
Ležáky *PL* 144 C5
Lézan *F* 35 B7
Lézat-sur-Lèze *F* 33 D8
Lezay *F* 28 C5
Lezhë *AL* 163 F8
Lézignan-Corbières *F* 34 D4
Lezoux *F* 30 D3
Lezuza *E* 55 B8
Lgota Górna *PL* 143 E7
Lgota Wielka *PL* 143 D7
Lhenice *CZ* 76 D6
Lherm *F* 33 D8
L'Hermenault *F* 28 B4
L'Honor-de-Cos *F* 33 B8
L'Hôpital *F* 186 C2
L'Horme *F* 30 E6
Lhospitalet *F* 33 B8
L'Hospitalet de l'Infant *E* 42 F5
L'Hospitalet de Llobregat *E* 43 E8
L'Hospitalet-du-Larzac *F* 34 C5
L'Hospitalet-près-l'Andorre *F* 33 E9
L'Hostal del Alls *E* 42 F5
L'Huisserie *F* 23 D10
Lhuître *F* 25 C11
Lia *N* 111 C14
Liabøen *N* 104 E4
Liancourt *F* 18 F5
Liapades *GR* 168 E2
Liart *F* 19 E9
Liarvåg *N* 94 D3
Liatorp *S* 88 B6
Liatroim *IRL* 6 D6
Libáň *CZ* 77 C11
Libatse *EST* 131 D8
Libberton *UK* 5 D9
Libčany *CZ* 77 B10
Libčeves *CZ* 76 B5
Liberadz *PL* 139 D9
Liberec *CZ* 81 E8
Liběšice *CZ* 80 E6
Libiąż *PL* 143 F7
Libice nad Cidlinou *CZ* 77 B10
Libina *CZ* 77 C12
Libochovice *CZ* 76 B6
Libofshë *AL* 168 C2
Libohovë *AL* 168 D3
Libouchec *CZ* 80 E6
Libourne *F* 28 F5
Libramont *B* 19 E11
Librazhd *AL* 168 B3
Librilla *E* 55 D10
Librós *E* 47 D10
Licata *I* 58 E4
Licciana Nardi *I* 69 E9
Licenza *I* 62 C3
Liceras *E* 40 F5
Lich *D* 21 C11
Lichfield *UK* 11 F8
Lichnov *CZ* 142 F4
Lichnowy *PL* 138 B6
Lichtaart *B* 182 C5
Lichte *D* 75 B9
Lichtenau *D* 17 E11
Lichtenau *D* 27 C9
Lichtenau im Waldviertel *A* 77 E8
Lichtenberg *D* 75 B9
Lichtenstein *D* 79 E12
Lichtentanne *D* 79 E11
Lichtenvoorde *NL* 17 E7
Lichtenwörth *A* 77 G10
Lichterfelde *D* 84 E5
Lichtervelde *B* 19 B7
Ličí *LV* 135 B9
Ličí *LV* 135 D13
Lička Jesenica *HR* 156 C3
Lički Osik *HR* 156 C3
Licko Lešće *HR* 156 C3
Licodia Eubea *I* 59 E6
Licques *F* 15 F12
Ličupe *LV* 135 C10
Licurici *RO* 160 D3
Lida *BY* 137 F11
Lidečko *CZ* 146 C5
Liden *S* 103 A12
Lidhult *S* 87 B12
Lidingö *S* 99 D10
Lidköping *S* 91 B13
Lido *I* 66 B5
Lido Adriano *I* 66 D5
Lido di Classe *I* 66 D5
Lido di Foce Verde *I* 62 E3
Lido di Jesolo *I* 66 A6
Lido di Metaponto *I* 61 C7
Lido di Ostia *I* 62 D2
Lido di Siponto *I* 63 D9
Lido di Spina *I* 66 C5
Lidoriki *GR* 174 B5
Lidsälen *S* 102 C8
Lidsjöberg *S* 106 C8
Lidzbark *PL* 139 D8
Lidzbark Warmiński *PL* 136 E2
Liebenau *D* 17 C12
Liebenau *D* 17 F12
Liebenburg *D* 79 B7
Liebenfels *A* 73 C9
Liebenwalde *D* 84 E4
Lieberose *D* 80 C6
Liebertwolkwitz *D* 79 D11
Liebling *RO* 159 B7
Lieboch *A* 148 C4
Liedakkala *FIN* 119 C13
Liédena *E* 32 E3
Liedenpohja *FIN* 123 F11
Liège *B* 19 C12
Liegi *LV* 134 C2
Liehittäjä *S* 117 E11
Liekokylä *FIN* 119 C16
Lieksa *FIN* 125 D14
Lielklapari *LV* 135 D11
Lielmēmele *LV* 135 D10
Lielvārde *LV* 135 C9

Lielvircava *LV* 134 C7
Liempde *NL* 183 B6
Lien *S* 105 D15
Lien *S* 107 E10
Lienden *NL* 183 B7
Lienen *D* 17 D9
Lienz *A* 73 C6
Liepa *LV* 135 B10
Liepāja *LV* 134 C2
Liepāre *LV* 134 C7
Liepas *LV* 135 B10
Liepe *D* 84 C4
Liepen *D* 84 C4
Liepene *LV* 134 B3
Liepgarten *D* 84 C6
Liepimäjärvi *FIN* 117 B11
Lieplaukalė *LT* 134 E4
Lieplaukė *LT* 134 E4
Liepna *LV* 133 B2
Lieponys *LT* 137 E10
Liepupe *LV* 135 B8
Lier *B* 19 B10
Lier *N* 95 C12
Lierbyen *N* 95 C12
Liernais *F* 25 F11
Lierneux *B* 19 D12
Liesek *SK* 147 C9
Lieser *D* 21 E8
Lieshout *NL* 183 B6
Lieso *FIN* 127 C12
Liessel *NL* 16 F5
Liestal *CH* 27 F8
Liešťany *SK* 146 D6
Liešti *RO* 161 B11
Lietavská Lúčka *SK* 147 C7
Lietekylä *FIN* 121 E11
Lietor *E* 55 B9
Lievestuore *FIN* 123 F16
Lievoperä *FIN* 119 F16
Liezen *A* 73 A9
Liezère *LV* 135 B12
Liffol-le-Grand *F* 26 D4
Lifford *IRL* 4 F2
Liffré *F* 23 D8
Lifton *UK* 12 D6
Ligardes *F* 33 B6
Lìgatne *LV* 135 B10
Ligist *A* 73 B10
Lignano Pineta *I* 73 E7
Lignano Sabbiadoro *I* 73 E7
Lignan-sur-Orb *F* 34 D5
Ligné *F* 23 F9
Ligneuville *B* 20 D6
Lignières *F* 29 B10
Ligny-en-Barrois *F* 26 D3
Ligny-le-Châtel *F* 25 E10
Ligonchio *I* 66 D1
Ligota *PL* 142 C4
Ligowo *PL* 139 E7
Ligueil *F* 29 A7
Ligugé *F* 29 B6
Lihme *DK* 86 B3
Lihula *EST* 131 D7
Liigvalla *EST* 131 C12
Liikasenvaara *FIN* 121 B14
Liikavaara *S* 116 D6
Liimattala *FIN* 123 E15
Liimattala *FIN* 123 E17
Liipantönkkä *FIN* 123 E16
Liittoperä *FIN* 123 C16
Liiva *EST* 130 D7
Lijar *E* 55 E8
Liješće *BIH* 157 B9
Liješka *BIH* 158 F3
Liljeva Rijeka *MNE* 163 D7
Likavka *SK* 147 C8
Likenäs *S* 102 E5
Līksna *LV* 135 E12
Lilaia *GR* 175 B5
Liland *N* 108 B8
Liland *N* 111 D12
Lilaste *LV* 135 B8
Lilla Edet *S* 91 C11
Lillåfors *S* 118 B6
Lilla Tjärby *S* 87 B12
Lillby *FIN* 123 C10
Lille *B* 16 F3
Lille *F* 19 C7
Lillebo *N* 102 C3
Lillebonne *F* 18 E2
Lillehammer *N* 101 D12
Lillers *F* 18 C5
Lillesand *N* 90 C3
Lilleshall *UK* 11 F7
Lillestrøm *N* 95 C14
Lille Værløse *DK* 87 D10
Lille Skensved *DK* 87 D10
Lillfjäten *S* 102 C7
Lillhamra *S* 102 C7
Lillhärdal *S* 102 C7
Lillholmsjö *S* 105 D16
Lillholmträsk *S* 107 B16
Lilliesleaf *UK* 5 D11
Lillkågeträsk *S* 118 C5
Lillkyrka *S* 98 C8
Lillmörtsjö *S* 103 B11
Lillo *E* 47 E6
Lillögda *S* 107 C13
Lillpite *S* 118 D6
Lillroten *S* 103 A14
Lillsaivis *S* 118 B7
Lillselet *S* 116 E9
Lillsjöhögen *S* 106 E8
Lillskog *S* 103 C9
Lillström *S* 103 B12
Lillviken *S* 109 C11
Lilyache *BG* 165 C8
Lilyanovo *BG* 169 A9
Lima *S* 102 D5
Limana *I* 72 D5
Limankövi *TR* 167 F10
Limanowa *PL* 144 D1
Limanu *RO* 155 F3
Limatola *I* 60 A2
Limavady *UK* 4 E3
Limay *F* 24 B6
Limbach *D* 21 F11
Limbach-Oberfrohna *D* 79 E12
Limbadi *I* 59 B8

Limbaži *LV* 135 A9
Limburg *B* 20 C5
Limburg an der Lahn *D* 21 D10
Limburgerhof *D* 187 C5
Lime *DK* 86 C6
Limedsforsen *S* 102 E5
Limenaria *GR* 171 C7
Limenas *GR* 177 C6
Limenas Chersonisou *GR* 178 E9
Limerick *IRL* 8 C5
Limingen *N* 105 B15
Limingoån *S* 116 E9
Liminka *FIN* 119 E14
Liminpuro *FIN* 120 E9
Limmared *S* 91 D14
Limmen *NL* 16 C3
Limnes *GR* 175 D6
Limni *GR* 175 B7
Limnochori *GR* 169 C6
Limoges *F* 29 D8
Limogne-en-Quercy *F* 33 B9
Limone Piemonte *I* 37 C7
Limonest *F* 30 D6
Limone sul Garda *I* 69 B10
Limosano *I* 63 D7
Limoux *F* 33 D10
Limpias *E* 40 B5
Lin *AL* 168 B4
Lina *S* 116 D4
Lina älv *S* 116 D4
Linakhamari *RUS* 114 D10
Linards *F* 29 D9
Linares *E* 54 C5
Linares de Mora *E* 48 D3
Linares de Riofrío *E* 45 C9
Linaria *GR* 175 A10
Lincent *B* 19 C11
Lincoln *UK* 11 E10
Lindås *N* 100 E2
Lindau *D* 79 B11
Lindau *D* 82 B6
Lindau (Bodensee) *D* 71 B9
Lindberg *S* 102 D4
Linde *DK* 86 C3
Lindelse *DK* 87 F7
Linden *D* 21 C11
Linden *D* 82 B6
Lindenberg *D* 80 B6
Lindenberg *D* 83 D12
Lindenberg im Allgäu *D* 71 B9
Lindenfels *D* 21 E11
Lindern (Oldenburg) *D* 17 C9
Lindesberg *S* 97 C13
Lindesnäs *S* 97 B12
Lindewitt *D* 82 A6
Lindfield *UK* 15 E8
Lindholmen *S* 99 C10
Lindi *EST* 131 E8
Lindknud *DK* 86 D4
Lindkoski *FIN* 127 D15
Lindlar *D* 21 B9
Lindome *S* 91 D11
Lindores *UK* 5 C10
Lindoso *P* 38 E3
Lindow *D* 84 E3
Lindsdal *S* 89 B10
Lindsjön *S* 103 B12
Lindstedt *D* 79 A10
Lindved *DK* 86 D5
Lindwedel *D* 78 A6
Linge *N* 100 B6
Lingen (Ems) *D* 17 C8
Lingenfeld *D* 21 F10
Lingfield *UK* 15 E8
Linghed *S* 103 B12
Linghem *S* 92 C7
Lingolsheim *F* 27 C8
Lingrasmo *N* 111 B18
Linguaglossa *I* 59 D7
Linguizzetta *F* 37 G10
Linhares *P* 44 C5
Linia *PL* 138 B4
Liniewo *PL* 138 B5
Linkmenys *LT* 135 F11
Linköping *S* 92 C7
Linksness *UK* 3 H10
Linkuva *LT* 135 D7
Linlithgow *UK* 5 D9
Linna *EST* 131 E11
Linnamäe *EST* 130 D7
Linnankylä *FIN* 123 F10
Linneryd *S* 89 B8
Linnich *D* 20 C6
Linow *D* 83 D13
Linsburg *D* 17 D12
Linsell *S* 102 B6
Linthal *CH* 71 D8
Linton *UK* 15 C9
Linum *D* 84 E3
Linxe *F* 32 C3
Linz am Rhein *D* 21 C8
Linz *A* 76 F6
Linyola *E* 42 D5
Lion-sur-Mer *F* 23 B11
Liorac-sur-Louyre *F* 29 F7
Lios Mór *IRL* 9 D7
Lios Póil *IRL* 8 D2
Lios Tuathail *IRL* 8 D4
Lipa *BIH* 157 D8
Lipa *PL* 139 D10
Lipany *SK* 145 E2
Lipar *SRB* 158 B4
Lipari *I* 59 C6
Lipawki *BY* 133 E6
Lipcani *MD* 153 A9
Lipce Reymontowskie *PL* 141 G1
Liperi *FIN* 125 E12
Liphook *UK* 15 E7
Lipiany *PL* 85 D7
Lipik *HR* 149 F8
Lipinki *PL* 144 D3
Lipinki Łużyckie *PL* 81 C9
Lipiny *PL* 81 C9
Lipjan *RKS* 164 D3
Lipka *PL* 85 D12
Lipki Wielkie *PL* 85 E9
Lipnica *PL* 85 C12
Lipnica *PL* 138 C5
Lipnica *PL* 143 E8
Lipnica Murowana *PL* 144 D1
Lipnik *PL* 143 E11

Lipník nad Bečvou *CZ* 146 B5
Lipnița *RO* 161 E11
Lipno *PL* 81 C11
Lipno *PL* 139 E7
Lipolist *SRB* 158 D4
Liposthey *F* 32 B4
Lipova *RO* 151 E8
Lipova *RO* 153 C10
Lipová *SK* 146 E6
Lipovac *SRB* 164 B4
Lipová-Lázně *CZ* 77 B12
Lipovăț *RO* 153 C11
Lipovec *CZ* 77 D11
Lipoveni *MD* 154 D3
Lipoveni *MD* 154 D3
Lipovljani *HR* 149 F7
Lipovo *RUS* 136 D5
Lipovo *RUS* 132 C6
Lipovu *RO* 160 E3
Lipowiec *PL* 139 D11
Lipowiec Kościelny *PL* 139 D9
Lipowina *PL* 139 B9
Lippoldsberg (Wahlsberg) *D* 78 C6
Lippstadt *D* 17 E10
Lipsk *PL* 140 C8
Lipsko *PL* 141 H5
Liptál *CZ* 146 C5
Liptovská Kokava *SK* 147 C9
Liptovská Lúžna *SK* 147 C8
Liptovská Osada *SK* 147 D8
Liptovská Teplička *SK* 147 D9
Liptovské Revúce *SK* 147 D8
Liptovský Hrádok *SK* 147 C9
Liptovský Mikuláš *SK* 147 C9
Lipusz *PL* 138 B5
Lipůvka *CZ* 77 D11
Liqenas *AL* 168 C4
Liré *F* 23 F9
Lis *AL* 163 F9
Lisa *RO* 152 F5
Lisa *RO* 160 F6
Lisac *BIH* 157 D8
Lisac *HR* 162 D4
Lisacul *IRL* 6 E5
Lisbane *UK* 7 C11
Lisbellaw *UK* 7 D7
Lisboa *P* 50 B1
Lisburn *UK* 7 C10
Liscannor *IRL* 8 C4
Liscarney *IRL* 6 E3
Liscarroll *IRL* 8 D5
Lisciano Niccone *I* 66 F5
Liscoteanca *RO* 161 C11
Lisdoonvarna *IRL* 6 F4
Liseleje *DK* 87 C10
Lisets *BG* 165 C10
Lisewo *PL* 138 D6
Lisgarode *IRL* 8 C6
Lisgoold *IRL* 8 E6
Lisia Góra *PL* 143 F11
Lisięcice *PL* 142 F4
Lisieux *F* 23 B12
Lisiy Nos *RUS* 129 E13
Lisjö *S* 98 C7
Liskeard *UK* 12 E6
Lisková *SK* 147 C8
Lisków *PL* 142 C5
Lisky *UA* 155 C4
Lisle *F* 4 F3
Lislea *UK* 4 F3
L'Isle-Adam *F* 25 B7
L'Isle-de-Noé *F* 33 D7
L'Isle-d'Espagnac *F* 29 D6
L'Isle-en-Dodon *F* 33 D7
L'Isle-Jourdain *F* 29 C7
L'Isle-Jourdain *F* 33 C8
L'Isle-sur-la-Sorgue *F* 35 C9
L'Isle-sur-le-Doubs *F* 26 F6
L'Isle-sur-Serein *F* 25 E11
Lisle-sur-Tarn *F* 33 C9
Lisma *FIN* 113 E15
Lismakin *IRL* 9 C7
Lismanaapa *FIN* 117 D17
Lismore *UK* 4 B6
Lisnagry *IRL* 8 C5
Lisnakill *IRL* 9 D8
Lisnarrick *UK* 7 D7
Lisnaskea *UK* 7 D8
Lisne *UA* 154 F5
Lišnja *BIH* 157 C8
Lišov *CZ* 77 D7
Lisów *PL* 142 E4
Lispole *IRL* 8 D2
Lisronagh *IRL* 9 D7
Lisryan *IRL* 7 E8
Liss *UK* 15 E7
Lissan *UK* 4 F3
Lisse *NL* 16 D3
Lissendorf *D* 185 D6
Lisskogsbrändan *S* 97 A10
Lissone *I* 69 B7
Lissycasey *IRL* 8 C4
List *D* 86 E2
Listed *DK* 89 E8
Listerby *S* 89 C8
Listerlin *IRL* 9 D8
Listowel *IRL* 8 D4
Listrac-Médoc *F* 28 E4
Listry *IRL* 8 D3
Liszki *PL* 143 F8
Liszkowo *PL* 85 D12
Lit *S* 106 E7
Lița *RO* 160 F5
Litakovo *BG* 165 D8
Litava *SK* 147 E8
Litene *LV* 133 B2
Liteni *RO* 153 B9
Litér *H* 149 B9
Lit-et-Mixe *F* 32 B3
Lith *NL* 183 B6
Lithakia *GR* 174 D2
Lithi *GR* 177 C7
Lithines *GR* 179 E11
Litija *SLO* 73 D10
Litke *H* 147 E9
Litlefjord *N* 113 B14
Litmalahti *FIN* 125 E9
Litmaniemi *FIN* 125 E10
Litochoro *GR* 169 D8
Litoměřice *CZ* 76 B6
Litomyšl *CZ* 77 C10
Litovel *CZ* 77 C12
Litschau *A* 77 E8
Littau *CH* 70 C6
Littleborough *UK* 11 D7
Littlehampton *UK* 15 F8
Littlemill *UK* 13 K9
Little Oakley *UK* 15 D11
Littleport *UK* 15 C9

Littlestone-on-Sea *UK* 15 F10
Littleton *IRL* 9 C7
Littoinen *FIN* 126 E7
Litva *BIH* 157 D10
Litvínov *CZ* 80 E5
Litzendorf *D* 75 C9
Liudvinavas *LT* 136 D6
Liukkunen *FIN* 125 D14
Liŭža *LV* 133 C3
Livada *RO* 145 H7
Livada *RO* 151 B10
Livadaki *GR* 174 D4
Livadeia *GR* 175 C6
Livaderó *GR* 169 D7
Livaderó *GR* 171 B6
Livadi *GR* 169 D7
Livadi *GR* 175 E10
Livadia *GR* 169 B9
Livadia *GR* 175 B7
Livanates *GR* 175 B7
Līvāni *LV* 135 D12
Livari *MNE* 163 E7
Livarot *F* 23 B12
Līvbērze *LV* 134 C6
Livernon *F* 33 B9
Liverpool *UK* 10 E6
Livezeni *RO* 152 D5
Livezi *RO* 160 D3
Livezi *RO* 153 D9
Livezile *RO* 152 C5
Livezile *RO* 159 D10
Līvi *LV* 135 B10
Livigno *I* 71 D10
Livingston *UK* 5 D9
Livno *BIH* 157 E7
Livo *FIN* 119 C17
Livold *SLO* 73 E10
Livonniska *FIN* 121 C10
Livorno *I* 66 E1
Livorno Ferraris *I* 68 C5
Livron-sur-Drôme *F* 30 F6
Liw *PL* 139 F12
Lixing-lès-St-Avold *F* 186 C2
Lixnaw *IRL* 8 D3
Lixouri *GR* 174 C1
Lizard *UK* 12 F4
Lizarraga *E* 32 E1
Lizdeni *LV* 131 F10
Lizums *LV* 135 B12
Lizy-sur-Ourcq *F* 25 B9
Lizzanello *I* 61 C10
Lizzano *I* 61 C9
Ljeskode Vode *BIH* 157 C8
Ljig *SRB* 158 E5
Ljønes *N* 108 B8
Ljørdal *N* 102 D4
Ljosland *N* 90 B1
Ljøsne *N* 100 D7
Ljuberađa *SRB* 164 C5
Ljubija *BIH* 156 C6
Ljubinje *BIH* 162 D5
Ljubiš *SRB* 158 F5
Ljubljana *SLO* 73 D10
Ljubogošta *BIH* 157 E10
Ljubovija *SRB* 158 E3
Ljubuški *BIH* 157 F8
Ljugarn *S* 93 E13
Ljung *S* 91 D13
Ljunga *S* 103 B11
Ljungaverk *S* 103 B11
Ljungby *S* 87 B13
Ljungby *S* 109 F15
Ljungbyhed *S* 87 C12
Ljungbyholm *S* 89 B10
Ljungdalen *S* 102 A4
Ljungsarp *S* 91 D14
Ljungsbro *S* 92 C7
Ljungskile *S* 91 C10
Ljupina *H* 149 F7
Ljusa *BIH* 157 D7
Ljušci Palanka *BIH* 156 C5
Ljusdal *S* 103 C11
Ljusfallshammar *S* 92 B7
Ljusne *S* 103 D13
Ljusnedal *S* 102 A4
Ljustorp *S* 103 A13
Ljusträsk *S* 118 C3
Ljusvattnet *S* 118 E6
Ljuti Dolac *BIH* 157 F8
Ljutomer *SLO* 148 C6
Llabjan *RKS* 164 D3
Llagostera *E* 43 D9
Llamas de la Ribera *E* 39 C8
Llanaelhaearn *UK* 12 A6
Llanarth *UK* 12 A6
Llanarthney *UK* 12 B6
Llanbadarn Fawr *UK* 12 A6
Llanbadrig *UK* 10 E3
Llanbedr *UK* 10 F3
Llanbedrog *UK* 10 F3
Llanberis *UK* 10 E3
Llanbister *UK* 13 A8
Llanblethian *UK* 13 C8
Llançà *E* 34 F5
Llandanwg *UK* 10 F3
Llandderfel *UK* 10 F4
Llandeilo *UK* 12 B7
Llandissilio *UK* 12 B5
Llandovery *UK* 13 B7
Llandrillo *UK* 10 F5
Llandrindod Wells *UK* 13 A8
Llandudno *UK* 10 E4
Llandwrog *UK* 10 F3
Llandysul *UK* 12 A6
Llanegwad *UK* 12 B6
Llaneilian *UK* 10 E3
Llanelli *UK* 12 B6
Llanelltyd *UK* 10 F4
Llanelwy *UK* 10 E5
Llanes *E* 39 B10
Llanfaelog *UK* 10 E3
Llanfair Caereinion *UK* 10 F5
Llanfairfechan *UK* 10 E4
Llanfair Talhaiarn *UK* 10 E4
Llanfairpwllgwyngyll *UK* 10 E4
Llanfihangel-ar-arth *UK* 12 A6
Llanfyllin *UK* 10 F5
Llangadfan *UK* 10 F5
Llangadog *UK* 13 B7
Llangefni *UK* 10 E3
Llangeler *UK* 12 A6
Llangelynin *UK* 10 F3
Llangollen *UK* 10 F5
Llangurig *UK* 13 A7
Llangynidr *UK* 13 B8
Llanidloes *UK* 13 A7
Llanilar *UK* 12 A6
Llanishen *UK* 13 B8

Llanllwchaiarn UK 10 F5
Llanllyfni UK 10 E3
Llannerch-y-medd UK 10 E3
Llan-non UK 12 A6
Llannon UK 12 B6
Llanrhaeadr-ym-Mochnant UK 10 F5
Llanrhidian UK 12 B6
Llanrhystud UK 12 A6
Llanrug UK 10 E3
Llanrumney UK 13 B8
Llanrwst UK 10 E4
Llansanffraid Glan Conwy UK 10 E4
Llansannan UK 10 E4
Llansawel UK 12 A6
Llantilio Pertholey UK 13 B8
Llantrisant UK 13 B8
Llantwit Major UK 13 C8
Llanuwchllyn UK 10 F4
Llanwddyn UK 10 F5
Llanwenog UK 12 A6
Llanwnda UK 10 E3
Llanwnog UK 10 F5
Llanwrtyd Wells UK 13 A7
Llanybydder UK 12 A6
Llapushnik RKS 163 D10
Llardecans E 42 E5
Llaurí E 48 F4
Llavorsí E 33 E8
Llay UK 10 E5
Lledrod UK 12 A7
Lleida E 42 D5
Llera E 51 C7
Llerena E 51 C7
Lliria E 48 E3
Llívia E 33 F9
Llodio E 40 B6
Llombai E 48 F3
Lloret de Mar E 43 D9
Lloseta E 49 E10
Llubí E 57 B11
Llucmajor E 57 C10
Lniano PL 138 C5
Lo B 18 C6
Loamneș RO 152 F4
Loano I 37 C8
Loarre E 32 F4
Löbau D 81 D7
Lobbæk DK 89 E7
Lobe LV 135 C10
Löbejün D 79 C10
Lobera de Onsella E 32 F3
Löbérği LV 131 F12
Łobez PL 85 C9
Lobith NL 183 B8
Löbnitz D 83 B13
Łobodno PL 143 E6
Lobón E 51 B6
Lobonäs S 103 C9
Loburg D 79 B11
Łobzenica PL 85 D12
Locana I 31 E11
Locarno CH 68 A6
Locate di Triulzi I 69 C7
Loccum (Rehburg-Loccum) D 17 D12
Loceri I 64 D4
Lochaline UK 4 B5
Lochau A 71 B9
Lochawe UK 4 C6
Lochboisdale UK 2 L2
Lochcarron UK 2 L5
Lochdon UK 4 C5
Lochearnhead UK 5 C8
Lochem NL 16 D6
Lochen A 76 F4
Lochend UK 3 L8
Loches F 24 F4
Loch Garman IRL 9 D10
Lochgelly UK 5 C10
Lochgilphead UK 4 C6
Lochgoilhead UK 4 C7
Lochinver UK 2 J6
Lochmaben UK 5 E10
Lochmaddy UK 2 K2
Lochovice CZ 76 C5
Łochów PL 139 E12
Lochranza UK 4 D6
Lochristi B 19 B8
Loch Sgioport UK 2 L2
Lociki LV 135 E13
Lockenhaus A 149 B6
Lockerbie UK 5 E10
Lockne S 106 E7
Löcknitz D 84 D6
Locks Heath UK 13 D12
Lockton UK 11 C10
Locmaria-Plouzané F 22 D2
Locmariaquer F 22 E6
Locminé F 22 E6
Locorotondo I 61 B8
Locquirec F 22 C4
Locri I 59 C9
Locronan F 22 D3
Loctudy F 22 E3
Loculi I 64 C4
Löddeköpinge S 87 D12
Lödderitz D 79 C10
Loddin D 84 B6
Lødding N 105 D10
Loddiswell UK 13 E7
Loddon UK 15 B11
Lodè I 64 B4
Lode LV 135 B10
Loděnice CZ 76 B6
Löderup S 88 E6
Lodève F 34 C5
Lodi I 69 C8
Løding N 108 B8
Lødingen N 111 D10
Lodi Vecchio I 69 C7
Lodosa E 32 F1
Lödöse S 91 C11
Łodygowice PL 147 B8
Łódź PL 143 C7
Loeches E 46 D6
Loenen NL 183 A8
Löf D 21 D8
Løfallstrand N 94 B4
Lofer A 73 A6
Löffingen D 27 E9
Lofos GR 169 D7
Lofsdalen S 102 B5
Loftahammar S 93 D9
Lofthus N 94 B5
Lofthus N 111 E10
Loftus UK 11 B10
Log SLO 73 D8
Loga N 94 F5
Logatec SLO 73 E9
Lögda S 107 C14

Lögdeå S 107 D16
Loggerheads UK 11 F7
Loghill IRL 8 C4
Logkanikos GR 174 E5
Logofteni MD 153 B11
Logreşti RO 160 D3
Logron F 24 D5
Logroño E 41 D7
Logrosán E 45 F10
Løgstør DK 86 B4
Løgstrup DK 86 B4
Løgten DK 86 C6
Løgumkloster DK 86 E3
Lohals DK 87 E7
Lohberg D 76 D4
Lohéac F 23 E8
Lohe-Rickelshof D 82 B6
Lohfelden D 78 D6
Lohijärvi FIN 119 B12
Lohiluoma FIN 129 B10
Lohiniva FIN 117 D13
Lohiranta FIN 121 B12
Lohja FIN 127 E11
Lohjan kunta FIN 127 E11
Lohme D 84 A5
Lohmen D 80 B6
Lohmen D 83 C12
Löhnberg D 21 C10
Löhne D 17 D11
Lohne (Oldenburg) D 17 C10
Lohra D 21 C11
Lohr am Main D 74 C6
Lohsa D 81 D7
Lohtaja FIN 123 B11
Lohusuu EST 131 D14
Loiano I 66 D3
Loigny-la-Bataille F 24 D6
Loimaa FIN 127 D9
Loimaan kunta FIN 127 D8
Loiré F 23 E10
Loiri-Porto San Paolo I 64 B3
Loiron F 23 D10
Loisy-sur-Marne F 25 C12
Loitz D 84 C4
Loivos P 38 E5
Loivos do Monte P 44 B5
Loja E 53 B8
Loja LV 135 B9
Løjt Kirkeby DK 86 E4
Loka brunn S 97 C11
Lokakylä FIN 123 D14
Lokalahti FIN 126 D5
Lokavec SLO 73 E8
Lokca SK 147 C8
Loke S 106 F7
Løken N 95 C14
Lokeren B 19 B9
Loket CZ 75 B12
Lokev SLO 73 E8
Lokka FIN 115 C3
Løkken DK 90 E6
Løkken N 104 E7
Løkkipärä FIN 123 C13
Lőkösháza H 151 E7
Lokrume S 93 D13
Loksa EST 131 B11
Løksa N 111 C14
Løksebotn N 111 C14
Lokuta EST 131 D9
Lokuti EST 131 C9
Lokve HR 67 B10
Lokve SLO 73 D8
Lokve SRB 159 C7
Løkvoll N 112 D6
Lolishniy Shepit UA 152 A6
Lollar D 21 C11
L'Ollería E 56 D3
Lom BG 159 F11
Lom N 101 C9
Łomazy PL 141 G8
Lombez F 33 D7
Lombheden S 119 B9
Lomborg DK 86 B2
Lomello I 68 C6
Lomen N 101 D9
Łomianki PL 139 F10
Lomma S 87 D12
Lommatzsch D 80 D4
Lomme F 18 C6
Lommel B 19 B11
Łomnica PL 81 B9
Lomnice CZ 77 D10
Lomnice nad Lužnicí CZ 77 D7
Lomnice nad Popelkou CZ 77 D7
Lomonosov RUS 129 F12
Lompolo FIN 117 B13
Lomsdalen N 101 E12
Lomsjö S 107 C11
Lomsti BG 166 C6
Lomträsk S 109 E18
Lomträsk S 118 C5
Łomża PL 139 D13
Lonato I 66 B1
Lønborg DK 86 D2
Lončari BIH 157 C10
Lončarica HR 149 E8
Londa I 66 E4
Londerzeel B 19 C9
Londinières F 18 E3
London UK 15 D8
Londonderry UK 4 F2
Lone LV 135 D10
Long F 18 D4
Longa GR 174 F4
Longages F 33 D8
Longares E 41 F9
Longarone I 72 D5
Long Ashton UK 13 C9
Long Bennington UK 11 F10
Longbridge Deverill UK 13 C10
Longchaumois F 31 C8
Long Compton UK 13 B11
Long Crendon UK 14 D6
Long Eaton UK 11 F9
Longeau-Percey F 26 E3
Longecourt-en-Plaine F 26 F3
Longerak N 90 B2
Longeville-en-Barrois F 26 F3
Longeville-lès-St-Avold F 186 C2
Longeville-sur-Mer F 28 C3
Longford IRL 7 E8
Longframlington UK 5 E13
Longhope UK 3 H10
Longhorsley UK 5 E13
Longhoughton UK 5 E13
Longi I 58 D5
Long Itchington UK 13 A12
Longlier B 19 E11

Long Melford UK 15 C10
Longmorn UK 3 K10
Longny-au-Perche F 24 C4
Longobardi I 60 E6
Longobucco I 61 E7
Longomel P 44 F5
Longos GR 168 B3
Longos Vales P 38 D3
Long Preston UK 11 C7
Longridge UK 10 D6
Longroiva P 45 C6
Long Stratton UK 15 B11
Long Sutton UK 11 F12
Longton UK 10 D6
Longtown UK 5 E11
Longueau F 18 E5
Longué-Jumelles F 23 F11
Longuenesse F 18 C5
Longueville F 25 C9
Longueville-sur-Scie F 18 E3
Longuyon F 19 F12
Longwood IRL 7 F9
Longwy F 19 E12
Lonigo I 66 B3
Lonin N 95 C10
Löningen D 17 C9
Łoniów PL 143 F12
Lonja HR 149 F7
Lonja HR 149 F8
Lonkan N 110 C9
Lonkka FIN 121 D13
Lonlay-l'Abbaye F 23 C10
Lönneberga S 92 D7
Lonneker NL 183 A9
Lonny F 184 E2
Lons F 32 D5
Lönsboda S 88 C6
Lonsee D 74 E6
Lons-le-Saunier F 31 B8
Lønstrup DK 90 E6
Lontzen B 183 D8
Lónya H 145 G5
Loo EST 131 C9
Loon op Zand NL 16 E4
Loos F 182 D2
Loosdorf A 77 F8
Loose UK 15 E10
Lopadea Nouă RO 152 E3
Lopar HR 67 C10
Lopătari RO 161 C9
Lopate NMK 164 E4
Lopatica NMK 168 B5
Lopatki PL 143 C7
Lopatnic MD 153 A10
Lopatovo RUS 132 F4
Lopcombe Corner UK 13 C11
Lopenka FIN 127 D8
Löpe EST 131 D7
Lopera E 53 A8
Lopik NL 182 B5
Łopiennik Górny PL 144 A7
Lopigna F 37 G9
Lopik NL 182 B5
Loppa N 112 C7
Loppersum NL 17 B7
Loppi FIN 127 D11
Lopra FO 2 C3
Łopuszno PL 143 E9
Lopyan BG 165 D9
Lora N 101 B9
Lora del Río E 51 D8
Loranca de Tajuña E 47 D6
Loràs S 106 E8
Lörby S 88 C7
Lorca E 55 D9
Lorch D 21 D9
Lorch D 74 E6
Lordelo P 38 F4
Lordosa P 44 C5
Lørenskog N 95 C13
Loreo I 66 B5
Loreto I 67 F8
Loreto Aprutino I 62 C5
Lorgues F 36 D4
Lorient F 22 E5
Lorignac F 28 E4
Loriguilla E 48 E3
Lőrinci H 150 B4
Loriol-sur-Drôme F 30 F6
Lormes F 25 F10
Loro Ciuffenna I 66 E4
Loro Piceno I 67 F7
Lorquin F 186 D2
Lorrez-le-Bocage-Préaux F 25 D8
Lorris F 25 E8
Lorsch D 21 E11
Lørslev DK 90 E7
Lörstrand S 103 C11
Lörudden S 103 B14
Lorup D 17 C9
Lorupe LV 135 B9
Los S 103 C9
Losacino E 39 E7
Losa del Obispo E 48 E3
Los Alcázares E 56 F3
Los Algezares E 56 F2
Los Arcos E 32 E1
Losar de la Vera E 45 D9
Los Arenales del Sol E 56 E4
Los Barrios E 53 D6
Los Barrios de Luna E 39 C8
Los Belones E 56 F3
Los Cantareros E 55 D10
Los Corrales E 53 B7
Los Corrales de Buelna E 40 B3
Loscos E 42 E1
Los Dolores E 56 F2
Løset N 101 D14
Los Gallardos E 55 E9
Losheim D 21 E7
Łosiów PL 142 E4
Łosina PL 141 F9
Łosiów PL 142 E4
Los Maldonados E 56 F2
Los Molinos E 46 C4
Lösnich D 21 D7
Losna N 101 D12
Losne F 26 F3
Losomäki FIN 125 D11
Losone CH 68 A6
Losova MD 154 C3
Losovaara FIN 121 E11
Los Palacios y Villafranca E 51 E8
Los Pandos E 40 B4

Los Pedrones E 47 F10
Los Pozuelos de Calatrava E 54 B5
Los Rábanos E 41 E9
Los Royos E 55 D9
Los Santos E 45 C9
Los Santos de Maimona E 51 C7
Loßburg D 27 D9
Losse F 32 B5
Losser NL 17 D7
Lossiemouth UK 3 K10
Lostallo CH 69 A7
Loštice CZ 77 C11
Los Tojos E 40 B3
Lostwithiel UK 12 E5
Los Villares E 53 A9
Los Yébenes E 46 E5
Lőt S 89 A11
Lote N 100 C4
Løten N 101 E14
Løten N 101 E14
Lotenhulle B 182 C2
Loth UK 3 J9
Lothmore UK 3 J9
Lotorp S 92 B7
Lotte D 17 D9
Lottefors S 103 D11
Löttorp S 89 A11
Lottstetten D 27 E10
Lottum NL 183 C8
Lotyń PL 85 C11
Lotzorai I 64 D4
Louannec F 22 C5
Loučeň CZ 77 B8
Loučná nad Desnou CZ 77 B12
Loučovice CZ 76 E5
Loudéac F 22 D6
Loudes F 30 E4
Loudun F 28 A6
Loué F 23 E11
Loue F 23 E11
Louejärvi FIN 119 B14
Louejoki FIN 119 B14
Loughborough UK 11 F9
Loughbrickland UK 7 D10
Lougher IRL 8 D3
Loughgall UK 7 D9
Loughglinn IRL 6 E5
Loughrea IRL 6 F5
Loughton UK 15 D9
Lougratte F 33 A7
Louhans F 31 B7
Louka CZ 146 D4
Loukas GR 174 D5
Loukisia GR 175 C7
Loukkojärvi FIN 119 D15
Loukusa FIN 121 C10
Loulans F 26 F5
Loulay F 28 C4
Loulé P 50 E3
Louny CZ 76 B5
Loupiac F 32 A5
Lourdes F 32 D5
Lourdoueix-St-Pierre F 29 C9
Louriçal P 44 D3
Lourinhã P 44 F2
Lourmarin F 35 C9
Louros GR 174 A2
Loury F 24 E7
Lousã P 44 D4
Lousa P 50 B1
Lousada P 38 F3
Louth UK 11 E11
Louth IRL 7 E9
Loutra GR 174 B3
Loutra GR 177 A8
Loutra Aidipsou GR 175 B7
Loutra Eleftheron GR 170 C6
Loutraki GR 174 B3
Loutraki GR 175 D6
Loutra Kyllinis GR 174 D3
Loutra Smokovou GR 174 A5
Loutra Ypatis GR 174 B5
Loutro GR 169 C7
Loutro GR 169 E6
Loutro GR 174 B3
Loutros GR 171 C10
Louvain B 19 C10
Louvain B 182 D5
Louveigné B 183 D7
Louverné F 23 D10
Louvie-Juzon F 32 D5
Louviers F 18 F3
Louvigné-du-Désert F 23 D9
Louvroil F 19 D8
Louze F 25 D12
Lov DK 87 E9
Lövånger S 118 F6
Lovas HR 157 B11
Lovasberény H 149 B11
Lovászi H 149 C7
Lovászpatona H 149 B9
Lovberg S 106 A7
Lövberga S 107 D9
Lovćenac SRB 158 B4
Lovčica-Trubín SK 147 D7
Løve DK 87 E8
Lovech BG 165 C10
Løvel DK 86 B4
Lovendegem B 19 B8
Lovere I 69 B9
Lovero I 69 A9
Lövestad S 88 D5
Loviisa FIN 127 E15
Lovik N 111 C10
Lovikka S 116 D9
Loviste HR 157 F7
Lovliden S 107 C10
Lövnäs S 102 D5
Lövnäs S 107 B9
Lövnäsvallen S 102 C4
Lovnidol BG 166 D4
Lövő H 149 A7
Lovosice CZ 76 B6
Lovraeid N 94 D4
Lovran HR 67 B9
Lovreč HR 67 B8
Lovreč HR 157 F7
Lovrenc SLO 148 C5
Lovrin RO 159 F6
Lövsele S 118 F6
Lövsjö S 106 B8
Lövstabruk S 99 B10
Lövstalöt S 99 C9
Lövstrand S 107 C10

Lövtjärn S 103 D11
Lovund N 108 D3
Lövvik S 103 A15
Lövvik S 107 C9
Łowcza PL 141 H8
Lowdham UK 11 E9
Löwenberg D 84 E4
Löwenstein D 27 B11
Lower Ballinderry UK 7 C10
Lower Cam UK 13 B10
Lower Kilchattan UK 4 C4
Lowestoft UK 15 C12
Lowick UK 5 D13
Łowicz PL 141 G1
Low Street UK 15 B11
Łowyń PL 81 A9
Loxstedt D 17 B11
Löyä FIN 123 E11
Loyettes F 31 D7
Loymola RUS 129 B16
Löytö FIN 128 B7
Löytökylä FIN 119 D16
Löytövaara FIN 121 C11
Lozarevo BG 167 D7
Lozen BG 165 F7
Lozen BG 166 C5
Lozenets BG 167 E9
Loženo D 17 E8
Lozna SRB 159 F6
Lozna RO 151 C11
Loznica BG 155 F1
Loznitsa BG 161 F9
Loznitsa BG 167 C12
Lozorno SK 77 F12
Lozovac HR 156 E4
Lozovik SRB 159 E7
Lozoya E 46 C5
Lozoyuela E 46 C5
Lozzo di Cadore I 72 D5
Lú IRL 7 E9
Luanco E 39 A8
Luarca E 39 A6
Lubaczów PL 144 C7
Lubań PL 81 D8
Lubāna LV 135 C13
Lubanowo PL 84 D6
Lübars D 79 B11
Lubartów PL 141 H7
Lubasz PL 85 E11
Lubawa PL 139 C8
Lubawka PL 81 E10
Lübbecke D 17 D11
Lübben D 80 C5
Lübbenau D 80 C5
Lübbow D 83 E10
Lubczyna PL 85 C7
Lube LV 134 B4
Lübeck D 83 C9
Ľubeľa SK 147 C8
Lubenec CZ 76 B4
Lubeník SK 145 F1
Lübersac F 29 E8
Lübesse D 83 D10
Lubezere LV 134 B5
Lubiai LT 134 E3
Lubián E 39 D6
Łubianka PL 138 D6
Lubichowo PL 138 C5
Lubicz Dolny PL 138 D6
Lubijcin PL 81 A9
Lubień PL 147 B9
Lubień Kujawski PL 139 F7
Lubieszewo PL 85 D9
Lubiń PL 81 C10
Lubina SK 146 D5
Lubiszyn PL 85 E7
Lubin PL 81 D10
Lublin PL 141 H7
Lubliniec PL 142 E6
Lubmin D 84 B5
Lubnia PL 138 C4
Lubnice PL 143 F11
Lubniewice PL 81 A8
Łubno PL 85 B12
Łubno PL 85 B10
Łubno PL 143 B7
Lubochnia PL 141 G2
Lubomierz PL 81 D9
Lubomino PL 139 B9
Luboń PL 81 B11
Luborzyca PL 143 F9
Łubotín SK 145 E3
Łubowidz PL 139 D8
Łubowo PL 85 B11
Lubraniec PL 138 E6
Lubrín E 55 E8
Lubrza PL 81 B8
Lubrza PL 142 F4
Lubsko PL 81 C7
Lübstorf D 83 C10
Lubstów PL 138 F5
Lubsza PL 142 E4
Lübtheen D 83 D10
Luby CZ 75 B11
Lübz D 83 D12
Luc F 33 B11
Luc F 35 A6
Lucainena de las Torres E 55 E8
Lucan IRL 7 F10
Lučani SRB 158 F5
Lúcar E 55 E8
Luçay-le-Mâle F 24 F5
Lucca I 66 E2
Lucca Sicula I 58 D3
Lucé F 24 D5
Luče SLO 73 D10
Lučenec SK 147 E9
Lucena del Cid E 48 D4
Lucenay-lès-Aix F 30 B3
Lucenay-l'Évêque F 25 F11
Luc-en-Diois F 35 A9
Luceni E 41 E9
Lucera I 63 D8
Lucéram F 37 D6
Lucey F 26 C4
Lucfalva H 147 E10
Luchapt F 29 C7
Luché-Pringé F 23 E12
Lucheux F 18 D5
Lüchow D 83 E10
Luchy F 18 E5
Luciana E 54 B4
Lucien PL 139 F7
Lucieni RO 161 D6

Lucignano I 66 F4
Lucija SLO 67 A8
Lucillo de Somoza E 39 D7
Lucito I 63 D7
Luciu RO 161 D10
Lucka D 79 D11
Luckau D 80 C5
Luckenwalde D 80 B4
Lucksta S 103 B13
Lückstedt D 83 E11
Lúčky SK 147 C8
Luco dei Marsi I 62 D4
Luçon F 28 C3
Lucq-de-Béarn F 32 D4
Luc-sur-Mer F 23 B11
Lucy-le-Bois F 25 E10
Ludag UK 2 L2
Ludanice SK 146 E6
Ludányhalászi H 147 E9
Ludbreg HR 149 D7
Lüdenscheid D 21 B9
Lüderitz D 79 A10
Lüdersdorf D 83 C9
Ludeşti RO 160 D6
Lüdinghausen D 17 E8
Ludlow UK 13 A9
Ludogortsi BG 161 F9
Ludoni RUS 132 E5
Ludoş RO 152 F3
Ludres F 186 D1
Ludoş RO 152 F3
Ludvigsborg S 87 D13
Ludvika S 97 B13
Ludwigsau-Friedlos D 78 E6
Ludwigsburg D 27 C11
Ludwigsfelde D 80 B4
Ludwigshafen D 21 F10
Ludwigshafen am Rhein D 21 F10
Ludwigslust D 83 D10
Ludwigsstadt D 75 B9
Ludwin PL 141 H7
Ludza LV 133 C3
Lüe F 32 B4
Luelmo E 39 F7
Lüerdissen D 78 C6
Luesia E 32 F3
Lueta RO 153 E6
Lug BIH 162 D5
Lug BIH 157 F9
Luga RUS 132 D6
Lugagnano Val d'Arda I 69 D8
Lugano CH 69 A6
Lüganuse EST 131 C14
Lugau D 79 E12
Lügde D 17 E12
Luglon F 32 B4
Lugnano in Teverina I 62 B2
Lugnäs S 91 B14
Lugnvik S 103 A14
Lugnvik S 106 E7
Lugny-lès-Charolles F 30 C5
Lugo E 38 B4
Lugo I 66 D4
Lugo-di-Nazza F 37 G10
Lugoj RO 159 B8
Lugones E 39 B8
Lugos F 32 B4
Lugrin F 31 C10
Lugros E 55 E6
Luhačovice CZ 146 C5
Luhalahti FIN 127 B9
Luhamaa EST 132 F1
Luhanka FIN 127 B14
Luhden D 17 D12
Luhe-Wildenau D 75 C11
Lühmannsdorf D 84 B5
Luhtapohja FIN 125 E15
Luhtikylä FIN 127 D13
Luhy UA 152 A4
Luhy UA 154 A5
Luica RO 161 E9
Luica RO 161 E9
Luik B 19 C12
Luik B 183 D7
Luikonlahti FIN 125 E11
Luimneach IRL 8 C5
Luino I 68 B6
Luintra E 38 D4
Luiro FIN 115 C4
Luiro FIN 115 D10
Luisant F 24 D5
Luizi Călugăra RO 153 D9
Luka BIH 157 F9
Luka SRB 159 E9
Lukácsháza H 149 B7
Luka nad Jihlavou CZ 77 D9
Lukavac BIH 157 C10
Lukavec CZ 77 C7
Lukavica BIH 157 C7
Łukawiec PL 144 C7
Luke NMK 164 E5
Lukeswell IRL 9 D8
Lüki BG 165 F10
Lukivtsi UA 152 A6
Lukkaroistenperä FIN 119 F13
Lukov CZ 146 C5
Lukovë AL 168 E2
Lukovit BG 165 C9
Lukovo HR 67 C10
Lukovo SRB 159 F8
Lukovo SRB 163 C11
Lukovo Šugare HR 67 D11
Lukovytsya UA 153 A8
Łuków PL 141 G6
Łukowa PL 144 C6
Łukowisko PL 141 F7
Lukšiai LT 136 D7
Lukštai LT 135 D11
Luky UA 145 D7
Lula I 64 C3
Luleå S 118 C8
Lüleburgaz TR 173 B7
Lüllemäe EST 131 F12
Lullymore IRL 7 F9
Lümanda EST 130 E4
Lumbier E 32 E3
Lumbrales E 45 C7
Lumbreras E 41 D7
Lumbres F 18 C5
Lumby DK 86 E6
Lumezzane I 69 B9
Lumijoki FIN 119 E14
Lumimetsä FIN 119 F14
Lumina RO 155 E3
Lumio F 37 F9
Lummelunda S 93 D12
Lummen B 19 C11
Lumparland FIN 99 B14

Lumphanan UK 3 L11
Lumpiaque E 41 E9
Lumpzig D 79 E11
Lumsheden S 103 E11
Luna E 41 D10
Luna RO 152 D3
Lunamatrona I 64 D2
Lunano I 66 E5
Lunas F 34 C5
Lunca MD 154 D2
Lunca RO 151 B9
Lunca RO 152 D5
Lunca RO 153 B9
Lunca RO 160 F5
Lunca Banului RO 154 D2
Lunca Bradului RO 152 D6
Lunca Cernii de Jos RO 159 B10
Lunca Corbului RO 160 D5
Lunca de Jos RO 153 D7
Lunca Ilvei RO 152 C5
Lunca de Sus RO 153 D7
Luncaviţa RO 153 D7
Luncaviţa RO 155 C2
Luncoiu de Jos RO 151 E10
Lund DK 86 D3
Lund N 105 B11
Lund N 110 E8
Lund N 111 C15
Lund S 87 D12
Lundamo N 104 E8
Lundbjörken S 102 E8
Lundby DK 86 F7
Lundby DK 87 E9
Lunde DK 86 D2
Lunde DK 86 E6
Lunde N 95 D10
Lunde N 111 C16
Lunde S 103 A14
Lundeborg DK 87 E7
Lundebyvollen N 102 E3
Lundegård N 94 F6
Lunden D 82 B6
Lunderskov DK 86 E4
Lundin Links UK 5 C11
Lundsbrunn S 91 C13
Lundsjön S 106 D7
Lünebach D 20 D6
Lüneburg D 83 D8
Lunel F 35 C7
Lünen D 17 E9
Lunery F 29 B10
Lunéville F 26 C5
Lungani RO 153 C10
Lunger CH 70 D6
Lungro I 60 D6
Lungsjön S 107 E10
Lungern CH 70 D6
Lunguleţu RO 161 D7
Lunino RUS 136 D5
Lunkkaus FIN 115 D3
Lunna BY 140 D11
Lunnäset S 102 A6
Lünne D 17 D8
Lunneborg N 111 B16
Lunow D 84 E6
Lunteren NL 183 A7
Luogosanto I 64 A3
Luohua FIN 119 E14
Luokė LT 134 E5
Luokkala FIN 121 D15
Luola-aapa FIN 119 C15
Luoma-aho FIN 123 D11
Luopa FIN 122 E9
Luopajärvi FIN 123 E10
Luopioinen FIN 127 C12
Luostari RUS 114 E10
Luosto FIN 115 D11
Luotolahti FIN 128 C8
Luovankylä FIN 122 F7
Luovttejohka N 113 C21
Lupac RO 159 C8
Lupandi LV 133 E3
Łupawa PL 85 B12
Lupeni RO 152 E6
Lupeni RO 159 C11
Lupiac F 33 C6
Lupiñén E 41 D10
Lupión E 53 A9
Łupków PL 145 E5
Lupoglav HR 67 B9
Łupowo PL 85 E8
Luppa D 80 D3
Lupopperä FIN 121 D9
Luppy F 26 C5
Lupşa RO 151 E11
Lupşanu RO 161 E9
Luque E 53 A8
Luras I 64 B3
Lurbe-St-Christau F 32 D4
Lurcy-Lévis F 30 B2
Lure F 26 E5
Lurgan UK 7 D10
Lurgan IRL 6 D6
Luri F 37 F10
Lurøy N 108 D4
Lurs F 35 C10
Lurudal N 105 C13
Lury-sur-Arnon F 24 F7
Lusca IRL 7 E10
Lusciano I 60 B2
Luserna San Giovanni I 31 F11
Lushnjë AL 168 C2
Lusi FIN 127 C15
Lusignan F 29 C6
Lusigny-sur-Barse F 25 D11
Lusk IRL 7 F10
Lus-la-Croix-Haute F 35 A10
Lusminki FIN 121 C14
Lusnić BIH 157 E6
Luso P 44 D4
Luspa FIN 116 B7
Luspebryggan S 109 B18
Lussac F 28 F5
Lussac-les-Châteaux F 29 C7
Lussac-les-Églises F 29 C8
Lussan F 35 B7
Lüssow D 83 C12
Lusta UK 2 K3
Lustad N 105 C12
Lustenau A 71 C9
Luster N 100 D6
Lustivere EST 131 D12
Łutcza PL 144 D5
Lutago I 72 C4
Lütau D 83 D8
Lütetsburg D 17 A8
Lutherstadt Wittenberg D 79 C12

Lütjenburg D 83 B9
Lutnes N 102 D4
Lutocin PL 139 E8
Lutomiersk PL 143 C7
Luton UK 15 D8
Lutowiska PL 145 E6
Lutrini LV 134 C4
Lutry PL 136 E2
Luttenberg NL 183 A8
Lutterbach F 27 E7
Lutterworth UK 13 A12
Lututów PL 142 D5
Lützelbach D 21 E12
Lützen D 79 D11
Lutzerath D 21 D8
Lutzingen D 75 E8
Lutzmannsburg A 149 B7
Lützow D 83 C10
Luua EST 131 D13
Luujoki FIN 119 C14
Luukkola FIN 129 C9
Luukkonen FIN 129 B9
Luumäen kk FIN 128 D8
Luumäki FIN 128 D8
Luunja EST 131 E13
Luupujoki FIN 123 C17
Luupuvesi FIN 123 C17
Luusniemi FIN 124 G7
Luusua FIN 115 F2
Luvia FIN 126 C6
Luvos S 109 C16
Lux F 26 F3
Luxembourg L 20 E6
Luxe-Sumberraute F 32 D3
Luxeuil-les-Bains F 26 E5
Luxey F 32 B4
Luyego de Somoza E 39 D7
Luyksgestel NL 183 C6
Luz P 50 E2
Luz P 50 E4
Luz P 51 C5
Luzaga E 47 C8
Lužani HR 157 B8
Luzarches F 25 B7
Luz-Ardiden F 32 E5
Luže CZ 77 C10
Luzech F 33 B8
Lužec nad Vltavou CZ 76 B6
Luzenac F 33 E9
Luzern CH 70 C6
Luzhany UA 153 A7
Luzhki BY 133 F3
Luzianes P 50 D3
Luz i Madh AL 168 B2
Luzino PL 138 A5
Luzmela E 40 B3
Łużna PL 144 D3
Lūžņa LV 130 F3
Lūznava LV 133 D2
Luzón E 47 B8
Luz-St-Sauveur F 32 E6
Luzy F 30 B4
Luzzara I 66 C2
Luzzi I 60 E6
L'viv UA 144 D9
Lwówek PL 81 B10
Lwówek Śląski PL 81 D9
Lyady RUS 132 D4
Lyaskelya RUS 129 B15
Lyaskovets BG 166 C5
Lyavoshki BY 133 E2
Lybokhora UA 145 F6
Lybster UK 3 J10
Lychen D 84 D4
Lycksaberg S 107 A14
Lycksele S 107 B15
Lydd UK 15 F10
Lyderslev DK 87 E10
Lydford UK 12 D6
Lydney UK 13 B9
Lyeninski BY 141 F10
Lyfjord N 111 A16
Lygna N 95 B13
Lykofi GR 171 B10
Lykoporia GR 175 C6
Lyly FIN 127 B11
Lylykylä FIN 121 E11
Lyman UA 155 B5
Lymans'ke UA 154 D5
Lymans'ke UA 155 C2
Lyme Regis UK 13 D9
Lymington UK 13 D11
Lymm UK 10 E7
Lynäs S 103 D12
Lyndhurst UK 13 D11
Lyne DK 86 D2
Lyneham UK 13 B11
Lynemore UK 3 L9
Lyness UK 3 H10
Lyngby DK 87 C7
Lyngdal N 94 F6
Lyngmoen N 112 D5
Lyngså S 86 A2
Lyngseidet N 111 A19
Lyngsnes N 105 B10
Lynmouth UK 13 C7
Lynton UK 13 C7
Lyntupy BY 137 C13
Lyökki FIN 126 D5
Lyon F 30 D6
Lyons-la-Forêt F 18 F3
Lyrestad S 91 B15
Lyrkeia GR 175 D6
Lysabild DK 86 F6
Lysá pod Makytou SK 146 C4
Łyse PL 139 D12
Lysebotn N 94 D5
Lysekil S 91 C9
Lyshchytsy BY 141 F9
Lysice CZ 77 D11
Lysnes N 111 B14
Łysomice PL 138 D6
Lysøysund N 104 D7
Lysroll N 111 D11
Lyss CH 31 A11
Lysvik S 97 B9
Łyszkowice PL 141 G1
Lytchett Minster UK 13 D10
Lytham St Anne's UK 10 D5
Lytovezh UA 144 B9
Lyubashivka UA 154 B6
Lyuben BG 165 E10
Lyubimets BG 166 F6
Lyublino RUS 136 D1
Lyubomyrka UA 154 B4
Lyubyntsi UA 145 E8
Lyulyakovo BG 167 D8

M

Maakeski FIN 127 C13

Maalahti FIN 122 E7
Maalismaa FIN 119 D15
Maam IRL 6 E3
Maaninka FIN 124 D8
Maaninkavaara FIN 115 E3
Maanselkä FIN 125 C10
Maaralanpera FIN 123 C16
Maardu EST 131 B10
Maarheeze NL 183 C7
Maaria FIN 126 D7
Maarianvaara FIN 125 E11
Maarn NL 183 A6
Maarssen NL 16 D4
Maarssenbroek NL 183 A6
Maas IRL 6 C6
Maasbracht NL 19 B12
Maasbree NL 16 F6
Maasdam NL 182 B5
Maaseik B 19 B12
Maaselkä FIN 121 F14
Maasen D 17 C11
Maasland NL 16 E2
Maasmechelen B 19 C12
Maassluis NL 16 E2
Maastricht NL 19 C12
Määttälä FIN 123 C12
Määttälänvaara FIN 121 B14
Maavesi FIN 124 F9
Mablethorpe UK 11 E12
Macael E 55 E8
Maçanet de Cabrenys E 34 F4
Maçanet de la Selva E 43 D9
Mação P 44 E5
Măcăreşti MD 153 C11
Macastre E 48 F3
Maccagno I 68 A6
Macchiagodena I 63 D6
Macclesfield UK 11 E7
Macduff UK 3 K12
Macea RO 151 E7
Maceda E 38 D4
Maceda P 44 C3
Macedo de Cavaleiros P 39 E6
Maceira P 44 C4
Maceira P 44 E3
Macelj HR 148 D5
Macerata I 67 F7
Macerata Feltria I 66 E5
Măceşu de Jos RO 160 F3
Măceşu de Sus RO 160 F3
Machados P 50 C5
Machairas GR 174 B3
Machault F 19 F10
Machecoul F 28 B2
Machelen B 182 D4
Machen UK 13 B8
Machern D 79 D12
Machliny PL 85 D10
Machov CZ 77 A10
Machowa PL 143 F11
Machrihanish UK 4 E5
Machynlleth UK 10 F4
Maciejowice PL 141 G5
Măcin RO 155 C2
Macinaggio I 37 F10
Măciuca RO 160 D4
Mačkatica SRB 164 D5
Mackenbach D 21 F9
Mackenrode D 79 C8
Mačkovci SLO 148 C6
Macomer I 64 C2
Mâcon F 30 C6
Macosquin UK 4 E3
Macotera E 45 C10
Macroom IRL 8 E5
Macugnaga I 68 B4
Mačvanska Mitrovica SRB 158 D4
Mačvanski Pričinović SRB 158 D4
Mád H 145 G3
Madan BG 165 B7
Madan BG 171 A7
Mädan S 103 A15
Madängsholm S 91 C14
Madara BG 167 C8
Madaras H 150 E3
Mădăraş RO 151 D8
Mădărjac RO 153 C10
Maddalena Spiaggia I 64 E3
Maddaloni I 60 A2
Made NL 16 E3
Madekoski FIN 119 E15
Madeley UK 10 F7
Mäder A 71 C9
Madetkoski FIN 115 C1
Madiran F 32 C5
Madley UK 13 A9
Madliena LV 135 C10
Madocsa H 150 D2
Madona LV 135 C12
Madonna di Campiglio I 69 A10
Madrid E 46 D5
Madridejos E 46 F5
Madrigal de las Altas Torres E 45 B11
Madrigal de la Vera E 45 D10
Madrigal del Monte E 40 D4
Madrigalejo E 45 F9
Madrigueras E 47 F9
Madroñera E 45 F9
Mădulari RO 160 D4
Madzharovo BG 171 A9
Mæl N 95 C9
Maël-Carhaix F 22 D5
Maella E 42 E4
Maello E 46 C3
Maenclochog UK 12 B5
Maenza I 62 D4
Mære N 105 D10
Măerişte RO 151 C10
Mäetaguse EST 131 C14
Maeztu E 41 C7
Mafra P 50 B1
Magacela E 51 B8
Magallón E 41 E9
Magalluf E 49 E10
Magaña E 41 E7
Magašici BIH 158 E3
Magaz E 40 E3
Magdala D 79 E9
Magdeburg D 79 B10
Magdeburgerforth D 79 B11
Magenta I 69 C6
Magescq F 32 C3
Mägeşti RO 151 C10
Maggia CH 68 A6
Maghera UK 4 F3
Magherafelt UK 4 F3
Magheralin UK 7 D10
Măgherani RO 152 D5
Maghery UK 7 C9
Maghull UK 10 D5

Magione I 66 F5
Măgiotsa EST 132 E1
Măgireşti RO 153 D9
Magisano I 59 A10
Maglaj BIH 157 C9
Magland F 31 C10
Maglavit RO 159 E11
Magliano de'Marsi I 62 C4
Magliano in Toscana I 65 B4
Magliano Sabina I 62 C2
Maglič SRB 158 F6
Maglie I 61 C10
Maglód H 150 C3
Magnac-Laval F 29 C8
Magné F 28 C4
Magnières F 26 D6
Magnor N 96 C7
Magnuszew PL 141 G4
Magny-Cours F 30 B3
Magny-en-Vexin F 24 B6
Mágocs H 149 D10
Magoula GR 174 E5
Magstrup DK 86 E4
Magueija P 44 B5
Maguilla E 51 C8
Maguiresbridge UK 7 D8
Măgura RO 153 D9
Măgura RO 160 E6
Măgura Ilvei RO 152 C5
Maguré RKS 164 D3
Măgurele RO 161 E8
Măgurele RO 161 E8
Măgureni RO 161 D7
Măguri-Răcătău RO 151 D11
Magy H 145 H4
Magyaralmás H 149 B10
Magyaratád H 149 D9
Magyarbánhegyes H 151 E6
Magyarbóly H 149 E10
Magyaregregy H 149 D10
Magyarhomorog H 151 C8
Magyarkeszi H 149 C10
Magyarnándor H 147 F8
Magyarpolány H 149 B9
Magyarszék H 149 D10
Mahala UA 153 A8
Maheriv UA 144 B8
Mahíde E 39 E7
Mahlberg D 186 E4
Mahlsdorf D 83 E10
Mahlu FIN 123 E14
Mahlwinkel D 79 B10
Mahmudia RO 155 C4
Mahmudiye TR 171 E11
Mahmutköy TR 173 C6
Mahón E 57 B13
Mahora E 47 F9
Mahovo HR 149 E6
Mähring D 75 C12
Mahtra EST 131 C10
Maia P 44 B3
Maials E 42 E5
Măicăneşti RO 161 C10
Maîche F 27 F6
Maida I 59 B9
Maiden Bradley UK 13 C10
Maidenhead UK 15 D7
Maiden Newton UK 13 D9
Maidens UK 4 E7
Maidstone UK 15 E10
Maienfeld CH 71 C9
Maierato I 59 B9
Maierhöfen D 71 B10
Maieru RO 152 C5
Măieruş RO 153 F7
Maigh Chromtha IRL 8 E5
Maigh Cuilinn IRL 6 F4
Maiglean Rátha IRL 7 F8
Maignelay-Montigny F 18 E6
Maijanen FIN 117 D14
Maikammer D 186 C5
Maillas F 32 B4
Maillebois F 24 C5
Maillezais F 28 C4
Mailly-le-Camp F 25 C11
Mailly-le-Château F 25 E10
Mailly-Maillet F 18 D6
Mailovac SRB 159 D7
Mainaschaff D 187 B7
Mainbernheim D 75 C7
Mainburg D 75 E10
Mainham IRL 7 F9
Mainhardt D 74 D6
Mainiemi FIN 127 B16
Mainistir Eimhín IRL 7 F8
Mainistir Fhear Maí IRL 8 D6
Mainistir Laoise IRL 9 C8
Mainistir na Búille IRL 6 E6
Mainistir na Corann IRL 8 E6
Mainistir na Feile IRL 8 D4
Mäinnikkö S 116 D8
Mainsat F 29 C10
Maintenon F 24 C6
Mainua FIN 121 F9
Mainvilliers F 24 D5
Mainz D 21 D10
Maiolati Spontini I 67 E7
Maiorca P 44 D3
Maiorga P 44 E3
Mairena del Alcor E 51 E8
Maisach D 75 F9
Maishofen A 73 B6
Mašiagala LT 137 D11
Maissau A 77 E9
Maisse F 25 D7
Maissin B 184 E3
Maivala FIN 121 B14
Maizières-lès-Metz F 186 C1
Majadahonda E 46 D5
Majadas de Tiétar E 45 E9
Majava FIN 121 C12
Majavatn N 105 A14
Majdan BIH 157 D7
Majdan Królewski PL 144 C4
Majdan Nepierski PL 144 C7
Majdanpek SRB 159 E8
Majovakylä FIN 119 C17
Majs H 149 E11
Majšperk SLO 148 D5
Majtum S 109 D18
Makád H 149 B11
Makarove UA 145 G6
Makarove UA 154 C4
Makarska HR 157 F7
Mäkelänranta FIN 119 C17
Makhnovka RUS 133 B6
Mäkikylä FIN 123 E12
Makkola FIN 125 G11
Makkoshotyka H 145 G4
Makkum NL 16 B4
Maklár H 147 F10
Makljenovac BIH 157 C9
Makó H 150 E5

Makoc RKS 164 D3
Mākoņkalns LV 133 D2
Mąkoszyce PL 142 E4
Makov SK 147 C7
Maków PL 141 G2
Maków PL 141 H4
Makowarsko PL 138 D4
Maków Mazowiecki PL 139 E11
Maków Podhalański PL 147 B9
Makrakomi GR 174 B5
Makresh BG 159 F10
Makri GR 171 C9
Makrisia GR 174 D4
Makrochori GR 169 C7
Makrychori GR 169 E7
Makrygialos GR 169 D8
Makrygialos GR 169 E7
Makrynitsa GR 169 F8
Makryrrachi GR 174 A5
Maksamaa FIN 122 D8
Maksniemi FIN 119 C14
Maksymilianowo PL 138 D5
Malá E 53 B9
Mala IRL 8 D5
Mala S 87 C13
Mălâ S 107 A15
Mala Bosna SRB 150 E4
Mala Čista HR 156 E4
Malacky SK 77 F12
Málaga E 53 C8
Malagón E 46 F5
Malahide IRL 7 F10
Malahvianvaara FIN 121 E14
Malaia RO 160 C4
Mălăieşti MD 154 D5
Mălainn Bhig IRL 6 C5
Mălainn Mhóir IRL 6 C5
Mala Kladuša BIH 156 B4
Malalbergo I 66 C4
Malá Lehota SK 147 D7
Malamocco I 66 B5
Malancourt F 19 F11
Malandrino GR 174 C5
Malanøw PL 142 C5
Malansac F 23 E7
Malaryta BY 141 G10
Mălăska FIN 119 F16
Mala Subotica HR 149 D7
Malaucène F 35 B9
Malaunay F 18 E3
Malaussanne F 32 C5
Mălăvaara FIN 115 E4
Mălăvānnäs S 107 A14
Mała Wieś PL 139 E9
Malax FIN 122 E7
Malaya Byerastavitsa BY 140 D9
Malbekkvatn N 114 C7
Malbork PL 138 B7
Malborn D 21 E7
Malbouzon F 34 A5
Malbuisson F 31 B9
Malcesine I 69 B10
Malchin D 83 C13
Malchow D 83 D12
Malcocinado E 51 C8
Malcov SK 145 E3
Malczyce PL 81 D10
Măldăeni RO 160 E5
Măldăreşti RO 160 C5
Maldegem B 19 B7
Malden NL 16 E5
Maldon UK 15 D10
Małdyty PL 139 C8
Malè I 71 E11
Malechowo PL 85 B11
Malejân E 41 E8
Maleme GR 178 E6
Malemort-du-Comtat F 35 B9
Malemort-sur-Corrèze F 29 E9
Malente D 83 B9
Mălera S 89 B9
Malesco I 68 A6
Malesherbes F 25 D7
Malesina GR 175 B7
Malestroit F 23 E7
Maletto I 59 D6
Malevo BG 166 F5
Malexander S 92 C6
Malfa I 59 C6
Malgersdorf D 75 E12
Malgovik S 107 B10
Malgrat de Mar E 43 D9
Malhadas P 39 E7
Malia GR 178 E9
Malicorne-sur-Sarthe F 23 E11
Maliena LV 133 B2
Mali Idoš SRB 158 B4
Malijai F 36 C4
Mălilla S 92 E7
Mali Lošinj HR 67 C9
Mali Mokryany UA 144 D7
Malin IRL 4 E2
Malin Beg IRL 6 C5
Málinec SK 147 D9
Malines B 19 B9
Malines B 182 C4
Mălini RO 153 C8
Malin More IRL 6 C5
Malinska HR 67 B10
Maliq AL 168 C4
Mālishevë RKS 163 E9
Maliskylä FIN 123 C14
Malissard F 30 F6
Maliuc RO 155 C4
Maljamerica N 111 C13
Mali Zvornik SRB 157 D11
Maljasalmi FIN 125 E11
Malkara TR 173 C6
Małkinia Górna PL 139 E13
Malko Gradishte BG 166 F5
Malko Tŭrnovo BG 167 F9
Mallaig UK 4 A5
Mâllångsbo S 103 D10
Mallén E 41 E9
Mallentin D 83 C10
Mallersdorf D 75 E11
Malles Venosta I 71 D11
Mallica UK 173 F7
Malling DK 86 C6
Mallow UK 8 D5
Mallwyd UK 10 F4
Malm N 105 C9
Malmån S 107 B12
Malmbäck S 92 D4
Malmberget S 116 D5
Malmby S 98 D9
Malmédy B 20 D6

Malmein N 94 E4
Malmesbury UK 13 B10
Malmköping S 93 A9
Malmö S 87 D12
Malmslätt S 92 C7
Malnaş RO 153 E7
Malnes N 110 C8
Malo I 69 B11
Malo Crniče SRB 159 D7
Makri GR 171 C9
Malo Konare BG 165 E9
Małomice PL 81 C8
Malomir BG 167 E7
Malomozhayskoye RUS 136 D3
Malón E 41 E8
Malona GR 181 D8
Malonno I 69 A9
Malorad BG 165 C8
Malošište SRB 164 C4
Malo Titovo BIH 156 D5
Malovăţ RO 159 D10
Mâløy N 100 C2
Maloye-Lugovoye RUS 136 D2
Maloye Sitna BY 133 E6
Malpartida E 45 F8
Malpartida de Cáceres E 45 F8
Malpartida de la Serena E 51 B8
Malpartida de Plasencia E 45 E8
Malpas UK 10 E6
Malpica E 38 B2
Malpica de Tajo E 46 E3
Malpica do Tejo P 44 E6
Mālpils LV 135 B9
Malsch D 27 C9
Mălsnes N 111 B16
Malsfeld D 78 D6
Malšice CZ 77 D7
Mâlsnes N 111 B16
Mălsryd S 91 D13
Mâlsta S 106 E7
Malta A 73 C8
Malta LV 133 D2
Malta P 45 C6
Maltas Trūpi LV 135 D13
Maltby UK 11 E9
Maltby le Marsh UK 11 E12
Maltepe TR 173 C8
Malterdingen D 27 D8
Malters CH 70 C6
Malton UK 11 C10
Malu cu Flori RO 160 C6
Maluenda E 41 F8
Mălūk Izvor BG 166 D5
Malu Mare RO 160 E3
Malung S 102 E6
Malungsfors S 102 E6
Mălupe LV 133 B2
Mălureni RO 160 C5
Măluşteni RO 153 E11
Malva E 39 E9
Malvaglia CH 71 E7
Malveira P 50 B1
Malvik N 105 E9
Malý Horeš SK 145 G4
Mályi H 145 G2
Malý Šariš SK 145 E3
Mally Bereznyy UA 145 F5
Mamaia RO 155 E3
Mamarrosa P 44 D3
Mambrilla de Castejón E 40 E4
Mamer L 20 E6
Mamers F 24 D3
Mamirolle F 26 F5
Mammaste EST 131 E14
Mammendorf D 75 F9
Mammola I 59 C9
Mamoiada I 64 C3
Mamone I 64 B3
Mamonovo RUS 139 B8
Mamuras AL 168 A2
Mamushë RKS 163 E10
Mana MD 154 C3
Maña SK 146 E6
Manacor E 57 B11
Manage B 19 D9
Manamansalo FIN 120 F9
Mañaria E 41 B6
Mânâs S 97 A9
Manasia RO 161 D9
Manasterz PL 144 C5
Manastir BG 165 F10
Mănăstirea RO 161 E9
Mănăstirea Caşin RO 153 E9
Mănăstirea Humorului RO 153 B7
Manastirica SRB 159 D7
Manastirsko BG 167 C7
Mănăştiur RO 151 F9
Mancera de Abajo E 45 C10
Mancha Real E 53 A9
Manchester UK 11 E7
Manching D 75 E9
Manchita E 51 B7
Manciano I 65 B5
Manciet F 33 C6
Mandal N 90 C1
Mândalen N 100 A6
Mandalo GR 169 C7
Mandas I 64 D3
Mandatoriccio I 61 E7
Mandayona E 47 C7
Mandelbachtal-Ormesheim D 186 C3
Mandelieu-la-Napoule F 36 D5
Mandello del Lario I 69 B7
Manderscheid D 21 D7
Mandeure F 27 F6
Mandino Selo BIH 157 E7
Mandli H 145 G5
Mandok H 145 G5
Mandra GR 175 C7
Mândra RO 152 F6
Mandraki GR 181 C6
Mandres-en-Barrois F 26 D3
Mandriko GR 181 D7
Mandritsa BG 171 B10
Manduria I 61 C9
Mane F 35 C10
Mane F 33 D7
Mâneciu RO 161 C7
Manerbio I 66 B1
Mañeru E 32 E2
Mănești RO 160 D6
Mănești RO 161 D7
Manětín CZ 76 C4

Manfredonia I 63 D9
Mangalia RO 155 F3
Manganeses de la Lampreana E 39 E8
Manganeses de la Polvorosa E 39 D8
Mângberg S 102 E8
Mângbyn S 118 F6
Manger N 100 E2
Mangiennes F 19 F12
Mangotsfield UK 13 C9
Mângsbodarna S 102 D6
Mangualde P 44 C5
Manhay B 19 D12
Manhuelles F 26 B4
Mani GR 171 B10
Maniace I 59 D6
Maniago I 73 D6
Maniakoi GR 168 D5
Manieczki PL 81 B11
Manilva E 53 D6
Manisa TR 177 B9
Manises E 48 E4
Manjärvträsk S 118 C4
Manjaur S 107 B16
Mank A 77 F8
Mankarbo S 99 B8
Mankila FIN 119 E15
Mannar B 19 D11
Manna GR 175 D6
Mânnamaa EST 130 D6
Mannersdorf an der Rabnitz A 149 B7
Mannheim D 21 F10
Männikuste EST 131 E8
Manningtree UK 15 D11
Manolada GR 174 C3
Manole BG 165 E10
Manoleasa RO 153 B10
Manoppello I 62 C6
Manorbier UK 12 B5
Manorhamilton IRL 6 D6
Manosque F 35 C10
Manowo PL 85 B10
Manresa E 43 D7
Mânsåsen S 105 E16
Manschnow D 81 A7
Mansfeld D 79 C9
Mansfield UK 11 E9
Mansfield Woodhouse UK 11 E9
Mansilla D 27 D8
Mansilla de las Mulas E 39 D9
Mansle F 29 D6
Manso F 37 G9
Mansonville F 33 B7
Manston UK 13 D10
Mânsträsk S 109 E18
Mantamados GR 171 F10
Mantasia GR 174 A5
Manteigas P 44 D5
Mantel D 75 C11
Manternach L 186 B1
Mantes-la-Jolie F 24 C6
Mantes-la-Ville F 24 C6
Manthelan F 24 F4
Mantila FIN 122 F9
Mantoche F 26 F4
Mantorp S 92 C6
Mantoudi GR 175 B7
Mantova I 66 B2
Mäntsälä FIN 127 D13
Mänttä FIN 123 E13
Mäntyharju FIN 128 C6
Mäntyjärvi FIN 117 D16
Mäntyjärvi FIN 121 B10
Mäntylahti FIN 124 D8
Mäntyluoto FIN 126 B5
Mäntyvaara S 116 E7
Manuel E 48 F3
Manulla IRL 6 E4
Mány H 149 A11
Manyas TR 173 D8
Manzac-sur-Vern F 29 E7
Mânzăleşti RO 161 C9
Manzanal de Arriba E 39 E7
Manzanal del Puerto E 39 C7
Manzanares E 55 A6
Manzanares el Real E 46 C5
Manzaneda E 38 D5
Manzaneque E 46 E5
Manzanilla E 51 E7
Manzano I 73 E7
Manzat F 30 D2
Maó E 57 B13
Mão EST 131 D11
Maoča BIH 157 C10
Maqellarë AL 168 A3
Maqueda E 46 D4
Mar P 38 E2
Mara I 64 C2
Marac F 26 E3
Maracalagonis I 64 E3
Marachkova BY 133 E6
Mărăcineni RO 160 D5
Mărăcineni RO 161 C9
Maradik SRB 158 C5
Maråker S 103 D13
Maranchón E 47 B8
Maranello I 66 D2
Marange-Silvange F 186 C1
Maranhão P 50 A4
Marano I 66 B2
Marano di Napoli I 60 B2
Marano sul Panaro I 66 D2
Marans F 28 C3
Maranville F 25 D12
Mărăşeşti RO 153 F10
Mărăşu RO 155 D1
Maratea I 60 C5
Marateca P 50 B2
Marathea GR 169 E6
Marathias GR 174 C5
Mârâthâ GR 181 C6
Marathokampos GR 177 D8
Marathonas GR 175 C8
Marathopoli GR 174 E4
Maraussan F 34 D5
Maraye-en-Othe F 25 D10
Marazion UK 12 E4
Marazlyiivka UA 154 C5
Marbach CH 70 D5
Marbach am Neckar D 27 C11
Marbäck S 91 D13
Marbella E 53 C7
Marboz F 31 C7
Mârbești RO 160 D6
Marburg an der Lahn D 21 C11
Marby S 105 E16
Marça E 42 E5

Marcali H 149 C8
Marcaltő H 149 B8
Mârčana HR 67 C8
Marcaria I 66 B2
Mărcăuţi MD 153 A10
Marcellina I 62 C3
Marcelová SK 149 A10
Marcenais F 28 E5
Marcenat F 30 E2
March UK 11 F12
Marchamalo E 47 C6
Marchaux F 26 F5
Marche-en-Famenne B 19 D11
Marchegg A 77 F11
Marchena E 51 E9
Marchenoir F 24 E5
Marcheprime F 28 F4
Marchiennes F 19 D7
Marchin B 19 D11
Marchtrenk A 76 F6
Marciac F 33 C6
Marciana I 65 B2
Marciana Marina I 65 B2
Marcianise I 60 A2
Marciano della Chiana I 66 F4
Marciena LV 135 C12
Marcigny F 30 C5
Marcillac-sur-Célé F 33 A9
Marcilla E 32 F2
Marcillac F 28 E4
Marcillac-la-Croisille F 29 E10
Marcillac-Vallon F 33 B10
Marcillat-en-Combraille F 29 C11
Marcilly-en-Gault F 24 F6
Marcilly-en-Villette F 24 E7
Marcilly-le-Hayer F 25 D10
Marcilly-sur-Eure F 24 C5
Marcinkonys LT 137 E9
Marcinkowice PL 85 D10
Marcinkowice PL 144 D2
Marcinowice PL 81 E11
Marciszów PL 81 E10
Marck F 18 C4
Marckolsheim F 27 D8
Marco de Canaveses P 44 B4
Marcoing F 19 D7
Marcòn I 66 A5
Marcoux F 36 C4
Marcq-en-Barœul F 19 C7
Mărculeşti MD 153 B12
Mârdaklev S 87 A11
Mardarivka UA 154 B5
Mardeuil F 25 B11
Mârdsele S 107 B16
Mårdsjö S 106 E9
Mårdsjö S 107 C11
Mårdudden S 118 B6
Marebbe I 72 C4
Marennes F 28 D3
Maresfield UK 15 F9
Mareuil F 29 E6
Mareuil-sur-Arnon F 29 B10
Mareuil-sur-Ay F 25 B11
Mareuil-sur-Lay-Dissais F 28 B3
Marey-sur-Tille F 26 E3
Marga RO 159 B9
Margarites GR 178 E8
Mărgăriteşti RO 160 D4
Margariti GR 168 F3
Margate UK 15 E11
Mârgâu RO 151 D10
Margecany SK 145 F3
Margerie-Hancourt F 25 C12
Margherita di Savoia I 60 A6
Marghita RO 151 F9
Margina RO 151 F9
Marginea RO 153 B7
Mărgineni RO 153 D9
Mărgineni RO 153 D9
Margionys LT 137 F9
Margone I 31 E11
Margonin PL 85 E12
Margraten NL 19 C12
Marguerittes F 35 C7
Margut F 19 E11
Marhaň SK 145 E3
Marià E 35 D8
Mariac F 30 F5
Mariager DK 86 B5
Maria Lankowitz A 73 B11
Maria Luggau A 73 C6
Marialva P 45 C6
Mariampole LV 133 D2
Mariana E 47 D8
Mariannelund S 92 D7
Mariano Comense I 69 B7
Marianopoli I 58 D4
Marianowo PL 85 D8
Mariánské Lázně CZ 75 C12
Mariapfarr A 73 B8
Maria Saal A 73 C9
Mariazell A 148 A4
Maribo DK 83 A10
Maribor SLO 148 C5
Marieberg S 92 A6
Mariefred S 98 D8
Mariehamn FIN 99 B13
Marieholm S 87 D12
Marieholm S 91 E14
Mariembourg B 19 D10
Marienberg D 80 E4
Marienhafe D 17 B8
Marienhagen D 78 B6
Marienheide D 21 B9
Mariental D 79 B8
Mariestad S 91 B14
Marifjøra N 100 D6
Marigliano I 60 B2
Marignana F 37 G9
Marignane F 35 D9
Marigné-Laillé F 24 E3
Marigny F 23 B9
Marigny-le-Châtel F 25 D10
Marijampolė LT 136 D7
Marikaj AL 168 B2
Marikostinovo BG 169 B9
Marín E 38 D2
Marina di Alberese I 65 B4
Marina di Amendolara I 61 D7
Marina di Arbus I 64 E1
Marina di Camerota I 60 C4
Marina di Campo I 65 B2
Marina di Carrara I 69 E8
Marina di Castagneto Donoratico I 66 F2
Marina di Cecina I 66 F1
Marina di Chieuti I 63 D8
Marina di Gioiosa Ionica I 59 C9
Marina di Grosseto I 65 B3
Marina di Leuca I 61 D10
Marina di Massa I 69 E8
Marina di Novaglie I 61 D10

Marina di Palma I 58 E4
Marina di Pulsano I 61 C8
Marina di Ragusa I 59 F6
Marina di Ravenna I 66 C5
Marinaleda E 53 B7
Marina Palmense I 67 F8
Marina Romea I 66 C5
Marina Schiavonea I 61 D7
Marinella I 58 D2
Marineo I 58 D3
Marines E 48 E3
Marines F 24 B6
Maringues F 30 D3
Marinha das Ondas P 44 D3
Marinha Grande P 44 E3
Marinhas P 38 E2
Marinići HR 67 B9
Marinka BG 167 E8
Marinkainen FIN 123 C10
Mārinkalns LV 133 B1
Marino I 62 D3
Mar'insko RUS 132 D4
Mariotto I 61 A7
Mărişelu RO 152 C5
Maritsa GR 181 D8
Marizy F 30 B5
Mārja EST 131 E13
Marjaliza E 46 E5
Mărjamaa EST 131 D8
Marjaniemi FIN 119 D13
Marjokylä FIN 121 E14
Marjonieemi FIN 127 C14
Marjovaara FIN 125 E15
Marjusaari FIN 123 D12
Mark UK 13 C9
Mark S 107 B10
Markaryd S 87 C13
Markaz H 147 F10
Markby FIN 122 D9
Markdorf D 27 E11
Markebäck S 92 B5
Markelo NL 17 D7
Market Deeping UK 11 F11
Market Drayton UK 10 F7
Market Harborough UK 15 C7
Markethill UK 7 D9
Market Rasen UK 11 E11
Market Warsop UK 11 E9
Market Weighton UK 11 D10
Markgröningen D 27 C11
Markhus N 94 C4
Marki PL 139 F11
Markina-Xemein E 41 B7
Markinch UK 5 C10
Märkisch Buchholz D 80 B5
Markitta S 116 D7
Markivka UA 154 D5
Markkina FIN 116 B8
Markkula FIN 123 C12
Marklohe D 17 C12
Marknesse NL 16 C5
Markneukirchen D 75 B11
Markopoulo GR 175 D8
Markovac SRB 159 C7
Markovac SRB 159 E7
Markovac Našićki HR 149 E10
Markovac Trojstveni HR 149 E12
Markowa PL 144 C5
Markranstädt D 79 D11
Marksuhl D 79 E7
Markt Allhau A 148 B6
Marktbergel D 75 D7
Markt Berolzheim D 75 D8
Markt Bibart D 75 C7
Marktbreit D 187 B9
Markt Erlbach D 75 D8
Markt Hartmannsdorf A 148 B5
Marktheidenfeld D 74 C6
Markt Indersdorf D 75 F9
Marktjärn S 103 A11
Marktl D 76 F3
Marktleugast D 75 B10
Marktoberdorf D 71 B11
Marktoffingen D 75 E7
Marktredwitz D 75 B11
Markt Rettenbach D 71 B10
Markt Sankt Martin A 149 A6
Markt Schwaben D 75 F10
Marktseft D 75 C7
Markt Wald D 71 A11
Markusevec HR 148 E6
Markušica HR 149 F11
Markušovce SK 145 F2
Markuszów PL 141 H6
Marl D 17 D10
Marl D 17 D12
Marlborough UK 13 C11
Marldon UK 13 E7
Marle F 19 E8
Marlengo I 72 C3
Marlenheim F 27 C8
Marlhes F 30 E5
Marlishausen D 79 E9
Marlow D 83 B13
Marlow UK 15 D7
Marly CH 31 B11
Marly F 19 D8
Marly F 26 B5
Marma S 99 B8
Marmagne F 25 F7
Marmagne F 30 B5
Marmande F 33 A6
Marmara GR 174 B5
Marmara TR 173 C8
Marmaracik TR 173 B8
Marmaraereğlisi TR 173 C8
Marmari GR 175 C9
Marmaris TR 181 C8
Marmaro P 38 E5
Marmelar P 50 C4
Marmeleira P 44 F3
Marmelete P 50 E2
Marmirolo I 66 B2
Marmolejo E 53 A8
Marmoutier F 27 C7
Marnardal N 90 C2
Marnäs S 103 E10
Marnay F 26 F5
Marne D 82 C6
Marne-la-Vallée F 25 C8
Marnes N 108 B7
Marnheim D 21 E10
Marnitz D 83 D11
Maroldsweisach D 75 B8
Marolles-les-Braults F 23 D12
Maromme I 18 E3
Maroneia GR 171 C9
Maroslele H 150 E5
Marostica I 72 E4
Maroué F 22 D6
Marousi GR 175 C8
Marpingen D 21 F8

Marpissa GR 176 E5
Marple UK 11 E7
Marpod RO 152 F5
Marquartstein D 72 A5
Marquion F 19 D7
Marquise F 15 F12
Marquixanes F 33 E10
Marradi I 66 D4
Marrasjärvi FIN 117 E14
Marraskoski FIN 117 E14
Marrubiu I 64 D2
Marrum NL 16 B5
Mârşa RO 161 E7
Mârşa RO 153 E7
Marsac-en-Livradois F 30 E4
Marsac-sur-Don F 23 E8
Marsaglia I 37 B10
Marsal F 26 C6
Marsala I 58 D1
Mârşani RO 160 E4
Marsanne F 35 A8
Marsberg D 17 F11
Marsciano I 62 B2
Marseillan F 35 D6
Marseille F 35 D9
Marseille-en-Beauvaisis F 18 E4
Marsico Nuovo I 60 C5
Marsico Vetere I 60 C5
Marsillargues F 35 C7
Marsjärn S 118 B8
Mårslet DK 86 C6
Mârsrëni AL 168 C2
Marssac-sur-Tarn F 33 C10
Märsta S 99 C9
Marstal DK 86 F6
Märstetten CH 27 E11
Marston UK 13 B12
Marstrand S 91 D10
Marstrup DK 86 E4
Märsylä FIN 123 C11
Marta I 62 B1
Mårtanberg S 103 E9
Martano I 61 C10
Martel F 29 E8
Martelange B 19 E12
Martellago I 66 A5
Martello I 71 D11
Mártély H 150 E5
Marten BG 161 F8
Martfeld D 17 C12
Martfö H 150 C5
Marthon F 29 D6
Martiago E 45 D8
Martignacco I 73 D7
Martigné-Briand F 23 F11
Martigné-Ferchaud F 23 E8
Martigné-sur-Mayenne F 23 D10
Martigny CH 31 C11
Martigny-le-Comte F 30 B5
Martigny-les-Bains F 26 D4
Martigny-les-Gerbonvaux F 26 D4
Martigues F 35 D9
Martiherrero E 46 C3
Martil MA 53 E6
Martim Longo P 50 E4
Martin SK 147 C7
Martina Franca I 61 B8
Martin Brod BIH 156 C5
Martinchel P 44 E4
Martinci SRB 158 C3
Martinci Čepinski HR 149 E10
Martín de la Jara E 53 B7
Martín del Río E 42 F2
Martín de Yeltes E 45 C8
Martinengo I 69 B8
Mârtineşti RO 151 F11
Martinet E 33 F9
Martinfeld D 79 D7
Martignana P 44 F3
Mârtiniş RO 152 E5
Martín Muñoz de las Posadas E 46 B3
Martinniemi FIN 119 D14
Martino UK 175 B7
Martinporra E 39 B8
Martinsberg A 77 F8
Martinščica HR 67 C9
Martinsicuro I 62 B5
Martis I 64 B2
Martizay F 29 B8
Martley UK 13 A10
Martna EST 130 D7
Martock UK 13 D9
Martonvaara FIN 125 D12
Martorell E 43 E7
Martos E 53 A9
Martres-Tolosane F 33 D8
Mårtsbo S 103 E13
Martti FIN 115 E3
Marttila FIN 127 D8
Marttila FIN 127 D13
Marttisenjärvi FIN 124 C7
Marugán E 46 C4
Maruggio I 61 C9
Marum NL 16 B6
Marum S 99 C12
Mărunţei RO 160 E4
Mărupe LV 135 C8
Maruszów PL 143 D12
Maruszów PL 144 B4
Marvão P 44 F6
Marvejols F 34 A5
Marville F 19 F11
Marxheim D 75 E8
Marxell D 27 C9
Mâry FIN 127 E9
Mar"yanivka UA 154 C6
Marykirk UK 5 B11
Marynnychi UA 152 A6
Marypark UK 3 L10
Maryport UK 5 F10
Marywell UK 5 A11
Marzabotto I 66 D3
Marzahna D 79 B12
Marzahne D 79 A12
Marzamemi I 59 F7
Marzan F 23 E7
Mârzăneşti RO 161 F6
Marzling D 75 F10
Masa E 40 C4
Masainas I 64 E2
Masalavés E 48 F3
Mas-Cabardès F 33 D10
Mascali I 59 D7
Mascalucia I 59 D7
Mascaraque E 46 E5
Mascarenhas P 38 E5
Maschito I 60 B5
Masclat F 29 F8
Mas de Barberáns E 42 F4

Mas de las Matas E 42 F3
Masegosa E 47 C8
Masegoso E 55 B8
Masegoso de Tajuña E 47 C7
Maselheim D 71 A9
Masera I 68 A5
Maserada sul Piave I 72 E5
Masevaux F 27 E6
Masfjorden N 100 E2
Masham UK 11 C8
Masi N 113 E12
Maside E 38 D3
Masi Torello I 66 C4
Maskalyanyaty BY 133 E7
Maskaur S 109 E14
Masku FIN 126 D7
Maslenika HR 156 D4
Maslinica HR 156 F5
Maşloc RO 151 E7
Maslovare BIH 157 C7
Masłówice PL 143 D8
Masone I 37 B9
Masquefa E 43 E7
Massa I 69 E9
Massa Fiscaglia I 66 C5
Massafra I 61 B8
Massa Lombarda I 66 D4
Massa Lubrense I 60 B2
Massamagrell E 48 E4
Massa Marittimo I 65 A3
Massa Martana I 62 B3
Massarosa I 66 E1
Massat F 33 E8
Massay F 24 F6
Maßbach D 75 B7
Masserett F 29 D9
Masseube F 33 D7
Massford UK 7 D10
Massiac F 30 E3
Massiaru EST 131 F9
Massimino I 37 C8
Massing D 75 F12
Mas-St-Chély F 34 B5
Maštácani RO 153 F9
Mastershausen D 185 D7
Masterud N 96 B7
Mastichari GR 177 F9
Mastrevik N 100 E1
Masua I 64 E1
Masugnsbyn S 116 D8
Masullas I 64 D2
Mäsvik N 112 D2
Masyevichy BY 141 G9
Maszewo PL 81 B7
Maszewo PL 85 B13
Maszewo PL 85 D8
Maszewo Duže PL 139 E8
Matabuena E 46 B5
Mataderecske H 147 F10
Matala FIN 119 C14
Matala GR 178 F8
Matalalahti FIN 124 C9
Matalascañas E 51 E6
Matalebreras E 41 E7
Matallana de Valmadrigal E 39 D9
Matamala de Almazán E 41 E6
Matamorosa E 40 C3
Mataporquera E 40 C3
Matapozuelos E 39 F10
Matara FIN 125 D12
Mataragka GR 169 F7
Mataragka GR 174 B3
Mataraselkä FIN 117 C17
Mataró E 43 E8
Mataruge MNE 163 C7
Mataruška Banja SRB 158 F6
Mătăsari RO 159 D11
Mătăsaru RO 161 D6
Mătăsvaara FIN 125 D13
Matca RO 153 F11
Mateeşti RO 160 C3
Matei RO 152 D4
Matelica I 67 F7
Matera I 61 B7
Materija SLO 73 E9
Mateševo MNE 163 D8
Matétszalka H 145 H5
Mateuşti MD 154 C3
Matfors S 103 B13
Matha F 28 D5
Mathay F 27 F6
Mathi I 68 C4
Matignon F 23 C7
Matigny F 18 E7
Matilda FIN 127 E8
Matinlompolo FIN 117 E12
Matino I 61 C10
Matiši LV 131 F10
Matkaniva FIN 119 F14
Matkavaara FIN 121 E12
Matku FIN 127 D10
Matkule LV 134 C5
Matlock UK 11 E8
Matosinhos P 44 B3
Matour F 30 C5
Matraballa H 147 F10
Mátramindszent H 147 F9
Mátraszele H 147 E9
Mátraszőlős H 147 F9
Mátraterenye H 147 F9
Mátraverebély H 147 F9
Matre N 94 C3
Matre N 100 E3
Matrei am Brenner A 72 B3
Matrei in Osttirol A 72 B6
Matsdal S 109 F10
Matsouki GR 174 B3
Matt CH 71 D8
Mattaincourt F 26 D5
Mattersburg A 149 A6
Mattighofen A 76 F4
Mattila S 119 C13
Mattinata I 63 D10
Mattinen FIN 119 B13
Mattisudden S 116 E3
Mattmar S 105 E15
Mattsmyra S 103 D9
Måttsund S 118 C8
Matulji LT 137 E10
Matušškovo SK 146 E5
Matyryna BY 133 F5
Maubert-Fontaine F 19 E9
Maubeuge F 19 D8
Maubourguet F 32 D6
Maubuisson F 28 E3
Mauchline UK 5 D8
Maud UK 3 K12

Mauerbach A 77 F10
Meddersheim D 21 E9
Meddo NL 183 A9
Mede I 68 C6
Medebach D 21 B11
Međeđa BIH 158 F1
Medeikiai LT 135 D9
Medelås S 107 B14
Medellín E 51 B8
Medemblik NL 16 C4
Medena-Selišta BIH 157 D6
Medeno Polje BIH 156 C5
Medenychi UA 145 E7
Međeš BIH 158 D3
Medesano I 66 C2
Medgidia RO 155 D2
Medgyesbodzás H 151 D6
Medgyesegyháza H 151 E7
Medhamn S 91 A14
Mediana E 41 F10
Mediaş RO 152 E4
Medicina I 66 D4
Medieşu Aurit RO 151 B11
Medinaceli E 47 B8
Medina de las Torres E 51 C7
Medina del Campo E 39 F10
Medina de Pomar E 40 C5
Medina de Rioseco E 39 E9
Medina-Sidonia E 52 D5
Medingënai LT 134 E4
Medinilla E 45 D9
Medininkai LT 137 D12
Medinos E 43 E7
Medis P 28 D4
Medkovets BG 159 F11
Medle S 118 E5
Medow CZ 77 C12
Medneva LV 133 B3
Medolla I 66 C3
Medousa GR 171 B8
Medovoye RUS 136 D1
Mjdrzechów PL 143 F10
Medsëdžiai LT 134 E3
Medskog S 103 E12
Medskogsbygget S 102 B5
Medstugan S 105 D12
Medugorje BIH 157 F8
Međuhana SRB 164 C3
Medulin HR 67 C8
Medumajdan HR 156 B5
Medumi LV 135 E12
Meduno I 73 D6
Meduređe SRB 163 B9
Medvedov SK 146 F5
Medveđa SRB 159 E7
Medveđa SRB 164 D4
Medviđa HR 156 D4
Medvode SLO 73 D9
Medyka PL 144 D6
Medze LV 134 C2
Medziibrod SK 147 D8
Medzilaborce SK 145 E4
Meenaclady IRL 6 B6
Meenavean IRL 6 C5
Meer B 182 C5
Meerane D 79 E11
Meerapalu EST 132 E1
Meerbeck D 17 D12
Meerbusch D 17 F7
Meerhout B 183 C6
Meerkerk NL 182 B5
Meerle B 182 C5
Meerlo NL 183 B8
Meersburg D 27 E11
Meerssen NL 19 C12
Meetkerke B 19 B12
Meeuwen B 19 B12
Mefjordvær N 111 A13
Mega Dereio GR 171 B10
Megala Kalyvia GR 169 F6
Megali Panagia GR 169 D10
Megali Volvi GR 169 C9
Megalochori GR 169 E6
Megalo Chorio GR 181 D6
Megalo Chorio GR 177 F9
Megalo Livadi GR 175 E9
Megalopoli GR 174 E5
Megara GR 175 C7
Megarfössta N 105 C13
Megen NL 183 B7
Megève F 31 D10
Megyaszó H 145 G3
Mehadia RO 159 D9
Mehamn N 113 A20
Mehedeby S 99 B8
Mehikoorma EST 132 E1
Méhkerék H 151 D7
Mehren D 21 D7
Mehring D 21 E7
Mehrstetten D 74 F6
Mehtäkylä FIN 119 F12
Mehun-sur-Yèvre F 25 F7
Meidrim UK 12 B6
Meigh UK 7 D10
Meigle UK 5 B10
Meijel NL 16 F5
Meikleour UK 5 B10
Meilen CH 27 F10
Meilhan F 32 C4
Meillac F 23 D8
Meillant F 29 B10
Meillerie F 31 C10
Meimoa P 45 D6
Meina I 68 B6
Meine D 79 B8
Meinersen D 79 B8
Meinerzhagen D 185 B8
Meiningen D 79 E7
Meira E 38 B5
Meiringen CH 70 D6
Meisdorf D 79 C9
Meise B 182 D4
Meisenheim D 21 E9
Meißen D 80 D4
Meißenheim D 186 E4
Meitingen D 75 E8
Meix-devant-Virton B 19 E11
Mejlby DK 86 C6
Mejorada E 46 D3
Mejorada del Campo E 46 D6
Mejrup DK 86 C3
Meka Gruda BIH 157 F9
Mekinje SLO 73 D10
Mel I 72 D5
Melalahti FIN 121 F10
Melampes GR 178 E8
Melanthi GR 168 D5
Melay F 25 F10
Melay F 26 E4
Melazzo I 37 B8

Melbä N 111 C11
Melbourn UK 15 C9
Melbu N 110 C8
Melby UK 3 E13
Melchow D 84 E5
Meldal N 104 E7
Meldola I 66 D5
Meldorf D 82 B6
Mele I 37 C9
Melegnano I 69 C7
Melenci SRB 158 B5
Melendugno I 61 C10
Melfi I 60 B5
Melfjordbotn N 108 C6
Melgaço P 38 D3
Melgar de Arriba E 39 D9
Melgar de Fernamental E 40 D3
Melgar de Tera E 39 E7
Melgven F 22 E4
Melhus N 104 E8
Melia GR 169 E7
Meliana E 48 E4
Melick NL 20 B6
Melicucco I 59 C9
Mélida E 32 F2
Melide E 38 C4
Melides P 50 C2
Meligalas GR 174 E4
Meliki GR 169 D7
Melilli I 59 E7
Melineşti RO 160 D3
Mélisey F 26 E6
Mélissa I 61 D10
Melissa I 61 B8
Melissano I 61 D10
Melissochori GR 175 C7
Melissopetra GR 168 D4
Melitaia GR 174 A5
Meliti GR 169 C6
Melito di Porto Salvo I 59 D8
Melivoia GR 169 E8
Melivoia GR 171 B7
Melk A 77 F8
Melkadalen N 111 D12
Melkefoss N 114 E7
Melksham UK 13 C10
Mellac F 22 E5
Mellajärvi FIN 119 B13
Mellakoski FIN 119 B13
Mellanfjärden S 103 C13
Mellansel S 107 E14
Mellansjö S 103 A10
Mellanström S 109 E15
Mellau A 71 C9
Mellbystrand S 87 B11
Melle B 19 C8
Melle D 17 D10
Melle F 28 C5
Mellen D 83 D11
Mellenbach-Glasbach D 79 E9
Mellendorf (Wedemark) D 78 A6
Mellensee D 80 B4
Mellerud S 91 B11
Mellerup DK 86 B6
Mellgård S 103 A9
Mellilä FIN 127 D8
Mellin D 79 A8
Mellingen D 79 E9
Mellösa S 93 A9
Mellrichstadt D 75 B7
Melmerby UK 5 F11
Melnbārži LV 135 B11
Melnica BG 166 E6
Melnica SRB 159 E7
Mělník CZ 77 B6
Mel'nikovo RUS 129 D12
Melnitsa BG 167 E7
Melnsils LV 130 F5
Melón E 38 D3
Meloy N 110 C9
Meløysund N 108 C6
Melrand F 22 E5
Melres P 44 B4
Melrose UK 5 D11
Mels CH 71 C8
Melsdorf D 83 B8
Melsungen D 78 D6
Melsvik N 112 C10
Meltaus FIN 117 E14
Meltham UK 11 D8
Melton UK 15 C11
Melton Mowbray UK 11 F10
Meltosjärvi FIN 117 E13
Melun F 25 C8
Melvaig UK 2 K5
Melvich UK 3 H9
Mélykút N 150 E3
Melzo I 69 C7
Memaliaj AL 168 D2
Membrey F 26 E5
Membribio E 55 B6
Membrío E 45 E7
Mëmele LV 135 D10
Memleben D 79 D9
Memmelsdorf D 75 C8
Memmingen D 71 B10
Memmingerberg D 71 B10
Memsie UK 3 K12
Menaggio I 69 A7
Menai Bridge UK 10 E3
Menaldum NL 16 B5
Menars F 24 E5
Menasalbas E 46 E4
Menat F 30 C2
Mendavia E 32 F1
Mende F 34 B5
Menden (Sauerland) D 17 F9
Mendenitsa GR 175 B6
Menderes TR 177 C9
Mendig D 21 D8
Mendiga P 44 E3
Mendigorría E 32 E2
Menditte F 32 D4
Mendrisio CH 69 B6
Ménéac F 22 D6
Menem TR 177 B9
Menen B 19 C7
Menesjärvi FIN 117 A16
Ménesplet F 29 A6
Ménessaire F 25 F11
Menet F 29 E11
Menetes GR 181 F6
Menetou-Salon F 25 F7
Menfi I 58 D2
Mengabril E 51 B8
Mengele LV 135 C10
Mengen D 27 E11
Mengerskirchen D 21 C10
Mengeš SLO 73 D10
Mengíbar E 53 A9
Menidi GR 174 A3

Ménigoute F 28 C5
Ménil-la-Tour F 26 C4
Ménil-sur-Belvitte F 26 D6
Menin B 19 C7
Menin B 182 D2
Menkijärvi FIN 123 E11
Menkulas AL 168 C4
Mennecy F 25 C7
Mennetou-sur-Cher F 24 F6
Menou F 25 F9
Mens F 31 F8
Mensignac F 29 E7
Menslage D 17 C9
Mentana I 62 C3
Menteroda D 79 D8
Menteşe TR 181 B8
Menton F 37 D7
Méntrida E 46 D4
Menznau CH 70 C6
Méobecq F 29 B8
Meolo I 72 E5
Méounes-les-Montrieux F 35 D10
Meppel NL 16 C6
Meppen D 17 C8
Mequinenza E 42 E4
Mer F 24 E5
Mera RO 153 F9
Merag HR 67 C9
Meråker N 105 E11
Meråkervollen N 105 E11
Merano I 72 C3
Méraq F 32 C5
Merasjärvi S 116 C9
Merasjärvi S 117 C10
Merate I 69 B7
Merbes-le-Château B 19 D9
Mercadillo E 40 B5
Mercatale I 66 F5
Mercatello sul Metauro I 66 E5
Mercatino Conca I 66 E5
Mercato San Severino I 60 B3
Mercato Saraceno I 66 E5
Mercëz RO 153 C11
Merching D 71 A11
Merchtern B 19 C9
Mercogliano I 60 B3
Mercuer F 29 E9
Mercuès F 33 B8
Merdrignac F 22 D7
Mèrdzene LV 133 C3
Mere B 19 C8
Mere UK 13 C10
Méreau F 24 F7
Merei RO 161 C9
Merelbeke B 19 C8
Meremäe EST 132 F3
Merenberg D 185 C9
Mereni MD 154 D3
Mereni RO 161 E7
Mereni RO 155 E2
Mereni RO 161 E7
Mérens-les-Vals F 33 E9
Mereşti RO 153 E6
Méréville F 24 D7
Merghindeal RO 152 F5
Mergozzo I 68 B5
Meria F 37 F10
Méribel-les-Allues F 31 E10
Meric TR 171 B10
Merichas GR 175 E9
Merichleri BG 166 E4
Mérida E 51 B7
Meriden UK 13 A11
Merijärvi FIN 119 F12
Merikarvia FIN 126 B6
Meriläinen FIN 123 D12
Merimasku FIN 126 E6
Mëřín CZ 77 D9
Mering D 71 A11
Merişani RO 160 D5
Mérk H 151 B9
Merkendorf A 148 C5
Merkendorf D 75 D8
Merkenes S 109 C11
Merkinė LT 137 E9
Merklín CZ 76 C4
Merklingen D 74 E6
Merksplas B 182 C5
Merlara I 66 B3
Merlevenez F 22 E5
Merlines F 29 D10
Mern DK 87 E10
Mērnieki LV 131 F8
Mernye H 149 C9
Merošina SRB 164 C5
Merrey F 26 D4
Mersch L 20 E6
Merseburg (Saale) D 79 D10
Mersinbeleni TR 177 D10
Mers-les-Bains F 18 D3
Mērsrags LV 134 B6
Mertajärvi S 116 B8
Mertert L 20 E6
Mertesdorf D 186 B2
Merthyr Tydfil UK 13 B8
Mertingen D 75 E8
Mertloch D 185 D7
Mértola P 50 D4
Mertzwiller F 186 D4
Méru F 18 F5
Mervans F 31 B7
Mervent F 28 B4
Merville F 18 C6
Merville F 33 C8
Méry-sur-Seine F 25 C10
Merzen D 17 D8
Merzig D 21 F7
Mesagne I 61 B9
Mesagros GR 175 D8
Mésanger F 23 F9
Mesão Frio P 44 B5
Mesaria GR 177 E6
Meschede D 17 F10
Meschers-sur-Gironde F 28 D4
Mešeišta NMK 168 B4
Mesocco CH 71 D8
Mesochora GR 168 F5
Mesochori GR 169 F6
Mesochori GR 181 E6
Mesola I 66 C5
Mesolongi GR 174 C3

Mesongi GR 168 F2
Mesopotamia GR 168 C5
Mesopotamos GR 168 F4
Mesoraca I 59 A10
Mesotopos GR 177 A7
Mespelbrunn D 187 B7
Mesquer F 22 F7
Messac F 23 E8
Messanges F 32 C3
Meßdorf D 83 E11
Messei F 23 C10
Messeix F 29 D11
Messejana P 50 D3
Messina I 59 C8
Messincourt F 19 E11
Messini GR 174 E5
Meßkirch D 27 E11
Messlingen S 102 A4
Meßstetten D 27 D10
Mesta BG 165 F8
Mesta GR 177 C6
Mestanza E 54 B4
Městec Králové CZ 77 B8
Mestervik N 111 B16
Mesti GR 171 C9
Mestlin D 83 C11
Město Albrechtice CZ 142 F4
Město Touškov CZ 76 C4
Mestre I 66 B5
Mesves-sur-Loire F 25 F8
Mesztegnyő H 149 C8
Meta I 60 B2
Métabief F 31 B9
Metagkitsi GR 169 D10
Metajna HR 67 C11
Metalliko GR 169 B8
Metallostroy RUS 129 F14
Metaxades GR 171 B10
Metelen D 17 D8
Meteş RO 151 E11
Metfield UK 15 C11
Methana GR 175 D7
Metheringham UK 11 E11
Methlick UK 3 L12
Methoni GR 178 B2
Methven UK 5 C9
Methwold UK 11 F13
Metković HR 157 F8
Metlika SLO 148 E4
Metnitz A 73 C9
Metochi GR 174 C3
Metochi GR 175 B8
Metovnica SRB 159 F9
Mētriena LV 135 C12
Metsäkansa FIN 127 C10
Metsäkylä FIN 121 D11
Metsäkylä FIN 120 D7
Metsälä FIN 126 B7
Metsämaa FIN 127 D9
Metschow D 84 C3
Metskula EST 130 D15
Metslawier NL 16 B6
Metsovo GR 168 E5
Mettäjärvi S 117 E11
Mettendorf D 20 E6
Mettenheim D 75 F11
Mettet B 19 D10
Mettingen D 17 D9
Mettlach D 21 F7
Mettmann D 17 F7
Mettray F 24 F4
Metz F 26 B5
Metzervisse F 20 F6
Metzingen D 27 C11
Meudt D 185 D8
Meulan F 24 B6
Meulebeke B 19 C7
Meung-sur-Loire F 24 E6
Meursault F 30 B6
Meuselwitz D 79 D11
Meuzac F 29 D8
Mevagissey UK 12 E5
Mexborough UK 11 E9
Meximieux F 31 D7
Mey UK 3 H10
Meyenburg D 83 D12
Meylan F 31 E8
Meymac F 29 D9
Meyrargues F 35 C10
Meyreuil F 35 D10
Meyronnes F 36 C5
Meyrueis F 34 B5
Meyssac F 29 E9
Meysse F 35 A8
Meythet F 31 D9
Mezapos GR 178 B3
Mežgare LV 135 C12
Mežciems LV 134 C7
Mezdra BG 165 C8
Mèze F 35 D6
Mézel F 36 C4
Mežgale LV 135 D11
Mezhdurech'ye RUS 136 D4
MezibořI CZ 80 E5
Mežica SLO 73 C10
Mézidon-Canon F 23 B11
Mezibořice CZ 80 D5
Mézières-en-Brenne F 29 B8
Mézières-sur-Issoire F 29 C7
Mézilhac F 30 F5
Mézilles F 25 E9
Mezinmeští CZ 81 E10
Mézin F 33 B6
Mezio P 44 C5
Mezőberény H 151 D7
Mezőcsát H 147 F11
Mezőcsokonya H 149 D9
Mezőfalva H 149 C11
Mezőgyán H 151 D8
Mezőhegyes H 150 E6
Mezőkeresztes H 145 H2
Mezőkovácsháza H 151 E6
Mezőkövesd H 147 F11
Mezőlak H 149 B8
Mezőnyárád H 145 H2
Mezőörs H 149 A9
Mézos F 32 B3
Mezőszemere H 150 B6
Mezőszentgyörgy H 149 B10
Mezőszilas H 149 C10
Mezőtárkány H 150 B5
Mezőtúr H 150 D6
Mezőzombor H 145 G3
Mezquita de Jarque E 42 F2
Mežvalde LV 134 C4
Mežvidi LV 133 C3
Mežvidi LV 134 C4
Mezzana I 69 A10
Mezzano I 37 C10
Mezzocorona I 69 A11

Mezzojuso I 58 D3
Mezzoldo I 69 A8
Mezzolombardo I 69 A11
Mga RUS 129 F15
Miączyn PL 144 B8
Miajadas E 45 F9
Mialet F 29 D7
Miały PL 85 E10
Miastczko Krajeńskie PL 85 D12
Miastko PL 85 B11
Miastków Kościelny PL 141 G5
Miastkowo PL 139 D12
Miavaig UK 2 J3
Mica RO 152 C3
Mica RO 152 C3
Micăsasa RO 152 E4
Miceşti RO 160 D5
Miceşti de Câmpie RO 152 D4
Michałany SK 145 F4
Michalová SK 147 D9
Michalovce SK 145 F4
Michałów PL 143 F9
Michałów Górny PL 141 G4
Michałowice PL 142 E3
Michałowice PL 143 F8
Michałowo PL 140 D9
Michelau in Oberfranken D 75 B9
Michelbach an der Bilz D 74 D6
Micheldorf A 73 C9
Micheldorf in Oberösterreich A 73 A9
Michelfeld D 74 D6
Michelstadt D 21 E12
Michendorf D 80 B4
Michorzewo PL 81 B10
Michów PL 141 G6
Mickelspiltom FIN 127 D15
Mickelsträsk S 118 F4
Mickleton UK 5 F12
Mickleton UK 13 A11
Micleşti RO 153 D11
Miclova MD 154 C3
Midbea UK 3 G11
Middagsbukt N 111 B17
Middelbeers NL 183 C6
Middelburg NL 16 E1
Middelfart DK 86 D5
Middelharnis NL 16 E2
Middelkerke B 18 B4
Middelstum NL 17 B7
Middenbeemster NL 16 C3
Middenmeer NL 16 C4
Middleham UK 11 C8
Middlemarsh UK 13 D10
Middlesbrough UK 11 B9
Middleton UK 11 D7
Middleton Cheney UK 13 A12
Middleton in Teesdale UK 5 F12
Middletown UK 7 D9
Middlewich UK 10 E7
Midhurst UK 15 F7
Midleton IRL 8 E6
Midlum D 17 A11
Midsund N 100 A5
Midtgård N 114 D8
Midtskogberget N 102 D3
MiŠvágur FO 2 A2
Midwolda NL 17 B8
Mid Yell UK 3 D14
Miechów PL 143 F9
Miechów Charsznica PL 143 F8
Miedes E 41 F9
Miedziana Góra PL 143 E10
Miedzichowo PL 81 B9
Miedzna PL 139 F13
Miedźna PL 143 G7
Miedźno PL 143 E6
Międzybórz PL 142 D4
Międzychód PL 81 A9
Międzylesie PL 77 B11
Międzyrzec Podlaski PL 141 G7
Międzyrzecz PL 81 B9
Międzyzdroje PL 84 C6
Miehikkälä FIN 128 D8
Miehlen D 185 D8
Miejsce Piastowe PL 145 D4
Miejska Górka PL 81 C11
Miękinia PL 81 D11
Miekojärvi S 119 B10
Mielan F 33 D6
Mielec PL 143 F11
Mielęcin PL 85 E12
Mielno PL 85 B10
Mieluskylä FIN 119 F14
Mielżyn PL 81 A8
Mieming A 71 C12
Mierasjärvi FIN 113 D19
Mieraslompolo FIN 113 D11
Miercurea-Ciuc RO 153 E7
Miercurea Nirajului RO 152 D5
Miercurea Sibiului RO 152 F3
Mieres E 39 B8
Mieres E 43 C9
Mierlo NL 16 F5
Mierojävri N 113 E11
Mieroszów PL 81 E10
Miers F 29 F9
Mierzęcice PL 143 F7
Mierzyn PL 84 D6
Miesau D 186 C3
Miesbach D 72 A4
Mieścisko PL 85 E12
Miesenbach D 186 C4
Miessaure S 116 E4
Mieste D 79 B9
Mieszków PL 142 B3
Mieszkowice PL 84 E6
Mietingen D 71 A9
Mietków PL 81 D11
Mietoinen FIN 126 D6
Miettilä FIN 129 C10
Mieussy F 31 C10
Mięzia E 45 B7
Migennes F 25 E10
Migliarino I 66 C4
Migliarino I 66 C4
Migliaro I 66 C4
Miglionica I 61 B6
Mignaloux-Beauvoir F 29 B6
Mignano Monte Lungo I 60 A1
Migné-Auxances F 29 B6
Mignovillard F 31 B9
Miguel Esteban E 47 E6
Miguelturra E 54 B5
Migushino RUS 133 E7
Mihăeşti RO 160 C4
Mihăeşti RO 160 C5
Mihăeşti RO 160 E5
Mihai Bravu RO 155 D3
Mihai Bravu RO 161 E8
Mihăileni RO 152 F4

Mihăileni RO 153 B8
Mihăileni RO 153 E7
Mihăilești RO 161 D7
Mihăilești RO 161 E7
Mihail Kogălniceanu RO 155 C3
Mihail Kogălniceanu RO 155 D3
Mihail Kogălniceanu RO 155 E2
Mihailovca MD 154 D3
Mihai Viteazu RO 152 D3
Mihai Viteazu RO 155 D3
Mihajlovac SRB 159 D6
Mihajlovac SRB 159 D6
Mihajlovo SRB 158 C5
Mihălășeni RO 153 B10
Miháld H 149 D8
Mihalţ RO 152 E3
Mihályi H 149 A8
Miheşu de Câmpie RO 152 D4
Mihla D 79 D7
Miiluranta FIN 123 C16
Mijares E 46 D3
Mijas E 53 C7
Mijdrecht NL 16 D3
Mijoska MNE 163 E7
Mikalayeva BY 133 F6
Mikepércs H 151 C8
Mikhalishki BY 137 D13
Mikhalkovo BG 165 F9
Mikhaltsi BG 166 C4
Mikhaylovo BG 165 B8
Mikhaylovo BG 166 C5
Mikitamäe EST 132 E2
Mikitsikha BY 133 F6
Mikkeli FIN 128 B7
Mikkelin mlk FIN 128 B7
Mikkelvik N 112 C3
Mikkola FIN 117 D16
Mikladalur FO 2 A3
Miklavž SLO 148 C5
Miklebostad N 111 C13
Mikleuš HR 149 E9
Mikofajki PL 136 F4
Mikofajki Pomorskie PL 139 C7
Mikofów PL 143 F6
Mikoszewo PL 138 B6
Mikre BG 165 C10
Mikri Volvi GR 169 C10
Mikro Dereio GR 171 B10
Mikrokambos GR 169 C8
Mikromilia GR 170 B6
Mikropoli GR 169 B9
Mikrothives GR 169 F8
Mikrovalto GR 169 D6
Mikstat PL 142 C4
Mikuláš CZ 77 C8
Mikulasovice CZ 80 E6
Mikulčice CZ 77 E12
Mikulov CZ 77 E11
Mikulovice CZ 142 F4
Mikušovce SK 146 C6
Miladinovci NMK 164 F4
Milagro E 32 F2
Milagros E 40 E4
Miłakowo PL 139 B9
Milano I 69 C7
Milano Marittima I 66 D5
Milanówek PL 141 F3
Milaş RO 152 D4
Milas TR 181 B7
Milatkoviče SRB 163 C10
Milatos GR 179 E10
Milazzo I 59 C7
Milborne Port UK 13 D10
Milcoiu RO 160 C4
Milcov RO 160 E4
Milcovul RO 161 B10
Mildenhall UK 15 C10
Mildstedt D 82 B6
Mileanca RO 153 A9
Milejczyce PL 141 E8
Milejewo PL 139 B8
Milejów PL 141 H7
Milena I 58 E4
Mileşti MD 153 C11
Milestone IRL 8 C6
Mileto I 59 B9
Milevsko CZ 76 D6
Milfield UK 5 D12
Milfontes P 50 D2
Milford IRL 7 B7
Milford IRL 8 D5
Milford Haven UK 12 B4
Milhão P 39 E6
Milhaud F 35 C7
Milici BIH 157 D11
Milicz PL 81 C12
Milies GR 169 F8
Milikov CZ 147 B7
Milín CZ 76 C6
Milina GR 175 A7
Milis I 64 C2
Militello in Val di Catania I 59 E6
Miliyewe UA 152 A6
Milizac F 22 D2
Miljana HR 148 D5
Miljeno BIH 157 E11
Miljevina BIH 157 E10
Milkel D 81 D6
Miłki PL 136 F4
Milkovitsa BG 160 F5
Milk'ovtsi BG 165 D9
Mill NL 16 E5
Millares E 48 F3
Millas F 34 E4
Millau F 34 B5
Millay F 30 B5
Millbrook UK 12 E6
Millbrook IRL 7 E8
Millesimo I 37 C8
Millevaches F 29 D10
Millford UK 7 D9
Millhouse UK 4 D6
Millières F 26 D3
Millingen aan de Rijn NL 183 B8
Millisle UK 7 C11
Millom UK 10 C5
Milloshevë RKS 164 D3
Millport UK 4 D7
Millstatt A 73 C8
Millstreet IRL 8 D4
Millstreet IRL 9 D7
Milltown IRL 6 E5
Milltown IRL 7 F9
Milltown IRL 8 D1
Milltown IRL 8 D3
Milltown Malbay IRL 8 C4
Milly-la-Forêt F 25 D7
Milmarcos E 47 B9
Milmersdorf D 84 D5
Milmort B 183 D7
Milna HR 156 F5

Milngavie UK 5 D8
Milnthorpe UK 10 C6
Milo I 59 D7
Miłocice PL 85 C11
Milohnić HR 67 B9
Miłomłyn PL 139 C8
Miloradz PL 138 B6
Miłosław PL 142 B3
Milot AL 163 F8
Milotice CZ 77 E12
Milovaig UK 2 L3
Milovice CZ 77 B7
Milow D 79 A11
Miłówka PL 147 B8
Miltach D 75 D12
Miltenberg D 21 E12
Miltini LV 134 C6
Milton UK 2 L8
Milton UK 5 B9
Milton Keynes UK 15 C7
Miltzow D 84 B4
Milutinovac SRB 159 F7
Milverton UK 13 C8
Milz D 75 B8
Mimetiz E 40 B5
Mimizan F 32 B3
Mimizan-Plage F 32 B3
Mimoň CZ 81 E7
Mina de São Domingos P 50 D5
Mín an Chladaigh IRL 6 B6
Minas de Riotinto E 51 D6
Minateda E 55 C9
Minaya E 47 E9
Minde P 44 E3
Mindelheim D 71 A10
Mindelstetten D 75 E10
Minden D 17 D11
Minderhout B 182 C5
Mindnes N 108 E3
Mîndrești RO 153 A10
Mindszent H 150 D5
Mindszentgodisa H 149 D10
Mindtangen N 108 E3
Mindūnai LT 135 F11
Mindya BG 166 C5
Minehead UK 13 C8
Mineo I 59 E6
Mineralni Bani BG 166 F4
Minerbio I 66 C4
Minervino Murge I 60 A6
Minfeld D 27 B9
Minger MD 154 D2
Minglanilla E 47 E9
Mingorría E 46 C3
Miniac-Morvan F 23 C8
Minićevo SRB 159 F9
Minné I 59 C7
Minne S 103 B9
Minnertsga NL 16 B5
Minnesund N 95 B14
Minniaff UK 4 F5
Minot F 25 E12
Mińsk Mazowiecki PL 141 F5
Minsterley UK 10 F6
Mintiloglio GR 174 C4
Mintiu Gherlii RO 152 D4
Mintlaw UK 3 K13
Mintraching D 75 E11
Minturno I 62 E5
Minucciano I 66 D1
Minusio CH 68 A6
Mionica SRB 158 E5
Mionnay F 30 D6
Mios F 32 A4
Mira E 47 E10
Mira I 66 B5
Mira P 44 D3
Mirabeau F 35 C10
Mirabel E 45 E8
Mirabel F 33 B8
Mirabel-aux-Baronnies F 35 B9
Mirabella Eclano I 60 A4
Mirabella Imbaccari I 58 E5
Miradoux F 33 C7
Miraflores de la Sierra E 46 C5
Miralcamp E 42 D5
Miramare I 66 D6
Miramas F 35 D9
Mirambeau F 28 E4
Mirambel E 42 F3
Miramont-de-Guyenne F 33 A6
Miranda de Arga E 32 E2
Miranda de Ebro E 40 C6
Miranda do Corvo P 44 D4
Miranda do Douro P 39 E7
Mirande F 33 D6
Mirandela P 38 F5
Mirandilla E 51 A7
Mirandola I 66 C3
Mirandol-Bourgnounac F 33 B10
Miranje HR 156 E4
Mirano I 66 B5
Miras AL 168 C4
Mirash RKS 164 E3
Mirăslău RO 152 E3
Miratovac SRB 164 E4
Miravci NMK 169 B7
Miravet E 42 E5
Miravete de la Sierra E 42 F2
Mircea Vodă RO 155 E2
Mircea Vodă RO 161 C10
Mircești RO 153 C9
Mircze PL 144 B8
Mirebeau F 26 F3
Mirebeau F 28 B5
Mirebel F 31 B8
Mirecourt F 26 D5
Mirepoix F 33 D9
Mireşu Mare RO 151 C11
Mireval F 35 C6
Miričina BIH 157 C9
Mirkovci HR 157 B8
Mirkovo BG 165 D9
Mirlović Zagora HR 156 E5
Mirna SLO 73 E11
Mirna Peč SLO 73 E11
Mirocin PL 144 C6
Mirocin Górny PL 81 C9
Mironeasa RO 153 D10
Miroševce SRB 164 D4
Miroşi RO 160 E5
Miroslav CZ 77 E10
Miroslávas LT 137 E8
Mirosławiec PL 85 D10
Miroslovești RO 153 C9
Mirošov CZ 76 C5
Mirotice CZ 76 D6
Mirovice CZ 76 C6

Mirovtsi BG 167 C8
Mirow D 83 D13
Mirşid RO 151 C11
Mirsk PL 81 E8
Mirto Crosia I 61 D7
Mirueña de los Infanzones E 45 C10
Mirzec PL 141 H4
Misano Adriatico I 67 E6
Misano RO 151 D8
Mişca RO 151 C9
Mischii RO 160 E3
Miserey-Salines F 26 F4
Mishnyevichy BY 133 F7
Misi FIN E17
Misilmeri I 58 C3
Mišinci BIH 157 C8
Miske H 145 G2
Miskolc H 145 G3
Mislina HR 162 D5
Mislinja SLO 73 D11
Mišnjak HR 67 C10
Mison F 35 B10
Missanello I 60 C6
Missenträsk S 107 A17
Missillac F 23 F7
Misso EST 131 F14
Mistelbach A 77 E11
Mistelgau D 75 C9
Misten N 108 B7
Misterbianco I 59 D7
Misterhult S 93 E9
Misterton UK 11 E10
Mistretta I 58 D5
Mistros GR 175 B8
Misurina I 72 C5
Misvær N 108 B9
Mitandersfors S 96 B7
Mitchelstown IRL 8 D6
Mithymna GR 171 F10
Mitoc RO 153 A10
Mitrašinci NMK 165 F6
Mitreni RO 161 E9
Mitropoli GR 169 F6
Mitrousi GR 169 B9
Mitrova Reka SRB 163 C9
Mitrovicë RKS 163 D10
Mitry-Mory F 25 C8
Mittådalen S 102 A6
Mittelberg A 71 C10
Mittelberg A 71 A9
Mittelbiberach D 71 A9
Mittelkalbach D 74 B6
Mittelsinn D 187 A8
Mittenwald D 72 B3
Mittenwalde D 80 B5
Mittenwalde D 84 D5
Mitterbach am Erlaufsee A 148 A4
Mitterding A 76 F4
Mitterdorf im Mürztal A 148 A5
Mittersheim F 27 C6
Mittersill A 72 B5
Mitterskirchen D 75 F12
Mitterteich D 75 C11
Mittet N 100 A7
Mittiliden S 105 B16
Mittweida D 80 E3
Mitwitz D 75 B9
Mizhhir''ya UA 145 F8
Mizil RO 161 D8
Miziya BG 160 F3
Mjäldrunga S 91 D13
Mjällby S 88 C7
Mjällom S 107 F14
Mjede AL 163 F8
Mjelde N 110 D9
Mjelde N 111 A15
Mjöbäck S 88 A3
Mjölby S 92 C6
Mjølfell N 100 D5
Mjolnarodden N 110 D5
Mjönäs S 97 C10
Mjønes N 108 B8
Mjösjöby S 107 D15
Mjösund FIN 126 E7
Mjöträsk S 119 B9
Mjøvatnet S 103 A14
Mjøvattnet N 108 D7
Mjøvattnet S 118 C5
Mladá Boleslav CZ 77 B8
Mladá Vožice CZ 77 C7
Mladé Buky CZ 81 E9
Mladenovac SRB 158 E6
Mladenovo SRB 158 C3
Mladikovine BIH 157 D8
Mladinovo BG 166 F6
Mlado Nagoričane NMK 164 E4
Mława PL 139 D9
Mlekarevo BG 166 E6
Mlinice BIH 157 D7
Mlinište BIH 157 D6
Młodzieszyn PL 141 F2
Młynary PL 139 B8
Młynarze PL 139 E11
Mlynys'ka UA 145 E9
Mnichovice CZ 77 C7
Mnichovo Hradiště CZ 77 A7
Mnichów PL 143 E9
Mniów PL 143 E9
Mníšek nad Hnilcom SK 145 F2
Mníšek pod Brdy CZ 76 C6
Mniszew PL 141 G4
Mniszków PL 141 H1
Mo N 108 E3
Mo S 103 D12
Mo S 103 E12
Mo S 107 D11
Mo S 107 D11
Moacşa RO 153 F7
Moaña E 38 D2
Moara RO 153 B8
Moara Domnească MD 153 B10
Moara Vlăsiei RO 161 D8
Moate IRL 7 F7
Mobberley UK 11 E7
Moça RO 153 D9
Moçarria P 44 F3
Mocejón E 46 E5
Močenok SK 146 E5
Mochales E 47 B8
Mochos GR 178 E9
Mochowo PL 139 E8
Mochy PL 81 B10
Mociu RO 152 D4
Möckern D 79 B10
Mockfjärd S 97 A12
Möckmühl D 27 B11
Mockrehna D 79 C12
Moclín E 53 B9
Moclinejo E 53 C8
Mo i Rana N 108 D7
Moià E 43 D8
Moiano I 60 A3
Moieciu RO 160 C6
Moimenta da Beira P 44 C5
Moineşti RO 153 E8
Moingt F 30 D5
Moira UK 7 C10
Moirans F 31 E8
Moirans-en-Montagne F 31 C8
Moirax F 33 B7
Moires GR 178 E8
Mõisaküla EST 131 E10
Moisburg D 82 D7
Moisdon-la-Rivière F 23 E9
Moisei RO 152 B5
Moisiovaara FIN 121 E13
Moislains F 18 E6
Moissac F 33 B8
Moissac-Bellevue F 36 D4
Moissey F 31 B7
Moïta F 37 G10
Moita P 44 D4
Moita P 44 F3
Moita P 50 B2
Moixent-Mogente E 56 D3
Mojácar E 55 E9
Mojados E 39 E10
Mojkovac MNE 163 D8
Mojmírovce SK 146 E5
Mojstrana SLO 73 D8
Mojzesovo SK 146 E5
Møkkelvik N 110 D6
Möklinta S 98 B7
Mokobody PL 141 F6
Mokrá Hora CZ 77 D11
Mokrance SK 145 F3
Mokre PL 143 F10
Mokren BG 167 D7
Mokresh BG 160 F2
Mokrievo NMK 169 B8
Mokrin BG 150 F5
Mokrin BIH 157 C10
Mokronog SLO 73 E11
Mokronoge BIH 156 D5
Mokronoge BIH 157 E7
Möksy FIN 123 D12
Mol B 183 C6
Mol SRB 150 F5
Mola di Bari I 61 A8
Molai GR 178 B4
Molare I 37 B9
Molares E 51 C7
Molbergen D 17 C9
Mölbling A 73 C9
Molby DK 86 E4
Mold UK 10 E5
Moldava nad Bodvou SK 145 F2
Molde N 100 A6
Moldjord N 108 B8
Moldova Nouă RO 159 D8
Moldova-Sulița RO 152 B6
Moldoveni RO 153 D9
Moldovița RO 153 B7

Møldrup DK 86 B5
Moldvik N 111 C11
Moledo P 38 E2
Moledo P 44 C5
Molelos P 44 C4
Molenbeek-St-Jean B 182 D4
Molenstede B 183 C6
Molescroft UK 11 D11
Molesmes F 25 E11
Molești MD 154 D3
Mõletai LT 135 F10
Molfetta I 61 A7
Molfsee D 83 B8
Moliden S 107 E14
Molières F 29 F9
Molières F 33 B8
Molières-sur-Cèze F 35 B7
Moliets-et-Maa F 32 C3
Molina Aterno I 62 C5
Molina de Aragón E 47 C9
Molina de Segura E 56 C2
Molina di Ledro I 69 B10
Molinara I 60 A3
Molinaseca E 39 C7
Molinella I 66 C4
Molines-en-Queyras F 31 F10
Molinet F 30 C4
Molinges F 31 C8
Molinicos E 55 C9
Molini di Tures I 72 C4
Molino de Villobas E 32 F5
Molinos E 42 F3
Molins de Rei E 43 E8
Moliterno I 60 C5
Molitg-les-Bains F 33 E10
Molkojärvi FIN 117 D16
Molkom S 97 C10
Mollas AL 168 C3
Mölle S 87 C11
Molledo E 40 B3
Möllenbeck D 84 D4
Möllenhagen D 84 C3
Mollerussa E 42 D5
Molles F 30 C3
Mollet del Vallès E 43 D8
Molliens-Dreuil F 18 E5
Mollis CH 71 C7
Mölln D 83 C9
Mölln D 84 C4
Molló E 33 F10
Mollösund S 91 C9
Mölltorp S 92 C4
Mölnbo S 93 A10
Mölnlycke S 91 D11
Molnytsya UA 153 A8
Moloha UA 154 E6
Molompize F 30 E3
Molos GR 175 B6
Moløy F 26 E2
Molpe FIN 122 E6
Molschleben D 79 D8
Molsheim F 186 D3
Molunat HR 162 E5
Molve HR 149 D8
Molveno I 69 A10
Molvízar E 53 C9
Mombaldone I 37 B8
Mombeltrán E 45 D10
Mombercelli I 37 B8
Mömbris D 21 D12
Mombuey E 39 D7
Mona FIN 122 D8
Monachil E 53 B9
Monacia-d'Aullène F 37 H10
Monaghan IRL 7 D9
Monamolin IRL 9 C10
Monäs FIN 122 D8
Monashi UA 154 E6
Monasterace I 59 C10
Monasterevin IRL 7 F8
Monastir I 64 E3
Monastiraki GR 174 B2
Monbahus F 33 A7
Monbazillac F 29 F6
Moncada E 48 E4
Moncalieri I 37 A7
Moncalvo I 68 C5
Monção P 38 D3
Moncarapacho P 50 E4
Moncaut F 33 B6
Moncel-sur-Seille F 26 C5
Mönchengladbach D 20 B6
Mönchhof A 77 G11
Monchio delle Corti I 66 D1
Monchique P 50 E2
Mönchsdeggingen D 75 E8
Monclar F 33 B7
Moncofa E 48 F4
Moncontour F 22 D6
Moncontour F 28 B5
Moncoutant F 28 B4
Moncrabeau F 33 B6
Monda E 53 C7
Mondariz E 38 D3
Mondariz-Balneario E 38 D3
Mondavezan F 33 D7
Mondavio I 67 E6
Mondéjar E 47 D6
Mondello I 58 C3
Mondeville F 23 B11
Mondolfo I 67 E7
Mondoñedo E 38 B5
Mondorf-les-Bains L 20 E6
Mondoubleau F 24 E4
Mondovì I 37 C7
Mondragon F 35 B8
Mondragone I 62 E5
Mondsee A 73 A7
Monea UK 7 D7
Moneasa RO 151 E9
Moneen IRL 6 F5
Moneglia I 37 C10
Monegrillo E 41 E11
Monein F 32 D4
Monemvasia GR 178 B5
Monesiglio I 37 C8
Monesterio E 51 C7
Monestier-de-Clermont F 31 F8
Monestiés F 33 B10
Monéteau F 25 E10
Moneygall IRL 9 C7
Moneymore UK 4 F3
Moneyneany UK 4 F3
Moneyreagh UK 7 C11
Monfalcone I 73 E8
Monfero E 38 B3
Monfecorvino Rovella I 60 B3
Monflanquin F 33 A7
Monfort D 73 C7
Monforte P 44 F6

Monforte da Beira P 45 E6
Monforte d'Alba I 37 B7
Monforte del Cid E 56 E3
Monforte de Lemos E 38 B5
Monforte de Moyuela E 42 E2
Monfortinho P 45 D7
Monghidoro I 66 D3
Mongrando I 68 B5
Mongstad N 100 E2
Monguelfo I 72 C5
Monheim D 75 E8
Moniaive UK 5 E9
Monieux F 35 B9
Monifieth UK 5 C11
Monilea IRL 7 E8
Mönistre EST 131 E13
Monistrol-d'Allier F 30 F4
Monistrol de Calders E 43 D8
Monistrol de Montserrat E 43 D7
Monistrol-sur-Loire F 30 E5
Monivea IRL 6 F5
Mönkeberg D 83 B8
Moñki PL 140 D7
Monkokehampton UK 12 D6
Monléon-Magnoac F 33 D7
Monmouth UK 13 B9
Monnaie F 24 E4
Mönni FIN 125 E14
Monninkylä FIN 127 E14
Monok H 145 G3
Monolithos GR 181 D7
Monopoli I 61 B8
Monor H 150 C3
Monor RO 152 D5
Monostorapáti H 149 C9
Monostorpályi H 151 C8
Monóvar E 56 E3
Monpazier F 33 A7
Monreal E 32 E3
Monreal del Campo E 47 C10
Monreale I 58 C3
Monreith UK 4 F7
Monroy E 45 E8
Monroyo E 42 F3
Mons B 19 D8
Mons F 34 C4
Mons F 36 D5
Monsampolo del Tronto I 62 B5
Monsaraz P 51 C5
Monschau D 20 C6
Monsec F 29 E7
Monségur F 33 A6
Monselice I 66 B4
Monsempron-Libos F 33 B7
Monserrat E 48 F3
Monsheim D 187 B5
Mönsheim D 187 D6
Mønsted DK 86 C4
Monster NL 16 D2
Mönsterås S 89 A10
Monsummano Terme I 66 E2
Monta I 37 B7
Montabaur D 21 D9
Montaberner E 56 D4
Montady F 34 D5
Montagnac F 34 D5
Montagnana I 66 B3
Montagne F 28 F5
Montagney F 26 F4
Montagnol F 34 C5
Montagrier F 29 E6
Montaigu F 28 B3
Montaigu-de-Quercy F 33 B8
Montaigut F 30 C2
Montaigut-sur-Save F 33 C8
Montaione I 66 E2
Montalbán E 42 F2
Montalbán de Córdoba E 53 A7
Montalbano Elicona I 59 C7
Montalbano Jonico I 61 C7
Montalbo E 47 E7
Montalcino I 65 A4
Montale I 66 E3
Montalegre P 38 E4
Montalieu-Vercieu F 31 D7
Montalivet-les-Bains F 28 E3
Montallegro I 58 E3
Montalto delle Marche I 62 B5
Montalto di Castro I 65 C5
Montalto Marina I 65 C5
Montalto Uffugo I 60 E6
Montalvão P 44 E5
Montamarta E 39 E8
Montana BG 165 C7
Montana CH 31 C11
Montanejos E 48 D3
Montaner F 32 D5
Montano Antilia I 60 C4
Montans F 33 C9
Montaquila I 62 D6
Montargil P 44 F4
Montargis F 25 E8
Montastruc-la-Conseillère F 33 C9
Montataire F 18 F5
Montauban F 33 B8
Montauban-de-Bretagne F 23 D7
Montaudin F 23 D10
Montauriol F 33 C10
Montauroux F 36 D5
Montaut F 32 C4
Montaut F 32 D5
Montaut F 33 D9
Montayral F 33 B7
Montazzoli I 63 D6
Montbard F 25 E11
Montbarrey F 31 A8
Montbazens F 33 B10
Montbazin F 34 C6
Montbazon F 24 F4
Montbéliard F 27 E6
Montbenoît F 31 B9
Montbeton F 33 B8
Montblanc E 42 E6
Montboucher-sur-Jabron F 35 A8
Montbozon F 26 F5
Montbrió del Camp E 42 E6
Montbrison F 30 D5
Montbron F 29 D7
Montbrun F 33 A9
Montbrun-les-Bains F 35 B9
Montcada i Reixac E 43 E8
Montcavrel F 15 F12
Montceau-les-Mines F 30 B5
Montcenis F 30 B5
Montchanin F 30 B5
Montcornet F 19 E9
Montcresson F 25 E8
Montcuq F 33 B8
Montcy-Notre-Dame F 19 E10

Montdardier F 35 C6
Mont-Dauphin F 36 B5
Mont-de-Marsan F 32 C5
Montdidier F 18 E6
Mont-Dore F 30 D2
Monteagudo E 41 E8
Monteagudo de las Salinas E 47 E9
Monteagudo de las Vicarías E 41 F7
Montealegre del Castillo E 55 B10
Montebello Ionico I 59 D8
Montebello Vicentino I 66 B3
Montebelluna I 72 E5
Montebourg F 23 B9
Montebruno I 37 B10
Montecalvo in Foglia I 66 E6
Montecalvo Irpino I 60 A4
Monte-Carlo MC 37 D6
Montecarotto I 67 E7
Montecassiano I 67 F7
Monte Castello di Vibio I 62 B2
Montecastrilli I 62 B2
Montecatini Terme I 66 E2
Montecatini Val di Cecina I 66 F2
Montecchio I 67 E6
Montecchio Emilia I 66 C1
Montecchio Maggiore I 66 A3
Montech F 33 C8
Montechiaro d'Asti I 37 A8
Montechiarugolo I 66 C1
Montecilfone I 63 D7
Montecorice I 60 C3
Montecosaro I 67 F8
Montecreto I 66 D2
Monte da Pedra P 44 F5
Monte das Flores P 50 B4
Monte de Procida I 60 B2
Monte do Trigo P 50 C4
Montefalco I 62 B2
Montefalcone di Val Fortore I 60 A4
Montefalcone nel Sannio I 63 D7
Montefano I 67 F7
Montefelcino I 67 E6
Montefiascone I 62 B2
Montefiore dell'Aso I 62 A5
Montefiorino I 66 D1
Montefortino I 62 B4
Montefranco I 62 B3
Montefrío E 53 B8
Montegiordano I 61 C7
Montegiorgio I 67 F8
Monte Gordo P 50 E5
Montegranaro I 67 F8
Montegrotto Terme I 66 B4
Montehermoso E 45 D8
Monteiasi I 61 C8
Monteils F 33 B9
Monteiros P 44 A5
Montejaque E 53 C6
Montejícar E 53 A10
Montejo de la Sierra E 46 B5
Montelabbate I 67 E6
Montelanico I 62 D4
Montelavar P 50 B1
Montel-de-Gelat F 29 D11
Monteleone di Puglia I 60 A4
Monteleone di Spoleto I 62 B3
Monteleone d'Orvieto I 62 B2
Monteleone Rocca Doria I 64 C2
Montelepre I 58 C3
Montelibretti I 62 C3
Montélier F 31 F7
Montélimar F 35 A8
Montella I 60 B4
Montellano E 51 E8
Montels F 33 D8
Montelupo Fiorentino I 66 E3
Montelupone I 67 F8
Montemaggiore Belsito I 58 D4
Montemagno I 37 B8
Montemarciano I 67 E7
Montemayor E 53 A7
Montemayor de Pililla E 40 E3
Montembœuf F 29 D7
Montemesola I 61 B8
Montemiletto I 60 A3
Montemilone I 60 A5
Montemolín E 51 C7
Montemonaco I 62 B4
Montemor-o-Novo P 50 B3
Montemurlo I 66 E3
Montemurro I 60 C5
Montenay F 23 D10
Montendre F 28 E5
Montenegro de Cameros E 40 D6
Montenero di Bisaccia I 63 D7
Montenerodomo I 63 C6
Monte Porzio I 67 E7
Monteprandone I 62 B5
Montepulciano I 62 A1
Monterblanc F 22 E6
Monterchi I 66 F5
Monte Real P 44 E3
Montereale I 62 B4
Montereau-fault-Yonne F 25 D8
Monte Redondo P 44 E3
Monterenzio I 66 D3
Monteriggioni I 66 F3
Monteroduni I 63 D6
Monte Romano I 62 C1
Monteroni d'Arbia I 66 F3
Monteroni di Lecce I 61 C10
Monterosso Almo I 59 E6
Monterosso Calabro I 59 B9
Monterotondo I 62 C3
Monterotondo Marittimo I 66 F3
Monterrei E 38 E5
Monterroso E 38 C4
Monterrubio de la Serena E 51 B9
Monterubbiano I 62 A5
Montesa E 56 D3
Monte San Biagio I 62 E4
Monte San Giovanni Campano I 62 D5
Montesano Salentino I 61 D10
Montesano sulla Marcellana I 60 C5
Monte San Savino I 66 F4
Monte Santa Maria Tiberina I 66 F5
Monte Sant'Angelo I 63 D9
Monte San Vito I 67 E7
Montesarchio I 60 A3

Montescaglioso I 61 B7
Montesclaros E 46 D3
Montescudaio I 66 F2
Montese I 66 D2
Montesilvano I 63 B6
Montespertoli I 66 E3
Montesquieu-Volvestre F 33 D8
Montesquiou F 33 C6
Montes Velhos P 50 D3
Monteux F 35 B8
Montevago I 58 D2
Montevarchi I 66 E4
Montevecchia I 64 D2
Monteverde I 60 A5
Montevil P 50 C2
Montfaucon F 23 F9
Montfaucon F 28 F5
Montfaucon-d'Argonne F 19 F11
Montfaucon-en-Velay F 30 E5
Montferran-Savès F 33 C7
Montferrat F 36 D4
Montferrier F 33 E9
Montfoort NL 182 A5
Montfort F 23 D8
Montfort F 32 C4
Montfort NL 183 C7
Montfort-l'Amaury F 24 C6
Montfort-le-Gesnois F 24 D3
Montfort-sur-Risle F 18 F2
Montgai E 42 D5
Montgaillard F 33 D6
Montgaillard F 33 D6
Montgenèvre F 31 F10
Montgeron F 25 C7
Montgiscard F 33 D9
Montgivray F 29 B9
Montgomery UK 10 F5
Montguyon F 28 E5
Monthermé F 19 E10
Monthey CH 31 C10
Monthois F 19 F10
Monthureux-sur-Saône F 26 D4
Monti I 64 B3
Monticelli d'Ongina I 69 C8
Monticello F 37 F9
Montichiari I 66 B1
Monticiano I 65 A4
Montiel E 56 D5
Montier-en-Der F 25 D12
Montieri I 65 A4
Montiers-sur-Saulx F 26 C3
Montiglio I 68 C5
Montignac F 29 E8
Montignies-le-Tilleul B 19 D9
Montignoso I 66 D1
Montigny F 27 C6
Montigny-la-Resle F 25 E10
Montigny-le-Roi F 26 D3
Montigny-lès-Metz F 26 B5
Montigny-Mornay-Villeneuve-sur-Vingeanne F 26 E4
Montigny-sur-Aube F 25 E12
Montijo E 51 B9
Montijo P 50 B2
Montilla E 53 A7
Montillana E 53 A9
Montivilliers F 23 A12
Montizón E 55 C6
Montjaux F 34 B4
Montjean F 23 D10
Montjovet I 68 C4
Montlaur F 34 D4
Montlieu-la-Garde F 28 E5
Mont-Louis F 33 F10
Montluçon F 29 C11
Montluel F 31 D7
Montmarault F 30 C2
Montmartin-sur-Mer F 23 C8
Montmédy F 19 E11
Montmélian F 31 D9
Montmeló E 43 D8
Montmeyran F 31 F6
Montmeyan F 36 D4
Montmirail F 25 C10
Montmirail F 25 D9
Montmirey-le-Château F 26 F4
Montmoreau-St-Cybard F 29 E6
Montmorillon F 29 C7
Montmorin F 35 B10
Montmort F 31 B8
Montmort-Lucy F 25 C10
Montoir-de-Bretagne F 23 F7
Montoire-sur-le-Loir F 24 E4
Montoison F 30 F6
Montoito P 50 B4
Montola FIN 124 F8
Montón E 47 B10
Montone I 66 F5
Montopoli di Sabina I 62 C3
Montorio al Vomano I 62 B5
Montoro E 53 A8
Montory F 32 D4
Montournais F 28 B4
Montpelier IRL 8 C6
Montpellier F 35 C6
Montpeyroux F 34 A4
Montpezat F 33 D7
Montpezat F 35 B7
Montpezat-de-Quercy F 33 B8
Montpezat-sous-Bauzon F 30 F5
Montpon-Ménestérol F 29 E6
Montpont-en-Bresse F 31 B7
Mont-ras E 43 D10
Montréal F 25 E11
Montréal F 32 A6
Montréal F 33 D10
Montredon-Labessonnié F 33 C10
Montregard F 30 E5
Montréjeau F 33 D7
Montrésor F 24 F5
Montresta I 64 C2
Montret F 31 B7
Montreuil F 15 G12
Montreuil-Bellay F 23 F11
Montreuil-Juigné F 23 E10
Montreux CH 31 C10
Montrevault F 23 F9
Montrevel-en-Bresse F 31 C7
Montrichard F 24 F5
Montricoux F 33 B9
Montrond F 31 C10
Mont-roig del Camp E 42 E5
Montrond F 31 B8
Montrond-les-Bains F 30 D5
Montrose UK 5 B12
Montroy E 48 F3
Monts F 24 F4
Montsalvy F 29 F10
Montsauche-les-Settons F 25 F11

Montségur F 33 E9
Montseny E 43 D8
Montsoué F 32 C5
Mont-sous-Vaudrey F 31 B8
Monts-sur-Guesnes F 29 B6
Mont-St-Aignan F 18 F3
Mont-St-Jean F 25 F11
Mont-St-Martin F 19 E12
Mont-St-Vincent F 30 B5
Montsûrs F 23 D10
Montsuzain F 25 D11
Möntu EST 130 F4
Montuïri E 57 B10
Montvalent F 29 F9
Montville F 18 E3
Montzen B 183 D7
Monza I 69 B7
Monzelfeld D 21 E8
Monzingen D 21 E9
Monzón E 42 D4
Monzón de Campos E 40 D3
Mook NL 183 C7
Moone IRL 7 G9
Moordorf (Südbrookmerland) D 17 B8
Moorends UK 11 D10
Moorenweis D 75 F9
Moorfields UK 4 F4
Moorrege D 82 C7
Moorslede B 19 C7
Moorweg D 17 B9
Moos D 76 E3
Moosbach D 75 E11
Moosburg D 73 C9
Moosburg an der Isar D 75 F10
Moosinning D 75 F10
Mooste EST 131 E14
Mór H 149 B10
Mora E 46 E5
Mora P 50 B3
Mora S 97 B14
Mora S 102 C3
Móra d'Ebre E 42 E5
Mora de Rubielos E 48 D3
Moradillo de Roa E 40 E4
Morąg PL 139 C8
Morais P 39 E6
Moraïtika GR 168 E2
Morakovo MNE 163 D7
Moral de Calatrava E 54 B5
Moraleda de Zafayona E 53 B9
Moraleja E 45 D7
Moraleja del Vino E 39 F8
Moraleja de Sayago E 45 B9
Morales de Campos E 39 E9
Morales del Vino E 39 F8
Morales de Toro E 39 E9
Morales de Valverde E 39 E8
Moralina E 39 F7
Moräng S 118 D3
Morano Calabro I 60 D6
Morano sul Po I 68 C5
Morar UK 4 B5
Morărești RO 160 C5
Morasverdes E 45 C8
Morata de Jalón E 41 F9
Morata de Tajuña E 46 D5
Moratalla E 55 C9
Morava BG 166 C4
Morava MD 154 F2
Morava SLO 73 E10
Moravany CZ 77 B10
Moravany CZ 77 D11
Moravany SK 145 F4
Moravče SLO 73 D10
Moravice HR 67 B11
Moraviţa RO 159 C7
Moravka BG 166 C6
Morávka CZ 147 B7
Moravská Třebová CZ 77 C11
Moravské Budějovice CZ 77 D9
Moravské Lieskové SK 146 D5
Moravský Svätý Ján SK 77 E12
Morawica PL 143 E10
Morbach D 21 E8
Morbegno I 69 A8
Morbier F 31 B9
Mörbisch am See A 149 A7
Mörby S 99 C10
Mörbylånga S 89 B10
Morcenx F 32 B4
Morciano di Leuca I 61 D10
Morciano di Romagna I 66 E6
Morcone I 60 A3
Mordelles F 23 D8
Mordoğan TR 177 B8
Mordy PL 141 F7
More LV 135 B10
Moréac F 22 E6
Moreanes P 50 D4
Morebattle UK 5 D12
Morecambe UK 10 C6
Moreda E 39 B8
Moreda E 55 E6
Morée F 24 E5
Mörel CH 70 E6
Morella E 42 F3
Moreni RO 161 D7
Morenish UK 5 C8
Morentin E 32 E1
Moreruela de Tábara E 39 E8
Mores I 64 B2
Morestel F 31 D7
Moretonhampstead UK 13 D7
Moreton-in-Marsh UK 13 B11
Moret-sur-Loing F 25 D8
Moretta I 37 B7
Moreuil F 18 E5
Morez F 31 B9
Morfa Nefyn UK 10 F2
Morfasso I 69 D8
Morfi GR 168 D5
Morfou CY 177 A8
Morfovouni GR 169 F6
Morgat F 22 D2
Morgedal N 95 D8
Mörhange F 26 C6
Morges CH 31 B10
Morgex I 31 D11
Mori I 69 B10
Moria GR 177 A8
Moriles E 53 B7
Morille E 45 C9
Morina AL 163 E10
Morinë RKS 163 E9

Moringen D 78 C6
Morino I 62 D4
Moritzburg D 80 D5
Morjärv S 118 B9
Mörk N 104 F3
Mørke DK 86 C6
Morkkaperä FIN 115 E1
Mørkøv DK 87 D9
Morkovice CZ 77 D12
Mörkret S 102 C4
Morlaàs F 32 D5
Morlaix F 22 C4
Morlanne F 32 C4
Morlanwelz B 19 D9
Mörlenbach D 21 E11
Morley F 26 C3
Morley UK 11 D8
Morley's Bridge IRL 8 E4
Mörlunda S 89 A9
Mormanno I 60 D6
Mormant F 25 C8
Mormoiron F 35 B9
Mornant F 30 D6
Mornas F 35 B8
Mornese I 37 B9
Moroeni RO 161 C6
Morolo I 62 D4
Morón de Almazán E 41 F7
Morón de la Frontera E 51 E9
Moros E 41 F8
Morosaglia F 37 G10
Morottaja FIN 115 E4
Morović SRB 158 C3
Morozova RUS 129 F14
Morozovo BG 166 E5
Morozzo I 37 C7
Morpeth UK 5 E13
Mørrevatnet N 104 D8
Morrfjord N 110 D8
Morriston UK 13 B7
Morrovalle I 67 F8
Mörrum S 89 C7
Morsbach D 21 C9
Morschen D 78 D6
Morshyn UA 145 E8
Mörsil S 105 E15
Morskoga S 97 C13
Morskogen N 95 B14
Morsum D 17 C12
Mørsvikbotn N 111 E10
Mortagne-au-Perche F 24 C4
Mortagne-sur-Gironde F 28 E4
Mortagne-sur-Sèvre F 28 A4
Mortágua P 44 C4
Mortain F 23 C10
Mortara I 68 C6
Morteau F 31 A10
Morteaux-Coulibœuf F 23 C11
Mörtebo S 103 E12
Mortegliano I 73 E7
Mortelle I 59 C8
Morteni RO 160 E6
Mortensnes N 112 C11
Mortensnes N 114 C6
Mortimer's Cross UK 13 A9
Morton UK 11 F11
Mortrée F 23 C12
Mörtschach A 73 C6
Mortsel B 19 B9
Mörtsjön S 106 D7
Mortsund N 110 D6
Morud DK 86 E6
Morunglav RO 160 E4
Morville UK 10 F7
Mor'ye RUS 129 E15
Moryń PL 84 E6
Morzeszczyn PL 138 C6
Morzine F 31 C10
Moșana MD 153 A11
Mosbach D 21 F12
Mosbjerg DK 90 D7
Mosborough UK 11 E9
Mosby N 90 C2
Moscavide P 50 B1
Moščenica HR 148 F5
Moščenička Draga HR 67 B9
Moschopotamos GR 169 D7
Mosciano Sant'Angelo I 62 B5
Mościcha PL 140 C8
Moscovei MD 154 E2
Moscow UK 5 D8
Moseby DK 86 A5
Mosédis LT 134 D3
Mosel D 79 E11
Möser D 79 B10
Mosina PL 81 B11
Mosjø S 107 E13
Mosjøen N 108 E5
Moskaret N 101 B12
Mosko BIH 162 D5
Moskorzew PL 143 E8
Moskosel S 109 E17
Moskuvaara FIN 115 C1
Moslavna Podravska HR 149 E9
Moșna RO 152 E4
Moșna RO 153 D11
Moșnița Nouă RO 159 B7
Moșoaia RO 160 D5
Moso in Passiria I 72 C3
Mosonmagyaróvár H 146 F4
Mosonszolnok H 146 F4
Mošovce SK 147 D7
Mosqueruela E 48 D3
Moss N 91 A14
Mossala FIN 126 E5
Mossat UK 3 L11
Mossbo S 103 D11
Mössingen D 27 D11
Mossley UK 4 F5
Mossley UK 11 D7
Moss-side UK 4 E4
Mosstakan S 96 C7
Most BG 166 F5
Most CZ 76 A5
Mostar BIH 157 F8
Mosteiro E 38 B5
Mosteiro P 44 E5
Mosteiro P 50 D3
Mostek CZ 77 B9
Mosterhamn N 94 C2
Most na Soči SLO 73 D8
Móstoles E 46 D5
Mostová SK 146 E5
Mostowo PL 85 B11
Mostys'ka UA 144 D7
Mosty u Jablunkova CZ 147 B7
Mosvik N 105 D10
Mosyr UA 144 A9

Moszczenica PL 143 C8
Mota del Cuervo E 47 F7
Mota del Marqués E 39 E9
Moțăieni RO 160 C6
Motala S 92 B6
Motarzyno PL 85 B12
Moțăței RO 159 E11
Moţca RO 153 C9
Motherwell UK 5 D9
Môtiers CH 31 B10
Motike BIH 156 D6
Motike BIH 157 C7
Motilla del Palancar E 47 E9
Motilleja E 47 F9
Motjärnshyttan S 97 C10
Motovun HR 67 B8
Motril E 53 C9
Motru RO 159 D10
Motta Montecorvino I 63 D8
Motta San Giovanni I 59 C8
Motta Visconti I 69 C6
Motten D 74 B6
Möttingen D 75 E8
Mottola I 61 B8
Möttönen FIN 123 D13
Mötz A 71 C11
Mou DK 86 B6
Moucha GR 174 A4
Mouchamps F 28 B3
Mouchan F 33 C6
Mouchard F 31 B8
Moudon CH 31 B10
Moudros GR 171 E8
Mougins F 36 D6
Mouhijärvi FIN 127 B9
Moulars... Mouláres F 33 B10
Mouliherne F 23 F12
Moulin-Neuf F 33 D9
Moulins F 30 B3
Moulins-Engilbert F 30 B4
Moulins-la-Marche F 24 C3
Moulis-en-Médoc F 28 E4
Moulismes F 29 C7
Moult F 23 B11
Moulton UK 11 F11
Moulton UK 15 C11
Mountbellew IRL 6 F6
Mount Bellew IRL 6 F6
Mountbenger UK 5 D10
Mountcharles IRL 6 C6
Mountcollins IRL 8 D4
Mount Hamilton UK 4 F2
Mountjoy UK 7 C9
Mountmellick IRL 7 F8
Mount Norris UK 7 D10
Mount Nugent IRL 7 E8
Mountrath IRL 7 F8
Mountshannon IRL 8 C6
Mountsorrel UK 11 F9
Moura P 50 C5
Mourão P 50 C5
Mourenx F 32 D4
Mouriès F 35 C8
Mouries GR 169 B8
Mouriki GR 175 C7
Mourisca do Vouga P 44 C4
Mouriscas P 44 F4
Mournies GR 178 E7
Mourujärvi FIN 121 B11
Mouscron B 19 C7
Moussac F 35 C7
Moussey F 27 D7
Moussoulens F 33 D10
Moussy F 25 F9
Moustéru F 22 C5
Moustey F 32 B4
Mousthéni GR 170 C6
Moustiers-Ste-Marie F 36 D4
Mouthier-Haute-Pierre F 31 A9
Mouthiers-sur-Boëme F 29 D6
Mouthoumet F 34 E4
Moutier CH 27 F7
Moutier-d'Ahun F 29 C10
Moûtiers F 31 E10
Moutiers-les-Mauxfaits F 28 C3
Moutnice CZ 77 D11
Moutsouna GR 177 E7
Moux F 34 D4
Moux-en-Morvan F 25 F11
Mouy F 18 F5
Mouzaki GR 169 F6
Mouzaki GR 174 D4
Mouzay F 19 E11
Mouzon F 19 E11
Moviken S 103 C12
Movila RO 155 D1
Movila Miresii RO 155 C1
Movileni RO 153 C10
Movileni RO 160 E5
Movileni RO 153 F10
Moviliţa RO 161 D10
Moviliţa RO 161 E9
Moville IRL 4 E2
Movollen N 105 F11
Mowtie UK 5 B12
Moy UK 3 L8
Moy UK 7 D9
Moya E 47 E10
Moyard IRL 6 E2
Moyasta IRL 8 C3
Moycullen IRL 6 F4
Moy-de-l'Aisne F 19 E7
Moyenmoutier F 27 D6
Moyenneville F 18 D4
Moygashel UK 7 D9
Moylough IRL 6 F5
Moymore IRL 8 C5
Moyne IRL 7 E7
Moyvalley IRL 7 F9
Moyvore IRL 7 E7
Mozac F 30 D3
Mozăceni RO 160 D6
Mozárbez E 45 C9
Mozelj SLO 73 E10
Mozirje SLO 73 D10
Mozoncillo E 46 B4
Mozsgó H 149 D9
Mozyr' RUS 136 E3
Mračaj BIH 157 E7
Mrągowo PL 136 E4
Mrákov CZ 76 D3
Mrakovica BIH 157 B6

Mramor BIH 157 F8
Mramorak SRB 159 D6
Mratinje BIH 157 F10
Mrčajevci SRB 158 F6
Mrežičko NMK 169 B6
Mrkalji BIH 157 D10
Mrkonjić-Grad BIH 157 D7
Mrkopalj HR 67 B10
Mrmoš SRB 164 B3
Mrocza PL 85 D13
Mroczeń PL 142 D4
Mroczków PL 141 H3
Mroczno PL 139 D8
Mrozy PL 141 F5
Mścice PL 85 B10
Mściwojów PL 81 D10
Mšené Lázně CZ 76 B6
Mšeno CZ 77 B7
Mshinskaya RUS 132 C6
Mstów PL 143 E7
Mszana PL 147 B7
Mszana Dolna PL 144 D1
Mszczonów PL 141 G3
Muccia I 62 A4
Much D 21 C8
Muchalls UK 5 A12
Mucharz PL 147 B9
Mücheln (Geiseltal) D 79 D10
Muchow D 83 D11
Muchówka PL 144 D1
Much Wenlock UK 10 F6
Mucientes E 39 E10
Mücka D 81 D7
Mücke Große-Eichen D 21 C12
Mücke-Nieder-Ohmen D 21 C12
Muckross IRL 8 D4
Múcsony H 145 G2
Mudanya TR 173 D10
Mudau D 187 B7
Müdelheim D 183 C9
Müden (Aller) D 79 A7
Müden (Örtze) D 83 E8
Mudersbach D 185 C8
Müdrets BG 166 E6
Muel E 41 E10
Muelas del Pan E 39 E8
Muff IRL 4 E2
Muga de Sayago E 39 F7
Mugardos E 38 B3
Muge P 44 F3
Mügeln D 80 C4
Mügeln D 80 D4
Mugeni RO 152 E6
Muggensturm D 187 D5
Muggia I 73 E8
Muğla TR 181 B8
Müglen BG 167 D8
Müglizh BG 166 D5
Mugron F 32 C4
Mühlacker D 27 C10
Mühlanger D 79 C12
Mühlbachl A 72 B3
Mühlberg D 79 E8
Mühlberg D 80 D4
Mühldorf A 73 C7
Mühldorf am Inn D 75 F12
Mühldorf bei Feldbach A 148 C5
Mühlen A 73 B10
Mühlenbeck D 84 E4
Mühlhausen D 21 F11
Mühlhausen D 75 D9
Mühlhausen (Thüringen) D 79 D7
Mühltroff D 75 A10
Muhola FIN 123 D14
Muhos FIN 119 E15
Muhr am See D 75 D8
Muhur AL 163 F9
Muineachán IRL 7 D9
Muine Bheag IRL 9 C9
Muiños E 38 E4
Muirdrum UK 5 B11
Muirhead UK 5 C10
Muirkirk UK 5 D8
Muir of Ord UK 2 K8
Muizon F 19 F8
Mujdić BIH 157 D7
Mujejärvi FIN 125 C12
Mukacheve UA 145 G6
Mukařov CZ 77 C7
Mukhavyets BY 141 F9
Mukhovo BG 165 E8
Mukkala FIN 115 D5
Mukkavaara FIN 115 C15
Mula E 55 C10
Mulbarton UK 15 B11
Muleby DK 89 E7
Mulešići BIH 157 C9
Mulfingen D 74 D6
Mülheim an der Ruhr D 17 F7
Mülheim-Kärlich D 185 D7
Mulhouse F 27 E7
Muljava SLO 73 E10
Mullach Íde IRL 7 F10
Mullagh IRL 7 E7
Mullagh IRL 7 F9
Mullaghroe IRL 6 E6
Mullan IRL 7 D9
Mullan IRL 7 D11
Mullartown UK 7 D11
Müllheim D 27 E8
Mullhyttan S 92 A5
Mullinavat IRL 9 D8
Mullingar IRL 7 E8
Mullion UK 12 F4
Müllrose D 80 B6
Mullsjö S 91 D14
Mulrany IRL 6 E3
Mulsanne F 23 E12
Mulseryd S 91 D14
Multia FIN 123 F13
Multiperä FIN 121 C15
Mümliswil CH 27 F8
Munakka FIN 122 E9
Muñana E 45 C10
Munapirtti FIN 128 E6
Müncheberg D 80 A6
München D 75 F9
Münchenbernsdorf D 79 E10
Münchenbuchsee CH 31 A11
Münchhausen D 21 C11
Münchsteinach D 75 C8
Münchweiler an der Rodalb D 186 C4
Münchwilen CH 27 F11
Mundaka E 41 B6
Munderkingen D 187 E8
Mundesley UK 15 B11
Mundford UK 15 B10
Mundheim N 94 B3
Mundolsheim F 27 C8
Munebrega E 41 F8
Munera E 55 A8

Mungia E 40 B6
Mungret IRL 8 C5
Muñico E 45 C10
Muniesa E 42 E2
Munilla E 41 D7
Munka-Ljungby S 87 C11
Munkbyn S 103 B11
Munkebakken N 114 D6
Munkebo DK 86 E7
Munkedal S 91 C10
Munken N 104 D6
Munkflohögen S 106 D7
Munkfors S 97 C10
Munklia N 111 D14
Munksund S 118 D7
Munktorp S 98 C6
Munkzwalm B 19 C8
Munne FIN 128 C7
Münnerstadt D 75 B7
Munningen D 75 E8
Muñogalindo E 46 C3
Munsala FIN 122 D8
Münsingen CH 31 B12
Münsingen D 74 F5
Münster A 72 B4
Münster CH 70 E6
Münster D 17 E9
Münster D 21 E11
Münster D 83 E8
Munster F 27 D7
Münsterdorf D 82 C7
Munstergeleen NL 183 D7
Münsterhausen D 75 F7
Münstermaifeld D 185 D7
Muntendam NL 17 B7
Munteni RO 153 F10
Munteni-Buzău RO 161 D9
Muntenii de Jos RO 153 D11
Münzenberg D 21 C11
Münzkirchen A 76 F5
Muodoslompolo S 117 C10
Muonio FIN 117 C11
Muonionalusta S 117 C11
Muotathal CH 71 D7
Muotkajärvi FIN 117 B10
Muotkavaara FIN 117 C12
Mur SRB 163 C9
Muradiye TR 173 D9
Muradiye TR 177 B9
Murakeresztúr H 149 D7
Murán SK 147 D10
Muras E 38 B4
Murasson F 34 C4
Muraste EST 131 B8
Muraszemenye H 149 D7
Murat F 30 E2
Muratlar TR 181 B9
Muratli TR 173 B9
Murato F 37 F10
Murat-sur-Vèbre F 34 C4
Murau A 73 B9
Muravera I 64 E4
Murazzano I 37 C8
Murça P 38 F5
Murchante E 41 D8
Mürchevo BG 165 B7
Murchin D 84 C5
Murcia E 56 F2
Murczyn PL 138 E4
Mur-de-Barrez F 29 F11
Mûr-de-Bretagne F 22 D6
Mur-de-Sologne F 24 F6
Mureck A 148 C5
Mürefte TR 173 C7
Muret F 33 D8
Murgeni RO 153 E12
Murgenthal CH 27 F8
Murgeşti RO 161 C9
Murgia E 40 C6
Muri CH 27 F9
Muri CH 31 B11
Murias de Paredes E 39 C7
Muriedas E 40 B4
Murighiol RO 155 C4
Murillo de Río Leza E 32 F1
Murillo el Fruto E 32 F3
Murino MNE 163 C8
Murisengo I 68 C5
Murjäni LV 135 B9
Murjek S 116 F5
Murley UK 7 D8
Murlo I 66 F3
Murmastiene LV 135 C13
Murnau am Staffelsee D 72 A3
Muro E 57 B11
Muro F 37 F9
Muro P 38 F2
Muro de Alcoy E 56 D4
Murol F 30 D2
Murole FIN 127 B10
Muro Lucano I 60 B4
Muron F 28 C4
Murony H 151 D7
Muros E 38 C1
Muros E 39 A7
Muros I 64 B2
Murovane UA 144 C9
Murów PL 142 E4
Murowana Goślina PL 81 A12
Murrë AL 168 A3
Murrhardt D 74 E6
Murronkylä FIN 119 E16
Murroogh IRL 6 F4
Mursalli TR 177 D10
Mûrs-Erigné F 23 F10
Murska Sobota SLO 148 C6
Mursko Središče HR 149 C6
Murtas E 55 F6
Murtede P 44 D4
Murten CH 31 B11
Murter HR 156 E4
Murtino NMK 169 B8
Murto FIN 119 E15
Murtolahti FIN 125 D9
Murtomäki FIN 124 B9
Murtovaara FIN 121 C13
Murumoen N 105 C16
Murvica FIN 156 D3
Murviel-lès-Béziers F 34 D5
Mürzsteg A 148 A5
Murzynowo PL 81 A8
Mürzzuschlag A 148 A5
Mūsa LV 135 D8
Musbury UK 13 D8
Müschenbach D 185 C8
Musei I 64 E2
Muselievo BG 160 F5
Mushtisht RKS 163 E10
Musile di Piave I 72 E6
Muskö S 93 B12
Mussalo FIN 128 D6
Musselburgh UK 5 D10
Musselkanaal NL 17 C8

Mussidan F 29 E6
Mussomeli I 58 D4
Musson B 19 E12
Mussy-sur-Seine F 25 E12
Mustafakemalpaşa TR 173 D9
Müstair CH 71 D10
Mustamaa FIN 119 F17
Mustamaa FIN 123 D10
Mustasaari FIN 122 D7
Mustavaara FIN 121 D12
Mustavaara FIN 121 F12
Mustinlahti FIN 125 E10
Mustjala EST 131 E11
Mustola FIN 114 F4
Mustolanmäki FIN 125 C10
Mustolanmutka FIN 125 B10
Mustvee EST 131 D13
Muszaki PL 139 D10
Muszyna PL 145 E2
Muta SLO 73 C11
Mutala FIN 127 B10
Mutalahti FIN 125 F16
Müttevelli TR 177 B10
Muthill UK 5 C9
Mutilva Baja E 32 E2
Mutné SK 147 C8
Mutriku E 32 D1
Mutterstadt D 21 F10
Mutxamel E 56 E4
Mutzig F 27 C7
Mutzschen D 80 D3
Muuga EST 131 C13
Muukajärvi S 116 E10
Muuksi EST 131 B11
Muurame FIN 123 F15
Muuratjärvi FIN 123 C14
Muurikkala FIN 128 D8
Muurla FIN 127 E9
Muurola FIN 119 B14
Muurola FIN 128 D8
Muuruvesi FIN 125 D10
Muxía E 38 B1
Muzillac F 22 E7
Mužla SK 149 A11
Myahuny BY 137 C14
Myakishevo RUS 133 C5
Myaretskiya BY 133 F3
Myazhany BY 135 E13
Mybster UK 3 J10
Myckelgensjö S 107 D13
Myckle S 118 E5
Myedna BY 141 G9
Myggenäs S 91 C10
Myggsjö S 102 C8
Myhinpää FIN 124 F7
Myjava SK 146 D5
Mykanów PL 143 E7
Mykhal'cha UA 153 A7
Mykhaylivka UA 154 F5
Myki GR 171 B7
Myklebostad N 110 E9
Mykolaiv UA 145 D8
Mykolayivka UA 154 F4
Mykolayivka-Novorosiys'ka
UA 154 E5
Mykonos GR 176 E5
Mykulychyn UA 152 A5
Mykytychi UA 144 B9
Myllykoski FIN 128 D6
Myllykylä FIN 122 E8
Myllykylä FIN 127 E10
Myllykylä FIN 128 D7
Myllylahti FIN 121 D13
Myllymäki FIN 123 E12
Myloi GR 175 D6
Mylopotamos GR 178 C4
Mynämäki FIN 126 D7
Mynttilä FIN 128 C8
Myon F 31 A8
Myory BY 133 E3
Myra GR 169 F8
Myrås S 109 E14
Myre N 110 D5
Myre N 111 B10
Myresjö S 92 E5
Myrhaug N 101 A14
Myrheden S 118 D4
Myrhult S 92 B4
Myrina GR 171 E8
Myriokefala GR 178 E7
Myrkky FIN 122 F7
Myrland N 110 D5
Myrland N 110 D9
Myrland N 111 C10
Myrlandshaugen N 111 C13
Myrmoen N 101 A15
Myrne UA 155 B4
Myrnes N 112 C9
Myrnopillya UA 154 E4
Myrsini GR 174 D3
Myrsini GR 178 B3
Myrskylä FIN 127 D14
Myrties GR 177 F8
Myrtos GR 179 E10
Myrviken S 105 E16
Mysen N 95 C14
Myshall UK 9 C9
Myślachowice PL 143 F7
Myślakowice PL 81 E9
Myślenice PL 147 B9
Myślibórz PL 85 E7
Myślice PL 139 C8
Mysłowice PL 143 F7
Mysovka RUS 134 F2
Myssjö S 102 A7
Mystegna GR 177 A7
Mystras GR 174 E5
Myszków PL 143 E7
Myszyniec PL 139 D11
Mytikas GR 174 B2
Mytilini GR 177 A8
Mytilinioi GR 177 D8
Mytna SK 147 E9
Mýto CZ 76 C5

N

Nå N 94 B5
Naaldwijk NL 16 E2
Naamankylä FIN 119 E17
Naamijoki FIN 117 E11
Naantali FIN 126 E6
Naarden NL 183 A6
Näärinki FIN 128 B8
Naapurinvaara FIN 121 F11
Naarn im Machlande A 77 F7
Naartijärvi S 119 C11
Naarva FIN 125 D16
Naas IRL 7 F9
Näätämö FIN 114 D6
Näätänmaa FIN 125 F10
Näätävaara FIN 121 E13

Nabburg D 75 D11
Nábrád H 145 G5
Na Cealla Beaga IRL 6 C6
Načeradec CZ 77 C7
Nacha BY 137 E10
Náchod CZ 77 B10
Nacina Ves SK 145 F4
Näckådalen S 102 D7
Nackel D 83 E13
Nackenheim D 185 E9
Näcksjö S 103 C12
Na Clocha Liathe IRL 7 F10
Nacpolsk PL 139 E9
Nad IRL 8 D5
Nadalj SRB 158 C4
Nadarzyce PL 85 D11
Nadarzyn PL 141 F3
Naddvik N 100 D7
Nadeş RO 152 E5
Nădlac RO 150 E6
Nădrag RO 159 B9
Nadrichne UA 154 E4
Nádudvar H 151 C7
Näeni RO 161 C8
Nærbø N 94 E3
Næroset N 101 E13
Nærsnes N 95 C13
Næsbjerg DK 86 D3
Næstved DK 87 E9
Näfels CH 27 F11
Nafferton UK 11 C11
Nafpaktos GR 174 C4
Nafplio GR 175 D6
Nagele NL 16 C5
Naggen S 103 C10
Naglarby S 97 B14
Na Gleannta IRL 6 C6
Nagli LV 133 C1
Nagłowice PL 143 E9
Nagold D 27 C10
Nagore E 32 E3
Nago-Torbole I 69 B10
Nagu FIN 126 E6
Nagyatád H 149 D8
Nagybajom H 149 D9
Nagybánhegyes H 151 E6
Nagybaracska H 150 E2
Nagybarca H 145 G2
Nagyberény H 149 C10
Nagyberki H 149 D9
Nagycenk H 149 A7
Nagycsécs H 145 H2
Nagycserkesz H 145 H4
Nagydobos H 145 G5
Nagydorog H 149 C11
Nagyecsed H 145 H5
Nagyfüged H 150 B5
Nagyhalász H 145 G4
Nagyharsány H 149 E10
Nagyhegyes H 151 B7
Nagyigmánd H 149 A10
Nagyiván H 151 C6
Nagykálló H 145 H4
Nagykanizsa H 149 D7
Nagykapornak H 149 C7
Nagykáta H 150 C4
Nagykereki H 151 C8
Nagykőnyi H 149 C9
Nagykőrös H 150 C4
Nagykőrő H 150 C5
Nagykovácsi H 149 A11
Nagylak H 150 E6
Nagylóc H 147 E9
Nagylók H 149 C11
Nagylózs H 149 A7
Nagymágocs H 150 D5
Nagymaros H 149 A11
Nagynyárád H 149 E11
Nagyoroszi H 147 E8
Nagyréde H 150 B4
Nagyrécse H 149 C8
Nagyszénás H 150 D6
Nagyszokoly H 149 C10
Nagytárcsa H 150 B3
Nagytőke H 150 D5
Nagyvarsány H 145 G5
Nagyvázsony H 149 C9
Nagyvisnyó H 145 G1
Naha EST 132 F1
Naharros E 47 D8
Nahe D 83 C8
Nahirne UA 155 C2
Nahrendorf D 83 D9
Naidäş RO 159 D8
Naila D 75 B10
Nailloux F 33 D9
Nailsworth UK 13 B10
Naimakka S 116 A7
Naintré F 29 B6
Naipköy TR 173 C7
Nairn UK 3 K9
Naivasta S 116 E16
Nakkerud N 95 B12
Nakkila FIN 126 C7
Nákło CZ 77 C12
Nakło E 143 E8
Nakło SLO 73 D9
Nakło nad Notecią PL 85 D13
Nakomiady PL 136 E3
Näkotne LV 134 C6
Nakovo SRB 150 F5
Nakskov DK 83 A10
Nalbach D 186 C2
Nalbant RO 155 C3
Nalda E 41 D7
Nálden S 105 E16
Nałęczów PL 141 H6
Nálepkovo SK 145 F2
Näljänkä FIN 121 D11
Nalkki FIN 121 E10
Nalliers F 28 C3
Nalžovské Hory CZ 76 D5
Namborn D 21 E8
Nambroca E 46 E5
Namdalseid N 105 C10
Náměšť nad Oslavou CZ 77 D10
Náměšť na Hané CZ 77 C12
Námestovo SK 147 C8
Nämpnäs FIN 122 E6
Namsos N 105 C11
Namsskogan N 105 B12
Namsvatn N 105 B15
Namur B 19 D10
Namysłów PL 142 D4
Nana RO 161 E9
Nána SK 147 F7
Nançay F 24 F7
Nanclares de la Oca E 40 C6
Nancy F 26 C5

Nandrin B 183 D6
Nănești RO 161 B10
Nangis F 25 C9
Nannestad N 95 B13
Nanov RO 160 F6
Nans-les-Pins F 35 D10
Nant F 34 B5
Nanterre F 25 C7
Nantes F 23 F8
Nanteuil-le-Haudouin F 25 B8
Nantiat F 29 C8
Nantua F 31 C8
Nantwich UK 10 E6
Naousa GR 169 C7
Naousa GR 176 E5
Napajedla CZ 146 C5
Napiwoda PL 139 D9
Napkor H 145 H4
Napola I 58 D2
Napoli I 60 B2
Napp N 110 D5
Näpradea RO 151 C11
Náquera E 48 E4
Når S 93 E13
Nåra N 100 D1
Nárai H 149 B7
Narberth UK 12 B5
Narbolia I 64 C2
Narbonne F 34 D5
Narbonne-Plage F 34 D5
Narborough UK 15 B10
Narbuvoll N 101 B14
Narcao I 64 E2
Narcy F 25 F9
Nardò I 61 C10
Narechenski Bani BG 165 F10
Narew PL 140 E9
Narewka PL 141 E9
Närhilä FIN 123 E16
Narin IRL 6 C6
Närinciems LV 134 B5
Narkaus FIN 119 B16
Narken S 116 E9
Narlidere TR 177 C9
Narni I 62 B3
Naro I 58 E4
Narol PL 144 C7
Närpes FIN 122 F6
Narrosse F 32 C3
Narta HR 149 E7
Nartë AL 168 D1
Nāruja RO 153 F9
Naruska FIN 115 D6
Naruszewo PL 139 E9
Narva EST 132 C3
Narva FIN 127 C10
Narva-Jõesuu EST 132 C3
Närvijoki FIN 122 E7
Narvik N 111 D13
Narzole I 37 B7
Narzym PL 139 D9
Näs FIN 99 B14
Näs S 90 A5
Näs S 93 E12
Näs S 97 B12
Näs S 102 A8
Näsåker S 107 E11
Näsåud RO 152 C4
Nasavrky CZ 77 C9
Näsberg S 103 C10
Nasbinals F 34 A5
Näs bruk S 98 B6
Näsby S 89 C10
Na Sceiri IRL 7 E10
Näset S 103 D9
Nashec RKS 163 E10
Našice HR 149 E10
Nasielsk PL 139 E10
Näske S 107 E15
Näsliden S 107 A14
Naso I 59 C6
Nassau D 21 D9
Nassereith A 71 C11
Nässja S 92 C5
Nässjö S 92 D5
Nässjö S 107 D10
Nassogne B 19 D11
Nästansjö S 107 B11
Nästätten D 185 D8
Nästeln S 102 A7
Nastola FIN 127 D14
Năsturelu RO 161 F6
Năsum S 88 C7
Nasutów PL 141 H6
Näsviken S 103 C12
Näsviken S 106 D9
Naszály H 149 A10
Natalinci SRB 159 E6
Nateby UK 11 C7
Naters CH 68 A4
Natters A 71 C11
Naturno I 71 D11
Naucelle F 33 B10
Naucelles F 29 F10
Naudaskalns LV 133 B2
Nauders A 71 D11
Naudīte LV 134 C6
Nauen D 79 A12
Nauendorf D 79 C10
Nauheim D 21 E10
Naujac-sur-Mer F 28 E3
Naujamiestis LT 135 E8
Naujasis Daugėliškis LT 135 F12
Naujoji Vilnia LT 137 D11
Naukšēni LV 131 F10
Naul IRL 7 E10
Naulaperä FIN 121 E10
Naulavaara FIN 125 C10
Naumburg (Hessen) D 17 F12
Naumburg (Saale) D 79 D10
Naundorf D 80 D4
Naundorf D 80 E4
Naunhof D 79 D12
Nauroth D 21 C9
Naustad N 108 B8
Naustbukta N 105 B11
Naustdal N 100 C3
Nauste N 101 A8
Nautijaur S 109 C17
Nautsi RUS 114 E6
Nautsund N 100 D2
Nava E 39 B9
Navacepeda de Tormes E 45 D10
Navaconcejo E 45 D9
Nava de Arévalo E 46 C3
Nava de la Asunción E 46 B4
Nava del Rey E 39 F9
Nava de Sotrobal E 45 C10

Nandrin B 183 D6
Navadrutsk BY 133 F2
Navafría E 46 B5
Navahermosa E 46 E4
Navajas E 48 E4
Naval E 42 C4
Navalagamella E 46 D4
Navalcaballo E 41 E6
Navalcán E 45 D10
Navalcarnero E 46 D4
Navalero E 40 E6
Navalmoral E 46 D3
Navalmoral de la Mata E 45 E9
Navalonguilla E 45 D10
Navalosa E 46 D3
Navalperal de Pinares E 46 C4
Navalpino E 46 F3
Navaluenga E 46 D3
Navalvillar de Ibor E 45 E10
Navalvillar de Pela E 45 F10
Navamorcuende E 46 D3
Navan IRL 7 E9
Navapolatsk BY 133 E5
Navarcles E 43 D7
Navardún E 32 E3
Navarredonda de la Rinconada E 45 D8
Navarrenx F 32 D4
Navarrés E 48 F3
Navarrete E 41 D6
Navarrevisca E 46 D3
Navàs E 43 D7
Navascués E 32 E3
Navas de Estrena E 46 E3
Navas de Jorquera E 47 F9
Navas del Madroño E 45 E7
Navas del Rey E 46 D4
Navas de Oro E 46 B4
Navas de San Juan E 55 C6
Navasfrías E 45 D7
Navata E 43 C9
Navatalgordo E 46 D3
Nave I 69 B9
Nave P 50 E2
Nave de Haver P 45 C7
Năvekvarn S 93 B9
Navelli I 62 C5
Náverdal N 101 A12
Näverede S 106 E8
Nave Redonda P 50 E3
Näverkärret S 97 C14
Näverrys FIN 129 C15
Naverstad S 91 B10
Navès E 43 D7
Naves F 29 E9
Navezuelas E 45 E10
Navia E 39 A6
Navilly F 31 B7
Navit N 112 D8
Năvodari RO 155 E3
Năvrăgol S 89 C9
Nawojowa PL 145 D2
Naxos GR 176 E5
Nay-Bourdettes F 32 D5
Nazaré P 44 E2
Nazza D 79 D7
Ndroq AL 168 B2
Nea Agathoupoli GR 169 D8
Nea Alikarnassos GR 178 E9
Nea Anchialos GR 169 F8
Nea Apollonia GR 169 C10
Nea Artaki GR 175 B8
Nea Efesos GR 169 D7
Nea Epidavros GR 175 D7
Nea Figaleia GR 174 E4
Nea Filadelfeia GR 175 C8
Nea Fokaia GR 169 D9
Nea Ionia GR 169 F8
Nea Iraklitsa GR 171 C6
Nea Kallikrateia GR 169 D9
Nea Karvali GR 171 C6
Nea Karya GR 171 C7
Nea Kerdylia GR 169 C10
Nea Kios GR 175 D6
Nea Koroni GR 174 F4
Nea Lampsakos GR 175 C8
Neale IRL 6 E4
Nea Liosia GR 175 C8
Nea Madytos GR 169 C10
Nea Makri GR 175 C8
Nea Malgara GR 169 C8
Nea Mesimvria GR 169 C8
Nea Michaniona GR 169 D8
Nea Moudania GR 169 D9
Nea Olynthos GR 169 D9
Nea Pella GR 169 C8
Nea Peramos GR 171 C6
Nea Peramos GR 175 C7
Nea Plagia GR 169 D9
Neapoli GR 168 D5
Neapoli GR 178 B5
Neapoli GR 179 E10
Nea Poteidaia GR 169 D9
Nea Roda GR 170 D5
Nea Santa GR 169 C8
Nea Santa GR 171 B9
Nea Silata GR 171 B9
Nea Styra GR 175 C9
Neath UK 13 B7
Nea Tirynitha GR 175 D6
Nea Triglia GR 169 D9
Neaua RO 152 E5
Nea Vravrona GR 175 C9
Nea Vyssa GR 171 A11
Nea Zichni GR 169 B10
Nebel D 82 A4
Nébias F 33 E10
Nebljusi HR 156 D4
Nebra (Unstrut) D 79 D10
Nebreda E 40 E4
Nechanice CZ 77 B9
Neckarbischofsheim D 187 C6
Neckargemünd D 21 F11
Neckargerach D 21 F11
Neckarsteinach D 21 F11
Neckarsulm D 21 F12
Neckartenzlingen D 27 C11
Necşeşti RO 160 E5
Necton UK 15 B10
Nečujam FIN 156 F5
Neda E 38 B3
Nedansjö S 103 B12
Nedashiv CZ 146 C6
Neddemin D 84 C4
Nedelino BG 171 B8
Nedelišće HR 149 D6
Nederby DK 86 B4
Nederhogdal S 102 B7
Nederhorst den Berg NL 183 A6
Neder Hvam DK 86 C4
Nederlangbroek NL 183 A6
Nedervetil FIN 123 C10

Neder Vindinge DK 87 E9
Nederweert NL 16 F5
Nedlitz D 79 B11
Nedožery-Brezany SK 147 D7
Nedrebø N 94 E4
Nedre Saxnäs S 109 F14
Nedre Soppero S 116 C7
Nedstrand N 94 D3
Nedvědice CZ 77 D10
Nedyalsko BG 167 E7
Njdza PL 142 F5
Neede NL 17 D7
Needham Market UK 15 C11
Neer NL 183 C7
Neerijnen NL 183 B6
Neermoor D 17 B8
Neeroeteren B 183 C7
Neerpelt B 19 B11
Neetze D 83 D9
Nefyn UK 10 F2
Negenborn D 78 C6
Negoi RO 160 F2
Negomir RO 159 D10
Negorci NMK 169 B7
Negoslavci HR 157 B11
Negotin SRB 159 E10
Negotino NMK 169 B7
Negotino NMK 169 B7
Negrar I 66 A2
Negraşi RO 160 D6
Negredo E 47 B7
Negreira E 38 C2
Nègrepelisse F 33 B9
Negreşti RO 153 C10
Negreşti-Oaş RO 145 H7
Negri RO 153 D9
Negru Vodă RO 155 F2
Nehoiu RO 161 C8
Neiden N 114 D6
Neidín IRL 8 E3
Neitakaite S 116 E8
Neitsuanto S 116 D5
Neittävä FIN 119 E16
Neive I 37 B8
Nejdek CZ 75 B12
Nekézseny H 145 G1
Nekla PL 81 B12
Nekşö DK 89 E8
Nelas P 44 C5
Nellim FIN 114 F4
Nellingen D 74 E6
Nelson UK 11 D7
Nematoruys LT 137 D9
Neman RUS 136 C5
Nemanjica NMK 164 F4
Nemanskoye RUS 136 C5
Nembro I 69 B8
Nemea GR 175 D6
Nemenčinė LT 137 D11
Nemesgulács H 149 C8
Nemesnádudvar H 150 E3
Nemesvámos H 149 B9
Nemesvid H 149 C8
Németkér H 149 C11
Nemežis LT 137 D11
Nemours F 25 D8
Nemsdorf-Göhrendorf D 79 D10
Nemšová SK 146 D6
Nemunaitis LT 137 E9
Nemunėlio Radviliškis LT 135 D9
Nemyriv UA 144 C7
Nenagh IRL 8 C6
Nendaz CH 31 C11
Nenince SK 147 E8
Nenita GR 177 C7
Nennhausen D 79 A12
Nennslingen D 75 D9
Nenonpelto FIN 124 F8
Nentershausen D 21 D9
Nentershausen D 78 D6
Nenthead UK 5 F8
Nenzing A 71 C9
Neo Agioneri GR 169 C8
Neochoraki GR 175 C7
Neochori GR 169 B9
Neochori GR 171 B10
Neochori GR 174 A3
Neochori GR 174 C3
Neo Erasmio GR 171 C7
Neoi Epivates GR 169 C8
Neo Monastiri GR 169 F7
Neoneli I 64 C2
Neo Petritsi GR 169 B9
Neorić HR 156 E6
Neos Kafkasos GR 168 C5
Neos Marmaras GR 169 D10
Neos Mylotopos GR 169 C7
Neo Souli GR 169 B10
Neos Pagontas GR 175 B8
Neos Pyrgos GR 175 B7
Neos Skopos GR 169 B10
Néoules F 36 E4
Nepi I 62 C2
Nepomuk CZ 76 D5
Nérac F 33 B6
Neratovice CZ 77 B7
Nerău RO 150 F6
Neravai LT 137 E9
Nerchau D 79 D12
Nercillac F 28 D5
Nerdal N 101 A9
Nerde Gårdsjö S 103 E9
Néré F 28 D5
Nereju RO 153 F9
Neresheim D 75 E7
Neresnica SRB 159 E8
Neresnytsya UA 145 G8
Nereta LV 135 D10
Nereto I 62 B5
Nerezine HR 67 C9
Neris LV 135 D10
Nerja E 53 C9
Nerkoo FIN 124 D8
Nerlia N 108 E7
Nerokouros GR 178 E7
Néronde F 30 D5
Nérondes F 30 B2
Neroth D 21 D7
Nerpio E 55 D8
Nersac F 28 D6
Nersingen D 75 F7
Nerskogen N 101 A11
Nerushay UA 155 B5
Nerva E 51 D6
Nervesa della Battaglia I 72 E5
Nervieux F 30 D5
Nes FO 2 A3
Nes N 95 D9
Nes N 96 A2
Nes N 110 D9

Nes N 111 D10
Nes NL 16 B5
Nesbyen N 101 E10
Neschwitz D 80 D6
Nesebūr BG 167 D9
Neset N 112 C7
Nes Flaten N 94 C5
Nesgrenda N 90 B5
Nesheim N 94 D3
Nesje N 110 D8
Nesjegjerde N 100 A6
Nesland N 110 D5
Neslandsvatn N 90 B5
Nesle F 18 E6
Nesna N 108 D5
Nesovice CZ 77 D12
Nessa F 37 F9
Nesse D 17 A8
Nesseby N 114 C6
Nesselwang D 71 B11
Nesslau D 27 F11
Nessodtangen N 95 C13
Nestani GR 175 D5
Nestby N 108 D6
Nesterov RUS 136 D3
Neston UK 10 E5
Nestorio GR 168 D5
Nestoyita UA 154 B4
Nesttun N 94 B2
Nesvady SK 146 F6
Nesvatnstemmen N 90 B3
Nesvik N 94 D4
Nethy Bridge UK 3 L9
Netolice CZ 76 D6
Netphen D 21 C10
Netra (Ringgau) D 79 D7
Netretić HR 148 E4
Netstal CH 71 C8
Nettancourt F 25 C12
Nettersheim D 21 D7
Nettetal D 16 F6
Nettuno I 62 E3
Netvořice CZ 77 C7
Neu-Anspach D 21 D11
Neubrandenburg N 90 B3
Neuberg an der Mürz A 148 A5
Neubeuern D 72 A5
Neubiberg D 75 F10
Neubrandenburg D 84 C4
Neubruchhausen D 17 C11
Neubrunn D 187 B8
Neubukow D 83 B11
Neubulach D 27 C10
Neuburg am Rhein D 187 D5
Neuburg an der Donau D 75 E9
Neuburg-Steinhausen D 83 C11
Neuburxdorf D 80 D4
Neuchâtel CH 31 B10
Neu Darchau D 83 D9
Neudietendorf D 79 E8
Neudorf A 164 F4
Neudrossenfeld D 75 B10
Neuenbürg D 27 C10
Neuendettelsau D 75 D8
Neuenhagen Berlin D 80 A5
Neuenhaus D 17 D7
Neuenhof D 21 C8
Neuenkirchen (Oldenburg) D 17 C10
Neuenkirchen-Seelscheid D 21 C8
Neuenrade D 185 B8
Neuenstadt am Kocher D 27 B11
Neuenstein D 187 D7
Neuenwalde D 17 A11
Neuerburg D 20 D6
Neufahrn bei Freising D 75 F10
Neufahrn in Niederbayern D 75 E11
Neufchâteau B 19 E11
Neufchâteau F 26 D4
Neufchâtel-en-Bray F 18 E3
Neufchâtel-Hardelot F 15 F12
Neufchâtel-sur-Aisne F 19 F9
Neufeld an der Leitha A 77 G10
Neuffen D 27 D11
Neufmanil F 184 E2
Neufra D 27 D11
Neugersdorf D 81 E7
Neuharlingersiel D 17 A9
Neuhaus A 73 A11
Neuhaus A 73 C10
Neuhaus (Oste) D 17 A12
Neuhaus am Inn D 76 F3
Neuhaus am Klausenbach A 148 C6
Neuhaus am Rennweg D 75 A9
Neuhaus an der Pegnitz D 75 C10
Neuhausen CH 27 E10
Neuhausen D 80 E4
Neuhausen D 187 D6
Neuhausen ob Eck D 27 E10
Neuhof D 74 B5
Neuhof an der Zenn D 75 D8
Neuhofen an der Krems A 76 F6
Neuillé-Pont-Pierre F 24 E4
Neuilly F 25 F10
Neuilly-en-Thelle F 18 F5
Neuilly-le-Réal F 30 C3
Neuilly-l'Évêque F 26 E3
Neuilly-St-Front F 25 B9
Neu-Isenburg D 187 A6
Neukalen D 83 C13
Neu Kaliß D 83 D10
Neukirch D 80 D6
Neukirch D 21 C12
Neukirchen A 148 B4
Neukirchen D 83 B10
Neukirchen D 83 B10
Neukirchen am Großvenediger A 72 B5
Neukirchen an der Enknach A 76 F4
Neukirchen an der Vöckla A 76 F5
Neukirchen-Balbini D 75 D11
Neukirchen beim Heiligen Blut D 76 D3
Neukirchen vorm Wald D 76 E4
Neukloster D 83 C11

Neulengbach A 77 F9
Neuler D 75 E7
Neulewin D 84 E6
Neulikko FIN 121 E10
Neulise F 30 D5
Neu Lübbenau D 80 B5
Neum BIH 162 D4
Neumagen D 185 E6
Neumark D 79 E11
Neumarkt am Wallersee A 73 A7
Neumarkt im Mühlkreis A 77 F6
Neumarkt in der Oberpfalz
D 75 D9
Neumarkt in Steiermark A 73 B9
Neumarkt-Sankt Veit D 75 F12
Neu Mukran D 84 B5
Neumünster D 83 B7
Neunburg vorm Wald D 75 D11
Neundorf D 75 A11
Neung-sur-Beuvron F 24 E6
Neunkirch CH 27 E10
Neunkirchen A 148 A6
Neunkirchen D 21 C10
Neunkirchen D 73 E11
Neunkirchen am Brand D 75 C9
Neunkirchen am Sand D 75 C9
Neuötting D 75 F12
Neupetershain D 80 C6
Neupölla A 77 E8
Neureichenau D 76 E5
Neuruppin D 83 E13
Neuschönau D 76 E4
Neusiedl am See A 77 G11
Neusorg D 75 C10
Neuss D 21 B7
Neussargues-Moissac F 30 E2
Neustadt D 27 E9
Neustadt D 79 D10
Neustadt D 83 E12
Neustadt (Harz) D 79 C8
Neustadt (Hessen) D 21 C12
Neustadt (Wied) D 21 C8
Neustadt am Kulm D 75 C10
Neustadt am Rübenberge
D 78 A5
Neustadt an der Aisch D 75 C8
Neustadt an der Donau
D 75 E10
Neustadt an der Waldnaab
D 75 C11
Neustadt an der Weinstraße
D 21 F10
Neustadt bei Coburg D 75 B9
Neustadt-Glewe D 83 D11
Neustadt in Holstein D 83 B9
Neustadt in Sachsen D 80 D6
Neustift im Stubaital A 72 B3
Neustrelitz D 84 D4
Neutraubling D 75 E11
Neutrebbin D 80 A6
Neu-Ulm D 74 F7
Neuvéglise F 30 F2
Neuves-Maisons F 186 D1
Neuvic F 29 E6
Neuvic F 29 E10
Neuville-aux-Bois F 24 D7
Neuville-de-Poitou F 29 B6
Neuville-les-Dames F 31 C7
Neuville-lès-Dieppe F 18 E3
Neuville-sur-Saône F 30 D6
Neuvilly-en-Argonne F 24 C7
Neuvy-Grandchamp F 30 B4
Neuvy-le-Roi F 24 E4
Neuvy-Pailloux F 29 B9
Neuvy-St-Sépulchre F 29 B9
Neuvy-sur-Barangeon F 25 F7
Neuweiler D 27 C10
Neuwied D 21 D8
Neuwittenbek D 83 B8
Neu Wulmstorf D 83 D7
Neu Zauche D 80 C6
Neuzelle D 81 B7
Neu Zittau D 80 B5
Névache F 31 E10
Nevarénai LT 134 D4
Neveja LV 130 F4
Neveklov CZ 77 C7
Nevele B 19 B8
Neverfjord N 113 C12
Nevernes N 108 F4
Neverness N 110 C9
Neveronys LT 137 D9
Nevers F 30 A3
Nevesinje BIH 157 F9
Nevestino BG 165 E6
Névez F 22 E4
Neviano I 61 C10
Néville F 18 E2
Nevlunghavn N 90 B6
Nevsha BG 167 C8
New Abbey UK 5 F9
New Aberdour UK 3 K12
New Alresford UK 13 C12
Newark-on-Trent UK 11 E10
Newbawn IRL 9 D9
Newbiggin-by-the-Sea UK 5 E13
Newbliss IRL 7 D8
Newborough UK 11 F11
Newbridge UK 13 B8
Newbridge IRL 7 F9
New Buildings UK 4 F2
Newburgh UK 3 L12
Newburgh UK 5 C10
Newbury UK 13 C12
Newby Bridge UK 10 C6
Newcastle UK 7 D11
Newcastle UK 13 A8
Newcastle IRL 7 F10
Newcastle Emlyn UK 12 A6
Newcastleton UK 5 E11
Newcastle-under-Lyme
UK 11 E7
Newcastle upon Tyne UK 5 F13
Newcastle West IRL 8 D4
New Cumnock UK 5 E8
New Deer UK 3 K12
Newel D 21 E7
Newent UK 13 B10
New Galloway UK 5 E8
New Inn IRL 6 F6
New Inn IRL 7 E8
Newinn IRL 9 D7
New Kildimo IRL 8 C5
Newmarket UK 2 J4
Newmarket UK 15 C9
Newmarket IRL 8 D5
Newmarket UK 9 C8
Newmarket-on-Fergus IRL 8 C5
Newmill UK 3 K11
New Milton UK 13 D11
Newnham UK 13 B10

New Pitsligo UK 3 K12
Newport UK 3 J10
Newport UK 11 F7
Newport UK 12 A5
Newport UK 13 B9
Newport UK 13 C12
Newport UK 15 D9
Newport IRL 6 E3
Newport IRL 8 C6
Newport-on-Tay UK 5 C11
Newport Pagnell UK 15 C7
Newport Trench UK 4 F3
New Quay UK 12 A6
Newquay UK 12 E4
New Radnor UK 13 A8
New Romney UK 15 F10
New Ross IRL 9 D9
Newry UK 7 D10
Newton UK 4 C6
Newton UK 10 D7
Newton Abbot UK 13 D7
Newton Aycliffe UK 5 F13
Newton Ferrers UK 12 E6
Newtonhill UK 5 A12
Newton-le-Willows UK 10 E6
Newton Mearns UK 5 D8
Newtonmore UK 5 A8
Newton Stewart UK 4 F8
Newtown UK 10 F5
Newtown UK 13 A9
Newtown IRL 6 F6
Newtown IRL 8 B5
Newtown IRL 8 D6
Newtown IRL 9 C8
Newtownabbey UK 4 F5
Newtownards UK 7 C11
Newtownbarry UK 9 C9
Newtownbutler UK 7 D8
Newtown Crommelin UK 4 F4
Newtown Forbes IRL 7 E7
Newtown Mount Kennedy
IRL 7 F10
Newtown St Boswells UK 5 D11
Newtownshandrum IRL 8 D5
Newtownstewart UK 4 F2
Nexon F 29 D8
Neyland UK 12 B5
Nezamyslice CZ 77 D12
Nezavertailovca MD 154 D5
Nézsa H 147 F8
Nezvěstice CZ 76 C5
Nianfors S 103 C12
Niata GR 175 F6
Nibbiano I 37 B10
Nibe DK 86 B5
Nīca LV 134 D2
Nicastro I 59 B9
Nice F 37 D6
Ničgale LV 135 D12
Nichelino I 37 A7
Nickelsdorf A 77 G12
Nicolae Bălcescu RO 153 B9
Nicolae Bălcescu RO 153 E9
Nicolae Bălcescu RO 155 D3
Nicolae Bălcescu RO 155 E2
Nicolae Bălcescu RO 160 D4
Nicolae Bălcescu RO 161 E9
Nicolae Titulescu RO 160 E5
Nicolaevca MD 154 B2
Nicolosi I 59 D7
Nicoreni MD 153 B11
Nicorești RO 153 F10
Nicosia I 58 D5
Nicotera I 59 B8
Nicșeni RO 153 B9
Niculești RO 161 D7
Niculițel RO 155 C2
Nida LT 134 F1
Nidau CH 27 F7
Nidda D 21 D12
Nidzica PL 139 D9
Niebla E 51 E6
Nieborów PL 141 F2
Niebüll D 82 A5
Niebylec PL 144 D4
Niechanowo PL 138 F4
Niechcice PL 143 D8
Niechłonin PL 139 D9
Niechlów PL 81 C10
Niechorze PL 85 B8
Niedeeraichbach D 75 E11
Niederanven L 20 E6
Niederau D 80 D5
Niederaula D 78 E6
Niederbipp CH 27 F8
Niederbrechen D 21 D10
Niederbreitbach D 185 C7
Niederbronn-les-Bains F 27 C8
Niederdorf D 84 E5
Niederfischbach D 185 C8
Niederflinow D 84 E5
Niedergörsdorf D 80 C3
Niederkassel D 21 C8
Niederkirchen D 21 E9
Niederkrüchten D 20 B6
Niederndorf A 72 A5
Niederneisen D 21 D10
Niedernhall D 74 D6
Niedernhausen D 21 D10
Niederoderwitz D 81 E7
Nieder-Olm D 185 E9
Nieder-Roденbach D 21 D12
Niederroßla D 79 D9
Niedersachswerfen D 79 C8
Niederselters D 21 D10
Niederstetten D 74 D6
Niederurnen CH 27 F11
Niederviehbach D 75 E11
Niederwerrn D 75 B7
Niederwörresbach D 186 B3
Niederzissen D 21 D8
Niedrzwica Duża PL 141 H6
Niedźwiada PL 141 G7
Niedźwiadna PL 139 D12
Niefern-Öschelbronn D 27 C10
Niegosław PL 85 E8
Niegosławice PL 81 C9
Niegowa PL 143 E7
Niegripp D 79 B10
Nieheim D 17 E12
Niekerk NL 16 B6
Niekłań Wielki PL 141 H3
Niekursko PL 85 D10
Niel B 182 C4
Nielisz PL 144 B7
Niemce PL 141 H7
Niemcza PL 81 E11
Niemegk D 79 B12
Niemelä FIN 113 C20
Niemelä FIN 115 E5
Niemenkylä FIN 119 F12
Niemenkylä FIN 122 F7
Niemenkylä FIN 126 C6
Niemenkylä FIN 126 C6
Niemenpää FIN 119 B11

Niemis S 119 B11
Niemisel S 118 B7
Niemisjärvi FIN 123 F16
Niemisjärvi FIN 117 E11
Niemiskylä FIN 124 E8
Niemiskylä FIN 123 C17
Niemodlin PL 142 E4
Nienadówka PL 144 C5
Nienburg (Saale) D 79 C10
Nienburg (Weser) D 17 C12
Niepars D 84 B3
Niepołomice PL 143 F9
Nieporęt PL 139 F11
Nierstein D 21 E10
Niesa FIN 117 D11
Niesi FIN 117 D15
Niesky D 81 D7
Nieświń PL 141 H2
Nieszawa PL 138 E6
Nietsak S 109 C15
Nietulisko Duże PL 143 E11
Nieul F 29 D7
Nieuw-Amsterdam NL 17 C7
Nieuw-Bergen NL 16 E6
Nieuwegein NL 16 D4
Nieuwe-Niedorp NL 16 C3
Nieuwe Pekela NL 17 B7
Nieuwerkerk NL 16 E2
Nieuwerkerk aan de IJssel
NL 16 E3
Nieuwerkerken B 19 C11
Nieuwe-Tonge NL 16 E2
Nieuw-Heeten NL 16 D6
Nieuwkoop NL 16 D3
Nieuw-Loosdrecht NL 183 A6
Nieuw-Milligen NL 183 A7
Nieuw-Namen NL 182 C4
Nieuwolda NL 17 B7
Nieuwpoort B 18 B6
Nieuwveen NL 182 A5
Nieuw-Vennep NL 16 D3
Nieuw-Vossemeer NL 182 B4
Nieuw-Weerdinge NL 17 C7
Nievern D 185 D8
Niewęgłosz PL 141 G7
Niezabyszewo PL 85 B12
Nigrán E 38 D2
Nigrande LV 134 D4
Nigrita GR 169 C10
Nigüelas E 53 C9
Niherne F 29 B9
Niinilahti FIN 123 E15
Niinimaa FIN 123 E10
Niinimäki FIN 123 E16
Niinisalo FIN 126 B8
Niinivaara FIN 125 D11
Niinivesi FIN 123 E16
Niirokumpu FIN 121 B11
Níjar E 55 F8
Nijemci HR 157 B11
Nijkerk NL 16 D4
Nijlen B 19 B10
Nijmegen NL 16 E5
Nijverdal NL 17 D6
Nikaia GR 169 F7
Nikaranperä FIN 123 E14
Nikel' RUS 114 E8
Niki GR 168 C5
Nikisiani GR 170 C6
Nikiti GR 169 D10
Nikkala S 119 C11
Nikkaluokta S 111 E17
Nikkaroinen FIN 127 C14
Nikkeby N 112 C6
Nikodin NMK 169 B6
Nikokleia GR 169 C9
Nikolaevo BG 165 D10
Nikolaevo BG 166 D5
Nikola-Kozlevo BG 161 F10
Nikolovo BG 161 F5
Nikopol BG 160 F5
Nikopoli GR 174 A2
Nikšić MNE 163 D6
Nikyup BG 166 C5
Nilivaara FIN 117 C14
Nilivaara S 116 D7
Nilsiä FIN 125 D10
Nilvange F 20 F6
Nim DK 86 D5
Nîmes F 35 C7
Nimigea RO 152 C4
Nimis I 73 D7
Nimtofte DK 86 C7
Nin HR 67 D11
Nina EST 131 D14
Nindorf D 82 B6
Ninemile Bar UK 5 E9
Ninemilehouse IRL 9 D8
Ninove B 19 C9
Niort F 28 C5
Nirza LV 133 D3
Niš SRB 164 C4
Nisa P 44 E5
Nisbet UK 5 D11
Niscemi I 58 E5
Niška Banja SRB 164 C5
Niskankorpi FIN 123 C11
Niskanpera FIN 119 B15
Nisko PL 144 B5
Niskos FIN 123 F10
Nismes B 19 D10
Nispen NL 16 F2
Nisporeni MD 154 C2
Nissafors S 91 E14
Nissan-lez-Enserune F 34 D5
Nissilä FIN 123 C11
Nissinvaara FIN 121 B11
Nissoria I 58 D5
Nissumby DK 86 B2
Nissum Seminarieby DK 86 B2
Nistelrode NL 16 E5
Nistorești RO 153 F9
Nitaure LV 135 B10
Nitra SK 146 E6
Nitrianske Pravno SK 147 D7
Nitrianske Rudno SK 146 D6
Nitrianske Sučany SK 146 D6
Nitry F 25 E11
Nitta S 91 D13
Nittedal N 95 B13
Nittel D 20 E6
Nittenau D 75 D11
Nittendorf D 75 D10
Nittorp S 91 E14
Niukkala FIN 129 B12
Nivå DK 87 D11
Niva FIN 121 F14

Nivala FIN 123 C13
Nivankylä FIN 117 E15
Nivanpää FIN 117 E11
Nivelles B 19 C9
Nivenskoye RUS 136 D2
Nivillac F 23 E7
Nivillers F 18 F5
Nivnice CZ 146 D5
Nivyanin BG 165 C8
Niwiska PL 143 F12
Nižbor CZ 76 C5
Nižná SK 147 C9
Nižná Slaná SK 145 F2
Nižný Hrabovec SK 145 F4
Nižný Hrušov SK 145 F4
Nižný Šipov SK 145 F4
Nizza di Sicilia I 59 D7
Nizza Monferrato I 37 B8
Njavve S 109 C15
Njegovuđa MNE 163 C7
Njetjavare S 116 E4
Njivice HR 67 B10
Njurundabommen S 103 B13
Njutånger S 103 C13
No DK 86 C3
Noaillan F 32 B5
Noailles F 18 F5
Noain E 32 E2
Noale I 66 A5
Noalejo E 53 B9
Noasca I 31 E11
Nöbbele S 89 B8
Nobber IRL 7 E9
Nobitz D 79 E11
Noblejas E 46 E6
Nočaj SRB 158 D4
Nocé F 24 D4
Nocera Inferiore I 60 B3
Nocera Terinese I 59 A9
Nocera Umbra I 62 A3
Noceto I 66 C1
Noci I 61 B8
Nociglia I 61 C10
Nociūnai LT 135 F8
Nocrich RO 152 F5
Nødebo DK 87 D10
Nodeland N 90 C2
Nödinge S 91 D11
Nodland N 94 F4
Nods F 26 F5
Noé F 33 D8
Noepoli I 61 C6
Noer D 83 B8
Nœux-les-Mines LT 18 D6
Noez E 46 E4
Nofuentes E 40 C5
Nogales E 51 B6
Nogara I 66 B3
Nogaro F 32 C5
Nogent F 26 D3
Nogent-le-Bernard F 24 D3
Nogent-le-Roi F 24 D5
Nogent-le-Rotrou F 24 D4
Nogent-sur-Aube F 25 D11
Nogent-sur-Oise F 18 F5
Nogent-sur-Seine F 25 D9
Nogent-sur-Vernisson F 25 E8
Nogersund S 89 C7
Nógrád H 147 F8
Nógrádmegyer H 147 E9
Nógrádsáp H 147 F8
Nograles E 40 F6
Noguera de Albarracín E 47 D9
Noguères F 32 D4
Noguerueelas E 48 D3
Nohant-Vic F 29 B9
Nohfelden D 21 E8
Nohic F 33 C8
Noia E 38 C2
Noicattaro I 61 A7
Noidans-lès-Vesoul F 26 E5
Noilhan F 33 C7
Noirétable F 30 D4
Noirmoutier-en-l'Île F 28 A1
Noisseville F 26 B5
Noja E 40 B4
Nojorid RO 151 C8
Nokia FIN 127 C10
Nol S 91 D11
Nolay F 30 B6
Noli I 37 C8
Nolimo FIN 121 B11
Nolmyra S 99 B8
Nólsoy FO 2 A3
Nombela E 46 D4
Nomeland N 90 A2
Nomeny F 26 C5
Nomexy F 26 D5
Nomia GR 178 B5
Nonancourt F 24 C5
Nonantola I 66 C3
Nonaspe E 42 E4
None I 37 B7
Nonnenweier D 186 E4
Nonnweiler D 21 E7
Nontron F 29 D7
Nonza F 37 F10
Nõo EST 131 E13
Noordwijk aan Zee NL 182 A4
Noordwijk-Binnen NL 16 D3
Noordwijkerhout NL 16 D3
Noordwolde NL 16 C6
Noormarkku FIN 126 B6
Nootdorp NL 182 A4
Nopankylä FIN 122 E8
Noppikoski S 102 D8
Nor S 103 C11
Nora S 103 A12
Nørager DK 86 B5
Noragugume I 64 C2
Norberg S 97 B14
Norcia I 62 B4
Nordagutu N 95 D10
Nordanå S 103 A13
Nordanås S 107 C14
Nordanås S 107 B13
Nordanås S 109 E16
Nordanås S 109 C14
Nordande S 103 A11
Nordankäl S 107 D10
Nordanede S 103 A11
Nordanälden S 105 E16
Nordborg DK 86 E5
Nordbotn N 112 D9
Nordby DK 86 D7
Nordby DK 86 E5
Nordby DK 86 D7
Norddeich D 17 A8
Norddorf D 82 A4
Norddyrøy N 104 D5

North Grimston UK 11 C10
North Hykeham UK 11 E10
North Leigh UK 13 B12
North Shields UK 5 E14
North Somercotes UK 11 E12
North Sunderland UK 5 D13
North Tidworth UK 13 C11
Northton UK 2 K2
North Walsham UK 15 B11
Northwich UK 10 E6
Nortmoor D 17 B9
Norton UK 11 C10
Norton UK 15 C10
Norton Fitzwarren UK 13 C8
Nortorf D 83 B7
Nortrup D 17 C9
Nort-sur-Erdre F 23 F8
Norup DK 86 B6
Norvasalmi FIN 117 E15
Norwich UK 15 B11
Norwick UK 3 D15
Noşlac RO 152 E3
Nøss N 111 B10
Nossa Senhora da Boa Fé
P 50 B3
Nossa Senhora da Graça de
Póvoa e Meadas P 44 E5
Nossa Senhora da Graça do
Divor P 50 B4
Nossa Senhora da Graça dos
Degolados P 45 F3
Nossa Senhora das Neves
P 50 C4
Nossa Senhora de Machede
P 50 B4
Nossebro S 91 C12
Nossen D 80 D4
Nossendorf D 84 C4
Nossentiner Hütte D 83 C12
Noszlop H 149 B8
Noszvaj H 145 H1
Notaresco I 62 B5
Notia GR 169 B7
Nøtincs H 147 E8
Nötö FIN 126 F6
Noto I 59 F7
Notodden N 95 C10
Notre-Dame-de-Bellecombe
F 31 D10
Notre-Dame-de-Gravenchon
F 18 F2
Notre-Dame-de-Monts F 28 B1
Notre-Dame-d'Oé F 24 F4
Nötsch im Gailtal A 73 C8
Nottensdorf D 82 D7
Nottingham UK 11 F9
Nottuln D 17 E8
Notviken S 118 C8
Nouan-le-Fuzelier F 24 E7
Nouans-les-Fontaines F 24 F5
Nouart F 19 F11
Nouzilly F 24 E4
Nouzonville F 19 E10
Nõva EST 130 C7
Nova H 149 C7
Nová Baňa SK 147 E7
Nova Breznica NMK 164 F3
Nova Bukovica HR 149 E9
Nová Bystřice CZ 77 D8
Nová Cerekev CZ 77 D8
Novachene BG 165 C9
Novachene BG 165 B10
Novaci NMK 169 B6
Novaci RO 160 C3
Nova Crnja SRB 158 B6
Novadnieki LV 134 C4
Nová Dubnica SK 146 D6
Novafeltria I 66 E5
Nova Gorica SLO 73 E8
Nova Gradiška HR 157 B7
Nova Ivanivka UA 154 F4
Novaj H 145 H1
Novajidrány H 145 G3
Nova Kamena BG 161 F10
Nova Kasaba BIH 157 D11
Novaki HR 148 E5
Novakovo BG 166 F4
Nováky SK 147 D7
Novales E 40 B3
Novales E 41 D11
Nova Levante I 72 D4
Novalja HR 67 C10
Novallas E 41 E8
Nová Ľubovňa SK 145 E2
Nova Makhala BG 165 F9
Nova Nadezhda BG 166 E5
Nova Nekrasivka UA 155 C3
Nová Paka CZ 77 B9
Nova Pazova SRB 158 D5
Nova Pokrovka UA 155 B4
Novara I 68 C6
Novara di Sicilia I 59 C7
Nová Role CZ 75 B12
Nova Sela HR 157 F8
Nova Siri I 61 C7
Nova Siri Scalo I 61 C7
Novate Mezzola I 69 A7
Nova Topola BIH 157 B7
Nova Varoš SRB 163 C8
Nova Vas SLO 73 E9
Nová Včelnice CZ 77 B8
Nová Ves I 72 D4
Nová Ves nad Žitavou SK 146 E6
Nova Zagora BG 166 E6
Nové Hrady CZ 77 E7
Novelda E 56 E3
Novellara I 66 C2
Nové Město nad Metuj CZ
77 B10
Nové Mesto nad Váhom
SK 146 D5
Nové Město na Moravě CZ
77 C10
Nové Město pod Smrkem
CZ 81 E8
Nove Misto UA 145 D6
Noventa di Piave I 73 E6
Noventa Vicentina I 66 B4
Novés F 36 C4
Noves F 35 C8
Nové Veselí CZ 77 D9
Nové Zámky SK 146 F6
Novgorodka RUS 133 B5
Novgrad BG 161 F7
Novi Banovci SRB 158 D5
Novi Bečej SRB 158 B5
Novi di Modena I 66 C2

Novi Dojran NMK 169 B8
Noviergas E 41 E7
Novi Grad BIH 157 B9
Novigrad HR 67 B8
Novigrad HR 156 D4
Novigrad Podravski HR 149 D7
Novi Iskŭr BG 165 D7
Novi Karlovci SRB 158 C5
Novi Khan BG 165 D8
Novi Kneževac SRB 150 E5
Novi Kozarci SRB 150 F6
Novi Ligure I 37 B9
Novillars F 26 F5
Noville B 182 D5
Novi Marof HR 149 D6
Novion-Porcien F 19 E9
Novi Pazar BG 167 C8
Novi Pazar SRB 163 C10
Novi Sad SRB 158 C4
Novi Šeher BIH 157 C9
Novi Slankamen SRB 158 C5
Novi Travnik BIH 157 D8
Novi Vinodolski HR 67 B10
Novo Beograd SRB 158 D5
Novobonyivka UA 154 C5
Novob-čić F 37 B9
Novo Delchevo BG 169 B9
Novokhovansk RUS 133 C7
Novo Korito SRB 159 F9
Novo Mesto SLO 73 E11
Novo Miloševo SRB 158 B5
Novomoskovskiy RUS 139 A9
Novomykolayivka UA 155 B5
Novo Orahovo SRB 150 F4
Novo Oryakhovo BG 167 D9
Novopetrivka UA 154 C5
Novorzhev RUS 133 B6
Novosamarka UA 154 B5
Novosedly CZ 77 E11
Novoselė AL 163 F10
Novoselë AL 168 C1
Novoselec FIN 127 C10
Novoselets BG 166 E6
Novoselija BIH 157 C7
Novoselivka UA 154 B5
Novoselivka UA 154 D5
Novo Selo BG 159 E10
Novo Selo BG 161 F8
Novo Selo BG 165 E6
Novo Selo BG 166 F4
Novo Selo BIH 157 B8
Novo Selo BIH 157 C7
Novo Selo NMK 169 B8
Novo Selo SRB 159 F6
Novoselovo RUS 139 B9
Novoseltsi BG 167 D8
Nosel'ye RUS 132 C2
Novoselytsya UA 154 A6
Novoselytsya UA 154 A6
Novosil's'ke UA 155 C3
Novostroyevo RUS 136 E4
Novot SK 147 C8
Novo Virje HR 149 D8
Novovolyns'k UA 144 B9
Novoyavorivs'ke UA 144 D8
Novska PL 143 D7
Novský Bor CZ 81 E7
Nový Bydžov CZ 77 B8
Nový-Chevrières F 19 E9
Nový Dvor BY 137 F10
Nový Hrozenkov CZ 146 C6
Nový Jičín CZ 146 B6
Nový Knín CZ 76 C6
Nový Malín CZ 77 C12
Nový Pahost BY 133 F2
Nový Rychnov CZ 77 D8
Novyya Kruki BY 133 E3
Novyy Rozdil UA 145 E9
Nový Šivot SK 146 E4
Nowa Brzeźnica PL 143 D7
Nowa Cerekwia PL 142 F4
Nowa Chodorówka PL 140 D8
Nowa Dęba PL 143 F12
Nowa Karczma PL 138 B5
Nowa Ruda PL 81 E11
Nowa Sarzyna PL 144 C5
Nowa Słupia PL 143 E11
Nowa Sól PL 81 C9
Nowa Sucha PL 141 F2
Nowa Wieś Ełcka PL 140 C6
Nowa Wieś Lęborskie PL 85 A13
Nowa Wieś Wielka PL 138 E5
Nowa Wola PL 140 D9
Nowa Wola Gołębiowska PL
141 H4
Nowe PL 138 C6
Nowe Brusno PL 144 C7
Nowe Brzesko PL 143 F9
Nowe Czarnowo PL 84 D6
Nowe Miasteczko PL 81 C9
Nowe Miasto PL 139 E10
Nowe Miasto Lubawskie
PL 139 D8
Nowe Miasto nad Pilicą PL
141 G3
Nowe Miasto nad Wartą PL
81 B12
Nowe Ostrowy PL 143 B7
Nowe Piekuty PL 141 E7
Nowe Skalmierzyce PL 142 C4
Nowe Warpno PL 84 C6
Nowinka PL 136 F6
Nowogard PL 85 D9
Nowogród PL 139 D12
Nowogród Bobrzański PL 81 C8
Nowogródek Pomorski PL 85 E8
Nowogrodziec PL 81 D8
Nowosady PL 141 E9
Nowosielce PL 145 D5
Nowosolna PL 143 C8
Nowotaniec PL 145 D5
Nowowola PL 140 D8
Nowy Bartków PL 141 F7
Nowy Duninów PL 139 E7
Nowy Dwór PL 138 D4
Nowy Dwór PL 140 D9
Nowy Dwór Gdański PL 138 B7
Nowy Dwór Mazowiecki PL
139 F10
Nowy Kawęczyn PL 141 G2
Nowy Korczyn PL 143 F10
Nowy Lubliniec PL 144 C6
Nowy Sącz PL 145 D2
Nowy Staw PL 138 B7
Nowy Targ PL 147 C10
Nowy Tomyśl PL 81 B10
Nowy Wiśnicz PL 144 D1
Nowy Żmigród PL 145 D4
Noyal-Muzillac F 22 E7
Noyalo F 22 E6
Noyal-Pontivy F 22 D6
Noyant F 23 E12

Noyarey F 31 E8
Noyen-sur-Sarthe F 23 E11
Noyers F 25 E10
Noyers-sur-Cher F 24 F5
Noyers-sur-Jabron F 35 B10
Noyon F 18 E6
Nozay F 23 E8
Nozdrzec PL 144 D5
Nozeroy F 31 B9
Nuaillé-d'Aunis F 28 C4
Nuasjärvi FIN 117 E13
Nubledo E 39 A8
Nucet RO 151 E10
Nuci RO 161 D8
Nucșoara RO 160 C5
Nudersdorf D 79 C12
Nüdlingen D 75 B7
Nudyzhe UA 141 H10
Nueil-les-Aubiers F 28 B4
Nuenen NL 16 F5
Nueno E 41 D11
Nueva E 39 B10
Nueva-Carteya E 53 A8
Nueva Jarilla E 52 C4
Nuez de Ebro E 41 E10
Nufăru RO 155 C3
Nughedu di San Nicolò I 64 B3
Nuijamaa FIN 129 D10
Nuillé-sur-Vicoin F 23 E10
Nuits F 25 E11
Nuits-St-Georges F 26 F2
Nukari FIN 127 D12
Nukši LV 133 D3
Nuland NL 16 E4
Nule I 64 C3
Nules E 48 E4
Nulvi I 64 B2
Numana I 67 E8
Numansdorp NL 16 E2
Nummela FIN 127 E11
Nummi FIN 127 E10
Nummijärvi FIN 122 F8
Nummikoski FIN 122 F9
Nünchritz D 80 D4
Nuneaton UK 11 F9
Nunkirchen D 186 C2
Nunnanen FIN 117 B12
Nunnanlahti FIN 125 D12
Nuñomoral E 45 D8
Nunsdorf D 80 B4
Nunspeet NL 16 D5
Nuojua FIN 119 E17
Nuoksujärvi S 116 E9
Nuolijärvi FIN 125 C11
Nuoramoinen FIN 127 C14
Nuorgam FIN 113 C20
Nuoritta FIN 119 D16
Nuoro I 64 C3
Nuorunka FIN 120 C9
Nuottavaara FIN 117 D12
Nuottikylä FIN 121 E12
Nur PL 141 E6
Nuragus I 64 D3
Nurallao I 64 D3
Nuraminis I 64 E3
Nureci I 64 D2
Nuriye TR 177 B10
Nurmaa FIN 128 C6
Nurmes FIN 125 C12
Nurmesperä FIN 123 C15
Nurmijärvi FIN 125 C13
Nurmijärvi FIN 127 E12
Nurmo FIN 123 E9
Nürnberg D 75 D9
Nurney IRL 7 F9
Nurri I 64 D3
Nurste EST 130 D4
Nürtingen D 27 C11
Nurzec-Stacja PL 141 F8
Nus I 31 D11
Nusco I 60 B4
Nușeni RO 152 C4
Nușfalău RO 151 C10
Nusfjord N 110 D5
Nusnäs S 102 E8
Nusplingen D 27 D10
Nußbach A 76 G6
Nußdorf D 73 A6
Nußdorf am Inn D 72 A5
Nuštar HR 149 F11
Nustrup DK 86 E4
Nuth NL 19 C12
Nutheim N 95 C9
Nuttuperä FIN 123 C15
Nuuksujärvi S 116 C8
Nuupas FIN 119 B16
Nuutajärvi FIN 127 C9
Nuutila FIN 119 F16
Nuutilanmaki FIN 128 B8
Nuvsvåg N 112 C9
Nuvvus FIN 113 D17
Nuxis I 64 E2
Nüziders A 71 C9
Nya Bastuselet S 109 F16
Nyåker S 107 D13
Nyåker S 107 D16
Nyárád H 149 B8
Nyáregyháza H 150 C4
Nyárlőrinc H 150 D4
Nyársapát H 150 C4
Nybble S 91 A15
Nybergsund N 102 D3
Nyborg DK 87 E7
Nyborg N 114 C5
Nyborg S 119 C10
Nybro S 89 B9
Nybrostrand S 88 E5
Nyby FIN 122 E6
Nyby N 113 C15
Nyby S 106 E8
Nybyn S 103 C9
Nybyn S 107 E14
Nybyn S 118 D6
Nýdek CZ 147 B7
Nydri GR 174 B2
Nye S 89 A8
Nyékládháza H 145 H2
Nyelv N 114 C5
Nyergesújfalu H 149 A11
Nyhammar S 97 B12
Nyhem S 103 A10
Nyhem S 109 E14
Ny Højen DK 86 D5
Nyhus N 111 B15
Nyhyttan S 97 C12
Nyírábrány H 151 B9
Nyíracsád H 151 B8
Nyírád H 149 B8
Nyíradony H 151 B8
Nyírbátor H 151 B9
Nyírbéltek H 151 B9
Nyírbogát H 151 B9
Nyírbogdány H 145 G4
Nyíregyháza H 145 H4
Nyírgelse H 151 B8
Nyírgyulaj H 145 H5
Nyíribrony H 145 G4
Nyírkáta H 145 H5
Nyírmada H 145 G5
Nyírmeggyes H 145 H5
Nyírmihálydi H 151 B8
Nyírpazony H 145 H4
Nyírtass H 145 G5
Nyírtelek H 145 G4
Nyírtét H 145 G4
Nyírtura H 145 G4
Nyírvasvári H 151 B9
Nykarleby FIN 122 C9
Nyker DK 89 E7
Nykil S 92 C6
Nykøbing DK 83 A11
Nykøbing Mors DK 86 B3
Nykøbing Sjælland DK 87 D9
Nyköping S 93 B10
Nykrogen S 98 B6
Nykroppa S 97 C11
Nyksund N 110 C9
Nykvåg N 110 C7
Nykvarn S 93 A10
Nykyrke S 92 B5
Nyland S 107 D16
Nyland S 107 E13
Nyland S 122 C3
Nylars DK 89 E7
Nyliden S 107 D15
Nymburk CZ 77 B8
Nymfes GR 168 E2
Nymindegab DK 86 D2
Nymoen N 112 C6
Nynäshamn S 93 B11
Nyneset N 105 C13
Ny Nørup DK 86 D4
Nyoiseau F 23 E10
Nyon CH 31 C9
Nyons F 35 B9
Nyråd DK 87 E9
Nýřany CZ 76 C4
Nýrsko CZ 76 D4
Nyrud N 114 E6
Nysa PL 142 F3
Nysäter S 97 D8
Nysätern S 102 A5
Nysättra S 99 C11
Nysted DK 83 A11
Nysted N 111 C15
Nystrand S 118 C5
Nyträsk S 118 E4
Nytrøa N 111 B16
Nyúl H 149 A9
Nyvoll N 113 C11
Nyzhankovychi UA 144 D6
Nyzhni Petrivtsi UA 153 A7
Nyzhni Vorota UA 145 F7
Nyzhniy Bystryy UA 145 G8
Nyzhnya Vysots'ke UA 145 E7
Nyzhnya Yablun'ka UA 145 E6

O

Oadby UK 11 F9
Oakengates UK 10 F7
Oakham UK 11 F10
Oakley UK 13 B12
Oakley UK 13 C12
Oakley UK 15 C7
Oancea RO 154 F2
Oandu EST 131 C13
Oarja RO 160 D5
O Arrabal E 38 D2
Oarța de Jos RO 151 C11
Obal' BY 133 E7
Obal' BY 133 F6
Obalj BIH 157 F9
Oban UK 4 C6
O Barco E 39 D6
Obârșia RO 160 F4
Obârșia-Cloșani RO 159 C10
Obârșia de Câmp RO 159 E11
Obbola S 122 C4
Obdach A 73 B10
Obecnice CZ 76 C5
Obedinenie BG 166 C5
Obejo E 54 C3
Obeliai LT 135 E11
Oberaich A 73 B11
Oberalm A 73 A7
Oberammergau D 71 B12
Oberasbach D 75 D8
Oberau D 72 A3
Oberaudorf D 72 A5
Obercunnersdorf D 81 D7
Oberderdingen D 27 B10
Oberding D 75 F10
Oberdorla D 79 D7
Oberdrauburg A 73 C6
Oberegg CH 71 C9
Oberelsbach D 75 B7
Oberfell D 185 D7
Obergebra D 79 D8
Obergösgen CH 27 F8
Ober-Grafendorf A 77 F9
Obergriesbach D 75 F9
Obergünzburg D 71 B10
Obergurgl A 72 C3
Oberguriq D 80 D6
Oberhaag A 148 C4
Oberhaid D 75 C8
Oberharmersbach D 27 D9
Oberhausen D 17 F7
Oberhausen D 75 E9
Oberhausen-Rheinhausen D 187 C5
Oberheldrungen D 79 D9
Oberhof D 79 E8
Oberhofen CH 70 D5
Oberhoffen-sur-Moder F 186 D4
Oberkirch D 27 C9
Oberkochen D 75 E7
Oberkotzau D 75 B10
Oberlangen D 17 C8
Oberlungwitz D 79 E12
Obermarchtal D 71 A9
Obermaßfeld-Grimmenthal D 75 A7
Obermoschel D 21 E9
Obernai F 27 D7
Obernberg am Inn A 76 F4
Obernburg am Main D 187 B7
Oberndorf D 17 A12
Oberndorf am Lech D 75 E8
Oberndorf am Neckar D 27 D10
Oberndorf bei Salzburg A 73 A6
Oberneukirchen A 76 F6
Obernfeld D 79 C7
Obernheim D 27 D10
Obernheim-Kirchenarnbach D 186 C4
Obernkirchen D 17 D12
Obernzell D 76 E5
Ober-Olm D 185 E9
Oberpullendorf A 149 A7
Oberried D 27 E8
Oberrieden D 71 A10
Oberriet CH 71 C9
Ober-Roden D 21 E11
Oberrot D 187 C8
Oberrotweil D 27 D8
Oberschneiding D 75 E12
Oberschützen A 148 B6
Obersiebenbrunn A 77 F11
Obersinn D 74 B6
Obersontheim D 187 C8
Oberspier D 79 D8
Oberstadion D 71 A9
Oberstaufen D 71 B10
Oberstdorf D 71 C10
Oberstenfeld D 27 B11
Oberthal D 21 E8
Oberthulba D 187 A8
Obertilliach A 73 C6
Obertraubling D 75 E11
Obertrubach D 75 C9
Obertshausen D 21 D11
Oberursel (Taunus) D 21 D11
Obervellach A 73 C7
Oberviechtach D 75 D11
Oberwald CH 70 D6
Oberwart A 148 B6
Oberwesel D 21 D9
Oberwolfach D 187 E5
Oberwölz A 73 B9
Óbidos P 44 F2
Obiliq RKS 164 D3
Obing D 75 F11
Obinitsa EST 132 F1
Obitel BG 167 C7
Objat F 29 E8
Objazda PL 85 A12
Öblarn A 73 B8
Oblešewo NMK 164 F5
Oblikë e Madhe AL 163 E7
Obnova BG 165 C10
Obodivka UA 154 A4
Oboga RO 160 E4
O Bolo E 38 D5
Obón E 42 F2
Oborci BIH 157 D7
Oborín SK 145 F4
Oborishte BG 165 D9
Oborniki PL 81 A11
Oborniki Śląskie PL 81 D11
Obrazów PL 143 E12
Obreja RO 159 C9
Obrenovac SRB 158 D5
Obretenik BG 166 B5
Obrež SRB 158 D4
Obrež SRB 159 F7
Obrigheim D 21 F12
Obrigheim (Pfalz) D 187 B5
Obrnice CZ 76 A5
Obrochishte BG 167 C10
Obrov SLO 67 A9
Obrovac HR 156 D4
Obrovac SRB 158 C3
Obrowo PL 138 E6
Obrtići BIH 157 E10
Obruchishte BG 166 E5
Obryte PL 139 E11
Obrzycko PL 85 E11
Obsza PL 144 C6
Obudovac BIH 157 C10
Obyce SK 146 E6
Obzor BG 167 D9
O Cádavo E 38 B5
Ocaklar TR 173 D8
O Campo da Feira E 38 B3
Ocaña E 46 E6
Ocana F 37 H9
O Carballiño E 38 D3
O Castelo E 38 D3
O Castro E 38 C3
O Castro de Ferreira E 38 C4
Occhiobello I 66 C4
Occimiano I 68 C6
Očevlja BIH 157 D9
Ochagavía E 32 E3
O Chao E 38 B4
Ochiltree UK 5 E8
Ochla PL 81 C8
Ochodnica SK 147 C7
Ocholt D 17 B9
Ochsenfurt D 75 C7
Ochsenhausen D 71 A9
Ochtrup D 17 D8
Ocke S 105 E15
Ockelbo S 103 E12
Öckerö S 91 D10
Ockholm D 82 A5
Ocksjön S 102 A8
Ocland RO 152 E6
Ocna de Fier RO 159 C8
Ocna Mureș RO 152 E3
Ocna Sibiului RO 152 F4
Ocna Șugatag RO 152 B3
Ocnele Mari RO 160 C4
Ocnița RO 161 D7
Ocolina MD 154 A2
Ocoliș RO 151 E11
O Convento E 38 D2
O Corgo E 38 C5
Ocrkavlje BIH 157 E10
Öcsa H 150 C3
Öcsény H 149 D11
Öcsöd H 150 D5
Octeville-sur-Mer F 23 A12
Ocypel PL 138 C5
Odäile RO 161 C9
Ödåkra S 87 C11
Odda N 94 B5
Odden N 112 D5
Oddense DK 86 B3
Odder DK 86 D6
Oddsta UK 3 D15
Ödeborg S 91 B10
Odeceixe P 50 E2
Odeleite P 50 E5
Odelouca P 50 E3
Odelzhausen D 75 F9
Odemira P 50 D2
Ödena E 43 D7
Odensbacken S 92 A7
Odensberg S 91 C13
Odense DK 86 E6
Odensjö S 92 D4
Odensvi S 93 D8
Oderberg D 84 E6
Oderin D 80 B5
Odernheim am Glan D 21 E9
Oderzo I 72 E5
Ödeshog S 92 C5
Odiáxere P 50 E2
Odiham UK 15 E7
Ødis DK 86 E4
Odivelas P 50 B1
Odivelas P 50 C3
Ödkarby FIN 99 B13
Odobești RO 153 F10
Odobești RO 161 D7
Odolanów PL 142 C4
Odolena Voda CZ 76 B6
Odón E 47 C9
Odoorn NL 17 C7
Odorheiu Secuiesc RO 152 E6
Odry CZ 146 B5
Odrzywół PL 141 G3
Ödsmål S 91 C10
Ødsted DK 86 D4
Odŭrne BG 165 C10
Odžaci BIH 157 E9
Odžaci SRB 158 B3
Odžak BIH 157 B9
Odžak BIH 157 E6
Odžak BIH 157 F9
Odžak MNE 163 C7
Odziena LV 135 C11
Oebisfelde D 79 B8
Oedelem B 182 C2
Oedelsheim (Oberweser) D 78 C6
Oederan D 80 E4
Oeffelt NL 16 E5
Oegstgeest NL 16 D3
Oeiras P 50 B1
Oelde D 17 E10
Oelixdorf D 82 C7
Oelsnitz D 75 B11
Oelsnitz D 79 E12
Oene NL 183 A8
Oenkerk NL 16 B5
Oensingen CH 27 F8
Oerel D 17 B12
Oering D 83 C8
Oerlenbach D 75 B7
Oerlinghausen D 17 E11
Oestrich-Winkel D 21 D9
Oettersdorf D 79 E10
Oettingen in Bayern D 75 E8
Oetz A 71 C11
Oetzen D 83 D9
Oeversee D 82 A6
Œyreluy F 32 C3
Ofatinți MD 154 B4
Ófehértó H 145 H5
Ofena I 62 C5
Offanengo I 69 C8
Offemont F 27 E6
Offenbach am Main D 21 D11
Offenbach an der Queich D 187 C5
Offenberg D 76 E3
Offenburg D 27 D8
Offerdal S 105 E16
Offersøy N 111 D10
Offida I 62 B5
Offingen D 75 F7
Offranville F 18 E3
O Forte E 38 C3
Ofte N 95 D8
Ofterdingen D 27 D11
Oftersheim D 21 F11
Ogenbargen D 17 A9
Oger F 25 C11
Ogeu-les-Bains F 32 D4
Ogéviller F 26 C6
Oggevatn N 90 C3
Ogliastro F 37 F10
Ogliastro Cilento I 60 C4
Ogmore UK 13 C7
Ognyanovo BG 169 A10
Ogonelloe IRL 8 C6
Ogoya BG 165 D8
Ogra RO 152 E4
Ogre LV 135 C9
Ogren AL 168 D3
Ogresgals LV 135 C9
Ogrezeni RO 161 E7
Ogrodniki PL 139 E12
Ogrodzieniec PL 143 F8
Ogrosen D 80 C6
O Grove E 38 D2
Ogulin HR 67 B11
Oğulpaşa TR 172 A6
Ohaba RO 152 E3
Ohaba Lungă RO 151 F8
Ohanes E 55 E7
Ohey B 19 D11
Ohkola FIN 127 D13
Ohlsbach D 27 D8
Ohlstadt D 72 A3
Ohne D 17 D8
Ohorn D 80 D6
Ohrady SK 146 E5
Ohrdruf D 79 E8
Ohrid NMK 168 B4
Öhringen D 27 B11
Ohtaanniemi FIN 125 E11
Ohtanajärvi S 116 E10
Ohukotsu EST 131 C9
Oia GR 179 C9
Oiã P 44 C3
Oiartzun E 32 D2
Oichalia GR 169 E6
Øie N 105 B12
Oignies F 18 D6
O Igrexario E 38 D3
Oijärvi FIN 119 C15
Oijen NL 183 B6
Oijusluoma FIN 121 C13
Oikarainen FIN 119 B16
Oileán Ciarraí IRL 8 D4
Oilgate IRL 9 D9
Oímbra E 38 E5
Oinacu RO 161 F8
Oinas FIN 115 E2
Oinasjärvi FIN 124 C9
Oinofyta GR 175 C8
Oinoi GR 175 C7
Oion E 41 C7
O Irixo E 38 C3
Oiron F 28 B5
Oirschot NL 183 B6
Oiselay-et-Grachaux F 26 F4
Oisemont F 18 E4
Oisseau F 23 D10
Oissel F 18 F3
Oisterwijk NL 183 B6
Oisu EST 131 D11
Õisu EST 131 E11
Oitti FIN 127 D13
Oituz RO 153 E9
Oitylo GR 178 B3
Oivanki FIN 121 B13
Oizon F 25 F8
Öja FIN 123 C9
Öja S 93 E12
Ojakkala FIN 127 E11
Ojakylä FIN 119 D13
Ojakylä FIN 119 D14
Ojakylä FIN 119 F15
Ojanperä FIN 120 F8
Öjarn S 106 D8
Ojdula RO 153 F8
Öje S 102 E6
Öjebyn S 118 D6
Öjeforsen S 103 B9
Ojén E 53 C7
Öjingsvallen S 103 C8
Ojos Negros E 47 C10
Ojrzeń PL 139 E10
Öjung S 103 C10
Okalewo PL 139 D8
Okány H 151 D7
Okçular TR 181 C9
Økdal N 101 A12
Okehampton UK 12 D6
Okhotnoye RUS 136 D4
Oklaj HR 156 E5
Okně S 89 A11
Okoč SK 146 F5
Okonek PL 85 C11
Okopy PL 141 H9
Ököritófülpös H 145 H6
Okorsh BG 161 F10
Okříšky CZ 77 D9
Okrouhlice CZ 77 C8
Okrúhle SK 145 E4
Oksa PL 143 E9
Öksajärvi S 116 C8
Oksakoski FIN 123 D12
Oksava FIN 123 C14
Oksbøl DK 86 D2
Oksbøl DK 86 E5
Øksfjord N 112 C9
Øksnes N 110 C8
Øksneshamn N 110 D9
Oksvoll N 104 D7
Oktonia GR 175 B9
Okučani HR 157 B7
Ola E 38 D2
Olague E 32 E2
Olaine LV 135 C7
Olalhas P 44 E4
Oland N 90 B3
Olănești MD 154 E5
Olanu RO 160 D4
Olargues F 34 C4
Olari RO 151 E8
Olaszliszka H 145 G3
Olave E 32 E2
Oława PL 81 E12
Olazti E 32 E1
Olba E 48 D3
Olbendorf A 148 B6
Olbernhau D 80 E4
Olbersdorf D 81 E7
Olbersleben D 79 D9
Olbia I 64 B3
Olbicin PL 144 B5
Olbramovice CZ 77 C7
Olcea RO 151 D8
Oldcastle IRL 7 E8
Old Dailly UK 4 E7
Oldebroek NL 16 D5
Oldedalen N 100 C5
Oldehove NL 16 B6
Oldeide N 100 C2
Oldemarkt NL 16 C5
Olden N 100 C5
Olden S 105 D15
Oldenbrok D 17 B10
Oldenburg D 17 B10
Oldenburg in Holstein D 83 B9
Oldendorf D 17 A12
Oldenswort D 82 B5
Oldenzaal NL 17 D7
Olderdalen N 112 D6
Oldereid N 108 B8
Olderfjord N 113 C15
Oldernes N 113 C13
Oldervik N 108 C5
Oldervik N 111 A18
Oldervik N 113 B13
Oldham UK 11 D7
Old Head IRL 8 E5
Oldisleben D 79 D9
Old Leake UK 11 E12
Oldmeldrum UK 3 L12
Oldsum D 82 A4
Oldtown IRL 7 E10
Oleby S 97 B9
Olecko PL 136 E6
Oleggio I 68 B6
Oleiros P 44 E5
Oleksandrivka UA 154 C6
Olekseyivka UA 154 B3
Olemps F 33 B11
Olen B 19 B10
Ølen N 94 C3
Olesa de Montserrat E 43 D7
Oleśnica PL 142 D3
Oleśnica PL 143 F11
Olešnice CZ 77 C10
Olesno PL 142 E5
Olesno PL 143 F10
Oleszyce PL 144 C7
Olette F 33 E10
Olevano Romano I 62 D4
Olfen D 17 E8
Olgina EST 132 C3
Olginate I 69 B7
Ølgod DK 86 D3
Olgrinmore UK 3 J9
Olhalvo P 44 F2
Olhão P 50 E4
Olhava FIN 119 D14
Ølholm DK 86 D5
Olho Marinho P 44 F2
Ol'hopil' UA 154 A5
Oliana E 43 C6
Olías del Rey E 46 E5
Oliena I 64 C3
Oliete E 42 E2
Oligastro Marina I 60 C3
Olingdal S 102 C7
Olingsjövallen S 102 C7
Oliola E 42 D6
Olite E 32 F2
Oliva E 56 D4
Oliva de la Frontera E 51 C6
Oliva de Mérida E 51 B7
Olival P 44 E3
Olivares E 51 E7
Olivares de Júcar E 47 E8
Oliveira de Azeméis P 44 C4
Oliveira de Frades P 44 C4
Oliveira do Arda P 44 B4
Oliveira do Bairro P 44 C4
Oliveira do Conde P 44 D5
Oliveira do Douro P 44 B3
Oliveira do Hospital P 44 D5
Olivenza E 51 B5
Oliveri I 59 C7
Olivet F 24 E6
Oliveto Citra I 60 B4
Oliveto Lucano I 60 B6
Olivone CH 71 D7
Ol'ka SK 145 E4
Olkijoki FIN 119 E13
Olkiluoto FIN 126 C5
Olkkajärvi FIN 117 E16
Olkusz PL 143 F8
Ollaberry UK 3 D14
Ollerton UK 11 E9
Ollerup DK 86 E7
Olliergues F 30 D4
Ollila FIN 121 B13
Ollila FIN 127 D8
Ollilanniemi FIN 121 E12
Ollioules F 35 D10
Ollo E 32 E2
Öllölä FIN 125 F15
Ollolai I 64 C3
Ollon CH 31 C10
Ollsta S 106 E8
Ölmbrotorp S 97 D13
Ölme S 97 D11
Olmedilla de Roa E 40 E4
Olmedo E 46 B3
Olmedo I 64 B1
Olmeta-di-Tuda F 37 F10
Olmeto F 37 H9
Olmos de Ojeda E 40 C3
Olney UK 15 C7
Olocau E 48 E3
Olocau del Rey E 42 F3
Olofsfors S 107 D16
Olofstorp S 91 D11
Olofström S 88 C7
Olombrada E 40 F3
Olomouc CZ 146 B4
Olonne-sur-Mer F 28 B2
Olonzac F 34 D4
Oloron-Ste-Marie F 32 D4
Olost E 43 D8
Olot E 43 C8
Olovi CZ 75 B12
Olovo BIH 157 D10
Olpe D 21 B9
Olřišov CZ 142 G4
Olšany CZ 77 C11
Olšany u Prostějova CZ 77 C12
Olsätter S 97 C10
Olsbenning S 97 B15
Olsberg D 17 F10
Olsberg N 111 B16
Ölsboda S 92 A4
Olsfors S 91 D12
Olshammar S 92 B5
Olsøy N 104 D8
Olst NL 16 D6
Ølsted DK 86 D5
Ølsted DK 87 D10
Ølstykke DK 87 D10
Olszanica PL 145 E5
Olszanka PL 142 E3
Olszany PL 144 D6
Olszewo-Borki PL 139 D12
Olsztyn PL 136 F1
Olsztyn PL 143 E7
Olsztynek PL 139 C9
Olszyn PL 141 F8
Olszyna PL 81 D8
Olszyny PL 144 D2
Oltedal N 94 E4
Olten CH 27 F8
Oltenești RO 153 D11
Olteni RO 160 E6
Oltenița RO 161 E9
Oltina RO 155 E1
Olula del Río E 55 E8
Olustvere EST 131 D11
Olvan E 43 C7
Olvasjärvi FIN 120 D9
Ölvega E 41 E8
Olvena E 42 C4
Olvera E 51 F9
Olympiada GR 169 C10
Olympiada GR 169 E7
Olympos GR 181 E6
Olzai I 64 C3
Oma N 94 B3
Omagh UK 4 F2
Omalos GR 178 E6
Omarchevo BG 166 E6
Omarska BIH 157 C6
Ombersley UK 13 A10
Omedu EST 131 D14
Omegna I 68 B5
Ömerköy TR 173 E9
Ömerli TR 173 D9
O Mesón do Vento E 38 B3
Omessa F 37 G10
Omiš HR 157 F6
Omišalj HR 67 B10
Ommen NL 16 C6
Omne S 107 F14
Omolio GR 169 E8
Omoljica SRB 159 D6
Omont F 19 E10
Ömossa FIN 122 F7
O Mosteiro E 38 C2
Omurtag BG 167 C6
Omvriaki GR 174 A5
Oña E 40 C5
Ona N 100 A5
Onani I 64 C3
Onano I 62 B1
Oñati E 32 D1
Oncala E 41 E7
Oncești RO 153 E10
Onchan GBM 10 C3
Onda E 48 E4
Ondara E 56 D5
Ondarroa E 32 D1
Ondřejov CZ 77 C7
Ondres F 32 C3
Oneglia I 37 D8
Onesse-et-Laharie F 32 B3
Onești RO 153 E9
Onet-le-Château F 33 B11
Ongles F 35 B10
Oniceni RO 153 D10
Onich UK 4 B6
Onifai I 64 C4
Oniferi I 64 C3
Onițcani MD 154 C4
Onkamaa FIN 128 D7
Onkamo FIN 115 E6
Onkamo FIN 119 D15
Onkamo FIN 125 F14
Onkiniemi FIN 127 C15
Önnestad S 88 C6
Önningby FIN 99 B14
Önod H 145 G2
Onøya N 108 D4
Onsala S 91 E11
Onsbjerg DK 86 D7
Ønslev DK 83 A11
Onslunda S 88 D6
Onstmettingen D 27 D10
Onstwedde NL 17 B8
Ontiñena E 42 D4
Ontinyent E 56 D3
Ontojoki FIN 125 B11
Ontronvara RUS 125 E16
Onttola FIN 125 E13
Ontur E 55 B10
Onuškis LT 135 D11
Onuškis LT 137 E10
Onzain F 24 E5
Onzonilla E 39 C8
Oola IRL 8 C6
Ooltgensplaat NL 16 E2
Oonurme EST 131 C13
Oostakker B 19 B8
Oostburg NL 19 B7
Oostende B 18 B6
Oostendorp NL 16 D5
Oosterbeek NL 183 B7
Oosterend NL 16 B3
Oosterhesselen NL 17 C7
Oosterhout NL 16 E3
Oosterland NL 16 E2
Oosterwolde NL 16 C6
Oosterzele B 19 C8
Oostham B 183 C6
Oosthuizen NL 16 C3
Oostkamp B 19 B7
Oostkapelle NL 16 E1
Oostmalle B 182 C5
Oost-Souburg NL 16 F1
Oostvleteren B 18 C6
Oost-Vlieland NL 16 B4
Oostvoorne NL 182 B4
Ootmarsum NL 17 D7
Opaka BG 166 C6
Opalenica PL 81 B10
Opalenie PL 138 C6
Ópályi H 145 H5
Opan BG 166 E5
Opařany CZ 77 D6
Opatija HR 67 B9
Opatov CZ 77 C10
Opatovice nad Labem CZ 77 B9
Opatów PL 142 D5
Opatów PL 142 E6
Opatów PL 143 E11
Opatówek PL 142 C5
Opatowiec PL 143 F10
Opava CZ 146 B5
O Pazo E 38 D3
O Pazo de Irixoa E 38 B3
Ope S 106 E7
O Pedrouzo E 38 C3
Opeinde NL 16 B6
Opfenbach D 71 B9
Opglabbeek B 183 C7
Opheusden NL 183 B7
Opi I 62 D5
Opitter B 183 C7
Oploo NL 183 B7
Oplotnica SLO 148 D4
Opmeer NL 16 C3
Opochka RUS 133 C5
Opočno CZ 77 B10
Opoczno PL 141 H2
Opoeteren B 183 C7
Opole PL 142 E4
Opole Lubelskie PL 141 H5
Oporelu RO 160 D4
Oporets' UA 145 F7
O Porriño E 38 D2
Opoul-Périllos F 34 E4
Opovo SRB 158 C5
Oppach D 81 D7
Oppala S 103 E13
Oppdal N 101 A11
Oppeano I 66 B3
Oppeby S 92 C7
Oppedal N 100 D3
Oppegård N 95 C13
Oppenau D 27 D9
Oppenheim D 21 E10
Oppenweiler D 187 D7
Opphaug N 104 D7
Opphus N 101 D14
Oppido Lucano I 60 B5
Oppido Mamertina I 59 C8
Oppin D 79 C11
Opponitz A 73 A10
Oppurg D 79 E10
Oprisavci HR 157 B9
Oprișor RO 159 E11
Oprtalj HR 67 B8
Opsa BY 135 E13
Opsaheden S 97 B10
Optași-Măgura RO 160 D5
Ópusztaszer H 150 E5
Opuzen HR 157 F8
Opwijk B 19 C9
Őr H 145 H5
Ora I 72 D3
Øra N 112 C8
Orada P 50 B5
Oradea RO 151 C8
Oradour-sur-Glane F 29 D8
Oradour-sur-Vayres F 29 D7
Orah BIH 157 F7
Orah BIH 162 D5
Orahovičko Polje BIH 157 D8
Orahova BIH 157 B7
Orahov Do BIH 162 D4
Orahovica HR 149 E9
Oraio GR 171 B7
Oraiokastro GR 169 C8
Oraison F 35 C10
Orajärvi FIN 117 E12
Orakylä FIN 115 D1
Orange F 35 B8
Orani I 64 C3
Oranienburg D 84 E4
Órán Mór IRL 6 F5
Oranmore IRL 6 F5
Orašac BIH 156 C5
Orašac HR 162 D5
Orašac SRB 164 C5
Orašje BIH 157 B10
Orăștie RO 151 F11
Orăștioara de Sus RO 151 F11
Orașu Nou RO 145 H7
Örträsk S 103 C10

Orava EST 132 F1
Orava FIN 123 D10
Oravainen FIN 122 D8
Oravais FIN 122 D8
Oravala FIN 128 D6
Oravan S 107 B14
Oravikoski FIN 124 E9
Oravisalo FIN 125 F13
Oravița RO 159 C8
Oravivaara FIN 121 E11
Oravská Polhora SK 147 B8
Oravské Veselé SK 147 C8
Oravský Podzámok SK 147 C8
Orba E 56 D4
Orbacém P 38 E2
Ørbæk DK 86 E7
Orbais-l'Abbaye F 25 C10
Orbara E 32 E3
Orbassano I 37 A7
Orbe CH 31 B10
Orbeasca RO 160 E6
Orbec F 24 B3
Orbeni RO 153 E10
Orbetello I 65 C4
Ørbyhus S 99 B9
Orca P 44 D6
Orce E 55 D8
Orcera E 55 C7
Orchamps-Vennes F 26 F6
Orchies F 19 D7
Orchomenos GR 175 C6
Orchów PL 143 C7
Orchowo PL 138 E5
Orciano di Pesaro I 67 E6
Orcières F 36 B4
Orcival F 30 D2
Ordan-Larroque F 33 C6
Ordes E 38 B3
Ordizia E 32 D1
Ordona I 60 A5
Ordzhonikidze UA 154 C6
Orea E 47 C9
O Real E 38 B3
Örebäcken S 102 C4
Orebić HR 157 G7
Örebro S 97 D13
Ořechov CZ 77 D11
Öregcsertő H 150 D3
Öreglak H 149 C9
Öregrund S 99 B10
Orehoved DK 87 F9
Oreini GR 169 B10
Orekhovitsa BG 165 B9
Orellana de la Sierra E 51 A9
Orellana la Vieja E 51 A8
Ören TR 177 C10
Ören TR 181 B7
Orenhofen D 185 E6
Oreoi GR 175 B7
Orés E 32 F3
Oresh BG 166 B4
Oreshak BG 165 D10
Orestiada GR 171 B11
Öreström S 107 C16
Öretjänälen S 103 A10
Oreye B 19 C11
Orezu RO 161 D9
Orford UK 15 C12
Organi GR 171 B9
Organyà E 43 C6
Orgaz E 46 E5
Orgelet F 31 B8
Ørgenvika N 95 B11
Orgères-en-Beauce F 24 D6
Orgita EST 131 D8
Orgiva E 53 C10
Orgon F 35 C9
Orgosolo I 64 C3
Orhaneli TR 173 E10
Orhaniye TR 171 C10
Orhaniye TR 181 C8
Orhanlar TR 173 D8
Orhei MD 154 C3
Oria E 55 E8
Oria I 61 C9
O Rial E 38 D2
Origny-Ste-Benoîte F 19 E7
Orihuela E 56 E3
Orihuela del Tremedal E 47 C9
Orikhivka UA 154 F3
Orikum AL 168 D1
Orimattila FIN 127 D14
Oriniemi FIN 127 C9
Orio E 32 D1
Oriola P 50 C4
Oriolo I 61 C6
Oriolo Romano I 62 C2
Oripää FIN 126 D8
Orissaare EST 130 D6
Oristà E 43 D8
Oristano I 64 D2
Oristown IRL 7 E9
Öriszentpéter H 149 C6
Oriv UA 145 E8
Orivesi FIN 127 B11
Orizare BG 167 D9
Orizari NMK 164 F5
Ørjavik N 104 F2
Ørje N 96 D6
Orkanger N 104 E7
Orkelljunga S 87 C12
Örkény H 150 C3
Orla PL 141 E8
Orlamünde D 79 E10
Orlat RO 152 F3
Orlea RO 160 F4
Orléans F 24 E6
Orleşti RO 160 D4
Orlivka UA 155 C2
Orllan RKS 164 D3
Orlová CZ 146 B6
Orlov Dol BG 166 E6
Orlovets BG 166 C5
Orly PL 144 D6
Orlyak BG 161 F10
Orlyane BG 165 C10
Orma GR 169 C6
Ormanli TR 173 B9
Ormaryd S 92 D5
Ormea I 37 C7
Örményes H 150 C6
Örménykút H 150 D6
Ormos GR 176 D4
Ormos Panormou GR 176 D5
Ormos Prinou GR 171 C7
Ormož SLO 148 D6

Ormskirk UK 10 D6
Ormylia GR 169 D10
Ornaisons F 34 D4
Ornans F 26 F5
Örnäs S 97 A14
Örnäsudden S 109 E13
Ornavasso I 68 B5
Ornbau D 75 D8
Ornes N 105 B15
Ørnes N 108 C6
Orneta PL 139 B9
Ørnhøj DK 86 C3
Ornö S 93 A12
Ornontowice PL 142 F6
Örnsköldsvik S 107 E15
Orodel RO 159 E11
Orolik HR 157 B11
Orom SRB 150 F4
Oron-la-Ville CH 31 B10
Oronoz E 32 D2
Oroński PL 141 H3
Oropa I 68 B4
Oropesa E 45 E10
Oropesa del Mar E 48 D5
Ororbia E 32 E2
Orosei I 64 C4
Orosháza I 150 D6
Oroslavje HR 148 E5
Oroszlány H 149 B10
Orpierre F 35 B10
Orreaga E 32 D3
Orrefors S 89 B9
Orrios E 42 F2
Orrmo S 102 C7
Orroli I 64 D3
Orrviken S 106 E6
Orsa S 102 D8
Orsara di Puglia I 60 A4
Orsay F 24 C7
Örsbäck S 107 D17
Orsennes F 29 C9
Örserum S 92 D6
Orsières CH 31 C11
Örsjö S 89 B9
Orslev DK 87 E9
Ørsnes N 100 B6
Orsogna I 63 C6
Orsomarso I 60 D5
Orşova RO 159 D9
Ørsta N 100 B6
Ørsted DK 86 C6
Örsundsbro S 99 C8
Ortaca TR 181 C9
Ortacesus I 64 D3
Ortakent TR 177 E9
Ortaklar TR 177 D8
Ortaköy TR 173 B9
Orta Nova I 60 A5
Orte I 62 C2
Orten N 100 A5
Ortenberg D 21 D12
Ortenberg D 27 D8
Ortenburg D 76 E4
Orth an der Donau A 77 F11
Orthez F 32 D4
Ortholmen S 102 B6
Orthovouni GR 168 E5
Ortigosa E 41 D6
Ortigueira E 38 A4
Ørting DK 86 D6
Ortisei I 72 C4
Ortişoara RO 151 F7
Ortnevik N 100 D4
Orton UK 10 C6
Ortona I 63 C6
Ortovera I 37 C8
Ortrand D 80 D5
Örträsk S 107 C15
Ortueri I 64 C2
Örtülüce TR 173 D7
Ørum DK 86 C3
Ørum DK 86 C7
Orune I 64 C3
Orusco F 29 B10
Orval F 29 B10
Orvalho P 44 D5
Orvault F 23 F8
Orvieto I 62 B2
Örviken S 118 E6
Orvinio I 62 C3
Oryakhovo BG 160 F3
Orzesze PL 142 F6
Orzinuovi I 69 C8
Orzyny PL 139 C11
Orzysz PL 136 F4
Os N 101 B14
Osa N 100 E6
Osa de Vega E 47 E7
Ošani LV 135 D11
Osaonica SRB 163 C9
Osbaldwick UK 11 D9
Os Blancos E 38 D4
Osburg D 186 B2
Øsby DK 86 E5
Osby S 88 C5
Osbyholm S 87 D13
Oščadnica SK 147 C7
Oschatz D 80 D4
Oschersleben (Bode) D 79 B9
Oschiri I 64 B3
Ościsłowo PL 139 E9
Os Dices E 38 C2
Osdorf D 83 B8
Osečina SRB 158 E4
O Seixo E 38 E2
Oseja de Sajambre E 39 B9
Osek CZ 76 C5
Osek CZ 80 E5
Osen N 105 C9
Osen N 108 D6
Osenets BG 166 B6
Osenieki LV 134 C4
Osera E 41 E10
Oseşti RO 153 D10
Oset N 101 D10
Osetno PL 81 C10
Osi H 149 B10
Osica de Sus RO 160 E4
Osidda I 64 B3
Osie PL 138 C5
Osijciny PL 138 E6
Osieck PL 141 G4
Osieczna PL 81 C11
Osieczna PL 138 C5
Osiecznica PL 81 D8
Osiek PL 138 C5
Osiek PL 138 D3
Osiek PL 143 E11
Osiek PL 147 B8
Osiek Jasielski PL 145 D3
Osiek Mały PL 142 B6
Osiek nad Notecią PL 85 D12

Osielsko PL 138 D5
Osiglia I 37 C8
Osijek HR 149 E11
Osikovitsa BG 165 D9
Osilo I 64 B2
Osimo I 67 F7
Osina PL 85 C8
Osini I 64 D3
Osiny PL 141 H6
Osio Sotto I 69 B8
Osipaonica SRB 159 D7
Osjaków PL 142 D6
Oskar S 89 B9
Oskar-Fredriksborg S 99 D10
Oskarshamn S 93 E8
Oskarström S 87 B11
Oskava CZ 77 C12
Oskořínek CZ 77 B8
Osľany SK 146 D6
Oslättfors S 103 E12
Osli H 149 A8
Oslo N 95 C13
Osloß D 79 B8
Osma E 40 C5
Osma FIN 117 D15
Osmancali TR 177 B9
Osmancik TR 173 A7
Osmangazi TR 173 D11
Osmaniye TR 173 E10
Osmaniye TR 173 E10
Osmanki FIN 123 C16
Osmanli TR 173 A6
Os'mino RUS 132 C5
Osmolin PL 141 F1
Osmoloda UA 145 F9
Osnabrück D 17 D10
Osno Lubuskie PL 81 B7
Osny F 24 B7
Osoblaha CZ 142 F4
Osogna CH 69 A6
Osojnik HR 162 D5
Osoppo I 73 D7
Osor E 43 D9
Osor HR 67 C9
Osøren RO 151 C9
Osorno E 40 D3
Osowa PL 136 A6
Osowa Sień PL 81 C10
Osøyri N 94 B2
Ospitaletto I 69 B9
Oss NL 16 E5
Ossa GR 169 C9
Ossa de Montiel E 55 B7
Ossana I 69 A10
Osséja F 33 F9
Ossendrecht NL 182 C4
Osses F 32 D3
Ossiach A 73 C8
Oßmannstedt D 79 D9
Ossun F 32 D5
Östa S 98 B7
Östanå S 88 C6
Östanfjärden S 119 C10
Östansjö S 92 A5
Östansjö S 109 E16
Östanskär S 103 A13
Östanvik S 103 D9
Oštarije HR 156 B3
Ostaszewo PL 138 B6
Östavall S 103 B9
Ostbevern D 17 D9
Østbirk DK 86 D5
Østbjörka S 103 E9
Østby N 94 A9
Østby N 102 D4
Østby S 107 D13
Osted DK 87 D9
Ostellato I 66 C4
Osten D 17 A12
Ostend B 18 B6
Ostend B 182 C1
Ostenfeld (Husum) D 82 B6
Østerås S 107 E12
Østerby DK 86 C5
Øster Bjerregrav DK 86 C5
Øster Brønderslev DK 86 A5
Osterburg (Altmark) D 83 E11
Osterburken D 27 B11
Osterby D 82 B7
Østerby DK 86 A5
Østerbybruk S 99 B9
Østerbymo S 92 D6
Østerede S 107 E11
Österfärnebo S 98 B7
Osterfeld D 79 D10
Österforse S 107 E12
Östergarn S 93 E13
Östergraninge S 107 F12
Österhankmo FIN 122 D7
Osterhever D 82 B5
Osterhofen D 76 E4
Øster Højst DK 86 E4
Osterholz-Scharmbeck D 17 B11
Øster Hornum DK 86 B5
Øster Hurup DK 86 B6
Østerild DK 86 A3
Øster Jølby DK 86 B3
Österjörn S 118 D4
Østerlars DK 89 E7
Øster Lindet DK 86 E4
Österlisa S 99 C11
Östermarie DK 89 E8
Östermark FIN 126 D8
Ostermiething A 76 F3
Ostermundigen CH 31 B11
Östernorret S 107 C12
Östero FIN 122 D8
Osterode am Harz D 79 C7
Osterrönfeld D 82 B7
Österskucku S 102 A8
Östersund S 106 E7
Östersundom FIN 127 E13
Øster Tørslev DK 86 B6
Øster Ulslev DK 83 A11
Östervåla S 98 B8
Øster Vedsted DK 86 E3
Østervrå DK 90 E7
Øster Vrøgum DK 86 D2
Osterwieck D 79 C8
Østese N 94 B4
Ostfildern D 187 D7
Østofte DK 87 F10
Osthammar S 99 B10
Ostheim vor der Rhön D 75 B7
Osthofen D 21 E10
Ostia I 62 D2
Ostiano I 66 B1
Ostiglia I 66 B3
Ostiz E 32 E2
Östlöning S 103 A13
Östmark S 97 B8
Östmarkum S 107 E14

Östnor S 102 D7
Ostojićevo SRB 150 F5
Ostoros H 145 H1
Ostra I 67 E7
Ostra RO 153 C7
Östra Åliden S 118 D4
Ostrach D 27 E11
Östra Ed S 93 C9
Östra Frölunda S 91 E13
Östra Granberg S 118 C4
Östra Grevie S 87 E12
Östra Husby S 93 B9
Östra Ljungby S 87 C12
Östra Lovsjön S 106 D7
Östra Ormsjö S 107 C10
Östra Ryd S 93 C8
Östra Skrämträsk S 118 E5
Östra Sönnarslöv S 88 D6
Östra Stugusjö S 103 A9
Ostrau D 79 C11
Ostrau D 80 D4
Östra Vemmerlöv S 88 D6
Östra Yttermark FIN 122 E6
Oštrelj BIH 156 D5
Ostren AL 168 B3
Ostřetín CZ 77 B10
Østrhauderfehn D 17 B9
Ostricourt F 182 E2
Östringen D 21 F11
Ostritsa BG 166 B5
Ostritz D 81 D7
Ostróda PL 139 C8
Ostrołęka PL 139 D11
Ostromecko PL 138 D5
Ostroměř CZ 77 B9
Ostrorog PL 81 A10
Ostrov BG 160 F4
Ostrov CZ 76 B3
Ostrov RO 155 D2
Ostrov RO 161 E10
Ostrov RUS 133 C4
Ostrov SK 146 D5
Ostroveni RO 160 F3
Ostrove E 40 D3
Osowa S 136 E6
Ostrów u Macochy CZ 77 D11
Ostrów PL 143 F12
Ostrówek PL 141 G7
Ostrówek PL 142 D6
Ostrowice PL 85 C9
Ostrowiec PL 85 B11
Ostrowiec Świętokrzyski PL 143 E11
Ostrowite PL 138 F5
Ostrowite PL 139 D7
Ostrów Lubelski PL 141 H7
Ostrów Mazowiecka PL 139 E12
Ostrowo PL 138 E5
Ostrów Wielkopolski PL 142 C4
Ostrowy nad Okszą PL 143 E7
Ostrožac BIH 156 C5
Ostrožac BIH 157 E8
Ostrożeń PL 141 G5
Østrup DK 86 B4
Ostrzeszów PL 142 D4
Ostuni I 61 B9
Ostvik S 118 E6
Ostwald F 186 D4
Osula EST 131 F13
Osuna E 53 B6
Ošupe LV 135 C13
Osvallen S 102 A4
Osvica BIH 157 C8
Oswaldkirk UK 11 C9
Oswestry UK 10 F5
Oświęcim PL 143 F7
Ota F 37 G9
Otaci MD 154 A1
Otalampi FIN 127 E12
Otanki LV 134 C2
Otanmäki FIN 120 F9
Otaslavice CZ 77 D11
Otava FIN 128 B7
Otavice HR 156 E5
Oteiza E 32 E2
Oţeleni RO 153 C10
Oţelu Roşu RO 159 B9
Otepää EST 131 E12
Oteren N 111 B18
Oterma FIN 120 E9
Otero de Bodas E 39 E7
Otervik N 105 A11
Oteşani RO 160 C4
Oteševo NMK 168 C4
Otfinów PL 143 F10
Othem S 93 D13
Otigheim D 27 C10
Otišheim D 27 C10
Otišić HR 156 E5
Otívar E 53 C9
Otley UK 11 D8
Otley UK 15 C11
Otmuchów PL 77 B12
Otnes N 101 C14
Otočac HR 156 C3
Otok HR 157 B10
Otok HR 157 E6
Otoka BIH 156 C5
Otopeni RO 161 D8
Otorowo PL 81 A10
O Toural E 38 D2
Otovica NMK 164 E4
Otradnoye RUS 129 F14
Otranto I 61 C10
Otricoli I 62 B2
Otrokovice CZ 146 C5
Otta N 101 C11
Ottana I 64 C3
Ottaviano I 60 B2
Ottenby S 89 C10
Ottendorf-Okrilla D 80 D5
Ottenheim D 186 E4
Ottenhöfen im Schwarzwald D 27 C9
Ottenschlag A 77 F8
Ottensheim A 76 F6
Ottenstein D 17 D12
Otterbach D 186 C4
Otterbäcken S 91 B15
Otterberg D 21 E9
Otterburn UK 5 E12
Otter Ferry UK 4 C6
Otterfing D 72 A4
Otterlo NL 16 D5
Otterndorf D 17 A11
Ottersberg D 17 B12
Ottersweier D 27 C9
Otterup DK 86 D6

Otterwisch D 79 D12
Öttevény H 149 A9
Ottignies B 19 C10
Ottmarsheim F 27 E8
Ottobeuren D 71 B10
Ottobrunn D 75 F10
Öttömös H 150 E4
Ottone I 37 B10
Ottrau D 21 C12
Ottsjö S 105 E14
Ottsjön S 106 D7
Ottweiler D 21 F8
Otwock PL 141 F4
Otxandio E 41 B6
Otyń PL 81 C9
Otzing D 76 E3
Ouanne F 25 E9
Ouarville F 24 D6
Ouca P 44 C4
Oucques F 24 E5
Oud-Beijerland NL 16 E2
Ouddorp NL 182 B3
Oudehaske NL 16 C5
Oudemirdum NL 16 C5
Oudenaarde B 19 C8
Oudenbosch NL 16 E3
Oudenburg B 18 B7
Oudeschild NL 16 B3
Oude-Tonge NL 182 B4
Oudewater NL 182 A5
Oudon F 23 F9
Oud-Turnhout B 16 F3
Oud-Vossemeer NL 16 E2
Oudzele B 182 C2
Oued Laou MA 53 F6
Ouffet B 19 D11
Oughterard IRL 6 F4
Ougney F 26 F4
Ouguela P 45 F6
Ouistreham F 23 B11
Oulainen FIN 119 F13
Oulanka FIN 115 F5
Oulchy-le-Château F 25 B9
Oulder B 20 D5
Oullins F 30 D6
Oulton UK 15 C12
Oulu FIN 119 D14
Oulunsalo FIN 119 E14
Oulx I 31 E7
Oundle UK 15 C8
Oupeye B 19 C12
Ouranoupoli GR 170 D5
Oure DK 87 E7
Ourém P 44 E3
Ourense E 38 D4
Ourique P 50 D3
Ourol E 38 A4
Ouroux-en-Morvan F 25 F10
Ouroux-sur-Saône F 31 B6
Ourville-en-Caux F 18 E2
Oust F 33 E8
Outakoski FIN 113 D16
Outarville F 24 D7
Outeiro P 38 E2
Outeiro P 39 E6
Outeiro de Rei P 38 B4
Outeiro Seco P 38 E4
Outokumpu FIN 125 E12
Outomuro E 38 D4
Outreau F 15 F12
Outwell UK 11 F12
Ouveillan F 34 D4
Ouzouer-le-Marché F 24 E6
Ouzouer-sur-Loire F 25 E7
Ovada I 37 B9
Ovanåker S 103 D10
Ovar P 44 C3
Ovaro I 73 C6
Ovča SRB 158 D6
Ovcha Mogila BG 166 C4
Ovcharovo BG 166 E6
Ovchepolsi BG 165 E9
Ove DK 86 B5
Ovelgönne D 17 B10
Överammer S 107 E9
Överäng S 105 D14
Överberg S 102 B8
Overbister UK 3 G11
Øverbygd N 111 C17
Överbyn S 103 C12
Overdinkel NL 183 A10
Over Feldborg DK 86 C3
Övergård N 111 B18
Overhalla N 105 B11
Overhogdal S 102 B8
Överhörnäs S 107 E15
Over Hornbæk DK 86 C5
Overijse B 19 C10
Överissjö S 107 C13
Over Jerstal DK 86 E4
Överkalix S 119 B9
Overlade DK 86 B4
Överlännäs S 107 E13
Överlida S 91 E12
Overloon NL 16 E5
Overton UK 10 F6
Overton UK 13 C12
Övertorneå S 119 B11
Överturingen S 102 B8
Överum S 93 D8
Ovezande NL 16 F1
O Vicedo E 38 A4
Ovidiu RO 155 E3
Oviedo E 39 B8
Oviglio I 37 B8
Oviken S 105 E16
Ovindoli I 62 C5
Oviši LV 130 F3
Övodda I 64 C3
Øvra S 107 D11
Øvre Alta N 113 D11
Øvre Årdal N 100 D7
Øvre Bredåker S 118 C6
Øvre Flåsjön S 118 B7
Øvre Kildal N 112 D7
Øvre-Konäs S 105 D14
Øvrella N 95 C10

Øvre Rendal N 101 C14
Øvre Soppero S 116 B7
Øvre Tväråsel S 118 C5
Ovria GR 174 C4
Øvsjöbyn S 107 E9
Ovtrup DK 86 D2
Owen D 27 E11
Owingen D 27 E11
Owińska PL 81 A11
Owschlag D 82 B7
Oxberg S 102 D7
Oxelösund S 93 B10
Oxenhope UK 11 D8
Oxentea MD 154 C4
Oxford UK 13 B12
Oxhalsö S 99 C11
Oxie S 87 E12
Oxkangar FIN 122 D8
Oxshott UK 15 E8
Oxted UK 15 E9
Oxton UK 5 D11
Oxylithos GR 175 B9
Øyangen N 104 E3
Øydegarden N 104 E4
Øyenkilen N 91 A8
Øyer N 101 D12
Øyeren N 96 B7
Øyjord N 108 B9
Oy-Mittelberg D 71 B10
Øynes N 108 B9
Øynes N 111 C11
Oyonnax F 31 C8
Øysletta N 90 C2
Oyten D 17 B12
Øyvatnet N 111 C12
Oza E 38 B3
Ozaeta E 41 C7
Ozalj HR 148 E4
Ožarów PL 143 E12
Ożarów Mazowiecki PL 141 F3
Ožbalt SLO 148 C4
Özbaşı TR 177 D9
Özbek TR 177 C8
Ózd H 145 G1
Oždany SK 147 E9
Özdere TR 177 D9
Ozenna PL 145 E3
Ozersk RUS 136 E5
Ozieri I 64 B3
Ozimek PL 142 E5
Ozimica BIH 157 D9
Özlüce TR 181 B9
Ozoir-la-Ferrière F 25 C8
Ozolaine LV 133 D2
Ozoli LV 131 F9
Ozoli LV 131 F9
Ozoli LV 135 C12
Ozolmuiža LV 133 D2
Ozolnieki LV 134 C7
Ozora I 149 C10
Ozorków PL 143 C7
Ozun RO 153 F7
Ozzano dell'Emilia I 66 D3
Ozzano Monferrato I 68 C5

P

Pääaho FIN 121 D10
Pääjärvi FIN 123 E13
Paakinmäki FIN 121 F11
Paakkila FIN 125 E11
Paakkola FIN 119 C13
Paal B 183 C6
Paalasmaa FIN 125 D12
Paaso FIN 127 C12
Paasvere EST 131 C13
Paatela FIN 129 B9
Paattinen FIN 126 D7
Paatus FIN 113 D17
Paavola FIN 119 E14
Pabaiskas LT 135 F9
Paberžė LT 137 D11
Pabianice PL 143 C7
Pabillonis I 64 D2
Pabiržė LT 135 D9
Pabneukirchen A 77 F7
Pabrade LT 137 D12
Pabu F 22 C5
Pacanow PL 143 F11
Paceco I 58 D2
Pacheia Ammos GR 179 E10
Pachino I 59 F7
Pachni GR 171 B7
Paciano I 62 B2
Pácin H 145 G4
Pačir SRB 150 F3
Pack A 73 C10
Pačlavice CZ 77 D12
Pacos de Ferreira P 38 F3
Pacov CZ 77 D8
Pacsa H 149 C8
Păcureţi RO 161 C8
Pacyna PL 143 B8
Pacy-sur-Eure F 24 B5
Paczków PL 77 B12
Padankoski FIN 127 C13
Padborg DK 82 A6
Padbury UK 14 D7
Padej SRB 150 F5
Padene HR 156 D5
Paderborn D 17 E11
Paderne P 38 D3
Paderne P 50 E3
Paderne de Allariz E 38 D4
Padeş RO 159 D10
Padesh BG 165 F7
Padew Narodowa PL 143 F12
Padežine BIH 157 F8
Padiham UK 11 D7
Pădina RO 159 E11
Padina RO 161 D10
Padina SRB 159 C6
Padinska Skela SRB 158 D5
Padirac F 29 F9
Padise EST 131 C8
Padoby RO 133 F2
Padoux F 26 D6
Padova I 66 B4
Padra I 64 C2
Padrón E 38 C2
Padru I 64 B4
Padure LV 134 C3
Pădureni RO 153 D11
Paesana I 37 B6
Paese I 72 E5
Pag HR 67 D11
Pagani I 60 B3

Paganica I 62 C4
Paganico I 65 B4
Pagégiai LT 134 F3
Pagiriai LT 135 F8
Pagiriai LT 137 D11
Paglieta I 63 C6
Pagny-sur-Moselle F 26 C5
Pagondas GR 177 D8
Pagouria GR 171 B8
Pagramantis LT 134 F4
Paharova S 116 E8
Páhi H 150 D3
Pahkakoski FIN 119 D16
Pahkakumpu FIN 115 E3
Pahkakumpu FIN 121 C12
Pahkala FIN 119 F11
Pahkamäki FIN 123 D16
Pähl D 72 A3
Pahlen D 82 B6
Pahranichny BY 140 D9
Pahtaoja FIN 119 C14
Paião P 44 D3
Paide EST 131 D11
Paignton UK 13 E7
Paihola FIN 125 E13
Päijälä FIN 127 B12
Paikuse EST 131 E9
Pailhès F 33 D8
Paillet F 32 A5
Paimbœuf F 23 F7
Paimela FIN 127 C14
Paimio FIN 126 E8
Paimpol F 22 C5
Paimpont F 23 D7
Painswick UK 13 B10
Painten D 75 E10
Paipis FIN 127 E13
Paisley UK 5 D8
Paistu EST 131 E11
Paisua FIN 124 C9
Päiväjoki FIN 115 F2
Pajala S 117 D10
Pajares de la Lampreana E 39 E8
Pajarón E 47 E9
Pajęczno PL 143 D6
Pajukoski FIN 125 C11
Pajukoste FIN 113 C20
Pajūris LT 134 F4
Pajuskylä FIN 124 D7
Pajuvaara FIN 121 D14
Paka H 149 C7
Pakaa FIN 127 D14
Pakalné LT 134 F2
Pakalniai LT 135 F11
Pakapė LT 134 E6
Pakarila FIN 123 E17
Pakkala FIN 127 C11
Pakod H 149 C8
Pakość PL 138 E5
Pakosław PL 81 C12
Pakoštane HR 156 E4
Pákozd H 149 B11
Pakrac HR 149 F8
Pakruojis LT 134 E7
Paksuniemi S 116 C5
Pala EST 131 D14
Palacios del Sil E 39 C7
Palacios de Sanabria E 39 D6
Palaciosrubios E 45 B10
Palade EST 130 D5
Palafrugell E 43 D10
Palagianello I 61 B7
Palagiano I 61 B8
Palagonia I 59 E6
Palaia I 66 E2
Palaia Fokaia GR 175 D8
Palaiochora GR 169 C9
Palaiochora GR 178 E6
Palaiochori GR 169 D6
Palaiochori GR 169 D6
Palaiochori GR 169 D6
Palaiochori GR 175 C7
Palaiokastritsa GR 168 E2
Palaiokastro GR 177 D9
Palaiokipos GR 177 A7
Palaiokomi GR 170 C5
Palaiomonastiro GR 169 F6
Palaiopoli GR 176 D4
Palaiopyrgos GR 169 E6
Palaiopyrgos GR 169 E8
Palaiopyrgos GR 174 C4
Palaiovracha GR 174 B5
Palairos GR 174 B2
Palaiseau F 25 C7
Palamas GR 169 F7
Palamós E 43 D10
Palamuse EST 131 D13
Palanca RO 153 D8
Palanga LT 134 F2
Pålänge S 119 C9
Palanzano I 66 D1
Palárikovo SK 146 E6
Palas de Rei E 38 C4
Palata BY 133 E6
Palata I 63 D7
Pălatca RO 152 D4
Palau I 64 A3
Palavas-les-Flots F 35 C6
Palazzo Adriano I 58 D3
Palazzolo Acreide I 59 E6
Palazzolo sull'Oglio I 69 B8
Palazzo San Gervasio I 60 B6
Paldiski EST 131 C8
Pale BIH 157 E10
Pāle LV 131 F9
Paleičiai LT 134 F2
Palena I 63 D6
Palencia E 40 D3
Palenciana E 53 B7
Palenzuela E 40 D3
Palermo AL 168 D2
Palermo I 58 C3
Palešnica PL 144 D2
Palestrina I 62 D3
Pālèveniēlē LT 135 E10
Palež BIH 158 E3
Palhaça P 44 C4
Pálháza H 145 G4
Palia Kavala GR 171 C6
Paliano I 62 D4
Palić SRB 150 E4
Palinges F 30 B5
Palinuro I 60 C5
Paliouri GR 169 E10
Paliouria GR 169 E5
Palis F 25 D10
Paliseul B 19 E11
Palivere EST 131 D7
Palizzi I 59 D9
Paljakka FIN 121 B14
Paljakka FIN 121 E11

Pälkäne FIN 127 C11
Pålkem S 118 B7
Palkino RUS 132 F3
Palkisoja FIN 115 A2
Palkovice CZ 146 B6
Palladio GR 171 B8
Pallagorio I 61 E7
Pallarés E 51 C7
Pallas Green New IRL 8 C6
Pallas Green IRL 8 C6
Pälli FIN 119 E16
Palling D 75 F12
Palluau F 28 B2
Palluau-sur-Indre F 29 B8
Palma P 50 C2
Palma Campania I 60 B3
Palma del Río E 51 D9
Palma de Mallorca E 49 E10
Palma di Montechiaro I 58 E4
Palmadula I 64 B1
Palmanova E 49 E10
Palmanova I 73 E7
Palmaz P 44 C4
Palmeira P 38 E3
Palmela P 50 B2
Palmi I 59 C8
Pálmonostora H 150 D4
Palnackie UK 5 F9
Palneca F 37 H10
Palo del Colle I 61 A7
Palohuornas S 116 E6
Palojärvi FIN 115 E2
Palojärvi FIN 117 A10
Palojärvi FIN 117 E14
Palojoensuu FIN 116 B10
Palokki FIN 125 E11
Paloma FIN 113 E19
Palomäki FIN 125 D11
Palomar de Arroyos E 42 F2
Palomares del Río E 51 E7
Palomas E 51 B7
Palombara Sabina I 62 C3
Palombaro I 63 C6
Palomene LT 137 D9
Palomera E 47 C8
Palomeras del Campo E 47 E7
Palomonte I 60 B4
Palonoja FIN 115 B2
Palonselkä FIN 117 D13
Palonurmi FIN 125 D10
Paloperä FIN 115 F4
Palos de la Frontera E 51 E6
Palosenjärvi FIN 124 C8
Palotabozsok H 149 D11
Palotáshalom H 150 B4
Palovaara FIN 119 B16
Palovaara FIN 121 E13
Palovaara FIN 125 E14
Pals E 43 D10
Pålsboda S 92 A4
Palsmane LV 135 B12
Palsselkä FIN 117 D16
Pålsträsk S 118 C6
Paltamo FIN 121 F10
Paltanen FIN 124 F7
Paltaniemi FIN 121 F10
Paltin RO 153 F9
Pãltiniș RO 153 A9
Pãltiniș RO 159 C9
Pãltinoasa RO 153 B7
Paludi I 61 D7
Paluel F 18 E2
Paluknys LT 137 E10
Paluzza I 73 C7
Palyatskishki BY 137 E11
Palyesye BY 135 F12
Palzem D 20 E6
Pambukovica SRB 158 E4
Pameče SLO 73 C11
Pamfylla GR 177 A8
Pamhagen A 149 A7
Pamiers F 33 D9
Pampâji LV 134 C4
Pamparato I 37 C7
Pampelonne F 33 B10
Pampilhosa P 44 D4
Pampilhosa da Serra P 44 D5
Pampliega E 40 D4
Pamplona E 32 E2
Pampow D 83 C10
Pamproux F 28 C5
Pamukçu TR 173 E8
Panaci RO 152 C6
Panagia GR 171 C7
Panagia GR 171 E8
Panagia GR 178 E8
Panagyurishte BG 165 D9
Panagyurski Kolonii BG 165 D9
Panahor AL 168 C2
Panaitolio GR 174 B3
Panaja AL 168 C1
Pãnãșești MD 154 C3
Panassac F 33 D7
Pãnãtãu RO 161 C8
Panazol F 29 D8
Pancalieri I 37 B7
Pancar TR 177 C9
Pancarköy TR 173 B7
Pãncești RO 153 E10
Pancevo SRB 158 D6
Pancharevo BG 165 D7
Panciu RO 153 F10
Pancorbo E 40 C5
Pâncota RO 151 E8
Pancrudo E 42 F1
Pánd H 150 C4
Pandelys LT 135 E11
Pandino I 69 C8
Pandrup DK 86 A5
Pandy UK 13 B9
Panelia FIN 126 D6
Panemunė LT 136 C4
Panemunėlis LT 135 E10
Panes E 39 B10
Pãnet RO 152 D4
Panevėžys LT 135 E8
Panga EST 130 D4
Pângãrați RO 153 D8
Pange F 186 C1
Panicale I 62 A2
Panichkovo BG 166 F4
Panissières F 30 D5
Paniza E 41 F9
Panjas F 32 C5
Panjevac SRB 159 E8
Panjik BIH 157 C9
Panka FIN 124 D7
Pankajärvi FIN 125 D14
Pankakoski FIN 125 D14
Panker D 83 B9

Panki PL 142 E6
Pannarano I 60 A3
Pannes F 25 D8
Panni I 60 A4
Panningen NL 16 F5
Pannonhalma H 149 A9
Panóias P 50 D3
Panorama GR 169 C9
Panormos GR 178 E8
Panschwitz-Kuckau D 80 D6
Pantanassa GR 178 B4
Pãntãne FIN 122 F8
Pantelimon I 55 D2
Pantelimon RO 161 E8
Panticeu RO 152 C3
Pantoja E 46 D5
Panttikylä FIN 122 F8
Panttila FIN 122 F8
Pant-y-dwr UK 13 A8
Paola I 60 E6
Pápa H 149 B8
Papadianika GR 178 B4
Papasidero I 60 D5
Pápateszér H 149 B9
Papenburg D 17 B8
Papendorf D 83 B12
Papendrecht NL 16 E3
Papile LT 134 D5
Papilys LT 135 D10
Papín SK 145 E5
Papkeszi H 149 B10
Paplaka LV 134 D2
Papowo Biskupie PL 138 D6
Pappades GR 175 A7
Pappados GR 177 A7
Pappenheim D 75 E8
Papradno SK 146 C6
Paprotnia PL 141 F6
Par UK 12 E5
Parabita I 61 D10
Paracin SRB 159 E7
Paracuellos E 47 C8
Paracuellos de Jarama E 46 C5
Parád H 147 F10
Parada P 45 C6
Parada de Ester P 44 C4
Parada de Pinhão P 38 F4
Parada de Rubiales E 45 B10
Parada de Sil E 38 D4
Paradas E 51 E9
Paradeisi GR 181 D8
Paradeisia GR 174 D5
Paradeisos GR 171 B7
Paradela E 38 C5
Paradela P 38 E4
Paradela P 44 B5
Paradyż PL 141 H2
Parainen FIN 126 E7
Parakalamos GR 168 E4
Parakka S 116 D7
Parakoila GR 177 A7
Paralepa EST 130 C7
Paralia GR 174 C4
Paralia GR 175 F6
Paralia Avdiron GR 171 C7
Paralia Saranti GR 175 C7
Paralia Tyrou GR 175 E6
Paralio Astros GR 175 E6
Paramé F 23 C8
Parâmio P 39 E6
Páramo del Sil E 39 C7
Paramythia GR 168 F4
Paranesti GR 171 B6
Paranhos P 44 D7
Parantala FIN 123 E15
Parapotamos GR 168 E3
Paras N 111 B19
Pârâu RO 152 F6
Parava RO 153 E9
Paravola GR 174 B4
Paray-le-Monial F 30 C5
Parcani MD 154 B3
Parcent E 56 D4
Parcé-sur-Sarthe F 23 E11
Parchen D 79 B11
Parchim D 83 D11
Parchovany SK 145 F4
Parchów PL 81 D9
Parchowo PL 85 B13
Parciaki PL 139 D11
Parcoul F 28 E6
Parczew PL 141 G7
Pardais P 50 B5
Pardies F 32 D4
Pardilhó P 44 C3
Pardosi RO 161 C9
Pardubice CZ 77 B9
Parechcha BY 137 F9
Paredes de Coura P 38 E2
Paredes de Nava E 39 D10
Pareja E 47 C7
Parempuyre F 28 F4
Parenti I 61 E6
Parentis-en-Born F 32 B3
Parets del Vallès E 43 D8
Parey D 79 B10
Parga GR 168 F3
Pârgãrești RO 153 E9
Pargas FIN 126 E7
Parghelia I 59 B8
Pargny-sur-Saulx F 25 C12
Pargolovo RUS 129 E13
Parhalahti FIN 119 F12
Pãri EST 131 E11
Parigné-l'Évêque F 24 E3
Parikkala FIN 129 B9
Parincea RO 153 E10
Paris F 25 C7
Parisot F 33 B9
Parisot F 33 C9
Parissavaara FIN 125 E16
Pãrjãnsuo FIN 120 C9
Pârjol RO 153 D9
Park UK 4 F2
Parkajoki S 117 C10
Parkalompolo S 116 C9
Parkano FIN 123 F10
Parkila FIN 125 D11
Parkila FIN 123 C14
Parkila FIN 128 B7
Parkkima FIN 123 C15
Parkku FIN 127 B10
Parksepa EST 131 F14
Parkstein D 75 C11
Parkua FIN 125 C9
Parkumäki FIN 129 B10
Parla E 46 D5
Parlan F 29 F10
Parma I 66 C1
Parndorf A 77 G11
Pãrnu EST 131 E8

Pärnu-Jaagupi EST 131 D9
Paroikia GR 176 E5
Parola FIN 127 C11
Paron F 25 D9
Parowa PL 81 D8
Parrillas E 45 D10
Pârsama EST 130 D5
Parsau D 79 A8
Parsberg D 75 D10
Pârscov RO 161 C9
Pârșcoveni RO 160 E4
Parsęcko PL 85 C11
Parsteinsee D 84 E6
Partakko FIN 113 E20
Partaloa E 55 E8
Partanna I 58 D2
Parteboda S 103 A10
Partenen A 71 D10
Partenstein D 74 B6
Pârteștii de Jos RO 153 B7
Parthenay F 28 B5
Partheni GR 177 E8
Partinello F 37 G9
Partinico I 58 C3
Partizani GR 167 C8
Partizani SRB 158 E5
Partizánska Ľupča SK 147 C8
Partney UK 11 E12
Parton UK 4 E4
Partry IRL 6 E4
Partsi EST 131 E5
Parudaminys LT 137 D11
Pãrup DK 86 C4
Parva RO 152 C5
Parvianen S 119 C11
Pâryd S 89 B9
Parysów PL 141 G5
Parzęczew PL 143 C7
Pasai Donibane E 32 D2
Pașaçli TR 171 C10
Pașaköy TR 171 E10
Pașaköy TR 173 E8
Pașaköy TR 173 B7
Pasalankylä FIN 125 D10
Pașayiğit TR 172 C6
Pașcani RO 153 C9
Pasching A 76 F6
Paseka CZ 77 C12
Pas-en-Artois F 18 D5
Pasewalk D 84 C6
Pasi FIN 128 D7
Pasiàn di Prato I 73 D7
Pasiano di Pordenone I 73 E6
Pasieki PL 139 E12
Pasiene LV 133 D4
Pasikovci HR 149 F9
Paskalevets BG 166 C4
Paskalevo BG 155 F1
Pãskallavik S 89 A10
Pasłęk PL 139 B8
Pasmajärvi FIN 117 D12
Pašman HR 156 E3
Passage East IRL 9 D9
Passail A 148 B5
Passais F 23 D10
Passau D 76 E4
Passignano sul Trasimeno I 66 F5
Passow D 84 D6
Passy F 31 D10
Pastavy BY 135 F13
Pastende LV 134 B5
Pasto FIN 123 D12
Pastoriza E 38 B5
Pastrana E 47 D7
Pãstrãveni RO 153 C9
Pașuliene LT 135 E12
Pasvalys LT 135 D8
Pašvitinys LT 134 D7
Pasym PL 139 C10
Pasytsely UA 154 B5
Paszab H 145 G4
Paszowice PL 81 D10
Pásztó H 147 F9
Pasztowa Wola PL 141 H4
Pata SK 146 E5
Pataias F 44 F3
Patak H 147 E8
Patana FIN 123 D11
Pãtârlagele RO 161 C8
Patay F 24 D6
Patchway UK 13 B9
Pateley Bridge UK 11 C8
Pateniemi FIN 119 D14
Patergassen A 73 C8
Paterna E 48 E4
Paterna del Campo E 51 E7
Paterna del Madera E 55 C8
Paterna del Río E 55 E7
Paterna de Rivera E 52 C5
Paternion A 73 C8
Paternò I 59 D6
Paternò I 60 C5
Paternopoli I 60 B4
Patersdorf D 76 D3
Pãterud S 96 C7
Patiška Reka NMK 164 F4
Pátka H 149 B11
Patmos GR 177 E8
Patna UK 4 E7
Pãtnów PL 142 D6
Patokoski FIN 117 E14
Patoniva FIN 113 D19
Patos AL 168 C2
Patra GR 174 C4
Pãtrãuți RO 153 B8
Patrica I 62 D4
Patrick GBM 10 C2
Patrimonio F 37 F10
Patrington UK 11 D11
Pãtroha H 145 G4
Pattada I 64 B3
Pattensen D 78 B6
Patterdale UK 10 B6
Patti I 59 C6
Pãttikkã FIN 116 A9
Pãtulele RO 159 E10
Patumșiai LT 134 E5
Pãturages B 182 E3
Páty H 149 A11
Pau F 32 D5
Pãuca RO 152 E4
Paudorf A 77 F9
Pauillac F 28 E4
Pauillac F 33 C7
Paukarlahti FIN 124 E9
Paukkaja FIN 125 E14
Paukkeri FIN 120 D9
Paul P 44 D5
Paularo I 73 C7

Pãuleni-Ciuc RO 153 E7
Paulești MD 154 C2
Pãulești RO 151 B10
Paulhac-en-Margeride F 30 F3
Paulhan F 34 C5
Pãuliș RO 151 E8
Paulinenaue D 83 E13
Pãuliș RO 151 B10
Paulistrōm S 92 E7
Paûls E 42 F4
Paulx F 28 B2
Pãunești RO 153 E10
Pausa D 79 E11
Pãușești RO 160 C4
Pãușești-Mãglași RO 160 C4
Pauträsk S 107 B13
Pauvres F 19 F10
Pavel BG 166 C5
Pavel Banya BG 166 D4
Pavia I 69 C7
Pavia P 50 B3
Pavia di Udine I 73 E7
Pavias E 48 E4
Pavie F 33 C7
Pavilly F 18 E2
Pãvilosta LV 134 C2
Pavlenci NMK 164 E4
Pavliani GR 174 B5
Pavlikeni BG 166 C4
Pavliš SRB 159 C7
Pavlivka UA 154 C5
Pavlivka UA 154 F4
Pavlov CZ 77 D8
Pavlovce nad Uhom SK 145 F5
Pavullo nel Frignano I 66 D2
Pavy RUS 132 F6
Pãwesin D 79 A12
Pawłosiów PL 144 D6
Pawłów PL 141 H5
Pawłowice PL 81 C11
Pawłowiczki PL 142 F5
Pawonków PL 142 E6
Paxton UK 5 D12
Payerne CH 31 B10
Paymogo E 51 D5
Payrac F 29 F9
Payrin-Augmontel F 33 C10
Payzac F 29 E8
Pazardzhik BG 165 E9
Pazarič BIH 157 E9
Pazarköy TR 173 E7
Pazin HR 67 B8
Paziols F 34 E4
Pázmánd H 149 B11
Pazos E 38 D3
Pchelarovo BG 166 F4
Pchelin BG 165 E8
Pcim PL 147 B9
Peacehaven UK 15 F8
Peal de Becerro E 55 D6
Péaule F 23 E7
Pébrac F 30 E4
Peccia CH 71 E7
Peccioli I 66 E2
Pécel H 150 C3
Peceneaga RO 155 D2
Pechea RO 155 B1
Pechenga RUS 114 D10
Pechina E 55 F8
Pechory RUS 132 F2
Pęci BIH 156 D5
Pecica RO 151 E7
Pecigrad BIH 156 B4
Pecinci SRB 158 D4
Pecineaga RO 155 F2
Peciu Nou RO 159 B7
Pecka RO 151 F10
Pecka SRB 158 E4
Pečky CZ 77 B8
Pecolo CZ 77 B8
Pečovská Nová Ves SK 145 E3
Pečurice MNE 163 E7
Pedaso I 62 A5
Pedede LV 133 B2
Pederobba I 72 E4
Pedersker DK 89 E7
Pedersöre FIN 122 C9
Pedino GR 169 C6
Pedrafita do Cebreiro E 38 C5
Pedrajas de San Esteban E 40 F2
Pedralba E 48 E3
Pedralba de la Pradería E 39 D6
Pedraza de Campos E 39 E10
Pedrera E 53 B7
Pedro Abad E 53 A8
Pedro Bernardo E 46 D3
Pedroche E 54 C3
Pedrógão P 44 E3
Pedrógão P 45 D6
Pedrógão P 50 D4
Pedrógão Grande P 44 E4
Pedrógão Pequeno P 44 E4
Pedro-Martínez E 55 D6
Pedro Muñoz E 47 F7
Pedrosa E 40 B4
Pedrosa del Príncipe E 40 D3
Pedrosillo de los Aires E 45 C9
Pedroso E 41 D6
Pedroso P 44 B3
Pedrouzos E 38 C2
Peebles UK 5 D10
Peel GBM 10 C2
Peenemünde D 84 B5
Peer B 19 B11
Peera FIN 112 F7
Peetri EST 131 D11
Pefki GR 175 A7
Pefkofyto GR 168 C4
Pefkos GR 178 E4
Pega P 45 D6
Pegalajar E 53 A9
Pegau D 79 D11
Peggau A 148 B4
Pegli I 37 C9
Pegnitz D 75 C10
Pego E 56 D4
Pego P 44 F4
Pegões P 50 B2
Pjgów PL 81 D11
Peguera E 49 E9
Pehčevo NMK 165 F6
Pehkolanlahti FIN 120 F9
Pehlivanköy TR 173 B6
Peillac F 23 E7
Peille F 37 D6

Peine D 79 B7
Peio I 71 E11
Peipin F 35 B10
Peipohja FIN 126 C7
Peïra-Cava F 37 D6
Peiraias GR 175 D8
Peißen D 79 C10
Peißen D 79 C11
Peißenberg D 72 A3
Peiting D 71 B11
Peitz D 80 C6
Peize NL 17 B7
Pejë RKS 163 D9
Pekankylä FIN 121 E13
Pekanpää FIN 119 B11
Pekkala FIN 119 B11
Pekkilä FIN 121 F10
Pelago I 66 E4
Pelagićevo BIH 157 C10
Pelahustán E 46 D3
Pelarrodríguez E 45 C8
Pelasgia GR 175 B6
Pelčí LV 134 C3
Peleagonzalo E 39 F9
Pelēči LV 133 D2
Pelekas GR 168 E2
Peleta GR 175 E6
Pelhřimov CZ 77 D8
Pelinei MD 154 E2
Pelinia MD 153 B11
Pelishat BG 165 C10
Pélissanne F 35 C9
Pelitköy TR 173 E6
Pelkosenniemi FIN 115 D2
Pella GR 169 C7
Pellaro I 59 C8
Pellegrino Parmense I 69 D8
Pellegrue F 28 F5
Pellérd H 149 D10
Pellesmäki FIN 124 E9
Pellestrina I 66 B5
Pellevoisin F 29 B8
Pellinki FIN 127 E14
Pellizzano I 69 A10
Pello FIN 117 E11
Pello FIN 117 E11
Pélmonostor I 60 C4
Pelovo BG 165 C9
Pelplin PL 138 C6
Pelsin D 84 C5
Pelso FIN 119 E16
Peltokangas FIN 123 D12
Peltosalmi FIN 124 C8
Peltovuoma FIN 117 B12
Pélussin F 30 E6
Pelvoux F 31 F9
Pély H 150 C5
Pelynt UK 12 E5
Pembrey UK 12 B5
Pembridge UK 13 A9
Pembroke UK 12 B5
Pembroke Dock UK 12 B5
Pembury UK 15 E9
Pemfling D 75 D12
Pempelijärvi S 116 E8
Peñacerrada E 41 C6
Penacova P 44 D4
Peñafiel E 40 E3
Penafiel P 44 B4
Peñaflor E 51 D9
Peñaflor de Hornija E 39 E10
Penagos E 40 B4
Peñalba E 42 E3
Peñalén E 47 C8
Peñalsordo E 51 B9
Penalva do Castelo P 44 C5
Penamacor P 45 D6
Peñaranda de Bracamonte E 45 C10
Peñaranda de Duero E 40 E5
Peñarroya de Tastavins E 42 F4
Peñarroya-Pueblonuevo E 51 C9
Peñarrubia E 55 C10
Penarth UK 13 C8
Peñas de San Pedro E 55 B9
Peñausende E 39 F8
Penc H 150 B3
Pencader UK 12 A6
Pěnčín CZ 81 E8
Pendeen UK 12 E3
Pendilhe P 44 C5
Pendine UK 12 B5
Pendlebury UK 11 D7
Pendueles E 39 B10
Penela P 44 D4
Pénestin F 22 E7
Penészlek H 151 B9
Pengfors S 107 D17
Pengsjö S 107 C13
Pengsjö S 107 D17
Penha Garcia P 45 D6
Peniche P 44 F2
Penicuik UK 5 D10
Penikkajärvi FIN 121 C14
Peninki FIN 123 C9
Peninver UK 4 E5
Peñíscola E 48 D5
Penistone UK 11 D8
Penkridge UK 11 F7
Penkule LV 134 D6
Penkun D 84 D6
Penly F 18 E2
Penmarch F 22 E3
Pennabilli I 66 E5
Pennan UK 3 K12
Penna in Teverina I 62 C2
Pennapiedimonte I 63 C6
Pennautier F 33 D10
Penne F 33 B9
Penne I 62 C5
Penne-d'Agenais F 33 B7
Penrhiw-pâl UK 4 C4
Penrhyn Bay UK 10 E4
Penrith UK 5 F11
Penryn UK 12 E4
Pensala FIN 122 D8
Pentalofo GR 174 C3
Pentalofos GR 168 D5
Pentalofos GR 171 A10
Pentapoli GR 169 B10
Penteoria GR 174 C5
Pentinniemi FIN 119 C16
Pentir UK 10 E3
Pentone I 59 B10
Pentraeth UK 10 E3
Pentre UK 13 B7
Pentrefoelas UK 10 E4
Penttäjä S 117 E11

Péronne F 18 E6
Perosa Argentina I 31 F11
Pero Viseu P 44 D6
Perpignan F 34 E4
Perranporth UK 12 E4
Perrecy-les-Forges F 30 B5
Perreux F 30 C5
Perrigny F 31 B8
Perrogney-les-Fontaines F 26 E3
Perros-Guirec F 22 C5
Perrum-Âbmir FIN 113 E19
Persan F 25 B7
Persåsen S 102 A7
Persberg S 97 C11
Persbo S 97 B13
Persenbeug A 77 F8
Pershagen S 93 A11
Pershamawskaya BY 133 E4
Pershore UK 13 A10
Pershotravneve UA 155 C4
Pershyttan S 97 D13
Persnäs S 89 A11
Persön S 118 C6
Perstorp S 87 C12
Perth UK 5 C10
Perthes F 25 D8
Pertouli GR 168 E5
Perttaus FIN 117 D15
Pertteli FIN 127 E9
Perttula FIN 127 E12
Pertuis F 35 C10
Pertunmaa FIN 127 C15
Pertusa E 42 C3
Peruc CZ 76 B5
Peručica SRB 158 F3
Perugia I 66 F5
Perukka FIN 119 F17
Perunkajärvi FIN 117 E15
Perushtitsa BG 165 E10
Perušić HR 156 C3
Péruwelz B 19 C8
Pervalka LT 134 F2
Pervenchères F 24 D3
Pervomaisc MD 154 D5
Pervomayskoye RUS 129 E13
Perwez B 19 C10
Pesadas de Burgos E 40 C4
Pesaro I 67 E6
Pescaglia I 66 D1
Pescantina I 66 B2
Pescara I 63 C6
Pescasseroli I 62 D5
Pesceana RO 160 D4
Peschadoires F 30 D3
Peschici I 63 D10
Peschiera del Garda I 66 B2
Pescia I 66 D2
Pescina I 62 C5
Pescocostanzo I 62 D6
Pescolanciano I 63 D6
Pescopennataro I 63 D6
Pescorocchiano I 62 C4
Pesco Sannita I 60 A3
Peseux CH 31 B10
Peshkopi AL 163 F9
Peshtera BG 165 E9
Pesiökylä FIN 121 E12
Pesionranta FIN 121 E12
Pesmes F 26 F4
Pesnica SLO 148 C5
Pesochnyy RUS 129 E13
Peso da Régua P 44 B5
Pessac F 28 F4
Pessalompolo FIN 117 F12
Pessan F 33 C7
Pesse NL 17 C7
Peßnitz D 80 A3
Peștani NMK 168 B4
Peștera RO 155 E2
Peștișani RO 159 C10
Peștișu Mic RO 151 F10
Pesués E 40 B3
Pešurići BIH 157 E11
Petacciato I 63 C7
Petaïkkö FIN 119 E16
Petäikylä FIN 125 C12
Petäjäjärvi FIN 119 C17
Petäjäkangas FIN 119 D17
Petäjäkoski FIN 119 B14
Petäjäkoski FIN 119 F17
Pererätä FIN 123 F14
Petalax FIN 122 F6
Petalidi GR 174 F4
Pétange L 20 E5
Petas GR 174 A3
Petelea RO 152 D5
Peteranec HR 149 D8
Peterborough UK 11 F11
Peterculter UK 3 L12
Peterhead UK 3 K13
Péteri H 150 C3
Peterlee UK 5 F14
Petersberg D 74 A6
Petersdorf D 75 F9
Petersdorf auf Fehmarn D 83 B10
Petersfield UK 15 E7
Petershagen D 17 D11
Petershagen D 80 B6
Peterswell IRL 6 F5
Pétervására H 147 E10
Pethelinos GR 169 C10
Petilia Policastro I 59 A10
Petín E 38 D5
Petisträsk S 107 B17
Petite-Rosselle F 186 C2
Petitmont F 186 D2
Petit-Noir F 31 B7
Petko Karavelovo BG 166 C5
Petko Slaveykov BG 165 C10
Petkula FIN 115 C2
Petkus D 80 C4
Petlovača SRB 158 D3
Pet Mogili BG 166 E6
Petnehâza H 145 G5
Petöfibánya H 150 B4
Petra E 57 B11
Petra GR 171 D7
Petrãchioaia RO 161 D8
Petrades GR 172 B6
Petralia-Soprana I 58 D5
Petran AL 168 D3
Petrčane HR 156 D3
Petrella Salto I 62 C4
Petrella Tifernina I 63 D7
Petrer E 56 E3
Petreșți RO 151 B9
Petreșți RO 160 D6
Petreștii de Jos RO 152 D3

Petreto-Bicchisano F 37 H9
Petriano I 67 E6
Petricani RO 153 C8
Petrich BG 169 B9
Petrijevci HR 149 E11
Petrila RO 160 C2
Petrinja HR 148 F6
Petriş RO 151 E9
Petritoli I 62 A5
Petrivka UA 154 D4
Petrivs'k UA 154 E3
Petrochori GR 174 B4
Petrodvorets RUS 129 F12
Pétrola E 55 B9
Petromäki FIN 125 E9
Petronà I 59 A10
Petroşani RO 160 C2
Petrota GR 166 F6
Petroussa GR 170 B6
Petrov CZ 146 D4
Petrova RO 152 B4
Petrovac MNE 163 E6
Petrovac SRB 159 E7
Petrovany SK 145 F3
Petrovaradin SRB 158 C4
Petrovec NMK 164 F4
Petrovice CZ 76 C6
Petrovice u Karvine CZ 147 B7
Petroviči BIH 157 D10
Petrovići MNE 162 D6
Petrovo BG 169 B9
Petrovo RUS 136 D1
Petrovo Selo SRB 159 D9
Petruma FIN 125 F11
Petru Rareş RO 152 C4
Petruşeni MD 153 B10
Petřvald CZ 146 B6
Petřvald CZ 146 B6
Petsakoi GR 174 C5
Petsmo FIN 122 D7
Petten NL 16 C3
Pettigo UK 7 C7
Pettineo I 58 D5
Petting D 73 A6
Pettneu am Arlberg A 71 C10
Pettorano sul Gizio I 62 D5
Petürch BG 165 D7
Petworth UK 15 F7
Peuerbach A 76 F5
Peujard F 28 E5
Peura FIN 119 B14
Peurajärvi FIN 119 B16
Peurasuvanto FIN 115 C1
Pevensey UK 15 F9
Peveragno I 37 C7
Pewsey UK 13 C11
Pewsum (Krummhörn) D 17 B8
Pexonne F 27 D6
Peymeinade F 36 D5
Peynier F 35 D10
Peypin F 35 D10
Peyrat-le-Château F 29 D9
Peyrehorade F 32 C3
Peyriac-Minervois F 34 D4
Peyrieu F 31 D8
Peyrins F 31 E7
Peyrolles-en-Provence F 35 C10
Peyruis F 35 B10
Pézenas F 34 D5
Pjzino PL 85 D8
Pezinok SK 146 E4
Pezuls F 29 F7
Pfaffenberg D 75 E11
Pfaffendorf D 80 B6
Pfaffenhausen D 71 A10
Pfaffenhofen an der Ilm D 75 E10
Pfaffenhofen an der Roth D 187 E9
Pfaffenhoffen F 27 C8
Pfäffikon CH 27 F10
Pfaffing D 75 F11
Pfalzfeld D 185 D8
Pfalzgrafenweiler D 27 C10
Pfarrkirchen D 76 F3
Pfarrweisach D 75 B8
Pfarrwerfen A 73 B7
Pfedelbach D 27 B11
Pflach A 71 B11
Pfons A 72 B3
Pförring D 75 E10
Pforzen D 71 B11
Pforzheim D 27 C10
Pfreimd D 75 D11
Pfronstetten D 27 D11
Pfronten D 71 B11
Pfullendorf D 27 E11
Pfullingen D 27 D11
Pfunds A 71 D11
Pfungstadt D 21 E11
Pfyn CH 27 E10
Phalsbourg F 27 C7
Philippeville B 19 D10
Philippine NL 182 C3
Philippsburg D 187 C5
Philippsreut D 76 E5
Piacenza I 69 C8
Piadena I 66 B1
Piana F 37 G9
Piana GR 174 D5
Piana Crixia I 37 C8
Piana degli Albanesi I 58 D3
Piancastagnaio I 62 B1
Piandimeleto I 66 E5
Piàn di Scò I 66 E4
Pianella I 62 C6
Pianello Val Tidone I 37 B10
Piano del Voglio I 66 D3
Pianoro I 66 D3
Pianosa I 65 B2
Pianotolli-Caldarello F 37 J10
Pians A 71 C11
Piansano I 62 B1
Pianu RO 152 F2
Pias P 50 C5
Piaseczno PL 85 D7
Piaseczno PL 141 F4
Piasek PL 84 E6
Piaski PL 81 C12
Piaski PL 141 H7
Piastów PL 141 F3
Piątek PL 143 B7
Piątnica Poduchowna PL 139 D13
Piatra RO 160 F6
Piatra Neamţ RO 153 D8
Piatra Olt RO 160 E4
Piatra şoimului RO 153 D8
Piau-Engaly F 33 E6
Piazza al Serchio I 66 D1
Piazza Armerina I 58 E5
Piazza Brembana I 69 B8
Piazzatorre I 69 B8
Piazzola sul Brenta I 66 A4
Pibrac F 33 C8
Pićan HR 67 B9
Picar AL 168 D3
Picassent E 48 F4
Picauville F 23 B9
Picerno I 60 B5
Picher D 83 D10
Pichl bei Wels A 76 F5
Pickering UK 11 C10
Pico I 62 E5
Picón E 54 A4
Picoto P 44 B3
Picquigny F 18 E5
Pidbuzh UA 145 E7
Pidhorodtsi UA 145 E7
Pidlisne UA 153 A7
Piebalgas LV 135 C8
Piechcin PL 138 E5
Piechowice PL 81 E9
Piecki PL 139 C11
Piecnik PL 85 D10
Piedicorte-di-Gaggio F 37 G10
Piedicroce F 37 G10
Piedimonte Etneo I 59 D7
Piedimonte Matese I 60 A2
Piedimulera I 68 A5
Piedrabuena E 54 A4
Piedrafita de Babia E 39 C7
Piedrahita E 45 D10
Piedralaves E 46 D3
Piedras Albas E 45 E7
Piedras Blancas E 39 A8
Piedruja LV 133 E2
Piegaro I 62 B2
Piégut-Pluviers F 29 D7
Piehinki FIN 119 E12
Piekary Śląskie PL 143 F6
Piekielnik PL 147 C9
Piekoszów PL 143 E9
Pieksämäen mlk FIN 124 F8
Pieksämäki FIN 124 F8
Pielavesi FIN 123 D17
Pieleşti RO 160 E3
Pielgrzymka PL 81 D9
Pienava LV 134 C6
Pieniężnica PL 85 C11
Pieniężno PL 139 B9
Pieńkowo PL 85 B11
Piennes F 19 F12
Pieńsk PL 81 D8
Pienza I 62 A1
Piera E 43 D7
Pierowall UK 3 G11
Pierre-Buffière F 29 D8
Pierre-Châtel F 31 F8
Pierre-de-Bresse F 31 B7
Pierrefeu-du-Var F 36 E4
Pierrefitte-Nestalas F 32 E5
Pierrefitte-sur-Aire F 26 C3
Pierrefonds F 18 F6
Pierrefontaine-les-Varans F 26 F6
Pierrefort F 30 F2
Pierrelatte F 35 B8
Pierrepont F 19 F12
Pierres F 24 C6
Pierrevert F 35 C10
Piershil NL 182 B4
Piertinjaure S 109 C17
Pierzchnica PL 143 E10
Piesau D 75 A9
Pieścirogi PL 139 E10
Piesendorf A 73 B6
Pieski PL 81 B8
Piesport D 185 E6
Piešťany SK 146 D5
Pieszkowo PL 136 E2
Pieszyce PL 81 E11
Pietrabbondante I 63 D6
Pietracatella I 63 D7
Pietracorbara F 37 F10
Pietra-di-Verde F 37 G10
Pietragalla I 60 B5
Pietralba F 37 F10
Pietra Ligure I 37 C8
Pietralunga I 66 F5
Pietramelara I 60 A2
Pietramontecorvino I 63 D8
Pietraperzia I 58 E5
Pietraporzio I 36 C6
Pietrari RO 160 C4
Pietrasanta I 66 E1
Pietravairano I 60 A2
Pietroasa RO 151 D10
Pietroasele RO 161 C9
Pietroşani RO 161 F7
Pietrosella F 37 H9
Pietroşita RO 161 C6
Pietrowice Wielkie PL 142 F5
Pieve d'Alpago I 72 D5
Pieve del Cairo I 68 C6
Pieve di Bono I 69 B10
Pieve di Cadore I 72 D5
Pieve di Cento I 66 C3
Pieve di Soligo I 72 E5
Pieve di Teco I 37 C7
Pieve Fosciana I 66 D1
Pievepelago I 66 D2
Pieve Santo Stefano I 66 E5
Pieve Torina I 62 A4
Pieve Vergonte I 68 A5
Piffonds F 25 D9
Piges GR 174 E5
Pigi GR 169 E6
Pigi GR 177 A7
Piglio I 62 D4
Pigna I 37 D7
Pignans F 36 E4
Pignataro Interamna I 62 E5
Pignataro Maggiore I 60 A2
Pignola I 60 B5
Pihlajakoski FIN 127 B13
Pihlajalahti FIN 127 B10
Pihlajalahti FIN 129 B9
Pihlajavaara FIN 125 D16
Pihlajavesi FIN 123 F12
Pihlava FIN 126 B6
Pihtipudas FIN 123 D15
Piikkiö FIN 126 E8
Piilijärvi S 116 C6
Piiloperä FIN 121 B12
Piippola FIN 119 F15
Piipsjärvi FIN 119 F13
Piirsalu EST 131 C8
Piispa FIN 125 D11
Piispajärvi FIN 121 D13
Piittisjärvi FIN 119 B17
Pijnacker NL 16 D2
Pikkarala FIN 119 E15
Pikkula FIN 119 D17
Piktupėnai LT 134 F3
Pila I 66 C5
Piła PL 85 D11
Pilar de la Mola E 57 D8
Pilas E 51 E7
Piława PL 141 G5
Piława Górna PL 81 E11
Pilchowice PL 142 F6
Pilda LV 133 D3
Piles E 56 D4
Pilgrimstad S 103 A9
Pili GR 175 B8
Pilica PL 143 F8
Pilis H 150 C4
Piliscsaba H 149 A11
Pilisszántó H 149 A11
Pilisszentiván H 149 A11
Pilisvörösvár H 149 A11
Pill A 72 B4
Pilling UK 10 D6
Pilníkov CZ 77 A9
Pilpala FIN 127 D11
Pilsach D 75 D10
Pilskalns LV 135 B13
Pilsting D 75 E12
Piltene LV 134 B3
Pilträsk S 118 C4
Pilu RO 151 D7
Pilviškiai LT 136 D7
Pilzno PL 143 G11
Pimentel I 64 E3
Pimperne UK 13 D10
Pimpiö S 117 E10
Piña de Campos E 40 D3
Piña de Esgueva E 40 E3
Piñar E 53 B10
Pinarbaşi TR 171 E10
Pinarca TR 173 B9
Pinarejo E 47 E8
Pinarhisar TR 173 A8
Pinarköy TR 181 B7
Pinarlibelen TR 177 E10
Pinasca I 31 F11
Pincehely H 149 C10
Pinchbeck UK 11 F11
Pińczów PL 143 E10
Pindstrup DK 86 C6
Pineda de Cigüela E 47 D7
Pineda de la Sierra E 40 D5
Pineda de Mar E 43 D9
Pinela P 39 E6
Piñel de Abajo E 40 E3
Pinerolo I 31 F11
Pineto I 62 B6
Piney F 25 D11
Pinggau A 148 B6
Pinhal Novo P 50 B2
Pinhanços P 44 D5
Pinhão P 44 B5
Pinheiro P 50 C2
Pinheiro Grande P 44 F4
Pinhel P 45 C6
Pinhoe UK 13 D8
Pinilla de Molina E 47 C9
Pinilla de Toro E 39 E9
Pinkafeld A 148 B6
Pinneberg D 82 C7
Pinnow D 81 C7
Pino E 39 E7
Pino F 37 F10
Pino del Río E 39 C10
Pinofranqueado E 45 D8
Pinols F 30 E3
Piñor E 38 C3
Pinoso E 56 E2
Pinos-Puente E 53 B9
Pinsac F 29 F9
Pinsiö FIN 127 B9
Pinsoro E 41 D9
Pintamo FIN 121 D10
Pintano E 32 E3
Pinto E 46 D5
Pinwherry UK 4 E7
Pinzano al Tagliamento I 73 D6
Pinzio P 45 C6
Pinzolo I 69 A10
Piobbico I 66 E6
Piolenc F 35 B8
Pioltello I 69 C7
Piombino I 65 B3
Piombino Dese I 72 E4
Pionerskiy RUS 139 A9
Pionki PL 141 H4
Pionsat F 29 C11
Pioraco I 62 A4
Piornal E 45 D9
Piossasco I 31 F11
Piotrkowice PL 81 D11
Piotrków Kujawski PL 138 E5
Piotrków Trybunalski PL 143 D8
Piove di Sacco I 66 B5
Piovene Rocchette I 69 B11
Piperskärr S 93 D9
Pipirig RO 153 C8
Pipriac F 23 E8
Piqeras AL 168 D2
Pir RO 151 C9
Piraino I 59 C6
Piran SLO 67 A8
Pirčiupiai LT 137 E10
Pirdop BG 165 D9
Pirg AL 168 C4
Pirgovo BG 161 F7
Piriac-sur-Mer F 22 F6
Piricse H 151 B9
Pirin BG 169 A10
Pirinççi TR 173 B10
Pirjolteni MD 154 C2
Pirkkala FIN 127 C10
Pirlița MD 153 C11
Pirlița MD 154 A2
Pirmasens D 21 F9
Pirna D 80 E5
Pirnmill UK 4 D6
Pirot SRB 164 C6
Pirovac HR 156 E4
Pirtó H 150 E3
Pirttijoki FIN 121 E9
Pirttikoski FIN 119 B18
Pirttikoski FIN 119 F12
Pirttikoski FIN 127 C10
Pirttimäki FIN 123 C17
Pirttimäki FIN 124 E7
Pirttimäki FIN 125 C10
Pirttinen FIN 123 D10
Pirttivaara FIN 121 D14
Pirttivuopio S 111 E17
Pisa FIN 119 B14
Pisa I 66 E1
Pisanets BG 161 F8
Pisarovina HR 148 E5
Pisarovo BG 165 C9
Pischelsdorf in der Steiermark A 148 B5
Pişchia RO 151 F7
Pisciotta I 60 C4
Pişcolt RO 151 B9
Piscu RO 155 B1
Piscu Vechi RO 159 F11
Pišece SLO 148 D5
Písečná CZ 77 B12
Písek CZ 76 D6
Písek CZ 147 B7
Pishcha UA 141 G9
Pishchana UA 154 A5
Pishchanka UA 154 A3
Pisisaare EST 131 D11
Piskokefalo GR 179 E11
Piškorevci HR 157 B9
Pisoderi GR 168 C5
Pisogne I 69 B9
Pissonas GR 175 B8
Pissos F 32 B4
Pistiana GR 168 F5
Pisticci I 61 C7
Pisto FIN 121 D13
Pistoia I 66 E2
Pisz PL 139 C12
Piszczac PL 141 G8
Pitagowan UK 5 B9
Pitäjänmäki FIN 123 C15
Pitarque E 42 F2
Piteå S 118 D6
Piteşti RO 160 D5
Pithiviers F 25 D7
Pitigliano I 65 B5
Pitkäjärvi FIN 127 D9
Pitkälä FIN 129 B11
Pitkälahti FIN 124 E9
Pitkäsenkylä FIN 119 F12
Pitkyaranta RUS 129 B15
Pitlochry UK 5 B9
Pitomača HR 149 E8
Pitrags LV 130 F4
Pitres E 55 F6
Pîtres F 18 F3
Pitscottie UK 5 C11
Pitsinaiika GR 174 C4
Pitstone UK 15 D7
Pitsund S 118 D7
Pitt UK 13 C12
Pittem B 19 C7
Pitten A 148 A6
Pittentrail UK 3 K8
Pittenweem UK 5 C11
Pitvaros H 150 E6
Pivašiūnai LT 137 E9
Pivka SLO 73 E9
Pivnice SRB 158 C3
Piwniczna-Zdrój PL 145 E2
Pizarra E 53 C7
Pizzighettone I 69 C8
Pizzo I 59 B9
Pizzoferrato I 63 D6
Pizzoli I 62 C4
Pjedsted DK 86 D5
Pjelax FIN 122 F6
Pjenovac BIH 157 D10
Pjesker S 118 C3
Plaaz D 83 C12
Plabennec F 22 C3
Placencia de las Armas E 32 D1
Plachkovtsi BG 166 D4
Plācis LV 135 B9
Plaffeien CH 31 B11
Plagia GR 169 B8
Plaidt D 21 D8
Plăieşii de Jos RO 153 E8
Plaintel F 22 D6
Plaisance F 33 C6
Plaisance-du-Touch F 33 C8
Plaisir F 24 C6
Plaka GR 168 F5
Plaka GR 171 D8
Plaka GR 175 E6
Plaka GR 179 B8
Plakhtiyivka UA 154 E5
Plakias GR 178 E7
Plakovo BG 166 D5
Plana BIH 157 G9
Planá CZ 75 C12
Plana GR 169 D8
Planá nad Lužnicí CZ 77 D7
Plaňany CZ 77 B8
Plancher-Bas F 27 E6
Plancoët F 23 C7
Plancy-l'Abbaye F 25 C10
Plan-de-Baux F 31 F7
Plan-de-la-Tour F 36 E5
Plandište SRB 159 C7
Plan-d'Orgon F 35 C8
Planegg D 75 F9
Planès F 33 F10
Plāņi LV 131 F11
Plánice CZ 76 D4
Planina SLO 73 E9
Planina SLO 148 D4
Planina SRB 158 E3
Planinica SRB 159 F9
Planjane HR 156 E5
Plankenfels D 75 C9
Planos GR 174 D2
Plasencia E 45 D8
Plasencia del Monte E 41 D10
Plaški HR 156 B3
Plassen N 102 D4
Plášťovce SK 147 E7
Plasy CZ 76 C4
Plataies GR 175 C7
Platamona Lido I 64 B1
Platamonas GR 169 E8
Platamonas GR 171 B7
Platania I 59 A9
Platanias GR 175 A7
Platanistos GR 175 C10
Platanorrevma GR 169 D7
Platanos GR 174 D4
Platanos GR 175 A6
Platanos GR 178 E6
Platanos GR 178 E8
Platanovrysi GR 174 C4
Plătăreşti RO 161 E8
Plataria GR 168 F3
Plate D 83 C11
Plateliai LT 134 D3
Platerów PL 141 F7
Plati GR 171 A10
Platì I 59 C9
Platiana GR 174 D4
Platičevo SRB 158 D4
Platja d'Aro E 43 D10
Platja de Nules E 48 E4
Platone LV 134 C7
Plattling D 76 E3
Platy GR 169 C8
Platykampos GR 169 E8
Platys Gialos GR 176 E5
Platys Gialos GR 176 F4
Plau D 83 D12
Plaue D 79 E8
Plauen D 75 A11
Plav MNE 163 D8
Plavë RKS 163 E10
Plaveč SK 145 E2
Plavecký Štvrtok SK 77 F12
Plaviņas LV 135 C11
Plavna SRB 159 E9
Plavni UA 155 C3
Plavnica SK 145 E2
Plavno HR 156 D5
Plavy CZ 81 E8
Pławno PL 85 C10
Plaza E 40 B6
Plazac F 29 E8
Płazów PL 144 C7
Pleaux F 29 E10
Plech D 75 C9
Plecka Dąbrowa PL 143 B8
Pleine-Fougères F 23 C8
Pleikšņi LV 133 D2
Plélan-le-Grand F 23 D7
Plélo F 22 C6
Plémet F 22 D6
Plénée-Jugon F 23 D7
Pléneuf-Val-André F 22 C6
Pleniţa RO 159 E11
Plentzia E 40 B6
Plérin F 22 C6
Plescop F 22 E6
Pleşcuţa RO 151 E9
Pleşeni MD 154 E2
Plešivec SK 145 F1
Plesná CZ 75 B11
Pleśna PL 144 D2
Pleşoiu RO 160 E4
Plessa D 80 D5
Plessé F 23 E8
Plestin-les-Grèves F 22 C4
Pleszew PL 142 C4
Pleternica HR 157 B8
Plettenberg D 21 B9
Pleubian F 22 C5
Pleudihen-sur-Rance F 23 C8
Pleumartin F 29 B7
Pleumeur-Bodou F 22 C4
Pleurs F 25 C10
Pleuven F 22 E3
Pleven BG 165 C10
Plevník-Drienové SK 147 C7
Pleyben F 22 D4
Pleyber-Christ F 22 C4
Pliego E 55 D10
Plienciems LV 134 B6
Pliešovce SK 147 E8
Pliezhausen D 187 D7
Plikati GR 168 D4
Plisa BY 133 F3
Pliska BG 167 C8
Plitvica HR 156 C4
Pljevlja MNE 163 C7
Pllanejë RKS 163 E10
Ploaghe I 64 B2
Plobannalec F 22 E3
Plobsheim F 186 E4
Ploče HR 157 F7
Ploce LV 134 C2
Plochingen D 187 D7
Pločica SRB 159 D6
Płociczno PL 136 E6
Płock PL 139 E8
Plodovoye RUS 129 D13
Ploegsteert B 182 D1
Ploemeur F 22 E5
Ploeren F 22 E6
Ploërmel F 23 E7
Plœuc-sur-Lié F 22 D6
Plogonnec F 22 D3
Plogshagen D 84 A4
Ploieşti RO 161 D8
Plomari GR 177 B7
Plombières-les-Bains F 26 E5
Plomeur F 22 E3
Plomin HR 67 B9
Plomodiern F 22 D3
Plön D 83 B8
Plonéour-Lanvern F 22 E3
Plonévez-du-Faou F 22 D4
Plonevez-Porzay F 22 D3
Płońsk PL 139 E9
Plop MD 153 A11
Plopana RO 153 D10
Plopeni RO 161 C7
Plopi MD 154 B4
Plopii-Slăviteşti RO 160 F5
Plopşoru RO 160 D2
Plopu RO 161 C8
Plosca RO 160 E6
Ploscoş RO 152 D3
Ploska UA 152 B6
Płoskinia PL 139 B8
Płośnica PL 139 D9
Plößberg D 75 C11
Płoty PL 85 C8
Plötzin D 79 B12
Plötzky D 79 B10
Plou E 42 E2
Plouagat F 22 C6
Plouaret F 22 C4
Plouarzel F 22 D2
Plouay F 22 E5
Ploubalay F 23 C7
Ploubazlanec F 22 C5
Ploubezre F 22 C5
Ploudalmézeau F 22 C2
Ploudiry F 22 D3
Plouescat F 22 C3
Plouézec F 22 C6
Ploufragan F 22 D6
Plougasnou F 22 C4
Plougastel-Daoulas F 22 D3
Plougonven F 22 C4
Plougonver F 22 D5
Plougrescant F 22 C5
Plouguenast F 22 D6
Plouguerneau F 22 C2
Plouguernével F 22 D5
Plouguiel F 22 C5
Plouguin F 22 C2
Plouha F 22 C6
Plouharnel F 22 E5
Plouhinec F 22 D3
Plouhinec F 22 E5
Plouigneau F 22 C4
Ploumagoar F 22 C5
Ploumilliau F 22 C4
Plounévez-Moëdec F 22 C5
Plounévez-Quintin F 22 D5
Plouray F 22 D5
Plouvorn F 22 C3
Plouyé F 22 D4
Plouzané F 22 D2
Plouzévédé F 22 C3
Plovdiv BG 165 E10
Plozévet F 22 E3
Pluck IRL 7 C7
Plüderhausen D 187 D8
Plugari RO 153 C10
Plumbridge UK 4 F2
Plumelec F 22 E6
Pluméliau F 22 E6
Plumergat F 22 E6
Plumlov CZ 77 D12
Pluneret F 22 E6
Plungė LT 134 E3
Pluszkiejmy PL 136 E5
Pluvigner F 22 E5
Plužine BIH 157 F9
Plužine MNE 157 F10
Płużnica PL 138 D6
Pluzunet F 22 C5
Plwmp UK 12 A6
Plymouth UK 12 E6
Plympton UK 12 E6
Plymstock UK 12 E6
Plytra GR 178 B4
Plyussa RUS 132 E5
Plyusy BY 133 E2
Plzeň CZ 76 C4
Pniewo PL 85 D11
Pniewo PL 139 E11
Pniewo PL 143 B8
Pniewy PL 81 A10
Pniewy PL 141 G3
Poarta Albă RO 155 E2
Pobedím SK 146 D5
Pobedino RUS 136 D6
Poběžovice CZ 75 C12
Pobiedno PL 145 D5
Pobiedziska PL 81 B12
Pobierowo PL 85 B7
Pobikry PL 141 E7
Pobladura del Valle E 39 D8
Poblete E 54 B5
Pobłocie PL 85 A13
Poboleda E 42 E5
Poboru RO 160 D5
Pobožje NMK 164 E3
Počátky CZ 77 D8
Poceirão P 50 B2
Pöchlarn A 77 F8
Pociems LV 131 F9
Počitelj BIH 157 F8
Pociumbeni MD 153 B10
Pöcking D 75 G9
Pocking D 76 F4
Pocklington UK 11 D10
Pocola RO 151 D9
Pocrnje BIH 157 F9
Pocsaj H 151 C8
Pócspetri H 145 H4
Poczesna PL 143 E7
Podareš NMK 169 A8
Podari RO 160 E3
Podayva BG 161 F9
Podbiel SK 147 C8
Podbořany CZ 76 B4
Podborov'ye RUS 132 F4
Podbožur MNE 163 D6
Podbrdo BIH 157 D7
Podbrdo SLO 73 D8
Podbrezová SK 147 D9
Podčetrtek SLO 148 D5
Poddjbice PL 143 C6
Poděbrady CZ 77 B8
Podedwórze PL 141 G8
Podegrodzie PL 145 D2
Podelzig D 81 B7
Podem BG 165 B10
Podeni RO 159 D10
Podenii Noi RO 161 C8
Podensac F 32 A5
Podenzana I 69 E8
Podenzano I 37 B11
Podersdorf am See A 77 G11
Podgajci Posavski HR 157 C10
Podgaje PL 85 D11
Podgora HR 157 F7
Podgorač HR 149 F10
Podgorac SRB 159 F8
Podgoria RO 161 C10
Podgorica MNE 163 E7
Podgorica SLO 73 D10
Podgorie AL 168 C4
Podgoritsa BG 167 C6
Podgórze PL 139 D13
Podgrab BIH 157 E10
Podgrad SLO 67 A9
Podgrade BIH 157 E8
Podhorod SK 145 F5
Podhum BIH 157 E7
Podil's'k UA 154 B5
Podivín CZ 77 E11
Podkova BG 171 B8
Podkowa Leśna PL 141 F3
Podkrajewo PL 139 D9
Podkrepa BG 166 F5
Podkum SLO 73 D11
Podlapača HR 156 C4
Podlehnik SLO 148 D5
Podles NMK 169 A6
Podlipoglav SLO 73 D10
Podmilečje BIH 157 D7
Podnanos SLO 73 E8
Podnovlje BIH 157 C9
Podochori GR 170 C6
Podogora GR 174 B3
Podoima MD 154 B3
Podoleni RO 153 D9
Podolí CZ 77 D11
Podolie SK 146 D5
Podolínec SK 145 E2
Podorozhnye UA 145 E9
Podosoje BIH 162 D5
Podrašnica BIH 157 D6
Podravske Sesvete HR 149 D8
Podromanija BIH 157 E10
Podsreda SLO 148 D5
Podstrana HR 156 F6
Podtabor SLO 73 E10
Podturen HR 149 D7
Podu Iloaiei RO 153 C10
Podujevë RKS 164 D3
Poduri RO 153 E9
Podu Turcului RO 153 E10
Podvelež BIH 157 F8
Podvrška SRB 159 D9
Podvysoká SK 147 C7
Podwilk PL 147 B9
Poederlee B 182 C5
Poeni RO 160 E6
Poenița MD 153 A11
Pofi I 62 D4
Pogăceaua RO 152 D4
Pogana RO 153 E11
Pogar BIH 157 D9
Poggendorf D 84 B4
Poggiardo I 61 C10
Poggibonsi I 66 F3
Poggio Berni I 66 D5
Poggio Bustone I 62 B3
Poggio Catino I 62 C3
Poggiodomo I 62 B3
Poggio Imperiale I 63 D8
Poggio-Mezzana F 37 G10
Poggio Mirteto I 62 C3
Poggio Moiano I 62 C3
Poggio Picenze I 62 C5
Poggio Renatico I 66 C4
Poggiorsini I 60 B6
Poggio Rusco I 66 C3
Pöggstall A 77 F8
Pogny F 25 C11
Pogoanele RO 161 D9
Pogoniani GR 168 E3
Pogorzela PL 81 C12
Pogorzelice PL 85 B13
Pogradec AL 168 C4
Pogrodzie PL 139 B8
Pohja FIN 127 C12
Pohja FIN 127 E10
Pohja-Lankila FIN 129 C11
Pohjasenvaara FIN 117 D11
Pohjaslahti FIN 119 B17
Pohjaslahti FIN 123 F12
Pohjavaara FIN 121 F11
Pohjois-li FIN 119 D14
Pohjoisjärvi FIN 123 F13
Pohorelá SK 147 D10
Pohořelice CZ 77 E11
Pohronská Polhora SK 147 D9
Pohronský Ruskov SK 147 F7
Poian RO 153 E8
Poiana RO 161 D7
Poiana Blenchii RO 152 C3
Poiana Câmpina RO 161 C7
Poiana Cristei RO 161 B9
Poiana Lacului RO 160 D5
Poiana Mare RO 159 F11
Poiana Mărului RO 160 B6
Poiana Sibiului RO 152 F3
Poiana Stampei RO 152 C6
Poiana Teiului RO 153 C7
Poiana Vadului RO 151 E10
Poibrene BG 165 E8
Põide EST 130 D6
Poienari RO 153 D10
Poienarii Burchii RO 161 D8
Poienarii de Argeş RO 160 C5
Poieneşti RO 153 D11
Poieni RO 151 D10
Poienile de Sub Munte RO 152 B4
Poijula FIN 121 D10
Poikajärvi FIN 117 E15
Põikva EST 131 D10
Poinçon-lès-Larrey F 25 E11
Poing D 75 F10
Pointis-Inard F 33 D7
Poirino I 37 B7
Poiseux F 25 F9
Poissons F 26 D3
Poissy F 24 C7
Poitiers F 29 B6
Poix-de-Picardie F 18 E4
Poix-Terron F 19 E10
Pojan AL 168 C4
Pojanluoma FIN 122 E9
Pojatno HR 148 E5
Pojejena RO 159 D8
Pojo FIN 127 E10
Pojorâta RO 152 B6
Pókaszepetk H 149 C7
Poki LV 134 C7
Pokka FIN 117 B15
Pokleĉani BIH 157 E7
Pokoj BIH 156 C4
Pokój PL 142 E4
Pokrówka PL 141 H8
Pokupsko HR 148 F5
Polac RKS 163 D10
Polača HR 156 D5
Polače HR 162 D3
Pola de Allande E 39 B6
Pola de Laviana E 39 B8
Pola de Lena E 39 B8
Pola de Siero E 39 B8
Pola de Somiedo E 39 B7
Polaincourt-et-Clairefontaine F 26 E5
Połajewo PL 85 E11
Polán E 46 E4
Polanica-Zdrój PL 77 B10
Połaniec PL 143 F11
Polanów PL 85 B11
Polatsk BY 133 F5
Polch D 21 D8
Polcirkeln S 116 E6
Połczyno PL 138 A5
Połczyn Zdrój PL 85 C10
Polegate UK 15 F9
Polena BG 165 F7
Poleñino E 41 E11
Polepy CZ 76 A6
Polesella I 66 C4
Polessk RUS 136 D3
Polgár H 145 H3
Polgárdi H 149 B10
Polgaste EST 131 F13
Polia I 59 B9
Poliçan AL 168 C3
Poliçan AL 168 D3
Police PL 84 C7
Police nad Metuj CZ 81 E10
Polichni GR 169 C8
Polichnitos GR 177 A7
Polička CZ 77 C10
Policoro I 61 C7
Policzna PL 141 H5
Polientes E 40 C4
Polignano a Mare I 61 B8
Poligny F 31 B8
Polikrayshte BG 166 C5
Polisot F 25 D11
Polistena I 59 C9
Politika GR 175 B8
Pölitz D 83 C8
Polizzi Generosa I 58 D5
Põljä FIN 124 D9
Poljana SRB 159 D7
Poljanak HR 156 C4
Poljana Pakračka HR 149 F7
Poljane SLO 73 D9
Poljčane SLO 148 D5
Polje BIH 157 C8
Poljica HR 156 D3
Poljice BIH 157 D10
Poljice-Popovo BIH 162 D5

Polkowice PL 81 D10
Polla I 60 B4
Pöllakkä FIN 125 F11
Pöllau A 148 B5
Pöllauberg A 148 B5
Polle D 78 C5
Pollença E 57 B11
Pollenfeld D 75 E9
Pollenza I 67 F7
Pollfoss N 100 C7
Polliat F 31 C7
Pollica I 60 C4
Pollina I 58 D5
Pollitz D 83 E11
Pollos E 39 F9
Polmak N 113 C20
Polminhac F 29 F11
Polmont UK 5 D9
Polo FIN 121 C12
Polomka SK 147 D9
Polop E 56 D4
Pölöske H 149 C7
Połoski PL 141 G8
Polovragi RO 160 C3
Polperro UK 12 E5
Pöls A 73 B10
Polsbroek NL 182 B5
Polsingen D 75 E8
Polska Cerekiew PL 142 F5
Polski Gradets BG 166 E6
Polski Trŭmbesh BG 166 C5
Polsko Kosovo BG 166 C5
Polso FIN 123 D12
Poltár SK 147 E9
Põltsamaa EST 131 D11
Põlula EST 131 C13
Polumir SRB 163 B10
Põlva EST 131 E14
Polvela FIN 125 D12
Polvenkylä FIN 122 E8
Polverigi I 67 E7
Polvijärvi FIN 125 E12
Polyana UA 145 E6
Polyanets'ke UA 154 A6
Polyantho GR 171 B8
Polyany RUS 129 E11
Polyatiste BG 167 D8
Polydendro GR 169 D7
Polydrosos GR 175 B6
Polygyros GR 169 D9
Polykarpi GR 169 C7
Polykastano GR 168 D5
Polykastro GR 169 C8
Polymylos GR 169 D7
Polyneri GR 168 D5
Polypotamo GR 168 C5
Polzela SLO 73 D11
Pölzig D 79 E11
Pomaluengo E 40 B4
Pomarance I 66 F2
Pomarão P 50 D4
Pomarez F 32 C4
Pomarico I 61 B7
Pomarkku FIN 126 B7
Pomârla RO 153 A8
Pomáz H 150 B3
Pombal P 38 F5
Pombal P 44 E3
Pomer E 41 E8
Pomeroy UK 7 C9
Pomezeu RO 151 D9
Pomezí CZ 77 C10
Pomezia I 62 D3
Pomi RO 151 B11
Pomiechówek PL 139 F10
Pömiö FIN 119 C14
Pommard F 30 A6
Pommelsbrunn D 75 C10
Pommersfelden D 75 C8
Pomol BIH 157 D11
Pomorie BG 167 D9
Pomorsko PL 81 B8
Pompaire F 28 B4
Pompei I 60 B3
Pompey F 26 C5
Pompia GR 178 E8
Pompignan F 35 C6
Pomysk Mały PL 85 B13
Poncin F 31 C7
Ponferrada E 39 C6
Poniatowa PL 141 H6
Poniatowo PL 139 D8
Poniec PL 81 C11
Poniklá CZ 81 E8
Poniky SK 147 D8
Ponoarele RO 159 D10
Ponor RO 151 E11
Ponoshec RKS 163 E9
Pons F 28 D4
Ponsa FIN 127 B11
Ponsacco I 66 E2
Pont-à-Celles B 19 C9
Pontacq F 32 D5
Pontailler-sur-Saône F 26 F3
Pontaix F 31 F7
Pont-à-Marcq F 19 C7
Pont-à-Mousson F 26 C5
Pontardawe UK 13 B7
Pontarion F 29 D9
Pontarlier F 31 B9
Pontassieve I 66 E3
Pontaubault F 23 C9
Pont-Audemer F 18 F2
Pontault-Combault F 25 C8
Pontaumur F 29 D11
Pont-Aven F 22 E4
Pont-Canavese I 68 C4
Pontcharra F 31 E9
Pontchâteau F 23 F7
Pont-Croix F 22 D3
Pont-d'Ain F 31 C7
Pont-de-Buis-lès-Quimerch
    F 22 D3
Pont-de-Chéruy F 31 D7
Pont-de-Larn F 33 C10
Pont-de-l'Isère F 30 E6
Pont-de-Loup B 182 E5
Pont-de-Poitte F 31 B8
Pont-de-Roide F 27 F6
Pont-de-Salars F 34 B4
Pont de Suert E 33 F7
Pont-de-Veyle F 30 C6
Pont-d'Héraut F 35 B6
Pont-d'Ouilly F 23 C11
Pont-du-Casse F 33 B7
Pont-du-Château F 30 D3
Pont-du-Navoy F 31 B8
Ponte I 60 A3
Ponte Aranga E 38 B4
Ponteareas E 38 D3
Pontebba I 73 C7

Pontecagnano Faiano I 60 B3
Ponte Caldelas E 38 D3
Ponteceso E 38 B2
Pontechianale I 36 B6
Pontecorvo I 62 E5
Pontecurone I 37 B9
Ponte da Barca P 38 E3
Pontedassio I 37 D8
Pontedecimo I 37 B9
Ponte de Lima P 38 E2
Ponte dell'Olio I 37 B11
Pontedera I 66 E2
Ponte de Sor P 44 F4
Pontedeume E 38 B3
Ponte di Legno I 69 A10
Ponte di Piave I 72 E5
Ponte do Rol P 44 F2
Pontefract UK 11 D9
Ponte Gardena I 72 C4
Pontelagoscuro I 66 C4
Ponteland UK 5 E13
Pontelandolfo I 60 A3
Ponte-Leccia F 37 G10
Pontelongo I 66 B5
Ponte nelle Alpi I 72 D5
Ponte Nizza I 37 B10
Ponte Nossa I 69 A8
Ponte Nova I 72 D3
Pont-en-Royans F 31 E7
Pontenure I 69 C8
Pontenx-les-Forges F 32 B3
Ponte San Nicolò I 66 B4
Ponte San Pietro I 69 B8
Ponte Valga E 38 C2
Pontevedra E 38 D2
Pontével P 44 F3
Pont-Évêque F 30 D6
Pontevico I 66 B1
Pont-Farcy F 23 C9
Pontfaverger-Moronvilliers
    F 19 F9
Pontgibaud F 30 D2
Pont-Hébert F 23 B9
Pontinha P 50 B1
Pontinia I 62 E4
Pöntiö FIN 120 F2
Pontivy F 22 D6
Pont-l'Abbé F 22 E3
Pont-la-Ville F 25 D12
Pont-les-Moulins F 26 F5
Pont-l'Évêque F 23 B12
Pontlevoy F 24 F5
Pontoise F 24 B7
Pontokomi GR 169 D6
Pontones E 55 C7
Pontonnyy RUS 129 F14
Pontonx-sur-l'Adour F 32 C4
Pontoon IRL 6 E4
Pontorson F 23 C8
Pontpoint F 18 F6
Pontremoli I 69 E8
Pontresina CH 71 E9
Pontrhydfendigaid UK 13 A7
Pontrieux F 22 C5
Pontrilas UK 13 B9
Ponts E 42 D6
Pöntsö FIN 117 C12
Pont-Ste-Marie F 25 D11
Pont-Ste-Maxence F 18 F6
Pont-St-Esprit F 35 B8
Pont-St-Martin I 68 B4
Pont-sur-Yonne F 25 D9
Pontvallain F 23 E12
Pontyberem UK 12 B6
Pontycymer UK 13 B7
Pontypool UK 13 B8
Pontypridd UK 13 B8
Ponyativka UA 154 D6
Ponza I 62 F3
Ponzone I 37 B8
Poola FIN 122 E7
Poole UK 13 D11
Poolewe UK 2 K5
Pooley Bridge UK 5 F11
Pool of Muckhart UK 5 C9
Pootsi EST 131 E8
Pope LV 134 B3
Popeasca MD 154 D5
Poperinge B 18 C6
Popeşti RO 151 C9
Popeşti RO 153 C10
Popeşti RO 160 E6
Popeşti-Leordeni RO 161 E8
Popielów PL 142 E4
Popina BG 161 F1
Popintsi BG 165 E9
Popiţsa BG 165 C8
Poplaca RO 152 F4
Popoli I 62 C5
Popovača HR 149 E7
Popovica SRB 159 E9
Popovitsa BG 166 E4
Popovo BG 166 C6
Popów PL 141 F1
Popów PL 143 D6
Poppenhausen D 75 B7
Poppenricht D 75 D10
Poppi I 66 E4
Poprad SK 145 E1
Popricani RO 153 C11
Poproč SK 145 F2
Popsko BG 171 A9
Pópulo P 38 F4
Populonia I 65 B2
Porąbka PL 147 B8
Poraj PL 143 E7
Porcari I 66 E2
Porcia I 73 E6
Porcsalma H 145 H6
Porcuna E 53 A8
Pordenone I 73 E6
Pordim BG 165 C10
Porjeba PL 143 F7
Porjeba-Kocjbny PL 139 E12
Poreč HR 67 B8
Pori FIN 126 C6
Porice BIH 157 D7
Porjus S 116 E3
Porkhov RUS 132 F6
Porkkala FIN 127 F11
Porkuni EST 131 C12
Porlammi FIN 127 D15
Porlezza I 69 A7
Porlock UK 13 C7
Pornainen FIN 127 E13
Pornassio I 37 C7
Pornic F 23 F7
Pornichet F 23 F7
Poroina Mare RO 159 E10
Poros GR 174 C2
Poros GR 175 D7
Poroschia RO 160 F6

Poroshkove UA 145 F6
Poroszló H 150 B6
Porozina HR 67 B9
Porpi GR 171 B8
Porqueres E 43 C9
Porras FIN 127 D10
Porraskoski FIN 127 C12
Porrentruy CH 27 F7
Porreres E 57 B11
Porretta Terme I 66 D2
Porsangmoen N 113 D14
Pörsänmäki FIN 124 D8
Porsäsen N 102 E3
Porsgrunn N 90 A6
Porsi S 118 B5
Porspoder F 22 C2
Port N 113 D17
Porta F 33 E9
Portadown UK 7 D10
Portaferry UK 7 D11
Portaje E 45 E7
Portakallık TR 181 B8
Portalegre P 44 F6
Portals Vells E 49 F10
Portaria GR 169 D8
Portaria GR 169 F8
Portarlington IRL 7 F8
Port Askaig UK 4 D4
Portavogie UK 7 D12
Portbail F 23 B8
Port-Barcarès F 34 E5
Portbou E 34 F5
Port Brillet F 23 D10
Portchester UK 13 D12
Port d'Andratx E 49 E9
Port-de-Bouc F 35 D8
Port-de-Piles F 29 A7
Port de Pollença E 57 B11
Port de Sóller E 49 E10
Portel P 50 C4
Portel-des-Corbières F 34 D4
Portell de Morella E 42 F3
Port Ellen UK 4 D4
Port-en-Bessin-Huppain
    I 23 B10
Portencross UK 4 D7
Port Erin GBM 10 C2
Portes-lès-Valence F 30 F6
Portets F 28 F5
Port Eynon UK 12 B6
Port Glasgow UK 4 D7
Porth UK 13 B8
Porthcawl UK 13 C7
Port Henderson UK 2 K5
Porthleven UK 12 E4
Porthmadog UK 10 F3
Porticcio F 37 H9
Portici I 60 B2
Portieux F 26 D5
Portilla de la Reina E 39 B10
Portillo E 40 F2
Portillo de Toledo E 46 D4
Portimão P 50 E2
Portimo FIN 119 B16
Portimojärvi FIN 119 B11
Portishead UK 13 C9
Port-Joinville F 28 B1
Portknockie UK 3 K11
Port Láirge IRL 9 D8
Portland IRL 6 F6
Port-la-Nouvelle F 34 D5
Port Laoise IRL 7 F8
Portlaoise IRL 7 F8
Portloe UK 12 E5
Portlethen UK 5 A12
Port-Leucate F 34 E5
Port Logan UK 4 F7
Port-Louis F 22 E5
Portmagee IRL 8 E2
Portmahomack UK 3 K9
Portman E 56 F3
Portmarnock IRL 7 F10
Port Mearnóg IRL 7 F10
Portmuck UK 4 F5
Port-na-Con UK 2 H7
Portnaguran UK 2 J4
Portnahaven UK 4 D3
Portnalong UK 2 L4
Port-Navalo F 22 E6
Porto E 38 E5
Porto F 37 G9
Porto Alto P 50 B2
Porto Azzurro I 65 B2
Porto Botte I 64 E2
Portobravo E 38 C2
Porto Cervo I 64 A4
Porto Cesareo I 61 C9
Portocheli GR 175 E7
Portocolom E 57 C11
Porto Covo da Bandeira P
    50 D2
Porto Cristo E 57 B11
Porto d'Ascoli I 62 B5
Porto de Lagos P 50 E2
Porto de Mós P 44 E3
Porto do Barqueiro E 38 A4
Porto do Son E 38 C1
Porto Empedocle I 58 E4
Porto Ercole I 65 C4
Portoferraio I 65 B2
Portofino I 37 C10
Port of Ness UK 2 J4
Porto Garibaldi I 66 C5
Porto Germeno GR 175 C7
Portogruaro I 73 E6
Porto Koufo GR 170 E5
Porto Levante I 59 C6
Porto Levante I 66 B5
Pörtom FIN 122 E7
Portomaggiore I 66 C4
Portomarin E 38 C4
Port Omna IRL 6 F6
Portopalo di Capo Passero
    I 59 F7
Porto Petro E 57 C11
Portør N 90 B5
Porto Recanati I 67 F8
Porto Rotondo I 64 A4
Portorož SLO 67 A8
Porto San Giorgio I 67 F8
Porto Sant'Elpidio I 67 F8
Porto Santo Stefano I 65 C4
Portoscuso I 64 E1
Porto Tolle I 66 C5
Porto Torres I 64 B1
Porto-Vecchio F 37 H10
Portovenere I 69 E8
Porto Viro I 66 B5

Portpatrick UK 4 F6
Portrane IRL 7 F10
Port Reachrann IRL 7 F10
Portreath UK 12 E4
Portree UK 2 L4
Portroe IRL 8 C6
Portrush UK 4 E3
Portsalon IRL 4 E1
Portsmouth UK 13 D12
Portsoy UK 3 K11
Port-Ste-Marie F 33 B6
Portstewart UK 4 E3
Port-St-Louis-du-Rhône
    F 35 D8
Port St Mary GBM 10 C2
Port-St-Père F 23 F8
Port-sur-Saône F 26 E5
Port Talbot UK 13 B7
Portugalete E 40 B5
Porturna IRL 6 F6
Port-Vendres F 34 E5
Port William UK 4 F7
Poruba SK 147 D7
Porubne UA 153 A8
Porumbacu de Jos RO 152 F4
Porvoo FIN 127 E14
Porządzie PL 139 E11
Porzuna E 46 F4
Posada I 64 B4
Posada E 39 B8
Posada de Valdeón E 39 B10
Posadas E 53 A6
Posaga RO 151 E11
Poschiavo CH 69 A9
Posedarje HR 156 D3
Poseidonia GR 176 E4
Poseritz D 84 B4
Posina I 72 E3
Posio FIN 121 B11
Pösing D 75 D12
Possagno I 72 E4
Posseberg S 97 D11
Possendorf D 80 E5
Possesse F 25 C12
Pößneck D 79 E10
Posta I 62 B4
Poşta Câlnău RO 161 D9
Postal I 72 C3
Posta Piana I 60 A5
Postau D 75 E11
Posterholt NL 20 B6
Postiglione I 60 B4
Postioma I 72 E5
Postojna SLO 73 E9
Postoliska PL 139 F11
Postoloprty CZ 76 B5
Postomino PL 85 B11
Postřelmov CZ 77 C11
Postupice CZ 77 C7
Posušje BIH 157 F7
Poświętne PL 140 E7
Poświętne PL 141 G2
Potamia GR 171 C2
Potamia GR 171 C7
Potamochori GR 171 B8
Potamoi GR 170 B6
Potamos GR 168 E2
Potamos GR 178 C4
Potashnya UA 154 A5
Potcoava RO 160 E5
Potigowo PL 85 B12
Potenza I 60 B5
Potenza Picena I 67 F8
Potes E 39 B10
Potidania GR 174 C5
Potigny F 23 C11
Potkraj BIH 157 E7
Potku FIN 115 E3
Potlogi RO 161 D7
Potočac SRB 159 F7
Potočani BIH 157 E7
Potoci BIH 156 D6
Potoci BIH 157 F8
Potok Złoty PL 143 E7
Potsdam D 80 B4
Põtsõnvaara FIN 125 D16
Potštát CZ 146 B5
Pottendorf A 77 G10
Pottenstein A 77 F10
Pottenstein D 75 C9
Potterne UK 13 C10
Potters Bar UK 15 D8
Pöttmes D 75 E9
Potton UK 15 C8
Pöttsching A 146 F2
Potworów PL 141 G3
Pouancé F 23 E8
Pougny F 25 F9
Pougues-les-Eaux F 30 A3
Pougy F 25 D11
Pouillon F 32 C4
Pouilly-en-Auxois F 25 F12
Pouilly-sous-Charlieu F 30 C5
Pouilly-sur-Loire F 25 F8
Pouilly-sur-Saône F 31 A7
Poulaines F 24 F6
Pouldreuzic F 22 E3
Poulgorm Bridge IRL 8 E4
Pouligny-St-Pierre F 29 B8
Poulithra GR 175 E6
Poullaouen F 22 D4
Poulnamucky IRL 9 D7
Poulstrup DK 90 E7
Poulton-le-Fylde UK 10 D6
Pounta GR 176 E5
Poupas F 33 C7
Pourcieux F 35 D10
Pouri GR 169 F7
Pourrain F 25 E9
Pourrières F 35 C10
Pousada P 45 C6
Pousos P 44 E3
Poussan F 35 D6
Poussay F 26 D5
Poussu FIN 121 C13
Pouxeux F 26 D6
Pouyastruc F 32 C5
Pouydesseaux F 32 C5
Pouzac F 33 D6
Pouzauges F 28 B4
Pouzay F 24 F4
Povedilla E 55 B7
Poviglio I 66 B2
Povlja HR 157 F6
Póvoa de Lanhoso P 38 E3
Póvoa de São Miguel P 51 C5
Póvoa de Varzim P 38 F2
Póvoa do Concelho P 45 C6
Povoletto I 73 D7
Povorly CZ 80 E6
Powerscourt IRL 9 C10
Power's Cross IRL 6 F6
Powick UK 13 A10

Powidz PL 138 F4
Powmill UK 5 C9
Poyales del Hoyo E 45 D10
Poyatos E 47 D8
Poyntz Pass UK 7 D10
Poyrazcık TR 177 A9
Poysdorf A 77 E11
Pöytyä FIN 126 D8
Poza de la Sal E 40 C5
Pozaldez E 39 F10
Požarevac SRB 159 D7
Požarnica BIH 157 C10
Požega HR 149 F9
Požega SRB 158 F5
Pożeřě LT 134 E4
Pozezdrze PL 136 E4
Pozharan RKS 164 E3
Pozherevitsy RUS 133 D7
Poznań PL 81 B11
Pozo Alcón E 55 D7
Pozoamargo E 47 F8
Pozoantiguo E 39 E9
Pozoblanco E 54 C3
Pozo Cañada E 55 B9
Pozo de Guadalajara E 47 D6
Pozohondo E 55 B9
Pozo-Lorente E 47 F10
Pozondón E 47 D10
Pozořice CZ 77 D11
Pozorrubio E 47 E7
Poźrzadło Wielkie PL 85 D9
Pozuel del Campo E 47 C10
Pozuelo E 55 B8
Pozuelo de Alarcón E 46 D5
Pozuelo de Aragón E 41 E9
Pozuelo del Páramo E 39 D8
Pozuelo del Rey E 46 D6
Pozuelo de Zarzón E 45 D8
Pozuelos de Calatrava E 54 B4
Pozza di Fassa I 72 D4
Pozzallo I 59 F6
Pozzo I 59 F6
Pozzolo Formigaro I 37 B9
Pozzomaggiore I 64 C2
Pozzuoli I 60 B2
Pozzuolo del Friuli I 73 E7
Prabuty PL 139 C7
Prača BIH 157 E10
Prachatice CZ 76 D5
Pracht D 21 C9
Prackenbach D 76 D3
Prádanos de Ojeda E 40 C3
Pradejón E 41 D7
Pradelles F 30 F4
Prádena E 46 B5
Prades F 33 E9
Prades F 33 E9
Prades-d'Aubrac F 34 A4
Pradillo E 41 D6
Pradines F 33 B8
Prado E 38 D3
Prado de la Guzpeña E 39 C9
Prado del Rey E 53 C5
Pradoluengo E 40 C5
Præstbro DK 86 A6
Præstø DK 87 E10
Pragelato I 31 E9
Prägraten A 72 B5
Praha CZ 77 B6
Prahecq F 28 C5
Prahovo SRB 159 E10
Praia a Mare I 60 D5
Praia da Barra P 44 C3
Praia da Rocha P 50 E2
Praia da Tocha P 44 D3
Praia de Esmoriz P 44 C3
Praia de Mira P 44 D3
Praiano I 60 B3
Praid RO 152 D6
Prăjeni RO 153 B10
Prakovce SK 145 F2
Pralognan-la-Vanoise F 31 E10
Pralyetarsk BY 133 E8
Pramanta GR 168 E5
Prambachkirchen A 76 F5
Pranjani SRB 158 E5
Prapatnica HR 156 E5
Prašice SK 146 D6
Praszka PL 142 D5
Prata di Pordenone I 73 E6
Pratau D 79 C12
Prat de Comte E 42 F4
Pratdip E 42 E5
Pratella I 60 A2
Prato I 66 E3
Prato allo Stelvio I 71 D11
Pratola Peligna I 62 C5
Pratovecchio I 66 E4
Prats de Lluçanès E 43 C8
Prats-de-Mollo-la-Preste
    F 33 F10
Pratteln CH 27 E8
Praulienа LV 135 C12
Prauthoy F 26 E3
Pravda BG 161 F10
Pravdinsk RUS 136 E3
Pravets BG 165 D8
Pravia E 39 B7
Prčanj E 163 E6
Préchac F 32 B5
Préci I 62 B4
Précigné F 23 E11
Prečín SK 147 C7
Précy-sous-Thil F 25 F11
Predappio I 66 D4
Predazzo I 72 D4
Predeal RO 161 B7
Predeal-Sărari RO 161 C7
Predejane SRB 164 D5
Predeşti RO 160 E3
Preding A 148 C4
Predlitz A 73 B8
Predmeja SLO 73 E8
Předměřice nad Labem
    CZ 77 B9
Predosa I 37 B9
Pré-en-Pail F 23 D11
Prees UK 10 F6
Preetz D 83 B8
Préfailles F 23 F7
Preganziòl I 72 E5
Pregarten A 77 F7
Preignan F 33 C7
Preili LV 135 D13
Preitenegg A 73 C10
Preixan F 33 D10
Prejmer RO 153 F7
Prekaja BIH 156 D6

Preko HR 156 D3
Prekopa HR 148 F6
Preljina SRB 158 F5
Prelog HR 149 D7
Preločica SLO 148 D5
Prem SLO 73 E9
Premantura HR 67 C8
Prémery F 25 F9
Premia I 68 A5
Premià de Mar E 43 E8
Premilcuore I 66 E4
Prémilhat F 29 C11
Premnitz D 79 A11
Premosello Chiovenda I 68 A5
Prenzlau D 84 D5
Préporché F 30 B4
Přerov CZ 146 C4
Prerow D 83 B13
Preselentsi BG 155 F2
Preševo SRB 164 E4
Presicce I 61 D10
Prešov SK 145 F3
Pressac F 29 C7
Pressath D 75 C10
Pressbaum A 77 F10
Presseck D 75 B9
Pressel D 79 C12
Pressig D 75 B9
Prestatyn UK 10 E5
Prestbakken N 111 C15
Prestbury UK 13 B10
Pre-St-Didier I 31 D10
Prestebakke N 91 B10
Presteigne UK 13 A8
Prestfoss N 95 B11
Přeštice CZ 76 C4
Preston UK 5 D12
Preston UK 10 D6
Prestwick UK 4 E7
Prettin D 80 C3
Pretzier D 83 E10
Pretzsch D 79 C12
Pretzschendorf D 80 E5
Preuilly-sur-Claise F 29 B7
Preußisch Oldendorf D 17 D10
Prevalje SLO 73 C10
Prévenchères F 35 A6
Préveranges F 29 C10
Prevešt SRB 159 F7
Preveza GR 174 A2
Prevršac HR 156 C5
Prezid HR 73 E10
Prez-sous-Lafauche F 26 D3
Priaranza del Bierzo E 39 C6
Pribelja BIH 157 D6
Příběta SK 146 F6
Pribinić BIH 157 C8
Pribislavec HR 149 C6
Priboieni RO 160 D6
Priboj BIH 157 C10
Priboj SRB 163 C8
Přibor CZ 146 B6
Priborn D 83 D13
Priboy BG 165 D6
Příbram CZ 76 C5
Pribrezhnoye RUS 136 D1
Pribude HR 156 E5
Pribylina SK 147 C9
Pribyslav CZ 77 C9
Prichaly RUS 134 F2
Prichsenstadt D 75 C7
Priedaine LV 135 C9
Priedes LV 131 F9
Priego E 47 D8
Priego de Córdoba E 53 B8
Priekule LT 134 E2
Priekule LV 134 D2
Priekuļi LV 135 B11
Prienai LT 137 D8
Prienai LT 137 D8
Prien am Chiemsee D 72 A5
Prieros D 80 B5
Prieška SK 147 D7
Priestewitz D 80 D5
Prievidza SK 147 D7
Priežmale LV 133 D2
Prignac-et-Marcamps
    F 28 E4

Privlaka HR 67 D11
Privlaka HR 157 B10
Prizna HR 67 C10
Prizren RKS 163 E10
Prizzi I 58 D3
Prkosi BIH 156 C5
Prnjavor BIH 157 C8
Prnjavor SRB 158 D3
Proaza E 39 B7
Probištip NMK 164 E5
Probota RO 153 C10
Probsteierhagen D 83 B8
Probus UK 12 E5
Proceno I 62 B1
Prochoditsе RUS 134 F2
Prochowice PL 81 D10
Procida I 60 B2
Prodromos GR 175 C6
Produlești RO 161 D7
Proença-a-Nova P 44 E5
Proença-a-Velha P 45 D6
Profen D 79 D11
Profesor-Ishirkovo BG 161 F10
Profilia GR 181 D7
Profitis GR 169 C9
Profitis Ilias GR 178 E9
Profondeville B 19 D10
Progonat AL 168 D2
Prohn D 84 B4
Prokhladnoye RUS 134 F2
Prokopí GR 175 B7
Prokuplje SRB 164 C4
Prolaz BG 167 C6
Prolog BIH 157 E6
Promachoi GR 169 C6
Promachonas GR 169 B9
Promna PL 141 G3
Promyri GR 175 A7
Pronsfeld D 20 D6
Propriano F 37 H9
Proseč CZ 77 C10
Prosek AL 163 F8
Prosenik BG 167 D8
Prosenjakovci SLO 149 C6
Prosetín CZ 77 C9
Proshkava BY 133 F4
Proskynites GR 171 C8
Prosotsani GR 170 B5
Prosperous IRL 7 F9
Prossedi I 62 D4
Prostějov CZ 146 C4
Prostki PL 140 C6
Prostřední Bečva CZ 146 C6
Prószków PL 142 F4
Proszowice PL 143 F9
Proszówki PL 143 F9
Proti GR 170 C5
Protići BIH 157 D6
Protivín CZ 77 D11
Protivín CZ 76 D6
Protokklisi GR 171 B10
Protoria GR 178 E7
Prötzel D 80 A5
Provadiya BG 167 C8
Provatas GR 169 B9
Provenchères-sur-Fave F 27 D7
Provins F 25 C9
Provo SRB 158 D4
Prozor BIH 157 E8
Prozor HR 156 C3
Prnenjas AL 168 B4
Pruchnik PL 144 D6
Prudhoe UK 5 F13
Prudnik PL 142 F4
Prudziniki BY 133 E4
Prugos RKS 164 C3
Prügy H 145 G3
Pruna E 51 F9
Prundeni RO 160 D4
Prundu RO 161 E8
Prundu Bârgăului RO 152 C5
Prunières F 36 B4
Pruniers-en-Sologne F 24 F6
Prunişor RO 159 D10
Prusac BIH 157 D7
Prušánky CZ 77 E11
Prusice PL 81 D11
Prusinovice CZ 146 C5
Prüšiši LV 135 B13
Pruszcz PL 138 D4
Pruszcz PL 138 C5
Pruszcz Gdański PL 138 B6
Pruszków PL 141 F3
Prutz A 71 C11
Pružina SK 146 C6
Prvačina SLO 73 D8
Pryamobalka UA 154 F4
Prymors'ke UA 155 B5
Prymors'ke UA 155 B5
Pryozerne UA 155 B5
Pryputtya UA 153 A8
Przasnysz PL 139 D10
Przechlewo PL 85 C12
Przecieszyn PL 143 F7
Przecław PL 143 F11
Przedbórz PL 143 D8
Przedecz PL 138 F6
Przejazdowo PL 138 B6
Przelewice PL 85 D9
Przemęt PL 81 B10
Przemków PL 81 C9
Przemyśl PL 144 D6
Przerośl PL 136 E6
Przewale PL 144 B8
Przewłoka PL 141 G8
Przeworno PL 81 E12
Przeworsk PL 144 C5
Przewóz PL 81 D7
Przeździatka PL 141 F6
Przezdzik Wielki PL 139 D10
Przezmark PL 139 C7
Przine-Zdralovac BIH 156 D6
Przodkowo PL 138 B5
Przyborów PL 81 C9
Przyborowice PL 139 E9
Przybranowo PL 138 E6
Przygodzice PL 142 C4
Przykona PL 142 C5
Przyłęk PL 81 C3
Przyłęk PL 141 G4
Przylesie PL 142 E3
Przyrów PL 143 E8
Przysieki PL 144 D3
Przystajń PL 142 E6
Przystawice Duże PL 141 H3
Przysucha PL 141 H3
Przyszowa PL 144 S
Przytoczna PL 81 A9
Przytoczno PL 141 G6
Przytuły PL 140 D6
Przytyk PL 141 H3

Przywidz PL 138 B5
Psača NMK 164 E5
Psachna GR 175 B8
Psara GR 177 B6
Psarades GR 168 C5
Psari GR 174 E4
Psari GR 175 D6
Psathopyrgos GR 174 C4
Psinthos GR 181 D8
Pskov RUS 132 F3
Psychiko GR 169 B10
Psychro GR 178 E9
Pszczew PL 81 B9
Pszczółki PL 138 B6
Pszczonów PL 141 G1
Pszczyna PL 147 B7
Pteleos GR 175 A6
Pteri GR 174 C5
Ptolemaida GR 169 C6
Ptuj SLO 148 D5
Publier F 31 C10
Puchberg am Schneeberg A 146 F1
Pucheni RO 160 C6
Puchenii Mari RO 161 D8
Puchheim D 75 F9
Púchov SK 146 C6
Pucioasa RO 161 D7
Pučišće HR 157 F6
Puck PL 138 A5
Puckaun IRL 8 C6
Puçol E 48 E4
Pudasjärven kirkko FIN 119 D17
Pudasjärvi FIN 119 D17
Puddletown UK 13 D10
Puderbach D 185 C8
Pudinava LV 133 C3
Pudsey UK 11 D8
Puebla de Albortón E 41 F10
Puebla de Alcocer E 51 B9
Puebla de Alfindén E 41 E10
Puebla de Almenara E 47 E7
Puebla de Beleña E 47 C6
Puebla de Don Fadrique E 55 D8
Puebla de Don Rodrigo E 46 F3
Puebla de Guzmán E 51 D5
Puebla de la Calzada E 51 B6
Puebla de la Reina E 51 B7
Puebla de Lillo E 39 B9
Puebla del Maestre E 51 C7
Puebla del Príncipe E 55 B7
Puebla del Prior E 51 B7
Puebla del Salvador E 47 E9
Puebla de Obando E 45 C4
Puebla de Sanabria E 39 D6
Puebla de Sancho Pérez E 51 C7
Puebla de San Miguel E 47 D10
Puebla de Yeltes E 45 C8
Puente de Domingo Flórez E 39 D6
Puente de Génave E 55 C7
Puente del Congosto E 45 D9
Puente de Montañana E 42 C5
Puente de San Miguel E 40 B3
Puente-Genil E 53 B7
Puente la Reina E 32 E2
Puentenansa E 40 B4
Puente Viesgo E 40 B4
Puerto de Béjar E 45 D9
Puerto de Mazarrón E 56 F2
Puerto de San Vicente E 45 E10
Puerto Lápice E 46 F6
Puertollano E 54 B4
Puerto Lumbreras E 55 D9
Puertomingalvo E 48 D4
Puerto Real E 52 C4
Puerto Seguro E 45 C7
Puerto Serrano E 51 F8
Pueyo E 32 E2
Pueyo de Santa Cruz E 42 D4
Pufești RO 153 E10
Puget-Théniers F 36 D5
Puget-Ville F 36 E4
Pugnochiuso I 63 D10
Puhja EST 131 E12
Puhoi MD 154 D4
Puhos FIN 121 D10
Puhos FIN 125 F13
Puhovac BIH 157 D9
Pui RO 159 B11
Puiatu EST 131 E10
Puiești RO 153 E11
Puiești RO 161 C10
Puig E 48 E4
Puigcerdà E 33 F9
Puigpunyent E 49 E10
Puig-reig E 43 D7
Puikkola FIN 115 E3
Puikule LV 131 F9
Puimoisson F 36 D4
Puiseaux F 25 D7
Puisieux F 18 D6
Puisseguin F 28 F5
Puisserguier F 34 D5
Puivert F 33 E10
Pujaut F 35 B8
Pujols F 28 F5
Pujols F 33 B7
Puka EST 131 E12
Pukanec SK 147 E7
Pukaro FIN 127 D15
Pukavik S 88 C7
Pukë AL 163 E8
Pukiš BIH 157 C10
Pula HR 67 C9
Pula I 64 E3
Puławy PL 141 H5
Pulborough UK 15 F7
Pulfero I 73 D7
Pulgar E 46 E4
Pulheim D 21 B7
Pulju FIN 117 B13
Pułkau A 77 E9
Pulkkaviita FIN 115 D5
Pulkkila FIN 119 F15
Pulkkinen FIN 123 D11
Pulkonkoski FIN 124 D8
Pulpí E 55 E9
Pulsa FIN 128 D8
Pulsano I 61 C8
Pulsen D 80 D4
Pulskala FIN 115 D4
Pulsnitz D 80 D6
Pulsujärvi S 116 B6
Pułtusk PL 139 E11
Pulversheim F 27 E7
Pumpénai LT 135 E8
Pumpuri LV 134 C4
Pūņas LV 134 B4
Punat HR 67 B10

Pundsvik N 111 D12
Punduri LV 133 D3
Pungești RO 153 D10
Punghina RO 159 E10
Punia LT 137 D9
Punkaharju FIN 129 B11
Punkalaidun FIN 127 C9
Punkka FIN 128 C7
Punta Ala I 65 B3
Punta Križa HR 67 C9
Punta Sabbioni I 66 B5
Punta Umbría E 51 E6
Puokio FIN 120 E8
Puolakkavaara FIN 115 D1
Puolanka FIN 121 E10
Puoliväli FIN 124 D9
Puoltikasvaara S 116 D6
Puottaure S 118 B4
Pūpoli LV 133 D2
Puraći BIH 157 C8
Puračić BIH 157 C10
Purani RO 161 F6
Puras FIN 121 E12
Purbach am Neusiedler See A 77 G11
Purcari MD 154 D5
Purchena E 55 E8
Pûre LV 134 B5
Purgstall an der Erlauf A 77 F8
Purila EST 131 C9
Purkijaur S 109 C18
Pürksi EST 130 C7
Purley UK 15 E8
Purmerend NL 16 C3
Purmo FIN 123 C9
Purmojärvi FIN 123 D10
Purmsäti LV 134 D3
Purnu S 116 E4
Purnumukka FIN 115 B2
Purnuvaara FIN 121 C12
Purnuvaara S 116 D6
Purola FIN 128 E6
Purontaka FIN 123 C12
Purujärvi FIN 129 B12
Purujosa E 41 E8
Purullena E 55 E6
Pūrvenets BG 165 E10
Purviniškė LT 137 D7
Pūrvomay BG 166 E4
Pūrvomay BG 169 B9
Puša LV 133 D2
Pušalotas LT 135 E8
Puschendorf D 75 C8
Pushkinskiye Gory RUS 133 B5
Pušmucova LV 133 D3
Püspökladány H 151 C7
Püssi EST 131 C7
Pusterwald A 73 B9
Pusté Úľany SK 146 E5
Pustków PL 143 F11
Pustomyty UA 144 D8
Pustoshka RUS 133 D6
Pûstrovo BG 166 E4
Pusula FIN 127 E10
Puszcza Mariańska PL 141 G2
Puszczykowo PL 81 B11
Pusztaföldvár H 150 D6
Pusztakovácsi H 149 C9
Pusztamérges H 150 E4
Pusztamonostor H 150 B4
Pusztaszabolcs H 149 B11
Pusztaszer H 150 D4
Pusztavács H 150 C4
Pusztavám H 149 B10
Putaja FIN 126 C8
Putanges-Pont-Écrepin F 23 C11
Putbus D 84 B4
Putifigari I 64 B1
Putignano I 61 B8
Putikko FIN 129 B11
Putinci SRB 158 C4
Putineiu RO 160 F5
Putineiu RO 161 F7
Putkivaara FIN 119 B17
Putlitz D 83 D12
Putna RO 153 B8
Putnok H 145 G1
Putte B 19 B9
Putte NL 16 F2
Puttelange-aux-Lacs F 27 B6
Putten NL 16 D5
Puttgarden D 83 A10
Püttlingen D 21 F7
Putula FIN 127 C13
Putyla UA 152 B6
Putzar D 84 C5
Putzkau D 80 D6
Puukari FIN 125 C11
Puukkokumpu FIN 119 C14
Puumala FIN 129 B9
Puurmani EST 131 D12
Puurs B 19 B9
Puurtila FIN 125 F9
Puutossalmi FIN 124 E9
Puutturinjärvi FIN 119 E17
Puybrun F 29 F9
Puycasquier F 33 C7
Puygouzon F 33 C10
Puylaroque F 33 B9
Puylaurens F 33 C10
Puy-l'Évêque F 33 A8
Puymirol F 33 B7
Puymoyen F 29 D6
Puyôo F 32 C4
Pwllheli UK 10 F3
Pyaozerskiy RUS 121 C17
Pyatidorozhnoye RUS 139 B9
Pyatirech'ye RUS 129 D14
Pyelishcha BY 141 F11
Pyershamayski BY 137 F11
Pyhäjärvi FIN 113 F16
Pyhäjärvi FIN 115 D2
Pyhäjoki FIN 119 F12
Pyhäjoki FIN 126 C7
Pyhäjoki FIN 121 D12
Pyhäkylä FIN 121 E11
Pyhältö FIN 128 D7
Pyhämaa FIN 126 D5
Pyhänkoski FIN 119 F12
Pyhäntä FIN 119 F16
Pyhäntä FIN 121 F11
Pyhäntaka FIN 127 C14
Pyharanta FIN 126 D5
Pyhäsalmi FIN 123 C15
Pyhäselkä FIN 125 F13
Pyhe FIN 126 D6
Pyhra A 77 F9
Pyhtää FIN 128 E6
Pyle UK 13 C7
Pyles GR 181 E6
Pyli GR 169 F6
Pyli GR 175 C6
Pyli GR 177 F9

Pylkönmäki FIN 123 E13
Pylos GR 174 F4
Pylypets' UA 145 F7
Pyntäinen FIN 126 B7
Pyrbaum D 75 D9
Pyrgetos GR 169 E8
Pyrgi GR 177 C7
Pyrgiotika GR 175 D6
Pyrgoi GR 170 B6
Pyrgos GR 174 D3
Pyrgos GR 176 D5
Pyrgos GR 177 E8
Pyrgos GR 178 E9
Pyrgos Dirou GR 178 B3
Pyrrönperä FIN 123 B15
Pyrsogianni GR 168 D4
Pyrzyce PL 85 D7
Pyšely CZ 77 C7
Pyskowice PL 142 F6
Pyssyperä FIN 121 D10
Pysznica PL 144 B3
Pytalovo RUS 133 B3
Pythagoreio GR 177 D8
Pythio GR 169 D7
Pythio GR 172 B6
Pyykkölänvaara FIN 121 E12
Pyykköskylä FIN 121 D13
Pyyli FIN 125 F11

Qafzez AL 168 D4
Qelëz AL 163 E8
Qeparo AL 168 D2
Qerret AL 168 B2
Quadrazais P 45 D7
Quadri I 63 D6
Quakenbrück D 17 C9
Qualiano I 60 B2
Quaregnon B 182 E4
Quargnento I 37 B9
Quarona I 68 B5
Quarrata I 66 E2
Quarré-les-Tombes F 25 F10
Quarteira P 50 E3
Quartell E 48 E4
Quarten CH 27 F11
Quartu Sant'Elena I 64 E3
Quatre-Champs F 19 F10
Quatretonda E 56 D4
Quattro Castella I 66 C1
Quedlinburg D 79 C9
Queenborough UK 15 E10
Queensbury UK 11 D8
Queenstown IRL 8 E6
Queidersbach D 21 F9
Queige F 31 D9
Queijra P 44 C5
Quel E 41 D7
Quelaines-St-Gault F 23 E10
Quellendorf D 79 C11
Queluz P 50 B1
Quemada E 40 E4
Quend F 18 D4
Quendorf D 17 D8
Queralbs E 33 F10
Quercianella I 66 F1
Querenhorst D 79 B8
Querfurt D 79 D10
Quérigut F 33 F10
Quern D 82 A7
Quernheim D 17 D10
Quero E 47 E6
Quero I 72 E4
Querrieu F 18 E5
Quesa E 48 F3
Quesada E 55 D6
Quesnoy-sur-Deûle F 18 C7
Quessoy F 22 D6
Questembert F 22 E7
Quettehou F 23 A9
Quettreville-sur-Sienne F 23 C9
Quevauvillers F 18 E5
Queyrac F 28 E4
Quézac F 29 F10
Quiaios P 44 D3
Quiberon F 22 F5
Quickborn D 83 C7
Quierschied D 21 F8
Quiévrain B 19 D8
Quiévrechain F 182 E3
Quiliano I 37 C8
Quillan F 33 E10
Quillebeuf-sur-Seine F 18 F2
Quimper F 22 E3
Quimperlé F 22 E4
Quin IRL 8 C5
Quincinetto I 68 B4
Quincy-Voisins F 25 C8
Quingey F 26 F4
Quinson F 36 D4
Quinssaines F 29 C11
Quinta de la Serena E 51 B8
Quintana del Castillo E 39 C7
Quintana del Pino E 39 C7
Quintana del Puente E 40 D3
Quintana de Rueda E 39 C9
Quintana-Martín Galíndez E 40 C5
Quintanapalla E 40 D4
Quintanar de la Orden E 47 E6
Quintanar de la Sierra E 40 E5
Quintanar del Rey E 47 F9
Quintana Redonda E 41 E6
Quintanilla de Onésimo E 40 E3
Quintenas F 30 E6
Quintin F 22 D6
Quinto E 41 F11
Quinto CH 71 D7
Quintos P 50 D4
Quinzano d'Oglio I 69 C9
Quiroga E 38 D5
Quiruelas de Vidriales E 39 D8
Quismondo E 46 D4
Quissac F 35 C7
Quittebeuf F 24 B4
Qukës AL 168 B3

Raab A 76 F5
Raabs an der Thaya A 77 E8
Raahe FIN 119 E12
Raajärvi FIN 115 E1
Rääkkylä FIN 125 F13
Raalte NL 16 D6
Raamsdonksveer NL 16 E3
Raanujärvi FIN 117 E13
Raappananmäki FIN 121 E10
Raasiku EST 131 C10
Raate FIN 121 E14

Raattama FIN 117 B12
Rab HR C10
Rabac HR 67 B9
Rabaçal P 44 D4
Rabaçal P 45 C6
Rábade E 38 B4
Rábagani RO 151 D9
Rábahidvég H 149 B7
Rábakecöl H 149 B8
Rabanales E 39 E7
Rábapaty H 149 A8
Rabastens F 33 C9
Rabastens-de-Bigorre F 33 D6
Rábatamási H 149 A8
Raba Wyżna PL 147 B9
Rabča SK 147 C8
Rabčice SK 147 B9
Rabe SRB 150 E5
Rabenstein an der Pielach A 77 F8
Raben Steinfeld D 83 C11
Råberg S 107 B13
Rąbino PL 85 C9
Rabisha BG 159 E10
Rabivere EST 131 C9
Rąbka PL 147 B9
Rabrovo BG 159 E10
Rabrovo SRB 159 D8
Rača SRB 159 E6
Rača SRB 164 D3
Rǎcǎciuni RO 153 E9
Racale I 61 D10
Rácalmás H 149 B11
Racalmuto I 58 E4
Rǎcari RO 161 D7
Rǎcǎşdia RO 159 D8
Racconigi I 37 B7
Race SLO 148 D5
Rachanie PL 144 B8
Raches GR 175 B6
Raches GR 177 D7
Rǎchiţi RO 153 B9
Rǎchitoasa RO 153 E10
Rǎchitova RO 159 B10
Rachoni GR 171 C7
Raciąż PL 138 C4
Raciąż PL 139 E9
Racibórz PL 142 F5
Raciechowice PL 144 D1
Racimierz PL 85 C7
Rǎčinovci HR 157 C7
Rǎčišće HR 162 D3
Rǎčiu RO 152 D4
Ráckeve H 149 B11
Racksätter S 97 D14
Racksund S 109 D17
Racławice PL 143 F9
Rǎcoasa RO 153 E9
Racoş RO 152 E6
Racova RO 159 B8
Racovița RO 160 B6
Racovița RO 161 C10
Racoviţeni RO 161 C9
Raczki PL 136 F6
Rączki PL 139 D9
Rád H 150 B3
Råda S 91 C13
Råda S 97 B9
Råda S 97 B10
Rada de Haro E 47 E7
Radalj SRB 158 E3
Radanovo BG 166 C5
Rǎdǎşeni RO 153 C8
Rǎdǎuţi RO 153 B7
Rǎdǎuţi-Prut RO 153 A9
Raddestorf D 17 D11
Raddi in Chianti I 66 F3
Raddusa I 59 E6
Radë AL 168 B2
Rade N 95 D13
Radeberg D 80 D5
Radebeul D 80 D5
Radeburg D 80 D5
Radeče SLO 73 D11
Radečnica PL 144 B6
Radefeld D 79 D11
Radegast D 79 C11
Rădelsbräten A 77 D11
Radenci SLO 148 C6
Radenthein A 73 C8
Rădeşti RO 152 E3
Radevormwald D 21 B8
Radibor D 80 D6
Radičí BIH 157 C7
Radicofani I 62 B1
Radicondoli I 66 F3
Radilovo BG 165 E9
Radimĕř CZ 77 C10
Radis D 79 C12
Radizelj SLO 148 D5
Radko Dimitrievo BG 167 C8
Radków PL 77 A10
Radlje ob Dravi SLO 73 C11
Radłów PL 143 F10
Radmannsdorf = Radovljica
Radnejaur S 109 E15
Radnevo BG 166 E5
Radnice CZ 76 C5
Radohova BIH 157 D7
Radoč BIH 157 F9
Rādohova BG 165 C9
Rădoieşti RO 160 E6
Radojevo SRB 150 E5
Radolfzell am Bodensee D 27 E10
Radom PL 141 H4
Radom SRB 159 D7
Radomin PL 139 D7
Radomireşti RO 160 E5
Radomirtsi BG 165 C9
Radomsko PL 143 D7
Radomyšl CZ 76 D5
Radomyśl nad Sanem PL 144 B4
Radomyśl Wielki PL 143 F11
Radonice CZ 76 B4
Radopole LV 133 C1
Rǎdoşina SK 146 E5
Radošovce SK 146 D4
Radoszyce PL 143 D9
Radoszyn PL 81 B8
Radovan RO 160 E3
Radovanu RO 161 E9
Radovets BG 167 F6

Radovici MNE 163 E6
Radoviš NMK 169 A7
Radovljica SLO 73 D9
Radovnica SRB 164 E5
Radowo Małe PL 85 C8
Radożda NMK 168 B4
Radslavice CZ 146 C5
Radstadt A 73 B7
Radstock UK 13 C10
Rǎducǎneni RO 153 D11
Raducic HR 156 D5
Raduil BG 165 E8
Radujevac SRB 159 E10
Raduň BY 137 E10
Radvanice CZ 81 E10
Radvaň nad Laborcom SK 145 E4
Radviliškis LT 134 E7
Radwanice PL 81 C9
Radymno PL 144 D6
Radzanów PL 141 G3
Radzanowo PL 139 E8
Radzice Duże PL 141 H2
Radzicz PL 85 D12
Radzieje PL 136 E4
Radziejowice PL 141 F3
Radziemice PL 143 F9
Radzików PL 81 B7
Radziłów PL 140 D6
Radzymin PL 139 F11
Radzyń Chełmiński PL 138 D6
Radzyń Podlaski PL 141 G7
Raesfeld D 17 E7
Raffadali I 58 E4
Rafford UK 3 K9
Rafína GR 175 C9
Ráfov RO 161 D8
Rafsbotn N 113 C12
Raften N 110 D9
Raftsjöhöjden S 106 D8
Ragaciems LV 134 B6
Rágama E 45 C10
Ragana LV 135 B9
Rageliai LT 135 E11
Rägelin D 83 D13
Raglan UK 13 B9
Råglanda S 91 A13
Ragösen D 79 B12
Raguhn D 79 C11
Ragunda S 107 E10
Ragusa I 59 F6
Raguva LT 135 E9
Rahden D 17 D11
Rahkio FIN 126 D7
Rahkonen FIN 123 C12
Råholt N 95 B14
Rahman RO 155 D2
Rahovec RKS 163 E10
Rahumäe EST 132 E1
Raiano I 62 C5
Raikku FIN 125 F12
Räimä FIN 124 D9
Rain D 75 E8
Rain D 75 E11
Rainbach im Mühlkreis A 77 E6
Raippo FIN 129 D9
Räisälä FIN 115 E3
Raisdorf D 83 B8
Raisio FIN 126 E7
Raiskio FIN 119 C16
Raiskio FIN 125 D12
Raismes F 19 D7
Raistakka FIN 121 B11
Rait UK 5 C10
Raitajärvi S 119 B10
Raja EST 131 D13
Raja-Jooseppi FIN 115 B4
Rajala FIN 117 C16
Rajamäki FIN 127 D12
Rajastrand S 106 B8
Rajcza PL 147 B8
Rájec CZ 77 D11
Rajec SK 147 C7
Rajecká Lesná SK 147 C7
Rajgród PL 140 C7
Rajhrad CZ 77 D11
Rajince SRB 164 E4
Raka SLO 148 E4
Rakamaz H 145 G3
Rakek SLO 73 E9
Rakalj HR 67 C9
Rakek SLO 148 E4
Rakitna SLO 73 E9
Rakitovec SLO 67 B8
Rakitovo BG 165 F9
Rakke EST 131 D12
Rakkeby DK 90 E6
Rakkestad N 95 D14
Raklitsa BG 166 D6
Rákóczifalva H 150 C5
Rákócziújfalu H 150 C5
Rakoniewice PL 81 B10
Rakoshyn UA 145 G6
Rakova Bara SRB 159 D8
Rakovica HR 156 C4
Rakovinë RKS 163 E10
Rakovitsa BG 159 E9
Rakovník CZ 76 B5
Rakovo BG 165 E10
Rakovski BG 165 E10
Rakow D 84 B4
Raków PL 143 E11
Rakvere EST 131 C12
Ralingen D 21 E7
Ralja SRB 158 D6
Rälla S 89 B11
Ram SRB 159 D7
Ramacca I 59 E6
Rämälä FIN 123 E16
Ramales de la Victoria E 40 B5
Ramalhal P 44 F2
Ramatuelle F 36 E5
Ramberg N 110 D5
Rambervillers F 26 D6
Rambin D 84 B4
Rambouillet F 24 C6
Rambrouch L 19 E12
Rambucourt F 26 C4
Ramelton IRL 7 B7
Râmeţ RO 152 E3

Ramillies B 182 D5
Ramingstein A 73 B8
Ramirás E 38 D4
Ramløse DK 87 C10
Ramma S 102 C2
Ramme DK 86 C2
Rammen D 75 E8
Rammingen D 75 D8
Râmna RO 159 C8
Ramnäs S 98 C6
Ramnes N 95 D12
Ramsjö S 103 B10
Ramsau im Zillertal A 72 B4
Ramsberg S 97 C13
Ramsbottom UK 11 D7
Ramsele S 107 C12
Ramsele S 107 D10
Ramsen CH 27 E10
Ramsey UK 15 C8
Ramsey GBM 10 C3
Ramsey St Mary's UK 15 C8
Ramsgate UK 15 E11
Ramsi EST 131 E10
Ramsing DK 86 B3
Ramsjö S 103 B10
Ramsloh (Saterland) D 17 B9
Ramstadlandet N 105 B10
Ramstein D 21 F9
Ramsund N 111 D12
Ramučiai LT 134 F3
Ramundberget S 102 A3
Ramvik S 103 A14
Râmygala LT 135 E8
Rana N 111 D13
Rånäs S 99 C10
Rånäs S 102 C5
Rånäsfoss N 95 B14
Rance B 19 D9
Randaberg N 94 E3
Randalstown UK 4 F4
Randan F 30 C3
Randaträsk S 118 B6
Randazzo I 59 D6
Rånddalen S 102 B5
Randegg A 77 F7
Randen N 101 C10
Randen N 101 E8
Randers DK 86 C6
Randersacker D 187 B8
Randijaur S 109 C17
Randonnai F 24 C4
Randsjö S 102 A6
Randsverk N 101 C10
Råne N 95 B14
Rånea S 118 C8
Rånes F 23 C11
Rångedala S 91 D13
Rangendingen D 187 E6
Rangersdorf A 73 C6
Rangsby FIN 122 E6
Rangsdorf D 80 B4
Ranhados P 44 C5
Ranilović SRB 158 E6
Ranis D 79 E10
Ranizów PL 144 C4
Ranka LV 135 B12
Rankl LV 134 C3
Rankweil A 71 C9
Rannankylä FIN 123 C15
Rannankylä FIN 123 C16
Rannapungerja EST 131 D14
Rånnäväg S 91 D13
Ránnslöv S 87 C12
Ránnö S 103 B12
Rannsundet S 102 B5
Rannu EST 131 E12
Ranovac SRB 159 E7
Ransbach-Baumbach D 21 D9
Ransberg S 92 C4
Ransby S 102 E4
Ranskill UK 11 E9
Ranst B 182 C5
Ransta S 98 C7
Ranta FIN 113 F19
Rantajärvi S 117 E11
Rantakangas FIN 123 E11
Rantasalmen asema FIN 125 F10
Rantasalmi FIN 125 F10
Ranta-Töysä FIN 123 E10
Ranten A 73 B9
Rantsila FIN 119 E15
Ranttila FIN 113 E16
Rantum D 82 A4
Rantzausminde DK 86 E7
Ranua FIN 119 C17
Ranum DK 86 B4
Rånvassbotn N 111 D13
Raon-l'Étape F 27 D6
Raossi I 69 B11
Rapakkojoki FIN 123 C17
Rapallo I 37 C10
Raphoe IRL 4 F1
Rapice PL 81 B8
Răpina EST 132 E1
Rapla EST 131 C9
Rapolano Terme I 66 F4
Rapolla I 60 B5
Rapoltu Mare RO 151 F11
Raposa P 44 F4
Raposeira P 50 E2
Rapotín CZ 77 B12
Rapoula do Côa P 45 D6
Rappenau CH 27 F10
Räpplinge S 89 B11
Rapsani GR 169 E8
Raron CH 68 A4
Raša HR 67 B9
Rasavci BIH 156 D5
Raşca RO 153 C8
Rascafría E 46 C5
Râscǎieţi MD 154 D5
Raschau D 75 A12
Râşcov MD 154 B3
Rasdorf D 78 E6
Rasen I 72 C5
Rasharkin UK 4 F4
Rašica SLO 73 E10
Rasina EST 131 E14
Râşinari RO 160 B4

Rasines E 40 B5
Rasinja HR 149 D7
Rasinkylä FIN 121 E10
Rasinvaara FIN 125 F13
Räsjö S 103 B10
Raška SRB 163 C10
Rask Mølle DK 86 D5
Raškovice CZ 146 B6
Răsmireşti RO 161 F7
Rasno SRB 163 D9
Râşnov RO 161 B6
Rasony BY 133 E5
Rasovo BG 159 F11
Raspenava CZ 81 E8
Rasquera E 42 E5
Rast RO 160 F2
Raštani BIH 157 F8
Rastatt D 27 C9
Rastdorf D 17 C9
Rasteby N 111 D14
Råsted DK 86 B5
Rastede D 17 B10
Rastenfeld A 77 E8
Raštević HR 156 D4
Rasti FIN 117 C14
Rasti FIN 125 C12
Rastinkylä FIN 125 C13
Răstoliţa RO 152 D5
Rastošnica BIH 157 C10
Rastovac MNE 157 D9
Rastow D 83 D10
Råstrand S 107 A14
Răsuceni RO 161 F7
Rasueros E 45 B10
Rasvåg N 94 F5
Raszków PL 142 C4
Raszyn PL 141 F3
Rataje SRB 164 C3
Rätan S 102 B3
Ratan S 122 C5
Rätansbyn S 102 B3
Ratasjärvi FIN 117 E11
Ratasvuoma FIN 119 B12
Ratekau D 83 C9
Rateče SLO 73 C8
Rătesti RO 160 D6
Rathangan IRL 7 F9
Ráth Bhoth IRL 4 F1
Ráth Caola IRL 8 C5
Rathconrath IRL 7 E7
Rathcoole IRL 7 F10
Ráth Cúil IRL 7 F10
Ráth Domhnaigh IRL 9 C7
Rathdowney IRL 9 C7
Ráth Droma IRL 9 C10
Rathdrum IRL 9 C10
Rathenow D 79 A11
Rathfriland UK 7 D10
Rathgormuck IRL 9 D7
Ráth Iomgháin IRL 7 F9
Rathkeale IRL 8 C5
Rathkeevin IRL 9 D7
Rathlee IRL 6 D4
Rathluirc IRL 8 D5
Ráth Mealtain IRL 7 B7
Rathmolyon IRL 7 F9
Rathmore IRL 8 D4
Rathmullan IRL 4 E1
Ráth Naoi IRL 9 C10
Rathnew IRL 7 G10
Rathnure IRL 9 D9
Ratho UK 5 D10
Rathowen IRL 7 E7
Rathstock D 81 A7
Rathumney IRL 9 D9
Rathvilly IRL 9 C9
Ratiboř CZ 146 C5
Ratić BIH 158 D3
Ratingen D 17 F7
Ratíškovice CZ 77 E12
Rátka H 145 G3
Ratkovac RKS 163 E10
Ratkovo SRB 158 C3
Ratnieki LV 133 D2
Ratoath IRL 7 E10
Rattelsdorf D 75 B8
Ratten A 148 B5
Rattenberg A 72 B4
Rattosjärvi FIN 117 E13
Rattray UK 5 B10
Rättsel S 118 C3
Rättvik S 103 E9
Ratvika N 104 D8
Ratzeburg D 83 C9
Rätzlingen D 79 B9
Raubach D 185 C8
Rauda LV 135 C12
Raudāna N 90 B4
Raudanjoki FIN 117 D16
Raudaskylä FIN 123 B13
Raudeberg N 100 C2
Râu de Mori RO 159 C10
Raudėnai LT 134 D5
Raudondvaris LT 137 D8
Raudonė LT 136 C7
Rauen D 80 B6
Raufoss N 101 E13
Rauha FIN 129 C10
Rauhala FIN 117 C14
Raulhac F 29 F11
Rauma FIN 126 C6
Rauna LV 135 B11
Raunds UK 15 C7
Rauris A 73 B6
Råu Sadului RO 160 B4
Rauschenberg D 21 C11
Raussila FIN 128 D6
Rautalampi FIN 124 E7
Rautaperä FIN 114 E4
Rautas S 111 D18
Rautavaara FIN 125 C11
Rautenkranz D 75 B11
Rautila FIN 119 F11
Rautila FIN 125 E9
Rautio FIN 123 C12
Rautionmäki FIN 123 E16
Rautjärvi asema FIN 129 C11
Rautjärvi FIN 129 C11
Rautuskylä FIN 117 C13
Rauvanniemi FIN 129 B11
Rauzan F 28 F5
Ravanica BG 167 E9
Ravanusa I 58 F5
Rava-Rus'ka UA 144 C3
Ravasd H 149 A9
Ravča HR 157 F7
Ravda BG 167 D9
Ravels B 16 F3

Ravelsbach A 77 E9
Rävemåla S 89 B8
Ravenglass UK 10 C5
Ravenna I 66 D5
Ravensburg D 71 A9
Ravenstein NL 16 E5
Ravières F 25 E11
Ravijoki FIN 128 D8
Ravik N 108 B7
Rävlanda S 91 D12
Ravna Dubrava SRB 164 C5
Ravna Gora HR 67 B10
Ravna Reka SRB 159 E8
Ravne SLO 73 D11
Ravne na Koroškem SLO 73 C10
Ravnets BG 167 D8
Ravni BIH 157 F8
Ravnište SRB 164 C3
Ravnje SRB 158 D3
Ravnkilde DK 86 B5
Ravno BIH 157 E7
Ravno BIH 162 D4
Ravnogor BG 165 F9
Ravno Selo SRB 158 C4
Ravnshøj DK 90 E7
Ravnstrup DK 86 C4
Rävsön S 103 A15
Ravsted DK 86 E4
Rawa Mazowiecka PL 141 G2
Rawicz PL 81 C11
Rawmarsh UK 11 E7
Rawtenstall UK 11 D7
Raykovo BG 171 A7
Rayleigh UK 15 D10
Rayol-Canadel-sur-Mer F 36 E4
Räyrinki FIN 123 D11
Ražana SRB 158 E4
Ražanac HR 156 D3
Ražanj SRB 159 F8
Războieni RO 153 C9
Razboj BIH 157 B7
Razdelna BG 167 C9
Razdol BG 169 A9
Razdrto SLO 73 E8
Razès F 29 C8
Razgrad BG 160 F2
Razgrad BG 167 B7
Ražljevo BIH 157 C10
Razlog BG 165 F7
Razlovci NMK 165 F6
Ražňany SK 145 E3
Răzvad RO 161 D6
Reading UK 14 E7
Reaghstown IRL 7 E9
Real P 38 F3
Réalmont F 33 C10
Realmonte I 58 E3
Réalville F 33 B8
Rear Cross IRL 8 C6
Réaup F 33 B6
Reay UK 3 H9
Rebais F 25 C9
Rébénacq F 32 D5
Rebild DK 86 B5
Rebollosa de Jadraque E 47 B7
Reboly RUS 125 C15
Rebordelo E 38 B3
Rebordelo P 38 E5
Rebra RO 152 C4
Rebricea RO 153 D11
Rebrişoara RO 152 C4
Rebrovo BG 165 D7
Rebŭrkovo BG 165 D8
Reca SK 146 E4
Reçan RKS 163 E10
Recanati I 67 E8
Recaş RO 151 F8
Recco I 37 C10
Recea MD 153 B11
Recea MD 154 C3
Recea RO 151 B12
Recea RO 152 F5
Recea RO 160 D6
Recea-Cristur RO 152 C3
Recess IRL 6 F3
Recey-sur-Ource F 25 E12
Réchicourt-le-Château F 27 C6
Rechlin D 83 D13
Rechnitz A 149 B6
Recht B 20 D6
Rechtenbach D 74 C6
Reci RO 153 F7
Rečica SLO 73 D11
Rečice BIH 157 F8
Recke D 17 D9
Reckingen CH 70 E6
Recklinghausen D 17 E8
Recoaro Terme I 69 B11
Recoubeau-Jansac F 35 A9
Recsk H 147 B10
Recuerda E 40 F6
Recz PL 85 D9
Rjczno PL 141 H1
Reda PL 138 A5
Redalen N 101 E13
Redange L 20 E5
Redcar UK 11 B9
Redcastle IRL 4 E2
Redcross IRL 9 C10
Reddelich D 83 B11
Redditch UK 13 A11
Réde H 149 B9
Redea RO 160 E4
Redefin D 83 D10
Redhill UK 15 E8
Rédics H 149 C6
Réding F 27 C7
Redinha P 44 D3
Rediu RO 153 C11
Rediu RO 153 D8
Rediu RO 153 F11
Rediul Mare MD 153 A11
Rednitzhembach D 75 D9
Redon F 23 E7
Redondela E 38 D2
Redondelo P 38 E4
Redondo P 50 B4
Redován E 56 E3
Red Point UK 2 K5
Redruth UK 12 E4
Redsted DK 86 B3
Reduzum NL 16 B5
Rjdzikowo PL 85 B12
Rjdziny PL 143 E7
Reen IRL 8 E3
Reens IRL 8 C5
Reepham UK 15 B11
Rees D 16 E6
Reeßum D 17 B12

Reetz D 79 B11
Reetz D 83 D11
Reftele S 87 A13
Regalbuto I 59 D6
Regen D 76 E4
Regensburg D 75 D11
Regensdorf CH 27 F9
Regenstauf D 75 D11
Reggello I 66 E4
Reggio di Calabria I 59 C8
Reggiolo I 66 C2
Reggio nell'Emilia I 66 C2
Reghin RO 152 D5
Reghiu RO 153 F9
Regna S 92 B7
Regnitzlosau D 75 B11
Régny F 30 D5
Regöly H 149 C10
Regozero RUS 121 D17
Regstrup DK 87 D9
Reguengo E 38 D2
Reguengos de Monsaraz
P 50 C4
Rehau D 75 B11
Rehburg (Rehburg-Loccum)
D 17 D12
Rehden D 17 C10
Rehling D 75 F8
Rehlingen-Siersburg D 21 F7
Řehlovice CZ 80 E5
Rehmsdorf D 79 D11
Rehna D 83 C10
Rehula FIN 129 C9
Reibitz D 79 C11
Reichelsheim (Odenwald)
D 187 B6
Reichenau an der Rax A 148 A5
Reichenbach CH 70 D5
Reichenbach D 79 E11
Reichenbach D 187 B6
Reichenberg D 74 C6
Reichenfels A 73 B10
Reichenthal A 76 E6
Reichertsheim D 75 F11
Reichia GR 178 B5
Reichling D 71 B11
Reichmannsdorf D 75 A9
Reicholzheim D 74 C6
Reichraming A 73 A9
Reichshoffen F 27 C8
Reichstett F 186 D4
Reiden CH 27 F8
Reigada P 45 C7
Reigate UK 15 E8
Reignac F 28 E4
Reignier F 31 C9
Reil D 21 D8
Reilingen D 187 C6
Reillanne F 35 C10
Reillo E 47 E9
Reims F 19 F9
Reina E 51 C8
Reinach CH 27 F8
Reinach CH 27 F7
Reinbek D 83 C8
Reinberg D 84 B4
Reine N 110 E5
Reinfeld (Holstein) D 83 C8
Reinheim D 21 E11
Reinosa E 40 C3
Reinøysund N 114 D8
Reinsfeld D 21 E7
Reinskard N 112 D4
Reinskloster N 104 D7
Reinstad N 111 C10
Reinsvik N 104 E3
Reinsvoll N 101 E13
Reipa N 108 C5
Reisbach D 75 E12
Reischach D 75 F12
Reisjärvi FIN 123 C13
Reiskirchen D 21 C11
Reiss UK 3 J10
Reitan N 100 B8
Reitan N 101 A14
Reitano I 58 D5
Reith bei Seefeld A 72 B3
Reit im Winkl D 72 A5
Reittiö FIN 125 D7
Reivyčiai LT 134 D4
Rejmyre S 92 B7
Rejowiec PL 141 H8
Rejsby DK 86 E3
Reka HR 149 D7
Rekava LV 133 B3
Rekavice BIH 157 C7
Reken D 17 E8
Rekijoki FIN 127 E9
Rekkenbach D 21 D8
Rekken NL 17 D7
Reklynets' UA 144 C9
Rekovac SRB 159 F7
Rekowo PL 85 B12
Rekvik N 111 A15
Rekyva LT 134 E6
Reliquias P 50 D3
Relleu E 56 D4
Rellingen D 83 C7
Rém H 150 E3
Remagen D 21 C8
Rémalard F 24 D4
Rembercourt-Sommaisne
F 26 C3
Remda D 79 E9
Remels (Uplengen) D 17 B9
Remennikovo RUS 133 C5
Remeskylä FIN 123 C16
Remetea RO 151 D9
Remetea RO 152 D6
Remetea Chioarului RO 152 B3
Remetea Mare RO 151 F7
Remeţi RO 145 H8
Remetinec HR 149 D6
Remetské Hámre SK 145 F5
Remich L 20 F6
Remicourt B 183 D6
Remiremont F 26 D6
Remmam S 107 D14
Remmen S 102 B8
Remmet S 102 B7
Remnes N 108 E4
Remolinos E 41 E9
Remouchamps B 183 D7
Remoulins F 35 C8
Remplin D 83 C13
Remptendorf D 75 A10
Remscheid D 21 B8
Remte LV 134 C5
Remungol F 22 E6
Rémuzat F 35 B9
Rena N 101 D14
Renaison F 30 C4

Renålandet S 106 D8
Renazé F 23 E9
Rencēni LV 131 F10
Renchen D 27 C9
Renda UV 134 B4
Rende I 60 E6
Rendsburg D 82 B7
Renedo E 39 E10
Renedo I 66 B4
Renedo de la Vega E 39 D10
Renens CH 31 B10
Renfrew UK 5 D8
Renginio GR 175 B6
Rengsdorf D 21 C8
Rengsjö S 103 D12
Renholmen S 118 D6
Reni UA 155 C2
Renko FIN 127 D11
Renkomäki FIN 127 D14
Renkum NL 183 B7
Renndal N 104 E5
Rennerod D 21 C10
Rennertshofen D 75 E9
Rennes F 23 D8
Rennes-les-Bains F 33 E10
Renningen D 27 C11
Rennweg A 73 B8
Renon I 72 C3
Renov D 83 C10
Rens DK 86 F4
Rensjön S 111 D18
Reńska Wieś PL 142 F5
Renström S 118 E4
Renswoude NL 183 A7
Rentina GR 174 A4
Rentjärn S 107 A15
Rentweinsdorf D 75 B8
Renwez F 19 E10
Renzow D 83 C10
Repbäcken S 97 A13
Répcelak H 149 B8
Repedea RO 152 B5
Repino RUS 129 E12
Repki PL 141 F6
Replot FIN 122 D6
Repojoki FIN 117 B15
Repolka RUS 132 C6
Reposaari FIN 126 B5
Reppen D 108 C6
Reppenstedt D 83 D8
Reps AL 163 F9
Repton UK 11 F8
Repvåg N 113 B16
Requejo E 39 E6
Requena E 47 F10
Réquista F 33 B11
Rerik D 83 B11
Resana I 72 E4
Resarö S 99 D10
Resavica SRB 159 E8
Resele S 107 E12
Resen BG 166 C5
Resenbro DK 86 C5
Resende P 44 B5
Rešetari HR 157 B7
Reşiţa RO 159 C8
Resko PL 85 C8
Resna MNE 163 E6
Resolven UK 13 B7
Respenda de la Peña E 39 C10
Resse (Wedemark) D 78 A6
Ressons-sur-Matz F 18 E6
Restelicë RKS 163 F10
Restinga MA 53 E6
Reston UK 5 D12
Resuttano I 58 D5
Retamal E 51 B8
Retford UK 11 E10
Rethel F 19 E9
Rethem (Aller) D 17 C12
Rethymno GR 178 E7
Retie B 16 F4
Retiers F 23 E9
Retje SLO 73 E10
Retortillo E 45 C8
Retortillo de Soria E 40 F6
Retournac F 30 E5
Rétság H 147 F8
Retuerta del Bullaque E 46 F4
Retunen FIN 125 E11
Retz A 77 E9
Reuden D 79 B11
Reuilly F 24 F7
Reureith D 75 B8
Reus E 42 E6
Reusel NL 16 F4
Reute D 76 F3
Reute D 27 D8
Reuţel MD 153 B11
Reutlingen D 27 D11
Reutte A 71 C11
Reutuaapa FIN 119 B15
Reuver NL 16 F6
Revel F 33 D10
Revello I 37 B6
Revest-du-Bion F 35 B10
Révfülöp H 149 C9
Revholmen N 91 A8
Reviga RO 161 D10
Revigny-sur-Ornain F 26 C2
Revilla de Collazos E 40 C3
Revilla del Campo E 40 D4
Revin F 19 E10
Revine-Lago I 72 E5
Řevnice CZ 76 C6
Řevničov CZ 76 B5
Revò I 72 D3
Revonlahti FIN 119 E13
Revsnes N 100 D6
Revsnes N 111 C11
Revsund S 103 A9
Revúca SK 147 D10
Rewal PL 85 B8
Rexbo S 103 E9
Reyrieux F 30 D6
Rezé F 23 F8
Rēzekne LV 133 C2
Rezi H 149 C8
Rezina MD 154 B3
Rēzna LV 133 D2
Rezovo BG 167 F10
Rezzato I 66 A1
Rezzo I 37 C7
Rezzoaglio I 37 B10
Rhade D 17 D7
Rhauen D 21 E8
Rhayader UK 13 A7

Rheda-Wiedenbrück D 17 E10
Rhede D 17 E7
Rhede (Ems) D 17 B8
Rheden NL 183 A8
Rheinau D 27 C8
Rheinbach D 21 C7
Rheinberg D 17 E7
Rheinböllen D 185 E8
Rheinbreitbach D 21 C8
Rheine D 17 D8
Rheinfelden (Baden) D 27 E8
Rheinsberg D 84 D3
Rheinstetten D 27 C9
Rheinzabern D 187 C5
Rhêmes-Notre-Dame I 31 D11
Rhêmes-St-Georges I 31 D11
Rhenen NL 16 E5
Rhens D 185 D8
Rhiconich UK 2 J7
Rhinau F 27 D8
Rhinow D 83 E12
Rho I 69 B7
Rhode IRL 7 F8
Rhoden (Diemelstadt) D 17 F12
Rhoon NL 182 B4
Rhoose UK 13 C8
Rhoslanerchrugog UK 10 E4
Rhôs-on-Sea UK 10 E4
Rhossili UK 12 B6
Rhuddlan UK 10 E4
Rhydaman UK 12 B7
Rhyl UK 10 E5
Rhymney UK 13 B8
Riace I 59 C9
Riachos P 44 F3
Riaillé F 23 E9
Rialp E 33 F8
Riaño E 39 C10
Riano I 62 C3
Rians F 35 C10
Riantec F 22 E5
Rianxo E 38 C2
Riaz CH 31 B11
Riba E 40 B4
Ribadavia E 38 D3
Ribadelago E 39 D6
Riba de Mouro P 38 D3
Ribadeo E 38 A5
Riba de Saelices E 47 C8
Ribadesella E 39 B9
Ribaforada E 41 D8
Ribafrecha E 32 F1
Ribarci SRB 164 C3
Ribare SRB 164 C2
Ribari SRB 158 D3
Ribaritsa BG 165 D9
Riba-roja d'Ebre E 42 E4
Riba-roja de Turia E 48 E3
Ribbásen S 102 D7
Ribchester UK 10 D6
Ribe DK 86 E3
Ribeauvillé F 27 D7
Ribécourt-Dreslincourt F 18 E6
Ribeira E 38 C2
Ribeira P 38 E3
Ribeira de Pena P 38 E4
Ribemont F 19 E7
Ribera I 58 E3
Ribérac F 29 E6
Ribera del Fresno E 51 B7
Ribesalbes E 48 D4
Ribes de Freser E 33 F10
Ribiţa RO 151 E10
Ribnica BIH 157 D9
Ribnica SLO 73 E10
Ribnica SLO 148 C4
Ribnica SRB 158 F4
Ribnik HR 148 E4
Ribniţa MD 154 B4
Ribnitz-Damgarten D 83 B12
Ribnovo BG 165 F8
Ribota E 40 F5
Ricadi I 59 B8
Říčany CZ 77 C7
Říčany CZ 77 D10
Riccia I 63 E7
Riccio I 66 F5
Riccione I 66 D6
Riccò del Golfo di Spezia I 69 E8
Richardménil F 26 C5
Richelieu F 24 F4
Richhill UK 7 D9
Richka UA 152 A5
Richmond UK 11 C8
Richvald SK 145 E3
Rickeby S 103 D11
Rickenbach D 27 E8
Rickinghall UK 15 C10
Rickling D 83 B8
Rickmansworth UK 15 D8
Ricla E 41 E9
Ricse H 145 G4
Ridala EST 131 D8
Riddarhyttan S 97 C14
Ridderkerk NL 16 E3
Riddes CH 31 C11
Rīdzica SRB 150 F3
Riebiņi LV 135 D13
Riebnesluspen S 109 D13
Riec-sur-Belon F 22 E4
Ried CH 68 A5
Riede D 17 C11
Riedenburg D 75 E10
Rieder D 79 C9
Ried im Innkreis A 76 F4
Ried im Oberinntal A 71 C11
Ried im Zillertal A 72 B4
Ried in der Riedmark A 77 F7
Riedlingen D 71 A8
Riegelsberg D 21 F7
Riegersburg A 148 B5
Riego de la Vega E 39 D8
Riehe (Suthfeld) D 78 B5
Riehen CH 27 E8
Rielasingen-Worblingen
D 27 E10
Riello E 39 C8
Rielves E 46 E4
Riemst B 19 C12
Rieneck D 74 B6
Rieni RO 151 D9
Rieponlahti FIN 124 D7
Riepsdorf D 83 B9
Riesa D 80 D4
Rieseby D 83 A7
Riesi I 58 E5
Riestedt D 79 D9
Rietavas LT 134 E3
Rietberg D 17 E10
Rieth D 84 C6

Riethoven NL 183 C6
Rieti I 62 C3
Rietschen D 81 D7
Rieumes F 33 D8
Rieupeyroux F 33 B10
Rieutort-de-Randon F 34 A5
Rieux F 23 E7
Rieux F 33 D8
Riez F 36 D4
Rifiano I 72 C3
Rīga LV 135 C8
Rigaio GR 169 F8
Rigaud F 36 D5
Riggisberg CH 31 B11
Rignac F 33 B10
Rignano Flaminio I 62 C2
Rignano Garganico I 63 D9
Rignano sull'Arno I 66 E3
Rigny-le-Ferron F 25 D10
Rigny-sur-Arroux F 30 B5
Rigny-Ussé F 24 F3
Rigside UK 5 D9
Riihiniemi FIN 126 C5
Riihimäki FIN 127 D12
Riihivaara FIN 125 C14
Riikonkumpu FIN 117 C14
Riipi FIN 117 D16
Riippi FIN 122 F7
Riisipere EST 131 C8
Riistavesi FIN 125 D10
Riitiala FIN 127 B8
Riječa BIH 157 E10
Rijeka BIH 157 E10
Rijeka BIH 157 F11
Rijeka HR 67 B9
Rijeka Crnojevića MNE 163 E7
Rijen NL 16 E3
Rijkevorsel B 16 F3
Rijnsburg NL 16 D2
Rijsbergen NL 16 E3
Rijsel F 19 C7
Rijssen NL 17 D7
Rijswijk NL 16 D2
Rikava LV 133 C2
Riksgränsen S 111 D15
Rila BG 165 E7
Rilhac-Rancon F 29 D8
Rilland NL 182 C4
Rillé F 23 F12
Rillieux-la-Pape F 30 D6
Rillo E 42 F2
Rillo de Gallo E 47 C9
Rima S 103 C10
Rimavská Baňa SK 147 D9
Rimavská Seč SK 145 G1
Rimavská Sobota SK 147 E10
Rimavské Janovce SK 147 E10
Rimbach D 76 D3
Rimbach D 187 B6
Rimbo S 99 C10
Rimetea RO 152 E3
Rimforsa S 92 C7
Rimini I 66 D6
Rimjokk S 118 B4
Rimmilä FIN 127 D11
Rimóc H 147 E9
Rimogne F 184 E2
Rimont F 33 E8
Rimsbo S 103 D10
Rimše LT 135 E12
Rimšēnai LT 135 F12
Rimske Toplice SLO 73 D11
Rimsting D 72 A5
Rinchnach D 76 E4
Rincón de la Victoria E 53 C8
Rincón de Soto E 41 D8
Rinda LV 134 A3
Rindal N 104 E6
Rindsholm DK 86 C4
Rineia GR 176 E5
Rinella I 59 B6
Ringarum S 93 C8
Ringaudai LT 137 D8
Ringe D 17 C7
Ringe DK 86 E6
Ringebu N 101 C12
Ringelia N 101 E12
Ringen N 95 B12
Ringford UK 5 F8
Ringhals S 87 A10
Ringkøbing DK 86 C2
Ringleben D 79 D9
Ringsend UK 4 E3
Ringsta S 106 E7
Ringsted DK 87 E9
Ringvattnet S 106 C8
Ringwood UK 13 D11
Rinkaby S 88 D6
Rinkabyholm S 89 B10
Rinkenæs DK 86 F5
Rinkilä FIN 129 B10
Rinloan UK 5 A10
Rinn UK 5 B9
Rinneen IRL 8 C4
Rinøyvåg N 111 D10
Rintala FIN 123 D10
Rinteln D 17 D12
Rio GR 174 C4
Rio Caldo P 38 E3
Riebiņi LV 135 D13
Río de Mel P 44 C6
Rio de Moinhos P 50 B4
Rio de Moinhos P 50 B3
Rio de Onor P 39 E6
Rio di Pusteria I 72 C4
Riofrío E 46 C3
Riofrío de Aliste E 39 E7
Ríogordo E 53 C7
Rioja E 55 F8
Riola Sardo I 64 D2
Riolobos E 45 D8
Riolo Terme I 66 D4
Riols F 34 C4
Riom F 30 D3
Riomaggiore I 69 E8
Rio Maior P 44 F3
Rio Marina I 65 B3
Riom-ès-Montagnes F 29 E11
Rion-des-Landes F 32 C4
Rionegro del Puente E 39 D7
Rio nell'Elba I 65 B3
Rionero in Vulture I 60 B5
Rionero Sannitico I 63 D6
Rions F 32 A5
Riorges F 30 D5
Ríos E 38 E5
Rioseco de Tapia E 39 C8
Rio Tinto P 44 B3
Rio Torto P 38 E5
Rioz F 26 F5
Ripač BIH 156 C5

Ripacandida I 60 B5
Ripalimosano I 63 D7
Ripanj SRB 158 D6
Riparbella I 66 F2
Ripatransone I 62 B5
Ripe I 67 E7
Ripi I 62 D4
Ripiceni RO 153 B10
Ripley UK 11 E9
Ripley UK 11 E8
Ripoll E 43 C8
Ripon UK 11 C8
Riposto I 59 D7
Rips NL 183 B7
Riquewihr F 27 D7
Risan MNE 163 E6
Risarven S 103 C10
Risbäck S 106 B9
Risberg S 102 D6
Risca UK 13 B8
Rişca RO 151 D11
Rîşcani MD 153 B11
Riscle F 32 C5
Risdal N 90 B3
Risede S 106 C8
Risinge S 92 B7
Risisilden S 107 B16
Risnabben S 118 D3
Risnes N 94 E5
Risør N 90 B5
Risøyhamn N 111 C10
Rissa N 104 D7
Rissnaben S 118 D4
Risnes N 94 E5
Riste FIN 126 C5
Risteli FIN 125 B12
Risti EST 131 D8
Ristiina FIN 128 B7
Ristijärvi FIN 121 F11
Ristilä FIN 121 D11
Ristilampi FIN 117 E17
Ristinen FIN 124 D7
Ristinkylä FIN 125 E12
Ristioja FIN 117 E13
Ristonmännikkö FIN 117 D16
Riströsk S 107 B12
Risudden S 119 B11
Risum-Lindholm D 82 A5
Rĩtausmas LV 135 D8
Rite LV 135 D10
Ríteri LV 135 C11
Ritini GR 169 D7
Ritola FIN 123 E11
Ritterhude D 17 B11
Rittersdorf D 185 D5
Rittersgrün D 75 B12
Riudarenes E 43 D9
Riudecols E 42 E5
Riudoms E 42 E6
Riutta FIN 123 C12
Riutula FIN 117 B18
Rīva LV 134 C2
Riva del Garda I 69 B10
Riva di Solto I 69 B9
Riva di Tures I 72 C5
Rivanazzano I 37 B10
Rivarolo Canavese I 68 C4
Rivarolo Mantovana I 66 B1
Rivas-Vaciamadrid E 46 D6
Rive-de-Gier F 30 D6
Rivedoux-Plage F 28 C3
Rivello I 60 C5
Riverchapel IRL 9 C10
Rivergaro I 37 B10
Rivero E 40 B3
Riverstown IRL 7 F7
Riverstown IRL 8 E6
Rivery F 18 E5
Rives F 31 E8
Rivesaltes F 34 E4
Rivière-sur-Tarn F 34 B5
Rivignano I 73 E7
Rivinperä FIN 119 F16
Rivodutri I 62 B3
Rivoli I 68 C4
Rivolta d'Adda I 69 C8
Rixensart B 19 C10
Rixheim F 27 E7
Rixö S 91 C9
Riza GR 175 C6
Rizes GR 174 D5
Rizomata GR 169 D7
Rizomylos GR 169 F8
Rizziconi I 59 C8
Rjånes N 100 B3
Rjukan N 95 C9
Rø DK 89 E7
Ro I 66 C4
Rō S 103 A14
Roa E 40 E4
Roa N 95 B13
Roade UK 15 C7
Roadside UK 3 H10
Roadside of Kinneff UK 5 B12
Roager DK 86 E3
Roaillan F 32 B5
Roald N 100 A4
Roan N 104 C7
Roanne F 30 C5
Roata de Jos RO 161 E7
Roath UK 13 C8
Röbäck S 122 C4
Robănești RO 160 E4
Robbio I 68 C6
Robeasca RO 161 C10
Röbel D 83 D13
Robella I 37 B8
Robert-Espagne F 26 C3
Roberton UK 5 D9
Roberton UK 5 E11
Robertsfors S 118 F5
Robežnieki LV 133 E3
Robiac-Rochessadoule F 35 B7
Robilante I 37 C6
Robin Hood's Bay UK 11 C10
Robion F 35 C9
Robledo E 55 B8
Robledo de Chavela E 46 C4
Robledo del Mazo E 45 E10
Robledollano E 45 E9
Robles de la Valcueva E 39 C9
Robliza de Cojos E 45 C9
Robšce N 101 E11
Robregordo E 46 B5
Robres E 41 E11
Robres del Castillo E 32 F1
Roč HR 67 B8
Rocafort de Queralt E 43 E6
Rocamadour F 29 F9

Roca Vecchia I 61 C10
Roccabianca I 66 B1
Roccadaspide I 60 C4
Rocca d'Evandro I 60 A1
Rocca di Cambio I 62 C5
Rocca di Mezzo I 62 C5
Rocca di Neto I 61 E7
Rocca di Papa I 62 D3
Roccafranca I 69 C8
Roccagloriosa I 60 C4
Roccagorga I 62 D4
Rocca Grimalda I 37 B9
Rocca Imperiale I 61 C7
Roccalbegna I 65 B5
Roccalumera I 59 D7
Roccamandolfi I 63 D6
Rocca Massima I 62 D4
Roccamena I 58 D3
Roccamonfina I 60 A1
Roccamontepiano I 62 C6
Roccanova I 60 C6
Roccapalumba I 58 D4
Rocca Pia I 62 D5
Roccaraso I 62 D6
Rocca San Casciano I 66 D4
Rocca San Giovanni I 63 C6
Roccasecca I 62 D5
Roccasecca dei Volsci I 62 D4
Rocca Sinibalda I 62 C3
Roccastrada I 65 A4
Roccavione I 37 C6
Roccella Ionica I 59 C9
Rocchetta Sant'Antonio I 60 A4
Rochdale UK 11 D7
Roche UK 12 E5
Rochechouart F 29 D7
Rochefort B 19 D11
Rochefort F 28 D4
Rochefort-en-Terre F 23 E7
Rochefort-Montagne F 30 D2
Rochefort-sur-Nenon E 26 F4
Rochehaut B 184 E3
Roche-la-Molière F 30 E5
Rochemaure F 35 A8
Roches-Bettaincourt F 26 D3
Rocheservière F 28 B2
Rochester UK 5 E12
Rochester UK 15 E10
Rochetaillée F 26 E3
Rochford UK 15 D10
Rochfortbridge IRL 7 F8
Rochin F 19 C7
Rochlitz D 79 D12
Rociana del Condado E 51 E6
Ročinj SLO 73 D8
Rociu RO 160 D6
Rockanje NL 182 B4
Rockchapel IRL 8 D4
Rockcliffe UK 5 F9
Rockcorry IRL 7 D8
Rockenhausen D 21 E9
Rockesholm S 97 C12
Rockhammar S 97 C13
Rockhill IRL 8 D5
Rockingham UK 11 F10
Rockmills IRL 8 D6
Rockneby S 89 B10
Röcknitz D 79 D12
Rocourt-St-Martin F 25 B9
Rocroi F 19 E10
Roda de Bara E 43 E6
Roda de Ter E 43 D8
Rodalben D 21 F9
Rödalsel S 118 F3
Rodberg N 95 B9
Rødbergshamn N 111 B15
Rødby DK 83 A10
Rødbyhavn DK 83 A10
Rødding DK 86 B3
Rødding DK 86 E4
Rødding DK 86 E4
Rødeby S 89 C9
Rodeiro E 38 C4
Rødekro DK 86 E4
Rodel UK 2 K3
Rodellar E 32 F5
Rodelle F 34 A4
Roden NL 17 B6
Ródenas E 47 C10
Rodenkirchen (Stadland)
D 17 B10
Ródental D 75 B9
Rodewald D 82 E6
Rodewisch D 75 A11
Rodez F 33 B11
Rodi Garganico I 63 D9
Roding D 75 D12
Rodingträsk S 107 C14
Ródjebro S 98 B9
Rødkærsbro DK 86 C4
Rodleben D 79 C11
Rødlia N 108 E7
Rödmyra S 103 C10
Rodna RO 152 C5
Rododafni GR 174 C5
Rodolivos GR 170 C5
Rödön S 105 E16
Rodopoli GR 169 B9
Rodos GR 181 D8
Rødovre DK 87 D10
Rødsand N 111 B13
Rødsand N 114 C8
Rødseidet N 105 B11
Rödvattnet S 107 D13
Rødvig DK 87 F10
Roela EST 131 C13
Roermond NL 20 B5
Roeselare B 19 C7
Roeşti RO 160 D4
Roetgen D 20 C6
Röfors S 92 B5
Rofrano I 60 C4
Rogač HR 156 E5
Rogačica SRB 158 E4
Rogaška Slatina SLO 148 D5
Rogaszyce PL 142 D4
Rogatec SLO 148 D5
Rogatica BIH 157 E11
Rogătz D 79 B10
Roggel NL 16 F5
Roggenburg D 75 F7
Roggendorf D 83 C10
Roggentin D 84 D4
Roggiano Gravina I 60 D6
Roghudi I 59 D8
Rogienice Wielkie PL 139 D13
Rogil P 50 E2
Rogliano F 37 F10
Rogliano I 61 E6
Rognac F 35 D9

Rognan N 108 B9
Rognes F 35 C9
Rognes N 104 E8
Rognonas F 35 C8
Rogova RO 159 E10
Rogovë RKS 163 E10
Rogovka LV 133 C2
Rogów PL 141 G1
Rogowo PL 138 E4
Rogowo PL 139 E7
Rogozen BG 165 B8
Rogoznica HR 156 E4
Rogoźnica PL 81 D10
Rogoźniczka PL 141 F7
Rogoźno PL 85 E12
Rogslösa S 92 C5
Rogsta S 102 A8
Rogsta S 103 C13
Roguszyn PL 139 F12
Rohan F 22 D6
Röhlingen D 75 E7
Rohlsdorf D 83 D11
Rohlsdorf D 83 D12
Rohod H 145 G5
Rohovce SK 146 E4
Rohožník SK 77 F12
Rohr D 79 E8
Rohrau A 77 F11
Rohrbach D 75 E10
Rohrbach in Oberösterreich A 76 E5
Rohrbach-lès-Bitche F 27 B7
Rohrberg D 83 E10
Rohr in Niederbayern D 75 E10
Röhrmoos D 75 F9
Röhrnbach D 76 E5
Rohrsen D 17 C12
Rohuküla EST 130 D6
Roiffieux F 30 E6
Roisel F 19 E7
Roismala FIN 127 C8
Roivainen FIN 115 B2
Roiz E 40 B3
Roja LV 134 A5
Rojales E 56 E3
Röjan S 102 B7
Röjdåfors S 97 B8
Rojewo PL 138 E5
Rökå S 107 A15
Rokai LT 137 D8
Rokiciny PL 143 C8
Rokietnica PL 81 A11
Rokietnica PL 144 D6
Rokiškis LT 135 E11
Rokitno PL 141 F8
Rokksøy N 111 C10
Røkkum N 101 A8
Røkland N 108 C9
Roklum D 79 B8
Roknäs S 118 D6
Rokycany CZ 76 C5
Rokytnice CZ 146 C4
Rokytnice v Orlických Horách CZ 77 B10
Rolampont F 26 E3
Rold DK 86 B5
Røldal N 94 C5
Rolde NL 17 C7
Rolfs S 119 C10
Rolfstorp S 87 A10
Rollag N 95 B10
Rollán E 45 C9
Rolle CH 31 C9
Rolsted DK 86 E7
Rolvåg N 108 D3
Rolvsnes N 94 C2
Rolvsøy N 95 D14
Rom D 83 D11
Rom F 29 C6
Roma I 62 D3
Roma RO 153 B9
Roma S 93 D12
Romagnano Sesia I 68 B5
Romagné F 23 D9
Romainmôtier CH 31 B9
Romakkajärvi FIN 117 E13
Roman BG 165 C8
Roman RO 153 D9
Romana I 64 C2
Românașl RO 151 C11
Românești RO 153 B10
Români RO 153 D9
Romanija BIH 157 E10
Romano d'Ezzelino I 72 E4
Romano di Lombardia I 69 B8
Romanones E 47 C7
Romanovce NMK 164 E4
Romanovo RUS 136 D1
Romanowo PL 85 E11
Romanshorn CH 27 E11
Romans-sur-Isère F 31 E7
Romanu RO 155 C1
Romazy F 23 D9
Rombas F 20 F6
Rombiolo I 59 B9
Romeira P 44 F3
Rometta I 59 C7
Romeu P 38 E5
Romford UK 15 D9
Romhány H 147 F8
Römhild D 75 B8
Romillé F 23 D8
Romilly-sur-Seine F 25 C10
Rommerskirchen D 21 B7
Romont CH 31 B10
Rømonysæter N 102 C3
Romorantin-Lanthenay F 24 F6
Romos RO 151 F11
Romppala FIN 125 E13
Romrod D 21 C12
Romsey UK 13 D12
Rømskog N 96 C6
Romsley UK 13 A10
Romstad N 105 C11
Romuli RO 152 B4
Rona de Jos RO 145 H9
Rona de Sus RO 145 H9
Rönäs S 108 E8
Rønbjerg DK 86 B3
Roncade I 72 E5
Roncadelle I 66 A1
Roncal E 32 E4
Roncegno I 69 A11
Ronce-les-Bains F 28 D3
Ronchamp F 26 E6
Ronchi dei Legionari I 73 E7
Ronciglione I 62 C2
Ronco Canavese I 68 C4
Roncone I 69 B10
Ronco Scrivia I 37 B9
Ronda E 53 C6
Rønde DK 86 C6
Rondissone I 68 C4
Rone S 93 E12
Ronehamn S 93 E12
Rong N 94 A1
Rõngu EST 131 E12
Rönnäng S 91 D10
Rönnäs S 107 B10
Rönnbacken S 103 E11
Rönnberg S 109 E17
Rönnberget S 118 D6
Rønne DK 88 E7
Rønnebæk DK 87 E9
Ronneburg D 79 E11
Ronneby S 89 C8
Ronnebyhamn S 89 C8
Rønnede DK 87 E10
Ronnenberg D 78 B6
Rönneshytta S 92 B6
Rönnholm S 107 D16
Rønningen N 111 C15
Rönnliden S 109 F17
Rönnöfors S 105 D15
Rönö S 93 C9
Ronov nad Doubravou CZ 77 C9
Ronsberg D 71 B10
Ronse B 19 C8
Ronshausen D 78 E6
Ronvik N 108 B7
Rooaun IRL 6 F6
Roodeschool NL 17 B7
Rookchapel IRL 8 D4
Roosendaal NL 16 E2
Roosinpohja FIN 123 F13
Roosky IRL 7 E7
Roosna-Alliku EST 131 C11
Ropa PL 145 D3
Ropažl LV 135 C9
Ropcha UA 153 A7
Ropczyce PL 143 F12
Ropeid N 94 D4
Roperuelos del Páramo E 39 D8
Ropienka PL 145 D5
Ropinsalmi FIN 116 A7
Ropotovo NMK 168 B5
Roquebillière F 37 C6
Roquebrun F 34 C5
Roquebrune-Cap-Martin F 37 D6
Roquebrune-sur-Argens F 36 E5
Roquecor F 33 B7
Roquecourbe F 33 C10
Roquefort F 32 B5
Roquemaure F 35 B8
Roquesteron F 36 D6
Roquetas E 42 F5
Roquetas de Mar E 55 F7
Roquevaire F 35 D10
Rørbakken N 111 C14
Rörberg S 103 E12
Rørby DK 87 D8
Rore BIH 157 D6
Røros N 101 A14
Rorschach CH 71 C9
Rörvattnet S 105 D16
Rørvig DK 87 D9
Rørvik N 94 F5
Rørvik N 105 B10
Rørvik N 110 D7
Rörvik S 88 A7
Rosà I 72 E4
Rosala FIN 126 F7
Rosal de la Frontera E 51 D5
Rosans F 35 B9
Rosapenna IRL 7 B7
Rosário P 50 D3
Rosarno I 59 C8
Rosavci BIH 157 C6
Rosbach vor der Höhe D 21 D11
Roscanvel F 22 D2
Ros Cathail IRL 6 F4
Rosche D 83 E9
Roscigno I 60 C4
Rościszewo PL 139 E8
Roscoff F 22 C4
Ros Comáin IRL 6 E6
Roscommon IRL 6 E6
Ros Cré IRL 9 C7
Roscrea IRL 9 C7
Rosdorf D 78 D6
Rose I 60 E6
Rose MNE 162 E6
Rosée B 184 D2
Rosehearty UK 3 K12
Rosemarkie UK 3 K8
Rosen BG 167 E9
Rosenallis IRL 7 F8
Rosendal N 94 C4
Rosenfeld D 27 D10
Rosenfors S 92 E7
Rosengarten D 83 D7
Rosenheim D 72 A5
Rosenlund S 91 D13
Rosenow D 84 C4
Rosenthal D 21 C11
Roserberg S 99 C9
Roses E 43 C10
Rosețl RO 161 E10
Roseto Capo Spulico I 61 D7
Roseto degli Abruzzi I 62 B6
Roseto Valfortore I 60 A4
Roshchino RUS 129 E12
Rosheim F 27 C7
Roshven UK 4 B5
Roșia RO 151 D9
Roșia RO 152 F4
Roșia de Amaradia RO 160 C3
Roșia de Secaș RO 152 E3
Roșia Montană RO 151 E11
Rosica LV 133 D3
Rosice CZ 77 C9
Rosice CZ 77 D10
Rosières F 30 E4
Rosières-en-Santerre F 18 E6
Roșiești RO 153 E11
Rosignano Marittimo I 66 F1
Roșiile RO 160 D3
Roșiori RO 153 D10
Roșiori RO 161 D10
Roșiori de Vede RO 160 E6
Rositsa BG 155 F1
Rositsa BY 133 E3
Rositz D 79 D11
Rosiyanivka UA 154 C5
Roskhill UK 2 L3
Roskilde DK 87 D10
Rosko PL 85 E10
Roskovec AL 168 C2
Roskow D 79 B12
Röslau D 75 B10
Roslev DK 86 B3
Rosmalen NL 16 E4
Rosmaninhal P 45 E6
Ros Mhic Thriúin IRL 9 D9
Rosnowo PL 85 B10
Rosolina I 66 B5
Rosolina Mare I 66 B5
Rosolini I 59 F6
Rosoman NMK 169 A6
Rosoy F 25 D9
Rosporden F 22 E4
Rösrath D 21 C8
Rossano I 61 D7
Rossano Veneto I 72 E4
Roßbach D 76 E3
Rossbol S 106 E7
Rosscahill IRL 6 F4
Ross Carbery IRL 8 E4
Rosscarbery IRL 8 E4
Rosscor UK 6 D6
Roßdorf D 21 E11
Rossell E 42 F4
Rosselló E 42 D5
Rosses Point IRL 6 D5
Rossett UK 10 E6
Rossfjord N 111 B15
Roßhaupten D 71 B11
Rossiglione I 37 B9
Rossignol B 19 E11
Rössing (Nordstemmen) D 78 B6
Rossington UK 11 E9
Rossio ao Sul do Tejo P 44 F4
Roßla D 79 D9
Rosslare IRL 9 D10
Rosslare Harbour IRL 9 D10
Roßlau D 79 C11
Rosslea UK 7 D8
Roßleithen A 73 A9
Rossnowlagh IRL 6 C6
Rossön S 107 D10
Ross-on-Wye UK 13 B9
Rossosz PL 141 G8
Rossoszyca PL 142 C6
Rossow D 83 D13
Roßtal D 75 D8
Rossum NL 183 B6
Røssvassbukta N 108 E7
Rossvik N 105 A11
Rossvoll N 104 E4
Rossvoll N 111 B15
Roßwein D 80 D4
Røst N 108 A3
Röstabo S 103 D11
Röstånga S 87 C12
Rostarzewo PL 81 B10
Röste S 103 D11
Rostellan IRL 8 E6
Rostock D 83 B12
Rostrenen F 22 D5
Rostrevor UK 7 D10
Röström S 107 C10
Rostrup DK 86 B5
Rostundelva N 112 D6
Rosturk IRL 6 E3
Rostuša NMK 168 A4
Røstvollen N 102 B3
Røsvik N 109 B9
Rosvik S 118 D7
Röszke H 150 E5
Rot S 102 D7
Rota E 52 C4
Rota N 111 E10
Rota Greca I 60 E6
Rot am See D 74 D7
Rot an der Rot D 71 A10
Rotava CZ 75 B12
Rotberget N 97 A8
Roteberg S 103 D10
Rotello I 63 D7
Rotenburg (Wümme) D 17 B12
Rotenburg an der Fulda D 78 D6
Rötgesbüttel D 79 B8
Rotgülden A 73 B7
Roth D 75 D9
Rothbury UK 5 E13
Röthenbach an der Pegnitz D 75 D9
Rothenberg D 21 E11
Rothenbuch D 74 C5
Rothenburg (Oberlausitz) D 81 D7
Rothenburg ob der Tauber D 75 D7
Rothéneuf F 23 C8
Rothenfels D 74 C6
Rothenschirmbach D 79 D10
Rothenstein D 79 E10
Rotherham UK 11 E9
Rothes UK 3 K10
Rothesay UK 4 D6
Rotheux-Rimière B 19 C11
Rothiesholm UK 3 G11
Rothley UK 11 F9
Rothrist CH 27 F8
Rothwell UK 11 D9
Rothwell UK 15 C7
Rothwesten (Fuldatal) D 78 D6
Rotimlja BIH 157 F8
Rotimojoki FIN 123 C17
Rotnäset S 106 C9
Rotonda I 60 D6
Rotondella I 61 C7
Rotselaar B 19 C10
Rotsjö S 103 A10
Rotsjön S 118 F4
Rotsund N 112 D6
Rott D 71 B11
Rottach-Egern D 72 A4
Rott am Inn D 75 G11
Røttangen N 111 E10
Rottenacker D 71 A9
Röttenbach D 75 D9
Rottenbach D 79 E9
Rottenbuch D 71 B11
Rottenburg am Neckar D 27 D10
Rottenburg an der Laaber D 75 E11
Rottendorf D 74 C7
Rottenmann A 73 A9
Rotterdam NL 16 E3
Rotthalmünster D 76 F4
Röttingen D 74 C6
Rottleberode D 79 C8
Rottne S 89 A7
Rottnemon S 96 B8
Rottneros S 97 C9
Rottofreno I 69 C8
Rottweil D 27 D10
Rotunda RO 160 F4
Rötviken S 105 D16
Rötz D 75 D12
Rouans F 23 F8
Roubaix F 19 C7
Rouchovany CZ 77 D10
Rõude EST 131 D7
Roudnice nad Labem CZ 76 B6
Rouen F 18 F3
Rouffach F 27 E7
Rouffignac F 29 E7
Rõuge EST 131 F13
Rougé F 23 E9
Rougemont F 26 F5
Rougemont-le-Château F 27 E6
Roughton UK 15 B11
Rougnac F 29 D6
Rouillac F 28 D5
Rouillé F 28 C6
Roujan F 34 C5
Roukala FIN 119 F11
Roukalahti FIN 125 F13
Roulans F 26 F5
Roulers B 19 C7
Roulers B 182 D2
Roumazières-Loubert F 29 D7
Roumoules F 36 D4
Roundstone IRL 6 F3
Roundway UK 13 C11
Roundwood IRL 7 F10
Rouravaara FIN 117 C14
Roure I 31 E11
Rousínov CZ 77 D11
Rousky UK 4 F2
Rousset F 35 D10
Roussillon F 30 E6
Roussillon F 35 C9
Rousson F 35 B7
Routot F 18 F2
Rouvroy F 182 E1
Rouvroy-sur-Audry F 19 E10
Rouy F 30 A4
Rovala FIN 115 C5
Rovala FIN 115 E1
Rovaniemi FIN 117 F15
Rovanpää FIN 117 C15
Rovanpää FIN 117 E12
Rovapää FIN 115 D2
Rovastinaho FIN 119 C16
Rovato I 69 B9
Rovegno I 37 B10
Roverbella I 66 B2
Roveredo CH 69 A7
Rovereto I 69 B11
Rövershagen D 83 B12
Roverud N 96 B7
Roviano I 62 C3
Rovies GR 175 B7
Rovigo I 66 B4
Rovinari RO 159 D11
Rovine BIH 157 B7
Rovinj HR 67 B8
Rovinka SK 146 E4
Rovišće HR 149 E7
Rovisuvanto FIN 113 E16
Rovsättra S 99 B10
Rów PL 85 E7
Rowde UK 13 C10
Równa PL 142 C6
Rownaye BY 133 F6
Roxmo S 93 A8
Roxton UK 15 C8
Royal Wootton Bassett UK 13 B11
Royan F 28 D3
Royat F 30 D3
Roybon F 31 E7
Roybridge UK 4 B7
Roydon UK 15 C11
Roye F 18 E6
Royère-de-Vassivière F 29 D9
Røyken N 95 C12
Røykenes N 111 C10
Röykkä FIN 127 E12
Roylyanka UA 154 E5
Røyrvik N 105 B15
Røyse N 95 B12
Royston UK 15 C8
Royton UK 11 D7
Röyttä FIN 119 C12
Roytvollen N 105 A12
Royuela E 47 D10
Roza BG 166 E6
Rožaje MNE 163 D9
Rozalén del Monte E 47 E7
Rozalimas LT 134 E7
Różan PL 139 E11
Różanki PL 85 E8
Rozavlea RO 152 B4
Rozay-en-Brie F 25 C8
Rožďalovice CZ 77 B8
Rozdil UA 145 E9
Rozdil'na UA 154 D6
Rozdrażew PL 142 C4
Rozenburg NL 182 B4
Rozendaal NL 183 A7
Roženica HR 148 E5
Rozes LV 135 B11
Rozhniv UA 152 A6
Rozhnyativ UA 145 F9
Rozino BG 165 D10
Rožmitál pod Třemšínem CZ 76 C5
Rožňava SK 145 F2
Rożniatów PL 142 B6
Roznov RO 153 D9
Rožnov pod Radhoštěm CZ 146 C6
Roznowo PL 85 E11
Rozogi PL 139 D11
Rozovets BG 166 E4
Rozoy-sur-Serre F 19 E9
Rozprza PL 143 D8
Roztoka Wielka PL 145 E2
Roztoky CZ 76 B5
Roztoky UA 152 A6
Rozula LV 135 B9
Rožupe LV 135 D12
Roźwienica PL 144 D6
Rozzano I 69 C7
Rrëshen AL 163 F8
Rrogozhinë AL 168 B2
Rromanat AL 168 B2
Rtkovo SRB 159 D10
Rtyně v Podkrkonoší CZ 77 B10
Ru E 38 B3
Rua P 44 C5
Ruabon UK 10 F5
Ruaudin F 23 E12
Ruba LV 134 D5
Rúbaň SK 146 F6
Rubano I 66 B4
Rubashki BY 133 F3
Rubayo E 40 B4
Rubbestadneset N 94 C2
Rubena E 40 D4
Rübenau D 80 E4
Rubene LV 135 B10
Rubenl LV 135 D11
Rubí E 43 E8
Rubiá E 39 D6
Rubí de Bracamonte E 40 F2
Rubielos de la Cérida E 47 C10
Rubielos de Mora E 48 D3
Rubiera I 66 C2
Rubik AL 163 F8
Rublacedo de Abajo E 40 C5
Rucăr RO 160 C6
Rucava LV 134 D2
Ruciane-Nida PL 139 C12
Rückersdorf D 80 C5
Ruda PL 142 D6
Ruda S 89 A10
Rudabánya H 145 G2
Ruda Maleniecka PL 141 H2
Rudamina LT 137 D11
Rudamina LT 137 E7
Ruda Różaniecka PL 144 C7
Rudartsi BG 165 D7
Ruda Śląska PL 142 F6
Rudbāržl LV 134 C3
Ruddervoorde B 182 C2
Rude LV 134 B5
Rude LV 134 D2
Rudelzhausen D 75 E10
Rudersberg D 74 E6
Rudersdorf A 148 B6
Rüdersdorf Berlin D 80 B5
Rüdesheim D 185 E8
Rudice BIH 156 C5
Rudina SK 147 C7
Rūdiškės LT 137 D10
Rudka PL 141 E7
Rudkøbing DK 87 F7
Rudky UA 145 D7
Rudná CZ 76 B6
Rudna PL 81 C10
Rudna Glava SRB 159 E9
Rudňany SK 145 F2
Rudna Wielka PL 81 C11
Rudne UA 144 D8
Rudnica PL 81 A8
Rudnica SRB 163 C10
Rudnik BG 167 D9
Rudnik PL 142 F5
Rudnik SRB 158 E5
Rudniki PL 142 D6
Rudnik nad Sadem PL 144 C5
Rüdnitz D 84 E5
Rudno PL 141 G6
Rudno SRB 163 C9
Rudnyky UA 145 E8
Rudnytsya UA 154 A3
Rudo BIH 158 F3
Rudolfov CZ 77 E7
Rudolstadt D 79 E9
Rudozem BG 171 B7
Rudsgrendi N 95 C9
Rudsjön S 107 D10
Rudston UK 11 C11
Ruds Vedby DK 87 D8
Rudzāti LV 135 D13
Rudziczka PL 142 F4
Rudziniec PL 142 F5
Rue F 18 D4
Rueda E 39 F10
Rueda de Jalón E 41 E9
Rueil-Malmaison F 24 C7
Ruelle-sur-Touvre F 29 D6
Ruen BG 167 D8
Ruente E 40 B3
Ruffano I 61 D10
Ruffec F 29 B8
Ruffec F 29 C6
Ruffey-lès-Echirey F 26 F3
Ruffiac F 23 E7
Ruffieu F 31 C8
Ruffieux F 31 D8
Rufford UK 10 D6
Rufina I 66 E3
Rugāji LV 133 B2
Rugby UK 13 A12
Rugeley UK 11 F8
Rugge N 110 C8
Rugince NMK 164 E4
Ruginești RO 153 E10
Ruginoasa RO 153 C9
Rügland D 75 D8
Rugles F 24 C4
Rugvica HR 148 E6
Ruha FIN 123 E9
Ruhala FIN 124 G2
Ruhla D 79 E7
Ruhland D 80 D5
Ruhmannsfelden D 76 E3
Rühn D 83 C11
Ruhpolding D 73 A6
Ruhstorf an der Rott D 76 F4
Ruidera E 55 B7
Ruikka FIN 119 B14
Ruinas I 64 D2
Ruinen NL 16 C6
Ruinerwold NL 16 C6
Ruiselede B 19 B7
Ruismäki FIN 123 E10
Ruivães P 38 E3
Ruja PL 81 D10
Rūjiena LV 131 F10
Rujišta BIH 157 F8
Ruka FIN 121 B13
Rukajärvi FIN 121 B13
Rukkisperä FIN 119 E14
Rulbo S 102 C8
Rullbo S 102 E7
Rullnäs S 107 D15
Rully F 30 B6
Rülzheim D 27 B9
Rum A 72 B3
Rum BY 133 E6
Rum H 149 B7
Ruma SRB 158 C4
Rumboci BIH 157 E7
Rumburk CZ 81 E7
Rumelange L 20 F6
Rumelifeneri TR 173 B11
Rumenka SRB 158 C4
Rumes B 19 C7
Rumia PL 138 A5
Rumigny F 19 E9
Rumilly F 31 D8
Rummu EST 131 C8
Rummukkala FIN 125 F12
Rumo FIN 125 C11
Rumont F 26 C3
Rumšiškės LT 137 D9
Rumskulla S 92 D7
Rumst B 19 B9
Rumyantsevo BG 165 C9
Runaberg S 107 E11
Runcorn UK 10 E6
Runcu RO 155 D2
Runcu RO 160 C4
Runcu RO 160 C6
Runděni LV 133 D3
Rundfloen N 102 D4
Rundhaug N 111 B16
Rundmoen N 108 D8
Rundvik S 107 D16
Rungsted DK 87 D11
Runhällen S 98 B7
Runik RKS 163 D10
Runkel D 185 D9
Runni FIN 124 C7
Runović HR 157 F7
Runsten S 89 B11
Runtuna S 93 B10
Ruohokangas FIN 115 A3
Ruohola FIN 119 C15
Ruokojärvi FIN 117 D12
Ruokojärvi FIN 129 B11
Ruokojärvi S 117 E10
Ruokolahti FIN 129 C10
Ruokoniemi FIN 129 B10
Ruokotaipale FIN 128 C8
Ruolahti FIN 127 B13
Ruoms F 35 B7
Ruona FIN 119 B17
Ruona FIN 123 E11
Ruonajärvi FIN 117 D13
Ruopsa FIN 115 E2
Ruorasmäki FIN 127 B15
Ruotaanmäki FIN 124 D7
Ruoti I 60 B5
Ruotsalo FIN 123 C10
Ruotsinkylä FIN 127 D15
Ruotsinpyhtää FIN 127 D15
Ruottisenharju FIN 120 D9
Ruovesi FIN 123 G12
Rupa HR 67 B9
Rupea RO 152 E6
Ruppovaara FIN 125 F14
Rupt-sur-Moselle F 26 E6
Rus E 55 C6
Rus RO 152 C3
Rusånes N 108 C9
Rusănești RO 160 F5
Rusca Montană RO 159 B9
Ruscova RO 152 B4
Rusdal N 94 E4
Ruse BG 161 F7
Ruše SLO 148 C5
Rusele S 107 B14
Ruseni MD 153 A10
Rusețu RO 161 D10
Ruševo HR 149 F9
Rush IRL 7 E10
Rushden UK 15 C7
Rusiec PL 143 D6
Rușii-Munțl RO 152 D5
Rusinovo NMK 165 F6
Rusinów PL 141 H3
Rusinowo PL 85 D10
Ruskeala RUS 129 B14
Ruskele S 107 B15
Ruski Krstur SRB 158 B3
Rusko FIN 126 D7
Rusko Selo SRB 150 F6
Ruskov SK 145 F3
Ruskträsk S 107 B15
Rusnė LT 134 F2
Rusokastro BG 167 E8
Rušona LV 133 D1
Rüsselsheim D 21 D10
Russeluft N 113 C11
Russelv N 112 D5
Russhaugen N 111 D11
Russi I 66 D5
Rust A 77 G11
Rustad N 95 B13
Rustad N 101 D14
Rustrel F 35 C9
Ruswil CH 27 F9
Ruszów PL 81 D8
Rutalahti FIN 123 G15
Rute E 53 B8
Rute S 93 D13
Rutesheim D 187 D6
Rüthen D 17 F10
Ruthin UK 10 E5
Rüti CH 27 F10
Rutigliano I 61 A8
Rutino I 60 C4
Rutka-Tartak PL 136 E6
Rutki-Kossaki PL 140 D6
Rutledal N 100 D2
Rutošl SRB 163 B8
Rutten NL 16 C5
Rütten-Scheid D 183 C9
Rutvik S 118 C8
Ruuhensuo FIN 121 C9
Ruuhijärvi FIN 117 E13
Ruuhijärvi FIN 127 C15
Ruukki FIN 119 E14
Ruunaa FIN 125 D14
Ruurlo NL 17 D6
Ruusa EST 131 E14
Ruusmäe EST 131 F14
Ruutana FIN 123 C17
Ruutana FIN 127 B11
Ruuvaoja FIN 115 C5
Ruvanaho FIN 115 F6
Ruvaslahti FIN 125 E12
Ruvo del Monte I 60 B5
Ruvo di Puglia I 61 A6
Ruy F 31 D7
Ruynes-en-Margeride F 30 E3
Rŭzhevo Konare BG 165 E10
Ruzhintsi BG 159 F10
Ružina LV 133 D2
Ružindol SK 146 E5
Ružomberok SK 147 C8
Ruzsa H 150 E4
Ry DK 86 C5
Ryabovo RUS 129 E10
Ryakhovo BG 161 F8
Ryakhovtsite BG 165 C11
Rybachiy RUS 134 F1
Rybany SK 146 D6
Rybczewice Drugie PL 144 A6
Rybitví CZ 77 B9
Rybnik PL 142 F6
Rybno PL 139 C11
Rybno PL 139 D8
Rybno PL 141 F2
Ryboły PL 140 E8
Rychliki PL 139 C8
Rychnov nad Kněžnou CZ 77 B10
Rychnowo PL 139 C9
Rychtal PL 142 D4
Rychvald CZ 146 B6
Rychwał PL 142 B5
Ryczywół PL 85 E11
Ryczywół PL 141 G4
Ryd S 88 C7
Rydaholm S 88 B6
Rydal S 91 D12
Rydboholm S 91 D12
Ryde DK 86 C3
Ryde UK 13 D12
Rydet S 91 E11
Rydöbruk S 87 B12
Rydsgård S 87 E13
Rydsnäs S 92 D6
Rydułtowy PL 142 F5
Rydzyna PL 81 C11
Rye UK 15 F10
Ryen N 90 C3
Ryeng N 114 D8
Ryes F 23 B10
Rygge N 95 D13
Ryggesbro S 103 D10
Ryglice PL 144 D3
Rygnestad N 94 D6
Ryhälä FIN 129 B9
Ryhall UK 11 F11
Ryjewo PL 138 C6
Rykene N 90 C4
Ryki PL 141 G5
Rymań PL 85 C9
Rymanów PL 145 D4
Rýmařov CZ 81 G12
Rymättylä FIN 126 E6
Rymnio GR 169 D6
Ryn PL 136 F4
Rynarzewo PL 138 D4
Rynkeby DK 86 E7
Ryńsk PL 138 D6
Ryomgård DK 86 C7
Rypefjord N 113 B12
Rypin PL 139 D7
Rysjedal N 100 D2
Ryslinge DK 86 E7
Ryssby S 88 B6
Rysum (Krummhörn) D 17 B8
Rytel PL 138 C4
Rytilahti FIN 115 E3
Rytinki FIN 120 C9
Rytky FIN 123 C17
Rytkynkylä FIN 119 F14
Ryttylä FIN 127 D12
Rytwiany PL 143 E11
Ržanovo NMK 168 B4
Rząśnik PL 139 E11
Rzeczenica PL 85 C12
Rzeczyca PL 141 G2
Rzeczyca PL 142 C6
Rzįgnowo PL 139 D10
Rzejowice PL 143 D8
Rzekuń PL 139 D12
Rzepiennik Strzyżewski PL 144 D3
Rzepin PL 81 B7
Rzerzįczyce PL 143 E7
Rzesznikowo PL 85 C8
Rzeszów PL 144 C5
Rzgów PL 143 C8
Rzgów Pierwszy PL 142 B5
Rzuców PL 141 H3

S

Sääksjärvi FIN 123 D11
Sääksjärvi FIN 127 C10
Sääksjärvi FIN 127 D13
Sääksmäki FIN 127 C11
Saal D 83 B12
Saal an der Donau D 75 E10
Saalbach-Hinterglemm A 73 B6
Saalburg D 75 B10
Saales F 27 D7
Saalfeld D 79 E9
Saalfelden am Steinernen Meer A 73 B6
Saanen CH 31 C11
Saarbrücken D 21 F7
Saarburg D 21 E7
Sääre EST 130 F4
Saarela FIN 125 C13
Saaren kk FIN 129 B12
Saarenkylä FIN 117 E15
Saaresmäki FIN 124 B7
Saari FIN 129 B12
Saariharju FIN 120 C9
Saarijärvi FIN 123 E14
Saari-Kämä FIN 119 B16
Saarikoski FIN 112 F7
Saarikoski FIN 119 C15
Saarikoski FIN 119 E14
Saarikylä FIN 121 D13
Saario FIN 125 F14
Saaripudas FIN 117 D11
Saariselkä FIN 115 B2
Saarivaara FIN 121 E14
Saarivaara FIN 125 F15
Saarlouis D 21 F7
Saarwellingen D 21 F7
Saas CH 71 D9
Sääse EST 131 C12
Saas Fee CH 68 A4
Saas Grund CH 68 A4
Sääskjärvi FIN 127 D15
Säävälä FIN 119 D16
Šabac SRB 158 D4
Sabadell E 43 D8
Săbăoani RO 153 C9
Sabarat F 33 D8
Sabatynivka UA 154 A6
Sabaudia I 62 E4
Sabbionetta I 66 C1
Sabero E 39 C9
Sabile LV 134 B5
Sabiñánigo E 32 E5
Sabinov SK 145 E3
Sabiote E 55 C6
Sables-d'Or-les-Pins F 23 C7
Sablé-sur-Sarthe F 23 E11
Sablet F 35 B9
Sabnie PL 141 E6
Sabres F 32 B4
Sabro DK 86 C6
Sabrosa P 38 F4
Sabugal P 45 D6
Sabugueiro P 44 D5
Sabugueiro P 50 B3
Săcădat RO 151 C9
Săcălășeni RO 152 B3
Săcălaz RO 151 F7
Sacañet E 48 E3
Săcășeni RO 151 C10
Sacavém P 50 B1
Sacecorbo E 47 C8
Sacedón E 47 D7
Săcel RO 152 B4
Săcel RO 152 E5
Săcele RO 155 E3
Săcele RO 161 B7
Săcelu RO 160 C3
Săceni RO 160 E6
Saceruela E 54 B3
Sachseln CH 70 D6

Sachsenberg (Lichtenfels) D 21 B11
Sachsenbrunn D 75 B8
Sachsenburg A 73 C7
Sachsenhagen D 17 D12
Sachsenhausen (Waldeck) D 17 F12
Sachsenheim D 27 C11
Sacile I 72 E5
Sacoşu Turcesc RO 159 B7
Sacović BIH 156 E6
Sacquenay F 26 E3
Sacramenia E 40 E4
Săcueni RO 151 C9
Săcuieu RO 151 D10
Sačurov SK 145 F4
Sada E 38 B3
Sádaba E 32 F3
Sadala EST 131 D13
Sadali I 64 D3
Sadina BG 166 C6
Sadki PL 85 D12
Sadkowice PL 141 G3
Sadkowo PL 85 C10
Sadlinki PL 138 C6
Sadova MD 154 C2
Sadova RO 152 B6
Sadova RO 160 F3
Sadove UA 154 E4
Sadovets BG 165 C9
Sadovo BG 165 E10
Sadowie PL 143 E11
Sadowne PL 139 E12
Sadská CZ 77 B7
Sadu RO 160 B4
Sädvaluspen S 109 D12
Sæbø N 94 B6
Sæbø N 100 B4
Sæbøvik N 94 C3
Sæby DK 87 C3
Sæby DK 90 E8
Sæd DK 86 F3
Saelices E 47 E7
Saelices de la Sal E 47 C8
Saelices del Rio E 39 D9
Saelices de Mayorga E 39 D9
Saerbeck D 17 D9
Særslev DK 86 D6
Sæter N 104 C8
Sætra N 104 E6
Sætre N 95 C13
Saeul L 20 E5
Sævareid N 94 B3
Safaalan TR 173 B9
Safara P 51 C5
Säffle S 91 A12
Saffré F 23 E8
Saffron Walden UK 15 C9
Săg RO 151 C10
Săg RO 159 B7
Sagama I 64 C2
Sagard D 84 A5
Sage D 17 C10
Săgeata RO 161 C9
Sågen S 97 B11
Sagiada GR 168 E3
Sağırlar TR 173 F9
Sağlamtaş TR 173 C7
Sågmyra S 103 E9
Sagna RO 153 D10
Sagone F 37 G9
Sagres P 50 E2
Sagstua N 95 B15
Sagunto E 48 E4
Sagvåg N 94 C2
Ságvár H 149 C10
Sagy F 31 B7
Sahagún E 39 D9
Sahaidac MD 154 D3
Sahalahti FIN 127 C11
Sahankylä FIN 122 F8
Saharna Nouă MD 154 B3
Săhăteni RO 161 C8
Şahin TR 173 B6
Şahinli TR 172 D6
Sahl DK 86 C5
Sahrajärvi FIN 123 F14
Sahun E 33 E6
Sahune F 35 B9
Šahy SK 147 E7
Saiakopli EST 131 C12
Saighdinis UK 2 K2
Saija FIN 115 D5
Saija FIN 127 C10
Saikari FIN 124 E7
Saillagouse F 33 F10
Saillans F 35 A9
Sail-sous-Couzan F 30 D4
Saimaanharju FIN 129 C9
Säimen FIN 125 F12
Sains-Richaumont F 19 E8
St Abbs S 5 D12
St-Affrique F 34 C4
St-Agnan F 30 B4
St-Agnan-en-Vercors F 31 F7
St-Agnant F 28 D4
St-Agnant-de-Versillat F 29 C9
St Agnes UK 12 E4
St-Agrève F 30 E5
St-Aignan F 24 F5
St-Aignan-sur-Roë F 23 E9
St-Aigulin F 28 E5
St-Albain F 30 C6
St-Alban F 22 C6
St-Alban-Leysse F 31 D8
St Albans UK 15 D8
St-Alban-sur-Limagnole F 30 F3
St-Amand-en-Puisaye F 25 E9
St-Amand-les-Eaux F 19 D7
St-Amand-Longpré F 24 E5
St-Amand-Montrond F 29 B11
St-Amand-sur-Fion F 25 C12
St-Amans F 34 B3
St-Amans-des-Cots F 30 F2
St-Amans-Soult F 33 D10
St-Amant-de-Boixe F 29 D6
St-Amant-Roche-Savine F 30 D4
St-Amant-Tallende F 30 D3
St-Amarin F 27 E7
St-Ambroix F 35 B7
St-Amour F 31 C7
St-Andiol F 35 C8
St-André-de-Corcy F 31 D6
St-André-de-Cruzières F 35 B7
St-André-de-Cubzac F 28 F5
St-André-de-l'Eure F 24 C5
St-André-de-Sangonis F 34 C6
St-André-de-Valborgne F 35 B6
St-André-le-Gaz F 31 D8

St-André-les-Alpes F 36 D5
St-André-les-Vergers F 25 D11
St Andrews UK 5 C11
St-Angel F 29 D10
St Anne GBG 23 A7
St-Anthème F 30 D4
St-Antonin-Noble-Val F 33 B9
St-Août F 29 B9
St-Apollinaire F 26 F3
St-Arcons-d'Allier F 30 E4
St-Arnoult-en-Yvelines F 24 C6
St Asaph UK 10 E5
St-Astier F 29 E7
St-Astier F 29 F6
St Athan UK 13 C8
St-Auban F 36 D5
St-Auban-sur-l'Ouvèze F 35 B9
St-Aubin F 31 A7
St-Aubin-Château-Neuf F 25 E9
St-Aubin-d'Aubigné F 23 D8
St-Aubin-de-Blaye F 28 E4
St-Aubin-du-Cormier F 23 D9
St-Aubin-lès-Elbeuf F 18 F3
St-Aubin-sur-Mer F 23 B11
St-Aulaye F 29 E6
St Austell UK 12 E5
St-Avé F 22 E6
St-Avertin F 24 F4
St-Avold F 26 B6
St-Ay F 24 E6
St-Ayguif F 36 E5
St-Barthélemy-d'Agenais F 33 A6
St-Barthélemy-de-Vals F 30 E6
St-Bauzille-de-Putois F 35 C6
St-Béat F 33 E7
St-Beauzély F 34 B4
St Bees UK 10 C4
St-Benin-d'Azy F 30 A3
St-Benoît F 29 B6
St-Benoît F 33 D10
St-Benoît-du-Sault F 29 C8
St-Benoît-sur-Loire F 25 E7
St-Béron F 31 D8
St-Berthevin F 23 D10
St-Bertrand-de-Comminges F 33 D7
St-Blaise CH 31 A10
St-Blaise-la-Roche F 27 D7
St-Blin-Semilly F 26 D3
St-Boil F 30 B6
St-Bonnet-de-Bellac F 29 C7
St-Bonnet-de-Joux F 30 C5
St-Bonnet-en-Bresse F 31 B7
St-Bonnet-en-Champsaur F 36 B4
St-Bonnet-le-Château F 30 E5
St-Bonnet-le-Froid F 30 E5
St-Bonnet-sur-Gironde F 28 E4
St-Branchs F 24 F4
St Brelade GBJ 23 B7
St-Brevin-les-Pins F 23 F7
St-Briac-sur-Mer F 23 C7
St-Brice-en-Coglès F 23 D9
St Brides Major UK 13 C7
St-Brieuc F 22 C5
St-Bris-le-Vineux F 25 E10
St-Brisson F 25 F11
St-Broing-les-Moines F 25 E12
St Buryan UK 12 E3
St-Calais F 24 E4
St-Cannat F 35 C9
St-Céré F 29 F9
St-Cergue CH 31 C9
St-Cergues F 31 C9
St-Cernin F 29 E10
St-Chaffrey F 31 F10
St-Chamarand F 33 A8
St-Chamas F 35 C9
St-Chamond F 30 E6
St-Chaptes F 35 C7
St-Chef F 31 D7
St-Chély-d'Apcher F 30 F3
St-Chély-d'Aubrac F 34 A4
St-Chinian F 34 D4
St-Christol F 35 B9
St-Christol-lès-Alès F 35 B7
St-Christoly-Médoc F 28 E4
St-Christophe I 31 D11
St-Christophe-en-Bazelle F 24 F6
St-Christophe-en-Brionnais F 30 C5
St-Ciers-sur-Gironde F 28 E4
St-Cirq-Lapopie F 33 B9
St-Clair-du-Rhône F 30 E6
St-Clar F 33 C7
St-Claud F 29 D6
St-Claude F 31 C8
St Clears UK 12 B6
St-Clément F 25 D9
St-Clément F 26 E3
St-Clément F 29 E9
St Clement GBJ 23 B7
St-Clément-de-Rivière F 35 C6
St Columb Major UK 12 E5
St Combs UK 3 K13
St-Constant F 29 F10
St-Cosme-en-Vairais F 24 D3
St-Cricq-Chalosse F 32 C4
St-Cyprien F 29 F8
St-Cyprien F 33 B8
St-Cyprien F 34 E5
St-Cyr-sur-Loire F 24 F4
St-Cyr-sur-Mer F 35 D10
St Cyrus UK 5 B12
St David's UK 9 E12
St Day UK 12 E4
St-Denis F 25 C7
St-Denis-d'Anjou F 23 E11
St-Denis-de-Gastines F 23 D10
St-Denis-de-Jouhet F 29 B9
St-Denis-de-Pile F 28 F5
St-Denis-d'Oléron F 28 C3
St-Denis-en-Bugey F 31 D7
St-Denis-lès-Bourg F 31 C7
St Dennis UK 12 E5
St-Désert F 30 B6
St-Didier-en-Velay F 30 E5
St-Didier-sur-Chalaronne F 30 C6
St-Dié-d'Auvergne F 30 D3
St-Dizier F 25 C12
St-Dizier-Leyrenne F 29 C9
St-Dolay F 23 F7
St-Donat-sur-l'Herbasse F 31 E6
St-Doulchard F 25 F7
St-Adresse F 23 A12
Ste-Alvère F 29 F7
Ste-Bazeille F 33 A6
Ste-Cécile-les-Vignes F 35 B8

Ste-Croix CH 31 B10
Ste-Croix F 31 B7
Ste-Croix F 31 B7
Ste-Croix-Volvestre F 33 D8
Ste-Engrâce F 32 D4
Ste-Énimie F 34 B5
Ste-Eulalie d'Olt F 34 B4
Ste-Eulalie-en-Born F 32 B3
Ste-Feyre F 29 C9
Ste-Foy-de-Peyrolières F 33 D8
Ste-Foy-la-Grande F 29 F6
Ste-Foy-l'Argentière F 30 D5
Ste-Foy-lès-Lyon F 30 D6
Ste-Foy-Tarentaise F 31 D10
Ste-Geneviève F 18 F5
Ste-Geneviève-sur-Argence F 30 F2
Ste-Égrève F 31 E8
Ste-Hélène F 28 F4
Ste-Hermine F 28 B3
Ste-Livrade-sur-Lot F 33 B7
Ste-Lizaigne F 29 A10
Ste-Éloy-les-Mines F 30 C2
Ste-Lucie-de-Tallano F 37 H10
Ste-Marguerite F 186 E2
Ste-Marie F 34 E5
Ste-Marie-aux-Mines F 27 D7
Ste-Maure-de-Peyriac F 33 B6
Ste-Maure-de-Touraine F 24 F4
Ste-Maxime F 36 E5
Ste-Menehould F 25 B12
Ste-Mère-Église F 23 B9
St-Émiland F 30 B5
Ste-Orse F 29 E8
Ste-Pazanne F 23 F8
Ste-Radegonde F 28 B5
Ste-Erme-Outre-et-Ramecourt F 19 E8
St Erth UK 12 E4
Ste-Sabine F 25 F12
Ste-Savine F 25 D11
Ste-Sévère-sur-Indre F 29 C10
Ste-Suzanne F 23 D11
St-Étienne F 30 E5
St-Étienne-de-Baïgorry F 32 D3
St-Étienne-de-Fontbellon F 35 A7
St-Étienne-de-Fursac F 29 C9
St-Étienne-de-Montluc F 23 F8
St-Étienne-de-St-Geoirs F 31 E7
St-Étienne-de-Tinée F 36 C5
St-Étienne-du-Bois F 31 C7
St-Étienne-du-Rouvray F 18 F3
St-Étienne-en-Dévoluy F 36 A3
St-Étienne-les-Orgues F 35 B10
St-Étienne-lès-Remiremont F 26 D6
St-Étienne-Vallée-Française F 35 B6
Ste-Tulle F 35 C10
Ste-Vertu F 25 E10
St-Fargeau F 25 E9
St-Félicien F 30 E6
St-Félix-Lauragais F 33 D9
St Fergus UK 3 K13
St-Ferme F 28 F6
Saintfield UK 7 D11
St Fillans UK 5 C8
St-Firmin F 26 D5
St-Firmin F 31 F9
St-Flavy F 25 D10
St-Florent F 37 F10
St-Florent-des-Bois F 28 B3
St-Florent-le-Vieil F 23 F9
St-Florent-sur-Cher F 29 B10
St-Flour F 30 E3
St-Flovier F 29 B8
St-Fons F 30 D6
St-Fort-sur-Gironde F 28 E4
St-Frajou F 33 D7
St-François-Longchamp F 31 E9
St-Front-de-Pradoux F 29 E6
St-Fulgent F 28 B3
St-Galmier F 30 D5
St-Gaudens F 33 D7
St-Gaultier F 29 B8
St-Gein F 32 C5
St-Gély-du-Fesc F 35 C6
St-Genest-Malifaux F 30 E5
St-Geniez F 36 C4
St-Geniez-d'Olt F 34 B4
St-Genis-de-Saintonge F 28 E4
St-Genis-Laval F 30 D6
St-Genis-Pouilly F 31 C9
St-Genix-sur-Guiers F 31 D8
St-Genou F 29 B8
St-Geoire-en-Valdaine F 31 E8
St-Georges-Buttavent F 23 D10
St-Georges-d'Aurac F 30 E4
St-Georges-de-Commiers F 31 E8
St-Georges-de-Didonne F 28 D4
St-Georges-de-Luzençon F 34 B4
St-Georges-de-Mons F 30 D2
St-Georges-de-Reneins F 30 C6
St-Georges-de-Groseillers F 23 C10
St-Georges-d'Oléron F 28 D3
St-Georges-du-Vièvre F 18 F2
St-Georges-en-Couzan F 30 D4
St-Georges-lès-Baillargeaux F 29 B6
St-Georges-sur-Baulche F 25 E10
St-Georges-sur-Cher F 24 F5
St-Georges-sur-Loire F 23 F10
St-Geours-de-Maremne F 32 C3
St-Gérand-le-Puy F 30 C4
St-Germain-Chassenay F 30 B3
St-Germain-de-Calberte F 35 B6
St-Germain-de-la-Coudre F 24 D4
St-Germain-des-Fossés F 30 C3
St-Germain-d'Esteuil F 28 E4
St-Germain-du-Bel-Air F 33 A8
St-Germain-du-Bois F 31 B7
St-Germain-du-Corbéis F 23 D12
St-Germain-du-Plain F 31 B6
St-Germain-du-Puy F 25 F7
St-Germain-du-Teil F 34 B5
St-Germain-en-Laye F 24 C7
St-Germain-Laval F 30 D5

St-Germain-Lembron F 30 E3
St-Germain-les-Belles F 29 D8
St-Germain-les-Vergnes F 29 E9
St-Germain-l'Herm F 30 E4
St Germans UK 12 E6
St-Germé F 32 C5
St-Gervais F 28 B1
St-Gervais F 31 E7
St-Gervais-d'Auvergne F 30 C2
St-Gervais-la-Forêt F 24 E5
St-Gervais-les-Bains F 31 D10
St-Gervais-les-Trois-Clochers F 29 B6
St-Gervais-sur-Mare F 34 C5
St-Géry F 33 B9
St-Ghislain B 19 D8
St-Gildas-de-Brethmas F 35 B7
St-Gildas-des-Bois F 22 E7
St-Gilles F 35 C8
St-Gilles-Croix-de-Vie F 28 B2
St-Gingolph F 31 C10
St-Girons F 33 E8
St-Girons-Plage F 32 C3
St-Gobain F 19 E7
St-Guénolé F 22 E2
St-Guilhem-le-Désert F 35 C6
St-Haon-le-Châtel F 30 C4
St Helens UK 10 E6
St Helier GBJ 23 B7
St-Herblain F 23 F8
St-Hilaire F 33 D10
St-Hilaire-de-Brethmas F 35 B7
St-Hilaire-de-Riez F 28 B2
St-Hilaire-des-Loges F 28 C5
St-Hilaire-de-Villefranche F 28 D4
St-Hilaire-du-Harcouët F 23 C9
St-Hilaire-du-Rosier F 31 E7
St-Hilaire-Fontaine F 30 B4
St-Hilaire-le-Grand F 25 B11
St-Hilaire-St-Florent F 23 F11
St-Hippolyte F 27 D7
St-Hippolyte F 27 F6
St-Hippolyte-du-Fort F 35 C6
St-Honoré-les-Bains F 30 B4
St-Hostien F 30 E5
St-Hubert B 19 D11
St-Imier CH 27 F6
St-Ismier F 31 E8
St Ive UK 12 E6
St Ives UK 12 E4
St Ives UK 15 C8
St-Izaire F 34 C4
St-Jacques-de-la-Lande F 23 D8
St-James F 23 C9
St-Jean F 33 C8
St-Jean-Bonnefonds F 30 E5
St-Jean-Brévelay F 22 E6
St-Jean-d'Angély F 28 D4
St-Jean-d'Assé F 23 D12
St-Jean-de-Bournay F 31 D7
St-Jean-de-Braye F 24 E6
St-Jean-de-Daye F 23 B9
St-Jean-de-la-Ruelle F 24 E6
St-Jean-de-Losne F 26 F3
St-Jean-de-Luz F 32 D2
St-Jean-de-Maruéjols-et-Avéjan F 35 B7
St-Jean-de-Maurienne F 31 E9
St-Jean-de-Monts F 28 B1
St-Jean-de-Sixt F 31 D9
St-Jean-de-Védas F 35 C6
St-Jean-d'Illac F 28 F4
St-Jean-du-Bruel F 34 B5
St-Jean-du-Falga F 33 D9
St-Jean-du-Gard F 35 B6
St-Jean-le-Centenier F 35 A8
St-Jean-Pied-de-Port F 32 D3
St-Jean-Poutge F 33 C6
St-Jean-sur-Erve F 23 D11
St-Jeoire F 31 C9
St-Jeure-d'Ay F 30 E6
St-Jeures F 30 E5
St-Joachim F 23 F7
St John GBJ 23 B7
St John's Chapel UK 5 F12
St John's Town of Dalry UK 5 E8
St-Jores F 23 B9
St-Jorioz F 31 D9
St-Jory F 33 C8
St-Jouan-des-Guérets F 23 C8
St-Jouin-Bruneval F 23 A12
St-Jouin-de-Marnes F 28 B5
St-Julien F 31 C7
St-Julien F 33 D8
St-Julien-Beychevelle F 28 E4
St-Julien-Boutières F 30 F5
St-Julien-Chapteuil F 30 E5
St-Julien-de-Concelles F 23 F9
St-Julien-de-Vouvantes F 23 E9
St-Julien-du-Sault F 25 D9
St-Julien-du-Verdon F 36 D5
St-Julien-en-Beauchêne F 35 A10
St-Julien-en-Born F 32 B3
St-Julien-en-Genevois F 31 C9
St-Julien-l'Ars F 29 B7
St-Junien F 29 D7
St-Just F 35 B8
St Just UK 12 E3
St-Just-en-Chaussée F 18 E5
St-Just-en-Chevalet F 30 D4
St-Just-Ibarre F 32 D3
St-Just-Luzac F 28 D3
St-Just-Sauvage F 25 C10
St-Just-St-Rambert F 30 E5
St Keverne UK 12 E4
St-Lambert-des-Levées F 23 F11
St-Lary-Soulan F 33 E6
St-Laurent F 36 C5
St-Laurent-Bretagne F 32 D5
St-Laurent-d'Aigouze F 35 C7
St-Laurent-de-Carnols F 35 B8
St-Laurent-de-Cerdans F 34 F4
St-Laurent-de-Chamousset F 30 D5
St-Laurent-de-la-Cabrerisse F 34 D4
St-Laurent-de-la-Salanque F 34 E4
St-Laurent-de-Neste F 33 D6
St-Laurent-des-Autels F 23 F9
St-Laurent-du-Pont F 31 E8
St-Laurent-du-Var F 37 D6
St-Laurent-en-Caux F 18 E2
St-Laurent-en-Grandvaux F 31 B8
St-Laurent-les-Bains F 35 A6

St-Laurent-Médoc F 28 E4
St-Laurent-Nouan F 24 E6
St-Laurent-sur-Gorre F 29 D7
St-Laurent-sur-Sèvre F 28 B4
St-Léger B 19 E12
St-Léger-des-Vignes F 30 B3
St-Léger-sur-Yvelines F 24 C6
St-Léger-sous-Beuvray F 30 B5
St-Léonard F 26 D6
St-Léonard-de-Noblat F 29 D8
St Leonards UK 13 D11
St-Lizier F 33 D8
St-Lô F 23 B9
St-Lon-les-Mines F 32 C3
St-Loubès F 28 F5
St-Loup-Géanges F 30 B6
St-Loup-Lamairé F 28 B5
St-Loup-sur-Semouse F 26 E5
St-Lubin-des-Joncherets F 24 C5
St-Lunaire F 23 C7
St-Lupicin F 31 C8
St-Lyé F 25 D11
St-Lys F 33 C8
St-Macaire F 32 A5
St-Macaire-en-Mauges F 23 F10
St-Magne F 32 A4
St-Magne-de-Castillon F 28 F5
St-Maime F 35 C10
St-Maixent-l'École F 28 C5
St-Malo F 23 C7
St-Malo-de-la-Lande F 23 B8
St-Mamert-du-Gard F 35 C7
St-Marcel F 30 B6
St-Marcel F 29 B9
St-Marcel F 30 B6
St-Marcel-d'Ardèche F 35 B8
St-Marcel-lès-Annonay F 30 E6
St-Marcel-lès-Sauzet F 35 A8
St-Marcel-lès-Valence F 31 F6
St-Marcellin F 31 E7
St-Marc-sur-Seine F 25 E12
St-Mards-en-Othe F 25 D10
St Margaret's Hope UK 3 H11
St-Mars-d'Outillé F 24 E3
St-Mars-du-Désert F 23 F9
St-Mars-la-Brière F 24 D3
St-Mars-la-Jaille F 23 E9
St-Martial F 35 B6
St-Martial-de-Nabirat F 29 F8
St-Martial-de-Valette F 29 D7
St-Martin F 35 C6
St Martin GBG 22 B6
St Martin GBJ 23 B7
St-Martin-Boulogne F 15 F12
St-Martin-d'Ablois F 25 B10
St-Martin-d'Arrossa F 32 D3
St-Martin-d'Auxigny F 25 F7
St-Martin-de-Belleville F 31 E10
St-Martin-de-Castillon F 35 C10
St-Martin-de-Crau F 35 C8
St-Martin-de-Landelles F 23 C9
St-Martin-de-Londres F 35 C6
St-Martin-de-Ré F 28 C3
St-Martin-des-Besaces F 23 B10
St-Martin-des-Champs F 22 C4
St-Martin-d'Entraunes F 36 C5
St-Martin-de-Seignanx F 32 C3
St-Martin-de-Valamas F 30 F5
St-Martin-de-Valgalgues F 35 B7
St-Martin-d'Hères F 31 E8
St-Martin-d'Oney F 32 C4
St-Martin-du-Mont F 31 C7
St-Martin-du-Var F 37 D6
St-Martin-en-Bresse F 31 B7
St-Martin-le-Beau F 24 F4
St-Martin-sur-Ouanne F 25 E9
St-Martin-Valmeroux F 29 E10
St-Martin-Vésubie F 37 C6
St-Martory F 33 D7
St Mary's UK 3 H11
St-Mathieu F 29 D7
St-Mathurin F 28 B2
St-Maur F 29 B9
St-Maurice CH 31 C10
St-Maurice-de-Lignon F 30 E5
St-Maurice-des-Lions F 29 D7
St-Maurice-la-Souterraine F 29 C8
St-Maurice-l'Exil F 30 E6
St-Maurice-Navacelles F 35 C6
St-Maurin F 33 B7
St Mawes UK 12 E4
St-Max F 26 C5
St-Maximin-la-Ste-Baume F 35 D10
St-Médard-en-Jalles F 28 F4
St-Méen-le-Grand F 23 D7
St-Méloir-des-Ondes F 23 C8
St-Memmie F 25 C11
St-Menoux F 30 B3
St Merryn UK 12 D5
St-Mesmin F 25 D9
St-Mesmin F 29 E8
St-Michel F 19 E9
St-Michel F 28 D5
St-Michel-Chef-Chef F 23 F7
St-Michel-de-Castelnau F 32 B5
St-Michel-de-Maurienne F 31 E9
St-Michel-en-l'Herm F 28 C3
St-Michel-Mont-Mercure F 28 B4
St-Mihiel F 26 C4
St Monans UK 5 C11
St-Montant F 35 B8
St-Nabord F 26 D6
St-Nauphary F 33 C8
St-Nazaire F 23 F7
St-Nazaire-le-Désert F 35 A9
St-Nectaire F 30 D2
St Neots UK 15 C8
St-Nicolas B 183 D7
St-Nicolas F 34 C5
St-Nicolas-d'Aliermont F 18 E3
St-Nicolas-de-la-Grave F 33 B8
St-Nicolas-de-Port F 26 C5
St-Nicolas-de-Redon F 23 E7
St-Nicolas-du-Pélem F 22 D5
St-Oedenrode NL 16 E4
St-Omer F 15 C9
St Osyth UK 15 D11
St-Ouen F 18 D5
St-Ouen F 24 E5
St-Ouen-du-Tilleul F 18 F2
St-Ouen-sur-Iton F 24 C4
St-Ouen-Toits F 23 D10
St-Pair-sur-Mer F 23 C8
St-Palais F 32 D3
St-Palais-sur-Mer F 28 D3

St-Pal-de-Chalancon F 30 E4
St-Pal-de-Mons F 30 E5
St-Pantaléon F 30 B5
St-Pantaléon F 33 B8
St-Papoul F 33 D10
St-Pardoux-Isaac F 33 A6
St-Pardoux-la-Rivière F 29 E7
St-Parize-le-Châtel F 30 B3
St-Parres-lès-Vaudes F 25 D11
St-Paterne F 23 D12
St-Paterne-Racan F 24 E3
St-Paul F 36 E5
St-Paul-Cap-de-Joux F 33 C9
St-Paul-de-Fenouillet F 33 E11
St-Paul-de-Jarrat F 33 E9
St-Paul-en-Born F 32 B3
St-Paul-en-Forêt F 36 D5
St-Paul-et-Valmalle F 35 C6
St-Paulien F 30 E4
St-Paul-le-Jeune F 35 B7
St-Paul-lès-Dax F 32 C3
St-Paul-lès-Durance F 35 C10
St-Paul-Trois-Châteaux F 35 B8
St-Pé-de-Bigorre F 32 D5
St-Pée-sur-Nivelle F 32 D2
St-Péray F 30 F6
St-Père F 25 F10
St-Père-en-Retz F 23 F7
Peter in the Wood GBG 22 B6
St Peter Port GBG 22 B6
St-Phal F 25 D10
St-Philbert-de-Bouaine F 28 B2
St-Philbert-de-Grand-Lieu F 28 A2
St-Pierre F 31 D11
St-Pierre-d'Albigny F 31 D9
St-Pierre-de-Chignac F 29 E7
St-Pierre-de-Côle F 29 E7
St-Pierre-de-la-Fage F 34 C5
St-Pierre-de-Maillé F 29 B7
St-Pierre-de-Plesguen F 23 D8
St-Pierre-des-Champs F 34 D4
St-Pierre-des-Corps F 24 F4
St-Pierre-des-Échaubrognes F 28 A4
St-Pierre-des-Landes F 23 D9
St-Pierre-des-Nids F 23 D11
St-Pierre-de-Trivisy F 33 C10
St-Pierre-d'Irube F 32 D3
St-Pierre-d'Oléron F 28 D3
St-Pierre-du-Chemin F 28 B4
St-Pierre-du-Mont F 32 C4
St-Pierre-Église F 23 A9
St-Pierre-en-Faucigny F 31 C9
St-Pierre-le-Moûtier F 30 B3
St-Pierre-lès-Elbeuf F 18 F3
St-Pierre-lès-Nemours F 25 D8
St-Pierre-Montlimart F 23 F9
St-Pierre-Quiberon F 22 E5
St-Pierre-sur-Dives F 23 B11
St-Plancard F 33 D7
St-Pois F 23 C9
St-Poix F 23 E9
St-Pol-de-Léon F 22 C4
St-Pol-sur-Mer F 15 B9
St-Pol-sur-Ternoise F 18 D5
St-Pompont F 29 F8
St-Pons F 36 C5
St-Pons-de-Thomières F 34 D4
St-Porchaire F 28 D4
St-Pourçain-sur-Sioule F 30 C3
St-Prex CH 31 C9
St-Priest F 30 D6
St-Priest-de-Champs F 30 D2
St-Priest-Laprugne F 30 D4
St-Priest-Taurion F 29 D8
St-Privat F 29 E10
St-Privat-d'Allier F 30 F4
St-Prix F 30 B5
St-Projet F 33 B9
St-Puy F 33 C6
St-Quentin F 19 E7
St-Quentin-la-Poterie F 35 B7
St-Quirin F 27 C7
St-Rambert-d'Albon F 30 E6
St-Rambert-en-Bugey F 31 D7
St-Raphaël F 36 E5
St-Remèze F 35 B8
St-Rémy F 30 B6
St-Rémy-de-Provence F 35 C8
St-Remy-en-Bouzemont-St-Genest-et-Isson F 25 C12
St-Rémy-sur-Avre F 24 C5
St-Rémy-sur-Durolle F 30 D4
St-Renan F 22 D2
St-Révérien F 25 F10
St-Rhemy I 31 D11
St-Riquier F 18 D4
St-Romain-en-Gal F 30 D6
St-Romain-sur-Cher F 24 F5
St-Romans F 31 E7
St-Rome-de-Cernon F 34 B4
St-Rome-de-Tarn F 34 B4
St-Saëns F 18 E3
St Sampson GBG 22 B6
St-Saturnin-lès-Apt F 35 C9
St-Saud-Lacoussière F 29 D7
St-Saulge F 25 F10
St-Sauves-d'Auvergne F 29 D11
St-Sauveur F 32 D3
St-Sauveur F 26 E5
St-Sauveur-de-Montagut F 30 F6
St-Sauveur-en-Puisaye F 25 E9
St-Sauveur-Gouvernet F 35 B9
St-Sauveur-Lendelin F 23 B9
St-Sauveur-le-Vicomte F 23 B8
St-Sauveur-sur-Tinée F 36 C6
St-Sauvy F 33 C7
St-Savin F 28 E5
St-Savin F 29 B7
St-Savinien F 28 D4
St Saviour GBJ 23 B7
St-Sébastien-de-Morsent F 24 C5
St-Sébastien-sur-Loire F 23 F9
St-Seine-l'Abbaye F 25 E12
St-Sernin F 33 A7
St-Sernin-sur-Rance F 34 C4
St-Seurin-sur-l'Isle F 28 E5
St-Sever F 32 C4
St-Sever-Calvados F 23 C9
St-Siméon-de-Bressieux F 31 E7
St-Simon F 19 E7
St-Simon F 29 F11
St-Sorlin-d'Arves F 31 E9
St-Soupplets F 25 B8
St-Sulpice F 33 C9
St-Sulpice-Laurière F 29 C8
St-Sulpice-les-Champs F 29 C9
St-Sulpice-les-Feuilles F 29 C8
St-Sulpice-sur-Lèze F 33 D8

St-Sulpice-sur-Risle F 24 C4
St-Sylvain F 23 B11
St-Symphorien F 30 H4
St-Symphorien F 32 B5
St-Symphorien-de-Lay F 30 D5
St-Symphorien-sur-Coise F 30 D5
St Teath UK 12 D5
St-Thégonnec F 22 C4
St-Thibéry F 34 D5
St-Thiébault F 26 D4
St-Thurien F 22 E4
St-Trivier-de-Courtes F 31 C7
St-Trivier-sur-Moignans F 30 C6
St-Trojan-les-Bains F 28 D3
St-Tropez F 36 E5
St-Uze F 30 E6
St-Valérien F 25 D9
St-Valery-en-Caux F 18 E2
St-Valery-sur-Somme F 18 D4
St-Vallier F 30 B5
St-Vallier F 30 E6
St-Vallier-de-Thiey F 36 D5
St-Varent F 28 B5
St-Vaury F 29 C9
St-Victor F 30 E6
St-Victor-de-Cessieu F 31 D7
St-Victoret F 35 C9
St-Victor-la-Coste F 35 B8
St Vigeans UK 5 B11
St-Vigor-le-Grand F 23 B10
St Vincent I 68 B4
St-Vincent-de-Connezac F 29 E6
St-Vincent-de-Paul F 32 C4
St-Vincent-les-Forts F 36 C4
St-Vit F 26 F4
St-Vite F 33 B7
St-Vith B 20 D6
St-Vivien-de-Médoc F 28 E3
St-Xandre F 28 C3
St-Yan F 30 C5
St-Ybars F 33 D8
St-Yorre F 30 C3
St-Yrieix-la-Perche F 29 D8
St-Yrieix-sur-Charente F 29 D6
St-Yvy F 22 E4
St-Zacharie F 35 D10
Sainville F 24 D6
Saissac F 33 D10
Saittarova S 116 D8
Saïx F 33 C10
Sajaniemi FIN 127 D11
Šahyajince SRB 164 E5
Šajkaš SRB 158 C5
Sajóbábony H 145 G2
Sajókaza H 145 G2
Sajókeresztúr H 145 G2
Sajólád H 145 G2
Sajószentpéter H 145 G2
Sajószöged H 145 H3
Sajóvámos H 145 G2
Säjvis S 119 C11
Saka LV 134 C7
Saka LT 135 C11
Saka S 98 C7
Šaľa SK 146 E5
Salaca LV 131 F10
Šilacea RO 151 C9
Salacgrīva LV 131 F8
Sala Consilina I 60 C5
Salagnac F 29 E8
Salahmi FIN 124 C7
Salaise-sur-Sanne F 30 E6
Salakas LT 135 E12
Salakos GR 181 D7
Salakovac BIH 157 F8
Salamajärvi FIN 123 D13
Salamanca E 45 C9
Salamina GR 175 D7
Salandra I 61 B6
Salanki FIN 117 B13
Salantai LT 134 D3
Salar E 53 B8
Šālard RO 151 C9
Salardu E 33 E7
Salarli TR 172 B6
Salas E 39 B7
Salaš SRB 159 E9
Salas de los Infantes E 40 D5
Salash BG 164 B6
Salaspils LV 135 C8
Sălașu de Sus RO 159 C10
Sălățig RO 151 C11
Sălătrucel RO 160 C4
Sălătrucu RO 160 C5
Salaunes F 28 F4
Salberg S 107 D16
Salbertrand I 31 E10
Sălboda S 97 C8
Salbohed S 98 C6
Salbris F 24 F7
Salbu N 100 D2
Salcea RO 153 B8
Salching D 75 E12
Salcia RO 159 E10
Salcia RO 160 E5
Salcia RO 161 C3
Salcia Tudor RO 161 C10
Sălciile RO 161 D8
Šalčininkai LT 137 E11
Šalčininkėliai LT 137 E11
Sălciua RO 151 E11
Salcombe UK 13 E7
Sălcuța MD 154 D5
Sălcuța RO 160 E2
Saldaña E 39 C10
Saldenburg D 76 E4
Saldón E 47 D10
Salduero E 40 E6
Saldus LV 134 C4
Sale UK 11 E7
Sale I 37 B9
Saleby S 91 C13
Salem D 71 B9
Salem D 83 C9
Salemi I 58 D2

Salen UK 4 B5
Salen UK 4 B5
Sälen S 102 D5
Salernes F 36 D4
Salerno I 60 B3
Salers F 29 E10
Salettes F 30 F4
Saleux F 18 E5
Salford UK 11 E7
Şalgamli TR 173 B6
Salgótarján H 147 E9
Salgueiro P 44 E5
Salhus N 94 A2
Sali HR 156 E3
Salice Salentino I 61 C9
Saliceto I 37 C8
Saliena LV 135 C7
Saliena LV 135 E13
Salies-de-Béarn F 32 D4
Salies-du-Salat F 33 D7
Salignac-Eyvignes F 29 F8
Salillas de Jalón E 41 E9
Salinas E 39 A8
Salinas E 56 D3
Salinas del Manzano E 47 D9
Salinas de Pamplona E 32 E2
Salinas de Pisuerga E 40 C3
Salin-de-Giraud F 35 D8
Saline di Volterra I 66 F2
Sälinkää FIN 127 D13
Salins F 29 E10
Salins-les-Bains F 31 B8
Salir P 50 E3
Salisbury UK 13 C11
Sǎlişte RO 152 F3
Sǎlişte LV 135 E5
Sǎliştea de Sus RO 152 B4
Salka SK 147 F7
Sal'kove UA 154 A5
Sall DK 86 C5
Salla EST 131 D12
Salla FIN 115 E5
Sallanches F 31 D10
Sallent E 43 D7
Sallent de Gállego E 32 E5
Salles F 32 A4
Salles-Curan F 34 B4
Salles-d'Angles F 28 D5
Salles-la-Source F 33 B11
Salles-sur-l'Hers F 33 D9
Sallgast D 80 C5
Sälliku EST 131 C14
Sallinbjerg A 77 F8
Sallins IRL 7 F9
Sällsjö S 105 E15
Sallypark IRL 8 C6
Salme EST 130 E4
Salmerón E 47 C8
Salmeroncillos de Abajo E 47 C7
Salmi FIN 123 E10
Salmi S 119 B10
Salmijärvi FIN 121 D10
Salminen FIN 121 B12
Salminen FIN 124 B8
Salmivaara FIN 115 E4
Salmiyarvi RUS 114 E8
Salmoral E 45 C10
Salmtal D 21 E7
Salnava LV 133 C3
Salnö S 99 C11
Salo FIN 127 E9
Salò I 69 B10
Salobre E 55 B7
Salobreña E 53 C9
Saločiai LT 135 D8
Saloinen FIN 119 E12
Salon F 25 C11
Salon-de-Provence F 35 C9
Salonkylä FIN 123 C11
Salonpää FIN 119 E14
Salonta RO 151 D8
Salorino E 45 F6
Salornay-sur-Guye F 30 B6
Salorno I 69 A11
Salou E 42 E6
Salouël F 18 E5
Šalovci SLO 148 C6
Salsåker S 107 F14
Salsbruket N 105 B11
Salsburgh UK 5 D9
Salses-le-Château F 34 E4
Sälsig RO 151 B11
Salsomaggiore Terme I 69 D8
Salt E 43 D9
Saltara I 67 E6
Saltash UK 12 E6
Saltburn-by-the-Sea UK 11 B10
Saltcoats UK 4 D7
Salteras E 51 E7
Salthill IRL 6 F4
Salto P 38 E4
Saltoniškės LT 137 D11
Saltrød N 90 C4
Saltsjöbaden S 99 D10
Saltum DK 90 E6
Saltvik FIN 99 B14
Saltvik S 103 C13
Saludecio I 67 E6
Saluggia I 68 C5
Salur TR 173 D8
Salussola I 68 C5
Saluzzo I 37 B6
Salva RO 152 C4
Salvacañete E 47 D10
Salvagnac F 33 C9
Salvaleón E 51 B6
Salvaterra de Magos P 50 A2
Salvaterra do Extremo P 45 E7
Salvatierra E 32 E1
Salvatierra de los Barros E 51 C6
Salvatierra de Santiago E 45 F8
Salve I 61 D10
Salviac F 33 A8
Sály H 145 H2
Salzburg A 73 A7
Salzgitter D 79 B7
Salzhausen D 83 D8
Salzhemmendorf D 78 B6
Salzkotten D 17 E11
Salzmünde D 79 C10
Salzwedel D 83 E10
Salzweg D 76 E4
Samadet F 32 C5
Samaila SRB 158 F4
Samarate I 68 B6
Samarica HR 149 E7
Samarineşti RO 159 D11
Samassi I 64 E2
Samatan F 33 D7
Sambade P 39 F6
Sâmbăta RO 151 D9

Sambiase I 59 B9
Sambir UA 145 D7
Samboal E 40 F3
Samborzec PL 143 E12
Sambuca di Sicilia I 58 D3
Sambuca Pistoiese I 66 D3
Sambuco I 36 C6
Sâmbureşti RO 160 D4
Samedan CH 71 D9
Sameiro P 44 D6
Samer F 15 F12
Sames S 39 B9
Sami GR 174 C2
Samil P 39 E6
Samir de los Caños E 39 E7
Şamli TR 173 E8
Sammakko S 116 E7
Sammakkola FIN 125 C10
Sammaljoki FIN 127 C10
Sammatti FIN 127 E10
Sammichele di Bari I 61 B7
Samnaun CH 71 D10
Samobor HR 148 E5
Samoëns F 31 C10
Samões P 38 F5
Samokov BG 165 E8
Samokov NMK 164 F4
Samolaco I 69 A7
Samora Correia P 50 B2
Šamorín SK 146 E4
Samos E 38 C5
Samos GR 177 D8
Samoš SRB 159 C6
Samothraki GR 171 D9
Samovodene BG 166 C5
Samper de Calanda E 42 E3
Sampeyre I 37 C9
Sampieri I 59 F6
Sampigny F 26 C3
amşud RO 151 C10
Samswegen D 79 B9
Samtens D 84 B4
Samuelsberg N 112 D6
Samuil BG 167 B7
Samuilovo BG 166 E5
San Adrián E 32 F2
San Agustín de Guadalix E 46 C5
Sanaigmore UK 4 D5
San Amaro E 38 D3
Sânandrei RO 151 F7
San Andrés del Rabanedo E 39 C8
San Antolín E 39 B6
San Antonio E 47 E10
San Asensio E 40 C5
San Bartolomé de las Abiertas E 46 E3
San Bartolomé de la Torre E 51 E5
San Bartolomé de Pinares E 46 C3
San Bartolomeo al Mare I 37 D8
San Bartolomeo in Galdo I 60 A4
San Basilio I 64 D3
San Benedetto dei Marsi I 62 C5
San Benedetto del Tronto I 62 B5
San Benedetto Po I 66 B2
San Benito E 54 B3
San Benito de la Contienda E 51 B5
San Biagio di Callalta I 72 E5
San Biago Platani I 58 D4
San Bonifacio I 66 B3
San Buono I 63 D7
San Candido I 72 C5
San Carlos del Valle E 55 B6
San Casciano dei Bagni I 62 B1
San Casciano in Val di Pesa I 66 E3
San Cataldo I 58 E4
San Cataldo I 61 C10
San Cebrián de Castro E 39 E8
Sâncel RO 152 E3
Sancergues F 25 F8
Sancerre F 25 F8
San Cesario sul Panaro I 66 C3
Sancey-le-Grand F 26 F6
Sancheville F 24 D6
Sanchidrián E 46 C3
San Chirico Nuovo I 60 B6
San Chirico Raparo I 60 C6
San Cibrão das Viñas E 38 D4
San Cipirello I 58 D3
San Cipriano d'Aversa I 60 B2
San Clemente E 47 F8
San Clodio E 38 D5
Sancoins F 30 B2
San Colombano al Lambro I 69 C7
San Cosme E 38 A5
San Costantino Albanese I 61 C6
San Costanzo I 67 E7
Sâncrăieni RO 153 E7
Sâncraiu RO 151 D10
Sâncraiu de Mureş RO 152 D5
San Cristóbal de Entreviñas E 39 D8
San Cristóbal de la Vega E 46 B3
Sancti-Spíritus E 45 C8
Sancti-Spíritus E 51 B9
Sand N 94 D4
Sand N 95 B15
Sand N 110 C5
Sand (Bad Emstal) D 17 F12
Sanda S 93 E12
Sandager DK 86 E5
Sandamendi E 40 B5
San Damiano d'Asti I 37 B8
San Damiano Macra I 37 C6
Sandane N 100 C4
San Daniele del Friuli I 73 D7
San Daniele Po I 66 B2
Sandanski BG 169 A9
Sandared S 91 D12
Sandarne S 103 D13
Sandau D 83 E12
Sandbach UK 11 E7
Sandberg D 74 B7
Sandby DK 87 F8
Sande N 95 C12
Sande N 100 D5
Sande P 44 B4
Sandefjord N 90 A7
Sandeggen N 111 A17
Sandeid N 94 C3

Sandelva N 112 D7
San Demetrio Corone I 61 D6
San Demetrio ne Vestini I 62 C5
Sander N 96 B6
Sandersdorf D 79 C11
Sandershausen (Niestetal) D 78 D6
Sandersleben D 79 C10
Sandes N 110 C9
Sandes N 110 C9
Sandfjord N 114 B9
Sandfors S 118 E5
Sandgarth UK 3 G11
Sandhausen D 21 F11
Sandhead UK 4 F6
Sandhem S 91 D14
Sandhult S 91 D12
Sandhurst UK 15 E7
Sandiás E 38 D4
Sandkrug D 84 D5
Sandl A 77 E7
Sandland N 112 C8
Sandnäset S 102 A8
Sandnes N 90 A5
Sandnes N 94 E3
Sandnes N 108 B4
Sandneshamn N 111 A15
Sandness N 111 C11
Sandnessjøen N 108 D4
Sando E 45 C8
Sandomierz PL 143 E12
Sândominic RO 153 D7
San Donaci I 61 C9
San Donà di Piave I 72 E6
San Donato di Lecce I 61 C10
San Donato di Ninea I 60 D6
San Donato Milanese I 69 C7
San Donato Val di Comino I 62 D5
Sándorfalva H 150 E5
Sandown UK 13 D12
Sandøy N 100 A5
Sandplace UK 12 E6
Šandrovac HR 149 E8
Sandsele S 107 A13
Sandsend UK 11 B10
Sandsjö S 102 A8
Sandsjöfors S 92 E5
Sandsjönäs S 107 A13
Sandslån S 107 E13
Sandstad N 104 D6
Sandstedt D 17 B11
Sandstrak N 108 D5
Sandstrand N 111 C12
Sandtangen N 114 D7
Sandtorg N 111 C12
Sandträsk S 118 B6
Sânduleni RO 153 E9
Sânduleşti RO 152 D3
Sandvatn N 94 F5
Sandved DK 87 E9
Sandvik FO 2 B3
Sandvik N 101 D15
Sandvik N 108 B7
Sandvik N 111 A16
Sandvik N 111 B14
Sandvik N 113 B16
Sandvik S 103 C11
Sandvika N 95 C13
Sandvika N 105 D12
Sandviken S 103 E12
Sandviken S 107 E15
Sandviken S 107 F13
Sandviksjön S 106 C7
Sandvikvåg N 94 C2
Sandwich UK 15 E11
Sandwick UK 3 F14
Sandy UK 15 C8
Sanem L 20 E5
San Emiliano E 39 C8
San Esteban de Gormaz E 40 E5
San Esteban de la Sierra E 45 C9
San Esteban de Litera E 42 D4
San Esteban del Molar E 39 E8
San Esteban del Valle E 46 D3
San Fele I 60 B5
San Felice a Cancello I 60 A2
San Felice Circeo I 62 E4
San Felices de los Gallegos E 45 C7
San Felice sul Panaro I 66 C3
San Ferdinando I 59 C8
San Ferdinando di Puglia I 60 A6
San Fernando E 52 D4
San Fernando de Henares E 46 D5
San Fili I 60 E6
San Filippo del Mela I 59 C7
Sanfins do Douro P 38 F5
San Francisco Javier E 57 D7
San Fratello I 59 C6
Sanfront I 37 B6
Sânga S 107 E13
Sangarcía E 46 C3
Sangaste EST 131 F12
Sangatte F 15 F12
San Gavino Monreale I 64 D2
Sângeni I 62 B3
Sângeorgiu de Mureş RO 152 D5
Sângeorgiu de Pădure RO 152 E5
Sângeorz-Băi RO 152 C5
Sânger RO 152 D4
Sangerhausen D 79 D9
San Germano Chisone I 31 F11
Sângeru RO 161 C8
Sangijän S 119 C11
San Gimignano I 66 F3
San Ginesio I 62 A4
Sanginjoki FIN 119 E16
Sanginkylä FIN 119 E17
San Giorgio a Liri I 62 E5
San Giorgio della Richinvelda I 73 D6
San Giorgio del Sannio I 60 A3
San Giorgio di Lomellina I 68 C6
San Giorgio di Nogaro I 73 E7
San Giorgio di Piano I 66 C3
San Giorgio Ionico I 61 C8
San Giorgio la Molara I 60 A3
San Giorgio Lucano I 61 C7
San Giovanni a Piro I 60 C4
San Giovanni Bianco I 69 B8
San Giovanni d'Asso I 66 F4
San Giovanni Gemini I 58 D4
San Giovanni Incarico I 62 D4

San Giovanni in Croce I 66 B1
San Giovanni in Fiore I 61 E7
San Giovanni in Persiceto I 66 C3
San Giovanni Lupatoto I 66 B3
San Giovanni Rotondo I 63 D8
San Giovanni Suergiu I 64 E1
San Giovanni Teatino I 63 C6
San Giovanni Valdarno I 66 E4
Sangis S 119 C11
San Giuliano Terme I 66 E1
San Giuseppe Jato I 58 D3
San Giuseppe Vesuviano I 60 B3
San Giustino I 66 F5
San Godenzo I 66 E4
San Gregorio I 59 C8
San Gregorio Magno I 60 B4
San Gregorio Matese I 60 A2
Sangüesa E 32 E3
San Guiliano Milanese I 69 C7
San Guim de Freixenet E 43 D6
Sanguinet F 32 B3
Sanguinetto I 66 B3
Sani GR 169 D9
San Ildefonso E 46 C5
Sanislău RO 151 B9
Sanitz D 83 B12
San Javier E 56 F3
San Jordi E 42 G4
San Jorge de Alor E 51 B5
San José E 55 F8
San José del Valle E 52 C5
San José de Malcocinado E 52 D5
San Juan E 41 B6
San Juan de Alicante E 56 E4
San Juan de Aznalfarache E 51 E7
San Juan de la Nava E 46 D3
San Juan del Puerto E 51 E6
San Justo de la Vega E 39 D7
Sankt Aegyd am Neuwalde A 77 G9
Sankt Andrä A 73 C10
Sankt Andrä an Zicksee A 149 A7
Sankt Andreasberg D 79 C8
Sankt Anna S 93 C9
Sankt Anna am Aigen A 148 C5
Sankt Anton an der Jeßnitz A 77 G8
Sankt Augustin D 21 C8
Sankt Gallen A 73 A10
Sankt Gallen CH 27 F11
Sankt Gallenkirch A 71 C9
Sankt Gangloff D 79 E10
Sankt Georgen am Walde A 77 F7
Sankt Georgen im Schwarzwald D 27 D9
Sankt Gilgen A 73 A7
Sankt Goar D 21 D9
Sankt Goarshausen D 21 D9
Sankt Ingbert D 21 F8
Sankt Jakob im Rosental A 73 C9
Sankt Jakob im Walde A 148 B5
Sankt Jakob in Defereggen A 72 C5
Sankt Johann am Tauern A 73 B9
Sankt Johann im Pongau A 73 B7
Sankt Johann im Walde A 73 C6
Sankt Johann in Tirol A 72 A5
Sankt Julian D 21 E9
Sankt Katharinen D 185 C7
Sankt Lambrecht A 73 B9
Sankt Leonhard am Forst A 77 F8
Sankt Leonhard am Hornerwald A 77 F8
Sankt Leonhard im Pitztal A 71 C11
Sankt Lorenz A 73 A7
Sankt Lorenzen im Gitschtal A 73 C7
Sankt Lorenzen im Lesachtal A 73 C6
Sankt Lorenzen im Mürztal A 148 A4
Sankt Lorenzen ob Murau A 73 B9
Sankt Marein im Mürztal A 148 A4
Sankt Margarethen D 17 A12
Sankt Margarethen an der Raab A 148 B5
Sankt Margarethen bei Knittelfeld A 73 B10
Sankt Margarethen im Burgenland A 77 G11
Sankt Märgen D 27 D9
Sankt Martin A 73 B7
Sankt Martin A 77 E7
Sankt Martin im Mühlkreis A 76 F6
Sankt Michael im Burgenland A 148 B6
Sankt Michael im Lungau A 73 B8
Sankt Michael in Obersteiermark A 73 B11
Sankt Michaelisdonn D 82 C6
Sankt Moritz CH 71 D9
Sankt Nikolai im Saustal A 148 C4
Sankt Nikolai im Sölktal A 73 B9
Sankt Olof S 88 D6
Sankt Oswald bei Freistadt A 77 E7
Sankt Oswald ob Eibiswald A 73 C11
Sankt Pankraz A 73 A9
Sankt Paul im Lavanttal A 73 C10
Sankt Peter am Kammersberg A 73 B9
Sankt Peter am Ottersbach A 148 C5
Sankt-Peterburg RUS 129 F13
Sankt Peter-Freienstein A 73 B11
Sankt Peter-Ording D 82 B5
Sankt Pölten A 77 F9
Sankt Radegund A 76 F3
Sankt Ruprecht an der Raab A 148 B5
Sankt Stefan im Gailtal A 73 C8
Sankt Stefan ob Leoben A 73 B10
Sankt Stefan ob Stainz A 148 C4
Sankt Ulrich bei Steyr A 76 F6

Sankt Valentin A 77 F7
Sankt Veit am Vogau A 148 C5
Sankt Veit an der Glan A 73 C9
Sankt Veit an der Gölsen A 77 F9
Sankt Veit im Pongau A 73 B7
Sankt Veit in Defereggen A 72 C5
Sankt Wendel D 21 F8
Sankt Wolfgang D 75 F11
Sankt Wolfgang im Salzkammergut A 73 A7
San Lazzaro di Savena I 66 D3
San Leo I 66 E5
San Leonardo de Yagüe E 40 E5
San Leonardo in Passiria I 72 C3
San Lorenzo I 59 D8
San Lorenzo al Mare I 37 D7
San Lorenzo Bellizzi I 61 D6
San Lorenzo de Calatrava E 54 C5
San Lorenzo de El Escorial E 46 C4
San Lorenzo de la Parrilla E 47 E8
San Lorenzo di Sebato I 72 C4
San Lorenzo in Campo I 67 E6
San Lorenzo Nuovo I 62 B1
San Luca I 59 D8
Sanlúcar de Barrameda E 52 C4
Sanlúcar de Guadiana E 51 D5
Sanlúcar la Mayor E 51 E7
San Lucido I 60 E6
Sanluri I 64 D2
San Maddalena Vallalta I 72 C5
San Mamés de Campos E 40 D2
San Marcello I 62 A3
San Marcello Pistoiese I 66 D2
San Marco dei Cavoti I 60 A3
San Marco in Lamis I 63 D8
San Marino RSM 66 E5
San Martín E 32 C1
San Martín E 40 B6
Sânmartin RO 151 C8
Sânmartin RO 151 C8
Sânmartin RO 153 E7
San Martín de la Vega E 46 D5
San Martín de la Vega del Alberche E 45 D10
San Martín del Pimpollar E 45 D10
San Martín de Montalbán E 46 E4
San Martín de Pusa E 46 E3
San Martín de Unx E 32 E2
San Martín de Valdeiglesias E 46 D4
San Martino Buon Albergo I 66 B3
San Martino di Castrozza I 72 D4
San Martino di Lupari I 72 E4
San Martino di Venezze I 66 B4
San Martino in Badia I 72 C4
San Martino in Passiria I 72 C3
San Martino in Pensilis I 63 D8
San Mateo de Gállego E 41 E10
San Mauro Castelverde I 58 D5
San Mauro Forte I 60 C6
San Mauro Marchesato I 61 E7
San Mauro Pascoli I 66 D5
San Mauro Torinese I 68 C4
San Menaio I 63 D9
San Michele al Tagliamento I 73 E6
San Michele Mondovì I 37 C7
San Michele Salentino I 61 B9
San Miguel de Arroyo E 40 E3
San Miguel de Bernuy E 40 F4
San Miguel de Salinas E 56 F3
Sânmihaiu Almaşului RO 151 C11
Sânmihaiu de Câmpie RO 152 D4
Sânmihaiu Român RO 159 B7
San Millán de la Cogolla E 40 D6
San Miniato I 66 E2
Sänna S 92 B5
Sannahed S 92 A6
Sannazzaro de'Burgondi I 69 C6
Sannicandro di Bari I 61 B7
Sannicandro Garganico I 63 D9
Sannicola I 61 C10
San Nicola dell'Alto I 61 E7
San-Nicolao F 37 G10
San Nicolás del Puerto E 51 C8
Sânnicolau Mare RO 150 E6
San Nicolò I 66 C4
San Nicolò d'Arcidano I 64 D2
San Nicolò Gerrei I 64 E3
Sanniki PL 139 F8
Sanok PL 145 D5
San Pablo de los Montes E 46 E4
San Pancrazio I 72 C3
San Pancrazio Salentino I 61 C9
San Paolo di Civitate I 63 D8
Sânpaul RO 152 E4
San Pedro E 55 B8
San Pedro de Alcántara E 53 D7
San Pedro de Ceque E 39 D7
San Pedro del Arroyo E 46 C3
San Pedro de Latarce E 39 E9
San Pedro del Pinatar E 56 F3
San Pedro del Romeral E 40 B4
San Pedro de Rozados E 45 C9
San Pedro Manrique E 41 D7
San Pedro Palmiches E 47 D8
San Pellegrino Terme I 69 B8
Sânpetru RO 153 F7
Sânpetru de Câmpie RO 152 D4
Sânpetru Mare RO 150 E6
San Piero a Sieve I 66 E3
San Piero Patti I 59 C6
San Pietro I 59 D6
San Pietro di Cadore I 73 C6
San Pietro in Cariano I 66 B3
San Pietro in Casale I 66 C3
San Pietro in Guarano I 61 E6
San Pietro Vernotico I 61 C10
San Polo d'Enza I 66 C1
San Prospero I 66 C3
Sanquhar UK 5 E9
San Quirico d'Orcia I 65 A5
San Rafael del Río E 42 F4
San Remo I 37 D7
San Román E 38 B5
San Román de Cameros E 41 D7
San Román de la Cuba E 39 D10
San Román de los Montes E 46 D3

San Roque E 38 B2
San Roque E 38 D3
San Roque E 53 D7
San Rufo I 60 C4
San Salvador de Cantamunda E 40 C3
San Salvatore Monferrato I 37 B9
San Salvatore Telesino I 60 A2
San Salvo I 63 C7
San Sebastián de los Ballesteros E 53 A7
San Sebastián de los Reyes E 46 C5
San Secondo Parmense I 66 C1
Sansepolcro I 66 E5
San Severa I 62 C1
San Severino Lucano I 60 C6
San Severino Marche I 67 F7
San Severo I 63 D8
San Silvestre de Guzmán E 50 E5
Sânsimion RO 153 E7
Sanski Most BIH 157 C6
Sansol E 32 E1
San Sosti I 60 D6
San Sperate I 64 E3
San Spirito I 61 A7
Santa Amalia E 51 A7
Santa Ana E 55 B9
Santa Ana de Pusa E 46 E3
Santa Ana la Real E 51 D6
Santa Bárbara E 42 F5
Santa Bárbara de Casa E 51 D5
Santacara E 32 F2
Santa Catalina de Armada E 38 B2
Santa Catarina I 50 C3
Santa Caterina dello Ionio I 59 B10
Santa Caterina di Pittinuri I 64 C2
Santa Caterina Villarmosa I 58 D5
Santa Cesarea Terme I 61 C10
Santa Cilia de Jaca E 32 E4
Santa Clara-a-Nova P 50 E3
Santa Clara-a-Velha P 50 D3
Santa Clara de Louredo P 50 D4
Santa Coloma de Farners E 43 D9
Santa Coloma de Queralt E 43 D6
Santa Colomba de Somoza E 39 D7
Santa Columba de Curueño E 39 C8
Santa Comba Dão P 44 D4
Santa Comba de Rossas P 39 E6
Santa Cristina d'Aro E 43 D9
Santa Cristina de la Polvorosa E 39 D8
Santa Croce Camerina I 59 F6
Santa Croce del Sannio I 60 A3
Santa Croce di Magliano I 63 D8
Santa Croce sull'Arno I 66 E2
Santa Cruz E 43 D9
Santa Cruz da Tapa P 44 C4
Santa Cruz de Bezana E 40 B4
Santa Cruz de Campézo E 32 E1
Santa Cruz de la Serós E 32 E4
Santa Cruz de la Sierra E 45 F9
Santa Cruz de la Zarza E 47 E6
Santa Cruz de los Cáñamos E 55 B7
Santa Cruz del Retamar E 46 D4
Santa Cruz de Moya E 47 E10
Santa Cruz de Mudela E 55 B6
Santadi I 64 E2
Santa Domenica Talao I 60 D5
Santa Domenica Vittoria I 59 D6
Santa Elena E 55 C5
Santa Elena de Jamuz E 39 D8
Santa Elisabetta I 58 D4
Santaella I 53 A7
Santa Engracia E 32 F1
Santa Eufemia E 54 B3
Santa Eugènia E 49 E10
Santa Eulalia E 39 B9
Santa Eulalia E 47 C10
Santa Eulália P 51 A5
Santa Eulalia del Río E 57 D8
Santa Eulalia de Oscos E 38 B5
Santa Eulália de Riuprimer E 43 D8
Santa Fé E 53 B9
Santa Fiora I 65 B5
Sant'Agata de'Goti I 60 A3
Sant'Agata del Bianco I 59 C9
Sant'Agata di Esaro I 60 D5
Sant'Agata di Militello I 59 C6
Sant'Agata di Puglia I 60 A5
Sant'Agata Feltria I 66 E5
Santa Giusta I 64 D2
Santa Giustina I 72 D5
Sant'Agostino I 66 C3
Sant Agustí de Lluçanès E 43 C8
Santahamina FIN 127 E13
Santa Iria P 50 D4
Santa Justa P 50 A3
Sant'Alberto I 66 C5
Santalha P 38 E5
Santa Liestra y San Quílez E 33 F6
Santa Luce I 66 F2
Santa Lucia I 64 B4
Santa Lucia del Mela I 59 C7
Santa Lucía de Moraña E 38 C2
Santa Luzia P 50 E3
Santa Magdalena de Pulpís E 48 D5
Santa Mare RO 153 B10
Santa Margalida E 57 B11
Santa Margarida da Serra P 50 C2
Santa Margarida de Montbui E 43 D7
Santa Margarida do Sádão P 50 C3
Santa Margherita di Belice I 58 D3
Santa Margherita Ligure I 37 C10
Santa Maria CH 71 D10
Santa María E 32 A4

Santa Maria Capua Vetere I 60 A2
Santa Maria da Feira P 44 C3
Santa María de Cayón E 40 B4
Santa María de Corcó E 43 C8
Santa María de Huerta E 41 F7
Santa María del Berrocal E 45 C10
Santa María del Camí E 49 E10
Santa María del Campo E 40 D4
Santa María del Campo Rus E 47 E8
Santa María del Cedro I 60 D5
Santa Maria della Versa I 37 B10
Santa María de los Llanos E 47 F7
Santa María del Páramo E 39 D8
Santa María del Val E 47 C8
Santa María de Nieva E 55 E8
Santa María de Palautordera E 43 D8
Santa María di Castellabate I 60 C3
Santa María di Sala I 66 B5
Santa Maria la Real de Nieva E 46 B4
Santa Maria Maggiore I 68 A5
Santa María Navarrese I 64 D4
Santa Maria Nuova I 67 E7
Santa Maria Rezzonico I 69 A7
Santa-Maria-Sichè F 37 H9
Santa Marina I 60 C5
Santa Marina del Rey E 39 C8
Santa Marina Salina I 59 B6
Santa Marinella I 62 C1
Santa Marta E 51 B6
Santa Marta E 51 B6
Santa Marta de Penaguião P 44 B5
Santa Marta de Tormes E 45 C9
Sant'Ambroggio F 37 F9
Santana P 44 D3
Santana E 50 C1
Sântana RO 151 E8
Santana da Serra P 50 D3
Santana de Cambas P 50 D4
Santana do Mato P 50 B3
Sant'Anastasia I 60 B3
Sant'Anatolia di Narco I 62 B3
Santander E 40 B4
Sant'Andrea Apostolo dello Ionio I 59 B10
Sant'Andrea Frius I 64 E3
Sântandrei RO 151 C8
Sant'Angelo I 59 B9
Sant'Angelo a Fasanella I 60 C4
Sant'Angelo dei Lombardi I 60 B4
Sant'Angelo di Brolo I 59 C6
Sant'Angelo in Lizzola I 67 E6
Sant'Angelo in Vado I 66 E5
Sant'Angelo Lodigiano I 69 C7
Sant'Angelo Muxaro I 58 E4
Santa Ninfa I 58 D2
Sant'Anna Arresi I 64 E2
Sant'Antimo I 60 B2
Sant'Antioco I 64 E1
Sant Antoni de Portmany F 57 D7
Sant'Antonio Abate I 60 B3
Sant'Antonio di Gallura I 64 B3
Sant'Antonio di Santadi I 64 E2
Santanyí E 57 C11
Santa Olalla E 46 D4
Santa Oliva E 43 E7
Santa Ollala del Cala E 51 D7
Santa Pau E 43 C9
Santa Pola E 56 E3
Santar P 44 C5
Sant'Arcangelo I 60 C6
Santarcangelo di Romagna I 66 D5
Santarém P 44 F3
Sant'Arsenio I 60 C4
Santas Martas E 39 D8
Santa Severina I 61 E7
Santa Sofia I 66 E4
Santa Sofia P 50 B3
Santa Sofia d'Epiro I 61 D6
Santa Susana P 50 B3
Santa Susana P 50 C3
Santa Teresa di Gallura I 64 A3
Santa Teresa di Riva I 59 D7
Santău RO 151 C9
Santa Uxía E 38 C2
Santa Venerina I 59 D7
Santa Vitória P 50 D3
Santa Vitória do Ameixial P 50 B4
Sant Boi de Llobregat E 43 E8
Sant Carles de la Ràpita E 42 F5
Sant Celoni E 43 D8
Sant Cugat del Vallès E 43 E8
Sant'Egidio alla Vibrata I 62 B5
Sant'Elia Fiumerapido I 62 D5
Sant Elm E 49 E9
San Telmo E 51 D6
Sant'Elpidio a Mare I 67 F8
San Teodoro I 64 B4
Santeramo in Colle I 61 B7
Santervás de la Vega E 39 C10
Santes Creus E 43 E6
Sant Feliu de Guíxols E 43 D10
Sant Feliu de Pallerols E 43 C9
Sant Feliu Sasserra E 43 D8
Santhià E 68 C5
Sant Hilari Sacalm E 43 D9
Sant Hipòlit de Voltregà E 43 C8
Santiago de Alcántara E 45 E6
Santiago de Calatrava E 53 A8
Santiago de Compostela E 38 C2
Santiago de Covelo E 38 D3
Santiago de la Espada E 55 C7
Santiago de la Ribera E 56 F3
Santiago del Campo E 45 E8
Santiago do Escoural P 50 B3
Santiagomillas E 39 D7
Santibáñez de Béjar E 45 D9
Santibáñez de la Peña E 39 C10
Santibáñez de la Sierra E 45 C9
Santibáñez de Tera E 39 E8
Santibáñez de Vidriales E 39 D7
Santibáñez el Bajo E 45 D8
Santibáñez Zarzaguda E 40 C4
Sant'Ilario d'Enza I 66 C1
Santillana E 40 B3
Sântimbru RO 152 E3

Santiponce E 51 E7
Santisteban del Puerto E 55 C6
Santiuste de San Juan Bautista
  E 46 B1
Santiz E 45 B9
Sant Jaime Mediterráneo
  E 57 B13
Sant Jaume d'Enveja E 42 F5
Sant Joan de Labritja E 49 F8
Sant Joan de les Abadesses
  E 43 C8
Sant Joan de Vilatorrada
  E 43 D7
Sant Joan les Fonts E 43 C9
Sant Josep de sa Talaia E 57 D7
Sant Julià de Lòria AND 33 F9
Sant Llorenç de Morunys
  E 43 C7
Sant Llorenç des Cardassar
  E 57 B11
Sant Lluís E 57 B13
Sant Martí de Tous E 43 D7
Sant Martí Sarroca E 43 E7
Sant Mateu E 42 G4
Sant Miquel de Balansat E 57 C7
Santo Aleixo P 50 B5
Santo Aleixo da Restauração
  P 51 C5
Santo Amador P 51 C5
Santo André P 50 C2
Santo Antoniño E 38 C2
Santo António dos Cavaleiros
  P 50 B1
Santo Domingo de la Calzada
  E 40 D6
Santo Domingo de Silos E 40 E5
Santo Estêvão P 50 B2
Santo Estêvão P 50 E4
Santo Isidro de Pegões P 50 B2
Santok PL 85 E8
Santomera E 56 E2
Sant Omero I 62 B5
Santoña E 40 B5
Santo Pietro I 58 E6
Santo-Pietro-di-Tenda E 37 F10
Santo-Pietro-di-Venaco
  F 37 G10
Santo Stefano al Mare I 37 D7
Santo Stefano Belbo I 37 B8
Santo Stefano d'Aveto I 37 B10
Santo Stefano di Cadore I 72 C6
Santo Stefano di Camastra
  I 58 C5
Santo Stefano di Magra I 69 E8
Santo Stefano Quisquina
  I 58 D3
Santo Stino di Livenza I 73 E6
Santo Tirso P 38 F3
Santo Tomé E 55 C6
Santovenia E 39 E8
Santpedor E 43 D7
Sant Pere de Ribes E 43 E7
Sant Pere de Torelló E 43 C8
Sant Pere Pescador E 43 C10
Sant Pol de Mar E 43 D9
Santpoort NL 182 A5
Sant Privat d'en Bas E 43 C8
Sant Quintí de Mediona E 43 E7
Sant Quirze de Besora E 43 C8
Sant Sadurní d'Anoia E 43 E7
Santullano E 39 B8
Santu Lussurgiu I 64 C2
Santurtzi E 40 B5
Sant Vicenç dels Horts E 43 E8
Sant Vicenç de Castellet
  E 43 D7
San Venanzo I 62 B2
San Vendemiano I 72 E5
San Vero Milis I 64 C2
San Vicente de Alcántara
  E 45 F6
San Vicente de Arana E 32 E1
San Vicente de la Barquera
  E 40 B3
San Vicente de la Sonsierra
  E 40 C6
San Vicente del Raspeig E 56 E3
San Vicente de Palacio E 46 B3
San Vicente de Toranzo E 40 B4
San Vicenzo E 38 C2
San Vincenzo I 59 B7
San Vincenzo I 65 A3
San Vincenzo Valle Roveto
  I 62 D5
San Vitero E 39 E7
San Vito I 64 E4
San Vito al Tagliamento I 73 E6
San Vito Chietino I 63 C6
San Vito dei Normanni I 61 B9
San Vito di Cadore I 72 D5
San Vito lo Capo I 58 C2
San Vito Romano I 62 D3
San Vito sullo Ionio I 59 B9
San Vittoria in Matenano
  I 62 A4
Sanxenxo E 38 D2
Sanxhax AL 168 A2
Sanza I 60 C5
Sânzieni RO 153 E8
Sanzoles E 39 F8
São Barnabé P 50 E2
São Bartolomeu P 44 F5
São Bartolomeu da Serra
  P 50 C2
São Bartolomeu de Messines
  P 50 E3
São Bento do Cortiço P 50 B4
São Brás P 50 D4
São Brás de Alportel P 50 E4
São Brás do Regedouro P 50 C3
São Brissos P 50 C4
São Cosmado P 44 B5
São Domingos P 50 D2
São Facundo P 44 F4
São Francisco da Serra P 50 C2
São Geraldo P 50 B3
São Gregório P 50 B4
São Jacinto P 44 C3
São João da Madeira P 44 C4
São João da Pesqueira P 44 B6
São João da Venda P 50 E4
São João do Campo P 44 D4
São João dos Caldeireiros
  P 50 D4
São José da Lamarosa P 44 F4
São Lourenço de Mamporcão
  P 50 B4
São Luís P 50 D2
São Manços P 50 C4
São Marcos da Ataboeira
  P 50 D4
São Marcos da Serra P 50 E3
São Marcos do Campo P 50 C4

São Martinho da Cortiça
  P 44 D4
São Martinho das Amoreiras
  P 50 D3
São Martinho de Angueira P
  P 39 E7
São Martinho do Porto P 44 E2
São Matias P 50 C4
São Miguel de Acha P 45 D6
São Miguel de Machede P
  50 B4
São Miguel do Outeiro P 44 C4
São Miguel do Pinheiro P 50 D4
Saône F 26 F5
São Pedro da Cadeira P 44 F2
São Pedro de Muel P 44 E2
São Pedro de Solis P 50 E4
São Pedro do Sul P 44 C4
Saorge F 37 D7
São Romão P 51 B5
São Romão do Sado P 50 C3
São Sebastião dos Carros
  P 50 D4
São Teotónio P 50 D2
Saou F 35 A9
São Vicente P 38 E5
São Vicente da Beira P 44 D5
Sáp H 151 C7
Sapanca TR 173 B10
Săpânța RO 145 H8
Sapareva Banya BG 165 E7
Saparevo BG 165 E7
Sápata RO 160 D5
Sapes GR 171 B9
Sapna BIH 157 C10
La Pobla E 57 B11
Săpoca RO 161 C9
Sapotskin BY 137 F8
Sappada I 73 C6
Sap'parjākka N 113 D17
Sappee FIN 127 C11
Sappetan S 107 A13
Sappetsele S 107 A13
Sappisaasi S 116 C7
Sapri I 60 C5
Săpsalampi FIN 123 F11
Sapsoperä FIN 125 B11
Sara FIN 122 F9
Saraby N 113 C12
Saracena I 60 D6
Saračinec HR 148 D6
Sarafovo BG 167 D9
Saraiķi LV 134 C2
Saraisniemi FIN 119 F17
Saraiu RO 155 D2
Sarajärvi FIN 120 F3
Sarajärvi FIN 129 B11
Sarajevo BIH 157 E10
Saramo FIN 125 C12
Saramon F 33 C7
Saran F 24 E6
Sárand H 151 C8
Sarandë AL 168 E2
Sarantaporo GR 169 D7
Sarantsi BG 165 D8
Sa Ràpita E 57 C10
Sarasău RO 145 H8
Sarata UA 154 E5
Sărata Galbenă MD 154 D2
Sărata-Monteoru RO 161 C9
Sărata Nouă MD 154 E2
Sărăteni MD 154 D2
Sărăteni Vechi MD 154 B3
Saray TR 173 B8
Sarayakpınar TR 167 F7
Saraylar TR 173 C8
Sarbanovac SRB 159 F9
Sárbeni RO 161 E6
Sárbogárd H 149 C11
Sarcelles F 25 C7
Sarconi I 60 C5
Sardara I 64 D2
Sardice CZ 77 E12
Sardinia GR 174 B3
Sardoal P 44 E4
Sare F 32 D2
S'Arenal E 49 E10
Sarengrad HR 158 C3
Sarentino I 72 C3
Sárevere EST 131 D10
Sarezzo I 69 B9
Sargans CH 71 C8
Sarıca LT 137 C13
Saribeyler TR 173 F8
Saricaali TR 173 B7
Sariçam TR 177 B9
Sarichioi RO 155 D3
Saridanışment TR 167 F7
Sariegos E 39 C8
Sarıkemer TR 177 D9
Sariköy TR 173 D8
Sariñena E 42 D3
Šarišské Michaľany SK 145 E3
Sarıyer TR 173 B11
Sarjankylä FIN 123 B14
Sarkad H 151 D7
Sarkadkeresztúr H 151 D7
Särkelä FIN 121 C12
Särkeresztúr H 149 B11
Särkijärvi FIN 117 C15
Särkijärvi FIN 119 C15
Särkijärvi FIN 120 E9
Sarkikorpi FIN 125 D12
Särkilahti FIN 127 B14
Särkilahti FIN 129 B11
Särkiluoma FIN 121 B14
Särkimo FIN 122 D8
Särkisalmi FIN 129 B11
Särkisalo FIN 123 E16
Särkisalo FIN 127 E8
Şarköy TR 173 C7
Sârmaş RO 152 D4
Şărmaşu RO 152 D4
Sármellék H 149 C8
Sarmizegetusa RO 159 B10
Särna S 102 C5
Sarnadas do Ródão P 44 E5
Sarnaki PL 141 F7
Sarnano I 62 A4
Sárnate LV 134 B3
Sarnen CH 70 D6
Sarniç TR 177 D8
Sarnico I 69 B8
Sarno I 60 B3
Sarnowa PL 81 C11
Särnstugan S 102 C5

Särö S 91 D10
Sarochyna BY 133 F5
Saronno I 69 B7
Sárosd H 149 B11
Sárospatak H 145 G4
Šarovce SK 147 E7
Sarón E 39 C10
Sárpsborg N 95 D14
Sarracín E 40 D4
Sarral E 42 E6
Sarralbe F 27 C7
Sarrance F 32 D4
Sarrancolin F 33 E6
Sarraquinhos P 38 E4
Sarras F 30 E6
Sarre UK 15 E11
Sarre I 31 D11
Sarreaus E 38 D4
Sarrebourg F 27 C7
Sarreguemines F 27 B7
Sárrétudvari H 151 C7
Sarre-Union F 27 C7
Sarria E 38 C5
Sarrià de Ter E 43 C9
Sarrians F 35 B8
Sarrión E 48 D3
Sarroca de Lleida E 42 E5
Sarròch I 64 E3
Sarród H 149 A7
Sarrola-Carcopino F 37 G9
Sarron F 32 C5
Sarry F 25 C11
Sarsina I 66 E5
Sarstedt D 78 B6
Sárszentágota H 149 C11
Sárszentlőrinc H 149 C11
Sart B 183 D7
Sartaguda E 32 F1
Sarteano I 62 B1
Sartène F 37 H9
Sarti GR 170 E5
Sartilly F 23 C9
Sartininkai LT 134 F3
Sarud H 150 B6
Sarule I 64 C3
Sărulești RO 161 C9
Sărulești RO 161 E9
Sárvár H 149 B7
Sarvela FIN 122 F9
Sarvijoki FIN 122 E7
Sarvikumpu FIN 125 F12
Sarviluoma FIN 122 F8
Sarvinki FIN 125 E14
Sarvisé E 32 E5
Sarvisvaara S 116 E6
Särvsjön S 102 A5
Sar''ya BY 133 E3
Sarzana I 69 E8
Sarzeau F 22 E6
Sarzedas P 44 E5
Sarzedo P 44 B5
Sasa NMK 164 E6
Sasalli TR 177 C9
Sasamón E 40 D3
Sasbach D 27 C9
Sasbachwalden D 186 D5
Sasca Montană RO 159 D8
Saschiz RO 152 E5
Săsciori RO 152 F3
Sascut RO 153 E9
Sásd H 149 D10
Sasina BIH 157 C6
Sasnava LT 137 D7
Sassali FIN 117 D15
Sassano I 60 C5
Sassari I 64 B2
Sassello I 37 C8
Sassen D 84 B4
Sassenage F 31 E8
Sassenberg D 17 E10
Sassenheim NL 16 D3
Sassetta I 65 A3
Sassnitz D 84 A5
Sassocorvaro I 66 E5
Sassoferrato I 67 F6
Sasso Marconi I 66 D3
Sassuolo I 66 C2
Sástago E 41 F11
Satão P 44 C5
Satchinez RO 151 F7
Šateikiai LT 134 E3
Šatenäs villastad S 91 C12
Säter S 97 B14
Sätervallen S 102 A5
Säti LV 134 C5
Satıkli LV 135 C13
Sätila S 91 D11
Satillieu F 30 E6
Satkūnai LT 134 D7
Satnica Đakovačka HR 149 F10
Sätofta S 87 D13
Sátoraljaújhely H 145 G4
Satovcha BG 170 A5
Satow D 83 C11
Satriano di Lucania I 60 B5
Satrup D 82 A7
Sattajärvi FIN 119 B12
Sattajärvi S 117 D10
Sattanen FIN 117 D17
Satteins A 71 C9
Satteldorf D 75 D7
Satter S 116 E7
Sattledt A 76 F6
Satulung RO 151 B11
Satu Mare RO 151 B10
Saubusse F 32 C3
Saúca E 47 B7
Sauca MD 154 A1
Sǎuca RO 151 C9
Saucats F 32 A4
Săucești RO 153 D9
Sauclières F 34 C5
Sauda N 94 C4
Sauda N 60 D5
Saue EST 131 C9
Sauerlach D 75 G10
Sauga EST 131 E8
Saugnacq-et-Muret F 32 B4
Saugos LT 134 F2
Saugues F 30 F4
Saujon F 28 D4
Sauk AL 168 C3
Sauka LV 135 D10
Šaukėnai LT 134 E5
Saukko FIN 125 C11
Saukkoaapa FIN 115 D2
Saukkojärvi FIN 119 B16
Saukkola FIN 127 E10

Saukkoriipi FIN 117 E12
Saukonkylä FIN 123 E11
Sauland N 95 C9
Saulce-sur-Rhône F 35 A8
Sauldorf D 27 E11
Saulepi EST 130 E7
Săulești RO 160 D2
Saulgau D 71 A9
Saulgrub D 71 B12
Saulheim D 21 E10
Şăulia RO 152 D4
Şăuliu F 25 F11
Saulkalne LV 135 C8
Saulkrasti LV 135 B8
Sault F 35 B9
Sault-de-Navailles F 32 C4
Sault-lès-Rethel F 19 F9
Saulx F 26 E5
Saulxures-sur-Moselotte F 27 E6
Saulzais-le-Potier F 29 B11
Saumos F 28 F3
Saumur F 23 F11
Saunajärvi FIN 125 C13
Saunakylä FIN 123 D13
Saunavaara FIN 115 D3
Saundersfoot UK 12 B5
Saurat F 33 E9
Sauriešl LV 135 C8
Sauris I 73 C5
Saursfjord N 110 E9
Sausnēja LV 135 C11
Sausset-les-Pins F 35 D9
Saussy F 26 F2
Sautens A 71 C11
Sautiņi LV 135 C9
Sautron F 23 F8
Sautso Kraftverk N 113 D12
Sauvagnon F 32 D5
Sauve F 35 C6
Sauveterre-de-Béarn F 32 C4
Sauveterre-de-Guyenne F 28 F5
Sauveterre-la-Lémance F 33 A8
Sauviat-sur-Vige F 29 C9
Sauvo FIN 126 E8
Sauxillanges F 30 D3
Sauzet F 35 C10
Sauzon F 22 F5
Sauzé-Vaussais F 29 C6
Sava I 61 C9
Sava SLO 73 D10
Săvădisla RO 151 D11
Savalia RO 153 D8
Savaloja FIN 119 E15
Săvar S 122 C5
Săvârșin RO 151 E9
Săvăst S 118 C4
Săve S 91 D10
Săveni RO 153 B9
Săveni RO 161 D11
Saverdun F 33 D9
Saverna FIN 123 D15
Saverne F 27 C7
Săvi FIN 126 B8
Săviă FIN 123 D17
Saviaho FIN 125 C10
Savières F 25 D10
Savigliano I 37 B7
Savignac-les-Églises F 29 E7
Savignano Irpino I 60 A4
Savignano sul Rubicone I 66 D5
Savigné-l'Évêque F 23 D12
Savigneux F 30 D5
Savigny-en-Sancerre F 25 F8
Savigny-lès-Beaune F 25 F12
Savigny-sur-Braye F 24 E4
Savijärvi FIN 125 C11
Savikylä FIN 125 C11
Savilahti FIN 127 C10
Savimäki FIN 124 C7
Savines-le-Lac F 36 B4
Săvinești RO 153 D8
Savino Selo SRB 158 B4
Saviranta FIN 121 F10
Saviselkä FIN 123 B16
Savitaipale FIN 128 C8
Sävja S 99 C9
Šavnik MNE 157 G11
Savognin CH 71 D9
Savoisy F 25 E11
Savona I 37 C8
Savonlinna FIN 129 B10
Savonranta FIN 125 F12
Savran' UA 154 A6
Sävsjö S 92 E5
Sävsjön S 97 C12
Savudrija HR 67 B7
Savukoski FIN 115 D4
Sawin PL 141 H8
Sawston UK 15 C9
Sax E 56 D3
Saxdalen S 97 B12
Saxen A 77 F7
Saxhyttan S 97 C11
Saxilby UK 11 E10
Saxmundham UK 15 C9
Saxnäs S 106 B8
Saxon CH 31 C11
Saxthorpe UK 15 B11
Saxvallen S 105 D12
Sayalonga E 53 C8
Sayatón E 47 D7
Sayda D 80 E4
Säynäjä FIN 121 B13
Säynätsalo FIN 123 F15
Säyneinen FIN 125 D10
Sázava CZ 77 C7
Sazlı TR 171 E10
Sazli TR 177 D8
Sazoba TR 173 D7
Sazoba TR 177 B8
Scaër F 22 D4
Scăești RO 160 E3
Scafa I 62 C6
Scalasaig UK 4 C5
Scalby UK 11 C11
Scalea I 60 D5
Scaletta Zanclea I 59 C7
Scandale I 61 E7
Scandiano I 66 D2
Scandicci I 66 E3
Scandriglia I 62 C3
Scanno I 62 D5
Scano di Montiferro I 64 C2
Scansano I 65 B4
Scanzano Jonico I 61 C7
Scapa UK 3 H11

Scarborough UK 11 C11
Scardovari I 66 C5
Scarinish UK 4 B3
Scărişoara RO 151 E10
Scarlino I 65 B3
Scarperia I 66 E3
Scartaglen IRL 8 D4
Scartaglin IRL 8 D4
Ščenica-Bobani BIH 162 D5
Scenici I 62 C3
Scey-sur-Saône-et-St-Albin
  F 26 E4
Schaafheim D 187 B7
Schaan FL 71 C8
Schaarsbergen NL 183 A7
Schaerbeek B 19 C9
Schaesberg NL 20 C6
Schaffhausen CH 27 E10
Schafflund D 82 A6
Schafstädt D 79 D10
Schafstedt D 82 B6
Schäftlarn D 75 G9
Schagen NL 16 C3
Schaijk NL 183 B6
Schalchen A 76 F3
Schalkau D 75 B9
Schalkhaar NL 183 A8
Schalksmühle D 17 F9
Schänis CH 27 F11
Schapen D 17 D9
Schaprode D 84 A4
Scharbeutz D 83 B9
Schardenberg A 76 E4
Schärding A 76 F4
Scharendijke NL 16 E1
Scharnebeck D 83 D9
Scharnitz A 72 B3
Scharnstein A 73 A8
Scharrel (Oldenburg) D 17 B9
Scharwoude NL 16 C4
Schashagen D 83 B9
Schattendorf A 149 A7
Schebheim D 75 C7
Scheemda NL 17 B7
Scheer D 27 E11
Scheeßel D 82 D6
Schefflenz D 21 F12
Scheggia e Pascelupo I 67 F6
Scheggino I 62 B3
Scheia RO 153 C8
Scheia RO 153 E10
Scheibbs A 77 F8
Scheiblingkirchen A 148 A6
Scheidegg D 71 B9
Scheifling A 73 B9
Scheinfeld D 75 C7
Schela RO 155 C1
Schela RO 160 C2
Schelklingen D 74 F6
Schellerten D 79 B7
Schemmerhofen D 71 A9
Schenefeld D 82 B6
Schenefeld D 83 C7
Schenkenfelden A 76 E6
Schenkenzell D 187 E5
Schenklengsfeld D 78 E6
Schermbeck D 17 E7
Schermen D 79 B10
Schermerhorn NL 16 C3
Scherpenheuvel B 19 C10
Scherpenzeel NL 16 D4
Scherwiller F 27 D7
Scheßlitz D 75 C9
Scheveningen D 27 E11
Schezlitz D 75 C9
Schiavi di Abruzzo I 63 D6
Schiedam NL 16 E2
Schieder-Schwalenberg D 17 E12
Schieren L 20 E6
Schierling D 75 E11
Schiermonnikoog NL 16 B6
Schiers CH 71 D9
Schiffdorf D 17 A11
Schifferstadt D 21 F10
Schifflange L 20 E6
Schiffweiler D 186 C3
Schijndel NL 16 E4
Schilde B 16 F3
Schillingen D 186 B2
Schillingsfürst D 75 D7
Schilpario I 69 A9
Schiltach D 27 D9
Schiltigheim F 27 C8
Schimatari GR 175 C8
Schinnen NL 183 D7
Schinos GR 174 B3
Schinveld NL 20 C5
Schio I 69 B11
Schipkau D 80 C5
Schipluiden NL 182 B4
Schirmeck F 27 D7
Schirmitz D 75 C11
Schitu RO 160 E5
Schitu RO 161 E7
Schitu Duca RO 153 C11
Schitu Golești RO 160 C6
Schkeuditz D 79 D11
Schkölen D 79 D10
Schkölen D 79 D11
Schkopau D 79 D11
Schladen D 79 B8
Schladming A 73 B8
Schlangen D 17 E11
Schlangenbad D 185 D9
Schleching D 72 A5
Schleiden D 20 C6
Schleife D 81 C7
Schleinbach A 77 F10
Schleitheim CH 27 E9
Schleiz D 79 E11
Schlema D 79 E12
Schleppzig D 80 C5
Schleswig D 82 A7
Schleusingen D 75 A8
Schlieben D 80 C5
Schliengen D 27 E8
Schlier D 71 B9
Schlierbach A 73 A9
Schlieren CH 27 F9
Schliersee D 72 A4
Schlitters A 72 B4
Schlitz D 78 E6
Schlossberg A 148 C4
Schloss Holte-Stukenbrock
  D 17 E11
Schlossvippach D 79 D9
Schlosswil CH 70 D5
Schlotheim D 79 D7
Schluchsee D 27 E9
Schlüchtern D 74 B6
Schlüsselfeld D 75 C8
Schmallenberg D 21 B10

Schmelz D 21 F7
Schmidgaden D 75 D11
Schmidmühlen D 75 D10
Schmiedeberg D 80 E5
Schmölln D 79 E11
Schmölln D 84 D6
Schnabelwaid D 75 C10
Schnackenburg D 83 D11
Schnaittach D 75 C9
Schneeberg D 79 E12
Schneidlingen D 79 C9
Schneizlreuth D 73 A6
Schnelldorf D 75 D7
Schnellmannshausen D 79 D7
Schneverdingen D 82 D7
Schnürpflingen D 187 E8
Schobüll D 82 A6
Schödnia PL 142 E5
Schoenberg B 20 D6
Scholen D 17 C11
Schollene D 83 E12
Schöllkrippen D 187 A7
Schöllnach D 76 E4
Schömberg D 27 D10
Schömberg D 187 D6
Schönaich D 187 D7
Schönau D 76 F3
Schönau im Schwarzwald
  D 27 E8
Schönbach A 77 F8
Schönberg D 83 C9
Schönberg D 84 B3
Schönberg (Holstein) D 83 B8
Schönberg am Kamp A 77 E9
Schönberg im Stubaital A 72 B3
Schönborn D 80 C5
Schondra D 74 B6
Schönebeck D 83 D12
Schönebeck (Elbe) D 79 B10
Schöneck D 75 B11
Schönecken D 20 D6
Schönenberg-Kübelberg
  D 186 C3
Schönermark D 84 D5
Schönewalde D 80 C4
Schönewörde D 79 A8
Schongau D 71 B11
Schönhausen D 79 A11
Schöningen D 79 B8
Schönkirchen D 83 B8
Schönow D 84 E5
Schönsee D 75 C12
Schöntal D 74 D6
Schönthal D 75 D12
Schönwald D 80 C5
Schönwalde am Bungsberg
  D 83 B9
Schönwies A 71 C11
Schoondijke NL 16 F1
Schoonebeek NL 17 C7
Schoonhoven NL 16 E3
Schoorl NL 16 C3
Schopfheim D 27 E8
Schopfloch D 75 D7
Schöppenstedt D 79 B8
Schoppernau A 71 C10
Schöppingen D 17 D8
Schörfling am Attersee A 73 A8
Schorndorf D 27 D11
Schorndorf D 75 D12
Schortens D 17 A9
Schoten B 16 F3
Schötten D 21 C12
Schramberg D 27 D9
Schreckbach D 21 C12
Schrems A 77 E8
Schrepkow D 83 E12
Schriesheim D 21 F11
Schrobenhausen D 75 E9
Schröder D 83 A9
Schrozberg D 74 D6
Schruns A 71 C9
Schübelbach CH 27 F10
Schuby D 82 A6
Schulenberg im Oberharz
  D 79 C7
Schull IRL 8 E3
Schulzendorf D 84 E6
Schüpfheim CH 70 D6
Schuttertal D 186 E4
Schutterwald D 186 E4
Schüttorf D 17 D8
Schwaan D 83 C12
Schwabach D 75 D9
Schwäbisch Gmünd D 74 E6
Schwäbisch Hall D 74 D6
Schwabmünchen D 71 A11
Schwabstedt D 82 B6
Schwaförden D 17 C11
Schwaigern D 21 F11
Schwalbach D 21 F7
Schwalmstadt-Treysa D 21 C12
Schwalmstadt-Ziegenhain
  D 21 C12
Schwanberg A 73 C11
Schwanden CH 71 C8
Schwandorf D 75 D11
Schwanebeck D 79 C9
Schwanenstadt A 76 F5
Schwanewede D 17 B11
Schwangau D 71 B11
Schwanstetten D 75 D9
Schwarme D 17 C12
Schwarmstedt D 82 E7
Schwarz D 83 D13
Schwarza D 79 E8
Schwarzach D 76 E3
Schwarzau im Gebirge A 148 A5
Schwarzenau A 77 E8
Schwarzenbach A 149 A6
Schwarzenbach am Wald
  D 75 B10
Schwarzenbek D 83 C8
Schwarzenberg D 79 E12
Schwarzenborn D 78 E6
Schwarzenburg CH 31 B11
Schwarzenfeld D 75 D11
Schwarzheide D 80 D5
Schwaz D 72 B4
Schwechat A 77 F10
Schwedeneck D 83 B8
Schwedt an der Oder D 84 D6
Schwegenheim D 187 C5
Schwei (Stadland) D 17 A10
Schweiburg D 21 E7
Schweich D 21 E7
Schweigen-Rechtenbach
  D 27 B8
Schweighouse-sur-Moder
  F 186 D4
Schweinfurt D 75 B7
Schweinitz D 80 C4
Schweinrich D 83 D13

Schwelm D 17 F8
Schwendau D 72 B4
Schwendi D 71 A10
Schwenningen D 27 D10
Schwepnitz D 80 D5
Schwerin D 83 C10
Schweringen D 17 C12
Schwerte D 17 F9
Schwichtenberg D 84 C5
Schwieberdingen D 27 C11
Schwiesau D 79 A9
Schwindegg D 75 F11
Schwinkendorf D 83 C13
Schwoich A 72 A5
Schwyz CH 71 C7
Sciacca I 58 D3
Sciara I 58 D4
Scicli I 59 F6
Scigliano I 61 E6
Scilla I 59 C8
Ścinawa PL 81 D10
Scionzier F 31 C10
Scoarța RO 160 C2
Scobinți RO 153 C9
Scoglitti I 58 F5
Scolaticci I 59 C6
Scole UK 15 C11
Scone UK 5 C10
Sconser UK 2 L4
Scopello I 68 B5
Scoppito I 62 C4
Scorbé-Clairvaux F 29 B6
Scordia I 59 E6
Scornicești RO 160 D5
Scorrano I 61 C10
Scorțaru Nou RO 161 C11
Scorțeni MD 154 C3
Scorțeni RO 153 D9
Scorțeni RO 161 C7
Scorțoasa RO 161 C9
Scorton UK 11 C8
Scorzè I 66 A5
Scotch Corner UK 11 C8
Scotch Corner IRL 7 D9
Scotshouse IRL 7 D8
Scourie UK 2 J6
Scousburgh UK 3 F14
Scrabster UK 3 H9
Screeb IRL 6 F4
Screggan IRL 7 F7
Scremerston UK 5 D13
Scribbagh UK 6 D6
Scrioaștea RO 160 E5
Scriob IRL 6 F3
Sculeni MD 153 C11
Scundu RO 160 D4
Scunthorpe UK 11 D10
Scuol CH 71 D10
Scurcola Marsicana I 62 C4
Scurtu Mare RO 160 E6
Scutelnici RO 161 D9
Seaca RO 160 E5
Seaca RO 160 F6
Seaca de Pădure RO 160 E2
Seaford UK 15 F9
Seaham UK 5 F14
Seahouses UK 5 D13
Seamer UK 11 C11
Seapatrick UK 7 D11
Seara P 38 E2
Seascale UK 10 C5
Seaton UK 9 F9
Seaton UK 13 D8
Seaton Delaval UK 5 E13
Seaton Sluice UK 5 E14
Seaview UK 13 D12
Sebal P 44 D3
Sébazac-Concourès F 34 B4
Sebbersund DK 86 B5
Sebechleby SK 147 E7
Sebedražie SK 147 D7
Sebeș RO 152 F3
Šebetov CZ 77 C11
Sebezh RUS 133 D4
Sebiș RO 151 E9
Sebnitz D 80 E6
Sečanj SRB 159 C6
Seča Reka SRB 158 E4
Sećanja RO 161 C7
Secaș RO 151 F8
Sece LV 135 C10
Secemin PL 143 E8
Seckach D 27 B11
Seckau A 73 B10
Seclin F 29 7
Secondigny F 28 B5
Sečovce SK 145 F4
Sečovská Polianka SK 145 F4
Secu RO 160 E2
Secuieni RO 152 E5
Secuieni RO 153 D9
Secuieni RO 153 D10
Secusigiu RO 151 E6
Seda LT 134 D4
Seda LV 135 F11
Sedan F 19 E10
Sedano E 40 C4
Sedbergh UK 10 C6
Seddülbahir TR 171 D10
Seden DK 86 E6
Šédere LV 135 E12
Séderon F 35 B10
Sedgefield UK 5 F14
Sedico I 72 D5
Sedilo I 64 C2
Sedini I 64 B2
Sedlarica HR 149 E8
Sedlčany CZ 77 C6
Sedlec Prčice CZ 77 C7
Sedlice CZ 76 D5
Sedliská SK 145 F4
Sedlnice CZ 146 B6
Sedrina I 69 B8
Šeduva LT 134 E7
Sdziejowice PL 143 C7
Sjdzin PL 138 E6
Sdziszów PL 143 E9
Sdziszów Małopolski PL 143 F12
See A 71 C10
Seebach D 79 D8
Seebach D 187 D5
Seebergen D 79 E8
Seeboden A 73 C8
Seebruck D 72 A5
Seeburg D 79 C7
Seedorf D 83 C9
Seefeld D 75 F9
Seefeld D 80 A5
Seefeld (Stadland) D 17 B10

Seefeld in Tirol A 72 B3
Seeg D 71 B11
Seehausen D 79 B9
Seehausen D 84 C5
Seehausen (Altmark) D 83 E11
Seeheim-Jugenheim D 21 E11
Seekirchen am Wallersee A 73 A7
Seelow D 80 A6
Seelze D 78 B6
Seend UK 13 C10
Seeon D 72 A5
Sées F 23 C12
Seesen D 79 C7
Seeshaupt D 72 A3
Seethal A 73 B8
Seevetal D 83 D8
Seewiesen A 148 A4
Séez F 31 D10
Seferihisar TR 177 C8
Sefkerin SRB 158 C6
Sefrivatn N 105 A14
Seftigen CH 31 B12
Segalstad N 101 D12
Segarcea RO 160 E3
Segarcea-Vale RO 160 F5
Segeletz D 83 E13
Segersäng S 93 A11
Segersta S 103 D12
Segerstad S 89 C11
Segesd H 149 D8
Seggiano I 65 A6
Séglien F 22 D5
Seglinge FIN 99 B15
Seglingsberg S 98 C6
Seglodden N 114 C9
Seglvik N 112 C7
Segmon S 97 D9
Segni I 62 D4
Segonzac F 28 D5
Segorbe E 48 E4
Segovia E 46 C4
Segré F 23 E10
Ségrie F 23 D12
Ségur F 34 B4
Segura P 45 E7
Segura de la Sierra E 55 C7
Segura de León E 51 C6
Segura de los Baños E 42 F2
Segurilla E 46 D3
Sehlde D 79 B7
Sehma D 76 A3
Sehnde D 79 B6
Seia P 44 D5
Seica Mare RO 152 E4
Seica Micä RO 152 E4
Seiches-sur-le-Loir F 23 E11
Seiersberg A 148 B4
Seierstad N 105 B10
Seifhennersdorf D 81 E7
Seignelay F 25 E10
Seignelégier CH 27 F6
Seilhac F 29 E9
Seillans F 36 D5
Seilles B 183 E6
Seimeni RO 155 E2
Seinäjoki FIN 122 E9
Seini RO 151 B11
Seinsheim D 75 C7
Seipäjärvi FIN 117 D16
Seirijai LT 137 E8
Seissan F 33 D7
Seitajärvi FIN 115 C3
Seitin RO 150 E6
Seix F 33 E8
Seixal P 50 B1
Sejerby DK 87 D8
Sejerslev DK 86 B3
Sejet DK 86 D5
Sejny PL 136 E7
Sekkemo N 112 D8
Šekovici BIH 157 D10
Sjkowa PL 145 D3
Sjkowo PL 81 A10
Sel S 107 E10
Sel S 107 E13
Sela HR 149 E6
Selånger S 103 B13
Selanovtsi BG 160 F3
Selänpää FIN 128 C6
Selargius I 64 E3
Selaru RO 160 E6
Selattyn UK 10 F5
Selaya E 40 B4
Selb D 75 B11
Selbekken N 104 E7
Selbitz D 75 B10
Selbu N 105 E10
Selbustrand N 105 E9
Selby UK 11 D9
Selcë AL 163 D8
Selce HR 67 B10
Selce SK 147 D8
Selci Dakovacki HR 149 F10
Selçuk TR 177 D9
Selde DK 86 B4
Selegas I 64 E3
Selenicë AL 168 C2
Selent D 83 B8
Selero GR 171 B7
Šélestat F 27 D7
Selet S 118 C7
Selet S 122 B3
Seleus RO 151 E8
Seleuš SRB 159 C6
Selevac SRB 159 E6
Seleznevo RUS 129 D10
Selfjord N 110 D5
Sëll LV 135 C10
Selice SK 146 E5
Seligenstadt D 187 A6
Šelimbãr RO 152 F4
Selimiye TR 177 D10
Selimpaşa TR 173 B9
Selimşahlar TR 177 B9
Selishtë AL 168 A3
Selishte BG 165 E7
Selište SRB 159 F9
Selitë AL 168 C3
Selja S 102 D7
Selje N 100 B2
Seljeli N 108 D6
Seljelynes N 111 B17
Seljenes N 113 B15
Seljord N 95 D9
Selkajärvi S 116 E8
Selkälä FIN 115 D6
Selkirk UK 5 D11
Selkopp N 113 B13
Sella I 56 D4
Sellano I 62 B3
Sellasia GR 174 E5

Selles-St-Denis F 24 F6
Selles-sur-Cher F 24 F6
Sellia GR 178 E7
Sellia Marina I 59 B10
Sellières F 31 B8
Sellin D 84 B5
Sellindge UK 15 E10
Selling DK 86 C6
Sellingen NL 17 C8
Sellrain A 72 B3
Sellye H 149 E9
Selm D 17 E8
Selmes P 50 C4
Selmsdorf D 83 C9
Selnes N 104 D7
Selnes N 111 B17
Selnes N 111 C13
Selo BIH 157 D10
Selommes F 24 E5
Selongey F 26 E3
Selonnet F 36 C4
Selow D 83 C11
Selsey UK 15 F7
Selsingen D 17 B12
Selsjön S 107 E12
Selsøyvika N 108 C4
Sélsted DK 86 E4
Selters (Westerwald) D 21 C9
Seltjärn S 107 D13
Seltz F 27 C9
Selva E 57 B10
Selva N 104 D7
Selva dei Molini I 72 C4
Selva di Progno I 69 B11
Selva di Val Gardena I 72 C4
Selvar N 108 C3
Selvåg N 104 D6
Selvazzano Dentro I 66 B4
Selvino I 69 B8
Selyatyn UA 152 B6
Selzthal A 73 A9
Sem N 95 D12
Seman AL 168 C1
Sembadel F 30 E4
Sembrancher CH 31 C11
Sëme LV 134 B6
Séméac F 33 D6
Sémelay F 30 B4
Semeliškes LT 137 D10
Semeljci HR 149 F11
Semeni MD 153 C11
Semenivka UA 154 E6
Semerovo SK 146 E6
Semic SLO 73 E11
Semide F 44 D4
Semily CZ 81 E8
Seminara I 59 C8
Semizovac BIH 157 E9
Semlac RO 151 E6
Semlow D 83 B13
Semmenstedt D 79 B8
Semoutiers-Montsaon F 26 D3
Sempach CH 70 D5
Semproniano I 65 B5
Semriach A 148 B5
Šemša SK 145 F3
Šemševo NMK 164 E3
Semur-en-Auxois F 25 F11
Semur-en-Brionnais F 30 C5
Semussac F 28 D4
Sena E 42 D3
Seña SK 145 F3
Senales I 71 D11
Senarpont F 18 E4
Sénas F 35 C9
Sencelles E 57 B10
Sençur SLO 73 D9
Send UK 15 E7
Senden D 75 F7
Sendenhorst D 17 E9
Sendets F 32 D5
Sendim P 38 F3
Sendim P 39 F7
Şendreni RO 155 C1
Şendriceni RO 153 B8
Šenec SK 146 E4
Seneffe B 19 D9
Seneghe I 64 C2
Senes S 85 E8
Senez F 36 D4
Senftenberg A 77 F9
Senftenberg D 80 C6
Sengouagnet F 33 E7
Senhora do Rosário P 45 F6
Senica SK 146 D4
Senice na Hané CZ 77 C12
Senigallia I 67 E7
Senis I 64 D2
Senise I 60 C6
Senj HR 67 C10
Senje SRB 159 E7
Senjehesten N 111 B12
Senjehopen N 111 B13
Senjski Rudnik SRB 159 F8
Sen'kava BY 133 D3
Šenkvice SK 146 E4
Senlis F 25 B8
Sennan S 87 B11
Sennariolo I 64 C2
Sennecey-le-Grand F 30 B6
Sennels DK 86 B3
Sennen UK 12 E3
Sennik BG 165 D11
Sennori I 64 B2
Sennwald CH 71 C8
Sennybridge UK 13 B7
Senohrad SK 147 E8
Senokos BG 167 B10
Senonches F 24 C5
Senones F 27 D6
Senorbì I 64 E3
Senouillac F 33 C9
Senovo BG 161 F8
Senožece SLO 73 E9
Sens F 25 D9
Sens-de-Bretagne F 23 D8
Senta SRB 150 F5
Sentein F 33 E7
Senterada E 33 F8
Šentilj SLO 148 C5
Šentjernej SLO 148 E4
Šentjur pri Celju SLO 148 D4
Senuc F 19 F10
Šeon CH 27 F9
Šepak BIH 157 C11
Šepino I 60 A3
Sjpólno Krajeńskie PL 85 D13
Sjpólno Wielkie PL 85 C12
Seppois-le-Bas F 27 E7
Şepreuş RO 151 D8

Septèmes-les-Vallons F 35 D9
Septemvri BG 165 E9
Septemvriytsi BG 160 F2
Septeuil F 24 C6
Septfonds F 33 B9
Septmoncel F 31 C8
Sepulcro-Hilario E 45 C8
Sepúlveda E 40 F4
Sequals I 73 D6
Sequeros E 45 C8
Seraincourt F 19 E9
Seraing B 19 C12
Séranon F 36 D5
Seravezza I 66 E1
Šerbãneşti RO 160 E5
Šerbettar TR 173 B6
Šerboc RKS 163 D10
Sercaia RO 152 F6
Serdiana I 64 E3
Sered SK 146 E5
Seredka RUS 132 F3
Seredyne UA 145 F6
Seredžius LT 137 C7
Seregélyes H 149 B11
Seregno I 69 B7
Séreilhac F 29 D8
Šérene LV 135 C10
Sérent F 22 E6
Serényfalva H 145 G1
Serfaus A 71 C11
Sergen TR 167 F9
Serginès F 25 D9
Sergnano I 69 C8
Serhat TR 172 E6
Serhiyivka UA 154 B6
Seriate I 69 B8
Sérifontaine F 18 F4
Serifos GR 175 E10
Sérignan F 34 D5
Sérignan-du-Comtat F 35 B8
Serina I 69 B8
Serino I 60 B3
Serinyà E 43 C9
Sermaises F 25 D7
Sermano F 37 G10
Sermenin NMK 169 B7
Sermide I 66 C3
Sermite LV 134 C4
Sermoneta I 62 D4
Sermoyer F 31 B6
Sërmükšl LV 135 B11
Sernancelhe P 44 C6
Serno D 79 C11
Serock PL 138 D5
Serock PL 139 E11
Serokomla PL 141 G6
Serón E 55 E8
Serón de Nájima E 41 F7
Serooskerke NL 182 B3
Seròs E 42 E4
Serpa P 50 D4
Serpins F 44 D4
Serpneve UA 154 E4
Serqueux F 26 E4
Serquigny F 24 B4
Serra E 48 E4
Serracapriola I 63 D8
Serrada E 39 F10
Serra de'Conti I 67 E7
Serradifalco I 58 E4
Serra-di-Ferro F 37 H9
Serradilla I 45 E8
Serradilla del Arroyo E 45 C8
Serra-di-Scopamène F 37 H10
Serramanna I 64 E2
Serramazzoni I 66 D2
Serranillos I 46 D3
Serra Riccò I 37 B9
Serra San Bruno I 59 B9
Serrastretta I 59 A9
Serravalle RSM 66 E5
Serravalle di Chienti I 62 A3
Serravalle Scrivia I 37 B9
Serre I 60 B4
Serrejón E 45 E9
Serrenti I 64 E2
Serres F 35 B10
Serres GR 169 B10
Serres-Castet F 32 D5
Serri I 64 D3
Serrières F 37 G9
Serrières F 30 E6
Serrig D 21 E7
Serritslev DK 90 E6
Serrungarina I 67 E6
Sers F 29 D6
Sersale I 59 A10
Sertã P 44 E4
Sertolovo RUS 129 E13
Servance F 26 E6
Serverette F 30 F3
Serverette F 30 F3
Servia GR 169 D7
Servian F 34 D5
Serviana GR 168 E4
Servigliano I 62 A4
Servon-sur-Vilaine F 23 D9
Servoz F 31 D10
Sery F 19 E9
Serzedo P 44 B3
Sesa E 42 D3
Sesava LV 134 D7
Seseña E 46 D5
Seskarö S 119 C11
Seskinore UK 2 K5
Sesma E 32 F1
Sessa Aurunca I 60 A1
Sessa Cilento I 60 C4
Ses Salines E 57 C11
Sesta al Reghena I 73 D6
Sesto Calende I 68 B6
Sesto Campano I 60 A2
Sesto Fiorentino I 66 E3
Šestanovac HR 157 F6
Šeštokai LT 137 E7
Sestola I 66 D2
Sesto San Giovanni I 69 B7
Sestriere I 31 F10
Sestri Levante I 37 C10
Sestrino BG 165 E9
Sestroretsk RUS 129 E12
Sestu I 64 E3
Sešuoliai LT 135 F9
Sesvete HR 148 E6
Seta LV 135 B9
Šeta LT 135 F8
Sète F 35 D6

Setenil E 51 F9
Setermoen N 111 C15
Setermoen N 100 A6
Setiles E 47 C9
Setivoho BIH 158 F3
Šetonje SRB 159 E7
Setter UK 3 E14
Settimo San Pietro I 64 E3
Settimo Torinese I 68 C4
Settimo Vittone I 68 B4
Settle UK 11 C7
Setúbal P 50 B2
Seubersdorf in der Oberpfalz D 75 D10
Seui I 64 D3
Seuil-d'Argonne F 26 C3
Seulo I 64 D3
Seurre F 31 A7
Seva E 43 D8
Sevar BG 161 F9
Sevaster AL 168 D2
Sevel DK 86 C3
Sevenoaks UK 15 E9
Sevenum NL 16 F6
Sévérac-le-Château F 34 B5
Sever do Vouga P 44 C4
Severin HR 149 E7
Severinovca MD 154 A3
Severnyak BG 155 F1
Seveso I 69 B7
Ševětín CZ 77 D7
Sevettijärvi FIN 114 E5
Sevid HR 156 E5
Sévignacq F 32 D5
Sevilla E 51 E8
Sevilla la Nueva E 46 D4
Sevilleja de la Jara E 46 E3
Şevketiye TR 173 D8
Sevlievo BG 166 C4
Sevnica SLO 148 D4
Sevojno SRB 158 F4
Sévrier F 31 D9
Sexdrega S 91 D13
Seyches F 33 A6
Seyda D 79 C12
Seymen TR 173 B8
Seymen TR 173 B9
Seyne F 36 C4
Seynes F 35 B7
Seynod F 31 D9
Seyrek TR 177 B9
Seyssel F 31 D8
Seysses F 33 C8
Sežana SLO 73 E8
Sézanne F 25 C10
Sezemice CZ 77 B9
Sezimovo Ústí CZ 77 D7
Sezze I 62 E4
Sfaka GR 179 E10
Sfântu Gheorghe RO 153 F7
Sfântu Gheorghe RO 161 D9
Sfendami GR 169 D8
Sfinţeşti RO 160 E6
Sgiogarstaigh UK 2 J4
's-Gravendeel NL 16 E3
's-Gravenhage NL 16 D2
's-Gravenmoer NL 182 B5
's-Gravenpolder NL 16 F1
's-Gravenvoeren B 183 D7
's-Gravenzande NL 16 E2
Sgurgola I 62 D4
Shabla BG 155 F3
Shaftesbury UK 13 C10
Shalfleet UK 13 D12
Shalwy IRL 6 C5
Shanacrane IRL 8 E4
Shanagolden IRL 8 C4
Shanavogh IRL 8 C4
Shankill IRL 7 F10
Shanklin UK 13 D12
Shanlaragh IRL 8 E4
Shanvus IRL 6 E6
Shap UK 10 B6
Sharashova BY 141 E10
Sharavogue IRL 7 F7
Sharkawshchyna BY 133 F2
Sharkovo BG 167 E7
Shatilava BY 133 F6
Shats'k UA 141 H9
Shaverki RUS 129 D17
Shawbost UK 2 J3
Shchyrets' UA 145 D8
's-Heerenberg NL 16 E6
's-Heerenhoek NL 182 C3
Sheerness UK 15 E10
Sheffield UK 11 E9
Shefford UK 15 C8
Shelekhove UA 154 B6
Shëmri AL 163 E9
Shëngjergj AL 168 B3
Shëngjin AL 163 F8
Shepit UA 152 B6
Shepped UK 11 F9
Shepton Mallet UK 13 C9
Sherborne UK 13 D9
Sherborne St John UK 13 C12
Sherburn UK 11 C9
Sherburn in Elmet UK 11 D9
Shercock IRL 7 E9
Sheringham UK 15 B11
Sherston UK 13 B10
's-Hertogenbosch NL 16 E4
Shetaj AL 168 A2
Shevchenkove UA 155 B4
Shieldaig UK 2 K5
Shifnal UK 11 F7
Shijak AL 168 B2
Shildon UK 5 F13
Shillelagh IRL 9 C9
Shinfield UK 15 E7
Shinness Lodge UK 2 J8
Shinrone IRL 7 G7
Shipka BG 166 D4
Shipkovo BG 165 D10
Shipley UK 11 D8
Shipston on Stour UK 13 A11
Shipton UK 10 E6
Shiroka Lŭka BG 165 F10
Shirokë AL 163 E8
Shivachevo BG 166 D6
Shkodër AL 163 E8
Shkorpilovtsi BG 167 D9
Shlissel'burg RUS 129 F15
Shmoylovo RUS 132 F4
Sho BY 133 F4
Shoeburyness UK 15 D10
Shoreham-by-Sea UK 15 F8
Shotts UK 5 D9
Shrewsbury UK 10 F6
Shrewton UK 13 C11
Shrigley UK 7 D11
Shrivenham UK 13 B11
Shrule IRL 6 E4
Shtëpaj AL 168 C3

Shtërmen AL 168 C3
Shtërpcë RKS 164 E3
Shtime RKS 164 E3
Shtiqën AL 163 E9
Shtit BG 166 F6
Shtrŭklevo BG 161 F8
Shumen BG 167 C7
Shumilina BY 133 F6
Shushicë AL 168 B3
Shymkove UA 154 B6
Shyroke UA 145 G7
Shyroke UA 154 C5
Shyryayeve UA 154 C6
Siamanna I 64 D2
Sianów PL 85 B10
Siatista GR 169 D6
Šiaulėnai LT 134 E6
Šiauliai LT 134 E6
Sibari I 61 D6
Sibbhult S 88 C6
Sibbo FIN 127 E13
Sibenik HR 156 E4
Sibinj HR 157 B8
Sibiu RO 152 F4
Sible Hedingham UK 15 D10
Sibnica SRB 158 E5
Sibo S 103 D12
Šibot RO 151 F11
Sibsey UK 11 E12
Sic RO 152 D3
Sićevo SRB 164 C5
Sichevita RO 159 D8
Sicienko PL 138 D4
Siciny PL 81 C10
Sickte D 79 B8
Šicula RO 151 E8
Šiculeni RO 153 E7
Siculiana I 58 E3
Šid SK 147 E9
Šid SRB 158 C3
Sidari GR 168 E2
Sidbury UK 13 D8
Sidcup UK 15 E9
Sidrabini LV 135 C11
Siddi I 64 D2
Sideby FIN 122 F6
Sidensjö S 107 E14
Siderno I 59 C9
Sidgunda LV 135 C9
Sidmouth UK 13 D8
Sidirokastro GR 169 B9
Sidirokastro GR 174 E4
Sidironero GR 171 B6
Sidzina PL 142 E3
Siebe N 112 F11
Siebengewald NL 183 B8
Sieben lehn D 80 D4
Sieciebórzyce PL 81 C9
Sieciechów PL 141 G5
Siedlanka PL 143 F12
Siedlce PL 141 F6
Siedlec PL 81 B9
Siedlce PL 81 C8
Siedliska PL 144 D3
Siedlisko PL 81 C9
Siedlisko PL 85 E11
Siegburg D 21 C8
Siegen D 21 C10
Siegenburg D 75 E10
Sieggraben A 149 A6
Sieghartskirchen A 77 F10
Siegsdorf D 73 A6
Sieima N 112 E4
Siek D 83 C8
Siekasjärvi S 117 E10
Siekkinen FIN 121 C11
Sielec PL 144 A8
Sielnica SK 147 D8
Siemianowice Śląskie PL 143 F7
Siemianówka PL 141 E9
Siemiatycze PL 141 F7
Siemién PL 141 G7
Siemkowice PL 143 D6
Siemyśl PL 85 B9
Siena I 66 F3
Sieniawa PL 144 C6
Sieniec PL 142 D6
Siennica PL 141 F5
Siennica Różana PL 144 A7
Sienno PL 141 H5
Sieppijärvi FIN 117 D11
Siepraw PL 147 B9
Sieradz PL 142 C6
Sieraków PL 81 A9
Sieraków PL 142 E6
Sierakówek PL 139 F7
Sierakowice PL 138 B4
Sierck-les-Bains F 20 F6
Sierentz F 27 E7
Sierksdorf D 83 B9
Sierndorf A 77 F10
Sierning A 76 F6
Sieroszewice PL 142 C4
Sierpc PL 139 E8
Sierra de Fuentes E 45 F8
Sierra de Luna E 41 D10
Sierra de Yeguas E 53 B7
Sierre CH 31 C12
Sierro E 55 E8
Siershahn D 185 D8
Siersleben D 79 C10
Siete Iglesias de Trabancos E 39 F9
Sietow D 83 D13
Sieu RO 152 C5
Şieu-Măgheruş RO 152 C5
Şieu-Odorhei RO 152 C4
Sieverstedt D 82 A6
Sievi FIN 123 C13
Sievin asema eli Korhoskylä FIN 123 C13
Siewierz PL 143 F7
Sifferbo S 97 A13
Sifjord N 111 B13
Sig DK 86 D3
Sigean F 34 D4
Sigerfjord N 111 C10
Siggerud N 95 C13
Sighetu Marmaţiei RO 145 H8
Sighişoara RO 152 E5
Sigillo I 67 E6
Sigmaringen D 27 D11

Sigmaringendorf D 27 D11
Sigmen BG 167 D7
Sigmundsherberg A 77 E9
Signa I 66 E3
Signau CH 70 D5
Signes F 35 D10
Signy-l'Abbaye F 19 E9
Signy-le-Petit F 19 E9
Sigogne F 28 D5
Sigoulès F 29 F6
Sigri GR 177 A6
Sigtuna S 99 C9
Sigüeiro E 38 C3
Sigüenza E 47 B7
Sigüés E 32 E3
Sihlea RO 161 C10
Siguarda EST 130 D7
Sigulda LV 135 B9
Sihelné SK 147 B8
Sihtuuna FIN 119 B13
Sihva EST 131 E12
Siikainen FIN 126 B6
Siikajoki FIN 119 E13
Siika-Kämä FIN 119 B16
Siikakoski FIN 128 B8
Siilinjärvi FIN 124 D9
Siimusti EST 131 D12
Siironen FIN 119 F11
Siivikko FIN 121 D10
Sijarinska Banja SRB 164 D4
Sijekovac BIH 157 B8
Šijuelo B 182 C2
Sikås S 106 D8
Sikeå S 118 F5
Sikfors S 118 C6
Sikirevci HR 157 B9
Siklós H 149 E10
Siknäs S 118 C8
Sikole SRB 159 E9
Sikórz PL 139 E8
Sikovuono FIN 113 F18
Sikrags LV 130 F4
Siksele S 107 C14
Siksjö S 107 C13
Siksjön S 102 C8
Siksjönäs S 107 B10
Siksklni LV 134 C2
Sil S 107 D10
Šilajärvi LV 133 D1
Šilalė LT 134 F4
Silandro I 71 D11
Silanus I 64 C2
Silaveni LV 135 D10
Silba HR 67 D10
Silbaš SRB 158 C3
Silene LV 135 E13
Silenrieux B 19 D9
Siles E 55 C7
Silfiac F 22 D5
Siligo I 64 B2
Silinda RO 151 E8
Siliqua I 64 E2
Šilinis LT 134 F4
Silistra BG 161 E10
Silivaşu de Câmpie RO 152 D4
Silivri TR 173 B9
Siljan N 95 D11
Šiljansnäs S 102 E8
Siljeåsen S 106 C9
Silkeborg DK 86 C5
Silla E 48 F4
Sillamäe EST 132 C2
Sillano I 66 D1
Sillanpää FIN 123 D11
Sillans-la-Cascade F 36 D4
Sillebotten S 96 D7
Silleda E 38 C3
Sillé-le-Guillaume F 23 D11
Silleró S 97 A10
Sillian A 72 C5
Silloth UK 5 F10
Silnowo PL 85 C10
Silo HR 67 B10
Sils CH 71 E9
Sils E 43 D9
Silsjönäs S 107 D10
Siltaharju FIN 115 C11
Siltakylä FIN 128 E6
Silte S 93 E12
Siltukalns LV 135 C13
Šilutė LT 134 F3
Šiluva LT 134 E6
Silva LV 135 B11
Silvalen N 108 E3
Silvaplana CH 71 E9
Silvares P 44 D5
Silveiras P 50 B3
Silverdalen S 92 D7
Silvermines IRL 8 C6
Silverstone UK 13 A12
Silverton UK 13 D8
Silvi I 62 B6
Silz A 71 C11
Simala I 64 D2
Simancas E 39 E10
Šimand RO 151 E7
Šimandre F 31 B6
Šimanovci SRB 158 D5
Simat de la Valdigna I 56 C4
Simaxis I 64 D2
Simbach D 75 E12
Simbach am Inn D 76 F4
Simbario I 59 B9
Šimeria RO 151 F11
Simeri-Crichi I 59 A10
Simested DK 86 B5
Šimian RO 151 C9
Šimian RO 159 E9
Simiane-la-Rotonde F 35 C10
Simićevo SRB 159 D7
Siminicea RO 153 B8
Šimitli BG 165 F7
Šimleu Silvaniei RO 151 C10
Simmelsdorf D 75 C9
Simmerath D 20 C6
Simmerberg D 71 B9
Simmern (Hunsrück) D 21 E9
Simnas LT 137 E8
Simo FIN 119 C14
Simola FIN 129 D10
Simola de Sus RO 160 E3
Šimonešti RO 152 E6
Šimonys LT 135 E10

Simonkangas FIN 119 C17
Simonkylä FIN 119 C14
Simonsberg D 82 B5
Simonstorp S 93 B8
Simontornya H 149 C11
Simoulo GR 174 D4
Simorre F 33 D7
Simpelveld NL 20 C5
Simrishamn S 88 D6
Simskardet N 106 A5
Simuna EST 131 C12
Sina RO 161 C7
Sinabelkirchen A 148 B5
Sinaia RO 161 C7
Sinalepa EST 130 D7
Sinalunga I 66 F4
Sinanli TR 173 B7
Sinarades GR 168 E2
Sinarcas E 47 E10
Şinca RO 152 F6
Šindal DK 90 E7
Sindel BG 167 C9
Sindelfingen D 27 C10
Sindi EST 131 E9
Sindia I 64 C2
Sindos GR 169 C8
Sinekçi TR 173 D7
Sinekli TR 173 B9
Sinemorets BG 167 E9
Sines P 50 D2
Sinesti RO 153 C10
Sineşti RO 160 D3
Sineşti RO 161 D8
Sinetänsalmi FIN 117 E14
Sinetta FIN 117 E14
Sineu E 57 B11
Singen (Hohentwiel) D 27 E10
Sîngera MD 154 D3
Sîngerei MD 153 B12
Singhofen D 21 D9
Singsjön S 106 E8
Singureni RO 161 E7
Singusdal N 90 A4
Sinimäe EST 132 C3
Siniscola I 64 B4
Sinj HR 157 E6
Sinlabajos E 46 B3
Sin-le-Noble F 19 D7
Sinn D 21 C10
Sinnai I 64 E3
Sinnes N 94 E5
Sinoie RO 155 D3
Sinole LV 135 B12
Sinopoli I 59 C8
Sins CH 27 F9
Sinsheim D 21 F11
Sint Annaland NL 16 E2
Sint Annaparochie NL 16 B5
Sint Anthonis NL 16 E6
Sintautai LT 136 D7
Sîntava SK 146 E5
Sintea Mare RO 151 D8
Sintereag RO 152 C4
Sinteu RO 151 C10
Sint-Genesius-Rode B 182 D4
Sint-Gillis-Waas B 16 F2
Sint-Huibrechts-Lille B 183 C6
Sint Jansteen NL 182 C4
Sint-Katelijne-Waver B 19 B10
Sint-Laureins B 182 C3
Sint-Lenaarts B 182 C5
Sint Maartensdijk NL 16 E2
Sint-Margriete B 182 C3
Sint-Maria-Lierde B 19 C8
Sint-Martens-Latem B 19 B8
Sint Michielsgestel NL 183 B6
Sint Nicolaasga NL 16 C5
Sint-Niklaas B 19 B9
Sint Odilienberg NL 183 C8
Sint-Pauwels B 182 C4
Sint Philipsland NL 16 E2
Sint-Pieters-Leeuw B 19 C9
Sintra P 50 B1
Sint-Truiden B 19 C11
Sint Willebrord NL 16 E3
Sinzheim D 27 C9
Sinzig D 21 C8
Sióagárd H 149 D11
Siófok H 149 C10
Sion CH 31 C11
Sion RO 161 C8
Sion Mills UK 4 F1
Šipca MD 154 C2
Sipinen FIN 121 F12
Šipkovica NMK 163 E10
Sipola FIN 119 F15
Šipote RO 153 C10
Sipotina MD 154 C1
Šipovo BIH 157 D7
Sipplingen D 27 E11
Sippola FIN 128 D7
Sira N 94 F5
Širač HR 149 E8
Siracusa I 59 E7
Siran F 29 F10
Sirbmá N 113 C19
Siret RO 153 B8
Şireţel RO 153 C9
Sirevåg N 94 F5
Sirgala EST 132 C2
Širia RO 151 E8
Širig SRB 158 C4
Sirineasa RO 160 D4
Širitovci HR 156 E5
Siriu RO 161 C8
Sirkka FIN 117 C13
Sirkkakoski FIN 117 E12
Sirkkamäki FIN 123 E16
Sirmione I 66 B2
Širna RO 161 D7
Širnach CH 27 F11
Sirniö FIN 121 C11
Sirnitz A 73 C9
Sirod F 31 B9
Sirogojno SRB 158 F4
Sirok H 145 H1
Široki Brijeg BIH 157 F8
Sirolo I 67 E8
Sîrör F 31 B9
Sirotina BY 133 F4
Siruela E 54 B2
Sirutiškis LT 135 F7
Širvintos LT 137 C10
Sisak HR 149 F6
Sisante E 47 F8
Şişcani MD 154 D2
Šisco F 37 F10
Şişeşti RO 152 B3
Şişeşti RO 159 D10
Şişli TR 173 B10
Sissa I 66 C1

Sissach CH 27 F8
Sisses GR 178 E8
Sissonne F 19 E8
Şiştarovăţ RO 151 E8
Sisteron F 35 B10
Sistranda N 104 D5
Sistrans A 72 B3
Sita Buzăului RO 161 B8
Sitaniec PL 144 B7
Siteia GR 179 E11
Sitges E 43 E7
Sitkówka-Nowiny PL 143 E10
Sitnica BIH 157 C7
Sitochori GR 169 C10
Sitovo BG 161 E10
Sitsyenyets BY 133 E6
Sittard NL 19 C12
Sittensen D 82 D7
Sittersdorf A 73 C10
Sittingbourne UK 15 E10
Sitzendorf an der Schmida A 77 E9
Sitzenroda D 80 D3
Siulaisiadar UK 2 J4
Siuntio FIN 127 E11
Siuro FIN 127 C9
Siurua FIN 119 D16
Siurunmaa FIN 115 D1
Sivac SRB 158 B3
Sivakka FIN 125 C13
Sivakkajoki FIN 119 B13
Sivakkavaara FIN 125 E11
Siverić HR 156 E5
Siverskiy RUS 132 C7
Sivertgården N 108 E7
Sivry B 19 D9
Sivry-sur-Meuse F 19 F11
Sixarby S 99 B9
Six-Fours-les-Plages F 35 D10
Sixmilebridge IRL 8 C5
Sixmilecross UK 7 C8
Six Road Ends UK 4 F5
Sixt-Fer-à-Cheval F 31 C10
Sizun F 22 D3
Sjemeć BIH 158 F3
Sjenica SRB 163 C9
Sjetlina BIH 157 E10
Sjoa N 101 C11
Sjøåsen N 105 C10
Sjöbo S 87 D13
Sjöbotten S 118 E6
Sjöbrånet S 107 C17
Sjøholt N 100 B5
Sjølund DK 86 E5
Sjömarken S 91 D12
Sjonbotn N 108 D6
Sjørring DK 86 B3
Sjørslev DK 86 C4
Sjørup DK 86 C4
Sjösa S 93 B10
Sjösäter S 99 B11
Sjötofta S 91 E13
Sjötorp S 91 B14
Sjoutnäset S 106 B7
Sjøvassbotn N 111 B17
Sjøvegan N 111 C14
Sjövik S 91 D11
Sjulsåsen S 106 C7
Sjulsmark S 118 C7
Sjunnen S 92 E6
Sjuntorp S 91 C11
Sjursvik N 111 B12
Skademark S 107 E16
Skælsør DK 87 E8
Skærbæk DK 86 E3
Skævinge DK 87 D10
Skaftung FIN 122 F6
Skagen DK 90 D8
Skagersvik S 91 B15
Skäggebyn S 91 A12
Skagshamn S 107 E16
Skaidi N 113 C13
Skaidiškes LT 137 D11
Skaill UK 3 H11
Skaista LV 133 E2
Skaistgirisi LT 135 E8
Skaistgirys LT 134 D6
Skaistkalne LV 135 D9
Skala GR 174 C2
Skala GR 175 B9
Skala GR 175 F6
Skala GR 177 E8
Skała PL 143 F8
Skala Eresou GR 177 A6
Skala Kallonis GR 177 C4
Skala Marion GR 171 C7
Skålan S 102 A7
Skaland N 111 B13
Skala Oropou GR 175 C8
Skálavík FO 2 B3
Skalbmierz PL 143 F9
Skåle N 105 C15
Skålevik N 90 C3
Skälgården S 103 A12
Skáli FO 2 A3
Skalica SK 146 D4
Skalice CZ 81 E7
Skalité SK 147 C7
Skalitsa BG 166 E6
Skallelv N 114 C8
Skällinge S 87 A10
Skallvik S 93 C9
Skalmodal S 108 F8
Skalmsjö S 107 D13
Skalná CZ 75 B11
Skålö S 97 A11
Skaloti GR 171 B6
Skals DK 86 B4
Skålsjön S 103 D10
Skålsvik N 108 B7
Skalvallen S 103 C10
Skån S 103 B11
Skanderåsen S 102 A7
Skanderborg DK 86 C5
Skånes-Fagerhult S 87 C12
Skåne-Tranås S 88 D5
Skånevik N 94 C3
Skåningen N 112 C4
Skankalne LV 131 F10
Skänninge S 92 C6
Skanör med Falsterbo S 87 E11
Skansbacken S 97 B11
Skansen N 105 D9
Skansholm S 107 B10
Skansnäs S 107 A10
Skansnäs S 109 E13
Skansnäset S 106 C9
Skåpafors S 91 A11
Skape PL 81 B8
Skapiškis LT 135 E10
Skär N 94 D4
Skara S 91 C13

Skäran S 118 F6
Skarberget N 111 D11
Skärblacka S 92 B7
Skarda S 107 C15
Skardmodalen N 108 F7
Skardmunken N 111 A18
Skardstein N 111 B15
Skardsvåg N 113 A16
Skare N 94 C5
Skåre S 97 D9
Skärhamn S 91 D10
Skarkdalen S 102 A4
Skärkind S 92 C7
Skarnes N 96 B6
Skärplinge S 99 B9
Skarp Salling DK 86 B4
Skarrild DK 86 D3
Skärså S 103 D13
Skarsfjord N 112 D2
Skärsjövålen S 102 B5
Skarstad N 111 D11
Skärstad S 92 D4
Skarsvåg N 113 A15
Skarszewy PL 138 B5
Skårup DK 86 E7
Skarv N 112 C11
Skärvången S 105 D16
Skarvfjordhamn N 112 B11
Skarvsjöby S 107 B12
Skaryszew PL 141 H4
Skarżysko-Kamienna PL 141 H3
Skasenden N 96 B7
Skåstra S 103 C11
Skatamark S 118 C6
Skatan S 103 B13
Skattkärr S 97 D10
Skattungbyn S 102 D8
Skatvik N 111 B14
Skaudvile LT 134 F5
Skaugvoll N 108 B8
Skaulo S 116 D6
Šķaune LV 133 D3
Skavdal N 111 B10
Skave DK 86 C3
Skavnakk N 112 C7
Skawina PL 143 G8
Skebobruk S 99 C11
Skebokvarn S 93 A9
Skeda udde S 92 C7
Škēde LV 134 C4
Skede S 92 E6
Skedevi S 92 B7
Skedsmokorset N 95 B14
Skee S 91 B9
Skegness UK 11 E12
Skegrie S 87 E12
Skei N 100 C4
Skei N 105 A11
Skela SRB 158 D5
Skelby DK 87 E9
Skelde DK 82 A7
Skelhøje DK 86 C4
Skellefteå S 118 E5
Skellefteåhamn S 118 E6
Skelmersdale UK 10 D6
Skelton UK 11 B9
Šķeltova LV 133 D2
Skelund DK 86 B6
Skelwick UK 3 G11
Skėmiai LT 134 E7
Skenderaj RKS 163 D10
Skender Vakuf BIH 157 D7
Skenfrith UK 13 B9
Skepasto GR 174 C5
Skjpe PL 139 E7
Skepplanda S 91 D11
Skeppshamn S 103 B14
Skeppshult S 87 A12
Skeppsmalen S 107 E16
Skerries IRL 7 E10
Ski N 95 C13
Skiathos GR 175 A7
Skibbereen IRL 8 E4
Skibbild DK 86 C3
Skibby DK 87 D9
Skibe LV 134 C6
Skibinge DK 87 E10
Skibotn N 111 B19
Skidal' BY 140 C10
Skiemonys LT 135 F10
Skien N 90 A6
Šķieneri LV 135 B13
Skierbieszów PL 144 B7
Skierniewice PL 141 G2
Skiippagurra N 114 C4
Šķilbēni LV 133 B3
Skillebotn N 108 F3
Skillefjordnes N 113 C11
Skillingaryd S 92 E4
Skillinge S 88 E6
Skillvassbakk N 111 D10
Skinias GR 178 E9
Skinnarud N 101 E12
Skinnskatteberg S 97 C12
Skipmannvik N 109 B9
Skipness UK 4 D6
Skipsea UK 11 D11
Skipton UK 11 D7
Skiptvet N 95 D14
Skirlaugh UK 11 D11
Skitenelv N 111 A17
Skiti GR 169 E8
Skivarp S 87 E13
Skive DK 86 B4
Skivjan RKS 163 E9
Skivsjön S 107 C16
Skiwy Duże PL 141 E7
Skjærhalden N 91 A9
Skjåholmen N 113 B12
Skjånes S 113 B18
Skjånes N 113 B21
Skjåvika N 108 E6
Skjeberg N 91 A9
Skjeggedal N 90 B3
Skjelelv N 111 C13
Skjellbreid N 111 A17
Skjelman N 111 A17
Skjelnes N 111 A18
Skjelstad N 105 D10
Skjelvik N 108 B7
Skjern DK 86 D3
Skjern N 105 C9
Skjerstad N 108 B9
Skjervøy N 112 C6
Skjød DK 86 C5
Skjold N 111 B17
Skjoldastraumen N 94 D3
Skjolden N 100 D7
Skjombotn N 111 D13
Skjøtningberg N 113 A19
Skláby GR 175 C9
Skagi GR 169 B7
Skokloster S 99 C9
Škofja Loka SLO 73 D9
Škofljica SLO 73 E10
Skog S 103 D12
Skogaholm S 92 A6
Skoganvarri N 113 D15
Skoger N 95 C12
Skogfoss N 114 E7
Skoghall S 97 D9
Skogly N 114 E6
Skogmo N 105 B12
Skogn N 105 D10
Skogså S 118 C7
Skogsby S 89 B11
Skogsfjord N 112 C4
Skogshöjden S 91 C11
Skogstorp S 87 B10
Skogstorp S 98 D6
Skogstue N 112 D5
Skogum N 114 E6
Skoki PL 81 A12
Sköldinge S 93 A8
Skole UA 145 E8
Skollenborg N 95 C11
Sköllersta S 92 A6
Skoltevatn N 114 E7
Skomlin PL 142 D5
Skonseng N 108 D7
Skönvik S 103 C12
Skopelos GR 175 A8
Skopelos GR 177 A7
Skopi GR 179 E11
Skopje NMK 164 F3
Skopos GR 169 C6
Skopos GR 171 B7
Skopun FO 2 B3
Skórcz PL 138 C6
Skorica SRB 159 F8
Skorild N 104 E6
Skorogoszcz PL 142 E4
Skoroszyce PL 142 E3
Skorovatn N 105 B14
Skorped S 107 E13
Skorpetorp S 89 A10
Skørping DK 86 B5
Skorstad N 105 B10
Skórzec PL 141 F6
Skoteini GR 174 D5
Skotfoss N 90 A6
Skotina GR 169 D8
Skotoussa GR 169 B9
Skotselv N 95 C11
Skotterud N 96 C7
Skottsund S 103 B13
Skoura GR 175 E5
Skourta GR 175 C8
Skoutari GR 169 B10
Skoutari GR 178 B4
Skoutaros GR 171 F10
Skovby DK 86 C5
Skövde S 91 C14
Skoved S 107 E14
Skovlund DK 86 D3
Skovsgård DK 86 A4
Skra GR 169 B7
Skräddrabo S 103 D10
Skradin HR 156 E4
Skråmestø N 100 E1
Skranstad N 110 E9
Skravena BG 165 D7
Skrea S 87 B11
Skreia N 101 E13
Skriaudžiai LT 137 D8
Skrinyano BG 165 E6
Skřipov CZ 146 B5
Skřiveri LV 135 C10
Skröven S 116 E7
Skrunda LV 134 C4
Skrwilno PL 139 D8
Skrzatusz PL 85 D11
Skrzyńsko PL 141 H3
Skrzyszów PL 143 G11
Skucani BIH 157 E6
Skudeneshavn N 94 D2
Skuhrov nad Bělou CZ 77 B10
Skujene LV 135 B10
Skujetnieki LV 133 C2
Skuki LV 133 E3
Skulsk PL 138 F5
Skulte LV 135 B8
Skulte LV 135 C7
Skultorp S 91 C14
Skultuna S 98 C6
Skuodas LT 134 D3
Skurträsk S 107 C16
Skurup S 87 E13
Skuteč CZ 77 C9
Skutskär S 103 E13
Skutvik N 110 D8
Skwierzyna PL 81 A9
Skýcov SK 146 D6
Skydra GR 169 C7
Skyllberg S 92 B6
Skylnäs S 103 A9
Skyros GR 175 B9
Skyttmon S 106 E9
Skyttorp S 99 B9
Slabodka BY 133 E2
Słaboszów PL 143 F9
Sládkovičovo SK 146 E5
Slagavallen S 102 B5
Slagelse DK 87 E8
Slagnäs S 109 E15
Slaidburn UK 10 D7
Slaka S 92 C7
Slampe LV 134 C6
Slane IRL 7 E9
Slanec SK 145 F3
Slangerup DK 87 D10
Slănic RO 161 C7
Slănic Moldova RO 153 E8
Slano HR 162 D4
Slantsy RUS 132 C3
Slaný CZ 76 B6
Slap BIH 158 F3
Slap MNE 163 D7
Slobelevo BG 166 D4

Slap SLO 73 D8
Šlapaberžė LT 135 F7
Šlapanice CZ 77 D11
Släpträsk S 107 A15
Slate LV 135 D12
Slatina BIH 157 C7
Slatina BIH 157 C8
Slatina BIH 157 E8
Slatina BIH 157 E10
Slatina HR 149 E9
Slatina RO 160 E4
Slatina RO 153 C8
Slatina SRB 158 E4
Slatiňany CZ 77 C9
Slatina-Timiş RO 159 C9
Slatino NMK 168 B4
Slatinski Drenovac HR 149 E9
Slătioara RO 160 C3
Slătioara RO 160 E4
Slato BIH 157 F9
Slättberg S 102 D8
Slåttholmen N 110 D8
Slåttmon S 103 A13
Slattum N 95 C13
Slava Cerchezà RO 155 D3
Slava Rusă RO 155 D3
Slaveino BG 165 F10
Slavičín CZ 146 C5
Slavinja SRB 165 C6
Slavkov CZ 146 B5
Slavkovichi RUS 132 F5
Slavkov u Brna CZ 77 D11
Slavonice CZ 77 E8
Slavonski Brod HR 157 B9
Slavošovce SK 145 F1
Slavotin BG 165 B7
Slavovitsa BG 165 E9
Slavsk RUS 136 C4
Slavs'ke UA 145 F7
Slavyani BG 165 C10
Slavyanovo BG 165 C10
Slavyanovo BG 166 F5
Sława PL 81 C10
Sławatycze PL 141 G9
Sławęcin PL 141 F7
Sławków PL 143 F7
Sławno PL 85 B11
Sławoborze PL 85 C9
Sławsko PL 85 B10
Sleaford UK 11 F11
Sledmere UK 11 C10
Sleen NL 17 C7
Sleidinge B 182 C3
Sleights UK 11 C10
Slemmestad N 95 C12
Šlesin PL 138 D4
Šlesin PL 138 F5
Sletta N 112 C9
Slevik N 91 A8
Sliač SK 147 D8
Slidre N 101 D9
Sliedrecht NL 16 E3
Šlienava LT 137 D9
Sligachan UK 2 L4
Sligeach IRL 6 D6
Sligo IRL 6 D6
Slimminge DK 87 E9
Slimnic RO 152 F4
Slinfold UK 15 E8
Slipra N 105 D9
Slišane SRB 164 D4
Slite S 93 D13
Sliven BG 166 D6
Slivilești RO 159 D11
Slivnitsa BG 165 D7
Slivo Pole BG 161 F8
Sliwice PL 138 C5
Slatiné e Madhe RKS 164 D3
Slobidka UA 154 B4
Slobozia MD 154 D5
Slobozia RO 160 D6
Slobozia RO 161 D10
Slobozia RO 161 F7
Slobozia Bradului RO 161 C10
Slobozia Ciorăşti RO 161 B10
Slobozia Conachi RO 155 B1
Slobozia Mândra RO 160 F5
Slobozia Moară RO 161 D7
Slochteren NL 17 B7
Slöinge S 87 B11
Słomniki PL 143 F9
Słonowice PL 85 C9
Słońsk PL 81 A7
Slootdorp NL 16 C3
Slottsskogen S 99 C9
Slough UK 15 E7
Sloupnice CZ 77 B12
Sløvåg N 100 E2
Slovenj Gradec SLO 73 C11
Slovenská Bistrica SLO 148 D5
Slovenská Ľupča SK 147 D8
Slovenská Ves SK 145 E1
Slovenske Konjice SLO 148 D4
Slovenské Nové Mesto SK 145 G4
Slovenski Grob SK 146 E4
Slovinci HR 157 B6
Slovinky SK 145 F2
Słowra N 110 D8
Slov''yanoserbka UA 154 D5
Słowik PL 143 C7
Słubice PL 81 B7
Słubice PL 139 F8
Sluderno I 71 D11
Sluis NL 16 F1
Sluiskil NL 16 F1
Šluknov CZ 81 E6
Slunj HR 156 B4
Słupca PL 138 F4
Słupca PL 144 B4
Słupia PL 141 G1
Słupia PL 143 E9
Słupia PL 143 E8
Słupno PL 139 E8
Słupsk PL 85 B12
Slušovice CZ 146 C5
Slussfors S 109 F11
Slyuda RUS 115 D9
Smailholm UK 5 D11
Smålandsstenar S 87 A12
Smalåsen N 105 A14
Smalfjord N 113 C12
Smalininkai LT 136 C6
Smalvos LT 135 E13
Smârdan RO 155 C1
Smârdan RO 155 C2
Smărde LV 134 C6
Smârdioasa RO 161 F6

Smardzewice PL 141 H2
Smardzewo PL 81 B9
Smardzko PL 85 C9
Smarhon' BY 137 E13
Šmarje pri Jelšah SLO 148 D5
Šmarjeta SLO 148 E4
Šmartno SLO 73 D10
Šmartno SLO 73 D11
Smarves F 29 B6
Smedby S 89 B10
Smederevo SRB 159 D6
Smederevska Palanka SRB 159 E6
Smedjebacken S 97 B13
Smedsbyn S 118 C8
Smedvik N 110 D6
Smeeni RO 161 C9
Smelror N 114 C10
Smelteri LV 135 D13
Smigiel PL 81 B10
Šmihel BG 165 D7
Šofrincani MD 153 A10
Šofiko GR 175 D7
Śofovo BG 171 B11
Šofronievo BG 160 F3
Šoften DK 86 C6
Søgne N 90 A4
Sofular TR 181 C8
Sögel D 17 C9
Sogndalsfjøra N 100 D6
Søgne N 90 C2
Šoğucak TR 173 A8
Söğütalan TR 173 D10
Sofia BG 165 D7
Sofia MD 153 B11
Sofikó GR 171 B11
Sofiko GR 175 D7
Sofopolis RUS 121 C17
Smidary CZ 77 B8
Smidstrup DK 86 D5
Smidstrup DK 87 C10
Smilde NL 17 C6
Smilets BG 166 D6
Smilevo NMK 168 B5
Smilgiai LT 135 D9
Smilgiai LT 135 E8
Smilgiai LT 137 D9
Smilgynai LT 134 E7
Smilovci SRB 165 C6
Smiltene LV 135 B11
Smiltyně LT 134 E2
Smilyan BG 171 A7
Smines N 110 D6
Smiřice CZ 77 B9
Smögen S 91 C9
Smokvica HR 162 D2
Smokvica NMK 169 B7
Smołdzino PL 85 A12
Smolenice SK 146 D4
Smolice PL 81 C12
Smolmark S 96 C7
Smolnica RO 160 E3
Smolník SK 145 F2
Smolyan BG 171 A7
Smolyanovtsi BG 165 C9
Smørfjord N 113 B15
Smultį RO 155 C1
Smyadovo BG 167 C8
Smygehamn S 87 E12
Smykow PL 143 D9
Snagov RO 161 D8
Snainton UK 11 C10
Snaith UK 11 D9
Snålroa N 102 E2
Snappertuna FIN 127 E10
Snaptun DK 86 D6
Snarby N 111 A18
Snåre FIN 123 C10
Snartemo N 94 F6
Snåsa N 105 C12
Snave Bridge IRL 8 E4
Snedsted DK 86 B3
Sneek NL 16 B5
Sneem IRL 8 E3
Snejbjerg DK 86 C3
Snępele LV 134 C3
Snerta N 101 C15
Snertinge DK 87 D8
Snesslinge S 99 B9
Snesudden S 118 B4
Snettisham UK 11 F13
Śniadowo PL 139 D12
Snikere LV 134 D6
Snina SK 145 F5
Snjegotina Velika BIH 157 C8
Snøde DK 87 E7
Snøfjord N 113 B14
Snogebæk DK 89 E8
Snoghøj DK 86 D6
Snolldelev DK 87 D10
Soajo P 38 E3
Soars RO 152 F5
Soave I 66 B3
Søberg N 110 C7
Sobešlav CZ 77 D7
Sobienie-Jeziory PL 141 G4
Sobolew PL 141 G5
Sobota PL 143 B8
Soboth A 73 C11
Sobotín CZ 77 B12
Sobotište SK 146 D4
Sobotka CZ 77 B8
Sobótka PL 81 E11
Sobótka PL 142 C4
Sobótka PL 143 E12
Sobowidz PL 138 B6
Sobra HR 162 D4
Sobradelo E 39 D6
Sobradiel E 41 E9
Sobrado E 38 B3
Sobrado E 38 D3
Sobral da Adiça P 51 C5
Sobrance SK 145 F5
Sobreira Formosa P 44 E5
Søby DK 86 F6
Soča SLO 73 D8
Sočanica RKS 163 C10
Soçanicë RKS 163 C10
Socchieve I 73 D6
Sočerga SLO 67 A8
Sochaczew PL 141 F2
Sochocin PL 139 E9
Sochos GR 169 C9
Socodor RO 151 E6
Socol RO 159 D7
Socond RO 151 B10
Socovos E 55 C9
Socuéllamos E 47 F7
Sodankylä FIN 117 D17
Söderåkra S 89 C10
Söderala S 103 D12
Söderås S 103 E9
Söderbärke S 97 B14
Söderboda S 99 B10
Söderby-Karl S 99 C11
Söderbrunn S 91 C12
Söderfors S 98 B8
Söderköping S 93 C8
Söderhamn S 103 D13
Söderköpping S 93 C8
Söderkulla FIN 127 E13
Södersvik S 99 C11

Sóller E 49 E10
Sollerön S 102 E8
Sollested DK 83 A10
Sollichau D 79 C12
Sollières-Pont F 36 E4
Solliès-Toucas F 36 E4
Sollihøgda N 95 C12
Söllingen D 79 B8
Sollstedt D 79 D8
Solmaz TR 181 B9
Solms D 21 C10
Solnice CZ 77 B10
Solnik BG 167 D9
Solofra I 60 B4
Solojärvi FIN 113 F18
Solomiac F 33 C7
Solomos GR 175 D6
Solopaca I 60 A3
Solórzano E 40 B4
Solosancho E 46 C3
Solośnica SK 146 E4
Solothurn CH 27 F8
Solotvyna UA 145 H8
Soløy N 111 C14
Solre-le-Château F 19 D9
Solrød Strand DK 87 D10
Solsem N 105 A11
Solskjela N 104 E4
Sølsnes N 100 A6
Solsona E 43 D7
Solsvik N 94 B1
Solt H 150 D3
Soltau D 83 E7
Soltendieck D 83 E9
Sol'tsy RUS 132 E7
Soltszentimre H 150 D3
Soltvadkert H 150 D3
Solumshamn S 103 A14
Solva UK 9 E12
Solvalla S 99 C10
Solvarbo S 97 B14
Sølvesborg S 88 C7
Solvorn N 100 D6
Solymár H 149 A11
Soma TR 177 A10
Somain F 19 D7
Somberek H 149 D11
Somberg S 25 F12
Sombor SRB 150 F3
Sombreffe B 19 C10
Şomcuţa Mare RO 151 B11
Someren NL 16 F5
Somerniemi FIN 127 D10
Somero FIN 127 D10
Someronkylä FIN 119 F12
Somerovaara FIN 119 D16
Sömerpalu EST 131 F13
Somerton UK 13 C9
Šömeru EST 131 C12
Someş-Odorhei RO 151 C11
Somianka PL 139 E11
Sominy PL 138 B4
Somlóvásárhely H 149 B8
Sommacampagna I 66 B2
Somma Lombardo I 68 B6
Sommariva del Bosco I 37 B7
Sommarøy N 111 A16
Sommarøy N 111 A15
Sommarset N 109 A10
Sommatino I 58 E4
Somme-Leuze B 19 D11
Sommen S 92 C5
Sommepy-Tahure F 19 F10
Sömmerda D 79 D9
Sommerfeld D 84 E4
Sommerset DK 86 E4
Sommesous F 25 C11
Sommevoire F 25 D12
Sommières F 35 C7
Sommières-du-Clain F 29 C6
Somogyapáti H 149 D9
Somogyjád H 149 D9
Somogyszob H 149 D8
Somogyudvarhely H 149 D8
Somogyvár H 149 C9
Somonino PL 138 B5
Somontín E 55 E8
Somotor SK 145 G4
Somova RO 155 C3
Somovit BG 160 F5
Sompa EST 131 C14
Sompolno PL 138 F5
Sompujärvi FIN 113 G13
Son N 95 C13
Son NL 16 E4
Sona I 66 B2
Şona RO 152 E4
Şoncoboz CZ 47 F7
Soncillo E 40 C4
Soncino I 69 C8
Sonda EST 131 C13
Søndeled N 90 B5
Sønder Balling DK 86 B3
Sønder Bjerre DK 86 D5
Sønder Bjert DK 86 E5
Sønderby DK 86 F5
Sønder Dråby DK 86 B3
Sønder Felding DK 86 D3
Sønderho DK 86 A5
Sønder Hygum DK 86 E3
Sønder Nissum DK 86 C2
Sønder Omme DK 86 D3
Sønder Onsild DK 86 B5
Sønder Rubjerg DK 90 D6
Sondershausen D 79 D8
Søndersø DK 86 E6
Sønder Stenderup DK 86 E5
Sønder Vilstrup DK 86 E5
Sønder Vissing DK 86 C5
Sondheim DK 86 D2
Sondrio I 69 A8
Soneja E 48 E4
Songe N 90 B5
Songeons F 18 E4
Sonim P 38 E5
Sonka FIN 117 E14
Sonkajärvi FIN 124 C8
Sonkari FIN 123 C9
Sonkakoski FIN 124 C9
Sonkamuotka FIN 117 B10
Sonnberg D 75 B9
Sonneborn D 79 E8
Sonnefeld D 75 B9
Sonnewalde D 80 C5
Sonnino I 62 E4
Sonntag A 71 C9

Sonntagberg A 77 G7
Sonseca E 46 E5
Son Servera E 57 B11
Sońsk PL 139 E10
Sonstorp S 92 B7
Sonta SRB 157 A11
Sontheim an der Brenz D 75 E7
Sonthofen D 71 B10
Sontra D 78 D6
Soodla EST 131 C10
Söörmarkku FIN 126 B6
Soorts-Hossegor F 32 C3
Sopeira E 33 F7
Sopelana E 40 B6
Sopilja BIH 157 F9
Sopište NMK 164 F3
Soponya H 149 B10
Šoporňa SK 146 E5
Sopot BG 165 C9
Sopot BG 165 D10
Sopot PL 138 B6
Sopot RO 160 E3
Sopot SRB 158 D6
Sopotnica NMK 168 B5
Šopotu Nou RO 159 D8
Šoppela FIN 115 E3
Sopron H 149 A7
Sopronkövesd H 149 A7
Sora I 62 D5
Soraga I 72 D4
Soragna I 66 C1
Söråker S 103 A14
Söräng S 103 D11
Sorano I 62 B1
Søranøy N 108 B6
Sorbas E 55 E8
Sorbie UK 5 F8
Sorbiers F 30 E5
Sörbo S 103 D11
Sörböle S 103 B13
Sorbolo I 66 C1
Sörby S 97 C9
Sörbygden S 103 A11
Sørbymagle DK 87 E8
Sørbyn S 118 B7
Sörbyn S 118 E4
Sorcy-St-Martin F 26 C4
Sord IRL 7 F10
Sørdal N 109 C10
Sorde-l'Abbaye F 32 C3
Sore F 32 B4
Sørebø N 100 D3
Sören S 118 C9
Søreng N 111 B19
Soresina I 69 C8
Sorèze F 33 D10
Sörfjärden S 103 B13
Sørfjord N 108 D5
Sørfjord N 111 C14
Sørfjorden N 111 C10
Sørfjordmoen N 109 A10
Sörflärke S 107 E13
Sörfors S 103 B13
Sörfors S 122 C4
Sörforsa S 103 C12
Sorge D 79 C8
Sorges F 29 E7
Sorgono I 64 C3
Sorgues F 35 B8
Sørheim N 100 D6
Soria E 41 E7
Soriano Calabro I 59 B9
Soriano nel Cimino I 62 C2
Sorihuela E 45 D9
Sorihuela del Gaudalimar E 55 C6
Sorisdale UK 4 B4
Sørkjosen N 112 D6
Sørkjosen N 112 D6
Sorkwity PL 136 F3
Sørland N 110 E4
Sørlenangen N 111 A19
Sørli N 105 C15
Sørli N 111 B14
Sörmark S 97 B8
Sörmjöle S 122 C4
Sørmo N 111 C16
Sorn UK 5 D8
Sornac F 29 D10
Sørnoret S 107 C12
Sora DK 87 E9
Soroca MD 154 A2
Sorokino RUS 133 B7
Soroni GR 181 D7
Sorradile I 64 C2
Sørreisa N 111 B15
Sorrento I 60 B2
Sorring DK 86 C5
Sørrollnes N 111 C12
Sorsakoski FIN 124 F9
Sorsele S 109 E14
Sörsjön S 102 D5
Sorso I 64 B2
Sörstafors S 98 C6
Sørstraumen N 112 D8
Sort E 33 F8
Sortavala RUS 129 B14
Sortelha P 45 D6
Sortino I 59 E7
Sörtjärn S 102 B8
Sortland N 110 C9
Sør-Tverrfjord N 112 C8
Sõru EST 130 D7
Sørum N 101 E11
Sorumsand N 95 C14
Sorunda S 93 A11
Sörup D 82 A7
Sørvad DK 86 C3
Sørvær N 112 B8
Sörvåge S 107 C15
Sørvågen N 110 E5
Sørvágur FO 2 A2
Sörvattnet S 102 B4
Sørvik N 111 C13
Sörvik S 97 B13
Sørvika N 101 B15
Sörviken S 107 E9
Sorvilán E 55 F6
Sos F 33 B6
Sosandra GR 169 C7
Sösdala S 87 C13
Sos del Rey Católico E 32 E3
Sosedno RUS 132 E4
Soses E 42 D4
Soshe-Ostrivs'ke UA 154 C5
Sosnica PL 85 D10
Sośnie PL 142 D4
Sosno PL 138 D4
Sosnovka RUS 136 D3
Sosnovo RUS 129 D13

Sosnovyy Bor RUS 129 F11
Sosnowica PL 141 G8
Sosnowiec PL 143 F7
Sosnówka PL 141 G8
Soso FIN 119 E15
Sospel F 37 D6
Sossonniemi FIN 121 C14
Sost F 33 E7
Šoštanj SLO 73 D11
Sostís GR 171 B8
Sot N 96 C6
Șoțânga RO 160 D6
Sótés E 41 D6
Sotiel Coronada E 51 D6
Sotillo de la Adrada E 46 D3
Sotillo del Rincón E 41 E6
Sotin HR 157 B11
Sotkajärvi FIN 119 E16
Sotkamo FIN 121 F11
Soto E 39 A7
Soto de la Vega E 39 D8
Soto del Real E 46 C5
Soto de Ribera E 39 B8
Soto en Cameros E 41 D7
Sotopalacios E 40 D4
Sotos E 47 D8
Sotoserrano E 45 D8
Soto y Amío E 39 C8
Sotres E 39 B10
Sotrondio E 39 B8
Sotta F 37 H10
Sotteville-lès-Rouen F 18 F3
Sottomarina I 66 B5
Sottrum D 17 B12
Sottunga FIN 99 B15
Sotuélamos E 55 A7
Soual F 33 C10
Soubès F 34 C5
Soucy F 25 D9
Souda GR 178 E7
Soudan F 23 E9
Soueix F 33 E8
Souesmes F 24 F7
Soufflenheim F 27 C8
Sougia GR 178 E6
Souillac F 29 F8
Souilly F 26 B3
Souk el Had el Rharbia MA 52 E5
Souk-Khémis-des-Anjra MA 53 E5
Soukolojärvi S 119 B11
Souk Tleta Taghramet MA 53 E6
Soulac-sur-Mer F 28 D3
Soulaines-Dhuys F 25 D12
Soultgé F 33 E11
Souli GR 175 D6
Soullans F 28 B2
Soulom F 32 E5
Soumagne B 19 C12
Soumoulou F 32 C4
Souppes-sur-Loing F 25 D8
Souprosse F 32 C4
Sourdeval F 23 C10
Soure P 44 D3
Sournia F 33 E10
Souro Pires P 45 C6
Sourpi GR 175 A6
Sours F 24 D6
Sourzac F 29 E6
Sousceyrac F 29 F10
Sousel P 50 B4
Soustons F 32 C3
Southam UK 13 A12
Southampton UK 13 D12
South Anston UK 11 E9
South Bank UK 11 B9
Southborough UK 15 E9
South Cave UK 11 D10
South Chard UK 13 D9
Southend UK 4 E5
Southend-on-Sea UK 15 D10
Southery UK 11 F11
Southgate UK 15 D8
South Harting UK 15 F7
South Kelsey UK 11 E11
South Kirkby UK 11 D9
Southminster UK 15 D10
South Molton UK 13 C7
South Ockendon UK 15 D9
Southport UK 10 D5
South Queensferry UK 5 D10
South Shields UK 5 F14
Southwell UK 11 E10
Southwold UK 15 C12
South Woodham Ferrers UK 15 D10
Souto E 38 C4
Souto E 38 E2
Souto P 44 E4
Souto P 45 D7
Souto da Casa P 44 D5
Soutupera FIN 123 B14
Souvala GR 175 D8
Souvigny F 30 B3
Sovarna RO 159 D10
Sovata RO 152 D6
Soveja RO 153 E9
Soverato I 59 B10
Soveria I 59 A9
Soveria Mannelli I 59 A9
Sovetsk RUS 136 C4
Sovetskiy RUS 129 D10
Sovičl BIH 157 F7
Sovicille I 66 F3
Søvik N 100 A4
Søvind DK 86 D5
Sowno PL 85 B11
Sowno PL 85 D7
Soyaux F 29 D6
Soye F 26 F5
Soylu TR 173 B7
Soymy UA 145 F7
Soyons F 30 F6
Sozopol BG 167 E9
Spa B 19 D12
Spa IRL 8 D3
Spabrücken D 21 E9
Spačince SK 146 E5
Spačva HR 157 B10
Spadafora I 59 C7
Spaichingen D 27 D10
Spalding UK 11 F11
Spálené Poříčí CZ 76 C5
Spalt D 75 D8
Spangenberg D 78 D6
Spanish Point IRL 8 C4

Spantekow D 84 C5
Spanţov RO 161 E9
Sparanise I 60 A2
Sparbu N 105 D10
Spåre LV 134 B4
Sparreholm S 93 A9
Spartà I 59 C7
Sparti GR 174 E5
Sparto GR 174 B3
Spartylas GR 168 E2
Spasovo BG 155 F2
Spata GR 175 D8
Spay D 185 D8
Spean Bridge UK 4 B7
Specchia I 61 D10
Speen UK 13 C12
Speia MD 154 D4
Speicher D 21 E7
Speichersdorf D 75 C10
Spello I 62 B3
Spenge D 17 D10
Spennymoor UK 5 F13
Spentrup DK 86 B6
Spercheiada GR 174 B5
Sperlinga I 58 D5
Sperlonga I 62 E4
Spermezeu RO 152 C4
Spersboda S 99 C11
Spetalen N 95 D13
Spetses GR 175 E7
Spey Bay UK 3 K10
Speyer D 21 F10
Spezzano Albanese I 61 D6
Spiczyn PL 141 H7
Spiddal IRL 6 F4
Spielberg bei Knittelfeld A 73 B10
Spiere B 182 D2
Spiesen-Elversberg D 21 F8
Spiez CH 70 D5
Spigno Monferrato I 37 B8
Spigno Saturnia I 62 E5
Spiinigied'di N 113 E15
Spijk NL 17 B7
Spijkenisse NL 16 E2
Spikberg S 118 B5
Spilamberto I 66 C3
Spili GR 178 E7
Spilimbergo I 73 D6
Spilinga I 59 B8
Spillersboda S 99 C11
Spillum N 105 C11
Spilsby UK 11 E12
Spinazzola I 60 B6
Spincourt F 19 F12
Špindlerův Mlýn CZ 81 E9
Spinea I 66 B5
Spineni RO 160 D5
Spinetoli I 62 B5
Spink IRL 9 C8
Spinuş RO 151 C9
Spišská Belá SK 145 E1
Spišská Nová Ves SK 145 F2
Spišská Stará Ves SK 145 E1
Spišská Teplica SK 145 F1
Spišské Bystré SK 145 F1
Spišské Podhradie SK 145 F2
Spišské Vlachy SK 145 F2
Spišský Hrušov SK 145 F2
Spital am Pyhrn A 73 A9
Spital am Semmering A 148 A5
Spitsyno RUS 132 D2
Spittal UK 3 J10
Spittal an der Drau A 73 C7
Spittal of Glenshee UK 5 B10
Spitz A 77 F8
Spjald DK 86 C3
Spjelkavik N 100 B4
Spjutsbygd S 89 C9
Spjutsund FIN 127 E14
Split HR 156 F5
Spodnja Idrija SLO 73 D9
Spodnje Hoče SLO 148 C5
Spodsbjerg DK 87 F7
Spofforth UK 11 D9
Spoleto I 62 B3
Spoltore I 63 C6
Spondigna I 71 D11
Spontin B 19 D11
Spornitz D 83 D11
Spotorno I 37 C8
Spraitbach D 74 E6
Sprakensehl D 83 E9
Sprâncenata RO 160 E5
Sprängsviken S 103 A14
Spreenhagen D 80 B5
Spremberg D 80 C6
Sprendlingen D 185 D8
Spresiano I 72 E5
Sprimont B 19 C12
Spring RO 152 F3
Springe D 78 B6
Springholm UK 5 E9
Springliden S 107 A16
Sproatley UK 11 D11
Sprockhövel D 17 F8
Sproge S 93 E12
Sprowston UK 15 B11
Sproxton UK 11 C9
Spuņģēni LV 135 C11
Spuž MNE 163 D7
Spychowo PL 139 C11
Spydeberg N 95 C14
Spytihněv CZ 146 C5
Spytkowice PL 143 G7
Spytkowice PL 147 B9
Squillace I 59 B10
Squinzano I 61 C10
Sráid an Mhuilinn IRL 8 D4
Sraith Salach IRL 6 F3
Sranea IRL 6 D6
Srath an Urláir IRL 7 C7
Srbac BIH 157 B8
Srbobran SRB 158 C4
Srbovac RKS 163 D10
Srdević BIH 157 C10
Srebrenica BIH 158 E3
Srebrenik BIH 157 C10
Srebŭrna BG 161 E10
Sredets BG 166 E5
Sredets BG 167 E8
Središče SLO 148 D6
Srednje BIH 157 D10
Srednje PL 139 F10
Srednjevo SRB 159 D7
Sredno Gradishte BG 166 E4

Śrem PL 81 B12
Sremčica SRB 158 D5
Sremska Kamenica SRB 158 C4
Sremska Mitrovica SRB 158 D4
Sremski Karlovci SRB 158 C4
Srenica BIH 156 D6
Srnice BIH 157 C10
Srockowo PL 136 E4
Środa Śląska PL 81 D11
Środa Wielkopolska PL 81 B12
Srpska Crnja SRB 158 B6
Srpski Itebej SRB 159 B6
Srpski Miletić SRB 157 A11
Sta S 105 E13
Staatz A 77 E10
Stabbfors S 108 E8
Stabbursnes N 113 C14
Stabroek B 16 F2
Stablinieki LV 135 D13
Staburnäs S 107 A10
Staby DK 86 C2
Stachy CZ 76 D5
Stade D 17 A12
Staden B 18 C7
Stadl-Paura A 76 F5
Stadra S 97 C12
Stadsbygd N 104 E7
Stadskanaal NL 17 C7
Stadtallendorf D 21 C12
Stadtbergen D 75 F8
Stadthagen D 17 D12
Stadtilm D 79 E9
Stadtkyll D 21 D7
Stadtlauringen D 75 B7
Stadtlohn D 17 E7
Stadtroda D 79 E10
Stadtschlaining A 148 B6
Stadum D 82 A6
Stäfa CH 27 F10
Staffanstorp S 87 D12
Staffin UK 2 K4
Staffolo I 67 F7
Stafford UK 11 F7
Stahovica SLO 73 D10
Stai N 101 D14
Staicele LV 131 F9
Staig D 74 F6
Stainach A 73 A9
Staindrop UK 11 B8
Staines-upon-Thames UK 15 E7
Stainforth UK 11 C7
Stainforth UK 11 D9
Staintondale UK 11 C11
Stainville F 26 C3
Stainz A 148 C4
Stajan RO 151 D9
Staïti I 59 D9
Stakčín SK 145 E5
Stąkl LV 135 B13
Stakiai LT 134 F6
Stakkvik N 112 D4
Stakliškės LT 137 D9
Stakroge DK 86 D3
Stalać SRB 159 F7
Stalbe LV 135 B10
Stålbo S 98 B7
Stalbridge UK 13 D10
Stalden CH 68 A4
Staldzene LV 134 B3
Stalgėnai LT 134 E3
Stalgene LV 135 C7
Stalham UK 15 B12
Stalheim N 100 E5
Stalidzani LV 135 C13
Staliogargo N 113 B12
Stall A 73 C7
Stallarholmen S 98 D8
Ställberg S 97 C12
Ställdalen S 97 C12
Stallwang D 75 D12
Stalon S 107 B9
Stalowa Wola PL 144 B5
Stâlpeni RO 160 C5
Stålpu RO 161 C9
Stamboliyski BG 165 E10
Stambulčić BIH 157 E10
Stamford UK 11 F10
Stamford Bridge UK 11 D10
Stamfordham UK 5 E13
Stamnham D 75 E9
Stamna GR 174 B3
Stamnes A 71 C11
Stamsund N 110 D6
Stăncani RO 155 B2
Stăncuța RO 155 D1
Standard RO 160 D5
Standardbuiten NL 182 B5
Standish UK 10 D6
Stăneşti RO 159 C11
Stăneşti RO 161 F7
Stånga S 93 E12
Stångăceaua RO 160 D2
Stangnes N 111 C13
Stångviken S 105 D16
Stanhoe UK 15 B10
Stanhope UK 5 F12
Stănilesti RO 153 D12
Stanin PL 141 G6
Stânisesti RO 153 E10
Stanišić SRB 150 F3
Stanisławów PL 141 F5
Staniţa RO 153 D9
Stanišinci SRB 159 F9
Štanjel SLO 73 E8
Staňkov CZ 76 C4
Stankovany SK 147 C8
Stanley UK 5 C10
Stanley UK 5 F13
Stanley UK 11 D9
Stannington UK 5 E13
Stanomino PL 85 C10
Stanos GR 169 C10
Stanos GR 174 B3
Stan'ovsti BG 165 D6
Stans CH 71 D8
Stansted Mountfitchet UK 15 D9
Stanton UK 15 C10
Stanzach A 71 C11
Stanz im Mürztal A 148 B5
Stapar SRB 158 B3

Stará Huť CZ 76 C6
Stara Kamienica PL 81 E9
Stara Kamionka PL 140 C9
Stara Kiszewa PL 138 C5
Stara Kornica PL 141 F7
Stara Kuľna UA 154 B4
Stará Ľubovňa SK 145 E2
Stara Łubianka PL 85 D11
Stara Moravica SRB 150 F3
Stara Novalja HR 67 C10
Stara Pazova SRB 158 D5
Stara Plošćica HR 149 E7
Stara Reka BG 166 D6
Stara Rudnica PL 84 E6
Stara Sil' UA 145 E6
Stara Tsarychanka UA 154 D5
Stará Turá SK 146 D5
Stara vas-Bizeljsko SLO 148 E5
Stará Ves nad Ondřejnicí CZ 146 B6
Staravina NMK 169 B6
Stara Wieś PL 144 B6
Stara Zagora BG 166 E5
Starčevo SRB 158 D6
Starchiojd RO 161 C8
Starcross UK 13 D8
Stare Babice PL 141 F3
Stare Bogaczowice PL 81 E10
Stare Budkowice PL 142 E5
Stařeč CZ 77 D9
Stare Czarnowo PL 85 D7
Stare Dolistowo PL 140 C7
Stare Hołowczyce PL 141 F7
Stare Kurowo PL 85 E9
Staré Město CZ 77 B11
Staré Město u Uherského Hradiště CZ 146 C4
Stare Miasto PL 142 B5
Stare Pole PL 139 B7
Stare Puchały PL 141 E7
Stare Selo UA 144 D9
Stare Strącze PL 81 C10
Star Huta SK 147 E8
Stari LV 135 B13
Stari Banovci SRB 158 D5
Starigard HR 67 C10
Starigrad HR 156 D3
Starigrad HR 156 F6
Stari Gradac HR 149 E8
Stari Log SLO 73 E10
Stari Majdan BIH 156 C6
Stari Mikanovci HR 157 B10
Stari Trg SLO 67 B11
Stari Trg SLO 73 F10
Stari Troyany UA 155 B4
Starkenberg D 79 E11
Stärkesmark S 122 B3
Starnberg D 75 G9
Starogard PL 85 C9
Starogard Gdański PL 138 C6
Staroglavice BIH 158 E3
Starokozache UA 154 D5
Staro Oryakhovo BG 167 D9
Staropatitsa BG 159 F10
Staro Petrovo Selo HR 157 B8
Starosel BG 165 E10
Staro Selo BG 161 F9
Staro Selo BIH 157 D7
Staro Selo SRB 159 E7
Staroseltsi BG 165 B8
Starosiedle PL 81 C7
Starowa Góra PL 143 C7
Staroye Syalo BY 133 F7
Staro Zhelezare BG 165 E10
Starozřeby PL 139 E8
Startforth UK 11 B8
Starup DK 86 E5
Stary Brus PL 141 H8
Stary Cykarzew PL 143 E7
Stary Dzierzgoń PL 139 C7
Stary Dzikowiec PL 144 C4
Stary Kisielin PL 81 C8
Stary Kobrzyniec PL 139 E7
Stary Molín CZ 77 B9
Stary Majdan PL 144 C6
Starynovichy BY 133 F4
Stary Pahost BY 133 F3
Stary Plzenec CZ 76 C4
Stary Sącz PL 145 D1
Starý Starý SK 145 E1
Stary Sękôw PL 139 E11
Stary Targ PL 139 C7
Stary Uścimów PL 141 H7
Stary Zamość PL 144 B7
Starzyno PL 138 A5
Staškov SK 147 C7
Staßfurt D 79 C10
Staszów PL 143 E11
Stathelle N 90 A6
Statland N 105 C10
Statsås S 107 B11
Statzendorf A 77 F9
Stăuceni MD 154 C3
Stăuceni RO 153 B9
Staufen D 27 E8
Staufenberg D 80 C5
Staupitz D 80 C5
Stavajn SRB 163 C9
Stavang N 100 C2
Stavanger N 94 E3
Stavchany UA 153 A8
Stave N 111 B10
Stavelot B 20 D5
Stavenisse NL 16 E2
Stavern N 90 B7
Stavertsi BG 165 C8
Stavky UA 144 A9
Stavne UA 145 F6
Stavning DK 86 D3
Stavre S 103 A9
Stavre S 105 E16
Stavreviken S 103 A13
Stavrodromi GR 174 D4
Stavros GR 169 C9
Stavros GR 169 C10
Stavros GR 169 F7
Stavros GR 174 A4
Stavros GR 175 D7
Stavroupoli GR 171 B7
Stavrovoe UA 154 A6
Stavsätra S 103 C9
Stavsjö S 99 D11
Stavtrup DK 86 C6
Staw PL 85 E7

Stawiguda PL 139 C9
Stawiski PL 139 D13
Stawiszyn PL 142 C5
Steane N 95 C14
Stębark PL 139 C9
Steblevë AL 168 B3
Stebnyk UA 145 E8
Steccato I 61 F7
Štěchovice CZ 76 C6
Stechow D 79 A11
Steckborn CH 27 E10
Stede Broec NL 16 C4
Stedesdorf D 17 A9
Stedten D 79 D10
Steeg A 71 C10
Steenbergen NL 16 E2
Steenderen NL 183 A8
Steenvoorde F 18 C6
Steenwijk NL 16 C6
Steeton UK 11 D8
Stefanaconi I 59 B9
Ştefan cel Mare RO 153 D9
Ştefan cel Mare RO 153 D11
Ştefan cel Mare RO 153 E9
Ştefan cel Mare RO 155 D1
Ştefan cel Mare RO 160 D6
Ştefan cel Mare RO 160 F4
Ştefăneşti MD 154 C2
Ştefăneşti RO 153 B9
Ştefăneşti RO 160 D5
Ştefăneşti RO 160 F5
Ştefăneştii de Jos RO 161 D8
Ştefani GR 175 C8
Ştefanje HR 149 E7
Ştefan Karadzha BG 167 C8
Stefan-Karadzhovo BG 167 E7
Ştefanov SK 146 D4
Stefanovikeio GR 169 F8
Stefanovo BG 167 C9
Stefanovo BG 167 C9
Stefanovouno GR 169 E7
Ştefan Vodă MD 154 D5
Ştefan Vodă RO 161 E10
Ştefeşti RO 161 C7
Steffisburg CH 70 D5
Stegaurach D 75 C8
Stege DK 87 F10
Stegelitz D 79 B10
Stegersbach A 148 B6
Stegna PL 138 B7
Stegny PL 139 D8
Stehag S 87 D12
Steigen N 110 E8
Steigra S 79 D10
Steimbke D 17 C12
Stein D 75 D9
Stein NL 19 C12
Steinach A 72 B3
Steinach D 75 B9
Steinach D 186 E5
Steinach am Brenner A 72 B3
Stein am Rhein CH 27 E10
Steinau D 21 C11
Steinau an der Straße D 74 B5
Steinbach D 21 C11
Steinbach D 80 E4
Steinbach am Attersee A 73 A8
Steinbach am Wald D 75 B9
Steinbakk N 114 D7
Steinberg D 75 D11
Steine N 105 D11
Steine N 110 D7
Steinen D 27 E8
Steinfeld A 73 C7
Steinfeld D 186 C5
Steinfeld D 17 C10
Steinfeld (Oldenburg) D 17 C10
Steinfjord N 111 B13
Steinfort L 20 E5
Steinfurt D 17 D8
Steingaden D 71 B11
Steinhagen D 17 D10
Steinhagen D 84 B3
Steinheim D 17 E12
Steinheim N 111 B13
Steinheim am Albuch D 187 D6
Steinheim an der Murr D 187 D7
Steinhöring D 75 F11
Steinhorst D 83 E8
Steinigtwolmsdorf D 80 D6
Steinkirchen D 82 C7
Steinkjer N 105 C11
Steinland N 110 C11
Steinløysa N 100 A7
Steinsdorf D 81 B7
Steinsfeld D 75 D7
Steinshamn N 100 A5
Steinsholt N 95 C11
Steinskjærnes N 114 D7
Steinsland N 94 B2
Steinsvik N 100 C3
Steinwenden D 186 C4
Steinwiesen D 75 B9
Stejari RO 155 D3
Stejaru RO 160 E5
Stekene B 19 B9
Stelle D 83 D8
Stellendam NL 16 E2
Stelnica RO 155 E1
Stelpe LV 135 C9
Stemland N 108 B9
Stemmen D 82 D7
Stemnes N 111 D13
Stemnitsa GR 174 D5
Stemshorn D 17 D10
Stenåsa S 89 C11
Stenay F 19 F11
Stenbjerg DK 86 B2
Stenbo S 93 B8
Stendal D 79 A10
Stende LV 134 B4
Stenderup DK 86 D3
Steneset N 108 D4
Stengelse N 113 D11
Stenhammar S 93 B9
Stenhousemuir UK 5 C9
Stenico I 69 A10
Steninge S 87 B11
Stenis S 102 E7
Stenkulla S 93 D12
Stenlille DK 87 D9
Stenløse DK 87 D10
Stennäs S 107 D13
Steno GR 174 D5
Steno GR 175 E5
Stensån S 99 D11
Stensele S 107 A14

Stensjön S 92 D5
Stenskär S 99 B10
Stenstorp S 91 C14
Stensträsk S 107 A17
Stenstrup DK 86 E6
Stensund S 109 D15
Stensund S 109 F13
Stensved DK 87 F10
Stenton UK 5 D11
Stenträsk S 118 B3
Stenudden S 109 C14
Stenum DK 90 E6
Stenungsund S 91 C10
Stenvad DK 86 C7
Stenzharychi UA 144 B9
Stepanci NMK 164 F4
Stepanivka UA 154 D5
Ştěpánov CZ 77 C12
Stepen BIH 157 F10
Stepnica PL 84 C7
Stepojevac SRB 158 D5
Stepping DK 86 E4
Sterdyń-Osada PL 141 E6
Sterfawki-Wielkie PL 136 F4
Sterna GR 171 A10
Sterna GR 175 D6
Sterna HR 67 B8
Sternberg D 83 C11
Šternberk CZ 146 B4
Sternes GR 178 E7
Stes-Maries-de-la-Mer F 35 D7
Stężyca PL 81 A11
Stężyca PL 141 G5
Stia I 66 E4
Stibb Cross UK 12 D6
Stickney UK 11 E12
Stidsvig S 87 C12
Stiege D 79 C9
Stiens NL 16 B5
Stienta I 66 C4
Stigen S 91 B11
Stigliano I 60 C6
Stignano I 59 C9
Stigsjö S 103 A14
Stigtomta S 93 B9
Stijena BIH 156 C5
Stikli LV 134 B4
Stilligarry UK 2 L2
Stilling DK 86 C5
Stillington UK 11 C9
Stillorgan IRL 7 F10
Stilo I 59 C9
Stilton UK 11 G11
Stimpfach D 75 D7
Stintino I 64 B1
Stio I 60 C4
Štip NMK 164 F5
Stirfaka GR 174 B5
Stiring-Wendel F 21 F7
Stirling UK 5 C9
Štitar HR 157 B10
Štitar SRB 158 D4
Štitina CZ 146 B6
Štítnik SK 145 F1
Štity CZ 77 C11
Stiubieni RO 153 B9
Štiuca RO 159 B8
Ştivan HR 67 C9
Stjær DK 86 C5
Stjärnfors S 97 C13
Stjärnhov S 93 A9
Stjärnorp S 92 B7
Stjørdalshalsen N 105 E9
Støa N 101 E14
Stôa N 102 D4
Stobiecko Miejskie PL 143 D7
Stobreč HR 156 E6
Stoby S 87 C13
Stochov CZ 76 B5
Stocka S 103 C13
Stockach D 27 E11
Stockamöllan S 87 D12
Stockaryd S 88 A7
Stöcke S 122 C4
Stockelsdorf D 83 C9
Stockenboi A 73 C8
Stockerau A 77 F10
Stockheim D 75 B9
Stockholm S 99 D10
Stocking A 148 C5
Stockport UK 11 E7
Stocksfield UK 5 F13
Stockstadt am Rhein D 21 E10
Stockton-on-Tees UK 11 B9
Stockvik S 103 B13
Stoczek PL 139 E12
Stoczek Łukowski PL 141 G5
Stod CZ 76 C4
Stöde S 103 B12
Stoeneşti RO 160 C4
Stoeneşti RO 160 C6
Stoeneşti RO 160 E5
Stoeneşti RO 161 E7
Stoer UK 2 J6
Stoholm DK 86 C4
Stoianovca MD 154 E2
Stoicăneşti RO 160 D4
Stoilești RO 160 D4
Stoina RO 160 D2
Stojakovo NMK 169 B8
Stojdraga HR 148 E5
Stojnci SLO 148 D5
Stoke Ash UK 15 C11
Stoke-on-Trent UK 11 E7
Stokesay UK 13 A9
Stokesley UK 11 C9
Stokite BG 166 D4
Stokka N 108 E3
Stokkasjøen N 108 E4
Stokkdal N 111 D12
Stokke N 90 A7
Stokkemarke DK 83 A10
Stokkvågen N 108 D5
Stokmarknes N 110 C8
Štoky CZ 77 D9

Stolac BIH 157 F8
Stolberg (Harz) Kurort D 79 C8
Stolberg (Rheinland) D 20 C6
Stolczno PL 85 C12
Stolerova LV 133 D3
Stollberg D 79 F12
Stöllet S 97 B9
Stolmen N 94 B2
Stolniceni MD 154 D3
Stolniceni-Prăjescu RO 153 C9
Stolnici RO 160 D5
Stolnik BG 165 D8
Stolno PL 138 D6
Stolpe D 83 B8
Stolpe D 84 E6
Stolpen D 80 D6
Stolzenau D 17 C12
Stomio GR 169 E8
Stömne S 97 D8
Ston HR 162 D4
Stone UK 11 F7
Stone UK 13 B10
Stonehaven UK 5 B12
Stonehouse UK 5 D9
Stong N 94 D3
Stongfjorden N 100 D2
Stonglandseidet N 111 B13
Stonybreck UK 3 F13
Stoob A 149 A7
Stöpen S 91 C14
Stopiče SLO 148 E4
Stopki PL 136 E3
Stopnica PL 143 F10
Storå N 111 D11
Storå S 97 C13
Stora Åby S 92 C5
Storåbränna S 106 D7
Stora Höga S 91 C10
Stora Levene S 91 C12
Stora Melby S 91 C12
Stora Mellösa S 92 A6
Storarmsjö S 107 C16
Storås N 104 E7
Störåsen S 103 A10
Störåsen S 106 D7
Stora Skedvi S 97 B14
Stora Tallberg S 109 E17
Stora Vika S 93 B11
Storbäck S 106 B9
Storbacka FIN 123 D10
Storbäcken S 102 C4
Storberg S 109 E16
Stor-Blåsjön S 105 B16
Storbo S 102 C3
Storbogaren S 107 D14
Storbränna S 106 D7
Storbrännan S 118 E4
Storbukt N 113 B12
Storby FIN 99 B13
Stordal N 100 B6
Štore SLO 148 D4
Storebø N 94 B2
Storebro S 92 D7
Storebru N 100 C2
Store Damme DK 84 A2
Store Darum DK 86 E3
Store Heddinge DK 87 E10
Storeidet N 111 E10
Storekorsnes N 112 C11
Storelv N 112 B10
Store Merløse DK 87 D9
Store Molvik N 114 B5
Støren N 104 E8
Storenga N 112 D7
Store Rise DK 83 A8
Store Rørbæk DK 87 D9
Storfjäten S 102 C5
Storfjord N 111 B18
Storfjordbotn N 113 C18
Storfors S 97 C11
Storforshei N 108 D8
Storhågna S 102 A7
Storhallaren N 104 D5
Stor-Hallen S 102 A7
Storhogen S 106 E8
Stor-Hulljön S 103 A12
Štorje SLO 73 E8
Storjola S 106 B7
Storjord N 109 C9
Storjorda N 108 C6
Storjorda N 111 C11
Storkow D 80 B5
Storli N 101 A10
Storlien S 105 E12
Stormi FIN 127 D7
Stornara I 60 A5
Stornarella I 60 A5
Stornäs S 106 A8
Stornäs S 107 A11
Stornäset S 107 C10
Stornes N 111 C11
Stornoway UK 2 J4
Storo I 69 B10
Storobăneasa RO 161 F6
Storodden N 104 E6
Storöhamn S 119 C10
Storozhynets' UA 153 A7
Storrington UK 15 F8
Storrøsta N 101 B13
Storsand S 103 C13
Storsand S 118 B5
Storsandsjö S 107 C17
Störsätern S 102 B3
Storsävarträsk S 118 F3
Storsele S 107 D14
Storseleby S 107 B11
Storselø N 108 C4
Storsjö S 102 A5
Storskog S 109 E10
Storslett N 112 D6
Storsletta N 112 D6
Storsteinnes N 111 A18
Storsteinnes N 111 B17
Storsund S 118 C5
Storuman S 107 A12
Storvik N 108 C6
Storvik N 110 C9
Storvik N 112 C9
Storvik N 112 D7
Storvik S 103 A14
Storvorde DK 86 A6
Storvreta S 99 C9
Štós SK 145 F2
Stößen D 79 D10
Stössing A 77 F9
Stoszowice PL 81 E11
Støtt N 108 C5
Stötten am Auerberg D 71 B11
Stotternheim D 79 D9
Stouby DK 86 D5
Stoumont B 19 D12
Stourbridge UK 13 A10

Stourport-on-Severn UK 13 A10
Støvran N 108 B8
Støvring DK 86 B5
Støvset N 108 B8
Stow UK 5 D11
Stowijcino PL 85 A12
Stowmarket UK 15 C10
Stow-on-the-Wold UK 13 B11
Stoyan Mikhaylovski BG 167 C8
Stoyanovo BG 165 C10
Stoykite BG 165 F10
Stozher BG 167 C9
Stožne PL 136 E5
Stra I 66 B5
Straach D 79 C12
Strabane UK 4 F2
Strabla PL 141 E8
Strachan UK 5 A11
Strachówka PL 139 F12
Strachur UK 4 C6
Stracin NMK 164 E5
Stračiūnai LT 137 E8
Strackholt (Großefehn) D 17 B9
Strączno PL 85 D10
Stradalovo BG 165 E6
Stradbally IRL 7 F8
Stradbroke UK 15 C11
Stradella I 69 C7
Stradi LV 135 B13
Stradishall UK 15 C9
Stradone IRL 7 E8
Stradsett UK 11 F12
Straelen D 16 F6
Stragari SRB 158 E6
Strai N 90 C2
Straimont B 19 E11
Straja RO 153 B7
Strakhilovo BG 166 C5
Strakonice CZ 76 D5
Straldzha BG 167 D7
Stralki BY 133 E4
Straloch UK 5 B9
Strålsnäs S 92 C6
Stralsund D 84 B4
Strambino I 68 C4
Stramproy NL 19 B11
Strâmtura RO 152 B4
Stráňavy SK 147 C7
Strančice CZ 77 C7
Strand N 101 D15
Strand N 110 C9
Strand S 106 D9
Stranda N 100 B5
Strandå N 108 A8
Strandnorum S 91 C10
Strandval N 105 B10
Strangford UK 7 D11
Stranghele N 100 E3
Strängnäs S 98 D8
Strangolagalli I 62 D4
Strångsjö S 93 B8
Stráni CZ 146 D5
Strömstad S 91 B9
Strömsund S 106 D9
Strömsund S 107 A11
Stronachlachar UK 4 C7
Stroncone I 62 C3
Strongoli I 61 E8
Stronie Śląskie PL 77 B11
Stronsdorf A 77 E10
Strontian UK 4 B5
Stropi LV 135 E13
Stropkov SK 145 E4
Stroppiana I 68 C5
Stroppo I 36 B6
Stroud UK 13 B9
Strövelstorp S 87 C11
Strovija NMK 168 A5
Stróża PL 147 B9
Strpce RKS 164 E3
Strückingen (Saterland) D 17 B9
Struer DK 86 C3
Struga NMK 168 B4
Strugi-Krasnyye RUS 132 E5
Strullendorf D 75 C8
Strumë AL 168 C2
Strumica NMK 169 B7
Strumien PL 147 B7
Strumkivka UA 145 G5
Strumok UA 155 B3
Strumyani BG 169 A9
Strunkovice nad Blanicí
    CZ 76 D6
Strupii LV 133 C1
Struth D 79 D7
Struth UK 13 C2
Stružec HR 149 E7
Stryama BG 165 E10
Stryków PL 143 C8
Stryksele S 107 C16
Stryn N 100 C5
Stryy UA 145 E8
Strzała PL 141 F6
Strzałkowo PL 138 F5
Strzebień PL 143 E6
Strzegocin PL 139 D10
Strzegom PL 81 E10
Strzegowo PL 139 E9
Strzelce PL 143 B7
Strzelce Krajeńskie PL 85 E9
Strzelce Opolskie PL 142 E5
Strzelce Wielkie PL 143 D7
Strzeleczki PL 142 F4
Strzelin PL 81 E12
Strzelno PL 138 E5
Strzepcz PL 138 B5
Strzyżewice PL 144 A5
Strzyżów PL 144 D4
Sts-Geosmes F 26 E3
Stubbæk DK 86 F4
Stubbekøbing DK 84 A2
Stubben D 17 B11
Stubel BG 165 C7
Stubline SRB 158 D5
Stubno PL 144 D6
Stuchowo PL 85 C8
Studena BG 165 D6
Studena BG 166 F6
Studená CZ 77 D8
Studena UA 154 A3
Studençan RKS 163 E10
Studenci HR 157 E7
Studenec CZ 81 E8
Stüdenitz D 83 E12
Studénka CZ 146 B6
Studienka SK 77 E12
Studina RO 160 F4

Strelac SRB 164 D5
Strelcë AL 168 C4
Strelcha BG 165 D9
Strelli i Epërm RKS 163 D9
Strělínieki LV 135 C8
Strem A 149 B6
Strembo I 69 A10
Stremț RO 152 E3
Strenčí LV 131 F11
Strendur FO 2 A3
Strengberg A 77 F7
Strengelvåg N 110 C9
Strengen A 71 C10
Strengen N 95 D9
Stresa I 68 B6
Stretford UK 11 E7
Stretham UK 15 C9
Streufdorf D 75 B8
Strezimirovtsi BG 164 D5
Strib DK 86 D5
Striberg S 97 C12
Štříbro CZ 76 C3
Strichen UK 3 K12
Strigno I 72 D4
Štrigova HR 148 C5
Strijen NL 16 E3
Strimasund S 108 D8
Strimoniko GR 169 B9
Strinda N 111 E10
Strittjomvare S 109 E16
Strizivojna HR 157 B9
Strmec HR 148 E5
Strmen HR 149 F7
Strmica BIH 156 D5
Strmica HR 156 D5
Ströbeck D 79 C8
Strobl A 73 A7
Strøby DK 87 E10
Strøby Egede DK 87 E10
Stroeşti RO 160 C3
Stroevo BG 165 E10
Strofylia GR 175 B7
Ströhen D 17 C11
Stroieşti RO 153 B8
Strokestown IRL 6 E6
Ström S 108 E8
Strömbäck S 122 C4
Strömbacka S 103 C12
Stromberg D 185 E8
Stromeferry UK 2 L5
Strömfors S 118 E4
Strömholm S 109 C9
Stromiec PL 141 G4
Strommen N 95 C13
Strömmer S 96 C6
Strömnäs S 107 B10
Strömnäs S 118 D5
Stromness UK 3 H10
Strömsberg S 99 B9
Strömsbro S 103 E13
Strömsbruk S 103 C13
Strömsholden S 98 C6
Strømsmoen N 111 C16
Strömsnäs S 107 E19
Strömsnäsbruk S 87 B13

Studley UK 13 A11
Studsvik S 93 B10
Studsviken S 107 D14
Studzienice PL 85 B13
Studzieniec PL 85 E11
Stugubacken S 103 D11
Stugufläten N 100 B8
Stugun S 106 E9
Stukenjord N 109 C9
Stukmani LV 135 C11
Stülpe D 80 B4
Stulpicani RO 153 C7
Stumm A 72 B4
Stumsnäs S 102 E8
Stuorajavrre N 112 E10
Stuorragár'žl N 113 E16
Stupari BIH 157 D10
Stupava SK 77 F12
Stupsk PL 139 D9
Stupurai LT 134 D7
Sturefors S 92 C7
Stüri LV 134 C5
Sturko S 89 C9
Šturlić BIH 156 B4
Šturmai LT 134 E3
Sturminster Newton UK 13 D10
Sturno I 60 A4
Šturovo NMK 164 F5
Štúrovo SK 149 A11
Sturry UK 15 E11
Sturzelbronn F 27 B8
Stuttgart D 27 C11
Stützerbach D 79 E8
Styggedalen N 111 B19
Stylida GR 175 B6
Stypsi GR 171 F10
Styra GR 175 C9
Styrkesnes N 109 A9
Styrnäs S 107 E13
Styrrmanstø N 111 A18
Suances E 40 B3
Suaningi S 117 E10
Suatu RO 152 D3
Subačius LT 135 E9
Subaşı TR 171 B10
Subaşı TR 173 B9
Subaşı TR 177 C9
Subate LV 135 E11
Subbiano I 66 E4
Suben A 76 F4
Subiaco I 62 D4
Subkowy PL 138 B6
Subotica SRB 150 E4
Subotište SRB 158 D4
Subotniki BY 137 E12
Sučany SK 147 C7
Suceava RO 153 B8
Suceveni RO 153 F11
Sučevič HR 156 D5
Suceviţa RO 153 B7
Sucha Beskidzka PL 147 B9
Suchacz PL 139 B7
Suchá Hora SK 147 C9
Suchá Loz CZ 146 D5
Suchań PL 85 D8
Suchá nad Parnou SK 146 E4
Suchdol CZ 77 C8
Suchdol nad Lužnicí CZ 77 E7
Suchdol nad Odrou CZ 146 B5
Suchedniów PL 143 E10
Suchowola PL 140 C8
Suchożebry PL 141 F6
Suchy Dąb PL 138 B6
Suchy Las PL 81 B11
Sucina E 56 F3
Suciu de Sus RO 152 C4
Suckow D 83 D11
Sucleia MD 154 D5
Sučuraj HR 157 F7
Sudanell E 42 D5
Sudargas LT 136 D6
Sudbury UK 15 C10
Suddendorf D 17 D9
Suddesjaur S 109 E17
Süden D 82 B5
Süderbrarup D 82 A7
Süderhastedt D 82 B6
Süderlügum D 86 F3
Süderstapel D 82 B6
Sudervė LT 137 D11
Suditl RO 161 D11
Sudkov CZ 77 C11
Südlohn D 17 E7
Sudok S 118 B4
Sudoméřice CZ 146 D4
Sudova Vyshnya UA 144 D7
Sudovec HR 149 D6
Sudrabkalns LV 135 D9
Suđurađ HR 162 D4
Sudwalde D 17 C11
Sueca E 48 F4
Süedinenie BG 165 E10
Suelli I 64 D3
Suèvres F 24 E5
Şugag RO 152 F3
Suha BIH 157 D9
Suha BIH 157 F10
Suhaia RO 160 F6
Suharău RO 153 A8
Suharekë RKS 163 E10
Suhl D 79 E8
Suhlendorf D 83 E9
Suhmura FIN 125 F13
Suho Polje BIH 157 C11
Suhopolje HR 149 E8
Šuica BIH 157 E7
Šuici RO 160 C5
Šuijavaara S 116 B9
Suilly-la-Tour F 25 F9
Suinula FIN 127 B11
Suinula FIN 127 B12
Suippes F 25 B12
Sukeva FIN 124 C8
Sukhindol BG 166 C4
Sukhodil UA 145 E9
Sukošan HR 156 D3
Sükösd H 150 E2
Sukovo SRB 165 C6
Sukow D 83 C11
Sul N 105 D12
Suldrup DK 86 B5
Sulechów PL 81 B8
Sulęcin PL 81 B8
Sulęczyno PL 138 B4
Sulejów PL 141 H1
Sulejówek PL 141 F4
Suleskar N 94 C5
Sulesund N 100 B4
Suletea RO 153 E12
Sulina RO 155 D5

Süleymanli TR 177 B10
Sulików PL 81 D8
Sulingen D 17 C11
Sulişcazevo PL 85 D9
Suliţa RO 153 B9
Sulitjelma N 109 B11
Sulkava FIN 129 B9
Sulkava FIN 123 D16
Sulkavanjärvi FIN 123 D17
Sulkavanjärvi FIN 123 F11
Sułkowice PL 147 B9
Sullington UK 15 F8
Sully F 30 A5
Sully UK 13 C8
Sully-sur-Loire F 25 E7
Sulmierzyce PL 142 C4
Sulmierzyce PL 143 D7
Sulmona I 62 C5
Sülniac F 22 E6
Süloğlu TR 167 F7
Sułoszowa PL 143 F8
Surfleet UK 11 F11
Surgères F 28 C4
Surhuisterveen NL 16 B6
Súria E 43 D7
Surier I 31 D11
Surnadalsøra N 104 F5
Sürnevets BG 166 E4
Sürnevo BG 166 E5
Sürnitsa BG 165 F9
Sürovce SK 146 E5
Sursee CH 70 C6
Surte S 91 D11
Survilíškis LT 135 F8
Surwold D 17 C9
Sury-le-Comtal F 30 D5
Surzur F 22 E6
Susa I 31 E11
Suşani RO 160 D4
Sušara SRB 159 D7
Susch CH 71 D10
Susegana I 72 E5
Susek SRB 158 C4
Süsel D 83 B9
Suseni RO 152 D5
Suseni RO 153 D7
Suseni RO 160 D5
Sushitsa BG 166 C5
Sušice CZ 76 D5
Susiec PL 144 C7
Šuškova LV 133 D4
Susleni MD 154 C3
Susort N 94 D2
Susteren NL 19 B12
Sustinente I 66 B3
Sustrum D 17 C9
Susurluk TR 173 E9
Susuzmüsellim TR 173 B7
Susz PL 139 C7
Sutera I 58 D4
Suteşti RO 160 D4
Suteşti RO 161 C10
Sutina BIH 157 E7
Sutivan HR 156 F5
Sutjeska SRB 158 C6
Sütlüce TR 173 A8
Sutri I 62 C2
Sutterton UK 11 F11
Süttö H 149 A10
Sutton UK 15 C9
Sutton Bridge UK 11 F12
Sutton Coldfield UK 11 F8
Sutton in Ashfield UK 11 E9
Sutton-on-the-Forest UK 11 C9
Suure-Jaani EST 131 D10
Suurejõe EST 131 D10
Suuremõisa EST 130 D5
Suurimäki FIN 124 C8
Suurlahti FIN 129 B9
Suvainiškis LT 135 D10
Suvanto FIN 115 D2
Suvantokumpu FIN 117 C15
Suvereto I 65 A3
Suviekas LT 135 E12
Suvodol NMK 168 B5
Suvodol NMK 169 B6
Suvorov MD 155 B3
Suvorovo UA 155 B3
Suvorovo BG 167 C9
Suwałki PL 136 E6
Suystamo RUS 129 B15
Suze-la-Rousse F 35 B8
Suzzara I 66 C2
Svabensverk S 103 D10
Svalboviken S 93 A8
Svalenik BG 166 B6
Svallerup DK 87 D8
Svalöv S 87 D12
Svalsta S 93 B9
Svalyava UA 145 F6
Svaná S 98 C6
Svanabyn S 107 B12
Svanberga S 99 C11
Svaneke DK 89 E8
Svanesund S 91 C10
Svängsta S 89 C7
Svaningen S 106 C8
Svannäs S 109 D15
Svansele S 107 B17
Svanskog S 91 A12
Svanstein S 117 E11
Svanström S 118 C3
Svantjärnåsen S 102 B7
Svanträsk S 118 D3
Svanvik N 114 E8
Svanvik S 91 C10
Svappavaara S 116 C6
Svärdal N 111 A12
Svärdsjö S 103 E10
Svarinci LV 133 D4
Svarstad N 95 D11
Svartå S 92 A5
Svartbo S 103 D11
Svartbyn S 119 B9
Svarte S 87 E12
Svartemyr N 100 D3
Svärtinge S 92 B8
Svartisdalshytta N 108 D7
Svartlå S 118 B8
Svartnes S 108 B8
Svartnes N 111 B17
Svartnes N 114 B17
Svartöstaden S 118 C8
Svärtträsk S 107 A12
Svartvik S 103 A14

Suplac RO 152 E5
Suplacu de Barcău RO 151 C10
Suplai RO 152 C4
Supraśl PL 140 D8
Supru FIN 114 E4
Supur RO 151 C10
Súr H 149 B10
Surahammar S 98 C6
Suraia RO 153 F10
Şura Mare RO 152 F4
Şura Mică RO 152 F4
Surany SK 146 E6
Suraż PL 140 E7
Surbo I 61 C10
Surčin SRB 158 D5
Surdegis LT 135 E9
Surdila-Găiseanca RO 161 C10
Surdila-Greci RO 161 C10
Surduc RO 151 C11

Svartvik S 103 B13
Svartvik S 103 D13
Svatava CZ 75 B12
Svätuše SK 145 G4
Svätý Jur SK 146 E4
Svätý Peter SK 146 F6
Svebølle DK 87 D8
Svedala S 87 D12
Svédasai LT 135 E10
Svedja S 103 C13
Svedje S 106 C9
Svedje S 107 D15
Svedjebo S 103 C12
Svedlár SK 145 F2
Sveg S 102 B7
Sveggesundet N 104 E3
Šveicarija LT 137 C9
Sveindal N 90 C1
Sveio N 94 C2
Svejbæk DK 86 C5
Svekļ LV 135 B13
Švēkšna LT 134 E3
Svelgen N 100 C2
Svellingen N 104 D5
Svelvik N 95 C12
Švenčionėliai LT 135 F11
Švenčionys LT 135 F12
Svendborg DK 86 E7
Svene N 95 C11
Svenes N 90 B3
Svenljunga S 91 E13
Svenningdal N 108 F5
Svensbu N 111 A18
Svensbyn S 118 D6
Svenshögen S 91 C10
Svenstavik S 102 A7
Svenstrup DK 86 B5
Svenstrup DK 86 E5
Svenstrup DK 87 E8
Svente LV 133 E2
Šventeżeris LT 137 E8
Švēntoji LT 134 D2
Sverepec S 146 C6
Sveta Marija HR 149 D7
Světčiovem HR 131 F11
Svēte LV 134 C7
Sveti Nikola BG 167 C10
Sveti Nikola NMK 164 F4
Sveti Nikole NMK 164 F4
Sveti Rok HR 156 D4
Sveti Stefan MNE 163 E6
Světlá Hora CZ 142 F3
Světlá nad Sázavou CZ 77 C8
Svetlen BG 166 C6
Svetlogorsk RUS 139 A9
Svetlyy RUS 139 A9
Svetogorsk RUS 129 C10
Svetozar Miletić SRB 150 F3
Svetvinčenat HR 67 B8
Svezhen BG 165 D11
Švica HR 67 B8
Svidník SK 145 E4
Svihov CZ 76 D4
Sviiby CZ 77 C8
Svilajnac SRB 159 E7
Svilengrad BG 166 F6
Svindinge DK 86 E7
Svinesund S 91 A9
Svingen N 111 A18
Svingstad N 101 C12
Svingvoll N 101 D12
Svinhult S 92 D6
Svinia SK 145 E3
Sviniţa RO 159 D9
Svinná SK 146 D6
Svinnersta S 92 B5
Svinninge DK 87 D8
Svinninge S 99 D10
Svínoy FO 2 A4
Svir BY 137 D13
Svirkos LT 135 F12
Svishtov BG 160 F6
Svislach BY 140 D10
Svit SK 145 E1
Svitávka CZ 77 C10
Svitavy CZ 77 C10
Svitene LV 135 D8
Svitlodolyns'ke UA 154 E5
Svityaz' UA 141 H9
Svoboda BG 155 F1
Svoboda BG 166 F4
Svoboda RUS 136 D4
Svoboda nad Úpou CZ 81 E9
Svobodinovo BG 166 F4
Svode BG 165 C8
Svodna BIH 156 B6
Svoge BG 165 D7
Svogerslev DK 87 D10
Svolvær N 110 D4
Svorkmo N 104 E7
Svratka CZ 77 C10
Svrčinovec SK 147 C7
Svrljig SRB 164 C5
Svullrya N 96 B7
Swadlincote UK 11 F8
Swaffham UK 15 B10
Swalmen NL 20 B6
Swan IE 9 C8
Swanage UK 13 D11
Swanley UK 15 E9
Swanlinbar IRL 7 D7
Swansea UK 13 B7
Swanton Morley UK 15 B10
Swarożyn PL 138 B6
Swarzędz PL 81 B12
Swatragh UK 4 F3
Sway UK 13 D11
Świątki PL 139 C9
Świbno PL 138 B6
Świderki PL 141 G6
Świdnica PL 81 C8
Świdnica PL 81 E10
Świdnik PL 141 H7
Świdwin PL 85 C9
Świebodzice PL 81 E10
Świebodzin PL 81 B9
Świecie PL 138 D5
Święciechowa PL 81 C11
Świeciechów-Duży PL 144 B4
Świedziebnia PL 139 D8
Świekatowo PL 138 D5
Świeradów-Zdrój PL 81 E8
Świercze PL 139 E10
Świerczów PL 142 D4
Świerklaniec PL 143 E6
Świerklany Górne PL 142 F6
Świerzawa PL 81 E9
Świerże PL 141 H9
Świerzenko PL 85 B11
Świerzno PL 85 C7

Świeszyno PL 85 B10
Świątajno PL 136 E5
Świątajno PL 139 C11
Świątochłowice PL 143 F6
Świątoszów PL 81 D8
Świlcza PL 144 C4
Swindon UK 13 B11
Swineshead UK 11 F11
Swinford IRL 6 E5
Świnna PL 147 B8
Świnoujście PL 84 C6
Swinton UK 5 D12
Swinton UK 11 D7
Swobnica PL 84 D7
Swords IRL 7 F10
Swornegacie PL 85 C13
Syalyawshchyna BY 133 E5
Sycewice PL 85 B11
Syców PL 142 D4
Sydalen N 110 D7
Sydänmaa FIN 123 E10
Sydänmaankylä FIN 123 C16
Sydnes N 94 C3
Syelishcha BY 133 F4
Syfteland N 94 B2
Sykaminia GR 171 F10
Sykäräinen FIN 123 C12
Syke D 17 C11
Sykea GR 178 B4
Syki GR 169 E7
Sykia GR 169 E7
Sykia GR 170 D5
Sykies GR 169 F7
Sykkylven N 100 B5
Sykorrachi GR 171 C9
Sykourio GR 169 E8
Sylling N 95 C12
Sylte N 100 B6
Symbister UK 3 E14
Symi GR 181 C7
Synej AL 168 B2
Synevyr UA 145 G8
Synevyr'ka Polyana UA 145 F8
Synod Inn UK 12 A6
Synsiö FIN 123 F17
Synyak UA 145 F6
Syötekylä FIN 119 D17
Sypniewo PL 85 D11
Sypniewo PL 139 D11
Syrau D 75 A11
Syre UK 3 J8
Syrgenstein D 75 E7
Syri FIN 123 C13
Syrjä FIN 125 F10
Sysmä FIN 127 B14
Sysslebäck S 102 E4
Syston UK 11 F9
Syväjärvi FIN 117 D15
Syvänniemi FIN 124 E8
Syvärinpää FIN 125 D9
Syvde N 100 B3
Syvdsnes N 100 B3
Syvota GR 168 F3
Syvros GR 174 B2
Systen DK 90 E7
Syyspohja FIN 129 C10
Szabadegyháza H 149 B11
Szabadhidvég H 149 C10
Szabadkígyós H 151 D7
Szabadszállás H 150 D3
Szabolcsbáka H 145 G5
Szada H 150 B3
Szadek PL 143 C6
Szaflary PL 147 C9
Szajol H 150 C5
Szakály H 149 C10
Szakcs H 149 C10
Szakmár H 150 D3
Szakolý H 151 B8
Szakszend H 149 A10
Szalánta H 149 E10
Szalaszend H 145 G3
Szalkszentmárton H 150 D3
Szalonna H 145 G2
Szamocin PL 85 D12
Szamossezeg H 145 G5
Szamotuły PL 81 A11
Szank H 150 D4
Szany H 149 B8
Szár H 149 B10
Szarvas H 150 D6
Szászberek H 150 C5
Szatmárcseke H 145 G6
Szatymaz H 150 E5
Százhalombatta H 149 B11
Szczaniec PL 81 B9
Szczawin Kościelny PL 139 F8
Szczawne PL 145 E5
Szczawnica PL 145 E2
Szczawno-Zdrój PL 81 E10
Szczebrzeszyn PL 144 B6
Szczecin PL 84 D7
Szczecinek PL 85 C11
Szczekociny PL 143 E8
Szczepankowo PL 139 D12
Szczepanowo PL 138 E4
Szczerców PL 143 D7
Szczucin PL 143 F11
Szczuczyn PL 140 C6
Szczurkowo PL 136 C2
Szczurowa PL 143 F10
Szczutowo PL 139 E8
Szczyrk PL 147 B8
Szczytna PL 77 B10
Szczytniki PL 142 C5
Szczytno PL 139 C10
Szécsény H 147 E9
Szederkény H 149 E10
Szedres H 149 D11
Szeged H 150 E5
Szeghalom H 151 C7
Szegi H 145 G3
Szegvár H 150 D5
Székely H 145 G4
Székesfehérvár H 149 B10
Székkutas H 150 D6
Szekszárd H 149 D11
Szemud PL 138 B5
Szendehely H 147 F8
Szendrő H 145 G2
Szendrőlád H 145 G2
Szentendre H 150 B3
Szentes H 150 D5
Szentgál H 149 B9
Szentgotthárd H 148 C6
Szentistván H 150 B6
Szentkirály H 150 D4
Szentlászló H 149 D9
Szentlőrinc H 149 D9
Szentpéterfa H 149 B6
Szepetnek H 149 D7
Szepietowo PL 141 E7

Szerecseny H 149 B9
Szeremle H 149 D11
Szerencs H 145 G3
Szerep H 151 C7
Szerzyny PL 144 D3
Szestno PL 136 F3
Szigetbecse H 149 B11
Szigetcsép H 149 B11
Szigethalom H 150 C3
Szigetszentmiklós H 150 C3
Szigetvár H 149 D8
Szigliget H 149 C8
Szihalom H 150 B5
Szikszó H 145 G2
Szil H 149 A8
Szilvásvárad H 145 G1
Szirák H 147 F9
Szirmabesenyő H 145 G2
Szkaradowo PL 81 C12
Szklarska Poręba PL 81 E9
Szklary Górne PL 81 D10
Szlichtyngowa PL 81 C10
Szob H 147 F7
Sződ H 150 B3
Szőgliget H 145 F2
Szokolya H 147 F8
Szolnok H 150 C5
Szőlősgyörök H 149 C9
Szombathely H 149 B7
Szomód H 149 A10
Szomolya H 145 H1
Szówsko PL 144 C6
Szpetal Górny PL 81 C9
Szprotawa PL 81 C9
Szreńsk PL 139 D9
Szropy PL 138 C5
Sztabin PL 140 C8
Sztum PL 138 C7
Sztutowo PL 139 B7
Szubin PL 138 D4
Szűcsi H 150 B4
Szudziałowo PL 140 D9
Szügy H 147 E8
Szuhogy H 145 G2
Szulmierz PL 139 E9
Szumowo PL 141 E6
Szurdokpüspöki H 147 F9
Szydłów PL 143 E11
Szydłowo PL 85 D11
Szydłowo PL 139 D9
Szymanów PL 141 F2
Szymonka PL 136 F4
Szynkielów PL 142 D6
Szypliszki PL 136 E7

T

Taagepera EST 131 F11
Taalintehdas FIN 126 E8
Taapajärvi FIN 117 D13
Taastrup DK 87 D10
Tab H 149 C10
Tabajd H 149 B11
Tabaky UA 155 B3
Tabanera de Cerrato E 40 D3
Tábara E 39 E8
Tabarz D 79 E8
Tabasalu EST 131 C9
Tabdi H 150 D3
Taberg S 91 D15
Tabernas E 55 E8
Taberno E 55 E8
Tabivere EST 131 D13
Taboada E 38 C4
Taboadela E 38 D4
Tábor CZ 77 D7
Táborfalva H 150 C3
Tábua P 44 D4
Tabuaço P 44 B5
Tabuenca E 41 E8
Täby S 99 D10
Tác H 149 B10
Tacen SLO 73 D9
Tacherting D 75 F12
Tachov CZ 75 C12
Tackåsen S 103 C9
Tăcuta RO 153 D11
Tadasuni I 64 C2
Tadcaster UK 11 D9
Tadley UK 13 C12
Taebla EST 130 D7
Tælavåg N 94 B1
Taevaskoja EST 131 F14
Tafalla E 32 F2
Tafers CH 31 B11
Taff's Well UK 13 B8
Tafjord N 100 B6
Täfteå S 107 E15
Täfteå S 122 C4
Taga RO 152 D4
Tagaranna EST 130 D4
Taggia I 37 D7
Taghmon IRL 9 D9
Tagliacozzo I 62 C4
Taglio di Po I 66 B5
Taglio-Isolaccio F 37 G10
Tagnon F 19 F9
Tägsjöberg S 107 D13
Tahal E 55 E8
Tahemaa EST 131 E14
Tahilla IRL 8 E3
Tahitótfalu H 150 B3
Tähtelä FIN 117 D17
Taibon Agordino I 72 D5
Taíde P 38 E3
Tain UK 3 K8
Tainiemi FIN 119 C15
Tainijoki FIN 119 C15
Tain-l'Hermitage F 30 E6
Taintrux F 27 D6
Taio I 72 D3
Taipale FIN 119 B15
Taipale FIN 119 C14
Taipale FIN 123 D17
Taipale FIN 124 D8
Taipale FIN 126 B3
Taipaleenharju FIN 119 D17
Taipalsaari FIN 129 C10
Taipalus FIN 123 E10
Taivalkoski FIN 121 C14
Taivassalo FIN 126 D6
Taizé F 30 B6
Tajmište NMK 168 A4
Tajno Łanowe PL 140 C7
Takácsi H 149 B8
Takamaa FIN 127 B10
Takeley UK 15 D9
Takene S 97 C9
Takene S 97 D11
Taklax FIN 122 E6
Taksony H 150 C3
Taktakenéz H 145 G3

Taktaszada H 145 G3
Talais F 28 E3
Talamantes E 41 E8
Talamona I 69 A8
Talamone I 65 B4
Talange F 186 C1
Talant F 26 F3
Talarrubias E 51 A9
Talavera de la Reina E 46 E3
Talavera la Real E 51 B6
Talayuela E 45 E8
Talayuelas E 47 E10
Talea RO 161 C7
Talence F 28 F4
Talensac F 23 D8
Tales E 48 E4
Taleža BIH 162 D5
Talgarreg UK 12 A6
Talgarth UK 13 B8
Talhadas P 44 C4
Tali EST 131 E9
Taliard F 36 C4
Taliesin UK 10 F4
Táliga E 51 B5
Tälje S 103 A10
Talla I 66 E4
Tallås S 103 E11
Tallås S 109 E14
Tallåsen S 103 C11
Tallberg S 103 E8
Tallberg S 118 B8
Taller F 32 C3
Tallinn EST 131 C9
Talljärv S 118 B3
Talloires F 31 D9
Tallone F 37 G11
Tallós E 38 C2
Tallow IRL 8 D6
Tallsjö S 107 C14
Talluskylä FIN 124 D7
Tallvik S 119 B9
Tállya H 145 G3
Tălmaciu RO 160 B4
Talmay F 26 F3
Talmaz MD 154 D5
Talmont F 28 C4
Talmont-St-Hilaire F 28 C2
Talpa RO 160 E6
Talsano I 61 C8
Talsi LV 134 B5
Taluskylä FIN 119 F12
Talviainen FIN 127 B12
Talvik N 112 C10
Tama E 39 B10
Tămădău Mare RO 161 E9
Tamajón E 47 B6
Tamames E 45 C8
Tamarë AL 163 E8
Tamarino BG 167 E7
Tamarite de Litera E 42 D4
Tamashowka BY 141 G9
Tamási H 149 C10
Tamaşı RO 153 D10
Tambach-Dietharz D 79 E8
Tâmboeşti RO 161 B10
Tâme S 118 E6
Tamins CH 71 D8
Tamiş TR 171 E10
Tamm D 27 C11
Tammela FIN 121 C15
Tammela FIN 127 D10
Tammensiel D 82 A5
Tammijärvi FIN 127 B14
Tammiku EST 131 C12
Tammisaari FIN 127 F9
Tammispää EST 131 D14
Tâmna RO 159 D11
Tamnay-en-Bazois F 30 A4
Tamnes N 101 A15
Tamnič SRB 159 E9
Tampere FIN 127 C10
Tamsalu EST 131 C12
Tamsweg A 73 B8
Tamurejo E 54 B3
Tana Bru N 113 C21
Tanacu RO 153 D11
Tanaji LV 135 E13
Tananes N 114 B4
Tananger N 94 E3
Tãnãsoaia RO 153 E10
Tanaunella I 64 B4
Tancarville F 18 F1
Tandådalen S 102 D4
Țăndărei RO 161 D11
Tandö S 102 E5
Tandragee UK 7 D10
Tandsbyn S 106 E7
Tandsjöborg S 102 C8
Tang IRL 7 F7
Tångaberg S 87 A10
Tangavaene IRL 6 E6
Tãngböle S 105 E13
Tange DK 86 C5
Tangen N 95 A12
Tangen N 101 B14
Tangen N 111 B14
Tanger MA 52 E5
Tångeråsen S 105 D15
Tangerhütte D 79 B10
Tangermünde D 79 A10
Tängstamon S 107 E11
Tangstedt D 83 C8
Tängvattnet S 108 E8
Tanhua FIN 115 C3
Taninges F 31 C10
Tankapirtti FIN 115 B2
Tankavaara FIN 115 B2
Tanlay F 25 E11
Tann D 76 E3
Tann (Rhon) D 79 E7
Tanna D 75 B10
Tannadice UK 5 B11
Tännäs S 102 B4
Tannay F 19 E10
Tänndalen S 102 A3
Tannenbergsthal D 75 B11
Tännesberg D 75 C11
Tannhausen D 75 E7
Tannheim A 71 B11
Tannila FIN 119 D15
Tännö S 88 A6
Taninroda D 79 E9
Tanowo PL 84 C6
Tansa RO 153 D10
Tãntãreni RO 160 D2
Tantonville F 26 D5
Tanumshede S 91 B9
Tanvald CZ 81 E8

Taormina I 59 D7
Tapa EST 131 C11
Tapanikylä FIN 121 F13
Tapfheim D 75 E8
Tapia de Casariego E 39 A6
Tápióbicske H 150 C4
Tápiógyörgye H 150 C4
Tapionkylä FIN 117 E14
Tapionniemi FIN 115 E2
Tápióság H 150 C4
Tápiószecső H 150 C4
Tápiószele H 150 C4
Tápiószentmárton H 150 C4
Tápiószőlős H 150 C4
Táplánszentkereszt H 149 B7
Tapojärvi FIN 117 C11
Tapolca H 149 C8
Tappeluft N 112 C9
Tappenbeck D 79 B8
Tappernøje DK 87 E9
Taps DK 86 E4
Tapsony H 149 D8
Tar H 147 E9
Taracena E 47 C6
Taraclia MD 154 D4
Taraclia MD 155 B2
Taradeau F 36 E4
Taradell E 43 D8
Taramundi E 38 B4
Tarancón E 47 D6
Tarano I 62 C3
Taranto I 61 C8
Tarany H 149 D8
Tarare F 30 D5
Tarascon F 35 C8
Tarascon-sur-Ariège F 33 E9
Tarashany UA 153 A8
Tarazona E 41 E8
Tarazona de la Mancha E 47 F9
Tárbena E 56 D4
Tarbert UK 2 K3
Tarbert UK 4 D5
Tarbert UK 4 D5
Tarbert IRL 8 C4
Tarbes F 33 D6
Tarbja EST 131 D11
Tarbolton UK 4 D8
Tãrcaia RO 151 D9
Tarcal H 145 G3
Tarcău RO 153 D8
Tarcea RO 151 C9
Tarcento I 73 D7
Tarčin BIH 157 E9
Tarczyn PL 141 G3
Tard H 145 H2
Tardajos E 40 D4
Tardelcuende E 41 E6
Tardets-Sorholus F 32 D4
Tardienta E 41 E10
Tarendö S 116 D7
Țăreuca MD 154 B3
Târgale LV 134 B3
Targon F 28 F5
Târgovişte RO 161 D6
Targowa Górka PL 85 F12
Târgu Bujor RO 153 F10
Târgu Cărbuneşti RO 160 D3
Târgu Frumos RO 153 C9
Târgu Jiu RO 159 C11
Târgu Lăpuş RO 152 C3
Târgu Mureş RO 152 D5
Târgu Neamţ RO 153 C8
Târgu Ocna RO 153 E8
Târgu Secuiesc RO 153 E8
Târguşor RO 155 E2
Târgu Trotuş RO 153 E9
Tarhos H 151 D7
Tarifa E 53 D5
Tarigrad MD 153 A11
Tarján H 149 A9
Tárkány H 149 A9
Tarleton UK 11 F8
Târlişua RO 152 C4
Tarłów PL 143 E11
Târlungeni RO 161 B7
Tarm DK 86 D3
Tarmstedt D 17 B12
Târnaby S 108 E9
Tarnala FIN 129 B12
Tarnalelesz H 145 G1
Tarna Mare RO 145 G7
Tarnaméra H 150 B5
Târnamo S 108 E8
Tarnaörs H 150 B5
Tarnaszentmiklós H 150 B5
Târnãveni RO 152 E4
Tarnawa Duża PL 144 B6
Tarnawatka PL 144 B7
Tarnazsadány H 150 B5
Târnelv N 111 B15
Tårnes N 104 D7
Târnet N 114 D8
Tarnobrzeg PL 143 E12
Tarnogród PL 144 C6
Tarnos F 32 C3
Tarnoszyn PL 144 C8
Târnova RO 151 E8
Tarnow PL 85 C12
Tarnów PL 85 E7
Tarnów PL 143 F10
Tarnowiec PL 144 D4
Tarnówka PL 85 D11
Tarnowo Podgórne PL 81 B11
Tarnów Opolski PL 142 E5
Tarnowskie Góry PL 142 F6
Târnsjö S 98 B7
Tårnvik N 108 A9
Târnvik N 113 C18
Taron-Sadirac-Viellenave F 32 C5
Tarouca P 44 B5
Tarp D 82 A6
Tarp DK 86 D2
Tarpa H 145 G6
Tarprubežiai LT 136 E7
Tarquinia I 62 C1
Tarquinia Lido I 62 C1
Tarragona E 42 E6
Tàrrajaur S 109 D18
Tàrrega E 42 D6
Tarrel UK 3 K9
Tarrenz A 71 C11
Tarrio E 38 B3
Tárrega de Segarra E 42 D6
Tàrs DK 87 F8
Tårs DK 90 E7
Tarsele S 107 D11
Tarsia I 60 D6
Țânțăreni RO 160 D2
Țârşolţ RO 145 H7

Tartas F 32 C4
Tărtăşeşti RO 161 D7
Tartu EST 131 E13
Tărup DK 87 E7
Tarutyne UA 154 E5
Tarvaala FIN 123 E14
Tarvasjoki FIN 126 D8
Tarves UK 3 L12
Tarvisio I 73 C8
Tarvola RO 123 D11
Tasagăl TR 173 A7
Tașca RO 153 D8
Taşlıc MD 154 C4
Tâşnad RO 151 C10
Tass H 150 C3
Tassagh UK 7 D9
Tassin-la-Demi-Lune F 30 D6
Tászár H 149 D9
Tát H 149 A11
Tata H 149 A10
Tatabánya H 149 A10
Tataháza H 150 E3
Tătărani RO 153 D11
Tătârani RO 160 D6
Tătăranu RO 161 B10
Tătărăşti RO 153 D9
Tătărăştii de Jos RO 160 E6
Tătărăştii de Sus RO 160 E6
Tătărăuca MD 153 A12
Tatarbunary UA 154 F5
Tatarevo BG 166 E4
Tatarköy TR 173 B7
Tatarlar TR 167 F7
Tatárszentgyörgy H 150 C3
Tătranská Javorina SK 145 E1
Tattershall UK 11 E11
Tătuleşti RO 160 D5
Tau N 94 E3
Tauberbischofsheim D 74 C6
Taucha D 79 D11
Tauche D 80 B6
Tauer D 81 C6
Taufkirchen D 75 F10
Taufkirchen D 75 F11
Taufkirchen (Vils) D 75 F11
Taujėnai LT 135 E9
Țaul MD 153 A11
Taulé F 22 C4
Taulignan F 35 B8
Taunton UK 13 C8
Tauplitz A 73 A9
Tauragė LT 134 F4
Tauragnai LT 135 E11
Tauralaukis LT 134 E2
Taurasi I 60 A4
Taurene LV 135 B10
Tăureni RO 152 D4
Taurianova I 59 C9
Taurisano I 61 D10
Taurupe LV 135 C10
Tauste E 41 E9
Tăuţ RO 151 E8
Tăuteu RO 151 C9
Tăuţii-Măgherăuş RO 151 B10
Tauves F 29 D11
Tavankut SRB 150 E4
Tavannes CH 27 F7
Tavant F 24 F3
Tavarnelle Val di Pesa I 66 E3
Tavastila FIN 128 D6
Tavastkenkä FIN 119 F16
Tavaux F 31 A7
Tavel F 35 B8
Tavelsjö S 122 B4
Tavera F 37 G9
Taverham UK 15 B11
Taverna I 59 A10
Tavernay F 30 A5
Tavernelle I 62 A2
Tavernes F 36 D4
Tavernes de la Valldigna E 56 C4
Tavernspite UK 12 B5
Taverny F 25 B7
Tavertet E 43 D8
Taviano I 61 D10
Tavira P 50 E4
Tavistock UK 12 D6
Tavullia I 67 E6
Tawern D 186 C2
Taxenbach A 73 B7
Taxiarchis GR 169 D10
Taxobeni MD 153 C11
Tayfur TR 171 D10
Tayinloan UK 4 D5
Taylor's Cross IRL 7 F7
Taynuilt UK 4 C6
Tayvallich UK 4 C5
Tazlár H 150 D4
Tazlău RO 153 D9
Tazones E 39 A9
Tczew PL 138 C6
Tczów PL 141 H4
Teaca RO 152 D5
Teano I 60 A2
Tearce NMK 164 E3
Teasc RO 160 E3
Teba E 53 C7
Tebay UK 11 C7
Tebra E 38 D2
Tebstrup DK 86 D5
Techendorf A 73 C7
Techirghiol RO 155 E3
Tecuci RO 153 F10
Tedavnet IRL 7 D8
Teeranea IRL 6 F3
Teeriranta FIN 121 C14
Teerivaara FIN 120 B9
Tefeli GR 178 E9
Tegelen NL 16 F6
Teggiano I 60 C5
Téglás H 151 B8
Teglio I 69 A9
Tegnaset S 107 C16
Teichwolframsdorf D 79 E11
Teignmouth UK 13 D8
Teijo FIN 127 E9
Teillay F 23 E8
Teillet F 33 C11
Teisendorf D 73 A6

Tårstad N 111 D12
Teisnach D 76 D4
Teistungen D 79 D7
Teiu RO 160 D6
Teius RO 152 E3
Teixeiro E 38 B3
Teixoso P 44 D6
Tejadillos E 47 D9
Tejado E 41 E7
Tejeda de Tiétar E 45 D9
Tejn DK 89 E7
Tekeler TR 181 A7
Tekeli TR 177 C9
Tekeriš SRB 158 D4
Tekija BIH 157 D11
Tekija SRB 159 D9
Tekirdağ TR 173 C7
Tekovské Lužany SK 147 E7
Teksdal N 104 D7
Telaki PL 141 E6
Telatyn PL 144 B8
Telč CZ 77 D8
Telciu RO 152 C4
Teldau D 83 D9
Telečka SRB 150 F3
Telega RO 161 C7
Telekgerendás H 151 D6
Teleneşti MD 154 B2
Telerig BG 161 F11
Telese I 60 A3
Teleşti RO 159 C11
Telford UK 10 F7
Telfs A 72 B3
Telgárt SK 145 F1
Telgte D 17 E9
Telhado P 44 D5
Telish BG 165 C9
Teliu RO 161 B7
Teliucu Inferior RO 159 B10
Teljo FIN 125 C13
Telkkälä FIN 119 C17
Tellancourt F 19 E12
Tellejåkk S 118 B3
Tellin B 19 D11
Tellingstedt D 82 B6
Telnice CZ 77 D11
Telões P 38 F4
Telšiai LT 134 E4
Telti I 64 B3
Teltow D 80 B4
Tembleque E 46 E6
Temeleuţ MD 153 C12
Temeni GR 174 C5
Temerin SRB 158 C4
Temmes FIN 119 E15
Tempi GR 169 E8
Tempio Pausania I 64 B3
Temple Bar UK 12 A6
Templeboy IRL 6 D5
Temple Ewell UK 15 E11
Templemore IRL 9 C7
Temple Sowerby UK 5 F11
Templeton UK 12 B5
Templetuohy IRL 9 C7
Templeuve F 182 C2
Templin D 84 D5
Tempo UK 7 D8
Temse B 19 B9
Temska SRB 164 C6
Temü I 69 A9
Tenala FIN 127 E9
Ten Boer NL 17 B7
Tenbury Wells UK 13 A9
Tenby UK 12 B5
Tence F 30 E5
Tenda I 37 C7
Tendilla E 47 C7
Tendu F 29 B9
Teneniai LT 134 F4
Tenevo BG 167 E7
Tengelic H 149 C11
Tengen D 27 E10
Tengesdal N 94 C4
Tenhult S 92 D4
Teningen D 27 D8
Tenja HR 149 F11
Tenk H 150 B5
Tennänget S 102 D6
Tennenbronn D 27 D9
Tennevik N 111 C12
Tenneville B 19 D12
Tennevoll N 111 C14
Tennilä FIN 119 B16
Tennilä FIN 127 D13
Tennskjer N 111 B15
Tennstrand N 110 D9
Tenterden UK 15 E10
Tentúgal P 44 D3
Teora I 60 B4
Teovo NMK 169 A6
Tepasto FIN 117 C13
Tépe H 151 C7
Tepeboz TR 177 B7
Tepecik TR 173 B10
Tepecik TR 173 E10
Tepelenë AL 168 D3
Teplá CZ 76 C3
Teplice CZ 80 E5
Teplice nad Metují CZ 81 E10
Teplička SK 145 F2
Teplička nad Váhom SK 147 C7
Tepsa FIN 117 C15
Tepu RO 153 F10
Ter Aar NL 182 A5
Teramo I 62 B5
Tèrande LV 134 B3
Ter Apel NL 17 C8
Teratyn PL 144 B8
Tèrbaç AL 168 D2
Terborg-Silvolde NL 16 E6
Terchová SK 147 C8
Terebeşti RO 151 B10
Tereblya UA 145 G8
Teregova RO 159 C9
Teremia Mare RO 150 F6
Terena P 50 B4
Teresa E 48 E3
Teresa de Cofrentes E 47 F10
Teresin PL 141 F9
Terespol PL 141 F9
Teresva UA 145 G8
Terezín CZ 76 B6
Terezino Polje HR 149 E8
Tergnier F 19 E7
Tergu I 64 B2
Terikeste EST 131 E14
Terjärv FIN 123 C11
Terka PL 145 E5
Terlan I 72 C3

Těrlicko CZ 146 B6
Terlizzi I 61 A7
Terme Luigiane I 60 E5
Térmens E 42 D5
Termignon F 31 E10
Terminiers F 24 D6
Termini Imerese I 58 D4
Termoli I 63 D7
Termon IRL 7 B7
Termonfeckin IRL 7 E10
Ternaard NL 16 B5
Ternavka UA 153 A8
Ternberg A 76 G6
Terndrup DK 86 B6
Terneuzen NL 16 F1
Terni I 62 B3
Ternitz A 148 A6
Ternove UA 145 G8
Terovo GR 168 F4
Terpezita RO 160 E3
Terpni GR 169 C9
Terpsithea GR 169 E7
Terpyllos GR 169 B8
Terracina I 62 E4
Terrades E 34 F4
Terråk N 105 A12
Terralba I 64 D2
Terranova da Sibari I 61 D6
Terranova di Pollino I 60 D6
Terranuova Bracciolini I 66 E4
Terrasini I 58 C3
Terrassa E 43 D8
Terrasson-Lavilledieu F 29 E8
Terraube F 33 C7
Terravecchia I 61 E7
Terricciola I 66 E2
Terriente E 47 D10
Terrin IRL 7 E8
Terrinches E 55 B7
Terrou F 29 E9
Terrugem P 50 B3
Terrugem P 51 B5
Terryglass IRL 6 F6
Terslev DK 87 E9
Tërstenik RKS 164 D2
Tertenia I 64 D4
Teruel E 47 D10
Tervajoki FIN 122 D8
Tervakoski FIN 127 D12
Tervavaara FIN 121 D11
Tervel BG 161 F11
Terves F 28 B4
Tèrvete LV 134 D6
Tervo FIN 123 E17
Tervola FIN 119 B13
Tervuren B 19 C10
Tés H 149 B10
Tešanj BIH 157 C8
Těšany CZ 77 D11
Tešedíkovo SK 146 E5
Tesero I 72 D4
Tešica SRB 164 C4
Tesimo I 72 C3
Tesjoki FIN 127 E15
Teslić BIH 157 C8
Teslui RO 160 D4
Teslui RO 160 E4
Tespe D 83 D8
Tessenderlo B 19 B11
Tessin D 83 B12
Tesson F 28 D4
Tessy-sur-Vire F 23 C9
Testa dell'Acqua I 59 F6
Testa del Gargano I 63 D10
Testelt B 19 C10
Teşvikiye TR 173 C11
Tét H 149 A9
Tetbury UK 13 B10
Teţcani MD 153 A9
Teţchea RO 151 C9
Teteringen NL 182 B5
Teterow D 83 C13
Tetétlen H 151 C7
Teteven BG 165 D9
Teti I 64 C3
Tetney UK 11 E11
Tetoiu RO 160 D3
Tétouan MA 53 E6
Tetovo BG 161 F8
Tetovo NMK 163 E10
Tettens D 17 A9
Tettnang D 71 B9
Tetýň PL 85 D7
Teublitz D 75 D11
Teuchern D 79 D10
Teufen CH 27 F11
Teufenbach A 73 B9
Teulada E 56 D5
Teulada I 64 F2
Teunz D 75 D11
Teupitz D 80 B5
Teurajärvi FIN 117 D11
Teurajärvi S 116 E8
Teuro FIN 127 D10
Teuschnitz D 75 B9
Teutschenthal D 79 D10
Teuva FIN 122 F7
Tevaniemi FIN 127 B10
Tevansjö S 103 B9
Tevel H 149 D10
Tēvenāni LV 133 C1
Tevfikiye TR 171 E10
Teviothead UK 5 E11
Tewkesbury UK 13 B10
Teyssieu F 29 E9
Thal D 79 E7
Thale (Harz) D 79 C9
Thaleischweiler-Fröschen D 186 C4
Thalfang D 21 E7
Thalgau A 73 A7
Thalheim D 79 C11
Thalheim bei Wels A 76 F6
Thaljina BIH 157 F8
Thalmässing D 75 D9
Thalwil CH 27 F9
Thame UK 14 D7
Thann F 27 E7
Thannhausen D 75 F7
Thaon-les-Vosges F 26 D5
Tharandt D 80 E5
't Harde NL 16 D5
Tharsis E 51 D5
Thasos GR 171 C7
Thatcham UK 13 C12
Thaumiers F 29 B11
Thaxted UK 15 D9
Thaya A 77 E8
Thayngen CH 27 E10
Theale UK 13 C12
Thedinghausen D 17 C12
Theillay F 24 F7
Theißen D 79 D11

Theix F 22 E6
Them DK 86 C5
Themar D 75 A8
The Mumbles UK 12 B6
Thenay F 29 B8
Thénezay F 28 B5
Thenon F 29 E8
Theologos GR 171 C7
Théoule-sur-Mer F 36 D5
The Pike IRL 9 D7
Therma GR 171 D9
Thermi GR 169 C9
Thermisia GR 175 E7
Thermo GR 174 B4
Thermopyles GR 175 B6
Thérouanne F 18 C5
The Sheddings UK 4 F4
Thespies GR 175 C7
Thesprotiko GR 168 F4
Thessaloniki GR 169 C8
The Stocks UK 15 E10
Thetford UK 15 C10
Theth AL 163 E8
Theux B 19 C12
Thèze F 32 D5
Thèze F 35 B10
Thiaucourt-Regniéville F 26 C4
Thiberville F 24 B3
Thibie F 25 C11
Thiéblemont-Farémont F 25 C12
Thiendorf D 80 D5
Thiene I 72 E3
Thierhaupten D 75 E8
Thierrens CH 31 B10
Thiers F 30 D4
Thiersee A 72 A5
Thiersheim D 75 B11
Thiesi I 64 B2
Thießow D 84 B5
Thiézac F 29 E11
Thimert-Gâtelles F 24 C5
Thin-le-Moutier F 19 E10
Thionville F 20 F6
Thiron Gardais F 24 D4
Thirsk UK 11 C9
Thisted DK 86 B3
Thisvi GR 175 C6
Thiva GR 175 C7
Thivars F 24 D5
Thiviers F 29 E7
Thizy F 30 C5
Thoirette F 31 C8
Thoiry F 24 C6
Thoissey F 30 C6
Tholen NL 16 E2
Tholey D 21 F8
Thomastown IRL 9 C8
Thommen B 20 D6
Thônes F 31 D9
Thonnance-lès-Joinville F 26 D3
Thonon-les-Bains F 31 C9
Thorame-Haute F 36 C5
Thoras F 30 F4
Thoré-la-Rochette F 24 E4
Thorenc F 36 D5
Thorigny-sur-Oreuse F 25 D9
Thörl A 73 A11
Thorn NL 19 B12
Thornaby-on-Tees UK 11 B9
Thornbury UK 13 B9
Thorne UK 11 D10
Thorney UK 11 F11
Thornhill UK 5 E9
Thorning DK 86 C4
Thornton UK 10 D5
Thorpe-le-Soken UK 15 D11
Thorpe Market UK 15 B11
Thorpeness UK 15 C12
Thorsager DK 86 C6
Thorshøj DK 90 E7
Thorsø DK 86 C5
Thouarcé F 23 F11
Thouaré-sur-Loire F 23 F9
Thouars F 28 B5
Thouria GR 174 E5
Thourotte F 18 F6
Thrapston UK 15 C7
Threshfield UK 11 C7
Thropton UK 5 E13
Thrumster UK 3 J10
Thuès-entre-Valls F 33 E10
Thueyts F 35 A7
Thuin B 19 D9
Thuine D 17 D9
Thuir F 34 E4
Thum D 80 E3
Thun CH 70 D5
Thundersley UK 15 D10
Thüngen D 74 C6
Thüngersheim D 74 C6
Thuré F 29 B6
Thuret F 30 D3
Thurey F 31 B7
Thüringen A 71 C9
Thurins F 30 D6
Thürkow D 83 C13
Thurlby UK 11 F11
Thurles IRL 9 C7
Thurnau D 75 B9
Thursby UK 5 F10
Thurso UK 3 H9
Thury-Harcourt F 23 C11
Thusis CH 71 D8
Thwaite UK 11 C7
Thyborøn DK 86 B2
Thyez F 31 C10
Thymiana GR 177 C7
Thyregod DK 86 D4
Thyrnau D 76 E5
Tia Mare RO 160 F5
Tiana I 64 C3
Ţibana RO 153 D10
Ţibăneşti RO 153 D10
Tibble S 99 D9
Tiberget S 102 D6
Tibi E 56 D3
Tiboldaróc H 145 H2
Tibro S 92 C4
Ţibucani RO 153 C9
Tice BIH 156 D6
Ticehurst UK 15 E9
Ticha BG 167 D6
Tichá CZ 146 B6
Tichileşti RO 155 C1
Tičići BIH 157 D9
Ticknall UK 11 F9
Ţicleni RO 160 D2
Ţicuşu RO 152 F6
Ticvaniu Mare RO 159 C8
Tidaholm S 91 C14
Tidan S 91 B15
Tiddische D 79 A8
Tidenham UK 13 B9
Tidersrum S 92 D7
Tiebas E 32 E2
Tiedra E 39 E9
Tiefenbach D 75 D12
Tiefenbach D 76 E4
Tiefenbronn D 27 C10
Tiefencastel CH 71 D9
Tiefensee D 84 E5
Tiel NL 16 E4
Tielen B 182 C5
Tielt B 19 C7
Tiemassaari FIN 125 F10
Tienen B 19 C10
Tiengen D 27 E9
Tiercé F 23 E11
Tierga E 41 E8
Tierp S 99 B9
Tierzo E 47 C9
Ţifeşti RO 153 F10
Ţigănaşi RO 153 C10
Ţigăneşti RO 160 F6
Tigare BIH 158 E3
Tighina MD 154 D4
Ţighira MD 153 C11
Tighnabruaich UK 4 D6
Tignale I 69 B10
Tignes F 31 E10
Tigveni RO 160 C5
Tigy F 25 E7
Tiha Bârgăului RO 152 C5
Tihany H 149 C9
Tihemetsa EST 131 E10
Tihilä FIN 123 C16
Tihusniemi FIN 124 F9
Tiistenjoki FIN 123 E10
Tiitilänkylä FIN 123 E17
Tijesno HR 156 E4
Tijnje NL 16 B5
Tíjola E 55 E8
Tikkakoski FIN 123 F15
Tikkala FIN 123 F14
Tikkala FIN 125 F14
Tikkurila FIN 127 E13
Tikob DK 87 C10
Tilburg NL 16 E4
Tilbury UK 15 E9
Til-Châtel F 26 E3
Tildarg UK 4 F4
Tileagd RO 151 C9
Tilehurst UK 13 C12
Tilh F 32 C4
Tilişca RO 152 F3
Tillac F 33 D6
Tillberga S 98 C7
Tillicoultry UK 5 C9
Tillières-sur-Avre F 24 C5
Tilloy-et-Bellay F 25 B12
Tillyfourie UK 3 L11
Tilly-sur-Seulles F 23 B10
Tilvikai LT 134 E3
Tilža LV 133 C2
Tim DK 86 C2
Timahoe IRL 7 G8
Timár H 145 G3
Timau I 73 C7
Timelkam A 76 F5
Timiryazevo RUS 136 C4
Timişeşti RO 153 C9
Timişoara RO 151 F7
Timmele S 91 D13
Timmendorfer Strand D 83 C9
Timmernabben S 89 B10
Timmersdala S 91 B14
Timola FIN 125 F9
Timoleague IRL 8 E5
Timolin IRL 7 G9
Timoniemi FIN 121 F13
Timovaara FIN 125 D12
Timrå S 103 A13
Timring DK 86 C3
Timsgearraidh UK 2 J2
Tinahely IRL 9 C10
Tinajas E 47 D7
Tinalhas P 44 E5
Tinca RO 151 D8
Tinchebray F 23 C10
Tineo E 39 B7
Tiņģere LV 134 B5
Tinglev DK 86 F4
Tingsryd S 89 B7
Tingstad S 93 B8
Tingstäde S 93 D13
Tingvatn N 94 F6
Tingvoll N 100 A8
Tingwall UK 3 G10
Tinja BIH 157 C10
Tinjan HR 67 B8
Tinn N 95 C9
Tinnoset N 95 C10
Tinos GR 176 D5
Tiñosillos E 46 C3
Tinosu RO 161 D8
Tinqueux F 19 F8
Tintagel UK 12 D5
Tinténiac F 23 D8
Tintern Parva UK 13 B9
Ţinteşti RO 161 C9
Tintigny B 19 E12
Tīnūži LV 135 C9
Tiobraid Árann IRL 8 D6
Tione di Trento I 69 A10
Tipasoja FIN 125 B11
Tipperary IRL 8 D6
Tiptree UK 15 D10
Tipu EST 131 E10
Tira MD 154 B2
Tiranë AL 168 B2
Tiranges F 30 E4
Tirano I 69 A9
Tiraspol MD 154 D5
Tiraspolul Nou MD 154 D5
Tire TR 177 C10
Tiream RO 151 B9
Tīreļi LV 134 C7
Tirgo E 40 C6
Tiriez E 55 B8
Tirig E 48 D5
Tiriolo I 59 B10
Tirkšliai LT 134 D4
Tirnova MD 153 A11
Tirrenia I 66 E1
Tirro FIN 113 F18
Tirschenreuth D 75 C11
Tirstrup DK 86 C7
Tirteafuera E 54 B4
Tirza LV 135 B12
Tisău RO 161 C9
Tišća BIH 157 D10
Tishevitsa BG 165 C8
Tishono RUS 136 E2
Tišice CZ 77 B7
Tismana RO 159 C10
Tišnov CZ 77 D10
Tisovec SK 147 D9
Tistrup Stationsby DK 86 D3
Tisvilde DK 87 C10
Tiszaalpár H 150 D4
Tiszabecs H 145 G6
Tiszabezdéd H 145 G5
Tiszabő H 150 C5
Tiszabura H 150 C5
Tiszacsege H 151 B7
Tiszadada H 145 G3
Tiszaderzs H 150 B6
Tiszadob H 145 G3
Tiszaeszlár H 145 G3
Tiszaföldvár H 150 D5
Tiszafüred H 150 B6
Tiszagyenda H 150 C6
Tiszaigar H 150 B6
Tiszajenő H 150 C5
Tiszakanyár H 145 G4
Tiszakarád H 145 G4
Tiszakécske H 150 D5
Tiszakerecseny H 145 G5
Tiszakeszi H 151 B6
Tiszakürt H 150 D5
Tiszalök H 145 G3
Tiszalúc H 145 G3
Tiszanagyfalu H 145 G3
Tiszanána H 150 B6
Tiszaörs H 150 B6
Tiszapalkonya H 145 H3
Tiszapüspöki H 150 C5
Tiszaroff H 150 C5
Tiszasas H 150 D5
Tiszasüly H 150 C5
Tiszaszalka H 145 G5
Tiszaszentimre H 150 C6
Tiszasziget H 150 E5
Tiszatarján H 147 F12
Tiszatelek H 145 G4
Tiszatenyő H 150 C5
Tiszaug H 150 D5
Tiszaújváros H 145 H3
Tiszavárkony H 150 C5
Tiszavasvári H 145 H3
Titaguas E 47 E10
Titel SRB 158 C5
Ţiteşti RO 160 C6
Tithorea GR 175 B6
Tito I 60 B5
Titova Korenica HR 156 C4
Titov Drvar BIH 156 D5
Titran N 104 D4
Tittelsnes N 94 C3
Titting D 75 E9
Tittmoning D 76 F3
Titu RO 161 D7
Titulcia E 46 D5
Tiukuvaara FIN 117 C13
Tiurajärvi FIN 117 C12
Tivat MNE 163 E6
Tivenys E 42 F5
Tiverton UK 13 D8
Tivissa E 42 E5
Tivoli I 62 D3
Tizzano F 37 H9
Tjæreborg DK 86 E3
Tjäkkjokk S 109 E15
Tjällmo S 92 B6
Tjåmotis S 109 C16
Tjappsåive S 109 E17
Tjärmyrberget S 106 C9
Tjärn S 107 D13
Tjärnäs S 98 A6
Tjärnberg S 107 A15
Tjärstad S 92 C7
Tjäruträsk S 118 B8
Tjautas S 116 D5
Tjeldnes N 111 D11
Tjeldstø N 100 E1
Tjelle N 100 A7
Tjentište BIH 157 F10
Tjöck FIN 122 F6
Tjøme N 90 A7
Tjønnefoss N 90 B4
Tjorhom N 94 E5
Tjörnarp S 87 D13
Tjøtta N 108 E3
Tjuda FIN 126 E8
Tjuvskjær N 111 C13
Tkon HR 156 E3
Tleń PL 138 C5
Tlmače SK 147 E7
Tłuchowo PL 139 E7
Tlumačov CZ 146 C5
Tłuszcz PL 139 F11
Toab UK 3 F14
Toaca RO 152 D5
Tóalmás H 150 C4
Toano I 66 D2
Tobar an Choire IRL 6 D5
Tobarra E 55 B9
Tobercurry IRL 6 D5
Tobermore UK 4 F3
Tobermory UK 4 B4
Tobo S 99 B9
Tobyn S 97 C8
Tocane-St-Apre F 29 E6
Tocco da Casauria I 62 C5
Tocha P 44 D3
Töcksfors S 96 C6
Tocón E 53 B9
Todal N 104 E5
Toddington UK 13 B11
Todi I 62 B2
Todireni RO 153 B10
Todireşti RO 153 B8
Todireşti RO 153 C9
Todireşti RO 153 D10
Todmorden UK 11 D7
Todolella E 42 F3
Todorići BIH 157 D7
Todor-Ikonomovo BG 161 F10
Todorovo BG 161 F9
Todtmoos D 27 E8
Todtnau D 27 E8
Toén E 38 D4
Toft UK 3 E14
Toft N 108 F3
Tofta S 87 A10
Tofta S 93 D12
Tofte N 95 C13
Töftedal S 91 B10
Tofterup DK 86 D3
Toftir FO 2 A3
Toftlund DK 86 E4
Tofyeli BY 133 E5
Togher IRL 7 E10
Togher IRL 7 F7
Togher IRL 8 E4
Togston UK 5 E13
Tohmajärvi FIN 125 F14
Tohmo FIN 115 E2
Toholampi FIN 123 C12
Toija FIN 127 E9
Toijala FIN 127 C10
Toila EST 132 C2
Toirano I 37 C8
Toivakka FIN 119 B17
Toivakka FIN 123 F16
Toivala FIN 124 E9
Toivola FIN 128 C6
Tojaci NMK 169 B6
Töjby FIN 122 E6
Tők H 149 A11
Tokachka BG 171 B9
Tokaj H 145 G3
Tokarnia PL 143 E9
Tokarnia PL 147 B9
Tokod H 149 A11
Tököl H 149 B11
Tokrajärvi FIN 125 E15
Toksovo RUS 129 E14
Tolastadh Úr UK 2 J4
Tolbaños E 46 C3
Tolbert NL 16 B6
Tolcsva H 145 G3
Toledo E 46 E4
Tolentino I 67 F7
Tolfa I 62 C1
Tolga N 101 B14
Toliejai LT 135 F10
Tolja FIN 119 B17
Tolk D 82 A7
Tolkmicko PL 139 B8
Tollarp S 88 D5
Tollered S 91 D11
Tollesbury UK 15 D10
Tollo I 63 C6
Tølløse DK 87 D9
Töllsjö S 91 D12
Tolmachevo RUS 132 D6
Tolmezzo I 73 D7
Tolmin SLO 73 D8
Tolna H 149 D11
Tolnanémedi H 149 C10
Tolne DK 90 E7
Tolo GR 175 D6
Tolocăneşti MD 153 A11
Tolonen FIN 117 E14
Tolosa E 32 D1
Tolosa P 44 F5
Tolosenmäki FIN 125 F14
Tolox E 53 C7
Tolšići BIH 157 D10
Tolva E 42 C5
Tolva FIN 121 B12
Tolvayarvi RUS 125 F16
Tolve I 60 B6
Tomai MD 154 D2
Tomai MD 154 E3
Tomar P 44 E4
Tomares E 51 E7
Tomaševac SRB 158 C6
Tomaševo MNE 163 C8
Tomašica BIH 157 C6
Tomášikovo SK 146 E5
Tomášovce SK 147 E9
Tomaszów Lubelski PL 144 C7
Tomaszów Mazowiecki PL 141 G2
Tomatin UK 3 L9
Tombebœf F 33 A6
Tomelilla S 88 D5
Tomelloso E 47 F6
Tomeşti RO 151 E10
Tomeşti RO 151 F9
Tomeşti RO 153 C11
Tomice PL 147 B8
Tomiño E 38 E2
Tomintoul UK 3 L10
Tomislavgrad BIH 157 E7
Tømmerneset N 111 E10
Tommerup DK 86 E6
Tomnavoulin UK 3 L10
Tömörkény H 150 D5
Tompa H 150 E4
Tomra N 100 A5
Tomşani RO 161 D8
Tona E 43 D8
Tonara I 64 C3
Tonbridge UK 15 E9
Tondela P 44 C4
Tønder DK 86 F3
Tonezza del Cimone I 69 B11
Tongeren B 19 C11
Tongland UK 5 F8
Tongue UK 2 J8
Tönisvorst D 183 C8
Tonkopuro FIN 115 E4
Tonna UK 13 B7
Tonnay-Boutonne F 28 D4
Tonnay-Charente F 28 D4
Tonneins F 33 B6
Tonnerre F 25 E10
Tonnes N 108 C5
Tönning D 82 B5
Tønsberg N 95 D12
Tönsen S 103 D12
Tonstad N 94 E5
Tonsvik N 111 A17
Toombeola IRL 6 F3
Toomebridge UK 4 F4
Tootsi EST 131 D9
Topalu RO 155 D2
Topana RO 160 D5
Topares E 55 D8
Toparlar TR 181 C9
Topchii BG 161 F9
Töpchin D 80 B5
Topčic-Polje BIH 157 D8
Topcliffe UK 11 C9
Töpen D 75 B10
Topeno FIN 127 D11
Tophisar TR 173 D9
Topleţ RO 159 D9
Topliceni RO 161 C10
Topli Do SRB 165 C6
Topliţa RO 152 D6
Topliţa RO 159 B10
Töplitz D 79 B12
Topojë AL 168 C1
Topola SRB 158 E6
Topolčani NMK 168 B5
Topoľčany SK 146 D6
Topoľčianky SK 146 E6
Topolia GR 178 E6
Topólka PL 138 E6
Topolnica SRB 159 E9
Topoľníky SK 146 F5
Topolog RO 155 D2
Topolovac HR 149 F6
Topolovăţu Mare RO 151 F8
Topoloveni RO 160 D6
Topolovgrad BG 166 E6
Topolovnik SRB 159 D7
Topolovo BG 166 F4
Topolšica SLO 73 D11
Toponica SRB 158 F6
Toporec SK 145 E1
Toporivtsi UA 153 A8
Toporów PL 81 B8
Toporu RO 161 E7
Toporzyk PL 85 C10
Toppenstedt D 83 D8
Topraisar RO 155 E2
Topsham UK 13 D8
Topusko HR 156 B4
Torá E 43 D6
Toral de los Guzmanes E 39 D8
Toral de los Vados E 39 C6
Torano Castello I 60 E6
Torasalo FIN 125 F10
Toras-Sieppi FIN 117 C11
Torbalı TR 177 C9
Torbjörntorp S 91 C14
Torbygget S 102 A4
Torchiara I 60 C4
Torchiarolo I 61 C10
Torcy F 30 B5
Torda SRB 158 B5
Tordas H 149 B11
Tordehumos E 39 E9
Tordera E 43 D9
Tordesillas E 39 E10
Tordesilos E 47 C9
Töre S 118 C9
Töreboda S 91 B15
Toreby DK 83 A11
Torekov S 87 C11
Torella del Sannio I 63 D7
Torellano E 56 E3
Torelló E 43 C8
Toreno E 39 C6
Torestorp S 91 E12
Torgau D 80 C3
Torgelow D 84 C6
Torgiano I 62 A2
Torhamn S 89 C9
Torhout B 19 B7
Tori EST 131 E9
Torigni-sur-Vire F 23 B10
Torija E 47 C6
Torino I 68 C4
Toritto I 61 B7
Torkanivka UA 154 A4
Torkovichi RUS 132 D7
Torla E 32 E5
Torma EST 131 D13
Tormac RO 159 B7
Törmänen FIN 113 E19
Törmänen FIN 115 A2
Törmänki FIN 117 E13
Törmänmäki FIN 121 E10
Törmäsenvaara FIN 121 C13
Törmäsjärvi FIN 119 B12
Tormestorp S 87 C13
Tormón E 47 D10
Tormore UK 4 D6
Tornadizos de Ávila E 46 C3
Tornaľa SK 145 G1
Tornavacas E 45 D9
Tornby DK 90 D6
Tornemark DK 87 E9
Tornes N 100 A6
Tørnes N 111 D11
Tornesch D 82 C7
Torneträsk S 111 D18
Tornimäe EST 130 D6
Tornio FIN 119 C12
Tornjoš SRB 150 F4
Torno I 69 B7
Tornos E 47 C10
Tornow D 84 D4
Toro E 39 E9
Torö S 93 B11
Törökbálint H 149 B11
Törökszentmiklós H 150 C5
Torony H 149 B7
Toros BG 165 C9
Toroshino RUS 132 F4
Torp FIN 99 B13
Torpa S 92 D6
Torphins UK 3 L11
Torpo N 101 E9
Torpoint UK 12 E6
Torpsbruk S 88 A7
Torpshammar S 103 B11
Torquay UK 13 E7
Torquemada E 40 D3
Torralba de Calatrava E 54 A5
Torralba E 47 D8
Torralba I 64 B2
Torralba de Aragón E 41 E10
Torralba de El Burgo E 40 E6
Torralba de los Sisones E 47 C10
Torralba de Oropesa E 45 E10
Torrão P 50 C3
Torrböle S 107 D17
Torre-Alháquime E 51 F9
Torre Annunziata I 60 B2
Torrebaja E 47 D10
Torreblanca E 48 D5
Torreblascopedro E 53 A9
Torrebruna I 63 D7
Torrecaballeros E 46 C4
Torrecampo E 54 C3
Torre Canne I 61 B8
Torre-Cardela E 55 D6
Torrecilla de Alcañiz E 42 F3
Torrecilla de la Jara E 46 E3
Torrecilla de la Orden E 45 B10
Torrecilla del Rebollar E 42 F1
Torrecillas de la Tiesa E 45 E9
Torrecuso I 60 A3
Torre da Gadanha P 50 B3
Torre das Vargens P 44 F5
Torre de Coelheiros P 50 C4
Torre de Dona Chama P 38 E5
Torre de Embesora E 48 D4
Torredeita P 44 C4
Torre de Juan Abad E 55 B6
Torre del Bierzo E 39 C7
Torre del Burgo E 47 C6
Torre del Campo E 53 A9
Torre del Greco I 60 B2
Torre del Mar E 53 C8
Torredembarra E 43 E6
Torre de Miguel Sesmero
E 51 B6
Torre de Moncorvo P 45 B6
Torre de'Passeri I 62 C5
Torre de Santa María E 45 F8
Torredonjimeno E 53 A9
Torre do Terrenho P 44 C6
Torrefarrera E 42 D5
Torregamones E 39 F7
Torregrossa E 42 D5
Torreiglesias E 46 B4
Torreira P 44 C3
Torrejoncillo E 45 E8
Torrejoncillo del Rey E 47 D7
Torrejón de Ardoz E 46 D6
Torrejón del Rey E 46 C6
Torrejón el Rubio E 45 E9
Torrelacarcel E 47 C10
Torrelaguna E 46 C5
Torrelapaja E 41 E8
Torrelavega E 40 B3
Torrellas E 41 E8
Torrelles de Foix E 43 E7
Torrelobatón E 39 E9
Torrelodones E 46 C5
Torremaggiore I 63 D8
Torremanzanas-La Torre de les
Macanes E 56 D4
Torremayor E 51 B6
Torremegía E 51 B7
Torre Mileto I 63 D9
Torremocha E 45 F8
Torremocha de Jiloca E 47 C10
Torremolinos E 53 C7
Torrenostra E 48 D5
Torrent E 48 F4
Torrente del Cinca E 42 E4
Torrenueva E 55 B6
Torreorgaz E 45 F8
Torre Orsaia I 60 C4
Torre-Pacheco E 56 F3
Torre Pellice I 31 F11
Torreperogil E 55 C6
Torres E 53 A9
Torresandino E 40 E4
Torre San Giovanni I 61 D10
Torre Santa Susanna I 61 C9
Torres de Albánchez E 55 C7
Torres de Berrellén E 41 E9
Torres de la Alameda E 46 D6
Torres del Carrizal E 39 E8
Torresmenudas E 45 B9
Torres Novas P 44 F3
Torres Vedras P 44 F2
Torrevelilla E 42 F3
Torrevieja E 56 F3
Torrice I 62 D4
Torricella I 61 C9
Torricella in Sabina I 62 C3
Torricella Peligna I 63 C6
Torricella Sicura I 62 B5
Torricella Taverne CH 69 A6
Torrico E 45 E10
Torri del Benaco I 69 B10
Torridon UK 2 K5
Torriglia I 37 B10
Torrijas E 48 D3
Torrijo E 41 F8
Torrijo del Campo E 47 C10
Torrijos E 46 E4
Torrin UK 2 L4
Tørring DK 86 D4
Tørring N 105 C10
Torrita di Siena I 66 F4
Torroal P 50 C2
Torroella de Montgrí E 43 C10
Torrox E 53 C9
Torrubia del Campo E 47 E7
Torrubia de Soria E 41 E7
Tørrvika N 104 C7
Torsåker S 98 A6
Torsång S 97 B14
Torsås S 89 C10
Tørsbøl DK 86 F4
Torsböle S 107 E15
Torsborg S 102 A5
Torsby S 97 B9
Torsby S 97 C9
Torsebro S 88 C6
Torshälla S 98 D6
Tórshavn FO 2 A3
Torsholma FIN 126 E5
Torsken N 111 B13
Torslanda S 91 D10
Torsminde DK 86 C2
Torsö S 91 B14
Torsvåg N 112 C4
Törtel H 150 C4
Tortellà E 43 C9
Torteval GBG 22 B6
Torthorwald UK 5 E9
Tortinmäki FIN 126 D7
Tórtola de Henares E 47 C6
Tórtoles de Esgueva E 40 E3
Tortolì I 64 D4
Tortomanu RO 155 E2
Tortona I 37 B9
Tortora I 60 D5
Tortoreto I 62 B5
Tortorici I 59 C6
Tortosa E 42 F5
Tortozendo P 44 D5
Tortuera E 47 C9
Tortuna S 98 C7
Toruń PL 138 D6
Torun' UA 145 F8
Torup S 87 B12
Tõrva EST 131 E11
Tor Vaianica I 62 D2
Tõrvandi EST 131 E13
Torvenkylä FIN 119 F11
Torvik N 100 B3
Torvik N 104 F4
Torvikbukt N 104 F3
Torvikbygd N 94 B4
Torvinen FIN 117 D17
Torvizcón E 55 F6
Torvsjö S 107 C12
Torysa SK 145 E2
Torzym PL 81 B8
Tosbotn N 108 F4
Toscolano-Maderno I 69 B10
Tossa E 43 D9
Tossåsen S 102 A5
Tossåsen S 102 A7
Tossavanlahti FIN 123 D16
Tosse F 32 C3
Tösse S 91 B12
Tossicia I 62 B5
Tosside UK 11 C7
Tõstamaa EST 131 E7
Tostedt D 82 D7
Tószeg H 150 C5
Toszek PL 142 F6
Totana E 55 D10
Totebo S 93 D8
Tôtes F 18 E3
Toteşti RO 159 B10
Tótkomlós H 150 E6
Totland UK 13 D11
Totnes UK 13 E7
Totra S 103 E13
Tötszerdahely H 149 D7
Tøttdal N 105 C10
Tottijärvi FIN 127 C9
Totton UK 13 D12
Tótvázsony H 149 B9
Touça P 45 B6
Toucy F 25 E9
Touffailles F 33 B8
Touget F 33 C7
Toul F 26 C4
Toulon F 35 D10
Toulon-sur-Allier F 30 B3
Toulon-sur-Arroux F 30 B5
Toulouges F 34 E4
Toulouse F 33 C8
Tounj HR 156 B3
Touques F 23 B12
Tourch F 22 D4
Tourcoing F 19 C7
Tourlaville F 23 A8
Tournai B 19 C7
Tournan-en-Brie F 25 C8
Tournay F 33 D6
Tournecoupe F 33 C7
Tournefeuille F 33 C8
Tournon-d'Agenais F 33 B7
Tournon-St-Martin F 29 B7
Tournon-sur-Rhône F 30 E6
Tournus F 30 B6
Tourny F 24 B6
Tourouvre F 24 C4
Tours F 24 F4
Tourteron F 19 E10
Tourtoirac F 29 E8
Toury F 24 D6
Tous E 48 F3
Tõusi EST 131 D7
Touvois F 28 B2
Toužim CZ 76 B3
Tovačov CZ 146 C4
Tovariševo SRB 158 C3
Tovarné SK 145 F4
Tovarnik HR 157 B11
Toven N 108 D5
Tovrljane SRB 164 C3
Towcester UK 14 C7
Tower IRL 8 E5
Toymskardlia N 106 A5
Töysä FIN 123 E11
Traar D 183 C9
Trabada E 38 B5
Trabanca E 45 B8
Trabazos E 39 E7
Traben D 21 E8
Trąbki PL 144 D1
Trąbki Wielkie PL 138 B6
Traboch A 73 B10
Trabotivište NMK 165 F6
Traby BY 137 E12
Trachili GR 175 B9
Tradate I 69 B6
Trädet S 91 D14
Trædal N 111 D14
Trafrask IRL 8 E3
Tragacete E 47 D9
Tragana GR 175 B7
Tragano GR 174 D3
Tragjas AL 168 D2
Tragwein A 77 F7
Traian RO 153 D10
Traian RO 155 C1
Traian RO 155 E1
Traian RO 160 E4
Traian RO 160 F6
Traian Vuia RO 151 F9
Traid E 47 C9
Traiguera E 42 F4
Train D 75 E10
Traînel F 25 D9
Traînou F 24 E7
Traisen A 77 F9
Traiskirchen A 77 F10
Traismauer A 77 F9
Traitsching D 75 D12
Trakai LT 137 D10
Trakovice SK 146 E5
Trakšėdžiai LT 134 F2
Tralee IRL 8 D3
Trá Li IRL 8 D3
Tramacastilla E 47 D9
Tramagal P 44 F4
Tramariglio I 64 B1
Tramatza I 64 C2
Tramayes F 30 C6
Tramelan CH 27 F7
Trá Mhór IRL 9 D8
Tramonti di Sopra I 73 D6
Tramonti di Sotto I 73 D6
Tramore IRL 9 D8
Tramutola I 60 C5
Tranås S 92 C5
Tranbjerg DK 86 C6
Trancoso P 44 C5
Tranebjerg DK 86 D7
Tranemo S 91 E13
Trången S 105 D15
Trånghalla S 92 D4
Trängslet S 102 D6
Trångsviken S 105 E16
Trani I 61 A6
Trannes F 25 D12
Tranovalto GR 169 D6
Tranøy N 111 D10
Trans F 23 D8
Trans-en-Provence F 36 D4
Transtrand S 102 D5
Transtrand S 102 E5
Tranum DK 86 A4
Tranvik S 99 D11
Trapani I 58 C2
Trapene LV 135 B13
Traplice CZ 146 C4
Trappes F 24 C7
Trarbach D 185 E7
Traryd S 87 B13
Trasacco I 62 D5
Träskvik FIN 122 F7
Trasmiras E 38 D4
Trasobares E 41 E8
Tratalias I 64 E2
Traun A 76 F6
Traunreut D 73 A6
Traunstein A 77 F8
Traunstein D 73 A6
Traupis LT 135 E9
Trava SLO 73 E10
Trävad S 91 C13
Travagliato I 66 A1
Travanca do Mondego P 44 D4
Travassô P 44 C4
Travemünde D 83 C9
Travenbrück D 83 C8
Travers CH 31 B10
Traversetolo I 66 C1
Trávnica SK 146 E6
Travnik BIH 157 D8
Travo I 37 B11

Trawniki PL 141 H8
Trawsfynydd UK 10 F4
Trbovlje SLO 73 D11
Trbuk BIH 157 C9
Trebaseleghe I 72 E5
Trebatice SK 146 D5
Trebatsch D 80 B6
Třebechovice pod Orebem CZ 77 B9
Trebel D 83 E10
Treben D 79 D11
Trebenice CZ 76 B5
Trebenow D 84 D5
Trèbes F 33 D10
Trébeurden F 22 C4
Třebíč CZ 77 D9
Trebinje BIH 162 D5
Trebisacce I 61 D7
Trebišov SK 145 F4
Trebnje SLO 73 E11
Trebolle E 38 C4
Třeboň CZ 77 D7
Trebsen D 79 D12
Trebujena E 51 F7
Trebur D 21 E10
Trecastagni I 59 D7
Trecate I 68 C6
Trecenta I 66 B3
Tredegar UK 13 B8
Tredington UK 13 A11
Tredozio I 66 D4
Treehoo IRL 7 D8
Trefaldwyn UK 10 F5
Trefeglwys UK 10 F5
Treffen A 73 C8
Treffort-Cuisiat F 31 C7
Treffurt D 79 D7
Trefriw UK 10 E4
Trefynwy UK 13 B9
Tregaron UK 13 A7
Trégastel F 22 C4
Tregde N 90 C2
Tregnago I 66 A3
Tregony UK 12 E5
Trégueux F 22 D6
Tréguier F 22 C5
Trégunc F 22 E4
Tregynon UK 10 F5
Trehörna S 92 C5
Trehörningsjö S 107 D15
Treia D 82 A6
Treia I 67 F7
Treignac F 29 D9
Treigny F 25 E9
Treimani EST 131 F8
Treis D 185 D7
Trekanten S 89 B10
Treklyano BG 165 D6
Trélazé F 23 F11
Trelech UK 12 B6
Trélissac F 29 E7
Trelleborg S 87 E12
Trelleck UK 13 B9
Trélon F 19 D9
Tremadog UK 10 F3
Tremblay-les-Villages F 24 C5
Tremedal de Tormes E 45 B8
Tremelo B 19 C10
Trémentines F 23 F10
Tremês P 44 F3
Tremezzo I 69 B7
Trémolat F 29 F7
Třemošná CZ 76 C4
Trémouilles F 34 B4
Tremp E 42 C5
Tremance UK 12 E4
Trenčianska Turná SK 146 D6
Trenčianske Jastrabie SK 146 D6
Trenčianske Stankovce SK 146 D5
Trenčianske Teplice SK 146 D6
Trenčín SK 146 D6
Trendelburg D 21 A12
Trengereid N 94 B3
Trensacq F 32 B4
Trent D 84 A4
Trenta SLO 73 D8
Trentels F 33 B7
Trento I 69 A11
Tréon F 24 C5
Treorchy UK 13 B7
Trepča HR 148 F5
Treppeln D 81 B7
Trept F 31 D7
Trequanda I 66 F4
Tres Cantos E 46 C5
Trescléoux F 35 B10
Trescore Balneario I 69 B8
Tresenda I 69 A9
Tresfjord N 100 A6
Tresigallo I 66 C4
Tresjuncos E 47 E7
Treski EST 132 F2
Treskog S 97 C8
Trešnjevica SRB 158 F5
Tresnuraghes I 64 C2
Trespaderne E 40 C5
Tressait UK 5 B9
Třešť CZ 77 D8
Tretower UK 13 B8
Trets D 5 D10
Tretten N 101 D12
Treuchtlingen D 75 E8
Treuenbrietzen D 79 B12
Treungen N 90 A4
Trévé F 22 D6
Trevélez E 55 F6
Tréveray F 26 C3
Trèves F 34 B5
Trevi I 62 B3
Treviana E 40 C5
Trévières F 23 B10
Treviglio I 69 B8
Trevignano Romano I 62 C2
Trévillers F 27 F6
Treviño E 40 C6
Treviso I 72 E5
Trevllazër AL 168 C2
Trevões P 44 B6
Trévoux F 30 D6
Trézelles F 30 C4
Trezzo sull'Adda I 69 B8
Trgovište SRB 164 E5
Trhová Hradská SK 146 F5
Trhové Sviny CZ 77 E7
Trhoviště SK 145 F4
Triacastela E 38 C5
Triaize F 28 C3

Trianta GR 181 D8
Triantaros GR 176 D5
Tribalj HR 67 B10
Tribanj-Krušćica HR 156 D3
Tribehou F 23 B9
Tribsees D 83 B13
Tribunj HR 156 E4
Tricarico I 60 B6
Tricase I 61 D10
Tricesimo I 73 D7
Trichiana I 72 D5
Tricot F 18 E6
Trieben A 73 A9
Triebes D 79 E11
Trie-Château F 18 F4
Trier D 21 E7
Trierweiler D 186 B2
Trieste I 73 E8
Trie-sur-Baïse F 33 D6
Trieux F 20 F5
Trifeşti RO 153 C11
Trifeşti RO 153 D9
Triftern D 76 F4
Trigaches P 50 D4
Trigance F 36 D4
Triggiano I 61 A7
Triglitz D 83 D12
Trignac F 23 F7
Trigrad BG 165 F9
Triguères F 25 E8
Trigueros E 51 E6
Trigueros del Valle E 39 E10
Triigi EST 130 D5
Trikala GR 169 E6
Trikäta LV 131 F11
Trikeri GR 175 A7
Trilj HR 157 E6
Trillevallen S 105 E14
Trillick UK 7 D8
Trillo E 47 C7
Trilofo GR 174 E5
Trilofos GR 169 D8
Trilport F 25 C8
Trim IRL 7 E9
Trimdon UK 5 F14
Trimley St Mary UK 15 D11
Trimmis CH 71 D9
Trimsaran UK 12 B6
Trindade P 38 F5
Trindade P 50 D4
Tring UK 15 D7
Trinità d'Agultu I 64 B2
Trinitapoli I 60 A6
Trino I 68 C5
Trins A 72 B3
Trinta P 45 C6
Trinwillershagen D 83 B13
Triollo E 39 C10
Triora I 37 D7
Tripkau D 83 D10
Tripoli GR 174 D5
Tripotama GR 174 D4
Tripotamos GR 169 C7
Trippstadt D 21 F9
Triptis D 79 E9
Tritenii de Jos RO 152 D4
Trittau D 83 C8
Trittenheim D 21 E7
Trivalea-Moşteni RO 160 E6
Trivento I 63 D7
Trivero I 68 B5
Trivigno I 60 B5
Trizac F 29 E11
Trnakovac HR 157 B7
Trnava CZ 146 C5
Trnava SK 146 E5
Trnava SRB 158 F5
Trnavá Hora SK 147 D7
Trnjani BIH 157 C6
Trnovec nad Váhom SK 146 E5
Trnovica BIH 157 E9
Trnovo BG 165 C9
Tröbitz D 80 C4
Trochtelfingen D 27 D11
Tröbje S 103 E13
Troense DK 86 E7
Trofa P 38 F2
Trofaiach A 73 B11
Trofors N 108 E5
Trogir HR 156 E5
Troglan Bara SRB 159 F7
Troia I 60 A4
Tróia P 50 C2
Troianul RO 160 F6
Troina I 59 D6
Troisdorf D 21 C8
Troisfontaines F 186 D3
Trois-Ponts B 19 D12
Troisreux F 18 F5
Trois-Vèvres F 30 B3
Troisvierges L 20 D6
Troîtsoe MD 154 D4
Trojane SLO 73 D10
Trojanów PL 141 G5
Troldhede DK 86 D3
Trollbäcka N 113 C19
Trollhättan S 91 C11
Trollvik N 112 D6
Tromello I 69 C6
Tromsdalen N 111 A16
Tromsø N 111 A16
Tromvik N 111 A15
Tronco P 38 E5
Trondenes N 111 C12
Trondheim N 104 E8
Trones N 105 B13
Trongisvágur FO 2 B3
Tronnes N 108 E5
Trönninge S 87 B11
Trönninngeby S 87 A10
Trönö S 103 D12
Trontano I 68 A5
Tronvik N 105 D14
Tronzano Vercellese I 68 C5
Troo F 24 E4
Troon UK 4 D7
Trooz B 183 D7
Tropaia GR 174 D4
Tropea I 59 B8
Trory UK 7 D8
Trosa S 93 B11
Troškūnai LT 135 E9
Trossin D 79 C12
Trossingen D 27 D10

Tröstau D 75 B10
Trostberg D 75 F11
Trostyanets' UA 145 E9
Trostyanets' UA 153 A12
Troszyn PL 139 D12
Troubelovo F 23 B9
Troubsko CZ 77 D10
Troubki UK 10 C6
Troutbeck UK 10 C6
Trouville-sur-Mer F 23 B12
Trouy F 29 A10
Troviscal P 44 E4
Trowbridge UK 13 C10
Troyan BG 165 D10
Troyanovo BG 167 D8
Troyes F 25 D11
Troyits'ke UA 154 B6
Troyits'ke UA 154 D6
Troyon F 26 B3
Trpanj HR 157 F7
Trpejca NMK 168 C4
Trpezi MNE 163 D9
Trpinja HR 157 B11
Trsa MNE 157 F10
Trstená SK 147 C9
Trstené pri Hornáde SK 145 F3
Trstenik HR 162 D3
Trstenik SRB 163 B11
Trstice SK 146 E5
Trstín SK 146 D4
Trubar BIH 156 D5
Trubia E 39 B8
Truchas E 39 D7
Truchtersheim F 27 C8
Trud BG 165 E10
Trudovets BG 165 D8
Trudy BY 133 E6
Trujillanos E 51 B7
Trujillo E 45 F9
Trŭn BG 165 D6
Trun CH 71 D7
Trun F 23 C12
Trŭnchovitsa BG 165 C11
Truro UK 12 E4
Trușești RO 153 B10
Trusetal D 79 E7
Trush AL 163 F7
Truskava LT 135 F8
Trŭstenik BIH 157 C7
Trŭstenik BG 165 C9
Trustrup DK 87 C7
Trutnov CZ 81 E9
Try N 90 C2
Tryavna BG 166 D4
Trybusivka UA 154 A3
Tryfos GR 174 B3
Tryggelev DK 83 A9
Trygona GR 168 E5
Tryńcza PL 144 C6
Trysil N 102 D3
Tryškiai LT 134 D5
Trysunda S 107 E15
Tržac BIH 156 C4
Trzcianka PL 85 D10
Trzcianka PL 139 E12
Trzcianne PL 140 D7
Trzciel PL 81 B9
Trzebiatów PL 85 B8
Trzebiechów PL 81 B8
Trzebiel PL 81 C7
Trzebielino PL 85 B12
Trzebień PL 81 D8
Trzebieszów PL 141 G7
Trzebieszowice PL 77 B11
Trzebież PL 84 C7
Trzebinia PL 143 F7
Trzebiszewo PL 81 A8
Trzebnica PL 81 D12
Trzebownisko PL 144 C5
Trzeciewiec PL 138 D5
Trzemeszno PL 138 E4
Trzemeszno Lubuskie PL 81 B8
Trześcianka PL 140 E8
Trześniów PL 145 D5
Trzeszczany Pierwsze PL 144 B8
Trzič SLO 73 D9
Trzcińsko-Zdrój PL 84 E7
Trzyciąż PL 143 F8
Trzydnik Duży PL 144 B5
Tsagkaráda GR 169 F9
Tsakarisianos GR 174 C2
Tsalapitsa BG 165 E10
Tsamantas GR 168 E3
Tsaparevo BG 169 A9
Tsareva Livada BG 166 D4
Tsarev Brod BG 167 C8
Tsarevo BG 167 E9
Tsarimir BG 165 E10
Tsaritsani GR 169 E7
Tsar Kaloyan BG 161 F8
Tsar-Petrovo BG 159 F10
Tschagguns A 71 C9
Tschernitz D 81 C7
Tsebrykove UA 154 C6
Tselina BG 166 E4
Tsenovo BG 165 C7
Tserovo BG 165 C7
Tsiistre EST 131 F14
Tsirguliina EST 131 F12
Tsivaras GR 178 E7
Tsooru EST 131 F13
Tsotili GR 168 D5
Tsoukalades GR 174 B2
Tsŭrvaritsa BG 165 E6
Tsvetino BG 165 F8
Tsyatsyerki BY 133 E2
Tua N 105 B13
Tuaim Beola IRL 6 F3
Tuam IRL 6 E5
Tuamgraney IRL 8 C5
Tubbercurry IRL 6 D5
Tubbergen NL 17 D7
Tubilla del Agua E 40 C4
Tübingen D 27 C11
Tubize B 19 C9
Tubre I 71 D10
Tučepi HR 157 F7
Tuchan F 34 E4
Tüchen D 83 D12
Tuchheim D 79 B10
Turki LV 135 D12
Turková BY 133 B11
Tuchola PL 138 C4
Tuchomie PL 85 B12
Tuchów PL 144 D3
Tuchowicz PL 141 G6
Tuczna PL 141 G8
Tucznö PL 85 D10

Tudela de Duero E 40 E2
Tudora RO 153 B9
Tudor Vladimirescu RO 155 B1
Tudu EST 131 C13
Tudulinna EST 131 C14
Tudweiliog UK 10 F2
Tuejar E 47 E10
Tuenno I 72 D3
Tufeni RO 160 E5
Tufeşti RO 155 D1
Tuffé F 24 D4
Tufjord N 113 A12
Tuhkakylä FIN 125 D10
Tui E 38 D2
Tuili I 64 D2
Tuin NMK 168 A5
Tuiskula FIN 122 E8
Tüja LV 135 B8
Tukhkala RUS 121 C16
Tukhol'ka UA 145 F7
Tukums LV 134 C6
Tula I 64 D2
Tulach Mhór IRL 7 F8
Tulare SRB 164 D3
Tuławki PL 136 F2
Tulbing A 77 F10
Tulca RO 151 D8
Tulce PL 81 B12
Tulcea RO 155 C3
Tulčík SK 145 E3
Tulette F 35 B8
Tulghes RO 152 D5
Tuli BIH 162 D5
Tuliszków PL 142 B5
Tulje BIH 157 F8
Tulka S 99 B11
Tulla IRL 8 C5
Tullaghan IRL 6 D6
Tullamore IRL 7 F8
Tulle F 29 E9
Tullebølle DK 87 F7
Tulleng N 111 A15
Tulleråsen S 105 E16
Tullingsås S 106 D9
Tullins F 31 E7
Tulln A 77 F10
Tullow IRL 9 C9
Tully UK 7 D8
Tully UK 7 D8
Tullyallen IRL 7 E10
Tullyvin IRL 7 E8
Tulnici RO 153 F9
Tulos RUS 125 C14
Tulovo BG 166 D5
Tufowice PL 139 F9
Tufowice PL 142 E4
Tulppio FIN 115 C6
Tulsk IRL 6 E6
Tulstrup DK 87 D10
Tuluceşti RO 155 B2
Tumba S 93 A11
Tumbo S 98 D6
Tume LV 134 C6
Tummel Bridge UK 5 B8
Tumšupe LV 135 C9
Tun S 91 C13
Tunadal S 103 B13
Tuna-Hästberg S 97 B13
Tunari RO 161 D8
Tunel E 38 D2
Tunga UK 2 J4
Tungozero RUS 121 C17
Tunnerstad S 92 C4
Tunnhovd N 95 B9
Tunnsjø-Røyrvika N 105 B14
Tunsjön S 107 E12
Tuntall UK 15 C11
Tuntenhausen D 72 A5
Tunvägen S 102 A8
Tuohikotti FIN 128 D6
Tuohikylä FIN 115 E5
Tuolluvaara S 116 C4
Tuomioja FIN 119 E12
Tuorila FIN 126 B6
Tuoro sul Trasimeno I 66 F5
Tuovilanlahti FIN 124 D8
TupilaţI RO 153 C9
Tuplice PL 81 C7
Tuppurinmäki FIN 125 E10
Tura H 150 B4
Turanlar TR 177 D10
Turany SK 147 C8
Turawa PL 142 E5
Turba EST 131 C8
Turbe BIH 157 D8
Turbenthal CH 27 F10
Turburea RO 160 D2
Turceni RO 160 D2
Turcia E 39 C8
Turčianske Teplice SK 147 D7
Turcifal P 44 F2
Turcineşti RO 160 C2
Turcoaia RO 155 C2
Turda RO 152 D3
Turdaş RO 151 F11
Turégano E 46 B4
Turek PL 142 B5
Tureni RO 152 D3
Tureni RO 151 D12
Turew PL 81 B11
Turgeliai LT 137 E12
Tŭrgovishte BG 167 C7
Turgut TR 181 B8
Turgutalp TR 177 A10
Turgutbey TR 173 B7
Turgutlu TR 177 C10
Turgutreis TR 177 F9
Türi EST 131 D10
Turi I 61 B8
Turia RO 153 E8
Turić BIH 157 C10
Turie SK 147 C7
Turija BIH 157 C9
Turija SRB 158 B4
Turija SRB 159 D8
Turís E 48 F3
Turiya BG 166 D4
Tŭrje H 149 C8
Türka UA 145 E6
Türkeve H 150 C6
Türkgücü TR 173 B8
Turki LV 135 D12
Türkmenli TR 171 E10
Turkova BY 133 B11
Turkovići BIH 162 D5
Turku FIN 126 E7
Turlava LV 134 C3
Turleque E 46 E5
Turmantas LT 135 E12
Turmenti BIH 162 D5

Turna LV 131 F11
Tŭrnak BG 165 C10
Turňa nad Bodvou SK 145 F2
Tŭrnava BG 165 B8
Turnberry UK 4 E7
Turners Hill UK 15 E8
Turnišče SLO 149 C6
Turnov CZ 81 E8
Turnu Măgurele RO 160 F5
Turnu Roşu RO 160 B4
Turnu Ruieni RO 159 C9
Turobin PL 144 B6
Turośl PL 139 D12
Turośl PL 139 D12
Turów PL 141 G6
Turowo PL 85 C11
Turquel P 44 F2
Turrach A 73 C8
Turre E 55 E9
Turri I 64 D2
Turriff UK 3 K12
Tursi I 61 C6
Turţ RO 145 H7
Turtel NMK 164 F5
Turtola FIN 117 E12
Turulung RO 145 H7
Turup DK 86 E5
Tur"ya-Bystra UA 145 F6
Tur"ya-Polyana UA 145 F6
Turyatka UA 153 A8
Tur"ye UA 145 E7
Tur"yi Remety UA 145 F6
Turynka PL 144 B8
Turza Wielka PL 139 D9
Turzovka SK 147 C7
Tusa I 58 D5
Tuscania I 62 C1
Tuse DK 87 D9
Tuset N 105 D16
Tušilović HR 148 F5
Tuşnad RO 153 E7
Tussenhausen D 71 A11
Tüßling D 75 F12
Tuszów Narodowy PL 143 F11
Tuszyma PL 143 F12
Tuszyn PL 143 D8
Tutbury UK 11 F8
Tutjunniemi FIN 125 F13
Tuţora RO 153 C11
Tutova RO 153 E11
Tutow D 84 C4
Tutrakan BG 161 E9
Tuttlingen D 27 E10
Tütüncü TR 173 D8
Tutzing D 72 A3
Tützpatz D 84 C4
Tuudi EST 130 D7
Tuukkala FIN 127 B16
Tuuliharju FIN 117 D13
Tuulimäki FIN 121 F11
Tuulos FIN 127 C12
Tuuri FIN 123 E11
Tuusjärvi FIN 125 E10
Tuusniemi FIN 125 E10
Tuusula FIN 127 E12
Tuv N 101 E8
Tuv N 108 B8
Tuvattnett S 106 D7
Tuve S 91 D10
Tuvnes N 104 D5
Tuvträsk S 107 B15
Tuxford UK 11 E10
Tuzburgazi TR 177 D9
Tuzculu TR 177 B8
Tuzi MNE 163 E7
Tužina SK 147 D7
Tuzla BIH 157 C10
Tuzla RO 155 E3
Tuzla TR 171 E10
Tužno HR 148 D6
Tuzora MD 154 C2
Tuzsér H 145 G5
Tvååker S 87 A10
Tväråträsk S 107 A13
Tvardiţa MD 154 E3
Tärsele S 106 D9
Tärskog S 89 B10
Tvede DK 86 B6
Tvedestrand N 90 B4
Tveit N 90 C3
Tveit N 94 C3
Tverai LT 134 E4
Tverečius LT 135 F13
Tverråga N 108 D6
Tverråmo N 109 B10
Tverrelmo N 111 C18
Tversted DK 90 D7
Tving S 89 C8
Tvis DK 86 C3
Tvøroyri FO 2 B3
Tvorozhkovo RUS 132 E4
Tvrdošín SK 147 C9
Tvŭrditsa BG 166 D5
Tvŭrdošovce SK 146 E6
Twardogóra PL 142 D4
Twatt UK 3 G10
Tweedmouth UK 5 D12
Tweedsmuir UK 5 D10
Twello NL 16 D6
Twist D 17 C8
Twiste (Twistetal) D 17 F11
Twistringen D 17 C11
Two Mile Borris IRL 9 C7
Twomileborris IRL 9 C7
Two Mile Bridge IRL 7 F8
Tworóg PL 142 E6
Twyford UK 13 C12
Twyford UK 15 C7
Twyning UK 13 A10
Tyachiv UA 145 G8
Tychero GR 171 B10
Tychowo PL 85 C10
Tychowo PL 85 C10
Tychy PL 143 F7
Tydal N 105 E10
Tydavnet IRL 7 D8
Tyfors S 97 B11
Tygelsjö S 87 D11
Tyinkrysset N 101 D9
Tykocin PL 140 D7
Tylawa PL 145 D4
Tylissos GR 178 E9
Tyllinge S 93 B9
Tylmanowa PL 145 D1
Tylösand S 87 B11

Tylstrup DK 86 A5
Tymbark PL 144 D1
Tymień PL 85 B9
Tymowa PL 81 D10
Tyndaris UK 4 E7
Tynderö S 103 B14
Týnec nad Labem CZ 77 B8
Týnec nad Sázavou CZ 77 C7
Tynemouth UK 5 E14
Tyngsjö S 97 B12
Tyniewicze Małe PL 141 E8
Týniště nad Orlicí CZ 77 B10
Tynkä FIN 119 F12
Týn nad Vltavou CZ 77 D6
Tynset N 101 B13
Typpö FIN 119 F12
Typpyrä FIN 117 E14
Tyrämäki FIN 121 D12
Tyrävaara FIN 121 D12
Tyrawa Wołoska PL 145 D5
Tyresö S 99 D10
Tyringe S 87 C13
Tyristrand N 95 B12
Tyrjänsaari FIN 125 E15
Tyrnävä FIN 119 E15
Tyrnavos GR 169 E7
Tyrrellspass IRL 7 F8
Tyrväntö FIN 127 C11
Tysnes N 111 D10
Tysse N 94 B3
Tysse N 100 D2
Tyssebotnen N 100 E3
Tyssedal N 94 B5
Tystberga S 93 B10
Tyszowce PL 144 B8
Tytuvėnai LT 134 E6
Tyukod H 145 H6
Tyulenovo BG 167 B11
Tywyn UK 10 F3
Tzermiado GR 178 E9
Tzummarum NL 16 B5

U

Uachdar UK 2 L2
Uachtar Ard IRL 6 F4
Ub SRB 158 E5
Übach-Palenberg D 183 D8
Ubby DK 87 D8
Úbeda E 53 A10
Überherrn D 21 F7
Überlingen D 27 E11
Übersee D 72 A5
Ubli FIN 162 D2
Ubli MNE 163 D6
Ubli MNE 163 D7
Ubrique E 53 C6
Ubstadt-Weiher D 21 F11
Ucciani F 37 G9
Uccle B 19 C9
Ucea RO 152 F5
Uceda E 46 C6
Ucel F 35 A7
Ucero E 40 E5
Ucha P 38 E2
Uchanie PL 144 B8
Uchaux F 35 B8
Uchizy F 30 B6
Uchte D 17 D11
Üchtelhausen D 75 B7
Uchtspringe D 79 A10
Uckange F 186 C1
Uckfield UK 15 F9
Uckro D 80 C5
Uclés E 47 E7
Üçpinar TR 177 B9
Uda RO 160 D5
Udanin PL 81 D10
Udavské SK 145 F4
Udbina HR 156 D4
Udby DK 87 D9
Uddebo S 91 E13
Uddeholm S 97 B10
Uddel NL 183 A7
Uddevalla S 91 C10
Uddheden S 97 C8
Uddington UK 5 D9
Uden NL 16 E5
Udenhout NL 16 E4
Udenisht AL 168 C4
Uderns A 72 B4
Üdersdorf D 21 D7
Udeşti RO 153 B8
Udiča SK 146 C6
Udine I 73 D7
Udorpie PL 85 B13
Üdruma EST 131 D8
Udrycze PL 144 B7
Uebigau D 80 C4
Ueckermünde D 84 C6
Uehlfeld D 75 C8
Uelsen D 17 D7
Uelzen D 83 E9
Uetendorf CH 31 B12
Uetersen D 82 C7
Uettingen D 74 C6
Uetze D 79 B7
Uffculme UK 13 D8
Uffenheim D 75 C7
Uffing am Staffelsee D 72 A3
Ufhusen CH 70 C5
Ugåle LV 134 B4
Ugao SRB 163 C9
Ugarana E 40 B6
Uge DK 86 F4
Ugelbølle DK 86 C6
Ugento I 61 D10
Ugerløse DK 87 D9
Uggdal N 94 B2
Uggelheden S 102 E3
Uggelhuse DK 86 C6
Uggerby DK 90 D7
Uggerslev DK 86 D6
Uggiano la Chiesa I 61 C10
Ugijar E 55 F6
Ugine F 31 D9
Uglari HR 157 E6
Uglejevik BIH 157 C11
Ugod H 149 B9
Uğurlu TR 171 D9
Uğurlu TR 171 D9
Uharte-Arakil E 32 E1
Uherské Hradiště CZ 146 C4
Uherský Brod CZ 146 C5
Uhingen D 74 E6
Uhldingen D 27 E11
Uhlířské Janovice CZ 77 C8
Uhlstädt D 79 E9

Uhniv UA 144 C8
Uhtna EST 131 C13
Uhyst D 81 D7
Uig UK 2 K4
Uimaharju FIN 125 E14
Uimila FIN 127 C15
Uitgeest NL 16 C3
Uithoorn NL 16 D3
Uithuizen NL 17 B7
Uivar RO 159 B6
Ujazd PL 141 G1
Ujazd PL 142 F5
Ujezd CZ 77 C12
Ujezd CZ 77 C12
Ujezd CZ 146 C6
Újfehértó H 151 B8
Újhartyán H 150 C3
Újkér H 149 B7
Újkígyós H 151 D7
Újléta H 151 C8
Ujpetre H 149 E10
Újście PL 85 D11
Újsofy PL 147 C8
Újszalonta H 151 D7
Újszász H 150 C5
Újszentiván H 150 E5
Újszentmargita H 151 B7
Újszilvás H 150 C4
Újtikos H 145 H3
Újudvar H 149 C7
Ukhozhany UA 154 A5
Ukkola FIN 125 E14
Ukmergė LT 135 F9
Ukri LV 134 D6
Ula BY 133 F6
Ula TR 181 B8
Ulamiş TR 177 C8
Uland S 103 A14
Ulan-Majorat PL 141 G6
Ulanów PL 144 C5
Ulaş TR 173 B8
Ulassai I 64 D4
Ula Tirso I 64 C2
Ulbjerg DK 86 B4
Ulbroka LV 135 C9
Ulbster UK 3 J10
Ulcinj MNE 163 F7
Uldum DK 86 D5
Ulefoss N 95 D10
Uleila del Campo E 55 E8
Ulenurme EST 131 E13
Uléz AL 163 F8
Ulfborg DK 86 C2
Ulft NL 16 E6
Ulhówek PL 144 C8
Ulič SK 145 F5
Ulicoten NL 182 C5
Ulieş RO 152 E6
Ulieşti RO 161 D6
Ulila EST 131 E12
Uljanik HR 149 E8
Uljma SRB 159 C7
Ullånger S 107 E14
Ullapool UK 2 K6
Ullared S 87 A11
Ullatti S 116 D7
Ullava FIN 123 C11
Ullbergsträsk S 107 A17
Ulldecona E 42 F4
Ulldemolins E 42 E5
Ullerslev DK 86 E7
Ullervad S 91 B14
Ullés H 150 E4
Ulleskelf UK 11 D9
Ullisjaur S 107 A10
Ullits DK 86 B4
Üllő H 150 C3
Ullsfjord N 111 A18
Ulm D 74 E6
Ulma RO 152 B6
Ulmbach D 74 B5
Ulme P 44 F4
Ulmen D 21 D7
Ulmeni RO 151 C11
Ulmeni RO 161 C9
Ulmeni RO 161 E9
Ulmi RO 161 D6
Ulmu RO 161 D10
Ulmu RO 161 E9
Ulog BIH 157 F9
Uløybukta N 112 D6
Ulricehamn S 91 D13
Ulrichsberg A 76 E5
Ulrichstein D 21 C12
Ulrika S 92 C6
Ulriksberg S 97 B12
Ulriksfors S 106 C9
Ulrum NL 16 B6
Ulsberg N 101 A11
Ulsta UK 3 D14
Ulsted DK 86 A6
Ulsteinvik N 100 B3
Ulstrup DK 86 C5
Ulstrup DK 87 D10
Ulsvåg N 111 D10
Ultrå S 107 E16
Uluabat TR 173 D9
Ulucak TR 177 B9
Ulucak TR 177 C9
Ulukonak TR 181 A7
Ulväker S 91 C14
Ulvenhout NL 16 E3
Ulverston UK 10 C5
Ulvik N 94 B5
Ulvila FIN 126 C6
Ulvöhamn S 107 E15
Ulvsjön S 102 C7
Ulvsjön S 103 B11
Ulvwik S 103 A14
Ul'yanovka UA 154 A5
Ul'yanovo RUS 136 D5
Umag HR 67 B8
Umberleigh UK 12 D7
Umbertide I 66 F5
Umbrărești RO 153 F10
Umbriatico I 61 E7
Umbukta N 108 D4
Umčari SRB 158 D6
Umeå S 107 D17
Umfors S 108 D4
Umhausen A 71 C11
Umka SRB 158 D5
Umljanović BIH 156 E5
Ummeljoki FIN 128 D6
Ummern D 79 A7
Umnäs S 109 F11
Ümraniye TR 173 B11
Umurbey TR 172 D6
Umurga LV 135 A9
Una E 47 D9
Uña de Quintana E 39 D7

Unaja FIN 126 C6
Ünanov CZ 77 E10
Unapool UK 2 J6
Unari FIN 117 D15
Unbyn S 118 C7
Uncastillo E 32 F3
Undenäs S 92 B4
Undereidet N 112 D9
Underfossen N 113 C18
Undersåker S 105 E14
Undingen D 27 D11
Undløse DK 87 D9
Undva EST 130 D3
Undy UK 13 B9
Unelanperä FIN 120 F9
Ungerhausen D 71 A10
Ungheni MD 153 C11
Ungheni RO 152 E4
Ungheni RO 160 E5
Ungra RO 152 F6
Unguraşi RO 152 C4
Ungureni RO 153 B9
Ungureni RO 153 D10
Ungurpils LV 131 F9
Unhais da Serra P 44 D5
Unhais-o-Velho P 44 D5
Unhošť CZ 76 B6
Uničov CZ 77 C12
Uniejów PL 142 C6
Unieux F 30 E5
Unín SK 146 D4
Unirea RO 152 E3
Unirea RO 155 C1
Unirea RO 155 F3
Unirea RO 159 B10
Unirea RO 159 E11
Unirea RO 161 E11
Unisław PL 138 D5
Unkel D 21 C8
Unken A 73 A6
Unlingen D 71 A9
Unna D 17 E9
Unnaryd S 87 B13
Unnau D 185 C8
Unntorp S 102 D7
Unset N 101 C14
Unsholtet N 101 A14
Unstad N 110 D6
Untamala FIN 122 D9
Untamala FIN 126 D6
Unţeni RO 153 B9
Unterägeri CH 27 F10
Unterammergau D 71 B12
Unterdießen D 71 B11
Untergriesbach D 76 E5
Unterhaching D 75 F10
Unterkulm CH 27 F9
Unterlüß D 83 E8
Untermaßfeld D 75 A7
Untermerzbach D 75 B8
Untermünkheim D 74 D6
Unterneukirchen D 75 F12
Unterpleichfeld D 75 C7
Unterreit D 75 F11
Unterschächen CH 71 D7
Unterschleißheim D 75 F10
Untersiemau D 75 B8
Untersteinach D 75 B10
Unterweißenbach A 77 F7
Unterwössen D 72 A5
Unverre F 24 D5
Upavon UK 13 C11
Upenieki LV 134 C5
Upenieki LV 135 D12
Upesgrīva LV 134 B6
Upgant-Schott D 17 A8
Úpice CZ 77 A10
Upinniemi FIN 127 E11
Uplyme UK 13 D9
Upninkai LT 137 C10
Upper Knockando UK 3 L10
Upperlands UK 4 F3
Upphärad S 91 C11
Uppingham UK 11 F10
Upplanda S 99 B9
Upplands-Väsby S 99 C9
Uppsala S 99 C9
Uppsälje S 97 A11
Uppsete N 100 E5
Uppsjö S 103 C12
Upton upon Severn UK 13 A10
Upyna LT 134 E5
Upyna LT 134 F4
Upytė LT 135 E8
Urafirth UK 3 E14
Urago d'Oglio I 69 B8
Uraiújfalu H 149 B7
Uras I 64 D2
Ura Vajgurore AL 168 C2
Uraz PL 81 D11
Urbach D 21 C9
Urbania I 66 E6
Urbar D 185 D8
Urbe I 37 C9
Urberach D 187 B6
Urbino I 66 E6
Urbisaglia I 67 F7
Urbise F 30 C4
Určice CZ 77 D12
Urda E 46 F5
Urdari RO 160 D2
Urdax-Urdazuli E 32 D2
Urdorf CH 27 F9
Urdos F 32 E4
Urduña E 40 C6
Ure N 110 D6
Urecheni RO 153 C9
Urecheşti RO 153 E10
Urecheşti RO 161 B10
Urë e Shtrenjtë AL 163 E8
Urepel F 32 D3
Ureterp NL 16 B6
Urga LV 131 F9
Úrhida H 149 B10
Úri H 150 C4
Uri I 64 B2
Uricani RO 159 C11
Uriménil F 26 D5
Uringe S 93 A11
Uriu RO 152 C4
Urjala FIN 127 C10
Urk NL 16 C5
Ürkmez TR 177 C8
Úrkút H 149 B9
Urla TR 177 C8
Urlaţi RO 161 D8
Urlingford IRL 9 C7
Urmeniş RO 152 D4
Urmince SK 146 D6
Urnäsch CH 27 F11
Urnieta E 32 D2
Üröm H 150 B3
Urovica SRB 159 E9
Urrea de Gaén E 42 E3
Urrea de Jalón E 41 E9
Urretxu E 32 D1
Urriés E 32 E3
Urros P 45 B6
Urroz E 32 E3
Urrugne F 32 D2
Ursberg D 71 A10
Ursensollen D 75 D10
Urshult S 89 B7
Ursviken S 118 E6
Urszulin PL 141 H8
Urt F 32 D3
Urtenen CH 31 A11
Urtimjaur S 116 E5
Urueña E 39 E9
Ururi I 63 D8
Urville Nacqueville F 23 A8
Urzędów PL 144 B5
Urzica RO 160 F4
Urziceni RO 151 B9
Urziceni RO 161 D9
Urzicuţa RO 160 E3
Urzulei I 64 C4
Urzy F 30 A3
Usagre E 51 C7
Ušari BIH 157 C7
Ušće SRB 163 C10
Uschlag (Staufenberg) D 78 D6
Uście Gorlickie PL 145 D3
Uście Solne PL 143 F10
Uscio I 37 C10
Usedom D 84 C5
Usellus I 64 D2
Useras E 48 D4
Ushachy BY 133 F5
Ušī LV 130 F5
Usingen D 21 D11
Usini I 64 B2
Usk UK 13 B9
Uskali FIN 125 F14
Uskedal N 94 C3
Üsküdar TR 173 B11
Üsküp TR 167 F8
Uslar D 78 C6
Usma LV 134 B4
Úsov CZ 77 C12
Uspenivka UA 154 E5
Usquert NL 17 B7
Ussana I 64 E3
Ussassai I 64 D3
Usseglio I 31 E11
Ussel F 29 D10
Ussel F 30 E2
Usson-du-Poitou F 29 C7
Usson-en-Forez F 30 E4
Ustaoset N 95 B8
Ustaritz F 32 D3
Ust'-Chorna UA 145 G8
Ust'-Dolyssy RUS 133 D7
Úštěk CZ 80 E6
Uster CH 27 F10
Ustibar BIH 163 B7
Ustica I 58 B3
Ustikolina BIH 157 E10
Ústí nad Labem CZ 80 E6
Ústí nad Orlicí CZ 77 C10
Ustiprača BIH 157 E11
Ustirama BIH 157 E8
Ustka PL 85 A11
Ust'-Luga RUS 132 B3
Ustou F 33 E8
Ustovo BG 171 A7
Ustrem BG 167 E6
Ustroń PL 147 B7
Ustronie Morskie PL 85 B9
Ustrzyki Dolne PL 145 E6
Ustya UA 154 A5
Ustyluh UA 144 B9
Usurbil E 32 D1
Uszew PL 144 D2
Uszód H 149 C11
Utajärvi FIN 119 E16
Utåker N 94 C3
Utakleiv N 110 D6
Utanede S 103 A12
Utanen FIN 119 E16
Utansjö S 103 A14
Utanskog S 107 E14
Utarp D 17 A8
Utbjoa N 94 C3
Utebo E 41 E10
Utelle F 37 D6
Utena LT 135 F11
Utersum D 82 A4
Uthaug N 104 D7
Uthleben D 79 D8
Uthlede D 17 B11
Utiel E 47 E10
Utne N 94 B5
Utö S 93 B12
Utoropy UA 152 A6
Utrecht NL 16 D4
Utrera E 51 E8
Utrillas E 42 F2
Utrine SRB 150 F4
Utro N 104 C8
Utsjoki FIN 113 D18
Utskor N 110 C8
Uttendorf A 72 B6
Uttendorf A 76 F4
Uttenweiler D 71 A9
Utterbyn S 97 B9
Utterliden S 109 F17
Uttersberg S 97 C14
Uttersjö S 107 D14
Utterslev DK 87 F8
Utti FIN 128 D6
Utting am Ammersee D 71 A12
Uttoxeter UK 11 F8
Utula FIN 129 C9
Utvalnäs S 103 E13
Utvik N 100 C5
Utvorda N 105 B9
Utzedel D 84 C4
Uuemõisa EST 130 D7
Uukuniemi FIN 129 B13
Uulu EST 131 E9
Uura FIN 121 F10
Uurainen FIN 123 E14
Uuro FIN 122 F8
Uusikaarlepyy FIN 122 C9
Uusikartano FIN 126 D7
Uusikaupunki FIN 126 D5
Uusikylä FIN 123 D12
Uusikylä FIN 127 D15
Uusi-Värtsilä FIN 125 F14
Uva FIN 121 E11
Uvac BIH 158 F4
Uvåg N 110 C8
Úvaly CZ 77 B7
Uvanå S 97 B10
Uvdal N 95 B9
Üvecik TR 171 E10
Uvernet-Fours F 36 C5
Uv'jarätto N 113 D12
Uxbridge UK 15 D8
Uxeau F 30 B5
Üxheim D 21 D7
Uyeasound UK 3 D15
Uza F 32 B3
Užava LV 134 B2
Uzdin SRB 158 C6
Uzel F 22 D6
Uzer F 35 A7
Uzerche F 29 E9
Uzès F 35 B7
Uzhhorod UA 145 F5
Uzhok UA 145 F6
Užice SRB 158 F4
Uzlovoye RUS 136 D5
Uznovë AL 168 C2
Užpaliai LT 135 E11
Uzrechcha BY 133 F3
Uzundzhovo BG 166 F5
Uzunköprü TR 172 B6
Uzunkuyu TR 177 C8
Užventis LT 134 E5

V

Vaadinselkä FIN 115 E5
Vaajakoski FIN 123 F15
Vaajasalmi FIN 124 E7
Vääkiö FIN 121 D12
Vaala FIN 119 E17
Vaalajärvi FIN 117 D16
Vaale D 82 C6
Vaalimaa FIN 128 D8
Vaals NL 20 C6
Vaarakylä FIN 121 C11
Vaarankylä FIN 121 F10
Vaaranniva FIN 121 D11
Vaaraperä FIN 121 C13
Vaaraslahti FIN 123 D17
Väärinmaja FIN 124 G2
Vaas F 24 E3
Vaasa FIN 122 D7
Vaassen NL 16 D5
Väätäiskylä FIN 123 E13
Väätsa EST 131 D10
Vaattojärvi FIN 117 D12
Vabalninkas LT 135 E9
Vabole LV 135 D12
Vabre F 33 C10
Vabres-l'Abbaye F 34 C4
Vác H 150 B3
Văcăreşti RO 161 D6
Vaccarizzo Albanese I 61 D6
Váchartyán H 150 B3
Vacheresse F 31 C10
Vachlia GR 174 D4
Väckelsång S 89 B7
Vacov CZ 76 D5
Vacqueyras F 35 B8
Vácrátót H 150 B3
Văculeşti RO 153 B8
Vad RO 152 C3
Vad S 97 B14
Vadakste LV 134 D5
Vadaktai LT 135 E7
Vădastra RO 160 F4
Vădastriţa RO 160 F4
Vaddas gruver N 112 D8
Vădeni RO 155 C1
Väderstad S 92 C5
Vad Foss N 90 B5
Vadheim N 100 D3
Vadla N 94 D4
Vadocondes E 40 E4
Vadokliai LT 135 F8
Vado Ligure I 37 C8
Vadskinn N 111 C11
Vadsø N 114 C7
Vadstena S 92 C5
Vadu Crişului RO 151 D10
Vadu lui Isac MD 155 B2
Vadu Izei RO 145 H8
Vadul lui Vodă MD 154 C4
Vadul Turcului MD 154 B3
Vadum DK 86 A5
Vadu Moldovei RO 153 C8
Vadu Moţilor RO 151 E10
Vadu Paşii RO 161 C9
Vaduz FL 71 C9
Vadžgirys LT 134 F5
Væggerløse DK 83 A11
Vafaiika GR 171 B7
Vafiochori GR 169 B8
Våg N 94 D2
Vågaholmen N 108 C5
Vågåmo N 101 C10
Vagan BIH 157 D6
Vågan N 111 B14
Vågdalen S 106 D9
Våge N 94 B3
Våge N 94 G6
Vaggatem N 114 E6
Vaggeryd S 92 E4
Vågholmane N 100 A4
Vagia GR 175 C7
Văgiuleşti RO 159 D11
Vaglia I 66 E3
Vaglio Basilicata I 60 B5
Vagli Sotto I 66 D1
Vagney F 26 D6
Vagnhärad S 93 B11
Vagnsunda S 99 C11
Vagos P 44 C3
Vågseidet N 100 E2
Vågsele S 107 B14
Vågsodden N 108 E3
Vágur FO 2 C3
Vähäjoki FIN 119 B14
Vähäkangas FIN 123 B13
Vähäkyrö FIN 122 D8
Vähäniva FIN 116 B9
Vahanka FIN 123 E13
Vahastu EST 131 D10
Vahenurme EST 131 D8
Vähikkälä FIN 127 D12
Vahojärvi FIN 127 B9
Váhovce SK 146 E5
Vahterpää FIN 127 E15
Vahto FIN 126 D7
Vaiamonte P 44 F6
Vaiano I 66 E3
Vaickūniškės LT 137 D10
Vaida EST 131 C10
Vaidava LV 135 B10
Vaideeni RO 160 C3
Vaidotai LT 137 D11
Vaiges F 23 D11
Vaiguva LT 134 E5
Väike-Maarja EST 131 C12
Väike-Pungerja EST 131 C14
Vaikijaur S 116 E3
Vaikko FIN 125 D11
Vailly-sur-Aisne F 19 F8
Vailly-sur-Sauldre F 25 F8
Vaimastvere EST 131 D12
Väimela EST 131 F14
Vaimõisa EST 131 D8
Vainikkala FIN 129 D9
Vaiņode LV 134 D3
Vainotiškiai LT 134 F7
Vainupea EST 131 B12
Vairano Patenora I 60 A2
Vairano Scalo I 60 A2
Väisälä FIN 121 E11
Vaison-la-Romaine F 35 B9
Vaïssac F 33 B9
Vaišvydava LT 137 D9
Vaivio FIN 125 E12
Vaivre-et-Montoille F 26 E5
Vaja H 145 H5
Vajangu EST 131 C12
Vajdácska H 145 G4
Våje N 90 B4
Våje N 110 D5
Vajkal AL 168 A3
Vajmat S 109 C18
Vajska SRB 157 B11
Vajszló H 149 E9
Vajta H 149 C11
Vakarel BG 165 D8
Vakern S 97 B11
Vakiflar TR 173 B8
Vaklino BG 155 F2
Vaksdal N 94 B3
Vaksevo BG 165 E6
Vaksince NMK 164 E4
Vál H 149 B11
Valada P 44 F3
Vålådalen S 105 E13
Valadares P 38 D3
Valajanaapa FIN 119 C15
Valajaskoski FIN 119 B14
Valaliky SK 145 F3
Valand N 90 C2
Valandovo NMK 169 B8
Vålånger S 103 A14
Valanida GR 169 E7
Valanjou F 23 F10
Valareña E 41 D9
Vålåsjø N 101 B10
Valaská SK 147 D9
Valaská Belá SK 146 D6
Valašská Bystřice CZ 146 C6
Valašská Polanka CZ 146 C6
Valašské Klobouky CZ 146 C6
Valašské Meziříčí CZ 146 C5
Välax FIN 127 E14
Valberg F 36 C5
Valberg N 110 D6
Vålberg S 97 D9
Valbiska HR 67 B9
Valbo S 103 E13
Valbom P 44 B3
Valbona E 48 D3
Valbondione I 69 A9
Valbonë AL 163 E8
Valbonnais F 31 F8
Valbonne F 36 D6
Valbuena de Duero E 40 E3
Valča SK 147 C7
Valcabrère F 33 D7
Valcău de Jos RO 151 C10
Vâlcele RO 153 F7
Vâlcele RO 160 E5
Vâlcelele RO 161 C10
Vâlcelele RO 161 E10
Valdagno I 69 B11
Valdahon F 26 F5
Valdaora I 72 C5
Valddak N 113 C14
Valdealgorfa E 42 E3
Valdeblore F 37 C6
Valdecaballeros E 45 F10
Valdecañas de Tajo E 45 E9
Valdecarros E 45 C10
Valdecilla E 40 B4
Valdecuenca E 47 D10
Valdefuentes E 45 F8
Valdeganga E 47 F9
Valdeķi LV 134 C5
Valdelacasa E 45 C9
Valdelacasa de Tajo E 45 E10
Valdelamusa E 51 D6
Valdelinares E 48 D3
Valdemanco del Esteras E 54 B3
Valdemārpils LV 134 B5
Valdemarsvik S 93 C9
Valdemeca E 47 D9
Valdemorillo E 46 C4
Valdemoro E 46 D5
Valdemoro-Sierra E 47 D9
Valdenoches E 47 C6
Valdeobispo E 45 D8
Valdeolivas E 47 C8
Valdepeñas E 55 B6
Valdepeñas de Jaén E 53 A9
Valderas E 39 D9
Val-de-Reuil F 18 F3
Valderice I 58 C2
Valderiès F 33 B10
Valderøy N 100 A4
Valderrobres E 42 F4
Val de Santo Domingo E 46 D4
Valdestillas E 39 F10
Valdetormo E 42 F4
Valdetorres E 51 B7
Valdeverdeja E 45 E10
Valdevimbre E 39 D8
Valdgale LV 134 B5
Valdice CZ 77 B8
Valdidentro I 71 E10
Valdilecha E 47 D6
Val-d'Isère F 31 E10
Valdisotto I 71 E10
Valdivienne F 29 B7
Val-d'Izé F 23 D9
Valdobbiadene I 72 E4
Valdoie F 27 E6
Valdunquillo E 39 D9
Vale GBG 22 B6
Våle N 95 D12
Våle S 103 A8
Valea Adîncă MD 154 A3
Valea Argovei RO 161 E9
Valea Călugărească RO 161 D8
Valea Chioarului RO 151 C11
Valea Ciorii RO 161 D11
Valea Crişului RO 153 F7
Valea Danului RO 160 C5
Valea Doftanei RO 161 C7
Valea Dragului RO 161 E8
Valea Ierii RO 151 D11
Valea Largă RO 152 D4
Valea lui Mihai RO 151 B9
Valea Lungă RO 152 E4
Valea Lungă RO 161 C7
Valea Măcrişului RO 161 D9
Valea Mare MD 153 C11
Valea Mare RO 160 D3
Valea Mare RO 160 D6
Valea Mare RO 160 E4
Valea Mare-Pravăţ RO 160 C6
Valea Mărului RO 153 F11
Valea Moldovei RO 153 C8
Valea Nucarilor RO 155 C3
Valea Râmnicului RO 161 C10
Valea Salciei RO 161 C9
Valea Sării RO 153 F9
Valea Seacă RO 153 C9
Valea Seacă RO 153 E10
Valea Stanciului RO 160 F3
Valea Teilor RO 155 C3
Valea Ursului RO 153 D10
Valea Viilor RO 152 E4
Valea Vinului RO 151 B11
Vale da Rosa P 50 E4
Vale das Mós P 44 F4
Vale de Açor P 44 F5
Vale de Açor P 50 D4
Vale de Cambra P 44 C4
Vale de Cavalos P 44 F3
Vale de Espinho P 45 D7
Vale de Estrela P 45 C6
Vale de Figueira P 44 F3
Vale de Lobo P 50 E3
Vale de Prazeres P 44 D6
Vale de Reis P 50 C3
Vale de Salgueiro P 38 E5
Vale de Santarém P 44 F3
Vale do Peso P 44 F5
Válega P 44 C3
Valeggio sul Mincio I 66 B2
Valen N 94 C3
Valença P 38 D2
Valença do Douro P 44 B5
Valençay F 24 F6
Valence F 30 F6
Valence F 33 B7
Valence-d'Albigeois F 33 B10
Valence-sur-Baïse F 33 C6
Valencia E 48 F4
Valencia de Alcántara E 45 F6
Valencia de Don Juan E 39 D8
Valencia de las Torres E 51 C7
Valencia del Mombuey E 51 C5
Valencia del Ventoso E 51 C7
Valenciennes F 19 D8
Văleni RO 153 D11
Văleni RO 160 E5
Vălenii de Munte RO 161 C8
Valensole F 35 C10
Valentano I 62 B1
Valentigney F 27 F6
Valenza I 37 A9
Valenzano I 61 A7
Valenzuela E 53 A8
Valenzuela de Calatrava E 54 B5
Våler N 95 D13
Våler N 101 E15
Valera de Arriba E 47 E8
Valernes F 35 B10
Vales Mortos P 50 D5
Valestrand N 94 A2
Valevåg N 94 C2
Valfabbrica I 66 F6
Valfarta E 42 D3
Valfroicourt F 26 D5
Valfurva I 71 E10
Valga EST 131 F12
Valgalciems LV 134 B5
Valgale LV 134 B4
Valgrisenche I 31 D11
Valgu EST 131 D9
Valguarnera Caropepe I 58 E5
Valgunde LV 134 C7
Valhelhas P 44 D6
Valhermoso E 47 C9
Vålhovd N 101 E12
Välijoki FIN 119 B15
Välikangas FIN 119 B16
Väli-Kannus FIN 123 C11
Valikardhë AL 168 A3
Välikylä FIN 123 C11
Valimi GR 174 C5
Välitalo FIN 117 C16
Văliug RO 159 C9
Väli-Viirre FIN 123 C11
Valjala EST 130 E5
Valjevo SRB 158 E4
Valjok N 113 D16
Valjunquera E 42 F4
Valka LV 131 F12
Valkeajärvi FIN 123 F12
Valkeakoski FIN 127 C11
Valkeakoski S 117 E11
Valkeala FIN 128 D6
Valkeiskylä FIN 124 C8
Valkeiskylä FIN 125 D10
Valkenburg NL 19 C12
Valkenswaard NL 16 F4
Valkininkai LT 137 E10
Valkla EST 131 C10
Valko FIN 127 E15
Valkó H 150 B4
Valla S 93 A8
Valla S 107 E10
Valladolid E 39 E10
Valladolises E 56 F2
Vállaj H 151 B9
Vallåkra S 87 D11
Vallargärdet S 97 D10
Vallata I 60 A4
Vallauris F 36 D6
Vallberga S 87 C12
Vallbo S 105 E14
Vallbona d'Anoia E 43 D7
Vallda S 91 E10
Valldal N 100 B6
Vall d'Alba E 48 D4
Valldemossa E 49 E10
Valle E 40 B3
Valle LV 135 C10
Valle N 90 A2
Valle N 90 B6
Valle N 108 B6
Valle Castellana I 62 B5
Vallecorsa I 62 E4
Valle de Abdalajís E 53 C7
Valle de la Serena E 51 B8
Valle de Matamoros E 51 C6
Valle de Santa Ana E 51 C6
Valle di Cadore I 72 D5
Valledolmo I 58 D4
Valledoria I 64 B2
Valleiry F 31 C8
Vallelunga Pratameno I 58 D4
Valle Mosso I 68 B5
Vallen S 107 D11
Vallen S 118 F6
Vallenca E 47 D10
Vallendar D 185 D8
Vallentuna S 99 C10
Valleräs S 102 E6
Valleraugue F 35 B6
Vallermosa I 64 E2
Vallerotonda I 62 D5
Vallersund N 104 D7
Vallervatnet N 105 B15
Vallet F 23 F9
Valley D 72 A4
Valley UK 10 E2
Vallfogona de Riucorb E 42 D6
Vallières F 29 D10
Vallioniemi FIN 121 B12
Vallmoll E 42 E6
Valløby DK 87 E10
Vallo della Lucania I 60 C4
Vallo di Nera I 62 B3
Valloire F 31 E9
Vallombrosa I 66 E4
Vallon-en-Sully F 29 B11
Vallon-Pont-d'Arc F 35 B7
Vallorbe CH 31 B9
Vallorcine F 31 C10
Vallouise F 31 F9
Vallrun S 105 D16
Valls E 42 E6
Vallsbo S 103 E12
Vallsjön S 103 A11
Vallsta S 103 C11
Vallstena S 93 D13
Vallvik S 103 D13
Valmadrera I 69 B7
Valmadrid E 41 F10
Valmen N 101 D15
Valmiera LV 131 F10
Valmiermuiža LV 131 F10
Valmojado E 46 D4
Valmont F 18 E2
Valmontone I 62 D3
Valmorel F 31 E9
Valmy F 25 B12
Valnes N 108 B7
Valognes F 23 A9
Valongo P 44 B4
Valongo P 44 F5
Válor E 55 F6
Valoria la Buena E 40 E2
Valøy N 105 B9
Valøy N 105 C11
Valpaços P 38 E5
Valpalmas E 41 D10
Valpelline I 31 D11
Valperga I 68 C4
Valpovo HR 149 E10
Valpperi FIN 126 D7
Valras-Plage F 34 D5
Valréas F 35 B8
Valros F 34 D5
Vals CH 71 D8
Valsavarenche I 31 D11
Vålse DK 87 F9
Valseca E 46 B4
Valsequillo E 51 C9
Valsgård DK 86 B5
Valsgarth UK 3 D15
Valshed S 103 E9
Valsinni I 61 C6
Valsjöbyn S 105 C16
Valsjön S 103 B11
Valška SRB 158 E6
Valskog S 97 D14
Vals-les-Bains F 35 A7
Valsøybotn N 104 E5
Vålsta S 103 C13
Valstagna I 72 E4
Val-Suzon F 26 F2
Valtablado del Río E 47 C8
Valtero GR 169 B9
Valtesiniko GR 174 D5
Valtice CZ 77 E11
Valtiendas E 40 F4
Valtierra E 41 D8
Valtimo FIN 125 C11
Valtola FIN 128 C7
Valtopina I 62 A3
Valtos GR 171 A10
Valtotopi GR 169 C10
Valtournenche I 68 B4
Valtura HR 67 C8
Valu lui Traian RO 155 E2
Valun HR 67 C9
Väluste EST 131 E11
Valverde de Burguillos E 51 C6
Valverde de Júcar E 47 E8
Valverde de la Virgen E 39 C8
Valverde del Camino E 51 D6
Valverde de Leganés E 51 B6
Valverde del Fresno E 45 D7
Valverde de Llerena E 51 C8
Valverde del Majano E 46 C4
Valverde de Mérida E 51 B7
Valvika N 108 B8
Valvträsk S 118 B7
Valyra GR 174 E4
Vama RO 145 H7
Vama RO 153 B7
Vama Buzăului RO 161 B7
Vamberk CZ 77 B10
Vamdrup DK 86 E4
Våmhus S 102 D7
Vamlingbo S 93 F12
Vammala FIN 127 C8
Vammen DK 86 B5
Vamos GR 178 E7
Vámosmikola H 147 F7
Vámospércs H 151 B8
Vámosújfalu H 145 G3
Vampula FIN 126 C8
Vamvakofyto GR 169 B9
Vamvakou GR 169 F7
Vanaja FIN 127 D12
Vana-Koiola EST 131 F14
Vânători RO 152 E5
Vânători RO 153 C9
Vânători RO 153 F10
Vânători RO 155 B2
Vânători RO 159 E10
Vânătorii Mici RO 161 E7
Vânători-Neamţ RO 153 C8
Vanault-les-Dames F 25 C12
Vana-Vigala EST 131 D8
Vana-Võidu EST 131 E11
Váncsod H 151 C8
Vanda FIN 127 E12
Vandel DK 86 D4
Vandellós E 42 E5
Vandenesse F 30 B4
Vandenesse-en-Auxois F 25 F12
Vandœuvre-lès-Nancy F 186 D1
Vandoies I 72 C4
Vändra EST 131 D10
Vändträsk S 118 C6
Vandzene LV 134 B5
Vandžiogala LT 135 F7
Vāne LV 134 C5
Väne-Åsaka S 91 C11
Vänersborg S 91 C11
Vañes E 40 C3
Vang N 101 D9
Vånga S 92 B7
Vangaži LV 135 B9
Vänge S 93 E13
Vängel S 107 D10
Vangshamn N 111 B15
Vangshylla N 105 D10
Vangsnes N 100 D5
Vangsvik N 111 B14
Vanha-Kihlanki FIN 117 C10
Vanhakylä FIN 122 F7
Vänjaurbäck S 107 C15
Vänjaurträsk S 107 C15
Vânjuleţ RO 159 E10
Vânju Mare RO 159 E10
Vannareid N 112 C4
Vännäs S 122 C3
Vännäsberget S 118 B9
Vännäsby S 122 C3
Vannavalen N 112 C4
Vånne N 90 B2
Vannes F 22 E6
Vannvåg N 112 C4
Vannvikan N 104 D8
Vänö FIN 126 F7
Vansbro S 97 A11
Vanse N 94 F5
Vänsjö S 103 C9
Vantaa FIN 127 E12
Vanttausjärvi FIN 119 B17
Vanttauskoski FIN 119 B17
Vanvey F 25 E12
Vanyarc H 147 F8
Vanzone I 68 B5
Vaour F 33 B9
Vápenná CZ 77 B12
Vaplan S 105 E16
Vaqueiros P 50 E4
Vara EST 131 D13
Vara S 91 C12
Vara del Rey E 47 F8
Varades F 23 F9
Vărădia RO 159 C8
Vărădia de Mureş RO 151 E9
Varages F 35 C10
Varaire F 33 B9
Varajärvi FIN 119 B13
Varajoki FIN 121 F14
Varakļāni LV 135 C13
Varallo I 68 B5
Varangerbotn N 114 C5
Varano de'Melegari I 69 D8
Varapayeva BY 133 F2
Varapodio I 59 C8
Vărăşti RO 161 E8
Varatec RO 153 C8
Varaždin HR 149 D6
Varaždinske Toplice HR 149 D6
Varazze I 37 C9
Varberg S 87 A10
Vărbilău RO 161 C7
Varbla EST 130 E7
Varbó H 145 G2
Varbola EST 131 C8
Varces-Allières-et-Risset F 31 E8
Vârciorog RO 151 D9
Varda GR 174 C3
Vardali GR 174 A5
Varde DK 86 D2
Vardim BG 161 F6
Vardište BIH 158 F3
Vårdö FIN 99 B14
Vardø N 114 C10
Várdomb H 149 D11
Varejoki FIN 119 B13
Varekil S 91 C10
Varel D 17 B10
Varėna LT 137 E10
Varengeville-sur-Mer F 18 E2
Varenna I 69 A7
Varennes-en-Argonne F 19 F11
Varennes-St-Sauveur F 31 C7
Varennes-sur-Allier F 30 C3
Varennes-Vauzelles F 30 A3
Vareš BIH 157 D9
Varese I 69 B6
Varese Ligure I 37 C11
Varetz F 29 E8
Vârfu Câmpului RO 153 B8
Vârfuri RO 161 C7
Vârfurile RO 151 E10
Vårgårda S 91 C12
Vărgata RO 152 D5
Vârghiş RO 153 E7
Vargön S 91 C11
Vargträsk S 107 C15
Varhaug N 94 E3
Vari GR 175 D8
Vari GR 176 E4
Variaş RO 151 E6
Varik NL 183 B6
Variku EST 130 C7
Varilhes F 33 D9
Varimbombi GR 175 C8
Varín SK 147 C7
Våring S 91 B14
Variņi LV 135 B12
Variskylä FIN 121 F9
Varislahti FIN 125 E11
Varistaipale FIN 125 E11
Varjakka FIN 119 E14
Varjisträsk S 109 D18
Varkaliai LT 134 E3
Varkaus FIN 125 F9
Vārkava LV 135 D13
Varkhi BY 133 E7
Vârlezi RO 153 F11
Varlosen (Niemetal) D 78 D6
Vārme LV 134 C4
Värmlandsbro S 91 A13
Varna BG 167 C9
Varna I 72 C4
Varna SRB 158 D4
Värnamo S 88 A6
Värnäs S 97 B9
Varnavas GR 175 C8
Varnhem S 91 C14
Varniai LT 134 E4
Varnja EST 131 E14
Varnsdorf CZ 81 E7
Varntresken N 108 E7
Varnyany BY 137 D13
Väröbacka S 87 A10

Viişoara RO 153 A9
Viişoara RO 153 D8
Viişoara RO 153 E11
Viişoara RO 160 F6
Viitaila FIN 127 C13
Viitala FIN 123 E10
Viitamäki FIN 123 C16
Viitaniemi FIN 125 D10
Viitapohja FIN 127 B11
Viitaranta FIN 115 D3
Viitaranta FIN 121 B12
Viitasaari FIN 123 D15
Viitavaara FIN 121 E13
Viitavaara FIN 121 F12
Viitka EST 132 F1
Viitna EST 131 C12
Viivikonna EST 132 C2
Vijciems LV 131 F11
Vik N 90 C4
Vik N 94 B2
Vik N 108 B9
Vik N 108 F3
Vik N 110 D7
Vik S 88 D6
Vika FIN 117 E16
Vika N 101 A15
Vika S 97 A14
Vikajärvi FIN 117 E16
Vikan N 104 E5
Vikan N 105 B10
Vikarbyn S 97 B15
Vikarbyn S 103 E9
Vikartovce SK 145 F1
Vikbyn S 97 B15
Vike N 100 E3
Vike S 103 A12
Vikebukt N 100 A6
Vikedal N 94 D3
Viken S 87 C11
Viken S 103 B9
Viken S 103 B11
Vikersund N 95 C11
Vikeså N 94 E4
Vikevåg N 94 D3
Vikholmen N 108 D4
Viķi LV 131 F9
Vikingstad S 92 C6
Vikmanshyttan S 97 B14
Vikna N 105 B9
Vikoč BIH 157 F10
Vikøyri N 100 D5
Vikran N 111 A16
Vikran N 111 C12
Vikran N 113 A12
Viksberg S 97 B14
Viksjö S 103 A13
Viksjöfors S 103 D10
Viksmon S 107 E12
Viksna LV 133 B2
Viktarinas LT 137 E8
Vikten N 110 D5
Viktorivka UA 154 C6
Vikýřovice CZ 77 C11
Vila E 38 E3
Vila Boa P 45 D7
Vila Boa do Bispo P 44 B4
Vila Caiz P 38 F3
Vila Chã de Sá P 44 C5
Vila Cova da Lixa P 38 F3
Vilada E 43 C7
Viladamat E 43 C10
Vila da Ponte P 38 E4
Viladecans E 43 E8
Vila de Cruces E 38 C3
Vila de Frades P 50 C4
Vila de Rei P 44 E4
Vila do Bispo P 50 E2
Vila do Conde P 38 F2
Viladrau E 43 D8
Vilafamés E 48 D4
Vilafant E 43 C9
Vila Fernando P 51 B5
Vila Flor P 38 F5
Vila Franca das Naves P 45 C6
Vilafranca de Bonany E 57 B11
Vilafranca del Penedès E 43 E7
Vila Franca de Xira P 50 B2
Vilagarcía de Arousa E 38 C2
Vilajuïga E 34 F5
Vilaka LV 133 B3
Vilalba E 38 B4
Vilallonga E 33 F10
Vilamartín E 38 D5
Vilamarxant E 48 E3
Vilamoura P 50 E3
Vilāni LV 133 C1
Vila Nogueira de Azeitão P 50 B1
Vilanova E 38 B5
Vila Nova da Baronia P 50 C3
Vila Nova da Barquinha P 44 F4
Vilanova d'Alcolea E 48 D5
Vilanova de Arousa E 38 C2
Vila Nova de Cacela P 50 E4
Vila Nova de Famalicão P 38 F2
Vila Nova de Foz Coa P 45 B6
Vila Nova de Gaia P 44 B3
Vilanova de la Barca E 42 D5
Vilanova de L'Aguda E 42 D6
Vilanova de Meià E 42 D6
Vila Nova de Paiva P 44 C5
Vila Nova de Poiares P 44 D4
Vilanova de Prades E 42 E5
Vila Nova de São Bento P 50 D5
Vilanova de Sau E 43 D8
Vilanova i la Geltrú E 43 E7
Vila Pouca da Beira P 44 D5
Vila Pouca de Aguiar P 38 E4
Vila Praia de Âncora P 38 E2
Vilar P 38 E3
Vilar P 44 F2
Vilarandelo P 38 E5
Vilarchao E 38 D4
Vilar da Veiga P 38 E3
Vilar de Andorinho P 44 B3
Vilar de Barrio E 38 D4
Vilar de Santos E 38 D4
Vilardevós E 38 E5
Vila Real P 38 F4
Vila Real de Santo António P 50 E5
Vilarelho da Raia P 38 E5
Vilar Formoso P 45 C7
Vilarinho da Castanheira P 45 B6
Vilarinho do Bairro P 44 D3
Vilariño de Conso E 38 D5
Vila-rodona E 43 E6
Vila Ruiva P 50 C4
Vila Seca P 38 E2
Vilaseca de Solcina E 42 E6
Vilassar de Mar E 43 D8
Vilasund S 108 D8
Vila Velha de Ródão P 44 E5
Vila Verde P 38 E3
Vila Verde P 38 F4
Vila Verde P 44 D3
Vila Verde da Raia P 38 E5
Vila Verde de Ficalho P 51 D5
Vila Viçosa P 50 B5
Vilce LV 134 D7
Vilches E 55 C6
Vildbjerg DK 86 C3
Vilémov CZ 77 C9
Vilgāle LV 134 C3
Vilhelmina S 107 B11
Vilia GR 175 C7
Viljakkala FIN 127 B9
Viljandi EST 131 E11
Viljevo HR 149 E10
Viljolahti FIN 125 F10
Vilkaviškis LT 136 D7
Vilkene LV 131 F9
Vilkija LT 137 C8
Vilkjärvi FIN 128 D8
Vilkyškiai LT 134 F4
Villa Bartolomea I 66 B3
Villabate I 58 C3
Villablanca E 51 E5
Villablino E 39 C7
Villabona E 32 D1
Villabrágima E 39 E9
Villabuena del Puente E 39 F9
Villac F 29 E8
Villacañas E 46 E6
Villa Carcina I 69 B9
Villacarriedo E 40 B4
Villacarrillo E 55 C6
Villa Castelli I 61 B8
Villacastín E 46 C4
Villach A 73 C8
Villacidro I 64 E2
Villaciervos E 41 E6
Villaconejos E 46 D6
Villaconejos de Trabaque E 47 D8
Villada E 39 D10
Villa d'Almè I 69 B8
Villa del Prado E 46 D4
Villa del Río E 53 A8
Villadepera E 39 E7
Villa de Ves E 47 F10
Villadiego E 40 C3
Villadose I 66 B4
Villadossola I 68 A5
Villaeles de Valdavia E 39 C10
Villaescusa de Haro E 47 E7
Villaescusa la Sombría E 40 D5
Villafáfila E 39 E8
Villafeliche E 47 B10
Villaflores E 45 B10
Villafranca d'Asti I 37 B8
Villafranca de Córdoba E 53 A7
Villafranca de Ebro E 41 E10
Villafranca del Bierzo E 39 C6
Villafranca del Campo E 47 C10
Villafranca del Cid E 48 D4
Villafranca de los Barros E 51 B7
Villafranca de los Caballeros E 46 F6
Villafranca di Verona I 66 B2
Villafranca in Lunigiana I 69 E8
Villafranca-Montes de Oca E 40 D5
Villafranca Tirrena I 59 C7
Villafranco del Guadalquivir E 51 E7
Villafrati I 58 D3
Villafrechos E 39 E9
Villafruela E 40 E4
Villafuerte E 40 E3
Villagarcía de Campos E 39 E9
Villagarcía de la Torre E 51 C7
Villagarcía del Llano E 47 F9
Villagio Mancuso I 59 A10
Villagonzalo E 51 B7
Villagrande Strisaili I 64 D4
Villaharta E 54 C3
Villähde FIN 127 D14
Villahermosa E 55 B7
Villahermosa del Campo E 47 B10
Villahermosa del Río E 48 D4
Villaherreros E 40 D3
Villahizán E 40 D4
Villahoz E 40 D4
Villaines-en-Duesmois F 25 E12
Villaines-la-Juhel F 23 D11
Villajoyosa E 56 D4
Villala FIN 125 F13
Villala FIN 129 D9
Villa Lagarina I 69 B11
Villalago I 62 D5
Villalar de los Comuneros E 39 E9
Villa Latina I 62 D5
Villalba I 58 D4
Villalba de Duero E 40 E4
Villalba de Guardo E 39 C10
Villalba del Alcor E 51 E7
Villalba de la Sierra E 47 D8
Villalba de los Alcores E 39 E10
Villalba de los Barros E 51 B6
Villalba del Rey E 47 D7
Villalba dels Arcs E 42 E4
Villalba de Rioja E 40 C6
Villalcampo E 39 E7
Villalcázar de Sirga E 40 D2
Villalengua E 41 F8
Villalgordo del Júcar E 47 F8
Villa Literno I 60 A2
Villalobos E 39 E9
Villalón de Campos E 39 D9
Villalonga E 56 D4
Villalpando E 39 E9
Villalpardo E 47 F9
Villaluenga de la Sagra E 46 D5
Villalumbroso E 39 D10
Villamalea E 47 F9
Villamañán E 39 D8
Villamandos E 39 D8
Villamanrique E 55 B7
Villamanrique de la Condesa E 51 E7
Villamanta E 46 D4
Villamar I 64 D2
Villamartín E 51 F8
Villamartín de Campos E 39 D10
Villamassargia I 64 E2
Villamayor E 45 C9
Villamayor de Calatrava E 54 B4
Villamayor de Campos E 39 E9
Villamayor de Santiago E 47 E7
Villamayor de Treviño E 40 D3
Villamblard F 29 E7
Villamediana E 40 D3
Villamediana de Iregua E 32 F1
Villamejil E 39 C7
Villamesías E 45 F9
Villamiel E 45 D7
Villa Minozzo I 66 D1
Villamo FIN 122 F7
Villamor de los Escuderos E 39 F8
Villamuelas E 46 E5
Villamuriel de Cerrato E 40 E3
Villandraut F 32 B5
Villanova I 61 B9
Villanova d'Albenga I 37 C8
Villanova d'Asti I 37 B7
Villanova del Battista I 60 A4
Villanovafranca I 64 D3
Villanova Monferrato I 68 C5
Villanova Monteleone I 64 B1
Villanova Truschedu I 64 D2
Villanova Tulo I 64 D3
Villanterio I 69 C7
Villanúa E 32 E4
Villanubla E 39 E10
Villanueva de Alcardete E 47 E6
Villanueva de Alcorón E 47 C8
Villanueva de Algaidas E 53 B8
Villanueva de Argaño E 40 D4
Villanueva de Bogas E 46 E5
Villanueva de Cameros E 41 D6
Villanueva de Castellón E 56 C4
Villanueva de Córdoba E 54 C3
Villanueva de Gállego E 41 E10
Villanueva de Gómez E 46 C3
Villanueva de la Cañada E 46 D5
Villanueva de la Concepción E 53 C7
Villanueva de la Fuente E 55 B7
Villanueva de la Jara E 47 F9
Villanueva de la Reina E 53 A9
Villanueva del Arzobispo E 55 C6
Villanueva de las Cruces E 51 D5
Villanueva de la Serena E 51 B8
Villanueva de la Sierra E 45 D8
Villanueva de las Torres E 55 D6
Villanueva de la Vera E 45 D10
Villanueva del Campo E 39 E9
Villanueva del Duque E 54 C3
Villanueva del Fresno E 51 C5
Villanueva de los Castillejos E 51 D5
Villanueva de los Infantes E 55 B7
Villanueva del Rey E 51 C9
Villanueva del Río Segura E 55 C10
Villanueva del Río y Minas E 51 D8
Villanueva del Rosario E 53 C8
Villanueva del Trabuco E 53 B8
Villanueva de Mesía E 53 B8
Villanueva de San Carlos E 54 B5
Villanueva de San Juan E 51 E9
Villanueva de Tapia E 53 B8
Villanueva de Valdegovia E 40 C5
Villanuño de Valdavia E 40 C2
Villány H 149 E10
Villa Opicina I 73 E8
Villapalacios E 55 B7
Villaperuccio I 64 E2
Villapiana I 61 D6
Villapiana Lido I 61 D6
Villa Poma I 66 C3
Villapourçon F 30 B4
Villaputzu I 64 E4
Villaquejida E 39 D8
Villaquilambre E 39 C8
Villaralbo E 39 F8
Villaralto E 54 C3
Villarcayo E 40 C4
Villard-Bonnot F 31 E8
Villard-de-Lans F 31 E8
Villar de Cañas E 47 E7
Villar de Chinchilla E 55 B9
Villar de Ciervo E 45 C7
Villardeciervos E 39 E7
Villardefrades E 39 E9
Villar del Arzobispo E 48 E3
Villar de la Yegua E 45 C7
Villar del Buey E 39 F7
Villar del Cobo E 47 D9
Villar del Humo E 47 E9
Villar de los Barrios E 39 C6
Villar de los Navarros E 42 E1
Villar del Pedroso E 45 E10
Villar del Rey E 45 F7
Villar del Salz E 47 C10
Villar de Olalla E 47 D8
Villar de Peralonso E 45 B8
Villar de Rena E 45 F9
Villar de Torre E 40 D6
Villardompardo E 53 A8
Villareal E 48 E4
Villarejo de Fuentes E 47 E7
Villarejo de Montalbán E 46 E3
Villarejo de Orbigo E 39 D8
Villarejo de Salvanés E 47 D6
Villa Rendena I 69 A10
Villarente E 39 C9
Villares de la Reina E 45 B9
Villargordo E 53 A9
Villargordo del Cabriel E 47 E10
Villarino de los Aires E 39 F7
Villarluengo E 42 F2
Villarosa I 58 D5
Villarquemado E 47 C10
Villarramiel E 39 D10
Villarrasa E 51 E6
Villarreal de Huerva E 47 B10
Villarrín de Campos E 39 E8
Villarrobledo E 47 F7
Villarrodrigo E 55 C7
Villarroya de la Sierra E 41 F8
Villarroya de los Pinares E 42 F2
Villarrubia de los Ojos E 46 F5
Villarrubia de Santiago E 46 E6
Villarrubio E 47 E7
Villars F 29 E7
Villars F 31 B9
Villars-Colmars F 36 C5
Villars-du-Saz E 47 E8
Villars-les-Dombes F 31 D7
Villars-sur-Var F 36 D6
Villar-St-Pancrace F 31 F10
Villarta E 47 F9
Villarta de los Montes E 46 F3
Villarta de San Juan E 46 F6
Villasalto I 64 E3
Villasana de Mena E 40 B5
Villasandino E 40 D3
Villa San Giovanni I 59 C8
Villa San Pietro I 64 E3
Villa Santa Maria I 63 D6
Villasante de Montija E 40 B5
Villa Santina I 73 D6
Villasarracino E 40 D3
Villasavary F 33 D10
Villasayas E 41 F6
Villasbuenas E 45 B8
Villasdardo E 45 B8
Villaseca de Laciana E 39 C7
Villaseca de la Sagra E 46 E5
Villaseco de los Gamitos E 45 B8
Villaseco de los Reyes E 45 B8
Villaseco del Pan E 39 F8
Villasequilla de Yepes E 46 E5
Villasimius I 64 E4
Villasmundo I 59 E7
Villasor I 64 E2
Villaspeciosa I 64 E2
Villasrubias E 45 D7
Villastar E 47 D10
Villasur de Herreros E 40 D5
Villatobas E 46 E6
Villatoro E 45 C10
Villatoya E 47 F10
Villatuerta E 32 E2
Villaturiel E 39 C9
Villaumbrales E 39 D10
Villaurbana I 64 D2
Villava E 32 E2
Villavallelonga I 62 D5
Villavelayo E 40 D6
Villaverde de Guadalimar E 55 C7
Villaverde del Río E 51 D8
Villaverde y Pasaconsol E 47 E8
Villavernia I 37 B9
Villaviciosa E 39 B9
Villaviciosa de Córdoba E 54 C2
Villaviciosa de Odón E 46 D5
Villavieja E 48 E4
Villavieja de Yeltes E 45 C8
Villaviudas E 40 E3
Villayón E 39 B6
Villazanzo de Valderaduey E 39 C10
Villé F 27 D7
Villebois-Lavalette F 29 E6
Villebrumier F 33 C8
Villecomtal-sur-Arros F 33 D6
Villecomte F 26 E3
Villecroze F 36 D4
Villedaigne F 34 D4
Villedieu-la-Blouère F 23 F9
Villedieu-les-Poêles F 23 C9
Villedieu-sur-Indre F 29 B9
Ville-en-Tardenois F 20 F1
Villefagnan F 28 C6
Villefontaine F 31 D7
Villefort F 35 B6
Villefranche-d'Albigeois F 33 C10
Villefranche-d'Allier F 30 C2
Villefranche-de-Lauragais F 33 D9
Villefranche-de-Lonchat F 28 F6
Villefranche-de-Panat F 34 B4
Villefranche-de-Rouergue F 33 B10
Villefranche-du-Périgord F 33 A8
Villefranche-sur-Cher F 24 F6
Villefranche-sur-Mer F 37 D6
Villefranche-sur-Saône F 30 D6
Villefranque F 32 D3
Villegas E 40 D3
Villel E 47 D10
Villelaure F 35 C9
Villemandeur F 25 E8
Villemorien F 25 D11
Villemoustaussou F 33 D10
Villemur-sur-Tarn F 33 C9
Villena E 56 D3
Villenauxe-la-Grande F 25 C10
Villeneuve CH 31 C10
Villeneuve F 33 B10
Villeneuve F 35 C10
Villeneuve-au-Chemin F 25 D10
Villeneuve-d'Allier F 30 E3
Villeneuve-d'Ascq F 19 C7
Villeneuve-de-Berg F 35 A8
Villeneuve-de-Marsan F 32 C5
Villeneuve-de-Rivière F 33 D7
Villeneuve-la-Guyard F 25 D9
Villeneuve-l'Archevêque F 25 D10
Villeneuve-lès-Avignon F 35 C8
Villeneuve-lès-Béziers F 34 D5
Villeneuve-Loubet F 36 D6
Villeneuve-sur-Allier F 30 B3
Villeneuve-sur-Lot F 33 B7
Villeneuve-sur-Yonne F 25 D9
Villeneuve-Tolosane F 33 C8
Villepinte F 33 D10
Villercomtal F 33 A11
Villeréal F 33 A7
Villerest F 30 D5
Villerías E 39 E10
Villerouge-Termenès F 34 D4
Villers-Bocage F 18 E5
Villers-Bocage F 23 B10
Villers-Bretonneux F 18 E6
Villers-Carbonnel F 18 E6
Villers-Cotterêts F 19 F7
Villers-Écalles F 18 E2
Villers-en-Argonne F 25 B12
Villersexel F 26 E5
Villers-Farlay F 31 A8
Villers-le-Bouillet B 183 D6
Villers-le-lac F 31 A10
Villers-lès-Nancy F 26 C5
Villers-Outréaux F 19 D7
Villers-Semeuse F 19 E10
Villers-sur-Glâne CH 31 B11
Villers-sur-Mer F 23 B11
Villerville F 23 B12
Villery F 25 D11
Villeseneux F 25 C11
Ville-sur-Tourbe F 20 F3
Villetta Barrea I 62 D5
Villeurbanne F 30 D6
Villeveyrac F 35 C6
Villevocance F 30 E6
Villié-Morgon F 30 C6
Villiers-Charlemagne F 23 E10
Villiers-en-Lieu F 25 C12
Villiers-en-Plaine F 28 C4
Villiers-le-Sec F 26 D3
Villiers-St-Benoît F 25 E9
Villiers-St-Georges F 25 C9
Villieu-Loyes-Mollon F 31 D7
Villikkala FIN 128 D6
Villingen D 27 D9
Villingsberg S 97 D12
Villmar D 185 D9
Villoldo E 39 D10
Villora E 47 E9
Villorba I 72 E5
Villoria E 45 C10
Villoruela E 45 B10
Villotte-sur-Aire F 26 C3
Villshärad S 87 B11
Villvattnet S 118 E4
Vilmány H 145 G3
Vilnius LT 137 D11
Vilovi d'Onyar E 43 D9
Vilppula FIN 123 F12
Vilpulka LV 131 F10
Vils A 71 B11
Vils DK 86 B3
Vilsbiburg D 75 F11
Vilseck D 75 C10
Vil'shanka UA 154 A3
Vilshofen D 76 E4
Vilshult S 88 C7
Vilslev DK 86 E3
Vilsted DK 86 B4
Vilsund Vest DK 86 B3
Vilusi BIH 157 C7
Vilusi MNE 162 D6
Viluste EST 132 F1
Vilvestre E 45 B7
Vilvoorde B 19 C9
Vilzēni LV 131 F10
Vima Mică RO 152 C3
Vimbodí E 42 E6
Vimeiro P 44 F2
Vimercate I 69 B7
Vimianzo E 38 B1
Vimieiro P 50 B4
Vimioso P 39 E6
Vimmarvattnet S 107 D9
Vimmerby S 92 D7
Vimory F 25 E8
Vimoutiers F 23 C12
Vimpeli FIN 123 D11
Vimperk CZ 76 D5
Vimy F 18 D6
Vinac BIH 157 D7
Vinadio I 37 C6
Vinaixa E 42 E5
Vinařice CZ 76 B6
Vinaròs E 42 G4
Vinarsko BG 167 D8
Vinäs S 102 E8
Vinassan F 34 D5
Vinay F 31 E7
Vinberg S 87 B11
Vinça F 34 E4
Vinča SRB 158 D6
Vincey F 26 D5
Vinchiaturo I 63 E7
Vinci I 66 E2
Vind DK 86 C3
Vindblæs DK 86 B4
Vindeballe DK 86 F6
Vindeby DK 86 E7
Vindel-Ånäset S 118 F3
Vindelgransele S 107 A14
Vindeln S 107 C17
Vinderei RO 153 E11
Vinderslev DK 86 C4
Vinderup DK 86 C3
Vindhella N 100 D7
Vindinge DK 87 D10
Vindinge DK 87 E7
Vindsvik N 94 D4
Vinebre E 42 E5
Vineuil F 24 E5
Vineuil F 29 B9
Vinga RO 151 E7
Vingåker S 92 A7
Vingang S 102 E4
Vingrau F 34 E4
Vingrom N 101 D12
Vingsand N 104 C8
Vinhais P 39 E6
Vinhas P 39 E6
Vinica HR 148 D6
Vinica NMK 164 F5
Vinica SK 147 E8
Vinica SLO 67 B11
Viničné SK 146 E4
Viniegra de Arriba E 40 D6
Vinine BIH 157 G8
Vinišće HR 156 F5
Vinje N 94 C7
Vinje N 100 E4
Vinjeøra N 104 E5
Vinkovci HR 149 F11
Vinkt B 182 C2
Vinliden S 107 B13
Vinné SK 145 F4
Vinnes N 94 B3
Vinnesvåg N 94 B2
Vinni EST 131 C12
Vinninga S 91 C13
Vinningen D 27 B8
Vinodol SK 146 E6
Vinograd BG 166 C5
Vinogradets BG 165 E9
Vinon-sur-Verdon F 35 C10
Vinsa S 116 E9
Vinslöv S 87 C13
Vinsobres F 35 B9
Vinsternes N 104 E4
Vinstra N 101 C11
Vintervollen N 114 D8
Vintilă Vodă RO 161 C9
Vintjärn S 103 E11
Vintrosa S 97 D12
Vintturi FIN 123 C11
Vințu de Jos RO 152 E2
Viñuelas E 46 C6
Vinuesa E 40 E6
Vinzelberg D 79 A10
Viöl D 82 A6
Violay F 30 D5
Violès F 35 B8
Vipava SLO 73 E8
Vīpe LV 135 D12
Viperești RO 161 C8
Vipiteno I 72 C3
Vipperød DK 87 D9
Vipperow D 83 D13
Vir BIH 157 E7
Vir HR 67 D11
Vir SLO 73 D10
Vira S 93 B8
Virawlya BY 133 E7
Virazeil F 33 A6
Virbalis LT 136 D6
Virbi LV 134 B5
Virče NMK 165 F6
Vire F 23 C10
Virešl LV 135 B12
Vireux-Molhain F 184 D2
Vireux-Wallerand F 19 D10
Virey-sous-Bar F 25 D11
Virgen A 72 B5
Virginia IRL 7 E8
Viriat F 31 C7
Virieu-le-Grand F 31 D8
Virigneux F 30 D5
Virignin F 31 D8
Viriville F 31 E7
Virķēni LV 131 F10
Virkkala FIN 127 E11
Virkkula FIN 121 B13
Virkkunen FIN 121 C10
Virklund DK 86 C5
Virmaanpää FIN 124 E8
Virmaila FIN 127 C13
Virmutjoki FIN 129 C10
Virolahden kk FIN 128 D8
Virolahti FIN 128 D8
Virovitica HR 149 E8
Virpazar MNE 163 E7
Virpe LV 134 B4
Virrat FIN 123 F11
Virsbo S 97 C15
Virserum S 89 A9
Virtaniemi FIN 114 F4
Virtasalmi FIN 124 F8
Virton B 19 E12
Virtsu EST 130 D7
Virttaa FIN 126 D8
Viru-Jaagupi EST 131 C12
Viru-Nigula EST 131 C13
Viry F 31 C9
Vis HR 63 A10
Visaginas LT 135 E12
Višakio Rūda LT 137 D7
Visan F 35 B8
Vișani RO 161 C10
Visåsen S 103 A13
Visbek D 17 C10
Visborg DK 86 B6
Visby DK 86 E3
Visby S 93 D12
Visé B 19 C12
Višegrad BIH 158 F3
Visegrád H 149 A11
Viserba I 66 D6
Viseu P 44 C5
Vișeu de Jos RO 152 B4
Vișeu de Sus RO 152 B4
Vishnyeva BY 137 E13
Vishovgrad BG 166 C4
Visiedo E 42 F1
Visikums LV 133 B2
Vișina RO 160 D6
Vișina RO 160 F4
Vișinești MD 154 E2
Vișinești RO 161 C7
Visingsö S 92 C5
Visjövalen S 105 E12
Viskafors S 91 D12
Viskan S 103 B11
Viškl LV 135 D13
Vislanda S 88 B6
Visnes N 94 D2
Višnja Gora SLO 73 E10
Višnjićevo SRB 158 D3
Višňová CZ 81 E8
Višňové CZ 77 E10
Višňové SK 147 C7
Visoca MD 153 A11
Viso del Marqués E 55 B5
Visokë AL 168 C2
Visoko BIH 157 E9
Visone I 37 B9
Visonta H 150 B5
Visp CH 68 A4
Vissac-Auteyrac F 30 E4
Vissani GR 168 E4
Visselfjärda S 89 B9
Visselhövede D 82 E7
Vissenbjerg DK 86 E6
Visso I 62 B4
Vissoie CH 68 A4
Vist N 105 D10
Vistabella del Maestrazgo E 48 D4
Vistbäcken S 118 C5
Viștea RO 152 F5
Vistheden S 118 C5
Visthus N 108 E4
Viştytis LT 136 E6
Visuvesi FIN 123 F11
Visznek H 150 B5
Vita I 58 D2
Vitå S 118 C8
Vitaby S 88 D6
Vitåfors S 118 C8
Vitanová SK 147 C9
Vitanovac SRB 159 F6
Vitanovac SRB 164 C5
Vítaz SK 145 F2
Vitberget S 118 C4
Viterbo I 62 C2
Viterne F 26 C5
Vitez BIH 157 D8
Vitězná CZ 77 B9
Viti RKS 164 E3
Vitigudino E 45 B8
Vitina BIH 157 F8
Vītiņi LV 134 D5
Vitis A 77 E8
Vitkov CZ 146 B5
Vitkovići BIH 157 E10
Vītoliņi LV 134 C7
Vitolište NMK 169 B6
Vitomirești RO 160 D4
Vitomiricë RKS 163 D9
Vitorchiano I 62 C2
Vitoria-Gasteiz E 41 C6
Vitoševac SRB 159 F8
Vitovlje BIH 157 D8
Vitré F 23 D9
Vitrey-sur-Mance F 26 E4
Vitrolles F 35 D9
Vitry-en-Artois F 18 D6
Vitry-en-Perthois F 25 C12
Vitry-le-François F 25 C12
Vitry-sur-Loire F 30 B4
Vittangi S 116 C7
Vittaryd S 87 B13
Vitteaux F 25 F12
Vittel F 26 D4
Vittikko FIN 115 E4
Vittikko FIN 119 B13
Vittikkovuoma FIN 117 E12
Vittinge S 98 C8
Vittjärv S 118 C7
Vittoria I 59 F6
Vittorio Veneto I 72 E5
Vittsjö S 87 C13
Vitulano I 60 A3
Vitulazio I 60 A2
Vitvattnet S 102 B8
Vitvattnet S 107 D16
Vitvattnet S 119 B10
Vitzenburg D 79 D10
Viuruniemi FIN 125 E11
Vivario F 37 G10
Vivastbo S 98 B7
Viveiro E 38 A4
Vivel del Río Martín E 42 F2
Viver E 48 E3
Viverols F 30 E4
Viveros E 55 B7
Vivier-au-Court F 19 E10
Viviers F 35 B8
Viviez F 33 A10
Vivild DK 86 C6
Viv-le-Fesq F 35 C7
Vivonne F 29 C6
Vix F 25 E12
Vix F 28 C4
Vizantea-Livezi RO 153 F9
Vize TR 173 A8
Vizille F 31 E8
Vižinada HR 67 B8
Viziru RO 155 C1
Vizitsa BG 167 E9
Vizovice CZ 146 C5
Vizslás H 147 E9
Vizsoly H 145 G3
Vizzini I 59 E6
Vlaardingen NL 16 E2
Vlachata GR 174 C2
Vlacherna GR 174 D5
Vlachia GR 175 B8
Vlachiotis GR 175 F6
Vlachokerasia GR 174 E5
Vlachovice CZ 146 C5
Vlădaia RO 159 E11
Vladaya BG 165 D7
Vlădeni RO 153 B9
Vlădeni RO 153 C10
Vlădeni RO 155 D1
Vlădești RO 154 F2
Vlădești RO 160 C4
Vlădești RO 160 C5
Vladičin Han SRB 164 D5
Vlădila RO 160 E4
Vladimir MNE 163 E7
Vladimir RO 160 D3
Vladimirci SRB 158 D4
Vladimirescu RO 151 E7
Vladimirovac SRB 159 C6
Vladimirovo BG 161 F11
Vladimirovo BG 165 B7
Vladinya BG 165 C10
Vladislav CZ 77 D9
Vlad Țepeș RO 161 E10
Vladychen' UA 155 B3
Vlăhița RO 153 E7
Vlahovići BIH 157 F9
Vlanduk BIH 157 D8
Vlaole SRB 159 E8
Vlase SRB 164 D4
Vlasenica BIH 157 D10
Vlašici HR 156 D3
Vlašim CZ 77 C7
Vlasina Okruglica SRB 164 D5
Vlăsinești RO 153 B9
Vlasotince SRB 164 C5
Vlatkovići BIH 157 D7
Vlčany SK 146 E5
Vlčnov CZ 146 D5
Vledder NL 16 C6
Vleuten NL 16 D4
Vlijmen NL 16 E4
Vlissingen NL 16 F1
Vlochos GR 169 E7
Vlorë AL 168 D1
Vlotho D 17 D11
Vlycho GR 174 B2
Vnanje Gorice SLO 73 D9
Vnorovy CZ 146 D4
Voćin HR 149 E9
Vöcklabruck A 76 F5
Vöcklamarkt A 76 F4
Voden BG 161 F9
Voden BG 167 E7
Voderady SK 146 E5
Vodica BIH 157 D7
Vodice HR 67 B9
Vodice HR 156 E4
Vođinci HR 157 B10
Voditsa BG 166 C6
Vođjenica BIH 156 C5
Vodňany CZ 76 D6
Vodnjan HR 67 C8
Vodskov DK 86 A6
Vodstrup DK 86 B3
Voe UK 3 E14
Voel DK 86 C5
Voerde (Niederrhein) D 17 E7
Voerendaal NL 183 D7
Voerladegård DK 86 C5
Voerså DK 86 A6
Vogatsiko GR 168 D5
Vogelenzang NL 182 A5
Vogelgrun F 27 D8
Vogelsang D 84 D4
Vogelsdorf D 80 B5
Vogelweh D 21 F9
Voghera I 37 B10
Voghiera I 66 C4
Vognill N 101 A11
Vognsild DK 86 B4
Vogošća BIH 157 E9
Vogt D 71 B9
Vogtareuth D 72 A5
Vogüé F 35 A7
Vohburg an der Donau D 75 E10
Vohenstrauß D 75 C11
Vöhl D 21 B11
Võhma EST 130 D4
Võhma EST 131 D11
Vohonjoki FIN 120 B9
Vöhringen D 27 D10
Vöhringen D 71 A10
Voicești RO 160 D4
Void-Vacon F 26 C4
Voikkaa FIN 128 D6
Voikoski FIN 128 C6
Voila RO 152 F5
Voillecomte F 25 C12
Voineasa RO 160 C3
Voinești RO 153 C10
Voinești RO 160 C6
Voiron F 31 E8
Võiste EST 131 E8
Voiteg RO 159 C7
Voiteur F 31 B8
Voitsberg A 73 B11
Voivodeni RO 152 D5

Vojakkala FIN 119 C12
Vojakkala FIN 127 D11
Vojčice SK 145 F4
Vojens DK 86 E4
Vojka SRB 158 D5
Vojkovic BIH 157 E9
Vojnic HR 148 F5
Vojnik NMK 164 E4
Vojnik SLO 148 D4
Vojtjajaure S 108 E9
Voka EST 132 C2
Vokány H 149 E10
Vokil BG 161 F10
Voknavolok RUS 121 E16
Voksa N 100 B2
Volakas GR 170 B5
Volda N 100 B4
Voldby DK 86 C5
Voldby DK 87 C7
Voldum DK 86 C6
Volendam NL 16 D4
Volgsele S 107 B15
Volimes GR 174 D2
Volintiri MD 154 E5
Volissos GR 177 C6
Voljevac BIH 157 E8
Voljice BIH 157 E8
Volkach D 75 C7
Volkel NL 183 B7
Völkermarkt A 73 C10
Volketswil CH 27 F10
Völklingen D 21 F7
Volkmarsen D 17 F12
Volkovija NMK 163 F10
Voll N 101 E13
Völlen D 17 B8
Vollenhove NL 16 C5
Vollersode D 17 B11
Vollsjö S 87 D13
Volmunster F 27 B7
Voloiac RO 159 D11
Volonne F 35 B11
Volos GR 169 F8
Voloshcha UA 145 D8
Volosovo RUS 132 C5
Volosyanka UA 145 F6
Volovăţ RO 153 B7
Volovets' UA 145 F7
Volovo BG 166 B5
Volpago del Montello I 72 E5
Volpedo I 37 B9
Volpiano I 68 C4
Völs A 72 B3
Völschow D 84 C4
Voltaggio I 37 B9
Volta Mantovana I 66 B2
Volterra I 66 F2
Voltlage D 17 D9
Voltri I 37 C9
Voltti FIN 122 D9
Volturara Appula I 63 E8
Volturara Irpina I 60 B3
Volturino I 63 E8
Voluntari RO 161 E8
Volvic F 30 D3
Volx F 35 C10
Volya Arlamivs'ka UA 144 D7
Volyně CZ 76 D5
Vomp A 72 B4
Voneshta Voda BG 166 D5
Vonge DK 86 D4
Vonges F 26 F3
Voni GR 178 E9
Vonitsa GR 174 B2
Võnnu EST 131 E14
Võõpste EST 131 E14
Võõpsu EST 132 E2
Voorburg NL 16 D2
Voore EST 131 D13
Voorhout NL 182 A4
Voorschoten NL 16 D2
Voorthuizen NL 183 A7
Voose EST 131 C10
Võra FIN 122 D8
Voranava BY 137 E11
Vorau A 148 B5
Vorbasse DK 86 D4
Vorchdorf A 76 F5
Vorden NL 16 D6
Vordernberg A 73 A10
Vordorf D 79 B8
Vorë AL 168 B2
Voreppe F 31 E8
Vorey F 30 E4
Vorgod DK 86 C3
Vormsele S 107 B15
Vormstad N 104 E7
Vormsund N 95 B14
Vormträsk S 107 B15
Vorna FIN 119 F15
Vorniceni RO 153 B9
Voroi GR 178 E8
Vorokhta UA 152 A5
Vorona RO 153 B9
Vorra D 75 C9
Vorța RO 151 E10
Vorst B 19 B11
Vorwerk D 17 B12
Voskopojë AL 168 C4
Voşlăbeni RO 153 D7
Voss N 100 E4
Vosselaar B 16 F3
Voştane HR 157 E6
Võsu EST 131 B11
Votice CZ 77 C7
Votonosi GR 168 E5
Voudia GR 179 B8
Vougeot F 26 F2
Vouillé F 28 B4
Vouillé F 29 B6
Voukolies GR 178 E6
Voula GR 175 D8
Voulaines-les-Templiers F 25 E12
Vouliagmeni GR 175 D8
Voulpi GR 174 A4
Voulx F 25 D8
Vouneuil-sous-Biard F 29 B6
Vouneuil-sur-Vienne F 29 B7
Vourgareli GR 168 F5
Vourvourou GR 169 D10
Voutezac F 29 E8
Voutianoi GR 174 E5
Vouvant F 28 B4
Vouvray F 24 F4
Vouvry CH 31 C10
Vouzela P 44 C4

Vouzi GR 174 A5
Vouziers F 19 F10
Vouzon F 24 E7
Voves F 24 D6
Vowchyn BY 141 F8
Vowkawshchyna BY 133 E2
Voxna S 103 D10
Voynka BG 167 E7
Voynitsa RUS 121 D15
Voynovo BG 161 F10
Voynyagovo BG 165 D10
Vozmediano E 41 E8
Voznesenka-Persha UA 154 E4
Vrå DK 90 E6
Vrabevo BG 165 D10
Vráble SK 146 E6
Vračev Gaj SRB 159 D7
Vračevšnica SRB 158 E6
Vrachasi GR 179 E10
Vrachati GR 175 D6
Vrachesh BG 165 D8
Vrachnaiika GR 174 C4
Vracov CZ 146 D4
Vrádal N 95 D8
Vrakúň SK 146 F5
Vrani RO 159 C8
Vranino BG 167 B10
Vranisht AL 168 D2
Vranisht AL 168 D2
Vranjak SRB 158 E6
Vranje SRB 164 D4
Vranjska Banja SRB 164 D5
Vranov CZ 77 E11
Vranov nad Topľou SK 145 F4
Vransko SLO 73 D10
Vrapčište NMK 163 F10
Vrapičić BIH 157 F8
Vrasene B 182 C4
Vrasna GR 169 C10
Vratarnica SRB 159 F9
Vratimov CZ 146 B6
Vratnica NMK 164 E3
Vratsa BG 165 C8
Vrattsa BG 165 D10
Vražogrnac SRB 159 F9
Vrbanja SRB 158 E4
Vrbanja HR 157 C10
Vrbanjci BIH 157 C7
Vrbas SRB 158 B4
Vrbica SRB 150 E5
Vrbnik HR 67 B10
Vrbnik HR 156 D5
Vrbno pod Pradědem CZ 142 F3
Vrbovac SRB 159 E7
Vrbovce SK 146 D4
Vrbové SK 146 D5
Vrbovec HR 149 E6
Vrbovsko HR 67 B11
Vrchlabí CZ 81 E9
Vrčin SRB 158 D6
Vrdy CZ 77 C8
Vrebac HR 156 D4
Vrécourt F 26 D4
Vreden D 17 D7
Vreeland NL 183 A6
Vrees D 17 C9
Vrellë RKS 163 D9
Vrelo SRB 164 C4
Vrena S 93 B9
Vrensted DK 90 E6
Vŕesina CZ 146 B6
Vresse B 19 E10
Vresthena GR 175 E6
Vretstorp S 92 A5
Vrgorac HR 157 F7
Vrhnika SLO 73 E9
Vrhopolje BIH 157 C6
Vrhovine HR 156 C3
Vridsted DK 86 C4
Vries NL 17 B7
Vriezenveen NL 17 D7
Vrigstad S 92 E4
Vrinners DK 86 C6
Vrisa GR 177 A7
Vrizy F 19 F10
Vrlika HR 156 E5
Vrnjačka Banja SRB 163 B10
Vron F 18 D4
Vrontados GR 177 C7
Vrontero GR 168 C5
Vrontou GR 169 D7
Vroomshoop NL 17 D7
Vrosina GR 168 E4
Vrouchas GR 179 E10
Vroutek CZ 76 B4
Vrouwenpolder NL 16 E1
Vrpolje HR 156 E5
Vrpolje HR 157 B9
Vrrin AL 168 B1
Vršac SRB 159 C7
Vršani BIH 157 C11
Vrsar HR 67 B8
Vrsi HR 156 D3
Vrtoče BIH 156 C5
Vrtojba SLO 73 E8
Vrulja MNE 163 C7
Vrútky SK 147 C7
Vrutok NMK 163 F10
Vrũv BG 159 E10
Vrysia GR 169 E10
Vrysoula GR 168 F4
Všemina CZ 146 C5
Všeruby CZ 76 C4
Všetaty CZ 77 B7
Všestín CZ 146 C5
Vsevolozhsk RUS 129 E14
Vŭbel BG 160 F5
Vučitrn BIH 157 C8
Vučja Lokva SRB 163 C10
Vučje SRB 164 D4
Vught NL 16 E4
Vŭglevtsi BG 166 D5
Vuka HR 149 F11
Vukova Gorica HR 148 F4
Vukovar HR 157 B11
Vulcan RO 159 C11
Vulcan RO 160 B6
Vulcana-Băi RO 160 C6
Vŭlchedrŭm BG 160 F2
Vŭlchidol BG 167 C9
Vulpeni RO 160 E3
Vultureni RO 152 D3
Vultureni RO 152 D3
Vulturești RO 153 B8
Vulturești RO 153 D11
Vulturești RO 160 D4

Vulturu RO 155 D2
Vulturu RO 161 B10
Vuobmaved FIN 113 E16
Vuohijärvi FIN 128 C6
Vuohtomäki FIN 123 C16
Vuojärvi FIN 117 D17
Vuokatti FIN 121 F11
Vuokko FIN 125 D12
Vuolenkoski FIN 127 C15
Vuolijoki FIN 120 F8
Vuolle FIN 123 C11
Vuollerim S 118 B5
Vuonisjärvi FIN 125 D14
Vuonislahti FIN 125 D13
Vuonos FIN 125 E12
Vuontisjärvi FIN 117 B11
Vuorenmaa FIN 126 C8
Vuorenmaa FIN 126 C8
Vuoreslahti FIN 121 F9
Vuorimäki FIN 123 E11
Vuorimäki FIN 124 C9
Vuoriniemi FIN 129 B11
Vuosaari FIN 127 E13
Vuosanka FIN 121 F13
Vuostimo FIN 115 E2
Vuostimojärvi FIN 115 E2
Vuotso FIN 115 B2
Vuottolahti FIN 119 F17
Vuotunki FIN 121 B14
Vũrbitsa BG 166 E6
Vũrbitsa BG 167 D7
Vurpăr RO 152 F4
Vũrshets BG 165 C7
VurtNop FIN 123 E11
Vutcani RO 153 C11
Vuzlove UA 145 G5
Vvedenka UA 154 E4
Vyalets BY 133 F3
Vyalikaya Byerastavitsa BY 140 D10
Vyalikaya Stayki BY 133 F7
Vyalikaye Syalo BY 133 F3
Vyarkhovichy BY 141 F8
Vyartsilya RUS 125 F15
Vyaz RUS 133 C7
Vyborg RUS 129 D10
Vyčapy-Opatovce SK 146 E6
Východná SK 147 C9
Vydeniai LT 137 E10
Vydrany SK 146 E5
Vyerkhnyadzvinsk BY 133 F4
Vyetryna BY 133 F4
Vyhne SK 147 D7
Vyhoda UA 145 E8
Vy-lès-Lure F 26 E5
Vylkove UA 155 C5
Vylok UA 145 G6
Vynnyky UA 144 D9
Vynohradiv UA 145 G7
Vynohradivka UA 154 F4
Vyritsa RUS 132 C7
Vyronas GR 175 D8
Vyroneia GR 169 B9
Vyshgorodok RUS 133 B3
Vyshka UA 145 F6
Vyshkiv UA 145 E8
Vyshkove UA 145 G7
Vyshneve UA 145 F4
Vyshniv UA 141 H10
Vyskod' RUS 132 F7
Vyškov CZ 77 D12
Vyšné Ružbachy SK 145 E2
Vyšný Mirošov SK 145 E3
Vyšný Orlík SK 145 E4
Vyšný Šipov SK 145 E4
Vysoká nad Kysucou SK 147 C7
Vysokaye BY 141 F8
Vysoké Mýto CZ 77 C10
Vysokoye RUS 136 D4
Vytsotsk RUS 129 D10
Vyšší Brod CZ 76 E6
Vyssinia GR 168 C5
Vytina GR 174 D5
Vyzhnytsya UA 152 A6
Vyžuonos LT 135 E11
Vzmor'ye RUS 139 A9

W

Waabs D 83 A7
Waal D 71 B11
Waalre NL 16 F4
Waalwijk NL 16 E4
Waarschoot B 19 B8
Wabern D 21 F12
Wąbrzeźno PL 138 D6
Wachow D 79 A12
Wachtebeke B 19 B8
Wächtersbach D 21 D12
Wacken D 82 B6
Wackersdorf D 75 D11
Waddeweitz D 83 D9
Waddington UK 11 E10
Waddinxveen NL 16 D3
Wadebridge UK 12 D5
Wädenswil CH 27 F10
Wadern D 21 E7
Wadersloh D 17 E10
Wadgassen D 21 F7
Wadhurst UK 15 E9
Wadlew PL 143 D7
Wadowice PL 147 B9
Wadowice Górne PL 143 F11
Wądroże Wielkie PL 81 D10
Waganiec PL 138 E6
Wagenfeld D 17 C11
Wagenhoff D 79 A8
Wageningen NL 16 E5
Waghäusel D 21 F11
Waging am See D 73 A6
Wagna A 148 C5
Wagrain A 73 B7
Wągrowiec PL 85 E12
Wahlhausen D 79 D6
Wahlsdorf D 80 C4
Wahlstedt D 83 C8
Wahrenholz D 79 A8
Waiblingen D 27 C11
Waibstadt D 187 C6
Waidhaus D 75 C11
Waidhofen an der Thaya A 77 E8
Waidhofen an der Ybbs A 77 G7
Waigolshausen D 75 C7
Waimes B 20 D6
Wain D 71 A10
Wainfleet All Saints UK 11 E12
Wainhouse Corner UK 12 D5

Waischenfeld D 75 C9
Waizenkirchen A 76 F5
Wakefield UK 11 D8
Wakendorf II D 83 C8
Walberswick UK 15 C12
Walburg D 78 D6
Walchum D 17 C8
Walcourt B 19 D9
Wałcz PL 85 D10
Wald CH 27 F10
Wald D 27 E11
Wald D 75 D11
Waldachtal D 27 D10
Waldböckelheim D 185 E8
Waldbreitbach D 185 C7
Waldbröl D 21 C9
Waldbrunn-Lahr D 21 C10
Waldburg D 71 B9
Waldbüttelbrunn D 74 C6
Waldegg A 77 G10
Waldenbuch D 27 C11
Waldenburg CH 27 F8
Waldenburg D 74 D6
Walderbach D 75 D11
Waldershof D 75 C11
Waldesch D 185 D8
Waldfischbach-Burgalben D 21 F9
Waldhausen A 77 E8
Waldhausen im Strudengau A 77 F7
Waldheim D 80 D4
Walding A 76 F6
Waldkappel D 78 D6
Waldkirch D 27 D8
Waldkirchen D 76 E5
Waldkraiburg D 75 F11
Wald-Michelbach D 187 B6
Waldmohr D 21 F8
Waldmünchen D 75 D12
Wałdowo-Szlacheckie PL 138 D6
Waldrach D 21 E7
Waldsassen D 75 B11
Waldsee D 187 C5
Waldshut D 27 E9
Waldstatt CH 27 F11
Waldstetten D 75 F7
Waldstetten D 75 F7
Walenstadt CH 27 F11
Walferdange L 20 E6
Walim PL 81 E10
Walkendorf D 83 C13
Wallasey UK 10 E5
Walldorf D 21 F11
Walldorf D 79 E7
Walldürn D 27 A11
Wallenfels D 75 B9
Wallerfing D 76 E3
Wallern im Burgenland A 149 A7
Wallersdorf D 75 E12
Wallerstein D 75 E7
Wallgau D 72 A3
Wallhausen D 75 D7
Wallhausen D 185 E8
Wallisellen CH 27 F10
Walls UK 3 E11
Wallsbüll D 82 A6
Wallstawe D 83 E10
Walmer UK 15 E11
Wałowice PL 81 C7
Wals A 73 A6
Walsall UK 11 F8
Walschleben D 79 D8
Walsoken UK 11 F12
Walsrode D 82 E7
Waltenhofen D 71 B10
Waltershausen D 79 E8
Walsrode D 82 E7
Waltham on the Wolds UK 11 F10
Walton-on-Thames UK 15 E8
Walton on the Naze UK 15 D11
Wamba E 39 E10
Wamel NL 183 B6
Wanderup D 82 A6
Wandlitz D 84 E4
Wandre B 183 D7
Wang A 77 F8
Wangen D 79 C11
Wangen im Allgäu D 71 B9
Wängi CH 27 E10
Wankendorf D 83 B8
Wanna D 17 A11
Wanne-Eikel D 183 B10
Wanroij NL 183 B7
Wanssum NL 183 B8
Wanzleben D 79 B9
Wapenveld NL 16 D6
Wapiersk PL 139 D7
Waplewo PL 139 D9
Wapnica PL 85 E12
Wapno PL 85 E12
Warberg D 79 B8
Warboys UK 15 C8
Warburg D 17 F12
Warcop UK 11 B7
Warcq F 20 F5
Wardenburg D 17 B10
Wardington UK 13 A12
Ware UK 15 D8
Waregem B 19 C7
Wareham UK 13 D10
Waremme B 19 C11
Waren D 83 C13
Warendorf D 17 E9
Warffum D 17 B7
Warin D 83 C11
Waringstown UK 7 D10
Wark UK 5 E12
Warka PL 141 G4
Warkworth UK 5 E13
Warlingham UK 15 E8
Warloy-Baillon F 18 D6
Warlubie PL 138 C6
Warmeriville F 19 F9
Warmington UK 13 A12
Warminster UK 13 C10
Warmond NL 182 A5
Warmsen D 17 D11
Warnemünde D 83 B12
Warngau D 72 A4
Warnice PL 85 D7
Warnice PL 85 E7
Warnino PL 85 C10
Warniow D 79 E7
Warnsveld NL 16 D6
Warrenpoint UK 7 D10

Warrington UK 10 E6
Warsingsfehn D 17 B8
Warslow UK 11 E8
Warstein D 17 F10
Warszawa PL 141 F4
Warszkowo PL 85 B11
Warta PL 142 C6
Warta Bolesławiecka PL 81 D9
Wartberg an der Krems A 76 G6
Wartenberg D 75 F10
Wartenburg D 79 C11
Wartenberg-Angersbach D 78 E5
Warth A 71 C10
Wartin D 84 D6
Wartkowice PL 143 C7
Wartmannsroth D 74 B6
Warwick UK 13 A11
Warwick Bridge UK 5 F11
Warzyce PL 144 D4
Wasbek D 83 B7
Wasbister UK 3 G10
Wäschenbeuren D 74 E6
Washington UK 5 F13
Wasigny F 19 E9
Wasilków PL 140 D8
Waśniów PL 143 E11
Wąsosz PL 81 C11
Wąsosz PL 140 C6
Wąsowo PL 81 B10
Waspik NL 16 E3
Wasselonne F 27 C7
Wassen CH 71 D7
Wassenaar NL 16 D2
Wassenberg D 20 B6
Wasseralfingen D 75 E7
Wasserbillig D 186 B2
Wasserburg am Inn D 75 F11
Wasserliesch D 186 B2
Wasserlosen D 74 B7
Wassertrüdingen D 75 D8
Wassigny F 19 D8
Wassy F 26 C2
Wasungen D 79 E7
Watchet UK 13 C7
Watchgate UK 10 C6
Waterbeck UK 5 E10
Waterford IRL 9 D9
Watergrasshill IRL 8 D6
Wateringen NL 182 A4
Waterloo B 19 C9
Waterlooville UK 14 F6
Waterville IRL 8 E2
Watford UK 15 D8
Wathlingen D 79 A7
Watlington UK 14 D6
Watten F 18 C5
Watten UK 3 J10
Wattenbek D 83 B8
Wattens A 72 B4
Watten-Scheid D 183 C10
Wattenwil CH 31 B12
Wattlebridge UK 7 D8
Watton UK 15 B10
Wattwil CH 27 F11
Wavre B 19 C10
Wavrin F 18 C6
Wąwelno PL 85 D13
Wąwelno PL 142 E4
Wąwolnica PL 141 H6
Wawrów PL 85 E8
Wawrzeńczyce PL 143 F9
Waxweiler D 20 D6
Waxwelier D 20 D6
Waziers F 182 E2
Weaverham UK 10 E6
Wechmar D 79 E8
Wechselburg D 79 D12
Wedde NL 17 B8
Weddingstedt D 82 B6
Wedel (Holstein) D 82 C7
Wedemore UK 13 C9
Wednesbury UK 11 F7
Weede D 83 C8
Weedon Bec UK 13 A12
Weelde B 182 C5
Weem UK 5 B9
Weener D 17 B8
Weerberg A 72 B4
Weerselo NL 17 D7
Weert NL 16 F5
Weesen CH 27 F11
Weesp NL 16 D4
Weeze D 16 E6
Weferlingen D 79 B9
Wegberg D 20 B6
Wegenstedt D 79 B9
Weggis CH 71 C6
Wegscheid D 76 E5
Wehdel D 17 A11
Wehl NL 16 E6
Wehr D 27 E8
Wehrbleck D 17 C11
Wehringen D 71 A11
Weibern D 183 E10
Weibersbrunn D 74 C5
Weichering D 75 E9
Weichs D 75 F9
Weida D 79 E11
Weidenberg D 75 C10
Weiden in der Oberpfalz D 75 C11
Weidenstetten D 187 D8
Weidenthal D 21 F10
Weiersbach D 186 B3
Weigersdorf D 81 D7
Weihmichl D 75 E11
Weikersheim D 74 D6
Weil D 71 A11
Weil am Rhein D 27 E8
Weilbach D 21 E12
Weilburg D 21 D10
Weil der Stadt D 27 C10
Weilerbach D 186 C4
Weilerswist D 21 C7
Weilheim an der Teck D 74 E6
Weilheim in Oberbayern D 72 A3
Weilmünster D 21 D10
Weimar D 79 E9
Weinbach D 185 D9
Weinböhla D 80 D5
Weinfelden CH 27 E11
Weingarten (Baden) D 187 C6
Weingarten D 71 B9
Weinheim D 21 F11
Weinsberg D 21 F12
Weinsheim D 20 D6
Weinsheim D 21 E9
Weinstadt D 27 C11

Westerburg D 21 C9
Westerdale UK 3 J10
Westeregeln D 79 C9
Westerende-Kirchloog (Ihlow) D 17 B8
Westergate UK 15 F7
Westergellersen D 83 D8
Westerhaar NL 17 D7
Westerholt D 17 A8
Westerhorn D 82 C7
Westerhoven NL 183 C6
Westerland D 86 F2
Westerlo B 19 B10
Westerrönfeld D 82 B7
Westerstede D 17 B9
Westerstetten D 74 E6
Westervoort NL 16 E5
Westgate UK 5 F12
West Grinstead UK 15 F8
West Haddon UK 13 A12
Westhausen D 75 E7
Westhill UK 3 L12
Westhofen D 187 B5
Westhoffen F 186 D3
Westkapelle NL 19 A7
West Kilbride UK 4 D7
West Kirby UK 10 E5
West Knapton UK 11 C10
West Lavington UK 13 C11
Westleton UK 15 C12
West Linton UK 5 D10
West Looe UK 12 E6
West Lulworth UK 13 D10
Westmalle B 16 F3
West Malling UK 15 E9
West Meon Hut UK 13 C12
West Mersea UK 15 D11
West Moors UK 13 D11
Westness UK 3 G10
Westnewton UK 5 F10
Weston UK 11 F7
Weston D 13 D10
Weston-super-Mare UK 13 C9
Westonzoyland UK 13 C9
Westport IRL 6 E3
Westport Quay IRL 6 E3
West Somerton UK 15 B12
West Tarbert UK 4 D6
West-Terschelling NL 16 B4
Westward Ho! UK 12 C6
West Wellow UK 13 D11
West Winch UK 11 F12
Wethau D 79 D10
Wetheral UK 5 F11
Wetherby UK 11 D9
Wetlina PL 145 E5
Wetter (Hessen) D 21 C11
Wetter (Ruhr) D 185 B7
Wetteren B 19 B8
Wetterzeube D 79 D11
Wettin D 79 C10
Wettingen CH 27 F9
Wettrup D 17 C9
Wetzikon CH 27 F10
Wetzlar D 21 C11
Wevelgem B 19 C7
Wewelsfleth D 17 A12
Wexford IRL 9 D10
Weyarn D 72 A4
Weybridge UK 15 E8
Weyerbusch D 21 C9
Weyer Markt A 73 A10
Weyersheim F 27 C8
Weyhausen D 79 B8
Weyhe D 17 C11
Weymouth UK 13 D10
Whalton UK 5 E13
Whauphill UK 4 F8
Wheddon Cross UK 13 C7
Wherwell UK 13 C12
Whiddon Down UK 13 D7
Whitburn UK 5 D9
Whitby UK 11 C10
Whitchurch UK 10 F6
Whitchurch UK 13 B8
Whitchurch UK 13 D7
Whitecross UK 7 D10
Whitegate IRL 8 C6
Whitegate IRL 8 E6
Whitehall UK 3 G11
Whitehall UK 9 C8
Whitehaven UK 10 B4
Whitehead UK 4 F5
Whitehill UK 15 E7
Whitehills UK 3 K11
Whitfield UK 15 E7
Whithorn UK 5 F8
Whiting Bay UK 4 E6
Whitland UK 12 B5
Whitley Bay UK 5 E14
Whitstable UK 15 E11
Whittington UK 10 F5
Whittlesey UK 11 F11
Wiartel PL 139 C12
Wiatrowo PL 85 E12
Wiązów PL 81 E12
Wiązowna PL 141 F4
Wiązownica PL 144 C6
Wichelen B 19 B8
Wick UK 3 J10
Wick UK 13 C9
Wick UK 5 F7
Wickede (Ruhr) D 185 B8
Wickenrode (Helsa) D 78 D6
Wickford UK 15 D10
Wickham UK 13 D12
Wickham Market UK 15 C11
Wicklow IRL 7 G10
Wicko PL 85 A13
Widawa PL 143 D6
Widdern D 72 A3
Widecombe in the Moor UK 13 D7
Widnau CH 71 C9
Widnes UK 10 E6
Widuchowa PL 84 D6
Wijcbork PL 85 D12
Wieck am Darß D 83 B13
Wieczfnia Kościelna PL 139 D9
Wieda D 79 C7
Wiedensahl D 17 D12
Wiederau D 80 D3
Wiednitz D 80 D6
Wiefelstede D 17 B10
Wiehe D 79 D9
Wiehl D 21 C9
Wiek D 84 A4
Wijkszyce PL 142 F5
Wielbark PL 139 D10
Wieleń PL 85 E10
Wielenbach D 72 A3
Wielgie PL 139 E7
Wielgomłyny PL 143 D8

Wielichowo PL 81 B10
Wieliczka PL 143 G9
Wieliczki PL 136 F6
Wielka Wieś PL 143 F8
Wielkie Oczy PL 144 C7
Wielopole Skrzyńskie PL 143 G12
Wielowieś PL 142 E6
Wielsbeke B 182 D2
Wieluń PL 142 D6
Wiemersdorf D 83 C7
Wien A 77 F10
Wienerbruck A 77 G8
Wiener Neustadt A 77 G10
Wieniawa PL 141 H3
Wieniec PL 138 E6
Wienrode D 79 C8
Wiepke D 79 A9
Wieprz PL 147 B8
Wieprz PL 147 B8
Wiercień Duży PL 141 F7
Wierden NL 17 D7
Wieren D 83 E9
Wieringerwerf NL 16 C4
Wiernsheim D 187 D6
Wieruszów PL 142 D5
Wierzawice PL 144 C5
Wierzbica PL 141 H4
Wierzbica PL 141 H8
Wierzbica Górna PL 142 D4
Wierzbijcin PL 85 C8
Wierzbinek PL 138 F6
Wierzbnik PL 142 E3
Wierzbno PL 139 F12
Wierzbowa PL 81 D9
Wierzchlas PL 142 D6
Wierzchosławice PL 138 E5
Wierzchosławice PL 143 F10
Wierzchowo PL 85 C11
Wierzchowo PL 85 D10
Wierzchucino PL 138 A5
Wies A 148 C4
Wiesa D 80 E3
Wiesau D 75 C11
Wiesbaden D 21 D10
Wiesbaden A 77 F8
Wiesen A 149 A6
Wiesen CH 71 D9
Wiesen D 74 B5
Wiesenburg D 79 B11
Wiesenfelden D 75 D12
Wiesensteig D 187 D8
Wiesent D 75 D11
Wiesenthau D 75 C9
Wiesentheid D 75 C7
Wiesloch D 21 F11
Wiesmath A 148 A6
Wiesmoor D 17 B8
Wieszowa PL 142 F6
Wietmarschen D 17 C8
Wietze D 78 A6
Wietzen D 17 C12
Wietzendorf D 83 E7
Wiewiórczyn PL 143 C7
Wigan UK 10 D6
Wiggensbach D 71 B10
Wigierska Górka PL 147 B8
Wigliniec PL 81 D8
Wigmore UK 13 A9
Wigorzewo PL 136 E4
Wigorzyno PL 85 C9
Wigrów PL 139 F13
Wigrzynice PL 81 B8
Wigrzynowo PL 139 F9
Wigston UK 11 F9
Wigton UK 5 F10
Wigtown UK 5 F8
Wijchen NL 16 E5
Wijewo PL 81 C10
Wijhe NL 183 A8
Wijk aan Zee NL 16 D3
Wijk bij Duurstede NL 16 E4
Wijk en Aalburg NL 183 B6
Wil CH 27 F11
Wilamowice PL 147 B8
Wilczyța PL 139 B8
Wilczogóra PL 138 F5
Wilczyn PL 138 F5
Wildalpen A 73 A10
Wildau D 80 B5
Wildberg D 27 C10
Wildberg D 83 E13
Wildberg D 84 C4
Wildeck-Richelsdorf D 79 D7
Wildendürnbach A 77 E10
Wildenfels D 79 E12
Wildeshausen D 17 C10
Wildflecken D 74 B6
Wildon A 148 C5
Wildpoldsried D 71 B10
Wildshut A 76 F3
Wilga PL 141 G4
Wilhelminadorp NL 183 C6
Wilhelmsburg A 77 F9
Wilhelmsburg D 84 C5
Wilhelmsdorf D 27 E11
Wilhelmshaven D 17 A10
Wilkau-Haßlau D 79 E12
Wilkinstown IRL 7 E9
Wilkołaz Pierwszy PL 144 A5
Wilków PL 141 H5
Wilków PL 142 D4
Wilkowice PL 147 B8
Wilkowo PL 81 B8
Willand UK 13 D8
Willebadessen D 17 E12
Willebroek B 19 B9
Willemstad NL 16 E2
Willich D 17 F7
Willingdon UK 15 F9
Willingen (Upland) D 17 F11
Willingshausen D 21 C12
Willisau CH 27 F9
Williton UK 13 C8
Willstätt D 27 D8
Wilmersdorf D 84 D5
Wilmslow UK 11 E7
Wilnis NL 182 A5
Wilnsdorf D 21 C10
Wilp NL 183 A8
Wilsdruff D 80 D5
Wilsickow D 84 D5
Wilster D 82 C6
Wiltersdorf A 77 E11
Wiltingen D 21 E7
Wilton UK 13 C11
Wiltz L 20 E5
Wimblington UK 11 F12
Wimborne Minster UK 13 D11
Wimereux F 15 F12
Wimmelburg D 79 C10

Wimmis CH 70 D5
Wimpassing A 148 A6
Wincanton UK 13 C10
Winchburgh UK 5 D10
Winchelsea UK 15 F10
Wincheringen D 186 B1
Winchester UK 13 C12
Wincrange L 20 D5
Winda PL 136 E3
Windach D 71 A12
Windeby D 83 B7
Windermere UK 10 C6
Windesheim D 21 E9
Windhagen D 21 C9
Windigsteig A 77 E8
Windischeschenbach D 75 C11
Windischgarsten A 73 A9
Windorf D 76 E4
Windsbach D 75 D8
Windsor UK 15 E7
Wingate UK 5 F14
Wingene B 19 B7
Wingen-sur-Moder F 27 C7
Winklarn D 75 D11
Winkleigh UK 13 D7
Winklern A 73 C6
Winklern bei Oberwölz A 73 B9
Winnenden D 27 C11
Winnert D 82 B6
Winnica PL 139 E10
Winningen D 79 C9
Winningen D 185 D8
Winnweiler D 186 B4
Winschoten NL 17 B8
Winsen (Aller) D 83 E7
Winsen (Luhe) D 83 D8
Winsford UK 10 E6
Wińsko PL 81 D11
Winslow UK 15 D7
Winssen NL 183 B7
Winston UK 11 B8
Winsum NL 16 B5
Winterberg D 21 B11
Winterbourne Abbas UK 13 D9
Winterfeld D 83 E9
Winterswijk NL 17 E7
Winterthur CH 27 E10
Winterton UK 11 D10
Winton UK 11 C7
Wintrich D 185 E6
Wintzenheim F 27 D7
Winwick UK 15 C8
Winzenburg D 78 C6
Winzer D 76 E4
Wipperdorf D 79 D8
Wipperfürth D 21 B8
Wippra Kurort D 79 C9
Wirdumer (Wirdum) D 17 B8
Wirges D 185 D8
Wirksworth UK 11 E8
Wirsberg D 75 B10
Wisbech UK 11 F12
Wischhafen D 17 A12
Wishaw UK 5 D9
Wiskitki PL 141 F2
Wisła PL 147 B7
Wiślica PL 143 F10
Wismar D 83 C10
Wiśniew PL 141 F6
Wiśniowa PL 139 D9
Wiśniowa PL 144 D1
Wiśniowa PL 144 D4
Wissant F 15 F12
Wissembourg F 27 B8
Wissen D 21 C9
Wissenkerke NL 16 E1
Wistedt D 82 D7
Wisznia Mała PL 81 D12
Wisznice PL 141 G8
Witaszyce PL 142 C4
Witham UK 15 D10
Witheridge UK 13 D7
Withernsea UK 11 D12
Witkowo PL 85 D8
Witkowo PL 138 F4
Witley UK 15 E7
Witmarsum NL 16 B4
Witney UK 13 B12
Witnica PL 84 E6
Witnica PL 85 E7
Witonia PL 143 B7
Witosław PL 85 D12
Witry-lès-Reims F 19 F9
Wittdün D 82 A4
Witten D 17 F8
Wittenbach CH 27 F11
Wittenberge D 83 E11
Wittenburg D 83 C10
Wittenförden D 83 C10
Wittenhagen D 84 B4
Wittenheim F 27 E7
Wittibreut D 76 F3
Wittichenau D 80 D6
Wittighausen D 74 C6
Wittingen D 83 E9
Wittlich D 21 E7
Wittmar D 79 B8
Wittmund D 17 A9
Wittstock D 83 D12
Witzenhausen D 78 D6
Witzin D 83 C11
Witzwort D 82 B5
Wiveliscombe UK 13 C8
Wivenhoe UK 15 D11
Wiżajny PL 136 E6
Władysławów PL 142 B5
Władysławowo PL 138 A5
Wleń PL 81 E9
Włocławek PL 138 E7
Włodawa PL 141 G9
Włodzienin PL 142 F4
Włodzimierzów PL 143 D8
Włostów PL 143 E11
Włoszakowice PL 81 C10
Włoszczowa PL 143 E8
Wöbbelin D 83 D11
Woburn UK 15 D7
Wodynie PL 141 F5
Wodzierady PL 143 C7
Wodzisław PL 143 E9
Wodzisław Śląski PL 142 F5
Woensdrecht NL 182 C4
Wœrth F 27 C8
Wohlen CH 27 F9
Wohlen CH 31 B11
Wohlmirstedt D 79 D9
Wöhrden D 82 B6
Wohyń PL 141 G7
Woippy F 26 B5

Wojaszówka PL 144 D4
Wojcieszków PL 141 G6
Wojcieszów PL 81 E9
Wójcin PL 138 E5
Wojkowice PL 143 F7
Wojnicz PL 143 G10
Wojnowo PL 138 D4
Wojsławice PL 144 B8
Wola PL 138 E7
Wola Mołodycka PL 144 C6
Wola Mysłowska PL 141 G5
Wolanów PL 141 H3
Wola-Rjbkowska PL 141 G5
Wola Uhruska PL 141 H9
Wola Wierzbowska PL 139 E10
Wolbórz PL 141 H1
Wolbrom PL 143 F8
Wołczyn PL 142 D5
Woldegk D 84 D5
Woldendorp NL 17 B8
Wolfach D 27 D9
Wolfegg D 71 B9
Wolfen D 79 C11
Wolfenbüttel D 79 B8
Wolfenschiessen CH 71 D6
Wölfersheim D 21 D11
Wolfhagen D 17 F12
Wölfis D 79 E8
Wolfpassing A 77 F8
Wolfratshausen D 72 A3
Wolfsberg A 73 C10
Wolfsburg D 79 B8
Wolfstein D 21 E9
Wolfurt A 71 C9
Wolgast D 84 C5
Wolhusen CH 70 C6
Wolin PL 84 C7
Wólka PL 139 E7
Wólka PL 141 H7
Wólka Dobryńska PL 141 F8
Wolkersdorf A 77 F11
Wollbach D 75 B7
Wolleau CH 27 F8
Wolmirstedt D 79 B10
Wolnzach D 75 E10
Wołomin PL 139 F11
Woloskowola PL 141 H8
Wołów PL 81 D11
Wolpertshausen D 74 D6
Wolpertswende D 71 B9
Wolphaartsdijk NL 182 B3
Wölpinghausen D 17 D12
Wolsingham UK 5 F13
Wolsztyn PL 81 B10
Woltersdorf D 80 B4
Woltersdorf D 83 E13
Wolvega NL 16 C6
Wolverhampton UK 11 F7
Wombwell UK 11 D9
Wommelgem B 19 B10
Wommels NL 16 B5
Wonfurt D 75 B7
Wonsees D 75 C9
Woodbridge UK 15 C11
Woodcote UK 13 B12
Woodford IRL 6 F6
Woodhall Spa UK 11 E11
Woodland UK 5 F13
Woodley UK 15 E7
Woodstock UK 13 B12
Woodton UK 15 C11
Woodtown IRL 7 E9
Wool UK 13 D10
Wooler UK 5 D12
Wootton Bassett UK 13 B11
Worb CH 31 B12
Worbis D 79 D7
Worcester UK 13 A10
Wördern A 77 F10
Worfield UK 11 F7
Wörgl A 72 B5
Workington UK 5 F9
Worksop UK 11 E9
Workum NL 16 C4
Wörlitz D 79 C11
Wormeldange L 20 E6
Wormerveer NL 16 D3
Wormhout F 18 C5
Worms D 21 E10
Worpswede D 17 B11
Wörrstadt D 21 E10
Wörschach A 73 A9
Wortel B 182 C5
Wörth A 73 B6
Worth UK 15 E8
Wörth am Main D 187 B7
Wörth am Rhein D 27 B9
Worthing UK 15 F8
Woudenberg NL 183 A6
Woudrichem NL 16 E3
Woudsend NL 16 C5
Woustviller F 27 B7
Wouw NL 16 E2
Woźniki PL 143 E7
Wragby UK 11 E11
Wrangle UK 11 E12
Wrecsam UK 10 E6
Wrjczyca Wielka PL 143 E6
Wredenhagen D 83 D13
Wrelton UK 11 C10
Wremen D 17 A11
Wrestedt D 83 E9
Wrexham UK 10 E6
Wriedel D 83 D8
Wriezen D 84 E6
Wrist D 82 C7
Wróblew PL 142 C6
Wrocław PL 81 D12
Wrohm D 82 B6
Wronki PL 85 E10
Wronki PL 136 E5
Wrotnów PL 139 E7
Wrząielka PL 142 B6
Wrzelowiec PL 144 A4
Wrzeście PL 85 A12
Września PL 85 E11
Wrzosowo PL 85 B9
Wschowa PL 81 C10
Wulfen D 83 D12
Wülfershausen an der Saale D 75 B7
Wulfsen D 83 D8
Wulften D 79 C7
Wülknau D 83 C11
Wullersdorf A 77 E10
Wundschuh A 148 C4
Wünnenberg D 17 E11
Wünsdorf D 80 B4
Wunsiedel D 75 B11
Wunstorf D 78 B5

Wuppertal D 17 F8
Wurmlingen D 27 D10
Würselen D 20 C6
Wurzbach D 75 B10
Würzburg D 74 C6
Wurzen D 79 D12
Wüstenrot D 27 B11
Wusterhausen D 83 E12
Wusterhusen D 84 B5
Wustermark D 79 A12
Wusterwitz D 79 B11
Wustrau-Altfriesack D 84 E3
Wustrow D 83 E10
Wustrow D 84 D3
Wustrow D 83 B12
Wutha D 79 E7
Wutöschingen D 27 E9
Wuustwezel B 16 F3
Wyczechy PL 85 C13
Wydminy PL 136 F5
Wye UK 15 E10
Wygoda PL 143 E7
Wygon PL 85 D9
Wyhl D 27 D8
Wyk auf Föhr D 82 A5
Wylatowo PL 138 E4
Wylye UK 13 C11
Wymondham UK 15 B11
Wyre Piddle UK 13 A10
Wyrozby-Konaty PL 141 F6
Wyrzysk PL 85 D12
Wyśmierzyce PL 141 G3
Wysoka PL 85 D12
Wysoka PL 85 E8
Wysoka Kamieńska PL 85 C7
Wysokie PL 136 F6
Wysokie PL 144 B6
Wysokie Mazowieckie PL 141 E7
Wysowa PL 145 E3
Wyszanów PL 142 D5
Wyszki PL 141 E7
Wyszków PL 139 E11
Wyszogród PL 139 F9
Wyszyny PL 85 E11
Wythall UK 13 A11
Wytyczno PL 141 H8

X

Xanceda E 38 B3
Xanten D 17 E6
Xanthi GR 171 B7
Xarrë AL 168 E3
Xàtiva E 56 D3
Xeraco E 56 C4
Xeresa E 56 C4
Xermade E 38 B4
Xerta E 42 F5
Xertigny F 26 D5
Xifiani GR 169 C7
Xino Nero GR 169 C6
Xinzo de Limia E 38 D4
Xirivella E 48 F4
Xirochori GR 174 D4
Xirokampi GR 174 F5
Xirokampo GR 177 E8
Xiropotamos GR 170 B6
Xixón E 39 A8
Xonrupt-Longemer F 27 D6
Xove E 38 A4
Xunqueira de Ambía E 38 D4
Xylagani GR 171 C8
Xylokastro GR 175 C6
Xyloupoli GR 169 C9

Y

Yablanitsa BG 165 C9
Yablanovo BG 167 D7
Yabluniv UA 152 A5
Yabŭlchevo BG 167 D8
Yabŭlkovo BG 166 E4
Yağci TR 173 C7
Yağcilar TR 173 F9
Yağcili TR 173 F9
Yagoda BG 166 D5
Yakimovo BG 160 F2
Yakoruda BG 165 E8
Yakovo BG 169 A9
Yaliçiftlik TR 177 E9
Yalikavak TR 177 E9
Yalova TR 171 D10
Yambol BG 167 E7
Yamkino RUS 132 E7
Yamm RUS 132 E3
Yampil' UA 154 A2
Yangi TR 181 C9
Yanguas E 41 D7
Yanikağil TR 173 B8
Yaniskoski RUS 114 F5
Yankavichy BY 133 E5
Yankovo BG 167 C8
Yantarnyy RUS 139 A9
Yapildak TR 172 D6
Yaraş TR 181 B8
Yareva BY 135 F13
Yarlovo BG 165 E7
Yarm UK 11 B9
Yarmouth UK 13 D12
Yarnton UK 13 B12
Yarove UK 154 E4
Yarrow UK 5 D10
Yasen BG 165 C10
Yasen' UA 145 F9
Yasenkovo BG 167 B7
Yasenovets BG 161 F9
Yasinya UA 152 A4
Yasna Polyana BG 167 E9
Yasnoye RUS 134 F5
Yassiören TR 173 B10
Yasski RUS 132 F7
Yatağan TR 181 B8
Yate UK 13 B10
Yatton UK 13 C9
Yavora UA 145 E7
Yavoriv UA 144 D7
Yavoriv UA 152 A5
Yaxley UK 11 F11
Yaylagöne TR 172 C6
Yazibaşı TR 177 C9
Yaziköy TR 181 C6
Yazna BY 133 F7
Yazvina BY 133 F7
Ybbs an der Donau A 77 F8
Ybbsitz A 77 G7
Ychoux F 32 B4
Ydby DK 86 B2
Ydes F 29 E10

Ydra GR 175 E7
Ydrefors S 92 D7
Y Drenewydd UK 10 F5
Yeadon UK 11 D8
Yealmpton UK 12 E7
Yebra E 47 D7
Yebra de Basa E 32 F5
Yecla E 56 D2
Yecla de Yeltes E 45 C8
Yedy BY 133 F5
Yémeda E 47 E9
Yenice TR 171 C10
Yenice TR 173 C10
Yenice TR 173 E8
Yenice TR 173 E10
Yenice TR 181 B8
Yeniçiftlik TR 173 B8
Yeniçiftlik TR 173 D7
Yeniçiftlik TR 177 C10
Yenifoça TR 177 B8
Yenihisar TR 177 E9
Yeniköy TR 173 C7
Yeniköy TR 173 E6
Yeniköy TR 173 E8
Yeniköy TR 177 B8
Yeniköy TR 177 B10
Yeniköy TR 177 C10
Yeniköy TR 177 D10
Yeniköy TR 177 E7
Yeniköy TR 177 E9
Yeniköy TR 181 B8
Yenimuhacirköy TR 172 C6
Yenişakran TR 177 B9
Yenne F 31 D8
Yeovil UK 13 D9
Yepes E 46 E5
Yerkesik TR 181 B8
Yerseke NL 16 F2
Yershovo RUS 132 F3
Yerville F 18 E2
Yeşildere E 32 E13
Yeşilsirt TR 173 D7
Yeşilyurt TR 181 B8
Yeste E 55 C8
Yevtodiya UA 154 B4
Yezyaryshcha BY 133 F9
Y Fenni UK 13 B8
Ygos-St-Saturnin F 32 C4
Ygrande F 30 B2
Ykspihlaja FIN 123 C10
Ylakiai LT 134 D3
Ylåköngås FIN 113 D17
Ylä-Luosta FIN 125 E11
Ylämaa FIN 129 D9
Ylämylly FIN 125 E13
Yläne FIN 126 D7
Ylä-Valtimo FIN 125 C11
Ylihärmä FIN 122 D9
Yli-Ii FIN 119 D15
Ylijärvi FIN 128 D8
Yli-Kärppä FIN 119 C15
Ylikiiminki FIN 119 D16
Yli-Körkkö FIN 119 C12
Ylikulma FIN 127 E9
Yli-Kurki FIN 121 D10
Ylikylä FIN 115 E2
Ylikylä FIN 117 E15
Ylikylä FIN 119 B13
Ylikylä FIN 123 C10
Ylikylä FIN 125 C12
Yli-Kyrö FIN 117 B13
Yli-Lesti FIN 123 D13
Yli-Liakka FIN 119 C12
Yli-Livo FIN 120 C9
Yli-Muonio FIN 117 B10
Yli-Nampa FIN 119 B16
Ylinenjärvi FIN 119 B13
Yli-Olhava FIN 119 C15
Ylipää FIN 119 E12
Ylipää FIN 119 E13
Ylipää FIN 119 E13
Ylipää FIN 119 E14
Ylipää FIN 119 E13
Ylipää FIN 123 D11
Yli-Paakkola FIN 119 B13
Yli-Siurua FIN 119 C17
Ylistaro FIN 122 E9
Yli-Tannila FIN 119 D16
Ylitornio FIN 119 B12
Yli-Utos FIN 120 E8
Yli-Valli FIN 122 F8
Ylivieska FIN 123 B13
Yli-Vuotto FIN 119 E17
Ylläsjärvi FIN 117 C12
Ylöjärvi FIN 127 B10
Ylönkylä FIN 126 E8
Ylvingen N 108 E3
Ymonville F 24 D6
Yngsjö S 88 D6
Yogsjö S 87 E10
Yolageldi TR 173 A6
Yonkovo BG 161 F9
York UK 11 D9
Youghal IRL 9 E7
Yovkovo BG 155 F2
Yoxford UK 15 C12
Ypäjä FIN 127 D9
Ypäjänkylä FIN 127 D9
Ypati GR 174 B5
Yport F 18 E1
Yppäri FIN 119 F12
Ypsos GR 168 E2
Ypyä FIN 123 C13
Yrittäperä FIN 121 D10
Yrttivaara S 116 E7
Ysane S 88 C7
Ysbyty Ystwyth UK 13 A7
Yset N 101 A12
Ysselsteyn NL 183 C7
Yssingeaux F 30 E5
Ystad S 87 E13
Ystalyfera UK 13 B7
Ytre Arna N 94 B2
Ytre Ärnes N 111 B15
Ytre Billefjord N 113 C15
Ytre Kjæs N 113 B16
Ytre Sandvik N 113 C16
Ytterån S 105 E16
Ytterberg S 102 B7
Ytteresse FIN 123 C9
Ytterhogdal S 102 B8
Ytterjeppo FIN 122 D9
Yttermalung S 102 E6
Yttersta S 118 D6
Yttertträsk S 118 F4
Ytterturingen S 103 B9
Yttervik S 109 E18
Yttre Lansjärv S 116 E8
Yukaribey TR 177 B9
Yukarikizilca TR 177 C9

Yukhavichy BY 133 D5
Yuncler E 46 D5
Yuncos E 46 D5
Yundola BG 165 E8
Yunquera E 53 C7
Yunquera de Henares E 47 C6
Yuper BG 161 F8
Yuratsishki BY 137 E12
Yuravichy BY 133 F5
Yürüceker TR 173 E10
Yürük TR 173 C7
Yutz F 20 F6
Yuzhnyy RUS 136 D2
Yverdon CH 31 B10
Yvetot F 18 E2
Yvignac F 23 D7
Yvoir B 19 D10
Yvoire F 31 C9
Yvonand CH 31 B10
Yxnerum S 93 C8
Yxsjö S 107 C13
Yxsjöberg S 97 B12
Yzeure F 30 B3
Yzeures-sur-Creuse F 29 B7

Z

Zaamslag NL 182 C3
Zaandam NL 16 D3
Zaazyer"ye BY 133 E5
Zäbala SRB 158 C5
Zabalj SRB 158 C5
Zabari SRB 159 E7
Zabełtitz-Treugeböhla D 80 D5
Žabia Wola PL 141 F3
Zabica BIH 162 D5
Zabiče SLO 67 A9
Zabiele PL 139 D12
Zabierzów PL 143 F8
Ząbki PL 141 F4
Ząbkowice Śląskie PL 81 E11
Zablaće HR 156 E3
Žablak MNE 163 D7
Zabłudów PL 140 D8
Žabno HR 149 E7
Žabno HR 149 F10
Žabno PL 144 B6
Zabok HR 148 D5
Žabokreky SK 147 C7
Žabokreky nad Nitrou SK 146 D6
Zabolova LV 133 A2
Zabór PL 81 C9
Zabor"ye BY 133 E6
Zabów PL 85 D7
Zabowo PL 85 D8
Zăbrani RO 151 E8
Zabrđe BIH 157 C10
Zábřeh CZ 77 C11
Zăbriceni MD 153 A10
Zabrze PL 142 F6
Zaburdo BG 165 F10
Zabuzhzhya UA 141 H9
Zacharo GR 174 E4
Zaclǎu RO 155 C2
Žacléř CZ 81 E9
Zaczopki PL 141 F8
Zadar HR 156 D3
Zadunayivka UA 154 F4
Zádveřice CZ 146 C5
Zadzim PL 142 C6
Zadzyezhzha BY 133 E4
Zafarraya E 53 C8
Zafferana Etnea I 59 D7
Zafirovo BG 161 F9
Zafra E 51 C7
Zafra de Záncara E 47 E7
Zafrilla E 47 D9
Žaga SLO 73 D7
Zagar RO 152 E5
Žagarė LT 134 D6
Zagarise I 59 A10
Zaglik VR 67 E11
Zaglavlieri GR 169 C9
Žaglav HR 67 E11
Zagnańsk PL 143 E10
Zagon RO 153 F8
Zagorčane E 47 C6
Zagorë AL 163 E8
Zagorje SLO 73 D9
Zagorje ob Savi SLO 73 D10
Zagórów PL 142 B4
Zagortsi BG 167 E8
Zagórz PL 145 D5
Zagra E 53 B8
Zagra RO 152 C4
Zagrazhden BG 160 F5
Zagrazhden BG 166 F3
Zagreb HR 148 E5
Zagrodno PL 81 D9
Žagrović HR 156 D5
Zagvozd HR 157 F7
Žagubica SRB 159 E8
Zagvozd RO 153 D11
Zagvarékas H 150 C5
Zahara E 51 F9
Zahara de los Atunes E 52 D5
Zahinos E 51 C6
Zahna D 79 C12
Zahnitkiv UA 154 A3
Záhony H 145 G5
Záhorská Ves SK 77 F11
Zahor"ye BY 133 E9
Zăicani MD 153 B10
Zaiceva LV 133 B2
Zaidín E 42 D4
Žaiginys LT 134 F6
Zaim MD 154 D4
Zajas NMK 168 A4
Zaječar SRB 159 F9
Zaječí CZ 77 E11
Zajȩczów CZ 76 C5
Zajezierze PL 85 C9
Zákamenné SK 147 C8
Zákány H 149 D7
Zákányszek H 150 E4
Zakharnichy BY 133 E5
Zakliczyn PL 144 D1
Zakliców PL 144 B5
Zakłopača BIH 157 D11
Zakoma BIH 157 E10
Zakopane PL 147 C9
Zakros GR 179 E11
Zakrzew PL 141 H4
Zakrzewo PL 85 D12
Zakrzewo PL 85 E11
Zakrzówek-Wieś PL 144 B5
Zákupy CZ 81 E7
Zakynthos GR 174 D2
Zalaapáti H 149 C8
Zalaegerszeg H 149 C8
Zalakaros H 149 C8

Zalakomár H 149 C8
Zalalővő H 149 C7
Zalamea de la Serena E 51 B8
Zalamea la Real E 51 D6
Zalaszántó H 149 C8
Zalaszentbalázs H 149 C7
Zalaszentgrót H 149 C8
Zalaszentiván H 149 C7
Zalău RO 151 C11
Zalavár H 149 C8
Zalavas LT 137 D13
Załazy PL 141 H5
Zaldibar E 41 B6
Žalec SLO 73 D11
Zaļenieki LV 134 C7
Zalesie PL 85 E13
Zalesie PL 141 F8
Zalesie Śląskie PL 142 F5
Zalesje LV 133 C3
Zales'ye RUS 136 D4
Zaleszany PL 144 B4
Zalewo PL 139 C8
Zalha RO 152 C2
Zälīte LV 135 C8
Zalinovye RUS 136 D2
Zall-Dardhë AL 163 F9
Žalno PL 138 C4
Zalogovac SRB 159 F7
Zaltbommel NL 16 E4
Załuski PL 139 E9
Zalužje BIH 158 E3
Zalužnica HR 156 C3
Zalve LV 135 D10
Zalyessye BY 133 E13
Zalyessye BY 137 E13
Zam RO 151 E10
Zamárdi H 149 C9
Zamarte PL 85 C12
Žamberk CZ 77 B10
Zambrana E 40 C6
Zâmbreasca RO 160 E6
Zambrów PL 140 E6
Zambujal de Cima P 50 C1
Zambujeira do Mar P 50 D2
Zamcin PL 85 D9
Zamfirovo BG 165 C7
Zamogil'ye RUS 132 D4
Zámoly H 149 B10
Zamora E 39 E8
Zamość PL 139 C10
Zamość PL 144 B7
Zamoshsha BY 133 F2
Zamostea RO 153 B7
Zams A 71 C11
Zámutov SK 145 F4
Zandhoven B 19 B10
Žandov CZ 80 E6
Zandvliet B 16 F2
Zandvoort NL 16 D3
Zănești RO 153 D9
Zaniemyśl PL 81 B12
Zante LT 134 C5
Zaorejas E 47 C8
Zaovine SRB 158 F3
Zapałów PL 144 C6
Zaplanik BIH 162 D5
Zaplyus'ye RUS 132 E6
Zăpodeni RO 153 D11
Zapolice PL 142 C6
Zapolyarnyy RUS 114 E9
Zapol'ye RUS 132 E6
Zaporozhskoye RUS 129 D14
Zappeio GR 169 F7
Zapponeta I 60 A5
Zaprešić HR 148 E5
Zapruddzye BY 133 F3
Zaptyiv UA 154 B2
Zaragoza E 41 E10
Zărand RO 151 E8
Zarasko PL 85 C9
Zarasai LT 135 E12
Zaratamo E 40 B6
Zaratán E 39 E10
Zarautz E 32 D1
Zarcilla de Ramos E 55 D9
Zarjby PL 139 E13
Zarjby-Kościelne PL 139 E13
Zarechcha BY 133 E7
Žarėnai LT 134 E4
Žarki PL 143 E7
Žarki Wielkie PL 81 C7
Zărnești RO 160 B6
Zărnești RO 161 C9
Żarnovica SK 147 D7
Żarnów PL 141 H2
Żarnowiec PL 143 F8
Zaronava BY 133 F7
Zaros GR 178 E8
Žarošice CZ 77 D11
Żarów PL 81 E10
Zarpen D 83 C9
Zarra E 47 F10
Zarrentin D 83 C9
Zarszyn PL 145 D5
Żary PL 81 C8
Zarza Capilla E 51 B9
Zarza de Alange E 51 B7
Zarza de Granadilla E 45 D8
Zarza de Tajo E 47 D6
Zarzadilla de Totana E 55 D9
Zarza la Mayor E 45 E7
Zarzecze PL 144 D6
Zarzuela del Monte E 46 C4
Zarzuela del Pinar E 40 F3
Zas E 38 B2
Zasa LV 135 D11
Zaskarki BY 133 F5
Žaškov SK 147 C8
Žasliai LT 137 D10
Zásmuky CZ 77 C8
Zasów PL 143 F11
Zastražišće HR 157 F6
Žatec CZ 76 B5
Zaton HR 156 E4
Zaton HR 162 D5
Zátor CZ 142 F4
Zator PL 143 G7
Zatory PL 139 E11
Zătreni RO 160 D3
Zatyshshya UA 154 C5
Zaube LV 135 C10
Zau de Câmpie RO 152 D4
Zavadka UA 145 E7
Zavala BIH 162 D4
Zavalla HR 157 C6
Zavallya UA 153 A7
Zavallya UA 154 A3
Zavattarello I 37 B10

Závažná Poruba SK 147 C9
Zaventem B 19 C9
Zavet BG 161 F9
Zavidovič BIH 157 D9
Zavlaka SRB 158 E3
Závoaia RO 161 D10
Závod SK 77 E12
Závoi RO 159 B9
Zavoj SRB 165 C6
Zavutstsye BY 133 F4
Zavyachellye BY 133 F5
Zavydovychi UA 144 D8
Zawada PL 81 C9
Zawada PL 141 G1
Zawada PL 142 E4
Zawada PL 143 E7
Zawada PL 144 B7
Zawady PL 140 D7
Zawadzkie PL 142 E5
Zawichost PL 144 B4
Zawidów PL 81 D8
Zawidz Kościelny PL 139 E8
Zawiercie PL 143 F7
Zawoja PL 147 B8
Zawonia PL 81 D12
Zaytsevo RUS 132 F4
Žažina HR 148 E6
Zázrivá SK 147 C8
Žažvic HR 156 E4
Zbąszyn PL 81 B9
Zbąszynek PL 81 B9
Zbehy SK 146 E6
Zberoaia MD 153 D12
Zbiczno PL 139 D7
Zbiersk PL 142 C5
Zbiroh CZ 76 C5
Zblewo PL 138 C5
Zbludowice PL 143 F10
Zbójna PL 139 D12
Zbójno PL 139 D7
Zborov SK 145 E3
Zborovice CZ 146 C4
Zborov nad Bystricou SK 147 C7
Zbraslav CZ 77 D10
Zbraslavice CZ 77 C8
Zbrzeznica PL 140 D6
Zbuczyn Poduchowny PL 141 F6
Ždala HR 149 D8
Ždánice CZ 77 D12
Žďár CZ 77 C9
Žďár nad Sázavou CZ 77 C9
Zdenci HR 149 E9
Ždiar SK 145 E1
Zdice CZ 76 C5
Zdihovo HR 67 B11
Zďkov CZ 76 D5
Žďírec nad Doubravou CZ 77 C9
Zdounky CZ 146 C4
Ždralovac BIH 157 E6
Zdravets BG 167 C9
Zdravinje SRB 164 C3
Ždrelac HR 156 D3
Ždrelo SRB 159 E8
Zdunje NMK 164 F3
Zduńska Wola PL 143 C6
Zduny PL 81 D12
Zduny PL 141 F1
Zdynia PL 145 E3
Zdziarzec PL 144 B4
Zdziechowa PL 85 E13
Zdzieszowice PL 142 F5
Zdzitłowice PL 144 B6
Zjbowice PL 142 E5
Žebrák CZ 76 C5
Zebreira P 45 E6
Zebrene LV 134 C5
Zebrzydowa PL 81 D8

Zechlinerhütte D 84 D3
Zeddam NL 183 B8
Zeddiani I 64 D2
Zedelgem B 19 B7
Zederhaus A 73 B8
Žednik SRB 150 F4
Zjdowice PL 142 E6
Zeeland NL 16 E5
Zeebrugge B 19 B7
Zeewolde NL 183 A7
Zegama E 32 D1
Żegiestów PL 145 E2
Žegljane NMK 164 E4
Żegocina PL 144 D1
Zehdenick D 84 E4
Zehna D 83 C12
Zehren D 80 D4
Zeilarn D 76 F3
Žeimelis LT 135 D8
Žeimiai LT 135 F8
Zeiselmauer A 77 F10
Zeiskam D 187 C5
Zeist NL 16 D4
Zeithain D 80 D4
Zeitlofs D 74 B6
Zeitz D 79 D11
Zejmen AL 163 F8
Żelazków PL 142 C5
Zele B 19 B9
Želechlinek PL 141 G2
Żelechów PL 141 G5
Zelena UA 152 A5
Zelena UA 152 B5
Zelena UA 153 A9
Zeleneč SK 146 E5
Zeleni Jadar BIH 158 E3
Zelenikovo BIH 157 D9
Zelenikovo NMK 164 F4
Zelenogorsk RUS 129 E12
Zelenogradsk RUS 136 D1
Zelenohirs'ke UA 154 B6
Żeletava CZ 77 D9
Železná Ruda CZ 76 D4
Železné SK 147 D8
Železnice CZ 77 B8
Železniki SLO 73 D9
Železný Brod CZ 81 E8
Zelhem NL 16 D6
Želiezovce SK 147 E8
Zelina HR 148 E6
Zelinja BIH 157 C8
Želiv CZ 77 C8
Željuša BIH 157 F8
Żelków-Kolonia PL 141 F6
Zell D 75 B10
Zell (Mosel) D 21 D8
Zella-Mehlis D 79 E8
Zell am Harmersbach D 27 D9
Zell am Main D 187 B8
Zell am See A 73 B6
Zell am Ziller A 72 B4
Zell im Wiesental D 27 E8
Zellingen D 74 C6
Zell-Pfarre A 73 D9
Zelmeņi LV 134 D6
Želovce SK 147 E8
Zelów PL 143 D7
Zeltingen-Rachtig D 21 E8
Zeltiņi LV 135 B13
Zeltweg A 73 B10
Želva LT 135 F10
Zelzate B 19 B8
Žemaičių Naumiestis LT 134 F3
Žemberovce SK 147 E7
Zemblak AL 168 C4

Zembrów PL 141 E6
Zembrzyce PL 147 B9
Zemen BG 165 E6
Zemeno GR 175 C6
Zemeš RO 153 D8
Zemité LV 134 C5
Zemmer D 21 E7
Zemné SK 146 F6
Zemplénagárd H 145 G5
Zemplínske Hámre SK 145 F2
Zemst B 19 C9
Zemun SRB 158 D5
Zenica BIH 157 D8
Zennor UK 12 E3
Zentene LV 134 B5
Žepa BIH 157 E11
Žepče BIH 157 D9
Žeravice CZ 146 C4
Zerbst D 79 C11
Zerf D 21 E7
Zerind RO 151 D8
Żerków PL 142 B4
Zernez CH 71 D10
Zernien D 83 D9
Zernitz D 83 E12
Zero Branco I 72 E5
Zerpenschleuse D 84 E5
Zerrenthin D 84 D6
Zestoa E 32 D1
Žetale SLO 148 D5
Zeta RO 152 E6
Zetel D 17 B9
Zet'ovo BG 166 E4
Zeulenroda D 79 E10
Zeven D 17 B12
Zevenaar NL 16 E6
Zevenbergen NL 16 E3
Zevgolatio GR 175 D6
Zevio I 66 B3
Zeytinbaşı TR 181 C9
ZeytinbağI TR 177 A10
Zeytindağ TR 177 E9
Zeytineli TR 177 C8
Zeytinli TR 173 E6
Zeytinliova TR 177 B10
Zgierz PL 143 C7
Zgłobice PL 143 G10
Zgornje Bitnje SLO 73 D9
Zgornje Jezersko SLO 73 D9
Zgornji Duplek SLO 148 C5
Zgorzelec PL 81 D8
Zgropolci NMK 169 A6
Zgurita MD 153 A11
Zhabinka BY 141 F8
Zhabokrychka UA 154 A3
Zhdeniyevo UA 145 F7
Zhegėri RKS 164 E3
Zheleznodorozhnyy RUS 136 E3
Zhelyazkovo BG 167 E8
Zhelyu Voyvoda BG 167 D6
Zheravna BG 167 D6
Zhilino RUS 136 D4
Zhitkovo RUS 129 D11
Zhitnitsa BG 167 D6
Zhitom AL 168 C2
Zhodzishki BY 137 D13
Zhorany UA 141 H9
Zhovkva UA 144 C8
Zhovtantsi UA 144 D9
Zhovtneve UA 144 B9
Zhovtneve UA 155 B3
Zhovtyy Yar UA 154 F5
Zhuprany BY 137 E13
Zhur RKS 164 E2
Zhvyrka UA 144 C9

Zhydachiv UA 145 E9
Zhyrmuny BY 137 E11
Žiar nad Hronom SK 147 D7
Zibalai LT 137 C10
Zibello I 66 B1
Zibreira P 44 F3
Zicavo F 37 H10
Zichovice CZ 76 D5
Zickhusen D 83 C10
Zidani Most SLO 73 D11
Zidarovo BG 167 E8
Žĺdikai LT 134 D4
Žĺdlochovice CZ 77 D11
Ziduri RO 161 C10
Zijbice PL 81 E12
Ziedkalne LV 134 D6
Ziegelroda D 79 D9
Ziegendorf D 83 D11
Ziegenrück D 79 E10
Ziegra D 80 D4
Zieleniewo PL 85 B9
Zieleniewo PL 85 D9
Zielitz D 79 B10
Zielkowice PL 141 G2
Zielona Chocina PL 85 C12
Zielona Góra PL 81 C9
Zielona Góra PL 85 E11
Zielonka PL 139 F11
Zielonki PL 143 F8
Zieluń PL 139 D8
Ziemeri LV 135 C13
Ziemnice Wielkie PL 142 E4
Ziemupe LV 134 C2
Zierenberg D 17 F12
Zierikzee NL 16 E1
Zierzow D 83 D11
Ziesar D 79 B11
Žiežmariai LT 137 D9
Žigljen HR 67 C12
Žĺguri LV 133 B3
Žĺhárec SK 146 E5
Žĺhle CZ 76 B4
Zilaiskalns LV 131 F10
Žilina SK 147 C7
Žilinai LT 137 E10
Zillis CH 71 D8
Ziltendorf D 81 B7
Zilupe LV 133 C4
Zimandu Nou RO 151 E7
Zimbor RO 151 C11
Zimmersrode (Neuental) D 21 B12
Zimnicea RO 160 F6
Zimnitsa BG 167 D7
Žĺndaičiai LT 134 F5
Zingst D 83 B13
Zinkgruvan S 92 B6
Zinnowitz D 84 B6
Ziras LV 134 B3
Zirc H 149 B9
Zirchow D 84 C6
Žiri SLO 73 D9
Zirndorf D 75 D8
Žĺrnešti MD 154 E2
Zirņi LV 134 C4
Ziros GR 179 E11
Žirovnice CZ 77 D8
Zistersdorf A 77 E11
Žĺtište SRB 158 C6
Žitkovac SRB 164 C4
Žĺtni Potok SRB 164 C4
Žĺtomislič BIH 157 F8
Žĺtosřđa SRB 164 C4
Žĺtoše NMK 168 B5
Zitsa GR 168 E4

Zittau D 81 E7
Zitz D 79 B11
Živaja HR 157 B6
Živinice BIH 157 D10
Živogošče HR 157 F7
Zizers CH 71 D9
Zizurkil E 32 D1
Zlarin HR 156 E4
Zlata SRB 164 C4
Zlatar BG 167 C8
Zlatar H 148 D6
Zlatar-Bistrica HR 148 D6
Zlaté Hory CZ 142 G4
Zlaté Klasy SK 146 E4
Zlaté Moravce SK 146 E6
Zlaten Rog BG 159 E10
Zlatitsa BG 165 D9
Zlatna RO 151 E11
Zlatna Panega BG 165 C9
Zlatograd BG 171 B8
Zlatokop SRB 164 D4
Zławies Wielka PL 138 D5
Žĺebič SLO 73 E10
Žĺeby CZ 77 C8
Žĺēkas LV 134 B3
Zletovo NMK 164 F5
Žĺibinai LT 134 E4
Zlín CZ 146 C5
Zliv CZ 76 D5
Žĺjebovi BIH 157 D10
Žĺlakuqan RKS 164 D2
Žĺočieniec PL 85 C10
Złoczew PL 142 D5
Zlokuchene BG 165 E8
Zlonice CZ 76 B6
Zĺota PL 141 G2
Zĺota PL 143 F10
Zĺotniki Kujawskie PL 138 E5
Zĺotoryja PL 81 D9
Zĺotów PL 85 D12
Zĺoty Stok PL 77 B11
Zlozela BIH 157 D8
Žĺutice CZ 76 B4
Zmajevac BIH 156 C5
Zmajevac HR 149 E11
Zmajevo SRB 158 C4
Zmeyovo BG 166 E5
Žĺmigród PL 81 D10
Zmijavci HR 157 F7
Žĺminj HR 67 B8
Žĺmudź PL 144 A8
Znamensk RUS 136 D3
Žĺnin PL 138 E4
Znojmo CZ 77 E10
Zoagli I 37 C10
Zoblitz D 80 E4
Zoelen NL 183 B6
Zoersel B 16 F3
Zofingen CH 27 F8
Zogno I 69 B8
Zografou GR 175 D8
Zola Predosa I 66 C3
Zolder B 19 B11
Zoldo Alto I 72 D5
Žófkiewka-Osada PL 144 B6
Zölkow D 83 C11
Zollikofen CH 31 A11
Zollikon CH 27 F10
Zolotkovychi UA 144 D6
Zóftnica PL 85 C10
Žófynia PL 144 C5
Zomba H 149 D11
Zomergem B 19 B8
Zonhoven B 19 C11

Zoni GR 171 A10
Zoniana GR 178 E8
Zonnebeke B 18 C6
Zonza F 37 H10
Zörawina PL 81 E12
Zörbig D 79 C11
Zorge D 79 C8
Zorita E 45 F9
Zorita del Maestrazgo E 42 F3
Zorleni RO 153 E11
Zorlenţu Mare RO 159 C8
Zorneding D 75 F10
Zornheim D 21 E10
Zörnigal D 79 C12
Zornitsa BG 166 C5
Zornitsa BG 167 C9
Žory PL 142 F6
Zossen D 80 B4
Zottegem B 19 C8
Zoutkamp NL 16 B6
Zoutleeuw B 183 D6
Zovi Do BIH 157 F9
Zovka RUS 132 F3
Zrenjanin SRB 158 C5
Zrin HR 156 B5
Zrinski Topolovac HR 149 D7
Zrmanja Vrelo HR 156 D5
Zrnovci NMK 164 F5
Žĺruč nad Sázavou CZ 77 C8
Zsadány H 151 D7
Zsáka H 151 C7
Zsámbék H 149 A11
Zsámbok H 150 B4
Zsana H 150 E4
Zschaitz D 80 D4
Zscherben D 79 D10
Zschopau D 80 E4
Zschortau D 79 D11
Zsombó H 150 E4
Zuberec SK 147 C9
Zubia E 53 B9
Zubiaur E 40 B6
Zubiĺli BIH 157 D9
Zubiena I 68 C5
Zubieta E 32 D1
Zubieta E 32 D2
Zubin Potok RKS 164 D2
Zubiri E 32 E2
Zubřĺ CZ 146 B6
Zubrohlava SK 147 C9
Žĺubrów PL 81 B8
Žĺuć SRB 164 C3
Zucaina E 48 D4
Zuchwil CH 27 F8
Zudaire E 32 E1
Zudar D 84 B4
Zuera E 41 E10
Zufre E 51 D7
Zug CH 27 F10
Zuhatzu-Kuartango E 40 C6
Zuheros E 53 A8
Zuid-Beijerland NL 182 B4
Zuidhorn NL 16 B6
Zuidland NL 182 B4
Zuidlaren NL 17 B7
Zuidwolde NL 17 C6
Zuienkerke B 182 C2
Zújar E 55 D7
Žĺuków PL 141 G2
Žĺuków PL 141 G8
Žĺukowice PL 81 C9
Žĺukowo PL 138 B5
Žĺuljana HR 162 D3
Zulová CZ 77 B12
Zulte B 19 C7

Zumaia E 32 D1
Zumarraga E 32 D1
Zundert NL 16 F3
Zungri I 59 B8
Zunzarren E 32 E3
Zuoz CH 71 D9
Žĺupanja PL 157 B10
Žĺuras LV 134 B3
Zúrgena E 55 E8
Zürich CH 27 F10
Zurndorf A 77 G12
Žĺurnevo BG 161 F10
Žĺuromin PL 139 D8
Zurow D 83 C11
Zusmarshausen D 75 F8
Züsow D 83 C11
Züssow D 84 C5
Žĺutatlat LT 134 E2
Zutendaal B 19 C12
Zutphen NL 16 D6
Žĺužemberk SLO 73 E10
Žĺveçan RKS 164 D2
Zverino BG 165 C8
Zvezde AL 168 C4
Zvezdel BG 171 B8
Zvezdets BG 167 E8
Zvolen SK 147 D8
Zvolenská Slatina SK 147 D8
Zvonce SRB 164 D6
Zvorište RO 153 B8
Zwaagwesteinde D 75 F8
Zwartsluis NL 16 C5
Zweeloo NL 17 C7
Zweibrücken D 21 F8
Zweisimmen CH 31 B11
Zwenkau D 79 D11
Zwethau D 80 C4
Zwettl A 77 E8
Zwevegem B 19 C7
Zwevezele B 182 C2
Zwickau D 79 E12
Zwiefalten D 27 D11
Zwierzyn PL 85 E9
Zwierzyniec PL 144 B6
Zwiesel D 76 D4
Zwijndrecht B 19 B9
Zwijndrecht NL 16 E3
Zwinge D 79 C7
Zwingen CH 27 F8
Zwingenberg D 21 E11
Zwochau D 79 D11
Zwoleń PL 141 H5
Zwolle NL 16 C6
Zwönitz D 79 E12
Zwota D 75 B12
Zyabki BY 133 F4
Zyal'ki BY 133 E4
Zyalyonka BY 133 E5
Zychlin PL 143 B8
Žĺydowo PL 85 B11
Žĺydowo PL 85 F13
Žĺygaičiai LT 134 F4
Zygos GR 171 B6
Zygry PL 143 C6
Žĺyniai LT 134 E2
Žĺyraków PL 143 F11
Žĺyrardów PL 141 F2
Žĺyrzyn PL 141 H6
Žĺytkiejmy PL 136 E6
Žĺytniów PL 142 D6
Žĺytno PL 143 E8
Žĺywiec PL 147 B8
Žĺywocice PL 142 F4

Æ

Ærøskøbing DK 86 F6

Ø

Ødis DK 86 E4
Ødsted DK 86 D4
Øie N 105 B12
Økdal N 101 A12
Øksfjord N 112 C8
Øksnes N 110 C8
Øksneshamn N 110 D9
Ølen N 94 C3
Ølgod DK 86 D3
Ølholm DK 86 D5
Ølsted DK 86 D5
Ølsted DK 87 D10
Ølstykke DK 87 D10
Ønslev DK 83 A11
Øra N 112 C8
Ørbæk DK 86 E7
Ørgenvika N 95 B11
Ørjavik N 104 F2
Ørje N 96 D6
Ørnes N 108 C6
Ørnhøj DK 86 C3
Ørslev DK 87 E9
Ørsnes N 100 A5
Ørsta N 100 B4
Ørsted DK 86 C6
Ørting DK 86 D6
Ørum DK 86 C5
Ørum DK 86 C7
Øsby DK 86 E5
Østbirk DK 86 D5
Østby N 91 A9
Østby N 102 D4
Østengård DK 86 D4
Øster Assels DK 86 B3
Øster Bjerregrav DK 86 C5
Øster Brønderslev DK 86 A5
Østerby DK 86 A5
Øster Hornum DK 86 B5
Øster Hurup DK 86 B6
Øster Højst DK 86 E4
Østerild DK 86 A3
Øster Jølby DK 86 B3
Østerlars DK 89 E7
Øster Lindet DK 86 E4
Øster Vedsted DK 86 E3
Østervrå DK 90 E7
Øster Vrøgum DK 86 D2
Østese N 94 B4
Østrup DK 86 B4
Øverbygd N 111 C17
Øvergard N 111 B18
Øvre Alta N 113 D11

Øvre Kildal N 112 D7
Øvrella N 95 C10
Øvre Rendal N 101 C14
Øvre Årdal N 100 D7
Øvre Åstbru N 101 D13
Øyangen N 104 E7
Øydegarden N 104 D4
Øyenkilen N 91 A8
Øyer N 101 D12
Øyeren N 96 B7
Øyjord N 108 B9
Øynes N 108 B9
Øynes N 111 C11
Øyslebø N 90 C2
Øyvatnet N 111 C12

Å

Å N 104 F7
Å N 110 E4
Å N 111 B12
Å N 111 C13
Åberget S 109 E18
Åbo S 103 C10
Åbodarna S 107 E14
Åbogen N 96 B7
Åbosjö S 107 D13
Åby S 89 A7
Åby S 93 B8
Åbyen DK 90 D7
Åbyggeby S 103 E13
Åbyn S 118 D6
Åbytorp S 92 A6
Ådalsliden S 107 E11
Ådum DK 86 D3
Åfarnes N 100 A7
Åfjord N 104 D8
Åfoss N 90 A6
Ågerup DK 87 D10
Ågotnes N 94 B2
Ågskaret N 108 C5
Åheim N 100 B3
Åhus S 88 D6
Åkarp S 87 D12
Åkerbränna S 107 D11
Åkerby S 99 B9
Åkerholmen S 118 C6
Åkersberga S 99 D10
Åkers styckebruk S 98 D8
Åkerstrømmen N 101 C14
Åknes N 105 D12
Åkran N 105 D12
Åkrehamn N 94 E3
Åkullsjön S 118 F5
Åkvissan S 107 E13
Ål N 101 E9
Ålberga S 93 B9
Ålbo S 98 D7
Ålbæk DK 90 D7
Åle DK 86 D5
Åled S 87 B11

Ålem S 89 B10
Ålen N 101 A14
Ålesund N 100 B4
Ålgnäs S 103 D12
Ålgård N 94 E3
Ålhult S 92 D1
Ålloluokta S 109 B17
Ålmo N 104 E4
Ålsrode DK 87 C7
Ålstad N 110 E9
Ålund S 118 D6
Ålvik N 94 B4
Ålvund N 101 A9
Ålvundeid N 101 A9
Ålåsen S 106 D7
Åmdals Verk N 95 D8
Åminne FIN 122 E7
Åminne S 87 A14
Åmland N 94 F5
Åmli N 90 A3
Åmli N 90 B1
Åmmeberg S 92 B6
Åmot N 94 C7
Åmot N 95 B11
Åmot N 95 C8
Åmot N 95 C11
Åmot S 103 D11
Åmotfors S 96 C7
Åmsele S 107 B16
Åmsosen N 94 D3
Åmynnet S 107 E15
Åmål S 91 A12
Åmål S 91 B12
Åmøyhamn N 108 C5
Åna-Sira N 94 F4
Åndalsnes N 100 A7
Åneby N 95 B13
Ånes N 104 E4
Ånge S 103 A10
Ånge S 109 E14
Ångelsberg S 97 C15
Ångersjö S 102 C8
Ånn S 105 E13
Ånstad N 110 C8
Åpåsdalen N 109 D7
Åre S 105 E14
Årfor N 105 C10
Årgård N 105 C10
Århult S 89 A10
Årjäng S 96 C7
Årnäs S 92 A4
Årnes N 109 B7
Årnes N 104 F5
Årnes N 111 B15
Årnäs S 91 B14
Åros N 95 C12

Årosjåkk S 111 E17
Årre DK 86 D3
Årrenjarka S 109 C15
Årsandøy N 105 A12
Årsdale DK 89 E8
Årset N 105 A11
Årslev DK 86 E6
Årstein N 111 C14
Årsunda S 98 A7
Årvik N 100 B3
Årviksand N 112 C8
Årvågen N 104 E5
Åryd S 89 B7
Åryd S 89 C8
Årøybukta N 111 A19
Årøysund N 90 A7
Ås N 95 C13
Ås N 105 E11
Ås S 97 C12
Ås S 107 E11
Åsa S 91 E11
Åsan N 105 B12
Åsarna S 102 A7
Åsby S 97 A10
Åse N 111 B10
Åsebyn S 96 D7
Åseda S 89 A8
Åsegg N 105 C9
Åsele S 107 C12
Åselet S 118 D4
Åsen N 105 D10
Åsen S 102 D6
Åsen S 106 E7
Åsen S 109 E16
Åsenbruk S 91 B11
Åseral N 90 B1
Åshammar S 103 E12
Åskilje S 107 B13
Åskogen S 118 C7
Åsli N 101 E11
Åsljunga S 87 C13
Åsmansbo S 97 B13
Åsmarka N 101 D13
Åsskard N 104 E5
Åsta N 101 D14
Åstan N 104 E7
Åster S 105 E14
Åstorp S 87 C11
Åstrand N 95 B12
Återvänningen S 103 A13
Åtorp S 92 A4
Åträsk S 118 B7
Åträsk S 118 C5
Åttonträsk S 107 C13
Åtvidaberg S 92 C7
Åva FIN 126 E5
Åvestbo S 97 C14

Åvist FIN 122 D9

Ä

Aetsä FIN 126 C8
Ähtäri FIN 123 E12
Ähtärinranta FIN 123 E12
Äijäjoki FIN 116 B10
Äijälä FIN 123 E16
Äkäsjokisuu FIN 117 C9
Äkäslompolo FIN 117 C12
Ålandsbro S 103 A14
Älgarås S 103 B12
Älgered S 103 B12
Älghult S 89 A9
Älmeboda S 89 B9
Älmhult S 88 B6
Älmsta S 99 C11
Älta S 99 D10
Älvdalen S 102 D7
Älvho S 102 C6
Älvkarleby S 103 E13
Älvkarleö S 99 A8
Älvros S 102 B7
Älvsbyn S 118 C6
Älvsered S 87 A11
Älvängen S 91 D11
Ämmänkylä FIN 123 E9
Ämmänsaari FIN 121 E12
Åmådalen S 102 D8
Äng S 92 D5
Änge S 105 E16
Ängebo S 103 C11
Ängelholm S 87 C11
Ängersjö S 122 C3
Ängeslevä FIN 119 E15
Ängesträsk S 118 B8
Ängesån S 116 E8
Ångom S 103 B13
Äppelbo S 97 B11
Äran S 105 D9
Ärla S 98 D7
Ärnäs S 102 D5
Ärtled S 103 E9
Ärtrik S 107 E11
Äsarp S 91 C14
Äsbacka S 103 D11
Äsköping S 92 A8
Åssjö S 103 B12
Åtran S 87 A11
Äyskoski FIN 123 D17
Äystö FIN 122 F7
Äänekoski FIN 123 E15

Ö

Öckerö S 91 D10
Ödeborg S 91 B10
Ödeshog S 92 C5
Ödkarby FIN 99 B13
Ödsmål S 91 C10

Ödåkra S 87 C11
Öja S 103 C12
Öja S 93 E12
Öjarn S 106 D8
Öje S 102 E6
Öjebyn S 118 D6
Öjeforsen S 103 B9
Öjingsvallen S 103 C8
Öjung S 103 C10
Öksajärvi S 116 C8
Öllölä FIN 125 F15
Ölmbrotorp S 97 D13
Ölme S 97 D11
Ölsboda S 92 A4
Ömossa FIN 122 F7
Önnestad S 88 C6
Önningeby FIN 99 B14
Öratjärn S 103 C10
Öravan S 107 B14
Öravattnet S 106 E9
Örbyhus S 99 B9
Örbäck S 97 C15
Örebro S 97 D13
Örebäcken S 102 C4
Öregrund S 99 B10
Öreström S 107 C16
Öretjändalen S 103 A10
Örkelljunga S 87 C12
Örnsköldsvik S 107 E15
Örnäsudden S 109 E13
Örsbäck S 107 D17
Örserum S 92 C5
Örsjö S 89 B9
Örsundsbro S 99 C8
Örträsk S 107 C15
Örviken S 118 E6
Ösmo S 93 B11
Östa S 98 B7
Österfärnebo S 98 B7
Östergraninge S 107 F12
Österhankmo FIN 122 D7
Österjörn S 118 D4
Österlisa S 99 C10
Östermark FIN 126 C8
Östernoret S 107 D12
Östero FIN 122 D8
Österskucku S 102 A8

Östersund S 106 E7
Östersundom FIN 127 E13
Östervåla S 98 B8
Österås S 107 E12
Östhammar S 99 B10
Östloning S 103 A13
Östmark S 97 B8
Östmarkum S 107 E14
Östnor S 102 D7
Östra E 53 B9
Östra Frölunda S 91 E13
Östra Granberg S 118 C4
Östra Grevie S 87 E12
Östra Husby S 93 B9
Östra Ljungby S 87 C12
Östra Lovsjön S 106 D7
Östra Löa S 97 C13
Östra Ryd S 93 C8
Östra Ormsjö S 107 C10
Östra Skrämträsk S 118 E5
Östra Stugusjö S 103 A9
Östra Sönnarslöv S 88 D6
Östra Vemmerlöv S 88 D6
Östra Yttermark FIN 122 E6
Östra Åliden S 118 C4
Överammer S 107 E9
Överberg S 102 B7
Överbyn S 103 C12
Överhogdal S 102 B8
Överhörnäs S 107 E15
Överissjö S 107 C13
Överkalix S 119 B9
Överlannäs S 107 E13
Övermalax FIN 122 E7
Övermark FIN 122 E6
Övermorjärv S 118 B9
Övernäs S 109 D14
Överstbyn S 118 B7
Övertorneå S 119 B11
Överturingen S 102 B8
Övertänger S 103 C12
Överum S 93 D8
Överäng S 105 D14
Överö FIN 99 B15
Öv Långträsk S 109 E16
Övra S 107 D11
Övre Bredåker S 118 C6
Övre Flåsjön S 118 C7
Övre-Konäs S 105 D14
Övre Soppero S 116 B7
Övre Tväråsel S 118 C5
Övsjöbyn S 107 E9
Öxabäck S 91 E12

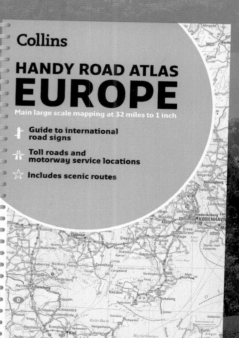